T0141319

About Island Press

Since 1984, the nonprofit organization Island Press has been stimulating, shaping, and communicating ideas that are essential for solving environmental problems worldwide. With more than 1,000 titles in print and some 30 new releases each year, we are the nation's leading publisher on environmental issues. We identify innovative thinkers and emerging trends in the environmental field. We work with world-renowned experts and authors to develop cross-disciplinary solutions to environmental challenges.

Island Press designs and executes educational campaigns in conjunction with our authors to communicate their critical messages in print, in person, and online using the latest technologies, innovative programs, and the media. Our goal is to reach targeted audiences—scientists, policymakers, environmental advocates, urban planners, the media, and concerned citizens—with information that can be used to create the framework for long-term ecological health and human well-being.

Island Press gratefully acknowledges major support of our work by The Agua Fund, The Andrew W. Mellon Foundation, The Bobolink Foundation, The Curtis and Edith Munson Foundation, Forrest C. and Frances H. Lattner Foundation, The JPB Foundation, The Kresge Foundation, The Oram Foundation, Inc., The Overbrook Foundation, The S.D. Bechtel, Jr. Foundation, The Summit Charitable Foundation, Inc., and many other generous supporters.

The opinions expressed in this book are those of the author(s) and do not necessarily reflect the views of our supporters.

GLOBAL ATLAS OF MARINE FISHERIES

About Sea Around Us

The Sea Around Us, named after the famous book by Rachel Carson, was launched in July 1999 as a joint research and extension project of the University of British Columbia in Vancouver, Canada, and the Philadelphia-based Pew Charitable Trusts, and has been funded since mid-2014 by the Paul G. Allen Family Foundation through Vulcan Philanthropic.

The Sea Around Us mission is to document and communicate fisheries impacts on marine ecosystems and to devise policies that can mitigate and reverse harmful trends while ensuring the social and economic benefits of sustainable fisheries. In the last 15 years, the Sea Around Us has worked toward these goals through a large number of hard-hitting publications in major scientific journals, and through an extensive outreach focusing on the environmental NGO community.

The Sea Around Us also makes key data available to fisheries scientists, managers, and to civil society in all countries of the world through its website (www.seaaroundus.org). This involves particularly detailed catch data from marine and estuaries fisheries, by country or territory, sector and species (groups), and catch-derived indicators of fisheries status.

These data, assembled as described in this Atlas, are corrected and updated at regular intervals, and thus its readers are invited to visit the Sea Around Us website for updates of, and details on, the data presented in this Atlas, and to contact its editors or authors.

The State of the World's Oceans Series; Daniel Pauly, series editor. Also in this series:

Five Easy Pieces: How Fishing Impacts Marine Ecosystems by Daniel Pauly

In a Perfect Ocean: Fisheries and Ecosystem in the North Atlantic by Daniel Pauly and Jay Maclean

GLOBAL ATLAS OF MARINE FISHERIES

A CRITICAL APPRAISAL OF CATCHES AND ECOSYSTEM IMPACTS

EDITED BY DANIEL PAULY
AND DIRK ZELLER

Library of Congress Control Number
2016941218

Design and typesetting by 2K/Denmark
Printed on recycled, acid-free paper

Manufactured in the United States of America
10 9 8 7 6 5 4 3 2 1

Keywords: catch reconstructions, Food and Agriculture Organization (FAO), high seas fisheries management, mariculture, marine biodiversity, marine reserves

CONTENTS

PART II. COUNTRIES & TERRITORIES ACCOUNTS

FOREWORD

Josh Reichert, Pew Charitable Trusts

In the autumn of 1997, I convened a small group of some of the world's most respected marine scientists to answer two questions:

- Is it possible to assess changes taking place in the world's oceans at regular intervals, looking at a broad number of factors that can provide a portrait of changing ocean health? If so, how would one go about it?
- Is it possible to determine with some degree of accuracy what is driving these changes, and what would be the outcome both for ocean life and human society if these causative factors were left unchecked?

With one exception, the members of the group said it was not doable without very large investments in monitoring technology, for which there was no recognizable donor other than governments, which would be unlikely to provide the funds needed. The one exception was Daniel Pauly, who, in his indomitable fashion, said it was possible to undertake this kind of analysis for ocean fisheries. Even for data-poor fisheries, of which there are many, he indicated that it is possible to reconstruct the past in order to compare it with the present and to infer likely future changes. Though not perfect, this kind of reconstruction and forecasting can provide a reasonable sense of how marine fisheries have changed over time, what the populations of specific species look like today, and where we are headed if our management of these resources does not change.

Late in the afternoon, the meeting came to an end, and the participants all left, with the exception of Daniel, whom I asked to stay for a brief discussion on some of the points he had made earlier in the day. That "brief discussion" turned into a professional relationship that has endured to this day. Daniel, and the *Sea Around Us* team, have produced some of the most groundbreaking fisheries science of the past 50 years and in the process have changed the way we think about the management of marine fish and the ocean systems on which they depend.

The goal of the *Sea Around Us* is to provide a portrait of the major changes that have taken place in populations of fish over time, primarily as a result of fishing, and to better understand the consequences of these changes to the broader ecosystems from which these fish are being taken.

The motivation behind the project was the absence of accurate and comprehensive information about the status of the world's ocean fisheries and the need for such information if we are to manage these resources prudently in the years ahead. Daniel and his colleagues at the Fisheries Centre of the University of British Columbia thought that the "official" effort to assess the state of the world's marine fisheries, which is undertaken by the Food and Agriculture Organization (FAO) of the United Nations, is flawed in very significant ways. First and foremost, the species that are the primary focus of FAO are caught overwhelmingly by large-scale, industrial fishers. However, this is only a fraction of the global marine catch. The FAO data do not include many of the recreational artisanal and subsistence fisheries, despite the fact that these "small-scale" fisheries make up about one quarter of the global marine catch and one third of the part of that catch that is destined for human consumption. Similarly, it does not include estimates of, or even placeholders for, illegally caught fish or fish that are not formally reported (estimated at one out of every five fish caught). Finally, it also fails to include discards, that is, fish that are thrown back into the sea, dead or dying, because they are not what the fishers are looking for. In short, the landing data sent to the FAO by its member countries, which form the base of its biannual *State of the World's Fisheries and Aquaculture* (SOFIA), tend to be strongly underreported.

The significance of this is profound for several reasons. First, these global catch statistics condition the way we look at and manage marine fisheries. For example, if we are not aware of the catches made by small-scale fisheries, we will underestimate the contribution of artisanal and subsistence fisheries to food security and of recreational fisheries to the tourism industry. This underestimation then justifies the neglect of these fisheries, despite their crucial contribution to local economies. Similarly, measures against illegal fishing are hard to justify—not least because of their cost—if the size of the illegal catches is not estimated.

The goal of the *Sea Around Us*, when it was initially launched, was to produce a more comprehensive and accurate historical portrait of the world's marine fisheries than that reported by the FAO and to put in place a system that would enable that portrait to be updated regularly. We underestimated the time needed for this project. It has taken longer than anticipated to gather the data, but like a jigsaw puzzle whose image gradually reveals itself, the project slowly began to unveil a crisis taking place in the world's oceans, one that can be met only by profound changes in the way we view and manage fisheries worldwide.

Daniel Pauly and his team of researchers have been able to quantify catches from the key fishing sectors worldwide: artisanal (small-scale fisheries), subsistence (small-scale fisheries), recreational (small-scale fisheries), and industrial (large-scale fisheries). The results of this analysis have fundamentally changed the way we define the scope of the fisheries crises and its solution.

What these data reveal is that developed countries underreport what is caught in their waters, often by as much as 50%, and developing countries underreport by 70%–200%. This profound difference is declining, however, as the landing statistics supplied to FAO by its member countries have improved over time. This is fortunate, because such discrepancies cannot be maintained for any length of time without introducing a grave disconnect between what we think we know and what really occurs in the water.

Recognizing these deficiencies enables us to overcome them. This upward reassessment of global catches also provides a sense of just how productive the oceans really are and a measure of optimism that we could enjoy high catches in the future if we rebuild depleted stocks and manage resources more prudently.

Daniel and his colleagues have produced the most comprehensive picture to date of the changing status of global fisheries. It is a sobering picture, but there is a silver lining. If we are less greedy and do not continuously overtax what they are capable of producing, the world's oceans have the potential to recover much of their former bounty.

Whether we take advantage of this opportunity is a political decision, far more than it is a technical one. For many years, we have failed to adequately measure what we are taking out of the world's oceans. We no longer have an excuse to do so. This global fisheries atlas tells a story of the decline of abundance and provides a series of concepts and tools that will help governments better measure the size of their catches and their impact on marine ecosystems. Thus, this atlas represents good science in the service of good conservation. It is a tool that can help us to better understand and document what is happening in the oceans so that we can manage marine fisheries in ways that will restore their productivity as opposed to accentuating their decline.

We would be wise to heed the insights contained in this atlas. Failure to do so will lead us further down a path we are now traveling, at an even faster pace. We know well where that path goes. The other road has a far better destination. It will take patience, discipline, political will, and short-term sacrifice. But it has a future that will provide for both people and nature, as opposed to the other path, which has none.

PREFACE

The atlas you are holding in your hands presents key results of the *Sea Around Us*, a research activity initiated at the University of British Columbia by the Pew Charitable Trusts, currently funded mainly by the Paul G. Allen Family Foundation, and devoted to studying and documenting human impacts on marine ecosystems, especially those caused by fisheries, and to propose policies to mitigate those impacts.

In the first 2 years of its existence, from mid-1999 to mid-2001, the *Sea Around Us* concentrated on the North Atlantic and attempted to answer the following six questions:

1 What are the total fisheries catches from the ecosystems, including reported landings, unreported landings, and discards at sea?
2 What are the biological impacts of these withdrawals of biomass for the remaining life in the ecosystems?
3 What would be the likely biological and economic impacts of continuing current fishing trends?
4 What were the former states of these ecosystems before the expansion of large-scale commercial fisheries?
5 How do the present ecosystems rate on a scale from "healthy" to "unhealthy"?
6 What specific policy changes and management measures should be implemented to avoid worsening of the present situation and improve ecosystems' health?

First, answers to these questions were published, for the North Atlantic, as a book in 2003,[1] and having passed this test, we began to apply the methods the *Sea Around Us* developed to answer these six questions for the global ocean.

In the process, we gradually realized that a key dataset we and most other researchers working on international fisheries were using—the global fisheries catch statistics assembled and disseminated annually since 1950 by the Food and Agriculture Organization of the United Nations (FAO)—was biased in a profound way. We hasten to add that FAO is not at fault: The bias, which works against small-scale fisheries (i.e., artisanal, subsistence, and recreational), is caused largely by most of its member countries not comprehensively including the catch of such fisheries in their annual data submission to FAO.

Thus, answering question 1 for the global ocean involved developing a global dataset that, in addition to including statistics that FAO provides, would also explicitly cover small-scale fisheries. Also, as appropriate for ecosystem-based management, we had to include in our database time series of fisheries discards, which FAO has documented globally in successive reports but kept outside its main database (which thus remains largely a database of "landings" rather than catches).

Our approach for answering question 1 was to perform (or encourage our colleagues throughout the world to perform) catch reconstructions whose scientific rationale and technical features are discussed in chapters 1 and 2 of this atlas, respectively. More than two hundred peer-reviewed journal articles, book chapters, reports, and working papers (all available at www.seaaroundus.org) that this work generated are summarized on pp. 185 to 457 of this atlas. These summaries describe, in the form of one-page accounts, the marine fisheries of the Exclusive Economic Zones (EEZs) of 273 countries (or parts thereof) and island territories, covering about 40% of the surface of the ocean, where about 95% (by weight) of the world marine catch is being taken. Complementing these summaries, chapter 3 describes the assembly of catch data documenting the industrial fisheries for large pelagics (mainly tuna), much of it on the high seas, outside the EEZs.

Jointly, these contributions demonstrate convincingly that the global marine fisheries catches are much higher than reported in official statistics; some of the implications for research and policy are briefly explored in chapter 14, which thus deals with question 6. Chapters 4 and 5 then cover various topics that the *Sea Around Us* worked on to address questions 2 to 5, some explicitly, others implicitly.

Because of its broad scope, we hope this atlas and its underlying data will be useful to researchers and students interested in comparative analyses of fisheries and marine biodiversity, and to the staff of international organizations, whether governmental or nongovernmental, with a stake in fisheries governance and marine conservation.

As for the *Sea Around Us*, our close association with the Pew Charitable Trusts ended in 2014, and we are now funded mainly by the Paul G. Allen Family Foundation. Also, the focus of the *Sea Around Us* has shifted from documenting fisheries' impacts on the oceans to mitigating these impacts, in collaboration with various governments and civil society. Information on our progress therein, along with the data underlying the graphs and analyses presented here, can be also found on our website (www.seaaroundus.org). Note, finally, that our home at the University of British Columbia changed from the Fisheries Centre, which regrettably ceased to exist in June 2015, to a new Institute for the Oceans and Fisheries.

Daniel Pauly
Dirk Zeller

NOTE

1. Pauly, D., and J. Maclean. 2003. *In a Perfect Ocean: Fisheries and Ecosystem in the North Atlantic Ocean*. Island Press, Washington, DC.

ACKNOWLEDGMENTS

We sincerely thank the Pew Charitable Trusts for supporting the *Sea Around Us* for more than 15 years, from mid-1999 to 2014. The fundamental trust that this support reflects was extremely valuable to us. It made us feel appreciated and resulted in more effective work. It enabled us to be creative and to think big, to tackle the long-term global fisheries issues that none of our colleagues could address, all without being monitored through short-term metrics.

We thank Ms. Rebecca Rimel, president and CEO of the Pew Charitable Trusts for her long-term support; Dr. Joshua Reichert for his inspiration and for formulating the six-point mission statement that has been our guiding star throughout; and Dr. Rebecca Goldburg for skillfully mediating between the different styles of an environmental advocacy organization and a university-based research group. We also thank the many dedicated Pew staffers with whom we established excellent relationships throughout the years and with whom we hope to continue collaborating in the future, if under different circumstances.

From mid-2014 on, the Paul G. Allen Family Foundation has provided the bulk of the support for the *Sea Around Us*, enabling a smooth transition for which we are extremely thankful. Also, the Paul G. Allen Family Foundation funded a complete overhaul of the *Sea Around Us* website (www.seaaroundus.org), implemented by outstanding staff at Vulcan Inc., enabling the visualization and effective delivery to users of the catch and related data generated by the *Sea Around Us* and featured in this atlas.

The work on this atlas received additional support from numerous foundations and other organizations, notably the Rockefeller Foundation, the Prince Albert II of Monaco Foundation, the Khaled bin Sultan Living Oceans Foundation, the MAVA Fondation pour la Nature, the Baltic 2020 Foundation, the National Geographic Society, the World Wildlife Fund for Nature, the Natural Resources Defense Council, Conservation International, the Bay of Bengal Large Marine Ecosystem Project, United Nations Environment Programme (UNEP), and the Intergovernmental Oceanographic Commission (IOC) of the United Nations Educational, Scientific and Cultural Organization (UNESCO), and several others.

A huge number of people were associated with the creation of this atlas, notably the 325 different authors of the national catch reconstructions on pp. 185 to 457 of this atlas. Unfortunately, we cannot do more here than thank them en bloc. Within the *Sea Around Us*, the following scientists, former graduate students of Daniel Pauly, research assistants, and volunteers, past and present, contributed extensively to the catch reconstructions: Dalal Al-Abdulrazzak, Melanie Ang, Andrea Au, Houman Azar, Sarah Bale, Milton Barbosa, Sebastian Baust, Dyhia Belhabib, Brajgeet Bhathal, Lea Boistol, Lisa Boonzaier, Shawn Booth, Ciara Brennan, Vania Budimartono, Elise Bultel, Annadel Cabanban, Devraj Chaitanya, William Cheung, Andrés Cisneros-Montemayor, Mathieu Colléter, Duncan Copeland, Kendyl Crawford, Ester Dividovich, Beau Doherty, Bridget Doyle, Leonie Färber, Katia Freire, Manuela Funes, Darah Gibson, Rhona Govender, Krista Greer, Andrea Haas, Sara Harper, Claire Hornby, Davis Iritani, Jennifer Jacquet, Boris Jovanović, Myriam Khalfallah, Kristin Kleisner, Danielle Knip, Daniel Kuo, Vicky Lam, Frédéric Le Manach, Alasdair Lindop, Stephanie Lingard, Jessica MacDonald, Ashley McCrea-Strub, Dana Miller, Elizabeth Mohammed, George Nguyen,

Devon O'Meara, Allan Padilla, Maria Lourdes Palomares, Lo Persson, Ciara Piroddi, Robin Ramdeen, Nazanin Roshan Moniri, Peter Rossing, Laurenne Schiller, Soohyun Shon, Patricia Sun, Wilf Swartz, Eric Sy, Louise Teh, Lydia Teh, Pablo Trujillo, Gordon Tsui, Aylin Ulman, Sadiq Vali, Liesbeth van der Meer, Liane Veitch, Nicolas Winkler, Yunlei Zhai, and Kyrstn Zylich. To them go our heartfelt thanks.

CHAPTER 1

ON THE IMPORTANCE OF FISHERIES CATCHES, WITH A RATIONALE FOR THEIR RECONSTRUCTION[1]

Daniel Pauly
Sea Around Us, Fisheries Centre, University of British Columbia, Vancouver, BC, Canada

Fishing must generate a catch, whether it is done by West African artisanal fishers supplying a teeming rural market, by the huge trawler fleets in Alaska that supply international seafood markets, by women gleaning on a reef flat in the Philippines to feed their families, or by an Australian angler bragging about it in a bar. Indeed, a fishery is defined by the amount and kind of fish caught and by their monetary value. This is how we judge a fishery's importance, compared with other fisheries and other sectors of the economy. It seems clear that the health of a fishery should by measured by changes in the magnitude and species composition of catches, along with other information, such as the growth and mortality of the fish that are exploited. Yet a debate has been recently raging about whether to use catch data to infer the status of fisheries, causing great confusion among fisheries scientists and managers. If the muddle continues, it could undermine the credibility of fisheries science.

The key role of catch data is the reason why the Food and Agriculture Organization (FAO) first began compiling fishery statistics soon after the agency was founded in 1945. A part of the United Nations' attempt to "quantify the world" (Ward 2004), these compendia turned, in 1950, into the much-appreciated FAO *Yearbook of Fisheries Catch and Landings*. The findings are based on annual data submissions by FAO member countries, vetted and harmonized by its staff. In contrast with the many international databases used to track major food crops (e.g., rice, wheat, maize), the *Yearbook* has been, to this day, the only global database of wild-caught fish and other marine species. As such, the *Yearbook* is widely cited as the major source for inferences on the status of fisheries in the world (Garibaldi 2012).

However, in many countries, particularly in the developing world, the government's role in monitoring their fisheries seems to end with the annual ritual of filling in catch report forms and sending them to FAO, as parodied in Marriott (1984). For others, mainly developed countries, collecting catch data from fishing ports and markets is only a start, and the bulk of their fishery-related research is in the form of "stock assessments." This term refers to a series of analytic procedures using a variety of data, often time series of commercial catch (figure 1.1), complemented by information on the age, size, or structure of the fish in that catch, tag and recapture data, stock abundances deduced through mathematical models or by fishery research vessels, and so on. The purpose is to infer the biomass of the populations or stocks that are being exploited and to propose levels of total allowable catch (TAC, or quota).

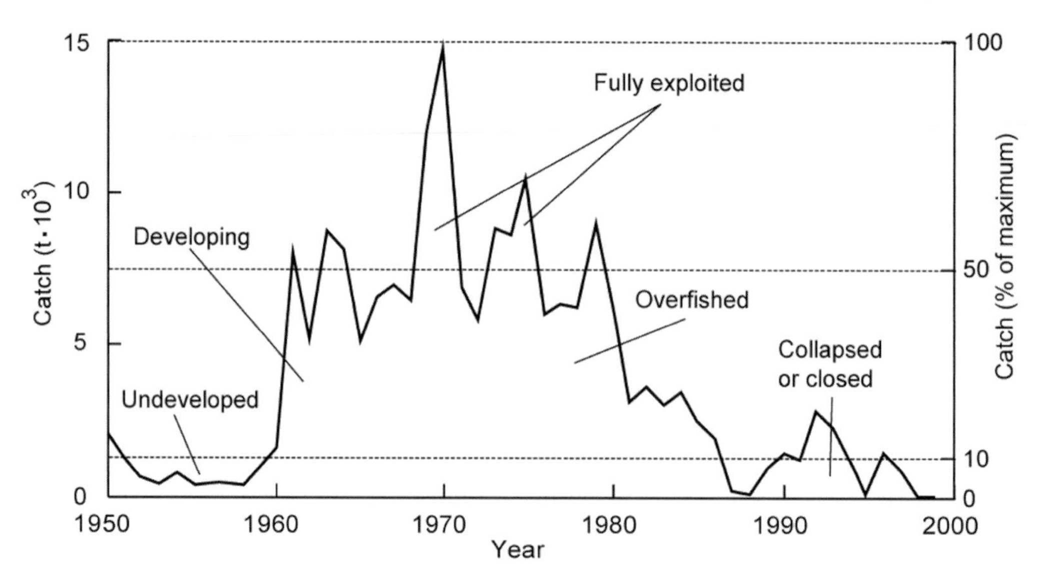

Figure 1.1. A typical catch time series, as can be used in conjunction with the method of Martell and Froese (2013) to perform a basic assessment of the fishery that generated that catch. The scale to the right defines the categories ("undeveloped," "developing," "fully exploited," etc.) used to describe the status of the underlying fisheries.

However, traditional stock assessments are extremely expensive, ranging from roughly US$50,000 per stock (assuming 6 months for experts to analyze existing data) to millions of dollars when fisheries-independent data are required (Pauly 2013). Along with a worldwide scarcity of expertise, this is why 20% at most of the more than 200 current maritime countries and associated island territories perform regular stock assessments. Moreover, these assessments deal only with the most abundant or most valuable species exploited. For some countries or territories, this may be one species, a dozen, or about two hundred, as in the United States. In all cases, this is only a small fraction of the number of species that are exploited, if only as unintended bycatch, which is often discarded.

Therefore, FAO always encouraged the development of methods that would allow scientists to infer the state of fisheries without stock assessments, or with limited ones (see Gulland 1969, 1971, 1983). This practice was driven by FAO's mission to inform policy makers about the state of fisheries in all countries of the world, including those without access to stock assessment expertise and the costly research vessels needed to collect fisheries-independent data.

To this end, FAO developed what are now called stock-status plots (SSPs; figure 1.2), which showed the status of the various fish stocks over time (Grainger and Garcia 1996). The status of each fishery was inferred from the shape of its catch time series. Essentially, increasing or stable catches meant fisheries were okay, and declining catches meant fisheries were in trouble. These SSPs or equivalent graphs were interpreted vertically, by comparing the percentage of stocks in a given state (e.g., "developing," "developed," "fully exploited," "overfished") in different years. The information was reported in press releases and in issues of the *State of the World Fisheries and Aquaculture* (SOFIA), a biannual narrative interpretation of the FAO fishery statistics. In successive SOFIAs (including the last available; see figure 13 in FAO 2014, p. 37), FAO notes that these percentages tend to get worse but does not analyze the SSPs further. However, much of what people throughout the world think they know about global fisheries originates from SSPs and similar approaches.

SSPs were also adopted and modified by researchers outside FAO, first by Rainer Froese of the GEOMAR Institute in Kiel, Germany (Froese and Kesner-Reyes 2002) and later by the group of which I am principal investigator, the *Sea Around Us*, whose work is featured in this

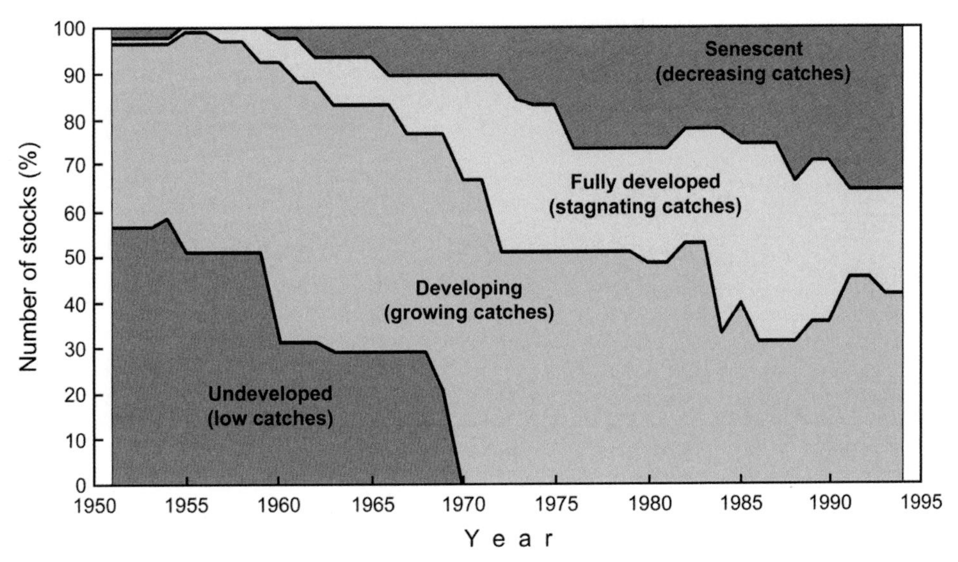

Figure 1.2. Interpretation of multiple catch time series (as in figure 1.1), pioneered by Grainger and Garcia (1996) and based on about 400 important stocks monitored by FAO. Such graphs, now called stock-status plots, are conventionally interpreted vertically, that is, by reading, for a given year, the percentage of stocks in the different categories. This is where the annually changing percentage of "overfished" or "collapsed" stocks that are communicated to the public originate. Note that these percentages are very sensitive to the details of definitions used for the different status categories. (Modified from Grainger and Garcia 1996.)

volume. Jointly, we demonstrated that an increased number of the stocks had "collapsed," meaning that catches were less than 10% of their historic maximum. Moreover, the transition from one state (e.g., "fully exploited") to another (e.g., "overexploited") was occurring at a faster rate than previously thought. These were dramatic findings, yet they generated little press and even less action.

The world finally started paying attention to the SSP findings in 2006, when Boris Worm and his colleagues published "Impacts of Biodiversity Loss on Ocean Ecosystem Services" in *Science* magazine (Worm et al. 2006). For the first time, the authors used these trends to project a date by which all stocks would "collapse": 2048.[2] The expert press release that accompanied the article (see Baron 2010) focused on this newsy aspect of what was a broad study, triggering an enormous amount of press coverage on all continents. The headlines were uniformly alarmist: "Fisheries collapse by 2048" (*The Economist*), "Seafood may be gone by 2048" (*National Geographic*), and "The end of fish, in one chart" (*The Washington Post*), among many more.

A strong pushback emerged, including wide and understandable criticism of the precise date, 2048, which mingled Mayan (2012) and Orwellian (1984) undertones. Stock assessment experts mocked the *projection*, which was mistaken for a *prediction*, with many arguing that scientists shouldn't extrapolate beyond the data. Yet good science always implies some inference beyond one's data; otherwise, it would consist only of *descriptions*. Moreover, most critiques overlooked the fact that "collapsed" stocks can continue to be fished. Indeed, this is what already occurs in vast areas of the ocean. Two notorious examples are the Swedish west coast, where a long-collapsed Atlantic cod stock continues to be exploited (Sterner and Svedäng 2005), and the Gulf of Thailand, where the demersal fish biomass was reduced in the 1990s to less than 10% of its value in the early 1960s, when trawling began, yet also continues to be exploited (Pauly and Chuenpagdee 2003). This is 2048—now.

Still, the criticism was so strong that several co-authors of the study opted not to defend it publicly. Consequently, fisheries scientists such as me, who are concerned with the state of global fisheries, had to either duck or defend the spirit of the 2048 projection, even if we did not agree with all its particularities.

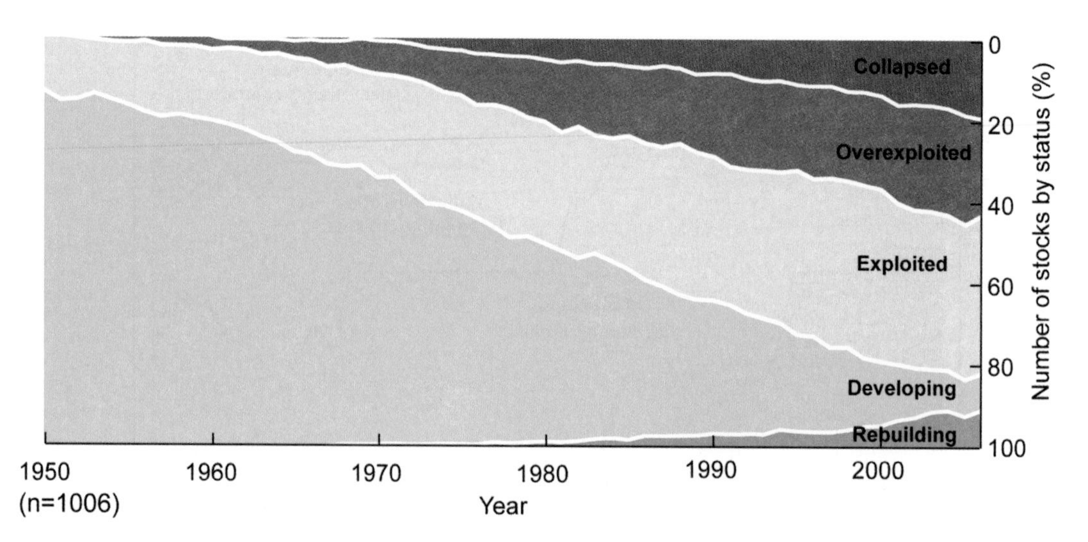

Figure 1.3. Stock-status plots (SSPs) based on more than 1,000 stocks worldwide, whose "developing" category combines the "undeveloped" and "developing" categories of figure 1.2 and which include a new category ("rebuilding"), which will hopefully increase with time (Kleisner et al. 2013). The 2048 projection mentioned in the text was derived by reading an SSP semihorizontally, that is, by extrapolation forward (and downward) from the line separating "overexploited" from "collapsed" stocks. Note the similarity of the trends suggested by this and figure 1.2, whose "senescent" or equivalent categories ("overexploited" and "collapsed") are clearly increasing. (Modified from Kleisner et al. 2013.)

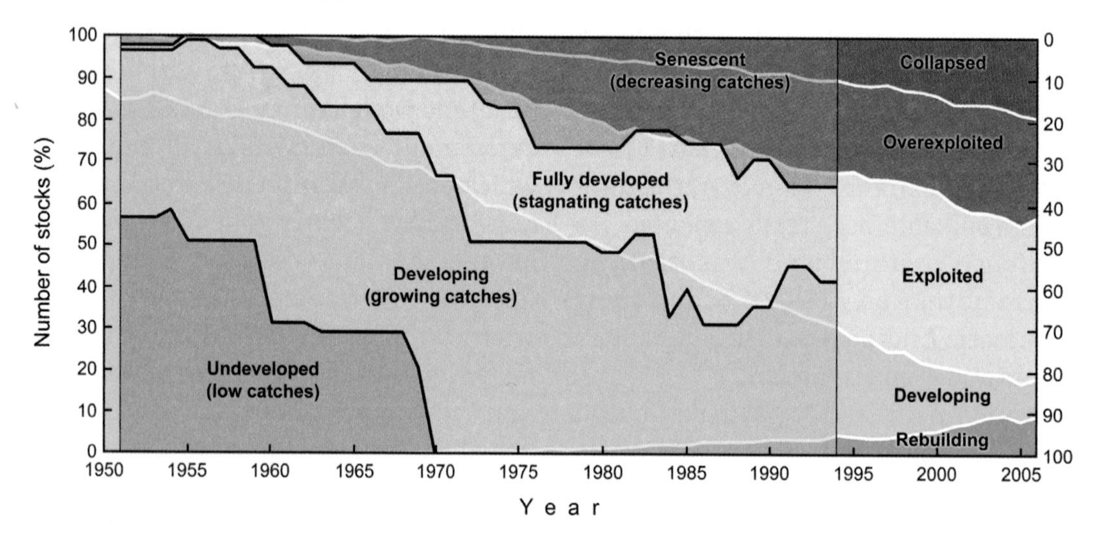

Figure 1.4. Superposing figures 1.2 and 1.3 shows that the trends suggested by these graphs are very robust, that is, they are not very sensitive to the details of the definitions of the categories (e.g., using 10% of the maximum historic catch for defining "collapsed" stocks). Also note that if Grainger and Garcia (1996) had used their graph (figure 1.2) for a 10-year prediction, it would be confirmed by this superposition. (Modified from Kleisner et al. 2013.)

To its credit, the projection was based on catch time series from virtually the entire world. The overwhelming majority showed that peak catches occurred several decades ago, with current catches increasingly derived from "overexploited" and "collapsed" stocks (figures 1.2 and 1.3). Although there is no way to *predict* where anything will be in 2048 or even 10 years from now, it would certainly be better if we could reverse current trends. So far we have not done so, even though some stocks are rebuilding (figures 1.3 and 1.4).

Before this defense could be mounted, the detractors began focusing on another criticism of the 2048 projection, claiming that catch data do not contain any information about stock status. In interviews, keynote lectures, and other outlets, they argued that full-fledged stock assessments are essential to understanding fisheries; without them, we are essentially left in the dark.

This is a case of allowing the perfect to become the enemy of the good. Even without perfect data, we can infer when fisheries are in serious trouble and make efforts to conserve them. After all, maintaining catches is the raison d'être of fisheries science. One can and should infer, at least tentatively, the status of fisheries from the catch data—*if* this is all we have (see figure 1.1; Froese et al. 2012, 2013; Kleisner et al. 2013). It is a mistake to assume that we must remain in Muggle-like ignorance unless we have access to the magic of stock assessments.

Accepting this doctrine would put us at the mercy of stock assessment models that can be fatally flawed. For example, the models used to study the Canadian northern cod fishery in the 1990s (Walters and Maguire 1996) were considered the best in the world. In fact, experts thought the models were so good that it was not necessary to consider the catch data from the coastal trap fisheries, which could not, like the trawlers, follow the cod to where they retreated as their numbers declined. Thus, the stock assessment experts were as surprised as the general public when the fishery had to be closed. The trawlers had decimated the stock under their noses, which they could have seen if they had analyzed the coastal trap data. Note that it is not even faulty stock assessments that are at issue here; it is the notion that one type of approach is so good that it makes all other approaches superfluous.

More importantly, this doctrine would discourage efforts to improve the quality of fisheries statistics worldwide, which is bemoaned by FAO in successive issues of SOFIA. It would also thwart attempts to manage, to the extent possible, the fisheries of developing countries. If leading fisheries scientists claim that catch data are useless, why would resource-starved governments invest in reforming and improving their statistical systems?

This flawed thinking would affect not only developing countries but also the community of stock assessment experts themselves. Without the collection of catch data, experts could end up either with beautiful stock assessment models applied to lousy data, as in the northern cod example above, or needing more of the costly fishery-independent data that can be used to correct for misreported commercial catch data (Beare et al. 2005).

We gain nothing from the notion that only a select group has the key to understanding fisheries, especially if that key cannot open any doors outside a small number of developed countries. Such claims undermine the credibility of the many fisheries scientists throughout the world who attempt to extract actionable insights from sparse data and to advise their governments on how to manage their fisheries even if they cannot afford formal stock assessments.

Fortunately, there is a solution: We all agree that many stocks need to be rebuilt and that doing so would lead to sustainable increases of catches and economic benefits (Sumaila et al. 2012). In fact, the more depleted the stocks currently are, the more is to be gained by rebuilding them.

Moreover, our systematic reevaluation of the FAO statistics suggests that developed countries tend to underreport their catches by about 30%–50% (Zeller et al. 2011), and many developing countries underreport by 100%–500% (Cisneros-Montemayor et al. 2013; Pauly and Zeller 2014; Zeller et al. 2007, 2015). (One notable exception is China, which overreports its catches because officials are rewarded for high yields.) This new perspective suggests that fisheries play a far more important role in the rural economy of developing countries than previously assumed and that rebuilding depleted fish populations on a grand scale would have greater benefits than so far imagined (other implications are presented in chapter 14).

Consequently, more attention should be given to the reliable collection of catch data throughout the world. In particular, we need to devise cost-effective systems to acquire accurate fisheries catch data, along with ancillary data on fishing effort, and its economic equivalent, catch value and fishing cost.

These ideas have been apparent to me since my first field experience in Ghana in

1971 and in Indonesia in 1974 and 1976. They were reinforced in 1979 when J. A. Gulland, a world-renowned scientist and senior staff member at FAO, commented that fisheries experts should emphasize three things: "the catch, the catch and the catch." Yet often catch data seem to be entirely missing from certain areas of countries or territories, particularly for informal, small-scale fisheries. In such cases, catch statistics can be reconstructed from other data.

The text below, slightly modified from an article I wrote in 1998, provides the rationale for such reconstructions.[3] It was inspired by discussions that took place at a conference held by the FishBase Project[4] in Trinidad in May and June 1998.

THE CATCH IN USING CATCH STATISTICS

It is widely recognized that catch statistics are crucial to fisheries management. However, the catch statistics routinely collected and published in most countries are deficient in numerous ways. This is particularly true of the national data summary sent annually by the statistical offices of various Caribbean and Pacific countries to the FAO for inclusion in their global statistics database (see Marriott 1984).

A common response to this situation has been to set up intensive but short-term projects devoted to improving national data reporting systems. Their key products are detailed statistics covering the (few) years of the project. However, without statistics from previous periods, these data are hard to interpret. This is a major drawback, because it is the changes in a dataset that demonstrate important trends.

Therefore, reconstructing past catches and catch compositions is a fundamental task for fisheries scientists and officers. In fact, it is necessary to fully interpret the data collected from current projects. For example, suppose that the fisheries department of Country A establishes, after a large and costly sampling project, that its reef fishery generated catches of 5 and 4 t/km^2/year for the years 1995 and 1996, respectively. The question now is, are these catch figures low values relative to the potential of the resource, thus allowing an intensification of the fishery, or high unsustainable values, indicative of an excessive level of effort?

Clearly, one approach would be to compare these figures with those of adjacent Countries B and C. However, these countries may lack precise statistics or have fisheries that use different gears. Furthermore, Country A's minister in charge of fisheries may be hesitant to accept conclusions based on comparative studies and may require local evidence before making important decisions affecting her country's fisheries. One approach to deal with this very legitimate requirement is to reconstruct and analyze time series, covering the years preceding the recent period for which detailed data are available and going as far back in time as possible (e.g., to the year 1950, when the aforementioned annual FAO statistics begin). Such data make it possible to quickly evaluate the status of fisheries and their supporting resources and to evaluate whether further increases in effort will be counterproductive (box 1.1).

BASIC METHODOLOGY FOR CATCH AND EFFORT RECONSTRUCTION

The key part of the methodology proposed here is psychological: One must overcome the notion that "no information is available," which is the wrong default setting when dealing with an industry such as fisheries. Rather, one must realize that fisheries are social activities, bound to throw large shadows onto the societies in which they are conducted. Therefore, records usually exist that document some aspects of these fisheries. All that is needed is to find them and to judiciously interpret the data they contain. Important sources for such undertaking are

1 Old files of the Department of Fisheries
2 Peer-reviewed journal articles
3 Theses and scientific and travel reports, accessible in departmental or local libraries or branches of the University of the West Indies or the University of the South Pacific, or through regional databases

BOX 1.1. QUICK INTERPRETATION OF CATCH AND EFFORT DATA

Daniel Pauly, *Sea Around Us*, University of British Columbia, Vancouver, Canada

There is a huge literature dealing with the fitting of surplus production models to time series of catch and effort data. Strangely enough, one rarely finds quick assessments based on the key properties of these models. A simple method for such assessments is presented here.

Two key predictions of the parabolic Schaefer model (Schaefer 1957; Ricker 1975) are that catch/effort (U) declines linearly with effort (f) and that a stock is biologically overfished if U, in the fishery exploiting that stock, has dropped to less than 50% of its level at the onset of the fishery.

Thus, with two estimates of U, a higher one pertaining to an early state of the fishery (U_{then}) and a lower one pertaining to a later state or to the present state (U_{now}), and the corresponding levels of effort (f_{then} and f_{now}), one can assess the present status of a fishery by first calculating

$$b = (U_{then} - U_{now})/(f_{now} - f_{then}), \text{ and } a = U_{then} + (bf_{then}).$$

Then, using a and b, one can calculate Maximum Sustainable Yield (MSY) and its associated level of effort (f_{MSY}) from $f_{MSY} = a/(2b)$, and $MSY = (af_{MSY}) - (bf_{MSY} \cdot f_{MSY})$.

In the Philippines, in about 1900, in the absence of industrial gear, 119,000 fishers reportedly caught 500,000 t/year (Anonymous 1905, p. 564), or 4.2 t per fisher per year. In 1977, 501,000 small-scale fishers reportedly caught 713,000 t (Smith et al. 1980), or 1.42 t per fisher per year. Inserted in the above equation, these numbers lead to $b = 0.0000073$, $a = 5.069$, MSY = 880,000 t/year and f_{MSY} 347,000 fishers.

The theory and applications of surplus production models have often been the subject of fierce debates, notably on how sustainable MSY really is. However, it is generally agreed that a reduction by 50% or more of initial catch/effort indicates overfishing in just about any stock, at least in economic terms. Therefore, the quick diagnostics suggested above should always be useful as a first approach.

REFERENCES

Anonymous. 1905. *Censo de las Islas Filipinas*. Vol. IV. *Agricultura, estadística social e industrial*. U.S. Census Office, Washington, DC.

Ricker, W. E. 1975. Computation and interpretation of biological statistics of fish populations. *Bulletin of the Fisheries Research Board of Canada* 191.

Schaefer, M. B. 1957. A study of the dynamics of the fishery for yellowfin tuna in the eastern tropical Pacific Ocean. *Inter-American Tropical Tuna Commission Bulletin* 2: 247–268.

Smith, I. R., M. Y. Puzon, and C. M. Vidal-Libuano. 1980. Philippine municipal fisheries: a review of resources, technology and socioeconomics. *ICLARM Studies and Reviews* 4.

4 Records from harbormasters and other maritime authorities with information on numbers of fishing craft (small boats by type, large boats by length class or engine power)
5 Records from the cooperative or private sectors (e.g., companies exporting fisheries products, processing plants, importers of fishing gear)
6 Old aerial photos from geographic or other surveys (to estimate numbers of boats on beaches and along piers)
7 Interviews with old fishers

ESTIMATING CATCHES

Analysis of the scattered data obtained from these sources should be based on the simple notion that catch in weight (Y) is the product of catch/effort (U, also known as CPUE) times effort (f), or

$$Y = Uf. \qquad (1.1)$$

This implies that one should obtain from sources 1–7 estimates of the effort (how many fishers, boats, or trips) of each gear type and multiply it by the mean catch/effort of that gear type (e.g., mean annual catch per fisher or mean catch per trip). Because the catch/effort of small boats and of individual fishers will differ substantially from that of the larger boats, it is best to estimate annual catches by sector, gear, or boat type, with the total catch estimates then obtained by summing over all gear or boat types.

Moreover, because CPUE usually varies with season, estimation of Y should preferably be done on a monthly basis, by applying equation 1.1 separately for every month, then adding the monthly catch values to obtain an annual sum. Alternatively, a seasonally averaged CPUE can be used. This should be repeated for every component of the fishery, such as the small-scale and industrial components.

Once all quantitative information has been extracted from the available records, linear interpolations can be used to fill in the years for which estimates are missing. For example, if one has estimated 1,000 t as annual reef catch for 1950 and 4,000 t for 1980, then it is legitimate to assume that the catches were about 2,000 t in 1960 and 3,000 t in 1970. This may appear too daring. However, the alternative to this is to leave blanks (so-called no data entries), which later will invariably be interpreted as catches of *zero*, which is a far worse estimate than interpolated values.

ESTIMATING CATCH COMPOSITION

Once catch time series have been established for distinct fisheries (e.g., near shore or reef, shelf, oceanic), the job is to split these catches into distinct species or species groups. Unfortunately, comprehensive information on catch composition is usually lacking. Therefore, the job of splitting the catches must be based on fragmentary information, such as the observed catch composition of a few, hopefully representative, fishing units. Still, combining all available anecdotal information on the catch composition of a fishery (i.e., observed composition of scattered samples) should create reasonable estimates of mean composition. Thus, a report stating that "the catches consisted of groupers, snappers, grunts, and other fish" can be turned into 25% groupers, 25% snappers, 25% grunts, and 25% other fish as a reasonable first approximation.

A number of such approximations of catch composition can then be averaged into a representative set of percentages, which can be applied to the catches of the relevant period. These percentage catch compositions can be interpolated in time, for example, as 1950–1954 with a composition of 40% groupers, 20% snappers, 10% grunts, and 30% other fish, and 10%, 10%, 20%, and 60%, respectively, for these same groups in 1960–1964. In this case, the values for the intermediate period (1955–1959) can be interpolated as 25% groupers, 15% snappers, 15% grunts, and 45% other fish.

CONCLUSIONS

Estimating catches from the catch/effort of selected gear and fishing effort is a standard method of fisheries management. Reconstruction of historic catches and catch compositions series may require interpolations and other bold assumptions, justified by the unacceptability of the alternative (i.e., accepting catches to be recorded as zero or otherwise known to be incompatible with empirical data and historic records).

There is obviously more to reconstructing catch time series than outlined here, and some of the available methods are rather sophisticated (see Zeller et al. 2015). The major impediment to applying this technique is that colleagues initially do not trust themselves to reconstruct unseen quantities such as historic catches or believe that they can judge the likely level of catches in the absence of "properly" collected data. Yet it is only by making bold assumptions that we can obtain the historic catches needed for comparisons with recent catch estimates and thus infer major trends in fisheries (see also box 1.1).

One example may be given here. The FAO catch statistics for Trinidad and Tobago for the years 1950–1959 start at 1,000 t (1950–1952), then gradually increase to 2,000 t in 1959. Of this, 500–800 t was contributed by "Osteichtyes," 300–500 t by "*Scomberomorus maculatus*" (now known as *S. brasiliensis*), 100–200 t by "*Penaeus* spp.," and 0–100 t by "Perciformes" (presumably reef fishes). On the other hand, the same statistics report, for 1950–1959, catches of zero for fish that are targeted by fishers in Trinidad and Tobago, such as *Caranx* spp., "Clupeoids," *Thunnus alalunga*, *T. albacares*, and *Katsuwonus pelamis*.

Despite their obvious deficiencies, these and similar data from other Caribbean countries are commonly used to illustrate fisheries trends from the region. Fortunately, it is very easy to improve on this method. Thus, Kenny (1955) estimated, based on detailed surveys at the major market (Port of Spain) and a few reasonable assumptions, that the total catch from the island of Trinidad was on the order of 13 million pounds (2,680 t) in 1954 and 1955, about two times the FAO estimate at this time for both Trinidad and Tobago. Moreover, King-Webster and Rajkumar (1958) provide details of the small-scale fisheries existing on Tobago, from which fishing effort and a substantial catch can be estimated, notably of "carite" (*Scomberomerus regalis*). Furthermore, both of these sources include detailed catch compositions as well, indicating that several of the categories with entries of zero in the FAO statistics (e.g., the clupeoids) generated substantial catches in the 1950s. Other early sources exist that can be used to corroborate this point. Similar datasets exist in other Caribbean countries.

The text of Pauly (1998) ended here, and this introductory chapter will also, because the 273 one-page accounts for countries or territories presented in the second part of this atlas summarize the analyses I had hoped would be done in the Caribbean and elsewhere. Additionally, chapter 14 presents a first summary of our reconstruction work and its consequences for fisheries research and management.

ACKNOWLEDGMENTS

I thank Ms. Lucy Odling-Smee for inspiring the first part of this chapter and, belatedly, the participants of the Africa Caribbean Pacific–European Union (ACP-EU) Course on Fisheries and Biodiversity Management, held in Port of Spain, Trinidad and Tobago, May 21 to June 3, 1998, for their interest in discussions that led to the second half of this chapter. This is a contribution of the *Sea Around Us*, a research activity at the University of British Columbia initiated and funded by the Pew Charitable Trusts from 1999 to 2014 and currently funded mainly by the Paul G. Allen Family Foundation.

REFERENCES

Baron, N. 2010. *Escape from the Ivory Tower: A Guide to Making Your Science Matter*. Island Press, Washington, DC.

Beare, D. J., C. L. Needle, F. Burns, and D. G. Reid. 2005. Using survey data independently from commercial data in stock assessment: an example using haddock in ICES Division Vi(a). *ICES Journal of Marine Science* 62(5): 996–1005.

Cisneros-Montemayor, A., M. A. Cisneros-Mata, S. Harper, and D. Pauly. 2013. Extent and implication of IUU catch in Mexico's marine fisheries. *Marine Policy* 39: 283–288.

FAO. 2014. *The State of World Fisheries and Aquaculture*. Food and Agriculture Organization of the United Nations, Rome.

Fox, A. 1995. *Linguistic Reconstruction: An Introduction to Theory and Methods*. Oxford University Press, Oxford, England.

Froese, R., and K. Kesner-Reyes. 2002. *Impact of Fishing on the Abundance of Marine Species*. ICES CM 2002/L:12, Copenhagen, Denmark.

Froese, R., D. Zeller, K. Kleisner, and D. Pauly. 2012. What catch data can tell us about the status of global fisheries. *Marine Biology* 159(6): 1283–1292.

Froese, R., D. Zeller, K. Kleisner, and D. Pauly. 2013. Worrisome trends in global stock status continue unabated: a response to a comment by R. M. Cook on "What catch data can tell us about the status of global fisheries." *Marine Biology* 160: 2531–2533.

Garibaldi, L. 2012. The FAO global capture production database: a six-decade effort to catch the trend. *Marine Policy* 36(3): 760–768.

Grainger, R. J. R., and S. M. Garcia. 1996. *Chronicles of marine fishery landings (1950–1994): trend analysis and fisheries potential*. FAO Fisheries Technical Paper 359.

Gulland, J. A. 1969. Manual of methods for fish stock assessment. Part 1: fish population analysis. *FAO Manuals in Fisheries Science* 4.

Gulland, J. A. 1971. *The Fish Resources of the Oceans*. FAO/Fishing New Books, Farnham, Surrey, UK.

Gulland, J. A. 1983. *Fish Stock Assessment: A Manual of Basic Methods*. John Wiley & Sons, New York.

Kenny, J. S. 1955. Statistics of the Port-of-Spain wholesale fish market. *Journal of the Agricultural Society* (June): 267–272.

King-Webster, W. A., and H. D. Rajkumar. 1958. *A preliminary survey of the fisheries of the island of Tobago. Caribbean Commission Central Secretariat, Port of Spain*. Unpublished ms.

Kleisner, D. Zeller, K., R. Froese, and D. Pauly. 2013. Using global catch data for inferences on the world's marine fisheries. *Fish and Fisheries* 14(3): 293–345.

Marriott, S. P. 1984. Notes on the completion of FAO Form Fishstat NS1 (national summary). *Fishbyte, Newsletter of the Network of Tropical Fisheries Scientists* 2(2): 7–8. Reprinted in Zylich, K., D. Zeller, M. Ang, and D. Pauly (eds.). 2014. Fisheries catch reconstructions: islands, part IV. *Fisheries Centre Research Reports* 22(2): 157, University of British Columbia, Vancouver.

Martell, S., and R. Froese. 2013. A simple method for estimating MSY from catch and resilience. *Fish and Fisheries* 14(4): 504–514.

Palomares, M. L. D., W. W. L. Cheung, V. W. Y. Lam, and D. Pauly. 2015. The distribution of exploited marine biodiversity. Pp. 46–58 in D. Pauly and D. Zeller (eds.), *Global Atlas of Marine Fisheries*. Island Press, Washington, DC.

Pauly, D. 1998. Rationale for reconstructing catch time series. *EC Fisheries Cooperation Bulletin* 11(2): 4–7. [Available in French as "Approche raisonné de la reconstruction des séries temporelles de prises," pp. 8–10.]

Pauly, D. 2013. Does catch reflect abundance? Yes, it is a crucial signal. *Nature* 494: 303–306.

Pauly, D., and R. Chuenpagdee. 2003. Development of fisheries in the Gulf of Thailand Large Marine Ecosystem: analysis of an unplanned experiment. Pp. 337–354 in G. Hempel and K. Sherman (eds.), *Large*

Marine Ecosystems of the World 12: Change and Sustainability. Elsevier Science, Amsterdam.
Pauly, D., and D. Zeller. 2014. Accurate catches and the sustainability of coral reef fisheries. *Current Opinion in Environmental Sustainability* 7: 44–51.
Sterner, T., and H. Svedäng. 2005. A net loss: policy instruments for commercial cod fishing in Sweden. *AMBIO: A Journal of the Human Environment* 34(2): 84–90.
Sumaila, U. R., W. W. L. Cheung, A. Dyck, K. M. Gueye, L. Huang, V. Lam, D. Pauly, U. T. Srinivasan, W. Swartz, R. Watson, and D. Zeller. 2012. Benefits of rebuilding global marine fisheries outweigh costs. *PLoS ONE* 7(7): e40542.
Walters, C. J., and J.-J. Maguire. 1996. Lessons for stock assessments from the northern cod collapse. *Review in Fish Biology and Fisheries* 6: 125–137.
Ward, M. 2004. *Quantifying the World: UN Ideas and Statistics. United Nations Intellectual History Project Series*. Indiana University Press, Bloomington.
Watkins, C. 2000. *The American Heritage Dictionary of Indo-European Roots*, 2nd ed. Houghton Mifflin, Boston.
Worm, B., E. B. Barbier, N. Beaumont, J. E. Duffy, C. Folke, B. S. Halpern, J. B. C. Jackson, H. K. Lotze, F. Micheli, S. R. Palumbi, E. Sala, K. A. Selkoe, J. J. Stachowicz, and R. Watson. 2006. Impacts of biodiversity loss on ocean ecosystem services. *Science* 314: 787–790.
Zeller, D., S. Booth, and D. Pauly. 2007. Fisheries contribution to GDP: underestimating small-scale fisheries in the Pacific. *Marine Resources Economics* 21: 355–374.
Zeller, D., S. Harper, K. Zylich, and D. Pauly. 2015. Synthesis of under-reported small-scale fisheries catch in Pacific-island waters. *Coral Reefs* 34(1): 25–39.
Zeller, D., P. Rossing, S. Harper, L. Persson, S. Booth, and D. Pauly. 2011. The Baltic Sea: estimates of total fisheries removals 1950–2007. *Fisheries Research* 108: 356–363.

NOTES

1. Cite as Pauly, D. 2015. On the importance of fisheries catches, with a rationale for their reconstruction. Pp. 1–11 in D. Pauly and D. Zeller (eds.), *Global Atlas of Marine Fisheries: A Critical Appraisal of Catches and Ecosystem Impacts*. Island Press, Washington, DC.
2. This analysis, by Large Marine Ecosystem (LME) using FAO catch data spatialized by the *Sea Around Us*, was performed by Dr. Reg Watson, then a member of the *Sea Around Us*, who thus became a co-author of Worm et al. (2006).
3. The word *reconstruction* is here taken over from historic linguistics (a field that I have an amateur's interest in), wherein extinct languages (Proto-Austronesian, Proto-Bantu, or Proto-Indo-European) are "reconstructed" from words in the daughter languages and rules about phonetic shifts (see Fox 1995; Watkins 2000). One is never sure about the final product but can still offer it to one's colleagues for further scrutiny. It is the same for reconstructed catches.
4. See Palomares et al. (2015).

CHAPTER 2

MARINE FISHERIES CATCH RECONSTRUCTION: DEFINITIONS, SOURCES, METHODS, AND CHALLENGES[1]

Dirk Zeller and Daniel Pauly
Sea Around Us, University of British Columbia, Vancouver, BC, Canada

It is now well established that official fisheries catch data, for perfectly legitimate reasons, have historically ignored or underreported certain sectors (e.g., the subsistence or recreational sectors) as well as fisheries discards, notably because landing data were collected in many cases for purposes of taxation or the management of a small number of target species.

Nowadays, however, when fisheries need to be managed in the context of the ecosystems in which they are embedded (Pikitch et al. 2004), less than full accounting for all withdrawals from marine ecosystems is insufficient. Therefore, this contribution is part of the effort documented in this atlas to provide a time series of all marine fisheries catches from 1950, the first year that the Food and Agriculture Organization of the United Nations (FAO) produced its annual compendium of global fisheries statistics to 2010, that is, 61 years with sharply contrasting economic, political, and environmental conditions.

What is covered here are catches in the waters within the Exclusive Economic Zones

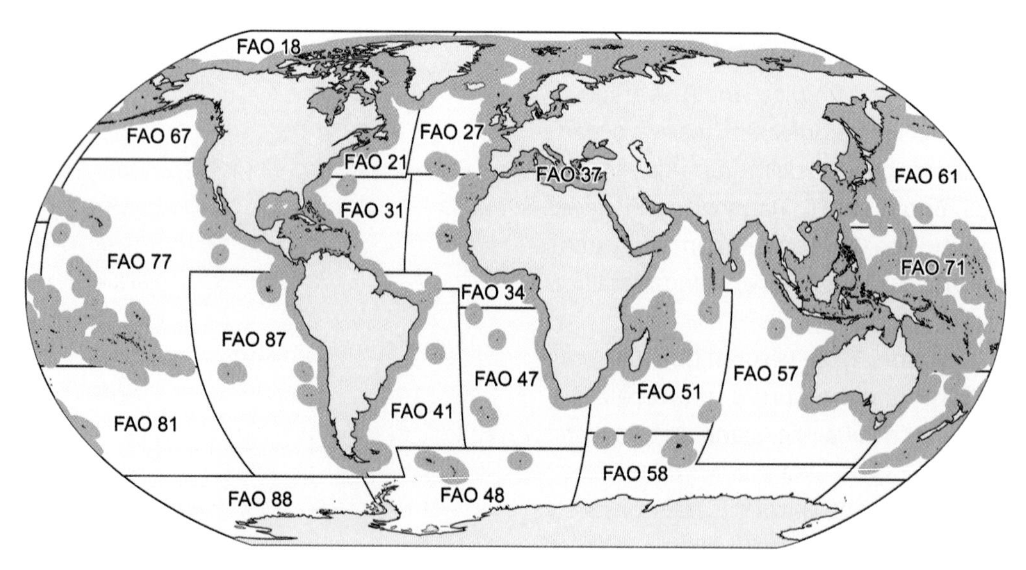

Figure 2.1. The extent and delimitation of countries' EEZs, as declared by individual countries or as defined by the *Sea Around Us* based on the fundamental principles outlined in UNCLOS (i.e., 200 nautical miles or midline rules), and the FAO statistical areas by which global catch statistics are reported. Note that for several FAO areas, some data exist by subareas as provided through regional organizations (e.g., International Council for the Exploration of the Sea [ICES] for FAO area 27). The *Sea Around Us* uses these spatially refined data to improve the spatial allocation of catch data, as described in chapter 5.

(EEZs, figure 2.1) that countries have claimed since they could do so under the United Nations Convention on the Law of the Sea (UNCLOS) or which they could claim under UNCLOS rules but have not (such as many countries around the Mediterranean). The delineations provided by the Flanders Marine Institute (see www.vliz.be and Claus et al. 2014) were used for our definitions of EEZs. Countries that have not formally claimed an EEZ were assigned EEZ-equivalent areas based on the basic principles of EEZs as outlined in UNCLOS (i.e., 200 nmi or midline rules). Note that we:

- Treat disputed zones (i.e., EEZ areas claimed by more than one country) as being owned by each claimant with respect to their fisheries catches, including the extravagant claims by one single country on large swaths of the open South China Sea.
- Treat EEZ areas before each country's year of EEZ declaration as "EEZ-equivalent waters" (with open access to all fishing countries during that time).

Therefore, this contribution deals with catches made in about 40% of the world ocean space, whereas the catches (mainly of tuna and other large pelagic fishes) made in the high seas, which cover the remaining 60%, are dealt with in chapter 3.

METHODS AND DEFINITIONS

The country-by-country fisheries catch reconstructions whose summaries form the core of this atlas are based on the rationale in Pauly (1998, and see chapter 1), as operationalized by Zeller et al. (2007, 2015). The former contribution asserted that there is no fishery with "no data" because fisheries, as social activities, throw a shadow onto the other sectors of the economy in which they are embedded, and it is always worse to put a value of zero (or the ubiquitous "no data" entry) for the catch of a poorly documented fishery than to estimate its catch, even roughly, because subsequent users of one's statistics will interpret the zeroes as "no catches" rather than "catches unknown" (see also Covey 2000).

Zeller et al. (2007) developed a six-step approach for implementing these concepts, as follows:

1. Identification, sourcing, and comparison of baseline reported catch time series, that is, FAO (or other international reporting entities) reported landings data by FAO statistical areas, taxon, and year; and national data series by area, taxon, and year.
2. Identification of sectors (e.g., subsistence, recreational), time periods, species, gears, and so on, not covered by (1), that is, missing data components. This is conducted via extensive literature searches and consultations with local experts.
3. Sourcing of available alternative information sources on missing data identified in (2), via extensive searches of the literature (peer-reviewed and gray, both online and in hard copies) and consultations with local experts. Information sources include social science studies (e.g., anthropology, economics), reports, colonial archives, datasets, and expert knowledge.
4. Development of data anchor points in time for each missing data component and expansion of anchor point data to country-wide catch estimates.
5. Interpolation for time periods between data anchor points, either linearly or assumption-based for commercial fisheries, and generally via per capita (or per fisher) catch rates for noncommercial sectors.
6. Estimation of total catch time series, combining reported catches (1) and interpolated, country-wide expanded missing data series (5).

Since these six points were originally proposed, a seventh point has come to the fore that cannot be ignored (Zeller et al. 2015):

7. Quantifying the uncertainty associated with each reconstruction.

The first part of this contribution expands on each of these seven reconstruction steps, based on the experience accumulated during the last decade, when completing or guiding the reconstructions that form the core of this atlas (see also figure 2.2 on next page).

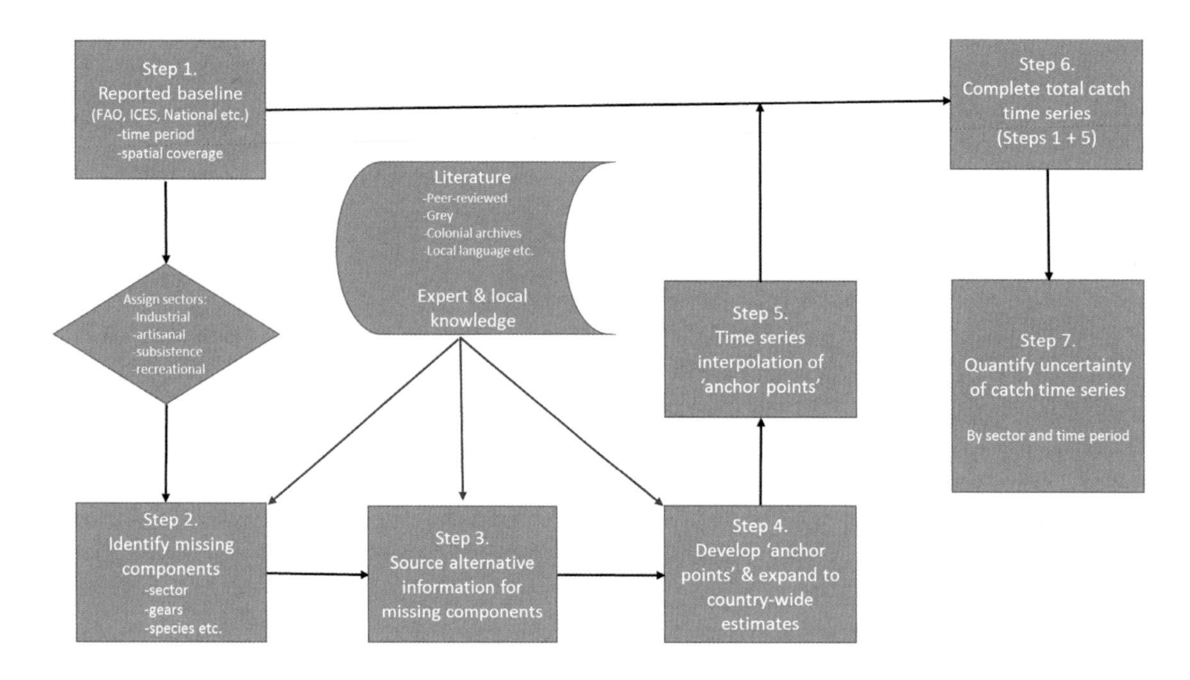

Figure 2.2. Conceptual representation of the 7-step catch reconstruction approach, as initially described in Zeller et al. (2007) and modified here and in Zeller et al. (2015).

Step 1: Identification, Sourcing, and Comparison of Existing, Reported Catch Time Series

Implicit in this first step is that the spatial entity that is to be reported on (e.g., EEZ of Germany in the Baltic Sea) be identified, delineated, and named, information that is not always obvious and posed serious problems to some of our external collaborators (box 2.1).

For most countries, the baseline data are the statistics reported by member countries to FAO (of whose existence a surprisingly large number of colleagues, especially in developing countries, are not aware). Whenever available, we also use data reported nationally for a first-order comparison with FAO data, which often assist in identifying catches probably taken in areas beyond national jurisdiction (i.e., either in EEZs of other countries or in high seas waters). The reason for this is that many national datasets do not include catches by national distant-water fleets fishing or landing catches elsewhere. As FAO assembles and harmonizes data from various sources, this first-order comparison enabled catches taken elsewhere to be identified and separated from truly domestic (national EEZ) fisheries (see chapter 5 for the spatial layering of reconstructed datasets).

For some countries, such as those resulting from the breakup of the USSR (Zeller and Rizzo 2007) and Yugoslavia (Rizzo and Zeller 2007), this involved using the post-breakup catch fractions to roughly split the pre-breakup catches reported for the USSR and Yugoslavia to generate approximations of the catch time series the newly emerged countries would have reported, had these countries already existed in 1950. In other words, we treat all countries recognized in 2010 (or acting like internationally recognized independent countries with regard to fisheries, e.g., the divided island of Cyprus; Ulman et al. 2015a) by the international community as having existed from 1950 to 2010. This was necessary, given our emphasis on "places," that is, on time series of catches taken from specific ecosystems. This also applies to "overseas territories," many of which were colonies and which have changed status and borders since 1950. Similar reassignments of former-USSR catches is also undertaken for the global tuna and large pelagic dataset (see chapter 3).

For several countries, the baseline was provided by other international bodies. In the case of EU countries, the baseline data originated from the International Council for

BOX 2.1. CATCH RECONSTRUCTIONS: THE CHALLENGES FOR AND OF LOCAL CO-AUTHORS

Aylin Ulman, *Sea Around Us*, University of British Columbia, Vancouver, Canada

Although we always intend for our catch reconstructions to assist them in their work, catch reconstructions are sometimes perceived as a challenge by the staff of national and international agencies that generate or report the catch data that are being thus corrected or complemented. These challenges often lead to noncooperation (e.g., on the part of ICES staff; see Zeller et al. 2011) and in several cases resulted in pressures on colleagues who started as co-authors but ended as anonymous informants (see Belhabib et al. 2012, Ulman et al. 2013).

In some other cases, it was necessary to make concessions regarding sensitive topics for colleagues to remain on board. Thus, the omission of an EEZ or "EEZ-equivalent waters" in the map on page 12 was not done inadvertently. Similarly, we had to create new geographic terms (e.g., "Central Morocco," "South Morocco"; see pp. 335 and 337, respectively), and negotiate regarding the island of Cyprus, separated into "north" and "south" sides since 1974 (see pp. 234 and 235). Two prominent scientists from each side helped in identifying acceptable names, and *north* and *south* were finally agreed upon. In the process, it appeared that even placing an article such as *the* before *north* and *south* or adding the suffix *ern* to the end of *north* and *south* would have been unacceptable. Goodwill triumphed here, as it did in the pair of cross-citing papers (Abudaya et al. 2013; Edelist et al. 2013) presenting reconstructions for the Gaza Strip and Israel (see pp. 273 and 300). Similarly, a Syrian co-author of Ulman et al. (2015) shared his data and profound knowledge on the marine fisheries of his country and remained optimistic throughout multiple exchanges of e-mails, all while his country was engulfed in a horrific civil war. Also, having a (South) Korean team member allowed the catch reconstructions for North and South Korea to rely on information available only in Korean (Shon et al. 2014a, 2014b).

Globally, there were few country reconstructions without a local first or co-author or a first or co-author with field experience in the countries in question. These were mostly small island developing states in the Pacific and the Caribbean (e.g., Ramdeen et al. 2013; Zeller et al. 2015), usually with small catches.

REFERENCES

Abudaya, M., S. Harper, A. Ulman, and D. Zeller. 2013. Correcting mis- and under-reported marine fisheries catches for the Gaza Strip: 1950–2010. *Acta Adriatica* 54(2): 241–252.

Belhabib, D., D. Pauly, S. Harper, and D. Zeller. 2012. Reconstruction of marine fisheries catches for Algeria, 1950–2010. Pp. 1–22 in D. Belhabib, D. Zeller, S. Harper, and D. Pauly (eds.), *Marine fisheries catches in West Africa, 1950–2010*, Part I. Fisheries Centre Research Reports 20(3), University of British Columbia, Vancouver, Canada.

Edelist, D., A. Scheinin, O. Sonin, J. Shapiro, P. Salameh, G. Rilov, Y. Benayhu, D. Schulz, and D. Zeller. 2013. Israel: Reconstructed estimates of total fisheries removals in the Mediterranean, 1950–2010. *Acta Adriatica* 54(2): 253–264.

Ramdeen, R., N. Smith, L. Frotté, S. Lingard, S. Harper, D. Zeller, and D. Pauly. 2013. Reconstructed total catches by the marine fisheries of countries of the wider Caribbean (1950–2010). Pp. 69–75 in *Proceedings of the 65th Meeting of the Gulf and Caribbean Fisheries Institute, Santa Marta, Colombia, 5–9 November 2012*, Vol. 65, Fort Pierce, FL.

Shon, S., S. Harper, and D. Zeller. 2014a. *Reconstruction of marine fisheries catches for the Republic of Korea (South Korea) from 1950–2010*. Fisheries Centre Working Paper #2014-19, University of British Columbia, Vancouver, Canada.

Shon, S., S. Harper, and D. Zeller. 2014b. *Reconstruction of marine fisheries catches for the Democratic People's Republic of Korea (North Korea) from 1950–2010*. Fisheries Centre Working Paper #2014-20, University of British Columbia, Vancouver, Canada.

Ulman, A., Ş. Bekişoğlu, M. Zengin, S. Knudsen, V. Ünal, C. Mathews, S. Harper, D. Zeller, and D. Pauly. 2013. From bonito to anchovy: a reconstruction of Turkey's marine fisheries catches (1950–2010). *Mediterranean Marine Science* 14(2): 309–342.

Ulman, A., A. Saad, K. Zylich, D. Pauly, and D. Zeller. 2015. *Reconstruction of Syria's fisheries catches from 1950–2010: signs of overexploitation*. Fisheries Centre Working Paper #2015-80, University of British Columbia, Vancouver, Canada.

Zeller, D., S. Harper, K. Zylich, and D. Pauly. 2015. Synthesis of under-reported small-scale fisheries catch in Pacific-island waters. *Coral Reefs* 34(1): 25–39.

Zeller, D., P. Rossing, S. Harper, L. Persson, S. Booth, and D. Pauly. 2011. The Baltic Sea: estimates of total fisheries removals 1950–2007. *Fisheries Research* 108: 356–363.

the Exploration of the Sea (ICES), which maintains fisheries statistics by smaller statistical areas, as required by the Common Fisheries Policy of the EU, which largely ignores EEZs. A similar area is the Antarctic continent and surrounding islands, whose fisheries are managed by the Commission for the Conservation of Antarctic Marine Living Resources (CCAMLR), where catches (including discards, a unique feature of CCAMLR) are available by smaller statistical areas (see Ainley and Pauly 2014).

When FAO data are used, care is taken to maintain their assignment to different FAO statistical areas for each country (figure 2.1). The point here is that the FAO statistical areas often distinguish between strongly different ecosystems, such as the Caribbean Sea from the coast of the Eastern Central Pacific in the case of Panama, Costa Rica, Nicaragua, Honduras, and Guatemala.

Step 2: Identification of Missing Sectors, Taxa, and Gear

This step is one where the contribution of local co-authors and experts is crucial (box 2.1). Four fisheries sectors potentially occur in the marine fisheries of a given coastal country, with the distinction between large-scale and small-scale being the most important point (Pauly and Charles 2015):

Industrial sector, consisting of large motorized vessels, requiring large sums for their construction, maintenance, and operation, either domestically, in the waters of other countries or the high seas, and landing a catch that is overwhelmingly sold commercially (as opposed to being consumed or given away by the crew). All gears that are dragged or towed across the seafloor or intensively through the water column (e.g., bottom- and mid-water trawls), no matter the size of the vessel deploying the gear, are here considered *industrial*, following Martín (2012), as are large pirogues (e.g., from Senegal; Belhabib et al. 2014) and "baby trawlers" (in the Philippines; Palomares and Pauly 2014) capable of long-distance fishing (i.e., in the EEZ of neighboring countries). Thus, the industrial sector can also be considered *large-scale and commercial* in nature.

Artisanal sector, consisting of small-scale (e.g., hand lines, gillnets) and fixed gears (e.g., weirs, traps) whose catch is predominantly sold commercially (notwithstanding a small fraction of this catch being consumed or given away by the crew). Our definition of artisanal fisheries relies also on adjacency: They are assumed to operate only in domestic waters (i.e., in their country's EEZ). Within their EEZ, they are further limited to a coastal area to a maximum of 50 km from the coast or to 200 m depth, whichever comes first. This is the area we call the Inshore Fishing Area (IFA; see Chuenpagdee et al. 2006). Note that the definition of an IFA assumes the existence of a small-scale fishery, and thus unpopulated islands (e.g., Kerguelen; Palomares and Pauly 2011) have no IFA, although they may have fisheries in their EEZ (which by our definition are *industrial*). The artisanal sector is thus defined as *small-scale and commercial*. The other small-scale sectors we recognize are subsistence and recreational fisheries, which overlap in many countries.

Subsistence sector, consisting of fisheries that often are conducted by women for consumption by their own families (box 2.2) or by indigenous groups, though with much overlap with artisanal fisheries (box 2.3). Note that we also count as subsistence catch the fraction of the yield of mainly artisanal boats that is given away to the crews' families or the local community (as occurs in the Red Sea fisheries; see Tesfamichael et al. 2012).

Recreational sector, consisting of fisheries conducted mainly for pleasure, although a fraction of the catch may end up being sold or consumed by the recreational fishers and their families and friends (Cisneros-Montemayor and Sumaila 2010). Unless data exist on catch-and-release mortalities in a given country, catch from recreational catch-and-release fisheries are not estimated. Often, fisheries that started out as subsistence (e.g., in the 1950s) changed progressively into recreational fisheries as economic development

BOX 2.2. WOMEN IN FISHERIES: THE GENDER DIMENSION TO RECONSTRUCTING TOTAL CATCHES

Sarah Harper, *Sea Around Us*, University of British Columbia, Vancouver, Canada

Both men and women participate in fishing activities, yet the contribution by women to the total catch is usually underestimated or ignored (Harper et al. 2013; Kleiber et al. 2015). Although it is mainly men who go out to sea to fish, women also fish, usually focusing on the gleaning and capture of invertebrates and small fish on reef flats and in shallow waters. The fishing activities of women, usually for subsistence purposes, are of crucial importance for household and communal food security (FAO 2013), yet they are the most neglected part of the sociopolitically marginalized small-scale fisheries sector (Pauly 2006). Moreover, in many countries women dominate postcapture activities, notably processing and marketing of fish and invertebrates, thus adding value to the catch. However, despite their substantial involvement in fisheries activities and contributions to the economy, women are often excluded from management and policy decisions of this resource (Bennett 2005).

This exclusion is intensified by the tendency of many countries to report to FAO and other international agencies only their catches of commercial and high-value species. Catch reconstructions, designed to correct for various kinds of bias in catch reporting, allow incorporation of the previously overlooked contributions by women, as in Senegal (Belhabib et al. 2014) or the Philippines (Palomares et al. 2014).

For most reconstructions, information on the contribution by women to the marine/estuarine catch in the countries in question was specifically sought, and quantitative estimates were included where available, or computed from proxies. This applied particularly to the small-island countries of the South Pacific (Zeller et al. 2015), where women are very involved in capture activities (Chapman 1987). Furthermore, the contribution by women was accounted for indirectly in the many instances where per capita seafood consumption rates were used in combination with coastal population data (including men, women, and children) to derive subsistence catch estimates. Although this inclusive approach probably accounts for some of the previously "invisible" catch components, the lack of reliable quantitative data on women in fisheries was a major limitation to being able to account fully and in all countries for their contribution to catches. Thus, the contribution by women to fisheries is probably substantial and plays a crucial role in food and income security in fishing communities the world over.

REFERENCES

Belhabib, D., V. Koutob, A. Sall, V. Lam, and D. Pauly. 2014. Fisheries catch misreporting and its implications: the case of Senegal. *Fisheries Research* 151: 1–11.

Bennett, E. 2005. Gender, fisheries and development. *Marine Policy* 29(5): 451–459.

Chapman, M. D. 1987. Women's fishing in Oceania. *Human Ecology* 15(3): 267–288.

FAO. 2013. *Good practice policies to eliminate gender inequalities in fish value chains*. Food and Agriculture Organization of the United Nations (FAO), Rome.

Harper, S., D. Zeller, M. Hauzer, D. Pauly, and U. R. Sumaila. 2013. Women and fisheries: contribution to food security and local economies. *Marine Policy* 39: 56–63.

Kleiber, D., L. M. Harris, and A. C. J. Vincent. 2015. Gender and small-scale fisheries: a case for counting women and beyond. *Fish and Fisheries* 16(4): 547–562.

Palomares, M. L. D., J. C. Espedido, V. A. Parducho, M. P. Saniano, L. P. Urriquia, and P. M. S. Yap. 2014. A short history of gleaning in Mabini, Batangas (Region IV, Subzone B, Philippines). Pp. 118–128 in M. L. D. Palomares and D. Pauly (eds.), *Philippine marine fisheries catches: a bottom-up reconstruction, 1950 to 2010*. Fisheries Centre Research Reports 22(1), University of British Columbia, Vancouver, Canada.

Pauly, D. 2006. Major trends in small-scale marine fisheries, with emphasis on developing countries, and some implications for the social sciences. *Maritime Studies (MAST)* 4(2): 7–22.

Zeller, D. S. Harper, K. Zylich, and D. Pauly. 2015. Synthesis of under-reported small-scale fisheries catch in Pacific-island waters. *Coral Reefs* 34(1): 25–39.

BOX 2.3. INDIGENOUS MARINE FISHERIES: A GLOBAL PERSPECTIVE

Andrés M. Cisneros-Montemayor[a,b] and Yoshitaka Ota[a]

[a]NEREUS Program, University of British Columbia, Vancouver, Canada

[b]Fisheries Economics Research Unit, University of British Columbia, Vancouver, Canada

Indigenous groups around the world include some 370 million people, 5% of the global population, but many live in precarious social and ecological conditions (UN-DESA 2009). For marine coastal indigenous groups, ocean resources are vital to their food security and the maintenance of their cultural heritage, and competition with other user groups has historically been a source of conflict. In the modern context, ecological and political pressures affect fisheries dynamics and governance among indigenous groups that are often already marginalized (see Pruner 2005) but exposed to global market and environmental changes. Moreover, the ongoing growth in users and alternative uses of ocean space has increased the literal and figurative confinement of coastal indigenous groups, jeopardizing their ability to continue fishing and, by proxy, maintaining their cultural identity.

Here, we briefly document an ongoing effort to provide a view of the global scale of the marine fisheries conducted by coastal indigenous people, such as the Haida of northern British Columbia and southern Alaska (Breinig 2001) or the Torres Strait Islanders, Australia (Johannes and MacFarlane 1991). We estimated their current number at around 29 million people in 1,800 distinct communities and 600 unique groups in 80 countries and all continents except Antarctica.

Based on fish catch and consumption data available for 12% of all groups, we estimated that the coastal indigenous peoples currently catch a total of around 1.6 million tonnes per year exclusively for subsistence needs, corresponding to 2% of the global officially reported catch (see chapter 14). The use of group-specific consumption data is crucial to an adequate estimation, because coastal indigenous groups on average consume more than three and up to twenty times as much fish as the other groups in their respective countries. Moreover, subsistence catch is only one component of indigenous fisheries; current data suggest that around 70% of total indigenous catch is sold commercially.

Details on fisheries by the coastal indigenous groups are available from www.seaaroundus.org. We hope that this information will contribute to a wider appreciation of the crucial role of fisheries in the food security and culture of coastal indigenous people.

REFERENCES

Breinig, J. C. 2001. Alaskan Haida narratives: maintaining cultural identity through subsistence. Pp. 19–28 in E. H. Nelson and M. A. Nelson (eds.), *Telling the Stories: Essays on American Indian Literatures and Cultures*. Peter Lang Publishing, New York.

Johannes, R. E., and J. W. MacFarlane. 1991. *Traditional fishing in the Torres Strait islands*. CSIRO Division of Fisheries, Marine Laboratories, Dickson, Australia.

Pruner, J. F. 2005. Aboriginal title and extinguishment not so clear and plain: a comparison of the current Maori and Haida experiences. *Pacific Rim Law and Policy Journal* 14(1): 253–288.

UN-DESA. 2009. *State of the World's Indigenous Peoples*, Vol. 9. United Nations Publications, New York.

increased in a given country and its cash economy grew.

Finally, for all countries and territories, we account for *discards*, here treated as "catch type" (and contrasted to "catch type" *retained landings*), which originate mainly from industrial fisheries for the years 1950 to 2010. Discarded fish and invertebrates are generally assumed to be dead, except for the U.S. fisheries, where the fraction of fish and invertebrates reportedly surviving is generally available on a per species basis (McCrea-Strub 2015). Because of a distinct lack of global coverage of information, we do not account for so-called underwater discards, or net-mortality of fishing gears (e.g., Rahikainen et al. 2004). We also do not address mortality caused by ghost-fishing of abandoned or lost fishing gear (Bullimore et al. 2001; He 2006; Renchen et al. 2010), even though it can be substantial, for example, about 4% of trap-caught crabs worldwide (Poon 2005).

Furthermore, we exclude from consideration all catches of marine mammals, reptiles, corals, sponges, and marine plants (the bulk of the plant material is primarily used not for human consumption but rather for cosmetic or pharmaceutical use). In addition, we do not estimate catches made for the aquarium trade, which can be substantial in some areas in terms of number of individuals but small in overall tonnage, because most aquarium fish are small or juvenile specimens (Rhyne et al. 2012). Note that at least one regional organization (the Secretariat of the Pacific Community, SPC) is coordinating the tracking of catches and exports of Pacific island countries involved in this trade (see Wabnitz and Nahacky 2014). Finally, we do not explicitly address catches destined for the Live Reef Fish Trade (LRFT; see Warren-Rhodes et al. 2003), although, given that these fisheries are often part of normal commercial operations, the catch tonnages of the LRFT are assumed to be addressed in our estimates of commercial catches. Our subsequent estimates of landed

BOX 2.4. WEIRS AND THE GROUND "TRUTH" IN THE PERSIAN GULF

Dalal Al-Abdulrazzak, *Sea Around Us*, University of British Columbia, Vancouver, Canada

Catch reconstruction requires accounting for all catches, by all gears. In the countries around the Persian Gulf, weirs (or stake nets; hadrah in Arabic) made of wooden poles, which are perpendicular to the coast, are generally not properly accounted for and their catch not included in official statistics. However, hadrah are visible from space and thus can be accounted for using Google Earth by combining counts (corrected for low visibility and occasional low map resolution) with scattered reports on their daily catches and fishing season duration to estimate their annual catch from the Gulf (Al-Abdulrazzak and Pauly 2014a). These results, which provided the first example of fisheries catch estimates from space, speak to the potential of satellite technologies for monitoring fisheries remotely, including illegal ones, as some observed hadrah were found operating in areas where they are banned, such as Qatar.

It is surprising that these obvious results should have inspired a critical response. However, Garibaldi et al. (2014) criticized the methods on which they were based and, in particular, attempted to show that the *hadrah* that were only partly visible (e.g., because of sea surface glare) were derelict and should not have been counted. Unfortunately for them, the two examples of "derelict *hadrah*" they provided were a *maskar* in one case [i.e., a different type of trap built of stones parallel to the coast and explicitly excluded from consideration by Al-Abdulrazzak and Pauly (2014a)] and an anchorage in the other case, built by the senior author's father in shallow waters in front of her family's house on the coast of Kuwait (Al-Abdulrazzak and Pauly 2014b).

Several lessons can be learned from this comedy of errors: Ground truthing requires nuanced expertise about a given area, which is not acquired simply by working for an organization with a mandate that happens to include the area in question; critiques alleging that things are "more complicated" (without quantifying the omitted factors and demonstrating that their omission distorts the results in question) do not advance science (see also Cheung et al. 2013); and using Google Earth can advance our knowledge of fisheries.

REFERENCES

Al-Abdulrazzak, D., and D. Pauly. 2014a. Managing fisheries from space: Google Earth improves estimates of distant fish catches. *ICES Journal of Marine Science* 71: 450–454.

Al-Abdulrazzak, D., and D. Pauly. 2014b. Ground-truthing the ground-truth: reply to Garibaldi et al.'s comment on "Managing fisheries from space: Google Earth improves estimates of distant fish catches." *ICES Journal of Marine Science* 71(7): 1927–1931.

Cheung, W. W. L., D. Pauly, and J. L. Sarmiento. 2013. How to make progress in projecting climate change impacts. *ICES Journal of Marine Science* 70: 1069–1074.

Garibaldi, L., J. Gee, T. Sachiko, P. Mannini, and D. Currie. 2014. Comment on: "Managing fisheries from space: Google Earth improves estimates of distant fish catches" by Al-Abdulrazzak and Pauly. *ICES Journal of Marine Science* 71(7): 1921–1926.

value of catches using the global ex-vessel fish price database (Sumaila et al. 2007; Swartz et al. 2013) will therefore undervalue the catch of any taxa destined directly to the LRFT. All the data omissions indicated above are additional factors why our reconstructed total catches are a conservative metric of the impacts of fishing on the world's marine ecosystems.

For any country or territory we check whether catches originating from the above four fishing sectors are included in the reported baseline of catch data, notably by examining their taxonomic composition and any metadata.

The absence of a taxon known to be caught in a country or territory from the baseline data (e.g., cockles gleaned by women on the shore of an estuary; see box 2.2) can also be used to identify a fishery that has been overlooked in the official data collection scheme, as can the absence of reef fishes in the coastal data of a Pacific island state (Zeller et al. 2015). However, to avoid double counting, tuna and other large pelagic fishes, unless known to be caught by a local small-scale fishery (and thus not likely to be reported to a regional fisheries management organization [RFMO]), are not included in this reconstruction step (industrial large pelagic catches are reconstructed using a global approach; see chapter 3).

Finally, if gears are identified in national data or information sources, but a gear known to exist in a given country is not included, then it can be assumed that its catch has been missed, as documented by Al-Abdulrazzak and Pauly (2013) for weirs in the Persian Gulf (see also box 2.4).

Step 3: Sourcing of Available Alternative Information Sources for Missing Data

The major initial source of information for catch reconstructions is governments' (and specifically their Department of Fisheries or equivalent agency) websites and publications, both online and in hard copies. Contrary to what could be expected, it is sometimes not the agency or staff responsible for collecting fisheries data that supplies the catch statistics to FAO but other agencies or staff, such as a statistical office or staff, with the result that much of the granularity of the original data (i.e., catch by sector, by species or by gear) is lost before it reaches Rome. Furthermore, the data request form sent by FAO each year to each country does not actively encourage improvements or changes in taxonomic composition, because the form (an Excel spreadsheet) contains the country's previous years' data in the same composition as submitted in earlier years and requests the most recent year's data. This encourages the pooling of detailed data at the national level into the taxonomic categories inherited through earlier (often decades old) FAO reporting schemes (e.g., Bermuda; Luckhurst et al. 2003). Thus, if we get back to the original data, much of the original granularity can be regained during reconstructions (e.g., Bermuda reconstruction; Teh et al. 2014). A second major source of information on national catches is international research organizations such as FAO, ICES, or SPC, an RFMO such as the Northwest Atlantic Fisheries Organization or CCAMLR (Cullis-Suzuki and Pauly 2010), or current or past regional fisheries development or management projects (many of them launched and supported by FAO), such as the Bay of Bengal Large Marine Ecosystem (BOBLME) Project. All these organizations and projects issue reports and publications describing, sometimes in great detail, the fisheries of their member countries. Another source of information is obviously the academic literature, now widely accessible through Google Scholar.

A good source of information for the earlier decades (especially the 1950s and 1960s) for countries that formerly were part of colonial empires (especially British or French) are the colonial archives in London (British Colonial Office), as well as the Archives Nationales d'Outre-Mer in Aix-en-Provence, and the publications of the Office de la Recherche Scientifique et Technique Outre-Mer (ORSTOM), for the former French colonies. A further good source of information and data is also

BOX 2.5. RECONSTRUCTING SUBSISTENCE FISH CATCH, WITH EMPHASIS ON SOUTHEAST ASIA

Lydia Teh, *Sea Around Us*, University of British Columbia, Vancouver, Canada

Subsistence fishing is fishing with the primary intent of meeting household nutritional needs, as opposed to artisanal (small-scale commercial) fishing, where the primary driver of fishing is the sale of catch for profit (Schumann and Macinko 2007). As it does in most of the world, subsistence fishing occurs throughout Southeast Asia, where it is a traditional source of animal protein for coastal communities (see Firth 1966). However, it tends to be unmonitored and unmanaged throughout most of the region. Consequently, subsistence catches are underrepresented in fisheries statistics kept by national agencies (de Graaf et al. 2011).

The estimation of subsistence catches, which is often gleaned by women and children (see box 2.2), commonly uses one of two methods. One is to estimate total fish consumption in coastal areas, on the assumption that subsistence fishers catch a given fraction of the seafood consumed by the coastal population. Temporal trends in coastal populations can be derived from international databases, whereas fish consumption rates rely on the availability of household socioeconomic surveys or nutritional studies, which are typically harder to find.

The other approach estimates subsistence catch on the basis of the fishing population and catch rates (see Palomares et al. 2014). Anthropological studies on fishing livelihoods can also provide valuable insight on who was fishing, the type of fish that fishers were catching, and the disposition of the catch (Firth 1966; Elliston 1967). They can also be used to infer the composition of reconstructed subsistence catches, which, besides small fishes, often consists of snails, octopus, crabs, and sea urchins and are hardly ever included in official fisheries statistics.

Understanding the social context of fishing is essential for reconstructing subsistence catches, because political, social, and economic conditions affect what and how subsistence fishers operate. Thus, in Vietnam and Cambodia, fishing activities, including for subsistence, were reduced in the 1970s due to war, whereas in Sabah, Malaysia, fishing intensity rose in the 1980s when refugees from the Philippines settled in Sabah's many remote islands and started to earn their livelihood from fishing (Teh et al. 2009). Although subsistence catch is sometimes small, it is important for food security at local scales and thus ought to be accounted for in national fisheries statistics.

REFERENCES

de Graaf, G. J., R. J. R. Grainger, L. Westlund, R. Willmann, D. Mills, K. Kelleher, and K. Koranteng. 2011. The status of routine fishery data collection in Southeast Asia, Central America, the South Pacific, and West Africa, with special reference to small-scale fisheries. *ICES Journal of Marine Science* 68: 1743–1750.

Elliston, G. R. 1967. *The Marine Fishing Industry of Sarawak*. University of Hull, UK.

Firth, R. 1966. *Malay Fishermen: Their Peasant Economy*. Archon Books, The Shoestring Press, Hamden, CT.

Palomares, M. L. D., J. C. Espedido, V. A. Parducho, M. P. Saniano, L. P. Urriquia, and P. M. S. Yap. 2014. A short history of gleaning in Mabini, Batangas (Region IV, Subzone B, Philippines). Pp. 118–128 in M. L. D. Palomares and D. Pauly (eds.), *Philippine marine fisheries catches: a bottom-up reconstruction, 1950 to 2010*. Fisheries Centre Research Reports 22(1), University of British Columbia, Vancouver, Canada.

Schumann, S., and S. Macinko. 2007. Subsistence in coastal fisheries policy: what's in a word? *Marine Policy* 31: 706–718.

Teh, L. S. L., L. C. L. Teh, D. Zeller, and A. Cabanban. 2009. Historical perspective of Sabah's marine fisheries. Pp. 77–98 in D. Zeller and S. Harper (eds.), *Fisheries catch reconstructions: islands*, Part I. Fisheries Centre Research Reports 17(5), University of British Columbia, Vancouver, Canada.

nonfishery sources, including household or nutritional surveys, which can be of great use for estimating unreported subsistence catches (box 2.5). We find the Aquatic Sciences and Fisheries Abstracts and the University of British Columbia library services (and especially its experienced librarians) and its Interlibrary Exchange invaluable for tracking and acquiring such older documents.

Our global network of local collaborators is also crucial in this respect, because they have access to key datasets, publications, and local

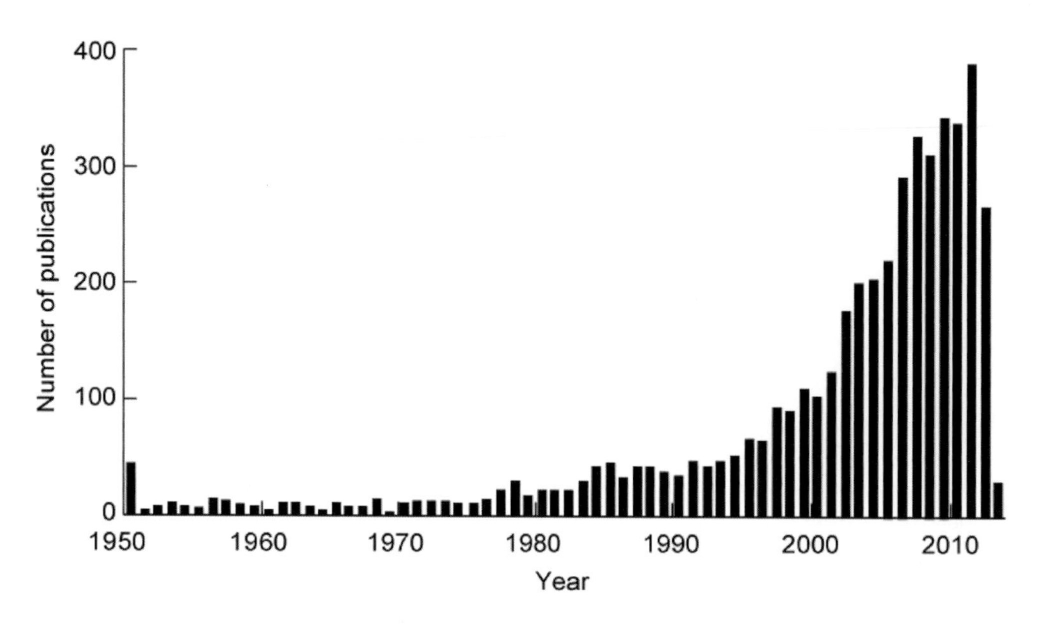

Figure 2.3. Number of publications (scientific and gray literature) and their publication dates used for 110 of the 270 country/territory catch reconstructions presented here. A total of 4,000 publications (excluding personal communications and online sources) were consulted for the 110 reconstructions, resulting in an average of 36 publications being used per reconstruction. The elevated number for 1950 is due to pooling of material dated before 1950 (as far back as the early twentieth or even late nineteenth century) that was used to inform 1950 anchor point information.

knowledge not available elsewhere (box 2.1), often in languages other than English (see below).

Figure 2.3 shows a plot of the publications used for and cited in 110 of the 273 country/territory catch reconstructions presented on pp. 185 to 457 of the atlas against their date of publication. Although recent publications predominate overall, older publications firmly anchor the 1950s catch estimates of many reconstructions. Note also that the data in figure 2.3 imply the use of, on average, about thirty-six publications per reconstruction (not counting online sources and personal communications, including orally provided information, e.g., via interviews or meetings, cited in the text of catch reconstruction reports). Although further details on sources are given in boxes 2.1–2.5, the reconstructions themselves should be consulted for fine-grained information on specific countries or territories. Every reconstruction we undertook is thoroughly documented and published, either in the peer-reviewed scientific literature (e.g., Zeller et al. 2011a, 2011b, 2015; Le Manach et al. 2012), as detailed technical reports in the publicly accessible and search engine–indexed Fisheries Centre Research Reports series (e.g., Zeller and Harper 2009; Harper and Zeller 2011; Harper et al. 2012) or the Fisheries Centre Working Paper series (e.g., Miller and Zeller 2013; Nunoo et al. 2014; Persson et al. 2014; Divovich et al. 2015a), or as reports issued by regional organizations (e.g., BOBLME 2011).

We use this opportunity to mention the issue of language. Some fisheries research groups behave as if something that is not published in English does not exist, and thus they add to the widespread misperception that "there are no data." We take this language bias very seriously in the *Sea Around Us*, which, besides team members who read Chinese, also has or had others who speak Arabic, Danish, Filipino/Tagalog, French, German, Hindi, Japanese, Portuguese, Russian, Spanish, Swedish, and Turkish. To deal with languages none of us master, we hired research assistants who spoke those languages, and we relied on our multilingual network of colleagues and friends throughout the world. Although it is true that English has become the undisputed language of science (Gordin 2015), other languages are used by billions of people, and one cannot assemble knowledge about the fisheries of the world without the willingness and capacity to explore the literature in languages other than English.

Step 4: Development and Expansion of Anchor Points

Anchor points are catch estimates usually pertaining to a single year and sector, often to an area not exactly matching the limits of the EEZ or IFA in question. Thus, an anchor point pertaining to a fraction of the coastline of a given country may need to be expanded to the entire coastline of a country, using fisher or population density, or relative IFA or shelf area as raising factor, as appropriate given the local conditions. In all cases, we are aware that case studies underlying or providing the anchor point data may have a case selection bias (e.g., representing an exceptionally good area or community for study, compared with other areas in the same country) and thus use any raising factors very conservatively. Hence, in many instances we may actually be underestimating any raised catches.

Step 5: Interpolation for Time Periods between Anchor Points

As a social activity involving multiple actors, fishing is very difficult to govern; in particular, fishing effort is difficult to reduce, at least in the short term. Thus, if anchor points are available for years separated by multiyear intervals, it will usually be more reasonable to assume that the underlying fishing activity went on in the intervening years with no data. Strangely enough, our continuity assumption is something that some colleagues are reluctant to make, which is why the catches of, say, small-scale fisheries monitored intermittently, often have jagged time series of reported catches. Exceptions to such continuity assumptions are obvious major environmental impacts such a hurricanes or tsunamis (e.g., cyclones Ofa and Val in 1990–1991 in Samoa; Lingard et al. 2012; hurricane Hugo in 1989 in Montserrat; Ramdeen et al. 2012) or major sociopolitical disturbances, such as military conflicts (e.g., 1989–2003 Liberian civil war; Belhabib et al. 2013), which we explicitly consider with regard to raising factors and the structure of time series. In such cases, our reconstructions mark the event through a temporary change (e.g., decline) in the catch time series (documented in the text of each catch reconstruction), if only to give pointers for future research on the relationship between fishery catches and natural catastrophes or conflicts. As an aside, we note here that the absence of such a signal in the officially reported catch statistics (e.g., a reduction in catch for a year or two) in countries having experienced a major event of this sort (e.g., cyclone Nargis in 2008 in Myanmar) is a sure sign that their official catch data are manufactured, without reference to what occurs on the ground (see also Jerven 2013). Overall, our reconstructions assume—when no information to the contrary is available—that commercial catches (i.e., industrial and artisanal) between anchor points can be linearly interpolated, whereas for noncommercial catches (i.e., subsistence and recreational), we generally use the fisher population trends over time to interpolate between anchor points (via per capita rates).

Radical and rapid effort reductions (or even their attempts) as a result of an intentional policy decision (and actual implementation) do not occur widely. One of the few exceptions that comes to mind is the trawl ban of 1980 in Western Indonesia, whose very partial implementation is discussed in Pauly and Budimartono (2015). The ban had little or no impact on official Indonesian fisheries statistics for Central and Western Indonesia, another indication that they, also, may have little to do with the realities on the ground. FAO (2014; pp. 10–11) hints at this being widespread in the Western Central Pacific and the Eastern Indian Ocean (incidentally the only FAO areas where reported catches appear to keep on increasing) when they note that "while some countries (i.e., the Russian Federation, India and Malaysia) have reported decreases in some years, marine catches submitted to FAO by Myanmar, Vietnam, Indonesia, and

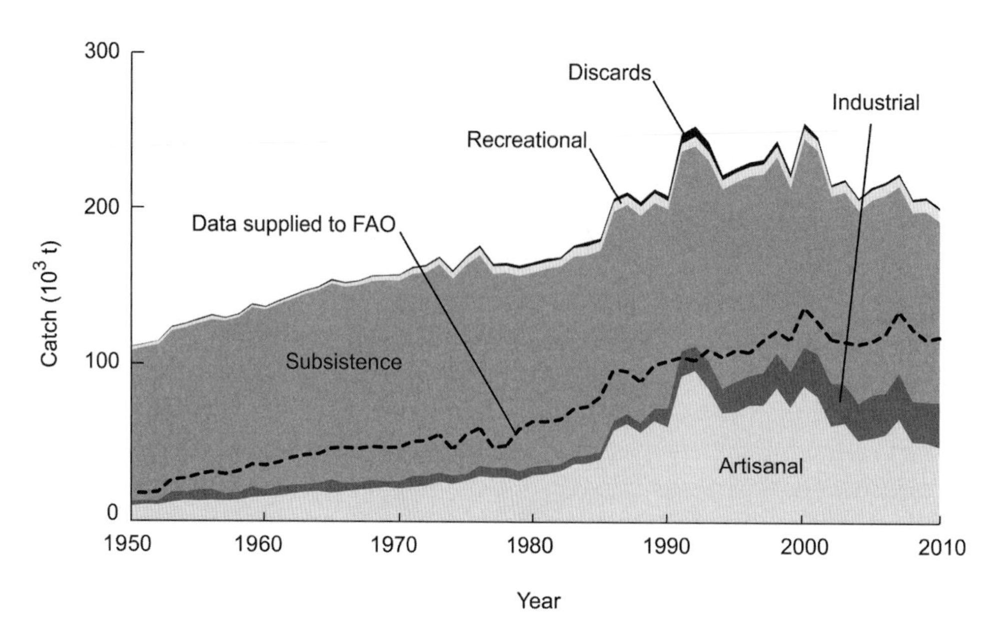

Figure 2.4. Catch reconstruction for the 25 Pacific island countries, states, and territories (Zeller et al. 2015) by the fishing sectors defined here: industrial (large-scale commercial), artisanal (small-scale commercial), subsistence (small-scale noncommercial), and recreational (small-scale noncommercial), with discards shown separately, and the official reported data as presented by FAO on behalf of these entities overlaid as a line graph. This clearly demonstrates the preponderance of commercial catch data in officially reported data as presented by FAO to the global community on behalf of countries. Note that industrial fisheries for large pelagic species (i.e., tuna and billfishes) are excluded from consideration by Zeller et al. (2015) unless they are conducted by truly domestic fleets. These industrial large pelagic fisheries and their catches are addressed separately using a global approach (see chapter 3). (Modified from Zeller et al. 2015.)

China show continuous growth, that is, in some cases resulting in an *astonishing* decadal increase (e.g., Myanmar up 121 percent, and Vietnam up 47 percent)" (emphasis added), which probably is FAO-diplomatic-speak for saying the officially reported data are made up.

Step 6: Estimation of Total Catch Time Series by Combining Steps 1 and 5

A reconstruction is completed when the estimated catch time series derived through steps 2–5 are combined and harmonized with the reported catch of step 1 (see figure 2.4 for a regional synthesis example). Generally, this will result in an increase of the overall catch, but the accounts on pp. 185–457 include several cases when the reconstructed total catch was lower than the reported catch. The best documented case of this situation is that of mainland China (Watson and Pauly 2001), whose overreported catches for local waters in the Northwest Pacific are inflated by underreported catches taken by Chinese distant water fleets, which in the 2000s operated in the EEZs of more than ninety countries, that is, most parts of the world's oceans (Pauly et al. 2014). The step of harmonizing reconstructed catches with the reported baselines obviously goes hand in hand with documenting the entire procedure, which is done via a text that is formally published in the scientific literature or, pending publication, is made available online as either a contribution in the Fisheries Centre Research Reports series (e.g., Harper et al. 2012) or as a Fisheries Centre Working Paper (e.g., Divovich et al. 2015a, 2015b). These documents (rather than only the summaries on pp. 185–457) should be consulted by anyone intending to work with our data. Both the data and the documentation associated with each reconstruction are available at www.seaaroundus.org.

Several reconstructions were performed in the mid- to late 2000s, when official data (i.e., FAO statistics or national data) were not available to 2010 (e.g., Zeller et al. 2006, 2007, 2008; Zeller and Harper 2009). All these cases were subsequently updated to 2010 through dedicated contributions (e.g., Divovich et al.

2015b) or forward-carry procedures (e.g., Zeller et al. 2015) adapted from the approach used for the estimation of missing catches for each country or territory.

Step 7: Quantifying the Uncertainty in Step 6

On several occasions, after having submitted reconstructions to peer-reviewed journals, we were surprised by the vehemence with which referees insisted on a quantification of the uncertainty involved in our reconstructions. We were surprised because catch data, in fisheries research, are never associated with a measure of uncertainty, at least not in the form of anything resembling confidence intervals. In most cases we pointed out that the issue at hand was not one of statistical *precision* (i.e., whether, upon reestimation, we could expect to produce similar results) but of statistical *accuracy*, that is, attempting to eliminate a systematic bias, a type of error that statistical theory does not really address.

However, this is an ultimately frustrating argument, as is the argument that officially reported catch data, despite being themselves sampled and estimated data (e.g., from commercial market sampling [Ulman et al. 2015b] or landing site sampling [Jacquet and Zeller 2007; Jacquet et al. 2010; McBride et al. 2013]) with unknown but potentially substantial margins of uncertainty, are never expected or thought to require measures of uncertainty. Therefore, we applied to all reconstructions summarized in this atlas the procedure in Zeller et al. (2015) for quantifying their uncertainly, which is inspired by the pedigrees' of Funtowicz and Ravetz (1990) and the approach used by the Intergovernmental Panel on Climate Change to quantify the uncertainty in its assessments (Mastrandrea et al. 2010).

This procedure consists of the authors of the reconstructions summarized in this atlas attributing to each catch estimate by fisheries sector (e.g., industrial, artisanal) in each of three periods (1950–1969, 1970–1989, and 1990–2010) a score expressing their evaluation of the quality of the time series: 1, *very low*; 2, *low*; 3, *high*; and 4, *very high*. Note the absence of a "medium" score, to avoid the nonchoice that this easy option would represent. Each of these scores corresponds to a percent range of uncertainty (table 2.1), adapted from Monte Carlo simulations in Ainsworth and Pitcher (2005) and Tesfamichael and Pitcher (2007).

The overall score for the reconstructed total catch of a sector or period can then be computed from the mean of the scores for each sectors, weighted by their catch, and similarly for the relative uncertainty. Alternatively, the percentage uncertainty for each sector and period can be used for a full Monte Carlo analysis (figure 2.5).

Note that this procedure was applied to countries' domestic catches (i.e., data "Layer 1"; see below and chapter 5) but not to foreign catches, whose uncertainty is generally very high and probably exceeds the ranges suggested in table 2.1.

FOREIGN AND ILLEGAL CATCHES

Foreign catches are catches taken by *industrial* vessels (by definition, all foreign fishing in the waters of another country is deemed to be industrial) of a coastal state in the EEZ or EEZ-equivalent waters of another coastal state. Because the high seas legally belong to no one (or to everybody, which is here equivalent), there can be no "foreign" catches in the high seas. Before UNCLOS and the declaration of EEZs by maritime countries, foreign catches were illegal only if conducted within the *territorial* waters of such countries (generally but not always 12 nmi). Since the declarations of EEZs by the overwhelming majority of maritime countries, foreign catches are considered illegal if conducted within the (usually 200 nmi) EEZ and without access agreement with the coastal state (except in the EU, whose waters are managed by a Common Fisheries Policy, which implies a multilateral access agreement).

Such agreements can be tacit and based on historic rights, or more commonly explicit and involving compensatory payment for the coastal state. The *Sea Around Us* has created a database of such access and agreements, which

Table 2.1. Scores for evaluating the quality of time series of reconstructed catches, with their approximate confidence intervals (IPCC criteria from figure 1 of Mastrandrea et al. 2010); the percentage intervals, here updated from Zeller et al. (2015), are adapted from Ainsworth and Pitcher (2005) and Tesfamichael and Pitcher (2007).

Score		± (%)	Corresponding IPCC criteria[a]
4	Very high	10	High agreement and robust evidence
3	High	20	High agreement and medium evidence ***or*** medium agreement and robust evidence
2	Low	30	High agreement and limited evidence ***or*** medium agreement and medium evidence ***or*** low agreement and robust evidence
1	Very low	50	Low agreement and low evidence

[a]Mastrandrea et al. (2010) note that "confidence increases" (and hence confidence intervals are reduced) "when there are multiple, consistent independent lines of high-quality evidence."

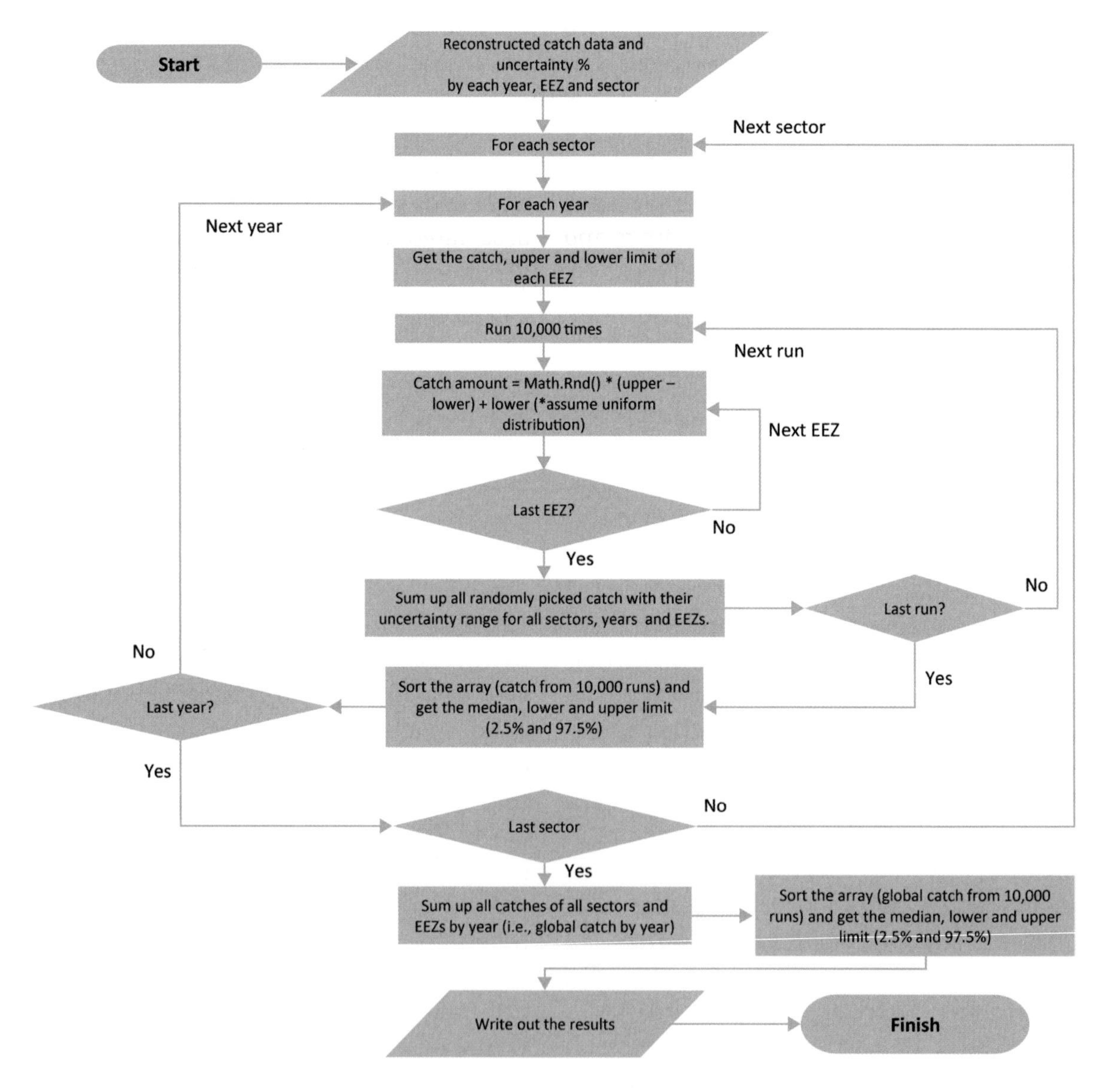

Figure 2.5. Flowchart detailing the manner in which the uncertainty scores (and the corresponding percentage intervals) in table 2.1 can be used to assess the uncertainty associated with the summed catches from a group of EEZs (e.g., those bordering the Mediterranean) or the entire world. MSY, maximum sustainable yield. (Computation and original graph courtesy of V. W. Y. Lam.)

is used to allocate the catches of distant-water fleets to the waters where they were taken (see chapter 5).

Suffice it here to say, therefore, that most catch reconstructions, in addition to identifying the catch of domestic fleets, often at least mention the foreign countries fishing in the waters of the country they cover (information we use in our access database), whereas other reconstructions explicitly quantify these catches (particularly in West Africa; see Belhabib et al. 2012).

This information is then combined and harmonized with the catches deemed to have been taken outside a country's EEZ, as derived in step 1 above and in chapter 5, and the landings of countries reported by FAO as fishing outside the FAO areas in which they are located (e.g., Spain in FAO area 27 reporting catches from area 51; figure 2.1), which always identifies these catches as distant-water landings and thus allows estimation of the catch by foreign fisheries in a given area and even EEZ. Conservative estimates of discards are then added to these foreign landings, estimated from the discarding rates of the fisheries operating in the countries or FAO areas in question. Ultimately, the total catch thus extracted from a given area (i.e., a chunk of EEZ or EEZ-equivalent waters, or high seas waters within a given FAO area) is then computed as the sum of three data layers: the reconstructed domestic catches (what the *Sea Around Us* calls "Layer 1" data, chapter 5), the inferred catches by foreign fleets ("Layer 2" data, chapter 5), and the tuna and other large pelagic fishes caught in the high seas and in EEZs, here treated separately from all other catches as "Layer 3" data (see chapter 3). Details of the harmonization and spatial allocation of these three data layers are presented in chapters 3 and 5.

Besides addressing foreign legal catches (including estimates of discards), we also examine illegal fishing. In line with INTERPOL, we believe that illegal fishing is a crime; here we define it as foreign fishing within the EEZ waters of another country without a formal or traditional permission to access. We do not treat domestic fisheries' violations of fishing regulations as illegal. In general, our reconstruction method cannot readily distinguish between legal and illegal foreign fishing (because we do not necessarily know about all access agreements). By default, our data pertain only to "reported" versus "unreported," irrespective of legal status of foreign fleets in a host country (Pauly et al. 2014; Belhabib et al. 2015). However, for about two dozen countries where the number of illegally operating vessels could be inferred (e.g., West Africa), the fleet size could be multiplied by appropriate catch per unit of effort rates, leading to an estimate of illegal catch, which was then harmonized with other layer 2 data.

In most cases, the magnitude of foreign catches will be highly uncertain (more so than domestic catches); however, the uncertainty associated with foreign catches was not assessed, contrary to domestic catches (see above, step 7).

CATCH COMPOSITION

The taxonomy of catches is what allows catches to be mapped in an ecosystem setting, because different taxa have different distribution ranges and habitat preferences (Close et al. 2006; chapter 4). Also, temporal changes in the relative contributions of different taxa in the catch data may also indicate changes in fishing operations or in dominance patterns in exploited ecosystems. Thus, various ecosystem state indicators can be derived from catch composition data, such as the "mean temperature of the catch," which tracks global warming (Cheung et al. 2013); "stock-status plots," which can provide a first-order assessment of the status of stocks (Kleisner et al. 2013; chapter 1); and the marine trophic index, which reveals instances of "fishing down marine food webs" (Pauly et al. 1998; Pauly and Watson 2005; Kleisner et al. 2014; see also www.fishingdown.org).

Most statistical systems in the world manage to present at least some of their catch in taxonomically disaggregated form (i.e., by

BOX 2.6. SPECIES COMPOSITION AND TAXONOMIC DISAGGREGATION AS CHALLENGES IN CARIBBEAN FISHERIES CATCH RECONSTRUCTIONS

Robin Ramdeen, *Sea Around Us*, University of British Columbia, Vancouver, Canada

For a catch reconstruction to be complete, estimates of catches have to be provided for each sector that exists in a given country (e.g., industrial, artisanal, subsistence, recreational), after which these catches must be split into distinct species or species groups (Pauly 1998; chapter 1). The latter task, which should also include discarded fish and is referred to as taxonomic disaggregation, presented a considerable challenge during the reconstructions of biodiverse tropical fisheries catches, such as those in the wider Caribbean region (e.g., Ramdeen et al. 2012, 2014).

Caribbean fisheries are usually small-scale coastal operations for which data are typically sparse, especially with regard to quantitative information (Salas et al. 2011). This is particularly true for information on the taxonomic composition of catches, which ideally is collected at landing sites through a labor-intensive process requiring fisheries officers well trained in fish identification, because the fisheries are usually multispecific, and land catches consisting of dozens of similar-looking species. Thus, given the scattered and often remote landing sites of Caribbean and other small island developing states, and fisheries departments with limited human resources, these catches are often reported as broad taxa, with the exception of unique and often exported taxa such as conch or spiny lobster (e.g., Ulman et al. 2015).

To taxonomically disaggregate the reconstructed catch estimates, we relied on more detailed catch studies from numerous short-term projects (1–3 years) conducted in the region. This allowed us to split larger groups into species. Also, qualitative information on seafood preferences (by locals and tourists) and export data, notably to the United States, proved useful for catch disaggregation. Finally, when such detailed catch studies related only to recent time periods, we included in a few cases an adjustment to reflect generally known changes in species availability. This applied especially to groupers (family Serranidae), which are highly targeted and highly vulnerable to the fishing techniques used on coral reefs and thus were far more common in catches in earlier time periods than currently.

REFERENCES

Pauly, D. 1998. Rationale for reconstructing catch time series. *EC Fisheries Cooperation Bulletin* 11(2): 4–7.

Ramdeen, R., D. Belhabib, S. Harper, and D. Zeller. 2012. Reconstruction of total marine fisheries catches for Haiti and Navassa Island (1950–2010). Pp. 37–45 in S. Harper, K. Zylich, L. Boonzaier, F. Le Manach, D. Pauly, and D. Zeller (eds.), *Fisheries catch reconstructions: islands*, Part III. Fisheries Centre Research Reports 20(5), University of British Columbia, Vancouver, Canada.

Ramdeen, R., K. Zylich, and D. Zeller. 2014. Reconstruction of total marine fisheries catches for St. Kitts and Nevis (1950–2010). Pp. 129–136 in K. Zylich, D. Zeller, M. Ang, and D. Pauly (eds.), *Fisheries catch reconstructions: islands*, Part IV. Fisheries Centre Research Reports 22(2), University of British Columbia, Vancouver, Canada.

Salas, S., R. Chuenpagdee, A. Charles, and J. C. Seijo (eds.). 2011. *Coastal fisheries of Latin America and the Caribbean*. FAO Fisheries and Aquaculture Technical Paper. No. 544. Food and Agriculture Organization, Rome.

Ulman, A., L. Burke, E. Hind, R. Ramdeen, and D. Zeller. 2015. *Reconstruction of total marine fisheries catches for the Turks and Caicos Islands (1950–2012)*. Fisheries Centre Working Paper #2015-63, University of British Columbia, Vancouver, Canada.

species), but many report a large fraction of their catch as overaggregated, uninformative categories such as "other fish" or "miscellaneous marine fishes" (or "marine fishes nei" [not elsewhere included] in FAO parlance). Interestingly, many official national datasets have better taxonomic resolution than the data reported to FAO by national authorities (see Luckhurst et al. 2003). It is highly likely that this is largely the result of the design of the data request form that FAO distributes to countries each year, which does not actively and easily encourage (even if accompanying instructional material suggests this option) that more detailed national taxonomic resolution data should be provided whenever possible. We aim to reduce the contribution of such overaggregated groups to a reconstruction by using the approach outlined in chapter 1, adapted to fit local conditions (see also box 2.6).

The species and higher taxa in the catch of a given country or territory can thus belong to one of three groups:

1. Species or higher taxa that were already included in the baseline reported data.
2. Species or higher taxa into which over-aggregated catches have been subdivided using two or more sets of catch composition data, such that the changing catch composition data reflect at least some of the observed changes of fishing operations or in the underlying ecosystem.
3. Species or higher taxa into which over-aggregated catches have been subdivided using only one set of catch composition data, which therefore cannot be expected to reflect changes in catch compositions caused by changes in fishing operations or in the underlying ecosystem. This group is also applied in cases where no local or national information on the taxonomic composition was available, and thus a taxonomic resolution from neighboring countries was applied.

We are labeling every taxon in the catch time series of every country with (1), (2), or (3) such that (3) and perhaps also (2) are not used to compute indicators such as outlined above (they would falsely suggest an absence of change), although we fear that this will still occur.

In summary, the approach we developed and used for undertaking the catch reconstructions for every maritime country/territory in the world (as summarized on pp. 185–457, and presented as a regional example in figure 2.4) consists of a well-structured system for using all available data sources and applying a conservative but comprehensive integration approach. With the addition of the recently developed estimation approach for uncertainty (step 7; Zeller et al. 2015), the approach presented here can provide a more nuanced view of fisheries catches (e.g., by fisheries sector; figure 2.4). Verifying and integrating these data into the global *Sea Around Us* database of fisheries catches, followed by spatial allocation of these catches in an ecosystem setting within given political constraints (i.e., EEZ access permissions), is the next step in using global reconstructed catch data. This process is described in chapter 5.

ACKNOWLEDGMENTS

This is a contribution of the *Sea Around Us*, a research activity at the University of British Columbia initiated and funded by the Pew Charitable Trusts from 1999 to 2014 and currently funded mainly by the Paul G. Allen Family Foundation. We are thankful to the many colleagues throughout the world, included Luca Garibaldi of FAO, for answering our questions. We acknowledge the steady support and assistance of the University of British Columbia library system and the associated interlibrary loan facility. Particularly, we thank librarian Sally Taylor for continuous support and expert assistance. Without such extensive library facilities, expertise, and services, this research would not have been feasible.

REFERENCES

Ainley, D., and D. Pauly. 2014. Fishing down the food web of the Antarctic continental shelf and slope. *Polar Record* 50(1): 92–107.

Ainsworth, C. H., and T. J. Pitcher. 2005. Estimating illegal, unreported and unregulated catch in British Columbia's marine fisheries. *Fisheries Research* 75(1–3): 40–55.

Al-Abdulrazzak, D., and D. Pauly. 2013. Managing fisheries from space: Google Earth improves estimates of distant fish catches. *ICES Journal of Marine Science*. 71(3): 450–455.

Belhabib, D., V. Koutob, A. Sall, V. Lam, and D. Pauly. 2014. Fisheries catch misreporting and its implications: the case of Senegal. *Fisheries Research* 151: 1–11.

Belhabib, D., Y. Subah, N. T. Broh, A. S. Jueseah, J. N. Nipey, W. Y. Boeh, D. Copeland, D. Zeller, and D. Pauly. 2013. *When "reality leaves a lot to the imagination": Liberian fisheries from 1950 to 2010*. Fisheries Centre Working Paper #2013-06, University of British Columbia, Vancouver.

Belhabib, D., U. R. Sumaila, V. Lam, E. A. Kane, D. Zeller, P. Le Billon, and D. Pauly. 2015. Euros vs. Yuan: a first attempt at comparing European and Chinese fishing access in West Africa. *PLOS One* 10(3): e0118351.

Belhabib, D., D. Zeller, S. Harper, and D. Pauly (eds.). 2012. *Marine fisheries catches in West Africa, 1950–2010*, Part I. Fisheries Centre Research Reports 20(3), University of British Columbia, Vancouver.

BOBLME. 2011. Fisheries catches for the Bay of Bengal Large Marine Ecosystem since 1950. Report prepared by S. Harper, D. O'Meara, S. Booth, D. Zeller, and D. Pauly (Sea Around Us Project). *Bay of Bengal Large Marine Ecosystem Project*, BOBLME-2011-Ecology-16, Phuket, Thailand.

Bullimore, B. A., P. B. Newman, M. J. Kaiser, S. E. Gilbert, and K. M. Lock. 2001. A study of catches in a fleet of "ghost-fishing" pots. *Fishery Bulletin* 99(2): 247–253.

Cheung, W. W. L., R. Watson, and D. Pauly. 2013. Signature of ocean warming in global fisheries catches. *Nature* 497: 365–368.

Chuenpagdee, R., L. Liguori, M. L. D. Palomares, and D. Pauly. 2006. *Bottom-up, global estimates of small-scale marine fisheries catches*. Fisheries Centre Research Reports 14(8), University of British Columbia, Vancouver.

Cisneros-Montemayor, A. M., and U. R. Sumaila. 2010. A global estimate of benefits from ecosystem-based marine recreation: potential impacts and implications for management. *Journal of Bioeconomics* 12: 245–268.

Claus, S., N. De Hauwere, B. Vanhoorne, P. Deckers, F. Souza Dias, F. Hernandez, and J. Mees. 2014. Marine regions: towards a global standard for georeferenced marine names and boundaries. *Marine Geodesy* 37(2): 99–125.

Close, C., W. W. L. Cheung, S. Hodgson, V. Lam, R. Watson, and D. Pauly. 2006. Distribution ranges of commercial fishes and invertebrates. Pp. 27–37 in M. L. D. Palomares, K. I. Stergiou, and D. Pauly (eds.), *Fishes in databases and ecosystems*. Fisheries Centre Research Reports 14(4), University of British Columbia, Vancouver.

Covey, C. 2000. Beware the elegance of the number zero. *Climatic Change* 44(4): 409–411.

Cullis-Suzuki, S., and D. Pauly. 2010. Failing the high seas: a global evaluation of regional fisheries management organizations. *Marine Policy* 34(5): 1036–1042.

Divovich, E., D. Belhabib, D. Zeller, and D. Pauly. 2015a. *Eastern Canada, "a fishery with no clean hands": marine fisheries catch reconstruction from 1950 to 2010*. Fisheries Centre Working Paper #2015-56, University of British Columbia, Vancouver.

Divovich, E., L. Färber, S. Shon, and K. Zylich. 2015b. *An updated catch reconstruction of the marine fisheries of Taiwan from 1950–2010*. Fisheries Centre Working Paper #2015-78, University of British Columbia, Vancouver.

FAO. 2014. *The State of World Fisheries and Aquaculture (SOFIA) 2014*. Food and Agriculture Organization, Rome.

Funtowicz, S. O., and J. R. Ravetz (eds.). 1990. *Uncertainty and Quality of Science for Policy*. Springer, Kluwer, Dordrecht, the Netherlands.

Gordin, M. D. 2015. *Scientific Babel: How Science Was Done Before and After Global English*. University of Chicago Press, Chicago.

Harper, S., and D. Zeller (eds.). 2011. *Fisheries catch reconstructions: islands*, Part II. Fisheries Centre Research Reports 19(4), University of British Columbia, Vancouver.

Harper, S., K. Zylich, L. Boonzaier, F. Le Manach, D. Pauly, and D. Zeller (eds.). 2012. *Fisheries catch reconstructions: islands*, Part III. Fisheries Centre Research Reports 20(5), University of British Columbia, Vancouver.

He, P. 2006. Gillnets: gear design, fishing performance and conservation challenges. *Marine Technology Society Journal* 40(3): 12.

Jacquet, J. L., H. Fox, H. Motta, A. Ngusaru, and D. Zeller. 2010 Few data but many fish: marine small-scale fisheries catches for Mozambique and Tanzania. *African Journal of Marine Science* 32(2): 197–206.

Jacquet, J. L., and D. Zeller. 2007. National conflict and fisheries: reconstructing marine fisheries catches for Mozambique. Pp. 35–47 in D. Zeller and D. Pauly (eds.), *Reconstruction of marine fisheries catches for key countries and regions (1950–2005)*. Fisheries Centre Research Reports 15(2). University of British Columbia, Vancouver.

Jerven, M. 2013. *Poor Numbers: How We Are Misled by African Development Statistics and What to Do about It*. Cornell University Press, Ithaca, NY.

Kleisner, K., H. Mansour, and D. Pauly. 2014. Region-based MTI: resolving geographic expansion in the Marine Trophic Index. *Marine Ecology Progress Series* 512: 185–199.

Kleisner, K., D. Zeller, R. Froese, and D. Pauly. 2013. Using global catch data for inferences on the world's marine fisheries. *Fish and Fisheries* 14(3): 293–311.

Le Manach, F., C. Gough, A. Harris, F. Humber, S. Harper, and D. Zeller. 2012. Unreported fishing, hungry people and political turmoil: the recipe for a food security crisis in Madagascar? *Marine Policy* 36: 218–225.

Lingard, S., S. Harper, and D. Zeller. 2012. Reconstructed catches for Samoa 1950–2010. Pp. 103–118 in S. Harper, K. Zylich, L. Boonzaier, F. Le Manach, D. Pauly, and D. Zeller (eds.), *Fisheries catch reconstructions: islands*, Part III. Fisheries Centre Research Reports 20(5), University of British Columbia, Vancouver.

Luckhurst, B., S. Booth, and D. Zeller. 2003. Brief history of Bermudian fisheries, and catch comparison between national sources and FAO records. Pp. 163–169 in D. Zeller, S. Booth, E. Mohammed, and D. Pauly (eds.), *From Mexico to Brazil: central Atlantic fisheries catch trends and ecosystem models*. Fisheries Centre Research Reports 11(6), University of British Columbia, Vancouver.

Martín, J. I. 2012. *The small-scale coastal fleet in the reform of the common fisheries policy*. Directorate-General for internal policies of the Union. Policy Department B: Structural and Cohesion Policies. European Parliament. IP/B/PECH/NT/2012_08, Brussels. www.europarl.europa.eu/studies.

Mastrandrea, M. D., C. B. Field, T. F. Stocker, O. Edenhofer, K. L. Ebi, D. J. Frame, H. Held, E. Kriegler, K. J. Mach, P. R. Matschoss, G.-K. Plattner, G. W. Yohe, and F. W. Zwiers. 2010. *Guidance Note for Lead Authors of the IPCC Fifth Assessment Report on Consistent Treatment of Uncertainties. Intergovernmental Panel on Climate Change (IPCC)*. www.ipcc.ch/pdf/supporting-material/uncertainty-guidance-note.pdf.

McBride, M. M., B. Doherty, A. J. Brito, F. Le Manach, L. Sousa, I. Chauca, and D. Zeller. 2013. *Taxonomic disaggregation and update to 2010 for marine fisheries catches in Mozambique*. Fisheries Centre Working Paper #2013-02, University of British Columbia, Vancouver.

McCrea-Strub, A. 2015. *Reconstruction of total catch by U.S. fisheries in the Atlantic and Gulf of Mexico: 1950–2010*. Fisheries Centre Working

Paper #2015-79, University of British Columbia, Vancouver.

Miller, D., and D. Zeller. 2013. *Reconstructing Ireland's marine fisheries catches: 1950–2010.* Fisheries Centre Working Paper #2013-10, University of British Columbia, Vancouver.

Nunoo, F. K. E., B. Asiedu, K. Amador, D. Belhabib, and D. Pauly. 2014. *Reconstruction of marine fisheries catches for Ghana, 1950–2010.* Fisheries Centre Working Paper #2014-13, University of British Columbia, Vancouver.

Palomares, M. L. D., and D. Pauly. 2011. A brief history of fishing in the Kerguelen Islands, France. Pp. 15–20 in S. Harper and D. Zeller (eds.), *Fisheries catch reconstruction: Islands, Part III.* Fisheries Centre Research Reports 19(4), University of British Columbia, Vancouver.

Palomares, M. L. D., and D. Pauly. 2014. Reconstructed marine fisheries catches of the Philippines, 1950–2010. Pp. 137–146 in M. L. D. Palomares and D. Pauly (eds.), *Philippine marine fisheries catches: a bottom-up reconstruction, 1950 to 2010.* Fisheries Centre Research Reports 22(1), University of British Columbia, Vancouver.

Pauly, D. 1998. Rationale for reconstructing catch time series. *EC Fisheries Cooperation Bulletin* 11(2): 4–10.

Pauly, D., D. Belhabib, R. Blomeyer, W. W. L. Cheung, A. Cisneros-Montemayor, D. Copeland, S. Harper, V. Lam, Y. Mai, F. Le Manach, H. Österblom, K. M. Mok, L. van der Meer, A. Sanz, S. Shon, U. R. Sumaila, W. Swartz, R. Watson, Y. Zhai, and D. Zeller. 2014. China's distant water fisheries in the 21st century. *Fish and Fisheries* 15: 474–488.

Pauly, D., and V. Budimartono (eds.). 2015. *Marine fisheries catches of Western, Central and Eastern Indonesia, 1950–2010.* Fisheries Centre Working Paper #2015-61, University of British Columbia, Vancouver.

Pauly, D., and T. Charles. 2015. Counting on small-scale fisheries. *Science* 347: 242–243.

Pauly, D., V. Christensen, J. Dalsgaard, R. Froese, and F. Torres. 1998. Fishing down marine food webs. *Science* 279: 860–863.

Pauly, D., and R. Watson. 2005. Background and interpretation of the "Marine Trophic Index" as a measure of biodiversity. *Philosophical Transactions of the Royal Society: Biological Sciences* 360: 415–423.

Persson, L., A. Lindop, S. Harper, K. Zylich, and D. Zeller. 2014. *Failed state: reconstruction of domestic fisheries catches in Somalia 1950–2010.* Fisheries Centre Working Paper #2014-10, University of British Columbia, Vancouver.

Pikitch, E. K., C. Santora, E. A. Babcock, A. Bakun, R. Bonfil, D. O. Conover, P. Dayton, P. Doukakis, D. L. Fluharty, B. Heneman, E. D. Houde, J. Link, P. A. Livingston, M. Mangel, M. K. McAllister, J. Pope, and K. J. Sainsbury. 2004. Ecosystem-based fishery management. *Science* 305: 346–347.

Poon, A. M.-Y. 2005. *Haunted waters: an estimate of ghost-fishing of crabs and lobsters by traps.* MSc thesis, University of British Columbia, Fisheries Centre, Vancouver.

Rahikainen, M., H. Peltonen, and J. Poenni. 2004. Unaccounted mortality in northern Baltic Sea herring fishery: magnitude and effects on estimates of stock dynamics. *Fisheries Research* 67(2): 111–127.

Ramdeen, R., A. Ponteen, S. Harper, and D. Zeller. 2012. Reconstruction of total marine fisheries catches for Montserrat (1950–2010). Pp. 69–76 in S. Harper, K. Zylich, L. Boonzaier, F. Le Manach, D. Pauly, and D. Zeller (eds.), *Fisheries catch reconstructions: islands,* Part III. Fisheries Centre Research Reports 20(5), University of British Columbia, Vancouver.

Renchen, G. F., S. Pittman, R. Clark, C. Caldow, and D. Olsen. 2010. Assessing the ecological and economic impact of derelict fish traps in the U.S. Virgin Islands. *Proceedings of the 63rd Gulf and Caribbean Fisheries Institute* 63: 41–42.

Rhyne, A. L., M. F. Tlusty, P. J. Schofield, L. Kaufman, J. A. Morris, and A. W. Bruckner. 2012. Revealing the appetite of the marine aquarium fish trade: the volume and biodiversity of fish imported into the United States. *PLoS ONE* 7(5):e35808.

Rizzo, Y., and D. Zeller. 2007. Country disaggregation of catches of former Yugoslavia. Pp. 149–156 in D. Zeller and D. Pauly (eds.),

Reconstruction of marine fisheries catches for key countries and regions (1950–2005). Fisheries Centre Research Reports 15(2), University of British Columbia, Vancouver.

Sumaila, U. R., A. D. Marsden, R. Watson, and D. Pauly. 2007. A global ex-vessel fish price database: construction and applications. *Journal of Bioeconomics* 9: 39–51.

Swartz, W., U. R. Sumaila, and R. Watson. 2013. Global ex-vessel fish price database revisited: a new approach for estimating "missing" prices. *Environmental Resource Economics* 56: 467–480.

Teh, L., K. Zylich, and D. Zeller. 2014. *Preliminary reconstruction of Bermuda's marine fisheries catches, 1950–2010*. Fisheries Centre Working Paper #2014-24, Fisheries Centre, University of British Columbia, Vancouver.

Tesfamichael, D., and D. Pauly (eds.). 2012. *Catch reconstruction for the Red Sea large marine ecosystem by countries (1950–2010)*. Fisheries Centre Research Reports 20(1), University of British Columbia, Vancouver.

Tesfamichael, D., and T. J. Pitcher. 2007. Estimating the unreported catch of Eritrean Red Sea fisheries. *African Journal of Marine Science* 29(1): 55–63.

Ulman, A., B. Çiçek, I. Salihoglu, A. Petrou, M. Patsalidou, D. Pauly, and D. Zeller. 2015a. Unifying the catch data of a divided island: Cyprus's marine fisheries catches, 1950–2010. *Environment, Development and Sustainability* 17(4): 801–820.

Ulman, A., A. Saad, K. Zylich, D. Pauly, and D. Zeller. 2015b. *Reconstruction of Syria's fisheries catches from 1950–2010: signs of overexploitation*. Fisheries Centre Working Paper #2015-80, University of British Columbia, Vancouver.

Wabnitz, C., and T. Nahacky. 2014. *Rapid aquarium fish stock assessment and evaluation of industry best practices in Kosrae*. Federated States of Micronesia. Secretariat of the Pacific Community, Noumea, New Caledonia.

Warren-Rhodes, K., Y. Sadovy, and H. Cesar. 2003. Marine ecosystem appropriation in the Indo-Pacific: a case study of the live reef fish food trade. *Ambio* 32(7): 481–488.

Watson, R., and D. Pauly. 2001. Systematic distortions in world fisheries catch trends. *Nature* 414: 534–536.

Zeller, D., S. Booth, P. Craig, and D. Pauly. 2006. Reconstruction of coral reef fisheries catches in American Samoa, 1950–2002. *Coral Reefs* 25: 144–152.

Zeller, D., S. Booth, G. Davis, and D. Pauly. 2007. Re-estimation of small-scale fishery catches for U.S. flag–associated island areas in the western Pacific: the last 50 years. *Fishery Bulletin* 105(2): 266–277.

Zeller, D., S. Booth, E. Pakhomov, W. Swartz, and D. Pauly. 2011a. Arctic fisheries catches in Russia, USA and Canada: baselines for neglected ecosystems. *Polar Biology* 34(7): 955–973.

Zeller, D., M. Darcy, S. Booth, M. K. Lowe, and S. J. Martell. 2008. What about recreational catch? Potential impact on stock assessment for Hawaii's bottomfish fisheries. *Fisheries Research* 91: 88–97.

Zeller, D., and S. Harper (eds.). 2009. *Fisheries catch reconstructions: islands*, Part I. Fisheries Centre Research Reports 17(5), University of British Columbia, Vancouver.

Zeller, D., S. Harper, K. Zylich, and D. Pauly. 2015. Synthesis of under-reported small-scale fisheries catch in Pacific-island waters. *Coral Reefs* 34(1): 25–39.

Zeller, D., and Y. Rizzo. 2007. Country disaggregation of catches of the former Soviet Union (USSR). Pp. 157–163 in D. Zeller and D. Pauly (eds.), *Reconstruction of marine fisheries catches for key countries and regions (1950–2005)*. Fisheries Centre Research Reports 15(2), University of British Columbia, Vancouver.

Zeller, D., P. Rossing, S. Harper, L. Persson, S. Booth, and D. Pauly. 2011b. The Baltic Sea: estimates of total fisheries removals 1950–2007. *Fisheries Research* 108: 356–363.

NOTE

1. Cite as Zeller, D., and D. Pauly. 2016. Marine fisheries catch reconstruction: definitions, sources, methods and challenges. Pp. 12–33 in D. Pauly and D. Zeller (eds.), *Global Atlas of Marine Fisheries: A Critical Appraisal of Catches and Ecosystem Impacts*. Island Press, Washington, DC.

CHAPTER 3

GLOBAL CATCHES OF LARGE PELAGIC FISHES, WITH EMPHASIS ON THE HIGH SEAS[1]

Frédéric Le Manach,[a,b] Pierre Chavance,[b] Andrés Cisneros-Montemayor,[c] Alasdair Lindop,[a] Allan Padilla,[a] Laurenne Schiller,[a] Dirk Zeller,[a] and Daniel Pauly[a]

[a]*Sea Around Us*, University of British Columbia, Vancouver, BC, Canada

[b]UMR 212 EME, Institut de Recherche pour le Développement, Sète, France

[c]Fisheries Economics Research Unit, University of British Columbia, Vancouver, BC, Canada

For many of the world's coastal regions, the exploitation of large pelagic fishes, particularly for tuna, has a long and significant history (Majkowski 2007). In the North Pacific, Japanese and Native Americans hunted Pacific bluefin tuna (*Thunnus orientalis*) more than five millennia ago (Anonymous 2013b), and people living in the Pacific Islands have fished for tuna at the subsistence level for centuries (SPC 2013). Similarly, in the Atlantic Ocean, especially in the Mediterranean Sea, the exploitation of various tuna species dates back to ancient times. Drawings of tuna have been discovered in the Grotta del Uzzo (c. 9000 BCE) and Genovese caves on Levanzo Island, near Sicily (c. 3000 BCE; Longo and Clarke 2012), and both the Phoenicians and the Romans used a variety of netting methods to capture these fish (Fromentin and Powers 2005; Di Natale 2010). Some of the earliest currencies of these societies depicted Atlantic bluefin tuna (*Thunnus thynnus*) on their coins (Di Natale 2011), and literature and artwork from ancient Greece attempted to portray the importance of these fish and the associated fisheries (Longo and Clarke 2012). Over time, fishing methods diversified, and by the eleventh century traditional trap fishing (*la tonnera*) in the Mediterranean had been established (Longo and Clarke 2012).

THE EXPANSION OF LARGE-SCALE TUNA FISHERIES

For centuries, tuna continued to be targeted by small, localized fisheries using traps and other artisanal methods. Commercial tuna fishing by longline and pole-and-line began around the Pacific Islands in the 1910s and 1920s. However, it was not until after World War II that industrial efforts began to intensify (Miyake 2005; Gillett 2007). Initially, smaller species (e.g., skipjack *Katsuwonus pelamis* and albacore tuna *Thunnus alalunga*) were sought for canning purposes and dried export by locally based but foreign-owned fleets (mainly from the United States and Japan). However, improvements in fishing vessel technology and shipping methods—as well as the development of flash freezing capabilities—precipitated a rapid expansion in the industry, not only in terms of the gears used but also with regard to the species targeted and the regions fished (Gillett 2007; Majkowski 2007).

Around the world, purse-seine catches quickly surpassed those of smaller pole-and-line operations, and a substantial shift in gear occurred in the 1970s. By the early 1980s the

European purse-seine fleet had expanded into the Indian Ocean (Stequert and Marsac 1983, 1989; Fonteneau 1996; Bayliff et al. 2005; Marsac et al. 2014), while purse-seine fleets in the western Pacific, which had simultaneously expanded outward from the Pacific islands and South American countries, began fishing in the eastern Pacific (Majkowski 2007). The next decade saw an even greater increase in purse-seine effort, often in combination with fish aggregating devices (FADs), by these fleets in all three oceans (Bromhead et al. 2003; Miyake 2005; Davies et al. 2014).

FADs exploit the propensity of tunas and other pelagic fishes to gather under floating objects (e.g., rafts formed by uprooted trees after typhoons) and follow them as they drift with the currents. The first stage for fisheries was to search for these "natural FADs"; this was quickly replaced by the use of artificial FADs, made of palm fronds and other natural products, which are either allowed to drift (d-FADs) until sufficient fish had gathered under them (at which point they are harvested) or are anchored (a-FADs) and harvested at set intervals (Floyd and Pauly 1984). The current stage of FAD development involves large concrete or metal contraptions, with electronics that monitor the biomass of fish under them and communicate the results via satellite to purse-seiners who harvest the FADs when the aggregated fish biomass is deemed sufficient (Dagorn et al. 2013).

In tandem with increasing management measures (primarily with regard to the implementation of regional fisheries management organizations [RFMOs]; United Nations 1995), the 1990s saw an increased prevalence of illegal fishing (Majkowski 2007). It was during this time that the use of flags of convenience intensified; also, many smaller coastal countries began chartering vessels from foreign fleets of longliners and purse-seiners (Majkowski 2007).

The high demand for sashimi-grade tuna also led to the onset of tuna "ranching," a capture-based husbandry practice whereby tuna (primarily bluefin but also yellowfin) are taken from wild schools and fattened in large ocean pens before being killed and exported, overwhelmingly to Japan (Miyake et al. 2010). These operations originated in the 1990s in the Mediterranean, where Croatia, Italy, Malta, and Spain are the main countries ranching Atlantic bluefin tuna, and southern Australia, where southern bluefin tuna (*Thunnus maccoyii*) is ranched. They later expanded to the Pacific coast of Mexico, where yellowfin tuna are fattened with locally caught sardine. By the end of the 2000s, around 12,000 t of tuna were ranched annually (FAO 2013b).

Despite tuna fisheries being among the most valuable in the world (FAO 2012), and despite the considerable interest by civil society in the management of large pelagic fishes, to date there are no global and comprehensive spatial datasets or atlases presenting the historical industrial catches of all the tuna and billfish species.

MATERIAL AND METHODS

The aim of this chapter is to present the methods used to produce this comprehensive atlas of the fisheries for large pelagic fishes and to discuss its output and implications.[2] To produce this atlas, we assembled various existing tuna datasets (table 3.1) and harmonized them using a rule-based approach.

For each ocean, the nominal catch data were spatialized according to reported proportions in the spatial data. For example, if France reported 100 t of yellowfin tuna in 1983 using longlines in the nominal dataset, but there were 85 t of yellowfin tuna reported spatially in 1983 by France using longlines, in four separate statistical cells (potentially of varying spatial size), the nominal 100 t for France were split up into those four spatial cells according to their reported proportion of total catch in the spatial dataset. This matching of the nominal and spatial records was done over a series of successive refinements, with the first being the best-case scenario, in which there were matching records for year, country, gear, and species. The last refinement was the worst-case scenario, in which there were no

Table 3.1. Overview of the various data sources used to create global catch maps of industrially caught tuna and other large pelagic fishes.

Ocean	RFMO	Sources		Spatial resolution	Countries/ gear/species
		Nominal catch	Spatialized catch		
Atlantic	ICCAT	ICCAT website	ICCAT website	1° × 1°, 5° × 5°, 5° × 10°, 10° × 10°, 10° × 20°, 20° × 20°	114/48/142
Indian	IOTC	IOTC website	IOTC website	1° × 1°, 5° × 5°, 10° × 10°, 10° × 20°, 20° × 20°	57/35/45
Eastern Pacific	IATTC	IATTC website	FAO Atlas of Tuna and Billfish Catches	5° × 5°[a]	28/11/19
Western Pacific	WCPFC	WCPFC website	WCPFC website	5° × 5°	41/9/9
Southern[b]	CCSBT[b]	Provided by CCSBT staff	CCSBT website	5° × 5°	11/8/1

[a]A number of these cells were straddling the Pacific and Atlantic Oceans. Their total catch was split into these two ocean basins, proportionately to the surface of the cell included in each ocean. This step was then corrected for biological distributions (based on www.fishbase.org); catches of both Atlantic bluefin tuna and Atlantic white marlin that were obviously wrongly allocated to the Pacific Ocean were reallocated back to the Atlantic Ocean.
[b]This RFMO covers all three oceans but deals only with southern bluefin tuna. Note that the other RFMOs also sometimes report this species (which we account for to avoid double-counting).
Acronyms: CCSBT, Commission for the Conservation of Southern Bluefin Tuna; FAO, Food and Agriculture Organization; IATTC, Inter-American Tropical Tuna Commission; ICCAT; International Commission for the Conservation of Atlantic Tunas; IOTC, Indian Ocean Tuna Commission; RFMO, regional fisheries management organization; WCPFC, Western and Central Pacific Fisheries Commission.

matching records except for the year of catch. For example, if France reported 100 t of yellowfin tuna caught in 1983 using longlines, but there were no spatial records for any country catching yellowfin tuna in 1983, the nominal 100 t for France was split up into spatial cells according to their reported proportion of total catch of any species and gear in 1983. After each successive refinement, the matched and nonmatched records were stored separately, so that at each new refinement, only the previous step's nonmatched records were used. The matched database was added to at the end of each step. The end result was a catch baseline database containing all matched and spatialized catches records, which sum up to the original nominal catch.

The catches thus assigned to the various sized tuna cells (1° × 1° to 20° × 20°; table 3.1) were then spatially reallocated to the standard 0.5° × 0.5° cells used by the *Sea Around Us* following the procedure described in chapter 5. All artisanal catches (i.e., any gear other than industrial-scale longlines, purse-seines, and pole-and-lines,[3] as well as offshore gillnets) were reallocated to the EEZs of origin of the fleet, as the *Sea Around Us* defines artisanal fleets as being restricted to domestic inshore areas (chapter 2). Here, only the industrial catches are presented.

Finally, a review of the literature was performed for each ocean to collect estimates of discards. Because of the limited amount of country- and fleet-specific data that this search yielded, it was decided that discard percentages should be averaged across the entire time period and applied to the region of origin of the fleet (e.g., East Asia or Western Europe), rather than the country of origin of the fleet. Similarly to the spatialization step described above, successive refinements were then performed to add discards to all reported catch.

For the Atlantic Ocean, we identified the lowest discard rate as 1.1% (longlines, North America; ICCAT 2009), the highest as 100% (longlines targeting swordfish; European Commission 2011), and the median discard rate as 10.7% (purse-seine, western and northern Europe; Amandè et al. 2011). For the Indian Ocean, we identified the lowest discard rate as 1.5% (gillnet, Asia; Shahifar 2012), the highest as 113% (longline; Alverson et al. 1994), and the median discard rate as 7.2% (longline, Asia; IOTC 2000). For the Pacific Ocean, we

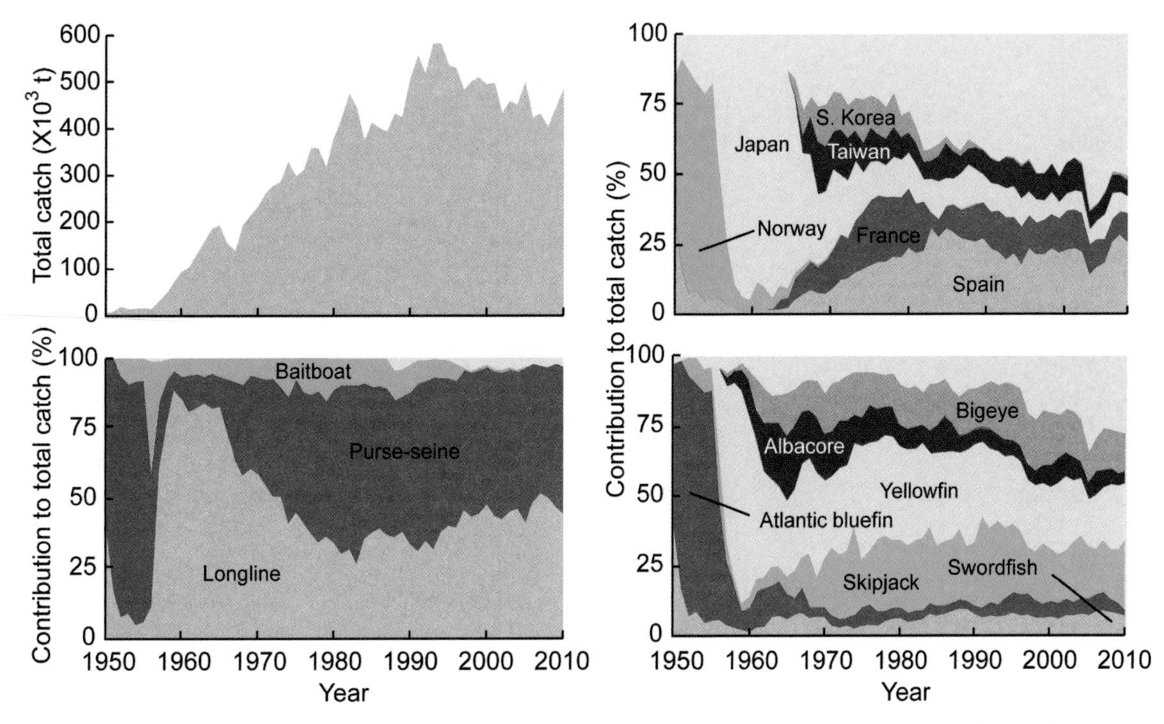

Figure 3.1. Industrial catch of large pelagic species in the Atlantic Ocean, 1950–2010. The top left panel shows nominal catches (without bycatch), whereas the top right, bottom left, and bottom right panels show this total catch as percentage contributions disaggregated by country, gear, and species, respectively. Gray areas are "other."

identified the lowest discard rate as 0.4% (pole-and-line; Kelleher 2005), the highest as 75% (longline, Federated States of Micronesia; Bailey et al. 1996), and the median discard rate as 20.5% (longline, Fiji, Hawaii; Lawson 1997; Kelleher 2005; WPRFMC 2013).

INDUSTRIAL TUNA FISHERIES OF THE ATLANTIC OCEAN

Industrial catches in the Atlantic Ocean steadily increased from very low values in 1950 to a high of almost 600,000 t/year in the mid-1990s (figure 3.1). They subsequently declined to around 400,000 t/year by the mid-2000s before rebounding beyond 500,000 t by 2012 (figure 3.1).

Longline catches became prevalent in the 1960s with the arrival of Japanese vessels in the Atlantic, but their contribution to the total catch decreased over time (figure 3.1). In the early 1980s, their contribution began to increase again to reach about 50% of the total catch by the mid-2000s. However, this relative increase was essentially caused by the migration of European purse-seiners to the Indian Ocean (Stequert and Marsac 1983, 1989; Fonteneau 1996; Bayliff et al. 2005; Marsac et al. 2014), as well as the overall decline of the EU fleet and number of fishing access agreements (Anonymous 2005, 2013a).

Purse-seiners (targeting skipjack and yellowfin tunas) are the second major gear in terms of catch (figure 3.1). Apart from a high contribution in the early 1950s (mostly from Norway), the purse-seine fleet only truly began expanding in the late 1960s and 1970s with the development of the French and Spanish fleets off West Africa, following the decreasing catches of albacore tuna in the Bay of Biscay (Chauveau 1989; Fonteneau et al. 1993; Sahastume 2002). It is noteworthy that the fleet of purse-seiners in the Atlantic Ocean has been increasing again since the late 2000s, with some vessels coming back from the Indian Ocean to avoid Somali piracy (Anonymous 2013a).

INDUSTRIAL TUNA FISHERIES OF THE INDIAN OCEAN

Industrial fisheries started in 1952 with the arrival of Japanese longliners (figure 3.2). By the mid-1950s, with Taiwan joining in,

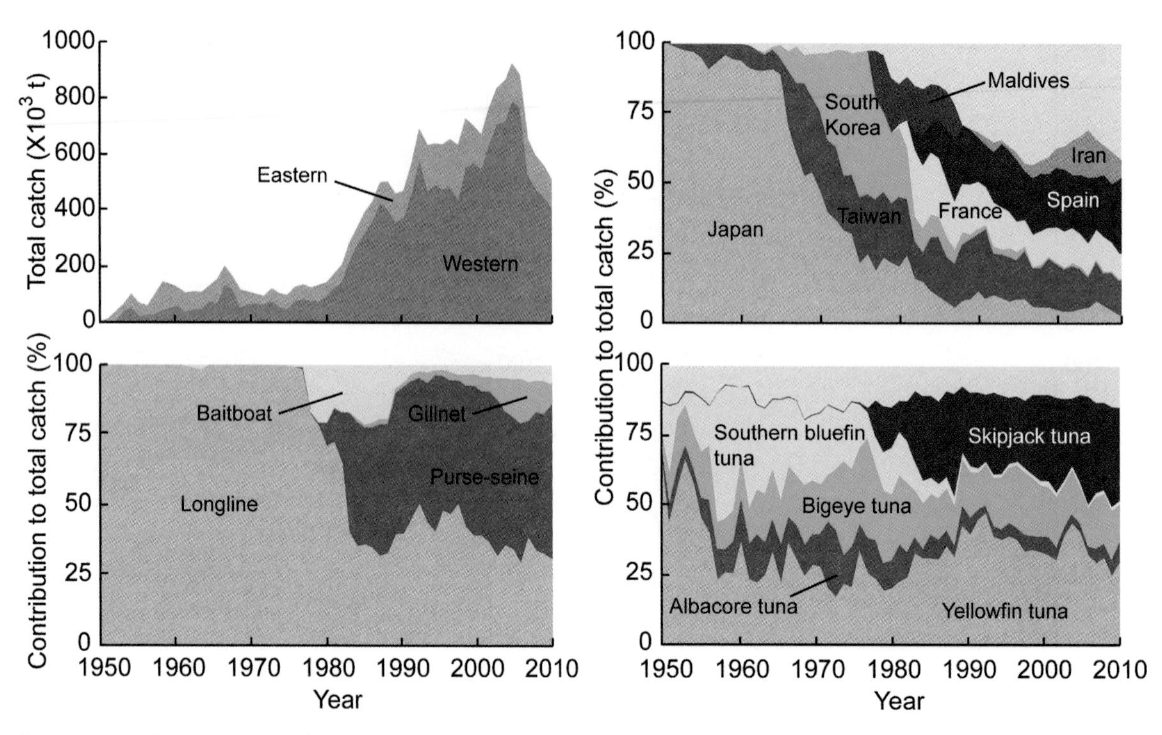

Figure 3.2. Industrial catch of large pelagic species in the Indian Ocean, 1950–2010. The top left panel shows nominal catches (without bycatch) by FAO area (51: western Indian Ocean; 57: eastern Indian Ocean), whereas the top right, bottom left, and bottom right panels show this total catch as percentage contributions disaggregated by country, gear, and species, respectively. Gray areas are "other."

catches reached 100,000 t/year. Until the migration of the European purse-seiners in the early 1980s, catches slightly fluctuated, but they always remained below 200,000 t/year (figure 3.2). Thereafter, catches increased to 600,000 t/year by the mid-1990s, and then again to more than 900,000 t/year by the mid-2000s (mostly because of the expansion of the Iranian gillnet fleet). Since then, industrial catches have steadily declined, at least in part because of the effects of Somali piracy in the region (Chassot et al. 2010; Martín 2011; Anonymous 2014).

With the expansion of the Japanese fleet in the 1950s through the 1970s, followed by the arrival of European purse-seiners, the contribution of the western Indian Ocean has consistently increased over time, from zero in 1950 to a stable level of about 80% after 1990 (figure 3.2). Focusing on the purse-seine fleets, FADs have been used since the beginning of this fishery in the Indian Ocean (Stequert and Marsac 1983, 1989; Davies et al. 2014). Drifting FADs, developed in the early 1990s (Bayliff et al. 2005), are now used predominantly in the northern area in the second half of the year but less in the Mozambique Channel because of the high occurrences of natural rafts (Davies et al. 2014).[4] About 55% of the total purse-seine catch was taken under d-FADs in the early 1980s, increasing to 74% in the late 2000s and early 2010s.

INDUSTRIAL TUNA FISHERIES OF THE PACIFIC OCEAN

The industrial tuna fisheries of the Pacific Ocean are currently among the most profitable fisheries in the world. Their main targets are skipjack, yellowfin, bigeye, and albacore tunas, but two species, yellowfin and skipjack, have contributed more than 79% of the catch by weight from 1950 to 2010 (figure 3.3). However, on a per-kilogram basis, the most valuable species caught in the Pacific are bigeye and bluefin (Majkowski 2007; Williams and Terawasi 2011).

At present, the majority of the catch is acquired through the use of purse-seines targeting yellowfin and skipjack for canning (Lehodey et al. 2011; Hall and Roman 2013;

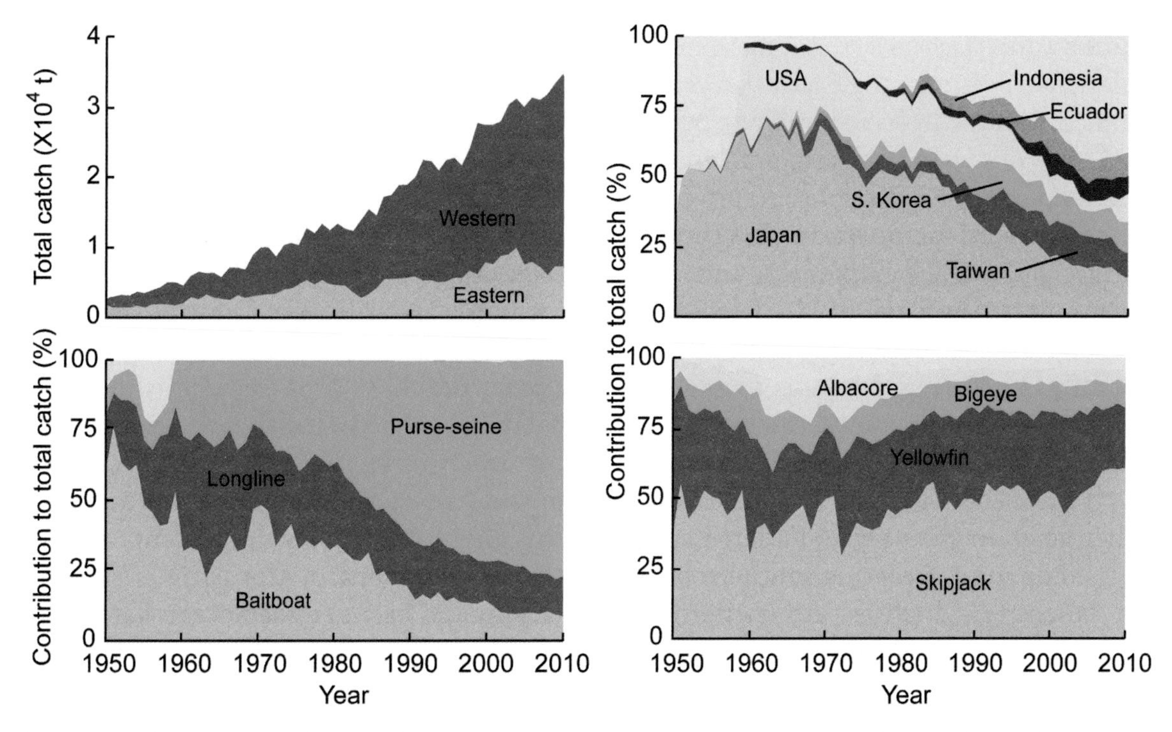

Figure 3.3. Industrial catch of large pelagic species in the Pacific Ocean, 1950–2010. The top left panel shows nominal catches (without bycatch) by area ("Western," WCPFC; "Eastern," IATTC), whereas the top right, bottom left, and bottom right panels show this total catch as percentage contributions disaggregated by country, gear, and species, respectively. Gray areas are "other."

Sumaila et al. 2014). Specifically, purse-seiners are responsible for more than 70% of the tuna caught in the Pacific Ocean (figure 3.3), and purse-seining with the use of d-FADs massively increased (Hall and Roman 2013; IATTC 2013b). Historically, distant-water fleets from Japan, Korea, Taiwan, and the United States were the main purse-seining operations in the Pacific Ocean, with fleets from, for example, China, Ecuador, El Salvador, New Zealand, and Spain becoming more prevalent in the region since the 2000s. In addition, since the late 1980s, Pacific island–based purse-seine fleets have steadily increased in number, and in 2010 there were seventy-eight locally based purse-seine vessels in the western part of the Pacific Ocean (Williams and Terawasi 2011).

Longlines are the second most common gear in the Pacific Ocean, contributing about 20% of the catch (figure 3.3). All longline fleets target primarily mature bigeye and yellowfin (which are flash frozen, then thawed for sale as fresh sashimi), as well as some albacore for canning (WCPFC 2011; Sumaila et al. 2014). Fleets from Asia, South America, and Spain have also targeted swordfish in the eastern part of the Pacific Ocean since the 1950s. Between 2000 and 2010, the total annual catch of this species averaged more than 800 t (1.8% of the total catch).

The prevalence of pole-and-line fishing (i.e., baitboat, figure 3.3) has decreased significantly over the last three decades (from more than 70% of the total catch in the early 1950s to about 10% in the 2000s), largely as a result of the expansion of purse-seining (Williams and Terawasi 2011). Nonetheless, this type of surface fishing remains a seasonal venture for Australia, Fiji, and Hawaii (domestic fleets), as well as Japan (both distant-water and domestic fleets) and a year-round fishery for domestic vessels from Indonesia, the Solomon Islands, and French Polynesia (Amoe 2005; Langley et al. 2010; WPRFMC 2013). Skipjack is the primary species landed by pole-and-line (81.4% of the catch).

THE SOUTHERN BLUEFIN TUNA FISHERY

Unlike the other tuna species, which form distinct populations in each ocean, southern bluefin tuna (*Thunnus maccoyii*) has a circumpolar distribution range between 8°S and 60°S, although it occurs mainly between 30°S and 50°S (www.fishbase.org). Initially, Australia began fishing for southern bluefin in the southern Pacific Ocean with the use of surface gears in the early 1950s, and in 1965 Japan entered the fishery with a longline fleet (CCSBT 2011; Polacheck 2012). Currently, nine countries (including the EU as a single entity) target southern bluefin; however, Japan and Australia are responsible for most of the catch of southern bluefin. Despite the use of other gears in the past and a large surface gear component targeting southern bluefin for ranching, longlines are currently the primary gear used to catch southern bluefin tuna (CCSBT 2011).

THE INDUSTRIAL CATCH OF LARGE PELAGIC FISHES: GLOBAL MAPS

The database described above allows mapping of the global catch of pelagic fishes (mainly tuna, as well as billfishes, and including bycatch and discards) by industrial fisheries, by year, from 1950 to 2010. As an example, we present a catch map representing this catch, averaged for the 11 years from 2000 to 2010 (figure 3.4).

CONCLUSION

This contribution introduced the first harmonized and spatially complete database of global large pelagic fisheries for all large pelagic taxa, including an estimate of discards. Until now, only regional (RFMO) or incomplete (e.g., the FAO Atlas of Tuna and Billfish Catches) databases existed, thus providing a truncated picture of these highly interconnected and global fisheries. Therefore, the interest of this new database lies in its ability to show the development of the various fisheries within and between each ocean basin (i.e., a clear advantage of scaling up), despite its preliminary nature. Notably, figures 3.1, 3.2, and 3.3, pertaining to the Atlantic, Indian, and Pacific Oceans, respectively, show an ominous progression, with catches peaking in the Atlantic in 1994 and in the mid-2000s in the Indian Ocean, but still increasing—until when?—in the Pacific Ocean.

Several points are perfectible, and the total catch is still thought to be incomplete and will have to be fixed in future iterations:

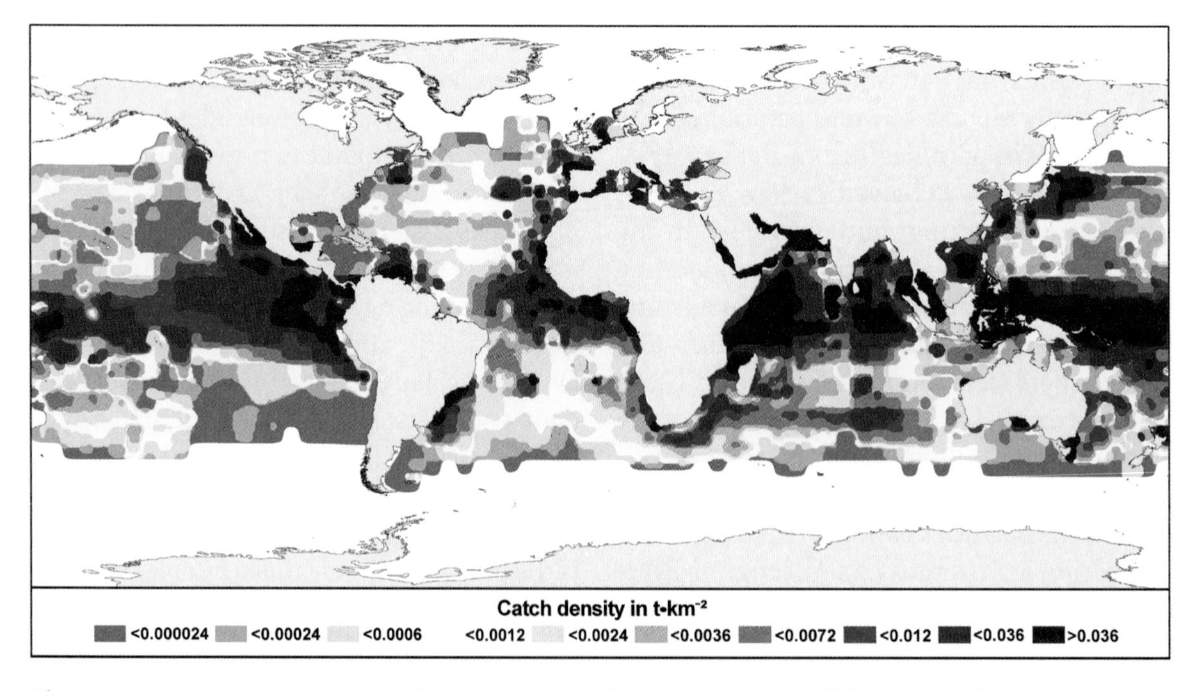

Figure 3.4. Average annual catches (t/km^2) of large pelagic species (tuna and billfishes, as well as associated bycatch and discards) for 2000–2010 as derived for the present database and spatially assigned to the *Sea Around Us* 0.5° × 0.5° cells.

- The Inter-American Tropical Tuna Commission (IATTC) posed some data problems by not releasing the spatialized catches that they were supposed to release in 2014. Our temporary solution was to apply the average spatial data of all other gears to the longline fleets, but this was only a stopgap measure. We hope that spatialized IATTC data will become available in the future, which will then improve mapping of tuna catches in the northeast Pacific.
- The ICCAT nominal catch database contains some qualitative geographic information (i.e., "subareas"), which are apparently not geographically defined. Thus, we could not use them to refine our coarse spatialization. If these subareas were to become geographically defined, we could force the allocation of the nominal catch to specific subareas rather than to the entire ICCAT area.
- Our FAD versus free-school breakdown could be improved, using the actual spatial and annual breakdown from the spatialized database rather than applying only the annual breakdown. This gear distinction is made only by the Indian Ocean Tuna Commission (IOTC) and ICCAT. We will later dedicate resources to this issue; in the meantime, our stopgap measure resulted in the same spatial allocation for both FAD and free-school catches.
- Discard rates presented here account for only a subset of the literature, and we encountered some initial difficulties in harmonizing all of them. Feedback from worldwide experts could allow us to refine these rates by integrating a rule-based approach by gear and country to our discard estimation.
- Finally, we also imagine that improved spatial assignment rules and other global databases such as www.fishbase.org could be used to refine our spatial distribution of the catch by, for example, restricting species to certain areas.

Thus, this chapter should be seen as version 1.0, and we hope it will trigger some interest in the community, ultimately resulting in the involvement of experts through feedback to improve our set of assumptions.

ACKNOWLEDGMENTS

This is a contribution of the *Sea Around Us*, a research activity at the University of British Columbia initiated and funded by the Pew Charitable Trusts from 1999 to 2014 and currently funded mainly by the Paul G. Allen Family Foundation. We are thankful to the many colleagues throughout the world for answering our questions and providing information. We acknowledge the steady support and assistance of the University of British Columbia library system and the associated interlibrary loan facility.

REFERENCES

Alverson, D. L., M. H. Freeberg J. G. Pope, and S. A. Murawski. 1994. *A global assessment of fisheries by-catch and discards*. FAO Fisheries Technical Papers T339, Rome.

Amandè, M. J., J. Ariz, E. Chassot, P. Chavance, A. Delgado de Molina, D. Gaertner, H. Murua, R. Pianet, and J. Ruiz. 2011. By-catch and discards of the European purse seine tuna fishery in the Atlantic Ocean: estimation and characteristics for 2008 and 2009. *Collective Volume of Scientific Papers: ICCAT* 65(5): 2113–2120.

Amoe, J. 2005. *Fiji tuna and billfish fisheries* [FR WP-12]. 1st Meeting of the Scientific Committee of the Western and Central Pacific Fisheries Commission WCPFC-SC1, Western and Central Pacific Fisheries Commission (WCPFC), Kolonia, Pohnpei State, Federated States of Micronesia, Noumea, New Caledonia.

Anonymous. 2005. *Specific Convention SC 12: the European tuna sector economic situation, prospects and analysis of the impact of the liberalisation of trade*. Framework Contract for performing evaluations, impact analyses and monitoring services in the context of Fisheries Partnership Agreements concluded between the community and non-member coastal states. Project FISH/2003/02, Oceanic Développement, Poseidon Aquatic Resource Management Ltd, and MegaPesca Lda, Brussels.

Anonymous. 2013a. *Contrat spécifique no. 5: revue des pêcheries thonières dans l'océan Atlantique Est. Rapport final*. Contrat cadre MARE/2011/01, lot 3: Évaluations rétrospectives et prospectives relatives à la dimension internationale de la politique commune des pêches. Cofrepeche, MRAG, Poseidon Aquatic Resource Management Ltd, NFDS, Brussels.

Anonymous. 2013b. *The Story of Pacific Bluefin*. The Pew Environment Group, Washington, DC.

Anonymous. 2014. *Specific contract no. 7: review of tuna fisheries in the western Indian Ocean*. Framework contract MARE/2011/01: Evaluation and impact assessment activities for DG MARE, lot 3: Retrospective and prospective evaluations on the international dimension of the Common Fisheries Policy: final report. COFREPECHE, MRAG, Poseidon Aquatic Resource Management Ltd, NFDS, Brussels.

Bailey, K., P. G. Williams, and D. Itano. 1996. *By-catch and discards in Western central Pacific tuna fisheries: a review of SPC data holdings and literature*. Oceanic Fisheries Programme Technical Report no. 34. Secretariat of the Pacific Community, Noumea, New Caledonia.

Bayliff, W. H., J. Ignacio de Leiva Moreno, and J. Majkowski. 2005. Management of tuna fishing capacity: conservation and socio-economics. Second Meeting of the Technical Advisory Committee of the FAO Project, Madrid (Spain), March 15–18, 2004. *FAO Fisheries Proceedings* 2, Food and Agriculture Organization of the United Nations (FAO), Rome.

Bromhead, D., J. Foster, R. Attard, J. Findlay, and J. Kalish. 2003. *A review of the impact of fish aggregating devices (FADs) on tuna fisheries: final report to Fisheries Resources Research Fund*. Commonwealth Department of Agriculture, Fisheries and Forestry, Bureau of Rural Sciences, Canberra, Australia.

CCSBT. 2011. *Report of the Sixteenth Meeting of the Scientific Committee*. July 19–28, 2011, Commission for the Conservation of Southern Bluefin Tuna, Bali, Indonesia.

Chassot, E., P. Dewals, L. Floch, V. Lucas, M. Morales-Vargas, and D. Kaplan. 2010. *Analysis of the effects of Somali piracy on the European tuna purse seine fisheries of the Indian Ocean*. IOTC-2010-SC-09, Indian Ocean Tuna Commission, Victoria, Seychelles.

Chauveau, J.-P. 1989. Histoire de la pêche industrielle au Sénégal et politiques d'industrialisation, 2ème partie: l'essor thonier et les limites d'une politique nationale d'industrialisation de la pêche (de 1955 aux premières années de l'Indépendance). *Cahiers des Sciences Humaines* 25(1–2): 259–275.

Dagorn, L., K. N. Holland, V. Restrepo, and G. Moreno. 2013. Is it good or bad to fish with FADs? What are the real impacts of the use of drifting FADs on pelagic marine ecosystems? *Fish and Fisheries* 14(3): 391–415.

Davies, T. K., C. C. Mees, and E. J. Milner-Gulland. 2014. The past, present and future use of drifting fish aggregating devices (FADs) in the Indian Ocean. *Marine Policy* 45: 163–170.

Di Natale, A. 2010. The eastern Atlantic bluefin tuna: entangled in a big mess, probably far from a conservation red alert. Some comments after the proposal to include bluefin tuna in CITES Appendix I. *Collective Volume of Scientific Papers: ICCAT* 65(3): 1004–1043.

Di Natale, A. 2011. *The iconography of tuna traps: an essential information for the understanding of the technological evolution of this ancient fishery*. ICCAT GBYP Symposium on Tuna Trap Fishery for Bluefin Tuna, May 23–25, 2011, Tangier, Morocco, International Commission for the Conservation of Atlantic Tunas (ICCAT), Madrid.

European Commission. 2011. *Impact assessment of discard reducing policies: EU discard annex*. Studies in the field of the Common Fisheries Policy and Maritime Affairs, lot 4: impact assessment studies related to the CFP, European Commission.

FAO. 2012. *The state of world fisheries and aquaculture*. Food and Agriculture Organization of the United Nations (FAO), Rome.

FAO. 2013a. *Atlas of tuna and billfish catches*. Fishery Statistical Collections (online; updated June 12, 2013; extracted June 27, 2013), Food and Agriculture Organization of the United Nations (FAO), Rome.

FAO. 2013b. FishStatJ: software for fishery statistical time series. V2.1.1. Food and Agriculture Organization of the United Nations (FAO), Rome.

Filmalter, J. D., M. Capello, J.-L. Deneubourg, P. D. Cowley, and L. Dagorn. 2013. Looking behind the curtain: quantifying massive shark mortality in fish aggregating devices. *Frontiers in Ecology and the Environment* 11(6): 291–296.

Floyd, J., and D. Pauly. 1984. Smaller size tuna around the Philippines: can fish aggregating devices be blamed? *Infofish Marketing Digest* 5(84): 25–27.

Fonteneau, A. 1996. Panorama de l'exploitation des thonidés dans l'océan Indien. Pp. 50–74 in P. Cayré and J. Y. Le Gall (eds.), *Le thon, enjeux et stratégies pour l'océan Indien*. Proceedings of the International Tuna Conference, November 27–29, 1996, Maurice. Indian Ocean Commission (IOC) and Office de la Recherche Scientifique et Technique Outre-Mer (ORSTOM), Maurice.

Fonteneau, A. 1997. *Atlas of tropical tuna fisheries: world catches and environment*. Office de la Recherche Scientifique et Technique Outre-Mer (ORSTOM), Paris.

Fonteneau, A. 2009. *Atlas of Atlantic Ocean Tuna Fisheries*. IRD Editions, Paris.

Fonteneau, A 2010. *Atlas of Indian Ocean Tuna Fisheries*. IRD Editions, Paris.

Fonteneau, A., T. Diouf, and M. Mensah. 1993. Tuna fisheries in the eastern tropical Atlantic. Pp. 31–103 in A. Fonteneau and J. Marcille (eds.), *Resources, fishing and biology of the tropical tunas of the Eastern Central Atlantic*. FAO Fisheries Technical Paper 292. Food and Agriculture Organization of the United Nations (FAO), Rome.

Fromentin, J.-M., and J. Powers. 2005. Atlantic bluefin tuna: population dynamics, ecology, fisheries and management. *Fish and Fisheries* 6: 281–306.

Gillett, R. 2007. *A short history of industrial fishing in the Pacific Islands*. Asia-Pacific Fishery Commission and Food and Agriculture Organization of the United Nations (FAO), Bangkok, Thailand.

Hall, M. A., and M. Roman. 2013. *Bycatch and non-tuna catch in the tropical tuna purse seine fisheries of the world*. FAO Fisheries Technical Paper 568, Food and Agriculture Organization of the United Nations (FAO), Rome.

Hallier, J.-P., and D. Gaertner. 2008. Drifting fish aggregation devices could act as an

ecological trap for tropical tuna species. *Marine Ecology Progress Series* 353: 255–264.
IATTC. 2013a. *Resolution C-13-05: data confidentiality policy and procedures*. 85th Meeting of the Inter-American Tropical Tuna Commission, June 10–14, 2013, Veracruz (Mexico). Inter-American Tropical Tuna Commission, La Jolla, CA. http://www.iattc.org/PDFFiles2/Resolutions/C-13-05-Procedures-for-confi dential-data.pdf. Accessed June 16, 2013.
IATTC. 2013b. *Tunas and billfishes in the eastern Pacific Ocean in 2012*. Fishery Status Report 11, Inter-American Tropical Tuna Commission (IATTC), La Jolla, CA.
ICCAT. 2009. Report of the 2008 ICCAT yellowfin and skipjack stock assessments meeting, July 21–29, 2008, Florianópolis (Brazil). SCRS/2008/016. *Collective Volume of Scientific Papers: ICCAT* 64: 669–927.
IOTC. 2000. *Indian Ocean tuna fisheries: data summary for 1989–1998*. IOTC Data Summary 20, Indian Ocean Tuna Commission (IOTC), Victoria, Seychelles.
Kelleher, K. 2005. *Discards in the world's marine fisheries. An update*. FAO Fisheries Technical Paper 470, Food and Agriculture Organization, Rome.
Langley, A., K. Uosaki, S. Hoyle, H. Shono, and M. Ogura. 2010. *A standardized CPUE analysis of the Japanese distant-water skipjack pole-and-line fishery in the western and central Pacific Ocean (WCPO), 1972–2009*. WCPFC Scientific Committee 6th Regular Session, August 10–19, 2010, Nuku'alofa, Tonga.
Lawson, T. A. 1997. *Estimation of bycatch and discards in central and western Pacific tuna fisheries: preliminary results*. Oceanic Fisheries Programme Internal Report No. 33, Secretariat of the Pacific Community, Noumea, New Caledonia.
Lehodey, P., J. Hampton, R. Brill, S. Nicol, I. Senina, B. Calmettes, H. Pörtner, L. Bopp, T. Ilyina, J. Bell, and J. Sibert. 2011. Vulnerability of oceanic fisheries in the tropical Pacific to climate change. Pp. 435–485 in J. Bell, J. Johnson, and A. Hobday (eds.), *Vulnerability of Tropical Pacific Fisheries and Aquaculture to Climate Change*. Secretariat of the Pacific Community (SPC), Noumea, New Caledonia.
Longo, S., and B. Clarke. 2012. The commodification of bluefin tuna: the historical transformation of the Mediterranean fishery. *Journal of Agrarian Change* 12: 204–226.
Majkowski, J. 2007. *Global fishery resources of tuna and tuna-like species*. FAO Fisheries Technical Paper 483, Food and Agriculture Organization of the United Nations (FAO), Rome.
Marsac, F., A. Fonteneau, and P. Michaud. 2014. *L'or Bleu des Seychelles: Histoire de la Pêche Industrielle au Thon dans l'Océan Indien*. IRD Editions, Marseille, France.
Martín, J. I. 2011. *Fisheries in the Seychelles and fisheries agreements with the EU*. Note IP/B/PECH/NT/2011_04, European Parliament, Directorate General for Internal Policies, Policy Department B: Structural and Cohesion Policies: Fisheries, Brussels.
Miyake, M. P. 2005. A brief history of the tuna fisheries of the world. Pp. 23–50 in W. H. Bayliff, J. Ignacio de Leiva Moreno, and J. Majkowski (eds.), *Management of tuna fishing capacity: conservation and socio-economics*. Second meeting of the Technical Advisory Committee of the FAO project, Madrid (Spain), March 15–18, 2004. FAO Fisheries Proceedings 2. Food and Agriculture Organization of the United Nations (FAO), Rome.
Miyake, M. P., P. Guillotreau, C.-H. Sun, and G. Ishimura. 2010. *Recent developments in the tuna industry: stocks, fisheries, management, processing, trade and markets*. FAO Fisheries and Aquaculture Technical Paper 543, Food and Agriculture Organization of the United Nations (FAO), Rome.
Polacheck, T. 2012. Assessment of IUU fishing for southern bluefin tuna. *Marine Policy* 36(5): 1150–1165.
Romanov, E. V. 2002. Bycatch in the tuna purse-seine fisheries of the western Indian Ocean. *Fishery Bulletin* 100(1): 90–105.
Sahastume, A. 2002. Thoniers et pêcheurs basques à Dakar: de la guerre au tournant de la senne (1945–1968). *Zainak* 21: 131–169.
Shahifar, R. 2012. *Estimation of bycatch and discard in Iranian fishing vessels (gillnets) in the IOTC*

area of competence during 2012. 8th Session of the IOTC Working Party on Ecosystems and Bycatch, IOTC-2012-WPEB08-42, Indian Ocean Tuna Commission (IOTC), Cape Town, South Africa.

SPC. 2013. *Balancing the needs: industrial versus artisanal tuna fisheries*. SPC Policy Brief 22/2013, Secretariat of the Pacific Community (SPC), Noumea, France.

Stequert, B., and F. Marsac. 1983. *Pêche thonière à la senne: evolution de la technique et bilan de dix années d'exploitation dans l'océan Indien*. Office de la Recherche Scientifique et Technique Outre-Mer (ORSTOM), Paris.

Stequert, B., and F. Marsac. 1989. *Tropical tuna: surface fisheries in the Indian Ocean*. Fisheries Technical Paper 282, Food and Agriculture Organization of the United Nations (FAO), Rome.

Sumaila, U. R., A. J. Dyck, and A. Baske. 2014. Subsidies to tuna fisheries in the western central Pacific Ocean. *Marine Policy* 43: 288–294.

United Nations. 1995. *Agreement for the implementation of the provisions of the United Nations Convention on the Law of the Sea of 10 December 1982 relating to the conservation and management of straddling fish stocks and highly migratory fish stocks*. Conference on Straddling Fish Stocks and Highly Migratory Fish Stocks, 6th session, July 24–August 4, 1995, United Nations, New York, NY.

WCPFC. 2011. *Tuna fishery yearbook 2010*. Western and Central Pacific Fisheries Commission (WCPFC), Noumea, New Caledonia.

Williams, P., and P. Terawasi. 2011. *Overview of tuna fisheries in the western and central Pacific Ocean, including economic conditions: 2010*. WCPFC Scientific Committee 7th Regular Session, August 9–17, 2011, Pohnpei, Federated States of Micronesia.

WPRFMC. 2013. *Pelagic fisheries of the western Pacific region: 2011 annual report*. Western Pacific Regional Fishery Management Council, Honolulu.

NOTES

1. Cite as Le Manach, F., P. Chavance, A. M. Cisneros-Montemayor, A. Lindop, A. Padilla, D. Zeller, L. Schiller, and D. Pauly. 2016. Global catches of large pelagic fishes, with emphasis on the high seas. Pp.34–45 in D. Pauly and D. Zeller (eds.), *Global Atlas of Marine Fisheries: A Critical Appraisal of Catches and Ecosystem Impacts*. Island Press, Washington, DC.
2. The Food and Agriculture Organization of the United Nations (FAO) has published a global, harmonized atlas, but it includes only the catch of twelve species of tuna and billfishes (i.e., albacore, Atlantic bluefin tuna, Atlantic white marlin, bigeye tuna, black marlin, blue marlin, Pacific bluefin tuna, skipjack tuna, southern bluefin tuna, striped marlin, swordfish, and yellowfin tuna; FAO 2013a). This atlas is available at www.fao.org/figis/geoserver/tunaatlas. For reasons of confidentiality of commercial interests, this dataset entirely lacks longline data for the eastern Pacific area after 1962, managed by the IATTC, although some data for the earlier time period have been published in aggregated form (Fonteneau 1997). A recent resolution on confidentiality rules may mean that these spatialized data may eventually become publicly available (IATTC 2013a). Fonteneau (1997) has also published a global atlas but did not estimate discards or scale up the spatialized data to 100% of the nominal catch. Updates were published later but at regional scales and without the Pacific Ocean (Fonteneau 2009, 2010).
3. Except when labeled "nonmechanized," "coastal," "small," or such that nonindustrial fishing can be inferred.
4. Note that the current intense use of d-FADs results in a fishing activity closer to "gathering" than "hunting," because it greatly reduces searching time and aggregates the fish in traceable locations. However, d-FADs are regularly criticized for generating high bycatch rates of juvenile yellowfin and bigeye tunas, reducing potential catches by the longline fleet (Bromhead et al. 2003; Miyake 2005), and nontarget species including at-risk shark species (Romanov 2002; Dagorn et al. 2013; Filmalter et al. 2013). They also may act as an "ecological trap" by tricking tunas into nonoptimal waters (Hallier and Gaertner 2008).

CHAPTER 4

THE DISTRIBUTION OF EXPLOITED MARINE BIODIVERSITY[1]

Maria Lourdes D. Palomares,[a]
William W. L. Cheung,[b] Vicky W. Y. Lam,[a]
and Daniel Pauly[a]
[a]*Sea Around Us*, University of British Columbia, Vancouver, BC, Canada
[b]Changing Oceans Research Unit and NF-UBC Nereus Program, University of British Columbia, Vancouver, BC, Canada

Ecosystem-based fisheries management (EBFM; Pikitch et al. 2004) must include a sense of place, where fisheries interact with the animals and plants of specific ecosystems. To assist researchers, managers, and policy makers attempting to implement EBFM schemes, the *Sea Around Us* presents biodiversity and fisheries data in spatial form onto a grid of about 180,000 half-degree latitude and longitude cells that can be regrouped into larger entities, such as the Exclusive Economic Zones (EEZs) of maritime countries (see chapter 2, and particularly figure 2.1), or the system of currently sixty-six Large Marine Ecosystems (LMEs) initiated by the National Oceanic and Atmospheric Administration (Sherman et al. 2007) and now used by practitioners throughout the world.

However, not all the marine biodiversity of the world can be mapped in this manner; thus, although FishBase (www.fishbase.org) includes all marine fishes described so far (more than 15,000 spp.), so little is known about the distribution of the majority of these species that they cannot be mapped in their entirety. The situation is even worse for marine invertebrates, despite huge efforts (see www.sealifebase.org). This also applies to commercially or otherwise exploited species of fishes and invertebrates, for many of which only rudimentary knowledge is available.

We define as "commercial" all marine fish or invertebrate species that are reported in the catch statistics of at least one of the member countries of the Food and Agriculture Organization of the United Nations (FAO) or are listed as part of commercial and noncommercial catches (retained as well as discarded) in country-specific catch reconstructions (see chapters 2 and 3). Fortunately, for most species occurring in the landings statistics of FAO, there were enough data in FishBase for at least tentatively mapping their distribution ranges. Similarly, most species of commercial invertebrates had enough information in SeaLifeBase for their approximate distribution range to be mapped. We discuss below the procedure we use for taxa that lacked sufficient data for mapping their distribution, which included few taxa in the FAO statistics and many from reconstructed catches, including discards. Note that the *Sea Around Us* works only on commercial species as defined above; the website of the AquaMap Project (www.aquamap.org) should be consulted for distribution range maps of other marine species of fish and invertebrates, and marine biodiversity in general. This contribution presents the methods (updated and improved from Close et al. 2006) by which all commercial species distribution ranges (totaling more than 2,500 for the 1950–2010 time period) were constructed or updated. It consists of a set of rigorously applied filters that will markedly improve the accuracy of the *Sea Around Us* maps and other products.

The filters used here are listed in the order in which they are applied. Before applying the filters presented below, the identity and nomenclature of each species are verified using FishBase (www.fishbase.org) or SeaLifeBase (www.sealifebase.org), two authoritative online encyclopedias covering the fishes of the world and marine nonfish animals, respectively, and their scientific and English common names are corrected if necessary. This information is then standardized throughout all *Sea Around Us* databases (see chapter 5). After the creation of all species-level distributions as described here, taxon distributions for higher taxonomic groups, such as genus and family, are generated by combining each taxon level's contributing components; for example, for the genus *Gadus*, all distributions of species within this genus are combined.

Note that the procedures presented here avoid the use of temperature and primary productivity to define or refine distribution ranges for any species, even though these factors strongly shape the distribution of marine fishes and invertebrates (Ekman 1967; Longhurst and Pauly 1987). This was done to allow subsequent analyses of distribution ranges to be legitimately performed using these variables, that is, to avoid circularity.

FILTER 1: FAO AREAS

The FAO has divided the world's oceans into nineteen statistical areas for reporting purposes (figure 2.1 in chapter 2). Information on the occurrence of commercial species within these areas is available primarily through FAO publications and the FAO website (www.fao.org), FishBase, and SeaLifeBase. Figures 4.1A and 4.2A illustrate the occurrence by FAO area of Florida pompano (*Trachinotus carolinus*) and silver hake (*Merluccius bilinearis*), examples representing pelagic and demersal species, respectively.

FILTER 2: LATITUDINAL RANGE

The second filter applied in this process is latitudinal ranges. After reviewing literature on the distribution of marine organisms, Charles Darwin concluded that "latitude is a more important element than longitude" (see Pauly 2004, p. 125, for the sources of this quote and one below). However, this does not mean that longitude and other factors do not play a role in determining a taxon's distribution. Still, in the following quote, Darwin illustrates how latitude provides the key to understanding the composition of certain fauna: "Sir J. Richardson says the Fish of the cooler temperate parts of the S. Hemisphere present a much stronger analogy to the fish of the same latitudes in the North, than do the strictly Arctic forms to the Antarctic."

The latitudinal range of a species is defined as the space between its northernmost and southernmost latitudes. This range can be found in FishBase for most fishes and in SeaLifeBase for many invertebrates. For fishes and invertebrates for which this information was lacking, latitudes were inferred from the latitudinal range of the EEZs of countries where they are reported to occur as endemic or native species, or from occurrence records in the Ocean Biogeographic Information System (OBIS) website (www.iobis.org). Note, however, that recent occurrence records (from the 1980s onward and known range extensions, e.g., of Lessepsian species) were not used to determine "normal" latitudinal ranges, because they tend to be affected by global warming (Cheung et al. 2009).

A species will not have the same probability of occurrence or relative abundance throughout its latitudinal range; it can be assumed to be most abundant at the center of its range (MacCall 1990). Defining the center of the latitudinal distribution range is done using the following assumptions:

1. For distributions confined to a hemisphere, a symmetrical triangular probability distribution is applied, which estimates the center of the latitudinal range as the average of the range: (northernmost + southernmost latitude)/2.
2. For distributions straddling the equator, the range is broken into three parts: the outer two thirds and the inner or middle third. If the equator falls in one of the outer thirds of the latitudinal range, then abundance is assumed to be the same

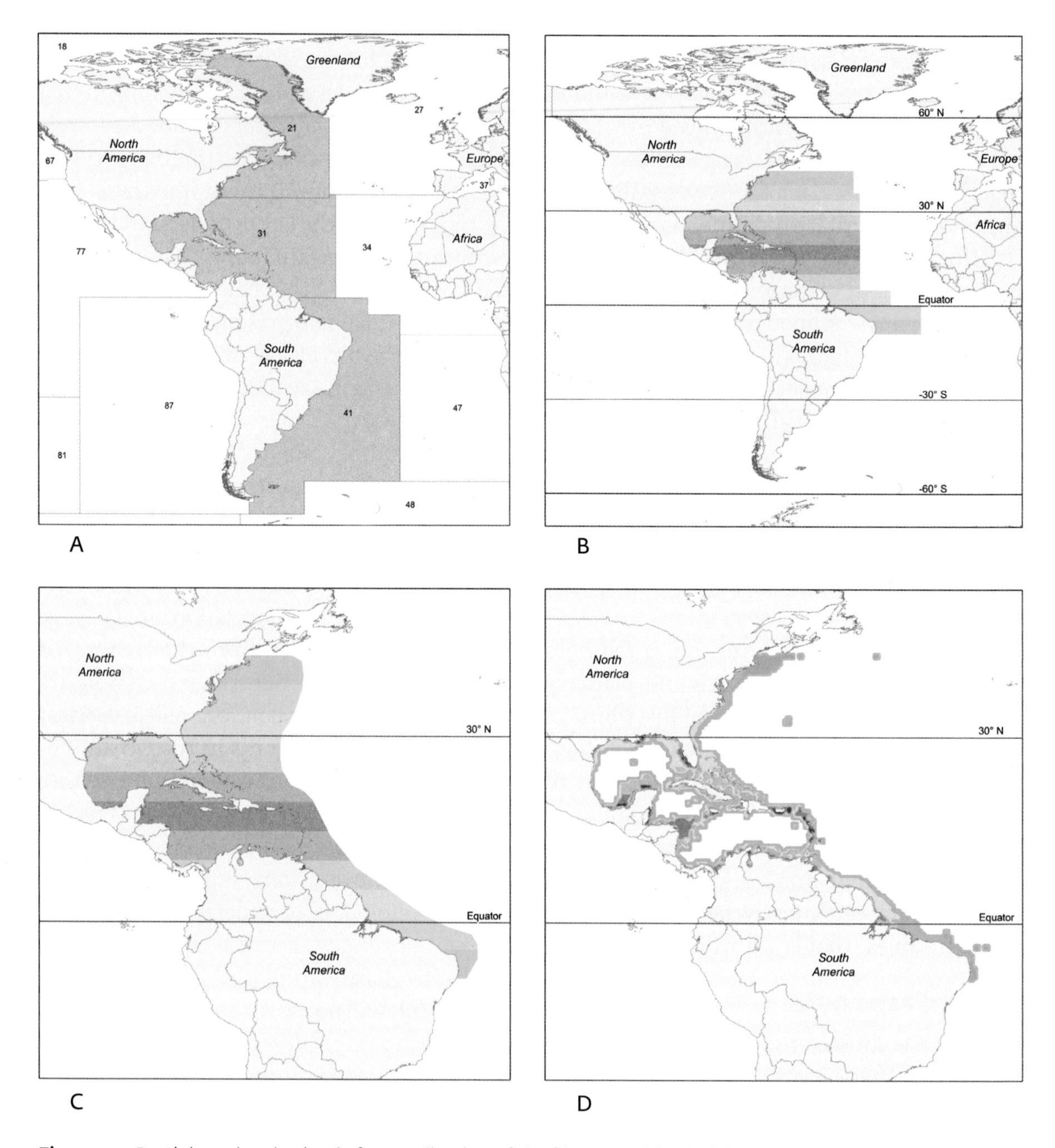

Figure 4.1. Partial results obtained after application of the filters used for deriving a species distribution range map for the Florida pompano (*Trachinotus carolinus*): (A) illustrates the Florida pompano's presence in FAO areas 21, 31, and 41; (B) illustrates the result of overlaying the latitudinal range (43°N to 9°S; see Smith 1997) over the map in (A). (C) shows the result of overlaying the (expert-reviewed) range-limiting polygon over (B). (D) illustrates the relative abundance of the Florida pompano resulting from the application of the depth range, habitat preference, and equatorial submergence filters on the map in (C). (Modified from Close et al. 2006.)

as in (1). However, if the equator falls in the middle third of the range, then abundance is assumed to be flat in the middle third and decreasing to the poles for the remainder of the range.

Figures 4.1B and 4.2B illustrate the result of the FAO and latitudinal filters combined. Both the Florida pompano and the silver hake follow symmetrical triangular distributions as mentioned in (1) above.

FILTER 3: RANGE-LIMITING POLYGON

Range-limiting polygons help confine species in areas where they are known to occur while preventing their occurrence in other areas where they could occur (because of environ-

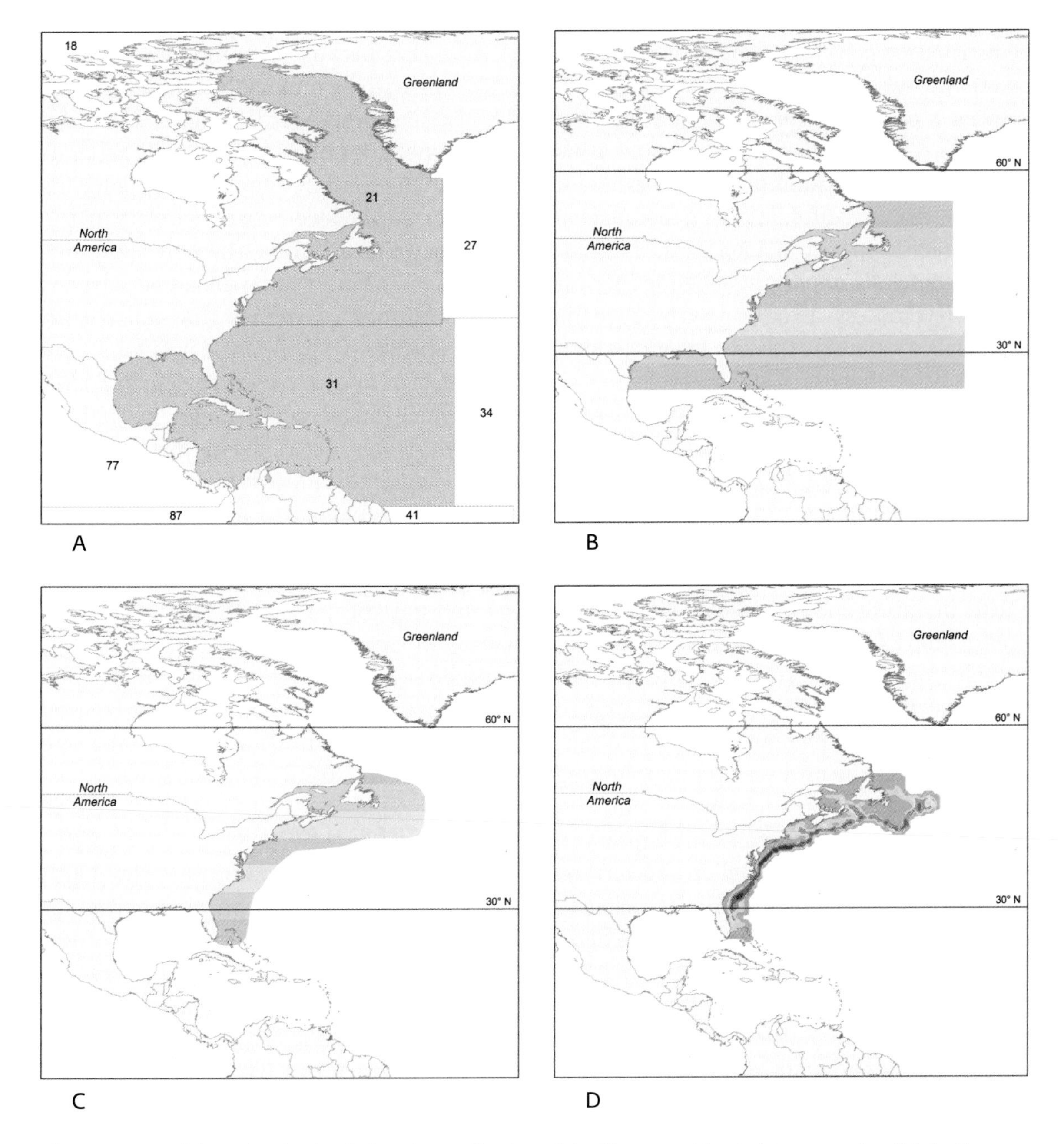

Figure 4.2. Partial results obtained after the application of the filters used for deriving a species distribution range map for the silver hake (*Merluccius bilinearis*): (A) illustrates the silver hake's presence in FAO areas 21 and 31. (B) illustrates the result of applying the FAO and latitudinal range (55°N to 24°N; see FAO-FIGIS 2001). (C) shows the result of overlaying the (expert-reviewed) range-limiting polygon over (B). (D) illustrates the silver hake's relative abundance resulting from the application of the depth range, habitat preference, and equatorial submergence filters on the map in (C). (Modified from Close et al. 2006.)

mental conditions) but do not. Distribution polygons for a vast number of species of commercial fish and invertebrates can be found in various publications, notably FAO's (species catalogues, species identification sheets, guides to the commercial species of various countries or regions), and in online resources, some of which were obtained from model predictions, such as AquaMaps (Kaschner et al. 2008; see also www.aquamaps.org). Such polygons are based mostly on observed species occurrences, which may or may not be representative of the actual distribution range of the species.

Occurrence records assume that the observer correctly identified the species being reported, which adds a level of uncertainty to the validity of distribution polygons. More often than not, taxonomic experts are required to review and validate a polygon before it is published, for example, in an FAO species catalogue. This review process is also important, notably for polygons that are automatically generated via model predictions such as AquaMaps. Note that for commercially important endemic species, this review process can be skipped because the polygon is restricted to the only known habitat and country where such species occur.

For species without published polygons, range maps are generated using the filter process described here and compared with the native distribution generated in AquaMaps. Differences between these two model-generated maps are verified using data from the scientific literature and OBIS/Global Biodiversity Information Facility (i.e., reported occurrences, notably from scientific surveys). Note that FAO statistics, in which countries report a given species in their catch, can be used as occurrence records, the only exception occurring when the species was caught by the country's distant-water fleet, as defined in chapter 2.

Polygons are drawn based on the verified map (i.e., with unverified occurrences deleted). Additionally, faunistic work covering the high-latitude end of continents or semienclosed coastal seas with depauperate faunas (e.g., Hudson Bay or the Baltic Sea) were used to avoid, where appropriate, distributions reaching into these extreme habitats. The result of this step (i.e., the information gathered from the verification of occurrences) is also provided to FishBase and SeaLifeBase to fill data gaps in both databases.

All polygons, whether available from a publication or newly drawn, were digitized with Esri's ArcGIS and were later used for inferences on equatorial submergence (see below). Figures 4.1C and 4.2C illustrate the result of the combination of the first three filters (i.e., FAO, latitude, and range-limiting polygons). These parameters and polygons will be revised periodically, as our knowledge of the species in question increases.

Note that because this mapping process deals only with commercially caught species, the distribution ranges for higher-level taxa (e.g., genera, families) were usually generated using the combination of range polygons from the lower-level taxa included in the higher-level taxon. Thus, the range polygons for genera were built using the range polygons of the commercial species that fall within them. Similarly, family-level polygons were generated from genus-level polygons, and so on. Latitude ranges, depth ranges, and habitat preferences were expanded in the same manner. Although this procedure will not produce the complete biological distribution of the genera and families in question, which usually consists of more species than are reported in catch statistics, it is likely that the generic names in catch statistics refer to the very commercial species that are used to generate the distribution ranges, because these taxa are often more abundant than the ones that are not reported in official catch statistics.

FILTER 4: DEPTH RANGE

Similar to the latitudinal range, the depth range (i.e., "[the] depth (in m) reported for juveniles and adults (but not larvae) from the most shallow to the deepest [waters]") is available from FishBase for most fish species and SeaLifeBase for many commercial invertebrates, along with their common depth, defined as the "[the] depth range (in m) where juveniles and adults are most often found. This range may be calculated as the depth range within which approximately 95% of the biomass occurs" (Froese et al. 2000). Given this, and based on Alverson et al. (1964), Pauly and Chua (1988), and Zeller and Pauly (2001), among others, the abundance of a species within the water column is assumed to follow a scalene triangular distribution, where maximum abundance occurs at the top one third of its depth range.

FILTER 5: HABITAT PREFERENCE

Habitat preference is an important factor affecting the distribution of marine species.

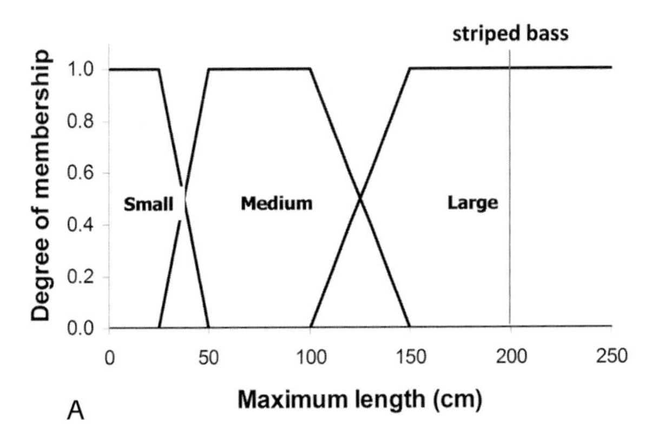

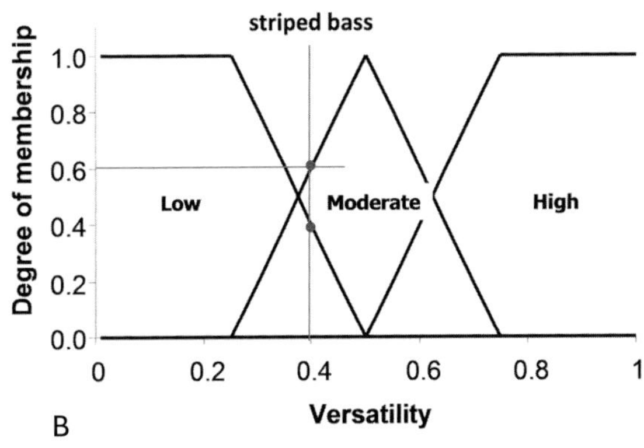

Figure 4.3. Fuzzy membership functions for the three categories of (A) maximum length and (B) habitat versatility of a species. Habitat versatility is defined as the ratio of the number of habitat types with which a species is associated to the total number of defined habitat types in table 4.1. For example, the striped bass (*Morone saxatilis*) grows to a maximum total length of 200 cm (large body size; degree of membership = 1). It occurs in estuaries and "other habitats" (2 of 5 defined habitats, i.e., versatility = 0.4, low to moderate; degree of membership = 0.4–0.6). (Modified from Close et al. 2006.)

Thus, the aim of this filter is to enhance the prediction of the probability that a species occurs in an area, based on its association with different habitats. Two assumptions are made here:

- That, other things being equal, the relative abundance of a species in a spatial half-degree cell is determined by a fraction derived from the number of habitats that a species associates with in that same cell and by how far the association effect will extend from that habitat.
- That the extent of this association is assumed to be a function of a species' maximum size (maximum length) and habitat versatility. Thus, a large species that inhabits a wide range of habitats is more likely to occur far from the habitats with which it is associated, whereas smaller species tend to have low habitat versatility (Kramer and Chapman 1999).

The maximum length and versatility of a species are classified into three categories, and it is assumed that a species can associate with one or more categories with different degrees of membership (0 to 1). A higher membership value means a higher probability that the species is associated with that particular category. The membership values are defined by a prespecified membership function for each of the length and versatility categories (figure 4.3). For example, the striped bass (*Morone saxatilis*) has a maximum length of 200 cm (total length). Based on the predefined membership function presented in figure 4.3A, the striped bass has a large body size with a membership of 1. Note that there are maximum length estimates for all the exploited species used by the *Sea Around Us*, derived from FishBase and SeaLifeBase.

The ability of a species to inhabit different habitat types, here referred to as "versatility," is defined as the ratio between the number of habitats with which a species is associated to the total number of habitats as defined in table 4.1. These habitats are categorized as "biophysical" (i.e., coral reef, estuary, sea grass, seamount, other habitats), "depth-related" (shelf, slope, or abyssal), and "distance from coast" (inshore or offshore). Because species are generally specialized toward "biophysical" habitats, this filter takes only those five habitats into consideration. Taking our example again, FishBase lists the following for striped bass: "Inhabit coastal waters and are commonly found in bays but may enter rivers in the spring to spawn" (Eschmeyer et al. 1983). This associates striped bass with estuaries and "other habitats" (i.e., when it enters rivers to spawn). Given that the total number of defined biophysical habitats is five, and the striped bass is associated with two of them, then the versatility of striped bass is

Table 4.1. Habitat categories used here, with some of the terms typically associated with them (in FishBase, SeaLifeBase and other sources).

Categories	Specifications of global map	Terms often used
Estuary	Alder (2003)	Estuaries, mangroves, river mouth
Coral	UNEP-WCMC et al. (2010)	Coral reef, coral, atoll, reef slope
Sea grass	Not yet available	Sea grass bed
Seamounts	Kitchingman and Lai (2004)	Seamounts
Other habitats	—	Muddy, sandy, or rocky bottom
Continental shelf	NOAA (2004)	Continental shelf, shelf
Continental slope	NOAA (2004)	Continental slope, upper or lower slope
Abyssal	NOAA (2004)	Away from shelf and slope
Inshore	NOAA (2004)	Shore, inshore, coastal, along shoreline
Offshore	NOAA (2004)	Offshore, oceanic

estimated to be 0.4 (i.e., 2/5). Finally, based on the defined membership functions shown in figure 4.3B, the versatility of striped bass is classified as "low" to "moderate," with a membership of approximately 0.4 and 0.6, respectively.

Determining Habitat Association

Qualitative descriptions relating the commonness of (or the preference of) a species to particular habitats (as defined in table 4.1) are given weighting factors as enumerated in table 4.2. Such descriptions are available from FishBase for most fishes and in SeaLifeBase for most commercially important invertebrates. Going back to our example, we thus know that striped bass occur in (and thus prefer) brackish water (i.e., estuaries) but enter freshwater (i.e., "other habitats") to spawn. Given the weighting system in table 4.2, estuaries is assigned a weight of 0.75 (*usually* occurs in) and "other habitats" is given a weight of 0.5 (assuming a *seasonal* spawning period).

Maximum Distance of Habitat Effect

Maximum distance of habitat effect (maximum effective distance) is the maximum distance from the nearest perimeter of the habitat that attracts a species to a particular habitat. This is defined by the maximum length and habitat versatility of the species using the heuristic rule matrix in table 4.3. Taking our example for the striped bass, with a "large" maximum length (membership = 1) and "low" to "moderate" versatility (membership values of 0.4 and 0.6), points to a "farthest" maximum effective distance in table 4.3. The degree of membership assigned to maximum effective distance is equal to the minimum membership value of the two predicates,[2] in this example, 1 vs. 0.4 = 0.4 and 1 vs. 0.6 = 0.6. When the same conclusion is reached from different rules, the final degree of membership equals the average membership value (in this example, [0.4 + 0.6]/2 = 0.50).

The maximum effective distance from the associated habitat can be estimated from the centroid value of each conclusion category, weighted by the degree of membership. The centroid values for "near," "far," and "farthest" maximum effective distances were defined as 1 km, 50 km, and 100 km, respectively. In our

Table 4.2. Common descriptions of relative abundance of species in habitats where they occur and their assigned weighting factors. The weighting factor for "other habitats" is assumed to be 0.1 when no further information is available.

Description	Weighting factor
Absent or rare	0.00
Occasionally, sometimes	0.25
Often, regularly, seasonally[a]	0.50
Usually, abundant in, prefer	0.75
Always, mostly, only occurs	1.00

[a]If a species occurs in a habitat but no indication of relative abundance is available, a default score of 0.5 is assumed.

Table 4.3. Heuristic rules that define the maximum effective distance from the habitat in which a species occurs. The columns in bold characters represent the predicates (categories of maximum body size and versatility), and those in italics represent the resulting categories of maximum effective distance.

	Maximum body size		
Versatility	**Small**	**Medium**	**Large**
Low	Near	Near	Near
Moderate	Far	Far	Farthest
High	Far	Farthest	Farthest

example, we obtained membership values of 0.4 for near (1 km) and 0.6 for farthest (100 km) maximum effective distance, respectively. This gives an estimate of (0.4 * 1 + 0 * 50 + 0.6 * 100)/(0.4 + 0 + 0.6) = 60.4 km (see figure 4.4).

Estimating Relative Abundance in a Spatial Cell

Several assumptions are made to simplify the computations. First, it is assumed that the habitat always occurs in the center of a cell and is circular in shape. Second, species density (per unit area) is assumed to be the same across any habitat type, and that density declines linearly from the habitat perimeter to its maximum effective distance. Given these assumptions, the total relative abundance of a species in a cell equals the sum of abundance on and around its associated habitat, expressed as:

$$B'T = [\alpha j + \alpha j + 1 \ldots (1 - \alpha j)], \ldots (1 - ed\, A), \quad (4.1)$$

where $B'T$ is the final abundance, αj, is the density away from the habitat from cell j, and A is the habitat area of the cell. The relative abundance resulting from the different habitat types is the sum of relative abundance and is weighted by their importance to the species.

Although these assumptions about the relationship between maximum length, habitat versatility, and maximum distance from the habitat may render predicted distributions uncertain at a fine spatial scale, this routine provides an explicit and consistent way to incorporate habitat considerations into distribution ranges.

FILTER 6: EQUATORIAL SUBMERGENCE

The equatorial submergence phenomenon was well known to Charles Darwin, who wrote that "we hear from Sir J. Richardson, that Arctic forms of fishes disappear in the seas of Japan & of northern China, are replaced by other assemblages in the warmer latitudes & reappear on the coast of Tasmania, southern New Zealand & the Antarctic islands" (Pauly 2004, p. 198).

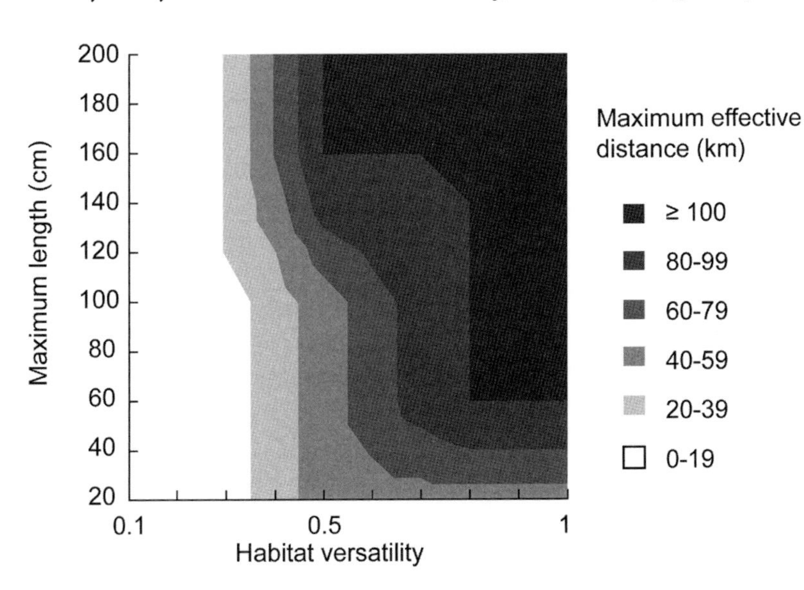

Figure 4.4. Maximum effective distance for striped bass (*Morone saxatilis*) estimated from the habitat versatility and maximum length of that species (see text). (Modified from Close et al. 2006.)

Eckman (1967) gives the current definition:

Animals which in higher latitudes live in shallow water seek in more southern regions archibenthal or purely abyssal waters. . . . This is a very common phenomenon and has been observed by several earlier investigators. We call it submergence after V. Haecker [1906–1908] who, in his studies on pelagic radiolarian, drew attention to it. In most cases, including those which interest us here, submergence increases towards the lower latitudes and therefore may be called equatorial submergence. Submergence is simply a consequence of the animal's reaction to temperature. Cold-water animals must seek colder, deeper water layers in regions with warm surface water if they are to inhabit such regions at all.

Equatorial submergence, indeed, is caused by the same physiological constraints that also determine the "normal" latitudinal range of species, as described above, and its shift due to global warming (chapter 8), that is, respiratory constraints fish and aquatic invertebrates experience at temperatures higher than that which they have evolved to prefer (Pauly 1998, 2010).

Modifying the distribution ranges to account for equatorial submergence requires accounting for two constraints: data scarcity and

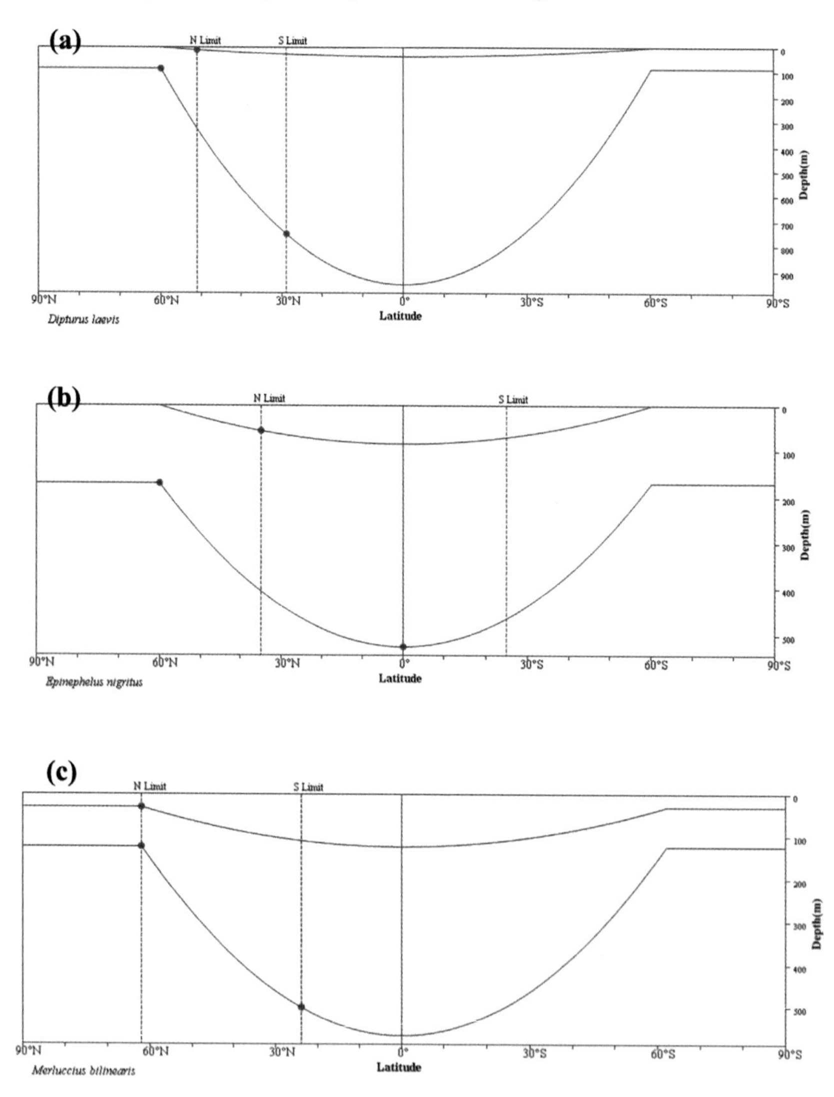

Figure 4.5. Shapes used to generate "equatorial submergence," given different depth and latitude data. (A) Case 1: barndoor skate (*Dipturus laevis*). When the distribution range of the species is at latitudes lower than 60°N or 60°S, the shallow parabola ($P_{shallow}$) is assumed to intercept zero at 60°N and 60°S. (B) Case 2: When the distribution range is spanning the northern and southern hemispheres, as in the case of the Warsaw grouper (*Epinephelus nigritus*), the deepest depth of the deep parabola (P_{deep}) is at the Equator. (C) Case 3: Silver hake (*Merluccius bilinearis*). The poleward limit of the latitudinal range (L_{high}) is at higher latitudes than 60°N or 60°S. (Modified from Pauly 2010.)

uneven distribution of environmental variables (e.g., temperature, light, food) with depth. FishBase and SeaLifeBase notwithstanding, there is little information on the depth distribution of most commercial species. However, in most cases, the following four data points are available for each species: the shallow end of the depth range ($D_{shallow}$), its deep end (D_{deep}), the poleward limit of the latitudinal range (L_{high}), and its low-latitude limit (L_{low}). If it is assumed that equatorial submergence is to occur, then it is logical to also assume that $D_{shallow}$ corresponds to L_{high} and that D_{deep} corresponds to L_{low}.

Also, we further mitigate data scarcity by assuming the shape of the function linking latitude and equatorial submergence. Here, two parabolas (P) are used (figure 4.5), one for the shallow limit of the distribution ($P_{shallow}$) and one for the deep limit (P_{deep}), with the assumption that both $P_{shallow}$ and P_{deep} are symmetrical about the Equator. In addition, maximum depths are assumed not to change poleward of 60°N and 60°S. The uneven distribution of the temperature gradient can be mimicked by constraining $P_{shallow}$ to be less concave than P_{deep} by setting the geometric mean (D_{gm}) of $D_{shallow}$ and D_{deep} as the deepest depth that $P_{shallow}$ can attain. Three points draw the parabolas. In most cases, $P_{shallow}$ is obtained with $D_{60°N} = 0$, $D_{60°S} = 0$, and $D_{Lhigh} = D_{shallow}$, and P_{deep} with $D_{60°N} = D_{gm}$, $D_{60°S} = D_{gm}$, and $D_{Llow} = D_{max}$. If L_{high} is in the northern hemisphere and L_{low} is in the south, P_{deep} is drawn with D_{deep} at the Equator and conversely for the southern hemisphere. Finally, it is assumed that if a computed $P_{shallow}$ intercepts zero depth at latitudes higher than 60°N or lower than 60°S, then $P_{shallow}$ is recomputed with $D_{60°N} = D_{shallow}$, $D_{60°S} = D_{shallow}$, and $D_{Lhigh} = 0$.

Figure 4.5 illustrates three cases of submergence based on different constraints. When this process is applied to a distribution based on latitudinal range and depth that did not account for submergence, these have the effect of "shaving off" parts of the shallow end of that distribution at low latitudes and, similarly, shaving off part of the deep end of the distribution at high latitudes. Also, besides leading to narrower and more realistic distribution ranges, this leads to narrowing the temperature ranges inhabited by the species in question, which is important for the estimation of their preferred temperature, as used when modeling global warming effects on marine biodiversity and fisheries (chapter 8).

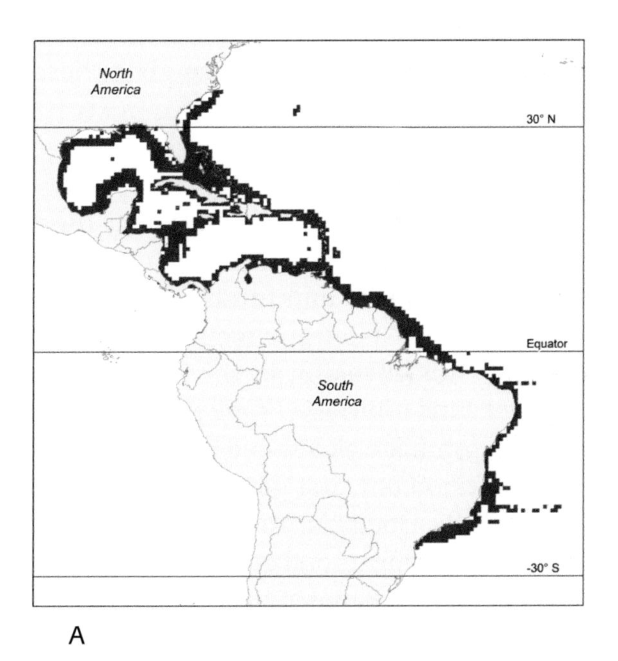

A

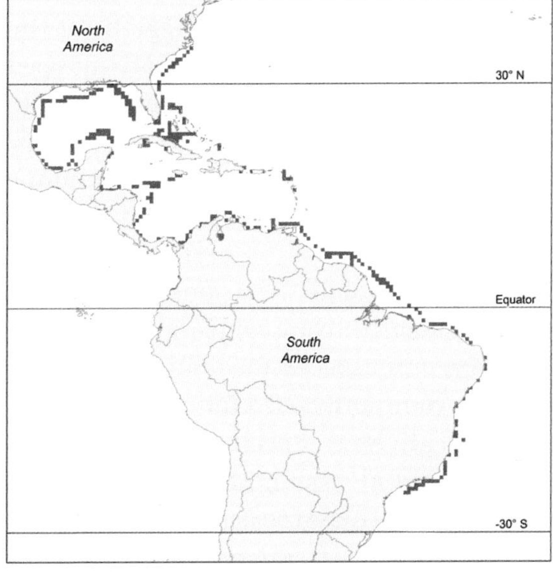

B

Figure 4.6. "Equatorial submergence" has the effect of "shaving off" areas from the distribution range of the Warsaw grouper, *Epinephelus nigritus*: (A) Original distribution; (B) Distribution adjusted for "equatorial submergence."

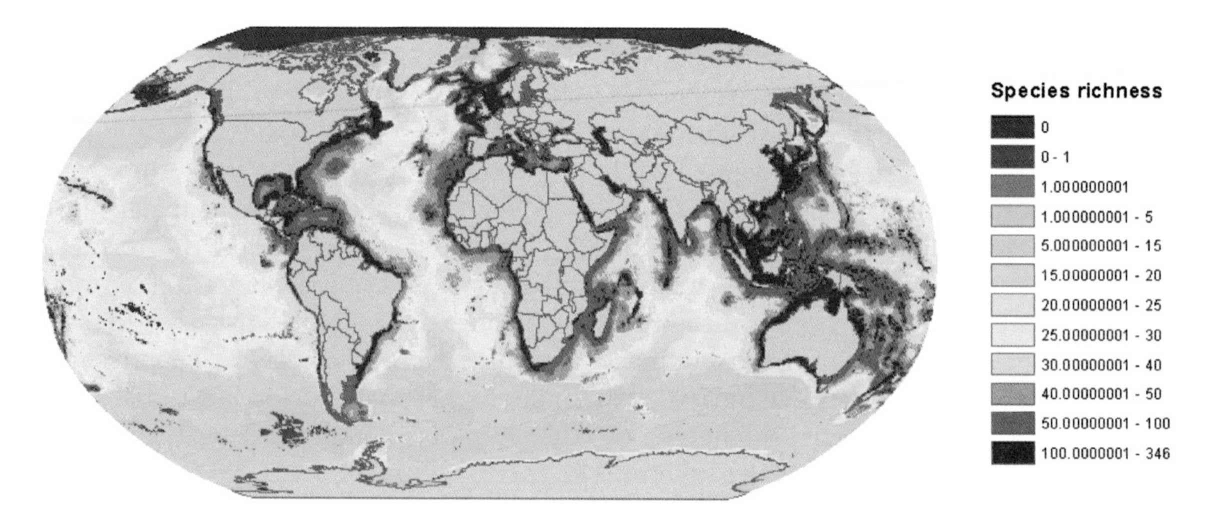

Figure 4.7. Distribution of the richness of "commercial" species (836 fish and 230 invertebrate spp.) created by the *Sea Around Us* by the mid-2000s for mapping fisheries and studying distribution shifts caused by global warming. The number of species available for such a map has increased since then by about 50%, as more species are included in reconstructed data and thus become "commercial" (see text).

CONCLUSION

The key outcome of the process described above consists of distribution ranges such as in figure 4.6 for more than 2,500 taxa, which can be viewed on the *Sea Around Us* website. They are also accessible via FishBase and SeaLifeBase (click "*Sea Around Us* distributions" under the "Internet sources" section of the species summary pages). These distribution ranges serve as basis for all spatial catch allocation done by the *Sea Around Us* (chapter 5).

The numbers of taxa used to spatialize fisheries catches in different regions of the globe (as described in chapter 5) are mapped in figure 4.7; these numbers, pertaining to the mid-2000s, have increased considerably since then, given the increasing taxonomic resolution of the reconstructed catch statistics used (chapter 2).

Predictions of distributions from the *Sea Around Us* algorithm are comparable in performance to other species modeling approaches that are commonly used for marine species (Jones et al. 2012). Specifically, AquaMaps (Kaschner et al. 2008), Maxent (Phillips et al. 2006), and the *Sea Around Us* algorithm are three approaches that have been applied to predict distributions of marine fishes and invertebrates. Jones et al. (2012) applied these three species distribution modeling methods to commercial fish in the North Sea and North Atlantic using data from FishBase and the Ocean OBIS. Comparing test statistics of model predictions with occurrence records suggest that each modeling method produced plausible predictions of range maps for each species. However, the pattern of predicted relative habitat suitability can differ substantially between models (Jones et al. 2013). Incorporation of expert knowledge, as discussed above with reference to Filter 3, generally improves predictions and therefore was given here particular attention.

ACKNOWLEDGMENTS

This is a contribution of the *Sea Around Us*, a research activity at the University of British Columbia initiated and funded by the Pew Charitable Trusts, from 1999 to 2014 and currently funded mainly by the Paul G. Allen Family Foundation. We thank Adrian Kitchingman, Ahmed Gelchu, and Reg Watson for their work on our first generation of range maps, Sally Hodgson and Chris Close for their work on the second generation, and the FIN team in Los Baños, Philippines, notably Elizabeth Bato, Jeniffer Espedido, Vina A. Parducho, Kathy Reyes, and Patricia M. S. Yap for the third generation of these maps.

REFERENCES

Alder, J. 2003. Putting the coast in the *Sea Around Us* project. *The Sea Around Us Newsletter* 15: 1–2.

Alverson, D. L., A. L. Pruter, and L. L. Ronholt. 1964. *A Study of Demersal Fishes and Fisheries of the Northeastern Pacific Ocean*. Institute of Fisheries, The University of British Columbia, Vancouver.

Cheung, W. W. L., V. W. Y. Lam, J. L. Sarmiento, K. Kearney, R. Watson, and D. Pauly. 2009. Projecting global marine biodiversity impacts under climate change scenarios. *Fish and Fisheries* 10: 235–251.

Close, C., W. W. L. Cheung, S. Hodgson, V. W. Y. Lam, R. Watson, and D. Pauly. 2006. Distribution ranges of commercial fishes and invertebrates. Pp. 27–37 in M. L. D. Palomares, K. I. Stergiou, and D. Pauly (eds.), *Fishes in databases and ecosystems*. Fisheries Centre Research Reports 14(4), University of British Columbia, Vancouver.

Ekman, S. 1967. *Zoogeography of the Sea*. Sidgwick & Jackson, London.

Eschmeyer, W. N., E. S. Herald, and H. Hammann. 1983. *A Field Guide to Pacific Coast Fishes of North America*. Houghton Mifflin, Boston, MA.

FAO-FIGIS. 2001. *A world overview of species of interest to fisheries*. Chapter: *Merluccius bilinearis*. Retrieved June 15, 2005, from www.fao.org/figis/servlet/species?fid=3026. 2p. FIGIS Species Fact Sheets. Species Identification and Data Programme-SIDP, FAO-FIGIS.

Froese, R., E. Capuli, C. Garilao, and D. Pauly. 2000. The SPECIES table. Pp. 76–85 in R. Froese and D. Pauly (eds.), *FishBase 2000: Concept, Design and Data Sources*. ICLARM, Los Baños, Laguna, Philippines.

Haecker, V. 1906–1908. *Tiefesee-Radiolarien. Wissenschaftliche Ergebnisse der Deutschen Tiefsee-Expedition auf dem Dampfer "Valdivia" 1898–1899* 14. Fischer, Jena, Germany.

Jones, M., S. Dye, J. Pinnegar, R. Warren, and W. W. L. Cheung. 2012. Modelling commercial fish distributions: prediction and assessment using different approaches. *Ecological Modelling* 225: 133–145.

Jones, M. C., S. R. Dye, J. K. Pinnegar, R. Warren, W. W. L. Cheung. 2013. Applying distribution model projections in the present for an uncertain future. *Aquatic Conservation: Marine and Freshwater Research* 23(5): 710–722.

Kaschner, K., J. S. Ready, E. Agbayani, J. Rius, K. Kesner-Reyes, P. D. Eastwood, A. B. South, S. O. Kullander, T. Rees, C. H. Close, R. Watson, D. Pauly, and R. Froese (eds.). 2008. AquaMaps Environmental Dataset: Half-Degree Cells Authority File (HCAF). www.aquamaps.org.

Kitchingman, A., and L. Lai. 2004. Inferences of potential seamount locations from mid-resolution bathymetric data. Pp. 7–12 in T. Morato and D. Pauly (eds.), *Seamounts: biodiversity and fisheries*. Fisheries Centre Research Reports 12(5), Fisheries Centre, University of British Columbia, Vancouver.

Kramer, D. L., and M. R. Chapman. 1999. Implications of fish home range size and relocation for marine reserve function. *Environmental Biology of Fishes* 55: 65–79.

Longhurst, A., and D. Pauly. 1987. *Ecology of Tropical Oceans*. Academic Press, San Diego, CA.

MacCall, A. 1990. *Dynamic Geography of Marine Fish Populations*. University of Washington Press, Seattle.

NOAA. 2004. ETOPO2: 2-Minute Gridded Global Relief Data. http://www.ngdc.noaa.gov/mgg/fliers/01mgg04.html.

Pauly, D. 1998. Tropical fishes: patterns and propensities. Pp. 1–17 in T. E. Langford, J. Langford, and J. E. Thorpe (eds.), Tropical fish biology. *Journal of Fish Biology* 53 (Supplement A).

Pauly, D. 2004. *Darwin's Fishes: An Encyclopedia of Ichthyology, Ecology and Evolution*. Cambridge University Press, Cambridge, England.

Pauly, D. 2010. *Gasping fish and panting squids: oxygen, temperature and the growth of water-breathing animals*. Excellence in Ecology (22), International Ecology Institute, Oldendorf/Luhe, Germany.

Pauly, D., and T.-E. Chua. 1988. The overfishing of marine resources: socioeconomic background in Southeast Asia. *AMBIO: A Journal of the Human Environment* 17(3): 200–206.

Phillips, S. J., R. P. Anderson, and R. E. Schapire. 2006. Maximum entropy modelling of species geographic distributions. *Ecological Modelling* 190: 231–259.

Pikitch, E. K., C. Santora, E. A. Babcock, A. Bakun, R. Bonfil, D. O. Conover, P. Dayton, P. Doukakis, D. L. Fluharty, B. Heneman, E. D. Houde, J. Link, P. A. Livingston, M. Mangel, M. K. McAllister, J. Pope, and K. J. Sainsbury. 2004. Ecosystem-based fishery management. *Science* 305: 346–347.

Sherman, K., M. C. Aquarone, and S. Adams. 2007. *Global applications of the Large Marine Ecosystem concept*. NOAA Technical Memorandum NMFS NE, 208.

Smith, C. L. 1997. *National Audubon Society Field Guide to Tropical Marine Fishes of the Caribbean, the Gulf of Mexico, Florida, the Bahamas, and Bermuda*. Alfred A. Knopf, Inc., New York.

UNEP-WCMC et al. 2010. *Global distribution of coral reefs, compiled from multiple sources*. UNEP World Conservation Monitoring Centre. http://data.unep-wcmc.org/datasets/1.

Zeller, D., and D. Pauly. 2001. Visualisation of standardized life history patterns. *Fish and Fisheries* 2(4): 344–355.

NOTES

1. Cite as Palomares, M. L. D., W. W. L. Cheung, V. W. Y. Lam, and D. Pauly. 2016. The distribution of exploited marine biodiversity. Pp. 46–58 in D. Pauly and D. Zeller (eds.), *Global Atlas of Marine Fisheries: A Critical Appraisal of Catches and Ecosystem Impacts*. Island Press, Washington, DC.
2. Predicate logic is a generic term for systems of abstract thought applied in fuzzy logic. In this example, the first-order logic predicate is "IF maximum weight is large," and the second-order logic predicate is "AND versatility is moderate." The resulting function, i.e., the conclusion category based on the predefined rules matrix in table 4.3, is "THEN maximum effective distance is farthest."

CHAPTER 5

THE *SEA AROUND US* CATCH DATABASE AND ITS SPATIAL EXPRESSION[1]

Vicky W. Y. Lam, Ar'ash Tavakolie, Maria L. D. Palomares, Daniel Pauly, and Dirk Zeller
Sea Around Us, University of British Columbia, Vancouver, BC, Canada

Although the fisheries catch reconstructions whose rationale and methods are described in chapters 1, 2, and 3 are all available online for checking (see www.seaaroundus.org), the taxonomically disaggregated time series of catch data they contain, covering 61 years (1950–2010), four fishing sectors (industrial, artisanal, subsistence, and recreational), two catch types (landed vs. discarded catch), and two types of reporting status (reported vs. unreported) for the Exclusive Economic Zones (EEZs) of all maritime countries and territories of the world or parts thereof (n = 273), are too large to be presented as flat tables in scientific or other articles, however detailed.

Therefore, the catch data generated by the reconstruction project of the *Sea Around Us* are stored in a dedicated catch reconstruction database, which interacts with the other databases of the *Sea Around Us* to generate various products, foremost among them spatially allocated fisheries catches to the 180,000 half-degree latitude and longitude cells covering the world's oceans. These products are freely available through our website and are likely to be widely used.

Because global catch maps and related products that are meaningful in terms of ecology as well as policy are one of the major outputs of the *Sea Around Us*, and because the spatial allocation process is closely tied to the catch reconstruction database, this database and the spatial allocation process are described together in this chapter.

CATCH RECONSTRUCTION DATABASE

The catch reconstruction database comprises all the catch reconstruction data for 1950 to 2010 by fishing country, taxon name, year of catch, catch amount, fishing sector, catch type, reporting status, input data source, and spatial location of catch, such as EEZ, FAO area, or other area designation (if applicable). The database is further subdivided into three different data layers, which include a layer with the catch taken by a fishing country in its own EEZ (called layer 1), the catch by each fishing country in other EEZs or the high seas (layer 2), and the catch of all tuna and large pelagic species caught by each fishing country's industrial fleet (layer 3). The basic structures of layers 1 and 2 are identical, and layer 3 differs slightly in structure because of the nature of the large pelagic input datasets (see chapter 3).

The process of data integration into the catch reconstruction database includes a data verification process, which is the first integration step undertaken after the original reconstruction dataset and associated reconstruction report are received after review by senior *Sea Around Us* staff. After the data verification process is completed for each country dataset, each record is allocated to one of the layers based on the taxon, sector, and area where the taxon was caught.

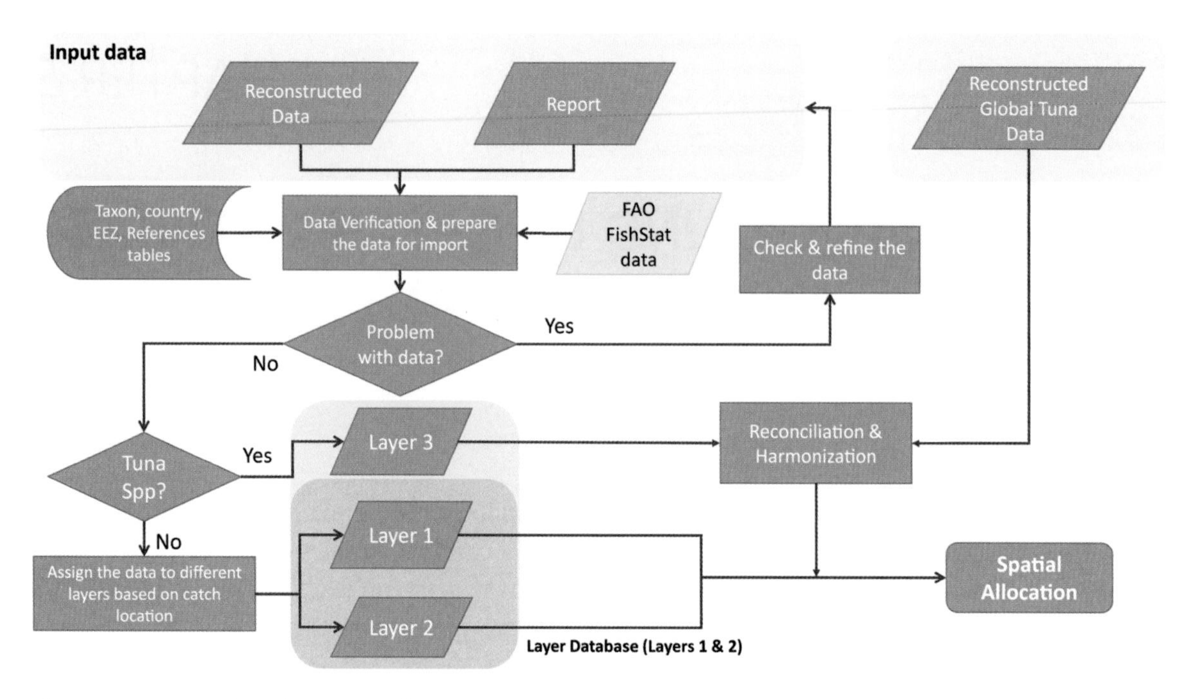

Figure 5.1. Data verification process for catch reconstruction data of the *Sea Around Us*. Details for the country- or territory-specific "Reconstructed Data" and "Report" are provided in chapters 1 and 2, and details for "Reconstructed Global Tuna Data" are described in chapter 3.

Data Verification Process

After initial, detailed review of each country or territory reconstruction dataset and associated technical report by senior Sea Around Us staff, the reconstruction dataset for each EEZ is further verified for accuracy and is formatted to fit the structure of the final database (see figure 5.1 for overview). For example, the total reported landings presented in the reconstruction dataset of each country or territory (which represent the catches landed and deemed reported to national authorities from within the EEZ of that country or territory) are compared with the reported data as present by FAO on behalf of the respective country or territory for each year. Any surplus FAO data are then considered to have been caught outside the EEZ of the given country or territory and thus are treated as part of layer 2 data. When an issue with the reconstructed catch data is identified, the issue is raised with the *Sea Around Us* catch reconstruction team and the original authors of the reconstruction for further checking and refining of the input data (figure 5.1). Additional data verification steps include harmonization of scientific taxon names in the reconstruction data with the official, recognized and standardized taxon names via the global taxonomic authorities of FishBase (www.fishbase.org) and SeaLifeBase (www.sealifebase.org; see chapter 4). Fishing country names and EEZ names are also checked and standardized against the Sea Around Us spatial databases. The fishing country and EEZ names allow us to link the catch data to the foreign fishing access database, which contains information on which fishing country can access the EEZ of another country (see "Foreign Fishing Access Database" section on p. 62).

Based on the location where each taxon was deemed to have been caught, each catch record is then assigned to a different layer (see "Structure of the Database" below). This includes a cross-checking process between the various reconstruction input datasets. For example, if an information source for Country A reported the landings of another country (B) in the EEZ of Country A, this catch of Country B is checked against the data in layer 2 of Country B, as provided through Country B reconstruction data. Emphasis is placed on avoiding double counting of catches.

Structure of the Database

As outlined above, the catch reconstruction database contains the catch data assigned to one of three layers.

Layer 1

This layer retains all the catches taken by a country within that country's own EEZ. It contains industrial, artisanal, subsistence, and recreational sector catches, subdivided by catch type (retained and landed vs. discarded catch) and reporting status (reported vs. unreported). However, this layer excludes all industrial catches of large pelagic fishes by a given country (see table 5.1 for the list of reported taxa excluded here), which are moved to layer 3 for later harmonization with the reconstructed global tuna data presented in chapter 3 (see also figure 5.1).

Layer 2

This layer contains data derived either directly from the reconstruction datasets and reconstruction technical reports (i.e., catches listed as being taken outside the country's own EEZ) or indirectly by subtracting the reconstructed catch identified as reported landings in a country's own EEZ (i.e., reported landings in layer 1) from the data reported by FAO on behalf of that country in the relevant (i.e., the "home" FAO area of a given fishing country) FAO area (excluding the taxa listed in table 5.1). Also, layer 2 includes catches by a given fishing country in all nonhome FAO areas (i.e., we refer to these catches as being taken by the distant-water fleets of the country in question). This layer includes only industrial fishing sector catches, as we define all fleets or gears that can operate outside of a given country's own EEZ waters as industrial (i.e., large-scale) in nature. The few documented cases where locally so-called artisanal fleets fish in neighboring EEZs, such as long-distance Senegalese pirogues operating in neighboring Mauritania or Guinea-Bissau (Belhabib et al. 2014), are internally reassigned to the industrial sector.

Table 5.1. Tuna and large pelagic taxa ($n = 29$) initially moved from country reconstruction datasets to layer 3 and later harmonized with industrially caught tuna and other large pelagic fishes (see also chapter 3).

Common name	Taxon name
Albacore	*Thunnus alalunga*
Atlantic bluefin tuna	*Thunnus thynnus*
Atlantic bonito	*Sarda sarda*
Atlantic sailfish	*Istiophorus albicans*
Atlantic white marlin	*Tetrapturus albidus*
Bigeye tuna	*Thunnus obesus*
Billfishes	Istiophoridae
Black marlin	*Makaira indica*
Black skipjack	*Euthynnus lineatus*
Blackfin tuna	*Thunnus atlanticus*
Blue marlin	*Makaira nigricans*
Bullet tuna	*Auxis rochei*
Indo-Pacific blue marlin	*Makaira mazara*
Indo-Pacific sailfish	*Istiophorus platypterus*
Kawakawa	*Euthynnus affinis*
Little tunny	*Euthynnus alletteratus*
Longbill spearfish	*Tetrapturus pfluegeri*
Longtail tuna	*Thunnus tonggol*
Mediterranean spearfish	*Tetrapturus belone*
Pacific bluefin tuna	*Thunnus orientalis*
Shortbill spearfish	*Tetrapturus angustirostris*
Skipjack tuna	*Katsuwonus pelamis*
Slender tuna	*Allothunnus fallai*
Southern bluefin tuna	*Thunnus maccoyii*
Striped marlin	*Kajikia audax*
Swordfish	*Xiphias gladius*
Tuna	*Thunnus* spp.
Wahoo	*Acanthocybium solandri*
Yellowfin tuna	*Thunnus albacares*

Layer 3

This layer initially included twenty-nine specific large pelagic taxa (table 5.1), whose reconstructed industrial catch data were moved to this layer for harmonization with the independently and globally reconstructed large pelagic dataset (chapter 3). The global tuna dataset combined taxonomically more diverse large pelagic catch datasets and added bycatch and discards associated with the global industrial tuna and large pelagic fisheries.

Thus, the final large pelagic dataset (harmonized layer 3, figure 5.1) contains about 140 taxa and their associated catches.

FOREIGN FISHING ACCESS DATABASE

The foreign fishing access database, initially derived from the fishing agreement database of FAO (1999), contains observed foreign fishing records and publicly accessible fishing agreements and treaties that were signed by fishing countries and the host countries in whose EEZs the foreign fleets were allowed to fish. In addition, the database also includes the start and end years of agreements or observed access. The type of access is also specified as "assumed unilateral," "assumed reciprocal," "unilateral," or "reciprocal." Also, the type of agreement is recorded in the database, and the agreement can be classified into bilateral agreements such as partnerships, multilateral agreements such as international conventions or agreements with regional fisheries organizations, private, licensing, or exploratory agreements. Additional information contained in this database relates to the type of taxa likely to be targeted by foreign fleets (e.g., tuna vs. demersal taxa), as well as any available data on fees paid or quotas included in the agreements.

Note that some agreements are not made public by either fishing country or host country. Thus, it is not readily possible to differentiate between foreign legal or illegal catches.

This database is used in conjunction with the catch reconstruction database and the taxon distribution database (see chapter 4) in the spatial allocation process that assigns catches to the global *Sea Around Us* half-degree latitude and longitude cell system.

SPATIAL ALLOCATION PROCEDURE

The spatial allocation procedure, although it relies on the same global *Sea Around Us* grid of half-degree cells that was used previously, is different from the approach used in the early phase of the *Sea Around Us* (until 2006) as described in Watson et al. (2004). In the earlier allocations, catches pertaining to large reporting areas (mainly FAO areas; see figure 2.1 in chapter 2) were allocated directly to the half-degree cells, subject only to constraints provided by initially derived taxon distributions for the various taxa (Close et al. 2006) and an earlier and more limited version of the fishing access database granting foreign fleets differential access to the EEZs of various countries (Watson et al. 2004). Thereafter, the catch by a given fishing country in a given EEZ was obtained by summing the catch that had been allocated to the cells making up the EEZ of that country (Watson et al. 2004). This process made the spatial allocation overly sensitive to the precise shape and cell probabilities of the taxon distribution maps and the precision of very problematic EEZ access rules for different countries. It often resulted in sudden and unrealistic shifts of allocated catches into and out of given EEZs purely through the lifting or imposing of EEZ access constraints. Attempts to improve the initial allocation procedure with more internal rules made it unwieldy, fragile, and extremely time consuming, and thus the *Sea Around Us* abandoned this approach in the mid-2000s.

The more structured allocation procedure that was devised as a replacement, and is described here (figure 5.2), relies on catch data that are spatially preassigned (through in-depth catch reconstructions; see chapter 2) to the EEZ or EEZ-equivalent waters (for years predating the declaration of individual EEZs) of a given maritime country or territory and, in the case of small-scale fisheries (i.e., the artisanal, subsistence, and recreational sectors), to their Inshore Fishing Areas (IFAs; Chuenpagdee et al. 2006). This radically reduces the number of access rules and constraints that the allocation procedure must consider, avoids fish catches showing up in the EEZs of the wrong country, and dramatically reduces the processing times of the allocation procedure.

Watson et al. (2004) also used the spatial allocation process to simultaneously disaggregate (i.e., taxonomically improve) uninfor-

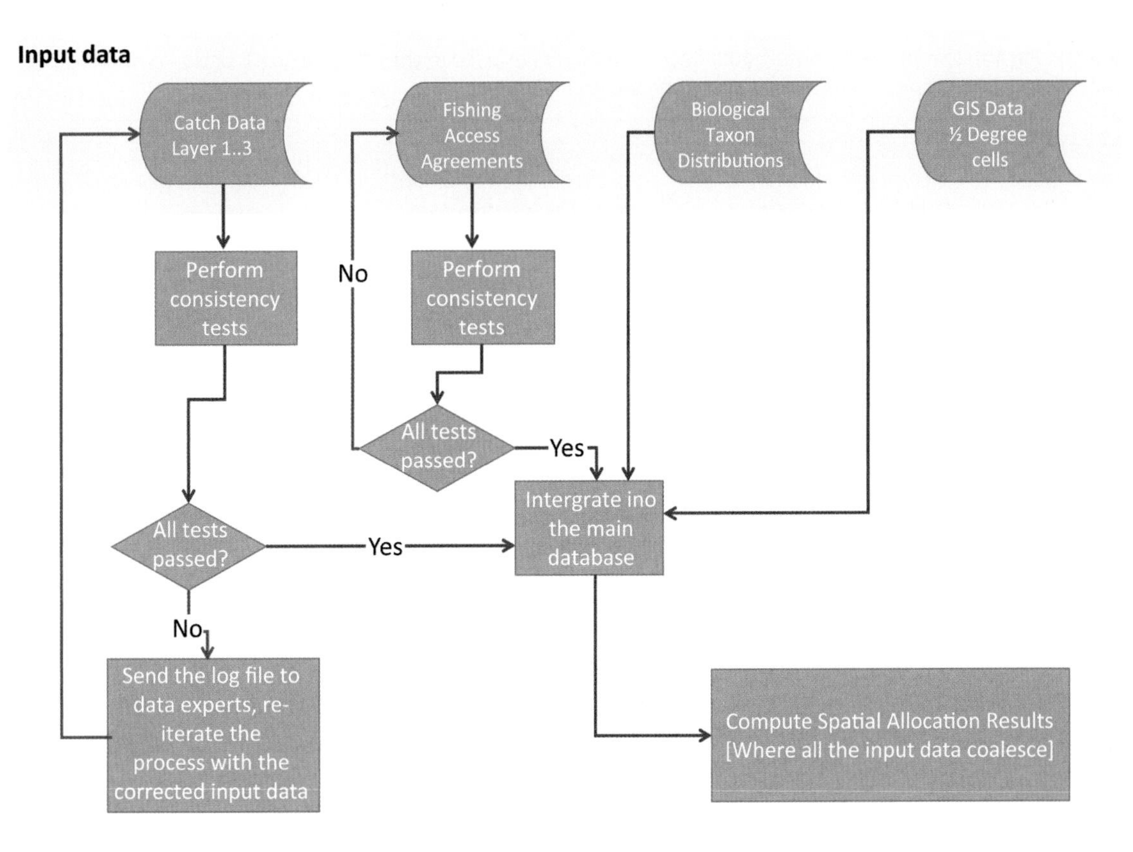

Figure 5.2. Spatial allocation procedure for catch reconstruction data of the *Sea Around Us*, resulting in the final half-degree allocated cell data. Details for the biological taxon distributions are provided in chapter 4.

mative taxa such as "miscellaneous marine fishes" (FAO term: "marine fishes nei") by relying on taxonomic information in neighboring half-degree cells. This further added to the complexity of the allocation procedure and increased the difficulty of tracing actual country, taxon, and catch entities through the process. This step was also discontinued in the new allocation approach. Instead, our new allocation procedure disaggregates the input catch data as part of the country-by-country catch reconstruction process (chapter 2), which therefore transparently documents the taxonomic changes in the associated technical report for each reconstruction. Within the catch reconstruction database, we keep track of the quality of the taxonomic disaggregation such that indicators sensitive to the quality of the disaggregation are not computed from inappropriate data (see "Catch Composition" in chapter 2).

These preallocation data processing modifications allow focusing on the truly spatial elements of the allocation, which are handled through a series of conceptual algorithmic steps. The general algorithm of spatial allocation of catches is harmonized for layers 1, 2, and 3 (table 5.2), which allows a better software flow while maintaining the conceptual differences in data layers. This starts with an overview of the new allocation process (figure 5.2), followed by a description of the unique features of each data layer and how it is handled, and ends with an overview of the general algorithm of the spatial allocation.

The spatial allocation of the catch is the process of computing the catch that can be allocated to each half-degree cell based on the overlap of three main components: the catch data, the fishing access observations and agreements, and the biological taxon distributions (figure 5.2; table 5.2). The overlap of these components is facilitated by a series of geographic information system (GIS) layers, which essentially bind them together.

Table 5.2. Parameters of the 3 spatial catch data input layers as used in the catch reconstruction database and in the spatial allocation to half-degree cells of the *Sea Around Us*.

Data layer	1	2	3
Taxa included	All except industrial large pelagics	All except industrial large pelagics	Large pelagics (n = 140+)
Spatial scope	Country's own EEZ	Other EEZs and high seas	Global tuna cells
Sectors	Industrial, artisanal, subsistence, recreational	Industrial	Industrial
Distributions	Biological	Biological	Biological
Fishing access	Automatically granted	Used	Used
Granularity of data	EEZ, IFA[a]	EEZ, high seas, ICES, CCAMLAR, NAFO, FAO, and other areas	Six types of tuna cells: 1 × 1, 5 × 5, 5 × 10, 10 × 10, 10 × 20, 20 × 20 degrees

[a]Inshore Fishing Area (IFA), defined as the area up to 50 km from shore or 200 m depth, whichever comes first (Chuenpagdee et al. 2006). Note that IFAs occur only along inhabited coastlines.

How Each Data Layer Is Conceptually Unique and How It Is Handled

In layer 1, the data come spatially organized by each fishing entity's EEZs. The allocation algorithm allocates the small-scale catches (i.e., artisanal, subsistence, and recreational) only to the cells associated with the IFA (Chuenpagdee et al. 2006) of that fishing entity's EEZs, while industrial catches can be allocated anywhere within that fishing entity's EEZs, as long as they remain compatible with the biological taxon distributions. Fishing access agreements are not applicable to this data layer, because a fishing entity (i.e., country) is always allowed to fish in its own EEZ waters.

To represent the historical expansion of industrial fishing since the 1950s in each country's waters, from more easily accessible areas closer to shore to the full extent of each country's EEZ, we use the depth adjustment function for domestic industrial catches described in Watson and Morato (2013). This function takes into account that, as domestic industrial catches increase over time, an increasing fraction are being taken progressively further offshore (but within EEZ boundaries).

In layer 2, the spatial granularity of the input catch data can be by EEZ, high seas, or any other form of regional reporting areas (i.e., ICES, CCAMLR, NAFO, or FAO statistical areas). However, in all cases it excludes the fishing entity's (fishing country's) own EEZ waters (which are treated in layer 1; table 5.2). In layer 2, the fishing access observations and agreements are used to delineate the areas that allow fishing for a particular fishing entity, year, and taxon. Once this area is computed, it is superimposed on the biological taxon distributions to derive the final layer 2 catch allocation.

In layer 3, which covers only industrial large pelagic fishes and their associated bycatch and discards, the input catch data are spatially organized by larger tuna cells, which range from 1 × 1 to 20 × 20 degrees (table 5.2; see also chapter 3). Similar to the region-specific areas in layer 2, these larger cells are intersected with all the EEZ boundaries to create a GIS layer that is suitable for use in the algorithm. Thereafter, the fishing access observations and agreements and taxon distributions are applied to calculate the final layer 3 catch allocation.

An Overview of the Algorithm

The spatial allocation algorithm consists of four main steps:

1. Validating and importing the fishing access observations and agreements database
2. Validating and importing the catch reconstruction database
3. Importing the biological taxon distributions
4. Computing the catch that can be allocated to each half-degree cell for each catch data layer in an iterative process

(allowing verifications and corrections to any of the input parameters)

1. Validating and Importing the Fishing Access Observations and Agreements Database

The fishing access observations and agreements are first verified using several consistency and matching tests (figure 5.2), and upon passing they are imported into the main allocation database. This fishing access information is subsequently used in two different processes: the verification process of the catch data (layers 1, 2, and 3) and the computing of the areas where a given fishing entity (i.e., country) is allowed to fish for a specific year and taxon.

2. Validating and Importing the Catch Reconstruction Database

The validating and importing of the catch data is a more complex process than the validating and importing process for the fishing access database. This process involves about twenty-two different preallocation data tests (figure 5.2), designed to make sure that the data are coherent from the standpoint of database logic. These tests range from simple tests such as "is the TaxonKey valid?" to more complex tests such as "validate if the given fishing entity has the required fishing access observations/agreements to fish in the given marine area." Every single row of catch data is examined via these tests, and if it passes all tests the data row in question is added to the main allocation database. If it fails *any* of the tests it is returned to the relevant *Sea Around Us* data experts for review, often involving the original authors of the catch reconstruction (figure 5.2). This process is repeated until all the data rows pass all the preallocation tests.

The process of importing the catch reconstruction database includes an important submodule for harmonizing the marine areas. This module is crucial, because the catch data come in a variety of different spatial reporting areas that are not globally homogeneous in GIS definitions (e.g., the EEZ of Albania is one entity, whereas the EEZs of India, Brazil, and the United States are subdivided into states or provinces; the northeast Atlantic uses ICES statistical areas). To harmonize these marine areas and make them accessible to the core allocation process, any given half-degree cell is split into its constituent countries' EEZs and high seas components. Then, the fishing access observations and agreements are applied to this layer to determine which of these "shards" of half-degree cells are allowing access to a given fishing entity. Once this is determined, these collections of "shards" are assigned to the given row of catch data; the result is a harmonized view of all the different marine areas. At present we have assigned more than 12,000 marine areas into their constituent "shards" of half-degree cells; these marine areas range from EEZs and LMEs to ICES, CCAMLR, NAFO, and FAO statistical areas. The procedure allows future marine areas to be readily assigned.

3. Importing Biological Taxon Distributions

Importing the biological taxon distributions is a fairly straightforward process. The more than 2,500 individual distributions (see chapter 4) are generated as individual text files (comma-separated values [csv] format) containing for each half-degree cell the specific taxon's probability of occurrence. These individual taxon distribution files are compiled into a database table for further use.

4. Computing and Allocating the Catch to Half-Degree Cells

Once steps 1, 2, and 3 are completed, we perform the computations that yield the final spatial half-degree allocation results. The catch of a given data row, TotalCatch, of taxon T is distributed among eligible half-degree cells, Cell 1 . . . *n*, using the following weighted average formula:

$$\text{Cell}_{i\,\text{AllocatedCatch}} = \text{Total Catch} \times \frac{\text{Cell}_{i\,\text{Surface Area}} \times \text{Cell}_{i\,\text{Relative Abundance of Taxon T}}}{\sum_{1}^{n} \text{Cell}_{i\,\text{Surface Area}} \times \text{Cell}_{i\,\text{Relative Abundance of Taxon T}}}$$

Throughout the allocation process, catch reconstruction parameters in addition to year and taxon, such as fishing sector, catch type, data layer, and reporting status, are preserved and carried over into the final half-degree allocated database.

Final Output

The final results of the intense and detailed database preparation and spatial allocation are time series of catches by half-degree cells that are ecologically reliable (i.e., taxa are caught where they occur and in relation to their relative abundance) and politically likely (e.g., by fishing country and within EEZ waters to which they have access), as documented in the summaries, p. 185 onward.

CONCLUSIONS

The previous global catch data spatialization of the *Sea Around Us* was akin to hacking a first path through the jungle of official catch data: One could go from place to place (i.e., EEZs) but with the dark suspicion that behind the data foliage, thousands of tonnes of misallocated data monsters lay hidden, ready to pounce. And pounce they did (i.e., colleagues complained of gross errors), which is why the new allocation system was developed.

This new system is more akin to a French garden, partitioned by clear lines (the EEZ boundaries), defining spaces into which known quantities (the reconstructed catches) are planted, with only the occasional poodle crossing the lines (i.e., errors will be small, easy to isolate and to correct). We invite and look forward to colleagues' assistance in tending this garden.

ACKNOWLEDGMENTS

We thank Reg Watson, whose pioneering work on spatializing the FAO catch database for the years 1950 to 2006 cleared the way for the reconceptualization presented here. We also thank Danielle Knip for initiating the reconstructed catch database. This is a contribution of the *Sea Around Us*, a research activity at the University of British Columbia initiated and funded by the Pew Charitable Trusts from 1999 to 2014 and currently funded mainly by the Paul G. Allen Family Foundation.

REFERENCES

Belhabib, D., V. Koutob, A. Sall, V. W. Y. Lam, and D. Pauly. 2014. Fisheries catch misreporting and its implications: the case of Senegal. *Fisheries Research* 151: 1–11.

Chuenpagdee, R., L. Liguori, M. L. D. Palomares, and D. Pauly. 2006. Bottom-up, global estimates of small-scale marine fisheries catches. *Fisheries Centre Research Reports* 14(8), University of British Columbia, Vancouver.

Close, C., W. W. L. Cheung, S. Hodgson, V. Lam, R. Watson, and D. Pauly. 2006. Distribution ranges of commercial fishes and invertebrates. Pp. 27–37 in M. L. D. Palomares, K. I. Stergiou, and D. Pauly (eds.), *Fishes in databases and ecosystems*. Fisheries Centre Research Reports 14(4), University of British Columbia, Vancouver.

FAO. 1999. FAO Fisheries Agreement Register (FARISIS). Committee on Fisheries, 23rd Session, February 15–19, 1999. COFI/99/ INF.9E.

Watson, R., A. Kitchingman, A. Gelchu, and D. Pauly. 2004. Mapping global fisheries: sharpening our focus. *Fish and Fisheries* 5: 168–177.

Watson, R., and T. Morato. 2013. Fishing down the deep: accounting for within-species changes in depth of fishing. *Fisheries Research* 140: 63–65.

NOTE

1. Cite as Lam, V. W. Y., A. Tavakolie, M. L. D. Palomares, D. Pauly, and D. Zeller. 2016. The *Sea Around Us* catch database and its spatial expression. Pp. 59–67 in D. Pauly and D. Zeller (eds.), *Global Atlas of Marine Fisheries: A Critical Appraisal of Catches and Ecosystem Impacts*. Island Press, Washington, DC.

CHAPTER 6

THE ECONOMICS OF GLOBAL MARINE FISHERIES[1]

Ussif Rashid Sumaila,[a] Andrés Cisneros-Montemayor,[a] Andrew J. Dyck,[a] Ahmed S. Khan,[a] Vicky W. Y. Lam,[b] Wilf Swartz,[a] and Lydia C. L. Teh[a]
[a]Fisheries Economic Research Unit, University of British Columbia, Vancouver, BC, Canada
[b]*Sea Around Us*, University of British Columbia, Vancouver, BC, Canada

Fishing is an economic activity, with global connections and linkages, including to other sectors of the global economy. Therefore, there is a need to study the economics of fisheries on a decidedly global basis and not generate pseudo-global coverage through case studies of dubious representativeness. However, this perspective is new to the study of fisheries, and important datasets were initially not available on a global basis to support such work. Therefore, the starting point for global fisheries economics work by the *Sea Around Us* (later in close collaboration with the Fisheries Economics Research Unit) was the creation, documentation, and preliminary analyses of several global databases. Over the last 15 years, several global databases were created and updated on

- ex-vessel fish prices (Sumaila et al. 2007; Swartz et al. 2013)
- cost of fishing (Sumaila et al. 2008; Lam et al. 2011)
- fisheries employment (Teh and Sumaila 2013)
- fisheries subsidies (Sumaila and Pauly 2006; Khan et al. 2006; Sumaila et al. 2010, 2013)
- recreational fisheries (Cisneros-Montemayor and Sumaila 2010)
- economic multipliers (Dyck and Sumaila 2010).

These studies were all conducted based on global marine fisheries landings data reported by FAO on behalf of its member countries. Because these official landings data are often much lower than the reconstructed total catches presented on pp. 185–457 of this atlas, it can be expected that the economic impact of fisheries is underestimated in several of the studies presented in this volume. Therefore, they will be gradually updated in the coming years. In the meantime, the database on which these studies were based will be made available through the website of the *Sea Around Us* (www.seaaroundus.org); they can be used to undertake large-scale bioeconomic analyses, as illustrated by Srinivasan et al. (2010, 2012) and Sumaila et al. (2012).

EX-VESSEL FISH PRICE DATABASE

The global Ex-vessel Fish Price Database described in Sumaila et al. (2007) was the first comprehensive database that presented average annual ex-vessel prices for all commercially exploited marine fish species and higher taxa, and does so by country. It contained more than 30,000 reported price items, covering the period from 1950 to the mid-2000s, and supplemented missing prices with estimates based on prices from a different year, species (group), or fleet nationality. The contribution of Sumaila et al. (2007) was updated and expanded by Swartz et al. (2013), who also

Table 6.1. Total fishing cost by type and by FAO region in 2003 (based on Lam et al. 2011).

	Fishing cost (US$ million)							
FAO region	**Fuel**	**Running**	**Repairs**	**Labor**	**Depreciation**	**Interest**	**Variable**	**Total**
Africa	1,739	1,218	683	1,511	888	597	5,151	6,636
Asia	8,272	7,396	4,372	14,968	5,067	3,142	35,008	43,217
Europe	2,340	3,594	2,896	6,767	2,163	774	15,597	18,534
North America	1,800	1,033	676	4,724	122	1,122	8,234	9,478
Oceania	588	865	640	1,383	260	158	3,475	3,893
Caribbean, South and Central America	1,779	1,200	1,061	1,650	516	301	5,690	6,507

revised the method for estimating missing prices. Key advantages of the new estimation approach are that it allows a larger number of observed prices to be used in the estimation of missing prices and better accounts for relative price differences between countries. This price database is linked to the *Sea Around Us* catch database (chapter 5) and thus allows estimation of landed value for any year and spatial area in the world.

These prices suggest that the worldwide marine fisheries catches, as reported by FAO (i.e., not yet incorporating the reconstructed catch data), had an ex-vessel value of US$100 billion in 2005, which is higher than the value of US$80–85 billion used in the studies documented further below. Also note that the landed values of reconstructed catches, which are generally higher than officially reported catches (see chapter 2), will be even higher than the US$100 billion estimate.

COST OF FISHING

The database of fishing cost (Lam et al. 2011) was the first and so far is the only global cost of fishing database to be documented in the primary literature. It provides crucial economic information that is needed for assessing the economics of global fisheries and should be useful for developing sustainable management scenarios. The database, which covers the 114 countries that landed approximately 98% of global marine fish catch in 2003 (based on FAO data), deals with two broad cost categories: variable and fixed costs. Variable costs include fuel cost, salaries for crew members, and repair and maintenance cost, whereas fixed costs include interest and depreciation costs of the invested capital (i.e., boats).

Costs varied between fishing gear types, with dredge and hook-and-line having the highest variable fishing costs and North America having the highest unit variable cost among regions (table 6.1). The global average variable cost per tonne of catch in 2003 was estimated to be US$1,125, and the corresponding total global variable fishing cost was US$86 billion. Given a global ex-vessel value of reported marine fisheries landings of about US$80–85 billion per year, this implies that the global fishing fleet is running at an annual operating loss of about US$1–6 billion per year without subsidies. Table 6.1 also provides summaries by regions of total variable fishing cost and average variable fishing cost per tonne of catch.

The updating of this database (beside the necessary harmonization with the reconstructed catches) will consist of structuring it, and the data it contains, in terms of fishing cost by commodity type (i.e., type of resources caught, e.g., small pelagic fishes, tuna, or demersal fishes) rather than by country only, because this will allow for finer-grained comparative analyses.

FISHERIES EMPLOYMENT

Marine fisheries contribute to the global economy in various ways, from the catching of fish to the provision of support services for the fishing industry. A general lack of detailed data and uncertainty about the level of employment in marine fisheries can lead to underestimation of fishing effort and hence overexploited fisheries or can result in inaccurate projections of economic and societal costs and benefits. To address this gap, a database of marine fisheries employment for 144 coastal countries was compiled. Gaps in employment data that emerged were filled using a Monte Carlo approach to estimate the number of direct and indirect fisheries jobs (Teh and Sumaila 2013).

This study emphasized the small-scale sectors, and more precisely artisanal fisheries, which globally provide the most jobs, and subsistence fisheries (for detailed definitions, see chapter 2). Around 260 ± 6 million people were found to be employed in global marine fisheries, encompassing full-time and part-time jobs in catching, processing, marketing, and otherwise handling fish, corresponding to 203 ± 5 million full-time equivalent jobs. Of these, 22 ± 0.45 million would be small-scale fishers, a figure similar to the estimate previously published by Chuenpagdee et al. (2006).

The results of this study can be used to improve management decision making and highlight the need to improve monitoring and reporting of the number of people employed in marine fisheries globally. Table 6.2 provides a summary of these employment estimates.

FISHERIES SUBSIDIES

Building on the publication of Munro and Sumaila (2002), a global database of subsidies provided to marine fisheries was developed and documented in Sumaila and Pauly (2006) and Khan et al. (2006), and updated in Sumaila et al. (2010, 2013). Therein, subsidies are grouped into three categories: "beneficial" ("good"), "capacity-enhancing" ("bad"), and "ambiguous" ("ugly"). The basis for this classification is the potential impact of given subsidy types on the sustainability of the fishery resource. Beneficial subsidies enhance the conservation of fish stocks over time; this includes subsidies that fund fisheries management and funds dispensed to establish (McCrea-Strub et al. 2011) and operate marine protected areas (Cullis-Suzuki and Pauly 2010). In contrast, capacity-enhancing subsidies, such as fuel subsidies, lead to overcapacity and ultimately overexploitation. Ambiguous subsidies can lead to either the conservation or overfishing of a given fish stock, such as buyback subsidies, which if not properly designed can lead to overcapacity (e.g., when a fisher can use the funds obtained from such schemes for the down payment on a new, more powerful vessel).

These estimates of global fisheries subsidies were presented to the World Trade Organization in Geneva and shaped numerous debates (Swartz and Sumaila 2013). The most recent update of these estimates (Sumaila et al. 2013), commissioned by and presented in October 2013 to the Fisheries Committee of the European Parliament in Brussels, are summarized in figure 6.1 by type and region. Figure 6.2 shows the similarity of the global sum to the estimates back to 2006 (once they are adjusted for inflation) and their differences from estimates published earlier by FAO (much higher) and the World Bank (lower).

Table 6.2. Marine fisheries employment worldwide in 2003 (thousands of jobs; adapted from Teh and Sumaila 2011).

Region	Direct	Indirect	Total employment
Europe	800 ± 71	1,700 ± 160	2,500 ± 230
Asia	40,000 ± 3,100	190,000 ± 15,000	230,000 ± 18,000
Africa	3,000 ± 150	14,000 ± 770	18,000 ± 910
South America	1,700 ± 330	3,900 ± 400	5,600 ± 710
Oceania	710 ± 120	160 ± 20	870 ± 130
North and Central America	3,000 ± 230	2,300 ± 170	5,400 ± 400

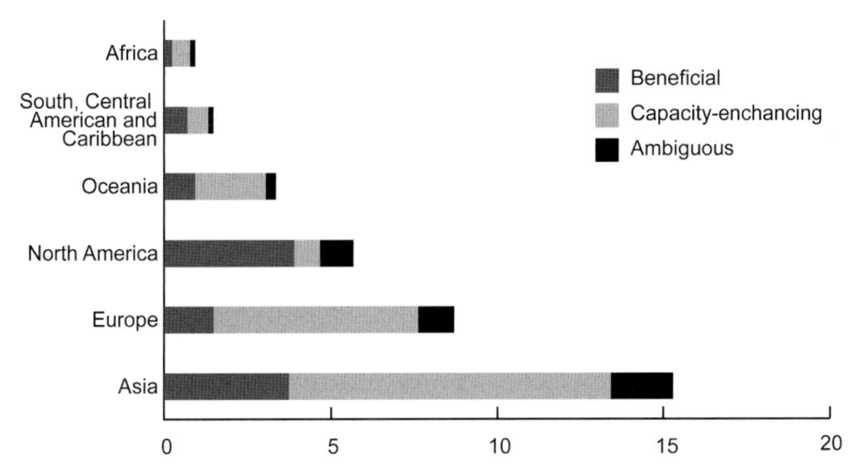

Figure 6.1. Government subsidies to marine fisheries by type and region. This shows that Asia is by far the greatest subsidizing region (43% of total), followed by Europe (25% of total) and North America (16% of total). For all regions, the amount of capacity-enhancing subsidies is higher than other categories, except for North and South America, which have higher beneficial subsidies. Adapted from FAO (1992), Milazzo (1998), Sumaila and Pauly (2006), and Sumaila et al. (2010).

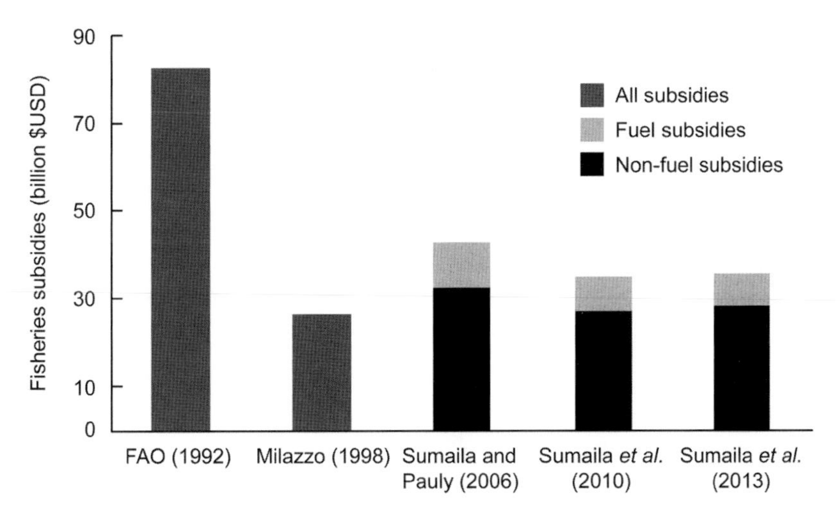

Figure 6.2. Successive global estimates of government subsidies. Adapted from FAO (1992), Milazzo (1998), Sumaila and Pauly (2006), and Sumaila et al. (2010, 2013). Note that the numbers in this graph were all converted via consumer price index to real 2009 US$, to make them mutually comparable.

Note that the current global estimates of about US$35 billion (in 2009 constant dollars) of government subsidies to marine fisheries may not need to be revised upward in consideration of higher reconstructed catches, even though low official catches were used for estimating subsidies for some countries, together with estimates of "subsidy intensity" from their neighbors (see Khan et al. 2006). The reason is that the estimates of subsidy intensity (i.e., subsidy per Catch Values for a given period), having been computed from catches that were too low, were themselves too high and thus, when multiplied with catches that were too low, should have yielded subsidy estimates that were not systematically biased one way or another.

ECOSYSTEM-BASED MARINE RECREATIONAL VALUES

Participation in ecosystem-based marine recreational activities (MRAs) has increased around the world, adding a new dimension to human uses of marine ecosystems and another good reason to implement effective management measures. A first step in studying the effects of MRAs at a global scale is to estimate their socioeconomic benefits, which are captured

Table 6.3. Summary of ecosystem-based marine recreation valuation results. All estimates are for 2003 (adapted from Cisneros-Montemayor and Sumaila 2010).

Item	Recreational fishing	Whale watching	Diving and snorkeling	Total
Expenditure (2003 US$; 10^9)	39.7	1.6	5.5	46.8
Participation (10^6)	58	13	50	121
Employment (10^3)	954	18	113	1,085

by three indicators: the amount of participation, employment, and direct expenditure by users (Cisneros-Montemayor and Sumaila 2010). A database of reported expenditure on MRAs was compiled for 144 coastal countries and a meta-analysis performed to calculate the yearly global benefits of MRAs in terms of expenditure, participation, and employment. It was estimated that more than 120 million people a year participate in MRAs, generating US$47 billion (2003) in expenditures and supporting one million jobs. The results of this study have several implications for resource managers and for the tourism industry. Aside from offering the first estimation of the global socioeconomic benefits of MRAs, this work provides insights into the drivers of participation and possible ecological impacts of these activities. Our results could also help direct efforts to promote adequate implementation of MRAs. Summaries of the numbers produced are provided in table 6.3.

MARINE FISH POPULATIONS' CONTRIBUTION TO THE WORLD ECONOMY

Although the gross revenue from marine capture fisheries is estimated to be US$80–85 billion annually (based on catches reported to FAO by member countries), a vast number of secondary economic activities, from boat building to running fish restaurants, are supported by the world's marine fisheries (see Christensen et al. 2014). Yet these related activities are rarely considered when evaluating the economic impact of fisheries. A study applying an input–output method was conducted to estimate the total direct, indirect, and induced impact of marine capture fisheries on the world economy (Dyck and Sumaila 2010). Specifically, the goal was to estimate the total output in an economy that is dependent (at least partially) on the output of marine fisheries. Herein, Leontief's "technological coefficients" at the catch levels reported by FAO for the early 2000s were used to estimate total output supported by marine fisheries throughout the economy (Dyck and Sumaila 2010). Although results suggest that there is a great deal of variation in fishing output multipliers between regions and countries (tables 6.4 and 6.5), the output multipliers suggest, at the global level, that the direct and indirect impacts of the marine fisheries sector are about three times the value of the landings at first sales (i.e., between US$225 and US$235 billion per year). In tables

Table 6.4. Output impacts in 2003 of the world's marine fisheries by region (values in US$ billion; from Dyck and Sumaila 2010), based on catch levels reported by FAO for the early 2000s.

Region	Landed value	Economic impact	Average multiplier
Africa	2.10	5.46	2.59
Asia	49.89	133.31	2.67
Europe	11.45	35.78	3.12
South America	7.20	14.78	2.05
North America	8.23	28.92	3.52
Oceania	5.22	17.06	3.27
World total	**84.10**	**235.31**	**2.80**

Table 6.5. Income impacts in 2003 of the world's marine fisheries by region (values in US$ billion; from Dyck and Sumaila 2010), based on catch levels reported by FAO for the early 2000.

Region	Landed value	Income effect	Average multiplier
Africa	2.10	1.30	0.62
Asia	49.89	35.30	0.71
Europe	11.45	8.70	0.76
South America	7.20	4.06	0.56
North America	8.23	10.06	1.22
Oceania	5.22	3.80	0.73
World total	**84.10**	**63.22**	**0.75**

6.4 and 6.5 we present output and income multipliers by region, respectively.

CONCLUSION

The databases described herein, and accessible via the *Sea Around Us*, should contribute to the gradual broadening of marine resource economics from the analysis of case studies and regional or national fisheries to the study and understanding of global marine populations and their use. One important advantage of global-scale studies is that they may reveal patterns, problems, and solutions that cannot be readily distinguished at smaller scales (Rosenthal and DiMatteo 2001). More importantly, however, we live in an increasingly globalized world, where many challenges raised by our use of marine resources are global. To deal with these challenges, we need global-level studies, and the contributions described here are important initial steps toward truly global fisheries economics.

ACKNOWLEDGMENTS

We thank the members of the Fisheries Economics Research Unit and the *Sea Around Us*, both at the University of British Columbia, for various inputs to this study. Thanks also to Sylvie Guénette and Patrizia Abdallah for translations from French and Spanish, respectively, and to Andrew Sharpless of Oceana for access to several sources of information. We also are grateful to Paulo Augusto Lourenço Dias Nunes, Ramiro Parrado, and especially Daniel Pauly for comments on earlier drafts of this contribution. We thank the *Sea Around Us*, a research activity at the University of British Columbia initiated and funded by the Pew Charitable Trusts from 1999 to 2014 and currently funded mainly by the Paul G. Allen Family Foundation. We are also grateful to the participants of the World Bank seminar on subsidies on October 30, 2006, particularly Bill Shrank and Matteo Milazzo.

REFERENCES

Christensen, V., S. de la Puente, J. C. Sueiro, J. Steenbeek, and P. Majluf. 2014. Valuing seafood: the Peruvian fisheries sector. *Marine Policy* 44: 302–311.

Chuenpagdee, R., L. Liguori, M. L. D. Palomares, and D. Pauly. 2006. *Bottom-up, global estimates of small-scale marine fisheries catches*. Fisheries Centre Research Reports 14(8), University of British Columbia, Vancouver, Canada.

Cisneros-Montemayor, A. M., and U. R. Sumaila. 2010. A global valuation of ecosystem-based marine recreation. *Journal of Bioeconomics* 12: 245–268.

Cullis-Suzuki, S., and D. Pauly. 2010. Marine Protected Area costs as "beneficial" fisheries subsidies: a global evaluation. *Coastal Management* 38(2): 113–121.

Dyck, A. J., and U. R. Sumaila. 2010. Economic impact of ocean fish populations in the global fishery. *Journal of Bioeconomics* 12: 227–243.

FAO. 1992. Marine fisheries and the Law of the Sea: a decade of change. Special chapter (revised) in *The State of Food and Agriculture* 1992, FAO Fisheries Circular 853. United Nations Food and Agriculture Organization, Rome.

Khan, A., U. R. Sumaila, R. Watson, G. Munro, and D. Pauly. 2006. The nature and magnitude of global non-fuel fisheries subsidies. Pp. 5–37 in U. R. Sumaila and D. Pauly (eds.), *Catching more bait: a bottom-up re-estimation of global fisheries subsidies*. Fisheries Centre Research Reports 14(6), University of British Columbia, Vancouver, Canada.

Lam, V. W. L., U. R. Sumaila, A. Dyck, D. Pauly, and R. Watson. 2011. Construction and potential applications of a global cost of fishing database. *ICES Journal of Marine Science* 68(9): 1–9.

McCrea-Strub, A., D. Zeller, U. R. Sumaila, J. Nelson, A. Balmford, and D. Pauly. 2011. Understanding the cost of establishing marine protected areas. *Marine Policy* 35: 1–9.

Milazzo, M. 1998. *Subsidies in world fisheries: a re-examination*. World Bank Technical Paper, Fisheries Series 406, The World Bank, Washington, DC.

Munro, G., and U. R. Sumaila. 2002. The impact of subsidies upon fisheries management and sustainability: the case of the North Atlantic. *Fish and Fisheries* 3: 233–250.

Rosenthal, R., and M. R. DiMatteo. 2001. Meta-analysis: recent developments in quantitative methods for literature reviews. *Annual Review of Psychology* 52: 59–82.

Srinivasan, U. T., W. Cheung, R. Watson, and U. R. Sumaila. 2010. Food security implications of global marine catch losses due to overfishing. *Journal of Bioeconomics* 12: 183–200.

Srinivasan, U. T., R. Watson, and U. R. Sumaila. 2012. Global fisheries losses at the Exclusive Economic Zone level, 1950 to present. *Marine Policy* 36: 544–549.

Sumaila, U. R., W. W. L. Cheung, A. Dyck, K. Gueye, L. Huang, V. W. Y. Lam, D. Pauly, T. Srinivasan, W. Swartz, D. Pauly, and D. Zeller. 2012. Benefits of rebuilding global marine fisheries outweigh costs. *PLoS ONE* 7(7): e40542.

Sumaila, U. R., A. Khan, A. Dyck, R. Watson, G. Munro, P. Tyedmers, and D. Pauly. 2010. A bottom-up re-estimation of global fisheries subsidies. *Journal of Bioeconomics* 12: 201–225.

Sumaila, U. R., V. Lam, F. Le Manach, W. Swartz, and D. Pauly. 2013. *Global Fisheries Subsidies* [EBook]. European Parliament, Brussels.

Sumaila, U. R., A. D. Marsden, R. Watson, and D. Pauly. 2007. Global Ex-vessel Fish Price Database: construction and applications. *Journal of Bioeconomics* 9: 39–51.

Sumaila, U. R., and D. Pauly. 2006. *Catching more bait: a bottom-up re-estimation of global fisheries subsidies*. Fisheries Centre Research Reports 14(6), University of British Columbia, Vancouver, Canada.

Sumaila, U. R., L. Teh, R. Watson, P. Tyedmers, and D. Pauly. 2008. Fuel price increase, subsidies, overcapacity, and resource sustainability. *ICES Journal of Marine Science* 65: 832.

Swartz, W., and U. R. Sumaila. 2013. Fisheries governance, subsidies and the World Trade Organization. Pp. 30–44 in I. Ekeland, D. Fessler, J. M. Lasry, and D. Lautier (eds.), *The Ocean as a Global System: Economics and Governance of Fisheries and Energy Resources*. ESKA Publishing, Portland, OR.

Swartz, W., R. Sumaila, and R. Watson. 2013. Global ex-vessel fish price database revisited: a new approach for estimating "missing" prices. *Environmental and Resource Economics* 56(4): 467–480.

Teh, L., and U. R. Sumaila. 2013. Contribution of marine fisheries to worldwide employment. *Fish and Fisheries* 14(1): 77–88.

NOTE

1. Cite as Sumaila, U. R., A. Cisneros-Montemayor, A. Dyck, A. Khan, V. W. Y. Lam, W. Swartz, and L. C. L. Teh. 2016. The economics of global marine fisheries. Pp. 68–75 in D. Pauly and D. Zeller (eds.), *Global Atlas of Marine Fisheries: A Critical Appraisal of Catches and Ecosystem Impacts*. Island Press, Washington, DC.

CHAPTER 7

GLOBAL EVALUATION OF HIGH SEAS FISHERY MANAGEMENT[1]

Sarika Cullis-Suzuki and Daniel Pauly
Sea Around Us, University of British Columbia, Vancouver, BC, Canada

Fishing is no longer a coastal phenomenon (Ban et al. 2013a; O'Leary et al. 2012). Over the last half century, advances in fishing technology coupled with coastal stock declines have prompted fisheries to expand beyond coastal waters and out into the high seas (Swartz et al. 2010). These previously inaccessible areas beyond national jurisdiction, that is, beyond the 200-nmi Exclusive Economic Zones (EEZs) of maritime countries, offered access to previously unexploited and extremely valuable fish stocks, especially of tuna. Global fish catch from the high seas thus increased tremendously (see chapter 3). However, limited regulation in these remote areas of ocean and inadequate management quickly led to severe stock declines (FAO 2009).

Regional fisheries management organizations (RFMOs) are intergovernmental bodies tasked with managing fish stocks found mostly in the high seas areas of the world ocean (figure 7.1; table 7.1).

Established by and made up of "member countries," often maritime countries around the part of the world ocean covered by the RFMO in question but also including any country with a "real interest" in the specific fishery, these members must manage, conserve, and ensure the long-term sustainability of the fisheries resources in their remit (UN 1982, 1995). This has proved to be a difficult task, and RFMOs face many challenges, from structural difficulties (e.g., catch allocation to new members; Munro 2007) to internal problems (e.g., data deficiencies; Collette et al. 2011; O'Leary et al. 2012), regional issues (including illegal fishing, corruption, and lack of enforcement; Pintassilgo et al. 2010; Sumaila et al. 2007), and broader problems associated with noncompliance with international treaties (Bjorndal and Munro 2003). These issues are not new and have been discussed for many years (Schiffman 2013), but their complexity has inhibited RFMO progress.

In 2010, in response to declining high seas stock trends and the observation that "RFMO performance has not lived up to expectation" (Lodge et al. 2007), a first global evaluation of the effectiveness of RFMOs was conducted (Cullis-Suzuki and Pauly 2010a, 2010b). Here, the key results of this analysis are summarized and updated, based on feedback from RFMO representatives, input from colleagues, and, where available, current data from recent stock assessments.

UPDATES TO GLOBAL RFMO DATABASE

There are currently nineteen marine RFMOs with management capacity (figure 7.1; table 7.1). Over the last decade, international calls for increasing RFMO coverage have been met (FAO 2012); today, almost every part of the global ocean is covered by at least one RFMO.

GLOBAL EVALUATION OF RFMO EFFECTIVENESS

2010 Study: Failing the High Seas

In 2010, a study was published that assessed the global effectiveness of nineteen current

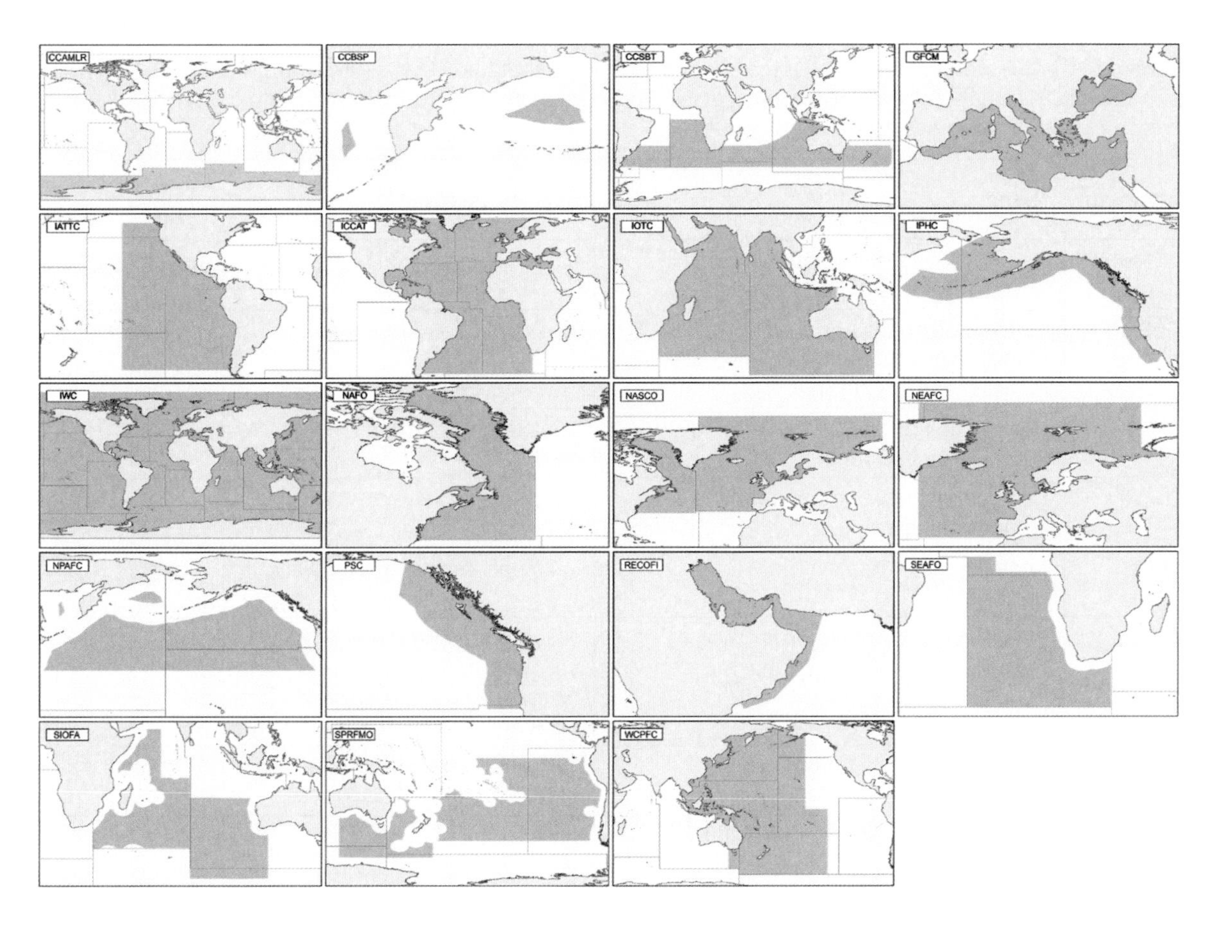

Figure 7.1. Maps representing each of the nineteen individual RFMOs and their designated convention areas. Thin red lines represent the border of FAO statistical areas.

RFMOs (table 7.1). This study, titled "Failing the High Seas: A Global Evaluation of Regional Fisheries Management Organizations" (Cullis-Suzuki and Pauly 2010b), assessed the overall performance of RFMOs as determined by how well they achieved management and conservation objectives mandated by international treaties (UN 1982, 1995). This was based on a two-tiered approach: assessing the effectiveness of RFMOs on paper and on the ground.

To assess RFMO effectiveness on paper, each RFMO was scored against a set of twenty-six best practice criteria developed from Lodge et al. (2007), where each criterion had ten possible scores, ranging from 1 to 10 (see also Alder et al. 2001). In addition to the eighteen RFMOs, two "outgroups" were also scored to test the criteria's discriminating ability: the World Wildlife Fund (an environmental nongovernment organization) and the U.S. National Marine Fisheries Service (a national fisheries management agency). A cluster analysis clearly identified the two non-RFMOs as outgroups, thus demonstrating that the criteria used in the study could distinguish between non-RFMOs and RFMOs (Cullis-Suzuki and Pauly 2010a). Across RFMOs, results revealed an average score of 57%, with a range of 43% to 74%. Out of five overarching categories, the highest scores were for "General information and organization," and the worst were for "Allocation" (Cullis-Suzuki and Pauly 2010a).

To assess RFMO effectiveness on the ground, stock assessments and scientific data were used to determine the state of stocks. Through plotting of relative fishing mortality and biomass data points, a score was obtained that reflected whether the stock was overfished or depleted (details below). Results showed that two thirds of fish stocks on the high seas and under RFMO management were either overfished or depleted, matching

Table 7.1. Average scores across RFMOs in 2010 and 2013. Note that five RFMOs (PSC, RECOFI, SEAFO, SIOFA, and SPRFMO) lacked sufficient data to be assessed. For supplementary information including score calculations and stock-specific data, see www.seaaroundus.org.

Acronym	Name	Species assessed	Mean score (%)	
			2010	2013
CCAMLR	Commission for the Conservation of Antarctic Marine Living Resources	Patagonian toothfish	100.0	100.0
CCBSP	Convention on the Conservation and Management of the Pollock Resources in the Central Bering Sea	Alaska pollock	33.3	33.3
CCSBT	Commission for the Conservation of Southern Bluefin Tuna	Southern bluefin tuna	0.0	33.3
GFCM	General Fisheries Commission for the Mediterranean	Sardine, anchovy	33.3	33.3
IATTC	Inter-American Tropical Tuna Commission	Yellowfin, bigeye, and skipjack tuna	33.3	77.8
ICCAT	International Commission for the Conservation of Atlantic Tunas	Bluefin tuna (West and East), yellowfin and skipjack tuna (West and East), bigeye and albacore tuna (North and South)	37.5	25.0
IOTC	Indian Ocean Tuna Commission	Yellowfin, albacore tuna, and bigeye tuna	77.8	77.8
IPHC	International Pacific Halibut Commission	Pacific halibut	33.3	33.3
IWC	International Whaling Commission	Fin, blue, sperm, right, sei, Bryde's, humpback, and minke whales (2 stocks)	33.3	33.3
NAFO	Northwest Atlantic Fisheries Organization	Redfish, cod (2 stocks), American plaice, Greenland halibut	41.7	20.0
NASCO	North Atlantic Salmon Conservation Organization	Atlantic salmon	33.3	33.3
NEAFC	North East Atlantic Fisheries Commission	Blue whiting, mackerel, golden redfish, herring	75.0	41.7
NPAFC	North Pacific Anadromous Fish Commission	Sockeye, chum, and pink salmon	77.8	77.8
PSC	Pacific Salmon Commission	—		—
RECOFI[a]	Regional Commission for Fisheries	—	N/Aa	—
SEAFO	South East Atlantic Fisheries Organization	—		—
SIOFA	South Indian Ocean Fisheries Agreement	—		—
SPRFMO	South Pacific Regional Fisheries Management Organization	—		—
WCPFC	Western and Central Pacific Fisheries Commission	Yellowfin, albacore, bigeye, and skipjack tuna	66.7	83.3
Total		**46 stocks**	**48.3**	**50.2**

[a]The RECOFI was not assessed in 2010; although RECOFI entered into force in 2001, it still does not provide enough information in its reports to assess the current state of stocks in its remit. RECOFI covers all marine organisms in waters of its member states: Bahrain, Iraq, Iran, Kuwait, Oman, Qatar, Saudi Arabia, and the United Arab Emirates.

estimates described by FAO (2009). The average score across RFMOs was 49%, ranging from 0% to 100% (table 5.1). There was no correlation between scores on paper and on the ground, suggesting a disconnect between RFMO intentions and actions.

Current Updated Evaluation

For this update, the focus is on recalculating RFMO effectiveness on the ground. Setbacks in determining RFMO effectiveness on paper centered mostly on data attributes. First, without standardization such data can be difficult to score (Kjartan Hoydal, NEAFC, personal communication, 2013). Furthermore, publicly accessible information can be limited or complicated to locate, or RFMOs can fail to provide information, resulting in a low score. Finally, high compliance does not always correlate with healthy fisheries, as suggested above and also shown in Alder et al. (2001). Thus by focusing on a quantitative and internationally recognized description of stock status, a framework is obtained that is more easily standardized (Froese and Proelss 2012).

To compute scores on the ground, forty-six fish stocks under current management across the fourteen different RFMOs with sufficient information for assessment were evaluated (table 7.1). Of the forty-eight stocks assessed in 2010, three were subsequently excluded after comments from RFMO managers (see Cullis-Suzuki and Pauly 2010a); three stocks were replaced with different stocks of the same species in response to data constraints and availability, and one new stock was added to the current study. Scores were calculated by plotting B/B_{MSY} against F/F_{MSY}, where B is the current stock biomass, F the current fishing mortality rate, and B_{MSY} and F_{MSY} generally accepted sustainability limits of biomass and fishing mortality rates, respectively (for scoring details see Q scores in Cullis-Suzuki and Pauly 2010b). Each plot had four quadrants; depending on which quadrant the data occupied, the stock was given a score of 0 (red quadrant: overfished and depleted, i.e., "threatened"), 1 (yellow quadrant: overfished or depleted, i.e., "at risk"), or 3 (green quadrant: not overfished or depleted; i.e., "stable"); see figure 7.2 for an example plot for the International Commission for the Conservation of Atlantic Tunas (ICCAT).

Since the 2010 evaluation, ten stocks have changed score (five have gone up, and five went down), and another two have moved from an "overfished" state to a "depleted" one, with no overall change to their score. The updated results reveal that currently, nearly three quarters of stocks are in poor condition, with 20% being threatened (i.e., overfished and depleted) and 52% being at risk (i.e., overfished or depleted). However, there has been a slight improvement of overall average stock scores across RFMOs, from 48.3% in 2010 to 50.2% (table 7.1).

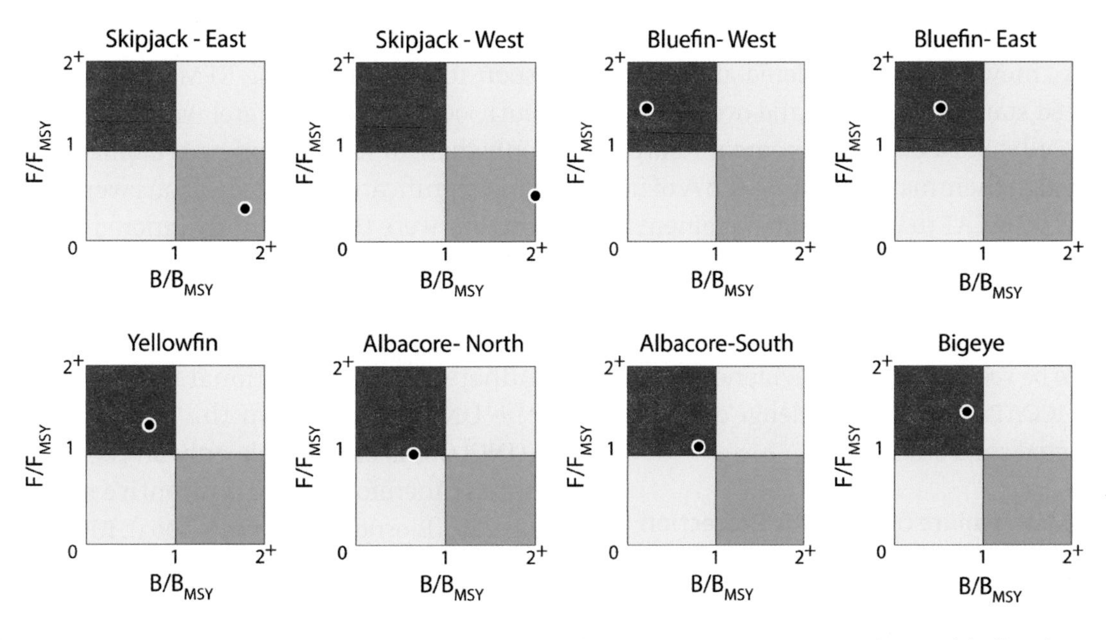

Figure 7.2. Current phase plots of eight principal tuna species under the management of ICCAT (similar phase plots for the other RFMOs with stock assessment results may be found at www.seaaroundus.org).

RECENT DEVELOPMENTS

Given that some RFMOs are doing better than others, and despite some steps toward progress (de Bruyn et al. 2013), it remains overwhelmingly clear that RFMOs are in need of improvement (Webster 2013). This is emphasized here through three important international events that have transpired over the last few years and reflect various aspects of the underwhelming performance of RFMOs.

ICCAT: Bluefin Tuna and CITES

In 2009, in response to the severely depleted state of Atlantic bluefin tuna (*Thunnus thynnus*), Monaco put forward the proposal to protect this species under the Convention on International Trade in Endangered Species of Wild Fauna and Flora (CITES; Nayar 2010). With the world's most profitable fish species on the line (Swing 2013), vocal tuna fishing countries fought the proposal, arguing that ICCAT, not CITES, was the appropriate regulating body to implement tuna management measures; the proposal was thus promptly defeated (FAO 2012). ICCAT, long criticized for its poor management record (Safina 1993), ranks here among the lowest scoring RFMOs (table 7.1) with six out of eight of their stocks qualifying as both overfished and depleted. Since 2010, none of ICCAT's stocks have improved: Five have stayed the same, and three decreased in score, moving from a depleted and underexploited state to a depleted and overfished one (see Cullis-Suzuki and Pauly 2010a; figure 7.2). Although there may have been a flurry of initial effort by ICCAT to improve management after the CITES ban failed (and their organization was brought under international scrutiny; Webster 2011), benefits on the ground have yet to be seen. There is no evidence to suggest that ICCAT is up to the challenge of rebuilding depleted tuna stocks.

CCAMLR: Failure of Antarctic Protection

With only one stock being assessed, CCAMLR scores the highest among RFMOs (table 7.1). Historically considered to be one of the better RFMOs in regard to conservation (Gilman et al. 2014), CCAMLR has garnered praise for their management contributions (Bodin and Österblom 2013). Over the last few years, however, CCAMLR has faced mounting criticism for failing to protect parts of Antarctica. The most recent attempt to protect the Ross Sea, supported by strong member countries such as the United States, New Zealand, and the EU, was defeated. This was the latest in a series of proposals calling for more protection in Antarctic waters; indeed, over the last year and a half, multiple proposals have been suggested, and all have been defeated (Cressey 2013). Although such a request for protection was supported by many countries, it was still vulnerable to being blocked by a few, exemplifying yet another fundamental problem associated with RFMOs: mandatory consensus for matters of import among members (Lodge et al. 2007).

SPRFMO: New RFMO, No More Fish

In 2012, after years of deliberation, SPRFMO finally came into force, and its establishment was deemed "a ground breaking development" (UN 2010). But in the time it took SPRFMO to be instituted, SPRFMO's dominant fishery, Chilean jack mackerel (*Trachurus murphyi*), suffered a decline of almost two thirds (ICIJ 2012). Indeed, over the course of 20 years, hastened by the impending regulatory input from the soon-to-be-functioning RFMO (UN 2010), by 2011, 90% of jack mackerel were gone. It was only then, at 10% of its original biomass levels, that significant catch reductions were agreed on (but were then promptly ignored the next year; Gjerde et al. 2013).

The jack mackerel story highlights a big challenge of the current RFMO framework: Adherence to international treaties such as the UN Convention on the Law of the Sea (UNCLOS) is enforceable only on cooperating states. Therefore, there is incentive not to cooperate (Bjorndal and Munro 2003). RFMOs can be seen as penalizing members by imposing catch restrictions; non-RFMO members, in contrast, are not bound by regulations, and thus the postponement of RFMO establishment becomes, in the short term, beneficial to fishing states (Gjerde et al. 2013). Despite advances in our understanding of these structural failures, it is disappointing to learn that such basic problems continue to plague even the newest of RFMOs.

THE ROAD AHEAD FOR RFMOS

Combining the results of Cullis-Suzuki and Pauly (2010b) with those of the present evaluation suggests that RFMOs are not effective management and conservation bodies on the high seas. Furthermore, they have not substantially improved over the last few years, as determined here by the state of the stocks in their remit. This is further supported by the recent rejection by RFMOs of conservation-based recommendations from the international community (Cressey 2013; UN 2010). Additionally, one of the biggest impediments to conducting such a study is dependence on available stock assessments (not to mention relevant reference points); these data are pronouncedly lacking in RFMOs and cause serious setbacks to stock evaluation (Froese and Proelss 2012; Powers and Medley 2013).

High seas management appears to be in a state of uncertainty: Recommended best practices have yet to be seriously implemented by RFMOs (Lodge et al. 2007), and strengthened international commitments under UN treaties still await consideration (Druel et al. 2012; UN 2012). Although many documents outline possible avenues for high seas improvement (Ban et al. 2013a; Clark 2011; Druel and Gjerde 2014; Englender et al. 2014; Pew Environment Group 2012; Veitch et al. 2012; Clark 2014), the high seas remain among the least understood and least protected ecosystems in the world (Ban et al. 2013b).

In May 2010, shortly after the contribution by Cullis-Suzuki and Pauly (2010b) was accepted for publication, the authors were invited to present their findings at the UN headquarters in New York during the UN Fish Stocks Agreement Review Conference,[2] and the first author attended (Cullis-Suzuki 2010). The turnout for this panel, which was organized by staff of the Pew Charitable Trusts and included two other marine scientists and a lawyer, was unexpectedly large and consisted mainly of RFMO delegates, many of whom expressed strong reservations and criticisms about the presentation on the results of the RFMO evaluation (Cullis-Suzuki 2010). Indeed, not only did they overwhelmingly reject its results, but many disagreed with the underlying data, although these data originated for the most part from the stock assessments the RFMOs themselves conducted and made available on their websites. An hour and a half of denunciations during the post-talk question period led to the extraordinary consensus among the delegates that RFMOs could not be the source of these unfavorable data and that any critique of RFMOs was unwarranted. Later on, as a follow-up to this presentation and to the supporting publication, the authors received e-mails from RFMO managers with detailed criticisms of the research presented in Cullis-Suzuki and Pauly (2010b). As a result of one such commentary, three stocks were eliminated from the initial Q score assessment; however, this did not affect the results (Cullis-Suzuki and Pauly, 2010a).

This and a similar but less intense experience of the second author at an event in early 2014 in Stockholm, Sweden, where the updated RFMO evaluation results were presented, exemplify what is perhaps the greatest obstacle to RFMOs achieving sustainably managed fisheries: RFMOs were created to allocate catch between competing fleets. RFMOs are fishery-oriented bodies first and conservation bodies second, if at all. RFMO delegates therefore represent fisheries' interests (Gjerde et al. 2013), and thus RFMO operations reflect their primary objective, which is to catch as much fish as possible, now (Webster 2013).

This basic organizational focus lies at the heart of failed management on the high seas. Moving away from allocation-based targets might mean recasting RFMOs as conservation bodies (Webster 2013), which could in turn lead to the start of the fundamental reform so urgently needed on the high seas (Gjerde et al. 2013). Actually, the more effective policy, as suggested by White and Costello (2014) and Sumaila et al. (2015), would be the globally more equitable policy to permanently close the entire high seas to fishing and to let maritime countries throughout the world benefit from the resulting resource recovery in their own EEZs.

ACKNOWLEDGMENTS

This is a contribution of the *Sea Around Us*, a research activity at the University of British Columbia initiated and funded by the Pew Charitable Trusts from 1999 to 2014 and currently funded mainly by the Paul G. Allen Family Foundation. We also thank Dr. Kjartan Hoydal of NEAFC for comments at two presentations on the data herein.

REFERENCES

Alder, J., G. Lugten, R. Kay, and B. Ferris. 2001. Compliance with international fisheries instruments. Pp. 55–80 in T. Pitcher, U. R. Sumaila, and D. Pauly (eds.), *Fisheries impacts on North Atlantic ecosystems: evaluations and policy exploration*. Fisheries Centre Research Reports 9(5). University of British Columbia, Vancouver, Canada.

Ban, N. C., N. J. Bax, K. M. Gjerde, R. Devillers, D. C. Dunn, P. K. Dunstan, A. J. Hobday, S. M. Maxwell, D. M. Kaplan, R. L. Pressey, J. A. Ardron, E. T. Game, and P. N. Halpin. 2013a. Systematic conservation planning: a better recipe for managing the high seas for biodiversity conservation and sustainable use. *Conservation Letters* 7(1): 41.

Ban, N. C., S. M. Maxwell, D. C. Dunn, A. J. Hobday, N. J. Bax, J. Ardron, K. M. Gjerde, E. T. Game, R. Devillers, D. M. Kaplan, P. K. Dunstan, P. N. Halpin, and R. L. Pressey. 2013b. Better integration of sectoral planning and management approaches for the interlinked ecology of the open oceans. *Marine Policy* 49: 127–136.

Bjorndal, T., and G. Munro. 2003. The management of high seas fisheries resources and the implementation of the UN fish stocks agreement of 1995. Pp. 1–35 in H. Folmer and T. Tietenberg (eds.), *The International Yearbook of Environmental and Resource Economics 2003/2004*. Edward Elgar Publishing, Cheltenham, UK.

Bodin, Ö., and H. Österblom. 2013. International fisheries regime effectiveness: activities and resources of key actors in the Southern Ocean. *Global Environmental Change* 23: 948–956.

Clark, E. A. 2011. Strengthening regional fisheries management: an analysis of the duty to cooperate. *New Zealand Journal of Public & International Law* 9(2): 223–246.

Clark, N. 2014. *An analysis of the transparency of marine governance organizations*. Master's thesis, Nicholas School of the Environment of Duke University, Durham, NC.

Collette, B. B., K. E. Carpenter, B. A. Polidoro, M. J. Juan-Jordá, A. Boustany, D. J. Die, C. Elfes, W. Fox, J. Graves, L. R. Harrison, R. McManus, C. V. Minte-Vera, R. Nelson, V. Restrepo, J. Schratwieser, C.-L. Sun, A. Amorim, M. Brick Peres, C. Canales, G. Cardenas, S.-K. Chang, W.-C. Chiang, N. de Oliveira Leite Jr., H. Harwell, R. Lessa, F. L. Fredou, H. A. Oxenford, R. Serra, K.-T. Shao, R. Sumaila, S.-P. Wang, R. Watson, and E. Yáñez. 2011. High value and long life—double jeopardy for tunas and billfishes. *Science* 333(6040): 291–292.

Cressey, D. 2013. Third time unlucky for Antarctic protection bid. *Nature News*, November 1, 2013. doi:10.1038/nature.2013.14085.

Cullis-Suzuki, S. 2010. The U.N. experience. *Sea Around Us Project Newsletter*, May/June, pp. 4–5.

Cullis-Suzuki, S., and D. Pauly. 2010a. *Evaluating global regional fisheries management organizations: methodology and scoring* (revised edition). Fisheries Centre Working Paper #2009-12, University of British Columbia, Vancouver, Canada.

Cullis-Suzuki, S., and D. Pauly. 2010b. Failing the high seas: a global evaluation of regional fisheries management organizations. *Marine Policy* 34(5): 1036–1042.

de Bruyn, P., H. Murua, and M. Aranda. 2013. The precautionary approach to fisheries management: how this is taken into account by tuna regional fisheries management organisations (RFMOs). *Marine Policy* 38: 397–406.

Druel, E., and K. M. Gjerde. 2014. Sustaining marine life beyond boundaries: options for an implementing agreement for marine biodiversity beyond national jurisdiction under the United Nations Convention on the Law of the Sea. *Marine Policy* 49: 90–97.

Druel, E., P. Ricard, J. Rochette, and C. Martinez. 2012. *Governance of marine biodiversity in areas beyond national jurisdiction at the regional level: filling the gaps and strengthening the framework for action. Case studies from the North-East Atlantic, Southern Ocean, Western Indian Ocean, South West Pacific*. IDDRI and AAMP, Paris.

Englender, D., J. Kirschey, A. Stöfen, and A. Zink. 2014. Cooperation and compliance control in areas beyond national jurisdiction. *Marine Policy* 49: 186–194.

FAO. 2009. *The state of the world fisheries and aquaculture 2008*. Food and Agriculture Organization of the UN, Rome.

FAO. 2012. *The state of world fisheries and aquaculture 2012*. Food and Agriculture Organization of the UN, Rome.

Froese, R., and A. Proelss. 2012. Evaluation and legal assessment of certified seafood. *Marine Policy* 36(6): 1284–1289.

Gilman, E., K. Passfield, and K. Nakamura. 2014. Performance of regional fisheries management organizations: ecosystem-based governance of bycatch and discards. *Fish and Fisheries* 15: 327–351.

Gjerde, K. M., D. Currie, K. Wowk, and K. Sack. 2013. Ocean in peril: reforming the management of global ocean living resources in areas beyond national jurisdiction. *Marine Pollution Bulletin* 74(2): 540–551.

ICIJ. 2012. *Looting the Seas*. The International Consortium of Investigative Journalists. Digital Newsbook, The Center for Public Integrity.

Lodge, M. W., D. Anderson, T. Lobach, G. Munro, K. Sainsbury, and A. Willock. 2007. *Recommended Best Practices for Regional Fisheries Management Organizations*. Chatham House, London.

Munro, G. R. 2007. Internationally shared fish stocks, the high seas, and property rights in fisheries. *Marine Resource Economics* 22: 425–443.

Nayar, A. 2010. Bad news for tuna is bad news for CITES. *Nature News*, March 23, 2010. doi:10.1038/news.2010.139.

O'Leary, B. C., R. L. Brown, D. E. Johnson, H. von Nordheim, J. Ardron, T. Packeiser, and C. M. Roberts. 2012. The first network of marine protected areas (MPAs) in the high seas: the process, the challenges and where next. *Marine Policy* 36(3): 598–605.

Pew Environment Group. 2012. *What states want from Rio+20: The ocean*. Available at http://www.pewtrusts.org/en/imported-old/other-resources/2011/12/13/what-states-want-from-rio20-the-ocean.

Pintassilgo, P., M. Finus, M. Lindroos, and G. Munro. 2010. Stability and success of regional fisheries management organizations. *Environmental and Resource Economics* 46(3): 377–402.

Powers, J. E., and P. A. H. Medley. 2013. *An evaluation of the sustainability of global tuna stocks relative to Marine Stewardship Council criteria*. ISSF Technical Report 2013-01A. International Seafood Sustainability Foundation, Washington, DC.

Safina, C. 1993. Bluefin tuna in the West Atlantic: negligent management and the making of an endangered species. *Conservation Biology* 7(2): 229–234.

Schiffman, H. S. 2013. The South Pacific Regional Fisheries Management Organization (SPRFMO): an improved model of decision-making for fisheries conservation? *Journal of Environmental Studies and Sciences* 3(2): 209–216.

Sumaila, U. R., V. Lam, D. Miller, L. Teh, R. Watson, D. Zeller, W. Cheung, I. Côté, A. Rogers, C. Roberts, E. Sala, and D. Pauly. 2015. Winners and losers in a world where the high seas is closed to fishing. *Scientific Reports* 5: 8481.

Sumaila, U. R., D. Zeller, R. Watson, J. Alder, and D. Pauly. 2007. Potential costs and benefits of marine reserves in the high seas. *Marine Ecology Progress Series* 345: 305–310.

Swartz, W., E. Sala, R. Watson, and D. Pauly. 2010. The spatial expansion and ecological footprint of fisheries (1950 to present). *PLoS ONE* 5(12): e15143.

Swing, K. 2013. Inertia is speeding fish-stock declines. *Nature* 494(7437): 314.

UN. 1982. *United Nations Convention on the Law of the Sea*. December 10, 1982, Montego Bay, Jamaica.

UN. 1995. *Provisions of the United Nations Convention on the Law of the Sea of 10 December 1982 relating to the conservation and management of straddling fish stocks and highly migratory fish stocks*. United Nations conference on straddling fish stocks and highly migratory fish stocks, August 4, 1995, New York.

UN. 2010. *Report of the resumed review conference on the agreement for the implementation of the provisions of the United Nations Convention on the Law of the Sea of 10 December 1982 relating to the conservation and management of straddling fish stocks and highly migratory fish stocks*. New York, NY, Vol. 46587.

UN. 2012. *The future we want*. United Nations General Assembly, A/Res/66/288, July 27, 2012.

Veitch, L., N. K. Dulvy, H. Koldewey, S. Lieberman, D. Pauly, C. M. Roberts, A. D. Rogers, and J. E. M. Baillie. 2012. Avoiding empty ocean commitments at Rio+20. *Science* 336: 1383–1385.

Webster, D. G. 2011. The irony and the exclusivity of Atlantic bluefin tuna management. *Marine Policy* 35: 249–251.

Webster, D. G. 2013. International fisheries: assessing the potential for ecosystem management. *Journal of Environmental Studies and Sciences* 3(2): 169–183.

White, C., and C. Costello. 2014. Close the high seas to fishing? *PLoS Biology* 12.3: e1001826.

NOTES

1. Cite as Cullis-Suzuki, S., and D. Pauly. 2016. Global evaluation of high seas fishery management. Pp. 76–85 in D. Pauly and D. Zeller (eds.), *Global Atlas of Marine Fisheries: A Critical Appraisal of Catches and Ecosystem Impacts*. Island Press, Washington, DC.
2. www.un.org/depts/los/convention_agreements/review_conf_fish_stocks.htm.

CHAPTER 8

GLOBAL-SCALE RESPONSES AND VULNERABILITY OF MARINE SPECIES AND FISHERIES TO CLIMATE CHANGE[1]

William W. L. Cheung[a] and **Daniel Pauly**[b]
[a]Changing Oceans Research Unit and NF-UBC Nereus Program, University of British Columbia, Vancouver, BC, Canada
[b]*Sea Around Us*, University of British Columbia, Vancouver, BC, Canada

Of the various ways humans affect marine ecosystems, climate change may be the most insidious and unrecognized. In fact, even if they believe that it is occurring, most people think climate change is going to affect us later, and thus there is no real urgency. As we show below, however, climate change has already begun to affect us in multiple ways, including through effects on the oceans and marine fisheries. This chapter is thus devoted to documenting some of the work through which scattered observations on the effect of climate change on marine organisms were generalized and, in the process, the first global maps of observed and predicted climate change impacts on marine biodiversity and fisheries were produced, thus complementing work performed in the terrestrial realm.

Climate change affects ocean properties including water temperature, oxygen level, and acidity. According to the Intergovernmental Panel on Climate Change (IPCC) 5th Assessment Report (AR5), there is compelling evidence that the heat content and the stratification of the ocean have been increasing in the twentieth century, while sea-ice and pH have been decreasing, and that these trends can be expected to continue in the next century under the climate change scenarios considered by the IPCC (2014). Also, available evidence indicates that climate change is expected to result in expansion of oxygen minimum zones, changes in primary productivity, changes in ocean circulation patterns, sea level rises, and increase in extreme weather events.

Marine fishery catches consist almost solely of fishes and invertebrates that are biologically sensitive to changes in temperature, oxygen level, and other ocean conditions; thus we expect that fisheries are being affected by climate change (CC) and ocean acidification (OA). In the ocean, physiological performance of aquatic and marine water-breathing organisms is strongly dependent on temperature and oxygen (Pauly 1981, 2010; Pörtner 2010). Because of the higher viscosity of water and because dissolved oxygen occurs in water in lower concentration than in the atmosphere, it is energetically costly for water-breathing organisms to obtain oxygen for respiration from water (Pauly 1981, 2010). Thus, changes in oxygen supply and demand are expected to have large implications for respiration and other body functions of fishes and invertebrates (i.e., water-breathing ectotherms). Although low oxygen tolerance thresholds vary across species and life stages, they tend to be highest for large water-breathing ectotherms. When temperature becomes either too high or too low, oxygen supply capacity decreases relative to oxygen demand and thus limits metabolism.

The ranges between the lowest and highest temperatures that are tolerated by organisms are generally consistent with the variability of environmental temperatures they are generally exposed to and can change during their life cycles. Also, smaller individuals are more heat tolerant than large ones, in line with observations of declining animal body sizes in warming oceans (Daufresne et al. 2009). Increases in CO_2 in the ocean may also have direct and indirect effects on growth, reproduction, and survivorship of fishes and invertebrates, particularly those that form calcium carbonate exoskeleton (Kroeker et al. 2013).

An understanding of the physiological sensitivity and responses to ocean temperature, oxygen, acidity, and other water properties allows us to develop hypotheses about how climate change and ocean acidification will affect exploited fish stocks and fisheries. Theses hypotheses include the following:

- Given ocean warming, fishes and invertebrates will be shifting their distributions, mainly to higher latitude and deeper water to maintain their thermal niche.
- In nontropical systems, warmer-water species will increase their contribution to local catches.
- Maximum body size of fishes will decrease as the oceans become warmer and less oxygenated.
- Global marine catches will decline, particularly in the tropics.

To examine these hypotheses at the global scale, we conducted a series of empirical and theoretical studies that made use of species' distribution ranges (see chapter 4) and spatially explicit global catch data from the *Sea Around Us* (see chapter 5) to evaluate the effects of climate change and ocean acidification on the distribution of exploited species, the species composition of catches, and projected fisheries catch potential.

As a conceptually first step (though one taken last in the sequence of studies described below), the signature of the effects of ocean temperature change on species composition of fisheries catches was studied, using a newly developed metric called mean temperature of the catch (MTC). The second step was to develop a species distribution model, called the dynamic bioclimate envelope model (DBEM), that predicts changes in the distribution ranges of exploited marine species and the patterns of species richness in response to changing ocean conditions. Once developed, the DBEM was modified so as to progressively account for an increasing number of features, such as the population dynamics of the species included, their dispersal modes, interactions with other species, association with different habitats, oxygen requirements, and resistance to low pH. The third step was to use macroecological theory to derive the theoretical relationship between net primary production (NPP), biogeography, and fisheries catch potential and to express that relationship in a single empirical equation. Then, with projected changes in NPP and species distributions combined, future changes in distribution of fisheries catch potential and the maximum body size of exploited species could be projected. Finally, by use of the DBEM and basic principles of geometry and physiology, the effects of ocean warming and deoxygenation on the maximum body size of exploited fishes could be projected.

Mean Temperature of the Catch

Marine fishes and invertebrates exhibit physiological thermal tolerances that constrain them to live within a certain range of temperatures. Thus, for example, the seasonal migration of fishes up and down along the coast of northwest Africa tracks the seasonal temperature oscillations along that same coast (figure 8.1). Similarly, as the oceans warm up, fishes and invertebrates have to shift their distributions in order to maintain themselves in habitats with their preferred temperature. This results (at locations outside of the tropics) in changes in species composition, as the taxa increase in abundance that are adapted to warmer waters. For example, warming in the eastern Mediterranean (i.e., in Greece) caused a reduction of bearded horse mussel *Modiulus*

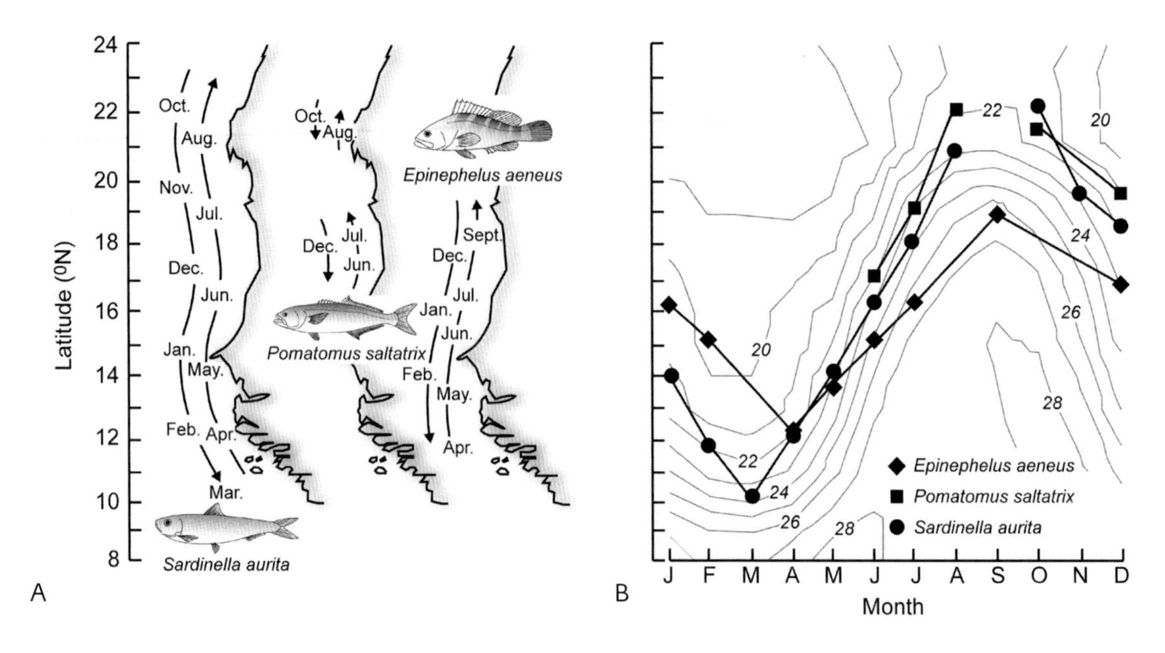

Figure 8.1. Seasonal latitudinal migrations of some northwest African fishes. (A) Summary of information on the occurrence in space (latitude) and time (month) of 3 species, *Sardinella aurita*, *Pomatomus saltator*, and *Epinephelus aeneus* (from Boëly et al. 1978, 1979; Champagnat and Domain 1978; Barry-Gérard 1994). (B) Same as in (A) but plotted against mean monthly temperature. Data from the Comprehensive Ocean–Atmosphere Data Set (COADS). The seasonal migrations result in the 3 species remaining in approximately the same temperature range (and hence having the same oxygen consumption) throughout the year. (Adapted from Pauly 1994.)

barbatus (Katsikatsou et al. 2011) and the establishment of the noxious silver-cheeked toadfish Lagocephalus sceleratus, a Red Sea or "Lessepsian" migrant (Kasapidis et al. 2007). Shifted species distribution ranges follow temperature clines from high to low, reflecting a lateral gradient at the basin scale (Pinsky et al. 2013; Poloczanska et al. 2013) or a vertical temperature gradient to deeper waters (Dulvy et al. 2008; Pauly 2010). However, the implications of such responses for global fisheries to ocean warming had not been empirically demonstrated.

The newly developed index, the MTC, shows that global catches are increasingly dominated by warmer-water species (Cheung et al. 2013c). The MTC is the weighted average of the preferred temperatures of the various fish and invertebrate species in the catch. The preferred temperature of each species (which is expected to be fairly stable in evolutionary time for most taxa) was predicted from overlaying the current distribution of the species, as mapped using the approach described in chapter 4 and sea surface temperature (SST). Therein, species that are distributed in warmer waters will have higher preferred temperature and vice versa. Thus, for example, if the catch of a small country in the temperate zone is increasingly dominated by warmer-water species, its MTC would increase.

Using the *Sea Around Us* catch data, the MTC was calculated for all the Large Marine Ecosystems (LMEs) of the world from 1970 to 2006. After the effects of fishing and large-scale oceanographic variability were accounted for, global MTC increased at a rate of 0.19°C per decade between 1970 and 2006, and nontropical MTC increased at a rate of 0.23°C per decade. In tropical areas, MTC increased initially because of the reduction in the proportion of subtropical species catches but subsequently stabilized as scope for further "tropicalization" of communities became limited (figure 8.2). By showing that changes in MTC are significantly related to changes in SST across LMEs, Cheung et al. (2013c) showed conclusively that ocean warming has already affected global fisheries catch composition in the past four decades. This is now being verified at smaller scales (Keskin and Pauly 2014; Tsikliras and Stergiou 2014).

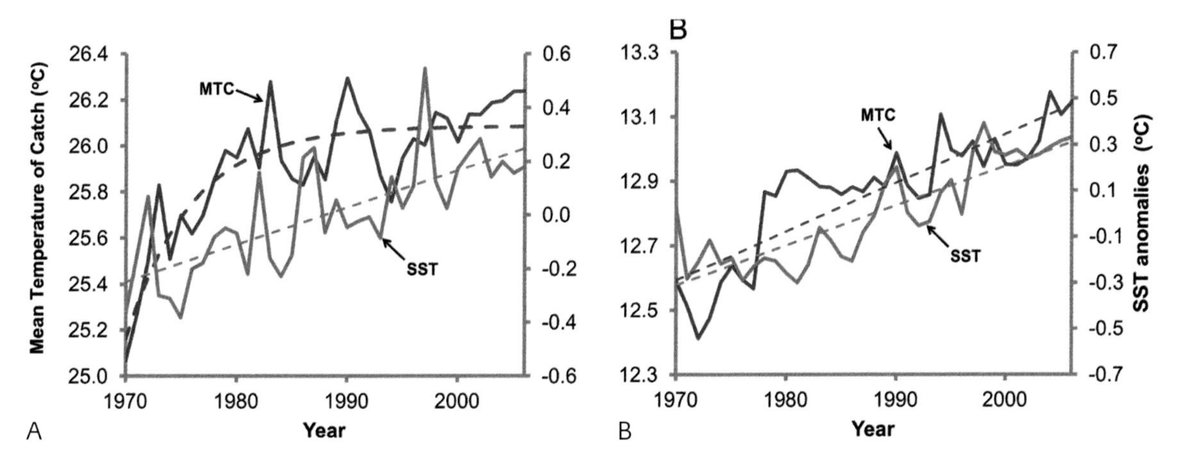

Figure 8.2. Observed trends in mean temperature of the catch (MTC) and sea surface temperature (SST) from (A) tropical and (B) nontropical Large Marine Ecosystems. (Adapted from Cheung et al. 2013c.)

Projecting Distribution Shifts of Exploited Species

Given that changes in the composition of fisheries catches are probably being driven by warming-induced biogeographic shifts, the next step was to investigate whether exploited species would continue to shift their biogeography in the future under climate change and how this would affect composition of the exploited species.

A DBEM was developed to project future distribution of more than 1,000 exploited fishes and invertebrates globally. The DBEM, described in Cheung et al. (2008b, 2009) and later in Cheung et al. (2011) and Fernandes et al. (2013), predicts a species' range (on a grid of about 180,000 cells of 30′ latitude by 30′ longitude representing the world ocean) based on the association between the modeled distributions and environmental variables. The current distribution of a taxon is predicted using a method developed by the *Sea Around Us* and described in chapter 4. Comparison with other species distribution modeling approaches that allow prediction of distributions for all the exploited species show that the DBEM performs equally well in terms of test statistics with observed occurrence data, for example, receiver operating characteristics (Jones et al. 2012). The DBEM infers preference profiles, defined as the suitability of different environmental conditions to the species covered, from their predicted current distributions, assuming that species' current distributions match their environmental preference. Distinguishing features of the DBEM relative to other species distribution models include the explicit representation of spatial population dynamics (Cheung et al. 2008b), ecophysiology (Cheung et al. 2011) and, in a new version of DBEM, trophic interactions (Fernandes et al. 2013).

We applied the DBEM to project future distributions of 1,066 species of exploited fishes and invertebrates under climate change scenarios developed by the IPCC. These species include the overwhelming majority of the taxa whose population is large enough to generate catches that are reported at the species level in the global fisheries statistics of the Food and Agriculture Organization of the United Nations (FAO) and thus represent a very large sample of marine macrofauna. The rate of range shift and the intensity of species invasion and local extinction in the global ocean by 2050 relative to the 2000s were then calculated.

The resulting projections show that climate change leads, overall, to range shifts to higher latitude and deeper waters (figure 8.3), although some species display range shift in opposite directions as they follow local rather than large-scale climate change gradients (Cheung et al. 2009; Pinsky et al. 2013). Thus, numerous local extinctions in the subpolar regions, the tropics, and semienclosed seas can be expected. Simultaneously, species invasions are projected to be most frequent in the Arctic and the Southern Ocean. Jointly,

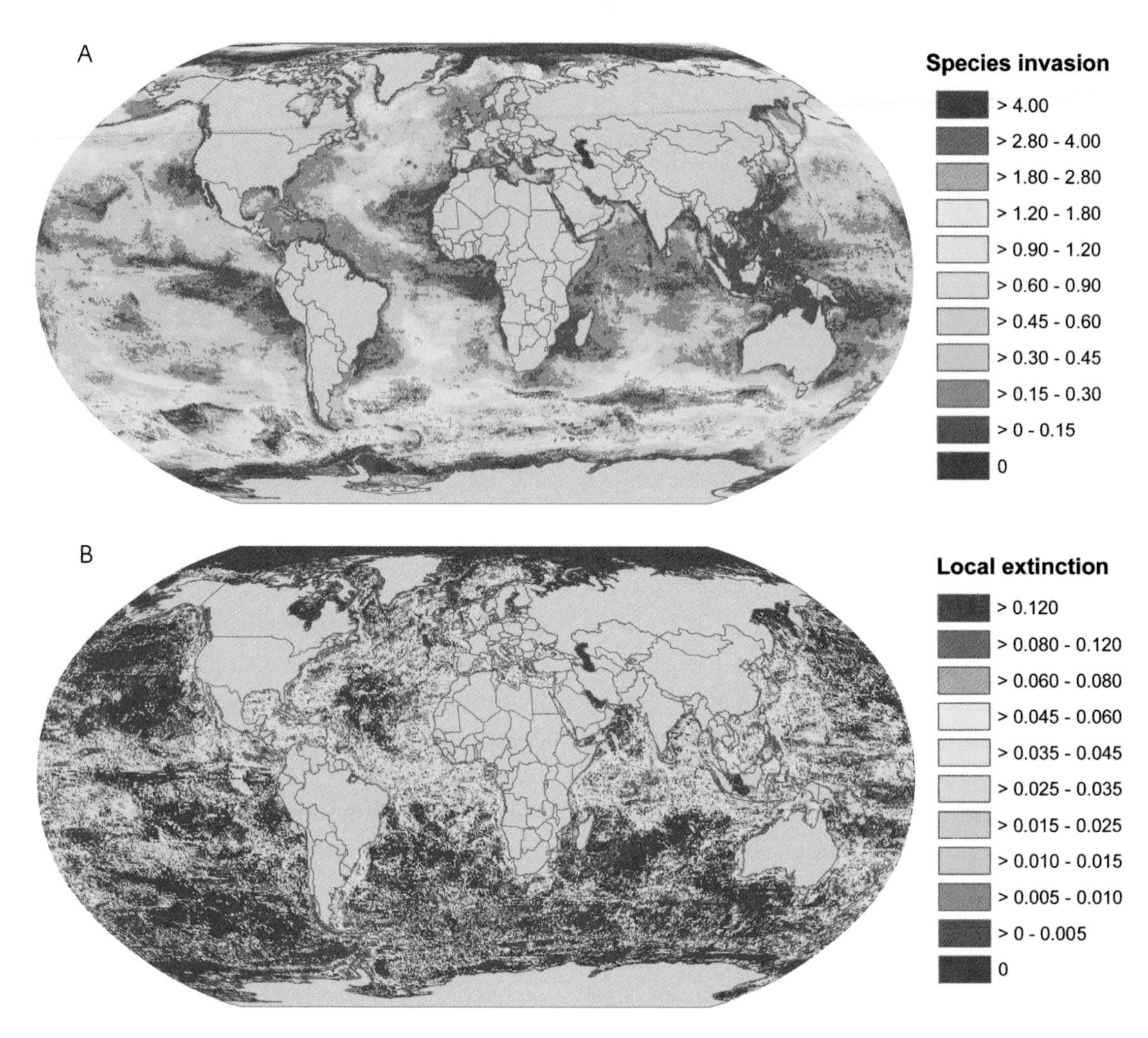

Figure 8.3. Projected intensity of (A) species invasion and (B) local extinctions by 2050 relative to 2000 (10-year average) under the IPCC Special Report on Emission Scenarios (SRES) A1B scenario. (Adapted from Cheung et al. 2009.)

these results, which are robust to the selection of species distribution models (Jones and Cheung 2015), suggest a dramatic turnover of current marine biodiversity, implying ecological disturbances that will massively disrupt the provision of ecosystem services. Moreover, these results support the hypothesis that the observed pattern of changes in species composition of catches, as indicated by the MTC introduced above, will continue in the future.

Relationship between NPP and Maximum Catch Potential

We developed an empirical relationship to predict maximum catch potential of a species based on NPP and the biogeography and ecology of marine water-breathers.

First, based on theories linking trophic energetics and allometric scaling of metabolism, a theoretical model was developed that relates the maximum catch potential of a species to its trophic level, geographic range, and mean NPP within the species' exploited range (Cheung et al. 2008a). Therein, the relationship between metabolic rate and body size of marine organisms, as quantified by an allometric equation and the energy available for a specific population of animals at trophic level (λ) in an ecosystem, was calculated based on a trophic transfer efficiency (TE) set at 10% (Pauly and Christensen 1995; Ware 2000). Finally, maximum sustainable yield (MSY) of the population was obtained from $rB_\infty/4$, where B_∞ is the biomass at carrying capacity and r is the intrinsic rate of population increase, as

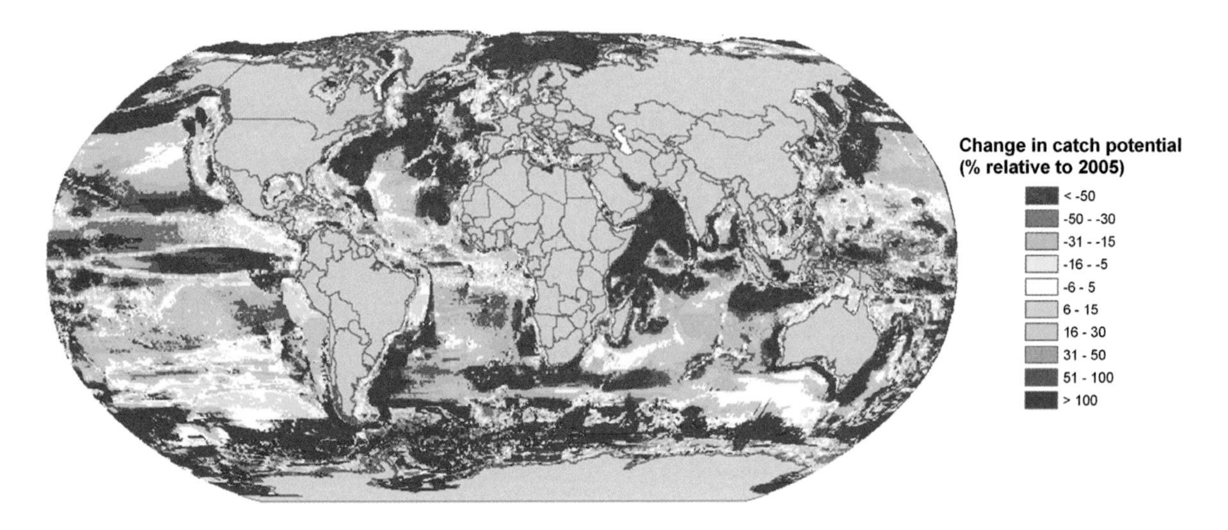

Figure 8.4. Projected change in maximum catch potential by 2050 relative to 2005 (10-year average) under the SRES A1B scenario. (Adapted from Cheung et al. 2010.)

implied in Schaefer (1954) and stated explicitly in Ricker (1975, p. 315) and Pauly (1984a, p. 140). This led to a theoretical relationship in which MSY was expressed as a function of NPP, λ, and TE.

This model was fitted to catch, ecological, and biogeographical data from 1,000 species of exploited marine fish and invertebrate species. This led to an empirical relationship (i.e., a multiple regression) between these species' approximated maximum catch potential and their ecology and biogeography. Therein, maximum catch potentials were assumed to be approximated by the average of their five highest annual catches (from 1950 and 2006), nearly the same assumption as made by Srinivasan et al. (2010). Additional variables were included in the empirical model to correct for biases resulting from the uncertainty inherent in the original catch data. The empirical model had a high explanatory power ($R^2 = 0.703$), and the signs and magnitudes of the partial slopes agreed with theoretical expectations. Friedland et al. 2012) suggest that chlorophyll A concentration is a better predictor of catch potential, but reanalysis of Cheung et al. (2008a) using chlorophyll A concentration instead of NPP does not result in a significantly better model. Thus, the empirical model can be combined with the DBEM to project the impacts of climate change on global marine fisheries, our next topic.

Projecting Future Catch Potential

By applying the empirical model described in the last section to projected future distributions derived from the DBEM, we could then estimate changes in global maximum catch potential for 1,066 species of exploited marine fish and invertebrates from 2005 to 2055 (Cheung et al. 2010). Species distribution projections were obtained from Sarmiento et al. (2004), working with the IPCC Special Report on Emission Scenarios (SRES) A1B and B2, along with projected changes in NPP. Once this information was incorporated into the empirical equation, it appeared that climate change would lead to a large-scale redistribution of global catch potential, with an average of 30%–70% increase in high-latitude regions and a drop of up to 40% in the tropics (figure 8.4). Moreover, predicted maximum catch potential declined in the lower-latitude margins of semienclosed seas, while it increased near the poleward tips of continental shelf margins, particularly in the Pacific Ocean. Among the twenty Exclusive Economic Zones (EEZs) with the highest landings according to the FAO, those with the highest increase in catch potential by 2055 belonged to Norway, Greenland, the United States (Alaska), and

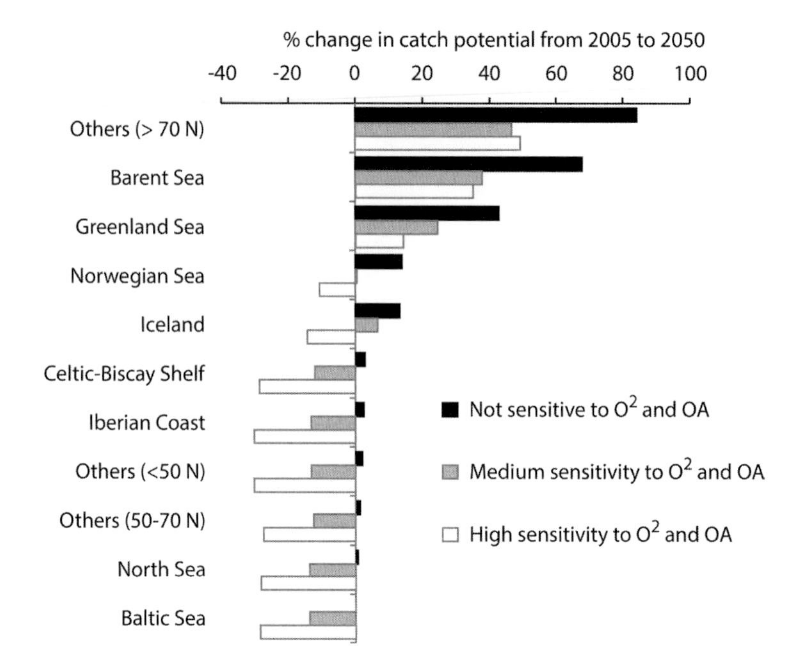

Figure 8.5. Projected change in maximum catch potential in Large Marine Ecosystems in the northeast Atlantic by 2050 relative to 2005 (10-year average) under the SRES A2 scenario. OA, ocean acidification. (Adapted from Cheung et al. 2011.)

the Russian Far East. In contrast, the catch potential of Indonesia, the contiguous United States, Chile, and China have EEZs whose catch potential was predicted to decline most strongly. These results highlight the need to develop adaptation and mitigations policies for climate change impacts on fisheries, particularly in the tropics.

Because the above studies did not account for the effects of changes in ocean biogeochemistry and phytoplankton community structure, a version of the DBEM was developed that incorporated these factors and was used to project the distributions and maximum catch potentials of 120 species of exploited demersal fish and invertebrates in the northeast Atlantic (Cheung et al. 2011). Using projections from the U.S. National Oceanic and Atmospheric Administration's Geophysical Fluid Dynamics Laboratory Earth System Model (ESM2.1) under the SRES A1B, we predicted ocean acidification and reduction in oxygen content to reduce growth performance, increase the rate of range shift, and lower the estimated catch potentials (10-year average of 2050 relative to 2005) by 20%–30% relative to simulations that did not consider these factors (figure 8.5). Consideration of changes in phytoplankton community structure may further reduce projected catch potentials by 10%. These results highlight the sensitivity of marine ecosystems to biogeochemical changes and the need to incorporate likely hypotheses of their biological and ecological effects in assessing climate change impacts.

Projecting Maximum Body Size of Fish and Invertebrates

Both theory and empirical observations support the hypothesis that warming and reduced oxygen will reduce body size of marine fishes (Pauly 1998a) and invertebrates (Pauly 1998b). Changes in temperature, oxygen content, and other ocean biogeochemical properties directly affect the ecophysiology of marine water-breathing organisms. Particularly, their physiological performance, including their growth rate and size at first reproduction, are strongly dependent on temperature and oxygen (Pauly 1981, 1984a, 1984b, 1998b, 2010; Pörtner and Farrell 2008). An organism's low oxygen tolerance threshold varies across species, body size, and life stage and is highest for large organisms. The oxygen tolerance threshold is set by the capacity of an organism's respiratory and circulatory systems to supply O_2 and cover demand. A corollary of the above is that distribution, growth, size at first reproduction, maximum body size, and survival of fishes are controlled

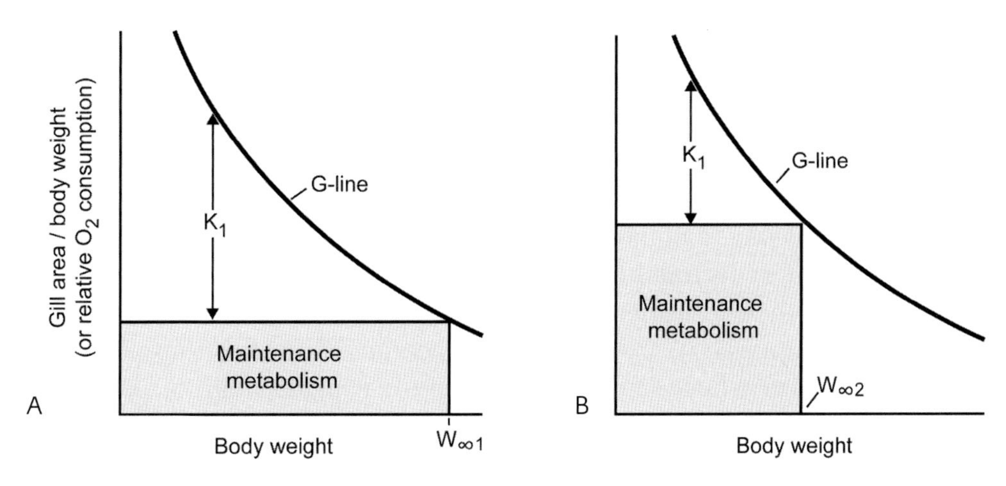

Figure 8.6. Diagram illustrating how maintenance metabolism determines asymptotic weight (W_∞), given a "G-line" defined by the growth of the gills relative to body weight, because at W_∞, relative gill area (and hence oxygen supply) is just enough for maintenance metabolism (shaded areas). (A) Fish exposed to a low level of stress (e.g., low temperature, abundant oxygen, abundant food). (B) Fish exposed to a higher level of stress (low oxygen concentration, high temperature, causing rapid denaturation of body protein, or low food density, requiring O_2 to be diverted to foraging rather than protein synthesis). Note that food conversion efficiency and hence also the scope for growth are directly related to the distance, in these graphs, between the G-line and the level of maintenance metabolism (see Pauly, 1981, 1984b, 2010). (Adapted from Pauly 2010.)

by the balance between oxygen supply and demand under different temperatures (Pütter 1920; Pauly 1981, 2010; Kolding et al. 2008). As fishes increase in size (weight), mass-specific oxygen demand increases more rapidly than oxygen supply (Pauly 1997). Thus, fish reach maximum body size when oxygen supply is balanced by oxygen demand (figure 8.6A). Moreover, the scope for aerobic respiration and growth decreases when size increases, that is, oxygen supply per unit body weight decreases (figure 8.6B). The decrease in food conversion efficiency that this implies decreases the biomass production of fish and invertebrate populations.

However, although the interrelationships of temperature, oxygen, and growth in water-breathers are well established in the laboratory (Pauly 2010), the extent to which maximum body size of fishes would be affected by projected changes in temperature and oxygen level in the oceans remained unexplored. The DBEM was thus used to examine the integrated biological responses of more

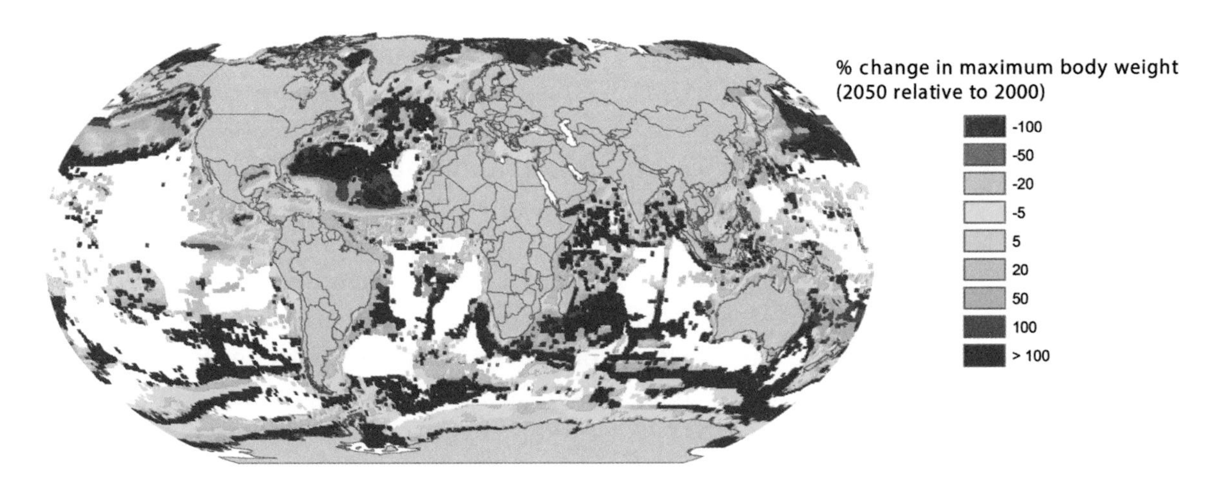

Figure 8.7. Projected change in maximum body weight of exploited fishes by 2050 relative to the 2000 period (20-year average) under the SRES A2 scenario. (Adapted from Cheung et al. 2013b.)

than 600 species of marine fishes to changes in distribution, abundance, and body size (Cheung et al. 2013a, 2013b), based on explicit representations of ecophysiology, dispersal, distribution, and population dynamics. The result was that assemblage-averaged maximum body weight is expected to decline by 14%–24% globally from 2000 to 2050 under a high-emission scenario (figure 8.7). The projected magnitude of decrease in body size is consistent with experimental (Forster et al. 2012; Cheung et al. 2013b) and field observations (Baudron et al. 2014). About half of this shrinkage is caused by changes in distribution and abundance, the remainder by changes in physiological performance. The tropical and intermediate latitudinal areas will be heavily affected, with an average reduction of more than 20%. Decreases in growth and body size should reduce the biomass production of fish populations, and hence fishery catches, and potentially alter trophic interactions.

CONCLUSION

In their series of studies, the authors and their collaborators detected a signature of ocean warming on global fisheries in the last four decades, and they also projected that such changes would continue over the next 40 years, leading to strong species turnover, redistribution of fisheries catch potential, and decreases in the maximum body sizes of exploited species of fishes and invertebrates. Results from these global-scale analyses highlighted the inequity of climate change impacts to different regions of the world. Specifically, the tropics will be affected by a high rate of local species extinction, a decrease in catch potential, and a larger decrease in body size of fishes. Many tropical communities are highly dependent on local fisheries for food and livelihood (e.g., Zeller et al. 2015), but their economic and societal capacity to adapt to climate change impacts on fisheries is often low. Thus, tropical fisheries are highly vulnerable to climate change, although tropical countries contribute little to the greenhouse gas emissions that cause climate change.

Future studies should address additional challenges to detecting, attributing, and projecting climate change and ocean acidification impacts on marine fisheries. First, the adaptive scope of exploited marine species and their fishers to impacts from climate change and ocean acidification should be evaluated. Second, different modeling approaches in projecting future seafood production under climate change and ocean acidification should be tested, to assess the utility of these approaches and quantify the level of uncertainty associated with the model projections. Third, the effects of multiple stressors (i.e., climate and nonclimate) and their interactions must be explored. In addition, more regional studies to downscale the global analyses must be conducted, through which the weaknesses associated with the coarse projections of ocean properties from global circulation models can be better addressed. Such regional-scale analyses are more useful for informing national fisheries and coastal management agencies, which will both be challenged by global warming in coming years.

ACKNOWLEDGMENTS

This is a contribution of the *Sea Around Us*, a research activity at the University of British Columbia initiated and funded by the Pew Charitable Trusts from 1999 to 2014 and currently funded mainly by the Paul G. Allen Family Foundation. W. W. L. Cheung also acknowledges the National Geographic Society, the Nippon Foundation–University of British Columbia Nereus Program, the Natural Sciences and Engineering Research Council of Canada, and the Centre for Fisheries and Aquaculture Sciences, which contributed funding to the research documented here.

REFERENCES

Barry-Gérard, M. 1994. Migration des poissons le long du littoral sénégalais. Pp. 215–234 in M. Barry-Gérard, T. Diouf, and A. Fonteneau (eds.), *L'évaluation des ressources exploitables par la pêche sénégalaise*. Édition ORSTOM, Paris.

Baudron, A. R., C. L. Needle, A. D. Rijnsdorp, and C. Tara Marshall. 2014. Warming temperatures and smaller body sizes: synchronous changes in growth of North Sea fishes. *Global Change Biology* 20: 1023–1031.

Boëly, T. 1979. Biologie de deux espèces de Sardinelles (*Sardinella aurita* Valenciennes 1887 et *Sardinella maderensis* Lowe 1841) des côtes sénégalaise. *Pêches Maritimes* July 1979: 426–430.

Boëly, T., J. Chabanne, and P. Fréon. 1978. Schéma migratoire de poissons pélagiques côtiers dans la zone sénégalo-mauritanienne. Pp. 63–70 in *Rapport du groupe de travail ad hoc sur les poissons pélagiques côtiers ouest africains de la Mauritanie au Libéria (26 °N à 5 °N)*. FAO/Comité des pêches pour l'Atlantique. Centre-Est: COPACE/RACE Sér. 78/10.

Champagnat, C., and F. Domain. 1978. Migrations des poissons démersaux le long des côtes ouest africaines de 10 °N à 24 °N. *Cahier ORSTOM, série Océanographie* 16(3–4): 239–261.

Cheung, W. W. L., C. Close, V. W. Y. Lam, R. Watson, and D. Pauly. 2008a. Application of macroecological theory to predict effects of climate change on global fisheries potential. *Marine Ecology Progress Series* 365: 187–197.

Cheung, W. W. L., J. Dunne, J. L. Sarmiento, and D. Pauly. 2011. Integrating ecophysiology and plankton dynamics into projected maximum fisheries catch potential under climate change in the northeast Atlantic. *ICES Journal of Marine Science* 68: 1008–1018.

Cheung, W. W. L., V. W. Y. Lam, and D. Pauly. 2008b. Dynamic bioclimate envelope model to predict climate-induced changes in distributions of marine fishes and invertebrates. Pp. 5–50 in W. W. L. Cheung, V. W. Y. Lam, and D. Pauly (eds.), *Modelling present and climate-shifted distribution of marine fishes and invertebrates*. Fisheries Centre Research Reports 16(3), University of British Columbia, Vancouver, Canada.

Cheung, W. W. L., V. W. Y. Lam, J. L. Sarmiento, K. Kearney, R. Watson, and D. Pauly. 2009. Projecting global marine biodiversity impacts under climate change scenarios. *Fish and Fisheries* 10: 235–251.

Cheung, W. W. L., V. W. Y. Lam, J. L. Sarmiento, K. Kearney, R. E. G. Watson, D. Zeller, and D. Pauly. 2010. Large-scale redistribution of maximum fisheries catch potential in the global ocean under climate change. *Global Change Biology* 16: 24–35.

Cheung, W. W. L., D. Pauly, and J. L. Sarmiento. 2013a. How to make progress in projecting climate change impacts. *ICES Journal of Marine Science* 70: 1069–1074.

Cheung, W. W. L., J. L. Sarmiento, J. Dunne, T. L. Frölicher, V. Lam, M. L. D. Palomares, R. Watson, and D. Pauly. 2013b. Shrinking of fishes exacerbates impacts of global ocean changes on marine ecosystems. *Nature Climate Change* 3: 254–258.

Cheung, W. W. L., R. Watson, and D. Pauly. 2013c. Signature of ocean warming in global fisheries catches. *Nature* 497: 365–368.

Daufresne, M., K. Lengfellner, and U. Sommer. 2009. Global warming benefits the small in aquatic ecosystems. *Proceedings of the National Academy of Sciences* 106: 12788–12793.

Dulvy, N. K., S. I. Rogers, S. Jennings, V. Stelzenmüller, S. R. Dye, and H. R. Skjoldal. 2008. Climate change and deepening of the North Sea fish assemblage: a biotic indicator of warming seas. *Journal of Applied Ecology* 45: 1029–1039.

Fernandes, J. A., W. W. Cheung, S. Jennings, M. Butenschön, L. Mora, T. L. Frölicher, M. Barange, and A. Grant. 2013. Modelling the effects of climate change on the distribution and production of marine fishes: accounting for trophic interactions in a

dynamic bioclimate envelope model. *Global Change Biology* 19(8): 2596–2607.
Forster, J., A. G. Hirst, and D. Atkinson. 2012. Warming-induced reductions in body size are greater in aquatic than terrestrial species. *Proceedings of the National Academy of Sciences* 109: 19310–19314.
Friedland, K. D., C. Stock, K. F. Drinkwater, J. S. Link, R. T. Leaf, B. V. Shank, J. M. Rose, C. H. Pilskaln, and M. J. Fogarty. 2012. Pathways between primary production and fisheries yields of large marine ecosystems. *PLoS ONE* 7: e28945.
IPCC. 2014. Summary for policymakers. In *Climate Change 2014: Impacts, Adaptation, and Vulnerability. Part A: Global and Sectoral Aspects*. Contribution of Working Group II to the Fifth Assessment Report of the Intergovernmental Panel on Climate Change [Field, C. B., V. R. Barros, D. J. Dokken, K. J. Mach, M. D. Mastrandrea, T. E. Bilir, M. Chatterjee, K. L. Ebi, Y. O. Estrada, R. C. Genova, B. Girma, E. S. Kissel, A. N. Levy, S. MacCracken, P. R. Mastrandrea, and L. L. White (eds.)]. Cambridge University Press, Cambridge, England.
Jones, M. C., and W. W. L. Cheung. 2015. Multi-model ensemble projections of climate change effects on global marine biodiversity. *ICES Journal of Marine Science* 72(3): 741–752.
Jones, M., S. Dye, J. Pinnegar, R. Warren, and W. W. L. Cheung. 2012. Modelling commercial fish distributions: prediction and assessment using different approaches. *Ecological Modelling* 225: 133–145.
Kasapidis, P., P. Peristeraki, G. Tserpes, and A. Magoulas. 2007. First record of the Lessepsian migrant *Lagocephalus sceleratus* (Gmelin 1789) (Osteichthyes: Tetraodontidae) in the Cretan Sea (Aegean, Greece). *Aquatic Invasions* 2(1): 71–73.
Katsikatsou, M., A. Anestis, H. O. Pörtner, T. Kampouris, and B. Michaelidis. 2011. Field studies on the relation between the accumulation of heavy metals and metabolic and HSR in the bearded horse mussel *Modiolus barbatus*. *Comparative Biochemistry and Physiology Part C: Toxicology & Pharmacology* 153: 133–140.
Keskin, Ç., and D. Pauly. 2014. Changes in the "mean temperature of the catch": application of a new concept to the north-eastern Aegean Sea. *Acta Adriatica* 55(2): 213–218.
Kolding, J., L. Haug, and S. Stefansson. 2008. Effect of ambient oxygen on growth and reproduction in Nile tilapia (*Oreochromis niloticus*). *Canadian Journal of Fisheries and Aquatic Science* 65: 1413–1424.
Kroeker, K. J., R. L. Kordas, R. Crim, I. E. Hendriks, L. Ramajo, G. S. Singh, C. M. Duarte, and J. P. Gattuso. 2013. Impacts of ocean acidification on marine organisms: quantifying sensitivities and interaction with warming. *Global Change Biology* 19: 1884–1896.
Pauly, D. 1981. The relationship between gill surface area and growth performance in fish: a generalization of von Bertalanffy's theory of growth. *Berichte der Deutschen Wissenschaftlichen Kommission für Meeresforschung* 28: 251–282.
Pauly, D. 1984a. *Fish population dynamics in tropical waters: a manual for use with programmable calculators*. ICLARM Studies and Reviews 8.
Pauly, D. 1984b. A mechanism for the juvenile-to-adult transition in fishes. *Journal du Conseil International pour l'Exploration de la Mer* 41: 280–284.
Pauly, D. 1994. Un mécanisme explicatif des migrations des poissons le long des côtes du Nord-Quest africain. Pp. 235–244 in M. Barry-Gérard, T. Diouf, and A. Fonteneau (eds.), *L'évaluation des ressources exploitables par la pêche artisanale sénégalaise*. Documents scientifiques présentés lors du Symposium, 8–13 février 1993, Dakar, Sénégal. ORSTOM Éditions, Paris.
Pauly, D. 1997. Geometrical constraints on body size. *Trends in Ecology and Evolution* 12: 442–443.
Pauly, D. 1998a. Tropical fishes: patterns and propensities. Pp. 1–17 in T. E. Langford, J. Langford, and J. E. Thorpe (eds.), Tropical fish biology. *Journal of Fish Biology* 53(Suppl. A).
Pauly, D. 1998b. Why squids, though not fish, may be better understood by pretending they are. Pp. 47–58 in A. I. L. Payne, M. R. Lipinski, M. R. Clarke, and M. A. C. Roeleveld

(eds.), Cephalopod biodiversity, ecology and evolution. *South African Journal of Marine Science* 20.

Pauly, D. 2010. *Gasping Fish and Panting Squids: Oxygen, Temperature and the Growth of Water-Breathing Animals*. Excellence in Ecology Series Vol. 22, International Ecology Institute, Oldendorf/Luhe, Germany.

Pauly, D., and V. Christensen. 1995. Primary production required to sustain global fisheries. *Nature* 374: 255–257.

Pinsky, M. L., B. Worm, M. J. Fogarty, J. L. Sarmiento, and S. A. Levin. 2013. Marine taxa track local climate velocities. *Science* 341: 1239–1242.

Poloczanska, E. S., C. J. Brown, W. J. Sydeman, W. Kiessling, D. S. Schoeman, P. J. Moore, K. Brander, J. F. Bruno, L. B. Buckley, M. T. Burrows, C. M. Duarte, B. S. Halpern, J. Holding, C. V. Kappel, M. I. O'Connor, J. M. Pandolfi, C. Parmesan, F. Schwing, S. A. Thompson, and A. J. Richardson. 2013. Global imprint of climate change on marine life. *Nature Climate Change* 3: 919–925.

Pörtner, H. O. 2010. Oxygen- and capacity-limitation of thermal tolerance: a matrix for integrating climate-related stressor effects in marine ecosystems. *Journal of Experimental Biology* 213: 881–893.

Pörtner, H. O., and A. P. Farrell. 2008. Physiology and climate change. *Science* 322: 690–692.

Pütter, A. 1920. Studien über physiologische Ähnlichkeit. VI. Wachstumsähnlichkeiten. *Pflüger's Archiv für die gesamte Physiologie* 180: 298–340.

Ricker, W. E. 1975. Computation and interpretation of biological statistics of fish populations. *Bulletin of the Fisheries Research Board of Canada* 191.

Sarmiento, J. L., R. Slater, R. Baber, L. Bopp, S. C. Doney, A. C. Hirst, J. Kleypas, R. Matear, U. Mikolajewicz, P. Monfray, V. Soldatov, S. A. Spall, and R. Stouffer. 2004. Response of ocean ecosystems to climate warming. *Global Biogeochemical Cycles* 18(3). doi:10.1029/2003GB002134.

Schaefer, M. B. 1954. Some aspects of the dynamics of populations important to the management of the commercial marine fisheries. *Bulletin of the InterAmerican Tropical Tuna Commission* 1: 27–56.

Srinivasan, U. T., W. W. L. Cheung, R. Watson, and U. R. Sumaila. 2010. Food security implications of global marine catch losses due to overfishing. *Journal of Bioeconomics* 12(3): 183–200.

Tsikliras, A. C., and K. I. Stergiou. 2014. Mean temperature of the catch increases quickly in the Mediterranean Sea. *Marine Ecology Progress Series* 515(2014): 281–284.

Ware, D. M. 2000. Aquatic ecosystems: properties and models. Pp. 161–194 in P. J. Harrison and T. R. Parsons (eds.), *Fisheries Oceanography: An Integrative Approach to Fisheries Ecology and Management*. Blackwell Science, Oxford.

Zeller, D., S. Harper, K. Zylich, and D. Pauly. 2015. Synthesis of under-reported small-scale fisheries catch in Pacific-island waters. *Coral Reefs* 34(1): 25–39.

NOTE

1. Cite as Cheung, W. W. L., and D. Pauly. 2015. Global-scale responses and vulnerability of marine species and fisheries to climate change. Pp. 86–97 in D. Pauly and D. Zeller (eds.), *Global Atlas of Marine Fisheries: A Critical Appraisal of Catches and Ecosystem Impacts*. Island Press, Washington, DC.

CHAPTER 9

MODELING THE GLOBAL OCEANS WITH THE ECOPATH SOFTWARE SUITE: A BRIEF REVIEW AND APPLICATION EXAMPLE[1]

Mathieu Colléter,[a,b] Audrey Valls,[c] Villy Christensen,[c] Marta Coll,[d,e] Didier Gascuel,[b] Jérôme Guitton,[b] Chiara Piroddi,[c] Jeroen Steenbeek,[c] Joe Buszowski,[c] and Daniel Pauly[a]
[a]*Sea Around Us*, University of British Columbia, Vancouver, BC, Canada
[b]Université Européenne de Bretagne, Agrocampus Ouest, Rennes, France
[c]Fisheries Centre, University of British Columbia, Vancouver, BC, Canada
[d]Institut de Recherche pour le Developpement, Sète, France
[e]Institut de Ciències del Mar (ICM-CSIC), Barcelona, Spain

The life sciences have reached a new era, that of the "big new biology" (Thessen and Patterson 2011). Ecology is following a similar path and has turned into a data-intensive science (Michener and Jones 2012). As demonstrated by this atlas and the mountain of data its completion entailed, this is also the case for marine biology and fisheries science. Indeed, our work is increasingly relying on large preexisting datasets, allowing new insights on phenomena visible mainly or only at global scales (e.g., Pauly 2007; Christensen et al. 2009; see also chapter 8).

However, data sharing in marine and fisheries biology is still not as extensive as in the historical "big" sciences, such as oceanography, meteorology, and astronomy, where massive data sharing is the norm (Pauly 1995; Edwards 2010; Thessen and Patterson 2011). The open-access principle of sharing information online for free has been increasingly applied to publications, but much less to data, mainly because of issues with recognition and sense of data ownership (Zeller et al. 2005; Vision 2010; Thessen and Patterson, 2011). Although incentives for digitization of nondigital materials have been growing, existing repositories have been estimated to represent less than 1% of the data in ecology (Reichman et al. 2011; Thessen and Patterson 2011).

GATHERING INFORMATION FOR AND FROM ECOPATH WITH ECOSIM MODELS

In aquatic ecology, the Ecopath with Ecosim (EwE) modeling approach has been widely applied to inform ecosystem-based management (e.g., Jarre-Teichmann 1998; Christensen and Walters 2004, 2011; Plaganyi and Butterworth 2004; Coll and Libralato 2012), since its original development in the early 1980s (Polovina 1984) and its relaunch in the early 1990s (Christensen and Pauly 1992). The EwE modeling approach was developed primarily as a tool to help fisheries management and answer "what if" questions about policy that could not be addressed with single-species assessment models (Pauly et al. 2000; Christensen and Walters 2004; 2011). The EwE software is user-friendly, free (under the terms of the GNU General Public License), and downloadable online (www.ecopath.org). Thus, hundreds of EwE models representing aquatic (and some

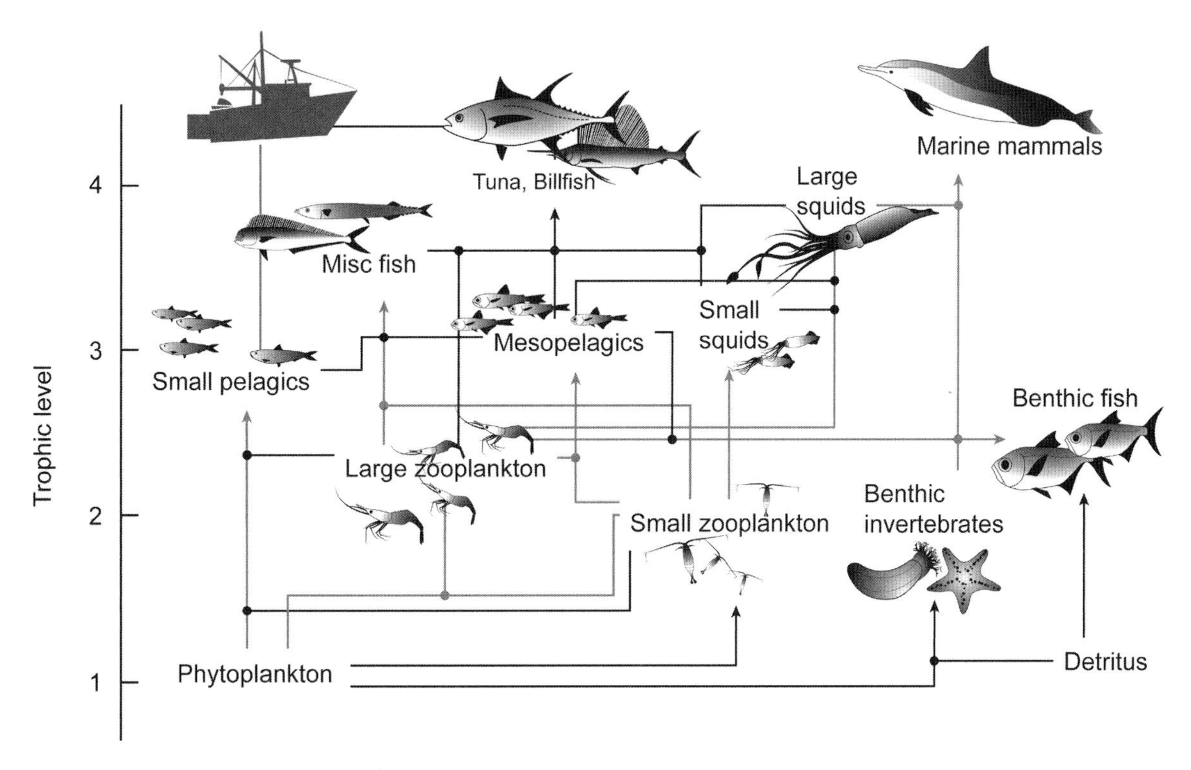

Figure 9.1. Flowchart of an Ecopath ecosystem model with a low number of groups. (Adapted from Pauly and Christensen 1993.)

terrestrial) ecosystems have been developed and published worldwide. The foundation of the EwE modeling approach is an Ecopath model, which creates a static mass-balanced snapshot of the resources in an ecosystem and their interactions, represented by trophically linked biomass "pools" (figure 9.1). The key principles and equations of EwE are presented in box 9.1, and more details can be found in the EwE online user guide (Christensen et al. 2008).

By formalizing available knowledge on a given ecosystem, EwE helps elucidate its structure and functioning and thus may be seen as an important source of mutually compatible data (Walters et al. 1997). Indeed, building an EwE model requires the collection, compilation, and harmonization of various types of information: descriptive data on species abundance, diet composition, and catch; computed data on species production and consumption; and the biomass trends resulting from various exploitation scenarios. Several meta-analyses based on smaller sets of EwE models have been performed, focusing either on theoretical ecology and ecological concepts (e.g., Christensen and Pauly 1993a; Pérez-España and Arreguín-Sánchez 1999, 2001; Gascuel et al. 2008; Arreguín-Sánchez 2011) or on ecosystems and species of particular interest (e.g., Pauly et al. 1999, 2009; Christensen et al. 2003a, 2003b). However, only few meta-analyses based on a large collection of EwE models have been published (e.g., Christensen 1995; Coll et al. 2012; Pikitch et al. 2012; Heymans et al. 2014).

Global Overview of EwE Applications and Presentation of a Meta-Analysis Case Study

EcoBase is an online information repository of EwE models published in the scientific literature, developed with the intention of making the models discoverable, accessible, and reusable to the scientific community (ecobase .ecopath.org). Details on the structure, usage, and capabilities of EcoBase can be found in the report introducing EcoBase (Colléter et al. 2013), which is available online. Colléter et al. (2013) first aimed to give a global overview of the applications of the Ecopath with Ecosim modeling approach in the scientific literature, using metadata gathered on the 435

BOX 9.1. ECOPATH KEY PRINCIPLES AND EQUATIONS

Mathieu Colléter and **Daniel Pauly**, *Sea Around Us*, University of British Columbia, Vancouver, Canada

The foundation of the EwE modeling approach is an Ecopath model (Polovina 1984; Christensen and Pauly 1992), which creates a static mass-balanced snapshot of the resources in an ecosystem and their interactions, represented by biomass "pools" linked by grazing or predation. The modeled food web is thus represented by "pools" or functional groups (i), which can be composed of species or groups of species that are ecologically similar or represent different age groups (or "stanzas") of a species. For each group, the Ecopath software solves two balancing equations: one to describe the production (equation 9.1), the other the energy balance (equation 9.2):

$$B_i \times \left(\frac{P}{B}\right)_i = \sum_{j=1}^{N} B_j \times \left(\frac{Q}{B}\right)_j \times DC_{ji} + \left(\frac{P}{B}\right)_i \times B_i \times (1 - EE_i) + Y_i + E_i + BA_i \quad (9.1)$$

$$Q_i = P_i + R_i + UA_i \quad (9.2)$$

where N is the number of groups in the model, B the biomass, P/B the production rate, Q/B the consumption rate, DC_{ji} the diet matrix representing the fraction of prey i in the diet of predator j, E the net migration rate, BA the biomass accumulation, Y the catches, EE the ecotrophic efficiency (i.e., the fraction of production that is used in the system), R the respiration, P the production, Q the consumption, and UA the unassimilated consumption because of egestion and excretion. The quantity (1 – EE) × P/B is the "other mortality" rate unexplained by the model.

Thus, the Ecopath model assumes the trophic network to be in a steady state during the reference period (usually 1 year), and consequently mass-balance occurs, where the production of the group is equal to the sum of all predation, nonpredatory losses, exports, biomass accumulations, and catches (see equation 9.1). Assuming there is no export and no biomass accumulation, and that the catches Y are known, only three of the four parameters *B*, *P/B*, *Q/B*, and *EE* have to be set initially for each group. The remaining parameter can be calculated by the software. The diet composition of each group is needed, that is, the percentage of the prey items in the diet of the group; rough initial diet composition estimates can be obtained from FishBase (www.fishbase.org) and SeaLifeBase (www.sealifebase.org).

REFERENCES

Christensen, V., and D. Pauly. 1992. The ECOPATH II: a software for balancing steady-state ecosystem models and calculating network characteristics. *Ecological Modelling* 61: 169–185.

Polovina, J. J. 1984. Model of a coral reef ecosystem, Part I: ECOPATH model and its application to French Frigate Shoals. *Coral Reefs* 3: 1–11.

EwE models registered in EcoBase. We focused on the objectives of the EwE-based studies, the complexity and scope of the models, and the general characteristics of the modeled ecosystems and noted the complementary use of EcoTroph in EwE models (box 9.2). Based on the year of publication of the models, we also analyzed the evolution of the EwE applications over the past 30 years.

We present an application example detailed in Christensen et al. (2014), based on 200 models and a method that has been previously applied to the North Atlantic, Southeast Asia, and West Africa (Christensen et al. 2003a, 2003b, 2004). Therein, the 200 EwE models in figure 9.2 were used to provide snapshots of how much life there was in the ocean at given points in time and space. Christensen et al. (2014) then evaluated how the environmental conditions at each point relate to environmental parameters, from which they developed a multiple regression model to predict biomass trends. Finally, they used global environmental databases to predict the spatial distribution of fish biomass. This allowed Christensen et al. (2014) to predict the biomass trends for

BOX 9.2. ECOTROPH: A TOOL TO ANALYZE FISHING IMPACTS ON AQUATIC FOOD WEBS

Didier Gascuel, Université Européenne de Bretagne, Agrocampus Ouest, Rennes, France

EcoTroph is a modeling approach articulated around the idea that an ecosystem can be represented by trophic spectra, representing the distribution across trophic levels, of biomass, production, catches, fishing mortalities, and so on. That is, ecosystems are viewed as biomass flowing from lower to higher trophic levels, through predation and ontogenetic processes (Gascuel 2005; Gascuel and Pauly 2009). Thus, the ecosystem biomass present at any given trophic level may be estimated from two simple equations, one describing biomass flow, the other their kinetics, which quantifies the velocity of biomass transfers toward top predators.

Modeling biomass flows as a quasiphysical process enables us to explore aspects of ecosystem functioning that complement EwE analysis. It provides users with simple tools to quantify fishing impacts at an ecosystem scale and a new way of looking at ecosystems. It also provides tools and indicators to analyze fishing fleets' interactions at the ecosystem level (Gasche and Gascuel 2013). EcoTroph can be used either in association with an existing Ecopath model or as a stand-alone application, especially in data-poor environments. It runs either as a plug-in of the EwE software or as an R-package (Colléter et al. 2013).

EcoTroph has also been used in specific case studies to assess the current fishing impacts at the ecosystem scale. High trophic levels have been found globally overexploited, for instance, in the Guinean EEZ (Gascuel et al. 2011), or the Bay of Biscay (Lassalle et al. 2012).

REFERENCES

Colléter, M., J. Guitton, and D. Gascuel. 2013. An introduction to the EcoTrophR package: analyzing aquatic ecosystem trophic network. *The R Journal* 5(1): 98–107.

Gasche, L., and D. Gascuel. 2013. EcoTroph: a simple model to assess fisheries interactions and their impacts on ecosystems. *ICES Journal of Marine Sciences* 70: 498–510.

Gascuel, D. 2005. The trophic-level based model: a theoretical approach of fishing effects on marine ecosystems. *Ecological Modelling* 189: 315–332.

Gascuel, D., S. Guénette, and D. Pauly. 2011. The trophic-level based ecosystem modelling approach: theoretical overview and practical uses. *ICES Journal of Marine Sciences* 68: 1403–1416.

Gascuel, D., and D. Pauly. 2009. EcoTroph: modelling marine ecosystem functioning and impact of fishing. *Ecological Modelling* 220: 2885–2898.

Lassalle, G., D. Gascuel, F. Le Loc'h, J. Lobry, G. Pierce, V. Ridoux, B. Santos, J. Spitz, and N. Niquil. 2012. An ecosystem approach for the assessment of fisheries impacts on marine top predators: the Bay of Biscay case study. *ICES Journal of Marine Sciences* 69: 925–938.

higher-trophic-level predatory fish (i.e., the larger predatory "table fish") and for the lower-trophic level prey fish, such as small pelagics (e.g., sardines, anchovies, capelins), which are used mainly for fishmeal and oil. Given the recent controversy over whether "fishing down the food web" is a phenomenon actually occurring in nature (Pauly et al. 1998; see also www.fishingdown.org) or a sampling artifact with no or little relation to the underlying ecosystem structure, we contribute here to this discussion by evaluating how the biomass of high-trophic-level species has changed relative to the biomass of low-trophic-level species.

APPLICATIONS OF THE EWE MODELING APPROACH

The 435 Ecopath or EwE models in Ecobase were used to tackle a wide range of ecological issues; notably, 87% of the models were developed to answer questions about ecosystem functioning, 64% to analyze fisheries, 34% to focus on particular species of interest, and 11% to consider environmental variability (the percentages add to more than 100 because models may have more than one purpose). Less than 10% of the models focused on marine protected areas (MPAs), pollution (e.g., Booth and Zeller 2005), or aquaculture. The model module that identifies the keystone species (or groups) in

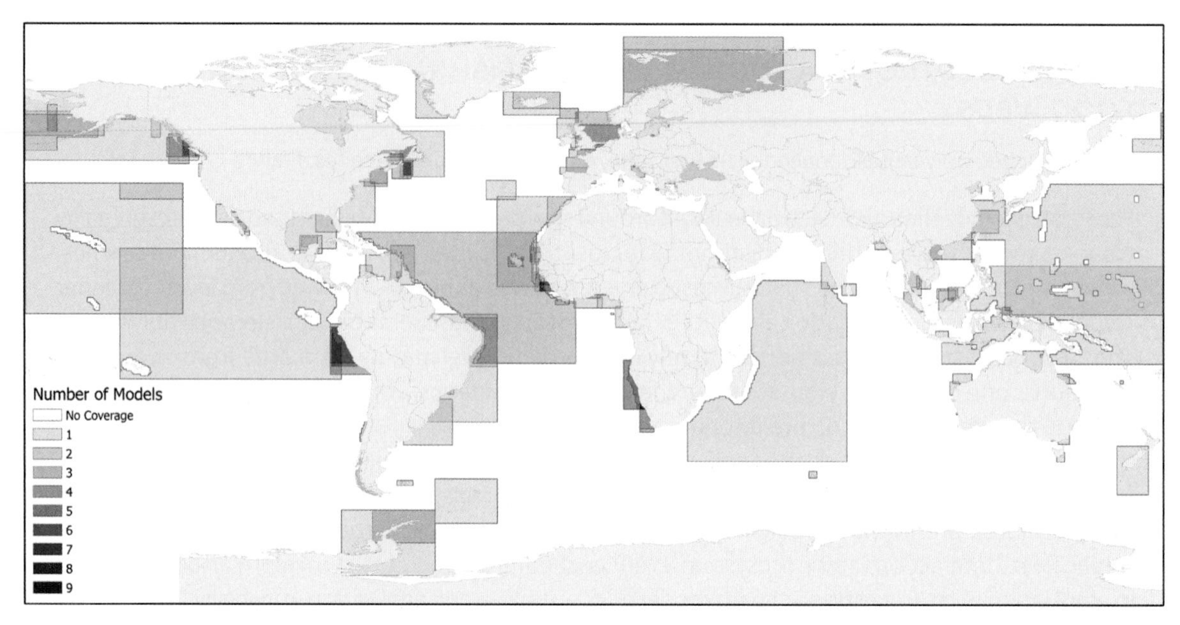

Figure 9.2. Areas covered by the 200 Ecopath models used by Christensen et al. (2014) to estimate historic changes in the biomass of fishes with high (≥3.5) and low (<3.5) trophic levels in the world ocean (see text).

ecosystems, based on Libralato et al. (2006), was used in 11% of the models, whereas the Ecotracer plug-in for tracking pollutants has been applied in less than 1% of the models (but see chapter 13).

The EcoTroph plug-in (see box 9.2 and figure 9.3) has also been little applied to date (2% of the models), but several of these applications focused on the effects of MPAs on the food web of ecosystems (Albouy et al. 2010; Colléter et al. 2012; Valls et al. 2012). In particular, Colléter et al. (2014) showed that the potential exports from small MPAs are of the same order of magnitude as the catch that could have been obtained inside the reserve, and Guénette et al. (2014), who used EcoTroph to assess the contribution of an MPA to the trophic functioning of a larger ecosystem, showed that the MPA of the Banc d'Arguin, in Mauritania, supports about 23% of the total production and 18% of the total catch of the Mauritanian shelf ecosystem, and up to 50% for coastal fish.

About three fourths of the 435 EwE models include between 10 and 40 functional (or taxonomic) groups, with 32% (141 models) including 10 to 20 groups. Overall, the numbers of groups range from 7 to 171 groups, but only 5 models include between 75 and 100 groups, and 2 models include more than 100 groups; 31% of

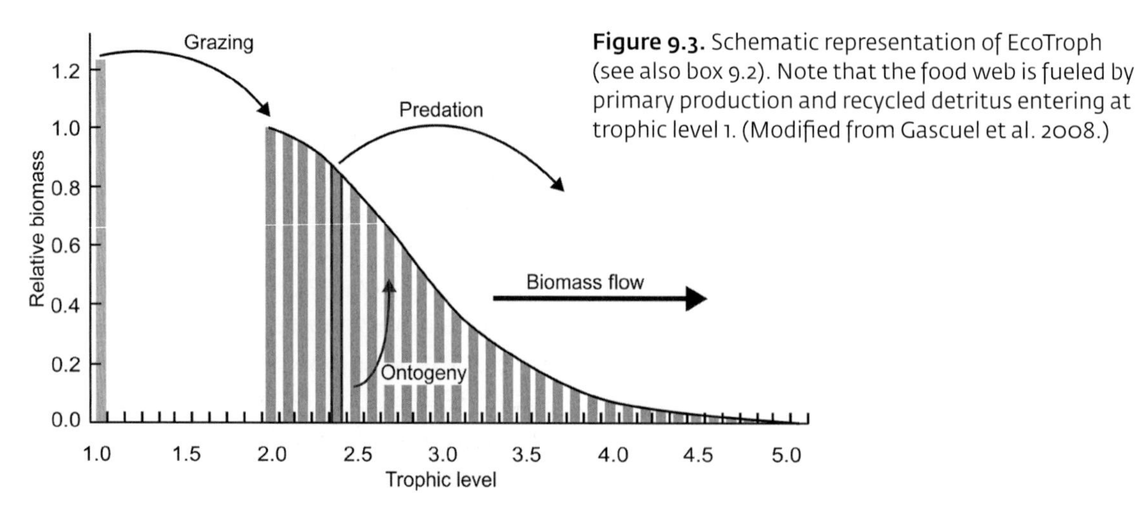

Figure 9.3. Schematic representation of EcoTroph (see also box 9.2). Note that the food web is fueled by primary production and recycled detritus entering at trophic level 1. (Modified from Gascuel et al. 2008.)

the models include groups corresponding to stanzas (age groups). Time dynamic (Ecosim) versions were developed for 41% of the models and spatially explicit (Ecospace) versions for 7% of the models.

About 70% of the models refer to a time period between 1980 and 2009, with 37% (159 models) corresponding to the 1990s. About three fourths of the models represent a time period lasting from 1 to 5 years, with 44% (192 models) corresponding to 1 year, which is the classical temporal scale of Ecopath models. The longest time period represented by a model is 40 years. The spatial extent covered by the models varies widely, from 0.005 km^2 to 34,640,000 km^2. However, model area does not exceed 1,000,000 km^2 for most models, and about half of the models cover an area ranging from 10,000 to 1,000,000 km^2. About 90% of models use wet weight as a unit (of which 88% express it in t/km^2), 5% carbon weight, and 4% dry weight. Only 4 models used calories (or joules) as a unit, and 1 used nitrogen (see chapter 13). Almost all models use year as a time unit, and only 10 models used day, month, or season.

The best-represented ecosystem types are continental shelves (32% of the models), bays and fjords (14%), open oceans (13%), and freshwater lakes (8%); 49% of the models are located in the tropics, 44% in temperate areas, and only 7% in high latitudes (see figure 9.2). EwE models have been developed to study aquatic ecosystem worldwide, with some regions better covered than others. Overall, the northern and central Atlantic Ocean are the regions with the highest proportion of EwE models. All FAO areas (figure 2.1) have at least one model, but five areas have about 40 models each: the northeast Atlantic and the eastern central Atlantic comprise 10% of the models each, and the western central Atlantic, the northwest Atlantic, and the Mediterranean and Black Sea comprise 9% of the models each. The Humboldt Current, the Gulf of Alaska, the Mediterranean, and the Guinea Current are the Large Marine Ecosystems (LMEs; see www.seaaroundus.org/data/#/lme/) with the highest number of models (at least 5% each). Three LMEs have no EwE model representing them: the Osyashio Current, the East Siberian Sea, and the Laptev Sea. Overall, few EwE models have been developed for the Indian Ocean and for Antarctic waters.

Recently developed models tend to be less aggregated and thus more complex, although highly aggregated models are still being published. In the first decade of the development of the EwE modeling approach, the total number of groups defined in the models ranged from seven to twenty-seven. Over time, the number of groups has expanded, up to sixty-seven groups in the past decade (excluding the few outlier models). The median is about fifteen groups between 1984 and 1993, and it is about thirty groups between 2004 and 2014. In contrast, the time period represented by the models tends to decrease over time; thus the median number of years represented by the models ranged from 3 years in 1984–1993 to 1 year in 2004–2014. The areas covered by the models have expanded toward very large areas, and the median area has shifted accordingly, from about 1,000 km^2 in 1984–1993 to about 100,000 km^2 in 1994–2014.

Fish Biomass in the World Ocean: A Century of Decline

Using 200 EwE models, each providing a snapshot of how much life there was in the ocean at given points in time and space, Christensen et al. (2014) evaluated trends in biomass of fish separately for higher-trophic-level predatory fish ("table fish") and for the lower-trophic-level prey fish. Their results suggest that the biomass of predatory fish has declined strongly (and significantly) over the last hundred years. For the 200 models, covering the period from 1880 to 2010, they evaluated how the conditions at each point relate to environmental parameters and other variables, and they obtained the multiple regression in table 9.1. The multiple coefficient of determination (R^2) they obtained is 0.70, indicating that the model they derived explains 70% of the variation in the dataset. The predictor variables are all highly significant apart from the factorial

Table 9.1. Parameter coefficients and associated test statistics for multiple linear regressions to predict the global marine biomass of predatory fishes. R^2 is 0.70; the t values are the ratios of an estimate and its standard error, and p (>|t|) indicates the probability of obtaining a larger t value; all but p values are significant ($\alpha = 0.001$).

Variable	Estimate	t value	p (>|t|)	Significant
Intercept	24.2500	54.8	<0.001	Yes
Year	−0.0151	−69.7	<0.001	Yes
log(distance)	−0.1008	−28.0	<0.001	Yes
log(prim. prod.)	1.1040	142.8	<0.001	Yes
Temperature	−0.0608	−69.6	<0.001	Yes
Upwelling index	0.0002	42.4	<0.001	Yes
FAO 18	0.0978	2.0	0.041	No
FAO 21	0.6361	19.9	<0.001	Yes
FAO 27	0.7966	28.4	<0.001	Yes
FAO 31	0.0605	1.7	0.091	No
FAO 34	−0.1952	−6.0	<0.001	Yes
FAO 37	−0.4279	−8.4	<0.001	Yes
FAO 41	1.0460	31.0	<0.001	Yes
FAO 47	0.6778	18.2	<0.001	Yes
FAO 48	1.1660	32.8	<0.001	Yes
FAO 57	1.1920	26.1	<0.001	Yes
FAO 61	1.1250	35.6	<0.001	Yes
FAO 67	1.5880	51.4	<0.001	Yes
FAO 71	1.2270	36.1	<0.001	Yes
FAO 77	0.4832	14.9	<0.001	Yes
FAO 87	0.3341	9.7	<0.001	Yes

variable for FAO areas 18 and 31 (representing the Amerasian Arctic and the Caribbean). The signs of the predictor variable coefficients all are as expected, that is, negative for biomass, distance, and temperature and positive for primary production and the upwelling index. The model suggested that we have lost 1.5% of the biomass of higher trophic level fish per year, on average, since the late nineteenth century.

Christensen et al. (2014) examined the relationship between observed and predicted values based on the coefficients in table 9.1 and observed that the multiple regression tends to overestimate abundance at low biomasses and underestimate it at high biomasses. This suggests that the model is conservative, that is, it does not overestimate changes in biomass over time. Such results also suggest that this multiple regression model would benefit from additional variables, such as "rugosity" (i.e., depth variability within spatial cells), substrate types, and fishing effort. However, Christensen et al. (2014) did not have access to global datasets of those variables and therefore had to ignore them for the time being. The implication is not that the present study is likely to be misleading but rather that better predictions will be obtained with additional predictor variables.

Using a resampling method, Christensen et al. (2014) randomly drew 30% of the 68,939 estimates of biomass over space and time, performed a multiple regression with each subsample, and obtained a distribution for each predictor variable. They then used each of the resampled regressions and the database of environmental parameters to predict

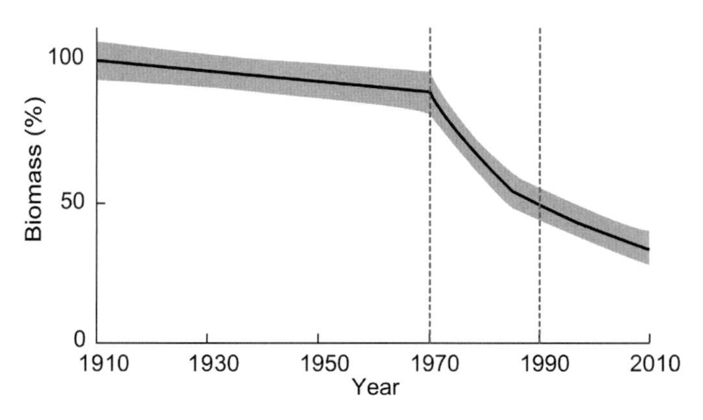

Figure 9.4. Global biomass trends for predatory fish in 1910–2010, as predicted based on 200 ecosystem models and 1,000 times random resampling of 30% of data points. The lines indicate median values and 95% confidence intervals. (Adapted from Christensen et al. 2014.)

global biomasses. Dividing the models into three time periods to increase the temporal resolution, and with the splits made in 1970 and 1990, they obtained multiple linear regressions similar to the regression reported above for the entire time period. Again the predictor variables were highly significant ($p < 10^{-16}$) and the regressions explained 66%–91% of the variability in the biomass data. Evaluating the time trends based on resampling the three regressions 1,000 times based on randomly selecting 30% of the biomass estimates for each case leads to figure 9.4.

Also, Christensen et al. (2014) estimated that the biomass of predatory fishes has declined by two thirds (66.4%, with 95% confidence intervals ranging from 60.2% to 71.2%) over the last hundred years. The decline was estimated to have been slow (10.8%, or 0.2%/year) up to 1970, then severe during 1970–1990 (41.6%, or 4.0%/year), and more slow since 1990 (14.0%, or 2.9%/year). Repeating the multiple linear regression for the entire time period and focusing on predicting the biomass of lower-trophic-level fish ($2.0 \leq TL < 3.0$) led to positive regression coefficients. The coefficient for the variable *year* (0.0085) suggested that the biomass of prey fish had been increasing over time by 0.85%/year. Over a hundred-year time period, this implies that there are now more than twice as many low-trophic-level (prey) fish in the global ocean than there were a century ago, an increase that may be caused by predation release.

CONCLUSION

The global overview of the EwE model applications showed that most models represented marine ecosystems, between 1980 and 2009, over a time period of 1 year and an area ranging from 10,000 to 1,000,000 km². The models generally include between ten and forty functional groups. Most models were built to analyze ecosystem functioning and inform fisheries management, principally in ecosystems located in the northern and central Atlantic Ocean. Half of the models were applied to tropical systems, and more than a third of the models were used to perform time dynamic simulations in Ecosim. Despite its complementarity with Ecopath, the EcoTroph plug-in has been applied to a few models only. However, EcoTroph is still a recent approach, and the development of the plug-in in R (see box 9.2) may allow a wider application.

In the first decade of its development (1984–1993), the EwE modeling approach essentially consisted of Ecopath models representing tropical marine systems, with a simple trophic structure. The initial emphasis on the tropics resulted from the development of EwE initially being centered at the International Center for Living Aquatic Resources Management (ICLARM, now WorldFish), then based in the Philippines, which was focused on developing methods for managing tropical ecosystems. In contrast, in the last two decades (1994–2014) EwE models were applied to a wider variety of ecosystems, including polar regions, and used to analyze a wider range of research topics, including pollution, aquaculture, and MPAs. The modeling practices have evolved over the past 30 years toward Ecopath models with larger spatial scales (up to 1,000,000 km^2), shorter temporal scales (typically 1 year for Ecopath), and more complex trophic structures (up to seventy functional groups). Although Ecosim has been used in a great proportion of the EwE models (41%), the same is not true for Ecospace (7%), which is surprising considering the insights Ecospace can provide (Pauly 2002).

We believe that the standardized metadata incorporated in EcoBase and used here will be valuable to perform meta-analyses based on EwE models. Indeed, the metadata may be used as selection criteria. By applying a scoring method on these criteria, a list of models of potential interest may be obtained. The pool of selected models may then be reused in EwE-based meta-analyses. Also, the metadata presented here may serve as a template of the necessary information that should be systematically provided when publishing EwE models. Lastly, the global and synthesized overview provided here may help elucidate the usage of and interest in the EwE modeling approach. Some regions and types of ecosystems have been widely analyzed with the EwE modeling approach, and others have remained poorly studied. Notably, modeling effort should be concentrated on the Indian Ocean.

The case study adapted from Christensen et al. (2014) found a major decline in the biomass of predatory fish, that is, of the larger fish that humans tend to eat, amounting to a reduction by two thirds over the last century, thus confirming earlier results of Tremblay-Boyer et al. (2011), obtained by applying the EcoTroph model (see box 9.2) to each of the 180,000 spatial cells also used here. Figure 9.4 shows that 55% of the decline noted by Christensen et al. (2014) occurred in the last 40 years, with the decline being strongest from 1970 to 1990, before it leveled off somewhat. However, this does not mean that conditions have started to improve globally. There may be regional improvements as reported by Worm et al. (2009); however, Christensen et al. (2014) did not see them at the global level.

The material summarized here contributes to the recent discussion on whether "fishing down the food web" is a sampling artifact or something that occurs in reality. Christensen et al. (2014) estimated that the predatory fish have declined by two thirds, whereas the prey fish may have more than doubled. Such doubling is likely to be linked to predation release, that is, the mechanism whereby reduction in predator populations leads to increases in prey abundance, as documented in Myers et al. (2007). Combined, the decrease of high-trophic-level fish and the increase of low-trophic-level fish strongly suggests that fishing down the marine food web occurs at the global scale and that this can be demonstrated through a method that is less dependent on fisheries catch estimates than previous "fishing down" studies, which itself has been central to the debate about "fishing down" (see also www.fishingdown.org).

In conclusion, we believe that open access is becoming mainstream in ecology, and we built the EcoBase repository as a contribution to this trend. This study was a first step toward a global integration of EwE-based metadata, and we hope that more meta-analyses can be facilitated by the availability of EcoBase (ecobase.ecopath.org).

ACKNOWLEDGMENTS

This is a joint contribution from the *Sea Around Us*, a research activity at the University of British Columbia initiated and funded by the Pew Charitable Trusts from 1999 to 2014 and currently funded mainly by the Paul G. Allen Family Foundation. M.C., D.G., and J.G. also acknowledge support from the French Ministry of Higher Education and Research and Agrocampus Ouest, and V.C. acknowledges support from the National Science and Engineering Research Council of Canada.

REFERENCES

Albouy, C., D. Mouillot, D. Rocklin, J. M. Culioli, and F. Le Loch. 2010. Simulation of the combined effects of artisanal and recreational fisheries on a Mediterranean MPA ecosystem using a trophic model. *Marine Ecology Progress Series* 412: 207–221.

Arreguín-Sánchez, F. 2011. Ecosystem dynamics under "top-down" and "bottom-up" control situations generated by intensive harvesting rates. *Hydrobiologica* 21: 323–332.

Booth, S., and D. Zeller. 2005. Mercury, food webs, and marine mammals: implications of diet and climate change for human health. *Environmental Health Perspectives* 113: 521–526.

Christensen, V. 1995. Ecosystem maturity: toward quantification. *Ecological Modelling* 77: 3–32.

Christensen, V., P. Amorim, I. Diallo, T. Diouf, S. Guénette, J. J. Heymans, A. Mendy, M. M. Ould Taleb Ould Sidi, M. L. D. Palomares, B. Samb, K. A. Stobberup, J. M. Vakily, M. Vasconcellos, R. Watson, and D. Pauly. 2004. Trends in fish biomass off northwest Africa, 1960–2000. Pp. 377–386 + plate VI in P. Chavance, M. Ba, D. Gascuel, M. Vakily, and D. Pauly (eds.), *Pêcheries maritimes, écosystèmes et sociétés en Afrique de l'Ouest: un demi-siècle de changement*. Collection des rapports de recherche halieutique ACP-UE 15.

Christensen, V., L. Garces, G. Silvestre, and D. Pauly. 2003a. Fisheries impact on the South China Sea Large Marine Ecosystem: a preliminary analysis using spatially-explicit methodology. Pp. 51–62 in G. Silvestre, L. Garces, I. Stobutzki, M. Ahmed, R. A. Valmonte-Santos, C. Luna, L. Lachica-Aliño, P. Munro, V. Christensen, and D. Pauly (eds.), *Assessment, management and future directions for coastal fisheries in Asian countries*. World-Fish Center Conference Proceedings 67.

Christensen, V., S. Guénette, J. J. Heymans, C. J. Walters, R. Watson, D. Zeller, and D. Pauly. 2003b. Hundred year decline of North Atlantic predatory fishes. *Fish and Fisheries* 4: 1–24.

Christensen, V., and D. Pauly. 1992. The ECOPATH II: a software for balancing steady-state ecosystem models and calculating network characteristics. *Ecological Modelling* 61: 169–185.

Christensen, V., and D. Pauly. 1993a. Flow characteristics of aquatic ecosystems. Pp. 338–352 in V. Christensen and D. Pauly (eds.), *Trophic models of aquatic ecosystems*, vol. 26. ICLARM Conference Proceedings, Manila, Philippines.

Christensen, V., and D. Pauly (eds.). 1993b. *Trophic models of aquatic ecosystems*. ICLARM Conference Proceedings, Manila, Philippines, 26.

Christensen, V., C. Piroddi, M. Coll, J. Steenbeek, J. Buszowski, and D. Pauly. 2014. A century of fish biomass decline in the ocean. *Marine Ecology Progress Series* 512: 155–166.

Christensen, V., and C. Walters. 2004. Ecopath with Ecosim: methods, capabilities and limitations. *Ecological Modelling* 172: 109–139.

Christensen, V., and C. J. Walters. 2011. Progress in the use of ecosystem modeling for fisheries management. Pp. 189–205 in V. Christensen and J. L. Maclean (eds.), *Ecosystem Approaches to Fisheries: A Global Perspective*. Cambridge University Press, Cambridge, England.

Christensen, V., C. J. Walters, R. Ahrens, J. Alder, J. Buszowski, L. B. Christensen, W. W. L. Cheung, J. Dunne, R. Froese, V. Karpouzi, K. Kaschner, K. Kearney, S. Lai, V. Lam, M. L. D. Palomares, A. Peters-Mason, C. Piroddi, J. L. Sarmiento, J. Steenbeek, U. R. Sumaila, R. Watson, D. Zeller, and D. Pauly. 2009. Database-driven models of the world's Large Marine Ecosystems. *Ecological Modelling* 220: 1984–1996.

Christensen, V., C. Walters, D. Pauly, and R. Forrest. 2008. *Ecopath with Ecosim Version 6. Users Guide*. Lenfest Ocean Futures Project 2008. Available at source.ecopath.org/trac/Ecopath/wiki/UsersGuide.

Coll, M., and S. Libralato. 2012. Contributions of food web modelling to the ecosystem approach to marine resource management in the Mediterranean Sea. *Fish and Fisheries* 13: 60–88.

Coll, M., J. Navarro, R. J. Olson, and V. Christensen. 2012. Assessing the trophic position and ecological role of squids in marine ecosystems by means of food-web models. *Deep Sea Research Part II: Topical Studies in Oceanography* 95: 21–36.

Colléter, M., D. Gascuel, C. Albouy, P. Francour, L. Tito De Morais, A. Valls, and F. Le Loch. 2014. Fishing inside or outside? A case studies analysis of potential spillover effect from marine protected areas, using food web models. *Journal of Marine Systems* 139: 383–395.

Colléter, M., D. Gascuel, J. Ecoutin, and L. T. De Morais. 2012. Modelling trophic flows in ecosystems to assess the efficiency of Marine Protected Area (MPA), a case study on the coast of Senegal. *Ecological Modelling* 232: 1–13.

Colléter, M., A. Valls, J. Guitton, L. Morissette, F. Arreguín-Sánchez, V. Christensen, D. Gascuel, and D. Pauly. 2013. *EcoBase: a repository solution to gather and communicate information from EwE models.* Fisheries Centre Research Reports 21(1), University of British Columbia, Vancouver, Canada.

Edwards, P. N. 2010. *A Vast Machine: Computer Models, Climate Data, and the Politics of Global Warming*. MIT Press, Boston.

Gascuel, D., L. Morissette, M. L. D. Palomares, and V. Christensen. 2008. Trophic flow kinetics in marine ecosystems: toward a theoretical approach to ecosystem functioning. *Ecological Modelling* 217: 33–47.

Guénette, S., B. Meissa, and D. Gascuel. 2014. Assessing the contribution of Marine Protected Areas to the trophic functioning of ecosystems: a model for the Banc d'Arguin and the Mauritanian Shelf. *PLoS ONE* 9(4): e94742.

Heymans, J. J., M. Coll, S. Libralato, L. Morissette, and V. Christensen. 2014. Global patterns in ecological indicators of marine food webs: a modelling approach. *PLoS ONE* 9: e95845.

Jarre-Teichmann, A. 1998. The potential role of mass balance models for the management of upwelling ecosystems. *Ecological Applications* 8: S93–S103.

Libralato, S., V. Christensen, and D. Pauly. 2006. A method for identifying keystone species in food web models. *Ecological Modelling* 195: 157–171.

Michener, W. H., and M. B. Jones. 2012. Ecoinformatics: supporting ecology as a data-intensive science. *Trends in Ecology and Evolution* 27: 85–93.

Myers, R. A., J. K. Baum, T. D. Shepherd, S. P. Powers, and C. H. Peterson. 2007. Cascading effects of the loss of apex predatory sharks from a coastal ocean. *Science* 315: 1846–1850.

Pauly, D. 1995. Anecdotes and the shifting baseline syndrome of fisheries. *Trends in Ecology & Evolution* 10: 430.

Pauly, D. 2002. Spatial modelling of trophic interactions and fisheries impacts in coastal ecosystems: a case study of Sakumo Lagoon, Ghana. Pp. xxxv, 289–296 in J. McGlade, P. Cury, K. A. Koranteng, and N. J. Hardman-Mountford (eds.), *The Gulf of Guinea Large Marine Ecosystem: Environmental Forcing and Sustainable Development of Marine Resources.* Elsevier Science, Amsterdam.

Pauly, D. 2007. The *Sea Around Us* Project: documenting and communicating global fisheries impacts on marine ecosystems. *AMBIO: A Journal of the Human Environment* 34(4): 290–295.

Pauly, D., F. Arreguin-Sanchez, J. Browder, V. Christensen, S. Manikchand-Heileman, E. Martinez, and L. Vidal. 1999. Toward a stratified mass-balance model of trophic fluxes in the Gulf of Mexico Large Marine Ecosystem. Pp. 278–293 in H. Kumpf, K. Steidinger, and K. Sherman (eds.), *The Gulf of Mexico Large Marine Ecosystem: Assessment, Sustainability, and Management*. Blackwell Science, Malden, MA.

Pauly, D., and V. Christensen. 1993. Stratified models of large marine ecosystems: a general approach, and an application to the

South China Sea. Pp. 148–174 in K. Sherman, L. M. Alexander, and B. D. Gold (eds.), *Stress, Mitigation and Sustainability of Large Marine Ecosystems*. American Association for the Advancement of Science, Washington, DC.

Pauly, D., V. Christensen, J. Dalsgaard, R. Froese, and F. Torres. 1998. Fishing down marine food webs. *Science* 279: 860–863.

Pauly, D., V. Christensen, and C. Walters. 2000. Ecopath, Ecosim and Ecospace as tools for evaluating ecosystem impact of fisheries. *ICES Journal of Marine Science* 57: 697–706.

Pauly, D., W. Graham, S. Libralato, L. Morissette, and M. L. D. Palomares. 2009. Jellyfish in ecosystems, online databases, and ecosystem models. *Hydrobiologia* 616: 67–85.

Pérez-España, H., and F. Arreguín-Sánchez. 1999. Complexity related to behavior of stability in modeled coastal zone ecosystems. *Aquatic Ecosystem Health and Management* 2: 129–135.

Pérez-España, H., and F. Arreguín-Sánchez. 2001. An inverse relationship between stability and maturity in models of aquatic ecosystems. *Ecological Modelling* 145: 189–196.

Pikitch, E. K., K. J. Rountos, T. E. Essington, C. Santora, D. Pauly, R. Watson, U. R. Sumaila, P. D. Boersma, I. L. Boyd, and D. O. Conover. 2012. The global contribution of forage fish to marine fisheries and ecosystems. *Fish and Fisheries* 15: 43–64.

Plaganyi, E., and D. Butterworth. 2004. A critical look at the potential of Ecopath with Ecosim to assist in practical fisheries management. *African Journal of Marine Science* 26: 261–287.

Polovina, J. J. 1984. Model of a coral reef ecosystem. *Coral Reefs* 3: 1–11.

Reichman, O., M. B. Jones, and M. P. Schildhauer. 2011. Challenges and opportunities of open data in ecology. *Science* 331: 703–705.

Thessen, A. E., and D. K. Patterson. 2011. Data issues in the life sciences. *ZooKeys* 150: 15–51.

Tremblay-Boyer, L., D. Gascuel, R. Watson, V. Christensen, and D. Pauly. 2011. Modelling the effects of fishing on the biomass of the world's oceans from 1950 to 2006. *Marine Ecology Progress Series* 442: 169–185.

Valls, A., D. Gascuel, S. Guénette, and P. Francour. 2012. Modeling trophic interactions to assess the potential effects of a marine protected area: case study in the NW Mediterranean Sea. *Marine Ecology Progress Series* 456: 201–214.

Vision, T. J. 2010. Open data and the social contract of scientific publishing. *BioScience* 60: 330–331.

Walters, C., V. Christensen, and D. Pauly. 1997. Structuring dynamic models of exploited ecosystems from trophic mass-balance assessments. *Reviews in Fish Biology and Fisheries* 7: 139–172.

Worm, B., R. Hilborn, J. K. Baum, T. Branch, J. S. Collie, C. Costello, M. J. Fogarty, E. A. Fulton, J. A. Hutchings, S. Jennings, O. P. Jensen, H. K. Lotze, P. M. Mace, T. R. McClanahan, C. Minto, S. R. Palumbi, A. M. Parma, D. Ricard, A. A. Rosenberg, R. Watson, and D. Zeller. 2009. Rebuilding global fisheries. *Science* 325: 578–585.

Zeller, D., R. Froese, and D. Pauly. 2005. On losing and recovering fisheries and marine science data. *Marine Policy* 29: 69–73.

NOTE

1. Cite as Colléter, M., A. Valls, V. Christensen, M. Coll, D. Gascuel, J. Guitton, C. Piroddi, J. Steenbeek, J. Buszowski, and D. Pauly. 2016. Modeling the global oceans with the Ecopath software suite: a brief review and application example. Pp. 98–109 in D. Pauly and D. Zeller (eds.), *Global Atlas of Marine Fisheries: A Critical Appraisal of Catches and Ecosystem Impacts*. Island Press, Washington, DC.

CHAPTER 10

JELLYFISH FISHERIES: A GLOBAL ASSESSMENT[1]

Lucas Brotz
Sea Around Us, University of British Columbia, Vancouver, BC, Canada

Jellyfish are considered traditional cuisine in China, where they have been eaten for more than 1,700 years (Omori and Nakano 2001; Li and Hsieh 2004). Other countries in Southeast Asia such as Thailand, Indonesia, and Malaysia have been catching jellyfish for decades, primarily for export to China and Japan. Despite this history, information on jellyfish fisheries is sparse and disaggregated. There are currently at least eighteen countries catching jellyfish for food (table 10.1), and an additional twelve are either exploring new fisheries or have been involved in jellyfish fisheries in the past (figure 10.1).

Many countries do not report their catches of jellyfish explicitly to the Food and Agriculture Organization of the United Nations (FAO), including them either as "miscellaneous marine invertebrates" (i.e., "marine invertebrates nei") or not at all. As a result, the current average annual catch of jellyfish reported by the FAO is approximately 350,000 t, but a global analysis reveals that the average annual catch during the period from 2000 to 2013 (figure 10.2) is at least 892,000 tonnes—more than 2.5 times previous estimates.

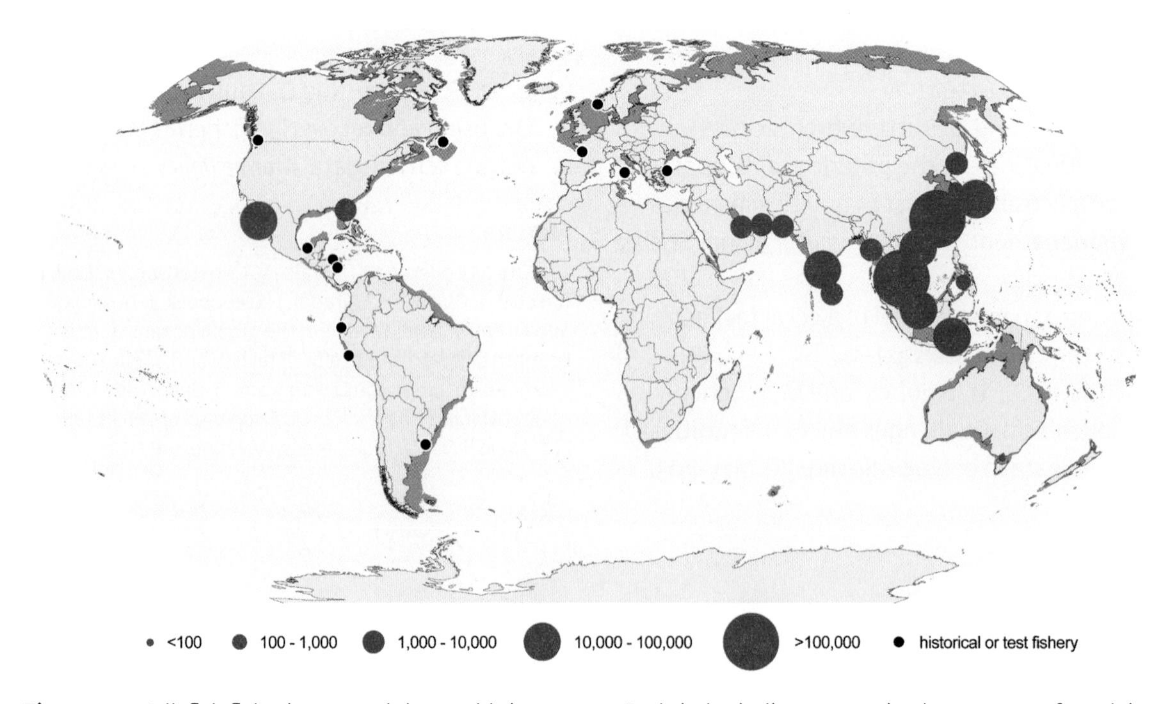

Figure 10.1. Jellyfish fisheries around the world since 2000. Red circles indicate magnitude category of catch in tonnes (see legend). Black circles indicate either historical fisheries (which no longer operate) or test fisheries (not yet established as commercial ventures). Circles are only approximately representative of catch locations.

Table 10.1. List of countries currently catching jellyfish for food with landings (mean of annual catches), as reported by FAO from 2000 to 2012, and as newly estimated for 2000 to 2013 (all values are in tonnes). Note that FAO also reports catches for more countries. However, they apparently refer to test fisheries (Nicaragua), past fisheries (Turkey), or discarded bycatch (Chile, Falkland Islands, Namibia, and the United Kingdom).

Country	Reported by FAO	New estimate	Sources
Australia	2	4	Fisheries Victoria and MAFRI (2002), Fisheries Victoria (2006)
Bahrain	2,378	2,737	Al-Abdulrazzak and Pauly (2013)
China	229,314	538,241	Li et al. (2014)
India	—	90,059	Anonymous (2005)
Indonesia	22,405	29,469	FAO (2014a)
Iran	3,451	5,968	Al-Abdulrazzak and Pauly (2013)
Japan	—	13,356	S. Uye, Hiroshima University (pers. comm.)
Korea (South)	—	457	Al-Abdulrazzak and Pauly (2013); FAO (2014a)
Malaysia	6,174	12,952	Teh and Teh (2014)
Mexico[a]	17,461	17,461	FAO (2014b)
Myanmar	2,175	2,183	SEAFDEC (2014)
Pakistan	—	3,000	Anonymous (2012); Khan (2012)
Philippines	18	856	FAO (2014a)
Russia	191	1,173	Yakovlev et al. (2005)
Sri Lanka	—	2,625	Perera (2008)
Thailand	71,736	141,864	SEAFDEC (2014); Teh et al. (2015)
United States	432	5,182	L. Brotz (unpublished data)
Vietnam	—	24,556	Nishikawa et al. (2008)
Total	**355,737**	**892,143**	Reconstructed total, probably an underestimate (see text)

[a]The value for Mexico is based on FAO data for 2011 and 2012 only. A more accurate estimate is approximately 9,315 t, based on the mean annual catch since 2000 reported by López-Martinez and Álvarez-Tello (2013).

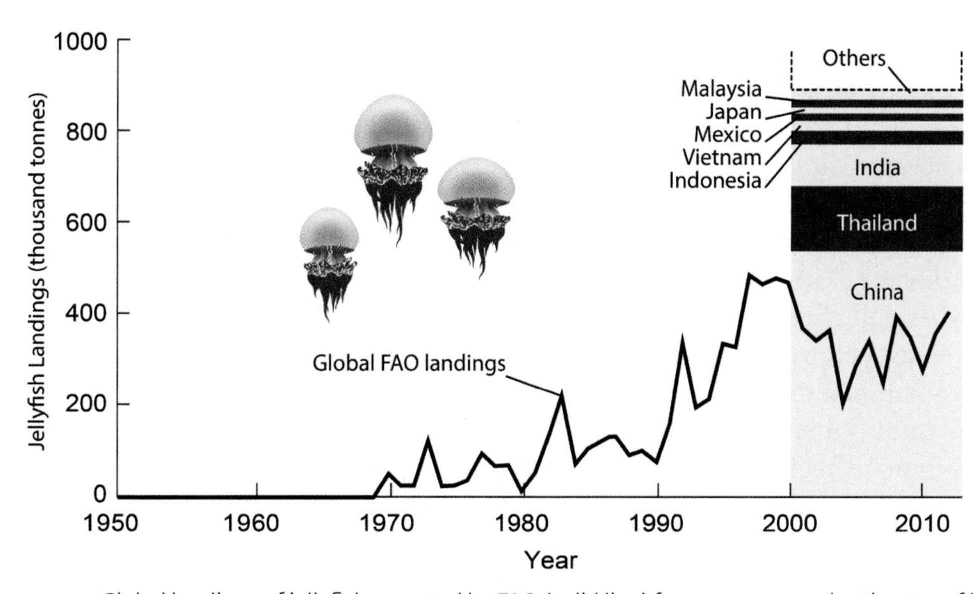

Figure 10.2. Global landings of jellyfish reported by FAO (solid line) for 1950–2012 and estimates of landings in this study for 2000–2013 (a likely underestimate; see text). The schematic representation of jellyfish is inspired by *Rhopilema esculentum*.

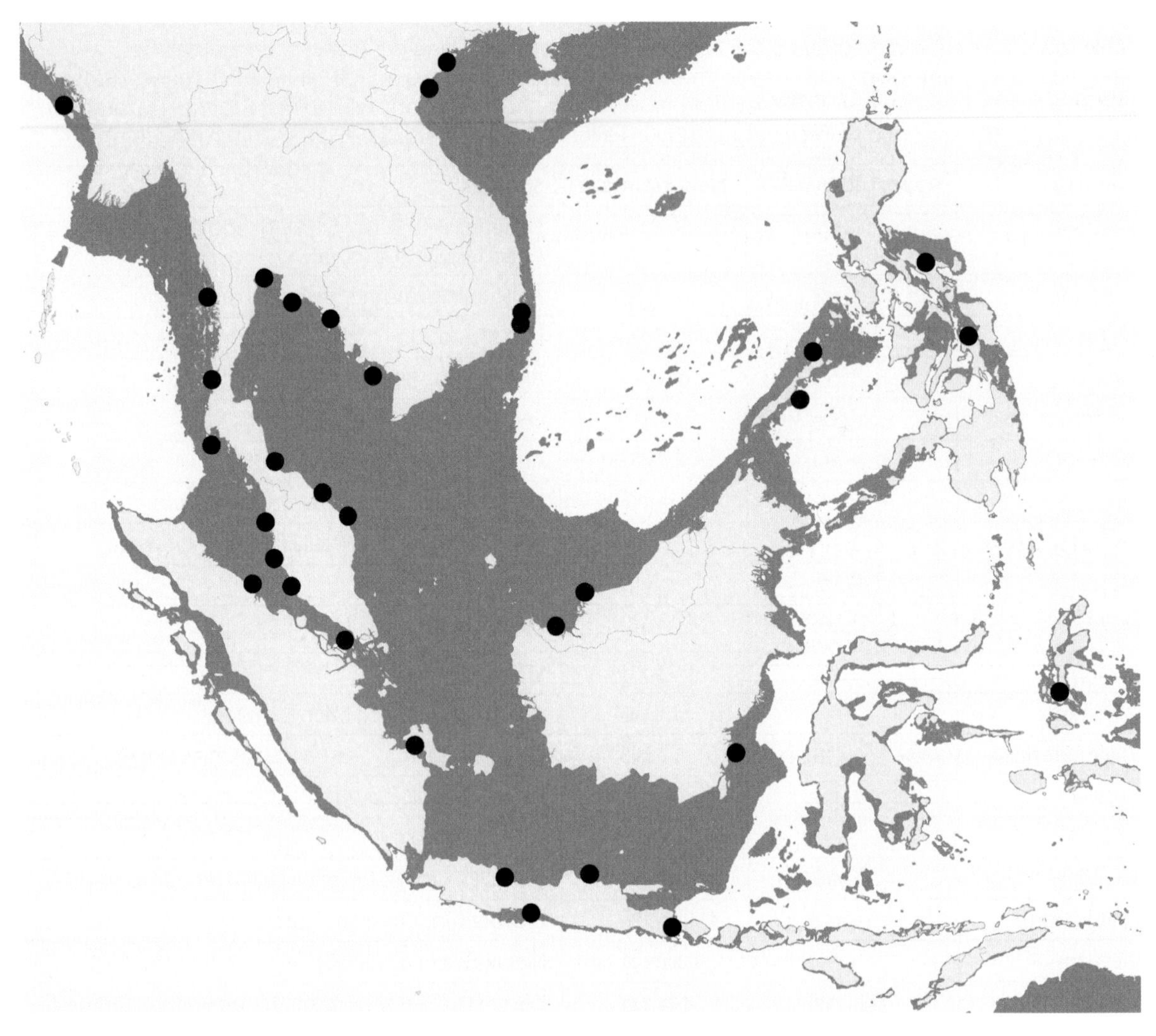

Figure 10.3. Locations of jellyfish fisheries in Southeast Asia; note that there are probably many additional locations that have yet to be documented. Based on Soonthonvipat (1976), Omori (1981), Rudloe (1992), Omori and Nakano (2001), Nishikawa et al. (2008), Kitamura and Omori (2010), and Thanh (2011).

Fisheries for jellyfish are usually characterized by short fishing seasons of a few months and dramatic interannual variations in catch (Omori 1978; Omori and Nakano 2001). In fact, rapid changes in exploitable biomass of jellyfish are probably more of a concern than for any other fishery (Kingsford et al. 2000). Combined with pollution from processing plants, as well as a conspicuous lack of research and regulation (Brotz and Pauly 2016), this has led to conflict and instability of jellyfish fisheries in many regions. Although the asexual reproductive phases of edible jellyfish are probably a buffer against overfishing, they do not appear to be a reliable safeguard, and overfishing of jellyfish stocks appears to have occurred in some locations, such as China (Dong et al. 2014) and the Salish Sea, in the Pacific Northwest (Mills 2001).

Jellyfish fisheries were historically concentrated in China, Japan, and Southeast Asia (figure 10.3). More recently, fisheries for jellyfish have expanded around the globe (figure 10.1), often driven by a combination of factors that may include a collapse of local fish stocks and increased interest from East Asian buyers. Established fishers, to whom jellyfish are often a costly nuisance, typically welcome these test fisheries enthusiastically. However, such exploratory jellyfish fisheries are often unsuccessful, potentially because of scant research, such as those in Canada (Sloan and Gunn 1985) or onerous regulations, as in Australia (Kingsford et al. 2000).

Table 10.2. Primary edible species of jellyfish in the order Rhizostomeae.

Species	Common name (Japan)	Country	Reference
Acromitus hardenbergi	"River type"	Malaysia, Indonesia, Thailand	Kitamura and Omori (2010)
Cassiopea ndrosia	Upside-down jellyfish	Philippines	Omori and Nakano (2001)
Catostylus mosaicus	Mosaic sea jelly, brown jelly blubber	Australia	Fisheries Victoria (2006)
Catostylus perezi	Banana jellyfish	Pakistan	Muhammed and Sultana (2008), Gul and Morandini (2013)
Cephea cephea	Crown, crowned, or cauliflower jellyfish	Thailand	Omori and Nakano (2001)
Crambione mastigophora	"Prigi type"	Indonesia	Omori and Nakano (2001), Kitamura and Omori (2010)
Crambionella annandalei	"Ball type," "sunflower type"	Myanmar	Kitamura and Omori (2010)
Crambionella helmbiru	"Cilacap type," "sunflower type"	Indonesia	Nishikawa et al. (2015)
Crambionella orsinia	—	India, Sri Lanka	Kuthalingam et al. (1989), NARA (2010)
Crambionella stuhlmanni	*Muttai chorri*	India	Kuthalingam et al. (1989), Mohan et al. (2011)
Lobonema smithi	—	India, Philippines	Kingsford et al. (2000), Murugan and Durgekar (2008)
Lobonemoides gracilis[b]	—	Philippines	Omori (1981), Kitamura and Omori (2010)
Lobonemoides robustus	"White type"	Indonesia, Myanmar, Vietnam, Thailand, Philippines	Kitamura and Omori (2010)
Lychnorhiza lucerna	—	Argentina	L. Brotz (unpublished data)
Nemopilema nomurai	Giant jellyfish, *echizen kurage*	China, Japan, Korea	Omori (1978), Morikawa (1984), Li et al. (2014)
Rhizostoma pulmo	Barrel jellyfish	Turkey, Italy	Ozer and Celikkale (2001), Armani et al. (2014)
Rhopilema esculentum	"Red type," flame jellyfish, *bizen kurage, aka kurage*	China, India, Indonesia, Japan, Korea, Malaysia, Thailand, Russia	Omori (1978), Morikawa (1984), Sloan (1986), Kingsford et al. (2000), Omori and Kitamura (2004), Yakovlev et al. (2005), Panda and Madhu (2009)
Rhopilema hispidium	"Sand type", *hizen kurage, shiro kurage*	China, Indonesia, Japan, Malaysia, Pakistan, Thailand, Vietnam	Kingsford et al. (2000), Omori and Kitamura (2004), Muhammed and Sultana (2008), Kitamura and Omori (2010), Gul and Morandini (2015)
Stomolophus meleagris	Cannonball jellyfish, jellyball	United States, Mexico	Hsieh et al. (2001), López-Martinez and Álvarez-Tello (2013)

[a]May be a synonym of *C. annandalei* (see Kitamura and Omori 2010).
[b]May be a synonym of *L. robustus* (see Kitamura and Omori 2010).

Table 10.3. Secondary or unconfirmed edible species of jellyfish.

Taxon	Common name	Country	Reference
Aurelia aurita	Moon jelly	Canada	DFA (2002a, 2002b)
Aurelia labiata	Moon jelly	Canada	Sloan and Gunn (1985)
Aurelia sp.	Moon jelly	India	Govindan (1984)
Carybdea rastoni	Box jellyfish	Taiwan	Purcell et al. (2007)
Chrysaora pacifica	Japanese sea nettle	Japan	Morikawa (1984), Huang et al. (1987)
Chrysaora plocamia	Humboldt jellyfish	Peru, Chile	L. Brotz (unpublished data)
Cotylorhiza tuberculata	Fried egg jellyfish	Italy	Armani et al. (2014)
Cyanea nozakii	Lion's mane jellyfish	China	Lu et al. (2003), Zhong et al. (2004), Dong et al. (2010)
Mastigias sp.	Spotted jellyfish	Thailand	Sloan and Gunn (1985)
Pelagia noctiluca	Mauve stinger	?	Armani et al. (2013)
Periphylla periphylla	Crown jellyfish, hat jellyfish	Norway	Wang (2007)
Phyllorhiza punctata	Brown jellyfish	Australia	Coleman et al. (1990), Kailola et al. (1993)
Rhizostoma octopus	Barrel jellyfish	Ireland	T. Doyle, National University of Ireland, Galway (pers. comm.)
Rhizostoma sp.	—	India	Chidambaram (1984)
Rhizostomatidae?	"Semi-China type"	Indonesia	Omori and Nakano (2001), Kitamura and Omori (2010)
Rhopilema nomadica	Nomad jellyfish	Turkey	Kingsford et al. (2000)
Rhopilema verrilli	Mushroom jellyfish	United States	Rudloe (1992), Kingsford et al. (2000)
Tamoya sp.	Box jellyfish	Kiribati	Shih (1977)

TARGET SPECIES

With the exception of Mexico, currently all catches of jellyfish reported by FAO are classified as *Rhopilema* spp., which is incorrect in many cases. The number of identified species of edible jellyfish worldwide is unclear and is typically underestimated (Omori 1981; Sloan 1986; Hsieh and Rudloe 1994; Omori and Nakano 2001; Armani et al. 2013), in part because of the taxonomy of edible jellyfish species being confused (Omori and Kitamura 2004). However, at least twenty different species of jellyfish have been identified as being consumed by humans (table 10.2).

Also, up to an additional sixteen edible species are either unconfirmed or under evaluation (table 10.3). Most edible species of jellyfish belong to the paraphyletic scyphozoan order Rhizostomeae. These jellyfish are typically large, with tough and rigid tissues. *Rhopilema esculentum* is the most valuable species and is currently the choice for hatchery and aquaculture operations in China (You et al. 2007; Dong et al. 2009). The giant jellyfish *Nemopilema nomurai* is also widely exploited in East Asia, in much larger quantities than have been reported until recently (see Li et al. 2014). There are reports that scyphomedusae from the order Semaeostomeae may also be consumed, such as *Aurelia*, *Chrysaora*, and *Cyanea*; however, it does not appear that any operations are currently targeting these less desirable species at commercial scales. There is also limited information to suggest cubozoans are consumed in some regions. Shih (1977) reported that the people of "Tarawa in the Pacific" (assumed to be Tarawa, Kiribati), consume freshly caught or sun-dried *Tamoya* sp. after boiling them. Purcell et al. (2007) reported that aboriginal peoples in Taitung, Taiwan eat cubomedusae. A number of jellyfish species have also been targeted for other reasons (e.g., nuisance, research, pharmaceuticals), which means that the total number of exploited jellyfish species is even higher (Kingsford et al. 2000).

Rhizostome jellyfish, which constitute the bulk of the edible species, have several life history characteristics that may help mitigate overfishing. Notably, they have a bipartite life cycle, consisting of a pelagic medusoid phase and a sessile polypoid phase. Female medusae are typically highly fecund, producing millions of eggs (Huang et al. 1985; Kikinger 1992). Fertilized planulae attach to hard substrates, which may be decreasing, as is the case with mangroves (Valiela et al. 2001), or increasing, as with anthropogenic substrates (Duarte et al. 2013). Polyps of many species may asexually bud additional polyps (Lucas et al. 2012) or transform into cysts capable of resisting harsh environmental conditions (Arai 2009). When conditions become favorable, polyps begin to segment and asexually release ephyrae through the process of strobilation. Each polyp may release numerous ephyrae and will often strobilate more than once in the same season. Ephyrae join the plankton and grow rapidly into medusae (Palomares and Pauly 2009), at which point they may be targeted for fisheries. This bipartite life cycle may provide a buffer against overfishing, because subsequent recruitment is possible even without spawning adults. Nonetheless, overfishing of jellyfish stocks is possible, as several case studies attest, and therefore management strategies for sustainable fisheries should be used.

CATCHING AND PROCESSING

Jellyfish are most often fished from small boats with crews of one to five fishers. Medusae are spotted from the surface and dip-netted individually. This method typically results in very low levels of bycatch. However, many juvenile fish use jellyfish as refugia (Purcell and Arai 2001), and numerous invertebrates are known to associate with jellyfish (Arai 2005), so concerns relating to bycatch cannot be eliminated entirely (Panda and Madhu 2009). Dip-nets, as well as nets with larger mesh sizes, may also facilitate the avoidance of smaller medusae, and several countries have implemented minimum size limits. A variety of other gears may also be used to catch jellyfish, including hooks, set-nets, gillnets, drift nets, purse seines, beach seines, weirs, and trawl nets.

Freshly caught jellyfish can spoil quickly, and therefore the catch is usually brought to a local processing facility within a few hours. Initial signs of spoilage include sliminess, color change, and unpleasant smell. Sometimes the catch is stored in containers with either seawater or a slurry of ice and alum—usually potassium aluminum sulfate, or $KAl(SO_4)_2$—on board the fishing vessels. Most edible jellyfish species have prominent oral arms rather than conspicuous tentacles. Depending on the species and the target market, the oral arms may be disposed of or processed separately. The bell is usually more expensive and was historically valued at more than twice as much as the oral arms (Hsieh and Rudloe 1994; Omori and Nakano 2001). However, demand for oral arms is increasing in China (Kitamura and Omori 2010), and there are recent reported cases where only the oral arms are targeted, with the bell being disposed of at sea (e.g., Mohan et al. 2011).

Processing facilities range from seaside tents and shacks to large, industrialized seafood processing factories. Facilities may employ dozens of laborers, including many women and children, who are sometimes the family members of fishers. Jellyfish are often scraped, sometimes with tools made of bamboo, to remove mucus or the surface "skin" if there are denticulations. The gastrovascular cavity and any developed gonads are typically removed (Chidambaram 1984), which may also be done near the end of processing (Rudloe 1992). The edges of the bell are sometimes trimmed and removed (Govindan 1984; Santhana-Krishnan 1984; Nishikawa et al. 2008). Jellyfish are usually rinsed with seawater, which appears to be especially important for species that produce a lot of mucus, such as cannonball jellyfish *Stomolophus meleagris* (Jones and Rudloe 1995).

Processing typically involves soaking the jellyfish in large vats or tanks containing different mixtures of salt and alum (which is granular, white, and odorless). These mixtures may be dry or in brine solution, and jellyfish are soaked in different mixtures for specified time periods. The salting process is intended to dehydrate and preserve the jellyfish without denaturing the collagen, resulting in a desirable firmness and crunchiness of the final product (Sloan and Gunn 1985). The alum will also reduce the pH and act as a disinfectant (Hsieh et al. 2001), and processing typically eliminates any remaining sting from nematocysts (Hsieh and Rudloe 1994). Using only salt or alum alone does not result in a satisfactory product (Wootton et al. 1982). Heat is avoided during processing, because it will quickly denature the collagen (Rigby and Hafey 1972). In some areas of Southeast Asia, such as Malaysia, Thailand, and the Philippines, a small amount of soda may be added to facilitate additional dehydration, thereby increasing the crispiness of the final product (Rudloe 1992; Hsieh et al. 2001; PCAMRD 2008). In certain locations, especially those exporting products to China, jellyfish are treated during processing with mixtures containing hydrogen peroxide in order to bleach the product white.

Jellyfish processing is stepwise and can take weeks before it is complete, although acceptable products have been produced in as little as 8 days using cannonball jellyfish (*S. meleagris*), which has a smaller maximum size than most edible species (Huang 1988). Mechanical drying has also been investigated to reduce processing time, but the resulting products were not satisfactory because of uneven dehydration (Wootton et al. 1982; Huang 1988; Hudson et al. 1997). Because processing is usually time-consuming and labor-intensive, it is a potential limiting economic factor in countries where labor costs are high. Processing techniques and formulas vary by region and species, and potentially even by batch, so many Asian processors employ "Jellyfish Masters" who make adjustments to obtain an acceptable product, often keeping their formulas as guarded secrets (Rudloe 1992; Jones and Rudloe 1995). Several different processing protocols based on Japanese and Thai preferences are outlined in Sloan and Gunn (1985), and numerous overviews of jellyfish processing are available (e.g., Chidambaram 1984; Govindan 1984; Santhana-Krishnan 1984; Rudloe 1992; Subasinghe 1992; Hsieh et al. 2001). After stepwise processing in salt–alum solutions, bells are often stacked in a pile and allowed to drain for several days (Subasinghe 1992). Salt may be sprinkled on the bells before stacking, and the bells may be rotated during dehydration to ensure proper drainage. In some locations, jellyfish may be consumed fresh or processed using oak leaves (Morikawa 1984); however, these are rarities in comparison with the vast quantities that are processed using salt and alum for mass consumption.

The final product can weigh anywhere from 7% to 30% of the raw wet weight, depending on the species in question and the specific processing formula used (Wootton et al. 1982; Huang 1986; Hsieh et al. 2001). Variation in final moisture content is typically dictated by different market preferences. For example, products bound for Japanese markets tend to be much crunchier than for Chinese markets. Products are often graded, and quality is determined based on size, texture, and color, with larger products that are whiter in color and of the preferred texture fetching the highest prices (Rudloe 1992). Color tends to turn from white to yellow to brown as the product ages, with a brown color being unacceptable. Jellyfish may be packed into crates or buckets for export, sent to local auctions, or sold in small retail packages. The shelf life of processed jellyfish is approximately 1 year at room temperature and may be increased to more than 2 years if kept cool (Hsieh et al. 2001). It is generally reported that the product will spoil if frozen (Huang 1986; Rudloe 1992; Subasinghe 1992; Hsieh et al. 2001); however, there are suggestions that the product can be stored at 0°C (Govindan 1984; Santhana-Krishnan 1984) or colder (Ozer and Celikkale 2001).

JELLYFISH AS FOOD

As stated by Hsieh et al. (2001), in China "jellyfish is more than a gourmet delicacy: it is a tradition." In many Asian countries, jellyfish is consumed in the home and at restaurants, ceremonies, and banquets. Although jellyfish are targeted and caught in numerous countries around the world, these countries export nearly all of their catch to Asia, usually after processing. Most consumption occurs in China, Japan, Malaysia, South Korea, Taiwan, Singapore, Thailand, and Hong Kong, with demand in Japan having increased especially in the 1970s (Omori and Nakano 2001). Smaller markets for jellyfish exist in many cities with inhabitants of Asian descent, who consume imported jellyfish. Although jellyfish as food continues to remain a novelty for most people in the western hemisphere, increasingly diversified urban populations have resulted in higher imports of jellyfish in many major cities. Combined with a growing and wealthier population in China, global demand for jellyfish is thus likely to continue to increase.

Both classically semidried and ready-to-eat jellyfish products are sold in a variety of different presentations and packaging. These may include bulk bins, plastic jars, soft plastic sealed bowls, and sealed plastic envelopes. Mislabeling of jellyfish products appears rampant (as it does for fish products; Jacquet and Pauly 2008) and may include a lack of adherence to local labeling regulations, a high rate of misidentifying species, and even labeling of the product as a vegetable, such as bamboo or tuber mustard (Armani et al. 2012, 2013).

Processed jellyfish has almost no intrinsic flavor and is therefore usually served with sauces that may include sesame oil, soy sauce, vinegar, or sugar. Edible jellyfish has a surprising crunch, and the sensation of biting into the crisp meat is sometimes referred to as the product's "sound." In preparation for consumption, processed jellyfish is usually soaked for several hours or overnight, sometimes with numerous water changes, in order to desalt and partially rehydrate the product. Traditional recipes may even call for jellyfish to be soaked for several days (Wootton et al. 1982), a process reminiscent of preparing salted cod. After soaking, the jellyfish may be scalded or blanched briefly in boiling water, which forms curls. Jellyfish is most often served at room temperature and sliced into thin strips. It may be served with sliced vegetables or sliced meat as an appetizer, salad, or soup. Many other Asian dishes include jellyfish as an ingredient. Ready-to-use or ready-to-eat jellyfish products have been developed more recently, which, as their name suggests, do not require soaking or scalding before eating. These products are sometimes packaged with spices or sauces as a ready-to-eat snack.

Rehydrated edible jellyfish are composed primarily of water, accounting for about 90% of the weight. The main organic component is collagen (Kimura et al. 1983), a connective tissue protein making up about 7% of the rehydrated product weight, a value that varies according to the species in question and the processing method used. Levels of fat, cholesterol, and carbohydrates are extremely low or undetectable. Lipid content may increase in specimens with well-developed gonads, but these are typically removed during processing. Tryptophan has been identified as the limiting amino acid at very low levels (Kimura et al. 1983), although this may vary by species (Khong et al. 2016). These characteristics have also led to edible jellyfish being declared as a natural diet food, with approximately 36 food calories per 100 g serving (USDA 2015).

A number of inorganic constituents are detectable in processed jellyfish, the most concerning of which is aluminum from the alum used in processing. Many salts and minerals may be removed through soaking and scalding; however, aluminum has been detected in the final edible product in significant quantities (Ogimoto et al. 2012). Ready-to-eat jellyfish products are not rinsed by the consumer before consumption and have also been shown to have high aluminum content (Wong et al. 2010; Armani et al. 2013). Consumption of aluminum has been linked to a number of negative health effects, including Alzheimer's disease (Perl and Brody 1980; Nayak 2002). The links between aluminum consumption and the associated negative health effects are still not well understood, and further research is needed. However, the development of a processing method that avoids the use of aluminum for edible jellyfish is desirable (Hsieh and Rudloe 1994). Additional additives may also be included in processed jellyfish, including monosodium glutamate or potassium sorbate (Armani et al. 2012).

In contrast to the negative effects associated with aluminum found in edible jellyfish products, there is also a long list of purported health benefits. Traditional Chinese medicine claims that eating jellyfish is beneficial for curing arthritis and gout, decreasing hypertension, treating bronchitis, alleviating back pain, curing ulcers and goiter, easing swelling, stimulating blood flow (especially during menstruation), remedying fatigue, softening skin, aiding weight loss, improving digestion, and treating cancer (Rudloe 1992; Hsieh and Rudloe 1994; Jones and Rudloe 1995; Hsieh et al. 2001). Australian Aborigines have also used dried jellyfish powder to treat burns (Hsieh and Rudloe 1994). Fishers in the Philippines believe that consumption of jellyfish will increase resistance to hypertension, back pain, arthritis, and malaria (PCAMRD 2008). Very few clinical trials have been conducted to test these claims, and although most remain neither proven nor disproven, many are unlikely. Hsieh et al. (2001) report the findings of a small study whereby several rats were fed jellyfish collagen after being injected with an arthritis-inducing reagent. Rats that were fed jellyfish collagen reportedly showed significantly reduced incidence, onset, and severity of arthritis in comparison to the control group. However, the details of the experiment are not presented, and numerous additional studies should be conducted before it can be claimed that consuming jellyfish is beneficial for sufferers of arthritis. Nonetheless, collagen has been shown to be beneficial in treating a wide variety of health conditions, including arthritis, so it remains plausible that consumption of jellyfish has some of the health benefits purported by traditional Chinese medicine (Hsieh and Rudloe 1994).

ESTIMATING THE CURRENT GLOBAL CATCH

A global estimate of current jellyfish landings was calculated by estimating the mean annual catch by country since the year 2000 (table 10.1; figure 10.2). Where possible, FAO catch statistics for jellyfish were verified from additional sources of data at the country or regional level. Some countries may report bycatch of jellyfish from other fisheries to FAO, regardless of whether it is landed. On one hand, it is positive that FAO reports these values, because they are part of the total catch. However, FAO makes no distinction between bycatch and targeted landings, which is problematic when interpreting the data. In this case, nations such as Namibia, the United Kingdom, and the Falkland Islands appear to have fisheries for jellyfish, when in fact these statistics are likely to indicate discarded bycatch. More detailed reporting by FAO and individual countries would be beneficial. In some cases (e.g., India), only production statistics were available, and a scaling factor of 4 was used to convert from semidried processed product weight back to wet weight. Processed jellyfish products can range anywhere from 7% to nearly 30% of the original wet weight, depending on the species and processing formula used. Because reported values are typically much less than 25% of the original wet weight (Omori 1981; Morikawa 1984; Huang 1986; Jones and Rudloe 1995; Fisheries Victoria and MAFRI 2002; Li et al. 2014), a scaling factor of 4 is conservative. In some cases, previously unreported landing estimates of jellyfish were added to reported FAO statistics, based on catch reconstructions performed as part of the *Sea Around Us*.

The global estimate of nearly 900,000 t/year is approximately 2.5 times larger than previous estimates (e.g., Omori and Nakano 2001) that were derived from FAO catch statistics.

Despite this difference, 900,000 t is probably an underestimate of the true global catch because of the conservative assumptions used and the fact that reporting of jellyfish catches is poor. For example, the estimate for India was calculated using mean production values between 2000 and 2003 (Anonymous 2005), which were reported by only one to three states depending on the year in question. However, at least six states in India are known to catch jellyfish. Therefore, the world catch of jellyfish for food probably exceeds 1 million t annually. There are also immense amounts of jellyfish that are caught for purposes other than human consumption. In China, large numbers of jellyfish are included in the vast quantities of trash fish from nontargeted fisheries that are delivered to factories to be turned into fishmeal (Cao et al. 2015). In addition, the global estimate presented here does not include any bycatch or discards of jellyfish, which can be huge, often resulting in losses to fishers of tens or hundreds of millions of dollars annually (Purcell et al. 2007; Uye 2008; Kim et al. 2012). In fact, the amount of discarded jellyfish bycatch is likely to exceed by far the landings of edible jellyfish, adding millions of tonnes to the world's marine catches.

THE FUTURE OF JELLYFISH FISHERIES

Jellyfish populations are increasing in many coastal ecosystems around the world (Brotz et al. 2012), often with severe impacts on human activities (Purcell et al. 2007). Eradication of some jellyfish populations has been successful in Hawaii (Hofmann and Hadfield 2002; Kelsey 2009); however, these isolated cases involved *Cassiopea* sp., which is not a strong natural disperser (Holland et al. 2004). Attempts to remove and thereby diminish jellyfish populations in other areas, such as parts of the Mediterranean Sea (Brotz and Pauly 2012), have been less successful. The development of jellyfish fisheries for food and medicine has been proposed as a possible strategy to deal with increasing jellyfish populations (e.g., Purcell et al. 2007; Richardson et al. 2009). Although such an approach may help humans cope with more jellyfish, it is not a true solution, for several reasons (Gibbons et al. 2016). First, only a small fraction of the more than one thousand species of jellyfish are preferred for consumption (see tables 10.2 and 10.3). Second, although overfishing does have the potential to reduce jellyfish populations, people involved in a profitable fishery desire sustainable catches and may attempt to increase catches through the use of hatchery programs, as is the case in China (Dong et al. 2009). Third, rapid development of jellyfish fisheries without sufficient research and management may lead to unintended consequences for ecosystems and contention among fishers, as is the case in India (Magesh and Coulthard 2004). In addition, pollution from jellyfish processing plants is a major concern and has prevented expansion of jellyfish fisheries in the United States and elsewhere. Finally, counting on jellyfish fisheries to resolve the problem of jellyfish blooms ignores the root causes of the problem. Increasing jellyfish blooms have been linked to numerous anthropogenic factors (Purcell et al. 2007; Richardson et al. 2009; Duarte et al. 2013). If we simply adapt to a new "normal" instead of addressing and correcting the underlying causes, our baseline will shift (Pauly 1995), and ultimately jellyfish may be the only seafood left.

ACKNOWLEDGMENTS

I thank the numerous fisheries scientists who corresponded with me about jellyfish fisheries around the world. I would especially like to thank Drs. Shin-ichi Uye and Miguel Cisneros Mata for hosting me and providing useful information on jellyfish fisheries in Japan and Mexico, respectively.

REFERENCES

Al-Abdulrazzak, D., and D. Pauly (eds.). 2013. *From dhows to trawlers: a recent history of fisheries in the Gulf countries, 1950 to 2010*. Fisheries Centre Research Reports 21(2), University of British Columbia, Vancouver, Canada.

Anonymous. 2005. *Handbook on Fisheries Statistics 2004*. Department of Animal Husbandry, Dairying and Fisheries, Ministry of Agriculture, Government of India, New Delhi, India.

Anonymous. 2012. *Fishing for jellyfish in Pakistan*. Dawn.com, Kerachi, Pakistan, edition of April 23.

Arai, M. N. 2005. Predation on pelagic coelenterates: a review. *Journal of the Marine Biological Association of the United Kingdom* 85(3): 523–536.

Arai, M. N. 2009. The potential importance of podocysts to the formation of scyphozoan blooms: a review. *Hydrobiologia* 616: 241–246.

Armani, A., P. D'Amico, L. Castigliego, G. Sheng, D. Gianfaldoni, and A. Guidi. 2012. Mislabeling of an "unlabelable" seafood sold on the European market: the jellyfish. *Food Control* 26: 247–251.

Armani, A., A. Giusti, L. Castigliego, A. Rossi, L. Tinacci, D. Gianfaldoni, and A. Guidi. 2014. Pentaplex PCR as screening assay for jellyfish species identification in food products. *Journal of Agricultural and Food Chemistry* 62(50): 12134–12143.

Armani, A., L. Tinacci, A. Giusti, L. Castigliego, D. Gianfaldoni, and A. Guidi. 2013. What is inside the jar? Forensically informative nucleotide sequencing (FINS) of a short mitochondrial *COI* gene fragment reveals a high percentage of mislabeling in jellyfish food products. *Food Research International* 54: 1383–1393.

Brotz, L., W. W. L. Cheung, K. Kleisner, E. Pakhomov, and D. Pauly. 2012. Increasing jellyfish populations: trends in Large Marine Ecosystems. *Hydrobiologia* 690(1): 3–20.

Brotz, L., and D. Pauly. 2012. Jellyfish populations in the Mediterranean Sea. *Acta Adriatica* 53(2): 211–230.

Brotz, L., and D. Pauly. 2016. Studying jellyfish fisheries: toward accurate national catch reports and appropriate methods for stock assessments. In G. L. Mariottini (ed.), *Jellyfish: Ecology, Distribution Patterns and Human Interactions*. Nova Publishers, Hauppauge, NY.

Cao, L., R. Naylor, P. Henriksson, D. Leadbitter, M. Metian, M. Troell, and W. Zhang. 2015. China's aquaculture and the world's wild fisheries. *Science* 347(6218): 133–135.

Chidambaram, L. 1984. *Export oriented processing of Indian jelly fish (Muttai Chori, Tamil) by Indonesian method at Pondicherry Region*. Technical and Extension Series (60). Marine Fisheries Information Service, Central Marine Fisheries Research Institute, Indian Council of Agricultural Research, Cochin, India.

Coleman, A., S. Micin, P. Mulvay, and R. Rippingale. 1990. *The brown jellyfish* (Phyllorhiza punctata) *in the Swan-Canning estuary*. Waterways Information No. 2. Swan River Trust, Perth, Australia.

DFA. 2002a. *Jellyfish exploratory survey, Trinity Bay*. Project summary: FDP 424-5, Emerging Fisheries Development, Fisheries Diversification Program. Department of Fisheries and Aquaculture, Government of Newfoundland and Labrador.

DFA. 2002b. *Production and marketing of jellyfish*. Project summary: FDP 421-3, Emerging Fisheries Development, Fisheries Diversification Program. Department of Fisheries and Aquaculture, Government of Newfoundland and Labrador.

Dong, J., L. X. Jiang, K. F. Tan, H. Y. Liu, J. E. Purcell, P. J. Li, and C. C. Ye. 2009. Stock enhancement of the edible jellyfish (*Rhopilema esculentum*) in Liaodong Bay, China: a review. *Hydrobiologia* 616: 113–118.

Dong, Z. J., D. Y. Liu, and J. K. Keesing. 2010. Jellyfish blooms in China: dominant species, causes and consequences. *Marine Pollution Bulletin* 60(7): 954–963.

Dong, Z., D. Liu, and J. K. Keesing. 2014. Contrasting trends in populations of *Rhopilema esculentum* and *Aurelia aurita* in Chinese waters. Pp. 207–218 in K. A. Pitt and C. H. Lucas (eds.), *Jellyfish Blooms*. Springer, Dordrecht the Netherlands.

Duarte, C. M., K. A. Pitt, C. H. Lucas, J. E. Purcell, S. Uye, K. Robinson, L. Brotz, M. B. Decker, K. R. Sutherland, A. Malej, L. Madin, H. Mianzan, J. M. Gili, V. Fuentes, D. Atienza, F. Pagès, D. Breitburg, J. Malek, W. M. Graham, and R. H. Condon. 2013. Is global ocean sprawl a cause of jellyfish blooms? *Frontiers in Ecology and the Environment* 11(2): 91–97.

FAO. 2014a. *Fishery commodities global production and trade*. Fishery Statistical Collections, Fisheries and Aquaculture Department, Food and Agriculture Organization of the United Nations, Rome, Italy. Available at http://www.fao.org/fishery/statistics/global-commodities-production/en.

FAO. 2014b. *Global capture production*. Fishery Statistical Collections, Fisheries and Aquaculture Department, Food and Agriculture Organization of the United Nations, Rome, Italy. Available at http://www.fao.org/fishery/statistics/global-capture-production/en.

Fisheries Victoria. 2006. *Statement of management arrangements for the Victorian developmental jellyfish fishery* (Catostylus mosaicus). Statement prepared for the Australian Department of Heritage and Environment. Fisheries Victoria Division of the Department of Primary Industries, Victoria, Australia.

Fisheries Victoria and MAFRI. 2002. *Developmental fisheries management plan, jellyfish* (Catostylus mosaicus), *2003–2005*. Fisheries Victoria and the Marine and Freshwater Resources Institute, Fisheries Division, Department of Natural Resources and Environment, Victoria, Australia.

Gibbons, M. J., F. Boero, and L. Brotz. 2016. We should not assume that fishing jellyfish will solve our jellyfish problem. *ICES Journal of Marine Science* 73(4): 102–1018.

Govindan, T. K. 1984. A novel marine animal with much export potential. *Seafood Export Journal* 16(7): 9–11.

Gul, S., and A. C. Morandini. 2013. New records of scyphomedusae from Pakistan coast: *Catostylus perezi* and *Pelagia* cf. *noctiluca*. *Marine Biodiversity Records* 6(e86): 1–6.

Gul, S., and A. C. Morandini. 2015. First record of the jellyfish *Rhopilema hispidum* from the coast of Pakistan. *Marine Biodiversity Records* 8(e30): 1–4.

Hofmann, D. K., and M. G. Hadfield. 2002. Hermaphroditism, gonochorism, and asexual reproduction in *Cassiopea* sp.: an immigrant in the islands of Hawai'i. *Invertebrate Reproduction and Development* 41(1–3): 215–221.

Holland, B. S., M. N. Dawson, G. L. Crow, and D. K. Hofmann. 2004. Global phylogeography of *Cassiopea*: molecular evidence for cryptic species and multiple invasions of the Hawaiian Islands. *Marine Biology* 145(6): 1119–1128.

Hsieh, Y. H. P., F. M. Leong, and J. Rudloe. 2001. Jellyfish as food. *Hydrobiologia* 451(1–3): 11–17.

Hsieh, Y. H. P., and J. Rudloe. 1994. Potential of utilizing jellyfish as food in Western countries. *Trends in Food Science and Technology* 5(7): 225–229.

Huang, M., J. Hu, Y. Wang, and Z. Chen. 1985. Preliminary study on the breeding habits of edible jellyfish in Hangzhou Wan Bay. *Journal of Fisheries of China* 9(3): 239–246 (in Chinese).

Huang, Y. 1986. *The processing of cannonball jellyfish* (Stomolophus meleagris) *and its utilization*. Proceedings of the 11th Annual Tropical and Subtropical Fisheries Technological Conference of the Americas, January 13–16, 1986. Tampa, FL.

Huang, Y. 1988. Cannonball jellyfish (*Stomolophus meleagris*) as a food resource. *Journal of Food Science* 53(2): 341–343.

Huang, Y., P. Christian, and D. Colson. 1987. *Harvest and preservation of a shrimp by-catch: cannonball jellyfish* (Stomolophus meleagris). Proceedings of the 12th Annual Tropical and Subtropical Fisheries Technological Conference of the Americas, November 9–11, 1987. Orlando, FL.

Hudson, R. J., N. F. Bridge, and T. I. Walker. 1997. *Feasibility study for establishment of a Victorian commercial jellyfish fishery*. Marine and Freshwater Resources Institute, Fisheries Research Development Corporation, Queenscliff, Australia.

Jacquet, J. L., and D. Pauly. 2008. Trade secrets: renaming and mislabeling of seafood. *Marine Policy* 32(3): 309–318.

Jones, R. P., and J. Rudloe. 1995. *Harvesting and processing of Florida cannonball jellyfish*. Report to the Florida International Affairs Commission, Florida Department of Commerce. Southeastern Fisheries Association.

Kailola, P. J., M. J. Williams, P. C. Stewart, R. E. Reichelt, A. McNee, and C. Grieve. 1993. *Australian fisheries resources*. Bureau of Resource Sciences and the Fisheries Research and Development Corporation, Canberra, Australia.

Kelsey, M. 2009. Stinging intruders: invasive jellyfish removed from Kaunakakai wharf. *The Molokai Dispatch*, Kaunakakai, Hawaii, June 23.

Khan, A. 2012. Jellyfish netting soars as its demand fetches good sum. *Business Recorder*, Karachi, Pakistan, June 17.

Khong, N. M. H., F. M. Yusoff, B. Jamilah, M. Basri, I. Maznah, K. W. Chan, and J. Nishikawa. 2016. Nutritional composition and total collagen content of three commercially important edible jellyfish. *Food Chemistry* 196: 953–960.

Kikinger, R. 1992. *Cotylorhiza tuberculata*: life history of a stationary population. *Marine Ecology* 13(4): 333–362.

Kim, D. H., J. N. Seo, W. D. Yoon, and Y. S. Suh. 2012. Estimating the economic damage caused by jellyfish to fisheries in Korea. *Fisheries Science* 78(5): 1147–1152.

Kimura, S., S. Miura, and Y. H. Park. 1983. Collagen as the major edible component of jellyfish (*Stomolophus nomurai*). *Journal of Food Science* 48(6): 1758–1760.

Kingsford, M. J., K. A. Pitt, and B. M. Gillanders. 2000. Management of jellyfish fisheries, with special reference to the order Rhizostomeae. *Oceanography and Marine Biology* 38: 85–156.

Kitamura, M., and M. Omori. 2010. Synopsis of edible jellyfishes collected from Southeast Asia, with notes on jellyfish fisheries. *Plankton and Benthos Research* 5(3): 106–118.

Kuthalingam, M. D. K., D. B. James, R. Sarvesan, P. Devadoss, S. Manivasagam, and P. Thirumilu. 1989. *A note on the processing of the jelly fish at Alambaraikuppam near Mahabalipuram*. Marine Fisheries Information Service, Technical and Extension Series (98). Central Marine Fisheries Research Institute, Indian Council of Agricultural Research. Cochin, India.

Li, J., J. Ling, and J. Cheng. 2014. On utilization of two edible macro-jellyfish and evaluation of the biomass of *Nemopilema nomurai* in China Sea. *Marine Fisheries* 36(3): 202–207 (in Chinese with English abstract).

Li, J. R., and Y. H. P. Hsieh. 2004. Traditional Chinese food technology and cuisine. *Asia Pacific Journal of Clinical Nutrition* 13(2): 147–155.

López-Martinez, J., and J. Álvarez-Tello. 2013. The jellyfish fishery in Mexico. *Agricultural Sciences* 4(6A): 57–61.

Lu, Z., Q. Dai, and Y. Yan. 2003. Fishery biology of *Cyanea nozakii* resources in the waters of Dongshan Island. *Chinese Journal of Applied Ecology* 14(6): 973–976 (in Chinese with English abstract).

Lucas, C. H., W. M. Graham, and C. Widmer. 2012. Jellyfish life histories: role of polyps in forming and maintaining scyphomedusa populations. *Advances in Marine Biology* 63: 133–196.

Magesh, S. J., and S. Coulthard. 2004. *Bloom or bust?* Samudra Report No. 39, November.

Mills, C. E. 2001. Jellyfish blooms: are populations increasing globally in response to changing ocean conditions? *Hydrobiologia* 451: 55–68.

Mohan, S., S. Rajapackiam, and S. Rajan. 2011. *Unusual heavy landings of jellyfish* Crambionella stuhlmanni *and processing methods at Pulicat landing centre, Chennai*. Marine Fisheries Information Service Technical and Extension Series (208). Central Marine Fisheries Research Institute, Indian Council of Agricultural Research, Cochin, India.

Morikawa, T. 1984. Jellyfish. *FAO INFOFISH Marketing Digest* 1(84): 37–39.

Muhammed, F., and R. Sultana. 2008. *New record of edible jellyfish,* Rhizostoma pulmo *from Pakistani waters. Marine Biodiversity Records* 1(e67): 1–3.

Murugan, A., and R. Durgekar. 2008. *Beyond the tsunami: status of fisheries in Tamil Nadu, India: a snapshot of present and long-term trends.* United Nations Development Programme and Ashoka Trust for Research in Ecology and the Environment, Bangalore, India.

NARA. 2010. Press statement on recent occurrence of jellyfish on south west coast. National Aquatic Resources Research and Development Agency, Sri Lanka.

Nayak, P. 2002. Aluminum: impacts and disease. *Environmental Research* 89(2): 101–115.

Nishikawa, J., S. Ohtsuka, N. Mulyadi, N. Mujiono, D. J. Lindsay, H. Miyamoto, and S. Nishida. 2015. A new species of the commercially harvested jellyfish *Crambionella* from central Java, Indonesia with remarks on the fisheries. *Journal of the Marine Biological Association of the United Kingdom* 95(3): 471–481.

Nishikawa, J., N. T. Thu, T. M. Ha, and P. T. Thu. 2008. Jellyfish fisheries in northern Vietnam. *Plankton and Benthos Research* 3(4): 227–234.

Ogimoto, M., K. Suzuki, J. Kabashima, M. Nakazato, and Y. Uematsu. 2012. Aluminium content in foods with aluminium-containing food additives. *Food Hygiene and Safety Science* 53(1): 57–62 (in Japanese with English abstract).

Omori, M. 1978. Zooplankton fisheries of the world: a review. *Marine Biology* 48: 199–205.

Omori, M. 1981. Edible jellyfish in the Far East waters: a brief review of the biology and fishery. *Bulletin of the Plankton Society of Japan* 28(1): 1–111 (in Japanese with English abstract).

Omori, M., and M. Kitamura. 2004. Taxonomic review of three Japanese species of edible jellyfish. *Plankton Biology and Ecology* 51(1): 36–51.

Omori, M., and E. Nakano. 2001. Jellyfish fisheries in southeast Asia. *Hydrobiologia* 451(1–3): 19–26.

Ozer, N. P., and M. S. Celikkale. 2001. Utilization possibilities of jellyfish *Rhizostoma pulmo*, as a food in the Black Sea. *Journal of Food Science and Technology* 38(2): 175–178.

Palomares, M. L. D., and D. Pauly. 2009. The growth of jellyfishes. *Hydrobiologia* 616: 11–21.

Panda, S. K., and V. R. Madhu. 2009. Studies on the preponderance of jellyfish in coastal waters of Veraval. *Fishery Technology* 46(2): 99–106.

Pauly, D. 1995. Anecdotes and the shifting baseline syndrome of fisheries. *Trends in Ecology and Evolution* 10(10): 430.

PCAMRD. 2008. Small fishermen in the Malampaya Sound of Palawan benefit from jellyfish. *The PCAMRD Waves*, Philippine Council for Aquatic and Marine Research and Development, Department of Science and Technology, January–March.

Perera, L. 2008. Jellyfish exports need checking. *The Sunday Times*, Sri Lanka, September 21.

Perl, D. P., and A. R. Brody. 1980. Alzheimer's disease: X-ray spectrometric evidence of aluminum accumulation in neurofibrillary tangle–bearing neurons. *Science* 208(4441): 297–299.

Purcell, J. E., and M. N. Arai. 2001. Interactions of pelagic cnidarians and ctenophores with fish: a review. *Hydrobiologia* 451(1–3): 27–44.

Purcell, J. E., S. Uye, and W. T. Lo. 2007. Anthropogenic causes of jellyfish blooms and their direct consequences for humans: a review. *Marine Ecology Progress Series* 350: 153–174.

Richardson, A. J., A. Bakun, G. C. Hays, and M. J. Gibbons. 2009. The jellyfish joyride: causes, consequences and management responses to a more gelatinous future. *Trends in Ecology and Evolution* 24(6): 312–322.

Rigby, R. J., and M. Hafey. 1972. Thermal properties of the collagen of jellyfish (*Aurella coerulea*) and their relation to its thermal behaviour. *Australian Journal of Biological Sciences* 25(6): 1361–1363.

Rudloe, J. 1992. *Jellyfish: a new fishery for the Florida Panhandle*. Report prepared for the Apalachee Regional Planning Council, Blounstown, Florida. U.S. Department of

Commerce, Economic Development Administration, EDA Project No. 04-06-03801.
Santhana-Krishnan, G. 1984. Salted jelly fish. *Seafood Export Journal* 16(7): 23–26.
SEAFDEC. 2014. Southeast Asian Fisheries Development Center, Fishery Statistical Bulletins, Bangkok, Thailand.
Shih, C. T. 1977. *A guide to the jellyfish of Canadian Atlantic waters*. Natural History Series (5), National Museum of Natural Sciences, National Museums of Canada, Ottawa, Canada.
Sloan, N. A. 1986. World jellyfish and tunicate fisheries, and the northeast Pacific echinoderm fishery. *Canadian Special Publication of Fisheries and Aquatic Sciences* 92: 23–33.
Sloan, N. A., and C. R. Gunn. 1985. Fishing, processing, and marketing of the jellyfish, *Aurelia aurita* [*labiata*], from southern British Columbia. *Canadian Industry Report of Fisheries and Aquatic Sciences* 157.
Soonthonvipat, U. 1976. Dried jelly fish. Pp. 149–151 in K. Tiews (ed.), *Fisheries Resources and Their Management in Southeast Asia*. Westkreuz-Druckerei, West Berlin, Germany.
Subasinghe, S. 1992. Jelly fish processing. *INFOFISH International* 4: 63–65.
Teh, L. C. L., and L. S. L. Teh. 2014. *Reconstructing the marine fisheries catch of peninsular Malaysia, Sarawak, and Sabah*, 1950–2010. Fisheries Centre Working Paper #2014-16, University of British Columbia, Vancouver, Canada.
Teh, L. C. L., D. Zeller, and D. Pauly. 2015. *Preliminary reconstruction of Thailand's fisheries catches:* 1950–2010. Fisheries Centre Working Paper #2015-01, University of British Columbia, Vancouver, Canada.
Thanh, C. 2011. Families ride crest of the wave in seafood. *Viet Nam News*, October 17.
USDA. 2015. National nutrient database for standard reference, release 28, United States Department of Agriculture, Agricultural Research Services, Nutrient Data Laboratory, Washington, DC. Accessed March 8, 2016.
Uye, S. 2008. Blooms of the giant jellyfish *Nemopilema nomurai*: a threat to the fisheries sustainability of the East Asian Marginal Seas. *Plankton and Benthos Research* 3(suppl. 3): 125–131.
Valiela, I., J. L. Bowen, and J. K. York. 2001. Mangrove forests: one of the world's threatened major tropical environments. *BioScience* 51(10): 807–815.
Wang, K. 2007. *The use of untraditional sea food: the commercialization of Norwegian jellyfish, red sea cucumber and whelk*. SINTEF Report, Trondheim, Norway.
Wong, W. W. K., S. W. C. Chung, K. P. Kwong, Y. Y. Ho, and Y. Xiao. 2010. Dietary exposure to aluminium of the Hong Kong population. *Food Additives and Contaminants Part A: Chemistry Analysis Control Exposure and Risk Assessment* 27(4): 457–463.
Wootton, M., K. A. Buckle, and D. Martin. 1982. Studies on the preservation of Australian jellyfish (*Catostylus* spp.). *Food Technology in Australia* 34(9): 398–400.
Yakovlev, Y. M., P. A. Borodin, and E. V. Osipov. 2005. The fishery of jellyfish *Rhopilema* in Peter the Great Bay. *Rybnoe Khozyaistvo* 5: 72–75 (in Russian).
You, K., C. H. Ma, H. W. Gao, F. Q. Li, M. Z. Zhang, Y. T. Qiu, and B. Wang. 2007. Research on the jellyfish (*Rhopilema esculentum*) and associated aquaculture techniques in China: current status. *Aquaculture International* 15(6): 479–488.
Zhong, X. M., J. H. Tang, and P. T. Liu. 2004. A study on the relationship between *Cyanea nozakii* breaking out and ocean ecosystem. *Modern Fisheries Information* 19(3): 15–17 (in Chinese).

NOTE

1. Cite as Brotz, L. 2016. Jellyfish fisheries: a global assessment. Pp. 110–124 in D. Pauly and D. Zeller (eds.), *Global Atlas of Marine Fisheries: A Critical Appraisal of Catches and Ecosystem Impacts*. Island Press, Washington, DC.

CHAPTER 11

GLOBAL SEABIRD POPULATIONS AND THEIR FOOD CONSUMPTION[1]

Michelle Paleczny,[a] Vasiliki Karpouzi,[a] Edd Hammill,[b] and Daniel Pauly[a]
[a]*Sea Around Us*, University of British Columbia, Vancouver, BC, Canada
[b]Spatial Community Ecology Laboratory, Utah State University, Logan, UT

Seabirds share the oceans with us, and they are strongly affected by fisheries both directly (e.g., as by-kills of longline fisheries) and indirectly, because fisheries reduce the abundance of their fish prey below crucial thresholds (Cury et al. 2011). Because they globally consume millions of tonnes of fish and marine invertebrates per year, seabirds play an integral role in the structure, function, and resilience of marine ecosystems. In this, they are similar to marine mammals, for which culling is often proposed, so we will have more fish to catch (Gerber et al. 2009; Morissette et al. 2012), or implemented (Pannozzo 2013), although at a smaller scale (see Hays 2015).

Because of seabirds' role in the functioning of marine ecosystems, the *Sea Around Us* has taken an early interest in mapping their worldwide distributions, such that they could be considered in global modeling efforts of the sort discussed in Christensen et al. (2009) and chapter 9. The first such product was the preliminary maps of Karpouzi et al. (2007), which came with a database of the estimates of abundance through time and the body sizes and diet compositions of 351 species of "seabirds" (Karpouzi 2005), now incorporated in SeaLifeBase (see www.sealifebase.org).

This database, which was used for a number of contributions on seabirds and their roles in marine ecosystems (Kaschner et al. 2006; Karpouzi and Pauly 2008), was extended by Paleczny (2012), and the number of species covered was reduced to true seabirds (i.e., benthic feeding ducks, scoters, eiders, and mergansers with little potential for overlap with fisheries were not considered further). Moreover, two distribution range maps (breeding and nonbreeding, i.e., foraging ranges) were generated for each of these 324 remaining species (see figure 11.1 for examples) and were then used to generate the global maps in this chapter and other products (Coll et al. 2012; Cheung et al. 2012; Paleczny and Pauly 2011).

SEABIRD BIODIVERSITY AND ECOLOGY

Seabirds are birds that have evolved to forage in the ocean but nest in colonies on islands and coastal cliffs. Brooke (2004) estimated 309 species of seabirds to have a cumulative population of 0.7 billion, and Paleczny (2012) estimated the current global seabird population at 0.77 billion (table 11.1). These populations belong to approximately 324 species (approximate because of recent taxonomic revisions, e.g., Rains et al. 2011), and four orders, Procellariiformes (i.e., petrels, diving petrels, storm petrels, and albatrosses), Charadriiformes (i.e., auks, terns, gulls, and skuas), Sphenisciformes (penguins), and Pelecaniformes (boobies, cormorants, frigatebirds, pelicans, and tropicbirds). Seabirds are unique among avian taxa for their relatively

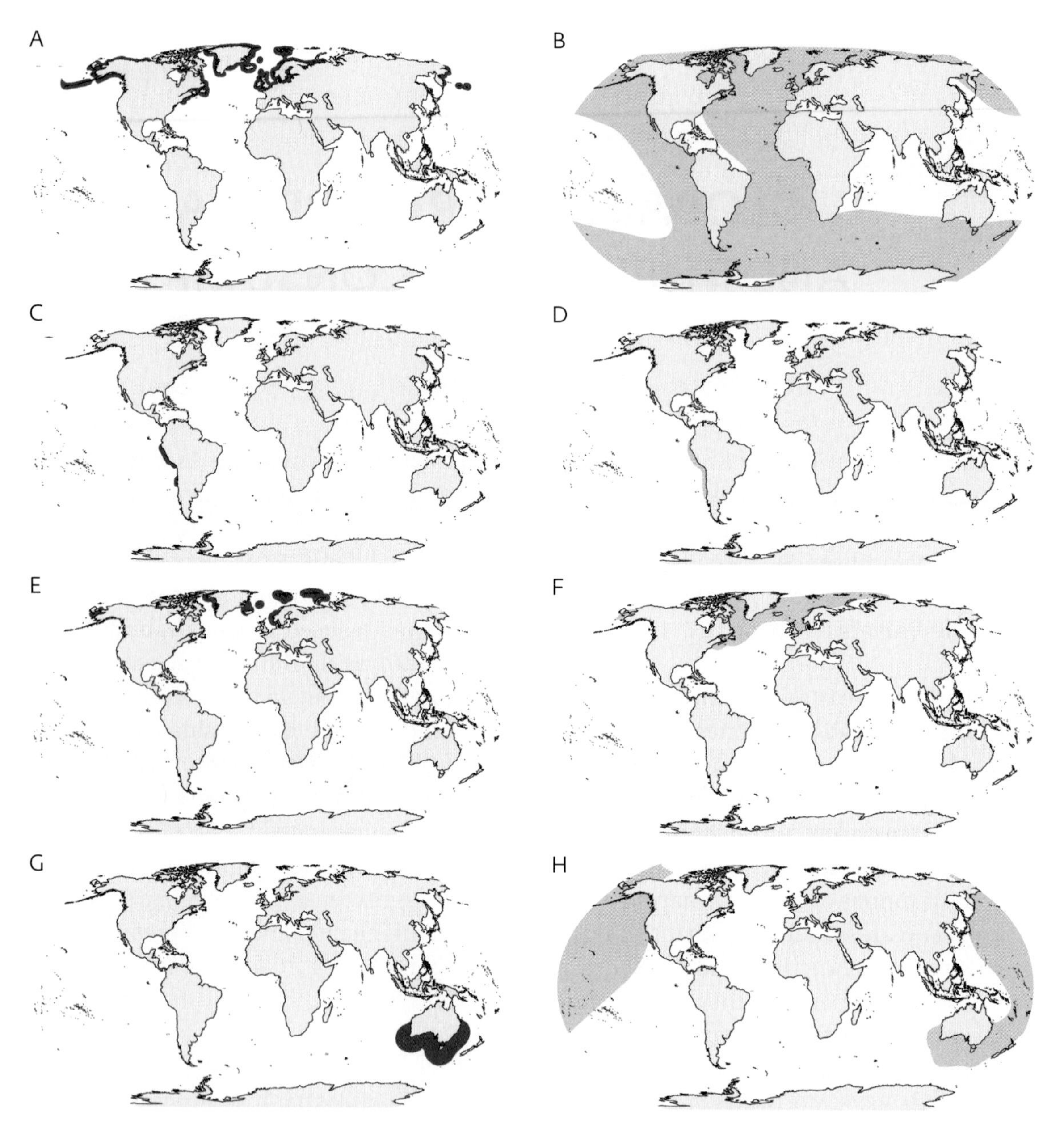

Figure 11.1. Examples of breeding season range (left) and total range (right) for four species of seabirds: (A, B) Arctic tern (*Sterna paradisaea*); (C, D) Peruvian booby (*Sula variegata*); (E, F) dovekie (*Alle alle*); and (G, H) short-tail shearwater (*Ardenna tenuirostris*). The breeding season ranges are estimated based on the maximum observed foraging radius from the colonies, and the foraging ranges are adapted from expert-derived range maps. (Adapted from Egevang et al. 2010; Gaston et al. 1998; Harrison 1987; Poole 2005.)

K-selected life history strategy (i.e., large body size, low population growth rate, and long lifespan) and ability to travel long distances to forage for prey (up to thousands of kilometers per foraging trip in some species). Jointly, the distribution of these 324 species covers the world's oceans, with species richness being highest in productive regions, particularly in the southern hemisphere (figure 11.2). The greater seabird endemism in the southern hemisphere may be a consequence of spatial isolation between breeding populations, as the distances between islands and continents supporting seabird colonies are greater than in the northern hemisphere.

The main prey of seabirds are krill, fish, and squid, and less commonly benthic crustaceans, other seabirds, marine mammal carrion, and jellyfish (see chapter 10). The relative importance of these diet items varies

Table 11.1. Minimum estimate of seabird population changes, 1950–2010 (from Paleczny 2012).

Family	Numbers (10^3) in year closest to 1950	Numbers (10^3) in year closest to 2010	Percentage remaining
Spheniscidae	102,397	65,283	64
Diomedeidae	5,498	5,170	94
Procellariidae	390,382	287,877	74
Hydrobatidae	32,695	46,065	141
Pelecanoididae	60,954	47,933	79
Phaethontidae	228	214	94
Pelecanidae	730	508	70
Phalacrocoracidae	17,945	6,082	34
Fregatidae	1,037	534	52
Sulidae	6,564	6,950	106
Stercorariidae	1,281	1,224	96
Laridae	39,957	34,779	87
Sternidae	139,430	53,797	39
Alcidae	196,787	212,058	108
Total	**995,887**	**768,472**	**77**

between seabird taxa, as well as regionally and seasonally. For example, seabirds may switch diets between breeding and nonbreeding season, with adults commonly provisioning high energy-density prey (e.g., forage fish) to their chicks. Seabirds are also prey in marine and coastal ecosystems, consumed by a variety of marine mammals (e.g., seals, sea lions, walrus, sea otters, killer whales, polar bears), sharks, coastal birds of prey (e.g., hawks, eagles), and other seabirds (see Hipfner et al. 2012). Seabirds share symbiotic foraging interactions with other marine fauna; for example, temperate foraging auks have mutualistic relationships with marine mammals (e.g., Anderwald et al. 2011), and most tropical sea-

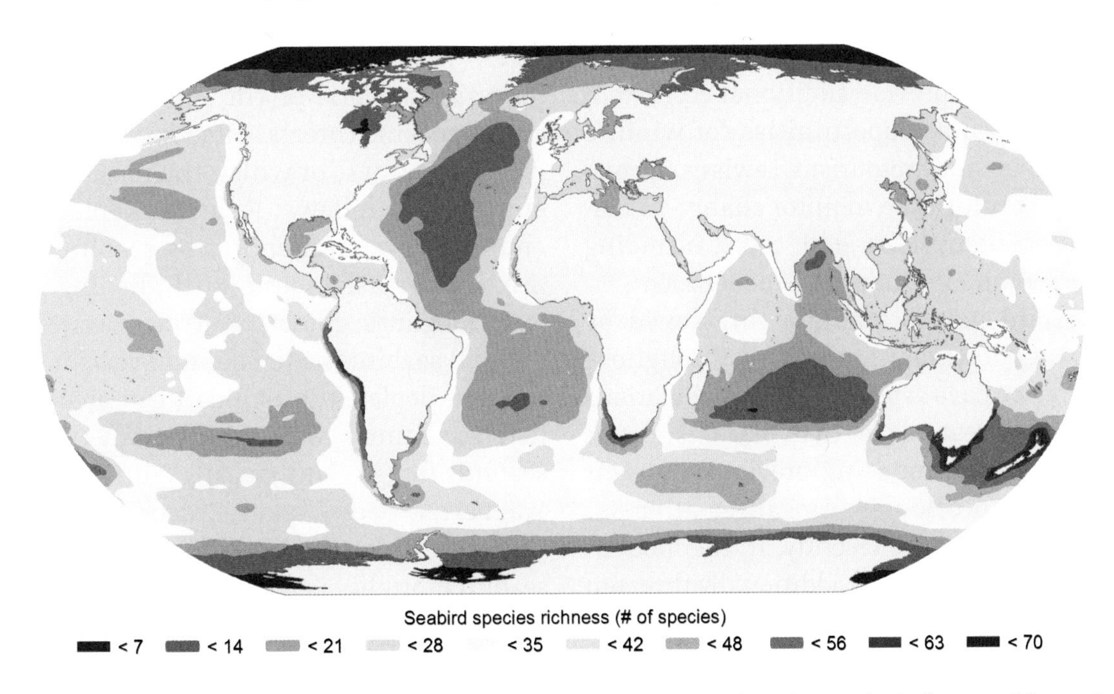

Figure 11.2. Global seabird species richness, as constructed using maps of total range (as in figure 11.1) for each of the world's 324 seabird species.

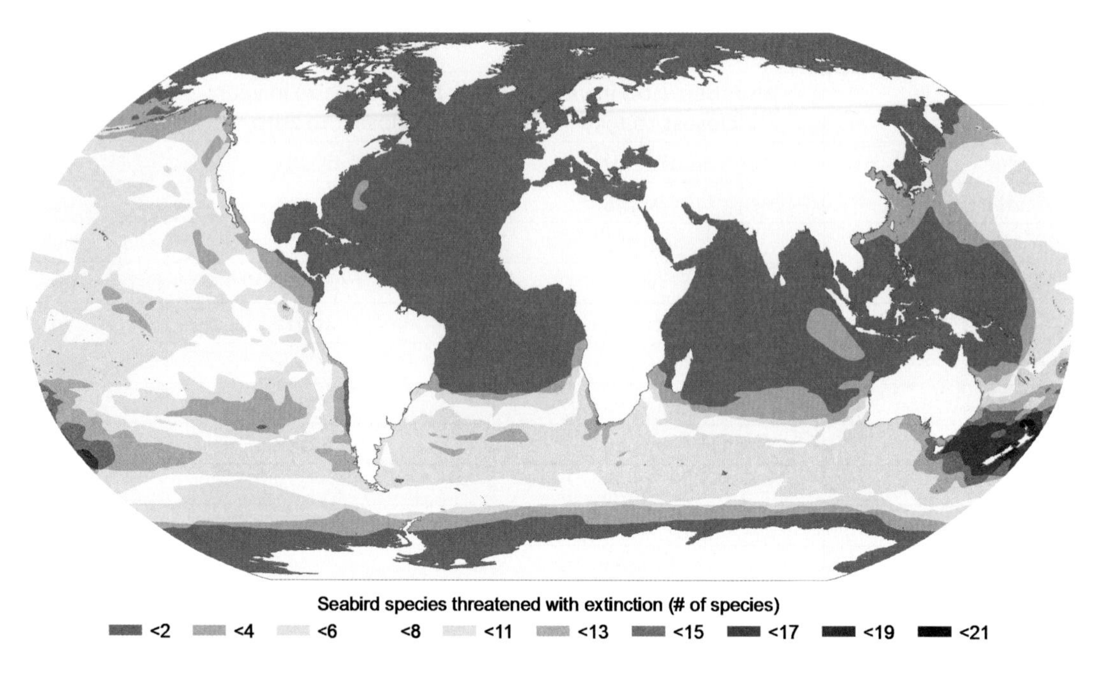

Figure 11.3. Seabird species considered threatened and on the IUCN Red List (IUCN 2012) (i.e., Vulnerable, Endangered, or Critically Endangered). Map constructed using foraging (nonbreeding) season range maps.

birds forage commensally with dolphins and tunas (Ballance and Pitman 1999). Seabirds are also important cross-ecosystem nutrient subsidizers, transporting nutrients via their guano to their breeding colonies, where they play a major role in enriching the productivity and biodiversity of the terrestrial and marine ecosystems surrounding their colonies (Croll et al. 2005). Because of their charismatic nature and accessibility at terrestrial breeding colonies, seabirds provide additional ecosystem services such as opportunities for wildlife interactions and ecotourism (Lewis et al. 2012) and opportunities to monitor change in marine ecosystems (Piatt et al. 2007), including fisheries-induced changes (Einoder 2009).

Seabird populations are threatened by humans, however (figure 11.3). Throughout history, we have depleted seabird populations by hunting seabirds for their feathers, meat, and oil and introducing previously absent predators to colonies (Croxall et al. 1984; Roberts 2007). More recently, in the modern industrial era, humans additionally threaten seabirds through coastal development, pollution, climate change, and fisheries (Croxall et al. 2012) and even through renewed targeted commercial harvesting for food (Grémillet et al. 2015). Seabird populations are particularly vulnerable to these threats because they have inherently low reproductive output and therefore slow population recovery rates (Russell 1999). Also, they range over large areas, which increases their probability of exposure to spatially heterogeneous anthropogenic threats (Jodice and Suryan 2010).

SEABIRDS AND FISHERIES

Fisheries are one of the greatest modern anthropogenic threats, affecting seabirds in numerous ways, of which the most threatening are as follows: Fisheries deplete the abundance or availability of seabird prey (i.e., forage fish, squid, krill); fishing gear (e.g., longlines, gillnets, and trawl nets) catch and kill seabirds as incidental bycatch; and fisheries deplete the abundance of predatory fish (e.g., tunas) with which many tropical seabirds forage commensally (Furness 2003; Wagner and Boersma 2011; Zydelis et al. 2009). Quantification of overall fisheries-induced seabird mortality is challenging because of a lack of reporting. However, fisheries-induced prey depletion has been associated with seabird breeding failures or population declines in most ecosystems supporting major forage fish

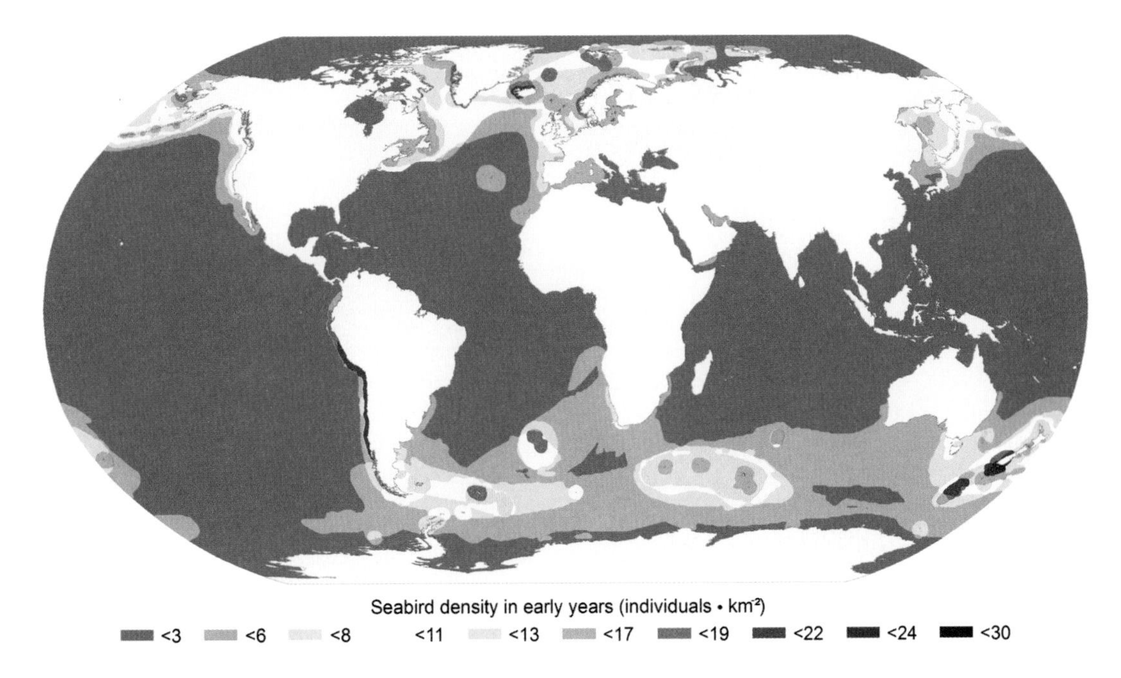

Figure 11.4. Global seabird density in the 1970s and 1980s. This map was constructed using the earliest available population size estimates for each of the world's 324 seabird species, mostly around 1980 (Paleczny 2012), converted to density by assuming that a breeding population of 70% of the total population is distributed evenly within the breeding season range for the duration of the breeding season and within the total range for the duration of the nonbreeding season, and a nonbreeding population of 30% of the total population is distributed evenly within the total range for the entire year.

fisheries (Montevecchi 2002), and hundreds of thousands of seabirds per year are reportedly killed as fisheries bycatch (Anderson et al. 2011; Moore and Zydelis 2008).

The effect of these various factors is that current seabird populations are depleted from their historical baseline, with data suggesting that the monitored portion of the global seabird population has declined by approximately 70% between 1950 and 2010 (Paleczny et al. 2015), with eleven of the fourteen seabird families experiencing a net decline (table 11.1). Additional but unquantified seabird declines occurred before seabird population monitoring, both before and during the modern industrial era. For example, hunting and introduced predators before population monitoring are estimated to have reduced tropical seabird populations to 1/100–1/1,000 of their historical abundance (Steadman 1997). According to the Red List of the International Union for Conservation of Nature (IUCN), one third of the world's seabird species are considered threatened with extinction and one half are known or suspected to be declining (Croxall et al. 2012).

Seabirds forage during the breeding season, within a few to thousands of kilometers from their breeding colonies, depending on the species. During the nonbreeding season, most seabirds (i.e., 220 of 324 species; Cox 2010) disperse or migrate to occupy larger foraging ranges (example in figure 11.1). The distribution of seabirds at sea generally reflects prey availability at large spatial scales (Shealer 2002). Globally, seabirds reach their highest densities in upwelling areas (i.e., Humboldt Current, Benguela Current, Eastern New Zealand) and in the northern and southern temperate regions (figure 11.4); these are regions that correspond to high ocean productivity (Longhurst et al. 1995).

Global maps of observed seabird population changes (figure 11.5) indicate that seabird density has declined over about 90% of the world's marine surface area, especially in the southern temperate regions, with stable populations observed only in some regions of the northern hemisphere. This pattern resembles the global distribution of seabird species threatened with extinction (figure 11.3).

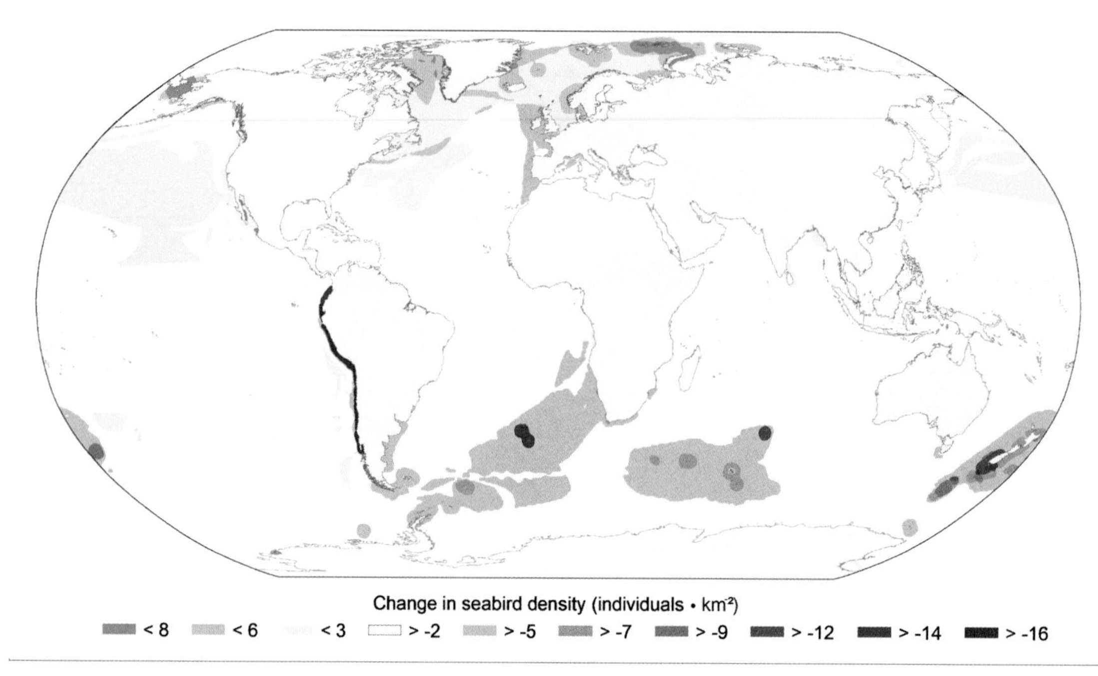

Figure 11.5. Observed change in global seabird density. Map constructed using the difference in density estimated using earliest and latest available population size records for each of the world's 324 seabird species (spanning an average of 16 years, and available for 86% of the global seabird population). The method used for figure 11.4 was used to estimate density.

When interpreting these observed patterns, one needs to consider that the overwhelming majority of initial estimates of seabird population size were made after the 1970s, and therefore population changes before and during the early modern industrial era are not captured by these maps. Thus, the northern hemisphere and the tropics (Roberts 2007) will have a low seabird baseline abundance. Observed increases in seabird populations in parts of the northern hemisphere can reflect population recovery after intense commercial hunting for feathers and oil (Colyer 2000; Grandgeorge et al. 2008), nesting habitat restoration projects (e.g., predator removal; Jones and Kress 2011), or higher availability to seabirds of fisheries discards as food (Kelleher 2005). Furthermore, early unobserved seabird population declines in the northern hemisphere may explain the low seabird density in some productive regions in the northern hemisphere, such as the California and Canary Currents (figure 11.4).

On the other hand, observed seabird declines in the southern hemisphere may best be explained by the southward expansion of fisheries since the 1970s (Swartz et al. 2010), within a period of increasing seabird population monitoring. Major fisheries-related threats in the southern hemisphere include bycatch by pelagic longlines (Anderson et al. 2011) and competition with the purse seine fisheries for Peruvian anchoveta (*Engraulis ringens*) and other forage fish fisheries in the Humboldt Current (see contributions in Pauly and Tsukayama 1987; Pikitch et al. 2014). There is also considerable evidence that climate change is contributing to widespread seabird population changes in the Antarctic (Croxall et al. 2002), notably affecting penguins (Forcada and Trathan 2009).

Mapping global seabird density and changes therein is inevitably limited by assumptions about seabird distribution, because plotting direct seabird observations is not possible at the global scale or over decades. Although these maps do account for the difference in seabird density (and hence prey consumption) between breeding and foraging areas, it is probable that they underestimate the importance of some foraging hotspots (e.g., shelf edges, frontal zones). However, it is impossible to accurately predict the locations of these foraging hotspots for all seabird species, given that the spatial distribution of

Table 11.2. Estimated food consumption per seabird family (t/year·10³) of 5 prey types in early and recent years. Annual consumption per family was calculated as the average contribution of diet items (based on all available diet data per species) to the estimated annual food intake per species (calculated using breeding and nonbreeding season duration and daily food intake as a function of the mean individuals in each species; Karpouzi 2005). The underlying data are maintained by and available from the *Sea Around Us*.

Order	Family	Estimated annual food consumption (t/year·10³) in early years/ in recent years					
		Krill, plankton	Other fish	Forage fish	Squid	Other[a]	Total
Charadrii-formes	Stercorariidae (skuas)	5/5	47/47	52/52	0/0	69/69	173/173
	Laridae (gulls)	419/389	1,394/1,295	2,070/ 1,923	167/155	867/805	4,917/4,567
	Sternidae (terns)	508/191	3,331/1,253	1,297/488	468/176	458/172	6,062/2,280
	Alcidae (auks)	2,906/ 2,494	3,869/ 3,320	7,389/ 6,341	479/411	202/174	14,845/ 12,740
Pelecani-formes	Pelecanidae (pelicans)	0/0	78/63	172/139	0/0	0/0	250/202
	Phaethontidae (tropicbirds)	0/0	11/10	6/6	3/3	0/0	20/19
	Sulidae (boobies)	0/0	467/560	564/677	50/60	1/1	1,082/1,298
	Phalacrocoracidae (cormorants)	6/2	1,888/680	1,172/422	34/12	68/25	3,168/1,141
	Fregatidae (frigatebirds)	1/1	91/48	65/34	40/21	0/0	197/104
Procellarii-formes	Procellariidae (petrels)	10,402/ 7,421	8,082/ 5,765	651/464	5,597/3,993	4,371/3,118	29,103/ 20,761
	Hydrobatidae (storm petrels)	199/270	202/274	8/11	49/67	19/26	477/648
	Diomedeidae (albatrosses)	158/142	684/614	36/32	715/641	158/142	1,751/1,571
	Pelecanoididae (diving petrels)	2,403/ 1,899	96/76	19/15	0/0	41/33	2,559/2,023
Sphenisci-formes	Spheniscidae (penguins)	20,788/ 14,474	13,558/ 9,440	1,869/1,301	2,528/1,760	551/384	39,294/ 27,359
Total		37,795/ 27,288	33,798/ 23,445	15,370/ 11,905	10,130/ 7,299	6,805/ 4,948	103,898/ 74,886

[a]Includes benthic crustaceans, other seabirds.

seabird foraging, especially at smaller spatial scales, varies seasonally and interannually (Weimerskirch 2007). Thus, the present maps were constructed using robust estimates of per species density, suitable for assessing large-scale patterns in seabird density and changes therein, but not more.

GLOBAL PREY CONSUMPTION BY SEABIRDS

The global consumption of prey by seabirds is estimated to have declined from approximately 104 million t/year in the 1970s and 1980s (figure 11.6) to 75 million t/year in 1990s and 2000s (see also table 11.2). Our modern estimate of global prey consumption is comparable to a previous but less detailed estimate of 70 million t/year (Brooke 2004), which it thus confirms. For comparison, it is estimated that the global marine mammal population consumes 168 million t/year (Kaschner et al. 2006).

The order of importance of prey types was consistent between early and recent years.

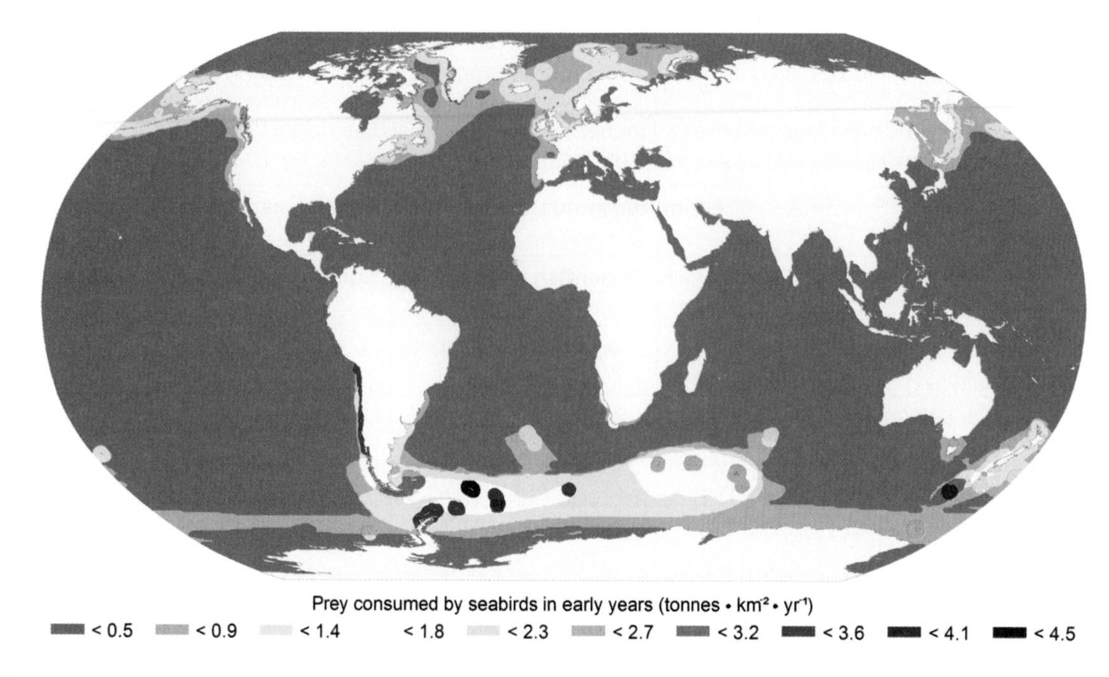

Figure 11.6. Annual food consumption by the global seabird population around the late 1970s and early 1980s. Map constructed using per species estimates of breeding and nonbreeding season density (from figure 11.5), multiplied by the daily food intake, estimated separately for the breeding and nonbreeding seasons.

Ordered from highest to lowest contribution to overall biomass, prey consumed were krill, fish, squid, and other diet items (table 11.2). Forage fish, an important commercial fish group, were estimated to make up 15%–16% (by mass) of all food consumed by seabirds and 31%–34% of the fish consumed by seabirds. It is important to note that although forage fish constitute a small percentage of the global food consumption by seabirds, they are of particular importance to the productivity of seabirds in upwelling ecosystems around the world (Cury et al. 2011).

Seabird prey consumption was historically highest in the temperate and upwelling regions, especially the Humboldt Current off Peru (figure 11.6) and where they declined most severely, mirroring the spatial distribution of global seabird population changes in figure 11.5.

It is important when interpreting these global estimates of prey consumption to be aware that food composition data are often biased toward the breeding season diet of seabirds, which may cause overestimation of the importance of fish in the diets of seabirds. On the other hand, calculating the relative contribution of diet items using fixed diet compositions does not account for the long-term change in seabird diets that has been observed in some seabirds as a result of long-term ecosystem changes, such as a decline in the trophic level of some species (Becker and Beissinger 2006).

The primary production required (PPR) to support seabirds changed from $0.79 \cdot 10^9$ t/year in the 1970s and 1980s to $0.63 \cdot 10^9$ t/year in the 1990s and 2000s, both estimates corresponding to approximately 1% of annual marine primary production. This is far less than the annual marine PPR to support marine fisheries, estimated at 8% by Pauly and Christensen (1995). Moreover, this estimate, being based on officially reported catch data (and not reconstructed total catches), probably underestimated the PPR of fisheries. This could make the actual PPR of seabirds' food consumption one order of magnitude smaller than that of fisheries catches. Such a result was to be expected: Although seabirds and marine fisheries take roughly similar amounts of biomass from the world's oceans, seabirds target much lower trophic levels than most fisheries.

CONCLUSIONS

This chapter demonstrates the spatial distribution of seabird density and food consumption, and their observed changes. Thus, it points to the potential effects of seabird decline on the marine ecosystems in which seabirds have strong effects on energy and nutrient fluxes. This summary of observed global seabird changes highlights the global importance of ecosystem-based management to protect seabirds and highlights key areas for seabird conservation, such as the temperate and upwelling regions of the southern hemisphere. Also, the spatially ubiquitous nature of global seabird decline lends support to the call for a large-scale approach to seabird conservation (Croxall et al. 2012; Jodice and Suryan 2010), and precautionary management in data-poor situations, an approach recommended for managing wildlife with uncertain population trends and uncertain or uncontrollable threats (Lauck et al. 1998). Marine no-take zones are a key large-scale and precautionary approach to seabird conservation and population recovery (Pichegru et al. 2010), protecting seabirds from direct and indirect fisheries-related threats and allowing sufficient prey to increase seabird productivity and in turn buffer additional mortality caused by other growing threats (e.g., pollution, coastal development, climate change).

ACKNOWLEDGMENTS

This is a contribution of the *Sea Around Us*, a research activity at the University of British Columbia initiated and funded by the Pew Charitable Trusts from 1999 to 2014 and currently funded mainly by the Paul G. Allen Family Foundation. We thank numerous colleagues for providing feedback or supplying unpublished datasets on bird distributions or abundance, notably David Grémillet, Bob Furness, Philippe Cury, Ian Boyd, Robert Crawford, James Mills, Henrik Österblom, John Piatt, Jean-Paul Roux, Lynne Shannon, William Sydeman, Matthieu Le Corre, Reg Watson, Kristin Kleisner, William Swartz, Patrick O'Hara, Louise Blight, and Joanna Smith.

REFERENCES

Anderson, O. R. J., C. J. Small, J. P. Croxall, E. K. Dunn, B. J. Sullivan, O. Yates, and A. Black. 2011. Global seabird bycatch in longline fisheries. *Endangered Species Research* 14: 91–106.

Anderwald, P., P. G. H. Evans, L. Gygax, and A. R. Hoelzel. 2011. Role of feeding strategies in seabird–minke whale associations. *Marine Ecology Progress Series* 424: 219–227.

Ballance, L. T., and R. L. Pitman. 1999. Foraging ecology of tropical seabirds. Pp. 2057–2071 in N. J. Adams and R. H. Slotow (eds.), *Proceedings from the 22nd International Ornithology Congress, Johannesburg*.

Becker, B. H., and S. R. Beissinger. 2006. Centennial decline in the trophic level of an endangered seabird after fisheries decline. *Conservation Biology* 20: 470–479.

Brooke, M. L. D. 2004. The food consumption of the world's seabirds. *Proceedings of the Royal Society of London. Series B: Biological Sciences* 271(suppl 4): S246–S248.

Cheung, W. W. L., D. Zeller, M. L. D. Palomares, D. Al-Abdulrazzak, L. Brotz, V. W. Y. Lam, M. Paleczny, and D. Pauly. 2012. *A preliminary assessment of climate change impacts on marine ecosystems and fisheries of the Arabian Gulf*. A Report to the Climate Change Research Group/LLC.

Christensen, V., C. J. Walters, R. Ahrens, J. Alder, J. Buszowski, L. B. Christensen, W. W. L. Cheung, J. Dunne, R. Froese, V. Karpouzi, K. Kaschner, K. Kearney, S. Lai, V. Lam, M. L. D. Palomares, A. Peters-Mason, C. Piroddi, J. L. Sarmiento, J. Steenbeek, U. R. Sumaila, R. Watson, D. Zeller, and D. Pauly. 2009. Database-driven models of the world's Large Marine Ecosystems. *Ecological Modelling* 220: 1984–1996.

Coll, M., C. Piroddi, C. Albouy, F. Ben Rais Lasram, W. W. L. Cheung, V. Christensen, V. S. Karpouzi, F. Guilhaumon, D. Mouillot, M. Paleczny, M. L. D. Palomares, D. Pauly, J. Steenbeek, P. Trujillo, and R. Watson. 2012. The Mediterranean Sea under siege: spatial overlap between marine biodiversity, cumulative threats and marine reserves. *Global Ecology and Biogeography* 21(4): 465–480.

Colyer, R. J. M. 2000. Feathered women and persecuted birds: the struggle against the plumage trade c. 1860–1922. *Rural History* 11: 57–73.

Cox, G. W. 2010. *Bird Migration and Global Change*. Island Press, Washington, DC.

Croll, D. A., J. L. Maron, J. A. Estes, E. M. Danner, and G. V. Byrd. 2005. Introduced predators transform subarctic islands from grassland to tundra. *Science* 307: 1959–1961.

Croxall, J. P., S. H. M. Butchart, B. Lascelles, A. J. Stattersfield, B. Sullivan, A. Symes, and P. Taylor. 2012. Seabird conservation status, threats and priority actions: a global assessment. *Bird Conservation International* 22: 1–34.

Croxall, J. P., P. G. H. Evans, and E. A. Schreiber (eds.). 1984. *Status and Conservation of the World's Seabirds*. Paston Press, Cambridge, England.

Croxall, J., P. Trathan, and E. Murphy. 2002. Environmental change and Antarctic seabird populations. *Science* 297: 1510–1514.

Cury, P. M., I. L. Boyd, S. Bonhommeau, T. Anker-Nilssen, R. J. M. Crawford, R. W. Furness, J. A. Mills, E. J. Murphy, H. Osterblom, M. Paleczny, J. F. Piatt, J.-P. Roux, L. Shannon, and W. J. Sydeman. 2011. Global seabird response to forage fish depletion: one-third for the birds. *Science* 334: 1703–1706.

Egevang, C., I. J. Stenhouse, R. A. Phillips, A. Petersen, J. W. Fox, J. R. D. Silk. 2010. Tracking of Arctic terns *Sterna paradisaea* reveals longest animal migration. *Proceedings of the National Academy of Sciences of the United States of America* 107: 2078–2081.

Einoder, L. D. 2009. A review of the use of seabirds as indicators in fisheries and ecosystem management. *Fisheries Research* 95: 6–13.

Forcada, J., and P. N. Trathan. 2009. Penguin responses to climate change in the Southern Ocean. *Global Change Biology* 15(7): 1618–1630.

Furness, R. W. 2003. Impacts of fisheries on seabird communities. *Scientia Marina* 67: 33–45.

Gaston, A. J., I. L. Jones, and I. Lewington. 1998. *The Auks: Alcidae*. Oxford University Press, Oxford, England.

Gerber, L., L. Morissette, K. Kaschner, and D. Pauly. 2009. Should whales be culled to increase fishery yields? *Science* 323: 880–881.

Grandgeorge, M., S. Wanless, T. Dunn, E. Timothy, M. Maumy, G. Beaugrand, and D. Gremillet. 2008. Resilience of the British and Irish seabird community in the twentieth century. *Aquatic Biology* 4(2): 187–199.

Grémillet, D., C. Peron, P. Provost, and A. Lescroel. 2015. Adult and juvenile European seabirds at risk from marine plundering off West Africa. *Biological Conservation* 182: 143–147.

Harrison, P. 1987. *Seabirds of the World: A Photographic Guide*. Princeton University Press, Princeton, NJ.

Hays, B. 2015. *Corps of Engineers to cull Oregon cormorants preying on endangered salmon*. Available at http://www.upi.com/Science_News/2015/02/09/Corps-of-Engineers-to-cull-Oregon-cormorants-preying-on-endangered-salmon/4501423520633/. Accessed April 19, 2015.

Hipfner, M. J., L. K. Blight, R. W. Lowe, S. I. Wilhelm, G. J. Robertson, R. T. Barrett, T. Anker-Nilssen, and T. P. Good. 2012. Unintended consequences: how the recovery of sea eagle *Haliaeetus* spp. populations in the northern hemisphere is affecting seabirds. *Marine Ornithology* 40: 39–52.

IUCN. 2012. International Union for Conservation of Nature Red List of Threatened Species, Version 2012.2: www.iucnredlist.org.

Jodice, P. G. R., and R. M. Suryan. 2010. The transboundary nature of seabird ecology. Pp. 139–165 in S. C. Trombulak and R. F. Baldwin (eds.), *Landscape-Scale Conservation Planning*. Springer, Amsterdam, the Netherlands.

Jones, H. P., and S. W. Kress. 2011. A review of the world's active seabird restoration projects. *The Journal of Wildlife Management* 76: 2–9.

Karpouzi, V. 2005. *Modelling and mapping trophic overlap between fisheries and the world's seabirds*. Department of Zoology, University of British Columbia, Vancouver, BC.

Karpouzi, V. S., and D. Pauly. 2008. A framework for evaluating national seabird conservation efforts. Pp. 62–70 in J. Alder and D. Pauly (eds.), *A comparative assessment of biodiversity, fisheries and aquaculture in 53 countries' Exclusive Economic Zones*. Fisheries Centre Research Reports 16(7), University of British Columbia, Vancouver, Canada.

Karpouzi, V. S., R. Watson, and D. Pauly. 2007. Modelling and mapping resource overlap between fisheries and seabirds on a global scale: a preliminary assessment. *Marine Ecology Progress Series* 343: 87–99.

Kaschner, K., V. Karpouzi, R. Watson, and D. Pauly. 2006. Forage fish consumption by marine mammals and seabirds. Pp. 33–46 in J. Alder and D. Pauly (eds.), *On the multiple uses of forage fish: from ecosystems to markets*. Fisheries Centre Research Reports 14(3), University of British Columbia, Vancouver, Canada.

Kelleher, K. 2005. *Discards in the world's marine fisheries. An update*. FAO Fisheries Technical Paper No. 470, Food and Agriculture Organization, Rome.

Lauck, T., C. W. Clark, M. Mangel, and G. R. Munro. 1998. Implementing the precautionary principle in fisheries management through marine reserves. *Ecological Applications* 8: S72–S78.

Lewis, S., J. Turpie, and P. Ryan. 2012. Are African penguins worth saving? The ecotourism value of the Boulders Beach colony. *African Journal of Marine Science* 1–8.

Longhurst, A., S. Sathyendranath, T. Platt, and C. Caverhill. 1995. An estimate of global primary production in the ocean from satellite radiometer data. *Journal of Plankton Research* 17: 1245–1271.

Montevecchi, W. A. 2002. Interactions between fisheries and seabirds. Pp. 527–557 in E. A. Schreiber and J. S. Burger (eds.), *Biology of Marine Birds*. CRC Press, Boca Raton, FL.

Moore, J. E., and R. Zydelis. 2008. Quantifying seabird bycatch: where do we go from here? *Animal Conservation* 11: 257–259.

Morissette, L., V. Christensen, and D. Pauly. 2012. Marine mammal impacts in exploited ecosystems: would large-scale culling benefit fisheries? *PLoS ONE* 7(9): e43966.

Paleczny, M. 2012. *An analysis of temporal and spatial patterns in global seabird abundance during the modern industrial era, 1950–2010, and the relationship between global seabird decline and marine fisheries catch*. MSc thesis, Department of Zoology, University of British Columbia, Vancouver.

Paleczny, M., E. Hammill, V. Karpouzi, and D. Pauly. 2015. Population trend of the world's monitored seabirds, 1950–2010. *PLoS ONE* 10(6): e0129342.

Paleczny, M., and D. Pauly. 2011. Seabirds in Canadian marine ecoregions: distribution and abundance. Pp. 41–46 in W. W. L. Cheung, D. Zeller, and D. Pauly. *Projected species shifts due to climate change in the Canadian Marine Ecoregions*. A report of the *Sea Around Us* Project to Environment Canada.

Pannozzo, L. 2013. *The Devil and the Deep Blue Sea: An Investigation into the Scapegoating of Canada' Grey Seal*. Fernwood Publishing, Black Point, Nova Scotia.

Pauly, D., and V. Christensen. 1995. Primary production required to sustain global fisheries. *Nature* 374: 255–257.

Pauly, D., and I. Tsukayama (eds.). 1987. *The Peruvian anchoveta and its upwelling ecosystem: three decades of change*. ICLARM Studies and Reviews 15.

Piatt, J., W. Sydeman, and F. Wiese. 2007. Introduction: a modern role for seabirds as indicators. *Marine Ecology Progress Series* 352: 199–204.

Pichegru, D., D. Grémillet, R. J. M. Crawford, and P. G. Ryan. 2010. Marine no-take zone rapidly benefits endangered penguin. *Biology Letters* 6(4): 498–501.

Pikitch, E. K., K. J. Rountos, T. E. Essington, C. Santora, D. Pauly, R. Watson, U. R. Sumaila, P. D. Boersma, I. L. Boyd, D. O. Conover, P. Cury, S. S. Heppell, E. D. Houde, M. Mangel, É. Plagányi, K. Sainsbury, R. S. Steneck, T. M. Geers, N. Gownaris, and S. B. Munch. 2014. The global contribution of forage fish to marine fisheries and ecosystems. *Fish and Fisheries* 15: 43–64.

Poole, A. (ed.). 2005. The Birds of North America Online. Available at http://bna.birds.cornell.edu/BNA/. Cornell Laboratory of Ornithology, Ithaca, NY.

Rains, D., H. Weimerskirch, and T. M. Burg. 2011. Piecing together the global population puzzle of the wandering albatrosses: genetic analysis of the Amsterdam albatross *Diomedea amsterdamensis*. *Avian Biology* 42: 69–79.

Roberts, C. 2007. *The Unnatural History of the Sea*. Island Press, Washington, DC.

Russell, R. W. 1999. Comparative demography and life history tactics of seabirds: implications for conservation and marine monitoring. *American Fisheries Society Symposium* 23: 51–76.

Shealer, D. 2002. Foraging behavior and food of seabirds. Pp. 137–177 in E. A. Schreiber and J. S. Burger (eds.), *Biology of Marine Birds*. CRC Press, Boca Raton, FL.

Steadman, D. W. 1997. Extinctions of Polynesian birds: reciprocal impacts of birds and people. Pp. 51–79 in P. V. Kirch and T. L. Hunt (eds.), *Historical Ecology in the Pacific Islands*. Yale University Press, New Haven, CT.

Swartz, W., E. Sala, S. Tracey, R. Watson, and D. Pauly. 2010. The spatial expansion and ecological footprint of fisheries (1950 to present). *PLoS ONE* 5(12): e15143.

Wagner, E. L., and P. D. Boersma. 2011. Effects of fisheries on seabird community ecology. *Reviews in Fisheries Science* 19: 157–167.

Weimerskirch, H. 2007. Are seabirds foraging for unpredictable resources? *Deep Sea Research Part II: Topical Studies in Oceanography* 54: 211–223.

Zydelis, R., J. Bellebaum, H. Österblom, M. Vetemaa, B. Schirmeister, A. Stipniece, M. Dagys, M. van Eerden, and S. Garthe. 2009. Bycatch in gillnet fisheries: an overlooked threat to waterbird populations. *Biological Conservation* 142: 1269–1281.

NOTE

1. Cite as Paleczny, M., V. S. Karpouzi, E. Hammill, and D. Pauly. 2016. Global seabird populations and their food consumption. Pp. 125–136 in D. Pauly and D. Zeller (eds.), *Global Atlas of Marine Fisheries: A Critical Appraisal of Catches and Ecosystem Impacts*. Island Press, Washington, DC.

CHAPTER 12

A GLOBAL ANALYSIS OF MARICULTURE PRODUCTION AND ITS SUSTAINABILITY, 1950–2030[1]

Brooke Campbell, Jackie Alder, Pablo Trujillo, and Daniel Pauly
Sea Around Us, University of British Columbia, Vancouver, BC, Canada

The biodiversity of the ecosystems in which our economy and culture are embedded provides us with food. However, we often act as though these environmental food services are somehow free and infinite. In reality, the scope and scale of our current human activities, and our tendency to rely on a short-term mindset, are damaging our environment and threatening this provisioning role of natural systems (Sumaila and Walters 2005).

This threat is evident in the evolution of global capture fisheries to their present state (Pauly and Zeller 2016). Indeed, capture fisheries alone are no longer expected to be capable of supplying the projected increases in the demand for food fish, a term used to collectively refer to finfish, molluscs, crustaceans, and other aquatic animals that are caught or farmed (FAO 2012b). Aquaculture is expected to both fill the supply gap and meet the growing worldwide consumption demand for fish (Ye 1999; FAO 2009a; FAO 2014).

Aquaculture is "the farming of aquatic organisms including fish, molluscs, crustaceans and aquatic plants, with some sort of intervention in the rearing process to enhance production, such as regular stocking, feeding, protection from predators, etc." (FAO 2008a), and its ability to provide fish for human consumption has changed dramatically since the first documented production of herbivorous pond fish in China more than 3,000 years ago (Ling 1977). Historically, aquaculture began as a low-intensity farming practice that applied basic rearing techniques to naturalized or native fish, primarily in freshwater pond environments. Today global-scale commercial aquaculture production across freshwater, brackish, and marine environments provides a large fraction of the fish consumed worldwide. Aquaculture therefore can also be expected to play a pivotal role in our attempt to meet the projected increases in global seafood demand.

Although the freshwater sector continues to be a very important contributor to global supplies of food fish, since 1970 there has been a reported threefold increase in the production and economic value of industrial-scale and intensively reared marine and brackish, or "mariculture," species (FAO 2009a), a trend that appears to be holding (FAO 2014). These species fetch a high price in international markets, but the effects of their rearing practices can be detrimental to the health of coastal ecosystems and their people (Trujillo 2007), and to fisheries as well (Goldburg 2008; Naylor et al. 1998, 2000; Pullin et al. 1992; Primavera 2006).

As part of its goal of improving understanding of the impact of fisheries on the world's marine ecosystems (Pauly 2007), the *Sea Around Us* supported research intended to improve understanding of global mariculture sector trends, linkages and processes, and their relationship to global fisheries, to people, and to the environment through time. This work led to a spatially and taxonomically disaggregated database of mariculture production from 1950 to 2010, an index of mariculture sustainability, and scenario-based simulations exploring how sustainable mariculture development policies might affect the long-term health and well-being of people and their environment vis-à-vis meeting the future demand for food fish in 2030. The following highlights parts of this work (see also Campbell and Pauly 2013).

GLOBAL ANALYSIS OF MARICULTURE PRODUCTION TRENDS SINCE 1950

In 1950, the Food and Agriculture Organization of the United Nations (FAO) began to disseminate annual worldwide statistics that, until 1985, combined fisheries landings with aquaculture production. Although many countries now publish their aquaculture production statistics online, the FAO remains the primary source of global aquaculture statistics and analyses. However, the growing interest in and need for reliable and increasingly detailed aquaculture statistics has highlighted the fact that global information systems for aquaculture lag behind systems for agriculture (FAO 2008b) and hinder understanding of aquaculture's role and status throughout the regions of the world.

To address some of these issues, a global database of marine and brackish aquaculture (or mariculture) production was constructed, using a bottom-up method and a detailed taxonomic and spatial resolution, as a mean of independently validating and further refining the existing FAO Global Aquaculture Production Database (FAO 2009a). This subnational-level Global Mariculture Database (GMD) covers the years 1950 to 2010 and was used to reanalyze reported trends in mariculture production and highlight ongoing issues chronic to the collection and interpretation of global aquaculture datasets.

ASSESSING THE ACCURACY OF GLOBAL MARICULTURE PRODUCTION DATA

To assess the accuracy of currently reported global mariculture production trends, the provincial-scale GMD data compiled by the

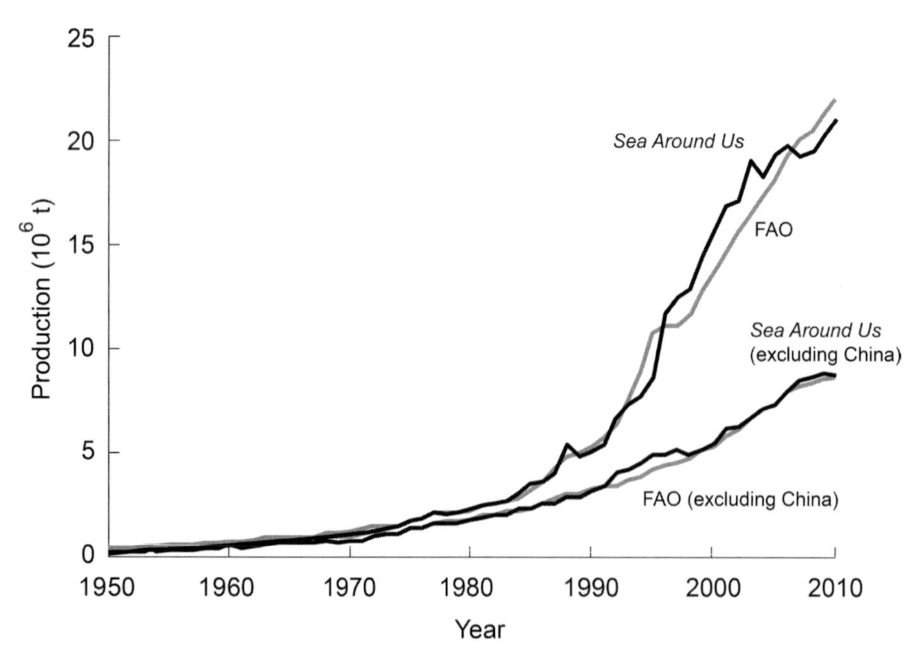

Figure 12.1. Comparison of global mariculture dataset trends between the FAO and the *Sea Around Us* GMD datasets, with and without China. Log-linear regressions of both datasets for the years 1970–2010 yielded R^2 values of 0.993 with China and 0.990 without China, indicating a strong match between datasets. The similarity in slopes during this same time period suggests a mean global rate of production increase of 7.7% per year with China and 5.5% per year without China. (Adapted from Campbell and Pauly 2013.)

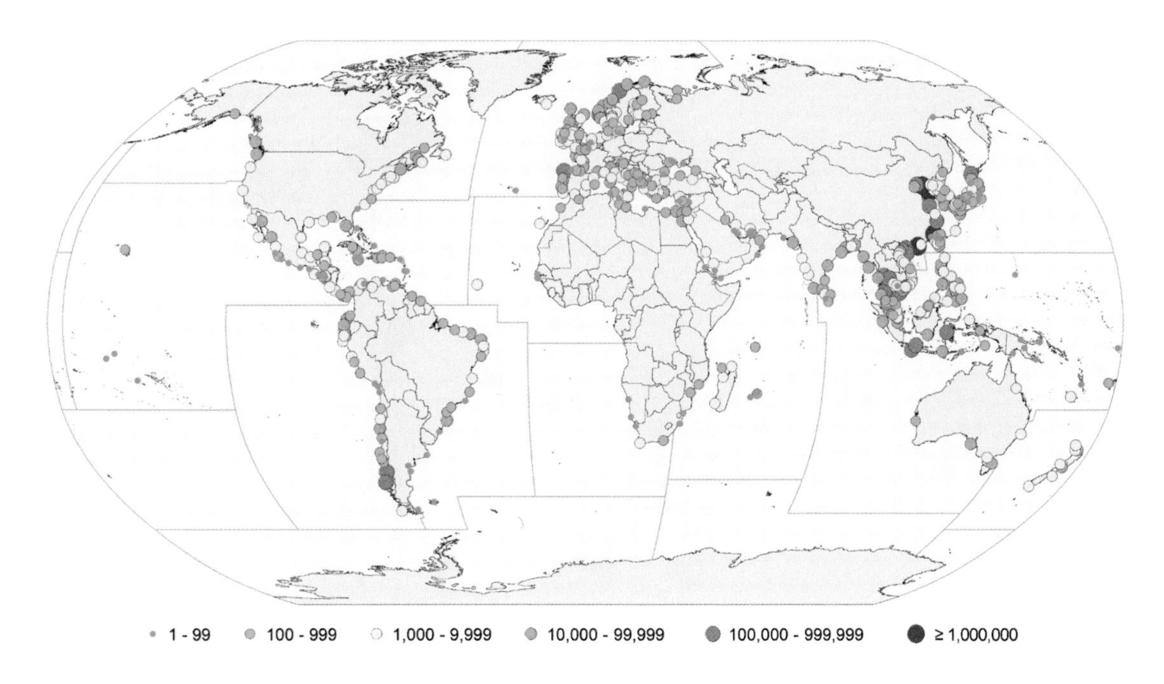

Figure 12.2. Average annual global mariculture production (t; all species combined) for 2000 to 2010, by coastal "province" (see text). (Adapted from Campbell and Pauly 2013.)

Sea Around Us were aggregated nationally and then globally. The resulting trend was then compared with the equivalent FAO FishStat Plus (v. 2.31) Global Aquaculture Production Database trend (FAO 2012a), whose content for 1950–1984 was derived by FAO through a post hoc disaggregation of their combined fisheries and aquaculture database, a necessary step, albeit fraught with uncertainties. Both datasets indicate a tripling in production between 1950 and 1970, as well as an overall similarity in total annual global production growth from 1970 to 2010 (i.e., just under three quarters of the compared annual production in these years is similar to within 10%). The similarity between the datasets increases to 80% when China is excluded from the analysis (figure 12.1).

The general resemblance between the datasets applies to all regions of the world except Africa, whose mariculture production continues to be negligible (see FAO 2014). The largest discrepancies between global datasets were found in the data-poor era before 1970, more than a decade before the establishment of FAO's aquaculture data repository (FAO 2009a). In these years, the *Sea Around Us* GMD provides the more conservative production estimate.

Note that we cannot rule out that the similarity between the FAO-reported mariculture statistic and the database presented here is due (at least in part) to the same bias (e.g., due to overreporting of provincial mariculture production from China). This possibility was hinted at in a previous issue of *State of the World's Fisheries and Aquaculture* (SOFIA; FAO 2012b), but although overreporting of fisheries catches was alluded to, the potential overreporting of mariculture production was not touched on in the last SOFIA (FAO 2014). Thus, whether or not China's mariculture data suffer from the same overreporting problems previously identified for China's wild capture fisheries (Watson and Pauly 2001) remains to be determined.

GEOGRAPHY OF GLOBAL MARICULTURE

The mariculture data in the *Sea Around Us* GMD were attributed to more than 600 different "provinces" (i.e., subnational entities) in 112 coastal countries and territories between 1950 and 2004, with an additional half-dozen countries initiating commercial production between 2005 and 2010 (figure 12.2). By comparison, the FAO distributes this historical production across a total of 21 large FAO areas (FAO 2012a).

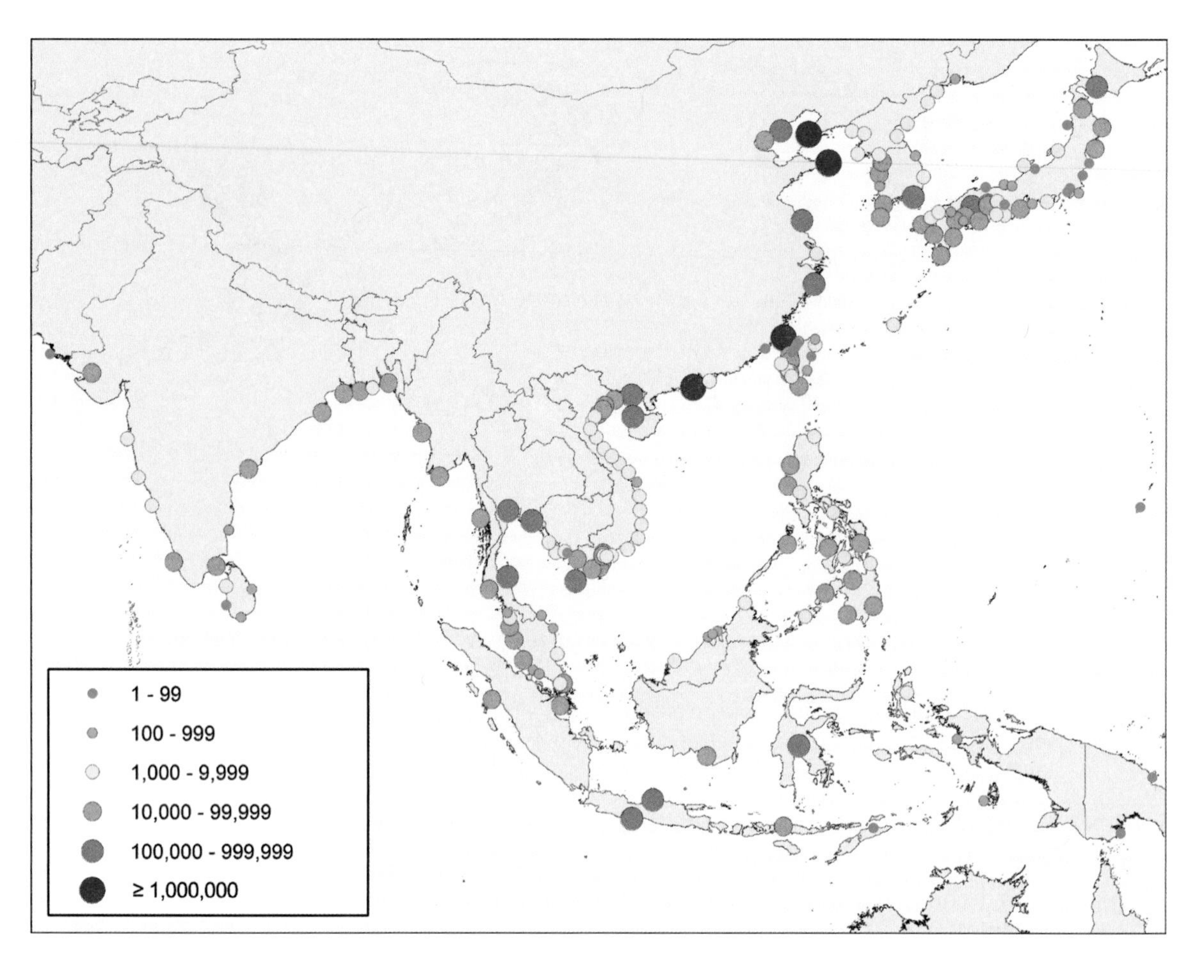

Figure 12.3. Average annual mariculture production (t; all species combined) in Asia for 2000 to 2010. Asia west of Pakistan is not shown here, because its mariculture production is negligible. (Adapted from Campbell and Pauly 2013.)

Figure 12.3 illustrates the average annual coastal mariculture production in Asia, distributed by province, between 2000 and 2010. Asia, both including and excluding China, has consistently produced the largest quantity of farmed marine and brackish species worldwide since 1950. Since 2000, China's top four mariculture-producing coastal provinces (Liaoning, Shandong, Fujian, and Guangdong) each produced more than any other maritime country, an annual average of more than 1 million t. Since 1980, three of these provinces experienced reported production increases of 1.5 to 3 million t, primarily bivalves such as Pacific cup oyster (*Crassostrea gigas*) and Manila clam (*Ruditapes philippinarum*). Note that although finfish and crustacean production is substantial in Asia, regional mariculture production is consistently dominated by bivalves.

FARMING UP THE MARINE FOOD WEB CONFIRMED

As the total number of farmed taxa has increased over time, so has the (production weighted) mean trophic level (TL) of the species produced (figure 12.4). Put differently, greater quantities of predator species are being farmed around the world. This phenomenon, previously observed in studies of FAO data that analyzed total global aquaculture (Pauly et al. 2001), as well as mariculture production in the Mediterranean (Stergiou et al. 2008; Tsirlikas et al. 2014), has been described as "farming up the food web" (Tacon et al. 2010). Farming up the food web is also apparent regionally (figure 12.4B). However, a decline in the mean TL of mariculture production occurred between 1980 and 2010 in Brazil, Denmark, Finland, Germany, Hong Kong, Nigeria, Norway, Peru, Singapore, and the United Kingdom; that is,

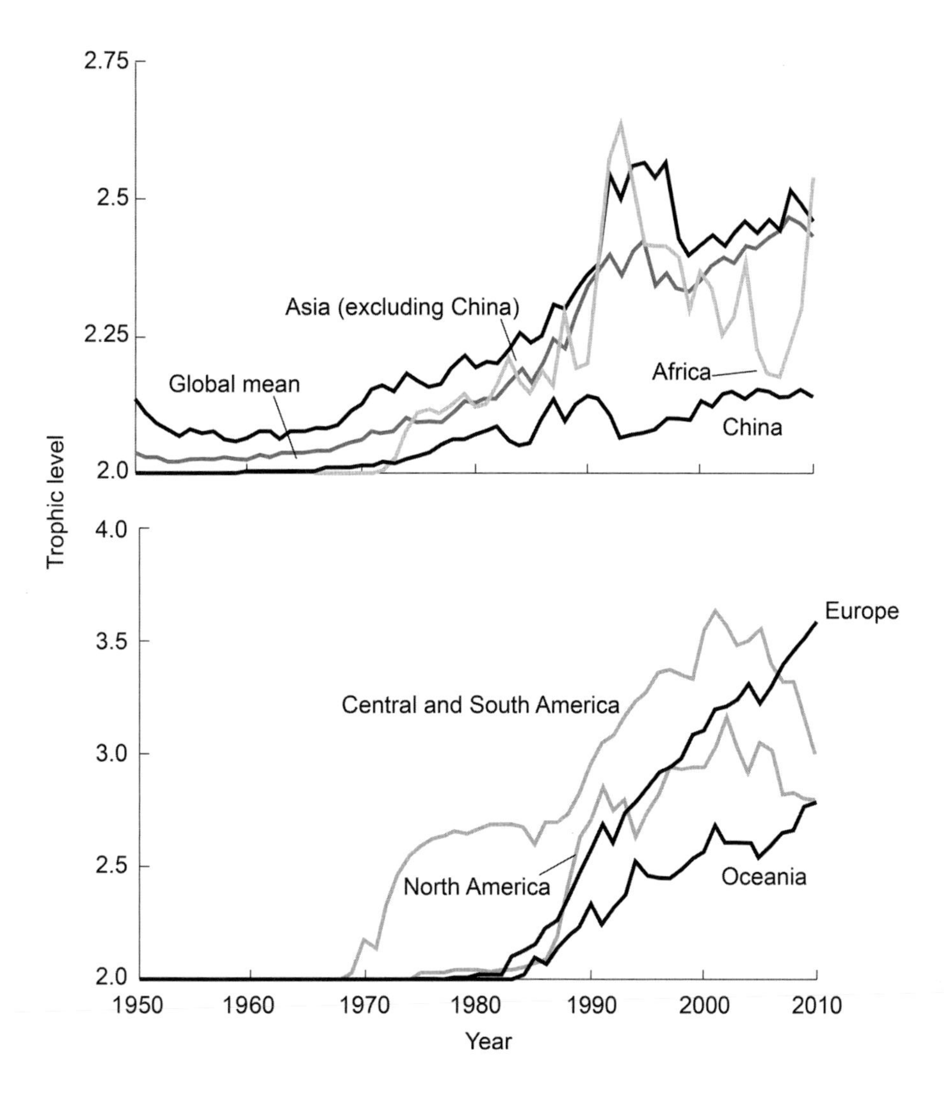

Figure 12.4. Change in the weighted mean trophic level (TL) of mariculture production in the GMD, demonstrating farming up the food web. (A) Change in the world on average and in regions where the weighted mean TL is stable or has decreased since the 1990s. (B) Change in regions with a marked increase in weighted mean TL since 1990. (Adapted from Campbell and Pauly 2013.)

these countries are currently producing greater quantities of lower-TL (herbivorous and omnivorous) species than they were in 1980. In contrast, China's weighted mean TL, with the majority of its production attributable to bivalves with a TL of 2.0, has remained stable since the mid-1980s. The large quantity of low-TL bivalves, brackish finfish, and crustaceans produced in China, and in Asia more broadly, are responsible for the low overall weighted mean global TL for mariculture.

GLOBAL ANALYSIS OF THE SUSTAINABILITY OF MARICULTURE

Currently, there are a number of codes of conduct and protocols for improving the sustainability of aquaculture (see the Aquaculture Stewardship Council at www.asc-aqua.org). Implementation of some of these codes is under way in a number of countries, with differing levels of commitment, especially in the developing world. However, given the growing consumer awareness of the long-term benefits of sustainable production (Costa-Pierce 2002), it might be useful to present here a set of explicit criteria that would allow consumers to determine whether the mariculture sector operates in a sustainable manner.

The *Sea Around Us* assessed the sustainability of mariculture in sixty-four major countries that involved eighty-six farmed finfish, crustacean, mollusc, and seaweed species and was

Table 12.1. Global summary statistics for the 13 indicators of mariculture sustainability (Trujillo 2008).

Indicator	Average	Standard deviation	Minimum	Maximum
Native/introduced	7.3	3.63	1	10
Export levels	4.5	1.87	1	10
Fishmeal use	5.2	3.62	1	10
Stocking intensity	4.8	2.63	1	10
Nutrition	7.6	2.46	2	10
Hatchery use	5.0	1.84	1	10
Antibiotic use	5.2	3.40	1	10
Habitat modification	5.0	2.04	1	10
Genetically modified organism use	6.4	1.78	2	10
Code of conduct compliance	5.2	1.44	1	9
Traceability	5.5	1.87	1	10
Employment	5.4	1.26	3	8
Waste management	5.4	3.08	1	10

performed based on thirteen indicators of performance covering ecological, economic, and social aspects of the industry (Trujillo 2007, 2008). This suite of indicators and the criteria that informed them were developed and adapted from peer-reviewed and industry literature and are roughly similar to the indicators and criteria developed by Volpe et al. (2010) for the twenty-two countries covered by the Global Aquaculture Sustainability Index. A single Mariculture Sustainability Index (MSI) was then derived by combining thirteen indicators, weighted by production to analyze differences between countries and species. An MSI score ranges between 1 and 10; a score of less than 6 is considered "unsustainable," between 6 and 8 is "approaching sustainability," and greater than 8 is "sustainable." For further detail on the scoring method and on the indicators and data sources used in the development of the assessment framework, see Trujillo (2007, 2008; Alder et al. 2010).

USING THE MSI TO ASSESS COUNTRIES' PERFORMANCE

Of the 64 countries assessed and the 361 country–species combinations generated, the lowest calculated weighted (by production) MSI score was 1.7. This was for whiteleg shrimp (*Litopenaeus vannamei*) farmed in Thailand, and the highest calculated weighted MSI score was 8.4, for seaweed farmed in Chile. Thirteen country–species combinations were greater than or equal to an MSI of 8 (sustainable), 112 cases were between 6 and 8 (approaching sustainability), and 236 cases were less than or equal to 6 (not sustainable). The average score for each indicator was between 4.5 and 7.6 (table 12.1).

Based on the weighted MSI and combined ecological and socioeconomic indicators, the ten highest-scoring countries (i.e., those with the most sustainable mariculture industries) were Germany, the Netherlands, Spain, Japan, Russian Federation, North Korea, South Korea, Ireland, France, and Argentina (table 12.2). Six of these top ten countries are developed and European, and three of the remaining countries are considered to be economies in transition.

There is no consistency between countries scoring high for the ecological indicators and countries scoring high for socioeconomic indicators. This is illustrated by Iceland, which was ranked thirteenth, with an MSI of 6.2 overall, but ranked twenty-second for its ecological score (5.4) and second for its socioeconomic score (7.1). The ten lowest-scoring countries (i.e., the least sustainable overall) were Guatemala, Cambodia, Bangladesh, Honduras,

Table 12.2. Rankings and mean weighted MSI scores of the top and bottom 10 countries, with ecological and socioeconomic scores by country; 10 indicates high sustainability and 1 indicates low sustainability of mariculture (Trujillo 2008).

Country	Rank			Score		
	Ecological	Socioeconomic	MSI	Ecological	Socioeconomic	MSI
Germany	1	1	1	9.0	7.1	8.0
Netherlands	2	2	2	9.0	7.1	8.0
Spain	3	3	3	8.7	7.1	7.9
Japan	5	6	4	7.5	6.5	7.0
Russian Federation	4	27	5	8.4	5.4	6.9
Korea, North	6	8	6	7.4	6.4	6.9
Korea, South	10	9	7	7.1	6.4	6.8
Ireland	7	12	8	7.4	6.1	6.8
France	14	5	9	6.4	7.0	6.7
Argentina	8	17	10	7.4	5.9	6.6
India	54	39	50	2.8	5.0	3.9
Faeroe Islands	40	56	51	4.5	3.0	3.8
Brazil	57	46	52	2.5	4.8	3.7
Norway	50	53	53	3.5	3.7	3.6
Chile	53	51	54	2.9	4.1	3.5
Belize	56	52	55	2.7	3.8	3.3
Myanmar	55	54	56	2.8	3.7	3.2
Honduras	52	57	57	3.2	3.0	3.1
Bangladesh	58	58	58	2.3	2.7	2.5
Cambodia	59	59	59	2.3	2.7	2.5
Guatemala	60	60	60	2.3	2.7	2.5

Myanmar, Belize, Chile, Norway, Brazil, and Faeroe Islands. Eight of the ten countries are developing and spread across Latin America and Asia. Most of these countries scored low for both ecological and socioeconomic indicators.

A principal component analysis undertaken after development of the indicators (Trujillo 2007) suggests that the indicators are a valid measure of overall sustainability in the mariculture industry, and the indicators selected in this analysis are capable of differentiating between high- and low-sustainability practices.

Countries that ranked high for overall mariculture sustainability are primarily from the developed world. In Europe, this high overall ranking is in part a reflection of consumer demand for sustainable seafood products and their desire for products free of contamination (Beardmore and Porte 2003; Volpe et al. 2010). In Japan and Korea, high scores reflect both a demand for high-quality seafood products (Bridger and Costa-Pierce 2002) and the production of substantial quantities of molluscs and plants. Most of the countries that scored high on the six ecological indicators did so because of their limited use of introduced species and fishmeal and their adequate treatment of waste and water. These countries also farm a high proportion of bivalves relative to their total production.

Mariculture in the lowest-ranking countries may not be sustainable; much of the production consists of semi-intensive to intensive production of crustaceans, in particular prawns, or carnivorous finfish, notably Atlantic salmon (*Salmo salar*). Production of these species relies heavily on the use of aquafeeds, which are rich in fishmeal and fish oil. Low scores for stocking density and insufficient waste treatment were also common among the lowest-scoring countries. Low ecological indicator scores suggest a higher risk of negative impacts to surrounding habitat, especially when these farms are open system cultures. The five lowest-scoring countries for socioeconomic indicators (i.e., Myanmar, Honduras, Bangladesh, Cambodia, and Guatemala) are all developing countries that intensively farm penaeid shrimps. Other low-scoring developed countries are European and produce large quantities of Atlantic salmon. The species–country combinations used in this assessment represent more than 95% of global mariculture production and thus represent the industry as a whole. Overall, most mariculture operations are not sustainable using current practices, in both developed and developing countries alike. Many policy makers promote the expansion of aquaculture for improving the economies of developing countries, including the creation of employment opportunities. However, this analysis suggests that this may not be a sustainable strategy because of the externalization of environmental costs. The future of the industry in developing countries in the short term (next 2 to 3 decades) will be a trade-off between short-term socioeconomic development and sustaining ecosystems. Overall, we hope with Pullin et al. (2007) "that use of broad biological, ecological and intersectional indicators will contribute to progress towards the sustainability of aquaculture."

MARICULTURE DEVELOPMENT SCENARIOS FOR THE NEXT DECADES

Aquaculture, particularly the mariculture subsector, is a growing contributor to global fish supply (FAO 2009a). This trend is anticipated to continue in the future as fish demand increases (FAO 2008c, 2010). This potential increase in global mariculture production has led to concerns about the sustainability of the sector (Pullin et al. 1992; Naylor et al. 1998, 2000; Naylor and Burke 2005; Pauly et al. 2002; Delgado et al. 2003; Primavera 2006; Goldburg 2008; Liu and Sumaila 2010). However, few forecasts and scenario exercises exist that explicitly examine the future of global aquaculture (Delgado et al. 2003; Brugère and Ridler 2004).

The UN Global Environmental Outlook (GEO) "story and simulation" scenarios assessment method is a departure from more traditional predictive models, which contain almost exclusively quantitative and price-mediated drivers of change. This is accomplished by providing both quantitative and replicable assessments of possible futures as well as a range of well-reasoned qualitative storylines (UNEP 2002; Ghosh 2007; Peterson et al. 2003; Pauly et al. 2003; Raskin 2005). The most comprehensive UN report on the environment and development to date, the GEO-4 "environment for development" assessment is primarily a capacity-building process (UNEP 2007). Its four overarching global development themes—*Markets First*, *Policy First*, *Security First*, and *Sustainability First*—and their underlying drivers, uncertainties, and critical assumptions were conceptualized and developed through a comprehensive process (UNEP 2007).

Key sources used in the construction of mariculture scenario storylines include the International Food Policy Research Institute's *Fish to 2020* (Delgado et al. 2003), *Global Aquaculture Outlook in the Next Decades* (Brugère and Ridler 2004), *State of World Aquaculture 2006* (FAO 2006), and *The State of World Fisheries and Aquaculture* (FAO 2010). Through qualitative narrative storylines and quantitative simulations of potential future production, the underlying purpose was to explore how the future of the global mariculture industry might unfold along four different possible development pathways to 2030. The analysis then relates the changes brought about by this development to the broader global seafood market.

- In the *Markets First* scenario, with a focus on the sustainability of markets rather than human–environment systems, the priority in commercial fisheries is maximizing profits.

- In the *Policy First* scenario, with a focus on the social and economic dimensions of development, the priority in fisheries is to find a balance between increasing profits, total catch, and jobs.
- In *Security First* total catch is emphasized.
- In *Sustainability First* the focus in fisheries is on ecosystem restoration.

However, emphasis is also given to increasing jobs and landings.

For further information on the GEO-4 scenarios work, refer to UNEP (2007). For the detailed mariculture scenarios, outcomes, and analysis, refer to Campbell (2010).

Markets First

In a *Markets First* world in 2030, key private sector actors, with active government support, are focused on improving the well-being of people and the environment through maximized economic growth and efficiency in the mariculture sector (UNEP 2007). This emphasis on economic drivers of sustainable development has led to an increased liberalization, strengthening, expansion, and creation of international and regional trade agreements, particularly within Asia but also between Asia and the rest of the world (UNEP 2007).

By 2030, the growing Indian and Chinese middle classes are a driving force behind increases in both total and per capita global demand for diversified and high-value marine seafood (Delgado et al. 2003; FAO 2009b). The widespread removal of trade barriers and technological constraints to increased production increases overall mariculture production more than the other global development scenarios. However, the overarching social priority of this scenario is to sustain profit rather than to sustain and improve the availability and accessibility of seafood for people (UNEP 2007). Therefore, seafood markets remain dictated by traditional supply and demand economics with few government controls (UNEP 2007), and the bulk of economic and social benefits derived from production still flows predominantly from poorer to richer countries and private entities (Kent 1997; Delgado et al. 2003).

Policy First

Under a *Policy First* scenario in 2030, government institutions worldwide, with active private and civil support, make efforts to resolve many of the issues facing humanity and the environment through top-down, policy-based reforms (UNEP 2007). Although economic growth remains a focal point for global mariculture development, it is acknowledged that such growth cannot be sustained without a stronger consideration of the negative social and environmental impacts that can accompany development. However, in practice most reform initiatives focus first and foremost on social considerations such as jobs and total production (UNEP 2007).

Policy reforms for mariculture are led by national governments and international institutions, including the FAO. These lead to improved resource sharing, a better alignment between social and political institutions, and greater political cohesion with international agreements such as the *Code of Conduct for Responsible Fishing* of the FAO. However, the slow pace of institutional reform and the inflexibility of a more centralized approach to implementing change mean that few major reforms to the mariculture industry are widely implemented by 2030 (Lake 1994; UNEP 2007).

Security First

In a world where security comes first, the benefits of mariculture production and development are available only to a privileged few (UNEP 2007). By 2030, to better control and monitor the movement of people, goods, and services within and across their respective borders, governments around the world, with support from powerful private actors, have implemented stronger restrictions on migration and trade.

The internal security focus of many government policies has led to a reduction in international cooperation and trade by 2030. Both Official Development Assistance for aquaculture extension activities and international trade in seafood are reduced, and what remains is strongly dependent on the interests of powerful governments, multinational corporations, and other powerful private interests (UNEP 2007). There is a growing distrust in the role and effectiveness of the United Nations and its specialized organizations such as the FAO, and these institutions are increasingly marginalized. The World Trade Organization becomes a leverage tool to gain more political and economic control (Smith 2006; Lynn 2010). As has occurred in capture fisheries (Alder and Watson 2007), countries unable to gain sufficient political and economic autonomy are strong-armed into expanding and intensifying mariculture production for export to economically developed foreign countries. The revenue from exported sales is brokered by, and primarily returned to, governments and private actors; poor and rural communities are marginalized.

Sustainability First

In a *Sustainability First* world in 2030, all government, private, and civil sector actors across all institutional levels are following through on their individual and collaborative commitments to address the most pressing social and environmental sustainability issues (UNEP 2007). In response to a growing social movement over the past 20 years that advocates for a more equitable treatment of social, economic, and environmental issues in development policies, both national and international institutions have collaboratively begun to rework their institutional and trade governance mandates to incorporate more than drivers of economic growth and efficiency (UNEP 2007). Globally, increases in jobs and total production are socially valued in the fisheries and aquaculture sectors, but only if the underlying marine ecosystem is maintained or restored (UNEP 2007). This new approach to governance increases the global focus on ecosystem restoration, includes a stronger emphasis on decision-making inputs from the private sector and civil society, and results in significant improvements to general cooperation and compliance in resource use issues worldwide (UNEP 2007).

Among wealthier major seafood consumers in the United States, Canada, and

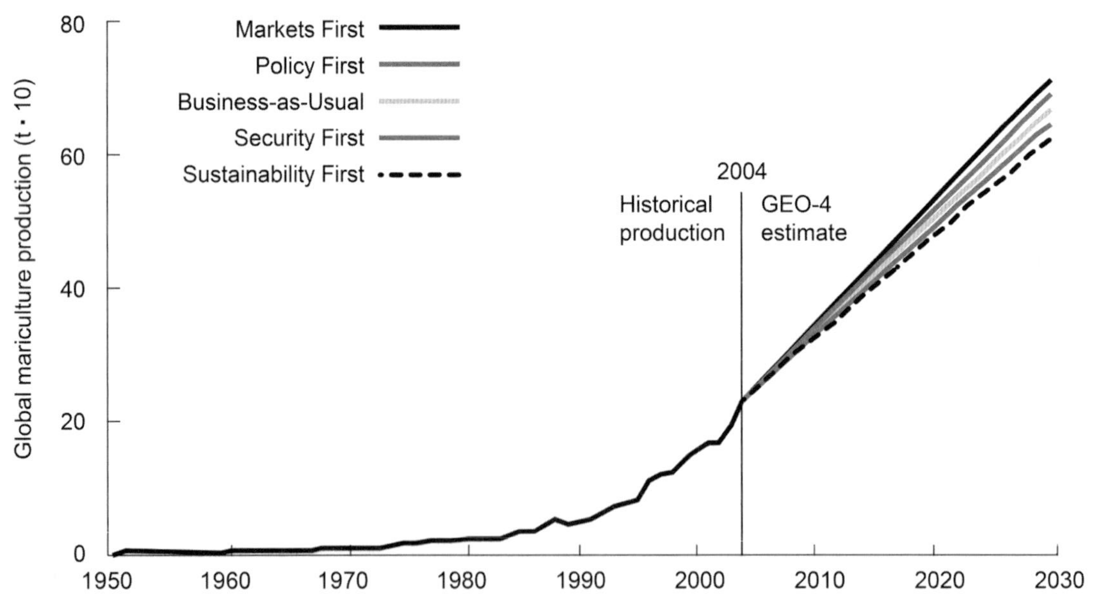

Figure 12.5. Extrapolation of global mariculture production trends, based on historical trends. Represented is a business-as-usual scenario, with all rate-changing factors held constant from 2004 forward, and the four GEO-4 scenarios, with production growth varying by increment rules described in Campbell (2010). Note that the extrapolation from the 2004 baseline (linear rather than exponential, to avoid absurdly high estimates) "predicts" the most recent estimate (FAO 2014), that is, a production of 40 million t in 2012 suggested by the *Market First* scenario. (Adapted from Campbell and Pauly 2013.)

the EU, there is an increasing growth and diversification in the demand for more responsible, ecologically sustainable, and ethically produced seafood products. This is a trend carried over and strengthened from previous decades (Lebel et al. 2002; Jansen and Vellema 2004).

PROJECTED MARICULTURE PRODUCTION UNDER DIFFERENT SCENARIOS

As a complement to the qualitative narrative storylines of possible production and sector futures, quantitative simulations of potential future production were generated to 2030, using past trends in mariculture production extrapolated forward (figure 12.5). As with other quantitative models developed in the GEO-4 assessment, this analysis uses historical time series data standardized to a common base year of 2000 (UNEP 2007).

If business-as-usual (BAU) rates of mariculture production continue onward from the 2004 baseline year (holding all else constant), by 2030 the quantity of farmed marine and brackish products worldwide could reach 67 million t, of which China might contribute nearly 70%. When this is compared with production simulations under the lowest growth rate scenario (*Sustainability First*) and the highest growth rate scenario (*Markets First*), the total difference in global mariculture tonnage is ±4.3 million t from the BAU baseline in 2030, with China contributing slightly less in the lowest-growth scenario. This implies an average increase in production of more than 1.5 million t per year.

When this simulated estimate of BAU mariculture production is combined with an estimate of freshwater aquaculture production that is based on total production proportions reported by the FAO between 2004 and 2006, total annual global aquaculture production from all production environments could be 172 million t of food fish by 2030. When total global aquaculture production estimates are then combined with the GEO-4 simulations of future marine fisheries landings (UNEP 2007), total simulated global fisheries and aquaculture production in 2030 range between 281 million t/year under the *Sustainability First* scenario and 317 million t/year of seafood under *Markets First*. These values seem exceedingly large.

CONCLUSIONS

As in the original GEO-4 assessment (UNEP 2007), a number of overarching policy messages can be summarized from the exploratory scenario outcomes. Notably, even under the overarching thematic influence of "environment for development," all but one of the development scenarios (*Sustainability First*) continue to prioritize a worldwide expansion, production increase, and intensification of high-value, high-environmental input, carnivorous marine finfish and crustacean species. Although market-driven choices are likely to increase total global mariculture production over the next two decades (as well as profits and some jobs), longer-term production growth may ultimately decrease in countries around the world because of rising environmental constraints. With many of the most serious negative ecological and social effects likely to be experienced by developing countries, the perceived benefits of market-driven pathways of action risk benefiting only a privileged few people.

However, if the global human population surpasses a projected 8.3 billion people by 2030 (UN 2009), an increase in the pressure on the world's marine resources under any scenario is inevitable. Furthermore, a *Sustainability First* approach to mariculture development does not overcome global inequities in the distribution of production and profit (UNEP 2007), nor will it eliminate the demand for high-value carnivorous species for consumption. Although a *Sustainability First* future may increase the total global production of higher-MSI bivalves (in the place of higher-trophic-level taxa with a lower MSI) and contribute to an increase in total global seafood tonnage, the actual availability of meat for consumption could be dramatically reduced because bivalve production is typically reported in shell weight (Ye 1999; Wijkstrom 2003). In addition, the lower comparative economic value of bivalves to finfish and crustaceans could mean that the overall profits derived from mariculture may decline in some countries even though production is increasing. Ultimately, this simulated variation highlights the uncertainty in dealing with the future, as well as the range of effects that individual and collective decisions can have on future global mariculture development.

ACKNOWLEDGMENTS

The is a contribution of the *Sea Around Us*, a research activity at the University of British Columbia initiated and funded by the Pew Charitable Trusts from 1999 to 2014 and currently funded mainly by the Paul G. Allen Family Foundation.

REFERENCES

Alder, J., S. Cullis-Suzuki, V. Karpouzi, K. Kaschner, S. Mondoux, W. Swartz, P. Trujillo, R. Watson, and D. Pauly. 2010. Aggregate performance in managing marine ecosystems in 53 maritime countries. *Marine Policy* 34: 468–476.

Alder, J., and R. Watson. 2007. Fisheries globalization: fair trade or piracy. Pp. 47–74 in W. Taylor, M. G. Schechter, and L. G. Wolfson (eds.), *Globalization: Effects on Fisheries Resources*. Cambridge University Press, Cambridge, England.

Beardmore, J. A., and J. S. Porte. 2003. *Genetically modified organisms and aquaculture*. FAO Fisheries Circular No. 989.

Bridger, C. J., and B. A. Costa-Pierce. 2002. Sustainable development of offshore aquaculture in the Gulf of Mexico. Pp. 255–265 in R. L. Creswell (ed.), *Proceedings of the fifty-third annual Gulf and Caribbean Fisheries Institute Meeting, November 2000*. Mississippi/Alabama Sea Grant Consortium, Fort Pierce, FL.

Brugère, C., and N. Ridler. 2004. *Global aquaculture outlook in the next decades: an analysis of national aquaculture production forecasts to 2030*. FAO Fisheries Circular. 0429–9329, Food and Agriculture Organization, Rome.

Campbell, B. 2010. *A global analysis of historical and projected mariculture production trends 1950–2030*. MSc thesis, Resource Management and Environmental Sciences, University of British Columbia, Vancouver, Canada.

Campbell, B., and D. Pauly. 2013. Mariculture: a global analysis of production trends since 1950. *Marine Policy* 39: 94–100.

Costa-Pierce, B. A. 2002. The Blue Revolution: aquaculture must go green. *World Aquaculture* 33(4): 4–5.

Delgado, C. L., N. Wada, M. W. Rosegrant, S. Meijer, and M. Ahmed. 2003. *Fish to 2020: supply and demand in changing markets*. International Food Policy Research Institute/ WorldFish Center, Washington, DC and Penang.

FAO. 2006. *State of world aquaculture 2006*. FAO Fisheries Technical Paper 500. FAO Inland Water Resources and Aquaculture Service, Fishery Resources Division, Rome.

FAO. 2008a. Crespi, V., and A. Cocher (comps.). *Glossary of aquaculture /Glossaire d'aquaculture/ Glosario de acuicultura*. FAO Fisheries and Aquaculture Department, Rome.

FAO. 2008b. *Strategy and outline plan for improving information on status and trends of aquaculture*. FAO, Rome.

FAO. 2008c. *Opportunities for addressing the challenges in meeting the rising global demand for food fish from aquaculture*. Committee on Fisheries, Sub-Committee on Aquaculture, 4th Session, Puerto Varas, Chile, October 6–10, 2008.

FAO. 2009a. *Fishery statistical collections: global aquaculture production*. FAO Fisheries and Aquaculture Department. Available at www.fao.org/fishery/statistics/global-aquaculture-production/en.

FAO. 2009b. *The state of world fisheries and aquaculture 2008*. FAO Fisheries and Aquaculture Department, Rome.

FAO. 2010. *The state of world fisheries and aquaculture 2010*. FAO Fisheries and Aquaculture Department, Rome.

FAO. 2012a. FishStatPlus (Version 2.31). Universal software for fishery statistical time series (Aquaculture Production: 1950–2010). FAO Fisheries Department, Fishery Information, Data, and Statistics Unit. Available at www.fao.org/fishery/statistics/.

FAO. 2012b. *The state of world fisheries and aquaculture 2012*. FAO Fisheries and Aquaculture Department, Rome.

FAO. 2014. *The state of world fisheries and aquaculture 2014*. FAO Fisheries and Aquaculture Department, Rome.

Ghosh, N. 2007. *A methodological framework of scenarios development: some reflections from the Global Environmental Outlook experiences*. Working Paper Series. Department of Policy Studies, TERI University, New Delhi.

Goldburg, R. 2008. Aquaculture, trade, and fisheries linkages: unexpected synergies. *Globalizations* 5(2): 183–194.

Jansen, K., and S. Vellema. (eds.). 2004. *Agribusiness and Society: Corporate Responses to Environmentalism, Market Opportunities, and Public Regulation*. Zed Books, London.

Kent, G. 1997. Fisheries, food security, and the poor. *Food Policy* 22(5): 393–404.

Lake, R. W. 1994. Central government limitations on local policy options for environmental protection. *The Professional Geographer* 46(2): 236–242.

Lebel, L., N. H. Tri, A. Saengnoree, S. Pasong, U. Buatama, and L. K. Thoa. 2002. Industrial transformation and shrimp aquaculture in Thailand and Vietnam: pathways to ecological, social, and economic sustainability. *AMBIO: A Journal of the Human Environment* 31(4): 311–323.

Ling, S. W. 1977. *Aquaculture in Southeast Asia: A Historical Overview*. University of Washington Press, Seattle.

Liu, Y., and U. R. Sumaila. 2010. Estimating pollution abatement costs of salmon aquaculture: a joint production approach. *Land Economics* 86(3): 569–584.

Lynn, J. 2010. WTO questions China's exports barriers. Reuters, Monday, May 31.

Naylor, R., and M. Burke. 2005. Aquaculture and ocean resources: raising tigers of the sea. *Annual Review of Environment and Resources* 30(1): 185–218.

Naylor, R. L., R. J. Goldburg, H. Mooney, M. Beveridge, J. Clay, C. Folke, N. Kautsky, J. Lubchenco, J. Primavera, and M. Williams. 1998. Nature's subsidies to shrimp and salmon farming. *Science* 282(5390): 883–884.

Naylor, R. L., R. J. Goldburg, J. H. Primavera, N. Kautsky, M. C. M. Beveridge, J. Clay, C. Folke, J. Lubchenco, H. Mooney, and M. Troell. 2000. Effect of aquaculture on world fish supplies. *Nature* 405(6790): 1017–1024.

Pauly, D. 2007. The *Sea Around Us* Project: documenting and communicating global fisheries impacts on marine ecosystems. *AMBIO: A Journal of the Human Environment* 36(4): 290–295.

Pauly, D., J. Alder, E. Bennett, V. Christensen, P. Tyedmers, and R. Watson. 2003. The future for fisheries. *Science* 302(5649): 1359–1361.

Pauly, D., V. Christensen, S. Guénette, T. J. Pitcher, U. R. Sumaila, C. J. Walters, R. Watson, and D. Zeller. 2002. Towards sustainability in world fisheries. *Nature* 418(6898): 689–695.

Pauly, D., P. Tyedmers, R. Froese, and L. Y. Liu. 2001. Fishing down and farming up the food web. *Conservation Biology in Practice* 2(4): 25.

Pauly, D., and D. Zeller. 2016. Catch reconstructions reveal that global marine fisheries catches are higher than reported and declining. *Nature Communications* 7: 10244.

Peterson, G. D., G. S. Cumming, and S. R. Carpenter. 2003. Scenario planning: a tool for conservation in an uncertain world. *The Journal of the Society for Conservation Biology* 17(2): 358–366.

Primavera, J. H. 2006. Overcoming the impacts of aquaculture on the coastal zone. *Ocean & Coastal Management* 49(9–10): 531–545.

Pullin, R. S. V., R. Froese, and D. Pauly. 2007. Indicators for the sustainability of aquaculture. Pp. 53–72 in T. M. Bert (ed.), *Ecological and Genetic Implications of Aquaculture Activities*. Kluwer Academic Publishers, Dordrecht, the Netherlands.

Pullin, R. S. V., H. Rosenthal, and J. Maclean (eds.). 1992. *Environment and aquaculture in developing countries*. Summary report of the Bellagio Conference on Environment and Aquaculture in Developing Countries, Bellagio, September 17–22, 1990. International Centre for Living Aquatic Resources Management Conference Proceedings 36.

Raskin, P. D. 2005. Global scenarios: background review for the Millennium Ecosystem Assessment. *Ecosystems* 8(2): 133–142.

Smith, J. M. 2006. Compliance bargaining in the WTO: Ecuador and the bananas dispute. Chapter 8, pp. 257–288 in J. S. Odell (ed.), *Negotiating Trade: Developing Countries in the WTO and NAFTA*. Cambridge University Press, Cambridge, England.

Stergiou, K. I., A. C. Tsikliras, and D. Pauly. 2008. Farming up Mediterranean food webs. *Conservation Biology* 23(1): 230–232.

Sumaila, U. R., and C. Walters. 2005. Intergenerational discounting: a new intuitive approach. *Ecological Economics* 52(2): 135–142.

Tacon, A. G. J., M. Metian, G. M. Turchini, and S. S. De Silva. 2010. Responsible aquaculture and trophic level implications to global fish supply. *Reviews in Fisheries Science* 18(1): 94–105.

Trujillo, P. 2007. *A global analysis of the sustainability of marine aquaculture*. Master of Science, Resource Management & Environmental Sciences Dissertation, University of British Columbia, Vancouver, Canada.

Trujillo, P. 2008. Using a mariculture sustainability index to rank countries' performance. Pp. 28–56 in J. Alder and D. Pauly (eds.), *A comparative assessment of biodiversity, fisheries and aquaculture in 53 countries' Exclusive Economic Zones*. Fisheries Centre Research Reports 16(7), University of British Columbia, Vancouver.

Tsirlikas, A., E. Mente, K. Stergiou, and D. Pauly. 2014. Shift in trophic level of Mediterranean mariculture species. *Conservation Biology* 28: 1124–1128.

UN. 2009. *World population prospects: the 2008 revision*. Population Division of the Department of Economic and Social Affairs of the United Nations Secretariat. Available at http://esa.un.org/unpp/.

UNEP. 2002. *Global Environmental Outlook (GEO-3)*. United Nations Environment Program Division of Early Warning and Assessment (DEWA), Nairobi.

UNEP. 2007. *Global Environmental Outlook: environment for development (GEO-4)*. United Nations Environment Program Division of Early Warning and Assessment (DEWA), Nairobi.

Volpe, J. P., M. Beck, V. Ethier, J. Gee, and A. Wison. 2010. *Global aquaculture performance index*. University of Victoria, Victoria, Canada.

Watson, R., and D. Pauly. 2001. Systematic distortions in world fisheries catch trends. *Nature* 414: 534–536.

Wijkstrom, U. N. 2003. Short- and long-term prospects for consumption of fish. *Veterinary Research Communications* 27(suppl. 1): 461–468.

Ye, Y. 1999. *Historical consumption and future demand for fish and fishery products: exploratory calculations for the years 2015/2030*. FAO Fisheries Circular. FAO, Rome.

NOTE

1. Cite as Campbell, B., J. Alder, P. Trujillo, and D. Pauly. 2016. A global analysis of mariculture production and its sustainability, 1950–2030. Pp. 137–151 in D. Pauly and D. Zeller (eds.), *Global Atlas of Marine Fisheries: A Critical Appraisal of Catches and Ecosystem Impacts*. Island Press, Washington, DC.

CHAPTER 13

POLLUTANTS IN THE SEAS AROUND US[1]

Shawn Booth,[a] William W. L. Cheung,[b] Andrea P. Coombs-Wallace,[a,c] Vicky W. Y. Lam,[a] Dirk Zeller,[a] Villy Christensen,[d] and Daniel Pauly[a]
[a]*Sea Around Us*, University of British Columbia, Vancouver, BC, Canada
[b]Changing Oceans Research Unit and NF-UBC Nereus Program, University of British Columbia, Vancouver, BC, Canada
[c]Frankfurt Zoological Society, Mpika, Zambia, and Imperial College, London
[d]Fisheries Centre, University of British Columbia, Vancouver, BC, Canada

The *Sea Around Us* was named after Rachel Carson's book of the same title (Carson 1951), and thus it is fitting that it should have undertaken various studies on the effects pollutants have on marine ecosystems, in part inspired by Carson's *Silent Spring* (Carson 1962). In this chapter, after a brief review of current issues in marine pollution, two vignettes are presented that examine pollutant dynamics in aquatic ecosystems: polychlorinated biphenyls (PCBs) in the northeast Pacific and methylmercury in the Faeroe Islands. Then, two global models are

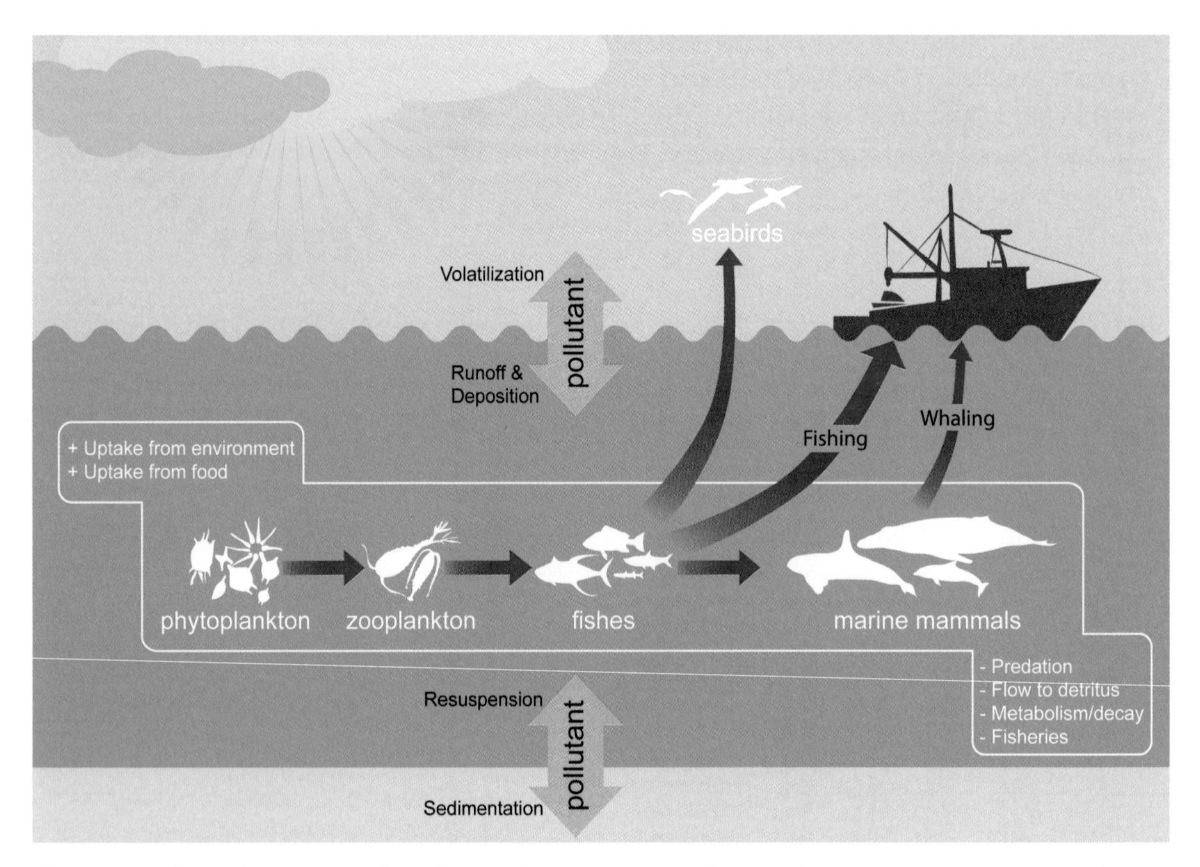

Figure 13.1. Schematic representation of how pollutants move within aquatic ecosystems; note that grazing and predation lead to bioaccumulation (see text). (Original figure by Ms. Evelyn Liu.)

BOX 13.1. THE ECOTRACER ROUTINE FOR TRACKING NUTRIENTS AND POLLUTANTS

Shawn Booth and Daniel Pauly, *Sea Around Us*, University of British Columbia, Vancouver, Canada

Ecotracer is a subroutine within the EwE modeling framework that allows the user to simulate the flow of pollutants (or nutrients such as nitrogen; see Kline and Pauly 1998) in an ecosystem by tracking the gains and losses in the functional groups and the environment. Ecotracer estimates pollutant concentrations in all functional groups, including those that may have no reported data available, and forcing functions can be used to simulate changes of pollutant inflow rates through time.

Four parameters describe the gains and losses to the environment: the initial concentration (g/km^2); the base inflow rate (g/km^2/year); the decay rate (per year), accounting for the pollutant being transformed to other substances; and the base volume exchange loss (per year). The initial environmental concentration is a snapshot affected by the base inflow rate, which can include atmospheric loading, runoff, volatilization, sedimentation, and resuspension.

Five parameters are used to describe the gains and losses in the biological groups: initial pollutant concentration per biomass (e.g., g/t), if known; direct absorption rates (e.g., grams of pollutant/tonne of biomass/environment concentration/year); decay rate (per year); assimilation as a proportion (0–1) of the exposure; and concentration of pollutant in immigrating biomass (e.g., g/t). One group must have a direct absorption rate in order for the pollutant to spread to other functional groups as determined from trophic interactions. If long-term equilibrium concentrations are sought, the initial pollutant concentration per biomass does not need to be known.

Pollutant concentrations in functional groups result from five processes: the predator–prey interactions as detailed in the EwE diet matrix; direct uptake from the environment for some organisms or diffusion across membranes; internal metabolism and decay processes, which transform the pollutant into other substances; migration effects; and fisheries, which can change the age structure of the population, resulting in changes in pollutant concentrations. Technical details are available in Christensen and Walters (2004).

REFERENCES

Christensen, V., and C. J. Walters. 2004. Ecopath with Ecosim: methods, capabilities and limitations. *Ecological Modelling* 172: 109–139.

Kline, T. C., and D. Pauly. 1998. Cross-validation of trophic level estimates from a mass-balance model of Prince William Sound using 15N/14N data. Pp. 693–702 in T. J. Quinn II, F. Funk, J. Heifetz, J. N. Ianelli, J. E. Powers, J. F. Schweigert, P. J. Sullivan, and C.-I. Zhang (eds.), *Proceedings of the International Symposium on Fishery Stock Assessment Models*. Alaska Sea Grant College Program Report No. 98-01.

presented, one on the bioaccumulation and concentration of dioxin in marine organisms as a result of atmospheric deposition, the other on the production, atmospheric transport, and deposition of dioxin on a global basis. Although pollution in marine waters has direct effects less often than in freshwaters, it often exacerbates the effects of other stressors, such as overfishing. Thus, the concepts, methods, and software presented herein may be useful to future students of marine pollution, especially in light of current concerns about the traceability and wholesomeness of seafood.

For a very long time it seemed true, especially for marine systems, that "the solution to pollution is dilution." However, biomagnification up the food web and bioaccumulation in long-lived organisms can effectively reverse the effect of dilution (figure 13.1), as can the sheer magnitude of the pollutant input. Thus, even the open oceans have now reached a stage where pollutants originating from various societal activities ranging from mining, manufacturing, and agriculture to consuming their products are reaching worrisome levels. Foremost is the thermal pollution and increased carbonic acid in the oceans, both the results of carbon dioxide emissions also responsible from global warming, which have profound effects on ocean life (chapter 8). Another kind of marine pollution is plastic pollution, much of it from plastic debris, some very small, which now contributes an increasing fraction of what fish (Moore 2008), marine turtles (Bugoni et al. 2001), and seabirds (chapter 10) mistakenly ingest instead of their natural prey.

The *Sea Around Us* used extensively, and further developed, ecosystem modeling tools, notably Ecopath with Ecosim (EwE; Pauly et al. 2000; Christensen and Walters 2004; chapter 9), which allow tracking of pollutants through and up a food web. This chapter thus briefly reviews the work of the *Sea Around Us* on pollutant tracking, including global models to simulate the oceanic dispersion and uptake of dioxin, whose global scale corresponds to the other issues in this volume.

Two vignettes are presented here that examine pollutant dynamics in aquatic ecosystems. The first vignette presents a model of bioaccumulation of PCBs in the eastern Bering Sea of Alaska (Coombs 2004), and the second deals with the bioaccumulation of mercury in the Faeroe Islands (Booth and Zeller 2005). The work on PCBs and mercury relied heavily on the Ecotracer routine of EwE (Christensen and Walters 2004; box 13.1).

We present two global models for dioxin, a group of chemicals that have seventeen forms with varying toxicological effects, with the most toxic form, 2,3,7,8-tetrachlorodibenzo-*p*-dioxin, being classified as a human carcinogen (IARC 1997). One examines the bioaccumulation and concentration of dioxin in marine organisms as a result of atmospheric deposition (Christensen and Booth 2006; Booth et al. 2013), and the other aims to simulate the production, atmospheric transport, and deposition of dioxin on a global basis. However, although the estimate of dioxin concentrations in marine organisms relied in part on the Ecotracer routine, the atmospheric transport model was developed outside the EwE framework.

PCBS IN ALASKA'S EASTERN BERING SEA

The eastern Bering Sea has undergone substantial ecological changes since 1950 that have affected the marine community's biomass and composition (NRC 1996). These changes have been attributed to the effects commercial whaling and fishing have had on the structure of the community (Trites et al. 1999) and to an oceanic regime shift in the late 1970s that favored the survival of one suite of species over another (Mantua et al. 1997). The area is also susceptible to depositions of pollutants that undergo long-range atmospheric transport, such as PCBs, because colder temperatures favor the entrapment of these pollutants in high-latitude areas.

PCBs are of concern because of their potential to decrease marine mammal populations in the eastern Bering Sea. Also, because many Alaskan native people consume marine mammals, they are also exposed to these toxins as part of their traditional diet. PCBs are lipophilic (fat soluble), resist degradation, persist in the environment, accumulate in marine mammal and fish tissue, are highly toxic, and can directly affect the health of animals and people. Toxic effects on marine mammals include immunosuppression, developmental abnormalities, carcinogenicity, endocrine disruption, reproductive impairment, neurotoxicity, skin disorders, tumors, and lipid degeneration (AMAP 1997; Kannan et al. 2000). Many of these effects are also observed in humans (Simmonds et al. 2002). Contaminant burdens may have greater impacts when people and animals are stressed, leading to compromised health, and in wild animals this may affect mortality levels indirectly (Loughlin and York 2000).

To gain a better understanding of the effects of these stressors, two ecosystem models were constructed using the EwE software that describe the eastern Bering Sea in the 1950s and the 1980s. Both model the accumulation of PCBs using the Ecotracer routine within EwE to assess the implications for marine mammals and people. In the Bering Sea, PCB levels rose steadily from the 1950s until the mid-1970s, when regulations restricting PCB use and production were implemented (Livingston and Low 1998), and this led to an initial reduction of PCB concentration in wildlife, but PCB concentrations stopped declining and have stabilized since the mid-1980s (Parker and Dasher 1999). This leveling off suggests that ongoing environmental cycling and continued leakage of PCBs are sufficient to maintain concentrations at a steady state (Aguilar et al. 2002).

The models describing the composition and flows of biomass in the ecosystem were compared with biomass trends taken from systematically collected assessment data in relation to fisheries and marine mammals (Trites et al. 1999). Similarly, concentration levels of PCBs were compared with reported concentrations. For marine mammals, simulated PCB concentrations in blubber were compared with reported blubber concentrations from male animals, because approximately 90% of their PCB load accumulates in blubber (Becker 2000), and male animals accumulate PCBs throughout their life, as compared to females, which can transfer some of their burden to their fetuses and also to calves when they are nursing. For other organisms, concentrations in muscle tissue were used because of its large contribution to body mass (Aguilar et al. 1999). Only total PCBs reported as the sum of individual congeners in a sample analyzed using more accurate, sensitive, and widely recognized gas chromatography methods were included (Valoppi et al. 1998).

The onset of significant PCB input occurred in sediment layers deposited after 1950 (Iwata et al. 1993). Accordingly, PCB concentrations predicted using the 1950s model depended on bottom-up processes. For the 1950s simulation, an environmental concentration of 9.2 g/km^2 (Kawano et al. 1986) and an inflow rate of 4.7 g/km^2/year (Iwata et al. 1994) were used to initialize the model. Time series data were also included in the 1950s model to account for changes in PCB input over time as a result of increasing global PCB production before the mid-1970s. Species and functional groups were considered to lack any PCBs at initialization so that the model estimated the bioaccumulation of PCBs from a starting point of zero. Plankton and large flatfish incorporated PCBs by direct uptake from the environment, and thus these groups were considered to be the trophic entry point of PCBs. These PCBs were then transferred to the remaining biological groups via trophic interactions. Many organisms have the ability to transform PCBs in their tissues, and thus, for most groups, a decay rate was used to match the predicted PCB concentrations to observed concentrations.

For the 1980s model, environmental concentrations were increased to 65 g/km^2 (Tysban 1999) and the inflow rate was increased to 6.3 g/km^2/year (Iwata et al. 1994) to reflect the increasing impact of PCBs on the ecosystem. Plankton groups were assigned uptake rates from the environment and, as in the 1950s model, most groups were assigned a decay rate. For both models, a simulation was run over a hundred-year time period to assess the movement of PCBs within the ecosystem.

Results from the 1950s model indicated that concentrations reached their peak in the mid-1980s. Concentrations then slowly declined in a delayed response to restrictions before reaching a steady state. The 1980s model was used to forecast future PCB concentrations in the eastern Bering Sea, and similar to the 1950s model, it also predicted that concentrations reached their peak in the mid-1980s, followed by a slow decline to a steady state.

PCB concentrations in Bering Sea seawater (0.00065 μg/L) and sediment (0.13 ppb, dry weight) are two orders of magnitude lower than guidelines set by the U.S. Environmental Protection Agency (EPA) and the National Oceanic and Atmospheric Administration, respectively. The leveling off of PCB concentrations from the 1980s model suggests that ongoing environmental cycling and continued input of PCBs into the Bering Sea is sufficient to maintain concentrations at current levels. Both models suggest that, with the exception of toothed whales, concentrations in marine mammals remained between 23% (beaked

Table 13.1. Estimated trophic level, highest concentration of PCB reached during the 1950s simulation, and the percentage difference between the range of the lower (7 ppm) and upper (15 ppm) threshold values established by Kannan et al. (2000) for the marine mammal groups modeled in the eastern Bering Sea model. Toothed whales are the only group that falls within immunosuppression threshold range.

Group	Trophic level	Concentration (ppm)	± Lower	± Upper
Sperm whales	4.75	1.9	−72.9	−87.3
Beaked whales	4.59	5.4	−22.9	−64.0
Toothed whales	4.36	11.3	+61.4	−24.7
Steller sea lions	4.30	1.9	−72.9	−87.3
Seals	4.01	1.2	−82.9	−92.0
Baleen whales	3.61	1.6	−77.1	−89.3
Walrus or bearded seals	3.52	0.6	−91.4	−96.0

whales) and 90% (walrus) lower than threshold levels throughout the time period from the 1950s to the present (table 13.1). However, the functional group "toothed whales" was within the 7 to 15 mg/kg wet weight threshold level established by Kannan et al. (2000).

Animals taken for subsistence from the eastern Bering Sea ecosystem are primarily from trophic levels 3 and 4. The results suggested that the greatest flow and concentrations of PCBs also occur at these trophic levels, with predicted concentrations in marine mammals and fish ranging from 0.005 to 2.15 ppm. Given that marine mammals and fish constitute a large proportion of the traditional Aleut diet, these concentrations, being well above EPA guidelines, represent a considerable contaminant risk for subsistence consumers. Perhaps the greatest contaminant concern for Alaskan natives is that PCB concentrations in their traditional foods are no longer declining, and the full extent of long-term chronic exposure for marine mammals and humans has yet to be realized (Carson 1962; Tanabe 2002).

MERCURY IN THE FAEROE ISLANDS

The Faeroe Islands are located in the North Atlantic Ocean southeast of Iceland, and their inhabitants have always relied heavily on marine resources for their livelihood and food security. In the early 2000s, fisheries accounted for 44.5% of gross domestic product (Zeller and Reinert 2004), and its traditional pilot whale hunt provides approximately 30% of total meat consumed on the islands (Faroe Government 2004). Although the whale hunts are important in terms of cultural identity and food security, there is also concern that the contaminant burden in whales could be detrimental to human health.

In the 1990s, it was found that children who were exposed prenatally to elevated levels of methylmercury suffered cognitive impairment (Grandjean et al. 1992, 1997). Subsequent studies also provided evidence of attenuated postnatal growth due to the transfer of methylmercury from mother to child during breastfeeding. In response to these findings, pregnant Faeroese women were reported to have decreased the amount of whale meat and fish in their diet (Weihe et al. 2003). Methylmercury exposure levels were assessed for the general population and for pregnant women, and these exposure levels were compared with the tolerable weekly intake levels advised by both the World Health Organization (WHO) and the EPA. Also of concern is the impact climate change may have on the accumulation of methylmercury: Increased temperatures lead to increased methylation rates, making increased amounts of methylmercury available. Therefore, increasing exposure levels would be expected to be found in human populations relying on marine resources as part of their diet.

Using EwE, an ecosystem model was constructed for the Faeroe Islands EEZ, and the flow of methylmercury was described via the Ecotracer routine (Booth and Zeller 2005).

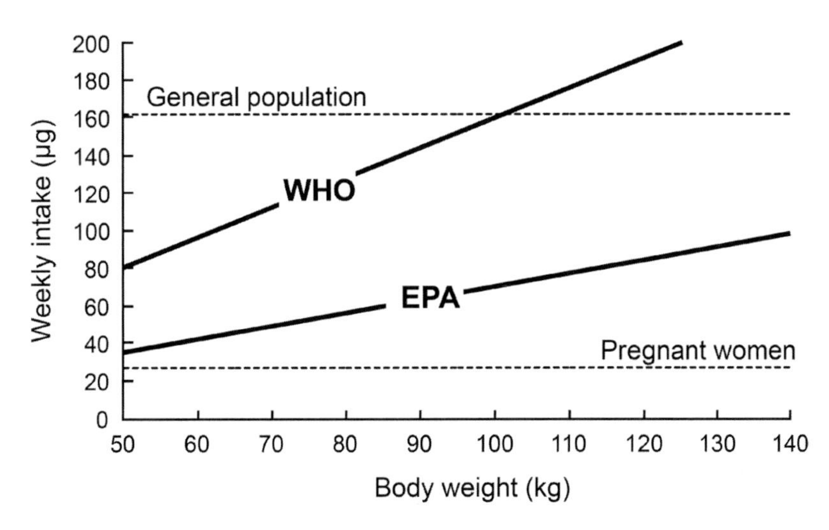

Figure 13.2. Weekly intake of methylmercury for the general population in the Faeroes and for pregnant women who lowered their intake of pilot whale and cod, the two marine species dominant in the traditional diet. People who weigh less than approximately 100 kg are above the WHO and EPA TWI limits, and pregnant women who lowered their weekly intake of pilot whale and cod are below the WHO and EPA TWI limits, irrespective of body weight. (Modified from Booth and Zeller 2005.)

The human exposure to methylmercury via consumption of cod and pilot whale meat was examined, the two main marine species providing most protein in the traditional diet. The environmental concentration was increased from 3.3 to 3.6 g/km² over 100 years to approximate the increase in the amount of methylmercury in the water column since the onset of the Industrial Revolution (Mason and Sheu 2002). The base inflow rate was considered to be only from atmospheric deposition and a value of 0.113 g/km²/year was used (Downs et al. 1998). In the first hundred years of the simulation, biotic groups and the environment were allowed to equilibrate to "present" conditions so that the predicted concentrations from the model simulation fell within the range of reported values. To assess the impacts of changes in fishing mortality rates and the effects of increasing methylation rates as a result of climate change on the ecosystem, a second simulation was undertaken from the present to 100 years in the future. We investigated the results of ocean temperature changes of 0.4°C and 1.0°C per century, the range expected by the International Panel on Climate Change (IPCC 2001). Changes in fishing and whaling mortality rates of 20% were also simulated on all targeted species, and the combined effects of climate change and changes in mortality rates were also investigated.

Methylmercury exposure levels based on the consumption of 12 g of pilot whale meat per week and 72 g of cod per week for the general population (Vestergaard and Zachariassen 1987) were compared with the tolerable weekly intake (TWI) limits set by the WHO of 1.6 µg/kg body weight and to the more risk-averse EPA value of 0.7 µg/kg body weight.[2] Weekly intake levels were also assessed for pregnant women who lowered their consumption to 1.45 g of pilot whale meat per week and 40.2 g of cod per week (Weihe et al. 2005). Model results indicate that women who changed their diets are below the limits set by the WHO and the EPA under both climate change scenarios and under the scenarios of changing fishing mortality rates. However, currently and under all scenarios the general population is above the methylmercury exposure levels set by the EPA, but for the WHO limits, the results depend on a person's body weight. Currently, people who weigh less than 102 kg and consume the reported average amount of pilot whale and cod are above the WHO limit, and under both climate change scenarios more people would be above the limit (figure 13.2).

Changes in the fishing and whaling mortality rates affected the concentration of methylmercury in the species and functional groups. Decreasing mortality rates on targeted species and functional groups led to an increase in biomass for the targeted species, but the prey items of these groups in the simulation showed a decrease in methylmercury concentrations. Decreasing the fishing and whaling mortality rates on targeted species, in the model simulation, leads to a decrease in the production/biomass (P/B) ratio, lowering the turnover rate of the population, and has the effect of extending longevity, allowing a longer time for accumulation to occur in the targeted species. However, simulating an increase in biomass for the targeted species also has the effect of decreasing the longevity of targeted species' prey items, thus lowering methylmercury concentrations in prey groups. Increasing mortality rates on targeted species had the opposite effect: lowered methylmercury concentrations in targeted species and groups, with consequential increases in concentrations of methylmercury in their prey items.

GLOBAL MODELS

Bioaccumulation of Dioxin

A global model involving spatial data was developed to describe the movement of dioxin up through the marine food web of the global oceans using the Ecospace and Ecotracer routines of the EwE modeling routine (Christensen and Booth 2006). It uses input data from a preliminary spatial dioxin ocean loading model as a result of atmospheric depositions (Zeller et al. 2006). Studies looking at the effects of dioxin in marine ecosystems are lacking, with most studies examining only concentrations in species of concern.

The underlying Ecopath model includes forty-two functional groups and is based on a modified model (Generic 37) developed for database-driven model construction distributed with the EwE software. The ecosystem is represented by a grid of 2° latitude × 2° longitude cells and extends from the equator to 70° latitude north and south. The oceans are divided into two depth zones (<200 m, >200 m), and most functional groups were assigned to both zones, except for small demersal fishes, reef fishes, seals, corals, and benthic plants, which were assigned to the shallower depth zone. Functional groups were assigned to all nineteen FAO statistical areas, and the groups were assigned primarily on estimates of primary production.

The modeling approach involved using predation, catches, and an assumed ecotrophic efficiency to estimate the biomass of each functional group. Default values for the Generic 37 model were maintained, but density levels were assigned for large sharks (0.1 t/km^2), jellyfish (0.1 t/km^2), seals and other pinnipeds (0.003 t/km^2), toothed whales (0.002 t/km^2), baleen whales (0.001 t/km^2), seabirds (0.001 t/km^2), macrobenthos and meiobenthos (1.5 and 2 t/km^2, respectively), corals (1 t/km^2), soft corals and sponges (2 t/km^2), and benthic plants (10 t/km^2). Catch data for each functional group were taken from the *Sea Around Us* and represent the catch taken in 2000. Spatially explicit data used in the Ecospace routine include primary production, biomass estimates for zooplankton, macrobenthos and meiobenthos, small and large mesopelagic fishes, and depth information for each half-degree spatial *Sea Around Us* cell.

Concentrations of dioxin in marine organisms were sourced mainly from the primary literature and represent reported values since 1990 in toxic equivalencies (TEQs). Seventeen congeners of dioxin have been reported as being toxic, and data reported as individual congeners were transformed into TEQs using the appropriate toxic equivalency factors (TEFs; Van den Berg et al. 1998). Concentrations in marine mammals were standardized to ng/kg lipid weight, and those for other organisms were standardized to µg/kg wet weight. Species with reported values were sorted into their respective functional groups and were placed within their representative region.

In the oceans, direct uptake rates of dioxin by primary producers and invertebrates are an important pathway for the transfer of pollutants up the food web. However, because of the

Table 13.2. Starting values used in the Ecotracer routine for the environment and biota.

Ecotracer inputs	Value
Initial concentration	0.1 t/km²
Base inflow rate	1.0 t/km²/year
Decay rate	1.0/year
Phytoplankton uptake rate	0.5 t dioxin/t in environment/t phytoplankton/year

lack of data on uptake rates, we assume that the dioxin is taken up only by phytoplankton once deposited to the ocean, and to prevent the accumulation of dioxin in the oceans, a decay rate was used for the environment (table 13.2). Thus, concentrations in the biota are a result of uptake of dioxin by phytoplankton and trophic transfer through the food web with no decay in biota. Under these initial conditions, the simulation was run for 22 years, and the results from the model were compared with the reported concentration values.

After the simulation was run, most groups reached equilibrium dioxin concentrations,

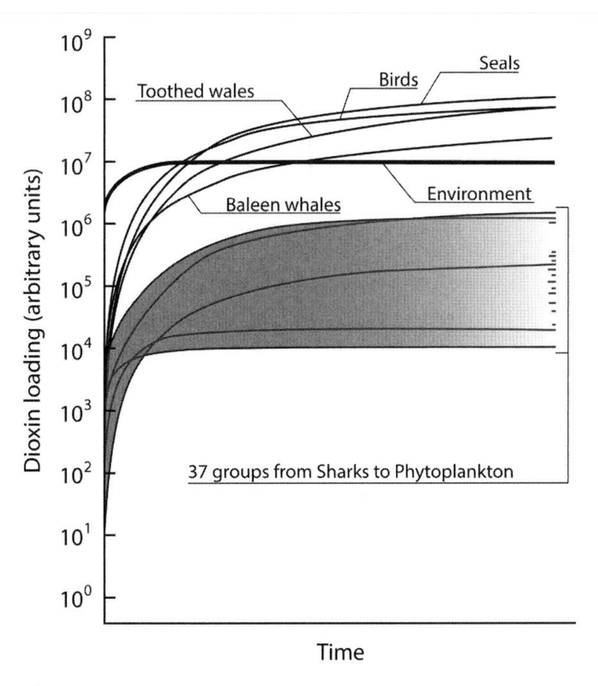

Figure 13.3. Predicted concentrations of dioxin in the global marine ecosystem model for each functional group, listed in order of highest to lowest values (top to bottom). At the end of the 22-year simulation, equilibrium had been reached in most functional groups, excluding whales (toothed and baleen), seals, and birds. (Adapted from Christensen and Booth 2006.)

excluding the whales (baleen and toothed), seals, and bird groups, which are long-lived (figure 13.3).

Excluding four outliers, the regression between predicted and observed dioxin concentrations explains 25% of the variation in the sample values ($p << 0.001$; figure 13.4) with a slope value of 0.84. Of the four outliers, two were associated with the polar regions, and the other two were associated with coastal areas in Asia. Predicted values for polar regions may be affected by both the duration of the atmospheric model (one year) and nonconsideration of reemission of dioxin from land back to the atmosphere caused by the "grasshopper effect" (Wania and Mackay 1996). Values for the coastal regions in Asia may also be affected by the input from coastal runoff and riverine outflow. The grasshopper effect and nonatmospheric inputs are not included in the preliminary atmospheric transport and deposition model of dioxin (but they were considered in the deposition models further below).

An important outcome of using a global EwE model with the Ecospace routine is that Ecotracer can predict the observed values of dioxin, although only within two orders of magnitude. The predictive power could be improved by having larger sample sizes for underrepresented areas. Because most concentration values are reported from coastal areas in developed countries (figure 13.5), samples from depths greater than 200 m and from developing areas could lead to a better fit. Improved fits may also be achieved by improving the atmospheric transport and deposition model to include the grasshopper effect, coastal runoff, and riverine inputs of dioxin, and these effects were considered in the updated atmospheric model for dioxin (see below).

Global Deposition of Atmospherically Released Dioxin

As a follow-up to the modeling work reported above, a global model of dioxin was developed that includes production, atmospheric diffusion and dispersion, transport from land to

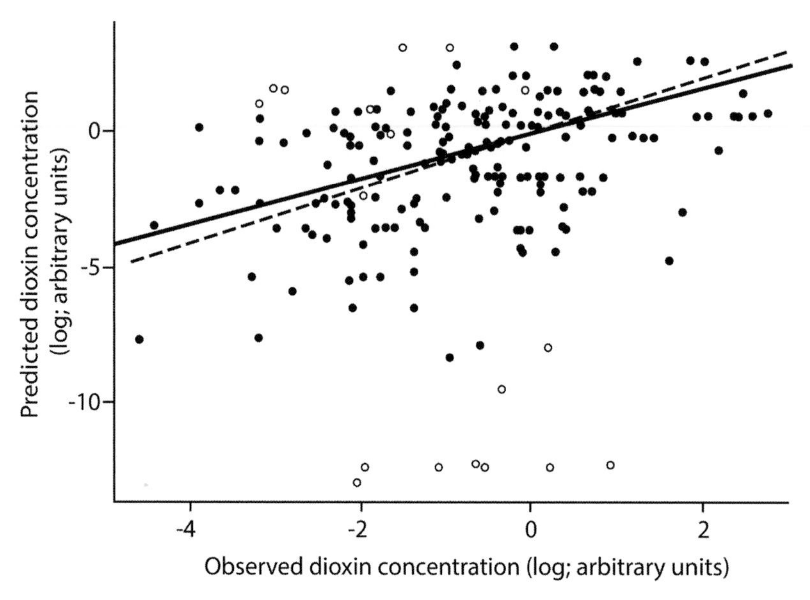

Figure 13.4. Predicted versus observed dioxin concentrations in various marine organisms (see text). The dashed line represents the case where the predicted values match the observed values, and the solid line is a fitted regression ($r^2 = 0.25$; $p < 1.7e^{-13}$). Open symbols are from 4 sample areas considered to be outliers and not included in the regression. (Adapted from Christensen and Booth 2006.)

coastal waters, and depositions to land and oceans (Booth et al. 2013). Its purpose was to highlight the deposition of dioxin in marine areas including countries' EEZs and the high seas to improve models such as that presented previously on the bioaccumulation of dioxin in marine ecosystems. Dioxin concentrations measured in marine organisms result from the input of dioxin to the oceans, but dioxin is not measured in the water column, and thus organism concentrations and sediment concentrations have served as a proxy to indicate areas that are more affected by dioxin than others. Monitoring programs for dioxin concentrations in organisms are sparse, and developing countries lack the resources to properly monitor the impacts. Thus, this model, of which only an outline is presented

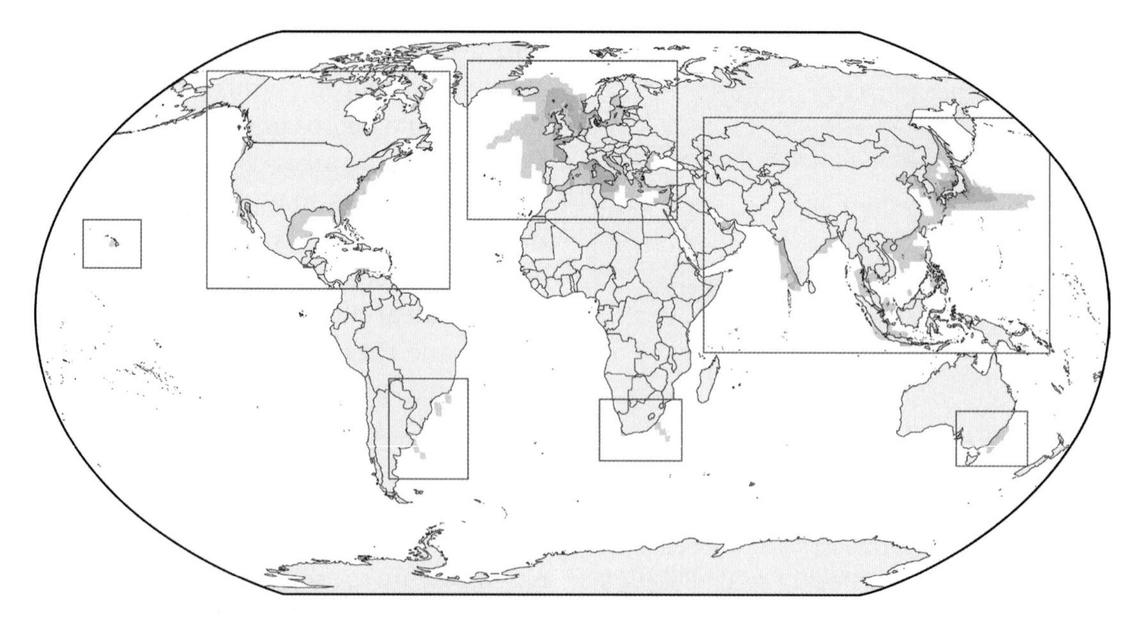

Figure 13.5. Predicted relative dioxin concentration in small pelagic fishes. Scale is from white to light blue to green, yellow, and orange. Note high concentrations in the waters around Asia, Europe, and North America, and lower concentrations in South America, Africa, and Australia. (Adapted from Christensen and Booth 2006.)

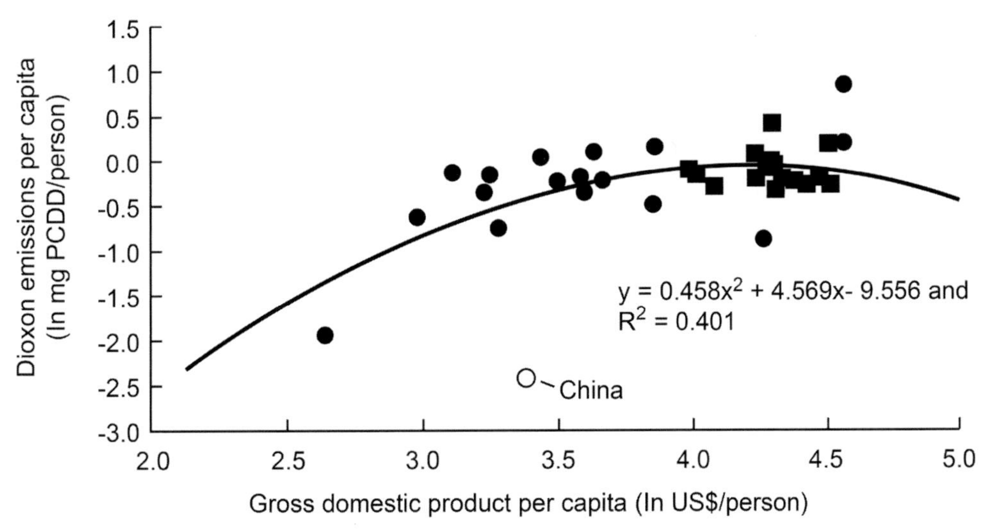

Figure 13.6. Environmental Kuznets curve used to estimate countries' per capita dioxin emissions. Original data used in Baker and Hites (2000) are shown in square symbols, new data are shown in circles, with China (open circle) omitted from analysis. (Adapted from Booth et al. 2013.)

here (but see Booth et al. 2013) identifies areas of potential concern, that is, where airborne dioxins are likely to be deposited.

A previous global mass balance of dioxin estimated global annual emissions of 13,100 kg ± 200 kg (Brzuzy and Hites 1996), which assumed that annual depositions to oceans contributed 5% to the global mass balance. In a preliminary run of our model we found that ocean depositions were approximately 38% and therefore increase global emissions to 17,226 kg. Mass balance studies have shown that depositions of dioxin are about ten times greater than reported emissions because of the formation of dioxin from pentachlorophenol, a common wood preservative, by photochemical transformation in the atmosphere (Baker and Hites 2000). Thus, each emission value was multiplied by 9.7 to account for this discrepancy, and the estimated annual emissions were considered to be released in weekly increments.

The production of dioxin for thirty-five countries was based on their reported inventories of annual atmospheric releases, with most of these countries having a single estimate between 1995 and 2002 (see Supplementary Online Material in Booth et al. 2013). The year 1998 was chosen as representative for the emission inventories, corresponding gross domestic product (GDP) data, and population data (see Chen 2004 for an example). Population data for each of the thirty-five countries were used to transform GDP and atmospheric releases of dioxin into per capita rates. The line of best fit through these points, representing an environmental Kuznets curve, was used to generate the atmospheric dioxin emissions for countries that have not completed a dioxin inventory (figure 13.6).

Within each country, we assigned dioxin emissions using spatial estimates of GDP. A global data set of spatialized estimates of GDP (Dilley et al. 2005) were mapped onto a global grid of half-degree cells, and dioxin emissions were then made directly proportional to the fraction that each land cell contributed to the country's total GDP per land area (i.e., GDP/km^2).

Dioxin dispersion in the atmosphere involved diffusion and the transport of dioxin with wind. Weekly releases of dioxin to the atmosphere were subjected to diffusion, using the diffusion constant for 2,3,7,8-TCDD dioxin of 4.86×10^{-6} m^2/s (Chiao et al. 1994), and to global wind patterns. Wind data consisted of global daily means of east–west and north–south wind components from the 40-year reanalysis dataset of the European Center for Medium-Range Weather Forecasts (2006). These data were further averaged over the 1991–2000 time period into weekly values.

The deposition of dioxin was simulated using the data in table 13.2 and the characteristic travel distance approach (CTD), which describes the distance an airborne semivolatile organic pollutant travels before reaching 1/*e* (i.e., ~37%) of its initial value (Bennett et al. 1998). The temperature-dependent CTD, which accounts for the grasshopper effect, for 2,3,7,8-TCDD dioxin was used to estimate the amount of dioxin deposited from the atmosphere to land and water (Beyer et al. 2000). Because the distance traveled depends on temperature and wind speeds, we used the temperature-dependent CTD for 2,3,7,8-TCDD dioxin at 5°C, 15°C, and 25°C (Klasmeier et al. 2004) to derive a temperature-dependent CTD for temperatures greater than 0°C.

The CTD is described by wind speed and the effective decay rate,

$$CTD = \mu / k_{eff}, \qquad (13.1)$$

where μ is the wind speed (m/s) and k_{eff} (per second) is the effective decay rate. The effective decay rate accounts for the transfer of dioxin from air to land and water surfaces and to biological components (e.g., plants). Because each geographic information system (GIS) cell has the wind speed and temperature as an attribute, a temperature-dependent CTD is defined, and therefore we rearrange the equation to solve for the effective decay rate for each GIS cell,

$$k_{eff} = CTD / \mu . \qquad (13.2)$$

Water basin transport of dioxin from land to coastal marine areas was also simulated using water runoff amounts as the driver. Data kindly supplied by Dr. C. J. Vörösmarty of the Water Systems Analysis Group at the University of New Hampshire (www.wsag.unh.edu) describe a global total of 6,031 basins, with 5,865 basins identified as ultimately exiting to coastal marine waters and 166 identified as landlocked. To identify coastal cells that receive dioxin via water basin transport, salinity gradient plots were used to determine whether a basin's outflow created a freshwater plume in marine waters. For discharge areas that had plumes, dioxin was deposited into the cells that created the plume; for river discharge areas that did not have identifiable plumes, we specified a single central coastal cell to receive the dioxin. For regions between 0° and 65° latitude N or S, we used a salinity threshold of 30 psu to identify freshwater plumes, whereas for regions located above 65° latitude N or S, we used a salinity threshold of 25 psu because of the large inputs of freshwater that remain in surface waters in polar areas. Salinity data were taken from the World Ocean Atlas 2005 (Antonov et al. 2006).

Water runoff was shown to be the dominant pathway of dioxin transport in a Japanese watershed (Kanematsu et al. 2009), accounting for more than 98% of total dioxin transport. We used a proportionality constant of 0.0004/year, derived from two studies examining dioxin transport (Vasquez et al. 2004; Kanematsu et al. 2009), in combination with water runoff data for each water basin (Fekete et al. 2000) to transport dioxin from water basins to coastal marine cells.

We simulated the dispersion and deposition of airborne dioxin by a two-dimensional diffusion–advection differential equation:

$$\frac{\partial A}{\partial t} = \frac{\partial}{\partial x}\left(D\frac{\partial A}{\partial x}\right) + \frac{\partial}{\partial y}\left(D\frac{\partial A}{\partial y}\right) - \frac{\partial}{\partial x}(\mu * A) - \frac{\partial}{\partial x}(v * A) - \lambda * A, \qquad (13.3)$$

where *A* is the amount of dioxin in a grid cell, D is the diffusion coefficient (4.86×10^{-6} m²/s), μ and v are the wind velocity components (in the N–S direction and E–W direction, respectively), and lambda (λ) is the decay rate (i.e., k_{eff}).

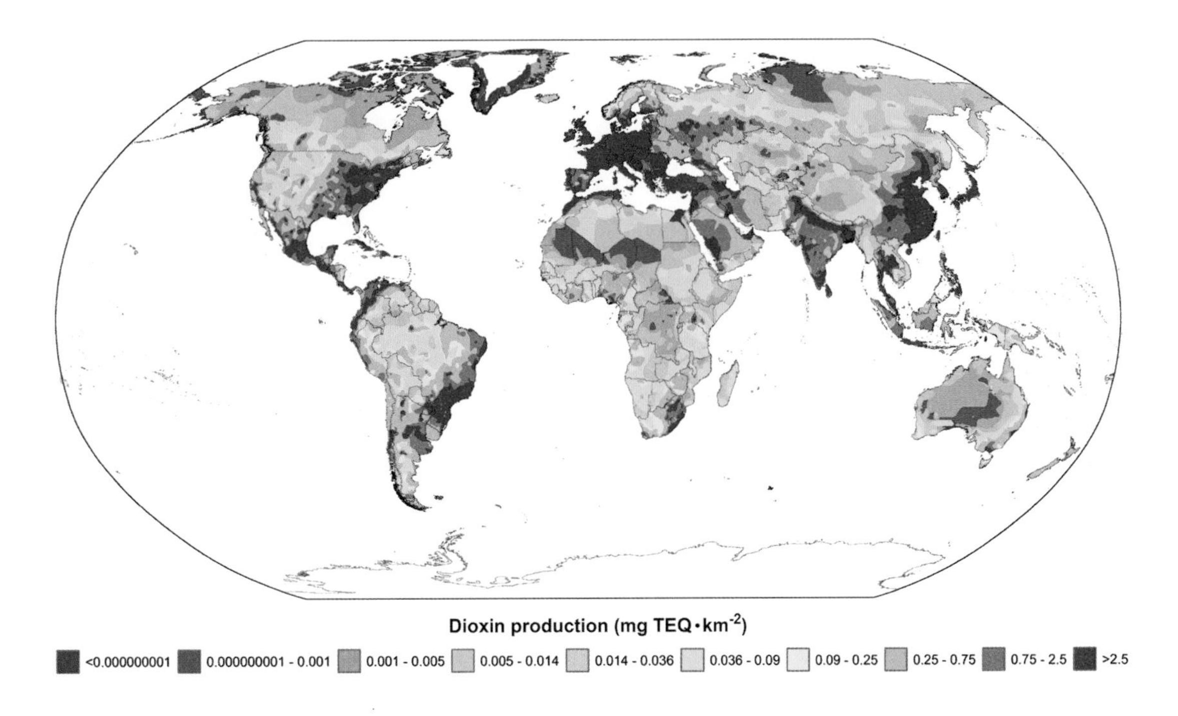

Figure 13.7. Global production of dioxin as toxic equivalents of 2,3,7,8-tetrachlorodibenzo-*p*-dioxin spatialized over the earth's surface, with emissions based on an environmental Kuznets curve. (Adapted from Booth et al. 2013.)

To numerically estimate the spatial and temporal amount of airborne dioxin above each cell, a finite difference technique using the alternating direction approach was used (Sibert and Fournier 1991). Briefly stated, the alternating direction approach requires that the amount of dioxin above each cell is determined by splitting each time step in half, determining the amount of dioxin at the end of the first half time step in a row-by-row fashion, and then determining the amount of dioxin at the end of each second half time step by column. Thus, we divided our 30-s time steps into two 15-s time steps and solved for the E-W and then the N-S movement. These two half time step processes were repeated for each time step in a week. To maintain mass balance, we redistributed any losses in the cumulative amount of dioxin in the system at the end of each time step, based on each cell's proportion of the overall total. The entire computation was repeated for each week of the year.

At the beginning of each week, each grid cell received its GDP-based share of global dioxin production, which was added to the amount that remained airborne at the end of the previous week. A circular boundary condition was also applied at all four edges of the global cell grid. In effect, this reconnected the cells in the rightmost column (i.e., 180°E longitude) to the cells of the same latitudes in the leftmost column (i.e., 180°W longitude). The application of the circular boundary condition also meant that the cells in the top row (i.e., 90°N latitude) were reconnected to those in the same row with a longitude difference of 180°. Similar reconnection of cells in the bottom row (i.e., 90°S latitude) also occurred.

For each cell, the amount of dioxin deposited to the earth's surface within each time step was determined as

$$\text{Dioxin deposited} = A * (1 - \exp[-(\omega/\text{CTD})(\text{ts})]), \quad (13.4)$$

where A is the amount of airborne dioxin above each cell, ω is the wind speed, CTD is the temperature-dependent characteristic travel distance, and ts is the time step (30 s).

The computer simulation of dioxin production suggested several areas of high local production of dioxin due to higher levels of

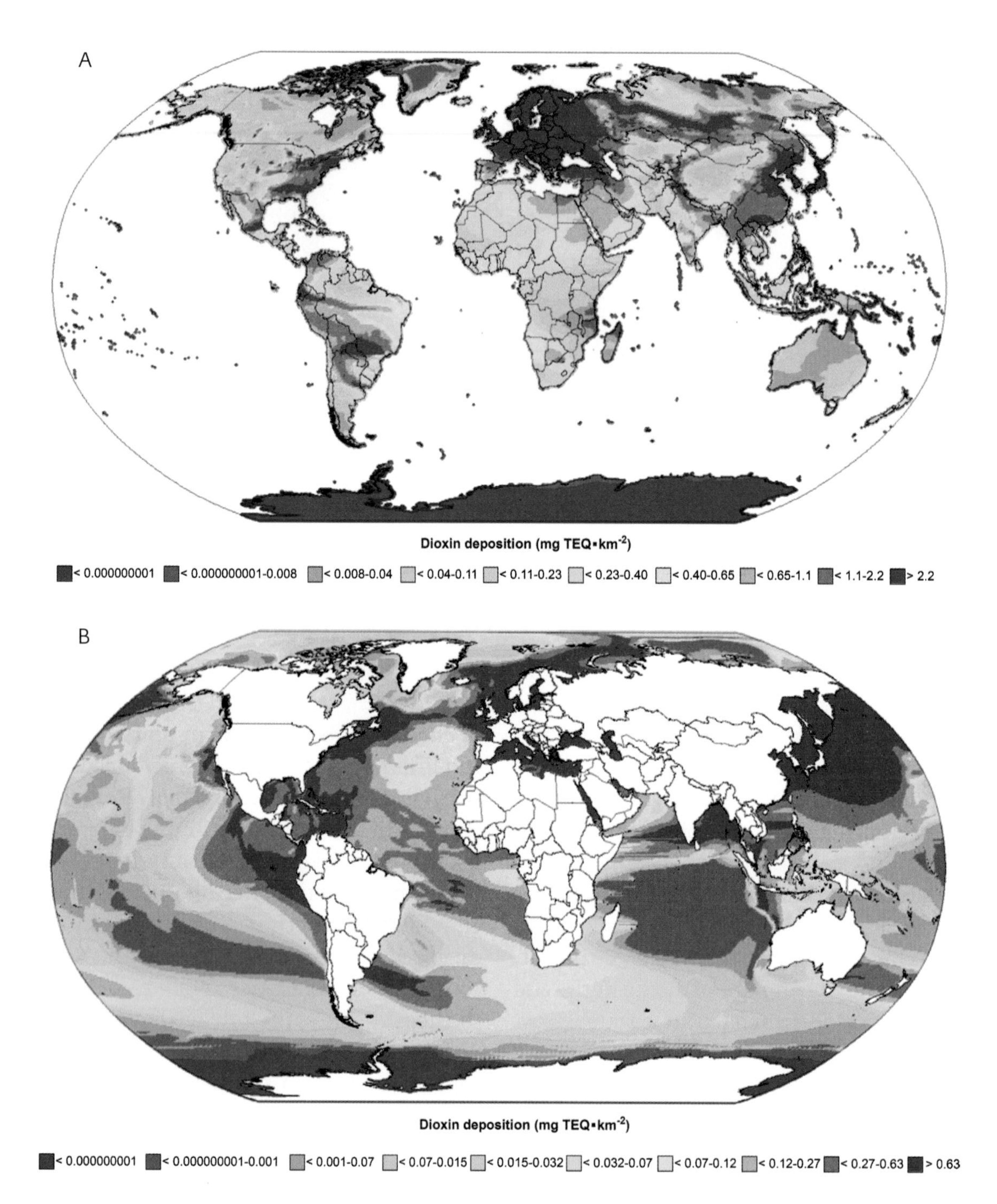

Figure 13.8. Deposits of dioxin to land (A) and to the world's oceans (B) presented as toxic equivalents of 2,3,7,8-tetrachlorodibenzo-*p*-dioxin after simulating 1 year of transport processes of global atmospheric emissions. (Modified from Booth et al. 2013.)

economic activity (figure 13.7). These were dominated by eastern North America, Europe, South Asia (particularly the Indian subcontinent), and East Asia (China, Japan, and South Korea). Countries belonging to the G20 account for more than 80% of the estimated annual emissions, with Japan, the United States, and China accounting for 30% of annual global emissions. However, it is smaller states such as Singapore and Malta that have the highest emissions per area.

After we ran the model to simulate 1 year's production, dispersion, deposition, and transport of dioxin, approximately 9 kg-TEQ (3%) of the annual dioxin production remained in the atmosphere. The model predicted that most of the annual production of dioxin, 163 kg-TEQ (57%), was deposited to land areas, and ocean waters received approximately 115 kg-TEQ (40%). Large parts of North America, most of central, northern, and Eastern Europe, as well as much of the Indian subcontinent and East Asia have high terrestrial depositions of dioxins (figure 13.8A). Dioxin depositions to land range from 1×10^{-8} mg-TEQ/km^2 to 146 mg-TEQ/km^2, with the lower values in the Antarctic and the highest values found in Europe and South Korea.

The model also suggested that ocean areas near the source emission areas also received high dioxin loads. These include the northeast and northwest Atlantic, Caribbean, Mediterranean, northern Indian Ocean, and large parts of the northwestern Pacific and South China Seas (figure 13.8B). However, several areas of low concentration of dioxins were also identified, specifically parts of the west coast of South America and northern parts of the west coast of North America. Marine deposited dioxin ranged from 1×10^{-8} mg-TEQ/km^2 to 33.5 mg-TEQ/km^2 and were similar to terrestrial deposits in that lower values were associated with the Antarctic, but the highest values were found in waters off Japan and South Korea. High dioxin depositions were also found in the marine waters of countries around Baltic and Mediterranean Seas.

Dioxin deposited to the oceans results from the production of dioxin on land, and thus the oceans can act as a sink for dioxin. The high seas receive the largest amount of dioxin (~36 kg-TEQ) as modeled here, and once standardized by area, depositions to the high seas are approximately 0.16 mg-TEQ/km^2. The most affected countries when comparing the ratio of deposits to emissions were found in Africa and Asia. Of the top twenty affected countries, eleven are located in Africa and six in Asia. The eleven African countries' per capita GDPs average less than US$250/person/year, and the Asian countries average less than US$450/person/year.

This work provides the opportunity to examine the impacts of dioxin on marine ecosystems. Coastal shelves provide most of the fish destined for human consumption (Pauly et al. 2002), and some coastal ecosystems (e.g., eastern North America, China, and Europe) receive much larger dioxin loads than other marine areas (e.g., most of South America and Australia). Past research has shown that dioxin levels in fish oils derived from forage fish around Europe and eastern North America have higher concentrations than those sourced from Peru (FAO 2002; Hites et al. 2004). We would expect a similar relationship for all ecosystem components, with higher concentrations found in places that receive higher inputs of dioxin.

Our model suggests that the oceans are strongly affected by dioxin. Previously, it was assumed that the ocean received only about 5% of the global annual production of dioxin (Baker and Hites 2000). Using spatial and temporal distributions of dioxin emissions in a kinematic model, we have shown that the oceans receive approximately 40% of the annual deposits. Although much of this is confined to coastal areas, the impacts on the high seas are not negligible and have consequences for food security. One concern is that dioxin is more likely to partition to plastic particles that are eaten by marine plankton and fishes, an entry point for accumulation (Rios et al. 2010). Thus, human populations with high seafood consumption levels may be exposed to higher levels of dioxin than previously thought.

This model does not account for direct (i.e., nonatmospheric) releases of dioxin to land or water, and this contributes to the lack of mass balance between emissions and depositions for dioxin. However, using the Kuznets curve for dioxin emissions lowers the mass balance discrepancy factor from 12.5 to 9.7. Polar regions receive little dioxin over the model simulation time of 1 year, but these areas may be affected by the accumulation of this toxin over longer time periods. Simulations that were run for a longer duration would show higher levels of dioxin in polar regions as dioxin migrated poleward as a result of the grasshopper effect.

CONCLUSIONS

The concepts, approaches, and results presented in this chapter illustrate that marine pollution studies, which are often performed in isolation from other ecosystem studies, can profitably be conducted with modifications of standard tools used by marine and fisheries scientists (e.g., the EwE and EcoTroph approach and software briefly described in chapter 9). If this is done systematically, it may lead to studies of ecosystem functioning and on the impact of fisheries, also providing results on the effects of persistent organic and other pollutants on ecosystems, biodiversity, fisheries, and hence public health. Integrative studies of this kind, which, as illustrated here, can be performed on a global basis, would thus greatly contribute to providing policy-relevant results.

ACKNOWLEDGMENTS

This work was undertaken as part of the *Sea Around Us*, a research activity at the University of British Columbia initiated and funded by the Pew Charitable Trusts from 1999 to 2014 and currently funded mainly by the Paul G. Allen Family Foundation. We thank Douw Steyn, Xu Wei, and Jordan Dawe of the Department of Earth and Ocean Sciences at the University of British Columbia and Zoreida Alajado, Sherman Lai, and Joe Hui, formerly with the *Sea Around Us*, for their input and advice. We also thank Dr. Charles Vörösmarty of the Water Systems Analysis Group at the University of New Hampshire for global data on water basins.

REFERENCES

Aguilar, A., A. Borrell, and T. Pastor. 1999. Biological factors affecting variability of persistent pollutant levels in cetaceans. *Journal of Cetacean Research and Management* (Special Issue 1): 83–116.

Aguilar, A., A. Borrell, and P. J. H. Reijnders. 2002. Geographical and temporal variation in levels of organochlorine contaminants in marine mammals. *Marine Environmental Research* 53(5): 425–452.

AMAP. 1997. *Arctic pollution issues: a state of the Arctic environment report*. Arctic Monitoring and Assessment Programme (AMAAP), Oslo, Norway.

Antonov, J. I., R. A. Locarnini, T. P. Boyer, A. V. Mishonov, and H. E. Garcia. 2006. World ocean atlas 2005, volume 2: Salinity. Levitus, S. (ed). *NOAA Atlas NESDIS* 62, U.S. Government Printing Office, Washington, DC. Available at www.nodc.noaa.gov. Accessed December 2007.

Baker, J. I., and R. A. Hites. 2000. Is combustion the major source of polychlorinated dibenzo-*p*-dioxins and dibenzofurans to the environment? A mass balance investigation. *Environmental Science and Technology* 34: 2879–2886.

Becker, P. R. 2000. Concentration of chlorinated hydrocarbons and heavy metals in Alaska arctic marine mammals. *Marine Pollution Bulletin* 40(10): 819–829.

Bennett, D. H., T. E. McKone, M. Matthies, and W. E. Kastenberg. 1998. General formulation of characteristic travel distance for semivolatile organic chemicals in a multimedia environment. *Environmental Science and Technology* 32: 4023–4030.

Beyer, A., D. Mackay, M. Matthies, F. Wania, and E. Webster. 2000. Assessing long-range transport potential of persistent organic pollutants. *Environmental Science and Technology* 30: 1797–1804.

Booth, S., J. Hui, Z. Alojado, V. Lam, W. W. L. Cheung, D. Zeller, D. Steyn, and D. Pauly. 2013. Global deposition of airborne dioxin. *Marine Pollution Bulletin* 75(1–2): 182–186.

Booth, S., and D. Zeller. 2005. Mercury, food webs, and marine mammals: implications of diet and climate change for human health. *Environmental Health Perspectives* 113: 521–526.

Brzuzy, L. P., and R. A. Hites. 1996. Global mass balance for polychlorinated dibenzo-*p*-dioxins and dibenzofurans. *Environmental Science and Technology* 30(6): 1797–1804.

Bugoni, L., L. Krause, and M. V. Petry. 2001. Marine debris and human impacts on sea turtles in southern Brazil. *Marine Pollution Bulletin* 42(12): 1330–1334.

Carson, R. L. 1951. *The Sea Around Us*. Oxford University Press, Oxford, England.

Carson, R. L. 1962. *Silent Spring*. Houghton Mifflin Company, Boston.

Chen, C.-M. 2004. The emission inventory of PCDD/PCDF in Taiwan. *Chemosphere* 54: 1413–1420.

Chiao, F. F., R. C. Currie, and T. E. McKone. 1994. *Final draft report: intermediate transfer factors for contaminants found in hazardous waste sites: 2,3,7,8-tetrachlorodibenzo-*p*-dioxin (TCDD)*. Risk Science Program (RSP), Department of Environmental Toxicology, University of California, Davis, CA.

Christensen, V., and S. Booth. 2006. Ecosystem modeling of dioxin distribution patterns in the marine environment. Pp. 83–102 in J. Alder and D. Pauly (eds.), *On the multiple uses of forage fish: from ecosystems to markets*. Fisheries Centre Research Reports 14(3), University of British Columbia, Vancouver, Canada.

Christensen, V., and C. J. Walters. 2004. Ecopath with Ecosim: methods, capabilities and limitations. *Ecological Modelling* 172: 109–139.

Coombs, A. P. 2004. *Marine mammals and human health in the eastern Bering Sea: using an ecosystem-based food web model to track PCBs*. MSc thesis, University of British Columbia, Resource Management and Environmental Studies, Vancouver, Canada.

Dilley, M., R. S. Chen, U. Deichmann, A. L. Lerner-Lam, and M. Arnold. 2005. *Natural disaster hotspots: a global risk analysis*. CIESIN and World Bank (unpublished data), Washington, DC.

Downs, S. G., C. L. MacLeod, and J. N. Lester. 1998. Mercury in precipitation and its relation to bioaccumulation in fish: a literature review. *Water Air Soil Pollution* 108: 149–187.

European Centre for Medium-Range Weather Forecasts. 2006. ECMWF 40-year reanalysis dataset. Available at http://data.ecmwf.int/data/index.html. Accessed February 2010.

FAO 2002. *Use of fishmeal and fish oil in aquafeeds: further thoughts on the fishmeal trap*, by M. B. New and U. N. Wijkström. FAO Fisheries Circular No. 975, Rome.

Faroe Government. 2004. *Whales and whaling in the Faroe Islands*. Available at: www.whaling.fo/index.htm (accessed June 2004).

Fekete, B., C. J. Vörösmarty, and W. Grabs. 2000. *Global, composite runoff fields based on observed river discharge and simulated water balances*. Available at www.grdc.sr.unh.edu. Accessed December 2007.

Grandjean, P., P. Weihe, P. J. Jørgensen, T. Clarkson, E. Cenichiari, and T. Viderø. 1992. Impact of maternal seafood diet on fetal exposure to mercury, selenium, and lead. *Archives of Environmental Health* 47(3): 185–195.

Grandjean, P., P. Weihe, R. F. White, F. Debes, S. Araki, K. Yokoyams, K. Murta, N. Sørensen, R. Dahl, and P. J. Jørgensen. 1997. Cognitive deficit in 7-year-old children with pre-natal exposure to methylmercury. *Neurotoxicology and Teratology* 19(6): 417–428.

Hites, R. A., J. A. Foran, D. O. Carpenter, M. C. Hamilton, B. A. Knuth, and S. J. Schwager. 2004. Global assessment of organic contaminants in farmed salmon. *Science* 303: 226–229.

IARC. 1997. *Polychlorinated dibenzo-p-dioxins and polychlorinated dibenzofurans: summary of data reported and evaluation*. International Agency for Research on Cancer (IARC) Monographs on the Evaluation of Carcinogenic Risks to Humans 69: 33.

IPCC. 2001. *Climate Change 2001: the scientific basis*. International Panel on Climate Change (IPCC), Cambridge University Press, Cambridge, UK. Available at www.grida.no/climate/ipcc_tar/wg1/htm. Accessed June 2004.

Iwata, H., S. Tanabe, N. Sakai, and R. Tatsukawa. 1993. Distribution of persistent organochlorines in the oceanic air and surface seawater and the role of ocean on their global transport and fate. *Environmental Science & Technology* 27(6): 1080–1098.

Iwata, H., S. Tanabe, M. Aramoto, N. Sakai, and R. Tatsukawa. 1994. Persistent organochlorine residues in sediments from the Chukchi Sea, Bering Sea and Gulf of Alaska. *Marine Pollution Bulletin* 28(12): 746–753.

Kanematsu, M., Y. Shimizu, K. Sato, S. Kim, T. Suzuki, B. Park, R. Saino, and M. Nakamura. 2009. Origins and transport of aquatic dioxins in the Japanese watershed: soil contamination, land use, and soil runoff events. *Environmental Science and Technology* 43: 4260–4266.

Kannan, K., A. L. Blankenship, P. D. Jones, and J. P. Giesy. 2000. Toxicity reference values for the toxic effects of polychlorinated biphenyls to aquatic mammals. *Human and Ecological Risk Assessment* 6(1): 181–201.

Kawano, M., S. Matsushita, T. Inoue, H. Tanaka, and R. Tatsukawa. 1986. Biological accumulation of chlordane compounds in marine organisms from the northern North Pacific and Bering Sea. *Marine Pollution Bulletin* 17(11): 512–516.

Klasmeier, J., A. Beyer, and M. Matthies. 2004. Screening for cold condensation potential of organic chemicals. *Organohalogen Compounds* 66: 2406–2411.

Livingston, P. A., and L. L. Low. 1998. *Bering Sea ecosystem research: planning, coordination, and communication*. Quarterly Report, Feature Article for Alaska Fisheries Science Center, National Marine Fisheries Service.

Loughlin, T. R., and A. E. York. 2000. An accounting of the sources of Steller sea lion, *Eumetopias jubatus*, mortality. *Marine Fisheries Review* 62(4): 40–45.

Mantua, N. J., S. J. Hare, Y. Zhang, J. M. Wallace, and R. C. Francis. 1997. A Pacific interdecadal climate oscillation with impacts on salmon production. *Bulletin of American Meteorological Society* 78: 1069–1079.

Mason, R. P., and G.-R. Sheu. 2002. Role of the ocean in the global mercury cycle. *Global Biogeochemical Cycles* 16(4): 40.1–40.14.

Moore, C. J. 2008. Synthetic polymers in the marine environment: a rapidly increasing, long-term threat. *Environmental Research* 108(2): 131–139.

NRC. 1996. *The Bering Sea Ecosystem*. National Academy Press, Washington, DC.

Parker, W., and D. Dasher. 1999. Pathways, sources and distribution of contaminants in the Arctic and Alaska. Pp. 5–48 in M. J. Bradley (ed.), *Alaska Pollution Issues*. Alaska Native Epidemiology Center, Anchorage, AK.

Pauly, D., V. Christensen, S. Guénette, T. J. Pitcher, U. R. Sumaila, C. J. Walters, R. Watson, and D. Zeller. 2002. Towards sustainability in world fisheries. *Nature* 418: 689–695.

Pauly, D., V. Christensen, and C. Walters. 2000. Ecopath, Ecosim, and Ecospace as tools for evaluating ecosystem impact of fisheries. *ICES Journal of Marine Science* 57: 697–706.

Rios, L. M., P. R. Jones, C. Moore, and U. V. Narayan. 2010. Quantitation of persistent organic pollutants adsorbed on plastic debris from the North Pacific gyre's "eastern garbage patch." *Journal of Environmental Monitoring* 12: 2189–2312.

Sibert, J. R., and D. A. Fournier. 1991. Evaluation of advection–diffusion equations for estimation of movement patterns from tag recapture data. Pp. 108–121 in R. S. Shomura, J. Majkowski, and S. Langi (eds.), *Interactions of Pacific tuna fisheries*. Proceedings of the first FAO expert consultation on interactions of Pacific tuna fisheries. FAO Fisheries Technical Paper No. 36, Volume 1, FAO, Rome.

Simmonds, M. P., K. Haraguchi, T. Endo, F. Cipriano, S. R. Palumbi, and G. M. Troisi. 2002. Human health significance of organochlorine and mercury contaminants in Japanese whale meat. *Journal of Toxicology and Environmental Health, Part A* 65(17): 1211–1235.

Tanabe, S. 2002. Contamination and toxic effects of persistent endocrine disrupters in marine mammals and birds. *Marine Pollution Bulletin* 45(1–12): 69–77.

Trites, A. W., P. A. Livingston, S. Mackinson, M. C. Vasconcellos, A. M. Springer, and D. Pauly. 1999. *Ecosystem change and the decline of marine mammals in the eastern Bering Sea: testing the ecosystem shift and commercial whaling hypotheses*. Fisheries Centre Research Reports 7(1), University of British Columbia, Vancouver, Canada.

Tysban, A. V. 1999. The BERPAC project: development and overview of ecological investigations in the Bering and Chukchi Seas. Pp. 713–731 in T. R. Loughlin and K. Ohtani (eds.), *Dynamics of the Bering Sea*. University of Alaska Sea Grant, Fairbanks, AK.

Valoppi, L., M. Petreas, R. Donohoe, L. Sullivan, and C. Callahan. 1998. *Use of PCV congener and homologue analysis in ecological risk assessments*. Biological Technical Advisory Group, U.S. Environmental Protection Agency, San Francisco, CA.

Van den Berg, M., L. Birnbaum, A. T. C. Bosveld, B. Brunström, P. Cook, M. Feeley, J. P. Giesy, A. Hanberg, R. Hasegawa, S. W. Kennedy, T. Kubiak, J. C. Larsen, F. X. R. van Leeuwen, A. K. D. Liem, C. Nolt, R. E. Peterson, L. Poellinger, S. Safe, D. Schrenk, D. Tillitt, M. Tysklind, M. Younes, F. Waern, and T. Zacharweski. 1998. Toxic equivalency factors (TEFs) for PCBs, PCDDs, PCDFs for humans and wildlife. *Environmental Health Perspectives* 106(12): 775–792.

Vasquez, A. P., J. L. Regens, and J. T. Gunter. 2004. Environmental persistence of 2,3,7,8-tetrachlorodibenzo-*p*-dioxin in soil around Hardstand 7 at Egline Air Force Base, Florida. *Journal of Soils and Sediments* 4(3): 151–156.

Vestergaard, T., and P. Zachariassen. 1987. Fødslukanning 1981–82. *Frødskaparrit* 33: 5–18.

Wania, F., and Mackay, D. 1996. Tracking the distribution of persistent organic pollutants. *Environmental Science and Technology* 30(9): 390A–396A.

Weihe, P., P. Grandjean, and P. J. Jørgensen. 2005. Application of hair-mercury analysis to determine the impact of a seafood advisory. *Environmental Research* 97: 201–208.

Weihe, P., U. Steurwald, S. Taheri, O. Faero, A. S. Veyhe, and D. Nicolajsen. 2003. The human health program in the Faroe Islands: 1985–2001. Pp. 194–198 in B. Deutch and J. C. Hansen (eds.), *AMAP Greenland and the Faroe Islands 1997–2001*. Arctic Monitoring and Assessment Programme, Oslo, Norway.

Zeller, D., S. Booth, V. Lam, C. Close, and D. Pauly. 2006. Global dispersion of dioxin: a spatial dynamic model with emphasis on ocean deposition. Pp. 67–82 in J. Alder and D. Pauly (eds.), *On the multiple uses of forage fish: from ecosystems to markets*. Fisheries Centre Research Reports 14(3), University of British Columbia, Vancouver, Canada.

Zeller, D., and J. Reinert. 2004. Modelling spatial closures and fishing effort restrictions in the Faroe Islands marine ecosystem. *Ecological Modelling* 172: 403–420.

NOTES

1. Cite as Booth, S., W. W. L. Cheung, A. P. Coombs-Wallace, V. W. Y. Lam, D. Zeller, V. Christensen, and D. Pauly. 2016. Pollutants in the seas around us. Pp. 152–170 in D. Pauly and D. Zeller (eds.), *Global Atlas of Marine Fisheries: A Critical Appraisal of Catches and Ecosystem Impacts*. Island Press, Washington, DC.
2. The U.S. EPA reference dose is 0.1 µg/kg body weight/day.

CHAPTER 14

TOWARD A COMPREHENSIVE ESTIMATE OF GLOBAL MARINE FISHERIES CATCHES[1]

Daniel Pauly and **Dirk Zeller**
Sea Around Us, University of British Columbia, Vancouver, BC, Canada

Marine fisheries are the chief contributors of wholesome seafood (finfish and marine invertebrates; here "fish"). In much of the global south, fish is the only or primary animal protein that rural people can afford (Mohan Dey et al. 2005); fish also are an important source of micronutrients essential to people with otherwise deficient nutrition (Kawarazuka and Béné 2011). However, the growing popularity of fish in countries with developed or rapidly developing economies created a demand that cannot be met by fish stocks in their own waters (e.g., the EU, the United States, China, and Japan).

These markets are increasingly supplied by fish imported from developing countries or caught in the waters of developing countries by various distant-water fleets (Swartz et al. 2010a, 2010b) with these consequences:

- Foreign or export-oriented domestic industrial fleets are increasingly fishing in the waters of developing countries (Le Manach et al. 2013; Belhabib et al. 2015a).
- Industrially caught fish has become a globalized commodity that is mostly traded between continents rather than consumed in the countries where it was caught (Alder and Sumaila 2004).
- The small-scale fisheries that traditionally supplied seafood to coastal rural communities and the interior of developing countries (notably in Africa; Belhabib et al. 2015b) are forced to compete with heavily subsidized foreign industrial fleets (Sumaila et al. 2008, 2010) without much support from their own governments (Jacquet and Pauly 2008).

The lack of attention that small-scale fisheries suffer in most parts of the world (Pauly 2006a) manifests itself also in the statistics that are submitted annually by many member countries to the Food and Agriculture Organization of the United Nations (FAO), which often omit or substantially underreport small-scale fisheries data (Zeller et al. 2015). FAO harmonizes the data submitted by its members, which then becomes the only global dataset of fisheries statistics in the world, widely used by policy makers and scholars (Garibaldi 2012).

However, this dataset not only underestimates artisanal (small-scale commercial) and subsistence fisheries but generally also omits the catch of recreational fisheries, discarded bycatch[2] (Zeller and Pauly 2005), and illegal and otherwise unreported catch, even when estimates are available (Zeller et al. 2011). Thus, except for a few obvious cases of overreporting (e.g., China; see Watson and Pauly 2001), the landings data updated and disseminated annually by the FAO grossly underestimate actual fisheries catch. Although the fact of this underestimation is widely known by FAO staff and among fisheries scientists working with FAO catch data, its global magnitude had not been explicitly documented until the publication of Pauly and Zeller (2016).

Here, we summarize the results of the catch reconstructions presented on pp. 185–457, in the hope that this may eventually lead to FAO member countries submitting to FAO fisheries statistics that are more realistic, which would facilitate the implementation of ecosystem-based fisheries management (Pikitch et al. 2004), a vital component of the *FAO Code of Conduct for Responsible Fisheries* (FAO 1995).

The catch reconstruction rests on two basic principles (Pauly 1998; chapter 1):

- When "no data are available" on the catch of a given fishery, it is not appropriate to enter "N.A." or "no data" into one's database. Such entries will later be turned into a zero, which is a bad estimate of the catch of a fishery known to exist (see also Covey 2000).
- Rather, an estimate should be inserted in all such cases, based on the fact that fishing is a social activity that is bound to throw a "shadow" on the society in which it is embedded and from which a rough (but better than zero) estimate of catch can be derived (e.g., from the seafood or the fuel consumed locally or the number of vessels engaged in fisheries and the average catch rate of vessels of this type).

Chapter 2 outlines the approach through which these principles were operationalized in the *Sea Around Us* (Zeller et al. 2007, 2015). Notably, when doing reconstructions, it quickly turned out that the perception of "no data" being available was not correct; the "social shadow" yielded hundreds of articles in the peer-reviewed and report literature with catch data, or data from which catch rates could be inferred, even for remote islands (Zeller et al. 2015). Also, it became obvious that many countries sent to FAO a stripped-down version of the catch data their fisheries research institutes actually possess and even publish on their websites.

Thus, it was straightforward, though tedious, to reconstruct time series of fisheries catch for all countries of the world from 1950, the first year FAO published its much-appreciated "Yearbook" of global fisheries statistics, to 2010. This was done by "sectors," defined as follows:

- *Industrial:* Large-scale vessels (trawlers, purse-seines, longliners), which may move fishing gear across the seafloor or through the water column (e.g., demersal and pelagic trawlers), irrespective of vessel size; this corresponds to the "commercial" sectors of countries such as the United States.
- *Artisanal:* Small-scale fisheries whose catch is predominantly sold (hence they are also "commercial fisheries").
- *Subsistence:* Small-scale operations whose catch is predominantly consumed by the people fishing it and their families.
- *Recreational:* Small-scale operations whose major purpose is enjoyment.

We used national or regional definitions of these sectors in each reconstruction, as a global definition is neither achievable nor useful. In addition to the reconstructions by sector, we also assigned catches to either "landings" (i.e., retained catch) or "discards" (i.e., discarded catch), labeled as either "reported" or "unreported" with regard to national and FAO data. Thus, reconstructions present "catch" as the sum of "landings" plus "discards." We note that FAO explicitly requests that countries not include discards in their FAO data submission. Thus, the FAO data represent a "landings" dataset.

The sum of the reconstructed catches of all sectors in all EEZs of the world (see pp. 185–457), plus the industrial catch of tuna and other large pelagic fishes in the high seas (see chapter 3), leads to two major observations. First, the trajectory of reconstructed catches differs distinctly from those reported by FAO on behalf of its member countries. The latter statistics suggest that, starting in 1950, the world catch increased steadily to around 86 million tonnes (mt) in 1996, stagnated, and then slowly declined to around 77 mt by 2010 (figure 14.1A).

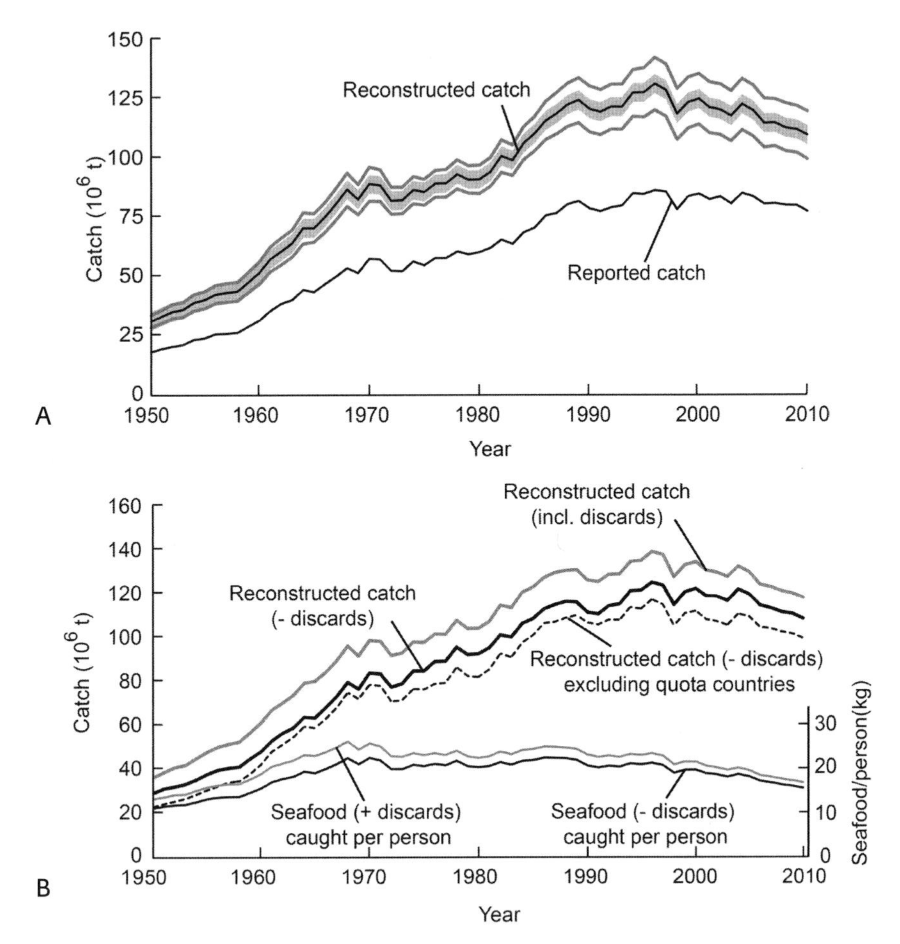

Figure 14.1. Trajectories of reported and reconstructed marine fisheries catches 1950–2010. (A) Note the difference between the world's marine fisheries catches, assembled by FAO from voluntary submissions of its member countries ("reported"), and the catch "reconstructed" to include all fisheries known to exist, in all countries and in the high seas ("reconstructed" = "reported" + estimates of "unreported"). See chapter 2 for details on catch reconstruction techniques and the Monte Carlo approach used to estimate the uncertainty associated with the reconstructed catches (gray area: uncertainty ranges as per table 2.1; gray lines: uncertainty ranges doubled). (B) Effect of removing the discards on estimates of global catch and of seafood caught per capita and of removing the catches of the major countries using quota management (i.e., United States, New Zealand, Australia, and northwestern Europe) on reconstructed total catches; the remaining catch still strongly declines. (Adapted from Pauly and Zeller 2016.)

In contrast, the reconstructed catch peaked at around 130 mt (±11 mt) in 1996 and strongly declined thereafter.[3] Overall, the reconstructed catches are 50% higher than the reported data. Second, since 1996 the reconstructed catch declined significantly, at a mean rate of 1.22 mt/year (±0.17 mt; $p < .01$), whereas FAO, at least until 2010, described the reported catch as characterized by "stability" (FAO 2011; Pauly and Froese 2012), although it exhibited a significant decline since 1996 (0.38 mt/year ± 0.13 mt; $p < .05$).

Note that the recent strong decline in catches is not caused by enlightened countries reducing catch quotas so that stocks can rebuild. Thus, a similar decline (1.01 mt/year ± 0.17; $p < .01$) is obtained when the catch from the United States, northwestern Europe, Australia, and New Zealand (i.e., countries where quota management predominates) is excluded (figure 14.1B). Note also that low quotas are not imposed when a stock is abundant; low quotas in fully developed fisheries are generally a management reaction to past or ongoing overfishing, just like declining catches in unmanaged, strongly exploited fisheries (Froese et al. 2012, 2013; Kleisner et al. 2013).

Closer examination of the reconstructed and reported catches in each of the 19 maritime

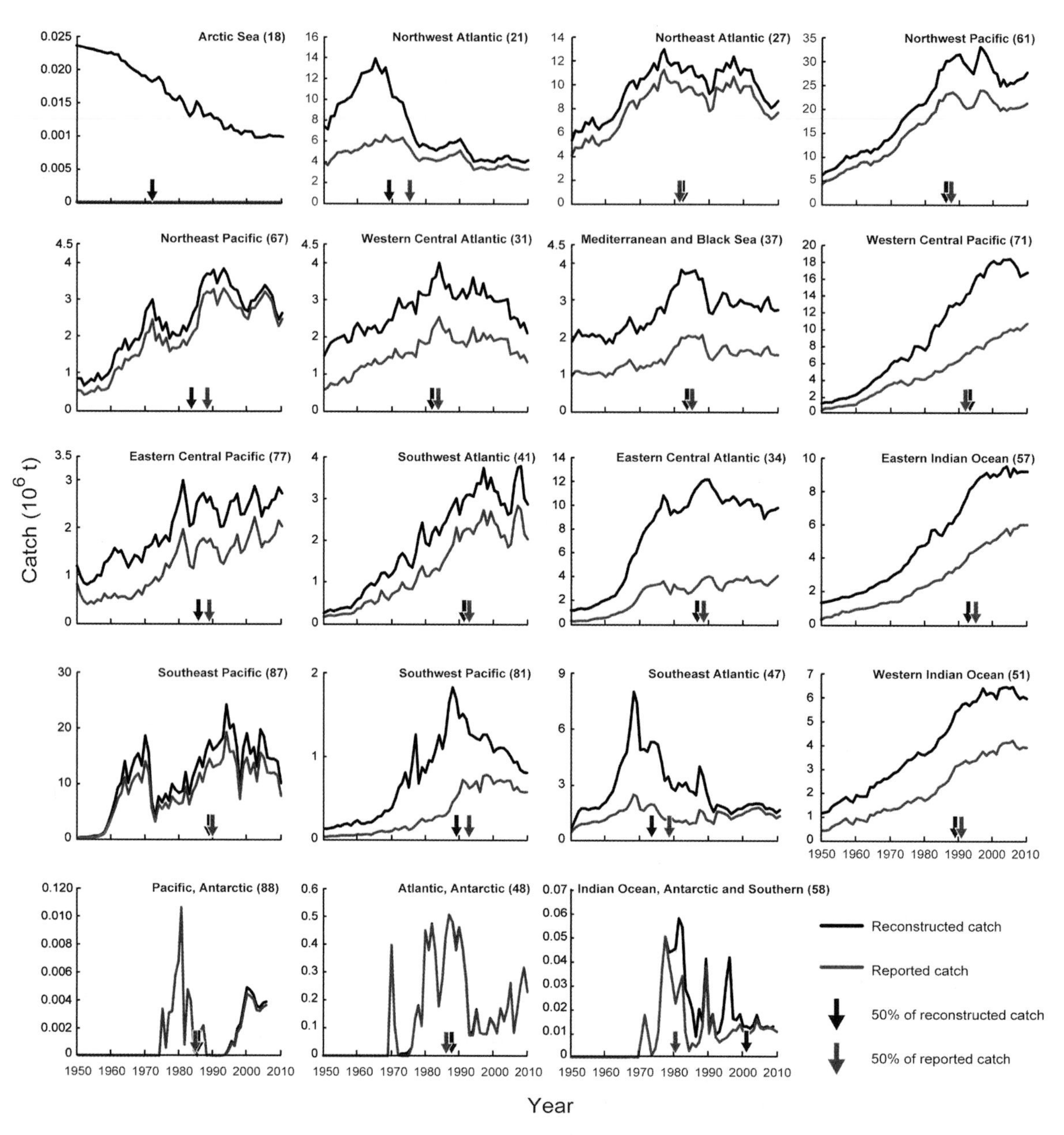

Figure 14.2. Reconstructed and reported catches in the 19 maritime "Statistical Areas" which FAO uses to roughly spatialize the world catch. Note that for Area 18 (Arctic), the reported catch by the United States and Canada was zero, and only Russia (former USSR) reported a minuscule catch in the late 1960s, even though the coastal fishes of the high Arctic are exploited by Inuit and others (Zeller et al. 2011). (Adapted from Pauly and Zeller 2016.)

FAO statistical areas suggests that the North Atlantic and North Pacific, where industrial fishing originated, were the first regions of the world to experience declining catches (figure 14.2). In contrast, lower-latitude areas experienced declines later, or still appear to have increasing catches. The increasing trend is most pronounced in the eastern Indian and western central Pacific oceans, where the reconstructed catches are very uncertain, because the statistics of various countries could only partially correct a regional tendency to exaggerate catches. FAO's eastern Indian and western central Pacific Ocean areas are also the only areas with an increasing reported catch, which, when added to that of other FAO areas, makes the world reported catch appear stable.

Catch reconstructions emphasize the assignment of all marine catches to fishing sectors (figure 14.3). Global marine catches

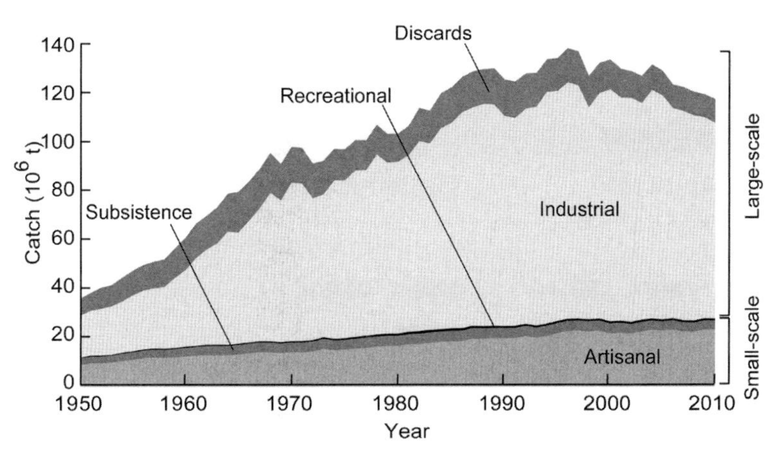

Figure 14.3. Reconstructed global catch by fisheries sectors. Based on domestic and foreign reconstructed catches in the EEZ of all countries and island territories of the world, plus high seas catches, by sectors, that is, large-scale (industrial) and small-scale (artisanal, subsistence, recreational), with discards (overwhelmingly from industrial fisheries) presented separately. (Adapted from Pauly and Zeller 2016.)

are dominated by industrial fisheries, which contributed 73 mt of landings in 2010, down from 87 mt in 2000. Thus, it is a declining industrial catch that is driving down global catches (assisted somewhat by declining discards; figure 14.3), in contrast to the artisanal sector, which generated a catch increasing from about 8 mt/year in the early 1950s to 22 mt/year in 2010 (figure 14.3). Although subsets of artisanal catches are sometimes included in official catch statistics provided to FAO, subsistence fisheries rarely are, even where they are crucial (Zeller et al. 2015). Worldwide, subsistence fisheries caught an estimated 3.8 mt/year between 2000 and 2010 (figure 14.3), a substantial fraction of it through gleaning by women in coastal ecosystems such as coral reef flats and estuaries (Harper et al. 2013). The importance of subsistence fishing for the food security of developing countries, particularly in the tropical Indo-Pacific, cannot be overemphasized (Chapman 1987; Palomares et al. 2014; Zeller et al. 2015).

Recreational fishing, whose current global estimate of just under 1 mt/year is rather imprecise, is declining in developed countries but increasing in developing countries. This activity, which generates an estimated US$40 billion/year of benefits, involves between 55 and 60 million people and generates about one million jobs worldwide (Cisneros-Montemayor and Sumaila 2010).

Discarded bycatch, generated mainly by industrial fishing, notably shrimp trawling (Andrew and Pepperell 1992), was estimated at 27 mt/year (±10 mt) and 7 mt/year (±0.7 mt) in global studies conducted for FAO in the early 1990s and 2000s, respectively (Alverson et al. 1994; Kelleher 2005). However, these point estimates were not incorporated into FAO's "catch" database, which thus consists only of landings.[3] Here, these studies were used, along with numerous other sources, to generate time series of discards (figure 14.3). They show that in 2000–2010, 10.3 mt/year of fish were discarded, a practice that was banned in Norway beginning in the late 1980s, and for the suppression of which the EU parliament passed legislation in 2014. Thus, discarded catches (until they are phased out) ought to be included in catch databases, if only to allow correct inferences on the state of the fisheries involved in this problematic practice.

Another way to present some of the major aspects of the catch reconstructions (and ancillary data) is in the form of an updated Thompson graph (Thompson 1988). This contrasts small-scale (artisanal and subsistence) and large-scale (industrial) fisheries (figure 14.4) such that the many social and ecological benefits of small-scale, particularly artisanal fisheries (see also Dioury 1985) become obvious.

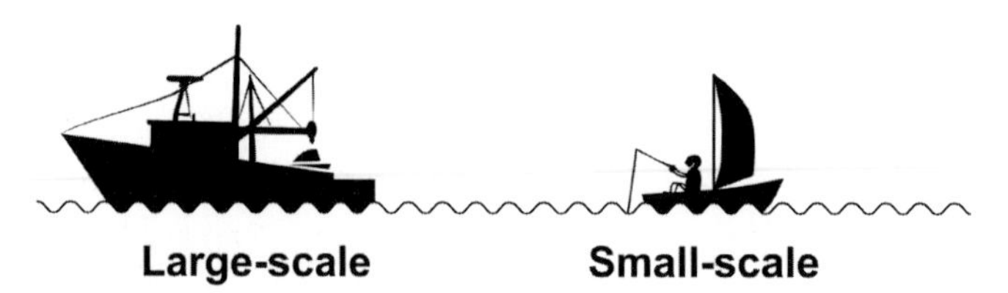

Fisheries Benefits	Large-scale	Small-scale
Annual landings for human consumption	40-45 million tonnes	25-30 million tonnes
Annual catch discarded at sea	10-12 million tonnes	about none
Annual catch for industrial reduction to fishmeal and oil, etc.	30-35 million tonnes	about none
Fuel used for tonne of fish for human consumption	5-20 tonnes	2-5 tonnes
Number of fishers employed	about 1/2 million	about 12 million
Government subsidies (billions of USD)	$$$$$ 25-30 billion USD	$ 5-7 billion USD

Figure 14.4. Comparing large- and small-scale fisheries during the period 2000–2010 through an updated Thompson graph (Thompson 1988; Pauly 2006b), which contrasts the performance of large-scale (industrial) and small-scale (artisanal and subsistence) fisheries on key criteria. The definitions of large-scale ("industrial," often mislabeled "commercial") and small-scale (often mislabeled "traditional") are those prevailing in each maritime country, yet they do not differ much (Chuenpagdee and Pauly 2008). The tonnage of fish reduced to fishmeal (Alder et al. 2008), and the fuel consumption figures (Tyedmers et al. 2005) were scaled up from nominal reported catches; while the numbers of fishers employed (Teh and Sumaila 2013) and the subsidies (chapter 6) were split into large- and small-scale (Jacquet and Pauly 2008).

SOME POLICY IMPLICATIONS

The reconstructed catch data presented here for all countries in the world allow us to formulate better policies for governing the world's marine fisheries, especially when combined with a similar reconstruction of the fishing effort to obtain this catch (e.g., by computing the engine power of the vessels involved therein; see Watson et al. 2013; Greer 2014). Also, the taxonomic composition of this catch, not presented here (but available by country and territory on pp. 185 to 457 and through the *Sea Around Us* website; see www.seaaroundus.org), will allow derivation of much more accurate catch-based indicators of fisheries status than has hitherto been possible.

A policy change that would be straightforward for FAO to implement is to request countries to submit their annual catch statistics separately for large-scale and small-scale fisheries, which would be an excellent contribution toward the implementation of the recently adopted *Voluntary Guidelines for Securing Sustainable Small-Scale Fisheries in the Context of Food Security and Poverty Eradication* (Pauly and Charles 2015). Given that FAO harmonizes all submitted data sources and even derives its own estimates for countries that do not submit data (Garibaldi 2012), FAO could readily derive sector assignments for countries that do not have such assignments.

The very high catches that were achieved in the 1990s were probably not sustainable. However, they do suggest that stock rebuilding, as successfully achieved in many U.S. fisheries, is a policy that needs wide implementation and would generate even higher sustained benefits than previously estimated from reported catches (Sumaila et al. 2012). On the other hand, the catch decline documented here is extremely worrisome in its implications for food security, as evidenced by the decline in per capita seafood availability (figure 14.1B). There is, moreover, little chance that aquaculture—at least the version of it that people in the West have in mind—can compensate for this decline, should it continue.

Most of what people in the West perceive as aquaculture is the farming of carnivorous fish such as salmon; however, such farming consumes more fish (anchovies, sardines, etc., in the form of fishmeal and oil) than it produces (Naylor et al. 2000; see also chapter 12). Hence, the more carnivorous fish we farm, the less fish we have for other purposes, including direct human consumption (figure 14.4; Pauly 2006b).

Furthermore, China, with 62% of the world's aquaculture production (FAO 2014), is now the world's largest importer of fishmeal (Cao et al. 2015), and its aquaculture industry has become increasingly dependent on wild capture fisheries. Thus, rather than a supplement or replacement to capture fisheries, aquaculture may become another driver for overfishing (Cao et al. 2015).

The last policy-relevant point to be made here transcends fisheries; it deals with the accuracy of the numbers used by the international community for decision making and for generating the factual knowledge needed for such decision making. After World War II, the newly created United Nations and its technical organizations, including the FAO, began a major project of "quantifying the world" (Ward 2004) to provide national and international agencies with the data on which to base their policies. As a result, large databases on agricultural crops and forest cover, for example, were created, and their accuracy is becoming increasingly important, given our propensity to push the exploitation of our natural ecosystems toward and beyond their limits (Rockström et al. 2009).

Therefore, we ought to check and validate these databases from time to time, lest they begin to produce "poor numbers" (Jerven 2013). For example, reports of member countries to FAO about their forest cover, when aggregated at the global level, suggested that the annual rate of forest loss between 1990/2000 and 2001/2005 was nearly halved, but the actual loss rate doubled when assessed by remote sensing and rigorous sampling (Lindquist et al. 2012). Here we show, similarly, that the main trend of the world marine fisheries catches is not one of "stability," as suggested by FAO (2011), but one of decline. Moreover, this decline, which began in the mid-1990s, started from a higher peak catch than suggested by the aggregate statistics supplied by FAO members, implying that we have more to lose if we let this decline continue. On the upside, this also means that there may be more to gain by rebuilding stocks.

A solution could therefore be for the global community to provide the FAO the funds to better and more intensively assist member countries in submitting better fishery statistics, especially statistics that cover all fisheries components, and to report data by sector (Pauly and Charles 2015). Such improved statistics can then lead to informed policy changes for rebuilding stocks and maintaining (sea) food security. Alternatively, FAO could team up with other groups (as was done for forestry statistics) to improve the fisheries statistics of member countries that often have fisheries departments with very limited human and financial resources.

Ultimately, the only database of international fisheries statistics that the world has (through FAO) should and can be improved. The rapid decline of fisheries catches clearly documented here is a good reason for this.

ACKNOWLEDGMENTS

The Pew Charitable Trusts, Philadelphia funded the *Sea Around Us* from 1999 to 2014, during which the bulk of the catch reconstruction work was performed. Since mid-2014, the *Sea Around Us* has been funded mainly by the Paul G. Allen Family Foundation and assisted by the staff of Vulcan, Inc., with additional funding from the Rockefeller, MAVA, and Prince Albert II Foundations. We also thank our many collaborators, the authors and co-authors of the catch reconstruction presented on pp. 185–457 in this volume, and Dr. Vicky Lam for the catch database and Monte Carlo simulation underlying figure 14.1.

REFERENCES

Alder, J., B. Campbell, V. Karpouzi, K. Kaschner, and D. Pauly. 2008. Forage fish: from ecosystems to markets. *Annual Reviews in Environment and Resources* 33: 153–166 [plus 8 pages of figures].

Alder, J., and U. R. Sumaila. 2004. Western Africa: the fish basket of Europe past and present. *The Journal of Environment and Development* 13: 156–178.

Alverson, D. L., M. H. Freeberg, J. G. Pope, and S. A. Murawski. 1994. *A global assessment of fisheries by-catch and discards*. FAO Fisheries Technical Papers T339, Rome.

Andrew, N. L., and J. G. Pepperell. 1992. The by-catch of shrimp trawl fisheries. *Oceanography and Marine Biology: An Annual Review* 30: 527–565.

Belhabib, D., U. R. Sumaila, V. W. Y. Lam., D. Zeller, P. Le Billon, E. A. Kane, and D. Pauly. 2015a. Euro vs. Yuan: comparing European and Chinese fishing access in West Africa. *PLoS One*. 10(3): e0118351.

Belhabib, D., U. R. Sumaila, and D. Pauly. 2015b. Feeding the poor: contribution of West African fisheries to employment and food security. *Ocean and Coastal Management* 10(3): e0118351.

Cao, L., R. Naylor, P. Henriksson, D. Leadbitter, M. Metian, M. Troell, and W. Zhang. 2015. China's aquaculture and the world's wild fisheries. *Science* 347: 133–135.

Chapman, M. D. 1987. Women's fishing in Oceania. *Human Ecology* 15: 267–288.

Chuenpagdee, R., and D. Pauly. 2008. Small is beautiful? A database approach for global assessment of small-scale fisheries: preliminary results and hypotheses. Pp. 575–584 in J. L. Nielsen, J. J. Dodson, K. Friedland, T. R. Hamon, J. Musick, and E. Vespoor (eds.), *Reconciling Fisheries with Conservation: Proceedings of the Fourth World Fisheries Congress*. American Fisheries Society, Symposium 49, Bethesda, MD.

Cisneros-Montemayor, A. M., and U. R. Sumaila. 2010. A global estimate of benefits from ecosystem-based marine recreation: potential impacts and implications for management. *Journal of Bioeconomics* 12: 245–268.

Covey, C. 2000. Beware the elegance of the number zero. *Climatic Change* 44(4): 409–411.

Dioury, F. 1985. Insights on the developmental aspect and future importance of artisanal fisheries. *ICLARM Newsletter* 8(3): 16–17.

FAO. 1995. *Code of Conduct for Responsible Fisheries*. Food and Agriculture Organization of the United Nations, Rome.

FAO. 2011. *The State of World Fisheries and Aquaculture (SOFIA) 2010*. Food and Agriculture Organization of the United Nations, Rome.

FAO. 2014. *State of World Fisheries and Aquaculture (SOFIA) 2014*. Food and Agriculture Organization of the United Nations, Rome.

Froese, R., D. Zeller, K. Kleisner, and D. Pauly. 2012. What catch data can tell us about the status of global fisheries. *Marine Biology* 159: 1283–1292.

Froese, R., D. Zeller, K. Kleisner, and D. Pauly. 2013. Worrisome trends in global stock status continue unabated: a response to a comment by R. M. Cook on "What catch data can tell us about the status of global fisheries." *Marine Biology* 160: 2531–2533.

Garibaldi, L. 2012. The FAO global capture production database: a six-decade effort to catch the trend. *Marine Policy* 36: 760–768.

Greer, K. 2014. *Considering the "effort factor" in fisheries: a methodology for reconstructing global fishing effort and CO_2 emissions, 1950–2010*. MSc thesis, University of British Columbia, Vancouver, Canada.

Harper, S., D. Zeller, M. Hauzer, U. R. Sumaila, and D. Pauly. 2013. Women and fisheries: contribution to food security and local economies. *Marine Policy* 39: 56–63.

Jacquet, J., and D. Pauly. 2008. Funding priorities: big barriers to small-scale fisheries. *Conservation Biology* 22: 832–835.

Jerven, M. 2013. *Poor Numbers: How We Are Misled by African Development Statistics and What to Do about It*. Cornell University Press, Ithaca, NY.

Kawarazuka, N., and C. Béné. 2011. The potential role of small fish species in improving micronutrient deficiencies in developing countries: building evidence. *Public Health and Nutrition* 14: 1927–1938.

Kelleher, K. 2005. *Discards in the world's marine fisheries. An update*. FAO Fisheries Technical Paper 470, Food and Agriculture Organization, Rome.

Kleisner, K., D. Zeller, R. Froese, and D. Pauly. 2013. Using global catch data for inferences on the world's marine fisheries. *Fish and Fisheries* 14: 293–311.

Le Manach, F., C. Chaboud, D. Copeland, P. Cury, D. Gascuel, K. M. Kleisner, A. Standing, U. R. Sumaila, D. Zeller, and D. Pauly. 2013. European Union's public fishing access agreements in developing countries. *PLOS One* 8(11): e79899.

Lindquist, E. J., R. D'Annunzio, A. Gerrand, K. MacDicken, F. Achard, R. Beuchle, A, Brink, H. D. Eva, P. Mayaux, J. San-Miguel-Ayanz, and H. J. Stibig. 2012. *FAO/JRC global forest land-use change from 1990 to 2005*. FAO Forestry Paper 169, Food and Agriculture Organization of the United Nations and European Commission Joint Research Center, Rome.

Mohan Dey, M., M. A. Rab, F. J. Paraguas, S. Piumsombun, R. Bhatta, M. F. Alam, and M. Ahmed. 2005. Fish consumption and food security: a disaggregated analysis by types of fish and classes of consumers in selected Asian countries. *Aquaculture Economics & Management* 9(1–2): 89–111.

Naylor, R. L., R. J. Goldburg, J. H. Primavera, N. Kautsky, M. C. M. Beveridge, J. Clay, C. Folke, J. Lubchenco, H. Mooney, and M. Troell. 2000. Effect of aquaculture on world fish supplies. *Nature* 405(6790): 1017–1024.

Palomares, M. L. D., J. C. Espedido, V. A. Parducho, M. P. Saniano, L. P. Urriquia, and P. M. S. Yap. 2014. A short history of gleaning in Mabini, Batangas (Region IV, Subzone B, Philippines). Pp. 118–128 in M. L. D. Palomares and D. Pauly (eds.), *Philippine marine fisheries catches: a bottom-up reconstruction, 1950 to 2010*. Fisheries Centre Research Reports 22(1), University of British Columbia, Vancouver, Canada.

Pauly, D. 1998. Rationale for reconstructing catch time series. *EC Fisheries Cooperation Bulletin* 11: 4–10.

Pauly, D. 2006a. Major trends in small-scale marine fisheries, with emphasis on developing countries, and some implications for the social sciences. *Maritime Studies (MAST)* 4(2): 7–22.

Pauly, D. 2006b. Babette's Feast in Lima. *Sea Around Us Project Newsletter* November/December (38): 1–2.

Pauly, D., D. Belhabib, R. Blomeyer, W. W. L. Cheung, A. Cisneros-Montemayor, D. Copeland, S. Harper, V. Lam, Y. Mai, F. Le Manach, H. Österblom, K. M. Mok, L. van der Meer, A. Sanz, S. Shon, U. R. Sumaila, W. Swartz, R. Watson, Y. Zhai, and D. Zeller. 2014. China's distant-water fisheries in the 21st century. *Fish and Fisheries* 15: 474–488.

Pauly, D., and T. Charles. 2015. Counting on small-scale fisheries. *Science* 347: 242–243.

Pauly, D., and R. Froese. 2012. Comments on FAO's State of Fisheries and Aquaculture, or "SOFIA 2010." *Marine Policy* 36: 746–752.

Pauly, D., and D. Zeller. 2016. Catch reconstructions reveal that global marine fisheries catches are higher than reported and declining. *Nature Communications* 7: 10244.

Pikitch, E. K., C. Santora, E. A. Babcock, A. Bakun, R. Bonfil, D. O. Conover, P. Dayton, P. Doukakis, D. L. Fluharty, B. Heneman, E. D. Houde, J. Link, P. A. Livingston, M. Mangel, M. K. McAllister, J. Pope, and K. J. Sainsbury. 2004. Ecosystem-based fishery management. *Science* 305: 346–347.

Rockström, J., W. Steffen, K. Noone, Å. Persson, F. S. Chapin, E. Lambin, T. M. Lenton, M. Scheffer, C. Folke, H. J. Schellnhuber, B. Nykvist, C. A. de Wit, T. Hughes, S. van der Leeuw, H. Rohde, S. Sörlin, P. K. Snyder, R. Costanza, U. Svedin, M. Falkenmark, L. Karlberg, R. W. Corell, V. J. Fabry, J. Hansen, B. Walker, D. Liverman, K. Richardson, P. Crutzen, and J. Foley. 2009. Planetary boundaries: exploring the safe operating space for humanity. *Nature* 461: 472–475.

Sumaila, U. R., W. W. L. Cheung, A. Dyck, K. M. Gueye, L. Huang, V. Lam, D. Pauly, U. T. Srinivasan, W. Swartz, R. Watson, and D. Zeller. 2012. Benefits of rebuilding global marine fisheries outweigh costs. *PLoS ONE* 7(7): e40542.

Sumaila, U. R., A. Khan, A. Dyck, R. Watson, R. Munro, P. Tyedmers, and D. Pauly. 2010. A bottom-up re-estimation of global fisheries subsidies. *Journal of Bioeconomics* 12: 201–225.

Sumaila, U. R., L. Teh, R. Watson, P. Tyedmers, and D. Pauly. 2008. Fuel price increase, subsidies, overcapacity and resource sustainability. *ICES Journal of Marine Science* 65: 832–840.

Swartz, W., E. Sala, S. Tracey, R. Watson, and D. Pauly. 2010a. The spatial expansion and ecological footprint of fisheries (1950 to present). *PLoS ONE* 5: e15143.

Swartz, W., U. R. Sumaila, R. Watson, and D. Pauly. 2010b. Sourcing seafood for the three major markets: the EU, Japan and the USA. *Marine Policy* 34: 1366–1373.

Teh, L., and U. R. Sumaila. 2013. Contribution of marine fisheries to worldwide employment. *Fish and Fisheries* 14: 77–88.

Thompson, D. 1988. The world's two marine fishing industries: how they compare. *Naga, the ICLARM Quarterly* 11: 17.

Tyedmers, P., R. Watson, and D. Pauly. 2005. Fueling global fishing fleets. *AMBIO: A Journal of the Human Environment* 34: 635–638.

Ward, M. 2004. *Quantifying the World: UN Ideas and Statistics*. Indiana University Press, Indianapolis.

Watson, R., W. W. L. Cheung, J. Anticamara, R. U. Sumaila, D. Zeller, and D. Pauly. 2013. Global marine yield halved as fishing intensity redoubles. *Fish and Fisheries* 14: 493–503.

Watson, R., and D. Pauly. 2001. Systematic distortions in world fisheries catch trends. *Nature* 414: 534–536.

Zeller, D., S. Booth, G. Davis, and D. Pauly. 2007. Re-estimation of small-scale fishery catches for U.S. flag–associated island areas in the western Pacific: the last 50 years. *Fishery Bulletin* 105: 266–277.

Zeller, D., S. Booth, E. Pakhomov, W. Swartz, and D. Pauly. 2011. Arctic fisheries catches in Russia, USA and Canada: baselines for neglected ecosystems. *Polar Biology* 34: 955–973.

Zeller, D., S. Harper, K. Zylich, and D. Pauly. 2015. Synthesis of under-reported small-scale fisheries catch in Pacific-island waters. *Coral Reefs* 34: 25–39.

Zeller, D., and D. Pauly. 2005. Good news, bad news: global fisheries discards are declining, but so are total catches. *Fish and Fisheries* 6: 156–159.

NOTE

1. Cite as Pauly, D., and D. Zeller. 2016. Toward a comprehensive estimate of global marine fisheries catches. Pp. 171–181 in D. Pauly and D. Zeller (eds.), *Global Atlas of Marine Fisheries: A Critical Appraisal of Catches and Ecosystem Impacts*. Island Press, Washington, DC.
2. However, FAO explicitly request that countries not include discards in their data submissions to FAO.
3. Note that in both cases, 1996 was not chosen visually (and arbitrarily) but was the output of the segmented regression analysis to which both the FAO and the reconstructed time series were subjected (see also Pauly and Zeller 2016).

FISHERIES BY COUNTRY AND TERRITORY, 1950–2010

The following 273 pages present single-page summaries of the fisheries in the Exclusive Economic Zones (EEZs) of maritime countries, or parts thereof, including the overseas territories of metropolitan countries. These summaries are arranged alphabetically, from Albania to Yemen (Red Sea), and the geographic index on pages 478–487 allows the reader to find a place whose summary may not be where it was expected, or which is mentioned in the summaries. The groups of EEZs presented in these summaries include a (coastal) ecosystem each, characterized by its geomorphological and oceanographic features, populated by a characteristic flora and fauna, and subjected to specific stresses from 1950 to 2010. Although the most significant impact is often fishing, hurricanes, tsunamis, and other major natural events are noted, along with wars, civil or not. Most of these summaries have the same authors as the reconstruction publications from which they are taken. All of the summaries have in common that they name and define the geographic entity in question (including its relationship to a larger entity, if appropriate), then present a small map ("figure 1") to define its EEZ or EEZ-equivalent marine area, as well as the continental shelf areas (to 200 m depth).

The catches in each country's or territory's EEZ are then described in the text and in "figure 2A" in terms of their magnitude in the 1950s and 2000s (taken to include the year 2010), with intermediate extremes (peaks or troughs) explicitly mentioned; the sectoral composition of this catch, that is, industrial and small-scale (artisanal, subsistence, and recreational); the disposition of the catch (landed or discarded); and the ratio of the reconstructed catch to the official landings reported by the FAO or other agencies on behalf of that country or territory. In each figure 2 caption, the uncertainty scores (as defined in table 2.1) associated with the (domestic) catch estimates are given by sector and disposition (Ind. = industrial; Art. = artisanal; Sub. = subsistence; Rec. = recreational, and Dis. = discards), for three periods (1950–1969, 1970–1989, 1990–2010). Figure 2B presents the catches by fishing country (domestic and foreign fleets) exploiting the EEZ in question. Note, however, that the estimates of foreign catches will be indicative only. Foreign catches, especially when they are made illegally (as is often the case), tend to be far more uncertain than domestic catches, with regard to the years such catches are taken, the quantities taken, and even the nationalities of the fleets that do the taking (or the nationality of the beneficial ownership of these fleets). The website of the *Sea Around Us* (www.seaaroundus.org) will present updated versions of these foreign catches, which may differ from those presented here. The website also includes an option for providing feedback, which can also be used by readers of this atlas.

Finally, figure 2C presents the taxonomic composition of the catch taken in these EEZ "chunks," that is, species or groups caught. Note that the catches refer to marine and brackish waters (estuaries and coastal lagoons) but not freshwater, that plants and vertebrates other than fishes (e.g., marine mammals, sea turtles) are not included, and that the products of sea farming or ranching, that is, mariculture, are not included either (but see chapter 12). The taxonomic resolution of figure 2C is limited, and readers wanting more details will find them at www.seaaroundus.org. Also note that, in a few cases, figure 2 may consist of only one or two panels.

For legibility's sake, and to save space, the scale label of the ordinate (*y*-scale) of figure 2 does not include trailing zeroes. Rather, the legend's scale is of the form $t \cdot 10^n$, where *n* indicates the number of trailing zeroes. Thus, *n* corresponds to the factor by which the tonnage along the *y*-scale should be multiplied: 1 = 10, 2 = 100, 3 = 1,000, etc.

We generally do not discuss the status of the fisheries, except in cases that are obvious and have been widely commented upon. After the description of the catches, the text often alludes to a major issue specific to the fisheries of the country or territory in question, or an interesting fact about it or the resources it exploits; the glossary on pp. 459 to 477 should be consulted for unfamiliar terms or acronyms. The references always include a citation to the articles or reports from which the summary was extracted (available from www.seaaroundus.org), to which references were added from the same or other sources, usually to reinforce a point unique to that summary.

Many of these reconstruction reports were made available, pending formal publication, via the Fisheries Centre Research Reports or the Fisheries Centre Working Papers series of the University of British Columbia (UBC), in Vancouver, Canada. To save space, these two series will be cited without reference to UBC; they can all be obtained from our website. The editors thank Mr. Chris Hoornaert for creating the maps cited as figure 1 in the 273 summaries that follow and Evelyn Liu for the time series graphs cited as figure 2.

ALBANIA[1]

Dimitrios K. Moutopoulos,[a] Brady Bradshaw,[b] and Daniel Pauly[c]

[a]Technological Educational Institute of Western Greece, Department of Aquaculture and Fisheries, Messolonghi, Greece
[b]Investigative Team, The Black Fish, Amsterdam, The Netherlands & Center for Marine Science, University of North Carolina, Wilmington [c]*Sea Around Us*, University of British Columbia, Vancouver, Canada

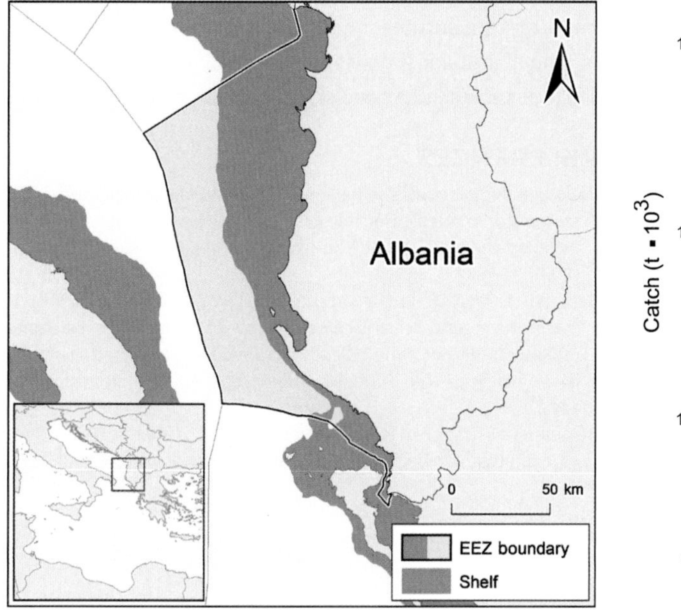

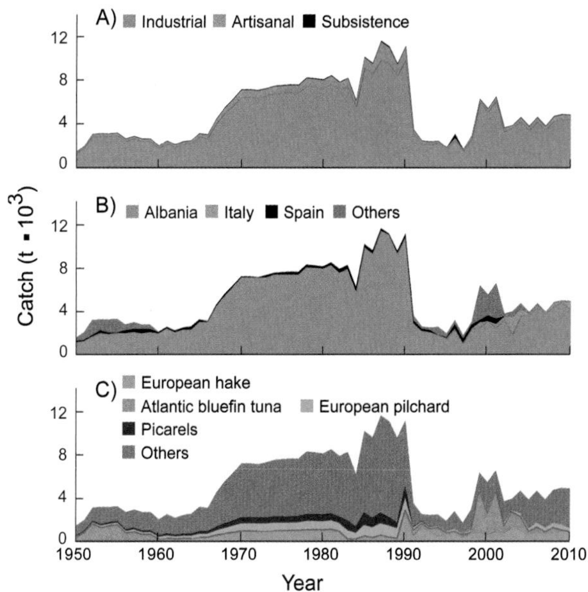

Figure 1. Albania's EEZ, of 11,100 km², including 6,090 km² of a shelf that is extensive and easy to trawl in the north and deep waters with a rocky seabed in the south. **Figure 2.** Domestic and foreign catches taken in the EEZ of Albania: (A) by sector (domestic scores: Ind. 2, 2, 3; Art. 2, 2, 3; Sub. 1, 1, 1; Dis. 1, 1, 2); (B) by fishing country (foreign catches are very uncertain); (C) by taxon.

Albania is a small country on the Adriatic Sea (figure 1) that, in 1991 and 1992, transitioned to an open democratic system after decades of isolation and backwardness. This account, based on the catch reconstruction of Moutopoulos et al. (2015), using only a handful of sources (e.g., Çobani 2005), consists of the official statistics reported by the FAO on behalf of Albania, complemented by estimates of discards and differentiating between industrial and small-scale gears. The majority of the total catches as allocated to Albania's EEZ are attributed to the industrial sector (figure 2A). Domestic fisheries accounted for the bulk of total catches in Albania's EEZ and were reconstructed to about 1,500 t/year in the 1950s, peaked at 8,000 t/year in the late 1980s, collapsed to early 1950s levels after the establishment of democracy, and appear to be rebuilding, reaching about 4,100 t in 2010 (figure 2B). Foreign fishing has been documented in Albania's waters since 1950, with the majority of catches being Italian (figure 2B). Marine fisheries provide direct employment to about 900 people, and more than 350 vessels are being deployed (some only occasionally), of which most are bottom trawlers operating in the north of the country's EEZ, and a few use (illegal) driftnets. The catch of the fisheries consists of European hake (*Merluccius merluccius*), Atlantic bluefin tuna (*Thunnus thynnus*), and European pilchard (*Sardina pilchardus*) (figure 2C).

REFERENCES

Çobani M (2005) Small-scale fisheries in Albania. Adriatic Sea Small-Scale Fisheries. Report of the AdriaMed Technical Consultation on Adriatic Sea Small-Scale Fisheries, FAO-MiPAF Scientific Cooperation to Support Responsible Fisheries in the Adriatic Sea. GCP/RER/010/ITA/TD15. AdriaMed Technical Documents, Split, Croatia. 15 p.

Moutopoulos D, Bradshaw B and Pauly D (2015) Reconstruction of Albania fishery catches by fishing gear (1950–2010). Fisheries Centre Working Paper #2015-12, 12 p.

1. Cite as Moutopoulos, D. K., B. Bradshaw, and D. Pauly. 2016. Albania. P. 185 in D. Pauly and D. Zeller (eds.) *Global Atlas of Marine Fisheries: A Critical Appraisal of Catches and Ecosystem Impacts*. Island Press, Washington, DC.

ALGERIA[1]

Dyhia Belhabib, Daniel Pauly, Sarah Harper, and Dirk Zeller

Sea Around Us, University of British Columbia, Vancouver, Canada

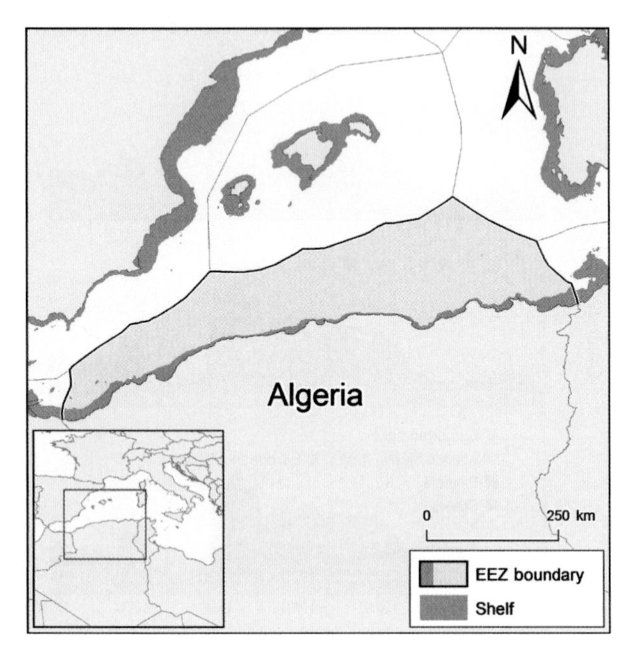

Figure 1. Algeria has an EEZ of about 129,000 km² and a narrow continental shelf of 10,500 km².

Algeria is the largest country in Africa, with a long coastline along the Mediterranean Sea (figure 1). Its population is based mainly in the north because of the arid Saharan climate, causing pressure on the coastal environment. The modernization process in the country began after independence from France in 1962 and was accelerated after the serious sociopolitical conflicts of the "black decade" (1991–2001), notably via massive subsidies to fisheries as an alternative livelihood for the country's youth. However, fisheries are both overexploited and undermonitored (Belouahem 2009), with symptoms of "fishing down" (Babouri et al. 2014). The total catches allocated to Algeria's EEZ were derived largely from the industrial and artisanal sectors (figure 2A). Belhabib et al. (2012) estimated Algeria's domestic marine fisheries catches including subsistence, artisanal, recreational, and industrial as more than 57,000 t in 1950, increasing to a peak of more than 270,000 t in 2007, before declining to about 210,000 t in 2010. Overall, reconstructed domestic catches were nearly twice as high as reported by the FAO on behalf of Algeria. Although most of this catch was domestic, some foreign fishing, for example by Spain and Japan, was also documented (figure 2B). The total catches consist mainly of European pilchards (*Sardina pilchardus*), an important substitute for nonmarine, more expensive sources of animal proteins (figure 2C); its declining catch is affecting the food security of Algerians.

REFERENCES

Babouri K, Pennino MG and Bellido JM (2014) A trophic indicator toolbox for implementing an ecosystem approach in data-poor fisheries: the Algerian and Bou-Ismael Bay example. *Scientia Marina* 78(S1): 37–51.

Belhabib D, Pauly D, Harper S and Zeller D (2012) Reconstruction of marine fisheries catches for Algeria, 1950–2010. Pp. 1–22 in Belhabib D, Pauly D, Harper S and Zeller D (eds.), *Marine fisheries catches in West Africa, 1950–2010, Part I*. Fisheries Centre Research Reports 20(3).

Belouahem S (2009) An ecosystemic approach to fisheries. The Algerian case. University of Montpellier II, France, 96 p.

1. Cite as Belhabib, D., D. Pauly, S. Harper, and D. Zeller. 2016. Algeria. P. 186 in D. Pauly and D. Zeller (eds.), *Global Atlas of Marine Fisheries: A Critical Appraisal of Catches and Ecosystem Impacts*. Island Press, Washington, DC.

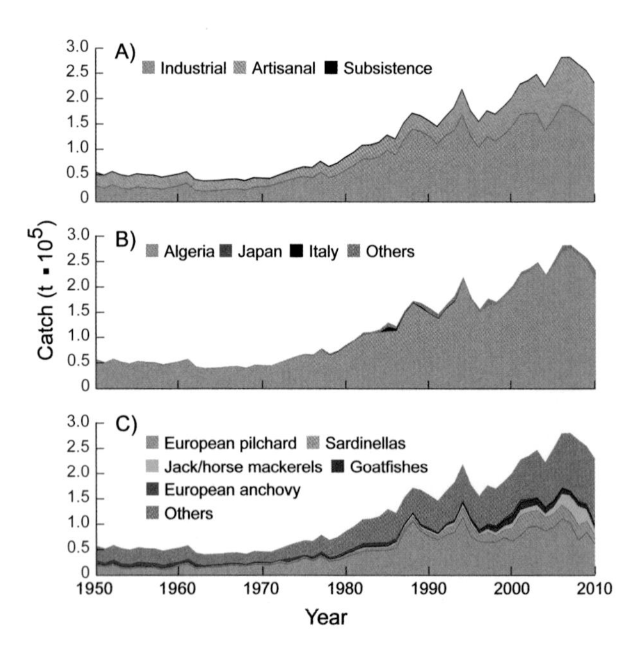

Figure 2. Domestic and foreign catches taken in the EEZ of Algeria: (A) by sector (domestic scores: Ind. 3, 3, 3; Art. 2, 2, 4; Sub. 3, 2, 2; Rec. –, 2, 2; Dis. 3, 3, 4); (B) by fishing country (foreign catches are very uncertain); (C) by taxon.

ANGOLA[1]

Dyhia Belhabib, Esther Divovich, and Daniel Pauly

Sea Around Us, University of British Columbia, Vancouver, Canada

Angola is located in southwest Africa and thus benefits from the upwelling that enriches the Benguela Current Large Marine Ecosystem (figure 1). Unlike in Namibia, Angola's shore does not flank a desert, so extensive small-scale fisheries compete with industrial fisheries all along the coast (Sowman and Cardoso 2010). However, their catch data are unreliable. The total allocated catch in the Angolan EEZ was attributed mainly to the industrial sector, although the artisanal fisheries are growing (figure 2A). Belhabib and Divovich (2015) reconstructed domestic catches of 180,000 t in 1950, a peak of 680,000 t in 1972, a decline to 130,000 t in 1976 after the departure of formerly "domestic" Portuguese fleets at independence, and a peak of 512,000 t in 2007. Domestic catches were, overall, 1.5 times those supplied to the FAO. Foreign fleets, such as that of Russia, have been documented in Angolan waters, many of them illegal, and were at their peak during the civil war (1975–2002, figure 2B). Catches consisted mostly of Cunene horse mackerel (*Trachurus trecae*), sardinellas (*Sardinella* spp.), and seabreams (*Dentex* spp.; figure 2C). Catches of Cunene horse mackerel declined over time. Although Angolan fisheries are overexploited, recreational fisheries are still perceived as having room for development (Potts et al. 2009) and could contribute significantly to the local economy.

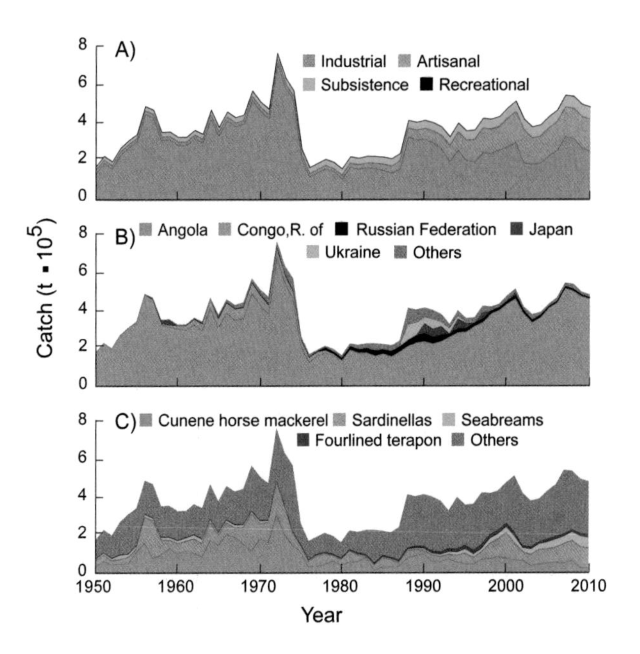

Figure 2. Domestic and foreign catches taken in the EEZ of Angola: (A) by sector (domestic scores: Ind. 3, 3, 2; Art. 1, 3, 3; Sub. 1, 3, 3; Rec. 2, 3, 3; Dis. 2, 2, 4); (B) by fishing country (foreign catches are very uncertain); (C) by taxon.

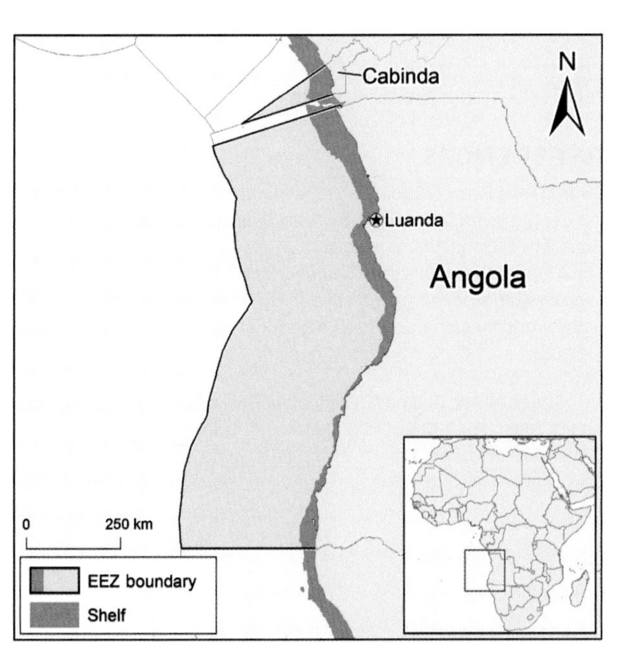

Figure 1. Angola (including the exclave of Cabinda) has a continental shelf of 50,900 km² and an EEZ of 491,000 km².

REFERENCES

Belhabib D and Divovich E (2015) Rich fisheries and poor data: a catch reconstruction for Angola, 1950–2010, an update of Belhabib and Divovich (2014). pp. 115–128 In Belhabib D and Pauly D (eds.), *Fisheries catch reconstructions: West Africa, Part II*. Fisheries Centre Research Reports 23(3).

Potts W, Childs AR, Sauer W and Duarte A (2009) Characteristics and economic contribution of a developing recreational fishery in southern Angola. *Fisheries Management and Ecology* 16(1): 14–20.

Sowman M and Cardoso P (2010) Small-scale fisheries and food security strategies in countries in the Benguela Current Large Marine Ecosystem (BCLME) region: Angola, Namibia and South Africa. *Marine Policy* 34(6): 1163–1170.

1. Cite as Belhabib, D., E. Divovich, and D. Pauly. 2016. Angola. P. 187 in D. Pauly and D. Zeller (eds.), *Global Atlas of Marine Fisheries: A Critical Appraisal of Catches and Ecosystem Impacts*. Island Press, Washington, DC.

ANTARCTICA[1]

David G. Ainley[a] and Daniel Pauly[b]

[a]H.T. Harvey & Associates, Los Gatos, CA [b]*Sea Around Us*, University of British Columbia, Vancouver, Canada

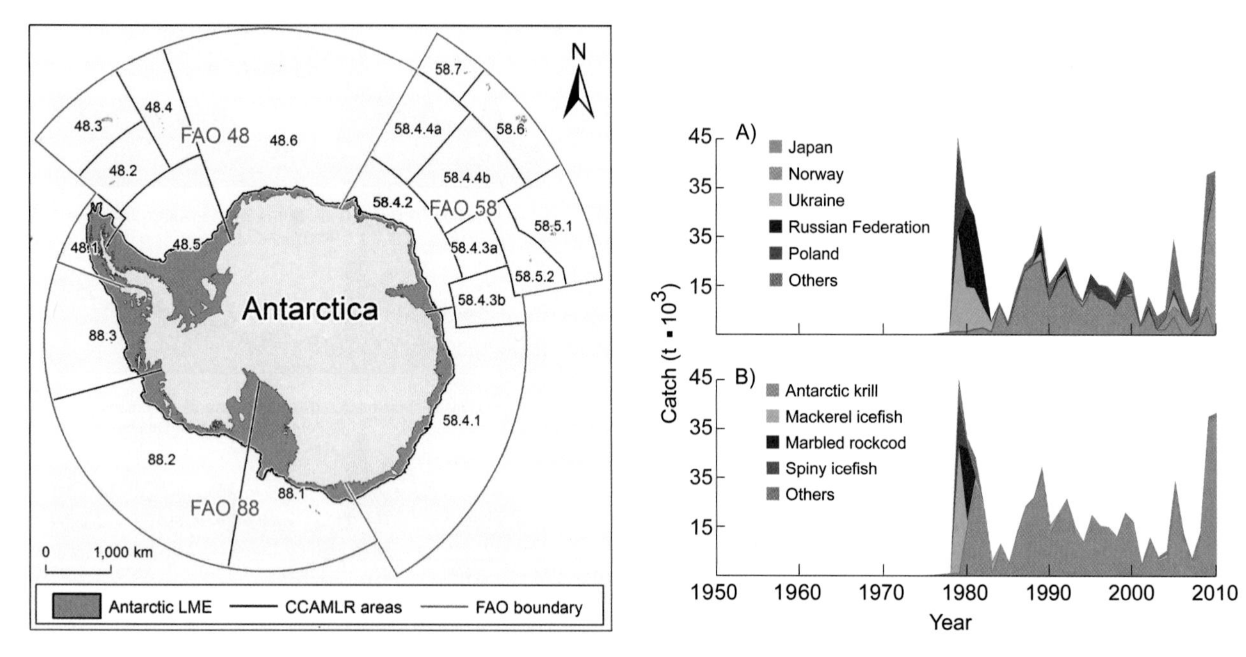

Figure 1. The Antarctic Large Marine Ecosystem (3.96 million km², with a shelf of 492,000 km²) and the corresponding FAO statistical areas, subareas and divisions. **Figure 2.** Domestic and foreign catches taken in the Antarctic Large Marine Ecosystem: (A) by fishing country (scores: Ind. –, 3, 3; Dis. –, 1, 2); (B) by taxon.

The biotic exploitation of the continental shelf and slope of the Antarctic continent, within the Antarctic Large Marine Ecosystem (Aquarone and Adams 2008), whose outer limits (figure 1) are used here in the absence of an EEZ, was one of serial depletion and substantially declined catches by a variety of countries (figure 2A). Along the Antarctic continent, marine mammals were decimated by the 1970s, and fishing for Antarctic krill (*Euphausia superba*) began upon the demise of groundfish in the 1980s and continued as the primary taxon targeted by the 2000s (figure 2B). The cetaceans and most groundfish stocks remain severely depressed, and their exploitation is now prohibited by the International Whaling Commission (IWC) and the Commission for the Conservation of Antarctic Marine Living Resources (CCAMLR). On the other hand, several countries now engage in krill fishing. CCAMLR is one of the few regional fisheries management organizations including bycatch and discards in their statistics. Thus, in the absence of hard data on illegal fishing, reconstructed and reported catches are similar (Ainley and Pauly 2014).

REFERENCES

Ainley D and Pauly D (2014) Fishing down the food web of the Antarctic continental shelf and slope. *Polar Record* 50: 92–107.

Aquarone MC and Adams S (2008) XVIII-57 Antarctic LME. pp. 764–773 In Sherman K and Hempel G (eds.), The UNEP large marine ecosystem report: A perspective on changing conditions in LMEs of the world's regional seas. UNEP Regional Seas Reports and Studies No. 182.

1. Cite as Ainley, D., and D. Pauly. 2016. Antarctica. P. 188 in D. Pauly and D. Zeller (eds.), *Global Atlas of Marine Fisheries: A Critical Appraisal of Catches and Ecosystem Impacts*. Island Press, Washington, DC.

ANTIGUA & BARBUDA[1]

Jeanel Georges,[a] Robin Ramdeen,[b] Kyrstn Zylich,[b] and Dirk Zeller[b]

[a]University of the West Indies, Cave Hill, St. Michael, Barbados
[b]*Sea Around Us*, University of British Columbia, Vancouver, Canada

Antigua & Barbuda are in the Lesser Antilles, between the Caribbean Sea and Atlantic Ocean (figure 1). Antigua is generally composed of limestone, with some volcanic rocks and hills, and Barbuda is a low-lying coral island. Antigua & Barbuda's fishery sector contributes to the food security of the islands and is important during downturns in tourism, the country's main economic sector. However, Carr and Heyman (2008) suggest that the fisheries of Antigua & Barbuda are at risk. The fisheries of these islands are mainly small-scale in nature, with the artisanal and subsistence sectors dominating the total allocated catch (figure 2A). A catch reconstruction undertaken by Georges et al. (2015), based on the scant fisheries literature on Antigua & Barbuda (e.g., Horsford 2002), produced an estimated total domestic catch of about 1,300 t/year in the 1950s, which increased to 3,000 t/year in the 2000s. This reconstructed domestic catch was 1.7 times the data reported by the FAO on behalf of Antigua & Barbuda and is probably an underestimate. Although the majority of the allocated catch for the Antigua & Barbuda EEZ came from the domestic fishery, foreign fishing by an unknown fishing entity and by Cuba and Venezuela, among others, was also documented (figure 2B). Figure 2C illustrates that catches consist mainly of reef species such as groupers (family Serranidae), snappers (family Lutjanidae), and grunts (family Haemulidae).

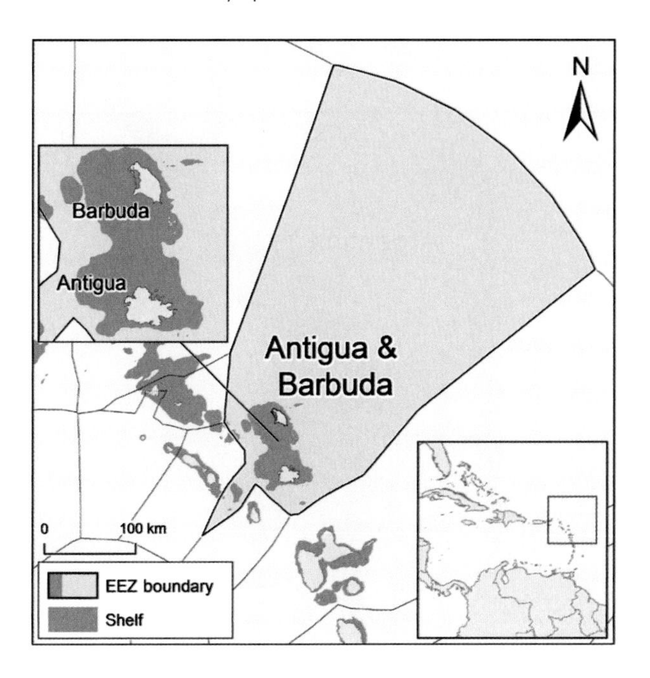

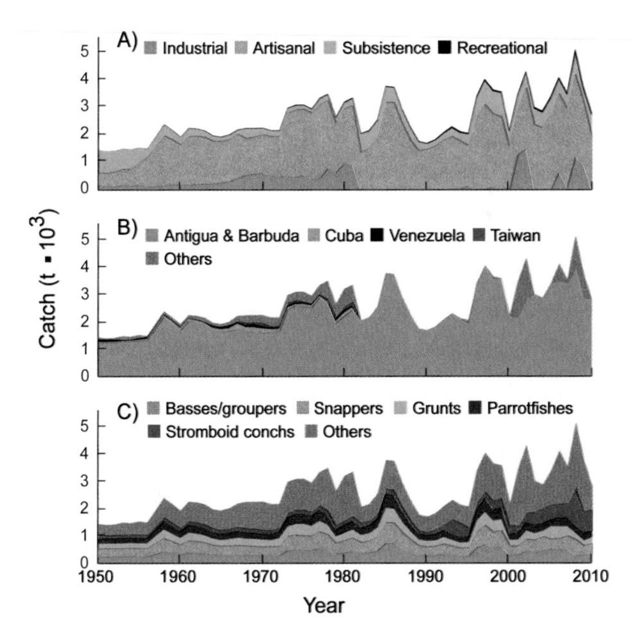

Figure 1. Antigua & Barbuda have a land area of 440 km², and their EEZ is 108,000 km². **Figure 2.** Domestic and foreign catches taken in the EEZ of Antigua & Barbuda: (A) by sector (domestic scores: Art. 2, 3, 3; Sub. 4, 3, 3; Rec. 2, 2, 2); (B) by fishing country (foreign catches are very uncertain); (C) by taxon.

REFERENCES

Carr LM and Heyman WD (2008) Jamaica bound? Marine resources and management at a crossroads in Antigua and Barbuda. *The Geographical Journal* 175(1): 17–38.

Georges J, Ramdeen R, Zylich K and Zeller D (2015) Reconstruction of total marine fisheries catch for Antigua and Barbuda (1950–2010). Fisheries Centre Working Paper #2015–13, 18 p.

Horsford I (2002) Economic viability of marine capture fisheries in Antigua & Barbuda: A case study. pp. 100–117 In FAO (ed.), Fisheries Report No. 640. Antigua & Barbuda Fisheries Division, St. Johns, Antigua.

1. Cite as Georges, J., R. Ramdeen, K. Zylich, and D. Zeller. 2016. Antigua & Barbuda. P. 189 in D. Pauly and D. Zeller (eds.), *Global Atlas of Marine Fisheries: A Critical Appraisal of Catches and Ecosystem Impacts*. Island Press, Washington, DC.

ARGENTINA[1]

Sebastian Villasante,[a,b] Gonzalo Macho,[b,c] Josu Isusu de Rivero,[a,b] Esther Divovich,[d] Kyrstn Zylich,[d] Dirk Zeller,[d] and Daniel Pauly[d]

[a]Department of Applied Economics, University of Santiago de Compostela, Spain
[b]Campus do Mar, International Campus of Excellence, Spain [c]Departamento de Ecoloxía e Bioloxía Animal, Universide de Vigo, Spain [d]*Sea Around Us*, University of British Columbia, Vancouver, Canada

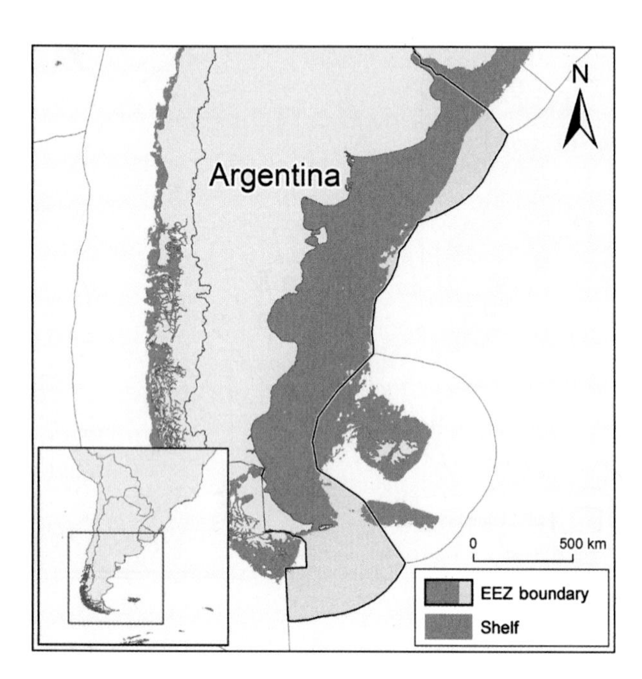

Figure 1. Argentina has a shelf of 796,000 km² and an EEZ of about 1.1 million km².

The main features of the massive shelf off Argentina, along the southeastern coast of South America (figure 1), is that it has gentle slopes and low relief, with soft sand bottom occurring on 65% of its surface, and rocky, hard bottom limited in extent (Bisbal 1995). This favored the development of industrial bottom trawl fisheries, notably targeting Argentinean hake *Merluccius hubbsi* (Dato et al. 2003). Total catches allocated to Argentina's EEZ were dominated by the industrial sector (figure 2A). Total domestic fisheries removals by Argentina's marine fisheries, as reconstructed by Villasante et al. (2015), were on the order of 110,000 t/year in the 1950s and increased almost exponentially to 2 million t in 1997 but declined thereafter to less than 1.4 million t/year in the 2000s. Domestic removals from the Argentinean EEZ were 1.55 times the data reported by the FAO, with the discrepancy caused mainly by unreported industrial catch and discards. Although catches were largely domestic, foreign catches, for example by China, South Korea, and Spain, were also documented (figure 2B). Hake accounted for a significant portion of total reconstructed catch, followed by Argentine shortfin squid (*Illex argentinus*), Patagonian grenadier (*Macroronus magellanicus*), and southern blue whiting (*Micromesistius australis*) (figure 2C). Although there are illegal forays by foreign fleets into Argentina's EEZ, the larger issue are the foreign fleets just outside the EEZ, potentially overexploiting straddling stocks.

REFERENCES

Bisbal GA (1995) The Southeast South American shelf large marine ecosystem: Evolution and components. *Marine Policy* 19(1): 21–38.

Dato CV, Villarino MF and Cañete GR (2003) Dinámica de la flota comercial argentina dirigida a la pesquería de merluza (*Merluccius hubbsi*) en el Mar Argentino. Período 1990–1997. INIDEP Informe Técnico 53, Instituto Nacional de Investigación y Desarrollo Pesquero, Mar del Plata, República Argentina, 25 p.

Villasante S, Macho G, Isusu de Rivero J, Divovich E, Zylich K, Harper S, Zeller D and Pauly D (2015) Reconstruction of Argentina's marine fisheries catches (1950–2010). Fisheries Centre Working Paper #2015-50, 6 p.

1. Cite as Villasante, S., G. Macho, J. Isusu de Rivero, E. Divovich, K. Zylich, D. Zeller, and D. Pauly. 2016. Argentina. P. 190 in D. Pauly and D. Zeller (eds.), *Global Atlas of Marine Fisheries: A Critical Appraisal of Catches and Ecosystem Impacts*. Island Press, Washington, DC.

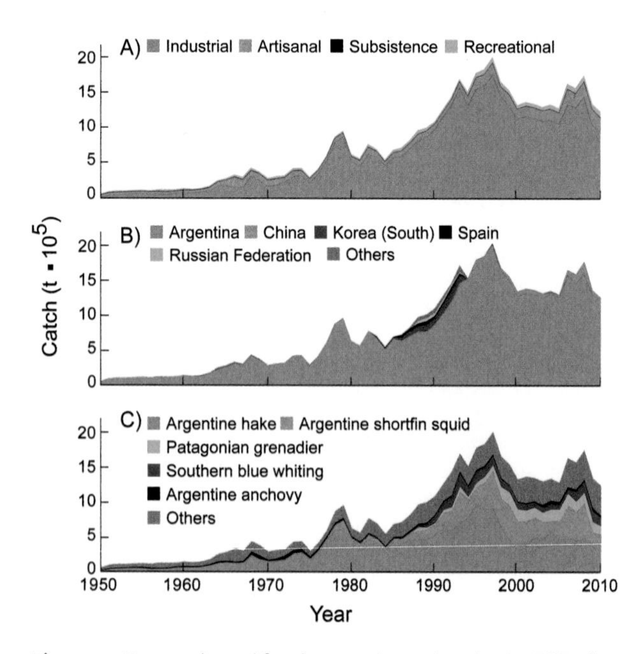

Figure 2. Domestic and foreign catches taken in the EEZ of Argentina: (A) by sector (domestic scores: Ind. 1, 1, 2; Art. 1, 1, 2; Sub. 1, 1, 2; Rec. 1, 1, 2; Dis. 1, 1, 2); (B) by fishing country (foreign catches are very uncertain); (C) by taxon.

AUSTRALIA[1]

Kristin M. Kleisner,[a] Ciara Brennan,[a] Anna Garland,[b] Stephanie Lingard,[a] Sean Tracey,[c] Paul Sahlqvist,[d] Angelo Tsolos,[e] Daniel Pauly,[a] and Dirk Zeller[a]

[a]*Sea Around Us*, University of British Columbia, Vancouver, Canada [b]Fisheries Queensland, Brisbane, Australia [c]University of Tasmania, Tasmania, Australia [d]Australian Bureau of Agricultural and Resource Economics and Sciences, Canberra, Australia [e]South Australia Research and Development Institute, Henley Beach, Australia

The Australian mainland (here including Tasmania but excluding other Australian islands; figure 1) has a diverse marine fauna, with more than 4,560 species of fish (411 endemic; www.fishbase.org, August 2015), and equally diverse fisheries. Reporting of landings from the commercial sector is generally accurate and taxonomically precise (Leatherbarrow et al. 2010). However, discards and landings from the recreational and indigenous sectors are not reported to the FAO. Industrial catches (69%), followed by the artisanal (17%) and recreational (13%) sectors, accounted for the majority of the total allocated catch in the Australian EEZ (figure 2A). Domestic landings from all sectors plus discards, as reconstructed by Kleisner et al. (2015), were about 75,000 t/year in the early 1950s, peaked at 400,000 t in 1991, and slowly declined to 200,000 t/year in the late 2000s. This domestic catch is nearly double the amount reported by Australia to the FAO. This difference is largely due to discards, which have declined substantially, and of which an unestimated but probably small fraction may survive after being discarded. Although most of the allocated catch was domestic, foreign fishing by Indonesia and Japan, among others, was also documented (figure 2B). Figure 2C presents a taxonomic breakdown of all catches, which include demersal species such as flatheads (family Platycephalidae) and smelt-whitings (Sillaginidae), as well as valuable species such as southern bluefin tuna (*Thunnus maccoyii*) and Australian spiny lobster (*Panulirus cygnus*).

REFERENCES

Kleisner KM, Brennan C, Garland A, Lingard S, Tracey S, Sahlqvist P, Tsolos A, Pauly D and Zeller D (2015) Australia: reconstructing estimates of total fisheries removals 1950–2010. Fisheries Centre Working Paper #2015–02, 26 p.

Leatherbarrow A, Sampaklis S and Mazur K (2010) Fishery status reports 2009: Status of fish stocks and fisheries managed by the Australian Government. Australian Bureau of Agricultural and Resource Economics, Bureau of Rural Sciences, Canberra. x+535 p.

1. Cite as Kleisner, K. M., C. Brennan, A. Garland, S. Lingard, S. Tracey, P. Sahlqvist, A. Tsolos, D. Pauly, and D. Zeller. 2016. Australia. P. 191 in D. Pauly and D. Zeller (eds.), *Global Atlas of Marine Fisheries: A Critical Appraisal of Catches and Ecosystem Impacts*. Island Press, Washington, DC.

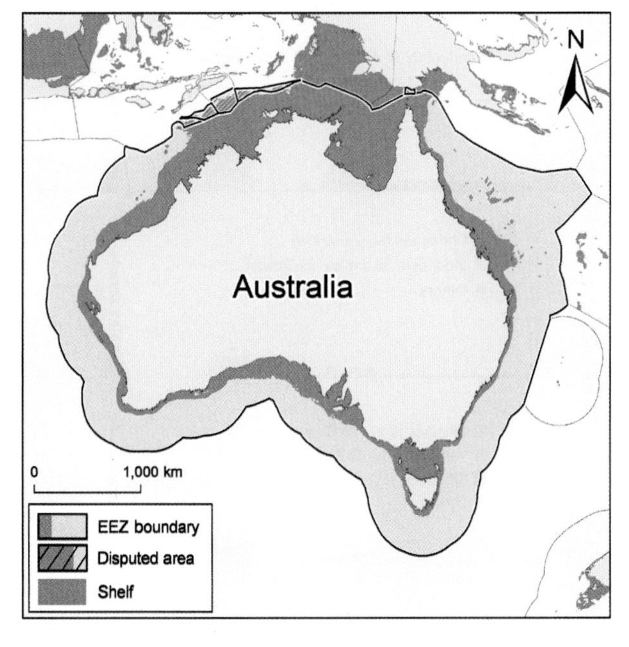

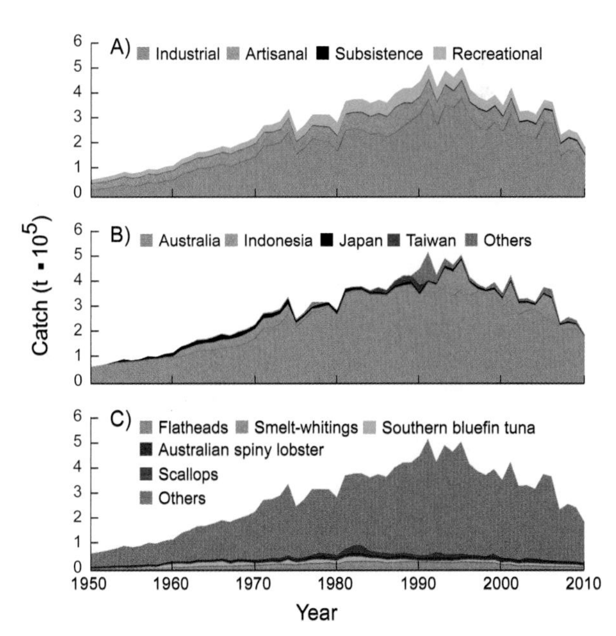

Figure 1. Australia has a shelf of 2.18 million km² and an EEZ of 6.37 million km². **Figure 2.** Domestic and foreign catches taken in the EEZ of Australia: (A) by sector (domestic scores: Ind. 2, 3, 4; Art. 2, 3, 4; Sub. 1, 1, 1; Rec. 1, 2, 3; Dis. 2, 2, 3); (B) by fishing country (foreign catches are very uncertain); (C) by taxon.

AUSTRALIA (CHRISTMAS ISLAND)[1]

Krista Greer, Sarah Harper, Dirk Zeller, and Daniel Pauly

Sea Around Us, University of British Columbia, Vancouver, Canada

Christmas Island is located 2,600 km northwest of Perth and 290 km south of Java, Indonesia (figure 1). Christmas Island remained uninhabited until the late 1800s, but its population, derived mostly from nearby Indonesia, grew after phosphate mining took off. In 1957, Christmas Island was transferred from Britain to Australia, and most of its land area has since been designated a national park. Four resorts currently form the backbone of a small tourism industry, devoted mainly to fishing and diving (Hourston 2010). The total allocated catch within Christmas Island's EEZ is predominantly from the industrial sector, with small catches, hardly visible in figure 2A, from the artisanal, subsistence, and recreational sectors. Reconstructed domestic catch, assembled by Greer et al. (2012, 2014) from fragmentary evidence, appears to have peaked at 72 t in 1966; from 1991 to the mid-1990s, it almost doubled, resulting in a secondary peak, caused by the introduction of large-scale fishing vessels in 1992. Then catches declined again, to 31 t/year by the late 2000s. Foreign catch were documented from Indonesia and probably other countries (figure 2B). From 1950 to 2010, the catch was dominated by skipjack tuna (*Katsuwonus pelamis*) and bigeye tuna (*Thunnus obesus*; figure 2C). The scarcity of information on Christmas Island is alarming, which also applies to illegal fishing in the EEZ surrounding the island, assumed to be substantial.

REFERENCES

Greer K, Harper S, Zeller D and Pauly D (2012) Cocos (Keeling) Islands and Christmas Island: Brief history of fishing and coastal catches (1950–2010). pp. 1–14 In Harper S, Zylich K, Boonzaier L, Le Manach F, Pauly D and Zeller D (eds.), *Fisheries catch reconstructions: Islands Part III*. Fisheries Centre Research Reports 20(5).

Greer K, Harper S, Zeller D and Pauly D (2014) Evidence for overfishing on pristine coral reefs: reconstructing coastal catches in the Australian Indian Ocean Territories. *Journal of the Indian Ocean Region* 10: 67–80.

Hourston M (2010) Review of the exploitation of marine resources of the Australian Indian Ocean Territories: The implications of biogeographic isolation for tropical island fisheries. Fisheries Research Report No. 208, Government of Western Australia, Department of Fisheries, Western Australia. iv + 45 p.

1. Cite as Greer, K., S. Harper, D. Zeller, and D. Pauly. 2016. Australia (Christmas Island). P. 192 in D. Pauly and D. Zeller (eds.), *Global Atlas of Marine Fisheries: A Critical Appraisal of Catches and Ecosystem Impacts*. Island Press, Washington, DC.

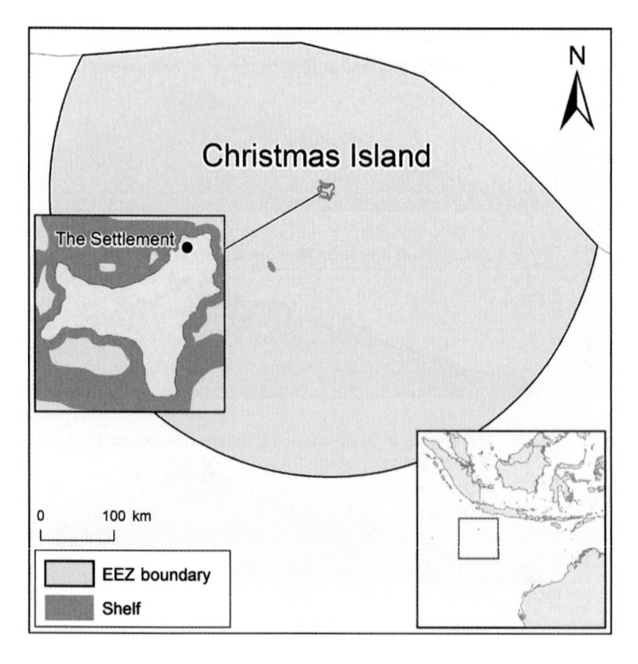

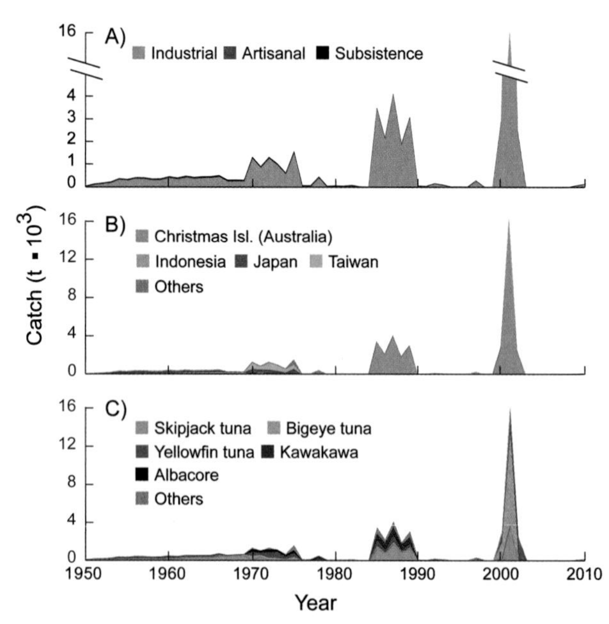

Figure 1. Christmas Island (Australia) has a land area of 139 km² and an EEZ of 328,000 km². **Figure 2.** Domestic and foreign catches taken in the EEZ of Christmas Island (Australia): (A) by sector (domestic scores: Ind. –, –, 2; Art. 1, 1, 1; Sub. 1, 1, 1; Rec. 1, 1, 1; Dis. 1, 1, 1); (B) by fishing country (foreign catches are very uncertain); (C) by taxon.

AUSTRALIA (COCOS [KEELING] ISLANDS)[1]

Krista Greer, Sarah Harper, Dirk Zeller, and Daniel Pauly

Sea Around Us, University of British Columbia, Vancouver, Canada

The Cocos (Keeling) Islands (figure 1), an atoll Charles Darwin famously visited in 1836 (Pauly 2004), are an Australian Indian Ocean territory where fisheries catch data are not routinely collected. Total catch allocated to the EEZ of Cocos (Keeling) Islands indicates that the bulk of removals are derived from the industrial sector, at least before the declaration of the EEZ in the late 1970s (figure 2A). An analysis of the extant literature combined with the reconstruction by Greer et al. (2012, 2014) suggests that the fisheries in the Cocos (Keeling) Island EEZ generated a catch of approximately 80 t/year in the 1950s (subsistence only), which increased, starting in the mid-1980s, to 250 t/year in recent years, mainly because of the introduction of recreational and later large-scale (industrial) fishing. Foreign fishing, largely by Japan and Taiwan, has been documented since the 1950s (figure 2B). Figure 2C illustrates that the total allocated catch is dominated by yellowfin tuna (*Thunnus albacares*), bigeye tuna (*Thunnus obesus*), and albacore tuna (*Thunnus alalungus*). Overfishing may have been occurring in the waters of the Cocos (Keeling) Islands since the late 1990s, as indicated by anecdotal evidence describing local depletions of less resilient species such as the groupers (family Serranidae; Hender et al. 2001). As part of a comprehensive coastal management plan, fisheries managers should therefore focus on determining primary target species and their vulnerability to overfishing.

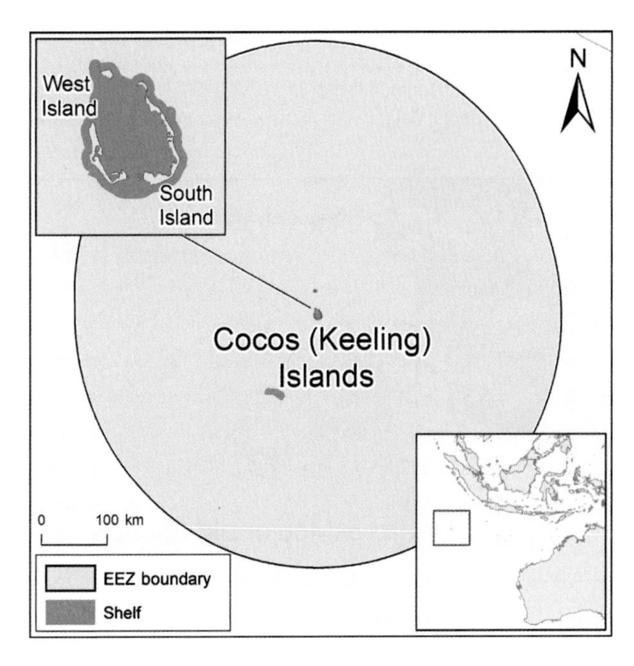

Figure 1. The Cocos (Keeling) Islands (Australia) have a land area of 24.0 km² and an EEZ of 467,000 km².

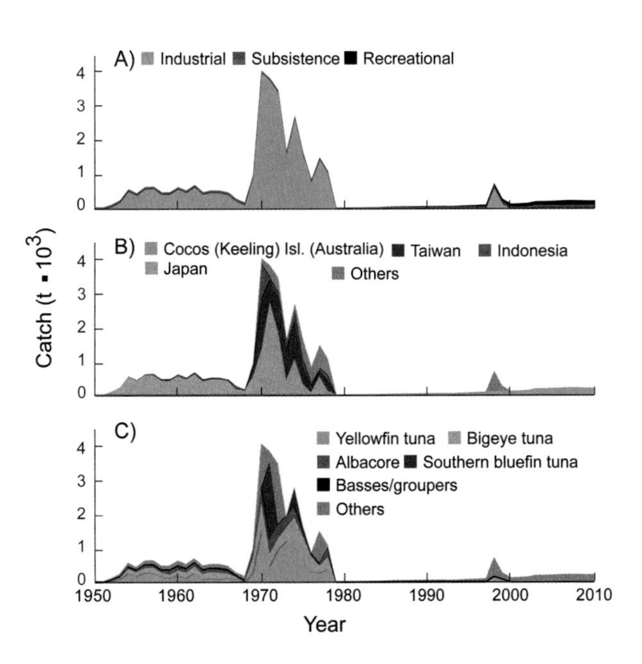

Figure 2. Domestic and foreign catches taken in the EEZ of Cocos (Keeling) Islands: (A) by sector (domestic scores: Ind. –, –, 3; Art. –, 1, 1; Sub. 1, 1, 2; Rec. –, 1, 2; (B) by fishing country (foreign catches are very uncertain); (C) by taxon.

REFERENCES

Greer K, Harper S, Zeller D and Pauly D (2012) Cocos (Keeling) Islands and Christmas Island: Brief history of fishing and coastal catches (1950–2010). pp. 1–14 In Harper S, Zylich K, Boonzaier L, Le Manach F, Pauly D and Zeller D (eds.), *Fisheries catch reconstructions: Islands part III*. Fisheries Centre Research Reports 20(5).

Greer K, Harper S, Zeller D and Pauly D (2014) Evidence for overfishing on pristine coral reefs: reconstructing coastal catches in the Australian Indian Ocean Territories. *Journal of the Indian Ocean Region* 10: 67–80.

Hender J, McDonald CA and Gilligan JJ (2001) Baseline surveys of the marine environments and stock size estimates of marine resources of the South Cocos (Keeling) Atoll (0–15m), eastern Indian Ocean. Marine Resources of the Cocos Atoll, Indian Ocean. iii + 69 p.

Pauly D (2004) Darwin's fishes: An encyclopedia of ichthyology, ecology and evolution. Cambridge University Press, UK. 340 p.

1. Cite as Greer, K., S. Harper, D. Zeller, and D. Pauly. 2016. Australia (Cocos [Keeling] Islands). P. 193 in D. Pauly and D. Zeller (eds.), *Global Atlas of Marine Fisheries: A Critical Appraisal of Catches and Ecosystem Impacts*. Island Press, Washington, DC.

AUSTRALIA (HEARD AND MCDONALD ISLANDS)[1]

Kristin M. Kleisner,[a] Ciara Brennan,[a] Anna Garland,[b] Stephanie Lingard,[a] Sean Tracey,[c] Paul Sahlqvist,[d] Angelo Tsolos,[e] Daniel Pauly,[a] and Dirk Zeller[a]

[a]*Sea Around Us*, University of British Columbia, Vancouver, Canada [b]Fisheries Queensland, Brisbane, Australia [c]University of Tasmania, Tasmania, Australia [d]Australian Bureau of Agricultural and Resource Economics and Sciences, Canberra, Australia [e]South Australia Research and Development Institute, Henley Beach, Australia

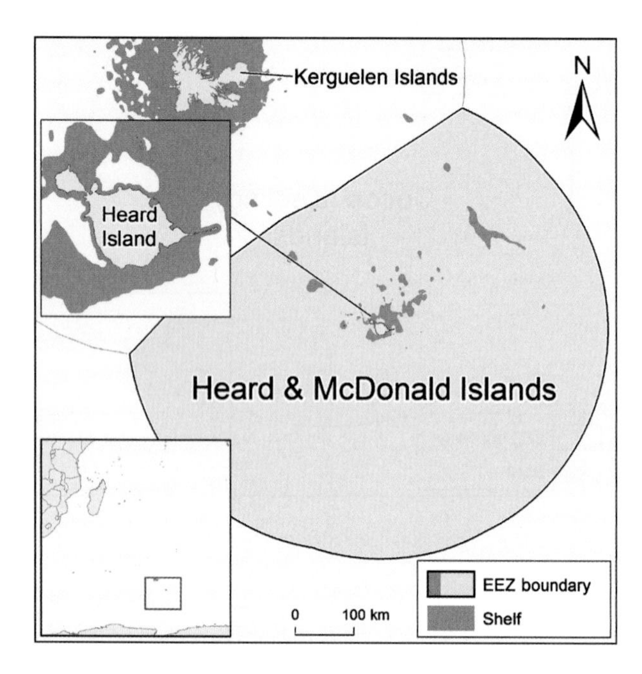

Figure 1. Heard and McDonald Islands have a land area of 363 km² and an EEZ of 417,000 km².

The Heard and McDonald Islands, part of Australia's external territories, are located in the Southern Ocean (figure 1). The fisheries around the islands are managed by the Australian Fisheries Management Authority (AFMA) and the Commission on the Conservation of Antarctic Marine Living Resources (CCAMLR). The majority of the total allocated catch within the Heard and McDonald Islands is from the domestic industrial fleets, with small removals from Japan and Ukraine (although not visible on graph) in the earlier time period (figure 2A). Total legal landings from the domestic fleet, as assembled by Kleisner et al. (2015) were 3 t/year in the early 1990s and 3,000 t/year in the late 2000s, with a peak in 2002 of approximately 4,500 t. Australian fishing within the Heard Island and McDonald Islands EEZ began in 1997 and targets Patagonian toothfish (*Dissostichus eleginoides*) and mackerel icefish (*Champsocephalus gunnari*) (figure 2B). The Heard Island and McDonald Islands Marine Reserve was established in 2002 and protects 65,000 km² of the EEZ; it is a strict nature reserve in which all fishing is prohibited. The AFMA patrols these waters quite effectively, as illustrated by the epic chase, all the way to South Africa, of the *Viarsa*, a pirate vessel that had been illegally fishing for Patagonian toothfish (a.k.a. "Chilean seabass") in the Heard and McDonald Island EEZ (Knecht 2006).

REFERENCES

Kleisner KM, Brennan C, Garland A, Lingard S, Tracey S, Sahlqvist P, Tsolos A, Pauly D and Zeller D (2015) Australia: reconstructing estimates of total fisheries removals 1950–2010. Fisheries Centre Working Paper #2015–02, 26 p.

Knecht, GB 2006. *Hooked: Pirates, Poaching, and the Perfect Fish*. Rodale, Emmaus, PA, 278 p.

1. Cite as Kleisner, K. M., C. Brennan, A. Garland, S. Lingard, S. Tracey, P. Sahlqvist, A. Tsolos, D. Pauly, and D. Zeller. 2016. Australia (Heard and McDonald Islands). P. 194 in D. Pauly and D. Zeller (eds.), *Global Atlas of Marine Fisheries: A Critical Appraisal of Catches and Ecosystem Impacts*. Island Press, Washington, DC.

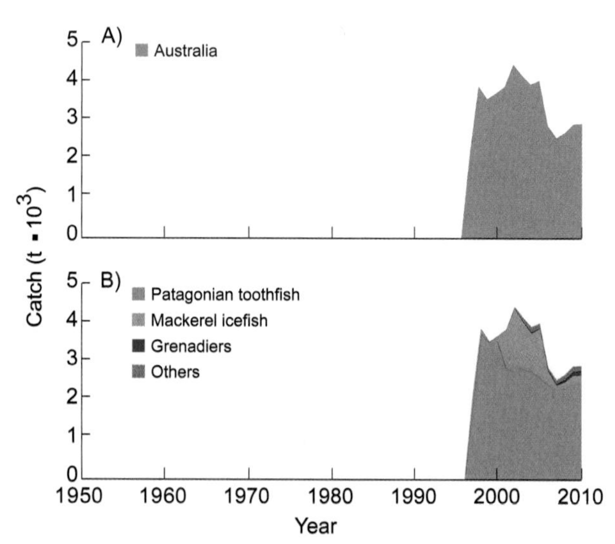

Figure 2. Domestic and foreign catches taken in the EEZ of Australia (Heard and McDonald Islands): (A) by fishing country (domestic scores: Ind. –, –, 4; Dis. 2, 2, 3; foreign catches are very uncertain); (B) by taxon.

AUSTRALIA (LORD HOWE ISLAND)[1]

Kristin M. Kleisner,[a] Ciara Brennan,[a] Anna Garland,[b] Stephanie Lingard,[a] Sean Tracey,[c] Paul Sahlqvist,[d] Angelo Tsolos,[e] Daniel Pauly,[a] and Dirk Zeller[a]

[a]*Sea Around Us*, University of British Columbia, Vancouver, Canada [b]Fisheries Queensland, Brisbane, Australia [c]University of Tasmania. Tasmania, Australia [d]Australian Bureau of Agricultural and Resource Economics and Sciences, Canberra, Australia [e]South Australia Research and Development Institute, Henley Beach, Australia

Lord Howe Island is in the South Pacific Ocean, approximately 700 km east of Sydney in FAO fishing area 81 (figure 1) and is under the jurisdiction of the Australian state of New South Wales. In 1982, the island was listed as a UNESCO World Heritage Area because of its high biodiversity. Also, in 2000 the waters between 2 and 12 nautical miles from shore were proclaimed a marine park under the National Parks and Wildlife Conservation Act of 1975 (Anonymous 2002). The total allocated catch is dominated by the industrial sector; additionally, there are local subsistence and recreational fisheries, though it was suggested that no more than 20 t/year is taken (Anonymous 2002). Kleisner et al. (2015) estimated that total landings from all sectors plus discards were 15 t/year in the early 1950s and have increased to an average of 263 t/year in the late 2000s. Removals within the Lord Howe Island EEZ are mainly from foreign vessels (figure 2A). Distant-water fisheries for tuna, billfish, and southern squid operate in waters adjacent to the Lord Howe Island EEZ, into which their vessels can be expected to launch occasional forays. Figure 2B presents a taxonomic breakdown of these catches, with squids, albacore tuna (*Thunnus alalunga*), and mackerels (*Scomberomorus* spp.) dominating.

REFERENCES

Anon (2002) Lord Howe Island Marine Park (Commonwealth waters). Management plan. Natural Heritage Trust, Environment Australia, Canberra. xii + 72 p.

Kleisner KM, Brennan C, Garland A, Lingard S, Tracey S, Sahlqvist P, Tsolos A, Pauly D and Zeller D (2015) Australia: reconstructing estimates of total fisheries removals 1950–2010. Fisheries Centre Working Paper #2015–02, 26 p.

1. Cite as Kleisner, K. M., C. Brennan, A. Garland, S. Lingard, S. Tracey, P. Sahlqvist, A. Tsolos, D. Pauly, and D. Zeller. 2016. Australia (Lord Howe Island). P. 195 in D. Pauly and D. Zeller (eds.), *Global Atlas of Marine Fisheries: A Critical Appraisal of Catches and Ecosystem Impacts*. Island Press, Washington, DC.

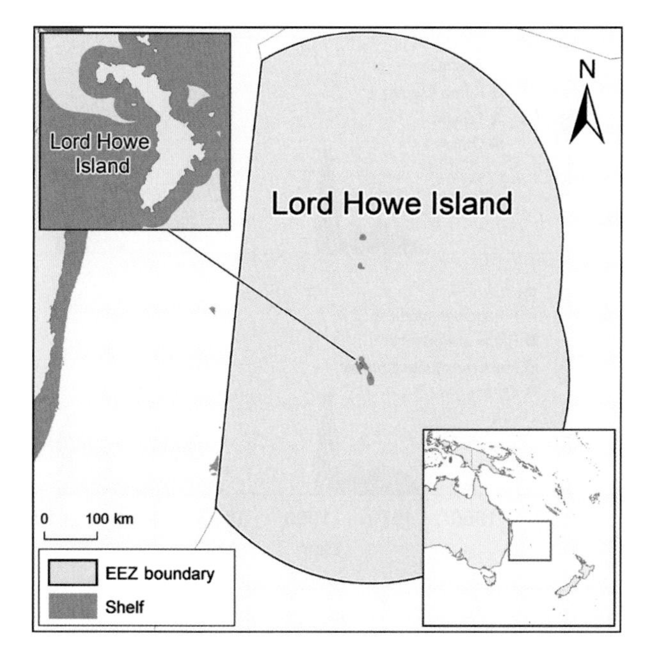

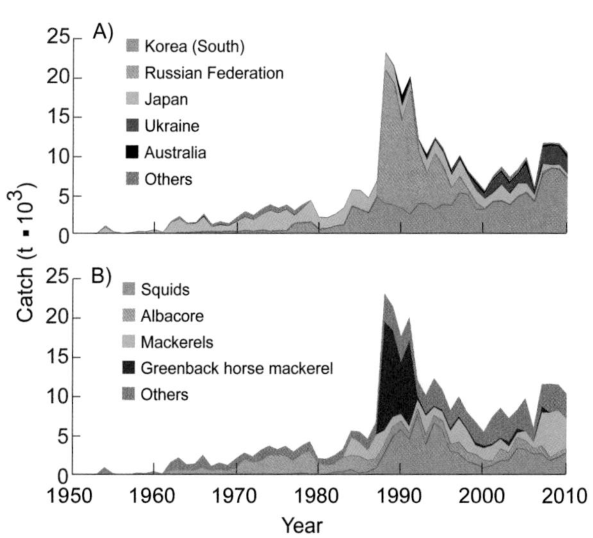

Figure 1. Lord Howe Island has a land area of 19.2 km² and an EEZ of 542,000 km². **Figure 2.** Domestic and foreign catches taken in the EEZ of Australia (Heard and McDonald): (A) by fishing country (domestic scores: Ind. –, –, –; Art. 2, 3, 4; Sub. 1, 1, 1; Dis. 2, 2, 3; foreign catches are very uncertain); (B) by taxon.

AUSTRALIA (MACQUARIE ISLAND)[1]

Kristin M. Kleisner,[a] Ciara Brennan,[a] Anna Garland,[b] Stephanie Lingard,[a] Daniel Pauly,[a] and Dirk Zeller[a]

[a]*Sea Around Us*, University of British Columbia, Vancouver, Canada [b]Fisheries Queensland, Brisbane, Australia
[c]University of Tasmania. Tasmania, Australia [d]Australian Bureau of Agricultural and Resource Economics and Sciences, Canberra, Australia [e]South Australia Research and Development Institute, Henley Beach, Australia

Macquarie Island is under the jurisdiction of the Australian state of Tasmania, and it is in the southwest Pacific Ocean, approximately 1,000 km southeast of Tasmania (figure 1). Macquarie Island became a wildlife sanctuary under the Tasmanian Animals and Birds Protection Act of 1928; in 1972 it also became a State Reserve under the Tasmanian National Parks and Wildlife Act of 1970. However, there are fisheries around Macquarie Island, and they are managed by the Australian Fisheries Management Authority (AFMA) and the Commission on the Conservation of Antarctic Marine Living Resources (CCAMLR). The total allocated catch was dominated by South Korea in the early time period and by the domestic fleet in the later time period (figure 2A). Total landings were about 530 t/year in the 1990s and 300 t/year in the late 2000s, with a peak in 1997 of approximately 850 t (Kleisner et al. 2015). Figure 2B presents a taxonomic breakdown of these catches, which consist mainly of Patagonian toothfish (*Dissostichus eleginoides*), and ridge scaled rattail (*Macrourus carinatus*). The feeding interactions within the marine ecosystem around Macquarie Island are being studied (see Goldsworthy et al. 2002; van den Hoff 2004), which, together with considering all fisheries withdrawals, should help foster an understanding of the functioning of this ecosystem and thus assist in the management of the fisheries embedded therein.

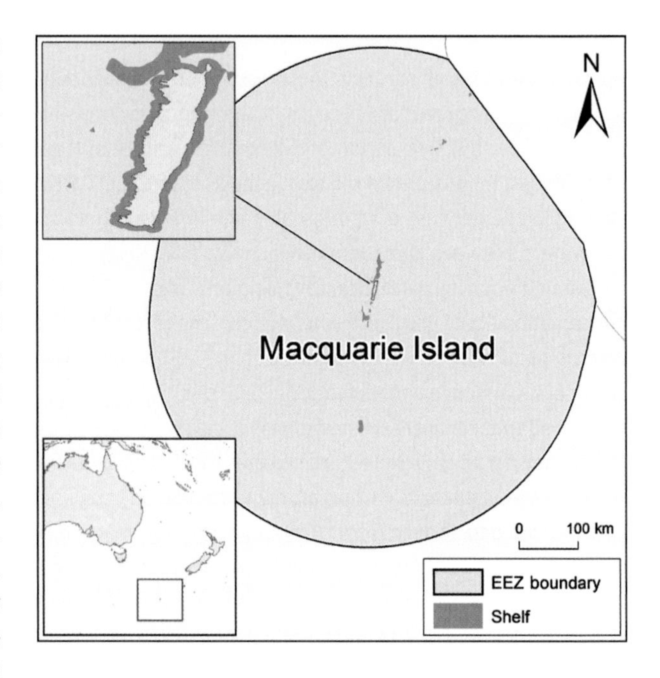

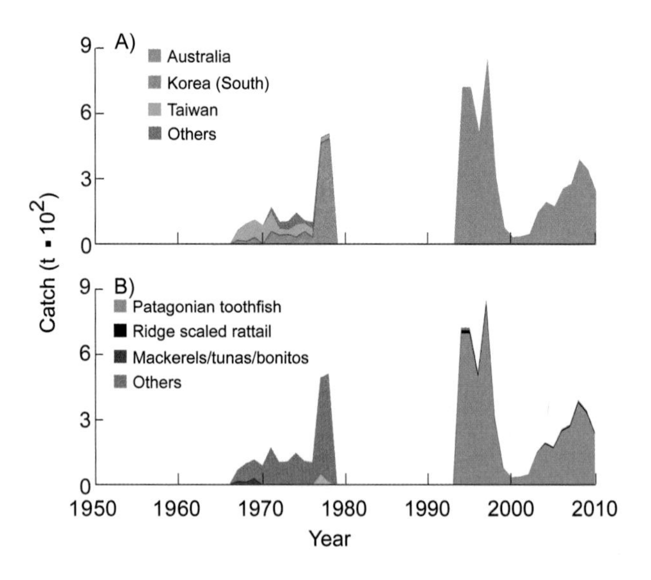

Figure 1. Macquarie Island has a land area of 134 km² and an EEZ of 477,000 km². **Figure 2.** Domestic and foreign catches taken in the EEZ of Australia (Macquarie Island): (A) by fishing country, all deemed foreign (foreign catches are very uncertain); (B) by taxon.

REFERENCES

Goldsworthy SD, Lewis M, Williams R, He X, Young J and van den Hoff J (2002) Diet of Patagonian toothfish (*Dissostichus eleginoides*) around Macquarie Island, South Pacific Ocean. *Marine and Freshwater Research* 53: 304–612.

Kleisner KM, Brennan C, Garland A, Lingard S, Tracey S, Sahlqvist P, Tsolos A, Pauly D and Zeller D (2015) Australia: reconstructing estimates of total fisheries removals 1950–2010. Fisheries Centre Working Paper #2015–02, 26 p.

van den Hoff J (2004) A comparative study of the cephalopod prey of Patagonian toothfish (*Dissostichus eleginoides*) and southern elephant seals (*Mirounga leonina*) near Macquarie Island. *Polar Biology* 27(10): 604–612.

1. Cite as Kleisner, K. M., C. Brennan, A. Garland, S. Lingard, S. Tracey, P. Sahlqvist, A. Tsolos, D. Pauly, and D. Zeller. 2016. Australia (Macquarie Island). P. 196 in D. Pauly and D. Zeller (eds.), *Global Atlas of Marine Fisheries: A Critical Appraisal of Catches and Ecosystem Impacts*. Island Press, Washington, DC.

AUSTRALIA (NORFOLK ISLAND)[1]

Kristin M. Kleisner,[a] Ciara Brennan,[a] Anna Garland,[b] Stephanie Lingard,[a] Sean Tracey,[c] Paul Sahlqvist,[d] Angelo Tsolos,[e] Dirk Zeller,[a] and Daniel Pauly[a]

[a]*Sea Around Us*, University of British Columbia, Vancouver, Canada [b]Fisheries Queensland, Brisbane, Australia [c]University of Tasmania, Tasmania, Australia [d]Australian Bureau of Agricultural and Resource Economics and Sciences, Canberra, Australia [e]South Australia Research and Development Institute, Henley Beach, Australia

Norfolk Island, one of Australia's external territories, is located in the South Pacific Ocean, approximately 1,500 km east of Brisbane in FAO fishing area 81 (figure 1). Norfolk Island is represented separately from Australia in the FAO database, and thus Kleisner et al. (2015) performed an independent catch reconstruction. Based on the reconstruction and literature sources, fishing activities around Norfolk Island EEZ were documented to be mostly industrial, with small removals by the inshore artisanal, recreational, and subsistence fisheries (Leatherbarrow et al. 2010; not all visible on figure 2A). Total reconstructed domestic catches for Norfolk Island were approximately 9 t/year in the 1950s, peaked at about 40 t in early 2000s, and dropped to 6 t/year in the late 2000s. The majority of removals were foreign fishing by Japan, among others, until the 1990s, when the domestic industrial fleets dominated the catch (figure 2B). Some of the catch was not identifiable by species, but albacore tuna (*Thunnus alalunga*), yellowfin tuna (*Thunnus albacares*), and bigeye tuna (*Thunnus obesus*) were caught throughout this period (figure 2C). It is likely that more species are being caught, given that 322 species of fish are reported from Norfolk Island (www.fishbase.org, Feb. 2014; Francis and Randall 1993); however, to remain conservative we relied here on AFMA (2010) for the primary species in the catch.

REFERENCES

AFMA (2010) Norfolk Island Inshore Fishery Data Summary 2006–2009. Australian Fisheries Management Authority Canberra. 20 p.

Francis MP and Randall JE (1993) Further additions to the fish faunas of Lord Howe and Norfolk Islands, Southwest Pacific Ocean. *Pacific Science* 47(2): 118–135.

Kleisner KM, Brennan C, Garland A, Lingard S, Tracey S, Sahlqvist P, Tsolos A, Pauly D and Zeller D (2015) Australia: reconstructing estimates of total fisheries removals 1950–2010. Fisheries Centre Working Paper #2015-02, 26 p.

Leatherbarrow A, Sampaklis S and Mazur K (2010) Fishery status reports 2009: Status of fish stocks and fisheries managed by the Australian Government. Australian Bureau of Agricultural and Resource Economics, Bureau of Rural Sciences, Canberra. x+535 p.

1. Cite as Kleisner, K. M., C. Brennan, A. Garland, S. Lingard, S. Tracey, P. Sahlqvist, A. Tsolos, D. Pauly, and D. Zeller. 2016. Australia (Norfolk Island). P. 197 in D. Pauly and D. Zeller (eds.), *Global Atlas of Marine Fisheries: A Critical Appraisal of Catches and Ecosystem Impacts*. Island Press, Washington, DC.

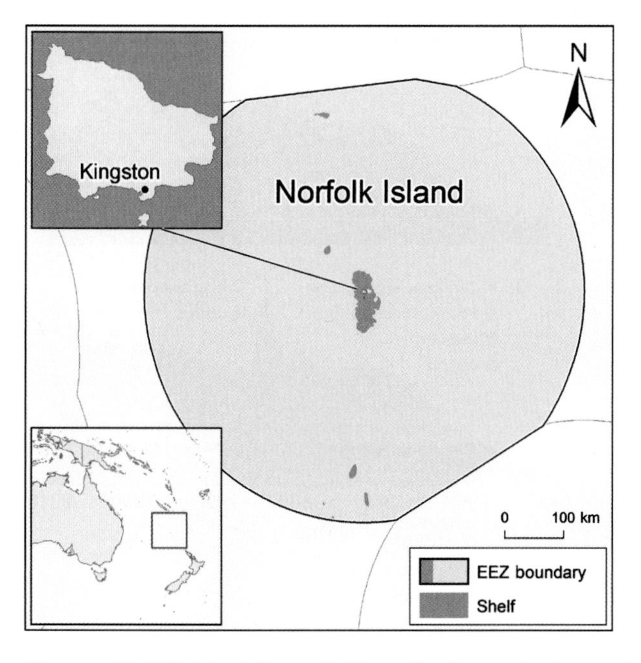

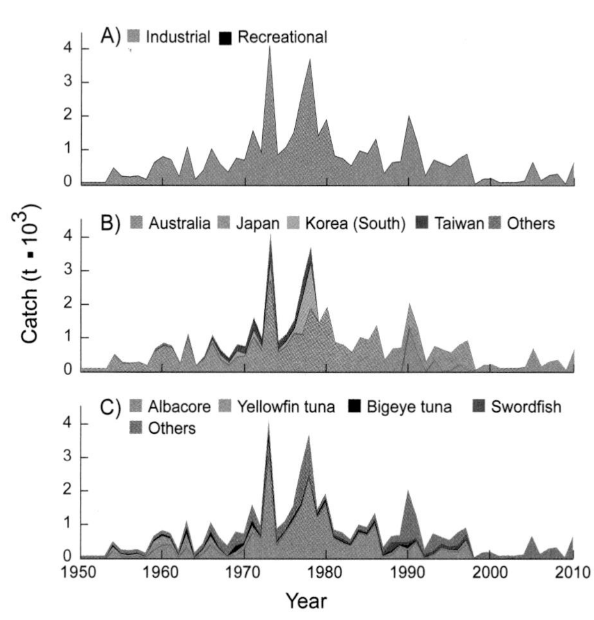

Figure 1. Norfolk Island (Australia) has a land area of 41 km² and an EEZ of 431,000 km². **Figure 2.** Domestic and foreign catches taken in the EEZ of Australia (Norfolk Island): (A) by sector (domestic scores: Ind. –, –, 4; Art. 2, 3, 4; Sub. 1, 1, 1; Rec. 1, 2, 3; Dis. 2, 2, 3); (B) by fishing country (foreign catches are very uncertain); (C) by taxon.

BAHAMAS[1]

Nicola S. Smith[a,b] and Dirk Zeller[c]

[a]Department of Marine Resources, Ministry of Agriculture, Marine Resources & Local Government, Nassau, Bahamas
[b]Earth to Ocean Research Group, Department of Biological Sciences, Simon Fraser University, Burnaby, Canada
[c]*Sea Around Us*, University of British Columbia, Vancouver, Canada

The Commonwealth of the Bahamas is an archipelago of more than 3,000 low-lying islands and cays located east of Florida in the United States (figure 1). Tourism is the primary industry, and since the 1970s the total number of annual visitors has outnumbered the resident population. Both tourists and residents expect to catch and eat local fish. The recreational sector dominates catches within the EEZ for the time period, although the industrial sector appears to have increasing catches (figure 2A). This account, based on a reconstruction by Smith and Zeller (2013, 2016), presents a comprehensive accounting of Bahamian catches, which ranged from about 2,000 t/year in the early 1950s to a peak of nearly 25,000 t in 1985 and 15,000–18,000 t/year in the late 2000s. Overall, the reconstructed catches were 2.6 times higher than the landings presented by the FAO on behalf of the Bahamas, mainly because of failure to report the catches from the recreational and subsistence fisheries. In particular, the former sector, described in Deleveaux and Higgs (1995), contributed 55% of reconstructed total catch yet remains unreported. The domestic sectors dominate catches; however, some foreign fishing, notably by Cuba, has been documented. These results provide a novel baseline for historic fisheries catches and their composition, notably Caribbean spiny lobster (*Panulirus argus*; figure 2B), which should be revised as better data become available.

REFERENCES

Deleveaux V and Higgs C (1995) A preliminary analysis of trends in the fisheries of the Bahamas based on the fisheries census (1995). Proceedings of the 48th Gulf and Caribbean Fisheries Institute 48: 353–359.

Smith NS and Zeller D (2013) Bahamas catch reconstruction: fisheries trends in a tourism-driven economy (1950–2010). Fisheries Centre Working Paper #2013–08, 28 p.

Smith NS and Zeller D (2016) Unreported catch and tourist demand on local fisheries of small island states: the case of the Bahamas, 1950–2010. *Fishery Bulletin* 114: 117–131.

1. Cite as Smith, N. S., and D. Zeller. 2016. Bahamas. P. 198 in D. Pauly and D. Zeller (eds.), *Global Atlas of Marine Fisheries: A Critical Appraisal of Catches and Ecosystem Impacts*. Island Press, Washington, DC.

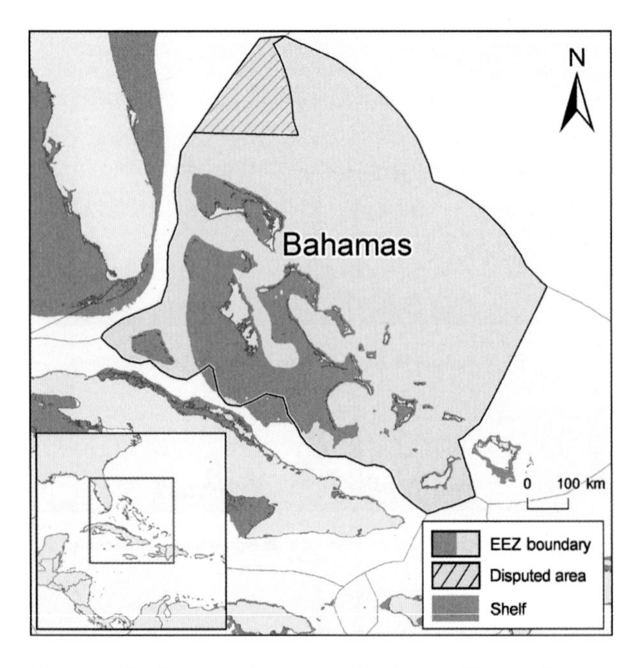

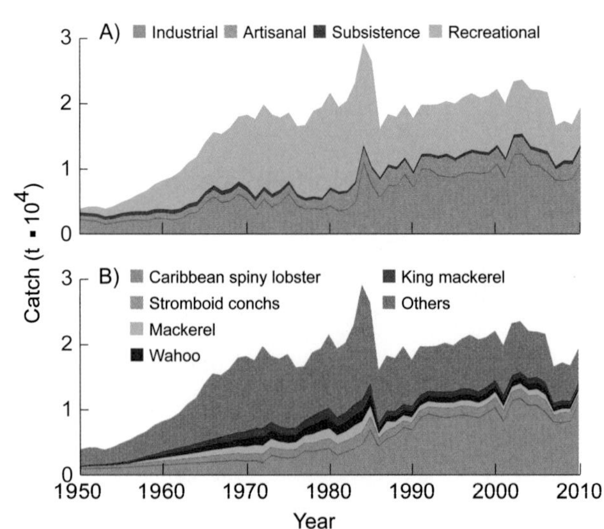

Figure 1. The Bahamas has a total land area of just under 14,000 km² and an EEZ of more than 628,000 km².
Figure 2. Fisheries catches in the EEZ of the Bahamas: (A) by sector (domestic scores: Ind. 2, 4, 4; Art. 3, 4, 4; Sub. 1, 1, 1; Rec. 1, 2, 3); (B) by taxon.

BAHRAIN[1]

Dalal Al-Abdulrazzak

Sea Around Us, University of British Columbia, Vancouver, Canada

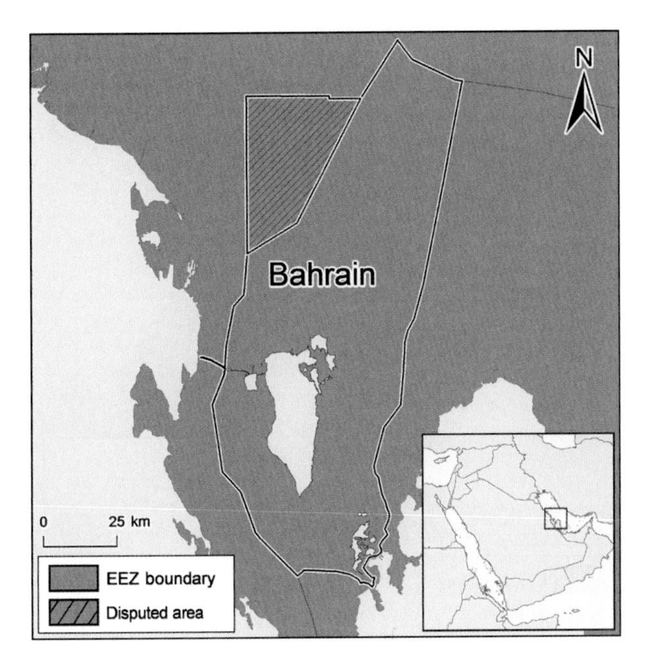

Figure 1. Bahrain (land area: 760 km²) and its EEZ (8,830 km², including the joint-regulation zone with Saudi Arabia).

Bahrain is the smallest of the Gulf States and the only island country in the region (figure 1). Bahrain has a rich maritime history that includes fishing and pearling. The fisheries consist of an important artisanal sector, whose catches dramatically increased in the 2000s, in contrast to industrial catches, which appeared to plummet (figure 2A). The reconstructed catch of Bahrain, based on Al-Abdulrazzak (2013), increased from 6,200 t in 1950 to 48,000 t in 1996, then declined and again increased to similar levels in the late 2000s, and corresponds to about 3.5 times the data reported by the FAO on behalf of Bahrain. The catch is overwhelmingly by the domestic fleet, although some foreign fishing does occur. The strong catch fluctuations are caused mainly by shrimp trawlers' discards (Abdulqader 2002) and not by the artisanal fishery, whose catches and growth were rather steady. Rabbitfishes (family Siganidae), swimming crab (Portunidae), herrings, shads, sardines and menhadens (Clupeidae), and fourlined terapon (*Pelates quadrilineatus*) were the major taxa in the catch, which also included a multitude of other species (figure 2B). There are growing concerns over Bahrain's fisheries (as for other Gulf fisheries; Al-Abdulrazzak et al. 2015), such as their continued use of illegal driftnets. Although catches appear to be increasing, it is likely that the decline of traditionally targeted taxa is masked by previously discarded species being retained for consumption by the increasing immigrant community in Bahrain.

REFERENCES

Abdulqader EAA (2002) The finfish bycatch of the Bahrain shrimp trawl fisheries. *Arab Gulf Journal of Scientific Research* 20(3): 165–174.

Al-Abdulrazzak D (2013) Missing sectors from Bahrain's reported fisheries catches: 1950–2010. pp. 1–6 In Al-Abdulrazzak D and Pauly D (eds.), *From dhows to trawlers: a recent history of fisheries in the Gulf countries, 1950 to 2010*. Fisheries Centre Research Reports 21(2).

Al-Abdulrazzak D, Zeller D, Belhabib D, Tesfamichael D, and Pauly D (2015) Total marine fisheries catches in the Persian/Arabian Gulf from 1950–2010. *Regional Studies in Marine Science* 2: 28–34.

1. Cite as Al-Abdulrazzak, D. 2016. Bahrain. P. 199 in D. Pauly and D. Zeller (eds.), *Global Atlas of Marine Fisheries: A Critical Appraisal of Catches and Ecosystem Impacts*. Island Press, Washington, DC.

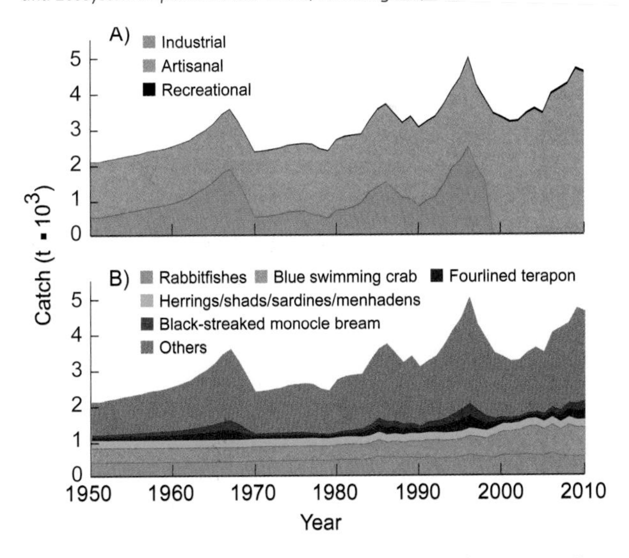

Figure 2. Domestic and foreign catches taken in the EEZ of Bahrain: (A) by sector (domestic scores: Ind. 1, 1, 1; Art. 3, 3, 3; Sub. 1, 1, 1; Rec. 2, 2, 2; Dis. 2, 2, 4); (B) by taxon.

BANGLADESH[1]

Hadayet Ullah,[a] Danielle Knip,[b] Darah Gibson,[b] Kyrstn Zylich,[b] and Dirk Zeller[b]

[a]WorldFish, Bangladesh Office, Dhaka, Bangladesh [b]*Sea Around Us*, University of British Columbia, Vancouver, Canada

Bangladesh opens to the Bay of Bengal (figure 1). In the absence of consistent catch data covering Bangladesh's fisheries (Islam 2003), Ullah et al. (2014) reconstructed these catches from 1950. FAO catch statistics were used as reported baseline, then adjusted using information from national reports, independent studies, local experts, and gray literature. The reconstruction yielded a catch of 220,000 t/year in the 1950s and 1 million t/year in the late 2000s, which was, overall, 2.5 times the landings reported by the FAO on behalf of Bangladesh. This discrepancy was caused mainly by unaccounted-for subsistence catches, underreported industrial catch, and discarded bycatch. Artisanal and subsistence fisheries dominate the catch (figure 2A). Although some foreign fishing, notably by Thailand, has been documented starting in the mid-1980s, the bulk of the catches are domestic. Hilsa shad (*Tenualosa ilisha*) was the largest single species-level contributor (figure 2B), and Bombay duck (*Harpadon nehereus*) was the most important species in the subsistence fishery. Islam (2003) reviewed the fisheries of Bangladesh and concluded that "the problems of the coastal and marine fisheries of Bangladesh are many and varied. The fisheries sector has been suffering from a chronic lack of well-planned management approaches, and persons with sufficient knowledge on the scientific basis of fisheries management and development are ignored from the higher level of policymaking."

REFERENCES

Islam MS (2003) Perspectives of the coastal and marine fisheries of the Bay of Bengal, Bangladesh. *Ocean & Coastal Management* 46: 763–796.

Ullah H, Gibson D, Knip D, Zylich K and Zeller D (2014) Reconstruction of total marine fisheries catches for Bangladesh: 1950–2010. Fisheries Centre Working Paper #2014-15, Vancouver. 10 p.

1. Cite as Ullah, H., D. Knip, D. Gibson, K. Zylich, and D. Zeller. 2016. Bangladesh. P. 200 in D. Pauly and D. Zeller (eds.), *Global Atlas of Marine Fisheries: A Critical Appraisal of Catches and Ecosystem Impacts*. Island Press, Washington, DC.

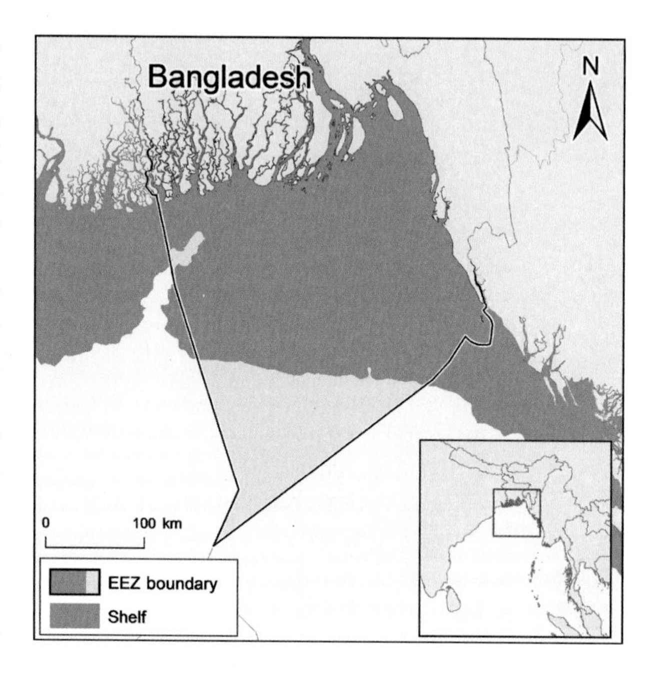

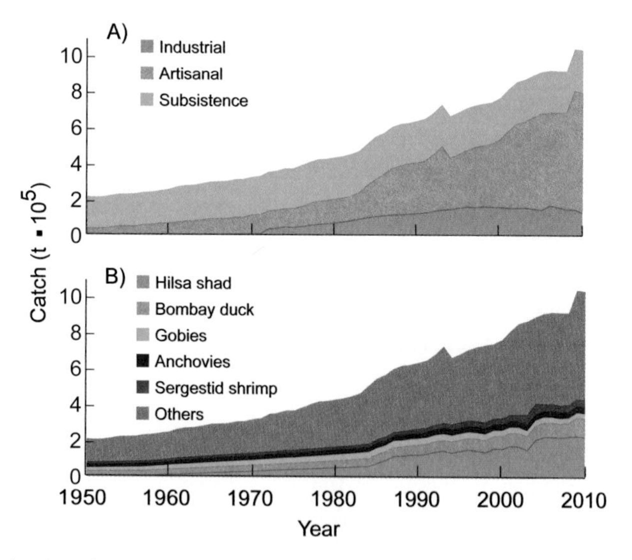

Figure 1. Bangladesh and its shelf (65,500 km²) and EEZ (84,800 km²).

Figure 2. Domestic and foreign catches taken in the EEZ of Bangladesh: (A) by sector (domestic scores: Ind. –, 2, 3; Art. 1, 2, 2; Sub. 2, 2, 2; Dis. 1, 2, 3); (B) by taxon.

BARBADOS[1]

Elizabeth Mohammed,[a] Alasdair Lindop,[b] Christopher Parker,[c] and Stephen Willoughby[c]

[a]Research and Resource Assessment, Caribbean Regional Fisheries Mechanism Secretariat, Eastern Caribbean Office, St. Vincent and the Grenadines [b]*Sea Around Us*, University of British Columbia, Vancouver, Canada
[c]Ministry of Agriculture, Food, Fisheries and Water Resource Management, Princess Alice Highway, Bridgetown, Barbados

Barbados is the most easterly country of the Lesser Antilles (figure 1). It has a narrow continental shelf and is surrounded by deep water. Fisheries are of high importance to this small island country, and within the Barbados EEZ, the artisanal sector dominates (figure 2A). There is also a growing industrial sector, following the introduction of ice boats in the mid-1970s. The reconstruction by Mohammed et al. (2015), which updated Mohammed et al. (2003), shows that domestic catches increased from 4,300 t in 1950 to 5,800 t in 2010. However, the trend was characterized by several large spikes, particularly in 1966 (9,200 t) and 1991 (8,500 t), where the catch quickly increased by more than 50% before declining just as swiftly over the following years. Overall, the reconstructed total domestic catch was 1.6 times that reported by the FAO on behalf of Barbados, with most of the disparity occurring in recent years, probably because of the exclusion of tertiary landing sites from the Fisheries Division's estimates of total landings. Foreign fishing appeared to have been stronger before the EEZ declaration in 1979 (figure 2B). Flyingfishes (family Exocoetidae), important in Barbados (see contributions in Oxenford et al. 2007) were the dominant taxon in the catch (figure 2C), with dolphinfish (*Coryphaena hippurus*) and Atlantic bonito (*Sarda sarda*) also major components.

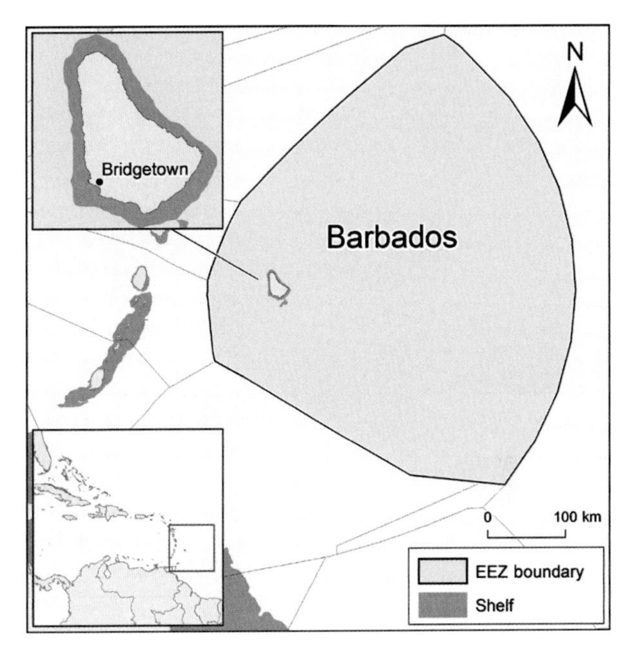

Figure 1. Barbados has a land area of 435 km² and an EEZ of 184,000 km².

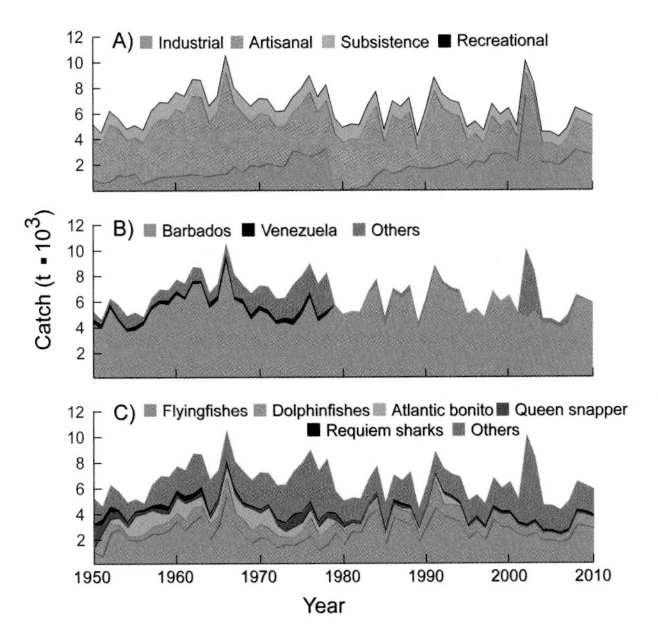

Figure 2. Domestic and foreign catches taken in the EEZ of Barbados: (A) by sector (domestic scores: Ind. –, 4, 4; Art. 2, 2, 2; Sub. 3, 3, 3; Rec. 3, 3, 3); (B) by fishing country (foreign catches are very uncertain); (C) by taxon.

REFERENCES

Mohammed E, Parker C and Willoughby S (2003) Barbados: reconstructed catches and fishing effort, 1940–2000. Pp. 45–66 In Zeller D, Booth S, Mohammed E and Pauly D (eds.), *From Mexico to Brazil: Central Atlantic fisheries catch trends and Ecosystem models*. Fisheries Centre Research Reports 11(6).

Mohammed E, Lindop A, Parker C and Willoughby (2015) Reconstructed fisheries catches of Barbados, 1950–2010. Fisheries Centre Working Paper #2015–16, 28 p.

Oxenford HA, Mahon R and Hunte W (2007) Biology and management of eastern Caribbean flyingfish. Centre for Resource Management and Environmental Studies, University of the West Indies, Barbados. 267 p.

1. Cite as Mohammed, E., A. Lindop, C. Parker, and S. Willoughby. 2016. Barbados. P. 201 in D. Pauly and D. Zeller (eds.), *Global Atlas of Marine Fisheries: A Critical Appraisal of Catches and Ecosystem Impacts*. Island Press, Washington, DC.

BELGIUM[1]

Ann-Katrien Lescrauwaet,[a] Els Torreele,[b] Magda Vincx,[c] Hans Polet,[b] and Jan Mees[a,c]

[a]Flanders Marine Institute (VLIZ), Oostende, Belgium [b]Institute for Agriculture and Fisheries Research (ILVO), Oostende, Belgium [c]Ghent University, MARBIOL, Gent, Belgium

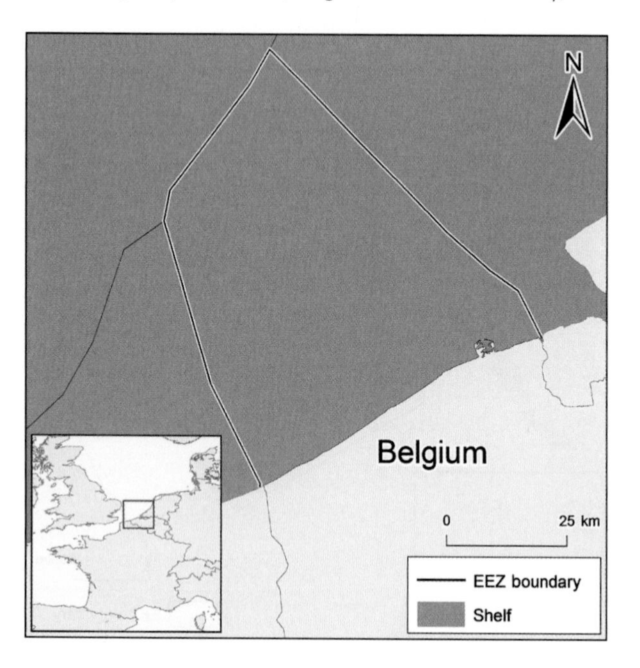

Belgium has a small EEZ in the North Sea (figure 1). Most of the catch within the EEZ of Belgium is derived from the industrial sector (figure 2A). Total domestic removals by Belgian fisheries were reconstructed by Lescrauwaet et al. (2015), including unreported and misreported landings of the industrial fleet, unreported landings by the recreational and artisanal/subsistence fisheries, and discards (Lescrauwaet et al. 2010, 2013). These removals from the Belgian EEZ were estimated at about 26,000 t/year in the early 1950s, which steadily declined to 5,000 t/year in the late 2000s, with total Belgian removals being 44% higher than officially reported over this period, with discards accounting for much of the discrepancy. The reconstruction also suggests that since the 2000s, approximately 50% of domestic catches in the Belgian EEZ are unreported landings and discards, which are increasingly taken by small-scale (<12 m) vessels not subject to reporting requirements and not taken into consideration in planning, monitoring, and enforcement. The total catches allocated to the Belgium EEZ were mainly domestic, although foreign fishing, by France and the United Kingdom, for example, also occurs (figure 2B). Figure 2C shows that common shrimp (*Crangon crangon*) and Atlantic herring (*Clupea harengus*) dominated catches. The support of small-scale fisheries is explicitly mentioned in the European Common Fisheries Policy; therefore, it is important to quantify recreational and noncommercial fishing activities relative to commercial coastal fisheries.

REFERENCES

Lescrauwaet AK, Debergh H, Vincx M and Mees J (2010) Historical marine fisheries data for Belgium: Data sources, data management and data integration related to the reconstruction of historical time-series of marine fisheries landings for Belgium. Fisheries Centre Working Paper #2010–08, 69 p.

Lescrauwaet AK, Fockedey N, Debergh H, Vincx M and Mees J (2013) Hundred and eighty years of fleet dynamics in the Belgian sea fisheries. *Reviews in Fish Biology and Fisheries* 23: 229–243.

Lescrauwaet AK, Torreele E, Vincx M, Polet H, Mees J, Lindop A, and Zylich K (2015) Invisible catch: a century of by-catch and unreported removals in sea fisheries, Belgium 1950–2010. Fisheries Centre Working Paper #2015–18, 18 p.

1. Cite as Lescrauwaet, A. K., E. Torreele, M. Vincx, H. Polet, J. Mees, A. Lindop, and K. Zylich. 2016. Belgium. P. 202 in D. Pauly and D. Zeller (eds.), *Global Atlas of Marine Fisheries: A Critical Appraisal of Catches and Ecosystem Impacts*. Island Press, Washington, DC.

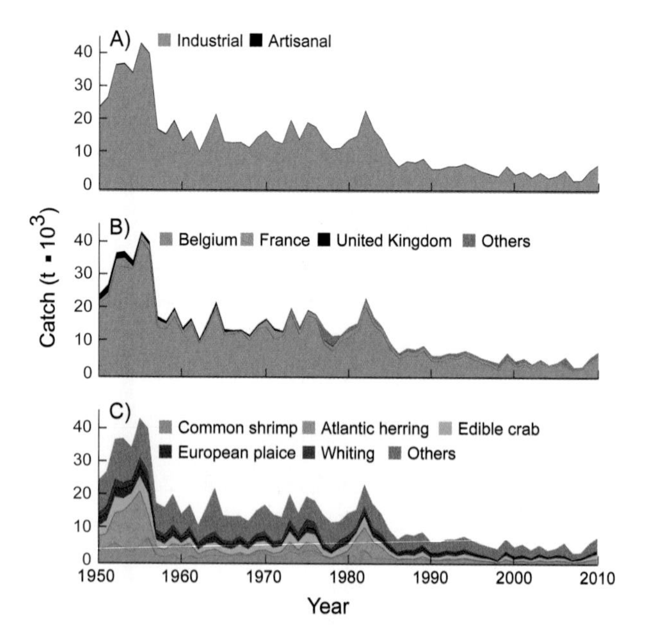

Figure 1. Belgium has a shelf of 3,480 km² and an EEZ of 3,480 km². **Figure 2.** Domestic and foreign catches taken in the EEZ of Belgium: (A) by sector (domestic scores: Ind. 4, 4, 4; Art. 2,–, –; Sub. 1, 1, 1; Rec. 1, 1, 1; Dis. 2, 2, 2); (B) by fishing country (foreign catches are very uncertain); (C) by taxon.

BELIZE[1]

Dirk Zeller,[a] Rachel Graham,[b] and Sarah Harper[a]

[a]*Sea Around Us*, University of British Columbia, Vancouver, Canada
[b]Ocean Giants Program, Wildlife Conservation Society, Punta Gorda, Belize

Belize, formerly British Honduras, is located on the east coast of Central America (figure 1), with Mexico to the north, while Belize, Guatemala, and Honduras share the Gulf of Honduras in the south (Perez 2009). The Belizean coastline is flanked by the second longest barrier reef in the world (Heyman and Kjerfve 2001), beyond which offshore areas drop off to between 500 and 1,000 m depth. The total catch allocated to the Belize EEZ is largely artisanal for the entire time period (figure 2A). The reconstruction of Zeller et al. (2011; updated to 2010) estimated total domestic marine fisheries catches of 4,100 t in 1950, peaking at 7,500 t in 1991, before declining to 6,200 t by 2010, 3.5 times the landings presented by the FAO on behalf of Belize. Although total removals were largely domestic, foreign catches by Cuba, among other countries, were also documented, although illegal foreign catches are largely missing (figure 2B). Stromboid conchs (*Strombus* spp.), yellowtail snapper (*Ocyurus chrysurus*), and lobster (*Panulirus argus*) made up almost a third of the total catch (figure 2C). Although recent plans for offshore oil drilling (Cisneros-Montemayor et al. 2013) have been shelved, coastal developments, excessive fishing effort, and difficult relations with Guatemala and Honduras in the south are threatening the future of Belize's fisheries.

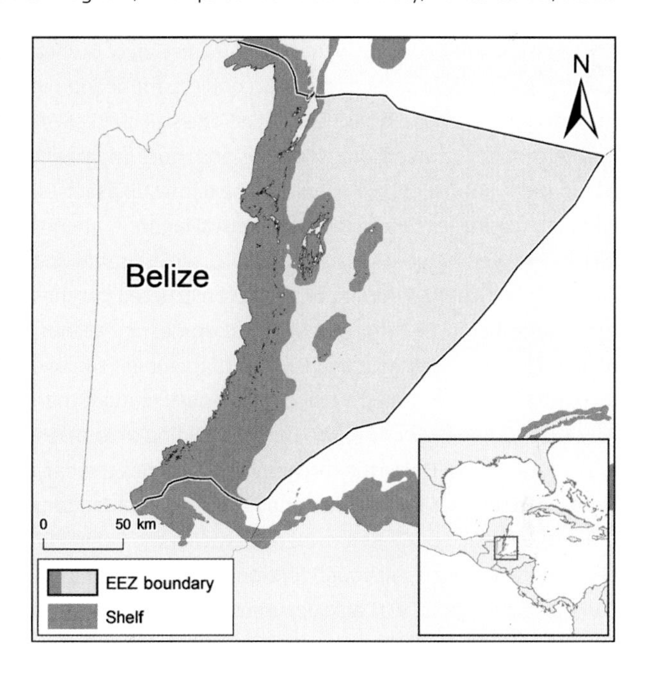

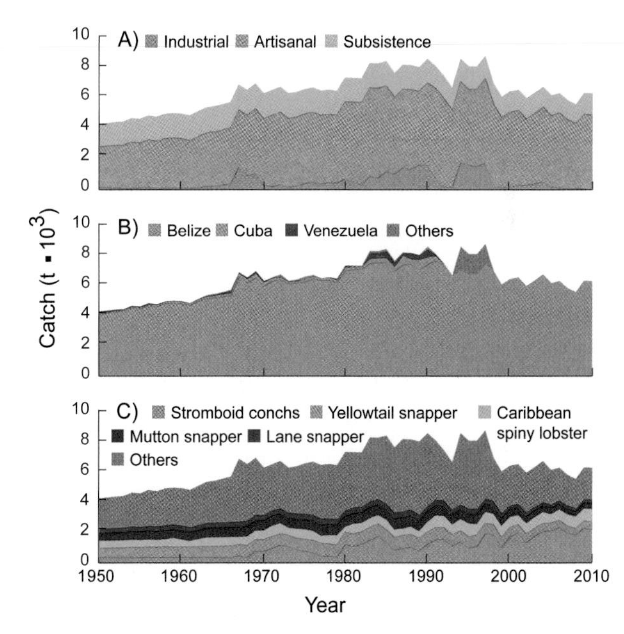

REFERENCES

Cisneros-Montemayor A, Kirkwood FG, Harper S, Zeller D, and Sumaila UR (2013) Economic use value of the Belize marine ecosystem: potential risks and benefits from offshore oil exploration. *Natural Resources Forum* 37(4): 221–230.

Heyman WD and Kjerfve B (2001) The Gulf of Honduras. pp. 17–32 In Seeliger U and Kjerfve B (eds.), *Coastal Marine Ecosystems of Latin America*. Springer-Verlag, Berlin, Heidelberg, New York.

Perez A (2009) Fisheries management at the tri-national border between Belize, Guatemala and Honduras. *Marine Policy* 33: 195–200.

Zeller D, Graham R and Harper S (2011) Reconstruction of total marine fisheries catches for Belize, 1950–2008. pp. 142–151 in Palomares MLD and Pauly D (eds.), *Too Precious to Drill: the Marine Biodiversity of Belize*, Fisheries Centre Research Reports 19(6).

1. Cite as Zeller, D., R. Graham, and S. Harper. 2016. Belize. P. 203 in D. Pauly and D. Zeller (eds.), *Global Atlas of Marine Fisheries: A Critical Appraisal of Catches and Ecosystem Impacts*. Island Press, Washington, DC.

Figure 1. Belize has a shelf of 10,500 km² and an EEZ of 36,200 km². **Figure 2.** Domestic and foreign catches taken in the EEZ of Belize: (A) by sector (domestic scores: Ind. 3, 3, 3; Art. 2, 2, 2; Sub. 1, 1, 1; Dis. 1, 2, 2); (B) by fishing country (foreign catches are very uncertain); (C) by taxon.

BENIN[1]

Dyhia Belhabib,[a] Maria C. Villanueva,[b] and Daniel Pauly[a]

[a]*Sea Around Us*, University of British Columbia, Vancouver, Canada [b]IFREMER, Centre de Bretagne, Sciences et Technologies Halieutiques, Plouzané, France

Benin is a small country in the West African Gulf of Guinea (figure 1); it is so beset by political conflicts and economic mismanagement that it has been called "the sick child of Africa." After agriculture, fishing is the second most important occupation, but fishers are poor, and the domestic fisheries, conducted mostly in brackish-water coastal lagoons, are not well monitored (Vogt et al. 2010). Small-scale artisanal and subsistence fisheries dominate the reconstructed catches within the Benin EEZ (figure 2A). Total domestic catches, reconstructed by Belhabib and Pauly (2015), appear to have increased from about 21,600 t in 1950 to a peak of more than 80,000 t/year in the early 1990s, before declining to 67,000 t in the late 2000s. Domestic catches were 2.3 times the data supplied to the FAO, 81% of which originated in coastal lagoons (Villanueva et al. 2006). Removals from within the Benin EEZ are mostly domestic, although foreign fishing, for example by France and China, was also documented and seems to be increasing (figure 2B). The composition of catches is uncertain, but perch-like fishes (Perciformes), West African ilisha (*Illisha africana*), croakers (*Pseudotolithus* spp.), and bonga shad (*Ethmalosa fimbriata*) are featured, as are invertebrates (figure 2C). In Benin, as elsewhere in West Africa, artisanal fishers and traders compete intensely with foreign industrial fleets, with deleterious economic and related health consequences for the local communities (Allison and Seeley 2004).

REFERENCES

Allison EH and Seeley JA (2004) HIV and AIDS among fisherfolk: a threat to "responsible fisheries"? *Fish and Fisheries* 5(3): 215–234.

Belhabib D and Pauly D (2015) Benin's fisheries: a catch reconstruction, 1950–2010. pp. 51–64 In Belhabib D and Pauly D (eds.), *Fisheries catch reconstructions: West Africa, Part II*. Fisheries Centre Research Reports 23(3).

Villanueva MC, Lalèyè P, Albaret JJ, Lae R, Tito de Morais L, and Moreau J (2006) Comparative analysis of trophic structure and interactions of two tropical lagoons. *Ecological Modelling* 197(3): 461–477.

Vogt J, Teka O, and Sturm U (2010) Modern issues facing coastal management of the fishery industry: A study of the effects of globalization in coastal Benin on the traditional fishery community. *Ocean & Coastal Management* 53(8): 428–438.

1. Cite as Belhabib, D., M. C. Villanueva, and D. Pauly. 2016. Benin. P. 204 in D. Pauly and D. Zeller (eds.), *Global Atlas of Marine Fisheries: A Critical Appraisal of Catches and Ecosystem Impacts*. Island Press, Washington, DC.

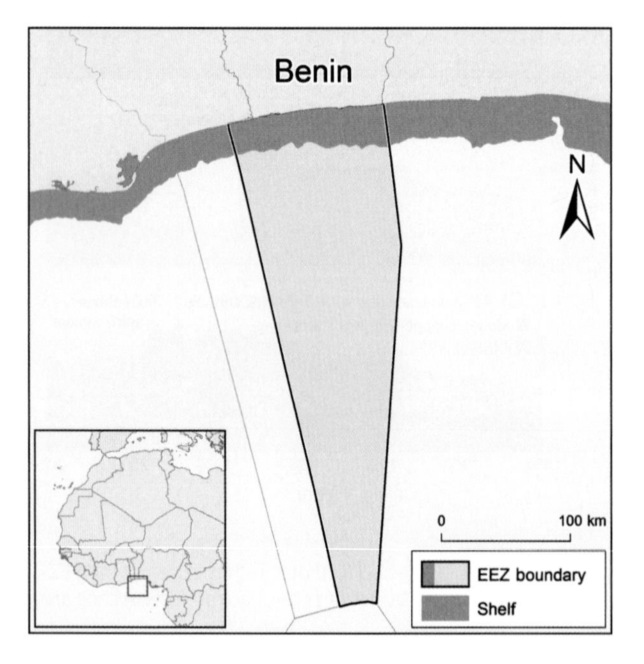

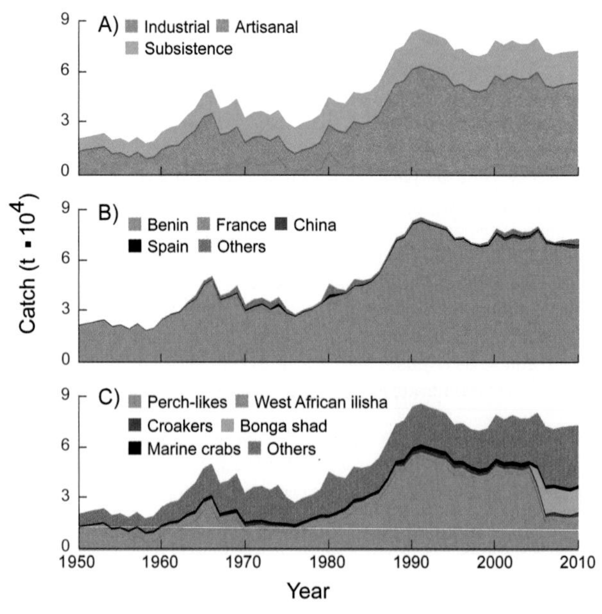

Figure 1. Benin has a continental shelf of 3,250 km² and an EEZ of about 30,300 km².

Figure 2. Domestic and foreign catches taken in the EEZ of Benin: (A) by sector (domestic scores: Ind. 3, 3, 3; Art. 2, 4, 4; Sub. 3, 3, 3; Dis. –, 1, 2); (B) by fishing country (foreign catches are very uncertain); (C) by taxon.

BOSNIA–HERZEGOVINA[1]

Davis Iritani, Kyrstn Zylich, and Dirk Zeller

Sea Around Us, University of British Columbia, Vancouver, Canada

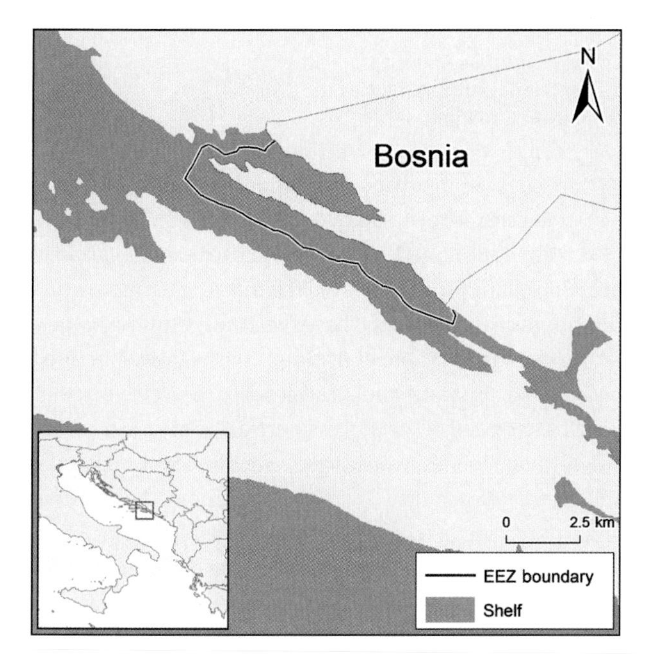

Figure 1. Bosnia–Herzegovina has a tiny shelf and EEZ (both 13.5 km²).

Bosnia–Herzegovina (here: "Bosnia") is small country in southeastern Europe with a very short coastline on the Adriatic Sea, and hence the Mediterranean (figure 1). The catches were based on officially reported landings as baseline data, including landings reported to the FAO by the former Yugoslavia, which were disaggregated to each member of the former Yugoslavia (Rizzo and Zeller 2007). The majority of the total catches allocated to the Bosnia EEZ are small, dominated by the subsistence sector (figure 2A), and as expected, fishing is essentially domestic (figure 2B). Domestic catches, reconstructed by Iritani et al. (2015), increased from 11 t in 1950 to a peak of about 80 t in 1990 and declined to about 65 t/year before increasing again to about 70 t by 2010. For the period from 1992 to 2010, when Bosnia began to independently report its catch to the FAO, the reconstructed domestic catches were 19 times the data presented by the FAO on behalf of Bosnia, with the difference being mainly due to unreported subsistence and recreational catches. The taxonomic composition of the domestic catches could not be ascertained from Bosnian data, and it was assumed similar, for each fishery sector, to the catch composition of the corresponding sector in nearby Croatia (figure 2C).

REFERENCES

Iritani D, Färber L, Zylich K, and Zeller D (2015) Reconstruction of fisheries catches for Bosnia–Herzegovina: 1950–2010. Fisheries Centre Working Paper #2015–15, 7 p.

Rizzo Y and Zeller D (2007) Country disaggregation of catches of former Yugoslavia. pp. 149–156 In Zeller D and Pauly D (eds.), *Reconstruction of marine fisheries catches for key countries and regions (1950–2005)*. Fisheries Centre Research Reports 15(2).

1. Cite as Iritani, D., L. Färber, K. Zylich, and D. Zeller. 2016. Bosnia–Herzegovina. P. 205 in D. Pauly and D. Zeller (eds.), *Global Atlas of Marine Fisheries: A Critical Appraisal of Catches and Ecosystem Impacts*. Island Press, Washington, DC.

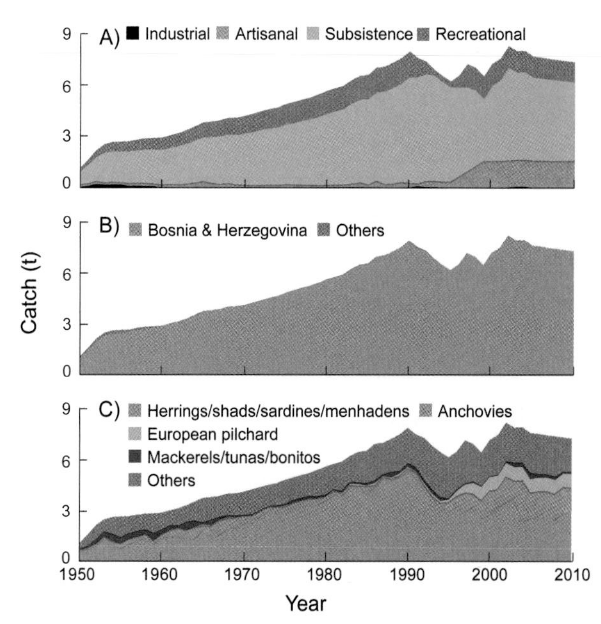

Figure 2. Domestic and foreign catches taken in the EEZ of Bosnia–Herzegovina: (A) by sector (domestic scores: Art. 2, 2, 3; Sub. 4, 4, 3; Rec. 1, 3, 3, Dis. 1, 1, 1); (B) by fishing country (foreign catches are very uncertain); (C) by taxon.

BRAZIL[1]

K. M. F. Freire,[a,all] J. A. N. Aragão,[CE] A. R. R. Araújo,[AP,PA,SE] A. O. Ávila-da-Silva,[SP] M. C. S. Bispo,[all] G. V. Canziani,[RS] M. H. Carneiro,[SP] F. D. S. Gonçalves,[PI,PB,BA] K. A. Keunecke,[RJ] J. T. Mendonça,[SP] P. S. Moro,[SP] F. S. Motta,[SP] G. Olavo,[BA] P. R. Pezzuto,[SC] R. F. Santana,[MA,ES,RJ] R. A. Santos,[PR] I. Trindade-Santos,[SC,RS] J. A. Vasconcelos,[RN] M. Vianna,[RJ] and E. Divovich[b]

[a]Departamento de Engenharia de Pesca e Aquicultura, Centro de Ciências Agrárias Aplicadas, Universidade Federal de Sergipe, Sergipe, Brazil (the abbreviations after the co-authors' names refer to figure 1). [b]*Sea Around Us*, University of British Columbia, Vancouver, Canada

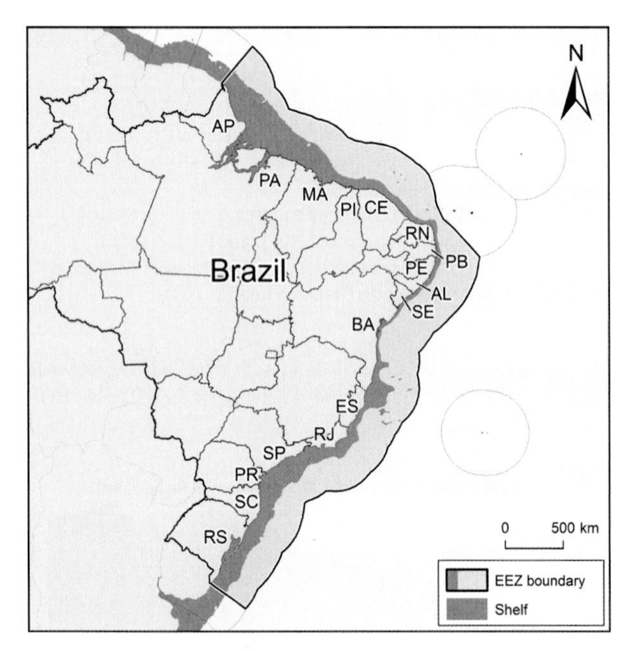

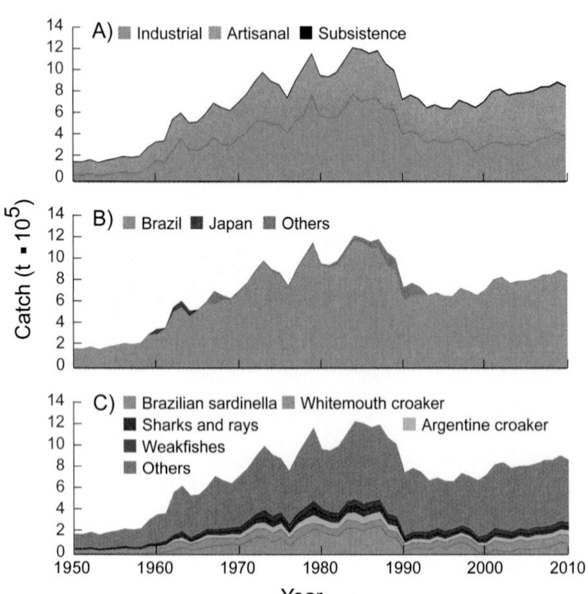

Figure 1. Mainland Brazil has an EEZ of 2.4 million km², of which 742,000 km² is shelf. **Figure 2.** Domestic and foreign catches taken in the mainland EEZ of Brazil: (A) by sector (domestic scores: Ind. 1, 3, 4; Art. 1, 3, 4; Sub. 1, 2, 3; Rec. 1, 2, 2; Dis. 1, 1, 1); (B) by fishing country (foreign catches are very uncertain); (C) by taxon.

The coast of mainland Brazil (figure 1) ranges from the tropical State of Amapá (AP; 4°N) to the temperate State of Rio Grande do Sul (RS; 34°S), and encompasses a wide range of ecologies and fisheries. The multiauthored catch reconstruction on which this account is based was thus performed largely at the state level (Freire et al. 2015). This was also required because the disparate statistical records (Dias Neto 2011) that were harmonized in the process were associated with about 446 common names, whose valid scientific counterparts are still not reliably identified (Freire and Pauly 2005). The industrial and artisanal sectors contributed the majority of the total catch allocated to the Brazil EEZ (figure 2A). The domestic catches (little foreign fishing seems to occur; figure 2B), aggregated to national level (but omitting Brazil's oceanic islands, covered in the next three accounts) were about 190,000 t/year in the early 1950s, peaked at nearly 1.2 million t in 1984, at the height of the industrial fishery for Brazilian "sardine" (*Sardinella brasiliensis*), and returned to lower levels after this fishery collapsed, with about 840,000 t/year in the late 2000s. The reconstructed domestic catches were 1.8 times the data reported by the FAO. Figure 2C shows that Brazilian sardinella (*Sardinella brasiliensis*) and croakers (*Micropogonias furnieri*, *Umbrina canosai*) were dominant within the Brazil EEZ.

REFERENCES

Dias Neto J (2011) Proposta de Plano Nacional de Gestão para o uso sustentável de camarões marinhos do Brasil. IBAMA, Brasília. 242 p.

Freire KMF and Pauly D (2005) Richness of common names of Brazilian marine fishes and its effect on catch statistics. *Journal of Ethnobiology* 25(2): 279–296.

Freire KMF et al. [see above] (2015) Reconstruction of catch statistics for Brazilian marine waters (1950–2010). pp. 3–30, In Freire KMF and Pauly D (eds.), *Fisheries catch reconstructions for Brazil's mainland and oceanic islands*. Fisheries Centre Research Reports 23(4).

1. Cite as Freire, K. M. F., J. A. N. Aragão, A. R. R. Araújo, A. O. Ávila-da-Silva, M. C. S. Bispo, G. V. Canziani, M. H. Carneiro, F. D. S. Gonçalves, K. A. Keunecke, J. T. Mendonça, P. S. Moro, F. S. Motta, G. Olavo, P. R. Pezzuto, R. F. Santana, R. A. Santos, I. Trindade-Santos, J. A. Vasconcelos, M. Vianna, and E. Divovich. 2016. Brazil. P. 206 in D. Pauly and D. Zeller (eds.), *Global Atlas of Marine Fisheries: A Critical Appraisal of Catches and Ecosystem Impacts*. Island Press, Washington, DC.

BRAZIL (FERNANDO DE NORONHA)[1]

Esther Divovich and Daniel Pauly

Sea Around Us, University of British Columbia, Vancouver, Canada

The archipelago of Fernando de Noronha consists of 1 larger island and 19 islets, 350 km east of the Brazilian coast at about 4°S, which Charles Darwin visited in February 1832 and later described (Armstrong 2004). It was originally a prison, and the first fishers were its captives, reportedly forced to swim from its shores and return with fish (IOPE 2010). The majority of the catches are from the industrial sector, although an artisanal fishery that developed after World War II supplied the resident population of Fernando de Noronha (figure 2A). In 1988, about 70% of the coastal waters to 50 m depth were declared part of an "Environmental Protection Area (EPA) designated for sustainable use" (Garla et al. 2006). Divovich and Pauly (2015) reconstructed catches for this island to be 209 t in 1950, increasing dramatically to a peak of 2,550 t in 1995 and subsequently declining to less than 100 t by 2010. The spatially allocated catches in this EEZ varied between 1950 and the early 1990s, in part because of foreign catches by South Korea and others (figure 2B), but they increased sharply in the mid-1990s, because of the domestic fleet, before declining in the late 2000s. Figure 2C illustrates that this catch consisted mainly of large pelagic taxa such as yellowfin tuna (*Thunnus albacares*), bigeye tuna (*Thunnus obesus*), and swordfish (*Xiphias gladius*).

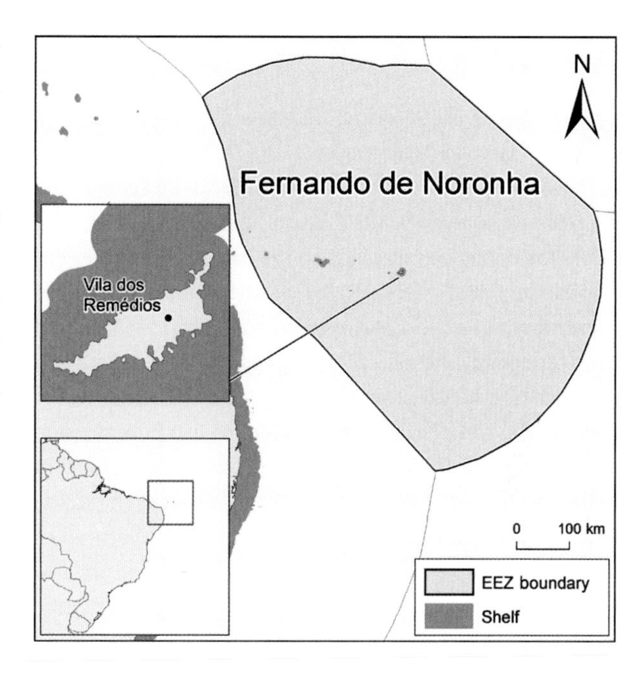

Figure 1. Fernando de Noronha (Brazil) has a land area of 23 km² and an EEZ of 363,000 km².

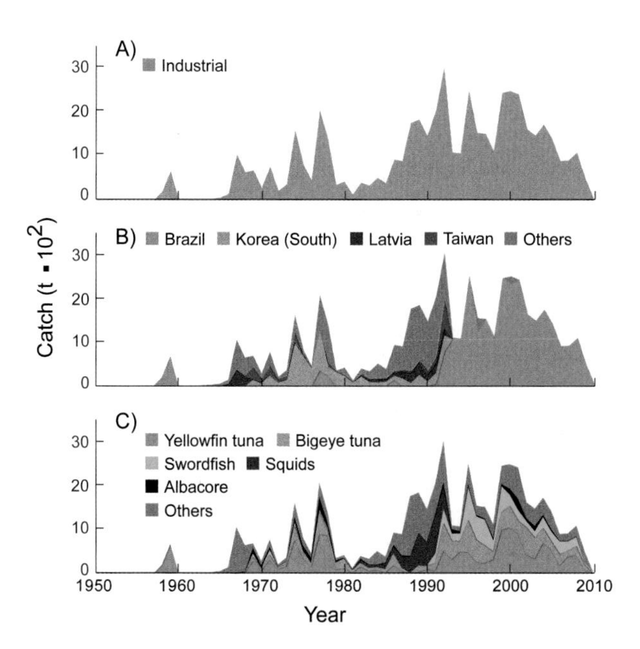

Figure 2. Domestic and foreign catches taken in the EEZ of Fernando de Noronha (Brazil): (A) by sector (domestic scores: Art. 2, 2, 2; Sub. 1, 1, 1; Dis. 1, 1, 1); (B) by fishing country (foreign catches are very uncertain); (C) by taxon.

REFERENCES

Armstrong P (2004) *Darwin's Other Islands*. Continuum, London & New York. 266 p.

Divovich E and Pauly D (2015) Oceanic islands of Brazil: catch reconstruction from 1950 to 2010). pp. 31–48 In Freire KMF and Pauly D (eds.), *Fisheries catch reconstructions for Brazil's mainland and oceanic islands*. Fisheries Centre Research Reports 23(4).

Garla R, Chapman D, Wetherbee B and Shivji M (2006) Movement patterns of young Caribbean reef sharks, *Carcharhinus perezi*, at Fernando de Noronha Archipelago, Brazil: the potential of marine protected areas for conservation of a nursery ground. *Marine Biology* 149(2): 189–199.

IOPE (2010) Diagnóstico socioeconômico da pesca artesanal na ilha de Fernando de Noronha. Diagnóstico socioeconômico da pesca artesanal do litoral de Pernambuco, p. 57–72, Vol. I, Instituto Oceanário de Pernambuco, Departamento de Pesca e Aqüicultura da UFRPE, Recife.

1. Cite as Divovich, E., and D. Pauly. 2016. Brazil (Fernando de Noronha). P. 207 in D. Pauly and D. Zeller (eds.), *Global Atlas of Marine Fisheries: A Critical Appraisal of Catches and Ecosystem Impacts*. Island Press, Washington, DC.

BRAZIL (ST. PETER AND ST. PAUL ARCHIPELAGO)[1]

Daniel Pauly,[a] José Airton Vasconcelos,[b] and Esther Divovich[a]

[a]*Sea Around Us*, University of British Columbia, Vancouver, Canada [b]IBAMA, Divisão de Controle, Monitoramento e Fiscalização Ambiental (DICAFI-Pesca), Natal, Rio Grande do Norte, Brazil

A group of islets on the Mid-Atlantic Ridge (Edwards and Lubbock 1980; figure 1), the St. Peter and St. Paul Archipelago (SPSPA) was visited by the *Beagle* on February 16, 1832. Charles Darwin wrote, "We only observed two kinds of birds. The booby lays her eggs on the bare rock; but the tern makes a very simple nest with sea-weed. By the side of many of these nests a small flying-fish was placed; which, I suppose, had been brought by the male bird for its partner. It was amusing to watch how quickly a large and active crab (*Graspus*), which inhabits the crevices of the rock, stole the fish from the side of the nest, as soon as we had disturbed the birds" (sources in Pauly 2004). Aside from the staff of a small research facility, SPSPA is uninhabited. Although the total catch allocated to the EEZ of SPSPA is mostly by domestic fleets, there are also non-Brazilian distant-water fleets, whose catches can be estimated from the broad pattern of pelagic fishing in that region (figure 2A). Divovich and Pauly (2015) report on the likely catch of a boat based in Pernanbuco State fishing in the SPSPA in the early 1970s and those of a fleet based in Natal and exploiting the SPSPA waters since 1988 and targeting large pelagics (i.e., tuna and swordfish) (Vaske et al. 2008; figure 2B).

REFERENCES

Divovich E and Pauly D (2015) Oceanic islands of Brazil: catch reconstruction from 1950 to 2010. pp. 31–48 In Freire KMF and Pauly D (eds.), *Fisheries catch reconstructions for Brazil's mainland and oceanic islands*. Fisheries Centre Research Reports 23(4).

Edwards A and Lubbock R (1980) Voyage to St. Paul's Rock. *Geographical Magazine* 52(8): 561–567.

Pauly D (2004) *Darwin's Fishes: An Encyclopedia of Ichthyology, Ecology and Evolution*. Cambridge University Press, Cambridge. xxv + 340 p.

Vaske Jr. T, Lessa R, Ribeiro A, Nóbrega M, Pereira A and Andrade C (2008) A pesca comercial de peixes pelágicos no arquipélago de São Pedro e São Paulo. *Brasil. Trop. Ocean* 34: 31–41.

1. Cite as Divovich, E., and D. Pauly. 2016. Brazil (St. Peter and St. Paul Archipelago). P. 208 in D. Pauly and D. Zeller (eds.), *Global Atlas of Marine Fisheries: A Critical Appraisal of Catches and Ecosystem Impacts*. Island Press, Washington, DC.

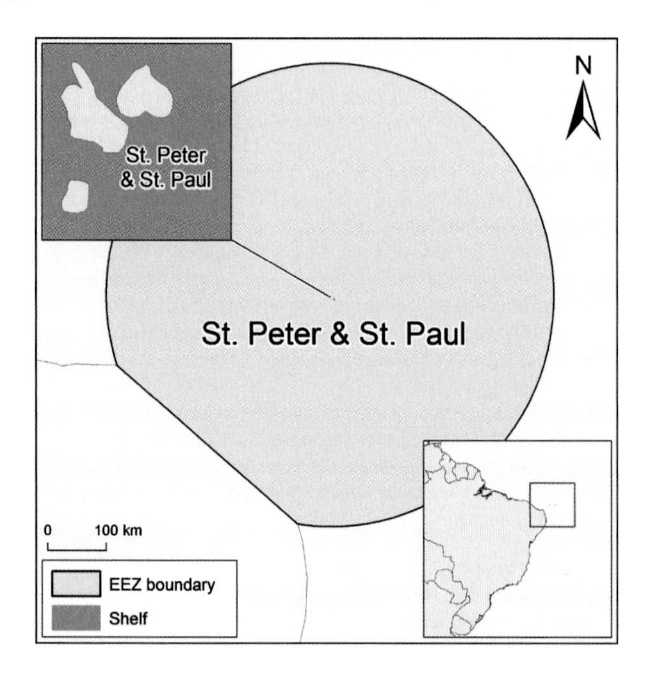

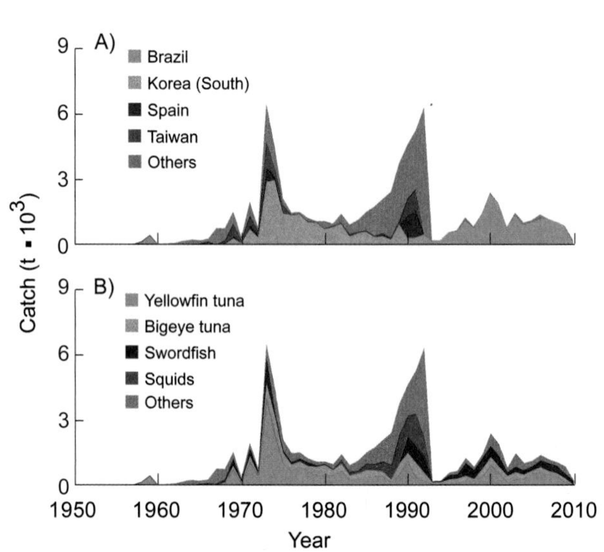

Figure 1. The St. Peter and St. Paul Archipelago has a land area of only 0.015 km² and an EEZ of 414,000 km².
Figure 2. Domestic and foreign catches taken in the EEZ of St. Peter and St. Paul Archipelago: (A) by fishing country (domestic scores: Ind. 2, 2, 2; Dis. –, 1, 1; foreign catches are very uncertain); (B) by taxon.

BRAZIL (TRINDADE & MARTIM VAZ ISLANDS)[1]

Esther Divovich and Daniel Pauly

Sea Around Us, University of British Columbia, Vancouver, Canada

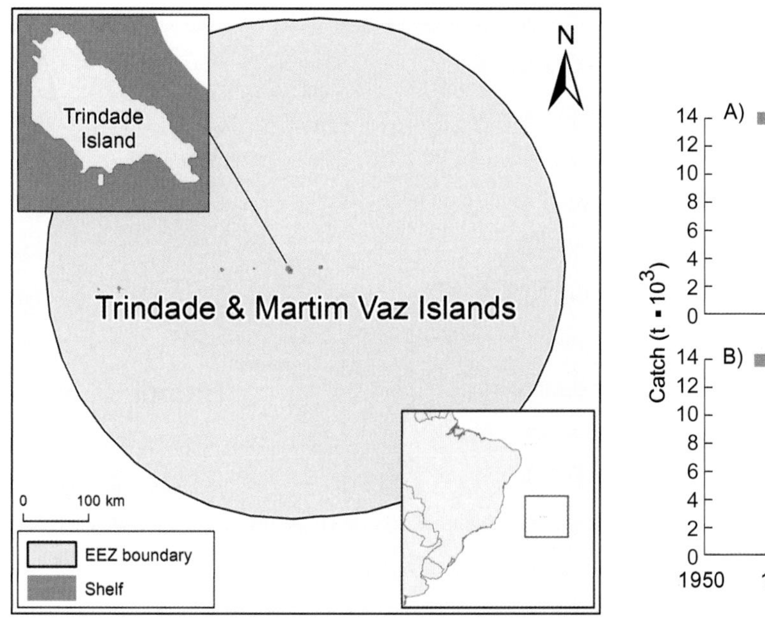

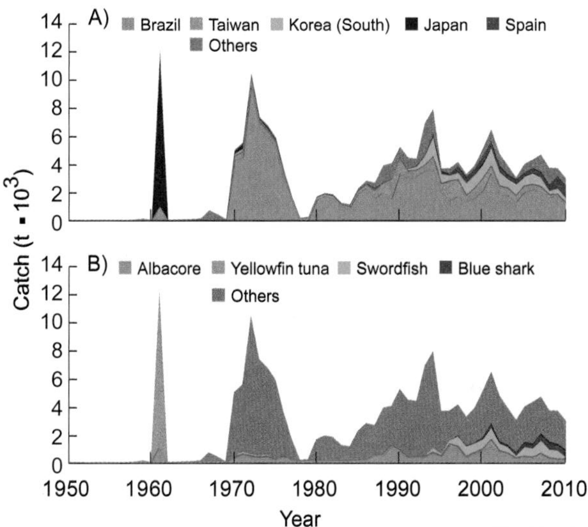

Figure 1. The Trindade and Martim Vaz Islands (Brazil) have a joint land area of about 20.7 km² and an EEZ of 470,000 km². **Figure 2.** Domestic and foreign catches taken in the EEZ of Trindade and Martim Vaz Islands (Brazil): (A) by fishing country (Brazil scores: Ind. 2, 2, 3; Dis. –, 1, 1; foreign catches are very uncertain); (B) by taxon.

The Brazilian islands of Trindade and Martim Vaz (TMVIG; figure 1) form an archipelago of five volcanic islands, about 1,200 km to the east of the Brazilian coast at 20°30′S. The islands are uninhabited except for a few dozen military personnel, but their coastal waters are fished from the mainland (Mazzoleni and Schwingel 2002; Pereira-Filho et al. 2011). Divovich and Pauly (2015) performed a reconstruction for Brazil's oceanic islands. These data, combined with global catches of large pelagics (see chapter 3), suggested total catches at about 3,400 t in 1990, declining to approximately 1,000 t by 2010. Although domestic catches composed a portion of the total removals from the EEZ of TMVIG, foreign catches by Taiwan, South Korea, and Spain, among others, dominated the allocated catch after 1986 (figure 2A). Taxonomically, the catches consist mainly of large pelagics such as albacore tuna (*Thunnus alalunga*), yellowfin tuna (*Thunnus albacares*), and swordfish (*Xiphias gladius*) (figure 2B). Since 1982, when Brazil declared its EEZ, the foreign catch has depended on access agreements it may have signed with distant-water fishing countries or on the latter's willingness to enter the EEZ without such agreement, which itself depends on the risk of getting caught. Interviews with Brazilian fishers operating around TMVIG reveal this is a real problem, because all of them "reported the presence of large Asian vessels operating clandestinely in Brazilian water" (Pinheiro et al. 2010).

REFERENCES

Divovich E and Pauly D (2015) Oceanic islands of Brazil: catch reconstruction from 1950 to 2010. pp. 31–48 In Freire KMF and Pauly D (eds.), *Fisheries catch reconstructions for Brazil's mainland and oceanic islands*. Fisheries Centre Research Reports 23(4).

Mazzoleni R and Schwingel P (2002) Biological aspects of pelagic longline caught species in south region of Trindade and Martin Vaz Islands in summer of 2001. *Brazilian Journal of Aquatic Science and Technology* 6(1): 51–57.

Pereira-Filho G, Menezes Amado-Filho G, Guimarães S, Moura R, Sumida P, Abrantes D, Bahia R, Güth A, Jorge R and Francini Filho R (2011) Reef fish and benthic assemblages of the Trindade and Martin Vaz island group, Southwestern Atlantic. *Brazilian Journal of Oceanography* 59(3): 201–212.

Pinheiro H, Martins A and Gasparini J (2010) Impact of commercial fishing on Trindade Island and Martin Vaz Archipelago, Brazil: characteristics, conservation status of the species involved and prospects for preservation. *Brazilian Archives of Biology and Technology* 53(6): 1417–1423.

1. Cite as Divovich, E., and D. Pauly. 2016. Brazil (Trindade and Martim Vaz Islands). P. 209 in D. Pauly and D. Zeller (eds.), *Global Atlas of Marine Fisheries: A Critical Appraisal of Catches and Ecosystem Impacts*. Island Press, Washington, DC.

BRUNEI[1]

Elviro A. Cinco,[a] Kyrstn Zylich,[b] Lydia C. L. Teh,[b] and Daniel Pauly[b]

[a]Fisheries Resources Evaluation Section, Department of Fisheries, Ministry of Industry and Primary Resources, Brunei Darussalam [b]*Sea Around Us*, University of British Columbia, Vancouver, Canada

The main sources of income for Brunei Darussalam, a small Islamic sultanate on the north coast of Borneo (or Kalimantan; figure 1), are (offshore) oil and gas. Agriculture and fisheries are also important. The total marine catch, as allocated to its EEZ, is dominated by the rapidly growing artisanal sector (figure 2A). This account, based on Cinco et al. (2015), presents the reconstructed domestic catches from Brunei's EEZ, which are 4 times as high as reported by the FAO on behalf of Brunei (there seems to be little foreign fishing; figure 2B). This catch was less than 3,000 t in the 1950s and was generated exclusively by small-scale fisheries (mainly artisanal, some subsistence, notably by the inhabitants of *kampong ayer*, the "water village" at the heart of Brunei Bay). An industrial trawl fishery started in the 1960s whose spatial dynamics are well documented (Pauly et al. 1997), and the total catch peaked at just under 20,000 t in 2010. Herrings, shads, and sardines (family Clupeidae), penaeid shrimps (Penaeidae), and mackerels, tunas, and bonitos (Scombridae) were the major components of the catch (figure 2C). Ebil et al. (2013) suggest that the changing catch composition, from demersal to pelagic dominance in the last decade, indicates worsening ecosystem health. Further improvement in the marine fisheries statistical systems should assist in mitigation efforts.

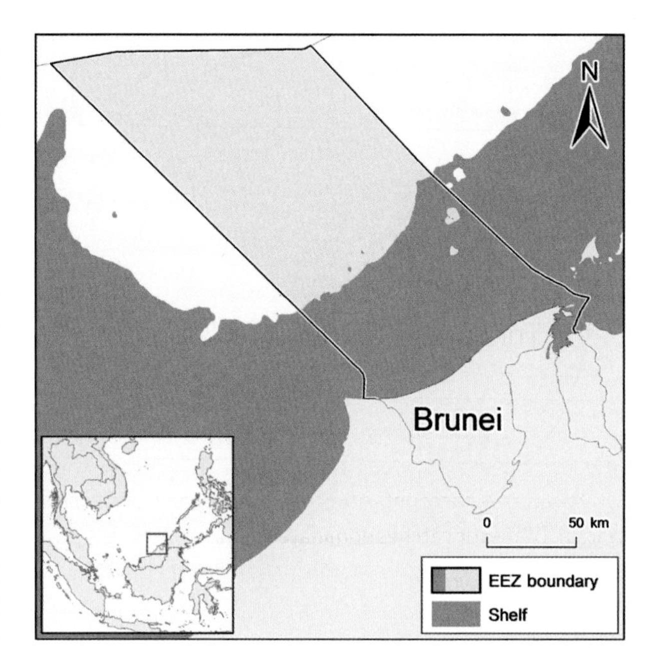

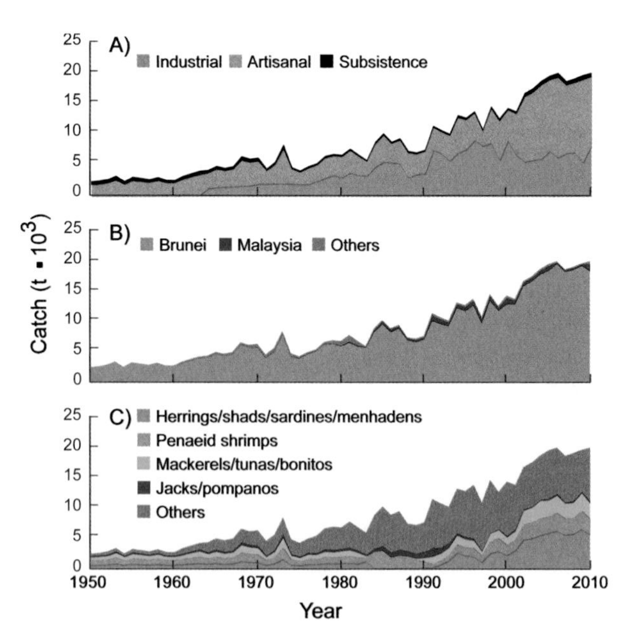

Figure 1. Brunei has a shelf of 8,920 km² and a small EEZ of 25,300 km². **Figure 2.** Domestic and foreign catches taken in the EEZ of Brunei: (A) by sector (domestic scores: Ind. 3, 4, 4; Art. 3, 4, 4; Sub. 3, 3, 4; Rec. 2, 3, 3; Dis. –, 3, 4); (B) by fishing country (foreign catches are very uncertain); (C) by taxon.

REFERENCES

Cinco EA, Zylich K and Teh LCL (2015) The marine and estuarine fisheries of Brunei Darussalam, 1950 to 2010. Fisheries Centre Working Paper #2015–29, 6 p.

Ebil S, Sheppard CRC, Wahab R, Price ARG and Bull JC (2013) Changes in community structure of finfish catches in Brunei Darussalam between 2000 and 2009. *Ocean and Coastal Management* 76: 45–51.

Pauly D, Gayanilo FC and Silvestre G (1997) A low-level geographic information system for coastal zone management, with application to Brunei Darussalam. Part 1: The concept and its design elements. *Naga, the ICLARM Quarterly* 20: 41–45.

1. Cite as Cinco, E. A., K. Zylich, L. C. L. Teh, and D. Pauly. 2016. Brunei. P. 210 in D. Pauly and D. Zeller (eds.), *Global Atlas of Marine Fisheries: A Critical Appraisal of Catches and Ecosystem Impacts*. Island Press, Washington, DC.

BULGARIA[1]

Çetin Keskin,[a] Aylin Ulman,[b] Violin Raykov,[c] Georgi M. Daskalov,[d] Kyrstn Zylich,[b] Daniel Pauly,[b] and Dirk Zeller[b]

[a]Faculty of Fisheries, University of Istanbul, Laleli, Istanbul, Turkey [b]*Sea Around Us*, University of British Columbia, Vancouver, Canada [c]Institute of Oceanography of Bulgarian Academy of Science, Varna, Bulgaria [d]Institute of Fisheries, Varna, Bulgaria

Bulgaria's EEZ in the Black Sea is small (figure 1). The total catch allocated to Bulgaria's EEZ is dominated by the industrial sector, with a small artisanal component (figure 2A). Total domestic fisheries removals from the EEZ by Bulgarian marine fisheries were estimated by Keskin et al. (2015) to have accounted for about 6,000 t/year through the 1950s and 1960s, increased to a peak above 20,000 t/year in the early 1980s, declined back to 5,000 t/year in the early 1990s, then experienced a second peak, reaching 20,000–30,000 t/year in the 2000s. The total reconstructed domestic catch of these fisheries for the 1950–2010 time period is 1.8 times the data FAO reports on behalf of Bulgaria, the discrepancy being mainly due to recent, unaccounted for catches of rapa whelk (*Rapana venosa*). Domestic catches dominated the total removals from Bulgaria's EEZ, although some foreign catches, mainly by Spain and Turkey, also occurred over the time period (figure 2B). As figure 2C reveals, the catch fluctuations reflected massive changes in the underlying fisheries, with a mix of large fish (e.g., bonito, *Sarda sarda*; Atlantic mackerel, *Scomber scombrus*) in the first two decades, being replaced later by sprat (*Sprattus spattus*), a small pelagic fish with high catches that contributed heavily to the first peak of the pelagic industrial fishery in the 1980s (Daskalov 2002).

REFERENCES

Daskalov GM (2002) Overfishing drives a trophic cascade in the Black Sea. *Marine Ecology Progress Series* 225: 53–63.

Keskin C, Ulman A, Raykov V, Daskalov G, Zylich K, Pauly D and Zeller D (2015) Reconstruction of fisheries catches for Bulgaria: 1950–2010. Fisheries Centre Working Paper #2015–20, 18 p.

1. Cite as Keskin, C., A. Ulman, V. Raykov, G. Daskalov, K. Zylich, D. Pauly, and D. Zeller. 2016. Bulgaria. P. 211 in D. Pauly and D. Zeller (eds.), *Global Atlas of Marine Fisheries: A Critical Appraisal of Catches and Ecosystem Impacts*. Island Press, Washington, DC.

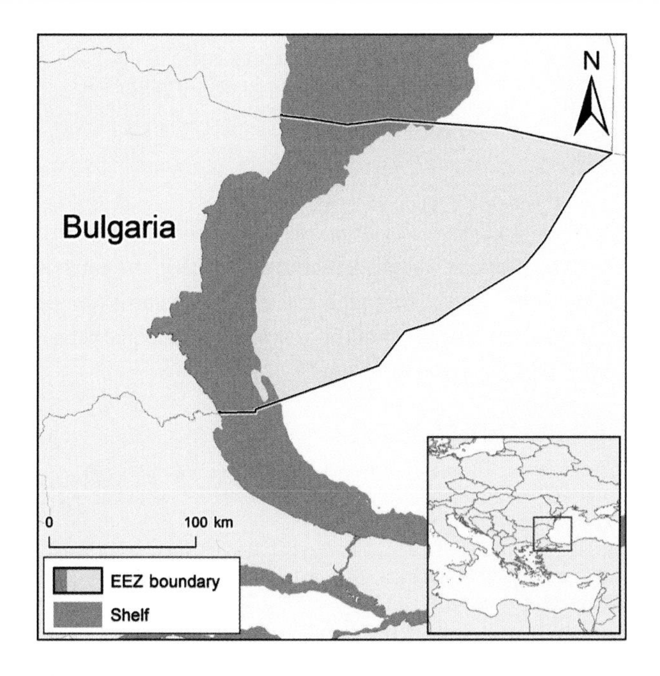

Figure 1. Bulgaria has a shelf of 11,900 km² and a small EEZ of 35,100 km².

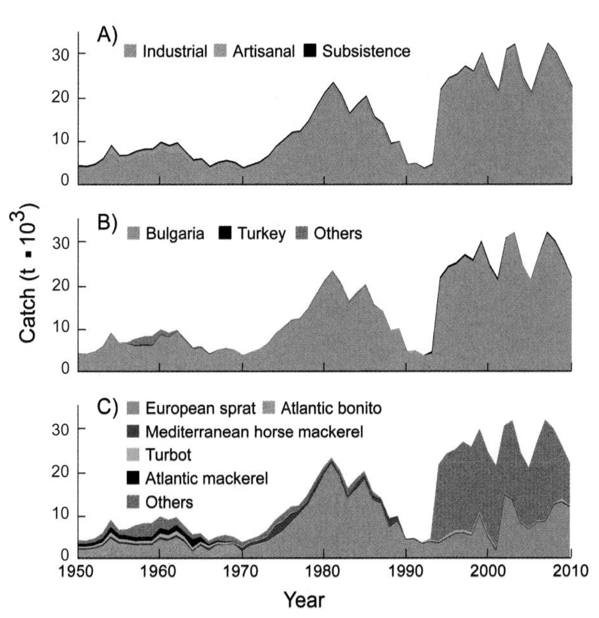

Figure 2. Domestic and foreign catches taken in the EEZ of Bulgaria: (A) by sector (domestic scores: Ind. 2, 3, 2; Art. 3, 3, 2; Sub 1, 1, 1; Rec. 1, 1, 1; Dis. 2, 2, 2); (B) by fishing country (foreign catches are very uncertain); (C) by taxon.

CAMBODIA[1]

Lydia C. L. Teh, Soohyun Shon, Kyrstn Zylich, and Dirk Zeller

Sea Around Us, University of British Columbia, Vancouver, Canada

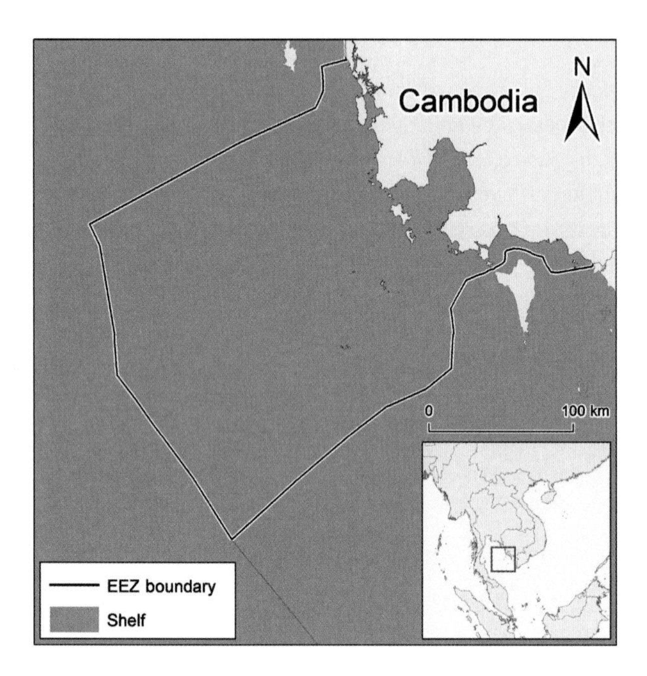

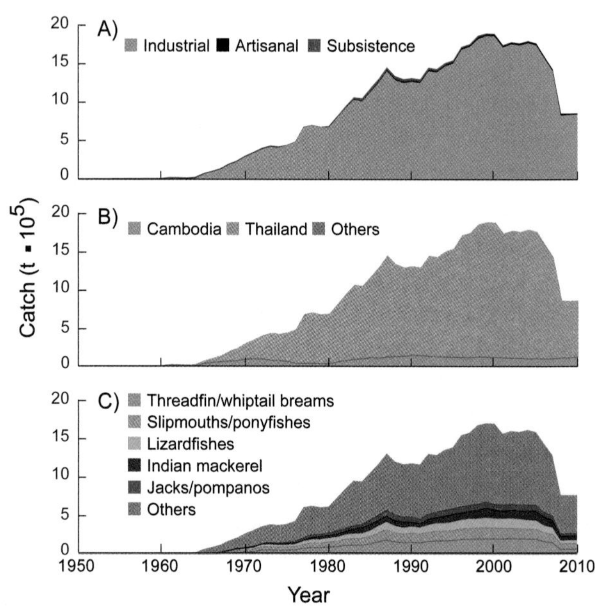

Figure 1. Cambodia has a shelf and EEZ of 48,000 km².
Figure 2. Domestic and foreign catches taken in the EEZ of Cambodia: (A) by sector (domestic scores: Ind. 1, 1, 1; Art. –, 2, 2; Sub. 1, 2, 2; Dis. 2, 2, 2); (B) by fishing country (foreign catches are very uncertain); (C) by taxon.

Cambodia is bordered by Thailand, Vietnam, Laos, and the Gulf of Thailand (figure 1). Cambodia's marine fisheries are small compared with its inland fisheries (Gillett 2004). Total marine catches allocated to Cambodia's EEZ were dominated by the industrial sector (figure 2A), although this was overwhelmingly dominated by foreign fleets from Thailand (figure 2B), with much of these catches being illegal. The reconstruction by Teh et al. (2014) estimated domestic catches in 1950 to be about 11,300 t. With minimal management, fishing effort in inshore areas rapidly intensified as small domestic trawlers raced to meet increasing demand for prawns and fish starting in the 1960s. Catch peaked at 140,000 t in 1990, before declining to 120,000 t by 2010. In 2010, the Cambodian industrial sector (i.e., vessels with engines >30 hp) contributed about 86% of the reconstructed total domestic catch of 120,000 t. Overall, the reconstructed domestic catches were 3 times the statistics reported by the FAO on behalf of Cambodia from 1950 to 2010, with fish sold at sea amounting to half the unreported catches. Given the overexploitation of the inshore resources, Cambodia's fisheries are expanding offshore, where they encounter legal and illegal foreign vessels from their neighbors to the north and east. Figure 2C suggests that taxonomically, the catch is mainly comprised of threadfin and whiptail breams (family Nemipteridae), slimy slipmouths or ponyfishes (Leiognathidae), and lizardfishes (Synodontidae).

REFERENCES

Gillett R (2004) The marine fisheries of Cambodia. FAO/FishCode Review, Food and Agriculture Organization of the United Nations (FAO), Rome (Italy). 57 p.

Teh LCL, Shon S, Zylich K and Zeller D (2014) Reconstructing Cambodia's marine fisheries catch, 1950–2010. Fisheries Centre Working Paper #2014–18, 2 p.

1. Cite as Teh, L. C. L., S. Shon, K. Zylich, and D. Zeller. 2016. Cambodia. P. 212 in D. Pauly and D. Zeller (eds.), *Global Atlas of Marine Fisheries: A Critical Appraisal of Catches and Ecosystem Impacts*. Island Press, Washington, DC.

CAMEROON[1]

Dyhia Belhabib and Daniel Pauly

Sea Around Us, University of British Columbia, Vancouver, Canada

Cameroon, named from the Portuguese Rio de Camarões ("Shrimp River"), is a country whose geographic location in the inner Gulf of Guinea results in its EEZ being the smallest in West Africa (figure 1). Despite political conflicts that heavily damaged its economy in the 1980s and 1990s, Cameroon partly recovered, although its people still suffer from food insecurity. Fisheries contribute more than 25% of animal protein consumption; however, inadequate governance resulted in severe overexploitation of the marine resources (ENVIREP-CAM 2011). The small-scale artisanal sector is the largest component of the fisheries in Cameroon's EEZ (figure 2A). Total domestic catches from the EEZ of Cameroon estimated by Belhabib and Pauly (2015) increased from 15,000 t in 1950 to a first peak of 89,000 t in 1977 and a second of more than 106,000 t in 2003, before declining to 70,000 t in 2010. Domestic catches overall were 1.5 times the data reported by the FAO on behalf of Cameroon, which trend upward while the reconstructed domestic catches trend down. Although catches were mainly domestic, foreign fishing, increasingly by China, had noticeable removals in the later time period (figure 2B). Figure 2C reveals that the bulk of the catch consisted of bonga shad (*Ethmalosa fimbriata*) and sardinellas (*Sardinella* spp.) among pelagics (Djama et al. 1990) and grunts (family Haemulidae) and croakers (family Sciaenidae; Djama and Pitcher 1989) among demersals.

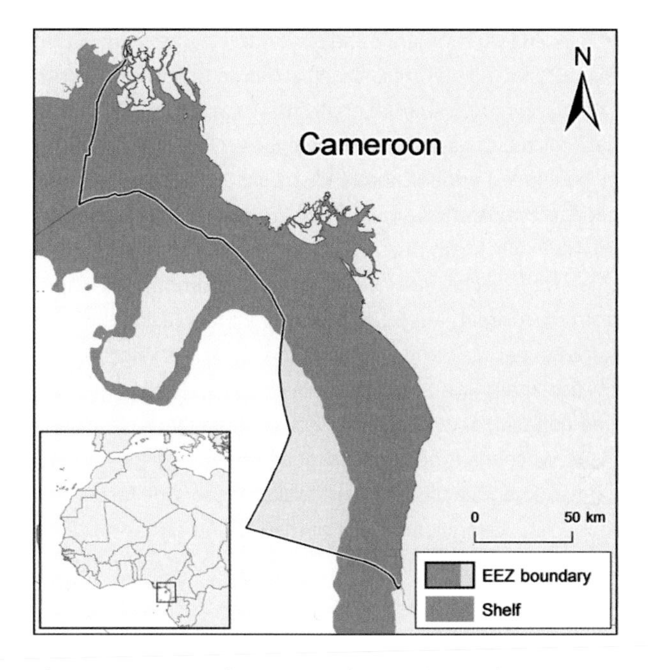

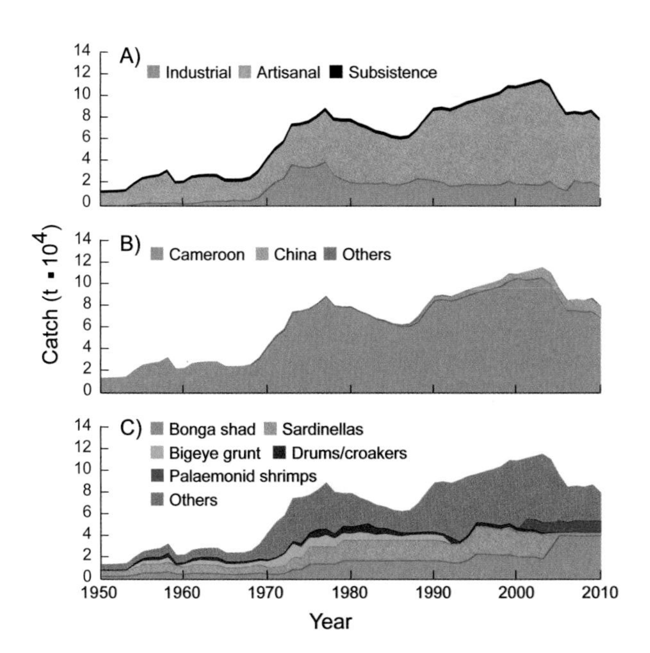

Figure 1. Cameroon has an EEZ of 14,700 km² and a continental shelf of 11,800 km². **Figure 2.** Domestic and foreign catches taken in the EEZ of Cameroon: (A) by sector (domestic scores: Ind. 2, 2, 2; Art. 1, 3, 3; Sub. 2, 2, 1; Dis. 2, 2, 3); (B) by fishing country (foreign catches are very uncertain); (C) by taxon.

REFERENCES

Belhabib D and Pauly D (2015) Reconstructing fisheries catches for Cameroon between 1950 and 2010. pp. 77–84 In Belhabib D and Pauly D (eds.), *Fisheries catch reconstructions: West Africa, Part II*. Fisheries Centre Research Reports 23(3).

Djama T and Pitcher T (1989) Comparative stock assessment of two sciaenid species, *P. typus* and *P. senegalensis* off Cameroon. *Fisheries Research* 7: 111–125.

Djama T, Nkumbe L and Ikome F (1990) Catch assessment of *Sardinella maderensis* in the Ocean Division, Cameroon. *Fishbyte: Newsletter of the Network of Tropical Fisheries Scientists* 18(1): 6–7.

ENVIREP-CAM (2011) Overview of Management and Exploitation of the Fisheries Resources of Cameroon, Central West Africa. Institut de Recherche Agricole pour le Developpement, Yaounde. 70 p.

1. Cite as Belhabib, D., and D. Pauly. 2016. Cameroon. P. 213 in D. Pauly and D. Zeller (eds.), *Global Atlas of Marine Fisheries: A Critical Appraisal of Catches and Ecosystem Impacts*. Island Press, Washington, DC.

CANADA (ARCTIC)[1]

Shawn Booth, Paul Watts, Lydia C. L. Teh, Kyrstn Zylich, and Dirk Zeller

Sea Around Us, University of British Columbia, Vancouver, Canada

Canada's arctic marine and estuarine fisheries occur within FAO statistical areas 18 and 21 (figure 1). Based on Booth and Watts (2007) and Teh et al. (2015), we present fish catch data for the Canadian Arctic (here for FAO area 18 only; area 21 is included under Canada East Coast). The total catches allocated to Arctic Canada's EEZ are primarily subsistence, possibly with a small artisanal component (figure 2A). The small-scale catches destined for human consumption increased from approximately 2,000 t in 1950 to 3,300 t/year in the early 1970s but declined to 900 t/year in the 2000s. Also included are the fish, mainly Arctic char (*Salvelinus alpinus*) used for feeding sled-dog teams, whose catch declined from 1960 to the early 1970s, when many dogs were shot or otherwise disposed of by a government eager for local populations to slide faster into modernity. Overall, none of the marine and estuarine catch in FAO area 18 was reported by the Canadian Department of Fisheries and Ocean to the FAO (Zeller et al. 2011). Figure 2B shows that this reconstructed catch consisted overwhelmingly of Arctic char, the rest consisting of more than fifteen other fish species, notably Atlantic salmon (*Salmo salar*), broad whitefish (*Coregonus nasus*), cods, and haddocks (family Gadidae).

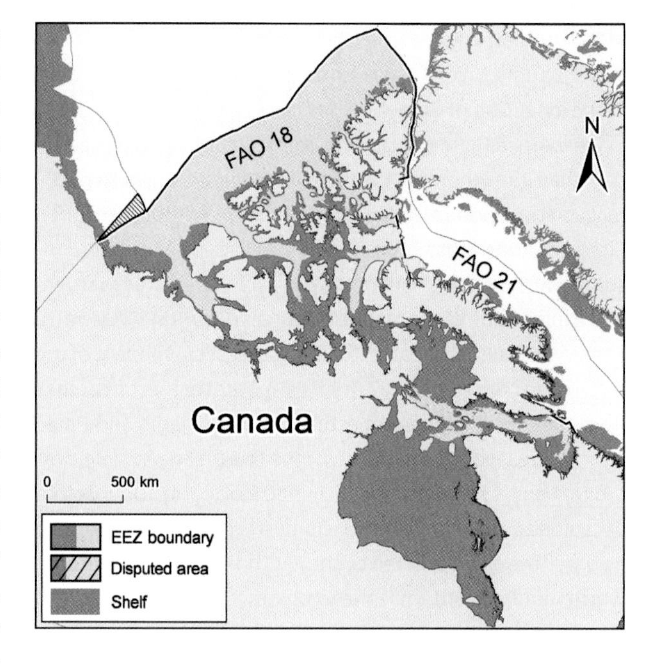

Figure 1. Map of Canada's shelf (1.83 million km²) and EEZ (3.02 million km²) in the Arctic (FAO area 18).

Figure 2. Domestic catches taken in the Arctic EEZ of Canada: (A) by sector (domestic scores: Art. 4, 4, 4; Sub. 3, 3, 3); (B) by taxon.

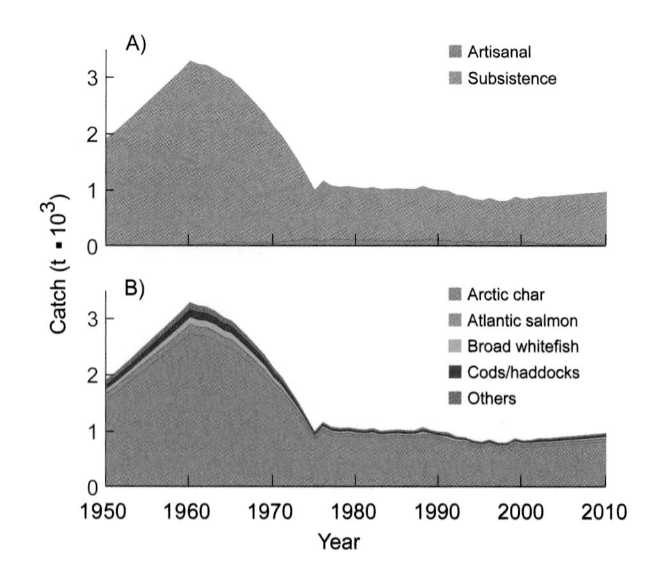

REFERENCES

Booth S and Watts P (2007) Canada's arctic marine fish catches. pp. 3–16 In Zeller D and Pauly D (eds.), *Reconstruction of marine fisheries catches for key countries and regions (1950–2005)*. Fisheries Centre Research Reports 15(5).

Teh LCL, Zylich K and Zeller D (2015) FAO area 18 (Arctic Sea): Catch data reconstruction extension of Zeller et al. (2011) to 2010. Fisheries Centre Working Paper #2015–14.

Zeller D, Booth S, Pakhomov E, Swartz W and Pauly D (2011) Arctic fisheries catches in Russia, USA and Canada: Baselines for neglected ecosystems. *Polar Biology* 34(7): 955–973.

1. Cite as Booth, S., P. Watts, L. C. L. Teh, K. Zylich, and D. Zeller. 2016. Canada (Arctic). P. 214 in D. Pauly and D. Zeller (eds.), *Global Atlas of Marine Fisheries: A Critical Appraisal of Catches and Ecosystem Impacts*. Island Press, Washington, DC.

CANADA (EAST COAST)[1]

Esther Divovich, Dyhia Belhabib, Dirk Zeller, and Daniel Pauly

Sea Around Us, University of British Columbia, Vancouver, Canada

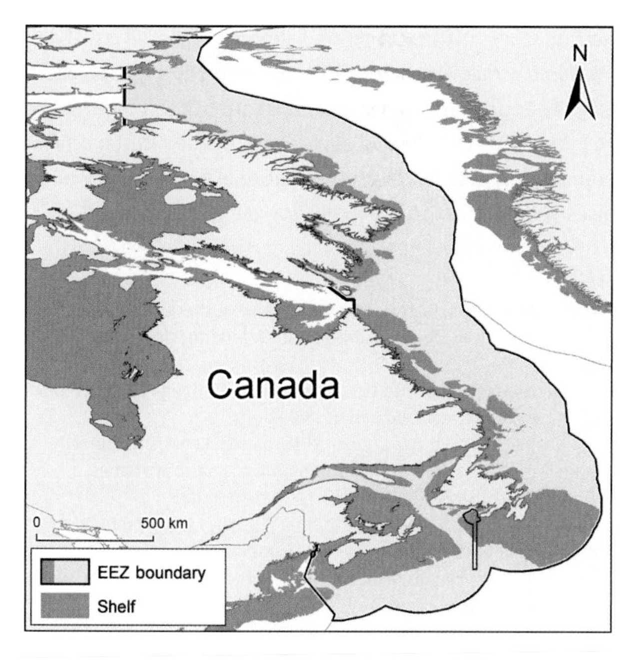

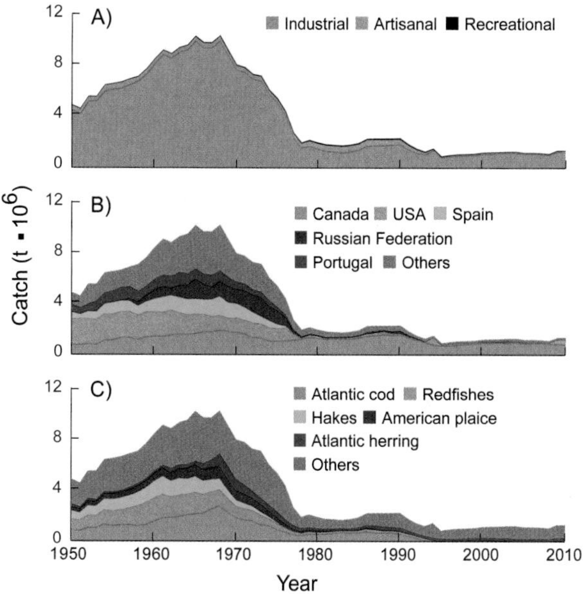

Figure 1. Canada's East Coast has an EEZ of 2.28 million km², of which 851,000 km² is shelf; some of the shelf reaches beyond the EEZ ("nose" and "tail"), another source of conflict. **Figure 2.** Domestic and foreign catches taken in the EEZ of Canada (East Coast): (A) by sector (domestic scores: Ind. 2, 2, 1; Art. 2, 2, 2; Sub. 1, 3, 1; Rec. 1, 2, 4; Dis. 2, 3, 3); (B) by fishing country (foreign catches are very uncertain); (C) by taxon.

Fisheries in eastern Canada (figure 1, including the Arctic part of FAO area 21) are thousands of years old. The industrial sector plummeted in the late 1960s but continued to dominate catches in Canada's East Coast EEZ throughout the time period (figure 2A). The reconstruction of Divovich et al. (2015) estimated domestic catches of about 800,000 t/year in the early 1950s, ascending to a peak of 2 million t in 1968, then slowly descending toward 900,000 t/year in the late 2000s. Overall, the domestic withdrawals from the waters of eastern Canada were 1.4 times the landings reported by the Canadian Department of Fisheries and Oceans to the FAO, although the bulk of unreported catch was taken before the EEZ declaration in 1977. Foreign vessels fishing in the EEZ were documented and classified as primarily the United States, Spain, Russia, and Portugal (figure 2B). As shown in figure 2C, Atlantic cod (*Gadus morhua*), a driver for European colonization and continued conflicts (Pélissac 2003), dominated the catch in the earlier time period. This conflict is ongoing, even after overfishing forced a moratorium on cod fishing in 1992 (Walters and Maguire 1996) and provided a stunning example of "fishing down" (Pauly et al. 2001). Although overall catches have declined since the moratorium, the small catches of cod are still high relative to its reduced biomass, which may explain its long-delayed, now tentative recovery.

REFERENCES

Divovich E, Belhabib D, Zeller D and Pauly D (2015) Eastern Canada, "a fishery with no clean hands": marine fisheries catch reconstruction from 1950 to 2010. Fisheries Centre Working Paper #2015–56, 37 p.

Pauly D, Palomares MLD, Froese R, Sa-a P, Vakily M, Preikshot D and Wallace S (2001) Fishing down Canadian aquatic food webs. *Canadian Journal of Fisheries and Aquatic Sciences* 58(1): 51–62.

Pélissac DS (2003) Une taupe chez les morues: halieuscopie d'un conflit. AnthropoMare, Mississauga, Canada, 293 p.

Walters C and Maguire JJ (1996) Lessons for stock assessment from the northern cod collapse. *Reviews in Fish Biology and Fisheries* 6(2): 125–137.

1. Cite as Divovich, E., D. Belhabib, D. Zeller, and D. Pauly. 2016. Canada (East Coast). P. 215 in D. Pauly and D. Zeller (eds.), *Global Atlas of Marine Fisheries: A Critical Appraisal of Catches and Ecosystem Impacts*. Island Press, Washington, DC.

CANADA (PACIFIC)[1]

Cameron H. Ainsworth

College of Marine Science, University of South Florida, St. Petersburg, FL

The Pacific coast of Canada consists exclusively of the coast of the province of British Columbia (BC), between the US states of Washington and Alaska (figure 1). The bulk of the catches allocated within the EEZ of Canada (Pacific) are industrial, with a slight recreational component that peaked in the 1990s (figure 2A). A catch reconstruction was performed by Ainsworth (2015, 2016), who estimated that total domestic withdrawals were on the order of 600,000 t/year in the 1950s, peaked twice, at about 700,000 t in 1963 and 610,000 t in 1991, and declined to 200,000 t/year in the 2000s. From 1950 to 2010, reconstructed domestic catches accounted for more than 24.3 million t, which is 1.8 times as high as the 13.2 million t officially reported by the FAO on behalf of Canada, a higher discrepancy than estimated for many other developed countries. Allocated catches within the EEZ is predominantly by the domestic fishery (figure 2B); however, foreign catches are here underestimated. Until the mid-1960s, catches were dominated by Pacific herring (*Clupea pallasii*) and only secondarily by several salmon species (*Oncorhynchus* spp.). After the decline of herring stocks in the mid-1960s, salmon dominated until hake (*Merluccius productus*) became prominent in the mid-late 1980s. Since the 2000s, hake dominates in terms of tonnage (figure 2C). More comprehensive coverage is needed in the official statistics of subsistence and recreational fisheries, as well as discards (Ainsworth and Pitcher 2005; Pauly et al. 2001).

REFERENCES

Ainsworth CH (2015) British Columbia Marine Fisheries Catch Reconstruction: 1873 to 2010. Fisheries Centre Working Paper #2015-62, 9 p.

Ainsworth CH (2016) British Columbia Marine Fisheries Catch Reconstruction: 1873 to 2011. *BC Studies* 188: 81–90.

Ainsworth CH and Pitcher TJ (2005) Estimating illegal, unreported and unregulated catch in British Columbia's marine fisheries. *Fisheries Research* 75(1–3): 40–55.

Pauly D, Palomares MLD, Froese R, Sa-a P, Vakily M, Preikshot D and Wallace S (2001) Fishing down Canadian aquatic food webs. *Canadian Journal of Fisheries and Aquatic Sciences* 58: 51–62.

1. Cite as Ainsworth, C. H. 2016. Canada (Pacific). P. 216 in D. Pauly and D. Zeller (eds.), *Global Atlas of Marine Fisheries: A Critical Appraisal of Catches and Ecosystem Impacts*. Island Press, Washington, DC.

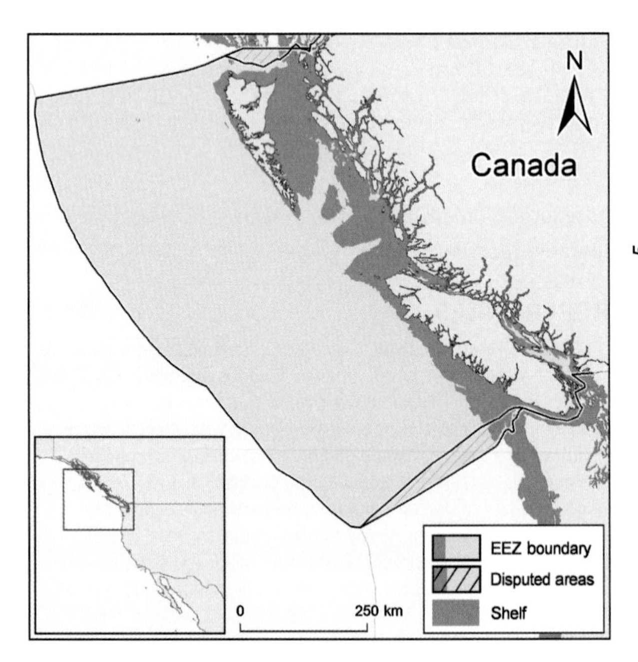

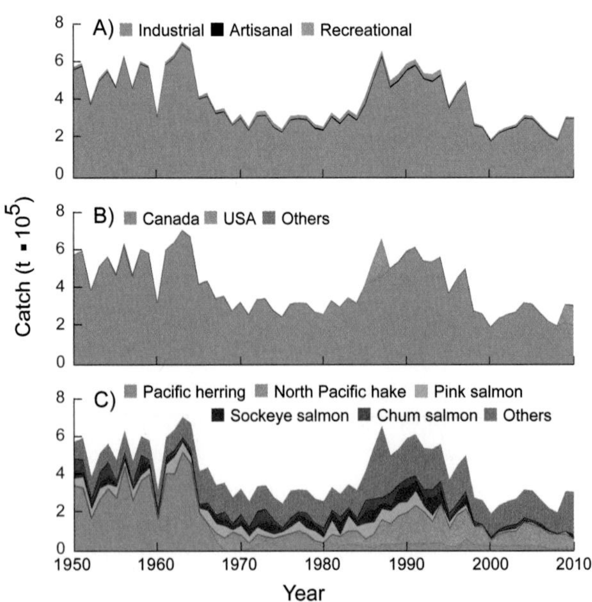

Figure 1. Canada's Pacific Coast (i.e., British Columbia) has an EEZ of 470,000 km², of which 87,400 km² is shelf.
Figure 2. Domestic and foreign catches taken in the EEZ of Canada (Pacific): (A) by sector (domestic scores: Ind. 4, 4, 4; Art. 1, 1, 2; Sub. 1, 1, 2; Rec. 3, 4, 4; Dis. 1, 1, 1); (B) by fishing country (foreign catches are very uncertain); (C) by taxon.

CAPE VERDE[1]

Dyhia Belhabib,[a] Carlos Alberto Monteiro,[b] Isaac Trindade Santos,[b]
Sarah Harper,[a] Kyrstn Zylich,[a] Dirk Zeller,[a] and Daniel Pauly[a]

[a]*Sea Around Us*, University of British Columbia, Vancouver, Canada [b]Instituto Nacional de Desenvolvimento das Pescas (INDP), Mindelo, Cabo Verde [c]Centro de Ciências Agrárias Aplicadas, Universidade Federal de Sergipe, Brazil

Cape Verde is an archipelagic country off northwest Africa (figure 1) with a dry climate that belies the country's name ("Green Cape"), and it imports most of its food. Fisheries are thus important to the 500,000 inhabitants and the growing number of tourists. However, the artisanal and industrial sectors are monitored at only 15% of the landing sites (Stobberup et al. 2005).

In the 1950s and 1960s, the artisanal sector dominated the total catches allocated to Cape Verde's EEZ; however, the industrial sector steadily increased and generated the bulk of the catch by the 2000s (figure 2A). Total domestic catches were estimated by Santos et al. (2012) at about 6,300 t/year between 1950 and 1963 and reached a peak of 30,000 t in 2005 (figure 2A). Overall, domestic catches were 1.9 times the landings supplied to the FAO; the small foreign catches were taken mainly by France and Spain (figure 2B). Catches consisted mostly of mackerel scad (*Decapterus marcellus*), yellowfin tuna (*Thunnus albacares*), and picarels (*Spicara* spp.), whose contribution recently increased (figure 2C). One particularly worrisome aspect of Cape Verdean fisheries is the use of dynamite to augment catches. Although suppressing this form of destructive fishing is difficult, it will have to be done, given the growing importance of local fish to Cape Verdeans.

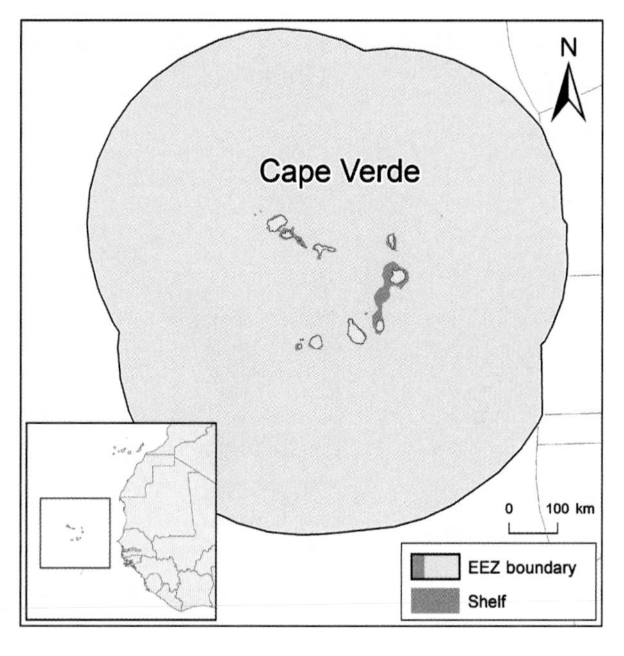

Figure 1. Cape Verde has a total land area of 4,000 km² and an EEZ of 797,000 km².

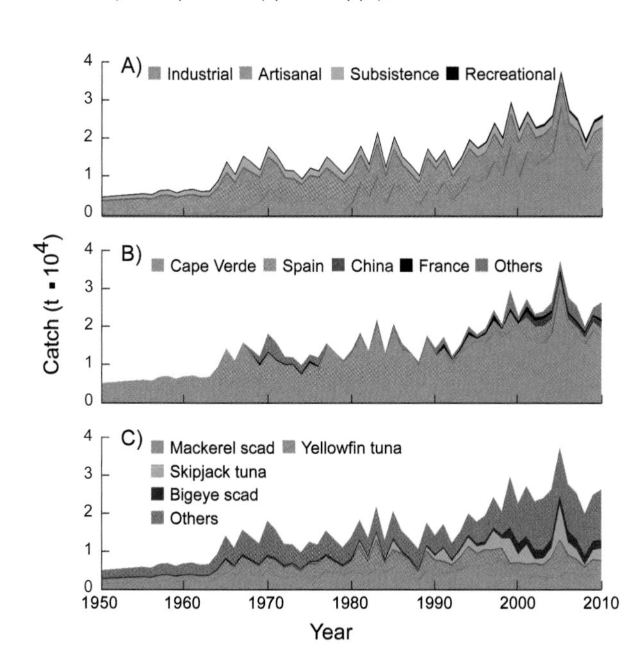

Figure 2. Domestic and foreign catches taken in the EEZ of Cape Verde: (A) by sector (domestic scores: Ind. 1, 1, 4; Art. 4, 4, 4; Sub. 2, 2, 2; Rec. 2, 2, 4; Dis. 2, 2, 2); (B) by fishing country (foreign catch are very uncertain); (C) by taxon.

REFERENCES

Santos I, Monteiro C, Harper S, Zeller D and Belhabib D (2012) Reconstruction of marine fisheries catches for the Republic of Cape Verde, 1950–2010. pp. 79–90 In Belhabib D, Zeller D, Harper S and Pauly D (eds.), *Marine fisheries catches in West Africa, Part 1*. Fisheries Centre Research Reports 20(3).

Stobberup KA, Amorim P and Monteiro VM (2005) Assessing the effects of fishing in Cape Verde and Guinea Bissau, Northwest Africa. p. 22 In Kruse GH, Gallucci VF, Hay DE, Perry RI, Peterman RM, Shirley TC, Spencer PD, Wilson B and Woodby D (eds.), Fisheries assessment and management in data-limited situations. University of Alaska Sea Grant College Program, Fairbanks.

1. Cite as Belhabib, D., C. Monteiro, I. T. Santos, S. Harper, K. Zylich, D. Zeller, and D. Pauly. 2016. Cape Verde. P. 217 in D. Pauly and D. Zeller (eds.), *Global Atlas of Marine Fisheries: A Critical Appraisal of Catches and Ecosystem Impacts*. Island Press, Washington, DC.

CHILE[1]

Liesbeth van der Meer,[a] Hugo Arancibia,[b] Kyrstn Zylich,[a] and Dirk Zeller[a]

[a]*Sea Around Us*, University of British Columbia, Vancouver, Canada [b]Unidad de Tecnología Pesquera, Universidad de Concepcion, Concepcion, Chile

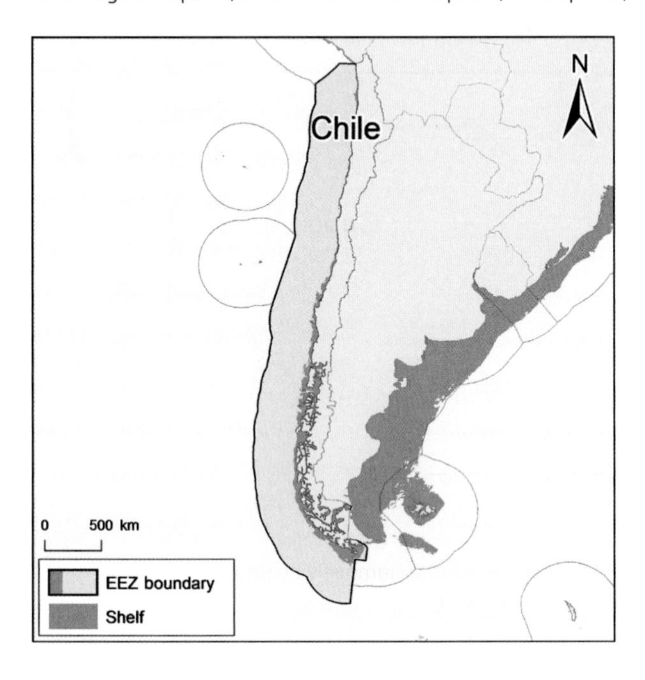

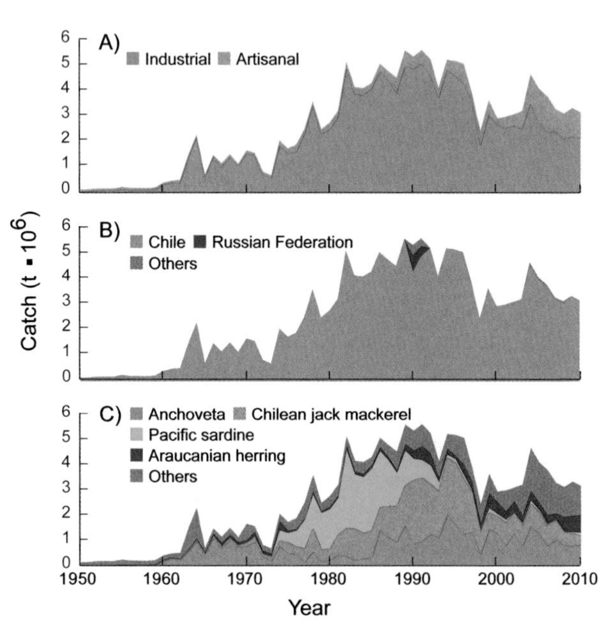

Figure 1. The Chilean mainland has a shelf of 228,000 km² and an EEZ of 1.98 million km². **Figure 2.** Domestic and foreign catches taken in the mainland EEZ of Chile: (A) by sector (domestic scores: Ind. 2, 2, 4; Art. 2, 2, 3; Dis. 1, 1, 1); (B) by fishing country (foreign catches are very uncertain); (C) by taxon.

The Republic of Chile is a narrow country bordering the southern Pacific and Antarctic Oceans to the west and south, respectively (figure 1), a geography that produced a population with strong ties to the sea. In addition to its mainland EEZ, Chile owns several oceanic islands, covered in the following pages. Although fishing accounts for only 0.4% of GDP and is dwarfed by mining, Chile's mainland landings in 2010 were the seventh largest in the world. The reason is that, except for its southernmost part, the Chilean EEZ overlaps with the southern half of the Humboldt Current Large Marine Ecosystem, the most productive marine ecosystem in the world. There, strong coastal winds drive water off the coast, resulting in upwelling of deep, nutrient-rich waters that allow an extremely high primary and secondary production, translated into abundant fish, seabirds, and marine mammals. Industrial domestic catches dominate the total removals from Chile's EEZ (figure 2A; figure 2B). However, the inshore catch of artisanal fishers, mainly benthic invertebrates (Castilla and Fernandez 1998), has been increasing over time. The domestic catch, reconstructed by van der Meer et al. (2015), started with approximately 170,000 t/year in the early 1950s and grew to more than 5.5 million t in 1989. The catches allocated to the EEZ consisted mainly of anchoveta (*Engraulis ringens*), Chilean jack mackerel (*Trachurus murphyi*), and Pacific sardine (*Sardinops sagax*; figure 2C).

REFERENCES

Castilla JC and Fernandez M (1998) Small-scale benthic fisheries in Chile: on co-management and sustainable use of benthic invertebrates. *Ecological Applications* sp1: 124–132.

van der Meer L, Arancibia H, Zylich K and Zeller D (2015) Reconstruction of total marine fisheries catches for mainland Chile (1950–2010). Fisheries Centre Working Paper #2015-91, 15 p.

1. Cite as van der Meer, L., H. Arancibia, K. Zylich, and D. Zeller. 2016. Chile. P. 218 in D. Pauly and D. Zeller (eds.), *Global Atlas of Marine Fisheries: A Critical Appraisal of Catches and Ecosystem Impacts*. Island Press, Washington, DC.

CHILE (EASTER ISLAND)[1]

Kyrstn Zylich,[a] Sarah Harper,[a] Roberto Licandeo,[b] Rodrigo Vega,[c] Dirk Zeller,[a] and Daniel Pauly[a]

[a]*Sea Around Us*, University of British Columbia, Vancouver, Canada [b]University of British Columbia, Vancouver, Canada [c]Global Ocean Legacy Project, The Pew Charitable Trusts, Santiago, Chile

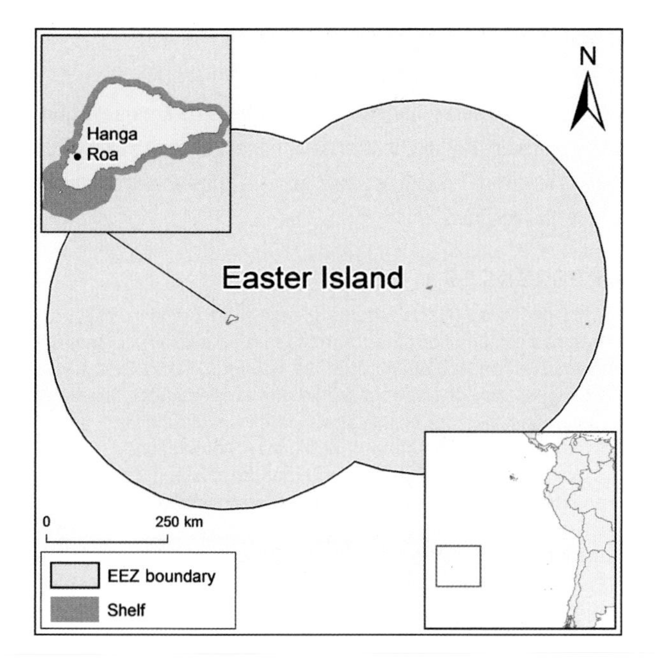

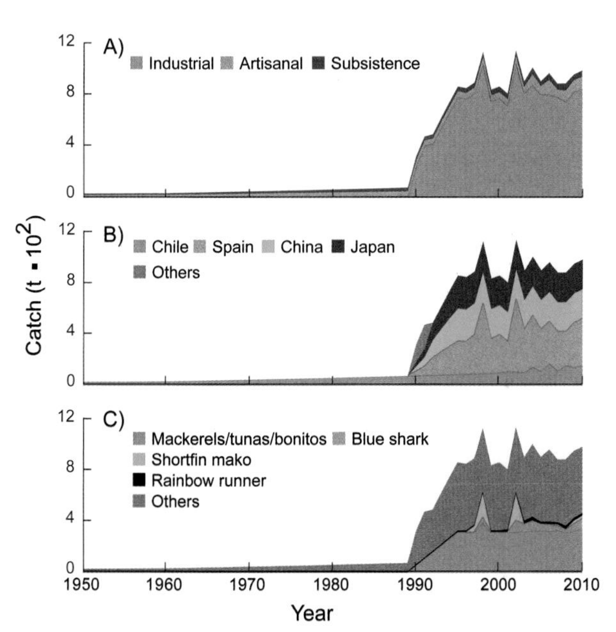

Figure 1. Easter Island Province, Chile, has a land area of 169 km² and an EEZ of 720,000 km². **Figure 2.** Domestic and foreign catches taken in the EEZ of Easter Island (Chile): (A) by sector (domestic scores: Ind. 1.3, 2.3, 3; Art. 1.3, 2.3, 3; Sub. 1.3, 2, 2); (B) by fishing country (foreign catches are very uncertain); (C) by taxon.

Easter Island, or Rapa Nui in the Polynesian language of its first inhabitants, is located 3,760 kilometers southwest of mainland Chile and forms, with the uninhabited and tiny island of Salas y Gómez, Chile's Easter Island Province (figure 1). This reconstruction focuses on Rapa Nui, because Salas y Gómez was never subjected to a sustained fishery, and most of its surrounding area became a marine reserve in 2010. The total catches allocated to Easter Island's EEZ are largely from the industrial sector and attributed to China (figure 2A; figure 2B). Small-scale subsistence and artisanal catch are important components of the domestic catch and were reconstructed by Zylich et al. (2014). Using fishers' interviews, various Chilean reports (e.g., Yáñez et al. 2007), and limited official data, it was estimated that domestic fisheries catches around Rapa Nui were about 30 t/year in the 1950s, which increased to about 130 t/year in the 2000s. Jumbo flying squid (*Dosidicus gigas*) and mackerel tuna and bonitos (family Scombridae) dominate the total removals within the EEZ (figure 2C). Satellite observation and radio communications also suggest that there is an illegal foreign tuna fishery in the EEZ of Easter Island Province, with catches very tentatively estimated at 630 t/year, which may contribute to the declining artisanal catch per effort experienced by Rapa Nui fishers (pers. comm. to R.V. and D.P.).

REFERENCES

Yáñez E, Silva C, Trujillo H, González E, Álvarez L, Manutomatoma L and Romero P (2007) Diagnóstico del Sector Pesquero de la Isla de Pascua. Pontificia Universidad Católica de Valparaíso, Valparaíso, Chile. 154 p.

Zylich K, Harper S, Licandeo R, Vega R, Zeller D and Pauly D (2014) Fishing in Easter Island: a recent history (1950–2010). *Latin American Journal of Aquatic Research* 24(2): 845–856.

1. Cite as Zylich, K., S. Harper, R. Licandeo, R. Vega, D. Zeller, and D. Pauly. 2016. Chile (Easter Island). P. 219 in D. Pauly and D. Zeller (eds.), *Global Atlas of Marine Fisheries: A Critical Appraisal of Catches and Ecosystem Impacts*. Island Press, Washington, DC.

CHILE (JUAN FERNANDEZ AND DESVENTURADAS ISLANDS)[1]

Kyrstn Zylich[a] and Liesbeth van der Meer[b]

[a]*Sea Around Us*, University of British Columbia, Vancouver, Canada [b]Oceana Latin American Office, Santiago, Chile

Chile controls three groups of oceanic islands, Easter Island (previous page) and the Juan Fernandez and Desventuradas Islands, closer to the Chilean mainland (figure 1). Because of their biological uniqueness, these islands were declared a National Park in 1935 and a UNESCO biosphere reserve in 1977. The Juan Fernandez Archipelago consists of 3 main volcanic islands: Robinson Crusoe, Alejandro Selkirk, and Santa Clara Island, which jointly had a population of 885 in 2012 and depend largely on tourism, industrial fishing from the mainland, and small-scale artisanal and subsistence fishing (figure 2A). In contrast, the Desventuradas Islands are uninhabited (except for a military base), and their coastal resources are exploited by Robinson Crusoe Island, whereas pelagic resources of their EEZ are exploited from the mainland industrial fleets, as are the pelagic resources of the Juan Fernandez EEZ. This justifies their treatment here in a single account, based on the reconstruction of Zylich and van der Meer (2015), which yielded a catch estimate of 640 t/year in the early 1950s, a peak of 1.2 million t in 1995, and about 300,000 t/year in the late 2000s. This erratic trajectory was due mainly to industrial fishing for Chilean jack mackerel (*Trachurus murphyi*) and pomfrets (Bramidae; figure 2B). The mainstay of the artisanal fishery is relatively stable (Arana and Toro 1985; Ernst et al. 2013), although the catch per effort has declined.

REFERENCES

Arana P and Toro C (1985) Distribucion del esfuerzo, rendimientos por trampa y composicion de las capturas en la pesqueria de la langosta de Juan Fernandez (*Jasus frontalis*). Pp. 157–185 in Arana P (ed.), Investigaciones Marinas en el Archipielago de Juan Fernandez. Escuela de Ciencias del Mar. Universidad Catolica de Valparaíso, Chile.

Ernst B, Chamorro JPM, Oresanz JM, Parma A, Porobic J and Roman C (2013) Sustainability of the Juan Fernandez lobster fishery (Chile) and the perils of generic science-based prescriptions. *Global Environmental Change* 23(6): 1381–1392.

Zylich K and van der Meer L (2015) Reconstruction of total marine fisheries catches for Juan Fernández Islands and the Desventuradas Islands (Chile) for 1950–2010. Fisheries Centre Working Paper #2015–92, 14 p.

1. Cite as Zylich, K., and L. van der Meer. 2016. Chile (Juan Fernandez and Desventuradas Islands). P. 220 in D. Pauly and D. Zeller (eds.), *Global Atlas of Marine Fisheries: A Critical Appraisal of Catches and Ecosystem Impacts*. Island Press, Washington, DC.

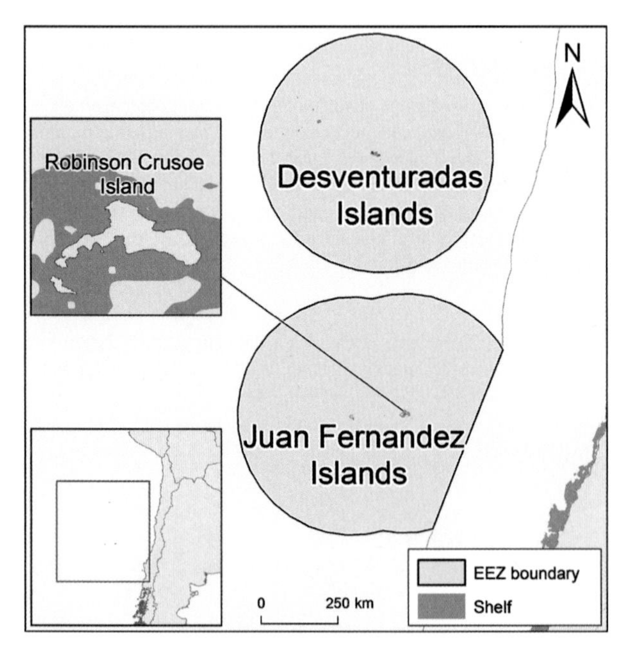

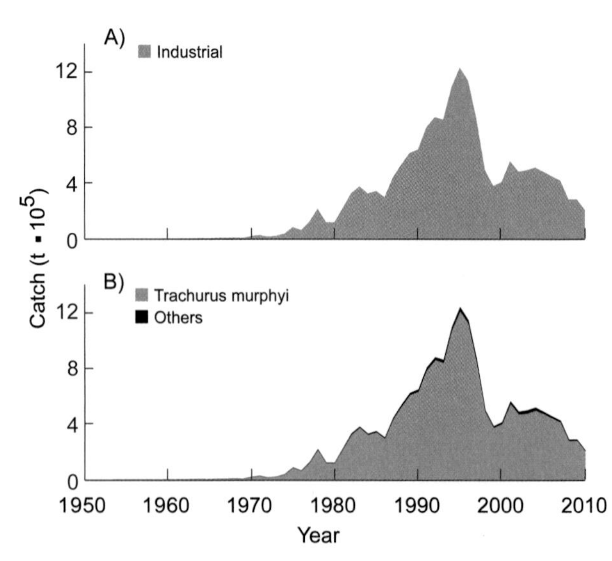

Figure 1. Juan Fernandez (land area 112 km²) and Desventuradas Islands (11 km²) have EEZs of 502,000 km² and 450,000 km², respectively. **Figure 2.** Domestic catches taken in the EEZ of Juan Fernandez and Desventuras: (A) by sector (scores: Ind. 2, 2, 4; Art. 4, 4, 4; Sub. 2, 2, 2; Rec. 2, 2, 1; Dis. 1, 1, 1); (B) by taxon.

CHINA[1]

Daniel Pauly and Frédéric Le Manach

Sea Around Us, University of British Columbia, Vancouver, Canada

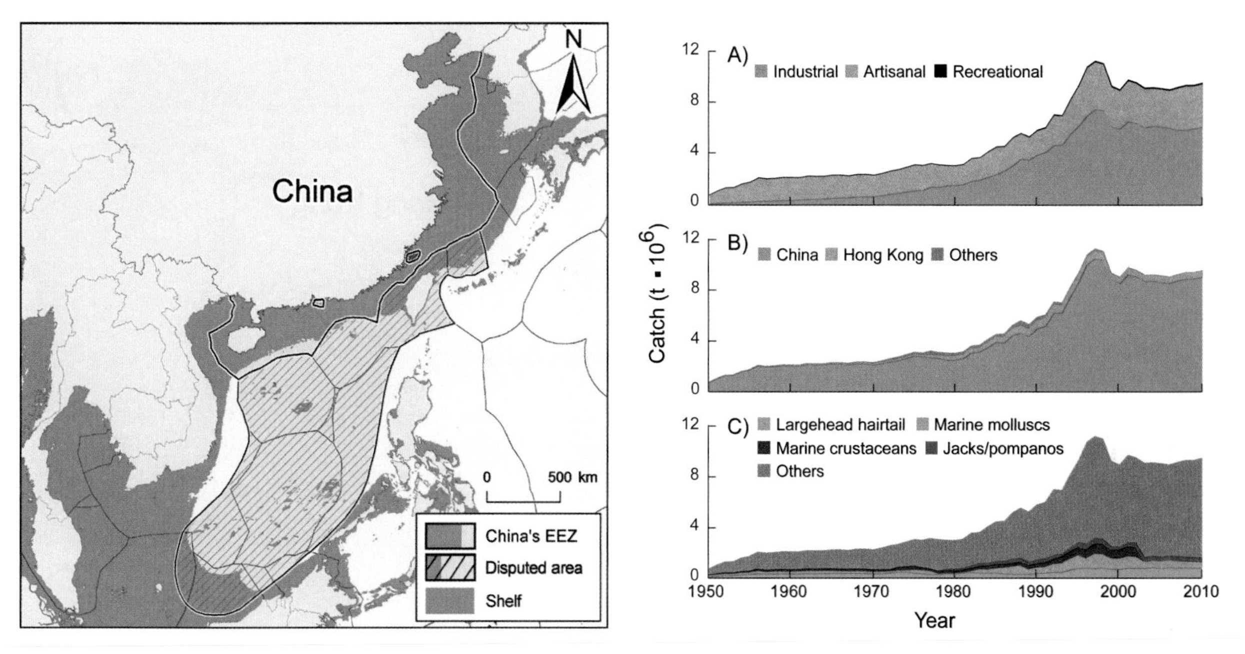

Figure 1. China's claimed EEZ, of 2.24 million km² (shelf area: 1.03 million km²), extends deep into the South China Sea (and is thus known as "cow tongue") and overlaps with the claims of all other riparian states. **Figure 2.** Domestic and foreign catches taken in EEZ of China: (A) by sector (domestic scores: Ind. 2, 3, 3; Art. 1, 2, 2; Sub. 2, 2, 2; Rec. –, 1, 1; Dis. 2, 2, 2); (B) by fishing country (foreign catches are very uncertain); (C) by taxon.

The People's Republic of China (here: China, excluding Hong Kong and Taiwan, but including Macau; figure 1) is reclaiming China's role as one of the leading cultures and economies of the world. However, China remains saddled with a system that generates fanciful production statistics. With regard to marine fisheries, this produces an inflated "domestic" catch (i.e., from the huge EEZ claimed by China; see figure 1), as demonstrated by Watson and Pauly (2001), and a large unreported catch by China's distant-water fleet (Pauly et al. 2014, Belhabib et al. 2015). The tentative reconstruction of China's domestic fisheries by Pauly and Le Manach (2015) documented the domestic and distant-water industrial sectors and the (domestic) small-scale sectors (artisanal, subsistence, and recreational). The total catches allocated to China's EEZ are mostly from the industrial and artisanal sectors (figure 2A). Although these catches were mainly domestic, foreign fishing by Hong Kong became more prevalent in the early 1980s (figure 2B). Figure 2C attempts to disaggregate reported Chinese catches, mostly consisting of uninformative "miscellaneous fishes nei," into species likely to be occurring in domestic catches, as inferred from the compositions of South Korea and Taiwan but also considering large (by)catches of what may be called "jellyfish sludge." We cannot overemphasize the tentative nature of the sectorial, spatial, and taxonomic disaggregations presented here.

REFERENCES

Belhabib D, Sumaila UR, Lam VWY, Kane EA, Zeller D, Le Billon P and Pauly D (2015) Euros vs. Yuan: A first attempt at comparing European and Chinese fishing access in West Africa. *PLoS ONE*, 10(3): e0118351.

Pauly D, Belhabib D, Blomeyer R, Cheung WWL, Cisneros-Montemayor AM, Copeland D, Harper S, Lam VWY, Mai Y, Le Manach F, Österblom H, Mok KM, van der Meer L, Sanz A, Shon S, Sumaila UR, Swartz W, Watson RA, Zhai Y and Zeller D (2014) China's distant-water fisheries in the 21st century. *Fish and Fisheries* 15: 474–488.

Pauly D and Le Manach F (2015) Tentative adjustments of China's marine fisheries catches (1950–2010). Fisheries Centre Working Paper #2015-28, 16 p.

Watson R and Pauly D (2001) Systematic distortions in world fisheries catch trends. *Nature* 414: 534–536.

1. Cite as Pauly, D., and F. Le Manach. 2016. China (Mainland). P. 221 in D. Pauly and D. Zeller (eds.), *Global Atlas of Marine Fisheries: A Critical Appraisal of Catches and Ecosystem Impacts*. Island Press, Washington, DC.

CHINA (HONG KONG)[1]

William W. L. Cheung

Changing Ocean Research Unit, University of British Columbia, Vancouver, Canada

Situated along the southeastern coast of China on the eastern shores of the Pearl River estuary (figure 1), Hong Kong was a British colony from the mid-19th century until 1997 and is currently a Special Administrative Region (SAR) in the People's Republic of China, with a high degree of autonomy except for foreign relations and defense. The industrial sector dominated the catch through the entire time period (figure 2A). Based on Cheung (2015), updating and expanding on Cheung and Sadovy (2004), reconstructed catches for the Hong Kong SAR are presented here. These were 5,000–8,000 t/year in the 1950s and early 1960s, which increased to a peak of 25,000–30,000 t/year in the 1970s to 1990s, then dropped to a plateau of about 13,000 t/year in the 2000s. The reconstructed domestic catches are 3 times the landings reported by the FAO and are assumed to have been taken in the Hong Kong EEZ. Although the bulk of the catches originate from the domestic fleet, small catch amounts by China and Taiwan were also documented (figure 2B). Figure 2C presents a taxonomic breakdown of these catches, which consisted of numerous invertebrates, especially "Natantian decapods" (i.e., shrimps) and clams, cockles, arkshells, as well as threadfins and whiptail breams (family Nemipteridae) among demersal fishes. Hong Kong's fisheries show many sign of severe overfishing.

REFERENCES

Cheung WWL (2015) Reconstructed catches in waters administrated by the Hong Kong Special Administrative Region. Fisheries Centre Working Paper #2015–93, 15 p.

Cheung WWL and Sadovy Y (2004) Retrospective evaluation of data-limited fisheries: a case from Hong Kong. *Reviews in Fish Biology and Fisheries* 14: 181–206.

1. Cite as Cheung, W. W. L. 2016 China (Hong Kong). P. 222 in D. Pauly and D. Zeller (eds.), *Global Atlas of Marine Fisheries: A Critical Appraisal of Catches and Ecosystem Impacts*. Island Press, Washington, DC.

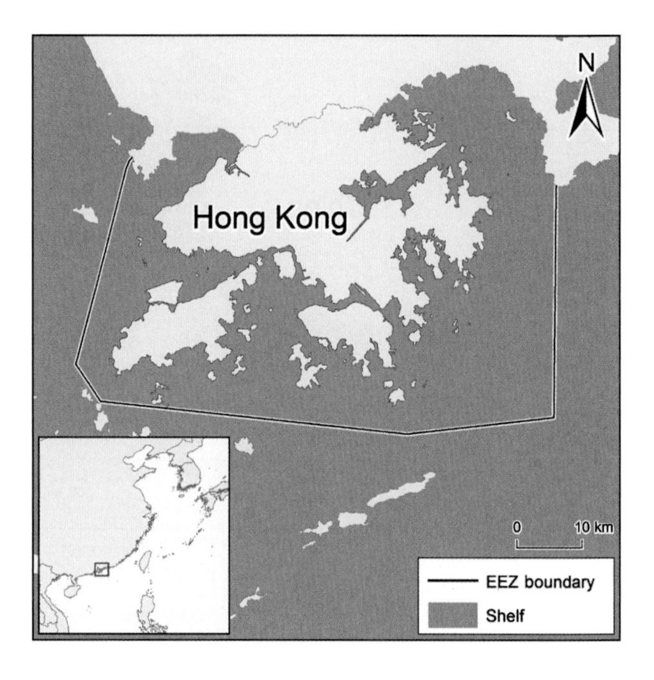

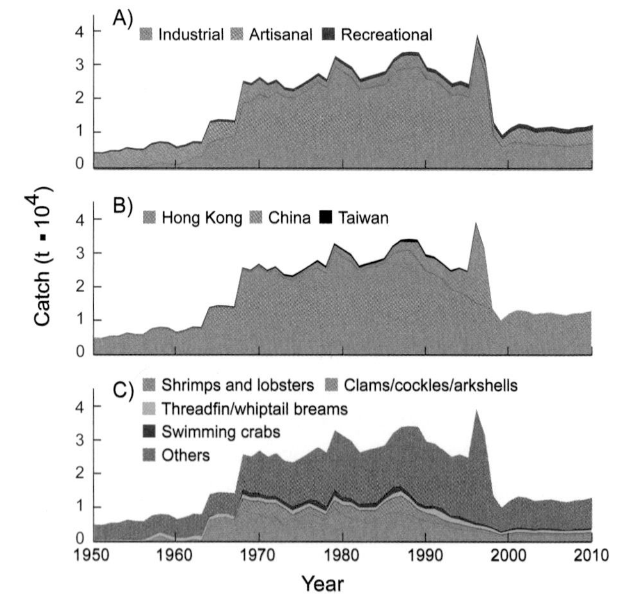

Figure 1. The waters administrated by Hong Kong (corresponding to an EEZ and shelf of about 2,000 km^2) are influenced by the outflow of freshwater from the Pearl River in the west and oceanic currents in the east.

Figure 2. Domestic and foreign catches taken in the EEZ of Hong Kong: (A) by sector (domestic scores: Ind. 3, 2, 2; Art. 2, 3, 3; Rec. 1, 1, 1; Dis. 1, 1, 1); (B) by fishing country (foreign catches are very uncertain); (C) by taxon.

COLOMBIA (CARIBBEAN)[1]

Jeffrey Wielgus,[a] Dalila Caicedo-Herrera,[b] Alasdair Lindop,[a] Tiffany Chen,[a] Kyrstn Zylich,[a] and Dirk Zeller[a]

[a]*Sea Around Us*, University of British Columbia, Vancouver, Canada [b]Fundación Omacha, Bogotá, Colombia

Colombia has coasts on both the Pacific and Atlantic Oceans, in the Caribbean (figure 1). The marine resources of the Colombian Caribbean are exploited by coastal communities (including of indigenous Wayuu), but industrial fisheries dominate the total catch allocated to the Caribbean EEZ of Colombia (figure 2A). The management of fisheries has been hampered by administrative changes, which led to the loss of statistical data. Wielgus et al. (2007, 2010) and Lindop et al. (2015) reconstructed the catches using secondary sources, including unique documents kindly made available by the Colombian Minister of Agriculture and Rural Development. The total catch allocated to the Caribbean EEZ was about 4,000 t/year in the early 1950s, increased rapidly to a peak of nearly 110,000 t in 1967, then declined to a low of about 13,000 t in 1976. After a secondary peak of 92,000 t in 1991, catches declined to about 15,000–16,000 t/year in the 2000s. From 1950 to 2010, the reconstructed catch was 3.4 times higher than the data in the Atlantic reported by the FAO on behalf of Colombia. Removals were predominantly domestic, although some catches by the United States were documented until 2000 (figure 2B). Figure 2C summarizes the taxonomic composition of the catch, which includes a wide range of demersal and pelagic fishes such as sea catfishes (family Ariidae), snooks (Centropomidae), mojarras (Gerreidae), and snappers (Lutjanidae).

REFERENCES

Lindop A, Chen T, Zylich K and Zeller D (2015) A reconstruction of Colombia's marine fisheries catches. Fisheries Centre Working Paper #2015–32, 15 p.

Wielgus J, Caicedo-Herrera D and Zeller D (2007) Reconstruction of Colombia's marine fisheries catches. pp. 69–79 In Zeller D and Pauly D (eds.), *Reconstruction of marine fisheries catches for key countries and regions* (1950–2005). Fisheries Centre Research Reports 15(2).

Wielgus J, Zeller D, Caicedo-Herrera D and Sumaila R (2010) Estimation of fisheries removals and primary economic impact of small-scale and industrial marine fishery in Colombia. *Marine Policy* 34: 506–513.

1. Cite as Wielgus, J., D. Caicedo-Herrera, A. Lindop, T. Chen, K. Zylich, and D. Zeller 2016. Colombia (Caribbean). P. 223 in D. Pauly and D. Zeller (eds.), *Global Atlas of Marine Fisheries: A Critical Appraisal of Catches and Ecosystem Impacts*. Island Press, Washington, DC.

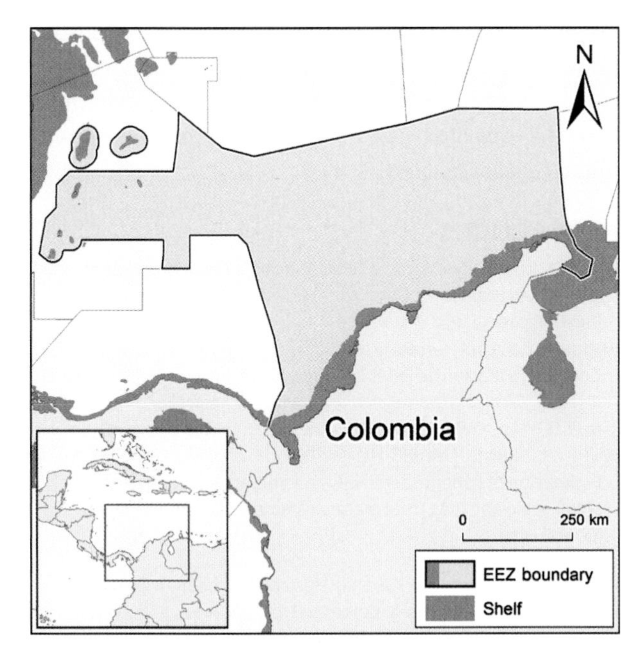

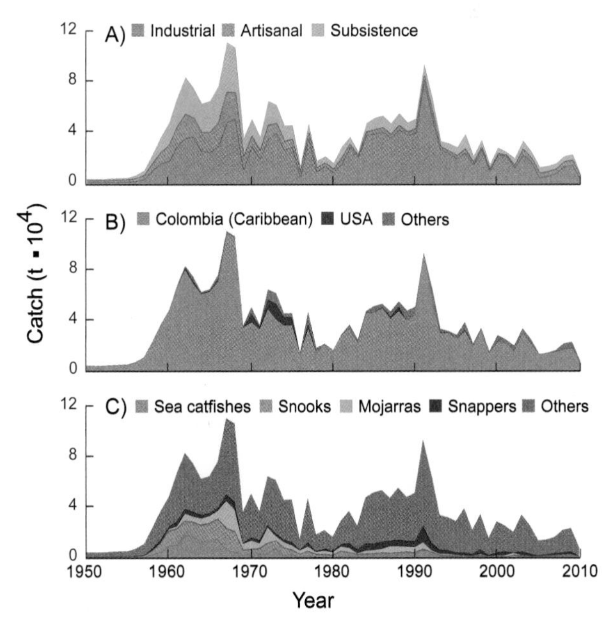

Figure 1. Colombia's EEZ in the Caribbean, including the San Andres Archipelago, is large (418,000 km², of which 33,200 km² is shelf). **Figure 2.** Domestic and foreign catches taken in the EEZ of Colombia (Caribbean): (A) by sector (domestic scores: Ind. 2, 2, 3; Art. 2, 2, 2; Sub. 2, 2, 2; Dis. 2, 2, 2); (B) by fishing country (foreign catches are very uncertain); (C) by taxon.

COLOMBIA (PACIFIC)[1]

Jeffrey Wielgus,[a] Dalila Caicedo-Herrera,[b] Alasdair Lindop,[a] Tiffany Chen,[a] Kyrstn Zylich,[a] and Dirk Zeller[a]

[a]*Sea Around Us*, University of British Columbia, Vancouver, Canada [b]Fundación Omacha, Bogotá, Colombia

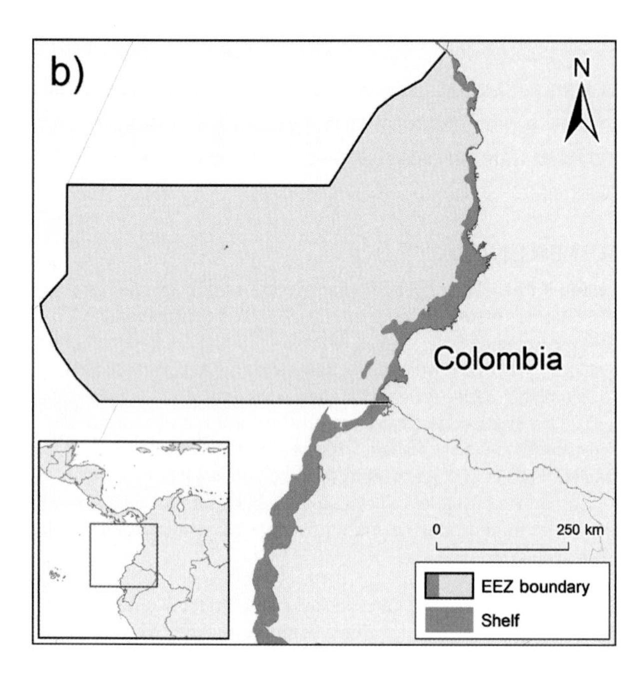

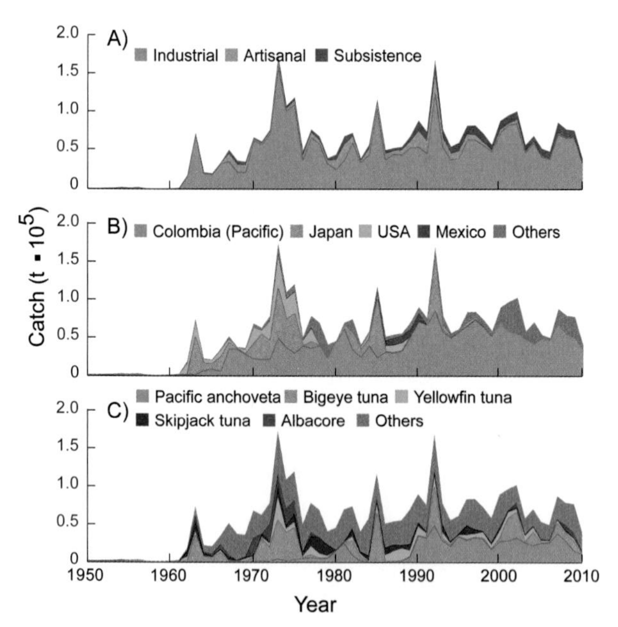

Figure 1. Colombia's EEZ in the Pacific covers 327,000 km², of which 16,500 km² is shelf. **Figure 2.** Domestic and foreign catches taken in the EEZ of Colombia (Pacific): (A) by sector (domestic scores: Ind. 2, 2, 3; Art. 2, 2, 2; Sub. 2, 2, 2; Dis. 2, 2, 2); (B) by fishing country (foreign catches are very uncertain); (C) by taxon.

Colombia's Pacific coast ranges from Panama in the north to Ecuador (figure 1). The fisheries resemble those in the Caribbean (see previous page), although they tend to fluctuate more because of environmental variations (Wielgus et al. 2010). The catch was driven largely by the industrial fishery. However, a declining artisanal sector over the last two decades, concomitant with a growth in noncommercial subsistence fishing, was also noted (figure 2A). The fisheries monitoring and management system is currently handled by the Colombian Institute for Rural Development. INCODER, its Spanish acronym, holds official landings data only from 1975, consisting of landings data for different number of taxa for different years. The catch reconstruction was undertaken by Wielgus et al. (2007, 2010) and updated by Lindop et al. (2015). The total allocated catch to Colombia Pacific's EEZ generated catch estimates of about 1,200 t/year in the 1950s, which grew to approximately 169,000 t in 1992, then declined to about 70,000 t/year in the late 2000s. Although the majority of the catch is domestic, a noticeable amount is by foreign vessels such as the United States, Japan, and Mexico, fishing in Colombia's EEZ (figure 2B). Figure 2C shows catches were driven largely by the industrial fishery for Pacific anchoveta (*Cetengraulis mysticetus*),which is used for producing fishmeal and fish oil, both of which are exported, as well as bigeye tuna (*Thunnus obesus*) and yellowfin tuna (*Thunnus albacares*).

REFERENCES

Lindop A, Chen T, Zylich K and Zeller D (2015) A reconstruction of Colombia's marine fisheries catches. Fisheries Centre Working Paper #2015–32, 15 p.

Wielgus J, Caicedo-Herrera D and Zeller D (2007) Reconstruction of Colombia's marine fisheries catches. pp. 69–79 In Zeller D and Pauly D (eds.), *Reconstruction of marine fisheries catches for key countries and regions (1950–2005)*. Fisheries Centre Research Reports 15(2).

Wielgus J, Zeller D, Caicedo-Herrera D and Sumaila UR (2010) Estimation of fisheries removals and primary economic impact of small-scale and industrial marine fishery in Colombia. *Marine Policy* 34: 506–513.

1. Cite as Wielgus, J., D. Caicedo-Herrera, A. Lindop, T. Chen, K. Zylich, and D. Zeller. 2016. Colombia (Pacific). P. 224 in D. Pauly and D. Zeller (eds.), *Global Atlas of Marine Fisheries: A Critical Appraisal of Catches and Ecosystem Impacts*. Island Press, Washington, DC.

COMOROS[1]

Beau Doherty,[a] Melissa Hauzer,[b] and Frédéric Le Manach[a]

[a]*Sea Around Us*, University of British Columbia, Vancouver, Canada
[b]Department of Geography, University of Victoria, Victoria, Canada

The Union of the Comoros is an archipelago in the western Indian Ocean, composed of three main islands (figure 1). Domestic fisheries consist of a small-scale boat fleet of pirogues and motor boats operated by men and shore-based fishing by women (Hauzer et al. 2013). Small-scale catches from the artisanal and subsistence sectors compose the bulk of the domestic catch in the Comoros EEZ (figure 2A). Doherty et al. (2015) reconstructed domestic catches of 1,200 t/year in the early 1950s, which increased to 10,000 t by the mid-1980s and about 18,500 t/year from 2005 to 2010. The rapid increase in later years was caused by the increasing number of motorized vessels, more efficient gear, and the use of fish aggregating devices (FADs) offshore. Overall, reconstructed catches are 1.4 times the data reported by the FAO for the Comoros; the discrepancy is caused mainly by an increase in catches since 1995, which is not reflected in the FAO data. Figure 2B shows that although domestic catches dominate the total removals, foreign fishing by Spain, France, and Japan also occurs. Figure 2C indicates that the reconstructed catch consists primarily of skipjack tuna (*Katsuwonus pelamis*) and yellowfin (*Thunnus albacares*) offshore and sardinellas (*Sardinella* spp.) and anchovies (family Engraulidae) closer inshore.

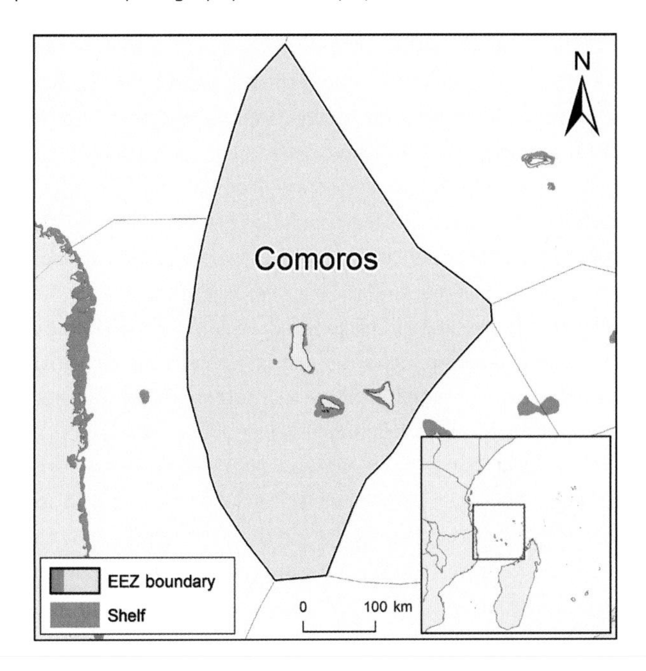

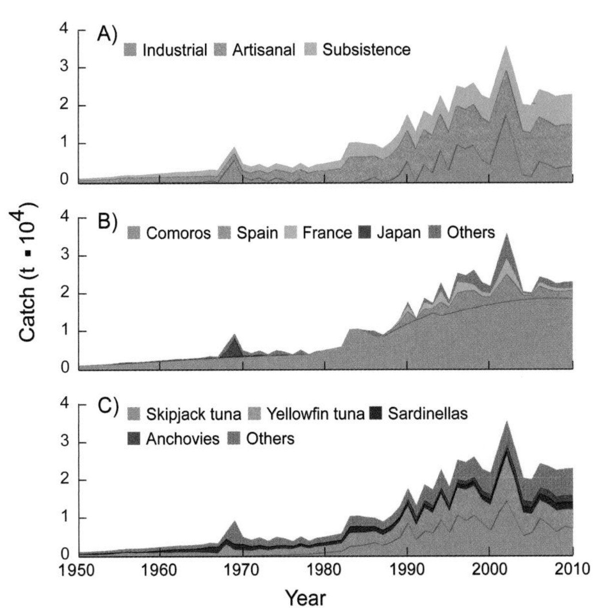

Figure 1. Comoros's three islands have a combined land area of 1,670 km² and an EEZ of 165,000 km². **Figure 2.** Domestic and foreign catches taken in the EEZ of Comoros: (A) by sector (domestic scores: Art. 1, 1, 1; Sub. 1, 1, 1); (B) by fishing country (foreign catches are very uncertain); (C) by taxon.

REFERENCES

Doherty B, Hauzer M and Le Manach F (2015) Reconstructing catches for the Union of the Comoros: uniting historical sources of catch data for Ngazidja, Ndzuwani and Mwali from 1950–2010. pp. 1–11 In Le Manach F and Pauly D (eds.), *Fisheries catch reconstructions in the Western Indian Ocean, 1950–2010*. Fisheries Centre Research Reports 23(2).

Hauzer M, Dearden P and Murray G (2013) The fisherwomen of Ngazidja island, Comoros: fisheries livelihoods, impacts, and implications for management. *Fisheries Research* 140: 28–35.

1. Cite as Doherty, B., M. Hauzer, and F. Le Manach. 2016. Comoros. P. 225 in D. Pauly and D. Zeller (eds.), *Global Atlas of Marine Fisheries: A Critical Appraisal of Catches and Ecosystem Impacts*. Island Press, Washington, DC.

CONGO (BRAZZAVILLE)[1]

Dyhia Belhabib and Daniel Pauly

Sea Around Us, University of British Columbia, Vancouver, Canada

The Republic of the Congo is located in the south of the Gulf of Guinea at the edge of the Benguela Current Large Marine Ecosystem in central West Africa, and its capital is Brazzaville (figure 1); it is here called Congo (Brazzaville) to distinguish it from the Democratic Republic of the Congo, formerly known as Zaïre. The population of the Congo depends greatly on fisheries as a source of income and food because of a high unemployment rate (Béné 2008). The total catch, as allocated to the EEZ of Congo (Brazzaville), is dominated by the rapidly growing industrial sector, whereas the artisanal sector is largely stagnant (figure 2A). Total domestic catches from within the EEZ, as estimated by Belhabib and Pauly (2015), were 7,000 t in 1950, increased to a peak of 99,000 t in 1977, declined in the 1980s, remained relatively constant at 30,000 t on average during the 1990s, and then increased slowly to 45,000 t in 2010. Overall, domestic catches were 2.8 times the data reported to the FAO. However, underreporting was strongest in the 1970s and 1980s, which significantly masked a tremendous decline in fisheries catches. Total removals were mainly domestic until 1999, as foreign fishing by China became prominent in the 2000s (figure 2B). Figure 2C shows that catches were dominated by Madeiran sardinella (*Sardinella maderensis*), shrimps and prawns, and sardinella (*Sardinella* spp., family Clupeidae).

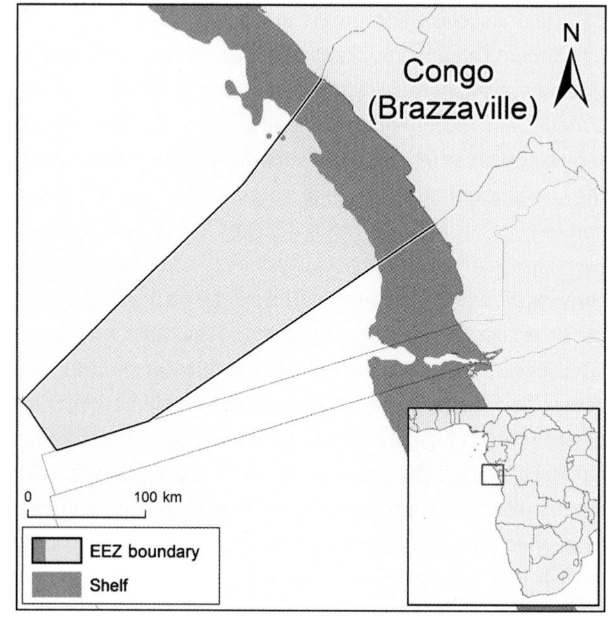

Figure 1. Congo (Brazzaville) has an EEZ of 39,600 km² and a continental shelf of 7,770 km².

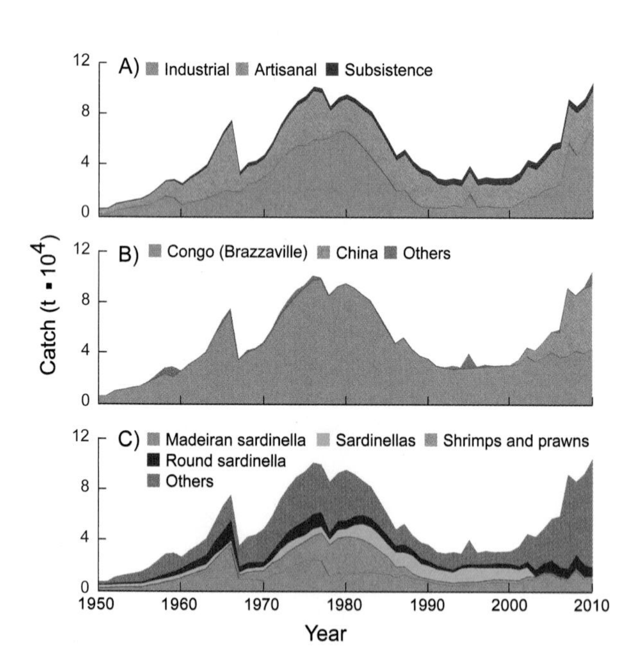

Figure 2. Domestic and foreign catches taken in the EEZ of Congo (Brazzaville): (A) by sector (domestic scores: Ind. 2, 4, 3; Art. 2, 4, 4; Sub. 2, 4, 4; Dis. 2, 2, 2); (B) by fishing country (foreign catches are very uncertain); (C) by taxon.

REFERENCES

Belhabib D and Pauly D (2015) The implications of misreporting on catch trends: a catch reconstruction for the People's Republic of the Congo, 1950–2010. pp. 95–106 In Belhabib D and Pauly D (eds.), *Fisheries catch reconstructions: West Africa, Part II*. Fisheries Centre Research Reports 23(3).

Béné C (2008) Contribution of Fishing to Households' Economy—Evidence From Fisher-Farmer Communities in Congo. WorldFish Center, Regional Offices for Africa, Penang, 11 p.

1. Cite as Belhabib, D., and D. Pauly. 2016. Congo (Brazzaville). P. 226 in D. Pauly and D. Zeller (eds.), *Global Atlas of Marine Fisheries: A Critical Appraisal of Catches and Ecosystem Impacts*. Island Press, Washington, DC.

CONGO (EX-ZAÏRE)[1]

Dyhia Belhabib, Sulan Ramdeen, Daniel Pauly

Sea Around Us, University of British Columbia, Vancouver, Canada

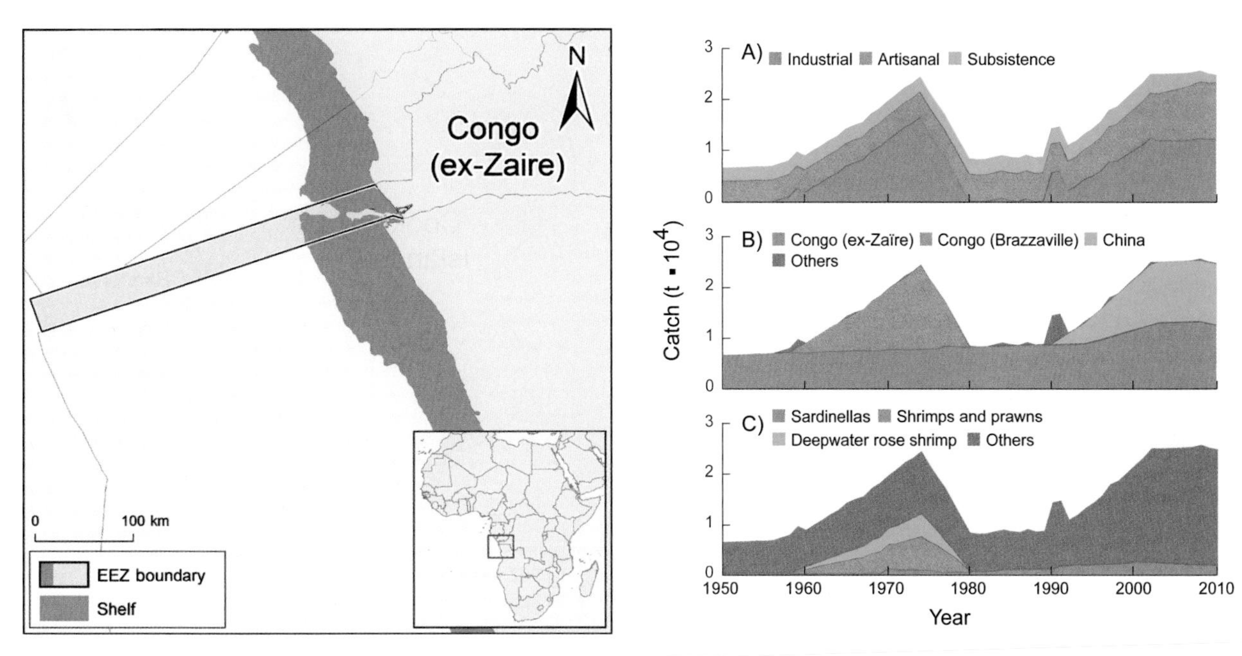

Figure 1. The EEZ of the Democratic Republic of the Congo covers 13,100 km², of which 2,690 km² is shelf. **Figure 2.** Domestic and foreign catches taken in the EEZ of the Democratic Republic of the Congo: (A) by sector (domestic scores: Ind. 2, 3, –; Art. 2, 3, 1; Sub. 1, 2, 2; Dis. 1, 1, –); (B) by fishing country (foreign catch is provided with higher uncertainty); (C) by taxon.

The Democratic Republic of Congo (formerly known as Zaïre) has an extremely short coastline between the coast of the Angolan exclave of Cabinda in the north and Angola proper in the south (figure 1). A succession of political disasters, notably the "Zaïrianization" of the country's economy and one of the deadliest wars since World War II, left the Congo's resources unmonitored and unmanaged and led to a tremendous growth of the informal economy, including fisheries (de Merode et al 2004). Catches, predominantly from the industrial sector, fluctuated throughout the time period, declining drastically in the 1980s, and the artisanal and subsistence sectors contributed stable amounts through the time period (figure 2A). Domestic catches, as reconstructed by Belhabib et al. (2015), were 6,600 t for 1950 and increased to about 9,000 t by the mid-1990s, then more rapidly to 12,000–13,000 t/year in the 2000s. Domestic catches were 4 times the data supplied to the FAO and generated consistent removals throughout the time period. Foreign fishing by Congo (Brazzaville) dominated in the 1960s and 1970s; however, foreign fishing by China had noticeable removals in the 1990s and 2000s (figure 2B). The taxonomic breakdown of catches shows that sardinellas (*Sardinella* spp.), shrimps and prawns, and deepwater rose shrimp (*Parapenaeus longirostris*) were the targeted species (figure 2C). The Congo (ex-Zaïre) provides an interesting example of the resilience of small-scale fisheries.

REFERENCES

Belhabib D, Ramdeen S and Pauly D (2015) An attempt at reconstructing the marine fisheries catches in the Congo (ex-Zaïre), 1950 to 2010. pp. 107–114 In Belhabib D and Pauly D (eds.), *Fisheries catch reconstructions: West Africa, Part II*. Fisheries Centre Research Reports 23(3).

de Merode E, Homewood K and Cowlishaw G (2004) The value of bushmeat and other wild foods to rural households living in extreme poverty in Democratic Republic of Congo. *Biological Conservation* 118(5): 573–581.

1. Cite as Belhabib, D., S. Ramdeen, and D. Pauly. 2016. Congo (ex-Zaïre). P. 227 in D. Pauly and D. Zeller (eds.), *Global Atlas of Marine Fisheries: A Critical Appraisal of Catches and Ecosystem Impacts*. Island Press, Washington, DC.

COOK ISLANDS[1]

Andrea Haas,[a] Teina Rongo,[b] Nicole Hefferman,[a] Sarah Harper,[a] and Dirk Zeller[a]

[a]*Sea Around Us*, University of British Columbia, Vancouver, Canada [b]Avarua, Rarotonga, Cook Islands

The Cook Islands, named after Captain Cook, who visited during his second (1773) and third (1777) voyages; (Kippis 1820), comprise 15 islands in the eastern central Pacific (figure 1). The catches, as reconstructed by Haas et al. (2012), include the large-scale sector, which is aimed at export, the reef-based subsistence sector, and the small-scale artisanal fishery (Hoffmann 2002), and are 2.4 times the data reported by the FAO on behalf of the Cook Islands (figure 2A). Overall, total domestic catches allocated to Cook Islands' EEZ increased from about 2,000 t in 1950 to a peak of 2,700 t in 1964 before declining to 1,200 t by 2010. Domestic catches, composing a small proportion of allocated catch, remained fairly constant through the time period, and foreign fishing was dominated by Japan in the 1960s and 1970s and by China in the 2000s (figure 2B). Albacore tuna (*Thunnus alalunga*), mackerels, tunas, bonitos (family Scombridae), and yellowfin tuna (*Thunnus albacares*) accounted for the overwhelming bulk of the total reconstructed catches (figure 2C). Illegal fishing is known to occur in the Cook Islands EEZ; however, patrolling and enforcing this immense expanse of ocean is a major challenge.

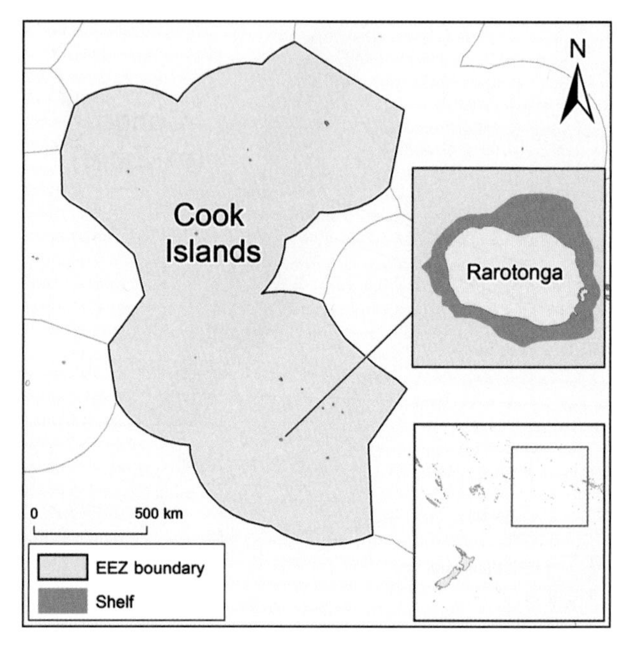

Figure 1. The Cook Islands have a land area of 261 km² and an EEZ of 1.96 million km². Most of the population (of about 20,000) lives on Rarotonga.

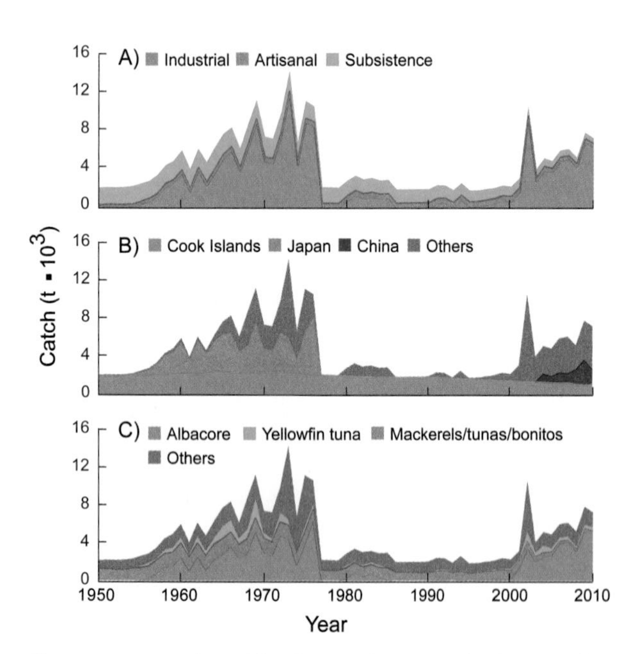

Figure 2. Domestic and foreign catches taken in the EEZ of Cook Islands: (A) by sector (domestic scores: Ind. –, –, 3; Art. 1, 2, 2; Sub. 1, 2, 2); (B) by fishing country (foreign catches are very uncertain); (C) by taxon.

REFERENCES

Haas A, Rongo T, Hefferman N, Harper S and Zeller D (2012) Reconstruction of the Cook Islands fisheries catches: 1950–2010. pp. 15–24 In Harper S, Zylich K, Boonzaier L, Le Manach F, Pauly D and Zeller D (eds.), *Fisheries catch reconstructions: Islands, Part III*. Fisheries Centre Research Reports 20(5).

Hoffmann TC (2002) The reimplementation of the Ra'ui: Coral reef management in Rarotonga, Cook Islands. *Coastal Management* 30: 401–418.

Kippis A (1820) A narrative of the voyages round the world performed by Captain James Cook with an account of his life in the intervening period. Chiswick Vol. I & II. C. and C. Whittingham, London.

1. Cite as Haas, A., T. Rongo, N. Hefferman, S. Harper, and D. Zeller. 2016. Cook Islands. P. 228 in D. Pauly and D. Zeller (eds.), *Global Atlas of Marine Fisheries: A Critical Appraisal of Catches and Ecosystem Impacts*. Island Press, Washington, DC.

COSTA RICA (CARIBBEAN)[1]

Pablo Trujillo,[a] Andrés M. Cisneros-Montemayor,[b] Sarah Harper,[a] Kyrstn Zylich,[a] and Dirk Zeller[a]

[a]*Sea Around Us*, University of British Columbia, Vancouver, Canada
[b]Fisheries Economics Research Unit, University of British Columbia, Vancouver, Canada

This account deals only with Costa Rica's comparatively featureless Caribbean coast (figure 1). Although total catch allocated to the Caribbean EEZ of Costa Rica fluctuates over time, it is dominated by the artisanal sector (figure 2A). The catch reconstruction of Trujillo et al. (2015) for this less productive coast led to allocated domestic catch estimates of 100 t/year for the 1950s, a peak of 2,600 t in 2000, and catches of about 1,400 t/year in the late 2000s, which corresponded, from 1950–2010, to 1.2 times the data reported by the FAO for Costa Rica in the Caribbean. The difference resulted largely from subsistence catches, neglected in official statistics. Fishing activities in the EEZ are mostly domestic, with some foreign fishing by Venezuela and Cuba, among other countries, in the earlier time period (figure 2B). The major fishery resource on Costa Rica's Caribbean coast is spiny lobster (*Panulirus argus*), targeted by artisanal fishers and sold to hotels or exporters. From the 1950s until 1998, lobster accounted for the majority of reported landings, followed by sea turtles (not included here) and fish, respectively. Since the onset of turtle protection, reported landings are dominated by lobster, followed by shrimps and fish. Figure 2C also indicates that sharks, ray, skates, and chimeras as well as penaeid shrimp (*Penaeus* spp.) are important components of the local catches, as in the West Indian topshell (Schmidt et al. 2002).

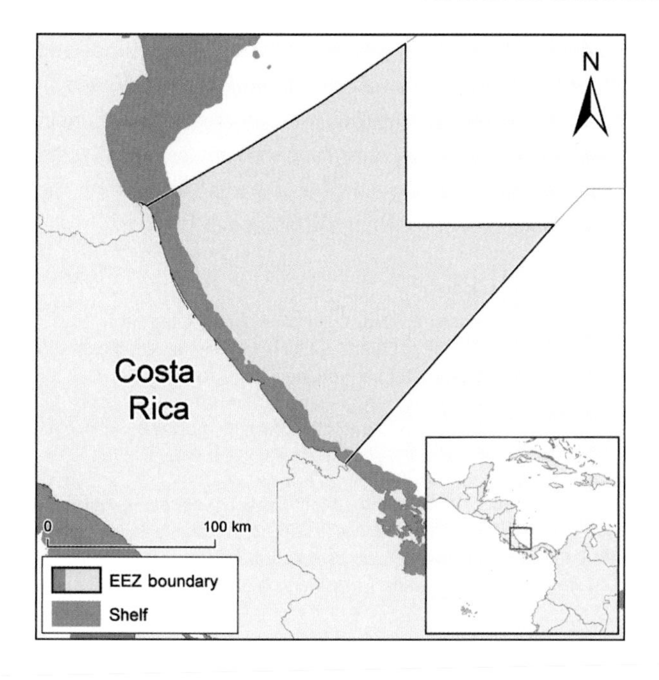

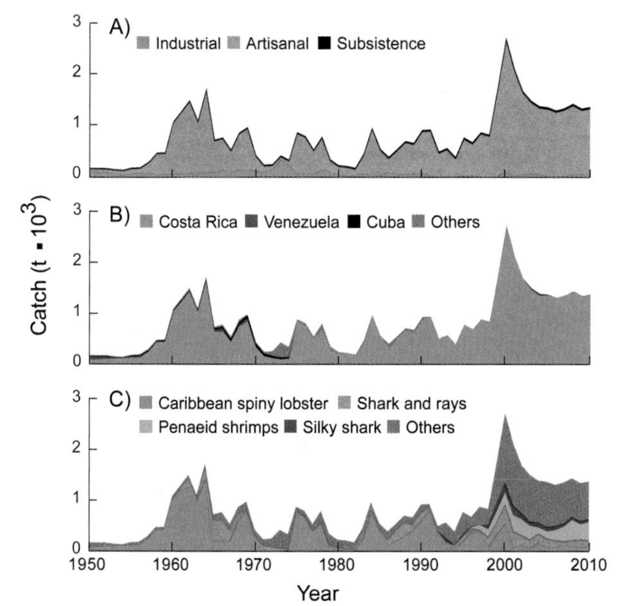

Figure 1. Costa Rica has, in the Caribbean, a shelf of 2,210 km² and an EEZ of 27,000 km². **Figure 2.** Domestic and foreign catches taken in the Caribbean EEZ of Costa Rica: (A) by sector (domestic scores: Art. 2, 3, 4; Sub. 2, 3, 3; Dis. 2, 3, 4); (B) by fishing country (foreign catches are very uncertain); (C) by taxon.

REFERENCES

Schmidt S, Wolff M and Vargas JA (2002) Population ecology and fishery of *Cittarium pica* (Gastropoda: Trochidae) on the Caribbean coast of Costa Rica. *Revista de Biología Tropical* 50: 3–4.

Trujillo P, Cisneros-Montemayor A, Harper S, Zylich K and Zeller D (2015) Reconstruction of Costa Rica's marine fisheries catches (1950–2010). Fisheries Centre Working Paper #2015–31, 16 p.

1. Cite as Trujillo, P., A. Cisneros-Montemayor, S. Harper, K. Zylich, and D. Zeller. 2016. Costa Rica (Caribbean). P. 229 in D. Pauly and D. Zeller (eds.), *Global Atlas of Marine Fisheries: A Critical Appraisal of Catches and Ecosystem Impacts*. Island Press, Washington, DC.

COSTA RICA (PACIFIC)[1]

Pablo Trujillo,[a] Andrés M. Cisneros-Montemayor,[b] Sarah Harper,[a] Kyrstn Zylich,[a] and Dirk Zeller[a]

[a]*Sea Around Us*, University of British Columbia, Vancouver, Canada
[b]Fisheries Economics Research Unit, University of British Columbia, Vancouver, Canada

The Republic of Costa Rica has coastlines on both the Atlantic and Pacific Oceans, but this account covers the Pacific coast and its major features: the gulfs of Papagayo, Nicoya, and Dulce and Coronado Bay (figure 1). The Gulf of Nicoya saw a fishery develop in the 1920s and is still the source of most of the country's catch. Another important feature of Costa Rica's waters (though more offshore) is the "dome," a permanent shallowing of the thermocline leading to a high oceanic productivity, including of tuna, the reason for Costa Rica's early membership in the Inter-American Tropical Tuna Commission. The industrial, domestic sector composed the bulk of the allocated total catch to the Pacific EEZ of Costa Rica (figure 2A; figure 2B). The reconstruction of Trujillo et al. (2015) yielded catches of 3,100 t in 1950, a peak of 174,000 t in 1986, and catches of 75,000 t/year in the late 2000s. Overall, reconstructed catches were 2.9 times the Pacific landings data supplied to the FAO by Costa Rica. Yellowfin tuna (*Thunnus albacares*), sharks, rays, skates, chimeras, and shrimps and prawns compose the bulk of catches within the EEZ (figure 2C). The marine realm has high touristic value (see contributions in Wehrtmann and Cortés 2009), and the composition of Costa Rica's fisheries catches reflects, if only to a small extent, the huge biodiversity on which the fisheries depend.

REFERENCES

Trujillo P, Cisneros-Montemayor A, Harper S, Zylich K and Zeller D (2015) Reconstruction of Costa Rica's marine fisheries catches (1950–2010). Fisheries Centre Working Paper #2015–31, 16 p.

Wehrtmann IS and Cortés J, editors (2009) Marine Biodiversity of Costa Rica, Central America Monographiae Biologicae 86. Springer and Business Media B.V., Berlin, Germany. 538 p + CD.

1. Cite as Trujillo, P., A. Cisneros-Montemayor, S. Harper, K. Zylich, and D. Zeller. 2016. Costa Rica (Pacific). P. 230 in D. Pauly and D. Zeller (eds.), *Global Atlas of Marine Fisheries: A Critical Appraisal of Catches and Ecosystem Impacts*. Island Press, Washington, DC.

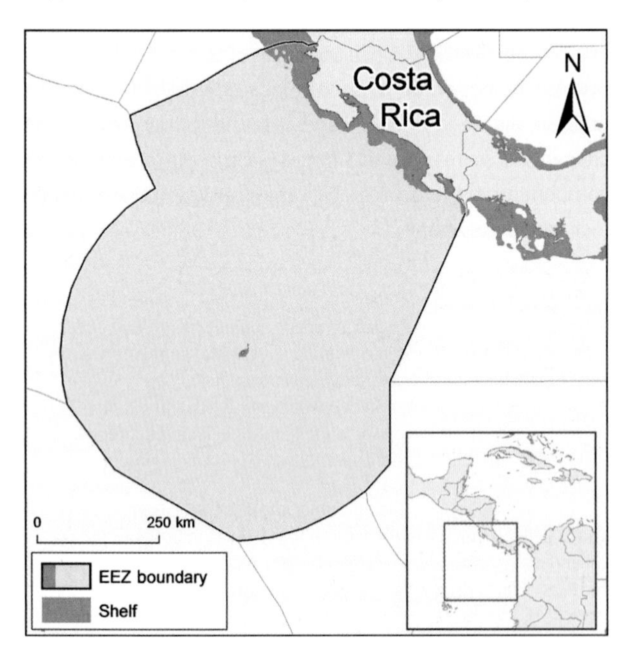

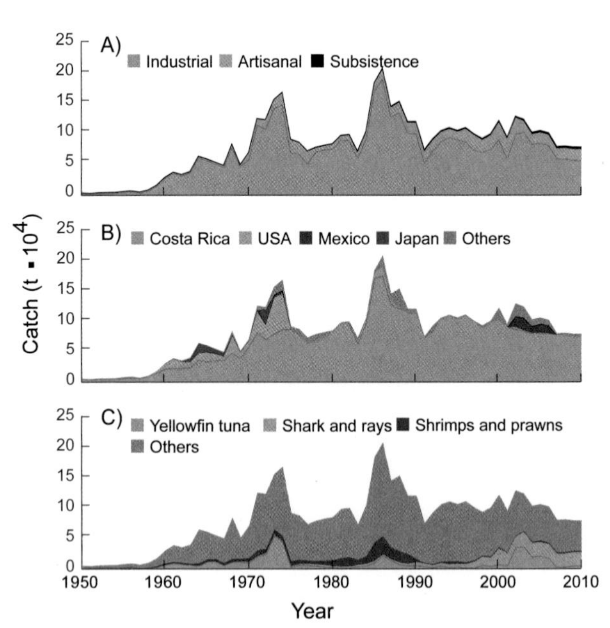

Figure 1. Costa Rica has, in the Pacific, a shelf of 17,100 km² and an EEZ of 545,000 km².

Figure 2. Domestic and foreign catches taken in the EEZ of Costa Rica (Pacific): (A) by sector (domestic scores: Ind. 2, 3, 4; Art. 2, 3, 4; Sub. 2, 3, 3; Rec. –, 1, 1; Dis. 2, 3, 4); (B) by fishing country (foreign catches are very uncertain); (C) by taxon.

CÔTE D'IVOIRE[1]

Dyhia Belhabib and Daniel Pauly

Sea Around Us, University of British Columbia, Vancouver, Canada

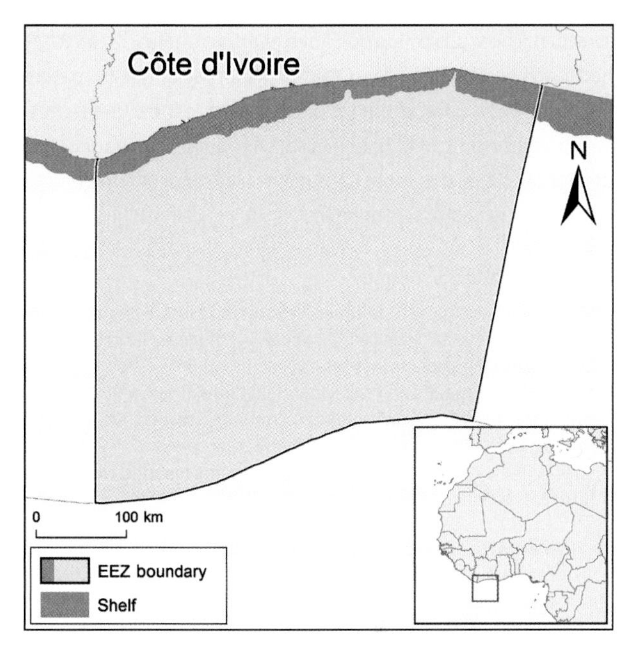

Figure 1. Côte d'Ivoire has a continental shelf of 12,300 km² and an EEZ of 174,000 km².

Côte d'Ivoire, located in Sub-Saharan West Africa, is one of the largest coastal countries of the Gulf of Guinea (figure 1). The economic prosperity inherited at independence in 1960 did not last; the cocoa-based economy collapsed, leading to an increase of the informal economy (including fisheries), notably for sardinella (Diaby 1996). The majority of the catch within the EEZ of Côte d'Ivoire is derived from the industrial sector, and catches were steadily increasing through the time period (figure 2A). Total domestic catches taken in the EEZ were estimated by Belhabib and Pauly (2015) at nearly 53,000 t in 1950, peaked at 118,000 t in 1966, and then gradually declined to 106,000 t/year by the late 1990s, before declining more strongly to about 60,000 t/year by the late 2000s. Reconstructed catches were about twice the data reported to the FAO after *faux poisson* (i.e., fish landed by foreign fleets in Abidjan that was caught outside the Ivoirian EEZ) was removed. Although catches were largely domestic before 1990, foreign catches, mainly by China, were documented in substantial amounts after 1990 (figure 2B). Catches consisted mostly of bonga shad (*Ethmalosa fimbriata*) and round sardinellas (*Sardinella aurita*; figure 2C). The coastal marine ecosystem is well documented (see contributions in Leloeuf et al. 1993), and therefore ecosystem-based management of the fisheries could be attempted to halt their decline; however, this would require that their problems be acknowledged.

REFERENCES

Belhabib D and Pauly D (2015) Côte d'Ivoire: fisheries catch reconstruction, 1950–2010. pp. 17–36 In Belhabib D and Pauly D (eds.), *Fisheries catch reconstructions: West Africa, Part II*. Fisheries Centre Research Reports 23(3).

Diaby S (1996) Economic impact analysis of the Ivoirian sardinella fishery. *Marine Resource Economics* 11(1996): 31–42.

Leloeuf P, Marchal E and Amon Kothia J (1993) Environnement et ressources aquatiques de Côte d'Ivoire. Tome 1-le Milieu Marin. ORSTOM, Paris. 589 p.

1. Cite as Belhabib, D., and D. Pauly. 2016 Côte d'Ivoire. P. 231 in D. Pauly and D. Zeller (eds.), *Global Atlas of Marine Fisheries: A Critical Appraisal of Catches and Ecosystem Impacts*. Island Press, Washington, DC.

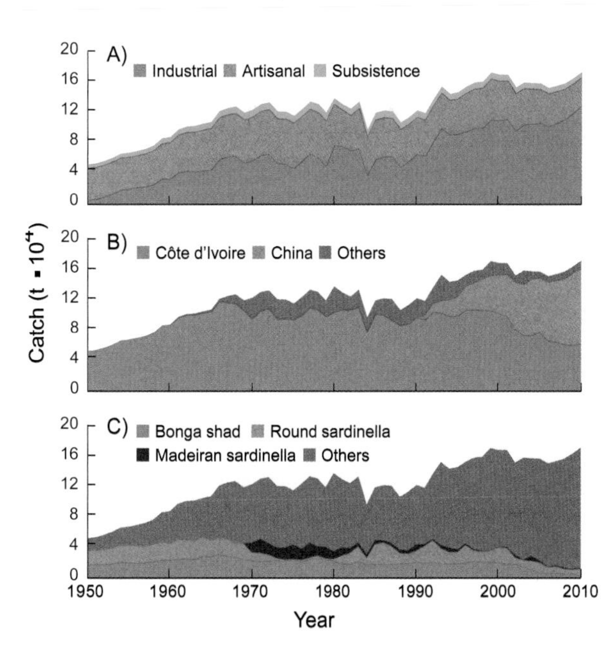

Figure 2. Domestic and foreign catches taken in the EEZ of Côte d'Ivoire: (A) by sector (domestic scores: Ind. 4, 4, 4; Art. 2, 3, 3; Sub. 2, 3, 2; Dis. 3, 3, 2); (B) by fishing country (foreign catches are very uncertain); (C) by taxon.

CROATIA[1]

Sanja Matić-Skoko,[a] Nika Stagličić,[a] Danijela Blažević,[a] Jasna Šiljić,[a] and Davis Iritani[b]

[a]Institute of Oceanography and Fisheries, Split, Croatia [b]*Sea Around Us*, University of British Columbia, Vancouver, Canada

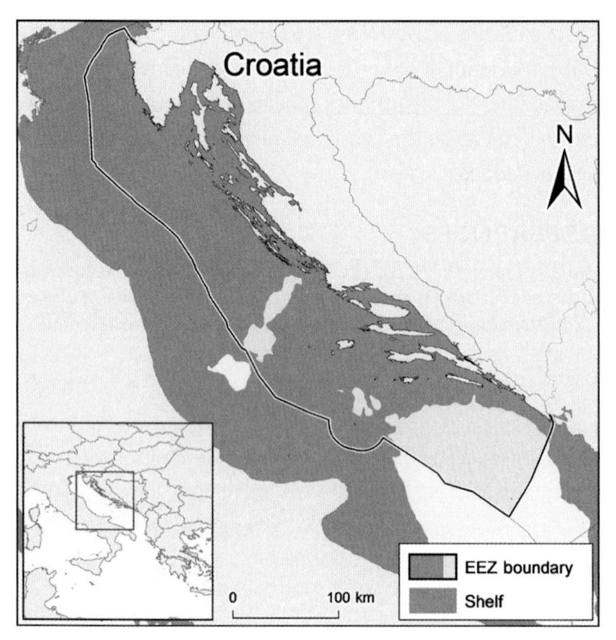

Figure 1. Croatia has a shelf of 44,800 km² and an EEZ of 56,000 km².

Croatia's coastline stretches along most of the eastern Adriatic Sea (figure 1), and its fisheries since ancient times have shaped both its people and the fisheries of this part of the Mediterranean (Matić-Skoko et al. 2011). As the 1990 breakup of the former Yugoslavia introduced a sharp discontinuity in the FAO fisheries database, Matić-Skoko et al. (2014) undertook a reconstruction of the Croatian domestic marine catch, which resulted in an estimate of about 25,000 t/year in the early 1950s, growing to a peak of 94,000 t in 1987, a decline to 35,000 t/year in the early 2000s, and an increase to 74,000 t/year in the late 2000s. The total catches allocated to Croatia's EEZ are primarily industrial, with a small artisanal component (figure 2A). Figure 2B shows that the bulk of catches are domestic; however, foreign fishing by Slovenia and Italy, among others, does seem to occur. Figure 2C presents the catch composition of the Croatian marine catch, which was always dominated by small pelagics: European pilchards (*Sardina pilchardus*), European anchovy (*Engraulis encrasicolus*), herrings, shads, sardines, menhadens (family Clupeidae), and European sprat (*Sprattus sprattus*). The study by Matić-Skoko et al. (2014) illustrated what had long been suspected by Croatian fishery scientists: Croatia's officially reported landings far underestimate the true catches. The current method of fisheries catch reporting is inadequate and incomplete, and more comprehensive reporting, including for all fishing sectors, is needed.

REFERENCES

Matić-Skoko S, Stagličić N, Blažević D, Šiljić J, and Iritani D (2014) Croatian marine fisheries (Adriatic Sea): 1950–2010. Fisheries Centre Working Paper #2014-26, 16 p.

Matić-Skoko S, Stagličić N, Kraljević M, Pallaoro A, Tutman P, Dragičević B, R. G and Dulčić J (2011) Croatian artisanal fisheries and the state of its littoral resources on the doorstep of entering the EU: effectiveness of conventional management and perspective for the future. *Acta Adriatica* 52(1): 87–100.

1. Cite as Matić-Skoko, S., N. Stagličić, D. Blažević, and D. Iritani. 2016. Croatia. P. 232 in D. Pauly and D. Zeller (eds.), *Global Atlas of Marine Fisheries: A Critical Appraisal of Catches and Ecosystem Impacts*. Island Press, Washington, DC.

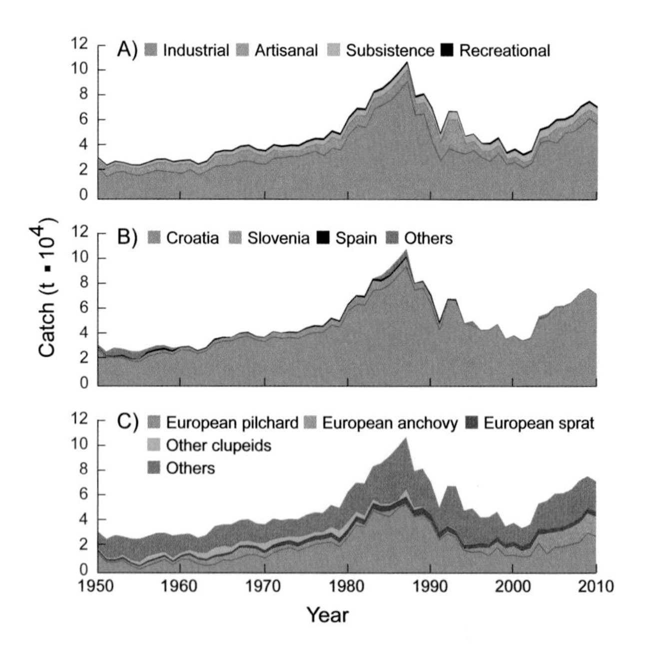

Figure 2. Domestic and foreign catches taken in the EEZ of Croatia: (A) by sector (domestic scores: Ind. 2, 3, 3; Art. 2, 3, 3; Sub. 1, 2, 2; Rec. 1, 1, 1; Dis. 2, 2, 2); (B) by fishing country (foreign catches are very uncertain); (C) by taxon.

CUBA[1]

Andrea Au, Kyrstn Zylich, and Dirk Zeller

Sea Around Us, University of British Columbia, Vancouver, Canada

The Cuban Archipelago is located in the northern Caribbean Sea, with the Gulf of Mexico to the west and the Atlantic Ocean to the east (figure 1). Although total removals are mostly artisanal, the industrial sector contributed a noticeable amount of the catch, notably between 1970 and 1990 (figure 2A). Au et al. (2014) updated and extended the preliminary reconstruction of Baisre et al. (2003), determining that the domestic catch from the Cuban EEZ increased from about 10,000 t in 1950 to a peak of just under 77,000 t in 1985, and then declined to approximately 28,500 t in 2010. Overall, this is 1.2 times the catch reported to and by the FAO. Catches are primarily domestic, and though not visible on the graph, small removals by foreign vessels did occur in the earlier time period (figure 2B). Caribbean spiny lobster (*Panulirus argus*) was the largest contributor, followed by lane snapper (*Lutjanus synagris*), grunts (family Haemulidae), sharks and rays, and northern pink shrimp (*Penaeus duorarum*) (figure 2C). The coastal ecosystems around Cuba were among the first, thanks to Baisre (2000, 2004) to have been diagnosed as unsustainably exploited according to indicators such as the mean trophic level and the mean maximum length of the fish in the catch, both of which strongly declined over his study period (1935–1995).

REFERENCES

Au A, Zylich K and Zeller D (2014) Reconstruction of total marine fisheries catches for Cuba (1950–2009). pp. 25–32 In Zylich K, Zeller D, Ang M and Pauly D (eds.), *Fisheries catch reconstructions: Islands, Part IV*. Fisheries Centre Research Reports 22(2).

Baisre J (2000) Chronicle of Cuban marine fisheries (1935–1995): trend analysis and fisheries potential. FAO Fisheries Technical Paper No. 394. 26 p.

Baisre J (2004) La pesca marítima en Cuba. Editorial Científico-Técnica, La Habana. 372 p.

Baisre JA, Booth S and Zeller D (2003) Cuban fisheries catches within FAO area 31 (Western Central Atlantic): 1950–1999. pp. 133–139 In Zeller D, Booth S, Mohammed E and Pauly D (eds.), *From Mexico to Brazil: Central Atlantic fisheries catch trends and ecosystem models*. Fisheries Centre Research Reports 11(6).

1. Cite as Au, A., K. Zylich, and D. Zeller. 2016. Cuba. P. 233 in D. Pauly and D. Zeller (eds.), *Global Atlas of Marine Fisheries: A Critical Appraisal of Catches and Ecosystem Impacts*. Island Press, Washington, DC.

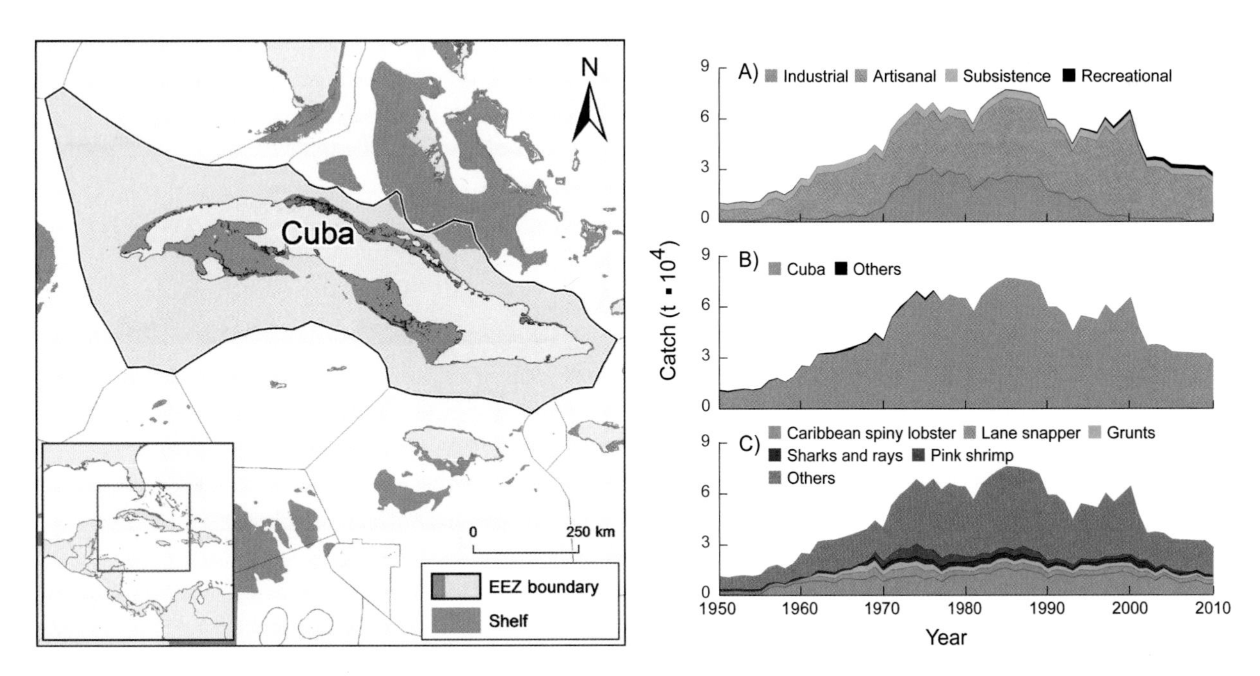

Figure 1. Cuba and its surrounding islands have a land area of 110,000 km² and an EEZ of 365,000 km². **Figure 2.** Domestic and foreign catches taken in the EEZ of Cuba: (A) by sector (domestic scores: Ind. 2, 3, 3; Art. 2, 3, 3; Sub. 1, 2, 2; Rec. 1, 1, 1; Dis. 2, 2, 2); (B) by fishing country (foreign catches are very uncertain); (C) by taxon.

CYPRUS (NORTH)[1]

Aylin Ulman,[a] Burak Ali Çiçek,[b] and Ilkay Salihoglu[c]

[a]*Sea Around Us*, University of British Columbia, Vancouver, Canada [b]Eastern Mediterranean University, Famagusta, TRNC [c]Near East University, Nicosia, TRNC

Cyprus is located in the eastern Mediterranean, and since 1974, a "Green Line" has divided the island into a Turkish north and a Greek south (figure 1). Catch reconstructions for both parts of Cyprus were performed by Ulman et al. (2013, 2015). The reconstruction for Cyprus (North) showed that from 1950 to 1973, domestic catches increased from 400 t/year to 1,300 t/year. Subsequently, catches dropped to 440 t in 1974, then increased, settling about 900 t/year in the late 2000s, after a brief period (1993–1997) during which five bottom trawlers generated catches of about 8,000 t/year. Also, a purse seiner operated in 2002; since then, industrial fishing has ceased. Total removals from within the EEZ of Cyprus (North) are mostly industrial for the time period, especially between 1992 and 1995, when the large-scale sector overwhelmed the small-scale sectors (figure 2A). Thus, for 1950 to 2010, the reconstructed catches were more than 10 times the amount reported by the FAO deemed to pertain to the northern part of Cyprus. Total allocated catches are primarily domestic; however, small removals by Spain have been documented in Cyprus (North)'s EEZ from 1950 to 2003 (figure 2B). Figure 2C shows that the fish caught have recently changed from groups typical of the eastern Mediterranean, picarel (*Spicara smaris*), red mullet (*Mullidus barbatus barbatus*), and common pandora (*Pagellus erythrinus*), to Lessepsian migrants.

REFERENCES

Ulman A, Çiçek B, Salihoglu I, Petrou A, Patsalidou M, Pauly D and Zeller D (2013) The reconstruction and unification of Cyprus' marine fisheries catch data, 1950–2010. Fisheries Centre Working Paper #2013–09, 72 p.

Ulman A, Çiçek B, Salihoglu I, Petrou A, Patsalidou M, Pauly D and Zeller D (2015) Unifying the catch data of a divided island: Cyprus' marine fisheries catches, 1950–2010. *Environment, Development and Sustainability* 17(4): 801–821.

1. Cite as Ulman, A., B. Çiçek, and I. Salihoglu. 2016. Cyprus (North). P. 234 in D. Pauly and D. Zeller (eds.), *Global Atlas of Marine Fisheries: A Critical Appraisal of Catches and Ecosystem Impacts*. Island Press, Washington, DC.

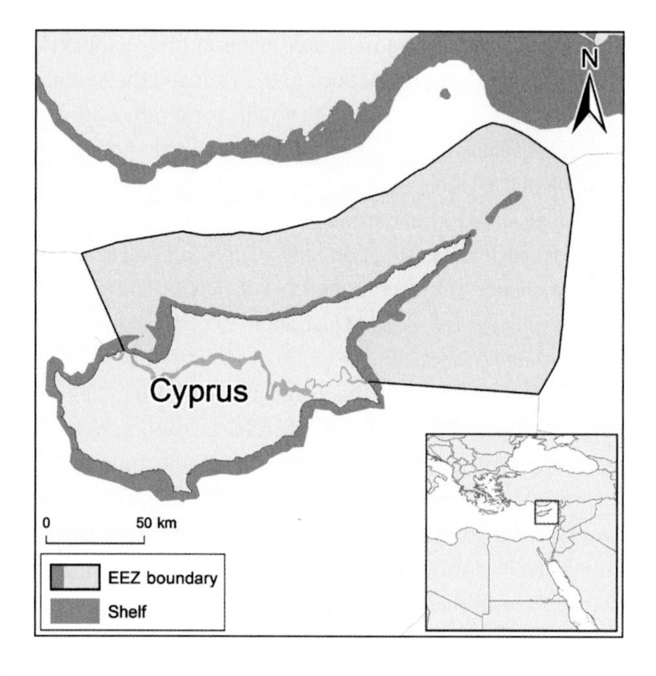

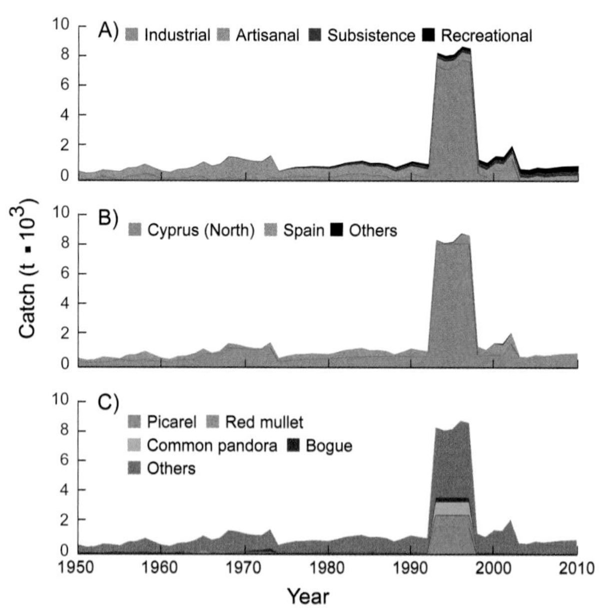

Figure 1. The island of Cyprus, showing the "Green Line" dividing north and south. The north's land area is 3,400 km²; the corresponding EEZ-equivalent area (17,700 km²) was drawn based on a nonbinding interpretation of UNCLOS principles. **Figure 2.** Domestic and foreign catches taken in the EEZ of Cyprus (North): (A) by sector (domestic scores: Ind. 1, –, 3; Art. 2, 2, 4; Sub. 1, 1, 1; Rec. 2, 1, 3; Dis. 2, 2, 1); (B) by fishing country (foreign catches are very uncertain); (C) by taxon.

CYPRUS (SOUTH)[1]

Aylin Ulman,[a] Antonis Petrou,[b] and Maria Patsalidou[b]

[a]*Sea Around Us*, University of British Columbia, Vancouver, Canada
[b]AP Marine Environmental Consultancy Ltd., Nicosia, Cyprus

Cyprus is the third largest Mediterranean island (figure 1). Since 1974, the catch data submitted to the FAO by "Cyprus" have originated only from the south. Thus, to obtain a consistent time series, Ulman et al. (2013) removed 40% of the annual reported catches from 1950–1973 from the reported total and allocated them to the north and supplemented the series with local data. The total catch allocated to Cyprus (South)'s EEZ is dominated by the industrial sector, with a noticeable, secondary artisanal component (figure 2A). Ulman et al. (2013, 2015) reconstructed the domestic catch, illustrating that in the early 1950s, the south averaged 600 t/year in removals, increased steadily, and peaked at just under 5,000 t in 1986. Catches declined to 2,800 t/year in the late 2000s, not least because of a reduction in industrial fishing. Overall, the reconstructed catches were 2 times the data reported by the FAO on behalf of (the whole of) Cyprus. Figure 2B shows that catches were primarily domestic; however, foreign fishing contributed a fair amount before the EEZ declaration in 2003. Figure 2C presents the catch composition that has recently shifted from groups typical of the eastern Mediterranean, such as picarel (*Spicara* spp.), bogue (*Boops boops*), and red mullet (*Mullus barbatus barbatus*), to Lessepsian (i.e., Red Sea) immigrants, notably the puffer (family Tetradontidae), which is discarded (and thus masks the decline in landings).

REFERENCES

Ulman A, Çiçek B, Salihoglu I, Petrou A, Patsalidou M, Pauly D and Zeller D 2013. The reconstructed and unified marine fisheries catches of Cyprus, 1950–2010. Fisheries Centre Working Paper #2013-09, 69 p.

Ulman A, Çiçek B, Salihoglu I, Petrou A, Patsalidou M, Pauly D and Zeller D (2015) Unifying the catch data of a divided island: Cyprus' marine fisheries catches, 1950–2010. *Environment, Development and Sustainability* 17(4): 801–821.

1. Cite as Ulman, A., A. Petrou, and M. Patsalido. 2016. Cyprus (South). P. 235 in D. Pauly and D. Zeller (eds.), *Global Atlas of Marine Fisheries: A Critical Appraisal of Catches and Ecosystem Impacts*. Island Press, Washington, DC.

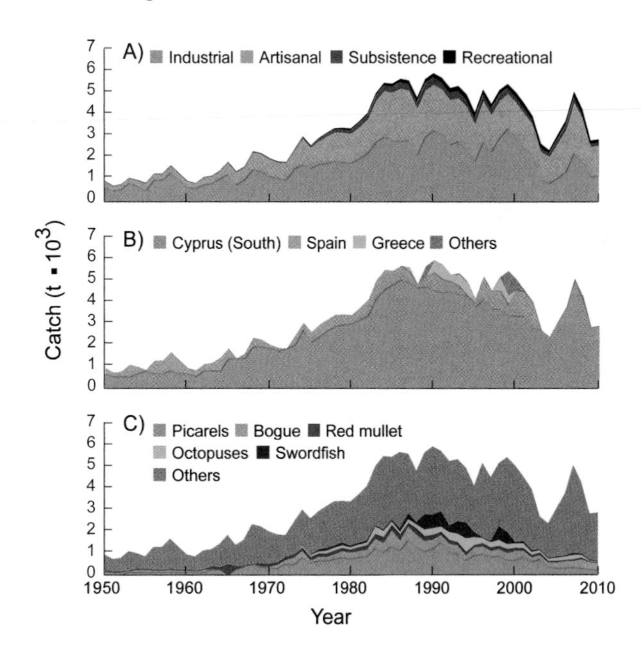

Figure 1. The island of Cyprus, showing the "Green Line" dividing north and south. The south's land area is 5,800 km², and the corresponding EEZ-equivalent area (80,400 km²) was drawn based on a nonbinding interpretation of UNCLOS principles. **Figure 2.** Domestic and foreign catches taken in the EEZ of Cyprus (South): (A) by sector (domestic scores: Ind. 1.5, 1.5, 2; Art. 2, 2, 2; Sub. 1, 1, 1; Rec. 2, 1, 2; Dis. 2, 2, 1); (B) by fishing country (foreign catches are very uncertain); (C) by taxon.

DENMARK (BALTIC SEA)[1]

Sarah Bale, Peter Rossing, Shawn Booth, and Dirk Zeller

Sea Around Us, University of British Columbia, Vancouver, Canada

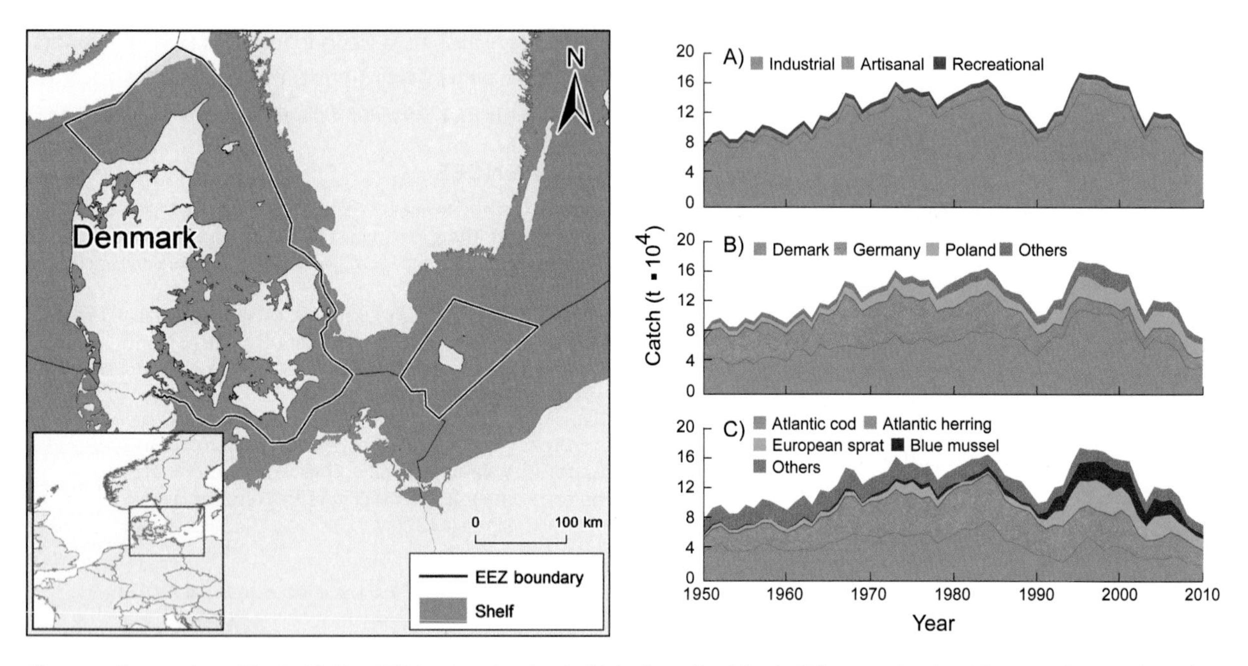

Figure 1. Denmark and its Baltic Sea EEZ (52,800 km²), of which virtually all is shelf (52,000 km²). **Figure 2.** Domestic and foreign catches taken in the Baltic EEZ of Denmark: (A) by sector (domestic scores: Ind. 3, 3, 3; Art. 2, 2, 2; Sub. 1, –, –; Rec. 2, 2, 2; Dis. 2, 2, 2); (B) by fishing country (foreign catches are very uncertain); (C) by taxon.

Denmark is located on the boundary of the Baltic and North Seas (figure 1); Jutland, the country's mainland peninsula, reaches northward to separate the North Sea and the Skagerrak from the Kattegat, which itself connects to the Baltic Sea through the Danish Sound and Belt. Although fisheries contribute only 0.5% of the GDP, they are integral to the livelihoods of many communities. Denmark is the EU's leading fishmeal-producing country, using both domestic catches and imported fish. The total catch, as allocated to its EEZ, is dominated by the fluctuations of the industrial sector, whereas artisanal and recreational catches remained fairly constant (figure 2A). Denmark's domestic fisheries catches, reconstructed by Bale et al. (2010), based mainly on data from the International Council for the Exploration of the Seas (ICES), were about 47,000 t in 1950, peaked at 110,000 t in 1995, then declined to about 35,000 t in the late 2000s. Overall, this was 1.5 times the data reported by ICES, with the discrepancy driven mainly by unreported industrial landings and discards (see also Zeller et al. 2011). Withdrawals are primarily domestic, but considerable catches are also taken by Germany and Poland (figure 2B). Figure 2C indicates that the major taxa caught by Denmark in the Baltic Sea are Atlantic cod (*Gadus morhua*), Atlantic herring (*Clupea harengus*), European sprat (*Sprattus sprattus*), and blue mussel (*Mytilus edulis*).

REFERENCES

Bale S, Rossing P, Booth S and Zeller D (2010) Denmark's marine fisheries catches in the Baltic Sea (1950–2007). pp. 39–62 In Rossing R, Booth S and Zeller D (eds.), *Total marine fisheries extractions by country in the Baltic Sea: 1950–present*. Fisheries Centre Research Reports 18(1). [updated to 2010]

Zeller D, Rossing P, Harper S, Persson L, Booth S and Pauly D (2011) The Baltic Sea: estimates of total fisheries removals 1950–2007. *Fisheries Research* 108: 356–363.

1. Cite as Bale, S., P. Rossing, S. Booth, and D. Zeller. 2016. Denmark (Baltic Sea). P. 236 in D. Pauly and D. Zeller (eds.), *Global Atlas of Marine Fisheries: A Critical Appraisal of Catches and Ecosystem Impacts*. Island Press, Washington, DC.

DENMARK (NORTH SEA)[1]

Darah Gibson,[a] Bernd Ueberschär,[b] Kyrstn Zylich,[a] and Dirk Zeller

[a]*Sea Around Us*, University of British Columbia, Vancouver, Canada
[b]GMA Association for Marine Aquaculture Ltd., Büsum, Germany

Denmark is a member of the European Union (EU) that borders the Baltic Sea, the Kattegat, Skagerrak, and North Sea (figure 1). Gibson et al. (2014) present a catch reconstruction for the fisheries in Denmark's North Sea EEZ using data from the International Council for the Exploration of the Seas (ICES) as a reporting baseline. This was improved upon using other data: ICES stock assessments, gray and peer-reviewed literature (e.g., Holm 2005), and local expert opinions. Catches were largely industrial, with a small and declining artisanal sector (figure 2A). The reconstructed domestic catch was 202,000 t in 1950, peaked at 1.53 million t in 1992, and declined to about 480,000 t/year in the late 2000s—overall 1.1 times the landings reported by ICES, driven mainly by unreported discards. Although it is mostly domestic in nature, slight catches by Germany and Sweden were also documented through the time period (figure 2B). Sand lances (family Ammodytidae) made up the largest catches from 1950–2010 and are the key species of the massive Danish industrial reduction fishery (i.e., for fishmeal). Herring (*Clupea harengus*) was important in the earlier years, and European sprat (*Sprattus sprattus*) and Atlantic cod (*Gadus morhua*) accounted for approximately similar amounts in the latter part of the period (figure 2C). Historically, a recreational fishery for Atlantic bluefin tuna (*Thunnus thynnus*) existed, which ended in 1964 because of the disappearance of the stock (MacKenzie and Ransom 2007).

REFERENCES

Gibson D, Ueberschär B, Zylich K and Zeller D (2014) Preliminary reconstruction of total marine fisheries catches for Denmark in the Kattegat, the Skagerrak and the North Sea (1950–2010). Fisheries Centre Working Paper #2014-25, 13 p.

Holm P (2005) Human impacts on fisheries resources and abundance in the Danish Wadden Sea, c1520 to the present. *Helgoland Marine Research* 59: 39–44.

MacKenzie BR and Ransom MA (2007) The development of the northern European fishery for north Atlantic bluefin tuna (*Thunnus thynnus*) during 1900–1950. *Fisheries Research* 87: 229–239.

1. Cite as Gibson, D., B. Ueberschär, K. Zylich, and D. Zeller. 2016. Denmark (North Sea). P. 237 in D. Pauly and D. Zeller (eds.), *Global Atlas of Marine Fisheries: A Critical Appraisal of Catches and Ecosystem Impacts*. Island Press, Washington, DC.

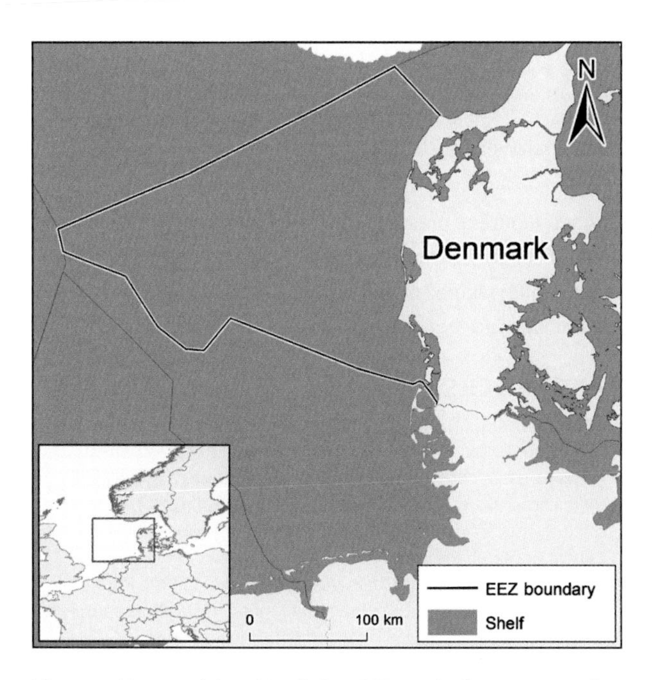

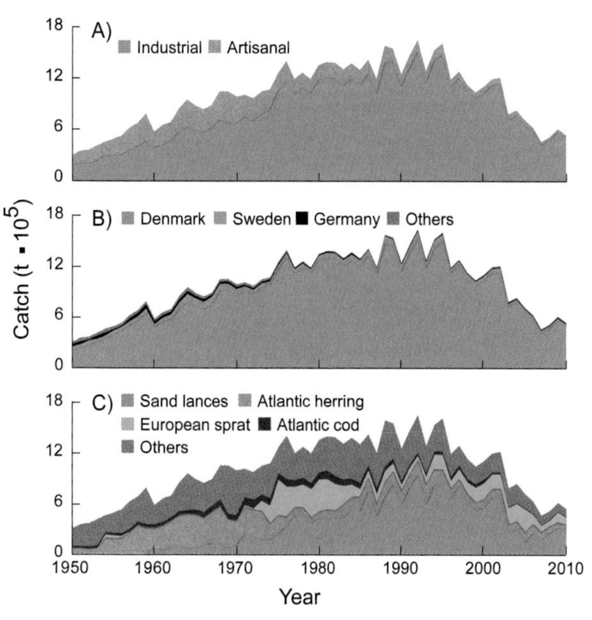

Figure 1. Denmark has North Sea EEZ-equivalent waters of 50,000 km² (incl. the Skagerrak and Kattegat), all of which is shelf.
Figure 2. Domestic and foreign catches taken in the North Sea EEZ of Denmark: (A) by sector (domestic scores: Ind. 3, 3, 4; Art. 3, 3, 4; Sub. 1, –, –; Rec. 1, 1, 2; Dis. 1, 1, 2); (B) by fishing country (foreign catches are very uncertain); (C) by taxon.

DJIBOUTI[1]

Mathieu Colléter,[a,b] Ahmed Darar Djibril,[c] Gilles Hosch,[d] Pierre Labrosse,[e]
Yann Yvergniaux,[f] Frédéric Le Manach,[a] and Daniel Pauly[a]

[a]*Sea Around Us*, University of British Columbia, Vancouver, Canada [b]Agrocampus Ouest, UMR985 Ecologie et Santé des Ecosystèmes, Rennes, France [c]Direction de la Pêche, Ministère de l'Agriculture, de l'Elevage et de la Mer, Djibouti [d]Fisheries Planning & Management, Luxembourg [e]Mission pour la Recherche et la Technologie, Haut-Commissariat de la République en Nouvelle-Calédonie, Nouméa [f]SmartFish Programme, Indian Ocean Commission, Ebène, Ile Maurice

The Republic of Djibouti is a small East African country north of Somalia (figure 1). A reconstruction of its marine fisheries catches was performed by Colléter et al. (2015) using a variety of historic records, consultant reports, and, for the last decades, data from Djibouti's Department of Fisheries. The industrial and artisanal sectors made up similar portions of the total catches allocated to Djibouti's EEZ (figure 2A). The domestic catch was estimated at about 140 t/year in the 1950s, which increased to about 1,100 t in 1990, before a decline caused by a conflict in the north of Djibouti. Catches then increased again, to about 2,000 t by 2010. Overall, the reconstructed catch of Djibouti for the 1950–2010 period was 1.5 times the adjusted data as reported by the FAO on behalf of Djibouti. Foreign catches by Yemen are comparable to domestic catches for the time period (figure 2B). For the postindependence period, the reconstruction was based mainly on locally available data, which differ from those published by the FAO on behalf of Djibouti yet probably better reflect the actual trend of artisanal catches. Figure 2C presents the composition of the catch based on field studies, including grouper species (e.g., *Epinephelus chlorostigma*) studied by Darar (1994) as well as snapper (family Lutjanidae) and mackerels (*Scomberomorus* spp.).

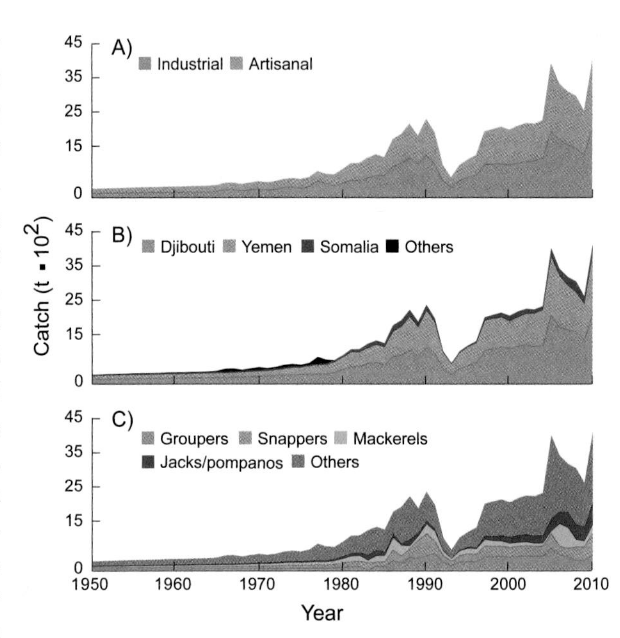

Figure 2. Domestic and foreign catches taken in the EEZ of Djibouti: (A) by sector (domestic scores: Art. 2, 3, 3; Sub. 2, 2, 2; Rec. 1, 1, 2; Dis. 2, 2, 2); (B) by fishing country (foreign catches are very uncertain); (C) by taxon.

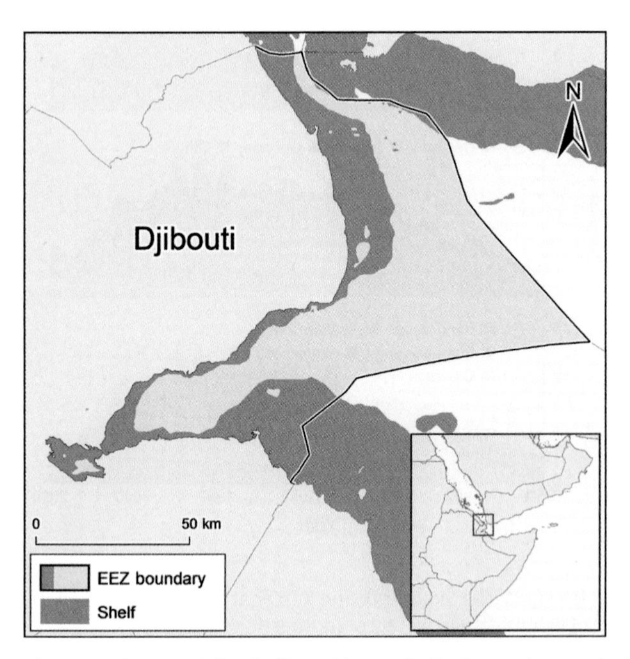

Figure 1. The Republic of Djibouti has a shelf of 2,340 km² and an EEZ of 7,040 km².

REFERENCES

Colléter M, Djibril AD, Hosch G, Labrosse P, Yvergniaux Y, Le Manach F and Pauly D (2015) Le Développement Soutenu de Pêcheries Artisanales: Reconstruction des Captures Marines à Djibouti de 1950 à 2010. pp. 13–25 In Le Manach F and Pauly D (eds.), *Fisheries catch reconstructions in the Western Indian Ocean, 1950–2010*. Fisheries Centre Research Reports 23(2).

Darar, A (1994) An account of fisheries development in the Republic of Djibouti with notes on the growth and mortality of three species of groupers. *Naga, the ICLARM Quarterly* 17(2): 30–32.

1. Cite as Colléter, M., A. D. Djibril, G. Hosch, P. Labrosse, Y. Yvergniaux, F. Le Manach, and D. Pauly. 2016. Djibouti. P. 238 in D. Pauly and D. Zeller (eds.), *Global Atlas of Marine Fisheries: A Critical Appraisal of Catches and Ecosystem Impacts*. Island Press, Washington, DC.

DOMINICA[1]

Robin Ramdeen, Sarah Harper, and Dirk Zeller

Sea Around Us, University of British Columbia, Vancouver, Canada

The Commonwealth of Dominica is an island of the Lesser Antilles, situated between the French islands of Guadeloupe in the north and Martinique in the south (figure 1). Dominica gained independence from Britain in 1978 and has since experienced a stable political history. Between 1886 and 1996, Dominica experienced 19 hurricanes, which caused extensive damage to many of Dominica's assets, including its fisheries sector. In 1979, Hurricane David almost demolished the island's entire fishing fleet; storms in 1996, 1997, and again in 1999 damaged coral reefs, seagrass beds, beach landing sites, and fisheries infrastructure. The fisheries were primarily subsistence, with an increasing artisanal and declining industrial component (figure 2A). The reconstruction by Ramdeen et al. (2014), based on numerous sources (e.g., Sebastien 2002), resulted in a catch of about 1,000 t in 1950, which increased to 1,700 t until Hurricane David struck in 1979. The catch since settled at about 1,500 t/year, 1.8 times the official reported data supplied to the FAO. Total removals allocated to Dominica's EEZ are mostly domestic. However, foreign fishing by neighboring and also Mediterranean countries generated noticeable removals from 1950 to the 1970s (figure 2B). Figure 2C documents the composition of the catch, which includes ballyhoo halfbeak (*Hemirhampus brasiliensis*), commonly used as bait, and a wide variety of reef-associated and pelagic fishes including Atlantic bonito (*Sarda sarda*) and common dolphinfish (*Coryphaena hippurus*).

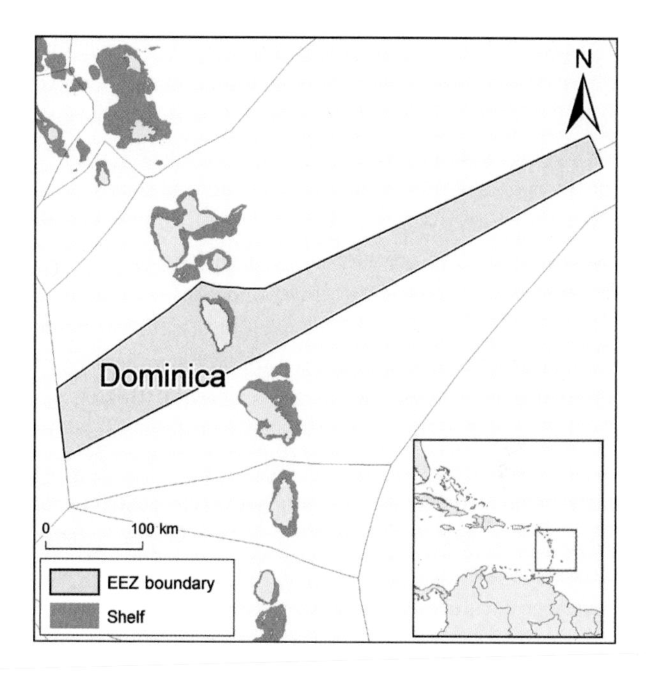

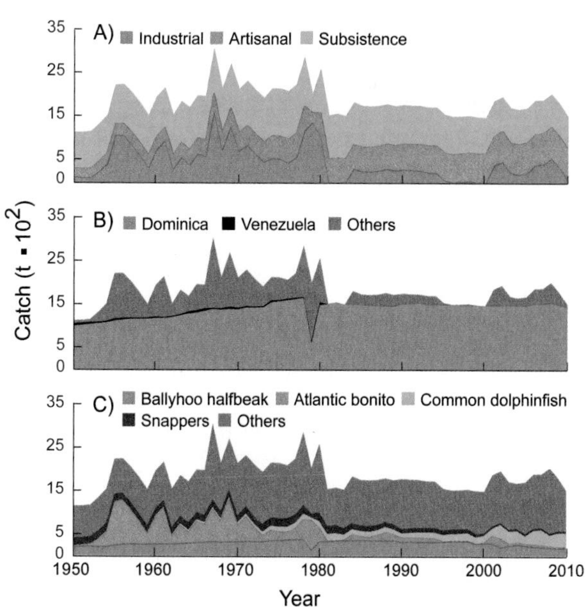

Figure 1. The Commonwealth of Dominica has a land area of 754 km² and an EEZ of 28,600 km². **Figure 2.** Domestic and foreign catches taken in the EEZ of Dominica: (A) by sector (domestic scores: Art. 2, 2, 2; Sub. 1, 2, 2); (B) by fishing country (foreign catches are very uncertain); (C) by taxon.

REFERENCES

Ramdeen R, Harper S and Zeller D (2014) Reconstruction of total marine fisheries catches for Dominica (1950–2010). pp. 33–42 In Zylich K, Zeller D, Ang M and Pauly D (eds.), *Fisheries catch reconstructions: Islands, Part IV*. Fisheries Centre Research Reports 22(2).

Sebastien RD (2002) National report of the Commonwealth of Dominica, pp. 27–34 In National reports and technical papers presented at the first meeting of the WECAFC Ad Hoc Working Group on the development of sustainable moored fish aggregating device fishing in the Lesser Antilles, October 8–11, 2001, Le Robert (Martinique).

1. Cite as Ramdeen, R., S. Harper, and D. Zeller. 2016. Dominica. P. 239 in D. Pauly and D. Zeller (eds.), *Global Atlas of Marine Fisheries: A Critical Appraisal of Catches and Ecosystem Impacts*. Island Press, Washington, DC.

DOMINICAN REPUBLIC[1]

Liesbeth van der Meer, Robin Ramdeen, Kyrstn Zylich, and Dirk Zeller

Sea Around Us, University of British Columbia, Vancouver, Canada

The Dominican Republic shares the island of Hispaniola with Haiti. This popular tourist destination has a north coast open to the Atlantic Ocean and a south coast opening to the Caribbean Sea (figure 1). Fishing has always been important for the people of the Dominican Republic, and the fisheries are mainly subsistence and artisanal (figure 2A). The catch reconstruction of van der Meer et al. (2014), based on a variety of sources, notably Herrera et al. (2011), yielded an estimate of about 17,000 t in 1950, which gradually increased, with some fluctuations, to a peak of 72,000 t in 1993 before settling at about 52,000 t/year in the late 2000s. Overall, this was 5.1 times the data reported by the FAO on behalf of the Dominican Republic, a discrepancy due to only about half of the artisanal catch being recorded and the large subsistence catch being completely missed. Though predominantly domestic, catches by foreign vessels such as Venezuela and Cuba, among others, have been documented in the EEZ (figure 2B). Major groups in the catch are snappers (family Lutjanidae), grunts (Haemulidae), mackerels, tunas, bonitos (Scombridae), jacks, pompanos (Carangidae), parrotfish (Scaridae), and porgies (Sparidae) (figure 2C). Also important is the spiny lobster (*Panulirus argus*). The growing population and the seafood demands of the tourism industry have increased pressure on the fisheries resources of the Dominican Republic to an unsustainable level.

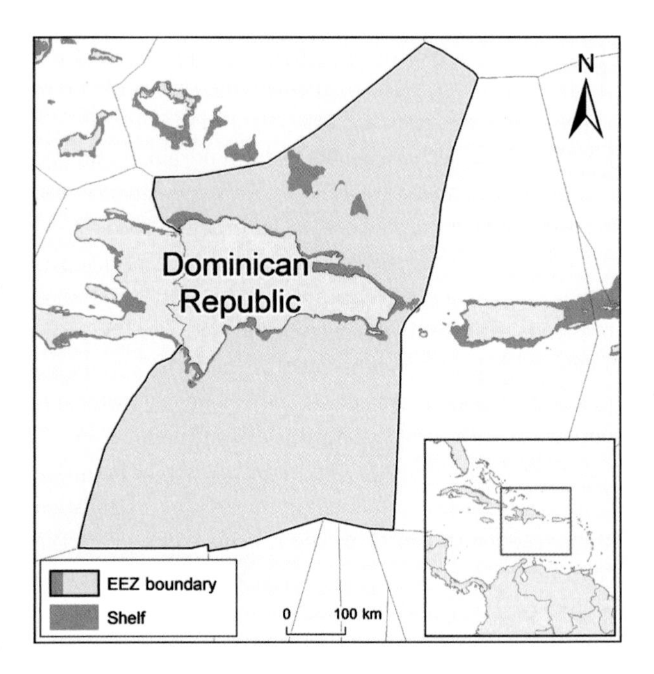

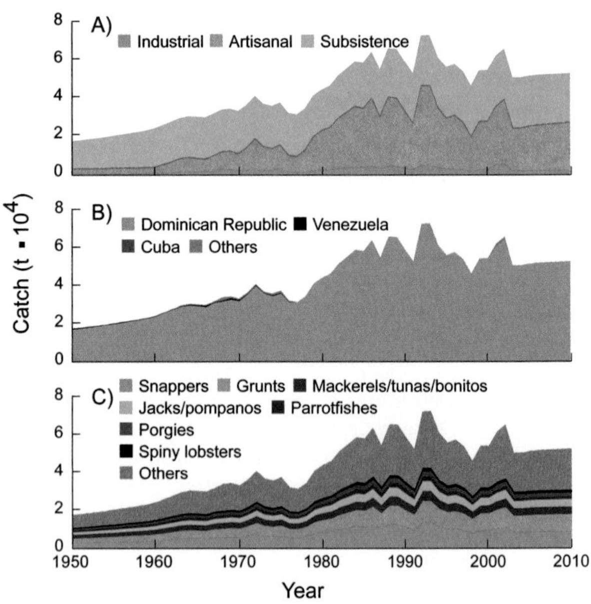

Figure 1. The Dominican Republic has a land area of 48,100 km² on the Island of Hispaniola and an EEZ of 269,000 km². **Figure 2.** Domestic and foreign catches taken in the EEZ of Dominican Republic: (A) by sector (domestic scores: Ind. 2, 2, 3; Art. 2, 4, 4; Sub. 2, 2, 3; Rec. 2, 2, 2); (B) by fishing country (foreign catches are very uncertain); (C) by taxon.

REFERENCES

Herrera A, Betancourt L, Silva M, Lamelas P and Melo A (2011) Coastal fisheries of the Dominican Republic. pp. 175–217 In Salas S, Chuenpagdee R, Charles A and Seijo JC (eds.), Coastal fisheries of Latin America and the Caribbean. FAO Fisheries and Aquaculture Technical Paper No. 544. FAO, Rome.

van der Meer L, Ramdeen R, Zylich K and Zeller D (2014) Reconstruction of total marine fisheries catches for the Dominican Republic (1950–2009). pp. 43–54 In Zylich K, Zeller D, Ang M and Pauly D (eds.), *Fisheries catch reconstructions: Islands, Part IV*. Fisheries Centre Research Reports 22(2).

1. Cite as van der Meer, L., R. Ramdeen, K. Zylich, and D. Zeller. 2016. Dominican Republic. P. 240 in D. Pauly and D. Zeller (eds.), *Global Atlas of Marine Fisheries: A Critical Appraisal of Catches and Ecosystem Impacts*. Island Press, Washington, DC.

ECUADOR[1]

Juan José Alava,[a,b] Alasdair Lindop,[c] and Jennifer Jacquet[c,d]

[a]School of Resource and Environmental Management, Simon Fraser University, Burnaby, Canada [b]Fundación Ecuatoriana para el Estudio de Mamíferos Marinos (FEMM), Guayaquil, Ecuador [c]*Sea Around Us*, University of British Columbia, Vancouver, Canada [d]Department of Environmental Studies, New York University, New York

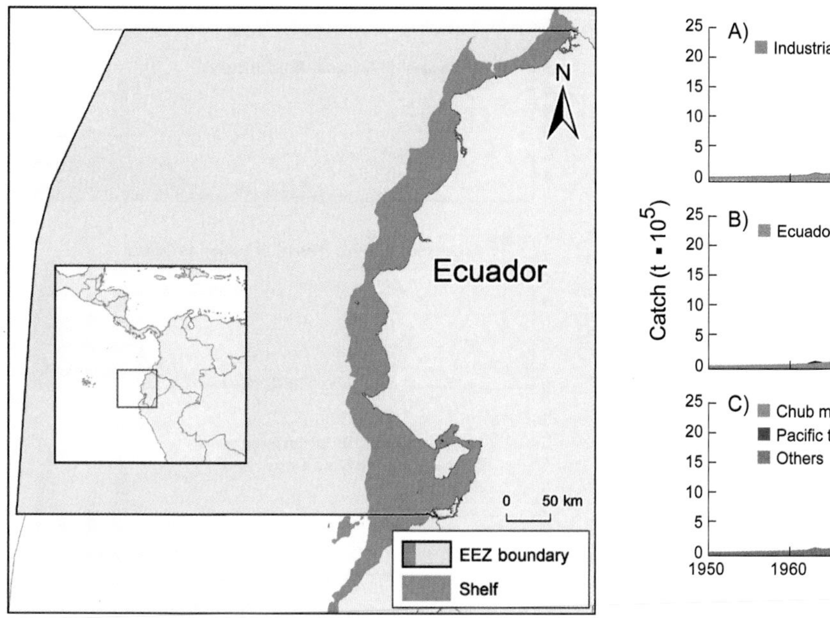

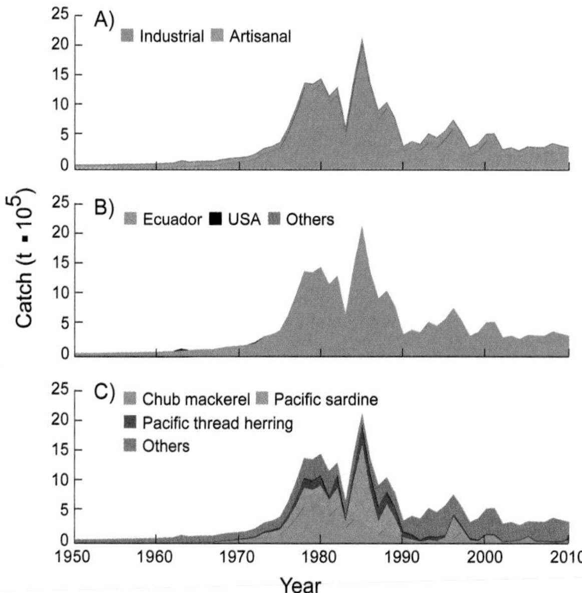

Figure 1. Ecuador has a shelf of 28,000 km² and an EEZ of 237,000 km². **Figure 2.** Domestic and foreign catches taken in the EEZ of Ecuador: (A) by sector (domestic scores: Ind. 2, 2, 2; Art. 3, 3, 3; Sub. 1, 1, 1; Dis. 1, 1, 2); (B) by fishing country (foreign catches are very uncertain); (C) by taxon.

Ecuador is located in the northwest of South America (figure 1) but has one oceanic province, the Galápagos (see next page). Ecuador is a low-income country faced with large political challenges, notably income and ethnic divides (Sachs 2005). Ecuador has sacrificed to the altar of shrimp culture a large fraction of its formerly abundant mangroves (Carvajal and Alava 2007), the nurseries of many of its marine species. Its catches are primarily industrial, with a small artisanal component that gradually increased (figure 2A). The catch reconstruction of Alava et al. (2015) shows that Ecuadorian fisheries, which yielded an exclusively small-scale catch of 60,000 t in 1950, were "developed" via shrimp trawlers and purse-seiners targeting small pelagics (used mainly for exported fishmeal). These drove total catches to a peak of 2.1 million t in 1985, which then declined to less than 400,000 t/year in the 2000s, despite continuing massive fishing effort, notably targeting sharks (Jacquet et al. 2008). Overall, fisheries catches from the Ecuadorian mainland EEZ were 1.9 times the data reported to the FAO, and mostly domestic, although some foreign fishing, by the United States occurred in the 1960s (figure 2B). The taxonomic composition is presented in figure 2C, where total catches are predominantly chub mackerel (*Scomber japonicus*) throughout the time period, despite the large amount of Pacific sardine (*Sardinops sagax*) during the peak in 1985.

REFERENCES

Alava JJ, Lindop A and Jacquet J (2015) Marine fisheries catch reconstructions for continental Ecuador: 1950–2010. Fisheries Centre Working Paper #2015-34, 25 p.

Carvajal R and Alava JJ (2007) Mangrove Wetlands Conservation Project and the shrimp farming industry in Ecuador: lessons learned. *World Aquaculture* 38(3): 14–17.

Jacquet J, Alava JJ, Pramod G, Henderson S and Zeller D (2008) In hot soup: sharks captured in Ecuador's waters. *Environmental Sciences* 5(4): 269–283.

Sachs J (2005) *The End of Poverty*. Penguin Books, London. xviii + 397 p.

1. Cite as Alava, J. J., A. Lindop, and J. Jacquet. 2016. Ecuador. P. 241 in D. Pauly and D. Zeller (eds.), *Global Atlas of Marine Fisheries: A Critical Appraisal of Catches and Ecosystem Impacts*. Island Press, Washington, DC.

ECUADOR (GALÁPAGOS)[1]

Laurenne Schiller,[a] Juan José Alava,[a,b,c] Jack Grove,[d] Günther Reck,[e] and Daniel Pauly[a]

[a]*Sea Around Us*, University of British Columbia, Vancouver, Canada [b]Fundación Ecuatoriana para el Estudio de Mamiferos Marino, Guayaquil, Ecuador [c]Resource and Environmental Management, Simon Fraser University, Burnaby, Canada [d]Section of Ichthyology, Natural History Museum of Los Angeles County, Los Angeles [e]Instituto de Ecología Aplicada, Universidad San Francisco de Quito, Quito, Ecuador

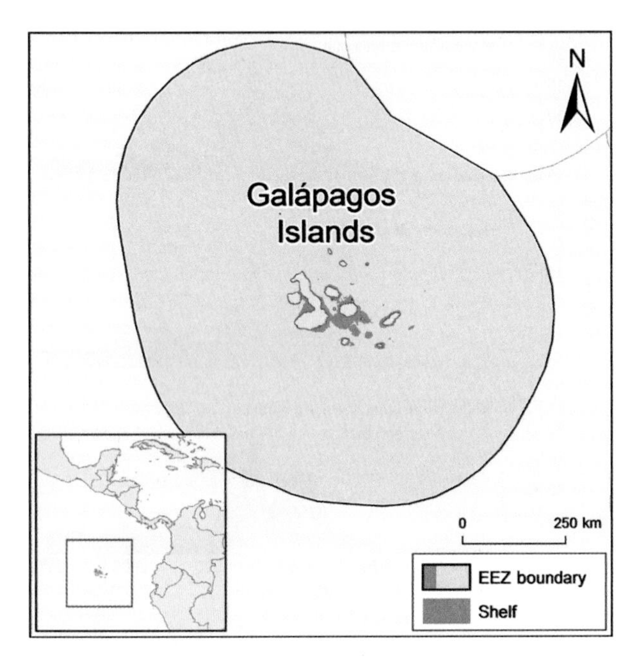

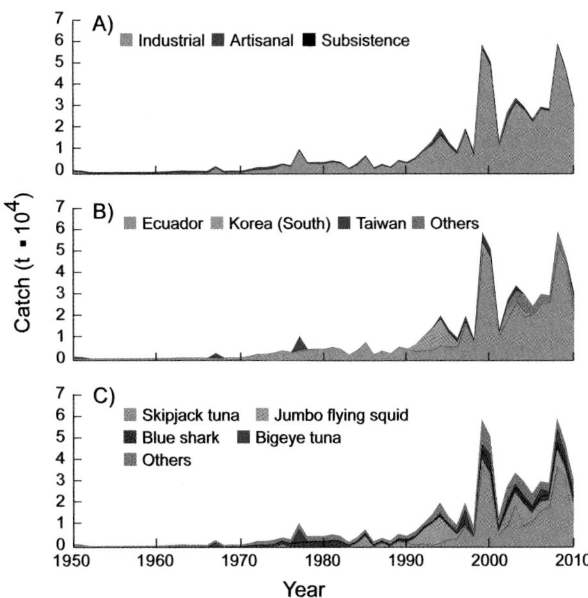

Figure 1. The Galápagos archipelago encompasses 13 islands (>10 km²) and more than 100 islets, for a total land area of 8,000 km², and is surrounded by an EEZ of about 836,000 km². **Figure 2.** Domestic and foreign catches taken in the EEZ of Ecuador (Galápagos): (A) by sector (domestic scores: Ind. 2, 3, 3; Art. 2, 4, 4; Sub. 1, 1, 2); (B) by fishing country (foreign catches are very uncertain); (C) by taxon.

The Galápagos Islands, an Ecuadorian province, are not a distinct entity in the catch statistics the FAO presents on behalf of Ecuador, although the islands are nearly 1,000 km to the west of the Ecuadorian mainland (figure 1). Catches within the Galápagos' EEZ were dominated by the mainland-based industrial fleet targeting tuna offshore, whereas the inshore, small-scale artisanal and subsistence fisheries of importance to Galápagos residents did not see their catch increase—with the exception of boom-and-bust runs on lobsters and sea cucumbers—despite expanding their effort and reach (figure 2A). The reconstruction by Schiller et al. (2013, 2015) led to a domestic catch estimate of about 600 t in the 1950s, increasing gradually to about 16,800 t in 1997, after which catches increased sharply and with substantial fluctuation to peak at 50,700 t in 2008 before declining to 18,100 t in 2010. Most catches are domestic; however, foreign fishing by South Korea in the early 1990s and Taiwan in the early 2000s was documented (figure 2B). Currently, illegal fishing is rampant, notably for shark fins, both within the 60-km marine reserve zone surrounding the Galápagos and in the rest of the 200-nm EEZ. Taxa exploited in the Galápagos' EEZ are mainly skipjack tuna (*Katsuwonus pelamis*), jumbo flying squid (*Dosidicus gigas*), blue shark (*Prionance glauca*), and bigeye tuna (*Thunnus obesus*; figure 2C).

REFERENCES

Schiller L, Alava J, Grove J, Reck G and Pauly D (2013) A reconstruction of fisheries catches for the Galápagos Islands, 1950–2010. Fisheries Centre Working Paper #2013-11, 38 p.

Schiller L, Alava JJ, Grove J, Reck G and Pauly D (2015) The demise of Darwin's fishes: evidence of fishing down and illegal shark finning in the Galápagos islands. *Aquatic Conservation: Freshwater and Marine Ecosystems* 25(3): 431–446.

1. Cite as Schiller, L., J. Alava, J. Grove, G. Reck, and D. Pauly. 2016. Ecuador (Galápagos). P. 242 in D. Pauly and D. Zeller (eds.), *Global Atlas of Marine Fisheries: A Critical Appraisal of Catches and Ecosystem Impacts*. Island Press, Washington, DC.

EGYPT (MEDITERRANEAN)[1]

Hatem Hanafy Mahmoud,[a] Lydia C. L. Teh,[b] Myriam Khalfallah,[b] and Daniel Pauly[b]

[a]College of Fisheries Technology & Aquaculture, Arab Academy for Science, Technology & Maritime Transport, Alexandria, Egypt [b]*Sea Around Us*, University of British Columbia, Vancouver, Canada

Egypt has coastlines on the Mediterranean and the Red Sea; this account deals with the former (figure 1). Total catches allocated to the Mediterranean EEZ of Egypt are primarily industrial; however, the artisanal and subsistence sectors steadily increased (figure 2A). As reconstructed by Mahmoud et al. (2015), Egypt's domestic catch in the Mediterranean averaged 35,000 t/year in the early 1950s, peaked at 150,000 t in 1999, and declined to 120,000 t/year in the late 2000s. Reconstructed catches were on average 1.9 times those reported by the FAO on behalf of Egypt; unreported catches were primarily from the industrial sector. Total allocated catch within the Mediterranean EEZ of Egypt were predominantly domestic, with little foreign fishing hardly visible in figure 2B. Round sardinella (*Sardinella aurita*) was the taxon contributing most to the reconstructed catch, followed by shrimps (figure 2C). Egypt's Mediterranean fisheries were affected by the construction of the Aswan High Dam, which, when closed in 1964, substantially reduced the flow of nutrient-rich Nile waters to the coastal zone (Nixon 2004). Fish landings decreased substantially in the mid-1960s and remained depressed until the late 1970s. Sardinella catches in particular declined sharply in the mid-1960s. Landings recovered only in the mid-1980s, attributed to nutrient loading from anthropogenic sources (Nixon 2003) due to increasing population pressure.

REFERENCES

Mahmoud HH, Teh LCL, Khalfallah M and Pauly D (2015) Reconstruction of marine fisheries statistics in the Egyptian Mediterranean Sea, 1950–2010. Fisheries Centre Working Paper #2015-85, 16 p.

Nixon S (2003) Replacing the Nile: are anthropogenic nutrients providing the fertility once brought to the Mediterranean by a great river? *AMBIO* 32(1): 30–39.

Nixon S (2004) The artificial Nile. *American Scientist* 92(2): 158–165.

1. Cite as Mahmoud, H. H., L. C. L. Teh, M. Khalfallah and D. Pauly. 2016. Egypt (Mediterranean). P. 243 in D. Pauly and D. Zeller (eds.), *Global Atlas of Marine Fisheries: A Critical Appraisal of Catches and Ecosystem Impacts*. Island Press, Washington, DC.

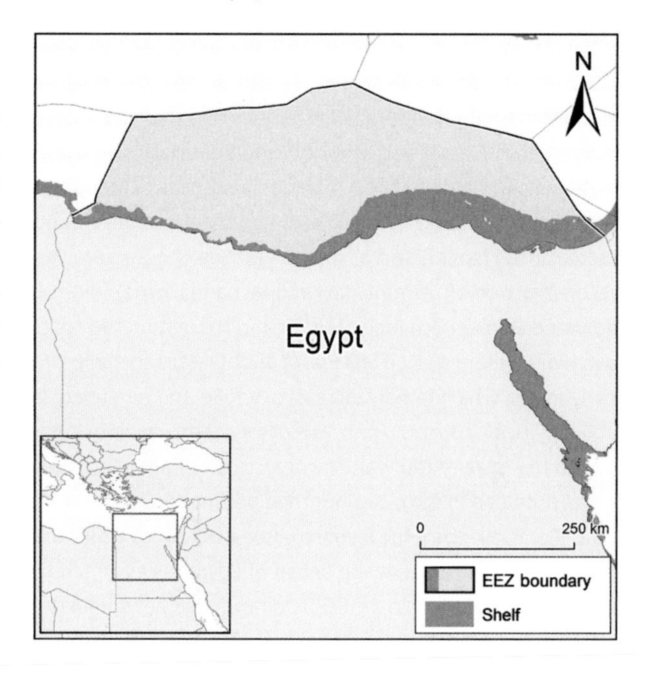

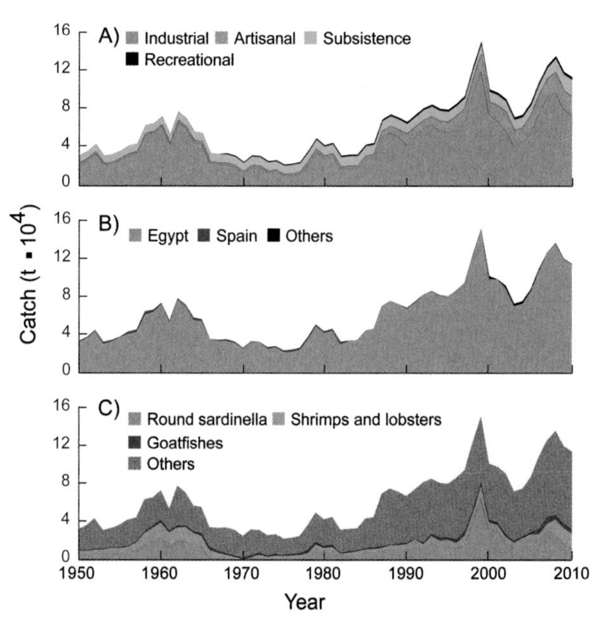

Figure 1. Egypt and its continental shelf (31,000 km²) and EEZ (170,000 km²) in the Mediterranean. **Figure 2.** Domestic and foreign catches taken in the Mediterranean EEZ of Egypt: (A) by sector (domestic scores: Ind. 2, 2, 2; Art. 2, 2, 2; Sub. 2, 2, 2; Rec. 1, 1, 1; Dis. 2, 3, 3); (B) by fishing country (foreign catches are very uncertain); (C) by taxon.

EGYPT (RED SEA)[1]

Dawit Tesfamichael[a,b] and Sahar Fahmy Mehanna[c]

[a]*Sea Around Us*, University of British Columbia, Vancouver, Canada [b]Department of Marine Sciences, University of Asmara, Eritrea [c]National Institute of Oceanography and Fisheries, Suez, Egypt

Egypt has coastlines on both the Red and the Mediterranean Sea, and this account deals with the former (figure 1). Egypt has highly developed fisheries in the Red Sea, where industrial purse seining and trawling dominate, although artisanal and subsistence fisheries also occur (Mehanna and El-Gammal 2007). Egypt also has the most developed recreational fishing sector in the Red Sea (figure 2A). The reconstruction of Tesfamichael and Mehanna (2012), slightly adjusted to meet *Sea Around Us* definitions, estimated total catch of Egypt in its Red Sea EEZ of about 5,600 t/year in the early 1950s, which rapidly increased in 1960 and remained at about 35,000–40,000 t/year, except for a sharp decline in 1973 due to the Israel–Arab war. The peak catch of about 50,000 t was obtained in 1993; catches then declined to 24,000 t by 2010. For 1950–2010, the reconstructed catch was 2 times the data reported by the FAO on behalf of Egypt (as assumed to have been caught within the EEZ). Domestic catches dominate, although some foreign catches seem to have occurred (figure 2B). Figure 2C presents the catch, dominated by pelagic fishes: jacks, pompanos (family Carangidae), red-eye round herrings (*Etrumeus sardina*), and scads (*Decapterus* spp.), but demersal taxa are also important.

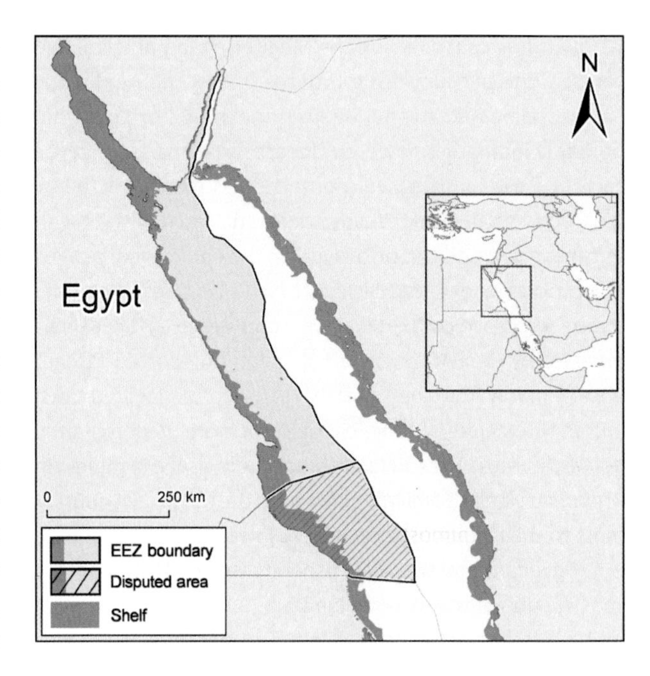

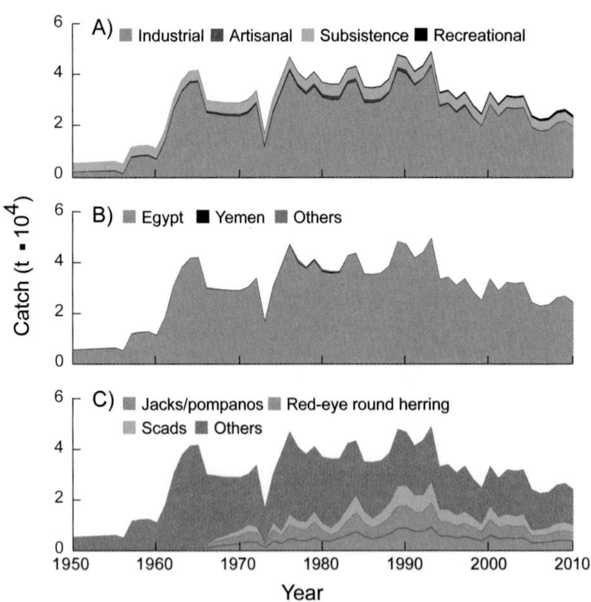

Figure 1. Egypt and its continental shelf (23,200 km²) and EEZ (91,300 km²) in the Red Sea. **Figure 2.** Domestic and foreign catches taken in the Red Sea EEZ of Egypt: (A) by sector (domestic scores: Ind. 2, 2, 2; Art. 2, 2, 2; Sub. 2, 2, 2; Rec. 1, 1, 1; Dis. 2, 3, 3); (B) by fishing country (foreign catches are very uncertain); (C) by taxon.

REFERENCES

Mehanna SF and El-Gammal FI (2007) Gulf of Suez fisheries: current status, assessment and management. *Journal of King Abdulaziz University (Marine Science)* 18: 3–18.

Tesfamichael D and Mehanna SF (2012) Reconstructing Red Sea fisheries of Egypt: Heavy investment and fisheries, pp. 23–50 In Tesfamichael D and Pauly D (eds.), *Catch reconstruction for the Red Sea Large Marine Ecosystem by countries (1950–2010)*. Fisheries Centre Research Reports 20(1).

1. Cite as Tesfamichael, D., and S. F. Mehanna. 2015. Egypt (Red Sea). P. 244 in D. Pauly and D. Zeller (eds.), *Global Atlas of Marine Fisheries: A Critical Appraisal of Catches and Ecosystem Impacts*. Island Press, Washington, DC.

EL SALVADOR[1]

Rodrigo Donadi,[a] Andrea Au,[b] Kyrstn Zylich,[b] Sarah Harper,[b] and Dirk Zeller[b]

[a]WWF Panama, Ciudad del Saber, Building 235, Clayton, Republic of Panama [b]*Sea Around Us*, University of British Columbia, Vancouver, Canada

El Salvador is a small country (figure 1) with a large population of about 7 million, resulting in the highest population density in Central America, particularly pronounced along its coast. The industrial domestic fishery dominates the total catch allocated to El Salvador's EEZ, but the artisanal sector has grown steadily (figure 2A; figure 2B). In contrast to the export-oriented industrial shrimp fisheries, the small-scale fisheries have barely been studied (Pauly and Agüero 1992), although they contribute directly to the livelihood of thousands of coastal people, as illustrated by Gammage (2004) in her study of inshore fishing by Salvadorian women. This imbalance can be overcome by a catch reconstruction, which gives equal emphasis to the various fishery sectors. Thus, the reconstruction by Donadi et al. (2015) yielded a domestic catch of 2,800 t/year in the early 1950s, generated almost exclusively by small-scale fisheries; a peak of 135,000 t (most of it industrially discarded fish) in 1966, when the shrimp trawl fishery was in full swing; and about 35,000 t/year in the late 2000s. Figure 2C presents shorthead lizardfish (*Synodus scituliceps*), pelagic red crab (*Pleuroncodes planipes*), and various shrimps as primary taxa. Although 6.9 times the data reported by the FAO on behalf of El Salvador, the reconstructed catch has a composition and a trajectory that identify this country as yet another victim of a faulty "fishery development" model.

REFERENCES

Donadi R, Au A, Zylich K, Harper S and Zeller D (2015) Reconstruction of marine fisheries in El Salvador, 1950–2010. Fisheries Centre Working Paper #2015–35, 22 p.

Gammage S (2004) The tattered net of statistics. pp. 36–40 In Kumar KG (ed.), Gender agenda—women in fisheries: a collection of articles from SAMUDRA Report. International Collective in Support of Fishworkers, India.

Pauly, D and Agüero M (1992) Small-scale fisheries in the neotropics: research and management issues, pp. 28–36 In M. Agüero (ed.), *Contribuciones para el estudio de la pesca artesanal en América Latina.* ICLARM Conference Proceedings 35.

1. Cite as Donadi, R., A. Au, K. Zylich, S. Harper, and D. Zeller. 2016. El Salvador. P. 245 in D. Pauly and D. Zeller (eds.), *Global Atlas of Marine Fisheries: A Critical Appraisal of Catches and Ecosystem Impacts*. Island Press, Washington, DC.

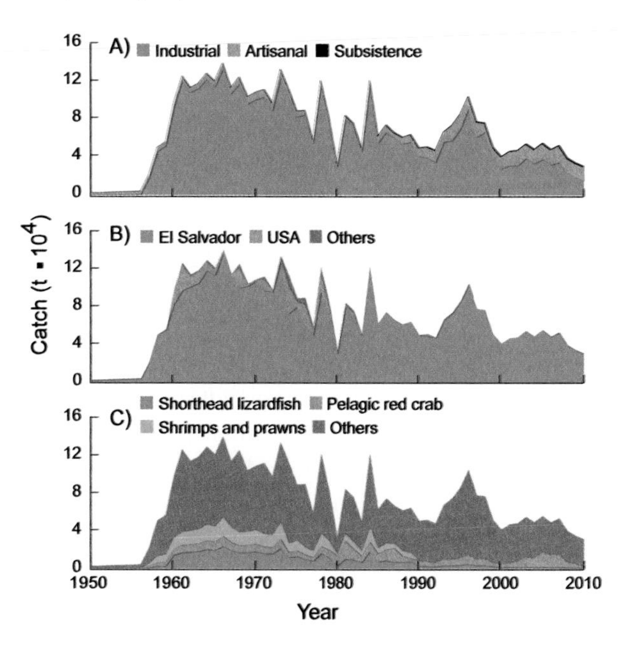

Figure 1. El Salvador, on the Pacific coast of Central America, has a shelf of 17,700 km² and an EEZ of about 93,700 km². **Figure 2.** Domestic and foreign catches taken in the EEZ of El Salvador: (A) by sector (domestic scores: Ind. 2, 3, 4; Art. 2, 3, 4; Sub. 2, 3, 1; Dis. 2, 3, 4); (B) by fishing country (foreign catches are very uncertain); (C) by taxon.

EQUATORIAL GUINEA[1]

Dyhia Belhabib,[a] Denis Hellebrandt,[b] Edward H. Allison,[b] and Daniel Pauly[a]

[a]*Sea Around Us*, University of British Columbia, Vancouver, Canada
[b]School of International Development, University of East Anglia, Norwich, UK

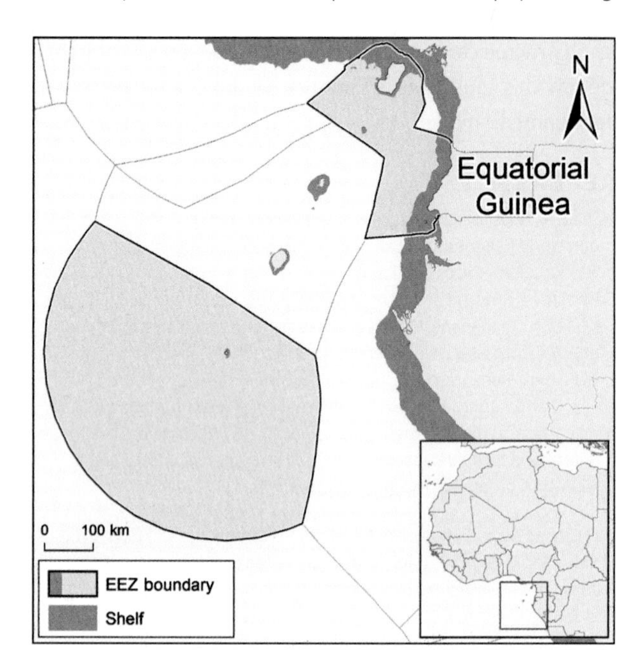

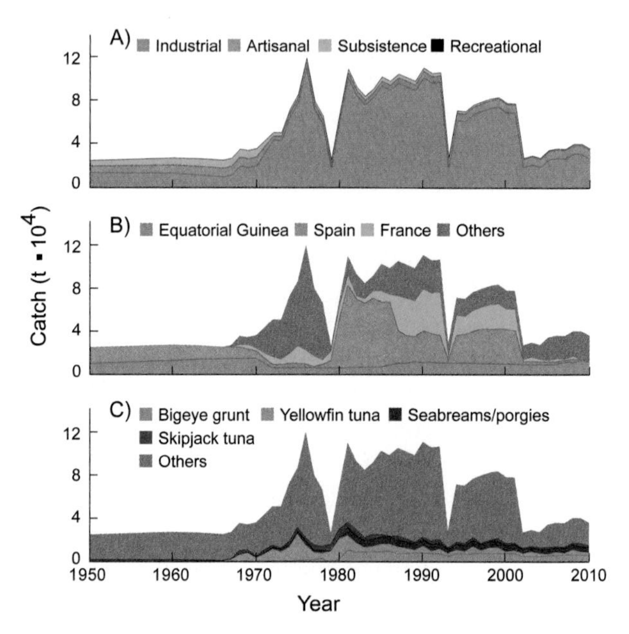

Figure 1. The EEZ of Equatorial Guinea covers 308,000 km², of which 12,000 km² is shelf. **Figure 2.** Domestic and foreign catches taken in the EEZ of Equatorial Guinea: (A) by sector (domestic scores: Ind. –, 3, 3; Art. 2, 2, 4; Sub. 2, 2, 4; Rec. –, –, 3; Dis. –, 2, 2); (B) by fishing country (foreign catches are very uncertain); (C) by taxon.

Equatorial Guinea, located in West Central Africa (figure 1), was the only Sub-Saharan African country colonized by Spain, with independence gained in 1968. The history of the country highlights the passage from one of the wealthiest economies of Africa to the most repressive dictatorship the continent has ever witnessed (Wood 2004). As a result of massive censorship, fisheries reports were scarce and the official data unrealistic. Total catches allocated to Equatorial Guinea's EEZ are primarily from foreign industrial fleets (figure 2A). The tentative catch reconstruction by Belhabib et al. (2015, 2016) suggested mean domestic catches of 13,000 t/year in the 1950s and 1960s, which declined dramatically to about 7,000 t/year during the reign of terror of the 1970s. Thereafter, catches increased again to average, with some fluctuation, just under 12,000 t/year in the late 2000s. Domestic catches were 4 times the data reported by the FAO on behalf of this country. Most of the removals were from foreign vessels, notably from Spain and France (figure 2B). Top taxa include bigeye grunt (*Brachydeuterus auritus*), seabreams, porgies (*Diplodus* spp.), and yellowfin tuna (*Thunnus albacares*; figure 2C). In contrast to what is suggested by official figures, domestic fisheries have been strongly affected by civil and political unrest. The noncommercial, subsistence sector declined because of civil and political conflicts, social and demographic changes, and the growing role of newly discovered oil resources.

REFERENCES

Belhabib D, Hellebrandt D, Allison E and Pauly D (2015) Equatorial Guinea: a catch reconstruction (1950–2010). Fisheries Centre Working Paper #2015–71, 24 p.

Belhabib D, Hellebrandt D, Allison EH, Zeller D and Pauly D (2016) Filling a blank on the map: 60 years of fisheries in Equatorial Guinea. *Fisheries Ecology and Management* 23: 119–132.

Wood G (2004) Business and politics in a criminal state: the case of Equatorial Guinea. *African Affairs* 103(413): 547–567.

1. Cite as Belhabib, D., D. Hellebrandt, E. Allison, and D. Pauly. 2016. Equatorial Guinea. P. 246 in D. Pauly and D. Zeller (eds.), *Global Atlas of Marine Fisheries: A Critical Appraisal of Catches and Ecosystem Impacts*. Island Press, Washington, DC.

ERITREA[1]

Dawit Tesfamichael[a,b] and Sammy Mohamud[c]

[a]*Sea Around Us*, University of British Columbia, Vancouver, Canada [b]Department of Marine Sciences, University of Asmara, Asmara, Eritrea [c]Ministry of Fisheries, Massawa, Eritrea

Eritrea borders the Red Sea (figure 1). Archaeological studies indicate that humans were exploiting near-shore marine organisms such as giant clams and other molluscs more than 100,000 years ago (Walter et al. 2000). Until 1991, Eritrea was the only marine province of Ethiopia, but the catch reconstruction of Tesfamichael and Mohamud (2012), slightly adjusted to meet *Sea Around Us* definitions, has treated Eritrea as an independent entity since 1950. Total removals were reconstructed to be mostly artisanal until the early 1970s, when the industrial sector surpassed the small-scale sectors and dominated the total catches allocated to Eritrea's EEZ (figure 2A). The domestic fisheries went through major shifts, yielding nearly 30,000 t/year, declining to catches of less than 2,500 t/year because of Eritrea's struggle for independence (i.e., from the mid-1970s to the early 1990s), and recovering and increasing again to about 4,000 t/year in the late 2000s. Overall, the reconstructed total catch for the period from 1950 to 2010 was 2.1 times the data reported by the FAO on behalf of Ethiopia and Eritrea from 1950 to 2010 (see also Tesfamichael and Pitcher 2007). Domestic fisheries contributed the bulk of catches until the mid-1970s, where foreign vessels from Egypt, among others, made large catches (figure 2B). Figure 2C shows that anchovies (family Engraulidae) and herrings, shads, and sardines (family Clupeidae) were important taxa in the catches from Eritrea's EEZ.

REFERENCES

Tesfamichael D and Mohamud S (2012) Reconstructing Red Sea fisheries catches of Eritrea: a case study of the relationship between political stability and fisheries development. pp. 51–70 In D. Tesfamichael and D. Pauly (eds.) *Catch reconstruction for the Red Sea ecosystem by countries (1950–2010)*. Fisheries Centre Research Reports 20(1).

Tesfamichael D and Pitcher TJ (2007) Estimating the unreported catch of Eritrean Red Sea fisheries. *African Journal of Marine Science* 29(1): 55–63.

Walter RC, Buffler RT, Bruggemann JH, Guillaume MMM, Berhe SM, Negassi B, Libsekal Y, Cheng H, Edwards RL, von Cosel R, Neraudeau D and Gagnon M (2000) Early human occupation of the Red Sea coast of Eritrea during the last interglacial. *Nature* 405(6782): 65–69.

1. Cite as Tesfamichael, D., and S. Mohamud. 2016. Eritrea. P. 247 in D. Pauly and D. Zeller (eds.), *Global Atlas of Marine Fisheries: A Critical Appraisal of Catches and Ecosystem Impacts*. Island Press, Washington, DC.

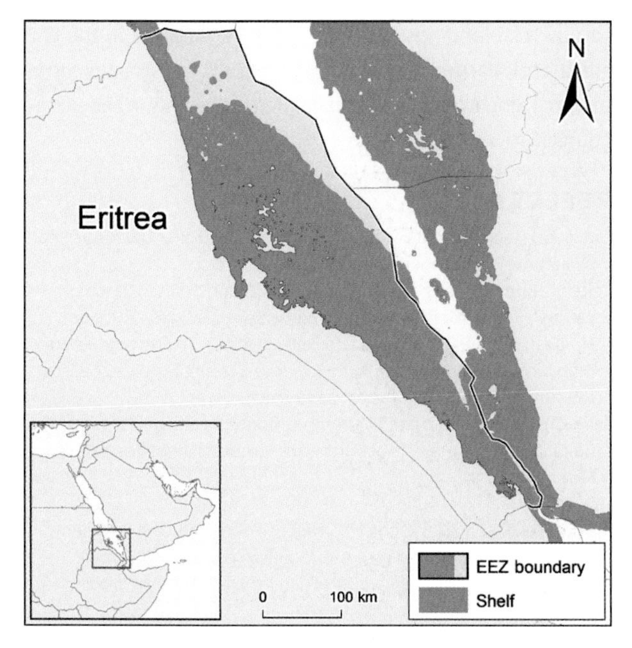

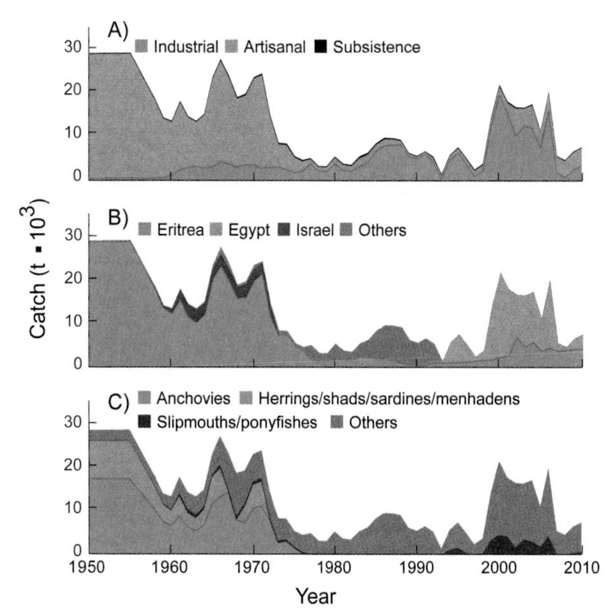

Figure 1. Eritrea has an EEZ of 78,400 km², of which 61,900 km² is shelf. **Figure 2.** Domestic and foreign catches taken in the EEZ of Eritrea: (A) by sector (domestic scores: Ind. 2, 3, 3; Art. 2, 3, 3; Sub. 2, 3, 3; Dis. 2, 2, –); (B) by fishing country (foreign catches are very uncertain); (C) by taxon.

ESTONIA[1]

Liane Veitch, Shawn Booth, Sarah Harper, Peter Rossing, and Dirk Zeller

Sea Around Us, University of British Columbia, Vancouver, Canada

Estonia is a small country on the eastern edge of the Baltic Sea, which became independent from the ex-USSR in 1991 (figure 1). Disaggregated Baltic catches of Estonia (extracted from statistics of the ex-USSR and other sources) were presented by Ojaveer (1999) and used (with data from the International Council for the Exploration of the Seas [ICES]) by Veitch et al. (2010) for their catch reconstruction (see also Zeller et al. 2011). Total catches allocated to Estonia's EEZ are mostly industrial, with a growing artisanal component (figure 2A). The reconstructed domestic catch from the Estonian EEZ, of about 10,000 t/year in the early 1950s, experienced a first peak of about 32,000 t/year in the mid-1970s, followed by one and a half decades of decline and a second peak of 40,000 t in 1997, with catches declining to 31,300 t in 2010. Approximately half of the total catches from the Estonian EEZ is from domestic fishing, and the other half is from foreign fishing by mainly neighboring countries (figure 2B). ICES reports on Estonia since 1991; from then to 2010, the reconstructed catches are 1.2 times the landing reported by ICES, mainly because of unreported commercial landings. The main species targeted in the EEZ are Atlantic herring (*Clupea harengus*), European sprat (*Sprattus sprattus*), and Atlantic cod (*Gadus morhua*; mainly in the 1980s; figure 2C).

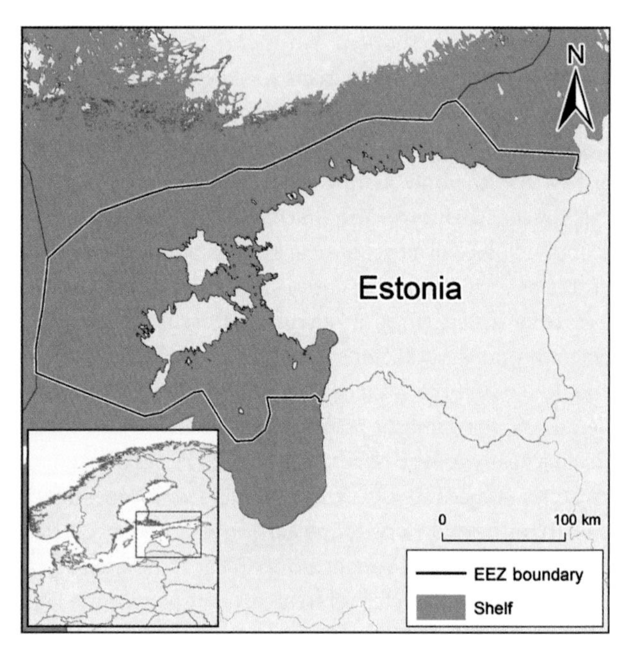

Figure 1. Estonia has an EEZ of 36,500 km², of which all is shelf.

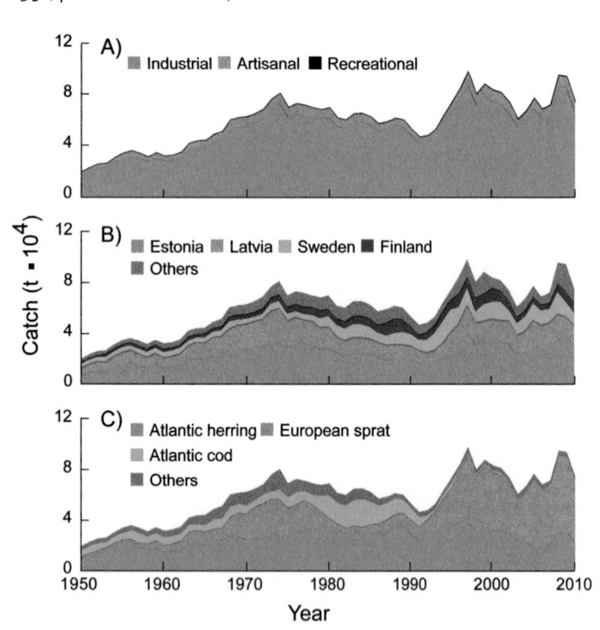

Figure 2. Domestic and foreign catches taken in the EEZ of Estonia: (A) by sector (domestic scores: Ind. 3, 3, 4; Art. 2, 2, 2; Rec. –, –, 2; Dis. 2, 2, 2); (B) by fishing country (foreign catches are very uncertain); (C) by taxon.

REFERENCES

Ojaveer H (1999) Exploitation of biological resources of the Baltic Sea by Estonia in 1928–1995. *Limnologica* 29: 224–226.

Veitch L, Booth S, Harper S, Rossing P and Zeller D (2010) Catch reconstruction for Estonia in the Baltic Sea from 1950–2007. pp. 63–84 In Rossing R, Booth S and Zeller D (eds.), *Total marine fisheries extractions by country in the Baltic Sea: 1950–present*. Fisheries Centre Research Reports 18(1). [updated to 2010]

Zeller D, Rossing P, Harper S, Persson L, Booth S and Pauly D (2011) The Baltic Sea: estimates of total fisheries removals 1950–2007. *Fisheries Research* 108: 356–363.

1. Cite as Veitch, L., S. Booth, S. Harper, P. Rossing, and D. Zeller. 2016. Estonia. P. 248 in D. Pauly and D. Zeller (eds.), *Global Atlas of Marine Fisheries: A Critical Appraisal of Catches and Ecosystem Impacts*. Island Press, Washington, DC.

FAEROE ISLANDS[1]

Darah Gibson, Kyrstn Zylich, and Dirk Zeller

Sea Around Us, University of British Columbia, Vancouver, BC, Canada

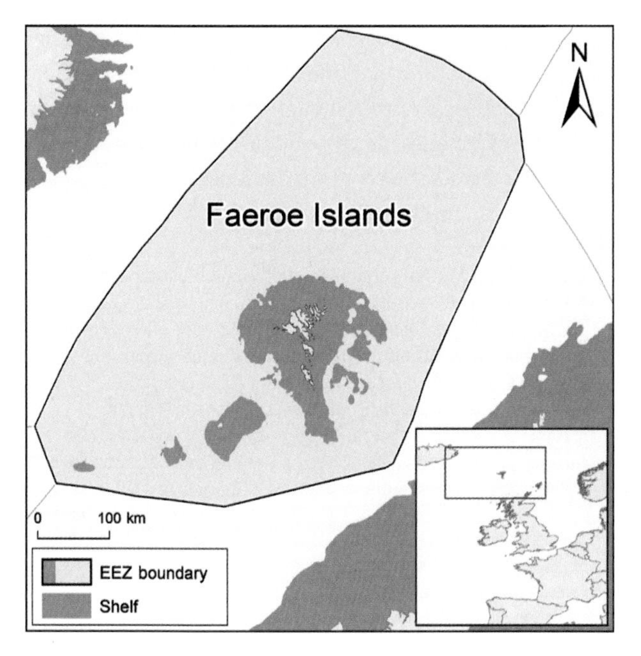

Figure 1. The Faeroes have a total land area of 1,460 km² and an EEZ of 268,000 km².

The Faeroe Islands (figure 1) belong to Denmark but enjoy considerable autonomy, notably in the management of their fisheries (Zeller and Reinert 2004). Most of the catch allocated to Faeroe Islands' EEZ are industrial, with a small but growing artisanal sector (figure 2A). Gibson et al. (2015) performed a reconstruction of Faeroese catches from 1950 to 2010, using statistics from the International Council for the Exploration of the Seas (ICES). Total domestic catches increased from an average of 42,000 t/year in the 1950s to 331,000 t/year in the 2000s, with a peak of 442,000 t in 2003, which is only 1.04 times the officially reported catch by ICES. Unreported catches seem low, because of the discard ban declared in 1994 (Reinert 2001). Foreign fishing by Norway made up the majority of the catch in the earlier time period, with domestic removals dominating catches in the Faeroe Islands EEZ by the end of the time period (figure 2B). Atlantic herring (*Clupea harengus*) and blue whiting (*Micromesistius poutassou*) made up a large fraction of the reconstructed catch (figure 2C). Atlantic cod (*Gadus morhua*), saithe (*Pollachius virens*), and haddock (*Melanogrammus aeglefinus*) remained prominent over the whole time series. The Faeroe Islands have an effective system for managing their commercial fisheries (Johnsen and Eliasen 2011) in consultation with their neighbors, but it will be severely taxed as temperatures rise, causing fish distributions to change.

REFERENCES

Gibson D, Zylich K and Zeller D (2015) Preliminary reconstruction of total marine fisheries catches for the Faeroe Islands in EEZ-equivalent waters (1950–2010). Fisheries Centre Working Paper #2015–36, 12 p.

Johnsen JP and Eliasen S (2011) Solving complex fisheries management problems: what the EU can learn from the Nordic experiences of reduction of discards. *Marine Policy* 35: 130–139.

Reinert J (2001) Faroese fisheries: discards and non-mandated catches. pp. 126–129 In Zeller D, Watson R and Pauly D (eds.), *Fisheries impacts on North Atlantic ecosystems: catch, effort and national/regional data sets*. Fisheries Centre Research Reports 9(3).

Zeller D and Reinert J (2004) Modelling spatial closures and fishing effort restrictions in the Faroe Islands marine ecosystem. *Ecological Modelling* 172: 403–420.

1. Cite as Gibson, D., K. Zylich, and D. Zeller. 2016. Faeroe Islands. P. 249 in D. Pauly and D. Zeller (eds.), *Global Atlas of Marine Fisheries: A Critical Appraisal of Catches and Ecosystem Impacts*. Island Press, Washington, DC.

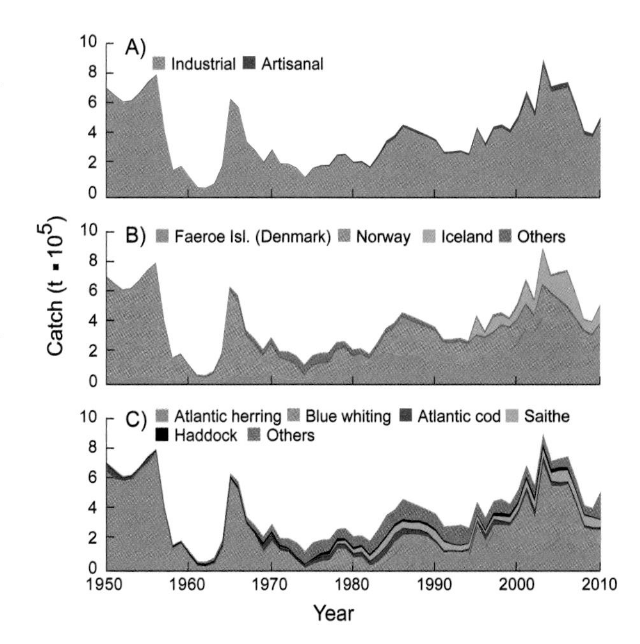

Figure 2. Domestic and foreign catches taken in the EEZ of Faeroe Islands: (A) by sector (domestic scores: Ind. 2, 3, 4; Art. 2, 3, 4; Sub. 1, 1, 1; Rec. 1, 1, 1; Dis. 1, 1, 1); (B) by fishing country (foreign catches are very uncertain); (C) by taxon.

FIJI[1]

Kyrstn Zylich,[a] Devon O'Meara,[a] Jennifer Jacquet,[a,b] Sarah Harper,[a] and Dirk Zeller[a]

[a]*Sea Around Us*, University of British Columbia, Vancouver, Canada [b]Department of Environmental Studies, New York University, New York, NY

The Republic of Fiji is an archipelago in the southwest Pacific Ocean (figure 1). Fiji's fisheries have undergone many changes in recent decades, and urbanization, technological innovations, and increased government incentives (e.g., subsidies, loans) have shaped their development. Although the subsistence sector dominates the total catch allocated to Fiji's EEZ, it has been gradually declining, while the industrial and artisanal sectors have increased (figure 2A). The total reconstructed domestic catch for Fiji, which was 40,000 t/year in the early 1950s, peaked at 48,000 t/year in the early 1970s, and was about 28,000 t/year in the 2000s, is 3 times the landings presented by the FAO on behalf of Fiji (Zylich et al. 2012). This discrepancy is due largely to much of the earlier subsistence catches being unreported. The subsistence catch, taken mainly by women (Vunisea 2005), has declined in recent times, with the reconstructed total catches being only 9% larger than the data reported since 2000 to the FAO. Although the catch is predominantly domestic, foreign vessels are fishing in Fiji's EEZ (figure 2B). The reconstructed catch is dominated by emperors and scavengers (family Lethrinidae), goatfishes (Mullidae), clams (Bivalvia), and parrotfishes (Scaridae; figure 2C). Surgeonfishes (Acanthuridae) are also prominent in the catch of reef fisheries, whose traditional management may need to be reestablished where it has faded away.

REFERENCES

Vunisea A (2005) Women's changing roles in the subsistence fishing sector in Fiji. pp. 89–105 In Novaczek I, Mitchell J and Vietayaki J (eds.), *Pacific Voices: Equity and Sustainability in Pacific Island Fisheries*. Institute of Pacific Studies, University of the South Pacific, Suva.

Zylich K, O'Meara D, Jacquet J, Harper S and Zeller D (2012) Reconstruction of marine fisheries catches for the Republic of Fiji (1950–2009). pp. 25–36 In Harper S, Zylich K, Boonzaier L, Le Manach F, Pauly D and Zeller D (eds.), *Fisheries catch reconstructions: Islands, Part III*. Fisheries Centre Research Reports 20(5).

1. Cite as Zylich, K., D. O'Meara , J. Jacquet, S. Harper, and D. Zeller. 2016. Fiji. P. 250 in D. Pauly and D. Zeller (eds.), *Global Atlas of Marine Fisheries: A Critical Appraisal of Catches and Ecosystem Impacts*. Island Press, Washington, DC.

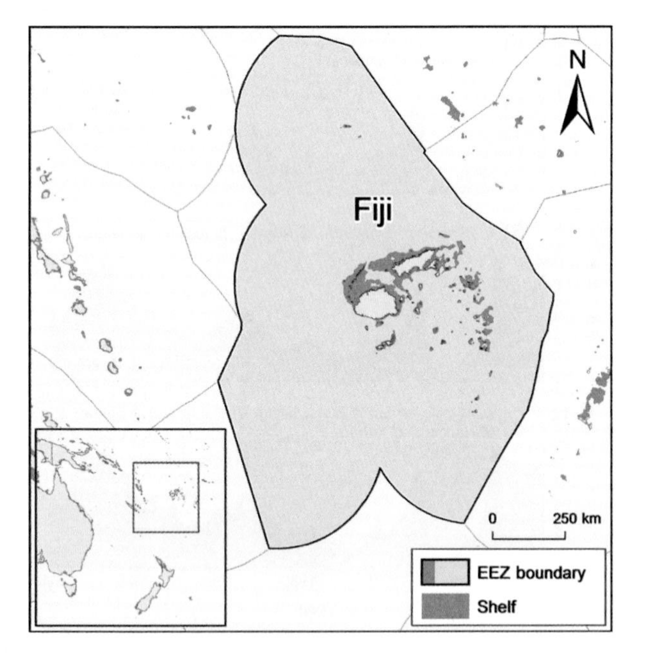

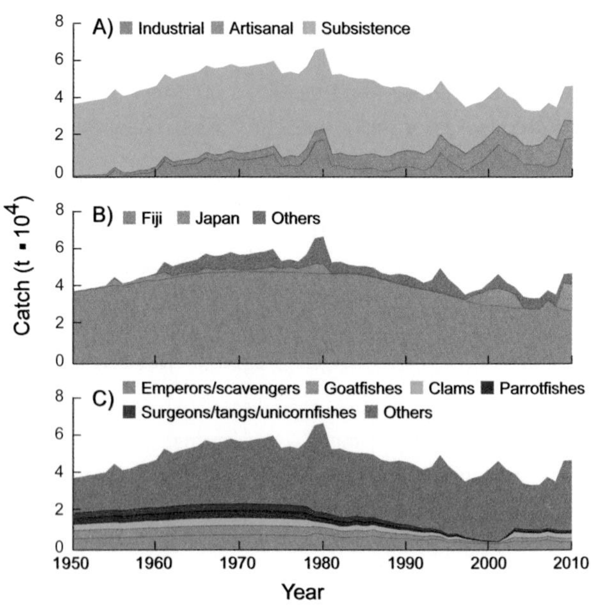

Figure 1. The Republic of Fiji has a land area of 18,000 km² and an EEZ of 1.28 million km², with the two largest islands, Viti Levu and Vanua Levu. **Figure 2.** Domestic and foreign catches taken in the EEZ of Fiji: (A) by sector (domestic scores: Art. 1, 1, 3; Sub. 2, 2, 3); (B) by fishing country (foreign catches are very uncertain); (C) by taxon.

FINLAND[1]

Peter Rossing, Sarah Bale, Sarah Harper, and Dirk Zeller

Sea Around Us, University of British Columbia, Vancouver, Canada

The Nordic Republic of Finland has coastal borders along the Gulf of Finland and the Baltic Sea proper (figure 1). The catches allocated to Finland's EEZ are predominantly from the industrial sector, which appears to have grown through the time period (figure 2A). The reconstruction of Finnish catches in their EEZ, based on Rossing et al. (2010), yielded estimates of total domestic withdrawals of 20,000 t/year in the early 1950s, a peak of 88,000 t in 1996, and catches of about 70,000 t/year in the late 2000s. These reconstructed catches, based on data from various sources, were 1.2 times the data officially reported by ICES of behalf of Finland (and deemed to have been taken within the Finish EEZ). Although the catch is mostly domestic, foreign fishing by other Baltic countries was noticeable toward the end of the period considered here (figure 2B). It should be mentioned that since 1953, Finland has included recreational catches in its official catch statistics, to our knowledge the only country in the world to do so. However, Finland could allocate more resources to decreasing the amount of unreported industrial catches and discards (Zeller et al. 2011). Figure 2C documents the taxonomic composition of the catch, in which Atlantic herring (*Clupea harengus*) and European sprat (*Sprattus sprattus*) dominate, reflecting the low species diversity of the northern Baltic.

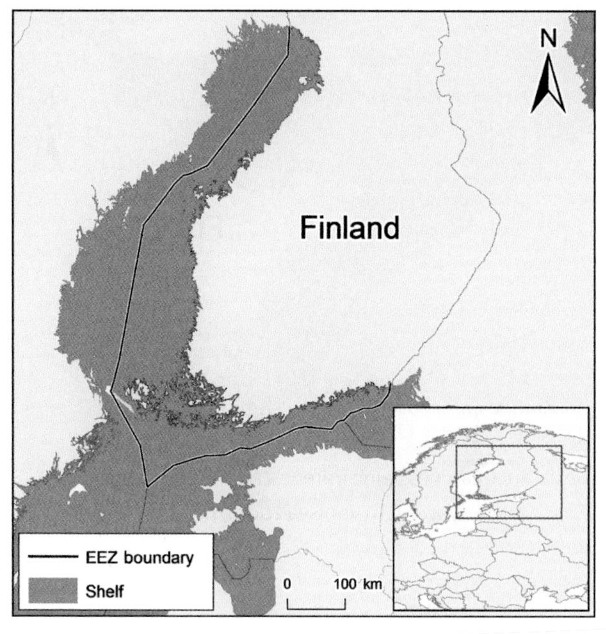

Figure 1. Finland and its Baltic Sea EEZ (81,500 km², of which nearly all is shelf).

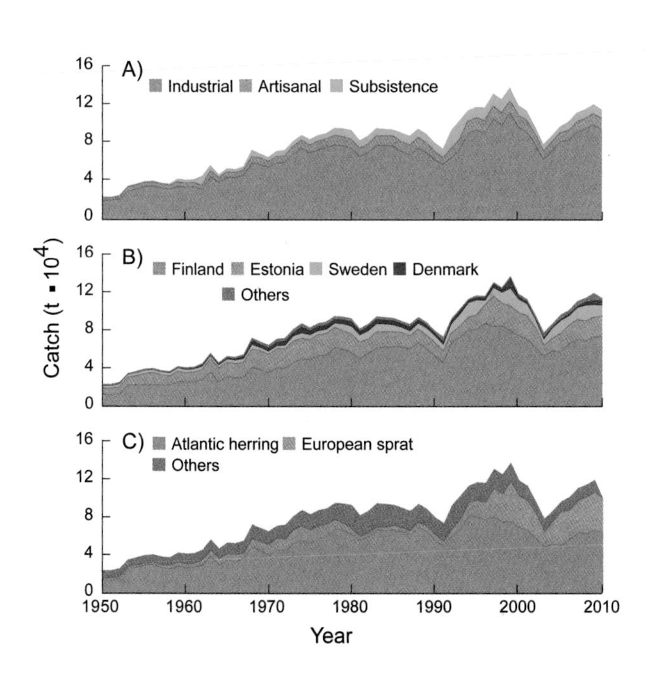

Figure 2. Domestic and foreign catches taken in the Baltic EEZ of Finland: (A) by sector (domestic scores: Ind. 2, 3, 3; Art. 2, 3, 3; Rec. 2, 3, 3; Dis. 2, 3, 3); (B) by fishing country (foreign catches are very uncertain); (C) by taxon.

REFERENCES

Rossing P, Bale S, Harper S and Zeller D (2010) Baltic Sea fisheries catches for Finland (1950–2007). pp. 85–106 In Rossing R, Booth S and Zeller D (eds.), *Total marine fisheries extractions by country in the Baltic Sea: 1950–present*. Fisheries Centre Research Reports 18(1). [Updated to 2010]

Zeller D, Rossing P, Harper S, Persson L, Booth S and Pauly D (2011) The Baltic Sea: estimates of total fisheries removals 1950–2007. *Fisheries Research* 108: 356–363.

1. Cite as Rossing, P., S. Bale, S. Harper, and D. Zeller. 2016. Finland. P. 251 in D. Pauly and D. Zeller (eds.), *Global Atlas of Marine Fisheries: A Critical Appraisal of Catches and Ecosystem Impacts*. Island Press, Washington, DC.

FRANCE (ATLANTIC COAST)[1]

Elise Bultel,[a] Didier Gascuel,[b] Frédéric Le Manach,[a] Daniel Pauly,[a] and Krystn Zylich[a]

[a]*Sea Around Us*, University of British Columbia, Vancouver, Canada [b]Ecologie et Santé des Ecosystèmes, Université Européenne de Bretagne, Rennes, France

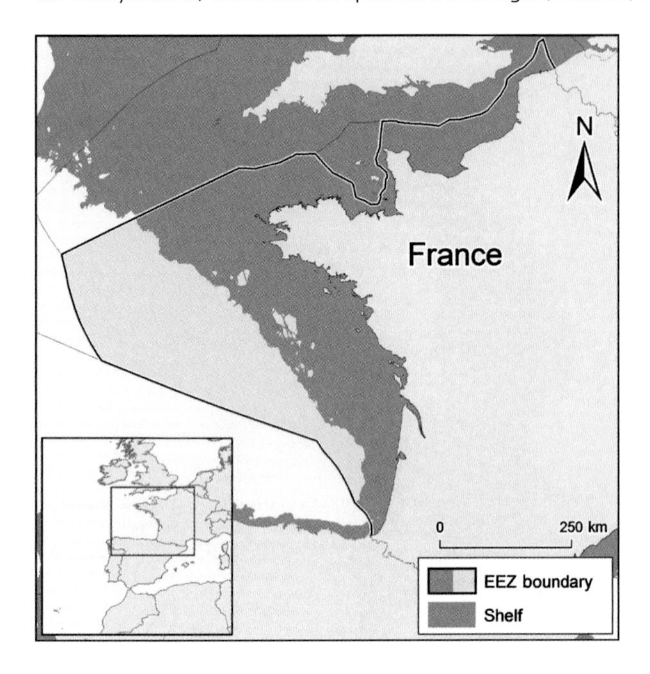

France has the third highest marine catch in Europe (Daurès et al. 2011) and an old fishing culture (particularly along its Atlantic coast; figure 1). However, there is no system in place to record all fisheries withdrawals from the French EEZ. Rather, industrial landings are emphasized, and much of the small-scale fishery catch remains outside the national recording system (Le Guilloux and Pauly 2010), as are discards. The major component of the total catch allocated to the Atlantic's EEZ of France are industrial landings and their discards, and the artisanal sector comes in second place (figure 2A). Bultel et al. (2015) reconstructed total French withdrawals from within the French Atlantic EEZ, which were about 5,200 t in the 1950s, then decreased to 17,000 t/year in the 1960s before rapidly increasing 240,000 t/year in the late 2000s, despite a substantial drop in 1981 and 1982 (to 100,000 t/year). It was possible to estimate that total catches from the French Atlantic EEZ, that is, by the industrial, artisanal, and recreational sectors and in the form of discards, were 1.5 times the official landings adjusted to the same area. Most of the catches are domestic; however, foreign fishing, by Denmark and the United Kingdom, for example, occurred throughout the time period (figure 2B). Figure 2C documents the gross composition of these catches, with cods and haddocks (family Gadidae), clams (Bivalvia), and herrings (Clupeiformes) being important.

REFERENCES

Bultel E, Gascuel D, Le Manach F, Pauly D and Zylich K (2015) Catch reconstruction for the French Atlantic coasts, 1950–2010. Fisheries Centre Working Paper #2015–37, 20 p.

Daurès F, Vignot C, Jacob C, Desbois Y, Le Grand C, Léonardi S, Guyader O, Macher C, Demanèche S, Leblond E and Berthou P (2011) Pêche professionnelle: V2bis. Analyse Economique et Sociale, IFREMER. 14 p.

Le Guilloux E and Pauly D (2010) Description synthétique des pêcheries françaises en 2007. Prepared for the French Ministry of Fisheries. Bloom Association and *Sea Around Us*, with the support of IFREMER, IRD and MNHN, Paris (France), 32 p.

1. Cite as Bultel, E., D. Gascuel, F. Le Manach, D. Pauly, and K. Zylich. 2016. France (Atlantic Coast). P. 252 in D. Pauly and D. Zeller (eds.), *Global Atlas of Marine Fisheries: A Critical Appraisal of Catches and Ecosystem Impacts*. Island Press, Washington, DC.

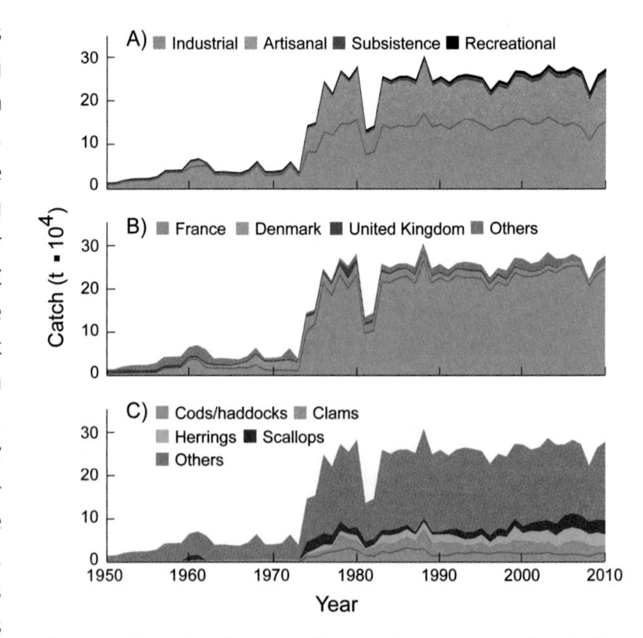

Figure 1. The Atlantic coast of France has an extensive shelf (143,000 km²) for an EEZ of 245,000 km². **Figure 2.** Domestic and foreign catches taken in the Atlantic EEZ of France: (A) by sector (domestic scores: Ind. 3, 3, 3; Art. 3, 3, 3; Sub. 1, 1, 1; Rec. 2, 2, 2; Dis. 2, 2, 2); (B) by fishing country (foreign catches are very uncertain); (C) by taxon.

FRANCE (CLIPPERTON ISLAND)[1]

Daniel Pauly

Sea Around Us, University of British Columbia, Vancouver, Canada

Clipperton Island, named after an English privateer, is a French territory in the eastern Central Pacific, about 1,300 km southwest of Acapulco, Mexico (figure 1). The geology and biology of Clipperton Island, including its coral reefs and their associated fauna, are well documented (see contributions in Charpy 2009), with the fish fauna showing a high degree of endemism (6.6%; Fourriére et al. 2014). Although there was an attempt at settlement, the island is uninhabited at present, and surveillance by the French Navy is scattered to nonexistent. The United States dominated total removals before EEZ declaration in 1978; thereafter, most of the catches were taken by vessels from Latin American countries (figure 2A). Diving for lobsters and shark finning also appear to occur near Clipperton Atoll, but the extent of these activities will probably remain highly uncertain. Thus, the catches made by various countries in the Clipperton Island EEZ must be estimated indirectly, that is, by "allowing" all countries that report catches of large pelagic fishes from the eastern Central Pacific (FAO area 77) to enter the Clipperton Island EEZ (Pauly 2009). Yellowfin tuna (*Thunnus albacares*) accounted for the majority of the catch from the EEZ. Skipjack tuna (*Katsuwonus pelamis*) and squids are also taken in the Clipperton area (figure 2B), but the catch of several taxa appears to have declined.

REFERENCES

Charpy L, editor (2009) Clipperton: Environnement et biodiversité d'un microcosme océanique. Muséum national d'histoire naturelle, Paris, and Institut de recherche pour le développement (IRD), Marseille (France). 420 p.

Fourriére M, Reyes-Bonilla H, Rodríguez-Zaragoza F and Crane N (2014) Fishes of Clipperton Atoll, Eastern Pacific: checklist, endemism, and analysis of completeness of the inventory. *Pacific Science* 68(3): 375–395.

Pauly D (2009) The fisheries resources of the Clipperton Island EEZ (France). pp. 35–37 In Zeller D and Harper S (eds.), *Fisheries catch reconstructions: Islands, Part I*. Fisheries Centre Research Reports 17(5).

1. Cite as Pauly, D. 2016. Clipperton Island. P. 253 in D. Pauly and D. Zeller (eds.), *Global Atlas of Marine Fisheries: A Critical Appraisal of Catches and Ecosystem Impacts*. Island Press, Washington, DC.

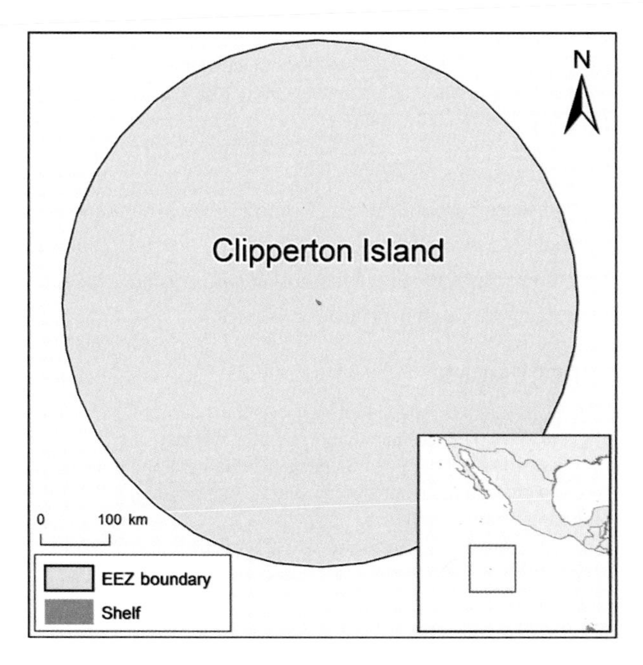

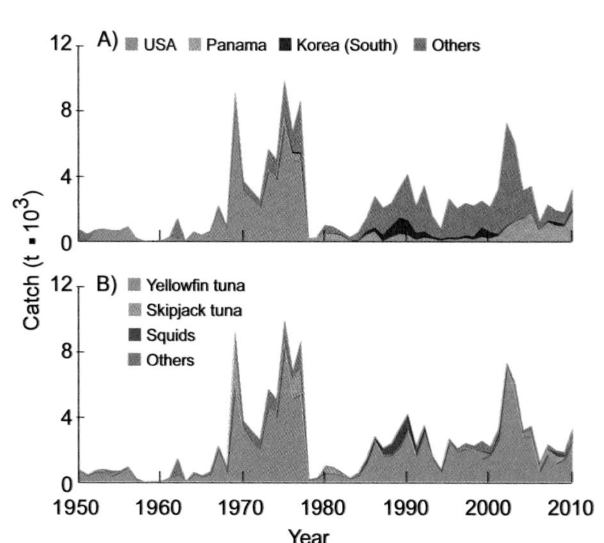

Figure 1. The land area of Clipperton Island is 8.8 km², and its EEZ has a surface area of 431,000 km² (www.clipperton.fr/).
Figure 2. Foreign catches taken in the EEZ of Clipperton Island: (A) by fishing country (foreign catches are very uncertain); (C) by taxon.

FRANCE (CORSICA)[1]

Frédéric Le Manach,[a] Delphine Dura,[a] Anthony Pere,[b,c] Jean-Jacques Riutort,[d] Pierre Lejeune,[b] Marie-Catherine Santoni,[e] Jean-Michel Culioli,[e] and Daniel Pauly[a]

[a]*Sea Around Us*, University of British Columbia, Vancouver, Canada [b]Station de Recherches Sous-marines et Océanographiques, Calvi, France [c]Faculté des Sciences et Techniques, Université de Corse, Corte, France [d]Bastia Offshore Fishing Association, Furiani, France [e]Office de l'Environnement Corse, Ajaccio France

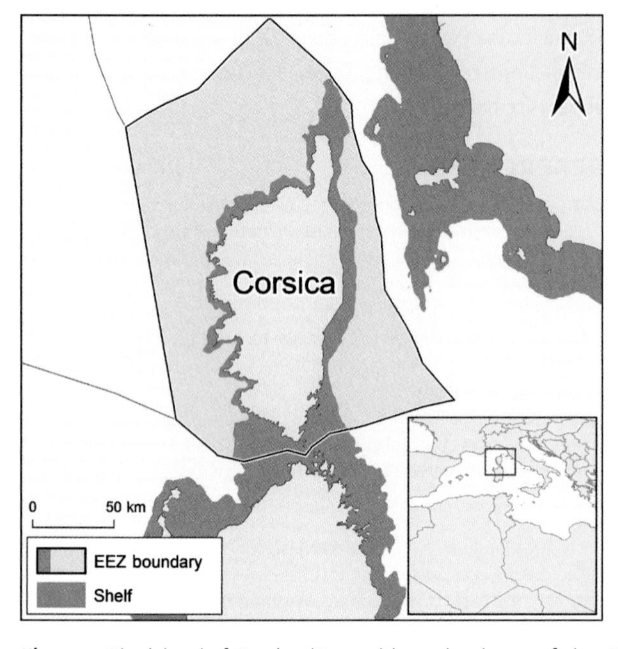

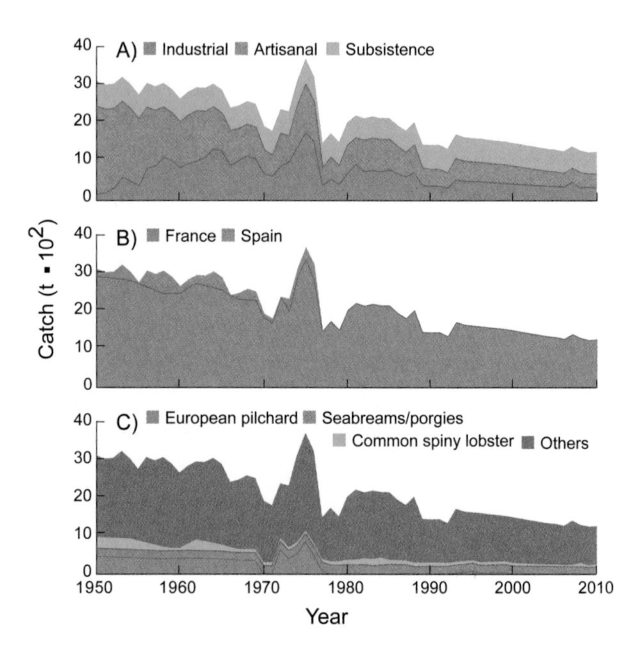

Figure 1. The island of Corsica (France) has a land area of about 8,700 km² and EEZ-equivalent waters of 24,000 km².
Figure 2. Domestic and foreign catches taken in the EEZ of Corsica (France): (A) by sector (domestic scores: Ind. 3, 4, 4; Art. 2, 3, 3; Rec. 2, 2, 2; Dis. 2, 3, 3); (B) by fishing country (foreign catches are very uncertain); (C) by taxon.

Corsica is located southeast of the French mainland and west of Italy (figure 1). The island, known as the birthplace of Napoléon Bonaparte, as well as for its historic and linguistic ties to Italy, is flanked by deep water along its west coast and a broad shelf along its east coast. Although the artisanal sector was predominant in the 1950s to 1960s, most catches from Corsica's EEZ are from the industrial sector (figure 2A). Other smaller and poorly documented subsistence and recreational fisheries also occur, but overall fishing pressure is low, and the number of full-time fishers is declining (Pere 2012). The total reconstructed catch from 1950 was about 2,800 t/year in the early 1950s but declined to about 1,300 t/year in the late 2000s, which corresponds, overall, to about 5 times the values reported by France to the FAO. The majority of the catch is domestic; however, foreign fishing was documented before the EEZ declaration in 1977 (figure 2B). Field investigations are needed to improve on these data, which were updated by Le Manach and Pauly (2015) from Le Manach et al. (2011) and presented here as the first approximation of total extractions from the waters around Corsica. Figure 2C shows that European pilchards (*Sardina pilchardus*), seabreams, porgies (*Diplodus* spp.), and common spiny lobster (*Palinurus elephas*) account for most of the catch from around Corsica.

REFERENCES

Le Manach F, Dura D, Pere A, Riutort JJ, Lejeune P, Santoni MC, Culioli JM and Pauly D (2011) Preliminary estimates of total fisheries catch in Corsica, France (1950–2008). pp. 3–14 In Harper S and Zeller D (eds.), *Fisheries catch reconstruction: Islands, Part II*. Fisheries Centre Research Reports 19(4).

Le Manach F and Pauly D (2015) Update of the fisheries catch reconstruction of Corsica (France), 1950–2010. Fisheries Centre Working Paper #2015-33, 5 p.

Pere, A (2012) Déclin des populations de langouste rouge et baisse de la ressource halieutique en Corse: causes et perspectives. PhD thesis, Université de Corse, Corte, France. 478 p.

1. Cite as Le Manach, F., D. Dura, A. Pere, J. J. Riutor, P. Lejeune, M. C. Santoni, J. M. Culioli, and D. Pauly. 2016. France (Corsica). P. 254 in D. Pauly and D. Zeller (eds.), *Global Atlas of Marine Fisheries: A Critical Appraisal of Catches and Ecosystem Impacts*. Island Press, Washington, DC.

FRANCE (CROZET ISLANDS)[1]

Patrice Pruvost,[a] Guy Duhamel,[a] Nicolas Gasco,[a] and Maria Lourdes D. Palomares[b]

[a]Muséum National d'Histoire Naturelle, Paris, France [b]*Sea Around Us*, University of British Columbia, Vancouver, Canada

The Crozet Islands are a small sub-Antarctic archipelago (figure 1), a part of the French Antarctic and sub-Antarctic Territories (TAAF), which also include the islands of Kerguelen, St. Paul, and Amsterdam (www.taaf.fr). The EEZ of the Crozet Islands is included in the purview of the Commission for the Conservation of Antarctic Marine Living Resources (CCAMLR) as part of its subarea 58.6, and its fish fauna is well documented (Duhamel et al. 2005). The fisheries resources of the uninhabited Crozet Islands were first explored in the mid-1960s, and distant-water fisheries began operating occasionally since the late 1970s, and more regularly since 1996. Catches from within the Crozet Islands EEZ were taken by domestic (French) vessels and South African fleets in the 2000s, although in 1978, some catch by Japan was documented (figure 2A). The catch data used for this account, based on Pruvost et al. (2015), indicate that in the last decade, domestic catches have been on the order of 1,000 t per season or year, with a declining trend, but were higher in the late 1990s, when the Crozet Island EEZ experienced higher levels of fishing pressure. Figure 2B shows that the target species of the longline fishing fleet is the Patagonian toothfish (*Dissotichus eleginoides*), and the bycatch, notably grenadiers (*Macrourus* spp.), previously discarded (but wisely recorded by CCAMLR) is increasingly retained, one sign of resource decline.

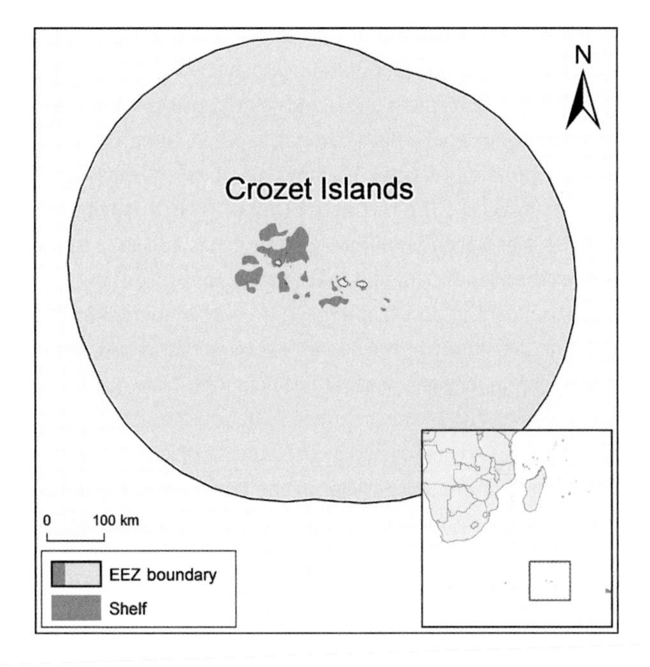

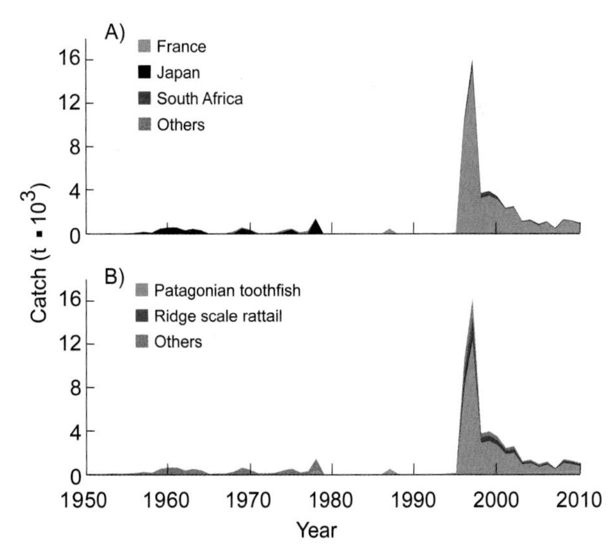

Figure 1. The Crozet Islands have a land area of 350 km² and an EEZ of 575,000 km². **Figure 2.** Domestic and foreign catches taken in the EEZ of Crozet Islands (France): (A) by fishing country (domestic scores: Ind. –, 3, 4; foreign catches are very uncertain); (B) by taxon.

REFERENCES

Duhamel G, Gasco N and Davaine P (2005) Poissons des îles Kerguelen et Crozet. Guide régional de l'océan Austral. Muséum national d'Histoire naturelle/Patrimoine naturels 63: 419 p.

Pruvost P, Duhamel G, Gasco N and Palomares MLD (2015) A short history of the fisheries of Crozet Islands. pp. 30–35 In Palomares MLD and Pauly D (eds.), *Marine fisheries catches of sub-Antarctic islands, 1950 to 2010*. Fisheries Centre Research Reports 23(1).

1. Cite as Pruvost, P., G. Duhamel, N. Gasco, and M. L. D. Palomares. 2016. France (Crozet Islands). P. 255 in D. Pauly and D. Zeller (eds.), *Global Atlas of Marine Fisheries: A Critical Appraisal of Catches and Ecosystem Impacts*. Island Press, Washington, DC.

FRANCE (FRENCH POLYNESIA)[1]

Sarah Bale,[a] Lou Frotté,[b] Sarah Harper,[a] and Dirk Zeller[a]

[a]*Sea Around Us*, University of British Columbia, Vancouver, Canada [b]Muséum National d'Histoire Naturelle, Station Marine de Concarneau, France

French Polynesia is a French overseas collectivity of more than 100 islands (population: 280,000 in 2010), has the largest EEZ among all Pacific islands, and consists of five main groups: the Society (with the capital, Papetee, comprising 70% of the population), Austral, Tuamotu, Gambier, and Marquesas islands (figure 1). The reconstruction of French Polynesia's catches was based on different approaches and data types in each island group (details in Bale et al. 2009). Considerable industrial fishing occurs in French Polynesia, in addition to growing artisanal and subsistence sectors (figure 2A). Most of the foreign catches, notably a peak in 1978, are by Japanese vessels (figure 2B). The reconstructed catches were about 5,000 t/year in the early 1950s and increased to 12,000 t/year in the early 1990s and 14,000 t/year in the late 2000s. Although catches originally were dominated by the subsistence sector, with about 6,000 t/year in the 2000s, commercial fisheries, propelled by a government initiative called the *pacte du progrès*, shot up in the 1990s, (figure 2A; Walker and Robinson 2009), mainly because of the development and offshore expansion of a domestic tuna fishery, to which the vast expanses of the EEZ of French Polynesia lend themselves (see also Zeller et al. 2015). Figure 2C presents the composition of the catch, which includes albacore tuna (*Thunnus alalunga*), bigeye tuna (*Thunnus obesus*), and bigeye scad (*Selar crumenophthalmus*).

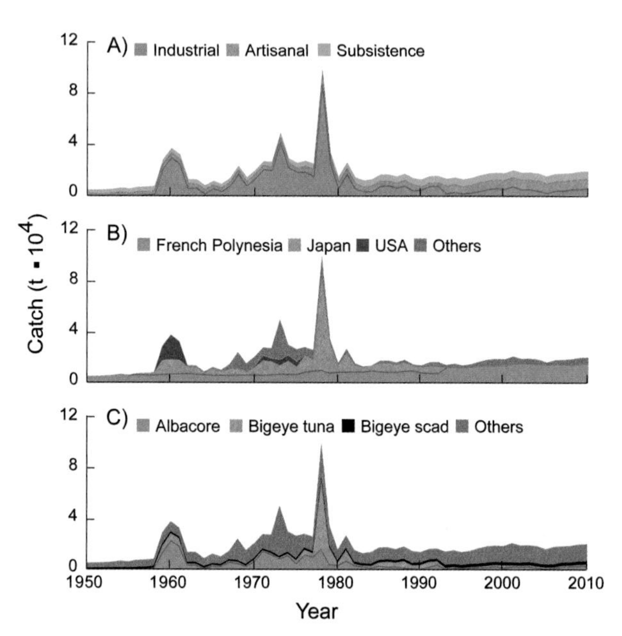

Figure 2. Domestic and foreign catches taken in the EEZ of French Polynesia: (A) by sector (domestic scores: Ind. –, 2, 3; Art. 2, 2, 3; Sub. 1, 1, 2); (B) by fishing country (foreign catches are very uncertain); (C) by taxon.

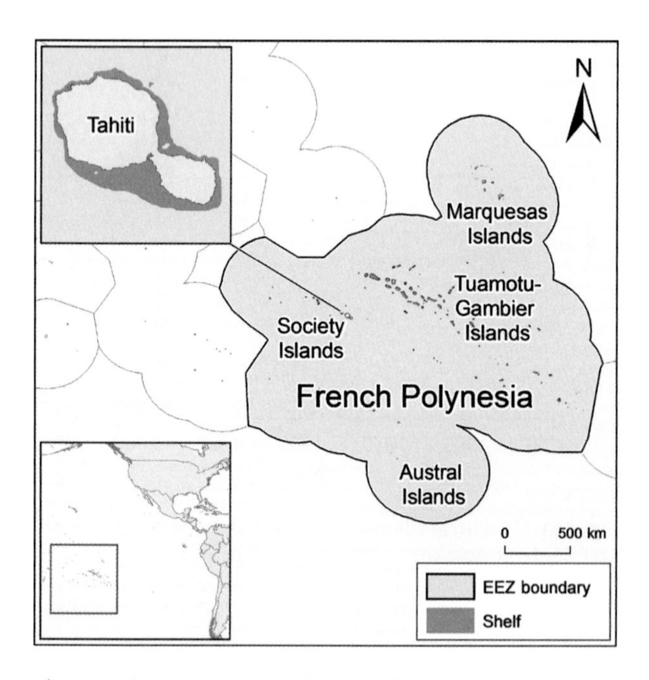

Figure 1. The more than 100 islands of French Polynesia have, jointly, a land area of 4,200 km² and an EEZ of nearly 4.8 million km².

REFERENCES

Bale S, Frotté L, Harper S and Zeller D (2009) Reconstruction of total marine fisheries catches for French Polynesia. pp. 53–65 In Zeller D and Harper S (eds.), *Fisheries catch reconstructions: Islands, Part I*. Fisheries Centre Research Reports 17(5).

Walker B and Robinson M (2009) Economic development, marine protected areas and gendered access to fishing resources in a Polynesian lagoon. *Gender, Place and Culture* 16(4): 467–484.

Zeller D, Harper S, Zylich K and Pauly D (2015) Synthesis of under-reported small-scale fisheries catch in Pacific-island waters. *Coral Reefs* 34(1): 25–39.

1. Cite as Bale, S., L. Frotté, S. Harper, and D. Zeller. 2016. France (French Polynesia). P. 256 in D. Pauly and D. Zeller (eds.), *Global Atlas of Marine Fisheries: A Critical Appraisal of Catches and Ecosystem Impacts*. Island Press, Washington, DC.

FRANCE (GUADELOUPE)[1]

Lou Frotté,[a] Sarah Harper,[b] Liane Veitch,[b] Shawn Booth,[b] and Dirk Zeller[b]

[a]Muséum National d'Histoire Naturelle, Station Marine de Concarneau, France [b]*Sea Around Us*, University of British Columbia, Vancouver, Canada

Guadeloupe is a French "Overseas Department" in the eastern Caribbean Sea, consisting of two main islands, Basse-Terre and Grande-Terre (figure 1). A large fraction of the catch is derived from the growing artisanal sector, and subsistence catches appear to be declining (figure 2A). A reconstruction of total domestic catches (Frotté et al. 2009; see also Desse 1989), combining commercial landings with unreported subsistence catches based on seafood consumption and trade data, and recreational catch estimates yielded catch estimates of 3,000 t/year in the early 1950s and about 12,000 t/year in the 2000s. This was 1.4 times the data reported by the FAO on behalf of Guadeloupe. However, our estimates of subsistence catches may be too low. Although catches are mostly domestic, foreign fishing has been documented to occur in Guadeloupe's EEZ, although the origin of the fleets generating these catches is largely unknown (figure 2B). Figure 2C presents a possible catch composition, adapted from that of nearby St. Lucia, which would make jacks (family Carangidae) the most important single taxon in the catch of Guadeloupe. Also, it is known that commercial vessels from other countries outside the Caribbean exploit fish stocks within the EEZs of this region, without the consent of the islands' governments. As local and tourism-based demand for fish is increasing rapidly, the research agencies concerned (e.g., IFREMER in Guadeloupe's case), have their work cut out for them.

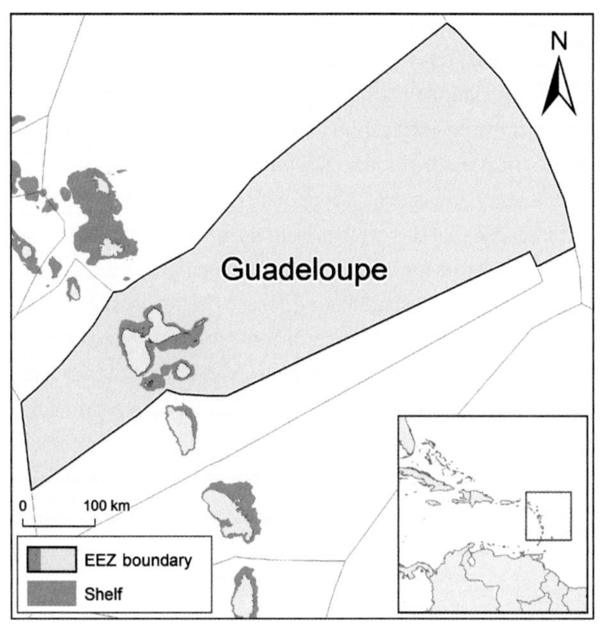

Figure 1. Guadeloupe has a combined land area of 1,600 km^2 and an EEZ of 90,600 km^2.

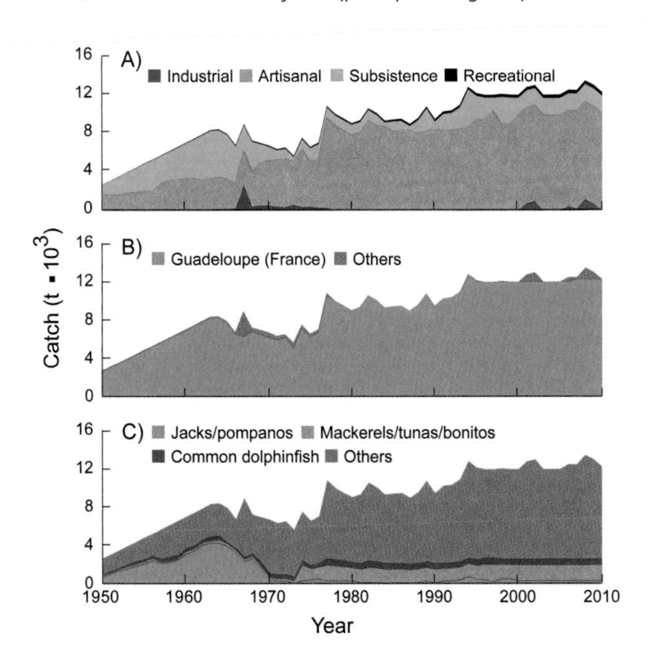

Figure 2. Domestic and foreign catches taken in the EEZ of Guadeloupe: (A) by sector (domestic scores: Art. 2, 3, 2; Sub. 1, 1, 1; Rec. 1, 1, 2); (B) by fishing country (foreign catches are very uncertain); (C) by taxon.

REFERENCES

Desse M (1989) Bilan des activités halieutiques dans les départements d'outre-mer de la Caraïbe. *La Pêche Maritime* 1328: 110–115.

Frotté L, Harper S, Veitch L, Booth S and Zeller D (2009) Reconstruction of marine fisheries catches for Guadeloupe from 1950–2007. pp. 13–19 In Zeller D and Harper S (eds.), *Fisheries catch reconstructions: Islands, part I*. Fisheries Centre Research Reports 17(5). [Updated to 2010]

1. Cite as Frotté, L., S. Harper, L. Veitch, S. Booth, and D. Zeller. 2016. France (Guadeloupe). P. 257 in D. Pauly and D. Zeller (eds.), *Global Atlas of Marine Fisheries: A Critical Appraisal of Catches and Ecosystem Impacts*. Island Press, Washington, DC.

FRANCE (GUIANA)[1]

Sarah Harper,[a] Lou Frotté,[b] Shawn Booth,[a] Liane Veitch,[a] and Dirk Zeller[a]

[a]*Sea Around Us*, Fisheries Centre, University of British Columbia, Vancouver, Canada [b]Muséum National d'Histoire Naturelle, Station Marine de Concarneau, France

French Guiana is a French overseas department on the northeast coast of South America (figure 1). It is an ideal location for launching the EU's geostationary satellites, and it is thus the home of "Europe's Spaceport." French Guiana has a continental shelf strongly affected by the outflow of the Amazon River (Artigas et al. 2003), which is exploited by massive fisheries. Most of the catch is from the growing industrial and artisanal sectors (figure 2A). Also, an informal sector operates near shore, for subsistence purposes. The reconstruction by Harper et al. (2015) yielded total domestic catch estimates of 600 t/year for the 1950s, a peak of 4,300 t in 1995, and 3,500 t/year for the late 2000s, which, overall, was 1.4 times the data reported by the FAO on behalf of French Guiana. These include substantial fishing by French vessels as well as foreign fishing by the United States and Venezuela (figure 2B). Figure 2C shows important catches of crustaceans, king weakfish (*Macrodon ancylodon*), as well as bycatch generated by the domestic shrimp trawl fishery, consisting of grunts (Haemulidae), croaker (Sciaenidae), and sea catfishes (Ariidae; figure 2C). Also, there is a prominent trawl fishery (Béné and Moguedet 1998) targeting mainly southern brown shrimp (*Farfantepenaeus subtilis*), a snapper fishery that targets red snapper (*Lutjanus purpureus*), and a coastal small-scale fishery that supplies local markets.

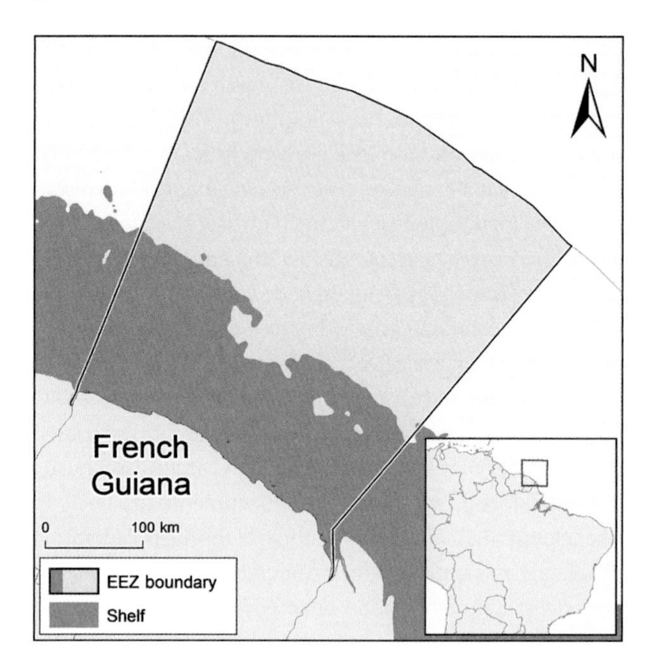

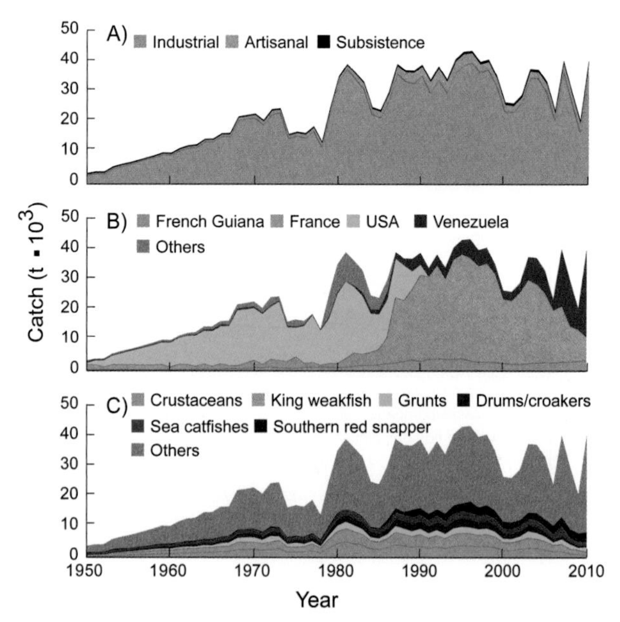

Figure 1. French Guiana has a shelf of 43,100 km² and an EEZ of 135,000 km². **Figure 2.** Domestic and foreign catches taken in the EEZ of French Guiana: (A) by sector (domestic scores: Art. 2, 3, 2; Sub. 1, 1, 1); (B) by fishing country (foreign catches are very uncertain); (C) by taxon.

REFERENCES

Artigas LF, Vendeville P, Leopold M, Guiral D and Ternon JF (2003) Marine biodiversity in French Guiana: Estuarine, coastal, and shelf ecosystems under the influence of Amazon waters. *Gayana* 67(2): 302–326.

Béné C and Moguedet P (1998) Global and local change: penaeid stocks in French Guyana, pp. 311–327 In M. H. Durand, P. Cury, R. Mendelssohn, A. Bakun, C. Roy and D. Pauly (eds.), *Local versus global changes in upwelling systems*. Séries Colloques et Séminaires, ORSTOM Editions, Paris.

Harper S, Frotté L, Booth S, Veitch L and Zeller D (2015) Reconstruction of marine fisheries catches for French Guiana from 1950–2010. Fisheries Centre Working Paper #2015-07, 11 p.

1. Cite as Harper, S., L. Frotté, S. Booth, L. Veitch, and D. Zeller. 2016. France (Guiana). P. 258 in D. Pauly and D. Zeller (eds.), *Global Atlas of Marine Fisheries: A Critical Appraisal of Catches and Ecosystem Impacts*. Island Press, Washington, DC.

FRANCE (ÎLES ÉPARSES)[1]

Frédéric Le Manach and Daniel Pauly

Sea Around Us, University of British Columbia, Vancouver, Canada

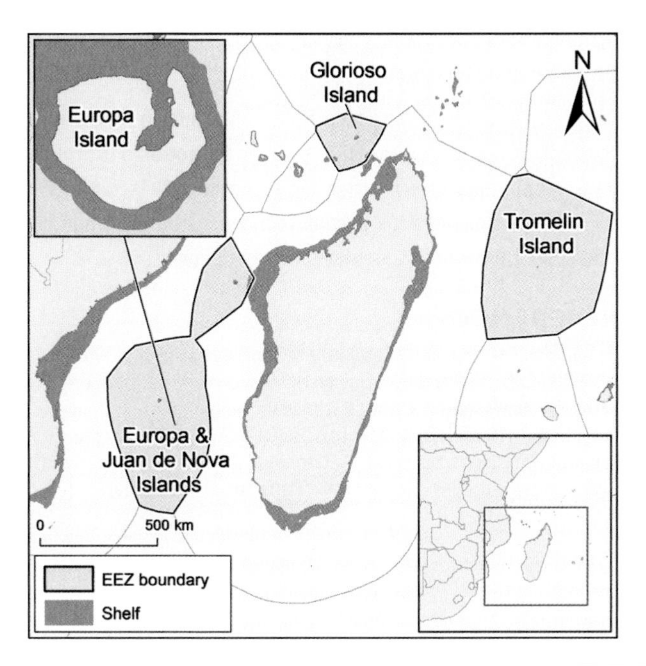

Figure 1. The Îles Éparses (with Tromelin, jointly managed by France and Mauritius) have a total land area of 38 km² and an EEZ of more than 640,000 km².

The Îles Éparses ("Scattered Islands") are located around Madagascar, in the western Indian Ocean. These islands are uninhabited and visited solely by scientists and the French Navy, which also patrols their waters. However, they are of prime importance to French and Spanish industrial fishing fleets targeting tuna and billfishes. They are also a biodiversity hotspot (particularly for birds and turtles, Le Corre and Safford 2001; as well as coral reef fishes, Durville et al. 2003). Catches are overwhelmingly from the domestic fleet, although illegal industrial fishing also occurs (figure 2A). Catches of tuna and billfishes, notably skipjack tuna (*Katsuwonus pelamis*) and yellowfin tuna (*Thunnus albacares*), are prominent in these waters (figure 2B). Since 1997, a small fleet of artisanal fishers from the Comoros Islands has been visiting the banks within the Glorieuses Islands EEZ, with annual catches estimated at less than 100 t in the late 2000s and declining. A surge in illegal activity has recently been observed around Bassas da India (recreational fishers from South Africa and Mozambique). The extent of these illegal activities remains poorly documented, although Le Manach and Pauly (2015) estimated that about 50 t of large pelagics are being caught annually, with an increasing trend. Since 2011, illegal fishers from Madagascar collecting holothurians for the Chinese market around the nominally protected Glorieuses Archipelago have been an increasing problem.

REFERENCES

Durville P, Chabanet P and Quod JP (2003) Visual census of the reef fishes in the natural reserve of the Glorieuses Islands (Western Indian Ocean). *Western Indian Ocean Journal of Marine Science* 2(2): 95–104.

Le Corre M and Safford RJ (2001) La Réunion and Îles Éparses. pp. 693–702 In Fishpool LDC and Evans MI (ed.) Important bird areas in Africa and associated islands: Priority sites for conservation. BirdLife Conservation Series 11. BirdLife International, Cambridge, UK.

Le Manach F and Pauly D (2015) First estimate of unreported catch in the French Îles Éparses, 1950–2010. pp. 27–35 In Le Manach F and Pauly D (eds.), *Fisheries catch reconstructions in the Western Indian Ocean, 1950–2010*. Fisheries Centre Research Reports 23(2).

1. Cite as Le Manach, F., and D. Pauly. 2016. France (Îles Éparses). P. 259 in D. Pauly and D. Zeller (eds.), *Global Atlas of Marine Fisheries: A Critical Appraisal of Catches and Ecosystem Impacts*. Island Press, Washington, DC.

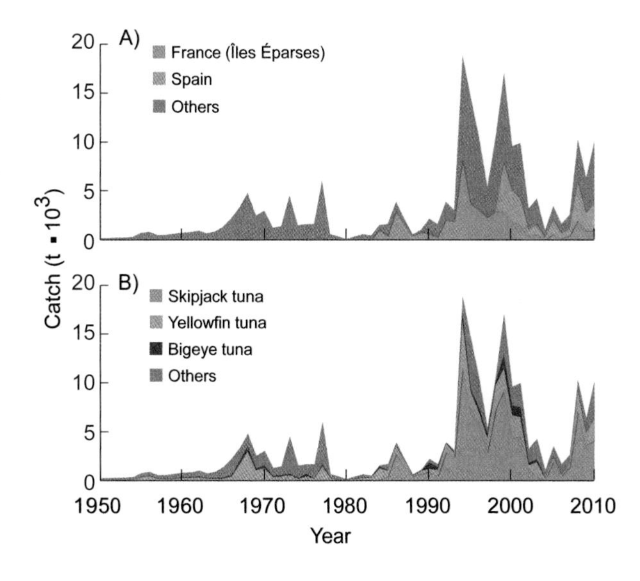

Figure 2. Domestic and foreign catches taken in the EEZ of Îles Éparses (with Tromelin): (A) by fishing country (foreign catches are very uncertain); (B) by taxon.

FRANCE (KERGUELEN ISLANDS)[1]

Maria Lourdes Deng Palomares and Daniel Pauly

Sea Around Us, University of British Columbia, Vancouver, Canada

The uninhabited Kerguelen Islands are part of the French Antarctic and sub-Antarctic Territories (www.taaf.fr) and consist of a main island ("Grande Terre") and a number of surrounding islets (figure 1). Statistics for the distant-water fisheries around the Kerguelen Islands, which began in 1970, were obtained from the Commission for the Conservation of Antarctic Marine Living Resources' *Statistical Bulletin* (area 58.5.1), complemented by statistics reported through the French KERPECHE program (Duhamel et al. 1997). Catches originally expressed in 6-month "seasons" (southern summer) were reexpressed as calendar years, which results in a slight between-season smoothing (Palomares and Pauly 2011). Research cruises in the 1960s, mostly by the former USSR (i.e., the present-day Russian Federation and Ukraine; figure 2A), led to the development of a fishery in the Kerguelen Islands starting in 1970, when 10 Russian bottom trawlers targeted mackerel icefish (*Champsocephalus gunnari*), gray rockcod (*Lepidonotothen squamifrons*), and marbled rockcod (*Notothenia rossii*) without management or control, which lasted until 1980, when France imposed a management scheme. Although illegal fishing continued (and is probably underestimated here), the fishery gradually shifted to a French longlining operation, concentrating, as shown in figure 2B, on Patagonian toothfish (*Dissostichus eleginoides*), with a large bycatch of skates and rays and ridge-scaled rattail (*Macrourous carinatus*), the last gradually replacing Patagonian toothfish as target species.

REFERENCES

Duhamel G, Pruvost P and Capdeville D (1997) By-catch of fish in longline catches off the Kerguelen Islands (Division 58.5.1) during the 1995/1996 season. *CCAMLR Science* 4: 175–193.

Palomares MLD and Pauly D (2011) A brief history of fishing in the Kerguelen Islands. pp. 15–20 In Harper S and Zeller D (eds.), *Fisheries catch reconstructions: Islands, Part II*. Fisheries Centre Research Reports 19(4).

1. Cite as Palomares, M. L. D., and D. Pauly. 2016. France (Kerguelen Islands). P. 260 in D. Pauly and D. Zeller (eds.), *Global Atlas of Marine Fisheries: A Critical Appraisal of Catches and Ecosystem Impacts*. Island Press, Washington, DC.

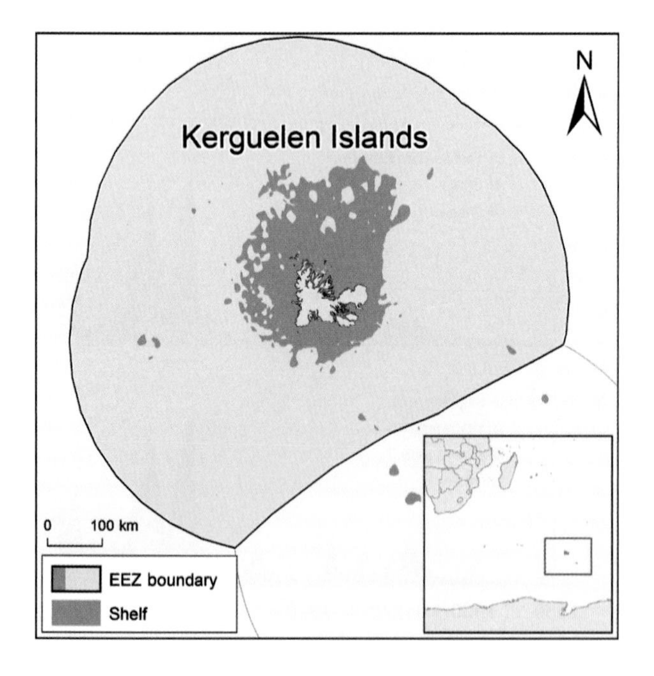

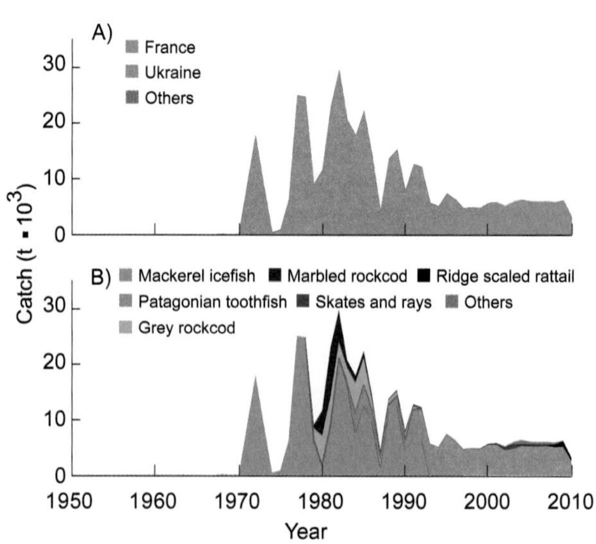

Figure 1. The Kerguelen Islands (France), showing the "Grande Terre" (land area 7,300 km²) and an EEZ of 568,000 km².
Figure 2. Domestic and foreign catches taken in the EEZ of Kerguelen Islands (France): (A) by fishing country (domestic scores: Ind. –, 3, 4; foreign catches are very uncertain); (B) by taxon.

FRANCE (MARTINIQUE)[1]

Lou Frotté,[a] Sarah Harper,[b] Liane Veitch,[b] Shawn Booth,[b] and Dirk Zeller[b]

[a]Muséum National d'Histoire Naturelle, Station Marine de Concarneau, France [b]*Sea Around Us*, Fisheries Centre, University of British Columbia, Vancouver, Canada

Martinique, in the eastern Caribbean Sea (figure 1), is one of four French "overseas departments." Most of the fisheries catch stems from the artisanal sector, with industrial catches plummeting after the declaration of an EEZ in 1978 and again in the mid-1990s (figure 2A). The reconstructed catch from Martinique, adapted from Frotté et al. (2009), was 3,100 t/year in the early 1950s, which increased to about 7,000 t/year in the 1990s and 8,000 t/year in the 2000s. This is 1.4 times the data reported by the FAO on behalf of Martinique and France, mainly because subsistence catches were not part of official statistics. Total allocated catches to the Martinique EEZ are primarily domestic, although foreign fishing by Mexico and the United States was documented before its EEZ declaration and by an unknown country after its EEZ declaration (figure 2B). The composition of this catch, adapted from Gobert (1990), is presented in figure 2C and includes Atlantic bonito (*Sarda sarda*), blackfin tuna (*Thunnus atlanticus*), and clupeoids such as herrings, sardines, and anchovies (figure 2C). Gobert (1994) summarizes this fishery thus: "Most of the catch is made up of species able to withstand high effort, with larger species having been eliminated through overfishing." To ensure that the fishery continues, the reefs must be protected (Bouchon et al. 2014), and the fisheries should be managed for sustainability, not encouraged to expand.

REFERENCES

Bouchon C, Bouchon-Navaro Y, Louis M, Mazeas F, Maréchal JP, Portillo P, Tregarot E and Reef Check (2014) French Antilles, pp. 233–237 In Jackson J, Donovan M, Kramer K and Lam VWY (eds.), *Status and trends of Caribbean coral reefs: 1970–2012*. Global Coral Reef Monitoring Network, IUCN, Gland, Switzerland.

Frotté L, Harper S, Veitch L, Booth S and Zeller D (2009) Reconstruction of marine fisheries catches for Martinique, 1950–2007. pp. 21–26 In Zeller D and Harper S (eds.), *Fisheries catch reconstructions: Islands, Part I*. Fisheries Centre Research Reports 17(5). [Updated to 2010]

Gobert B (1990) Production relative des pêcheries côtières en Martinique. *Aquatic Living Resources* 3: 181–191.

Gobert B (1994) Size structures of demersal catches in a multispecies multigear tropical fishery. *Fisheries Research* 19(1): 87–104.

1. Cite as Frotté, L., S. Harper, L. Veitch, S. Booth, and D. Zeller. 2016. Martinique. P. 261 in D. Pauly and D. Zeller (eds.), *Global Atlas of Marine Fisheries: A Critical Appraisal of Catches and Ecosystem Impacts*. Island Press, Washington, DC.

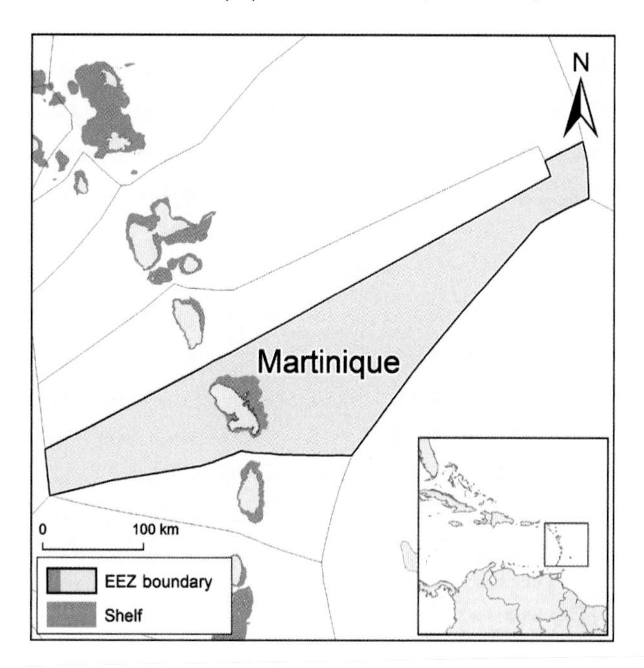

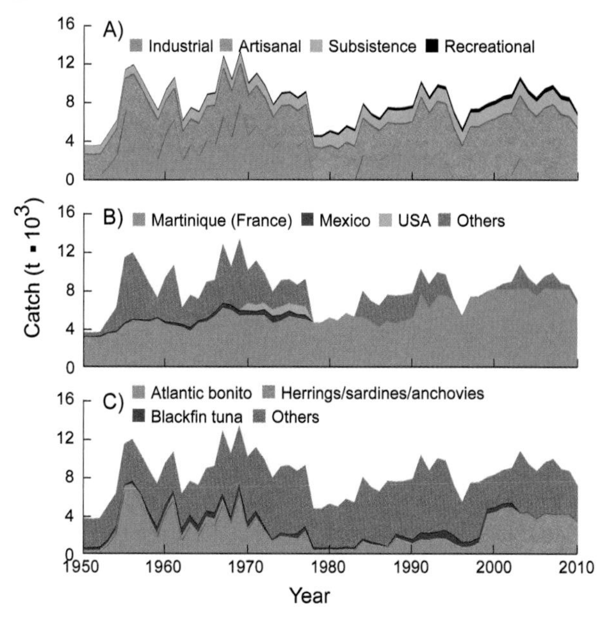

Figure 1. Martinique (land area 1,120 km²) has a small EEZ (47,400 km²) because of the proximity of other islands.
Figure 2. Domestic and foreign catches taken in the EEZ of Martinique: (A) by sector (domestic scores: Art. 2, 3, 2; Sub. 1, 2, 1; Rec. 1, 1, 2); (B) by fishing country (foreign catches are very uncertain); (C) by taxon.

FRANCE (MAYOTTE)[1]

Beau Doherty,[a] Frédéric Le Manach,[a] and Johanna Herfaut[b]

[a]*Sea Around Us*, University of British Columbia, Vancouver, Canada [b]Agence des Aires Marines Protégées, Parc Naturel Marin de Mayotte

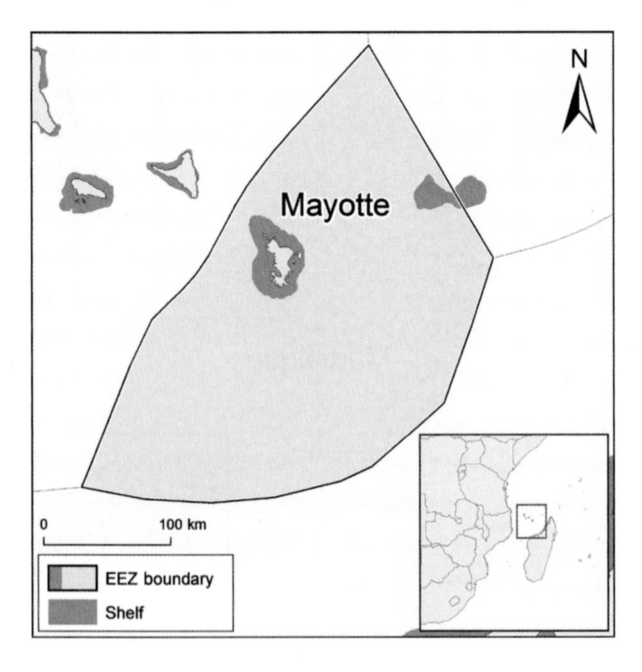

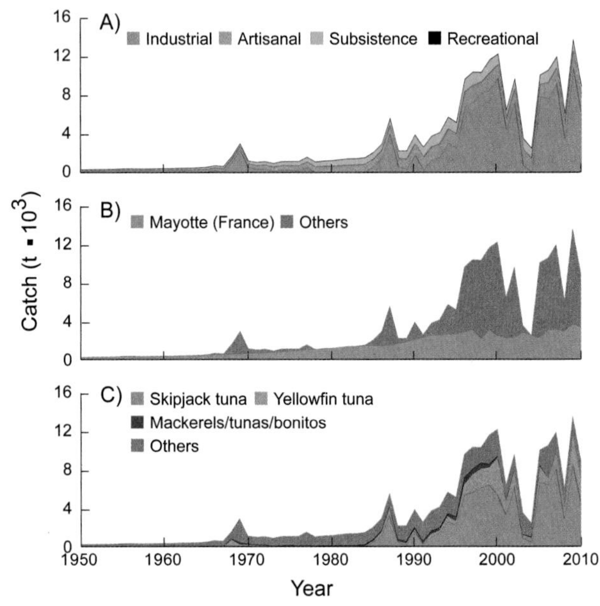

Figure 1. Mayotte has a land area of 393 km², and its EEZ is 63,000 km²; the islands of Anjouan and Mohéli to the west of Mayotte are part of the Comoros Islands. **Figure 2.** Domestic and foreign catches taken in the EEZ of Mayotte: (A) by sector (domestic scores: Art. 1, 1, 3; Sub. 1, 1, 3; Rec. –, 1, 1; Dis. –, –, –); (B) by fishing country (foreign catches are very uncertain); (C) by taxon.

Mayotte is a French overseas department in the Indian Ocean (figure 1). Domestic fisheries in Mayotte consist of shore-based subsistence and boat-based artisanal fisheries, both critical to local food security, and foreign industrial fishing was mostly by an unknown fishing country (figure 2A and 2B). The catches reported to the FAO, the Indian Ocean Tuna Commission (IOTC, for large pelagics), and national statistics were the primary sources used by Doherty et al. (2015) to reconstruct marine fisheries catches of Mayotte. Catches slowly increased from 300 t/year in the early 1950s to 2,000–3,000 t/year in the 1990s before settling to about 3,000 t/year in the late 2000s and were composed mainly of reef species because most of the effort occurred in Mayotte's lagoon (Maggiorani and Maggiorani 1990). Since then, catches have included more tuna (Herfaut 2004) because of the expansion of the domestic fleet fishing further offshore. Overall, the total reconstructed domestic catches from 1950–2010 are 1.9 times the domestic catches reported to and by the FAO (excluding catch from industrial tuna fisheries). Figure 2C, based on Maggiorani and Maggiorani (1990), suggests that catches from Mayotte's EEZ consist mainly of skipjack tuna (*Katsuwonus pelamis*), yellowfin tuna (*Thunnus albacares*), and mackerels, tunas, and bonitos (family Scombridae). The data in figure 2 should be useful as baseline for monitoring Mayotte's fisheries.

REFERENCES

Doherty B, Herfaut J, Le Manach F, Harper S and Zeller D (2015) Reconstructing domestic marine fisheries in Mayotte from 1950–2010. pp. 53–65 In Le Manach F and Pauly D (eds.), *Fisheries catch reconstructions in the Western Indian Ocean, 1950–2010*. Fisheries Centre Research Reports 23(2).

Herfaut J (2004) Suivi statistique de la pêcherie artisanale Mahoraise: évaluation de l'effort de pêche, des captures et des CPUE de 1997 à 2003. Service des Pêches de la Direction de l'Agriculture et de la Forêt. 31 p.

Maggiorani F and Maggiorani JM (1990) Enquête sur la pêche artisanale Mahoraise. Tome 2: Effort de pêche et captures. Collectivité Territoriale de Mayotte, Direction de l'Agriculture et de la Forêt, Service des Pêches. 75 p.

1. Cite as Doherty, B., F. Le Manach, and J. Herfaut. 2016. France (Mayotte). P. 262 in D. Pauly and D. Zeller (eds.), *Global Atlas of Marine Fisheries: A Critical Appraisal of Catches and Ecosystem Impacts*. Island Press, Washington, DC.

FRANCE (MEDITERRANEAN)[1]

Elise Bultel, Frédéric Le Manach, Aylin Ulman, and Dirk Zeller

Sea Around Us, University of British Columbia, Vancouver, Canada

France has a Mediterranean coast (figure 1) well known for its tourist sites, but its fisheries are not well documented. France's marine fisheries catches in the Mediterranean Sea (mainland EEZ only, excluding Corsica) were estimated using a catch reconstruction approach (Bultel et al. 2015). This produced an estimate of total fisheries removals (i.e., industrial, artisanal, recreational, and subsistence catches as well as discards) of about 30,000 t/year in the early 1950s, which peaked at about 100,000 t in 1972, 1992, and 2003 before declining to about 45,000 t/year in the late 2000s. This is 2.3 times the officially reported data, and the major reasons for this discrepancy are unreported artisanal (Guillou et al. 2002) and industrial catches. The main sectors were the industrial and artisanal fisheries, and the recreational sector appears to be growing (figure 2A). Figure 2B suggests that the majority of the catch is by domestic French vessels, although foreign fishing by Spain has been growing. Major landed taxa were European pilchard (*Sardina pilchardus*), European anchovy (*Engraulis encrasicolus*), and mullets (family Mugilidae; figure 2C). Elasmobranchs (sharks, rays, and skates) were the most frequently discarded, which is consistent with the risk of extinction several species seem to be facing, especially in the Mediterranean Sea (Bradai et al. 2012). Overall, French Mediterranean fisheries have been declining at a very rapid rate in recent years.

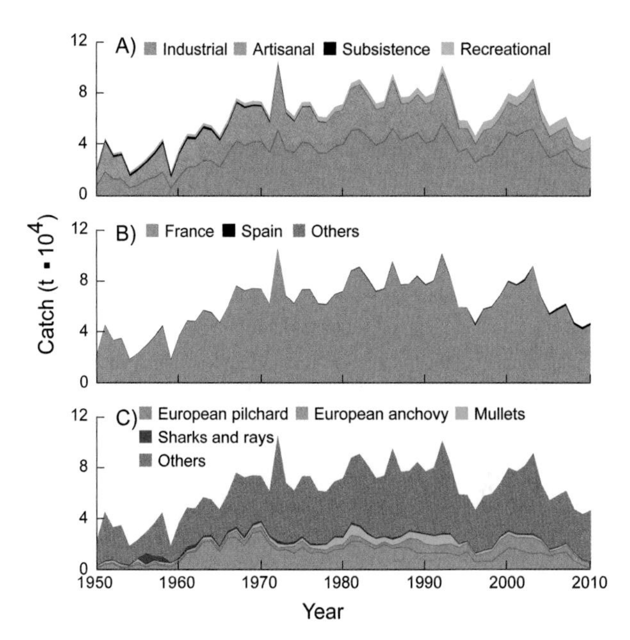

Figure 2. Domestic and foreign catches taken in the Mediterranean EEZ of France: (A) by sector (domestic scores: Ind. 3, 3, 3; Art. 3, 3, 3; Sub. 1, 1, 1; Rec. 2, 2, 2; Dis. 2, 2, 2); (B) by fishing country (foreign catches are very uncertain); (C) by taxon.

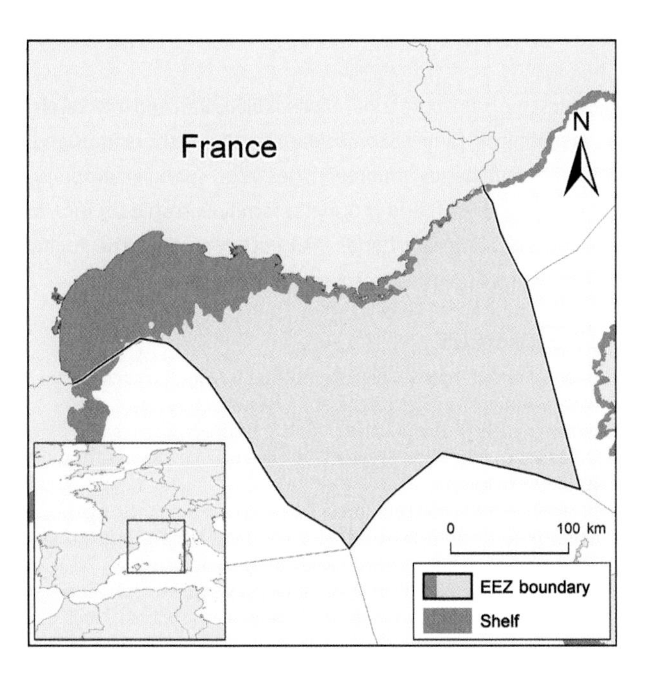

Figure 1. The Mediterranean coast of France has an EEZ of nearly 64,500 km² (excluding Corsica, but including Monaco, which has no fishery), of which 13,700 km² is shelf.

REFERENCES

Bradai MN, Saidi B and Enajjar S (2012) Elasmobranchs of the Mediterranean and Black Sea: status, ecology and biology: bibliographic analysis. FAO Studies and Reviews #91, 103 p.

Bultel E, Le Manach F, Ulman A and Zeller D (2015) Catch reconstruction for the French Mediterranean Sea, 1950–2010. Fisheries Centre Working Paper #2015–38, 20 p.

Guillou A, Lespagnol P and Ruchon F (2002) La pêche aux petits métiers en Languedoc-Roussillon en 2000–2001. Convention de participation au programme PESCA (PIC) DIRAM-IFREMER no. 00/3210040/YF, Convention de recherche Région Languedoc-Roussillon-IFREMER no. 00/1210041/YF, Sète.

1. Cite as Bultel, E., F. Le Manach, A. Ulman, and D. Zeller. 2016. France (Mediterranean). P. 263 in D. Pauly and D. Zeller (eds.), *Global Atlas of Marine Fisheries: A Critical Appraisal of Catches and Ecosystem Impacts*. Island Press, Washington, DC.

FRANCE (NEW CALEDONIA)[1]

Sarah Harper,[a] Lou Frotté,[b] Sarah Bale,[a] Shawn Booth,[a] and Dirk Zeller[a]

[a]*Sea Around Us*, University of British Columbia, Vancouver, Canada
[b]Muséum National d'Histoire Naturelle, Station Marine de Concarneau, France

New Caledonia, a French "special collectivity" in the southwestern Pacific, consists of a main island, "La Grande Terre," and several smaller islands (figure 1). Although the industrial sector dominates the total catch allocated to New Caledonia before EEZ declaration in 1978, it has been gradually declining, whereas the subsistence and recreational sectors have been increasing (figure 2A). The reconstruction of domestic catches in New Caledonia, based on Harper et al. (2009), consists of commercial data from government reports, to which catches from the subsistence and recreational sectors were added (e.g., Labrosse et al. 2006). Reconstructed domestic catches were about 2,500–3,000 t/year in the 1950s and 8,000–9,000 t/year in the 2000s. Catches were 5 times those presented by the FAO on behalf of New Caledonia. Most of this discrepancy was caused by nonconsideration of the recreational and subsistence sectors by the reporting agencies. Most removals were from domestic fishing; however, notable catches were taken by Japan and Taiwan (figure 2B). Figure 2C presents the taxonomic composition of these catches, which consisted mainly of large pelagics (e.g., albacore tuna, *Thunnus albacares*; and mackerels, tunas, bonitos, family Scombridae) for the offshore industrial and foreign fisheries, and reef fishes (e.g., spangled emperor, *Lethrinus nebulosus*; and groupers, family Serranidae) for the small-scale, inshore fisheries. As in other areas of the Pacific (Zeller et al. 2015), these resources are stressed.

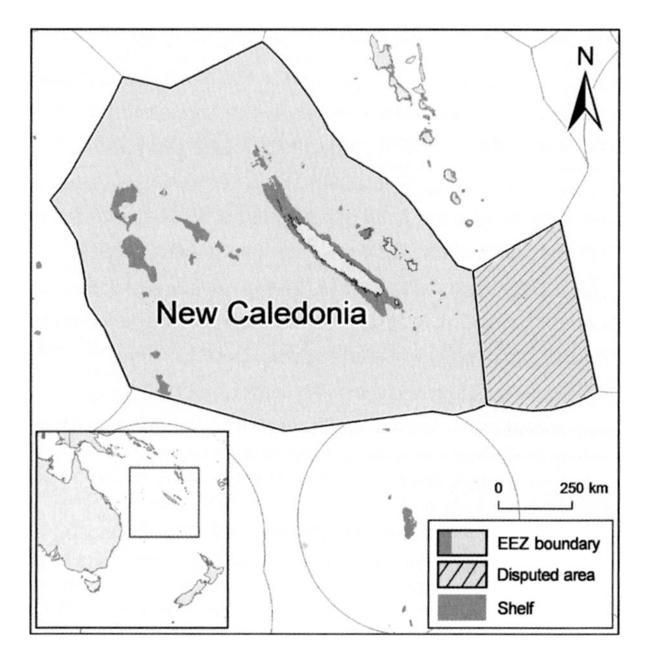

Figure 1. New Caledonia (land area: about 18,500 km²) has an EEZ of about 1.42 million km².

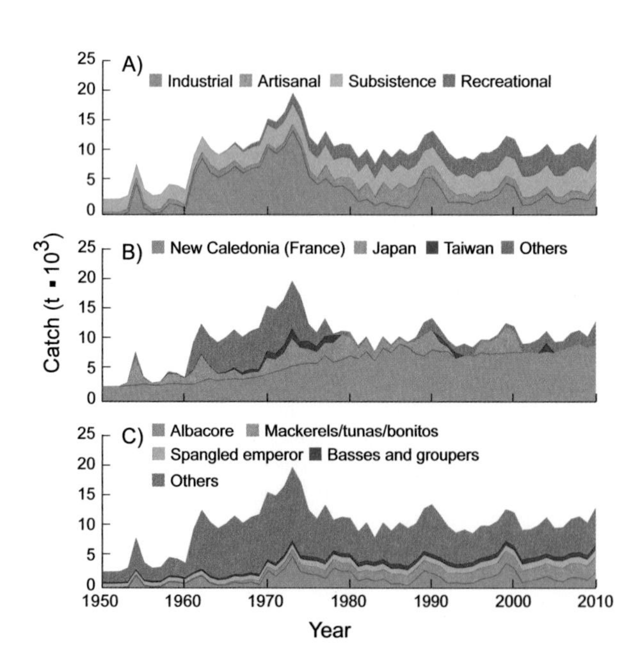

Figure 2. Domestic and foreign catches taken in the EEZ of New Caledonia: (A) by sector (domestic scores: Ind. –, 3, 3; Art. 2, 3, 3; Sub. 2, 2, 2; Rec. 1, 1, 1); (B) by fishing country (foreign catches are very uncertain); (C) by taxon.

REFERENCES

Harper S, Frotté L, Bale S, Booth S and Zeller D (2009) Reconstruction of total marine fisheries catches for New Caledonia (1950–2007). pp. 67–75 In Zeller D and Harper S (eds.), *Fisheries catch reconstructions: Islands, Part I*. Fisheries Centre Research Reports 17(5). [Updated to 2010]

Labrosse P, Ferrais J and Letourneur Y (2006) Assessing the sustainability of subsistence fisheries in the Pacific: The use of data on fish consumption. *Ocean & Coastal Management* 49: 203–221.

Zeller D, Harper S, Zylich K and Pauly D (2015) Synthesis of under-reported small-scale fisheries catch in Pacific-island waters. *Coral Reefs* 34(1): 25–39.

1. Cite as Harper, S., L. Frotté, S. Bale, S. Booth, and D. Zeller. 2016. New Caledonia. P. 264 in D. Pauly and D. Zeller (eds.), *Global Atlas of Marine Fisheries: A Critical Appraisal of Catches and Ecosystem Impacts*. Island Press, Washington, DC.

FRANCE (RÉUNION)[1]

Frédéric Le Manach,[a] Pascal Bach,[b] Léo Barret,[c] David Guyomard,[d] Pierre-Gildas Fleury,[e] Philippe S. Sabarros,[b] and Daniel Pauly[a]

[a]*Sea Around Us*, University of British Columbia, Vancouver, Canada [b]Institut de Recherche pour le Développement, Sète and Le Port, Réunion, France [c]Institut des Sciences de la Mer, Rimouski, Québec, Canada [d]Comité Régional des Pêches Maritimes et des Elevages Marins, Le Port, Réunion, France [e]Institut Français de Recherche pour l'Exploitation de la Mer, Le Port, Réunion, France

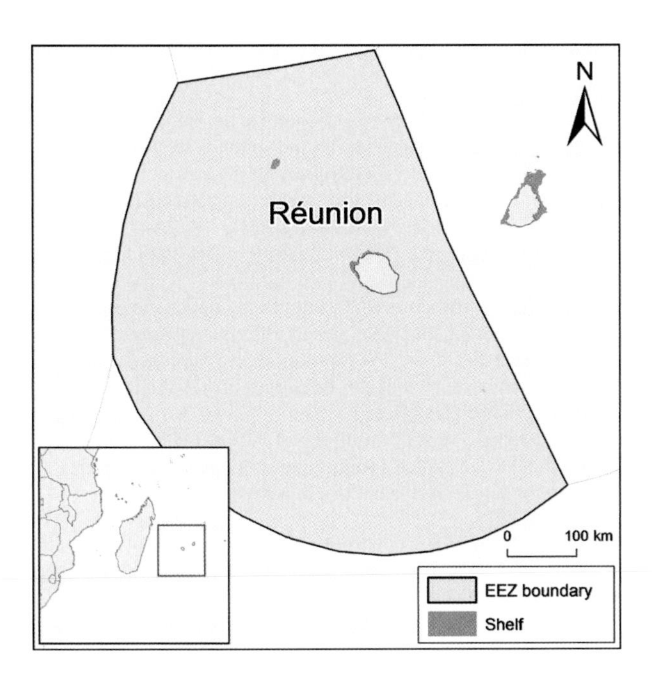

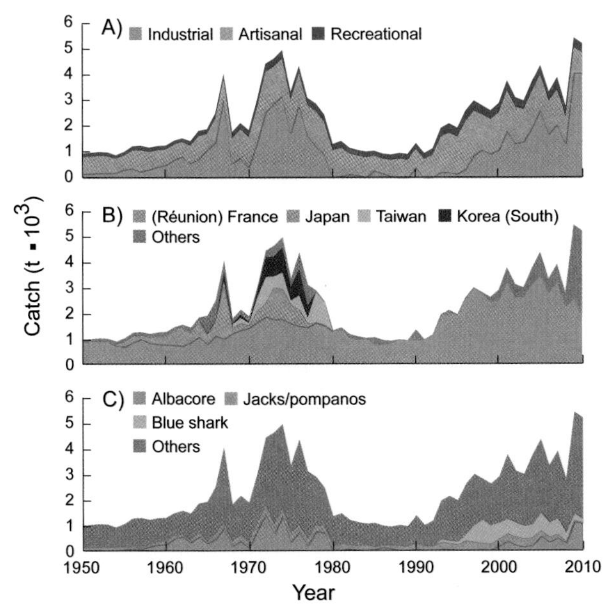

Figure 1. The island of Réunion (France) has a land area of 2,500 km² and a EEZ of 315,000 km². **Figure 2.** Domestic and foreign catches taken in the EEZ of Réunion (France): (A) by sector (domestic scores: Ind. 1, –, 4; Art. 3, 3, 3; Rec. 3, 3, 3; Dis. –, –, 4); (B) by fishing country (foreign catches are very uncertain); (C) by taxon.

Historically, the inhabitants of La Réunion (figure 1) have not been much of a fishing people, with rough sea conditions and a narrow shelf limiting opportunities. However, it has long served as a base for industrial fishing. Thus, vessels from La Réunion targeted lobsters around Saint Paul and Amsterdam Islands and cold-water fish species such as Patagonian toothfish around Kerguelen and Crozet Islands (Méralli-Ballou 2008). These two "distant-water" ventures are the most important fisheries contributors to the local economy. Overall, industrial catches are substantial but highly variable in these waters, in addition to the artisanal and recreational sectors (figure 2A). The domestic catches reconstructed by Le Manach et al. (2015) increased from about 850 t/year in the early 1950s to 2,800 t/year in the late 1990s, then decreased to 1,300 t by 2010. These catches were 2.4 times the data reported for Réunion. Catches within the EEZ are predominantly domestic, with foreign fishing mostly by Japan, Taiwan, and South Korea, before the EEZ declaration in 1978 and subsequently in the 2000s (figure 2B). Domestic fisheries expanded in the early 1990s, with a network of fish-aggregating devices around the island aimed at reducing the fishing pressure on coastal resources. Figure 2C presents the composition of the catch, in which pelagic fishes, notably albacore tuna (*Thunnus alalunga*), jacks and pompanos (family Carangidae), and blue shark (*Prionace glauca*) now figure most prominently.

REFERENCES

Le Manach F, Bach P, Barret L, Guyomard D, Fleury PG, Sabarros PS and Pauly D (2015) Reconstruction of the domestic and distant-water fisheries catch of La Réunion (France), 1950–2010. pp. 83–98 In Le Manach F and Pauly D (eds.), *Fisheries catch reconstructions in the Western Indian Ocean, 1950–2010*. Fisheries Centre Research Reports 23(2).

Méralli-Ballou P (2008) Homme libre, toujours. La pêche à l'île de La Réunion depuis le XVIIe siècle. Comité Régional des Pêches Maritimes et des Élevages Marins de La Réunion, Le Port, France. 111 p.

1. Cite as Le Manach, F., P. Bach, L. Barret, D. Guyomard, P. G. Fleury, P. S. Sabarros, and D. Pauly. 2016. France (La Réunion). P. 265 in D. Pauly and D. Zeller (eds.), *Global Atlas of Marine Fisheries: A Critical Appraisal of Catches and Ecosystem Impacts*. Island Press, Washington, DC.

FRANCE (SAINT BARTHÉLEMY)[1]

Elise Bultel, Alasdair Lindop, Robin Ramdeen, and Kyrstn Zylich

Sea Around Us, University of British Columbia, Vancouver, Canada

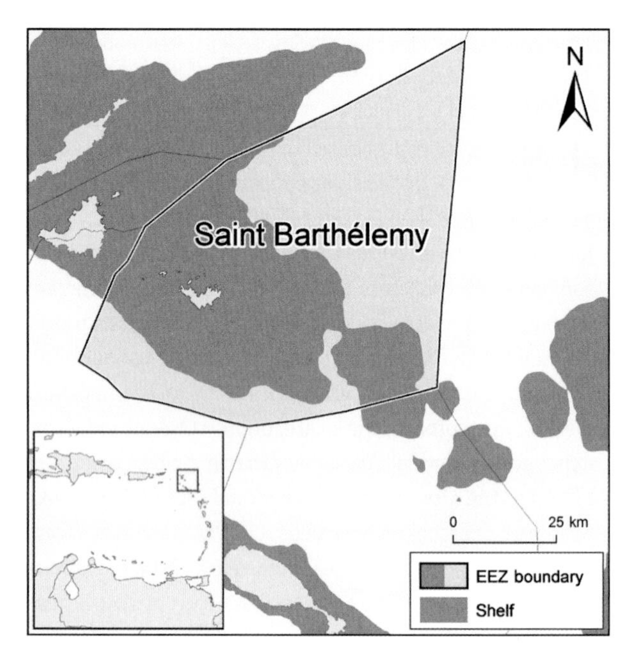

Figure 1. Saint Barthélemy (France) has a land area of 25 km² and an EEZ of 4,300 km².

Saint Barthélemy, also known as St. Bart's, is a French "overseas collectivity" in the Caribbean (figure 1) dependent on high-end tourism. Catches are predominantly from the small-scale sectors, artisanal and subsistence (figure 2A); however, some foreign industrial fishing by Cuba and Venezuela, among others, has been documented (figure 2B). The fisheries catches reconstructed by Bultel et al. (2015), based on Lorance (1989) and other reports, were estimated at 40 t/year in the 1950s, 200 t/year in the 1980s, and 270 t/year in the 2000s. Until 2007, Saint Barthélemy was administratively a part of Guadeloupe, and therefore the reconstructed catches cannot be easily compared with data reported to the FAO, although a preliminary attempt suggests that total catches might be 5.4 times the reported data deemed to relate to Saint Barthélemy. Figure 2C documents the composition of this catch to be mostly red hind (*Epinephelus guttatus*) and little tunny (*Euthynnus alletteratus*). One important feature of the fauna around Saint Barthélemy and neighboring Saint Martin is that many of its fishes, particularly the groupers (family Serranidae; Munro and Blok 2005), tend to be ciguatoxic (Bourdeau 1989). Ciguatera is a form of food poisoning caused by the ingested fish being contaminated by *Gambierdiscus toxicus*, a toxic alga. See www.fishbase.de/Topic/List.php?group=27 for a list of fish species and countries for which ciguatoxicity has been reported (see also Vakily et al. 2000).

REFERENCES

Bourdeau P (1989) Risk factors of ciguatera in the French West Indies in Saint-Barthélémy, Saint-Martin and Anguilla. *Revue d'élevage et de médecine vétérinaire des pays tropicaux* 42(3): 393–410.

Bultel E, Lindop A, Ramdeen R and Zylich K (2015) Reconstruction of marine fisheries catches for St. Barthélémy and St. Martin (French Caribbean, 1950–2010). Fisheries Centre Working Paper #2015-39, 9 p.

Lorance P (1989) Ressources démersales et description des pêcheries des bancs de Saint Martin et Saint-Barthélémy. IFREMER, 75 p.

Munro JL and Blok L (2005) The status of stocks of groupers and hinds in the Northeastern Caribbean. *Proceedings of the Gulf and Caribbean Fisheries Institute* 56: 283–294.

Vakily M, Pablico GT and Dalzell P (2000) The CIGUATERA database. pp. 277–281 In Froese R and Pauly D (eds.), *FishBase 2000: Concepts, Design and Data Sources*. ICLARM, Los Baños, Philippines.

1. Cite as Bultel, E., A. Lindop, R. Ramdeen, and K. Zylich. 2016. France (Saint Barthélemy). P. 266 in D. Pauly and D. Zeller (eds.), *Global Atlas of Marine Fisheries: A Critical Appraisal of Catches and Ecosystem Impacts*. Island Press, Washington, DC.

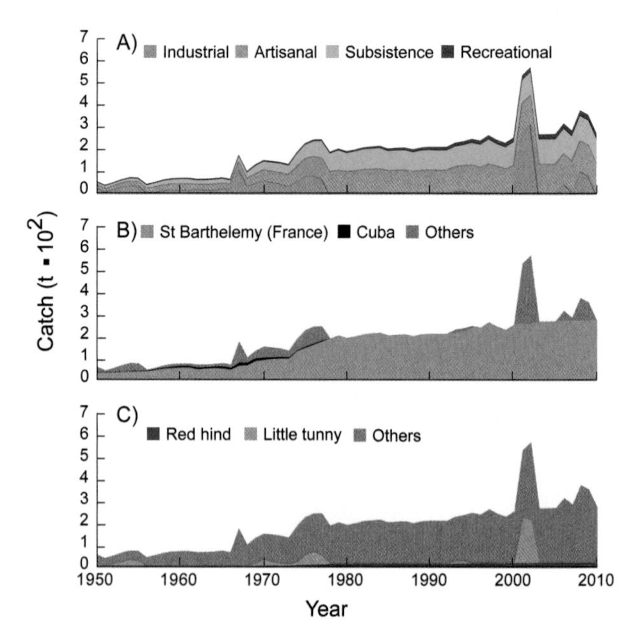

Figure 2. Domestic and foreign catches taken in the EEZ of Saint Barthélemy (France): (A) by sector (domestic scores: Art. 2, 2, 3; Sub. 1, 1, 1; Rec. 1, 1, 1; Dis. 1, 1, 2); (B) by fishing country (foreign catches are very uncertain); (C) by taxon.

FRANCE (SAINT MARTIN)[1]

Elise Bultel, Alasdair Lindop, Robin Ramdeen, and Kyrstn Zylich

Sea Around Us, University of British Columbia, Vancouver, Canada

Saint Martin is the French half of a Caribbean Island, whose other half is Dutch and helpfully called Sint Maarten (figure 1). Total removals from the EEZ of Saint Martin are predominantly due to artisanal and subsistence fisheries (figure 2A). These fisheries, whose catches were reconstructed by Bultel et al. (2015), based partly on Lorance (1989), yielded about 150 t/year in the 1950s, 830 t/year in the 1980s, and 1,100 t/year in the late 2000s. Total catches were 5.4 times the data deemed reported for Saint Martin, with the discrepancy caused by unreported subsistence and recreational catches and substantially underreported artisanal catches. Overall catches are mostly domestic, as shown in figure 2B. However, there is some foreign fishing, notably by Spain. Figure 2C, based on Lorance (1989) and other sources, suggests that catches are composed mainly of red hind (*Epinephelus guttatus*), stromboid conchs (*Strombus* spp.), grunts (family Haemulidae), and silk snappers (*Lutjanus vivanus*). Ciguatera is common around Saint Martin (Bourdeau 1989), as it is around Saint Barthélemy (see previous page), with no area in the Anguilla plateau apparently free from it. This may be the reason why fish imports into Saint Martin are high (Munro and Blok 2005) and may explain the absence of overexploitation, as seen around many otherwise similar Caribbean islands.

REFERENCES

Bourdeau P (1989) Risk factors of ciguatera in the French West Indies in Saint-Barthélémy, Saint-Martin and Anguilla. *Revue d'élevage et de médecine vétérinaire des pays tropicaux* 42(3): 393–410.

Bultel E, Lindop A, Ramdeen R and Zylich K (2015) Reconstruction of marine fisheries catches for St. Barthélémy and St. Martin (French Caribbean, 1950–2010). Fisheries Centre Working Paper #2015-39. 9 p.

Lorance P (1989) Ressources démersales et description des pêcheries des bancs de Saint Martin et Saint-Barthélémy. IFREMER. 75 p.

Munro JL and Blok L (2005) The status of stocks of groupers and hinds in the Northeastern Caribbean. *Proceedings of the Gulf and Caribbean Fisheries Institute* 56: 283–294.

1. Cite as Bultel, E., A. Lindop, R. Ramdeen, and K. Zylich. 2016. France (Saint Martin). P. 267 in D. Pauly and D. Zeller (eds.), *Global Atlas of Marine Fisheries: A Critical Appraisal of Catches and Ecosystem Impacts*. Island Press, Washington, DC.

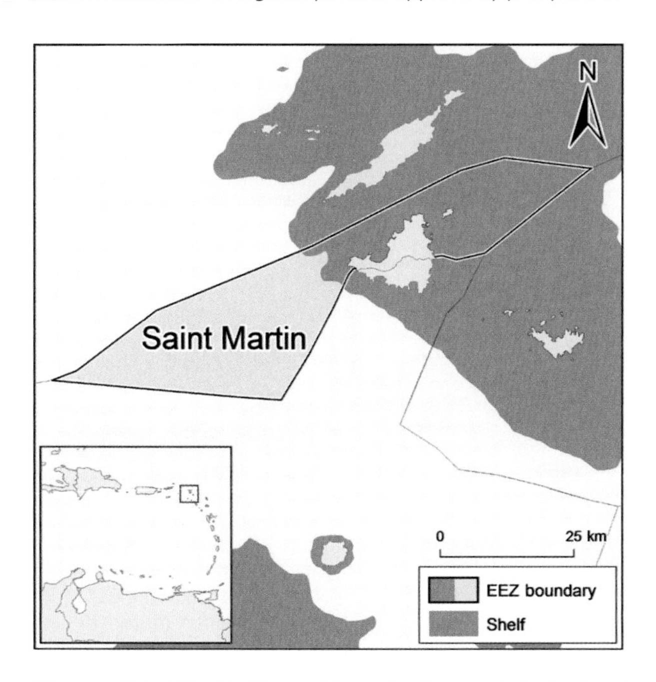

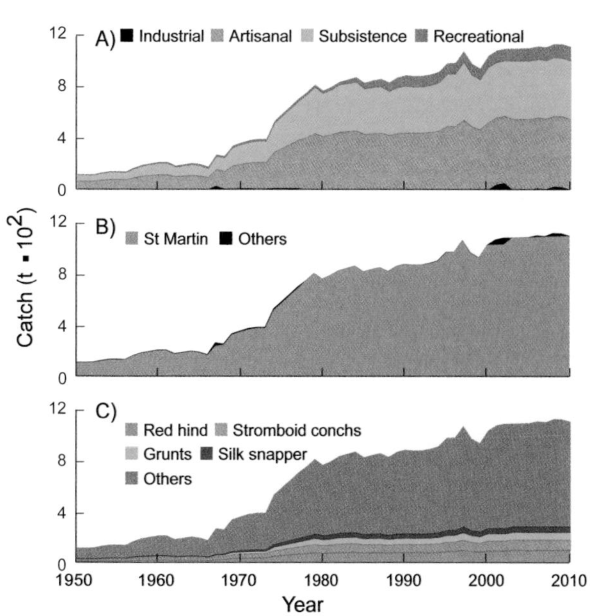

Figure 1. Saint Martin (France) has a land area of 56.2 km² and an EEZ of 1,070 km². **Figure 2.** Domestic and foreign catches taken in the EEZ of Saint Martin (France): (A) by sector (domestic scores: Art. 1, 1, 1; Sub. 1, 1, 1; Rec. 1, 1, 1; Dis. 1, 1, 1); (B) by fishing country (foreign catches are very uncertain); (C) by taxon.

FRANCE (ST. PAUL AND AMSTERDAM)[1]

Patrice Pruvost,[a] Guy Duhamel,[a] Frédéric Le Manach,[b] and Maria-Lourdes D. Palomares[b]

[a]Muséum National d'Histoire Naturelle, Paris, France [b]*Sea Around Us*, University of British Columbia, Vancouver, Canada

Saint Paul (6 km²) and Amsterdam (55 km²), which are about 90 km apart, are located in the southern Indian Ocean (figure 1) and are part of the French *Terres Australes et Antarctiques Françaises* (TAAF, www.taaf.fr), which also include the islands of Crozet and Kerguelen. The EEZ of St. Paul and Amsterdam, whose fish fauna has been well studied (Duhamel 1987, 1997), has been fished since 1949, with the exception of the 1957–1958 seasons. Pruvost et al. (2015), based on Duhamel (1987) and data reported to the Musée d'Histoire Naturelle (MNHN), reconstructed the catch as 1,000 t/year in the 1950s, which increased to a peak of 1,500 t in 1969 and decreased to 600 t/year in the late 2000s. Total removals from within the St. Paul and Amsterdam EEZ are solely domestic after the EEZ declaration in 1978; however, foreign fishing, notably by Japan, was noticeable before 1978 (figure 2A). Figure 2B presents the taxonomic composition of the catch, which includes species such as southern bluefin tuna (*Thunnus maccoyii*), albacore tuna (*Thunnus alalunga*), and yellowfin tuna (*Thunnus albacares*). Most important, however, is a French industrial trap fishery for Saint Paul rock lobster (*Jasus paulensis*), whose vessels are based in La Réunion and which is regulated by strict quotas set by the TAAF administration. The management of this resource includes onboard observers, who report to the MNHN in Paris.

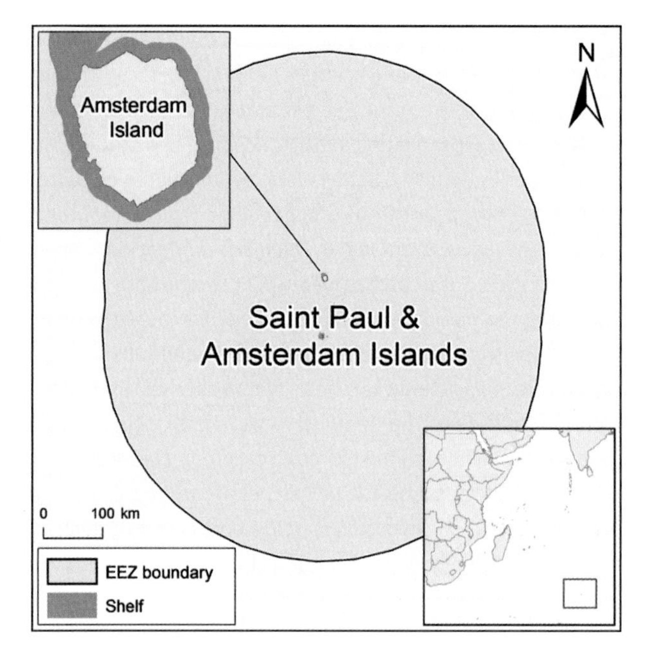

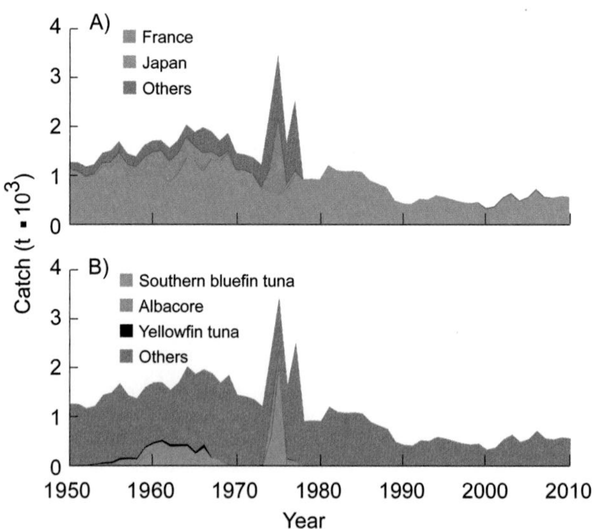

Figure 1. St. Paul and Amsterdam (France) have a joint land area of 72.9 km² and an EEZ of 509,000 km².

Figure 2. Domestic and foreign catches taken in the EEZ of St. Paul and Amsterdam: (A) by fishing country (domestic scores: Ind. 3, 3, 4; foreign catches are very uncertain); (B) by taxon.

REFERENCES

Duhamel G (1987) Ichthyofaune des secteurs indien occidental et atlantique oriental de l'océan austral: biogéographie, cycles biologiques et dynamique des populations. Thèse de Doctorat d'Etat, Université Pierre et Marie Curie, Paris, 687 p.

Duhamel G (1997) Deep-sea demersal ichthyofauna off the St-Paul and Amsterdam Islands (central southern Indian Ocean). pp. 185–194 In Séret B and Sire JY (eds.), Proceedings of the 5th Indo-Pacific Fish Conference, Nouméa, 3–8 November 1997, Paris.

Pruvost P, Duhamel G, Le Manach F and Palomares MLD (2015) A short history of the fisheries of Saint-Paul and Amsterdam Islands. pp. 37–45 In Palomares MLD and Pauly D (eds.), *Marine Fisheries Catches of Sub-Antarctic Islands, 1950 to 2010*. Fisheries Centre Research Reports 23(1).

1. Cite as Pruvost, P., G. Duhamel, F. Le Manach, and M. L. D. Palomares. 2016. France (St. Paul and Amsterdam). P. 268 in D. Pauly and D. Zeller (eds.), *Global Atlas of Marine Fisheries: A Critical Appraisal of Catches and Ecosystem Impacts*. Island Press, Washington, DC.

FRANCE (SAINT PIERRE ET MIQUELON)[1]

Elise Bultel and Kyrstn Zylich

Sea Around Us, University of British Columbia, Vancouver, Canada

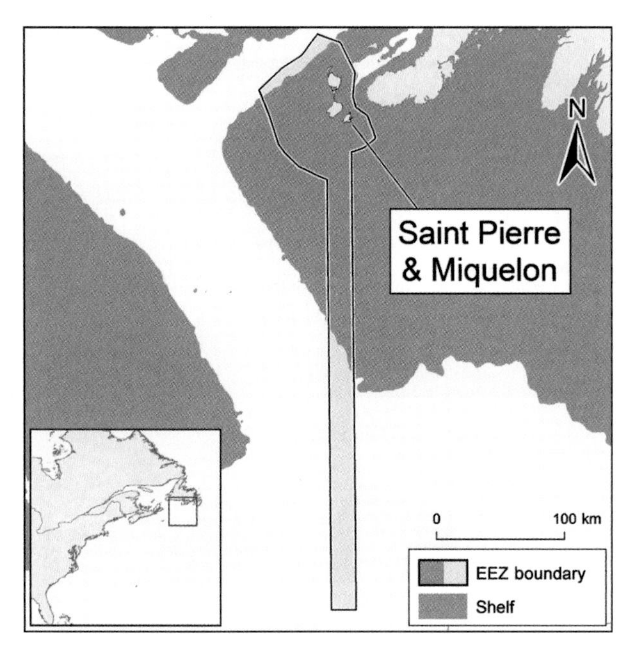

Figure 1. Saint Pierre et Miquelon (France; land area 242 km²), showing the three islands of St. Pierre, Miquelon, and Langlade, and their EEZ of 12,400 km².

Saint-Pierre and Miquelon (SPM), a French "overseas collectivity" and the last French foothold in North America, consist of three main islands: Saint-Pierre, Miquelon, and Langlade. The fisheries catches of SPM by sector are given in figure 2A, including the artisanal, motorized "doris" with a 2-person crew earlier exploiting Atlantic cod (*Gadus morhua*), along with some subsistence fishing in the early 1950s (Eynaud 1986). The cod stock and the fishery based theron declined, notably because of developments in Newfoundland, whose EEZ surrounds that of SPM. The decline intensified with the inception of stern trawling in SPM (Anonymous 1976), which competed for cod with the artisanal fishery. The reconstruction of Bultel and Zylich (2015) estimated catches of 3,100 t in 1950, of 12,200 t in 1987 and 1991, and 3,200 t in 2010. Overall, this is 1.4 times the catch reported by the FAO for SPM vessels, as assigned to the EEZ by Bultel and Zylich (2015), with the discrepancy being caused by discards and unreported landings. As shown in figure 2B, most of the catch is from the domestic Saint Pierre and Miquelon fleet, although foreign fishing such as by the United States and Canada occurs but plummeted by the late 1970 to 1980s. In addition to Atlantic cod, Atlantic salmon (*Salmo salar*), American sea scallop (*Placopecten magellanicus*), and American plaice (*Hippoglossoides platessoides*) are important species (figure 2C), although their stocks have also declined.

REFERENCES

Anonymous (1976) Rapport sur la pêche dans la région de Saint-Pierre et Miquelon. 22 p. Available at: http://archimer.ifremer.fr/doc/00046/15764/13166.pdf [Accessed: 21/07/2014].

Bultel E and Zylich K (2015) Fisheries catch reconstruction of the Western Atlantic French archipelago of Saint Pierre et Miquelon, 1950–2010. Fisheries Centre Working Paper #2015-42, 15 p.

Eynaud P (1986) Analyse du déclin de la pêche artisanale à St Pierre et Miquelon. 123 p. Available at: http://archimer.ifremer.fr/doc/00000/4088/ [Accessed: 05/05/2014].

1. Cite as Bultel, E., and K. Zylich. 2016. France (Saint Pierre et Miquelon). P. 269 in D. Pauly and D. Zeller (eds.), *Global Atlas of Marine Fisheries: A Critical Appraisal of Catches and Ecosystem Impacts*. Island Press, Washington, DC.

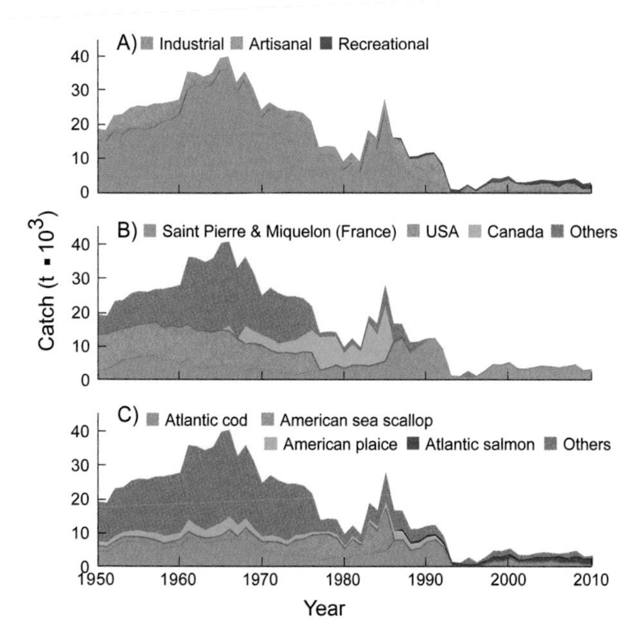

Figure 2. Domestic and foreign catches taken in the EEZ of Saint Pierre et Miquelon: (A) by sector (domestic scores: Ind. 3, 3, 4; Art. 3, 3, 4; Sub. 2, 2, 2; Rec. –, 1, 1; Dis. 2, 2, 2); (B) by fishing country (foreign catches are very uncertain); (C) by taxon.

FRANCE (WALLIS AND FUTUNA)[1]

Sarah Harper,[a] Lou Frotté,[b] Shawn Booth,[a] and Dirk Zeller[a]

[a]*Sea Around Us*, University of British Columbia, Vancouver, Canada [b]Muséum National d'Histoire Naturelle, Station Marine de Concarneau, France

The Wallis and Futuna Islands are a French Pacific territory located between Fiji and Samoa and consist of two island groups 200 km apart, Wallis and Futuna (plus Alofi; figure 1). The main island, Uvea (Wallis Island), has a lagoon protected by a coral reef. Futuna has neither but has a lot of earthquakes. The territory's economy is supplemented by remittances from expatriate workers in New Caledonia, French Polynesia, and France and from licensing distant-water fleets (mainly Japanese and South Korean) fishing for tuna (Kronen et al. 2008). The majority of the total removals allocated to the EEZ of Wallis and Futuna are from the foreign industrial sector, notably by Japan; however, domestic subsistence catches, though secondary, have been increasing through the time period (figure 2A, 2B). The catch reconstruction for these islands, based on Harper et al. (2009), yielded domestic catches of 400 t/year in the 1950s and 900–1,000 t/year in the 2000s. This consisted of a significant subsistence catch (much of it contributed by women; Kronen 2008) and an artisanal catch that started in 1970. The reconstructed catch for 1950–2010 is 4.6 times the data reported by the FAO, because until the early 2000s it reported only the commercial catch. Figure 2C show the gross composition of the catch from Wallis and Futuna to be albacore tuna (*Thunnus alalunga*), skipjack tuna (*Katsuwonus pelamis*), and yellowfin tuna (*Thunnus albacares*).

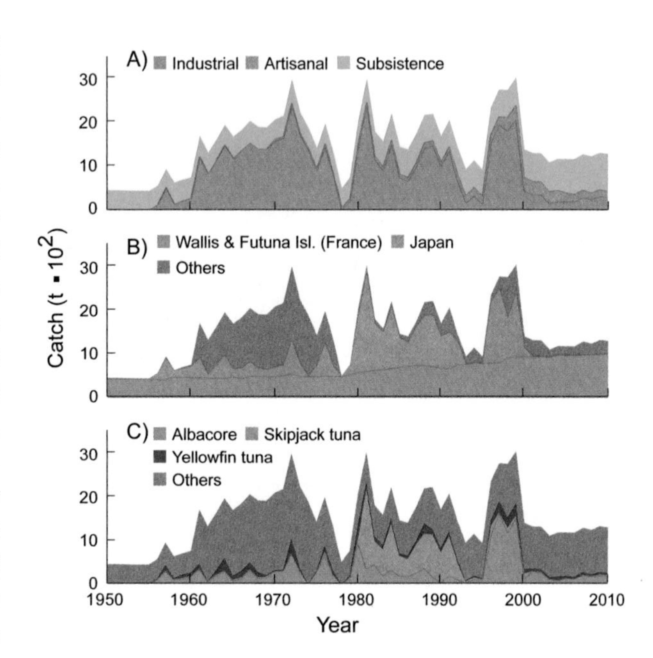

Figure 2. Domestic and foreign catches taken in the EEZ of Wallis and Futuna: (A) by sector (domestic scores: Art. 2, 2, 3; Sub. 1, 2, 3); (B) by fishing country (foreign catches are very uncertain); (C) by taxon.

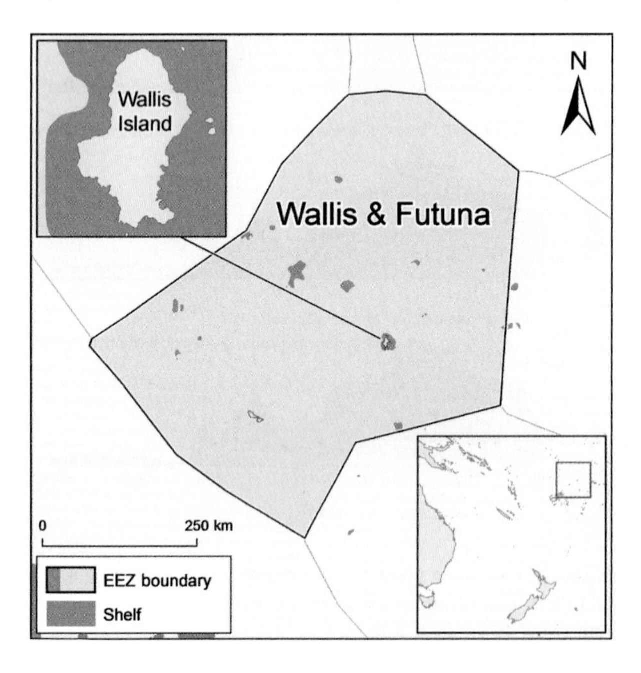

Figure 1. Wallis and Futuna (land area 177 km²), showing the three islands, Wallis, Futuna, and Alofi, and their EEZ of 258,000 km².

REFERENCES

Harper S, Frotté L, Booth S and Zeller D (2009) Reconstruction of marine fisheries catches for Wallis and Futuna Islands (1950–2007). pp. 99–104 In Harper S and Zeller D (eds.), *Fisheries catch reconstructions: Islands, Part I*. Fisheries Centre Research Reports 17(5). [Updated to 2010]

Kronen M (2008) Combining traditional and new fishing techniques: Fisherwomen in Niue, Papua New Guinea and Wallis and Futuna. *Women in Fisheries Information Bulletin* 18 (March): 11–15.

Kronen M, Tardy E, Boblin P, Chapman L, Lasi, Ferral, Pakoa K, Vigliola L, Friedman K, Magron F and Pinca S (2008) Wallis and Futuna country report: profiles and results from survey work at Vailal, Halalo, Leava and Vele. Pacific Regional Oceanic and Coastal Fisheries Development Programme, 333 p.

1. Cite as Harper, S., L. Frotté, S. Booth, and D. Zeller. 2016. France (Wallis and Futuna). P. 270 in D. Pauly and D. Zeller (eds.), *Global Atlas of Marine Fisheries: A Critical Appraisal of Catches and Ecosystem Impacts*. Island Press, Washington, DC.

GABON[1]

Dyhia Belhabib

Sea Around Us, University of British Columbia, Vancouver, Canada

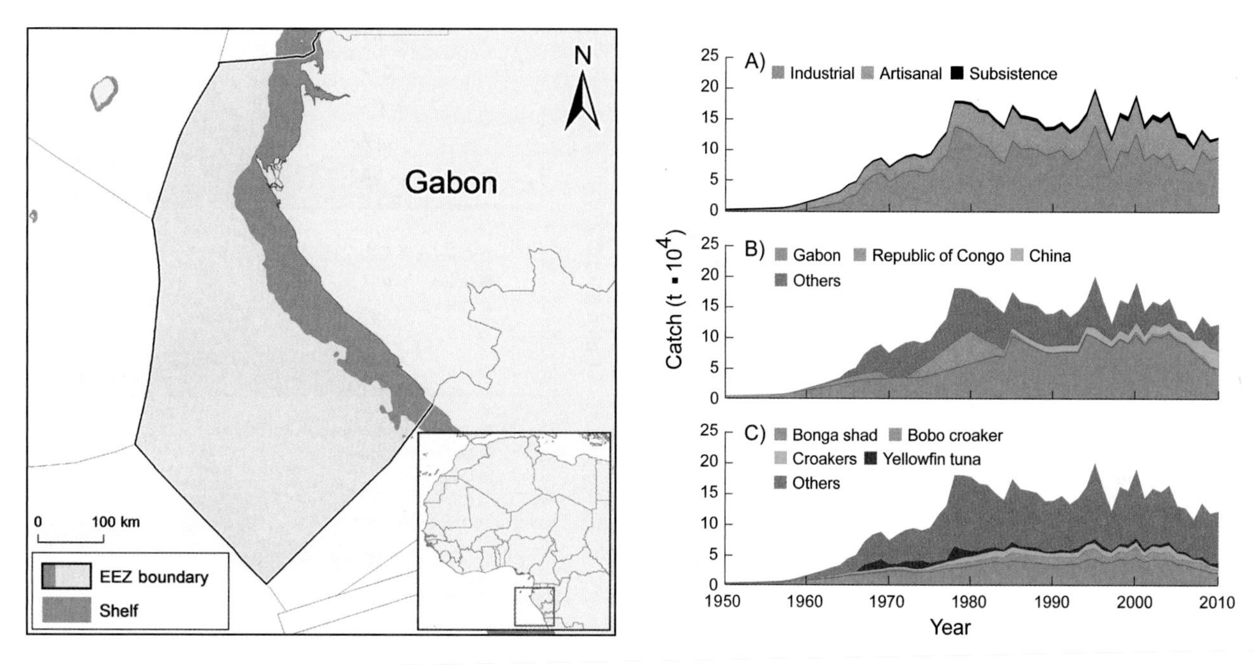

Figure 1. Gabon has an EEZ of 192,000 km² and a continental shelf of 36,600 km². **Figure 2.** Domestic and foreign catches taken in the EEZ of Gabon: (A) by sector (domestic scores: Ind. 3, 3, 4; Art. 2, 4, 4; Sub. 3, 2, 3; Dis. 2, 2, 3); (B) by fishing country (foreign catches are very uncertain); (C) by taxon.

Gabon is located on the Equator (figure 1) and was among the most prosperous economies of West Africa. After the oil price collapse in the 1970s, Gabon shifted to other extractive industries, such as fisheries, whose performance was then not well known (Bignouma 2011). The artisanal sector was dominant in the 1950s; however, by the mid-1960s, total catches allocated to the industrial sector exceeded small-scale catches, which remained the case for most of the remaining time period (figure 2A). Belhabib's (2015) reconstruction yielded domestic catches (excluding industrial tuna) of 5,000 t in early 1950s, which increased to a peak of 70,000 t in 1983 before declining to 36,000 t in 2010. Domestic catches were 2.5 times the data reported to the FAO, and underreporting was strongest in the 1980s. Most of the total removals allocated to Gabon's EEZ are domestic vessels; however, foreign catches by the Republic of Congo were prominent in the 1970s and 1980s and by China since the mid-1980s (figure 2B). Catches included more than 70 taxa, but most of it consisted of bonga shad (*Ethmalosa fimbriata*), Bobo croakers (*Pseudotolithus elongatus*), and other croaker (*Pseudotolithus* spp.) and yellowfin tuna (*Thunnus albacares*) (figure 2C). Léon (1993) describes a scheme to reduce sector conflicts wherein only Gabonese small-scale fishers are allowed to fish in the zone closest to the shore, non-Gabonese small-scale fishers in zones slightly further offshore, and foreign industrial vessels in the furthest offshore zone.

REFERENCES

Belhabib D (2015) Gabon fisheries between 1950 and 2010: a catch reconstruction. pp. 85–94 In Belhabib D and Pauly D (eds.), *Fisheries catch reconstructions: West Africa, Part II*. Fisheries Centre Research Report 23(3).

Bignouma G (2011) Le concept de périphérie appliqué à l'activité halieutique : impact sur l'aménagement des pêcheries maritimes artisanales au Gabon. *Geo-Eco-Trop* 2011(35): 33–40.

Léon MN (1993) Conflicts in coastal fisheries in Gabon. pp. 24–26 In Satia BP and Horemans B (eds.), Workshop on Conflicts in Coastal Fisheries in West Africa, 24–26 November 1993, Cotonou, Benin.

1. Cite as Belhabib, D. 2016. Gabon. P. 271 in D. Pauly and D. Zeller (eds.), *Global Atlas of Marine Fisheries: A Critical Appraisal of Catches and Ecosystem Impacts*. Island Press, Washington, DC.

GAMBIA[1]

Dyhia Belhabib,[a] Asberr Mendy,[b] Dirk Zeller,[a] and Daniel Pauly[a]

[a]*Sea Around Us*, University of British Columbia, Vancouver, Canada [b]Commission Sous-Régionale des Pêches, Dakar Sénégal

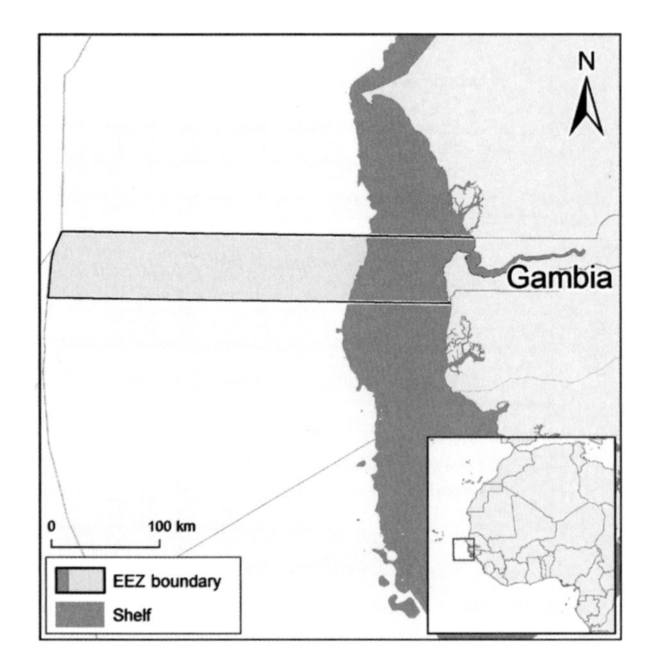

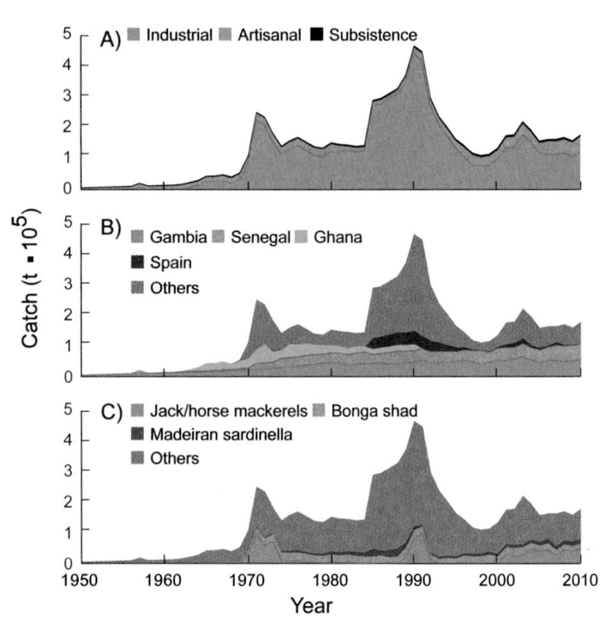

Figure 1. Gambia has an average width of 32 km, and its EEZ, of 22,700 km², is similarly narrow; its shelf covers 5,810 km².
Figure 2. Domestic and foreign catches taken in the EEZ of Gambia: (A) by sector (domestic scores: Ind. 3, 3, 3; Art. 2, 3, 3; Sub. 2, 3, 3; Rec. 2, 2, 3; Dis. 2, 3, 3); (B) by fishing country (foreign catches are very uncertain); (C) by taxon.

Gambia ("the Smiling Coast of Africa") is one of Africa's smallest coastal states. Despite the small size of the Gambian EEZ (figure 1), the catch of the legal and illegal foreign fisheries, reconstructed by Belhabib et al. (2013), is high. Industrial catches (including discards) contributed the bulk of the total reconstructed withdrawals, and the artisanal and subsistence fisheries (much of the latter involving women gathering shellfish along the mangrove-lined banks of the Gambia River) generated smaller contributions (figure 2A). The rest is contributed by a nascent recreational fishing industry catering to foreign tourists. Gambian domestic catches, on the other hand, increased from 5,700 t in 1950 to 47,000 t by 1980 and oscillated between 46,000 and 66,000 t/year in the 2000s. Overall, domestic catches are 2.5 times the data submitted to the FAO (see also Behabib et al. 2015). Total allocated catch in the EEZ is predominantly from foreign vessels such as those of Senegal, Ghana, and Spain (figure 2B). Despite a massive industrial trawl fishing effort (especially in the 1980s and 1990s), the key species in Gambia continue to be horse mackerels (*Trachurus* spp.) and bonga shad (*Ethmalosa fimbriata*), a locally popular species roughly similar to herring (figure 2C). However, its catch is declining, which, along with climate change (Jaiteh and Sarr 2011), further jeopardizes food security in Gambia, notably by increasing fish prices, and threatens the thousands of jobs these fisheries generate.

REFERENCES

Belhabib D, Mendy A, Zeller D and Pauly D (2013) Big fishing for small fishes: six decades of fisheries in The Gambia, "the smiling coast of Africa". Fisheries Centre Working Paper #2013-07, 20 p.

Belhabib D, Mendy A, Subah Y, Broh NT, Jueseah AS, Nipey N, Boeh WW, Willemse N, Zeller D and Pauly D (2015) Fisheries catch under-reporting in The Gambia, Liberia and Namibia, and the three Large Marine Ecosystems which they represent. *Environmental Development* doi:10.1016/j.envdev.2015.08.004.

Jaiteh M and Sarr B (2011) Climate change and development in The Gambia. Challenges to ecosystem goods and services. The Earth Institute, Columbia University, New York. 43 p.

1. Cite as Belhabib, D., A. Mendy, D. Zeller, and D. Pauly. 2016. Gambia. P. 272 in D. Pauly and D. Zeller (eds.), *Global Atlas of Marine Fisheries: A Critical Appraisal of Catches and Ecosystem Impacts*. Island Press, Washington, DC.

GAZA STRIP[1]

Mohammed Abudaya,[a] Sarah Harper,[b] Aylin Ulman,[b] and Dirk Zeller[b]

[a]University of Palestine, Faculty of Applied Engineering, Al Zahra, Gaza, Palestine [b]*Sea Around Us*, University of British Columbia, Vancouver, Canada

The Gaza Strip is a narrow stretch of land along the southwestern Palestinian coastal plains (figure 1). Total marine fisheries catches were estimated by Abudaya et al. (2013) from 1950 to 2010 by accounting for all fisheries sectors. Landings data have been reported by the FAO separately for the Gaza Strip since 1995 and represent the official records but cover only the larger-scale, industrial sector. Before 1995, FAO landings data for Gaza were assigned to Israel (Edelist et al. 2013). Here, these mis-assigned landings were reallocated to Gaza, showing that the artisanal sector dominates, with the industrial sector being secondary (figure 2A). Thus, the reconstructed total domestic catches for Gaza increased from 1,000 t in 1950 to about 5,000 t/year in the late 1970s, declined substantially in the mid-1980s, peaked in 1997 at 7,800 t, and averaged 3,500 t/year in the late 2000s. Fishing options are strongly affected by the spatial restrictions imposed by Israel. Total catches were 2.2 times the reported baseline (i.e., a portion of FAO landings data reported for Israel and those reported for Gaza). The majority of catches were domestic, with some foreign fishing being allocated to Gaza's EEZ (figure 2B). Catches were dominated by round sardinella (*Sardinella aurita*), seabreams (*Dentex* spp.), Atlantic mackerel (*Scomber scombrus*), red mullets (*Mugil* spp.), and a large "miscellaneous marine fishes" group (figure 2C).

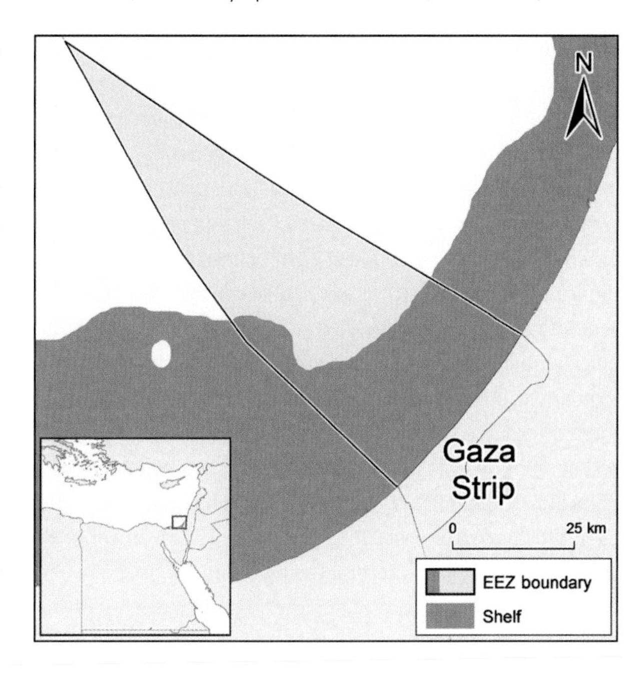

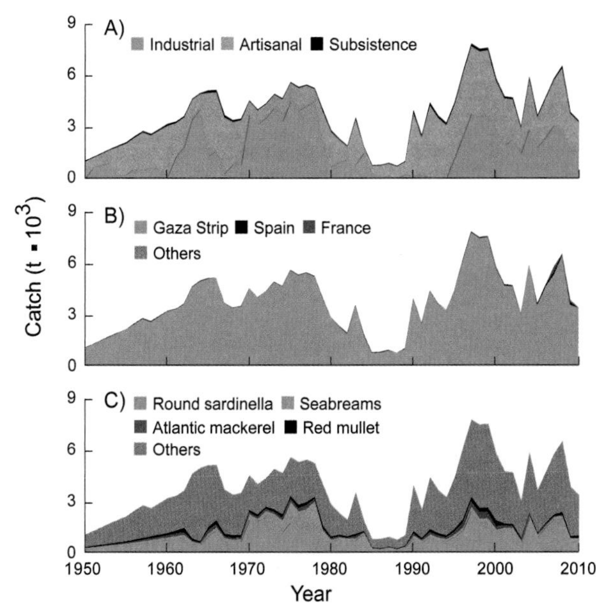

Figure 1. The Gaza Strip has a shelf area of 1,020 km² and an EEZ of 2,590 km². **Figure 2.** Domestic and foreign catches taken in the EEZ of the Gaza Strip: (A) by sector (domestic scores: Ind. 3, 4, 4; Art. 3, 3, 4; Sub. 3, 3, 3); (B) by fishing country (foreign catches are very uncertain); (C) by taxon.

REFERENCES

Abudaya M, Harper S, Ulman A and Zeller D (2013) Correcting mis- and under-reported marine fisheries catches for the Gaza Strip: 1950–2010. *Acta Adriatica* 54(2): 241–252.

Edelist D, Scheinin A, Sonin O, Shapiro J, Salameh P, Rilov G, Benayahu Y, Schulz D and Zeller D (2013) Israel: Reconstructed estimates of total fisheries removals in the Mediterranean, 1950–2010. *Acta Adriatica* 54(2): 253–264.

1. Cite as Abudaya, M., S. Harper, A. Ulman, and D. Zeller. 2016. Gaza Strip. P. 273 in D. Pauly and D. Zeller (eds.), *Global Atlas of Marine Fisheries: A Critical Appraisal of Catches and Ecosystem Impacts*. Island Press, Washington, DC.

GEORGIA[1]

Aylin Ulman and Esther Divovich

Sea Around Us, University of British Columbia, Vancouver, Canada

Georgia is located on the eastern Black Sea (figure 1). An ex-Soviet Republic, Georgia became independent in 1991. The catch reconstruction of Ulman and Divovich (2015) treated Georgia as if it had always been independent and also included Georgia's northern coast, the "faux-state" of Abkhazia, which, with Russia's help, declared itself independent in 1991. Catches from within Georgia's EEZ have been largely industrial (figure 2A). This reconstruction yielded estimates of 11,000 t/year in the 1950s, which peaked at 132,000 t in 1980 and then quickly crashed to just 12,000 t in 1991, before increasing again to nearly 100,000 t in 2010. For the period from 1988 to 2010, the reconstructed catch was 2.3 times that reported by the FAO on behalf of Georgia as a whole. Although catches are mostly domestic, foreign fishing by Turkey, Ukraine, and Russia has been allocated to Georgia's EEZ (figure 2B). The low taxonomic diversity of the catch (dominated by European anchovy, *Engraulis encrasicolus*; figure 2C) reflects the low diversity of the Black Sea. Abkhazia has waters in which Turkish (Oztürk et al. 2011) and other foreign vessels operate seemingly at will, notably to exploit the remnants of once abundant stocks of European anchovies. Overall, the work of Komakhidze et al. (2007) suggests that the Georgian coastal ecosystem is in as bad a shape as the politics on land.

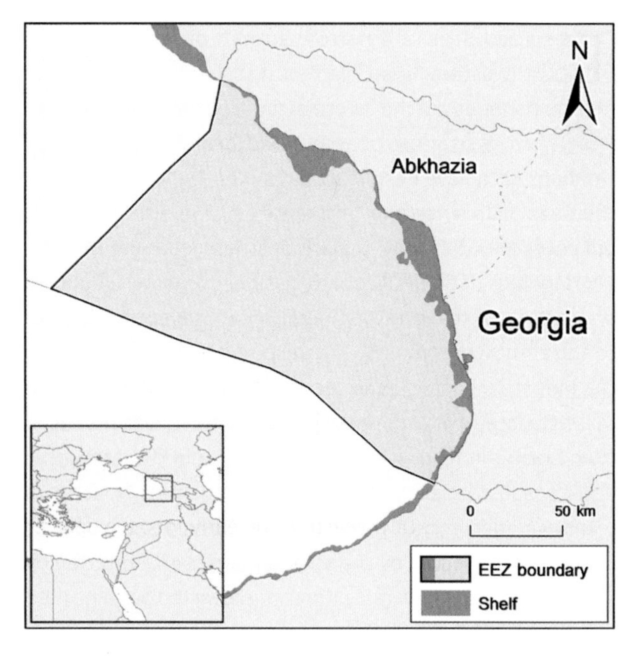

Figure 1. Georgia has a narrow continental shelf, 2,520 km², and an EEZ of 22,900 km². Note that in the Black Sea, waters deeper than 100–150 m are usually devoid of oxygen, and hence of fish.

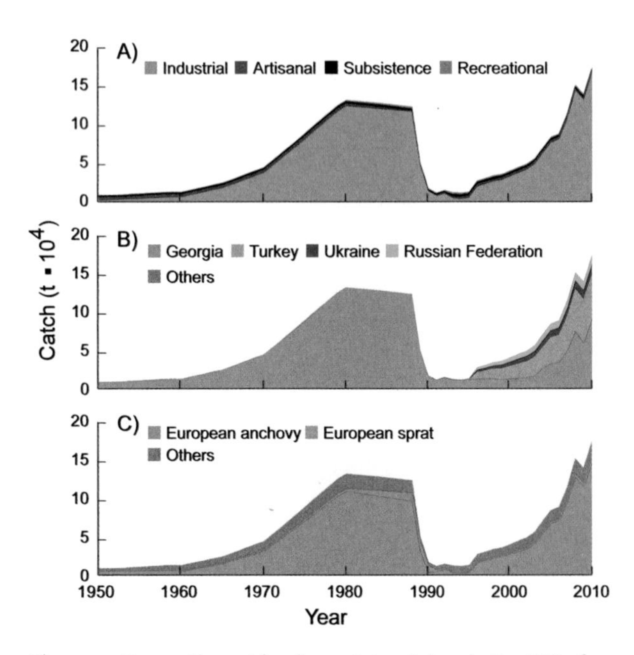

Figure 2. Domestic and foreign catches taken in the EEZ of Georgia: (A) by sector (domestic scores: Ind. 3, 3, 2; Art. 3, 4, 2; Sub. 1, 1, 2; Rec. 1, 1, 2; Dis. 3, 3, 3); (B) by fishing country (foreign catches are very uncertain); (C) by taxon.

REFERENCES

Komakhidze A, Goradze R, Diasamidze R, Mazmanidi N and Komakhidze G (2007) Fish, fisheries and dolphins as indicators of ecosystem health along the Georgian coast of the Black Sea. pp. 251–260 In Payne AIL, O'Brien CMS and Rogers SI (eds.), *Management of Shared Fish Stocks*. Blackwell Publishing Ltd, Cornwall, U.K.

Oztürk B, Keskin C and Engin S (2011) Some remarks on the catches of anchovy, *Engraulis encrasicolus* (Linnaeus, 1758), in Georgian waters by Turkish fleet between 2003 and 2009. *Journal of the Black Sea/Mediterranean Environment* 17(2): 145–158.

Ulman A and Divovich E (2015) The marine fishery catch of Georgia (including Abkhazia) 1950–2010. Fisheries Centre Working Paper #2015–88, 25 p.

1. Cite as Ulman, A., and E. Divovich. 2016. Georgia. P. 274 in D. Pauly and D. Zeller (eds.), *Global Atlas of Marine Fisheries: A Critical Appraisal of Catches and Ecosystem Impacts*. Island Press, Washington, DC.

GERMANY (BALTIC SEA)[1]

Peter Rossing,[a] Cornelius Hammer,[b] Sarah Bale,[a] Sarah Harper,[a] Shawn Booth,[a] Dirk Zeller,[a] and Rainer Froese[c]

[a]*Sea Around Us*, University of British Columbia, Vancouver, Canada [b]Institute for Baltic Sea Fisheries, Rostock, Germany [c]GEOMAR Helmholtz-Centre for Ocean Research, Germany

The Federal Republic of Germany has coastlines along both the North and Baltic Seas (figure 1), but much of the latter, until 1990, was part of the now defunct German Democratic Republic (East Germany). Total biomass withdrawals by Germany from the Baltic Sea were estimated by Rossing et al. (2010) from landing statistics of the International Council for the Exploration of the Sea (ICES), ICES stock assessment working group reports, and national data (including from the former East Germany) on discards and recreational fishing (Zeller et al. 2011). Catches in the Baltic EEZ are predominantly from the industrial sector, with declining catches by the artisanal fisheries; the recreational sector is hardly visible in figure 2A. This yielded reconstructed domestic catches from the German Baltic Sea EEZ of 36,000 t/year in the early 1950s, a peak of 65,000 t in 1977, a sudden decline in 1991 after reunification leading to a low of about 10,000 t/year in the early 2000s, and a partial recovery to about 15,000 t in 2010. The reconstructed catch from 1950–2010 was approximately 1.3 times the catches attributed to Germany by ICES. Most catches are domestic, but foreign fleets also take their share (figure 2B). Figure 2C presents only important taxa, which include Atlantic herring (*Clupea harengus*) and Atlantic cod (*Gadus morhua*), followed by European sprat (*Sprattus sprattus*) and blue mussel (*Mytilus edulis*).

REFERENCES

Rossing P, Hammer C, Bale S, Harper S, Booth S and Zeller D (2010) Germany's marine fisheries catches in the Baltic Sea (1950–2007). pp. 107–126 In Rossing R, Booth S and Zeller D (eds.), *Total marine fisheries extractions by country in the Baltic Sea: 1950–present*. Fisheries Centre Research Reports 18(1). [Updated to 2010]

Zeller D, Rossing P, Harper S, Persson L, Booth S and Pauly D (2011) The Baltic Sea: estimates of total fisheries removals 1950–2007. *Fisheries Research* 108: 356–363.

1. Cite as Rossing, P., C. Hammer, S. Bale, S. Harper, S. Booth, D. Zeller, and R. Froese. 2016. Germany (Baltic). P. 275 in D. Pauly and D. Zeller (eds.), *Global Atlas of Marine Fisheries: A Critical Appraisal of Catches and Ecosystem Impacts*. Island Press, Washington, DC.

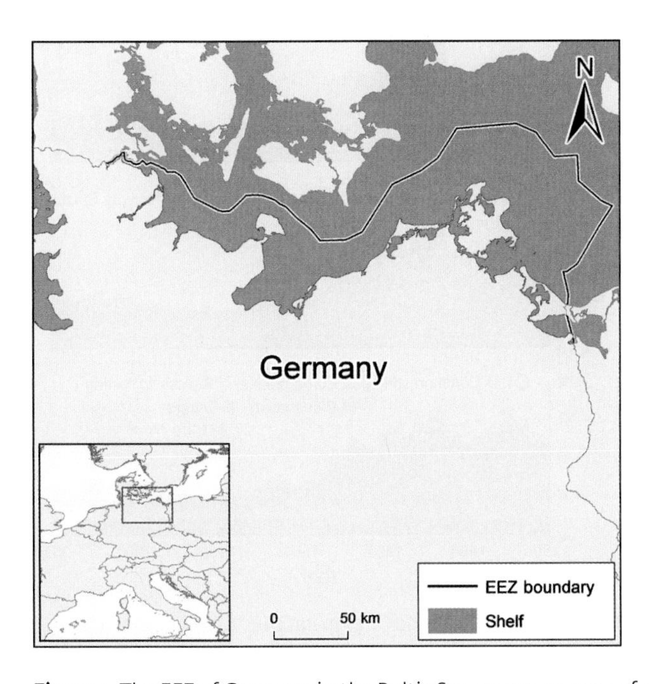

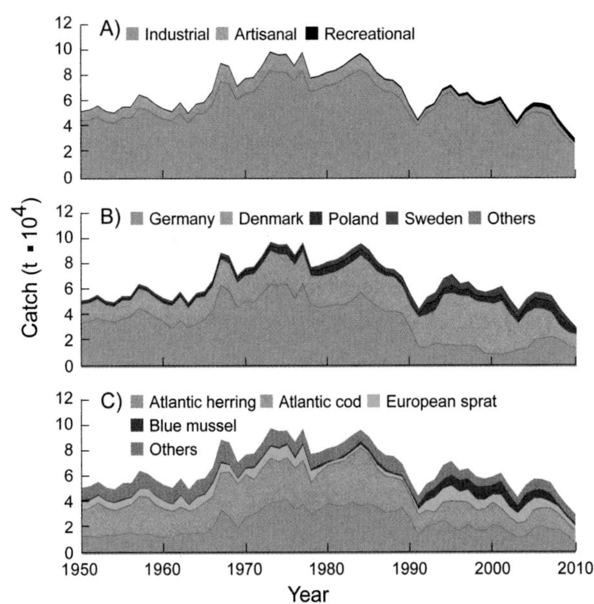

Figure 1. The EEZ of Germany in the Baltic Sea covers an area of 15,500 km², of which 15,500 km² is shelf. **Figure 2.** Domestic and foreign catches taken in the Baltic Sea EEZ of Germany: (A) by sector (domestic scores: Ind. 2, 2, 2; Art. 2, 2, 2; Rec. 1, 1, 1; Dis. 2, 2, 2); (B) by fishing country (foreign catches are very uncertain); (C) by taxon.

GERMANY (NORTH SEA)[1]

Darah Gibson,[a] Rainer Froese,[b] Bernd Ueberschär,[c] Kyrstn Zylich,[a] and Dirk Zeller[a]

[a]*Sea Around Us*, University of British Columbia, Vancouver, Canada [b]GEOMAR Helmholtz-Zentrum für Ozeanforschung, Kiel, Germany [c]GMA, Association for Marine Aquaculture Ltd., Büsum, Germany

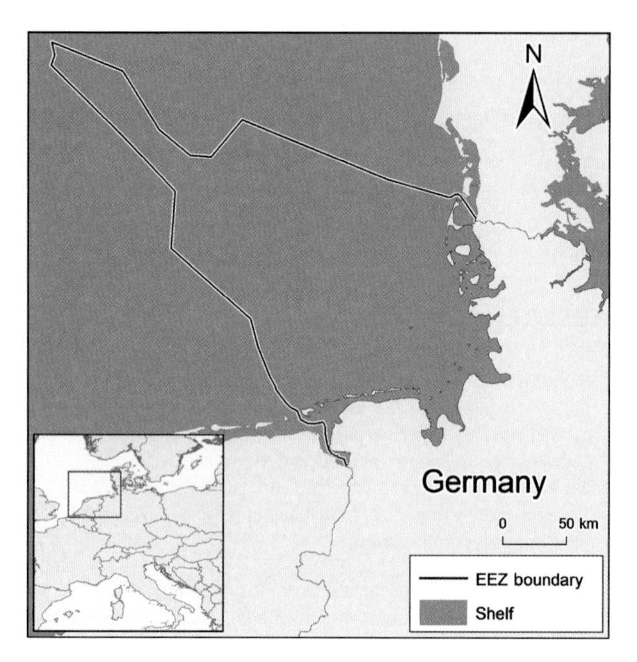

Figure 1. Germany's EEZ in the North Sea (41,000 km², all of which is shelf).

The Federal Republic of Germany has coastlines in the Baltic (see previous page) and North Seas. Gibson et al. (2015) present a reconstruction of German catches in the North Sea from 1950 to 2010, whose data were prorated via surface area to the German EEZ-equivalent area (figure 1). ICES data were used as reported baseline, which was then adjusted using ICES stock assessment working group reports, German data (Kaschner et al. 2001), and other data to estimate unreported landings, recreational and subsistence catches, and discards. The bulk of the catch is derived from the industrial sector, although the artisanal sector does exist. However, both appear to be have declined (figure 2A). The results were reconstructed catches taken within the EEZ, which averaged about 78,000 t/year in the 1950s, declined to 25,000 t in 1990, and averaged about 38,000 t/year in the late 2000s. Reconstructed catches were 1.9 times the data reported for Germany, largely because of unreported discards. Domestic catches dominate, but other North Sea countries' catches do occur (figure 2B). Removals in the EEZ are dominated by brown shrimp (*Crangon crangon*), European plaice (*Pleuronectes platessa*), Atlantic herring (*Clupea harengus*), and blue mussel (*Mytilus edulis*; figure 2C). Of major concern is the disappearance of North Sea cod from the German Bight (Froese and Quaas 2012).

REFERENCES

Froese R and Quaas M (2012) Mismanagement of the North Sea cod by the European Council. *Ocean and Coastal Management* 70: 54–58.

Gibson D, Froese R, Ueberschär B, Zylich K and Zeller D (2015) Reconstruction of total marine fisheries catches for Germany in the North Sea (1950–2010). Fisheries Centre Working Paper #2015-09, 11 p.

Kaschner K, Wolff G and Zeller D (2001) German fisheries: Institutional structure for reporting of catches and fleet statistics (1991–1999). pp. 130–134 In Zeller D, Watson R and Pauly D (eds.), *Fisheries impacts on North Atlantic ecosystems: Catches, effort and national/regional data sets*. Fisheries Centre Research Reports 9(3).

1. Cite as Gibson, D., R. Froese, B. Ueberschär, K. Zylich, and D. Zeller. 2016. Germany (North Sea). P. 276 in D. Pauly and D. Zeller (eds.), *Global Atlas of Marine Fisheries: A Critical Appraisal of Catches and Ecosystem Impacts*. Island Press, Washington, DC.

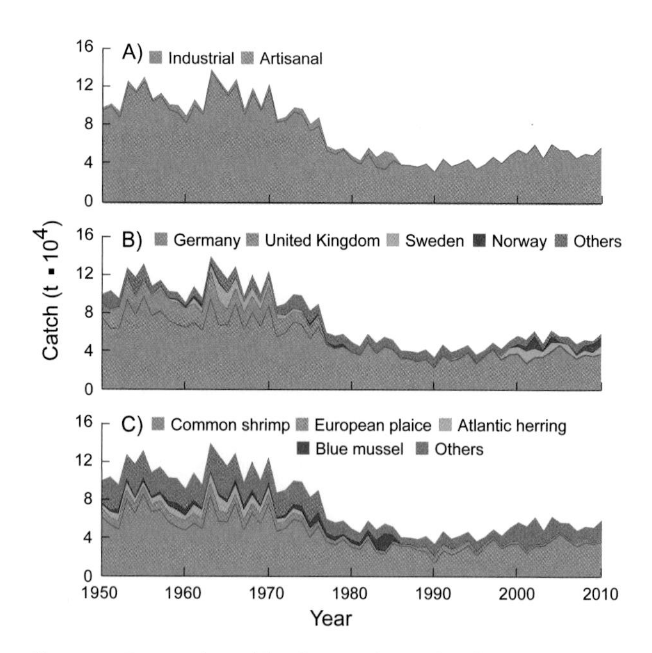

Figure 2. Domestic and foreign catches taken in the North Sea EEZ of Germany: (A) by sector (domestic scores: Ind. 3, 3, 4; Art. 3, 3, 4; Sub. 1, 1, 1; Rec. 1, 1, 2; Dis. 1, 1, 3); (B) by fishing country (foreign catches are very uncertain); (C) by taxon.

GHANA[1]

Francis K. E. Nunoo,[a] Berchie Asiedu,[a] Kofi Amador,[b] Dyhia Belhabib,[c] and Daniel Pauly[c]

[a]Department of Marine and Fisheries Science, University of Ghana, Accra, Ghana [b]Marine Fisheries Research Division, Ministry of Fisheries & Aquaculture Development, Tema, Ghana [c]*Sea Around Us*, University of British Columbia, Vancouver, Canada

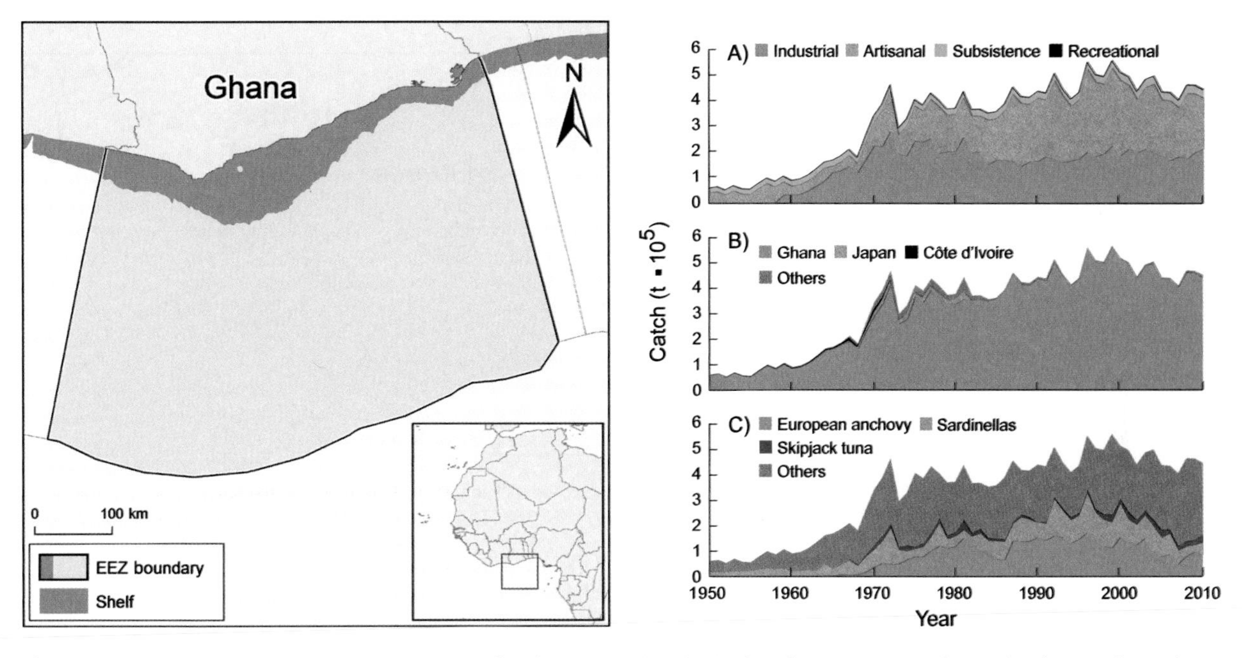

Figure 1. The EEZ of Ghana covers 225,000 km², of which 24,000 km² is shelf. **Figure 2.** Domestic and foreign catches taken in the EEZ of Ghana: (A) by sector (domestic scores: Ind. 4, 4, 4; Art. 3, 3, 3; Sub. 2, 2, 3; Rec. –, 2, 3; Dis. 2, 3, 3); (B) by fishing country (foreign catches are very uncertain); (C) by taxon.

Ghana, located in West Africa (figure 1), had one of the most vibrant fisheries in Africa. However, the fisheries sector nearly collapsed after its distant-water fleet had to return home because of the declaration of EEZs by West African countries (Atta-Mills et al. 2004). Although the Ghanaian fisheries are well documented, official data fail to include the catches of noncommercial sectors, especially the lagoon fisheries (Pauly 2002). Also, Ghana has few data on foreign legal and illegal fleets. The artisanal sector dominates catches, with the industrial sector contributing secondarily (figure 2A). Nunoo et al. (2015) estimated domestic catches as 63,000 t in 1950, which increased to a peak of 415,000 t in 1972 and then rapidly declined after the 1972 military coup to 260,000 t in 1973. Catches increased to 551,000 t in 1996 and then declined gradually to 440,000 t in 2010. Reconstructed domestic catches were 2 times the data supplied to the FAO. Some foreign fishing by Japan and Côte d'Ivoire exists, notably in the 1970–1980s (figure 2B). Major taxa include European anchovy (*Engraulis encrasicolus*), sardinellas (*Sardinella* spp.), specifically the round sardinella (*Sardinella aurita*), and tunas (figure 2C). Underestimating the catches of Ghana's fisheries highlights how reliance on underreported national data puts authorities at serious risk of allowing excessive fishing effort, underestimating the national fisheries, and mismanaging the marine ecosystems.

REFERENCES

Atta-Mills J, Alder J and Sumaila UR (2004) The unmaking of a regional fishing nation: The case of Ghana and West Africa. *Natural Resources Forum* 28: 13–21.

Nunoo FKE, Asiedu B, Amador K, Belhabib D and Pauly D (2015) Reconstruction of marine fisheries catches for Ghana, 1950–2010. Fisheries Centre Working Paper #2015-09, 11 p.

Pauly D (2002) Spatial modelling of trophic interactions and fisheries impacts in coastal ecosystems: a case study of Sakumo Lagoon, Ghana. pp. xxxv & 289–296. In J McGlade, P Cury, KA Koranteng and NJ Hardman-Mountford (eds.), *The Gulf of Guinea Large Marine Ecosystem: Environmental Forcing and Sustainable Development of Marine Resources*. Elsevier Science, Amsterdam.

1. Cite as Nunoo, F. K. E., B. Asiedu, K. Amador, D. Belhabib, and D. Pauly. 2016. Ghana. P. 277 in D. Pauly and D. Zeller (eds.), *Global Atlas of Marine Fisheries: A Critical Appraisal of Catches and Ecosystem Impacts*. Island Press, Washington, DC.

GREECE (EXCLUDING CRETE)[1]

Dimitrios K. Moutopoulos,[a] Athanassios C. Tsikliras,[b] and Konstantinos I. Stergiou[b]

[a]Technological Educational Institute of Western Greece, Department of Aquaculture and Fisheries, Mesolonghi, Greece [b]Laboratory of Ichthyology, Department of Zoology, School of Biology, Aristotle University of Thessaloniki, Thessaloniki, Greece [c]Institute of Marine Biological Resources and Inland Waters, Hellenic Centre for Marine Research, Anavyssos, Attiki, Greece

Greece has strong marine affinities, because of its geography (figure 1), and its fisheries have been exhaustively documented in their various aspects, some more strange than others (Moutopoulos and Koutsikopoulos 2014). Moutopoulos et al. (2015), based on Tsikliras et al. (2006), performed a catch reconstruction summarized here. Although total catches allocated to Greece's EEZ are largely from the artisanal and industrial sectors, a declining trend occurred after 1995 (figure 2A). Total domestic catches in Greek waters (excluding Crete, see next page) increased from 50,000 t/year in the 1950s to a peak of slightly more than 250,000 t in 1994, before declining to just under 160,000 t by 2010. These catches were 1.6 times the officially reported figures, mainly because of discards, the neglect of subsistence and recreational data, and the incomplete coverage of artisanal fisheries, still the mainstay of Greek fisheries. Figure 2B summarizes the total catch by country, illustrating total removals to be predominantly domestic. However, small foreign catches by various other fishing countries have occurred through the time period. Figure 2C presents the catch composition, in which European pilchard (*Sardina pilchardus*), European anchovy (*Engraulis encrasicolus*), and Mediterranean horse mackerel (*Trachurus mediterraneus*) are the most prevalent. Stergiou (2005) demonstrated that the Greek marine ecosystems exhibit symptoms of "fishing down" marine food webs.

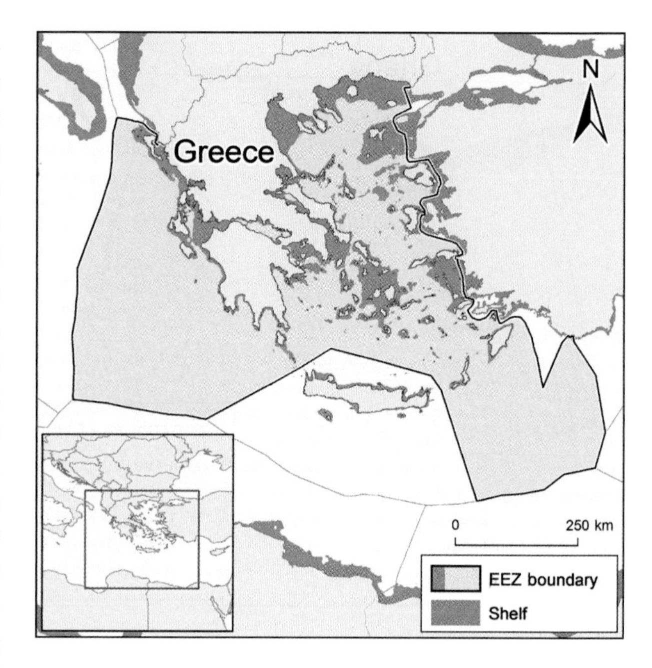

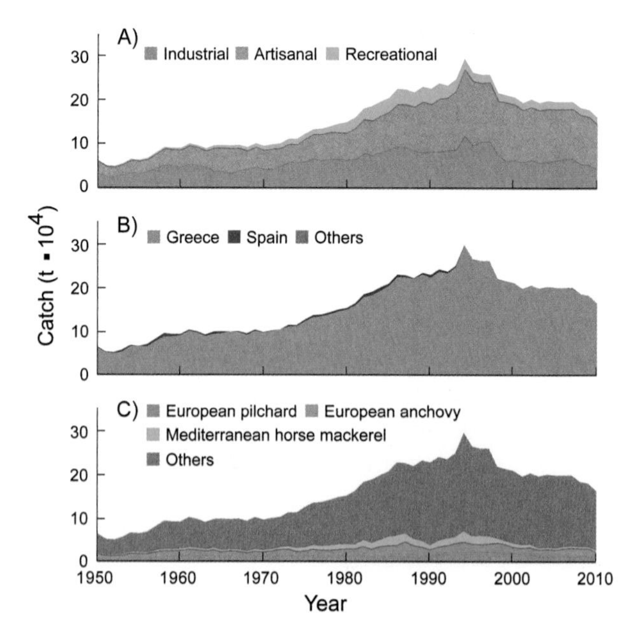

Figure 1. Greece (excluding Crete) has a shelf 68,500 km² and EEZ-equivalent waters of 397,000 km². **Figure 2.** Domestic and foreign catches taken in the EEZ of Greece (excluding Crete): (A) by sector (domestic scores: Ind. 3, 3, 3; Art. 2, 2, 3; Rec. 1, 1, 2; Dis. 1, 1, 3); (B) by fishing country (foreign catches are very uncertain); (C) by taxon.

REFERENCES

Moutopoulos DK and Koutsikopoulos C (2014) Fishing strange data in national fisheries statistics of Greece. *Marine Policy* 48: 114–122.

Moutopoulos DK, Tsikliras AC and Stergiou KI (2015) Reconstruction of Greek fishery catches by fishing gear and area (1950–2010). Fisheries Centre Working Paper #2015-11, 14 p.

Stergiou KI (2005) Fisheries impact on trophic levels: long-term trends in Hellenic waters. pp. 326–329 In Papathanassiou E and Zenetos A (eds.), *State of the Hellenic Marine Environment*. Hellenic Centre for Marine Research, Institute of Oceanography, Athens, Greece.

Tsikliras AC, Moutopoulos DK and Stergiou KI (2006) Reconstruction of Greek marine fisheries landings, and comparison of national with the FAO statistics. pp. 1–17 In Palomares MLD, Stergiou KI and Pauly D (eds.), *Fishes in databases and ecosystems*. Fisheries Centre Research Reports 14(4).

1. Cite as Moutopoulos, D. K., A. C. Tsikliras, and K. I. Stergiou. 2016. Greece (excluding Crete). P. 278 in D. Pauly and D. Zeller (eds.), *Global Atlas of Marine Fisheries: A Critical Appraisal of Catches and Ecosystem Impacts*. Island Press, Washington, DC.

GREECE (CRETE)[1]

Dimitris K. Moutopoulos,[a] Nicolas Bailly,[b] Athanassios C. Tsikliras,[c] and Konstantinos I. Stergiou[c]

[a]Technological Educational Institute of Mesolonghi, Department of Aquaculture and Fisheries Management, Mesolonghi, Greece [b]Hellenic Center for Marine Research, Crete, Greece [c]University of Thessaloniki, Thessaloniki, Greece

Crete is located in the southern part of the Aegean Sea (figure 1) and has a rich history marked by foreign conquests and the flowering of local cultures, both attested by a varied archeological heritage, which attracts numerous tourists (Duke 2007). Based on the reconstruction by Moutopoulos et al. (2015), the total marine catch by sector allocated to Crete for the period 1950–2010 has been largely commercial (figure 2A). More than half of the total catches were artisanal, with a declining industrial component. Recreational and subsistence catches contributed a small portion to the total catch, leading to a per capita consumption of fish (an essential component of the "Mediterranean diet," best exemplified in Crete) of 21 kg/year. Domestic catches increased from 1,700 t/year in the early 1950s, peaked at 18,000 t in 1992, then declined to 6,700 t/year in the late 2000s. Although the domestic fishery dominates, small catches were taken by Spain and other countries before the EEZ declaration in 1995 (figure 2B). Catches were dominated by picarel (*Spicara smaris*), bogue (*Boops boops*), large-eye dentex (*Dentex macrophthalmus*), and Mediterranean horse mackerel (*Trachurus mediterraneus*; figure 2C), as part of catches comprising more than 60 taxa. This has a price: Of 221 fish species occurring around Crete, as assessed by a thorough review of the literature by the second author, 26 are threatened and on the IUCN Red List (see also www.fishbase.org).

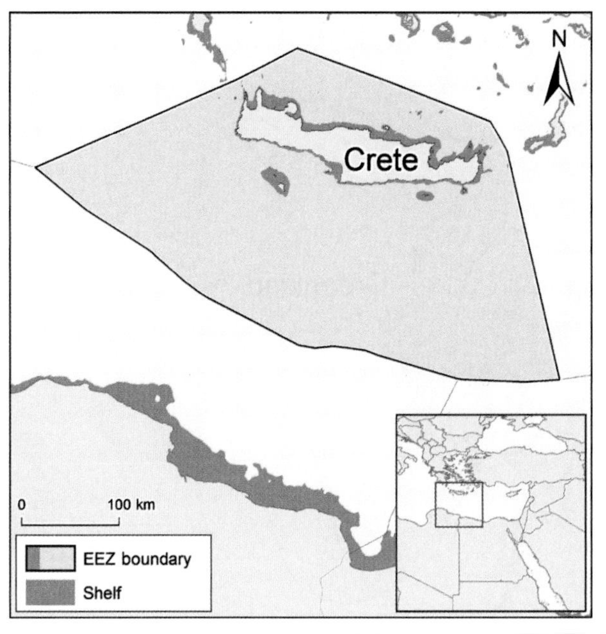

Figure 1. Crete (Greece) has a land area of 8,300 km² and an EEZ-equivalent area of 96,500 km².

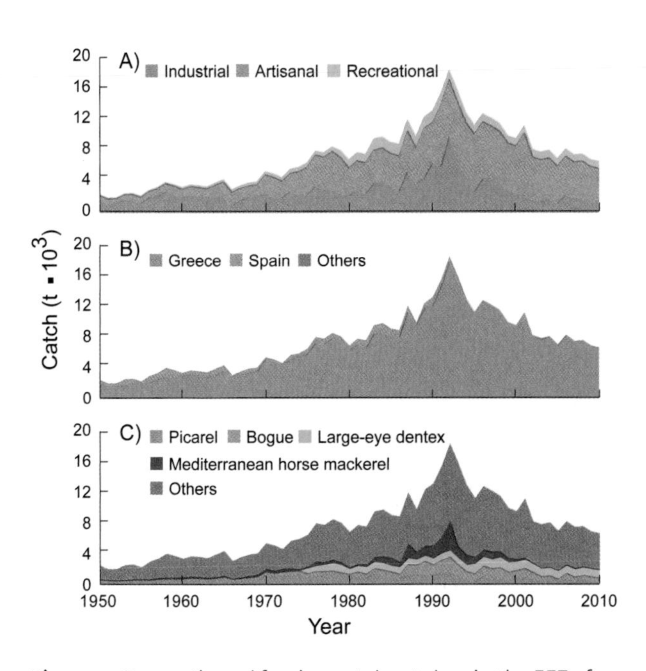

Figure 2. Domestic and foreign catches taken in the EEZ of Crete (Greece): (A) by sector (domestic scores: Ind. 3, 3, 3; Art. 2, 2, 3; Rec. 1, 1, 2; Dis. 1, 1, 3); (B) by fishing country (foreign catches are very uncertain); (C) by taxon.

REFERENCES

Duke, P.G. (2007). *The Tourists Gaze, the Cretans Glance: Archaeology and Tourism on a Greek Island*. Vol. 1. Left Coast Press, Santa Barbara.

Moutopoulos DK, Tsikliras AC and Stergiou KI (2015) Reconstruction of Greek fishery catches by fishing gear and area (1950–2010). Fisheries Centre Working Paper #2015–11, 14 p.

1. Cite as Moutopoulos, D. K., N. Bailly, A. C. Tsikliras, and K. I. Stergiou. 2016. Greece (Crete). P. 279 in D. Pauly and D. Zeller (eds.), *Global Atlas of Marine Fisheries: A Critical Appraisal of Catches and Ecosystem Impacts*. Island Press, Washington, DC.

GREENLAND[1]

Shawn Booth and Danielle Knip

Sea Around Us, University of British Columbia, Vancouver, Canada

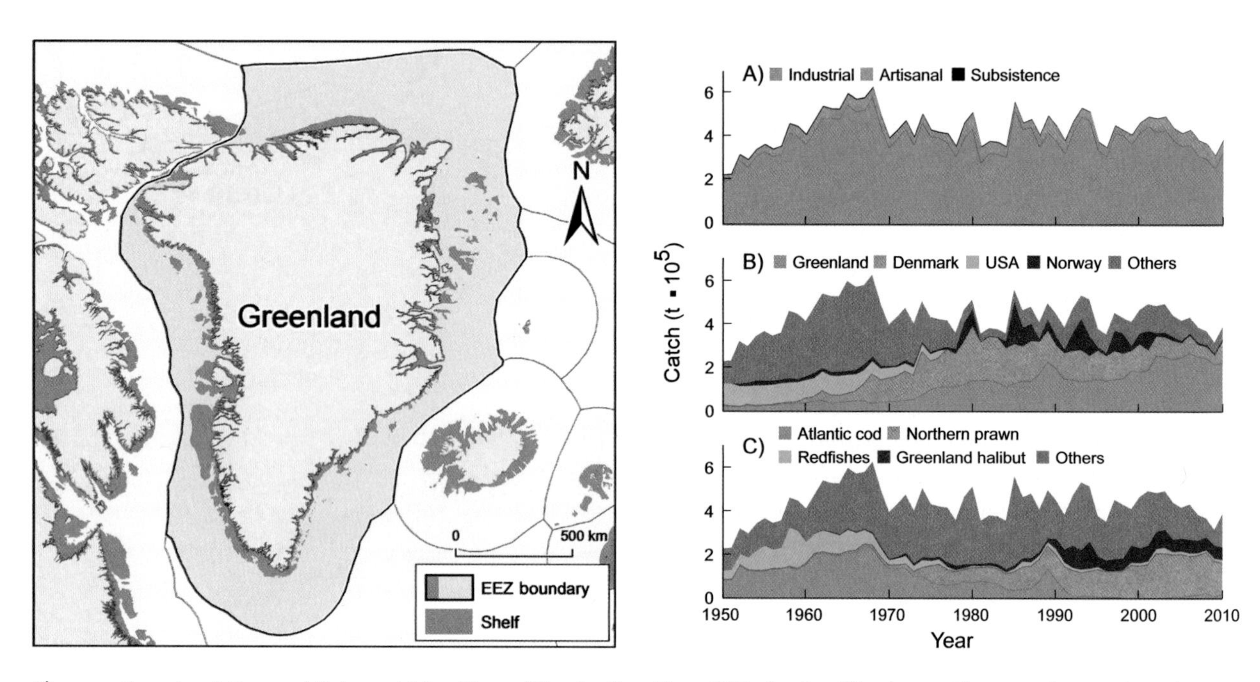

Figure 1. Greenland, the world's largest island (2.2 million km²), with an EEZ of 2.28 million km². **Figure 2.** Domestic and foreign catches taken in the EEZ of Greenland: (A) by sector (domestic scores: Ind. 2, 3, 3; Art. 2, 3, 3; Sub. 2, 2, 2; Dis. 3, 3, 3); (B) by fishing country (foreign catches are very uncertain); (C) by taxon.

Greenland, an autonomous territory within the Kingdom of Denmark, has coastlines along the Arctic Sea, the Northeast Atlantic, and the Northwest Atlantic (figure 1). Total catches of the marine fisheries in Greenland's waters from 1950 to 2010 were presented by Booth and Knip (2014). The bulk of the catch is generated from the industrial sector; however, there is a gradually increasing artisanal sector (figure 2A). The reconstructed total domestic catch of Greenland increased from 29,000 t in 1950 to 270,000 t in 2006, before declining to 225,000 t in 2010. However, although this is about 30% higher than the landings reported by the FAO and ICES on behalf of Greenland, this still only accounts for less than half the annual catch taken from Greenland waters, because of large catches by Danish and foreign vessels, such as from the United States and Norway (figure 2B). Thus, the discrepancy between the ICES and FAO data and the reconstructed catch is mostly caused by discards in the industrial fisheries. Atlantic cod (*Gadus morhua*) is the most important species in the subsistence sector, but northern prawn (*Pandalus borealis*) represents a substantial portion of the reconstructed total catch (figure 2C). The mean trophic level (TL) of the foreign and domestic catches declined strongly, by 0.26·TL units per decade, primarily because the fishery changed its main target from cod to shrimp (Hamilton et al. 2003), a classic case of "fishing down."

REFERENCES

Booth S and Knip D (2014) The catch of living marine resources around Greenland from 1950–2010. pp. 55–72 In Zylich K, Zeller D, Ang M and Pauly D (eds.), *Fisheries catch reconstructions: Islands, Part IV*. Fisheries Centre Research Reports 22(2).

Hamilton LC, Brown BC and Rasmussen RO (2003) West Greenland's cod-to-shrimp transition: Local dimensions of climate change. *Arctic* 56(3): 271–282.

1. Cite as Booth, S., and D. Knip. 2016. Greenland. P. 280 in D. Pauly and D. Zeller (eds.), *Global Atlas of Marine Fisheries: A Critical Appraisal of Catches and Ecosystem Impacts*. Island Press, Washington, DC.

GRENADA[1]

Elizabeth Mohammed[a] and Alasdair Lindop[b]

[a]Research and Resource Assessment, Caribbean Regional Fisheries Mechanism Secretariat, Eastern Caribbean Office, St. Vincent and the Grenadines [b]*Sea Around Us*, University of British Columbia, Vancouver, Canada

Grenada is a small country in the south eastern Caribbean Sea consisting of the primary island of Grenada and several islands at the southern end of the Grenadine chain (figure 1). The fisheries are largely foreign and industrial, and a small domestic component is derived from the artisanal sector (figure 2A, 2B), which for many years used traditional techniques (McConney and Baldeo 2007). A domestic industrial fishery began in the early 1990s, when 8 pelagic longliners with cold storage entered the fishery. The reconstruction of Mohammed and Lindop (2015), which updates earlier work by Mohammed and Rennie (2003), yielded a domestic catch estimate that, as a whole, and with some fluctuations, grew from 800 t/year in the early 1950s to a high of 3,100 t in 2003 and 2,500 t in 2010. Overall, the reconstructed catch of Grenada for 1950–2010 was 1.5 times the amount reported by the FAO on behalf of Grenada. Fish caught from around the island of Grenada made up 89% of the catch, with the rest from the southern end of the Grenadines. Atlantic bonito (*Sarda sarda*) was a primary taxon in the catch, particularly in the earlier years (figure 2C). King mackerel (*Scomberomorus cavalla*), bigeye scad (*Selar crumenophthalmus*), and yellowfin tuna (*Thunnus albacares*) were also important.

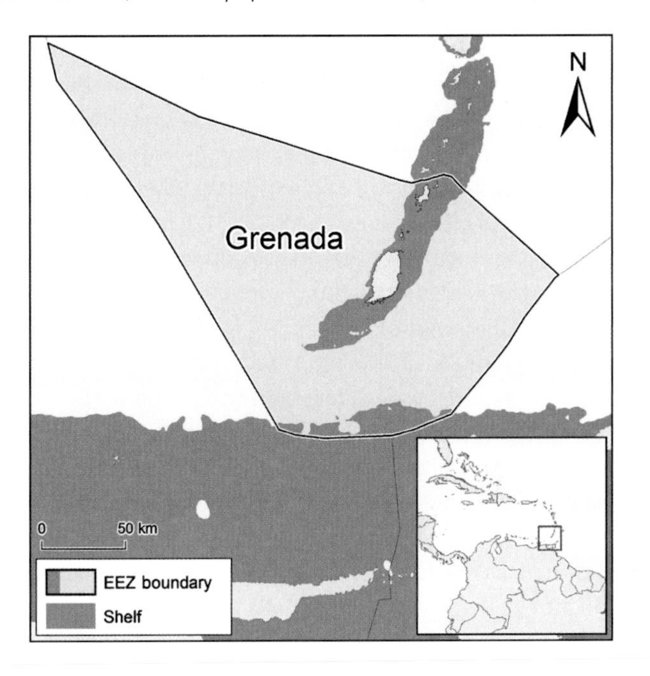

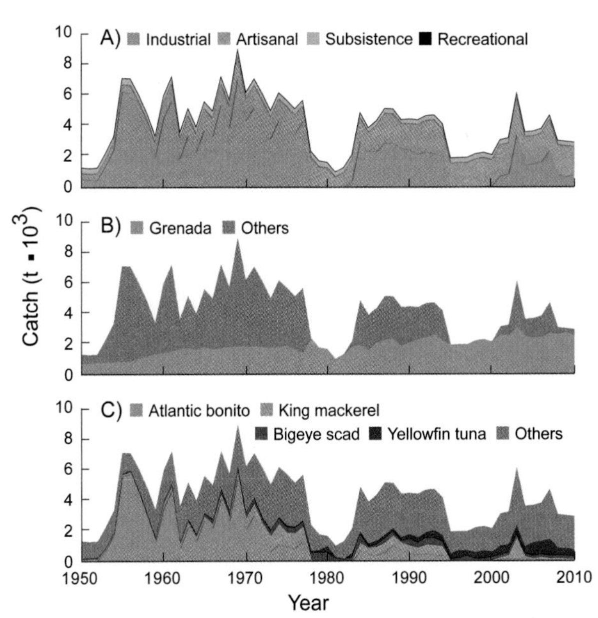

Figure 1. Grenada has a land area of 360 km² and an EEZ of 26,100 km². **Figure 2.** Domestic and foreign catches taken in the EEZ of Grenada: (A) by sector (domestic scores: Ind. –, –, 3; Art. 2, 3, 3; Sub. 2, 2, 2; Rec. 1, 1, 1); (B) by fishing country (foreign catches are very uncertain); (C) by taxon.

REFERENCES

McConney P and Baldeo R (2007) Lessons in co-management from beach seine and lobster fisheries in Grenada. *Fisheries Research* 87(1): 77–85.

Mohammed E and Lindop A (2015) Grenada: Reconstructed Fisheries Catches, 1950–2010. Fisheries Centre Working Paper #2015-40, 27 p.

Mohammed E and Rennie J (2003) Grenada and the Grenadines: Reconstructed fisheries catches and fishing effort, 1942–2001. pp. 67–94 In Zeller D, Booth S, Mohammed E and Pauly D (eds.), *From Mexico to Brazil: Central Atlantic fisheries catch trends and ecosystem models*. Fisheries Centre Research Reports 11(6).

1. Cite as Mohammed, E., and A. Lindop. 2016. Grenada. P. 281 in D. Pauly and D. Zeller (eds.), *Global Atlas of Marine Fisheries: A Critical Appraisal of Catches and Ecosystem Impacts*. Island Press, Washington, DC.

GUATEMALA (CARIBBEAN)[1]

Alasdair Lindop,[a] Marcelo Ixquiac-Cabrera,[b] Kyrstn Zylich,[a] and Dirk Zeller[a]

[a]*Sea Around Us*, University of British Columbia, Vancouver, Canada [b]Universidad de San Carlos de Guatemala, Centro de Estudios del Mar y Acuicultura, Ciudad Universitaria, Guatemala

Guatemala is an impoverished Central American country bordering the Pacific and Atlantic Oceans. In the Caribbean (figure 1), where its coast is squeezed between those of Belize and Honduras (Perez 2009), artisanal fishing dominates, because industrial fishing is banned from the Bay of Amatique and limited to offshore areas (figure 2A). The reconstruction by Lindop et al. (2015) led to tentative estimates of domestic catches that steadily increased from 350 t in 1950 to 3,800 t in 1994; later, the increase is accelerated, with catches reaching 6,700 t in 2001 and on average 6,100 t/year throughout the 2000s. Overall, this was 19 times the landings reported by the FAO on behalf of Guatemala, the discrepancy being caused mainly by nonreporting of the artisanal fishery and the underreporting of the industrial sector, which also discards about half of its catch. Total allocated catch within the Caribbean EEZ of Guatemala is predominantly domestic, with a few tonnes caught by foreign vessels from Cuba and other countries, although they are hardly visible on figure 2B. Anchovies (family Engraulidae) were overwhelmingly dominant in the catch, and mojarras (Gerreidae), cusk-eels (Ophidiidae), and penaeid shrimp (*Penaeus* spp.) also contributing (figure 2C). The artisanal fishery in the Bay of Amatique is well studied (Andrade and Midré 2011) and illustrates the "Malthusian overfishing" (Pauly 2006) that besets much of the developing world.

REFERENCES

Andrade H and Midré G (2011) The merits of consensus: Small-scale fisheries as a livelihood buffer in Livingston, Guatemala. pp. 427–448 In Jentoft S and Eide A (eds.), *Poverty Mosaics: Realities and Prospects in Small-Scale Fisheries*. Springer Netherlands.

Lindop A, Ixquiac-Cabrera M, Zylich K and Zeller D (2015) A reconstruction of marine fish catches in the Republic of Guatemala. Fisheries Centre Working Paper #2015–41, 17 p.

Pauly D (2006) Major trends in small-scale marine fisheries, with emphasis on developing countries, and some implications for the social sciences. *Maritime Studies (MAST)* 4(2): 7–22.

Perez A (2009) Fisheries management at the tri-national border between Belize, Guatemala and Honduras. *Marine Policy* 33(2): 195–200.

1. Cite as Lindop, A., M. Ixquiac-Cabrera, K. Zylich, and D. Zeller. 2016. Guatemala (Caribbean). P. 282 in D. Pauly and D. Zeller (eds.), *Global Atlas of Marine Fisheries: A Critical Appraisal of Catches and Ecosystem Impacts*. Island Press, Washington, DC.

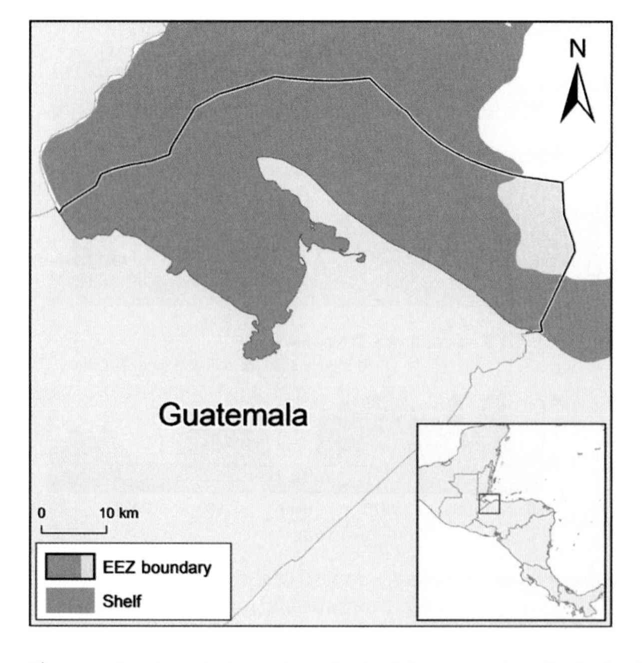

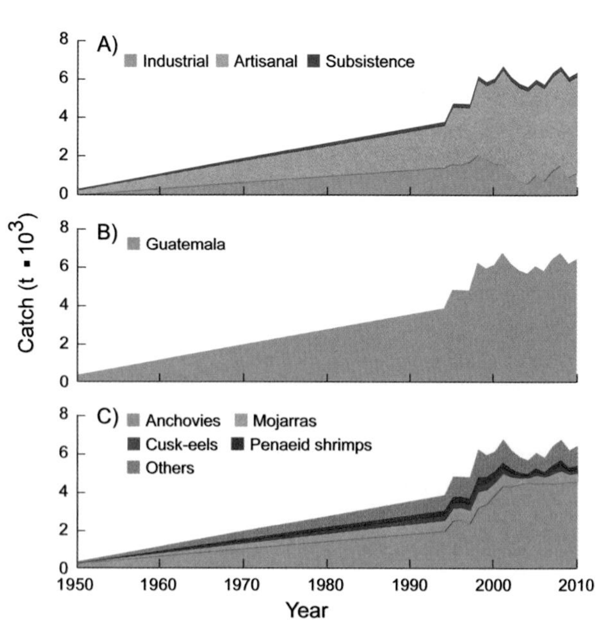

Figure 1. Guatemala has, along its Caribbean coast, a shelf of 1,480 km² and an EEZ of 1,600 km². **Figure 2.** Domestic and foreign catches taken in the Caribbean EEZ of Guatemala: (A) by sector (domestic scores: Ind. 2, 2, 2; Art. 2, 2, 2; Sub. 1, 1, 1; Dis. 1, 1, 1); (B) by fishing country (foreign catches are very uncertain); (C) by taxon.

GUATEMALA (PACIFIC)[1]

Alasdair Lindop,[a] Marcelo Ixquiac-Cabrera,[b] Kyrstn Zylich,[a] and Dirk Zeller[a]

[a]*Sea Around Us*, University of British Columbia, Vancouver, Canada [b]Universidad de San Carlos de Guatemala, Centro de Estudios del Mar y Acuicultura, Ciudad Universitaria, Guatemala

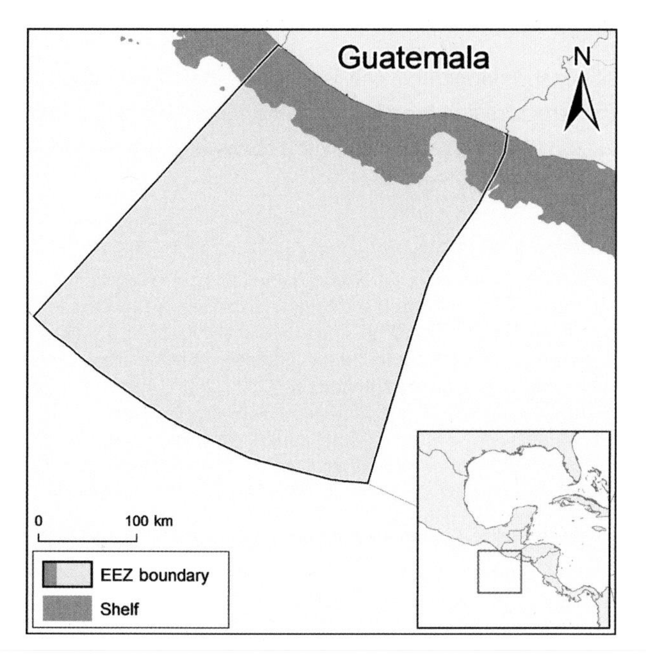

Figure 1. Guatemala has a shelf of 13,900 km² and an EEZ of 116,000 km² along its Pacific coast.

Guatemala is a very poor Central American country with coasts on the Caribbean Sea (see previous page) and the Pacific Ocean (figure 1), the latter being more important in terms of fisheries. The majority of the total allocated catch is derived from the industrial sector, which appears to be declining. The artisanal sector experienced accelerating growth in the late 1990s and has surpassed the industrial fishery to dominate catches (figure 2A). The total domestic catch from the Pacific coast of Guatemala, as reconstructed by Lindop et al. (2015), grew from 2,400 t in 1950 to a high of 69,000 t in 1996, and catches after 2005 decreased to 25,000 t in 2009. This amounted to 13 times the landings data reported by the FAO on behalf of Guatemala for the 1950–2010 period. The reasons for the substantial discrepancies are the underreporting of industrial landings, the nonreporting of the artisanal and subsistence fisheries, and the discards of the industrial (shrimp) fishery. Figure 2B summarizes catches by fishing country to be mostly domestic, although foreign fishing by the United States and Mexico, among others, has been documented mostly before the EEZ declaration in 1976. Total catches are dominated by drums, croakers (family Sciaenidae), and grunts (Haemulidae; figure 2C). The report by Alvarado and Mijango Lopez (1999) is one of the few studies of Pacific fisheries of Guatemala.

REFERENCES

Alvarado CR and Mijango Lopez N (1999) Estudio sobre la pesqueria de tiburon en Guatemala, Chapter 6. In Shotton R (ed.), *Case studies of management of elasmobranch fisheries*. FAO Technical Paper 378. (www.fao.org/docrep/003/x2097e/x2097E08.htm).

Lindop A, Ixquiac-Cabrera M, Zylich K and Zeller D (2015) A reconstruction of marine fish catches in the Republic of Guatemala. Fisheries Centre Working Paper #2015-41, 17 p.

1. Cite as Lindop, A., M. Ixquiac-Cabrera, K. Zylich, and D. Zeller. 2016. Guatemala (Pacific). P. 283 in D. Pauly and D. Zeller (eds.), *Global Atlas of Marine Fisheries: A Critical Appraisal of Catches and Ecosystem Impacts*. Island Press, Washington, DC.

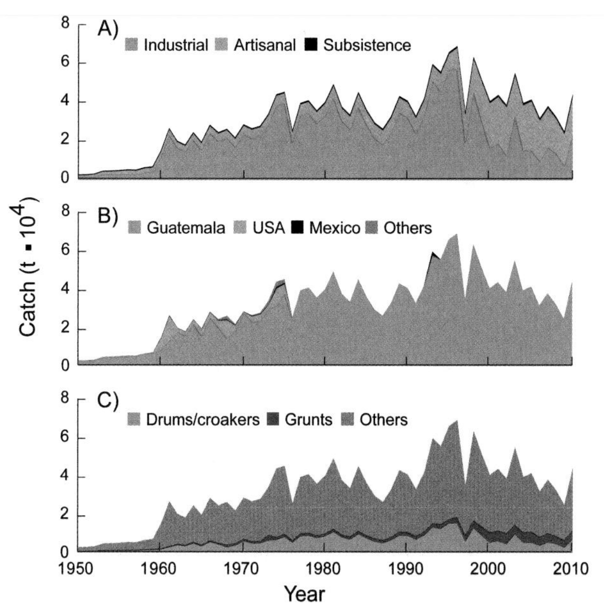

Figure 2. Domestic and foreign catches taken in the Pacific EEZ of Guatemala: (A) by sector (domestic scores: Ind. 2, 2, 2; Art. 2, 2, 2; Sub. 1, 1, 1; Dis. 1, 1, 1); (B) by fishing country (foreign catches are very uncertain); (C) by taxon.

GUINEA[1]

Dyhia Belhabib,[a] Alkaly Doumbouya,[b] Ibrahima Diallo,[b] Sory Traore,[b] Youssouf Camara,[b] Duncan Copeland,[a] Beatrice Gorez,[c] Sarah Harper,[a] Dirk Zeller,[a] and Daniel Pauly[a]

[a]*Sea Around Us*, University of British Columbia, Vancouver, Canada [b]Centre National des Sciences Halieutiques de Boussoura, Conakry, Guinea [c]Coalition for Fair Fishing Agreements, Bruxelles, Belgium

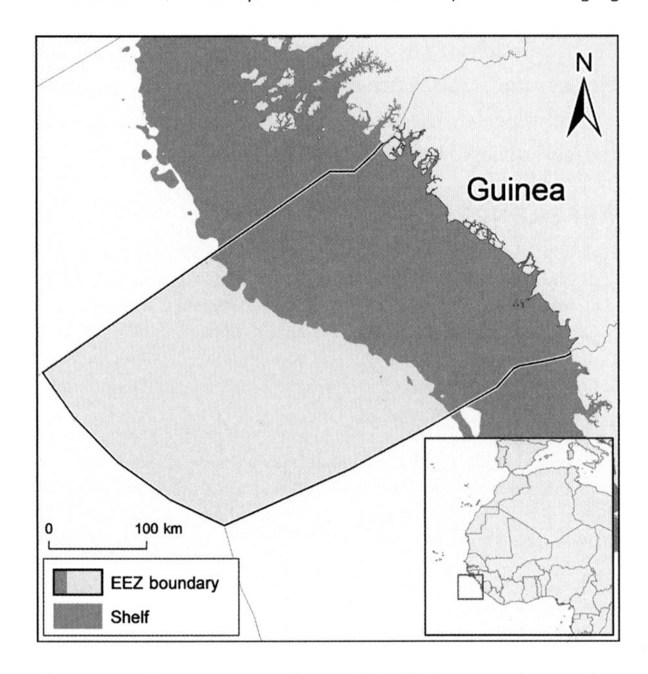

Figure 1. Guinea has a continental shelf of 50,000 km² and an EEZ of 109,000 km².

Guinea is a poor country in northwest Africa that has the largest continental shelf in the region (figure 1). However, only part of the artisanal and industrial domestic fisheries catches is officially reported (Domain 1999). Industrial, artisanal, and subsistence catches along with discards by the domestic and foreign legal and illegal fisheries were reconstructed by Belhabib et al. (2012). Industrial catches dominate total removals in Guinea's EEZ (figure 2A). The reconstructed domestic catch was 48,000 t in 1950, reached a peak of 178,000 t in 1985, and declined to 126,000 t in 1999, followed by an increase to about 230,000 t in 2010. Reconstructed domestic catches were 4 times the data supplied to the FAO, yet this was only a small component of the removals from Guinea's EEZ, at least when compared with the foreign fishing by, for example, Spain and Russia (figure 2B). Figure 2C summarizes catches by taxa, dominated by cephalopods (largely from foreign vessels), bonga shad (*Ethmalosa fimbriata*), and other small pelagic species (e.g., *Sardinella* spp.). Overexploitation and the lack of monitoring of foreign fleets have negative repercussions for local fishing communities, which are expanding their fishing grounds to catch fish whose individual size is shrinking. All of this is inducing massive changes in the structure and functioning of the ecosystem (Laurans et al. 2004), as indicated also by the occurrence of the "fishing down" phenomenon.

REFERENCES

Belhabib D, Doumbouya A, Diallo I, Traore S, Camara Y, Copeland D, Gorez B, Harper S, Zeller D and Pauly D (2012) Guinean fisheries, past, present and future? pp. 91–104 In Belhabib D, Zeller D, Harper S and Pauly D (eds.), *Marine fisheries catches in West Africa, Part I.* Fisheries Centre Research Reports 20(3).

Domain F (1999) Influence de la pêche et de l'hydroclimat sur l'évolution dans le temps du stock côtier (1985–1995). pp. 117–136 In Domain F, Chavance P and Diallo A (eds.), *La pêche côtière en Guinée: Ressources et exploitation.* Institut de Recherche pour le Développement (IRD), Conakry, Guinea.

Laurans M, Gascuel D, Chassot E and Thiam D (2004) Changes in the trophic structure of fish demersal communities in West Africa in the three last decades. *Aquatic Living Resources* 17(2): 163–173.

1. Cite as Belhabib, D., A. Doumbouya, I. Diallo, S. Traore, Y. Camara, D. Copeland, B. Gorez, S. Harper, D. Zeller, and D. Pauly. 2016. Guinea. P. 284 in D. Pauly and D. Zeller (eds.), *Global Atlas of Marine Fisheries: A Critical Appraisal of Catches and Ecosystem Impacts*. Island Press, Washington, DC.

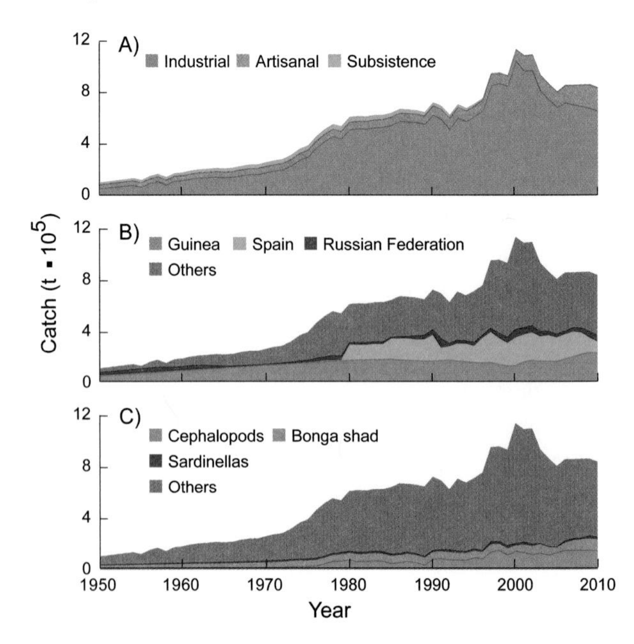

Figure 2. Domestic and foreign catches taken in the EEZ of Guinea: (A) by sector (domestic scores: Ind. 2, 4, 2; Art. 2, 3, 4; Sub. 2, 2, 2; Dis. 4, 4, 4); (B) by fishing country (foreign catches are very uncertain); (C) by taxon.

GUINEA-BISSAU[1]

Dyhia Belhabib and Daniel Pauly

Sea Around Us, Fisheries Centre, University of British Columbia, Vancouver, Canada

Guinea-Bissau, located in northwest Africa (figure 1), has a large continental shelf adjacent to an intricate lagoon system, whose productivity is enhanced by a coastal upwelling and riverine nutrients, respectively. These are optimal conditions for coastal fisheries. However, a difficult history, including a long war of liberation from Portuguese colonialism, had made the country one of the poorest in the world, with little monitoring of its fisheries, despite the strong dependence on fisheries for income and food security (Tvedten 1990). The main sector identified is the industrial sector, which includes both foreign and domestic vessels (figure 2A); the foreign component is represented by Russia, South Korea, China, and others (figure 2B). Domestic catches, as reconstructed by Belhabib and Pauly (2015), increased from 13,500 t in 1950 to 44,000 t in 2000 and averaged 33,400 t/year in the late 2000s. These reconstructed catches from Guinea-Bissau, including a growing woman-dominated subsistence sector, were estimated to be 10 times the data reported by the FAO on behalf of Guinea-Bissau. Figure 2C summarizes catches by taxon, with sea catfishes (family Ariidae), horse mackerels (*Trachurus* spp.), and tonguefishes (Cynoglossidae) being most important. The contribution of largely artisanally caught fish species such as bonga shad (*Ethmalosa fimbriata*) and croakers (Sciaenidae) to the total catch, not visible on the graph, appear to be decreasing over time, while subsistence catches of bivalves have increased.

REFERENCES

Belhabib D and Pauly D (2015) Fisheries in troubled waters: a catch reconstruction for Guinea-Bissau, 1950–2010. pp. 1–16. in D Belhabib and D Pauly (eds.). *Fisheries catch reconstructions: West Africa, Part II*. Fisheries Centre Research Reports 23(3).

Tvedten I (1990) The difficult transition from subsistence to commercial fishing. The case of the Bijagos of Guinea-Bissau. *Maritime Studies (MAST)* 3(1): 119–130.

1. Cite as Belhabib, D., and D. Pauly. 2016. Guinea-Bissau. P. 285 in D. Pauly and D. Zeller (eds.), *Global Atlas of Marine Fisheries: A Critical Appraisal of Catches and Ecosystem Impacts*. Island Press, Washington, DC.

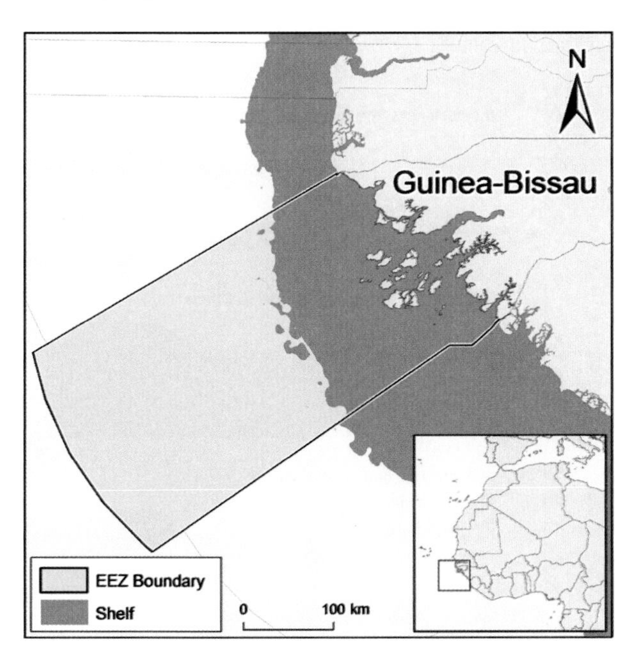

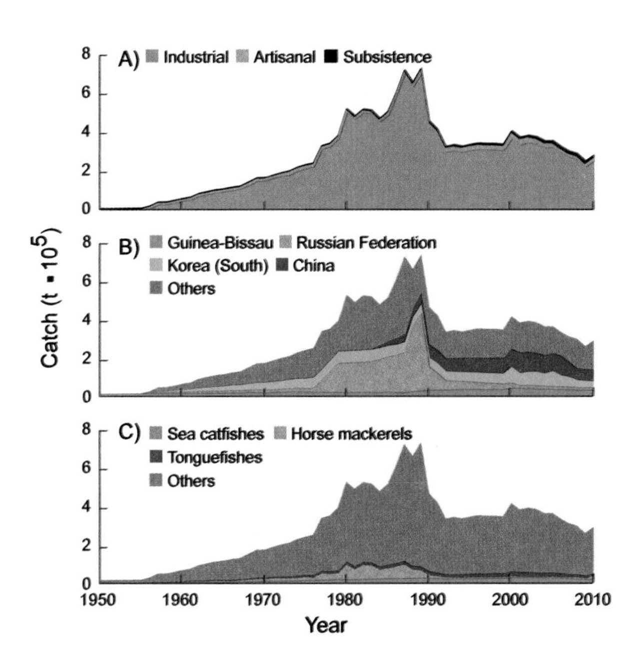

Figure 1. Guinea-Bissau (EEZ: 106,000 km²) has a continental shelf of 38,200 km². **Figure 2.** Domestic and foreign catches taken in the EEZ of Guinea-Bissau: (A) by sector (domestic scores: Art. 3, 3, 3; Sub. 2, 3, 3; Rec. –, 2, 4); (B) by fishing country (foreign catches are very uncertain); (C) by taxon.

GUYANA[1]

Jessica MacDonald, Sarah Harper, Shawn Booth, Dirk Zeller, and Daniel Pauly

Sea Around Us, University of British Columbia, Vancouver, Canada

Guyana, formerly known as British Guiana, is located on the northeast coast of South America (figure 1). Overall, total removals from Guyana's EEZ were predominantly artisanal, with a growing industrial sector (figure 2A). However, this trend masks the shift from nearly exclusive subsistence fisheries to a dominance of artisanal fisheries. Based on the reconstruction of MacDonald et al. (2015), the domestic marine fisheries within Guyana's EEZ had a catch of 11,000 t in 1950, which increased to a peak of 84,000 t in 2003, then declined to 65,000 t in 2010. Domestic catches make up a substantial portion of the total allocated catches, with some foreign fishing by Venezuela and the United States through the time period (figure 2B). Overall, domestic catches are 1.6 times the landings that FAO presents on behalf of Guyana, mainly because of unreported subsistence and underreported artisanal catches, as well as discards by the trawl fishery for Atlantic seabob (*Xiphopenaeus kroyeri*), the top taxon in Guyana's EEZ (figure 2C). These discards were mostly perfectly good fish to eat; Guyanese (and later Canadian) fisheries scientist Dr. Herbert Allsopp coined the term *bycatch* and ran a worldwide campaign for its retention and use (Pauly 2007; see also Allsopp 1982). The discarding bans and bycatch retention policies that are now widely being implemented or considered, for example in the EU, go back to this.

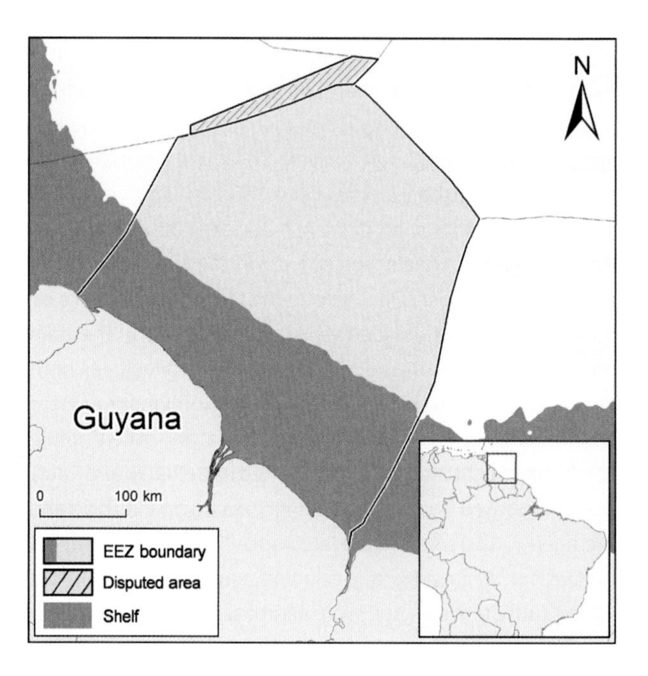

Figure 1. Guyana has a shelf of 50,500 km² and an EEZ of 140,000 km².

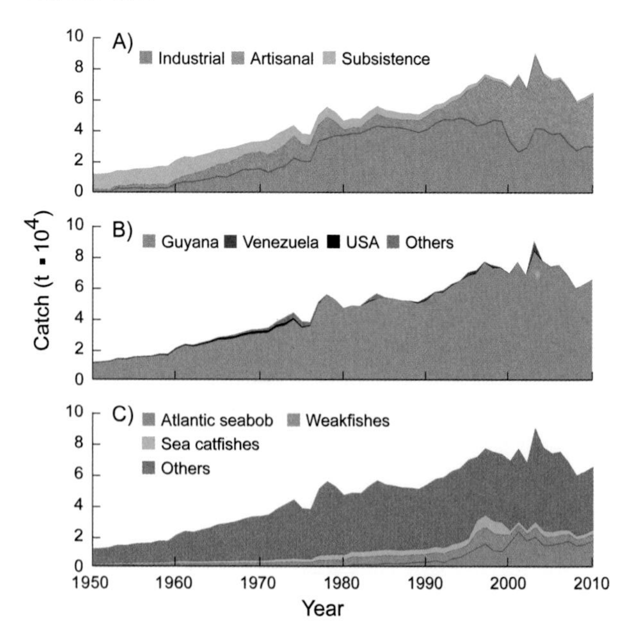

Figure 2. Domestic and foreign catches taken in the EEZ of Guyana: (A) by sector (domestic scores: Ind. 3, 3, 3; Art. 2, 2, 2; Sub. 1, 1, 1; Dis. 1, 2, 2); (B) by fishing country (foreign catches are very uncertain); (C) by taxon.

REFERENCES

Allsopp WHL (1982) Use of fish bycatch from shrimp trawling: future development. pp. 29–36 In Fish bycatch bonus from the sea. Report of the technical consultation on shrimp bycatch utilization, 27–30 October 1981, Georgetown, Guyana.

MacDonald J, Harper S, Booth S and Zeller D (2015) Guyana fisheries catch: 1950–2010. Fisheries Centre Working Paper #2015–21, 18 p.

Pauly D (2007) On bycatch, or how W.H.L. Allsopp coined a new word and created new insights. *Sea Around Us* Project Newsletter (44): 1–4.

1. Cite as MacDonald, J., S. Harper, S. Booth, D. Zeller and D. Pauly. 2016. Guyana. P. 286 in D. Pauly and D. Zeller (eds.), *Global Atlas of Marine Fisheries: A Critical Appraisal of Catches and Ecosystem Impacts*. Island Press, Washington, DC.

HAITI AND NAVASSA ISLAND[1]

Robin Ramdeen, Dyhia Belhabib, Sarah Harper, and Dirk Zeller

Sea Around Us, University of British Columbia, Vancouver, Canada

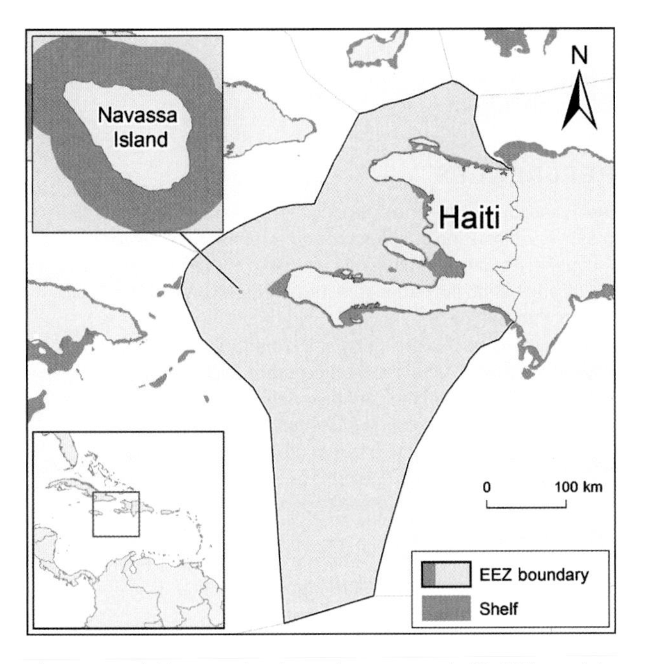

Figure 1. Haiti (27,000 km²), on the western half of Hispaniola Island, and its EEZ (124,000 km²), which here includes the U.S.-controlled Navassa Island, a marine reserve that is nevertheless visited by Haitian fishers.

The Republic of Haiti's EEZ (figure 1) consists of half of the mountainous, large island of Hispaniola and 6 small islands, with Navassa Island being a U.S.-run marine reserve (Wiener 2006), albeit one exploited by Haitian fishers. The small-scale artisanal and subsistence sectors contribute most catches from Haiti's EEZ and appear to be gradually increasing (figure 2A). These reconstructed estimates are derived from seafood consumption data combined with trade and aquaculture data (Ramdeen et al. 2012). Most of the catches allocated to the EEZ are domestic, with some foreign fishing by Cuba and other countries (figure 2B). The domestic catch for Haiti and Navassa was estimated at about 7,000 t/year in the early 1950s and 20,000–25,000 t/year since 2000, which is approximately 3 times the catch that FAO reports on behalf of Haiti. A large part of this discrepancy is caused by the inclusion of unreported subsistence catch and the improved accounting for artisanal fisheries catches of spiny lobster (*Panulirus argus*), shrimps, and stromboid conch (*Strombus* spp.), especially in recent decades. Aside from these invertebrates, fish such as wrasses (family Labridae), parrotfishes (Scaridae), and bar jack (*Caranx ruber*) are caught in the overfished coastal areas (Stickney and Kohler 1986) and large pelagic fishes further offshore (figure 2C). Haiti is a net importer of seafood, but the local fishing sector is important for domestic food security.

REFERENCES

Ramdeen R, Belhabib D, Harper S and Zeller D (2012) Reconstruction of total marine fisheries catches for Haiti and Navassa Island (1950–2010). pp. 37–45 In Harper S, Zylich K, Boonzaier L, Le Manach F, Pauly D and Zeller D (eds.), *Fisheries catch reconstructions: Islands, Part III*. Fisheries Centre Research Reports 20(5).

Stickney RR and Kohler CC (1986) Overfishing: the Haitian experience. *Naga, the ICLARM Quarterly*, 9(2): 5–7.

Wiener JW (2006) Oral history and contemporary assessment of Navassa Island fishermen. Report for the United States Department of Commerce, National Oceanic and Atmospheric Administration, National Marine Fisheries Service, and the NOAA Coral Reef Conservation Program, 35 p.

1. Cite as Ramdeen, R., D. Belhabib, S. Harper, and D. Zeller. 2016. Haiti and Navassa Island. P. 287 in D. Pauly and D. Zeller (eds.), *Global Atlas of Marine Fisheries: A Critical Appraisal of Catches and Ecosystem Impacts*. Island Press, Washington, DC.

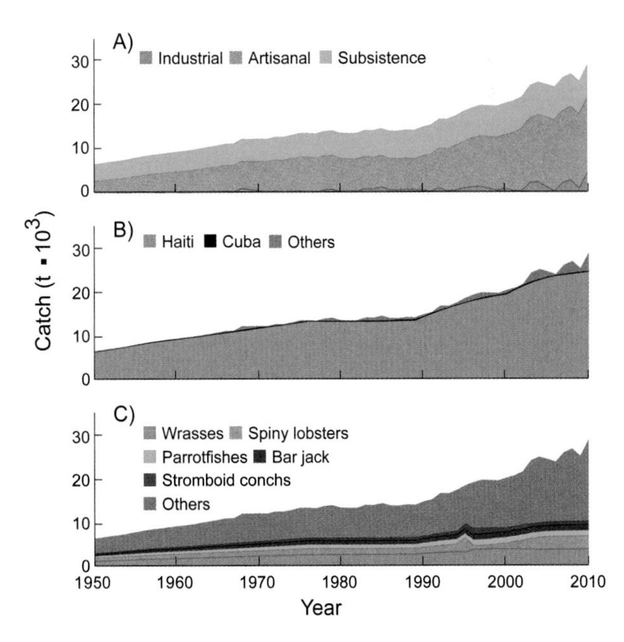

Figure 2. Domestic and foreign catches taken in the EEZ of Haiti and Navassa: (A) by sector (domestic scores: Art. 2, 2, 2; Sub. 1, 2, 2); (B) by fishing country (foreign catches are very uncertain); (C) by taxon.

HONDURAS (CARIBBEAN)[1]

Manuela Funes,[a] Kyrstn Zylich,[a] Esther Divovich,[a] Dirk Zeller,[a] Alasdair Lindop,[a] Daniel Pauly,[a] and Stephen Box[b]

[a]*Sea Around Us*, University of British Columbia, Vancouver, Canada [b]Centro de Ecología Marina, Edificio Florencia, Tegucigalpa, Honduras

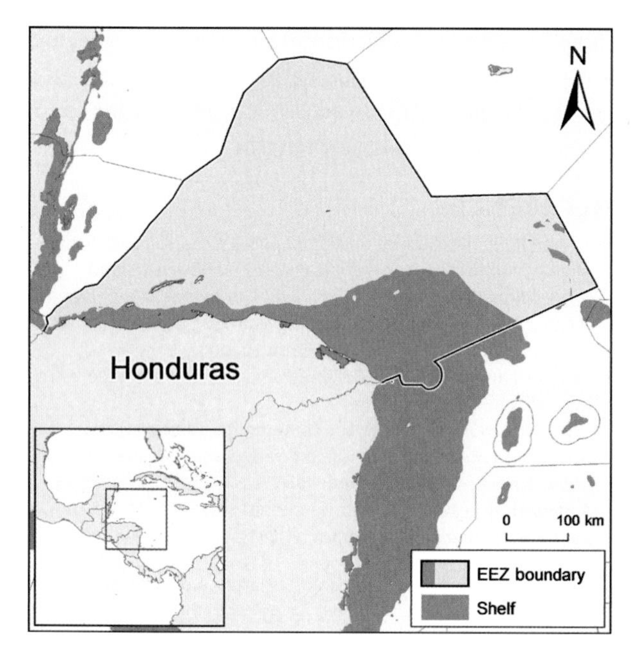

Figure 1. Honduras's Caribbean shelf is 60,300 km², and its EEZ is 218,000 km².

The Republic of Honduras is located in Central America and borders the Caribbean Sea in the north, Nicaragua in the southeast, El Salvador and the Pacific in the southwest, and Guatemala in the west (figure 1). Honduras has fisheries on both of its coasts, but this account, based on the catch reconstruction of Funes et al. (2015), deals only with the country's Caribbean EEZ. Honduras's main fisheries in the Atlantic are industrial but were later dominated by the artisanal sector (figure 2A). The reconstructed domestic catches of Honduras in its Caribbean EEZ were on average 2,600 t/year in the early 1950s, peaked at 25,000 t in 1988, and declined to about 9,000 t in 2010. Catches are mostly domestic, with foreign fishing by Cuba and Venezuela, among other countries through the time period (figure 2B). Figure 2C indicates that fisheries target mostly Caribbean spiny lobster (*Panulirus argus*), stromboid conchs (*Strombus* spp.), and penaeid shrimp (family Penaeidae; Mackenzie and Stehlik 1996), jacks (*Caranx* spp.), blue crab (*Callinectes sapidus*), and some finfish such as croakers, grunts, and groupers (Gobert et al. 2005). Shrimp trawl bycatch began to be exported for human consumption to Jamaica in 2010 rather than being discarded. As noted by Box and Canty (2011), the strong historic export orientation of Honduras's fisheries in the Caribbean may be a driver for overexploitation.

REFERENCES

Box SJ and Canty SWJ (2011) The long and short term economic drivers of overexploitation in Honduran coral reef fisheries due to their dependence on export markets. pp. 43–51 In Proceedings of the 63rd Gulf and Caribbean Fisheries Institute. November 1–5, 2010, San Juan, Puerto Rico.

Funes M, Zylich K, Divovich E, Zeller D, Lindop A, Pauly D and Box S (2015) Honduras, a fish exporting country: Preliminary reconstructed marine catches in the Caribbean Sea and the Gulf of Fonseca, 1950–2010. Fisheries Centre Working Paper #2015-90, 16 p.

Gobert B, Berthou P, Lopez E, Lespagnol P, Oqueli Turcios MD, Macabiau C and Portillo P (2005) Early stages of snapper–grouper exploitation in the Caribbean (Bay Islands, Honduras). *Fisheries Research* 73(1–2): 159–169.

Mackenzie CL and Stehlik LL (1996) The Crustacean and Molluscan Fisheries of Honduras. *Marine Fisheries Review* 58(3): 33–44.

1. Cite as Funes, M., K. Zylich, E. Divovich, D. Zeller, A. Lindop, D. Pauly, and S. Box. 2016. Honduras (Caribbean). P. 288 in D. Pauly and D. Zeller (eds.), *Global Atlas of Marine Fisheries: A Critical Appraisal of Catches and Ecosystem Impacts*. Island Press, Washington, DC.

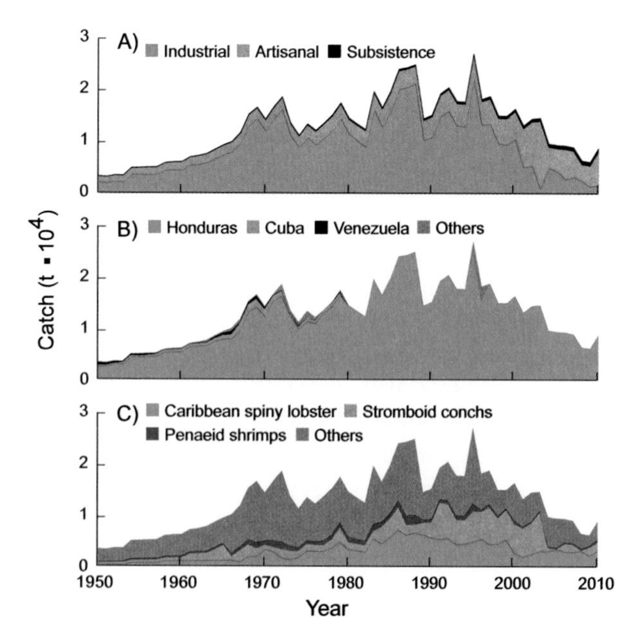

Figure 2. Domestic and foreign catches taken in the Caribbean EEZ of Honduras: (A) by sector (domestic scores: Ind. 2, 2, 2; Art. 2, 2, 2; Sub. 1, 1, 1; Dis. 2, 2, 2); (B) by fishing country (foreign catches are very uncertain); (C) by taxon.

HONDURAS (PACIFIC)[1]

Manuela Funes,[a] Kyrstn Zylich,[a] Esther Divovich,[a] Dirk Zeller,[a] Alasdair Lindop,[a] Daniel Pauly,[a] and Stephen Box[b]

[a]*Sea Around Us*, University of British Columbia, Vancouver, Canada [b]Centro de Ecología Marina, Tegucigalpa, Honduras, and Smithsonian Institution, Fort Pierce, Florida

The Republic of Honduras's Pacific coast is limited to the inner part of the Gulf of Fonseca (figure 1). The Gulf of Fonseca was surrounded by dense mangrove, much of which has been destroyed for shrimp mariculture (Stonich 1995; Thornton et al. 2003), whose growth is well documented. This is different for the fisheries, for which information is very scarce, especially for the earlier decades. Because the shallowness of the Fonseca Gulf does not allow for industrial vessels, fisheries are solely artisanal and subsistence (figure 2A). The reconstruction of Funes et al. (2015), based mainly on Soto (2012), suggests that domestic catches were 160–180 t/year in the 1950s and grew to 7,000 t/year in the 2000s. This was about 3 times the landings reported by the FAO on behalf of Honduras for the 1950–2010 period. The discrepancy between these estimates results from the local subsistence and artisanal fisheries remaining largely invisible to Honduran agencies, whose emphasis tended to be on shrimp, mariculture and Caribbean fisheries. Domestic fisheries are important in these waters; only some small foreign fishing by the United States, among other countries, has been documented (figure 2B). Figure 2C summarizes the approximate composition of the catch, in which shrimp (*Litopenaeus stylirostis*, *L. occidentalis*, and *L. vannamei*), weakfish (*Cynoscion squamipinnis*), and other croakers (*Ophioscion* spp.) were prominent.

REFERENCES

Funes M, Zylich K, Divovich E, Zeller D, Lindop A, Pauly D and Box S (2015) Honduras, a fish exporting country: Preliminary reconstructed marine catches in the Caribbean Sea and the Gulf of Fonseca, 1950–2010. Fisheries Centre Working Paper #2015–90, 16 p.

Soto L (2012) Informe de Evaluacion de la Actividad Pesquera en el Golfo de Fonseca, Honduras. Proyecto de Desarrollo Pesquero/ Actividades Acuicolas en el Golfo de Fonseca. Agencia Española de Cooperación Internacional para el Desarrollo, Tegucigalpa. 129 p.

Stonich SC (1995) The environmental quality and social justice implications of shrimp mariculture development in Honduras. *Human Ecology* 23(2): 143–168.

Thornton C, Shanahan M and Williams J (2003). From wetlands to wastelands: Impacts of shrimp farming. *The Society of Wetland Scientists Bulletin* 20(1): 48–53.

1. Cite as Funes, M., K. Zylich, E. Divovich, D. Zeller, A. Lindop, D. Pauly, and S. Box. 2016. Honduras (Pacific). P. 289 in D. Pauly and D. Zeller (eds.), *Global Atlas of Marine Fisheries: A Critical Appraisal of Catches and Ecosystem Impacts*. Island Press, Washington, DC.

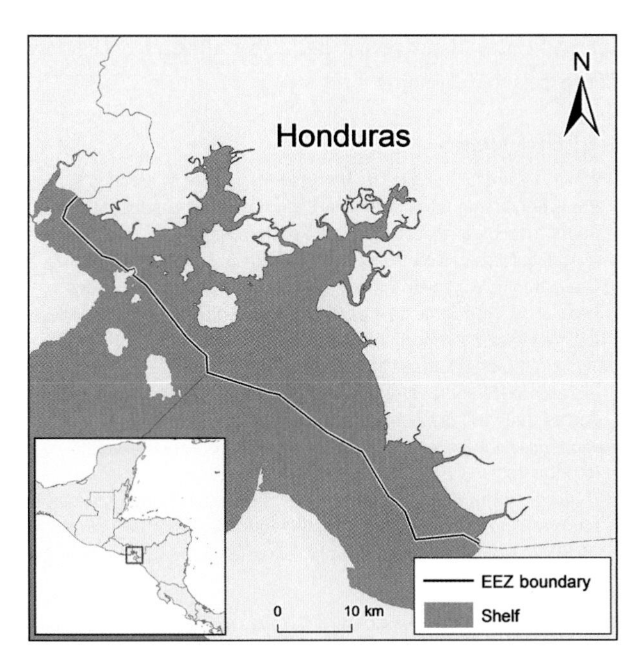

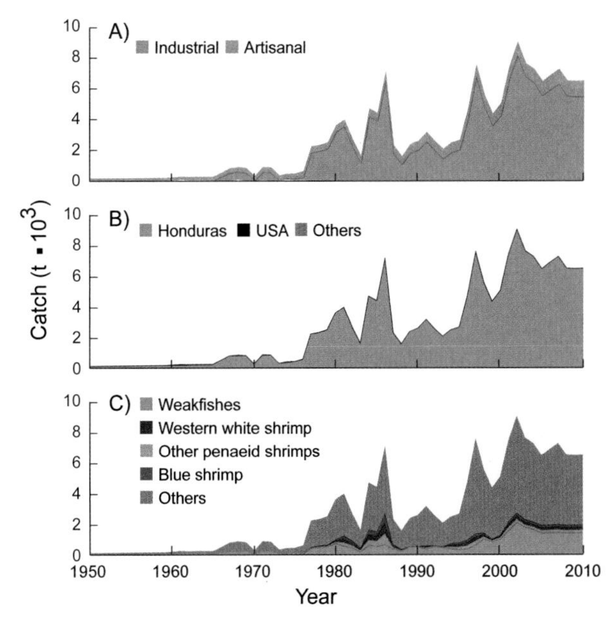

Figure 1. Honduras's EEZ in the Pacific is small (747 km²), consisting only of the inner Gulf of Fonseca, shared among El Salvador, Honduras, and Guatemala (all of it being shelf waters). **Figure 2.** Domestic and foreign catches taken in the Pacific EEZ of Honduras: (A) by sector (domestic scores: Art. 2, 2, 2; Sub. 1, 1, 1); (B) by fishing country (foreign catches are very uncertain); (C) by taxon.

ICELAND[1]

Hreiðar Þór Valtýsson

University of Akureyri, Borgir v/Nordurslod, Akureyri, Iceland

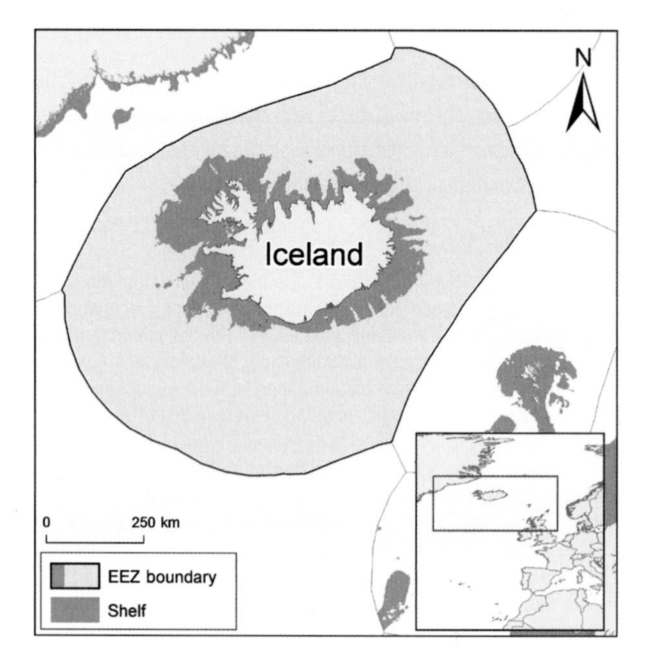

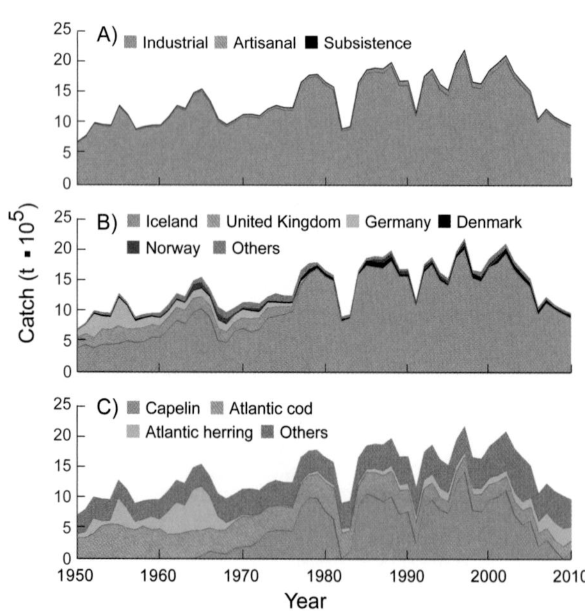

Figure 1. Iceland's land area is about 103,000 km², and its EEZ is 756,000 km². **Figure 2.** Domestic and foreign catches taken in the EEZ of Iceland: (A) by sector (domestic scores: Ind. 4, 4, 4; Art. 2, 3, 4; Sub. 1, 2, 3; Rec. –, –, 3; Dis. 1, 2, 3); (B) by fishing country (foreign catches are very uncertain); (C) by taxon.

Iceland is the second largest island in Europe and is located just below the Arctic Circle (figure 1). Marine resources have always been important, which is why Iceland fought hard to acquire an early EEZ (Bonfil et al. 1998). The economic importance of fisheries reached a zenith in 1949, when marine products contributed 97% of all exports. Valtýsson (2014), updating Valtýsson (2002), reconstructed withdrawals from Icelandic waters, including discards and unreported catch caught by Icelandic fleets. Although artisanal, subsistence, and recreational sectors do exist in Iceland's EEZ, most of the catches were from industrial fleets (figure 2A). This reconstructed catch ranged from 411,000 t in 1950 to a peak of 2 million t in 1997, before declining to 900,000 t by 2010 and, when cumulated over the entire time period, was 1.04 times the officially reported catch, based on a combination of ICES and Icelandic national data. Discards and unreported catches ranged from 2.7% to 6.8% of the reported landings and account for most of the difference between reported landings and reconstructed catches. Foreign catches were substantial, notably from Britain, Germany, and Norway (figure 2B). The most valuable species has been Atlantic cod (*Gadus morhua*), but the highest catches are capelin (*Mallotus villosus*; figure 2C). The fishing industry is more important than official statistics reveal, because it affects most sectors of the Icelandic economy.

REFERENCES

Bonfil R, Munro G, Sumaila UR, Valtysson H, Wright M, Pitcher T, Preikshot D, Haggan N and Pauly D (1998) Impacts of distant water fleets: an ecological, economic and social assessment. p. 111 In *The footprint of distant water fleet on world fisheries*. Endangered Seas Campaign. WWF International, Godalming, Surrey. [Also issued separately, with same title, as Bonfil et al. (Editors). 1998. Fisheries Centre Research Reports 6(6), 111 p.

Valtýsson H (2002) The sea around Icelanders: Catch history and discards in Icelandic waters. pp. 52–86 In Zeller D, Watson R and Pauly D (eds.), *Fisheries impacts on north Atlantic ecosystems: Catch, effort, and national/regional data sets*. Fisheries Centre Research Reports 9(3).

Valtýsson H (2014) Reconstructing Icelandic catches from 1950–2010. pp. 73–88 In Zylich K, Zeller D, Ang M and Pauly D (eds.), *Fisheries catch reconstructions: Islands, Part IV*. Fisheries Centre Research Reports 22(2).

1. Cite as Valtýsson, H. 2016. Iceland. P. 290 in D. Pauly and D. Zeller (eds.), *Global Atlas of Marine Fisheries: A Critical Appraisal of Catches and Ecosystem Impacts*. Island Press, Washington, DC.

INDIA[1]

Brajgeet Bhathal, Claire Hornby, Dirk Zeller, and Daniel Pauly

Sea Around Us, University of British Columbia, Vancouver, Canada

India (figure 1) is a huge country consisting of 28 states, governed locally (9 with access to the sea), and 7 Union Territories (UTs), governed from New Delhi. Two of the UTs are on the mainland, and two are archipelagos, one included here (the Lakshadweeps), the other on the next page. Although there are marked oceanographic differences between the Malabar Coast (along the Arabian Sea) and the Coromandel Coast (along the Bay of Bengal), both support extensive subsistence and artisanal fisheries, to which heavily subsidized industrial fisheries were added from the 1970s on (Bhathal 2014). Reconstructed catches (Hornby et al. 2015) increased from 1 million t/year in the early 1950s to about 3.8 million t/year in the late 2000s (figure 2A). However, catch per effort (here expressed as cumulative engine power and roughly proportional to resource abundance) plummeted from about 20–30 kg/kW/day to about 5 kg/kW/day in all maritime states and UTs (Bhathal 2014). Indian fisheries also suffer from a depletion of their larger fish species, as evidenced by the "fishing down" phenomenon (Bathal and Pauly 2008), although this is not visible in figure 2B. The reconstructed domestic catch, almost twice the landings reported by India to the FAO, and sustained mainly by the offshore expansion of the fisheries, is largely appropriated by industrial fleets, a perennial source of conflict with the beleaguered small-scale fishers.

REFERENCES

Bhathal B (2014) Government-led development of India's marine fisheries since 1950: catch and effort trends, and bioeconomic models for exploring alternative policies. PhD thesis, University of British Columbia, Department of Zoology, Vancouver. 333 p.

Bhathal B and Pauly D (2008) "Fishing down marine food webs" and spatial expansion of coastal fisheries in India, 1950–2000. *Fisheries Research* 91: 26–34.

Hornby C, Bhathal B, Pauly D and Zeller D (2015) Reconstruction of India's marine fish catch from 1950–2010. Fisheries Centre Working Paper #2015-77, 42 p.

1. Cite as Bhathal, B., C. Hornby, D. Zeller, and D. Pauly. 2016. India. P. 291 in D. Pauly and D. Zeller (eds.), *Global Atlas of Marine Fisheries: A Critical Appraisal of Catches and Ecosystem Impacts*. Island Press, Washington, DC.

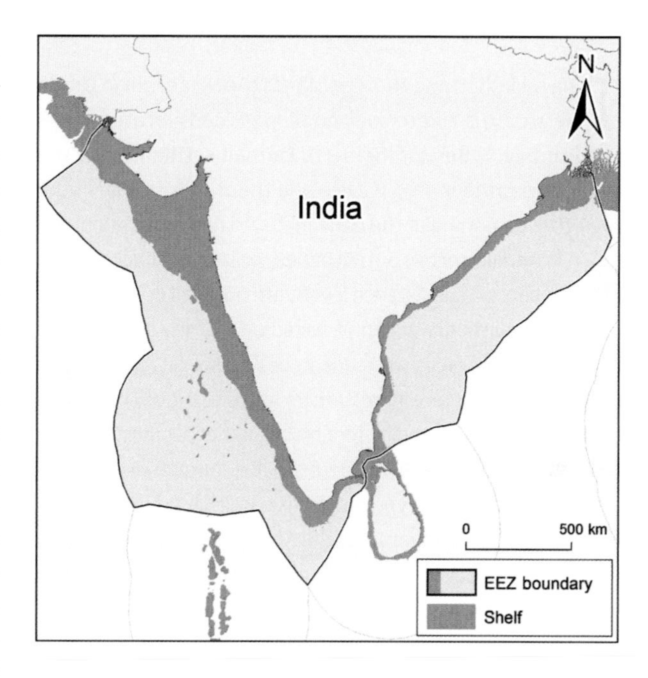

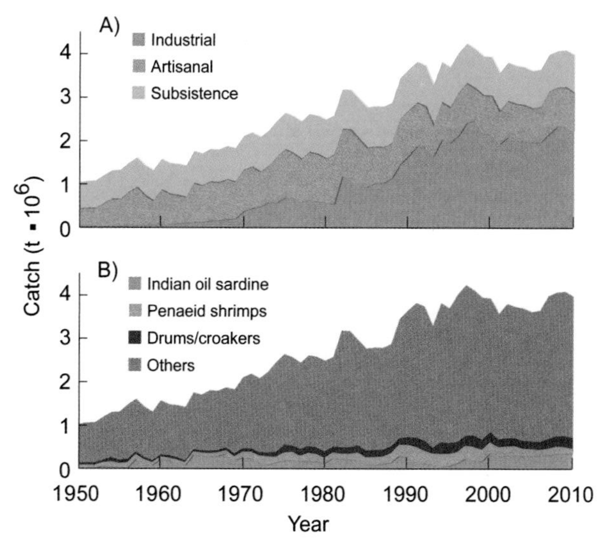

Figure 1. India's EEZ (excluding the Andaman and Nicobar Islands) covers 1.63 million km^2, of which 394,000 km^2 is shelf. **Figure 2.** Domestic catches in the Indian EEZ (excluding the Andaman and Nicobar Islands): (A) by sector (domestic scores: Ind. 2, 2, 2; Art. 2, 3, 3; Sub. 2, 2, 2; Dis. 1, 1, 1); (B) by taxon.

INDIA (ANDAMAN AND NICOBAR ISLANDS)[1]

Claire Hornby,[a] M. Arun Kumar,[b] Brajgeet Bhathal,[a] Sahir Advani,[c] Daniel Pauly,[a] and Dirk Zeller[a]

[a]*Sea Around Us*, University of British Columbia, Vancouver, Canada [b]Department of Ocean Studies and Marine Biology, Pondicherry University, Port Blair, Andaman Islands, India [c]Institute for the Oceans and Fisheries, University of British Columbia, Vancouver, Canada

The Andaman and Nicobar Islands (A&N), an Indian Union Territory, is an archipelago of 572 islands located in the Eastern Bay of Bengal (figure 1). Human settlement to the islands occurred in two waves, one thousands of years ago, the other beginning in the early 1950s, mainly from mainland India (Krishnakumar 2009). Fisheries, which have been slow to develop beyond subsistence levels, are not well documented, and this preliminary account, based on Hornby et al. (2015), documents their reconstructed catches. Although the domestic small-scale sectors are important for local livelihood (figure 2A), they are overshadowed by foreign industrial catches, notably from Thailand (figure 2B). Illegal fishing by Indonesian, Taiwanese, and other foreign vessels also is common in the EEZ, as is poaching in inshore waters of the A&N (Rajan 2003). Starting at 3,000–4,000 t/year in the early 1950s, domestic catches increased strongly in the late 1980s with the development of artisanal and industrial fisheries and reached about 111,000 t/year in the late 2000s. Overall, they were 3.8 times the data reported by the FAO on behalf of India/A&N. Although figure 2C presents the taxonomic composition to be dominated by threadfin and whiptail breams (family Nemipteridae), recent information pertaining to the marine export industry of these islands makes this estimate tentative. Notably, the nascent grouper fishery (Advani et al. 2013) will have to be considered, which is currently not accounted for.

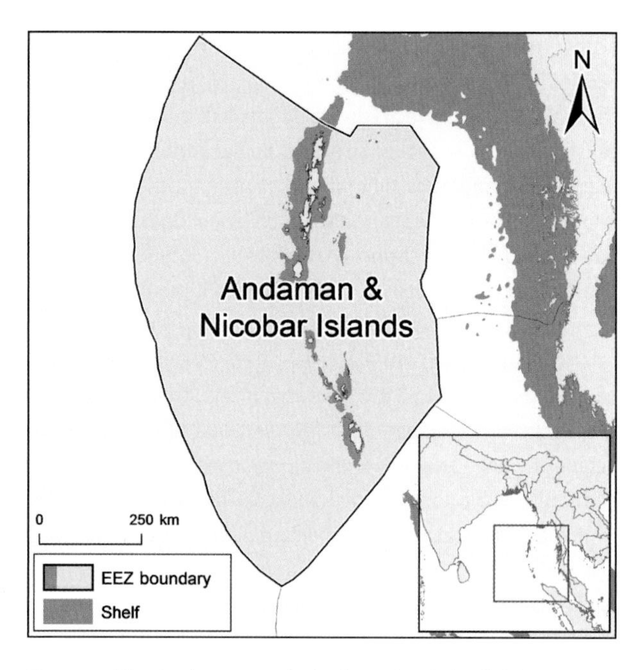

Figure 1. The Andaman and Nicobar Islands (India) have, collectively, a land area of 8,000 km² and an EEZ of 660,000 km².

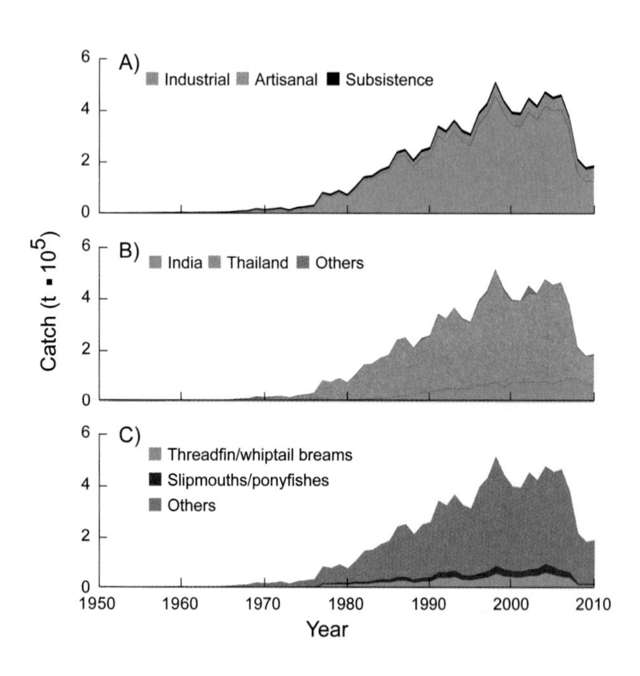

Figure 2. Domestic and foreign catches taken in the EEZ of Andaman and Nicobar Islands (India): (A) by sector (domestic scores: Ind. –, 2, 1; Art. 1, 2, 1; Sub. 1, 2, 1; Dis. 1, 1, 1); (B) by fishing country (foreign catches are very uncertain); (C) by taxon.

REFERENCES

Advani S, Sridhar A, Namboothri N, Chandi M and Oommen MA (2013) Emergence and transformation of marine fisheries in the Andaman Islands. Dakshin Foundation and ANET, 50 p.

Hornby C, Arun Kumar A, Bhathal B, Pauly D and Zeller D (2015) Reconstruction of the Andaman and Nicobar Islands marine fish catch from 1950–2010. Fisheries Centre Working Paper #2015-75, 27 p.

Krishnakumar MV (2009) Development or despoilation? The Andaman Island under colonial and postcolonial regimes. *Shima: The International Journal of Research into Island Cultures* 3(2): 104–117.

Rajan PT (2003) A field guide to marine food fishes of the Andaman and Nicobar Islands. Zoological Survey of India, Kolkata. 260 p.

1. Cite as Hornby, C., M. A. Kumar, B. Bhathal, S. Advani, D. Pauly, and D. Zeller. 2016. India (Andaman and Nicobar Islands). P. 292 in D. Pauly and D. Zeller (eds.), *Global Atlas of Marine Fisheries: A Critical Appraisal of Catches and Ecosystem Impacts*. Island Press, Washington, DC.

INDONESIA (CENTRAL)[1]

Vania Budimartono,[a] Muhammed Badrudin,[b] Esther Divovich,[a] and Daniel Pauly[a]

[a]*Sea Around Us*, University of British Columbia, Vancouver, Canada [b]Indonesia Marine and Climate Support Project, USAID, Jakarta, Indonesia

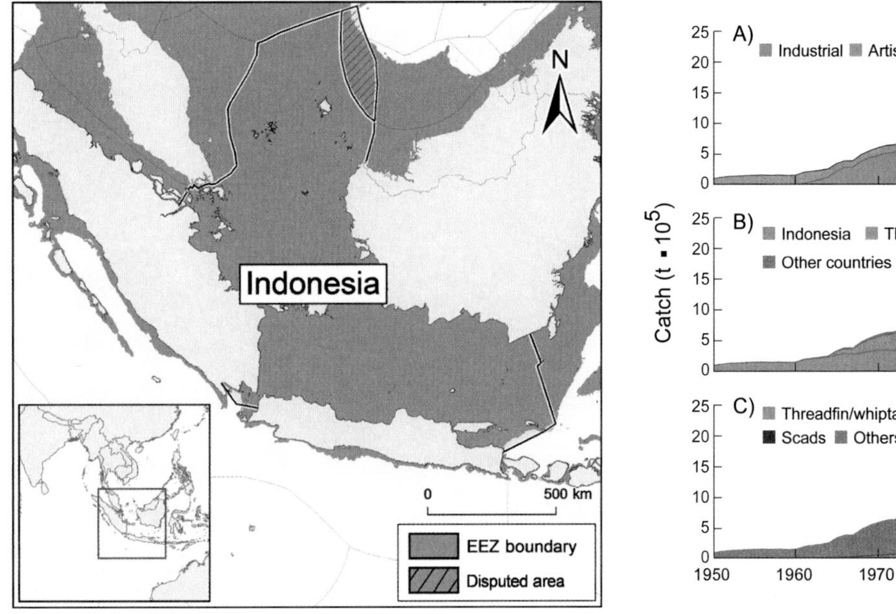

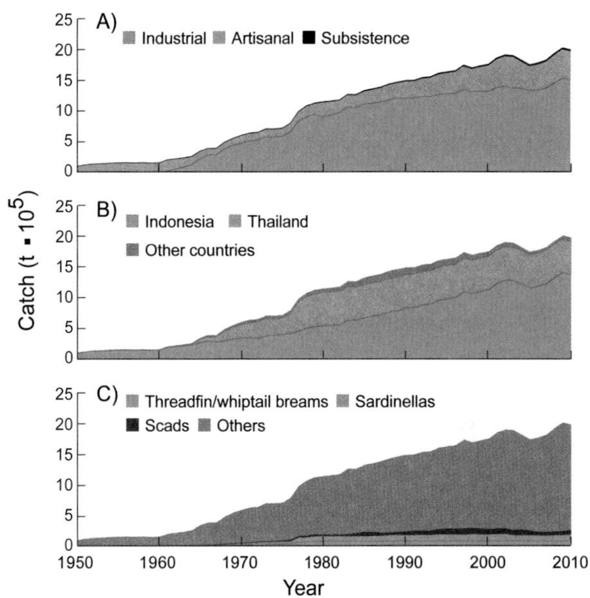

Figure 1. Central Indonesia as defined here has a shelf of 1 million km² and an EEZ of 1.02 million km². **Figure 2.** Domestic and foreign catches taken in the EEZ of Central Indonesia: (A) by sector (domestic scores: Ind. 1, 2, 1; Art. 1, 2, 1; Sub. 1, 2, 1; Rec. 1, 1, 1; Dis. 1, 1, 1); (B) by fishing country (foreign catches are very uncertain); (C) by taxon.

Indonesia ranges east to west as much as all of Southeast Asia, which is the reason why its EEZ is split here into three manageable subzones, with Central Indonesia defined, for the purposes of this account, as the Java Sea plus the southern end of the South China Sea (figure 1). This implied splitting the reported catch of provinces with coastlines in Central Indonesia and along the Indian Ocean, as documented in Budimartono et al. (2015), who also discuss how the changes resulting from the trawling ban of 1980 (Sarjono 1980) were taken into account. Most of the catches are from the industrial sector, in addition to the artisanal, subsistence, and recreational sectors and discards (figure 2A). Their reconstruction yielded catches of 120,000 t/year in the early 1950s, increasing to 1.3 million t/year in the late 2000s, although attempts were made to reduce the tendency of reported catches to increase exponentially, as reported time series have a fixed percentage added annually to the preceding value. The reconstructed domestic catch from Central Indonesia is 1.2 times the data officially reported for the pertinent provinces, to which a large foreign catch by vessels from Thailand is here added (figure 2B). Figure 2C presents its taxonomic composition to include threadfin breams (family Nemipteridae), sardinella (*Sardinella* spp.), and lizardfishes (Synodontidae). However, Indonesia harbors an immense marine biodiversity, only partly reflected in catches (Pauly and Martosubroto 1996).

REFERENCES

Budimartono V, Badrudin M, Divovich E and Pauly D (2015) A reconstruction of marine fisheries catches of Indonesia, with emphasis on Central and Eastern Indonesia, 1950–2010. pp. 2–26 In Pauly D and Budimartono V (eds.), *Marine Fisheries Catches of Western, Central and Eastern Indonesia*, 1950–2010. Fisheries Centre Working Paper #2015-61.

Pauly D and Martosubroto P, editors (1996) *Baseline studies in biodiversity: the fish resources of western Indonesia*. ICLARM Studies and Reviews 23, 390 p.

Sarjono I (1980) Trawlers banned in Indonesia. *ICLARM Newsletter* 3(4): 3.

1. Cite as Budimartono, V., M. Badrudin, E. Divovich, and D. Pauly. 2016. Indonesia (Central). P. 293 in D. Pauly and D. Zeller (eds.), *Global Atlas of Marine Fisheries: A Critical Appraisal of Catches and Ecosystem Impacts*. Island Press, Washington, DC.

INDONESIA (EASTERN)[1]

Vania Budimartono,[a] Muhammed Badrudin,[b] Esther Divovich,[a] and Daniel Pauly[a]

[a]*Sea Around Us*, University of British Columbia, Vancouver, Canada
[b]Indonesia Marine and Climate Support Project, USAID, Jakarta, Indonesia

In Eastern Indonesia (figure 1), corals and by extension the associated organisms are the most diverse in the world (Veron et al. 2009). The industrial fisheries sector, which has expanded considerably since 1970, dominates total removals from within these waters (figure 2A). Indonesia's enormous, less populated space (Dalzell and Pauly 1989) is difficult to cover through surveillance operations, and distant-water fleets (DWF) of Indonesia's northern neighbors roam at will, legally and illegally. However, total removals from within the waters still appear to be predominantly domestic (figure 2B). The reconstruction of Budimartono et al. (2015), which yielded a domestic catch of 120,000 t/year in the early 1950s, peaked at 2.5 million t in 1987, and, after a substantial decline in the early 1990s, increased again to 2.2 million t in 2010, was 1.7 times the officially reported data. The discrepancy was caused mainly by the high discards from shrimp trawlers rather than by the local coral reef fisheries. Top taxa include yellowstripe scad (*Selaroides leptolepis*), skipjack tuna (*Katsuwonus pelamis*), and scads (*Decapterus* spp.; figure 2C) The extraordinary level of illegal fishing in Eastern Indonesia was addressed in several of the sources cited in Budimartono et al. (2015), notably Wagey et al. (2009), who, using consultations with industry stakeholders, concluded that "the official statistics for the Arafura Sea fisheries are not in line with reality as regards both catch and effort in this fishery."

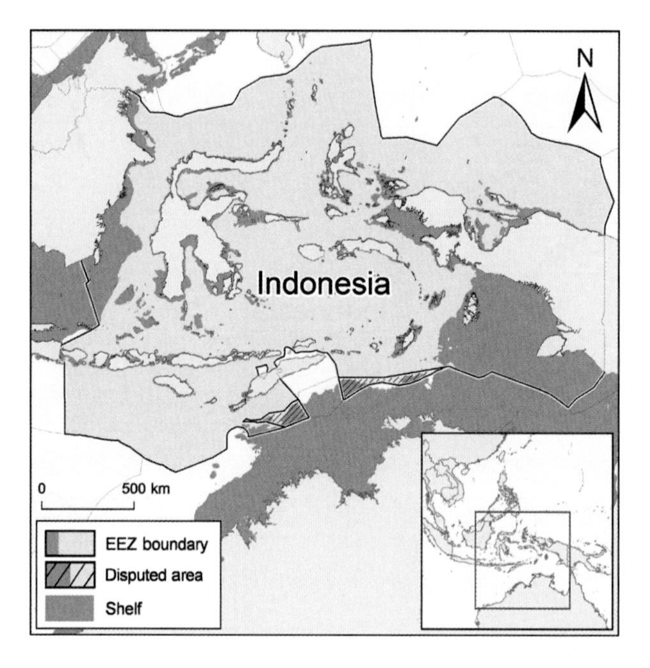

Figure 1. Eastern Indonesian as defined here has a shelf of 795,000 km² and an EEZ of 3.6 million km².

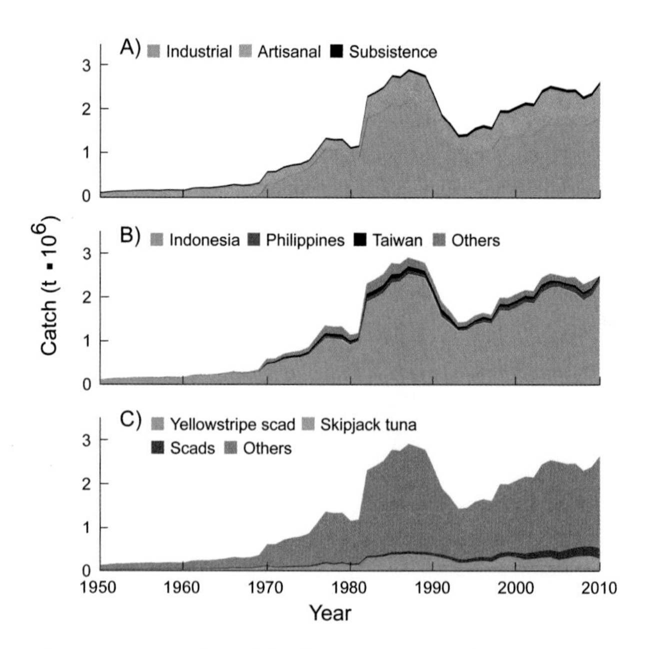

Figure 2. Domestic and foreign catches taken in the EEZ of Eastern Indonesia: (A) by sector (domestic scores: Ind. –, 2, 1; Art. 1, 2, 1; Sub. 1, 2, 1; Rec. 1, 1, 1; Dis. 1, 1, 1); (B) by fishing country (foreign catches are very uncertain); (C) by taxon.

REFERENCES

Budimartono V, Badrudin M, Divovich E and Pauly D (2015) A reconstruction of marine fisheries catches of Indonesia, with emphasis on Central and Eastern Indonesia, 1950–2010. pp. 2–26 In Pauly D and Budimartono V (eds.), *Marine Fisheries Catches of Western, Central and Eastern Indonesia, 1950–2010*. Fisheries Centre Working Paper #2015-61.

Dalzell P and Pauly D (1989) Assessment of the fish resources of Southeast Asia, with emphasis on the Banda and Arafura Seas. *Netherlands Journals of Sea Research* 24(4): 641–650.

Veron JEN, Devantier LM, Turak E, Green AL, Kininmonth S, Stafford-Smith M and Peterson N (2009) Delineating the coral triangle. *Galaxea, Journal of Coral Reef Studies* 11(2): 91–100.

Wagey GA, Nurhakim S, Nikijuluw VPH, Badrudin and Pitcher TJ (2009) A study of illegal, unreported, and unregulated (IUU) fishing in the Arafura Sea, Indonesia. Ministry of Marine Affairs and Fisheries, Jakarta, Indonesia.

1. Cite as Budimartono, V., M. Badrudin, E. Divovich, and D. Pauly. 2016. Indonesia (Eastern). P. 294 in D. Pauly and D. Zeller (eds.), *Global Atlas of Marine Fisheries: A Critical Appraisal of Catches and Ecosystem Impacts*. Island Press, Washington, DC.

INDONESIA (INDIAN OCEAN)[1]

Vania Budimartono,[a] Muhammed Badrudin,[b] and Daniel Pauly[a]

[a]*Sea Around Us*, University of British Columbia, Vancouver, Canada
[b]Indonesia Marine and Climate Support Project, USAID, Jakarta, Indonesia

In order to deal with the diversity of Indonesian fisheries, the EEZ was subdivided into three large subzones (see previous two pages). This subzone, which covers most of the Indonesian waters in the Indian Ocean (as defined by the limits of FAO area 57; figure 1), also includes the Indonesian part of the Bay of Bengal Large Marine Ecosystem (BOBLME) Project. The reported catch statistics of the 11 Indonesian provinces in that subzone (some of them in parts only) have numerous deficiencies, as assessed by the fact that they largely increase exponentially and too regularly, but apparently they do not reflect the 1980 trawling ban (although the impact of the tsunami of December 2004 may have been captured). The industrial sector dominates total allocated catches in these waters, although artisanal and subsistence do exist (figure 2A). A catch reconstruction partly accounting for these deficiencies (Budimartono et al. 2015) yielded catches of about 130,000 t/year in the 1950s and 1.2 million t/year in the 2000s, 1.4 times the landings reported by the FAO, as adjusted for the area in question. Although foreign fishing by Malaysia and Thailand, among other countries, does exist, the majority of the catch is domestic (figure 2B). The biodiversity of Indonesian waters, further detailed in Pauly and Martosubroto (1996), is reflected in figure 2C, which includes short mackerel (*Rastrelliger brachysoma*), scads (*Decapterus* spp.), and yellowstripe scads (*Selaroides leptolepis*).

REFERENCES

Budimartono V, Badrudin M and Pauly D (2015) Indonesian marine fisheries catches in the Western Indonesia (FAO Area 57) and in the Bay of Bengal large marine ecosystem project (BOBLME) area: a tentative reconstruction, 1950–2010. pp. 27–51 In Pauly D and Budimartono V (eds.), *Marine Fisheries Catches of Western, Central and Eastern Indonesia, 1950–2010*. Fisheries Centre Working Paper #2015–61.

Pauly D and Martosubroto P, editors (1996) Baseline studies in biodiversity: The fish resources of western Indonesia. International Center for Living Aquatic Resources Management (ICLARM), Makati City, Philippines. 390 p.

1. Cite as Budimartono, V., M. Badrudin, and D. Pauly. 2016. Indonesia (Indian Ocean). P. 295 in D. Pauly and D. Zeller (eds.), *Global Atlas of Marine Fisheries: A Critical Appraisal of Catches and Ecosystem Impacts*. Island Press, Washington, DC.

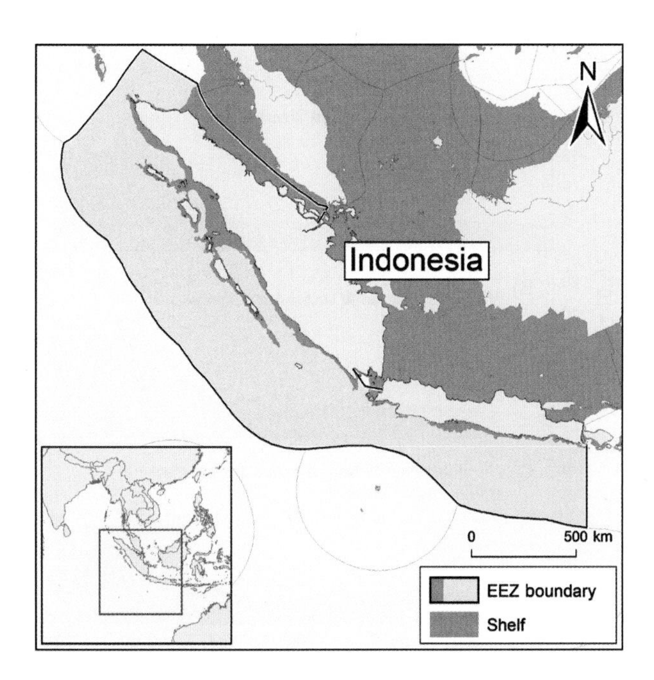

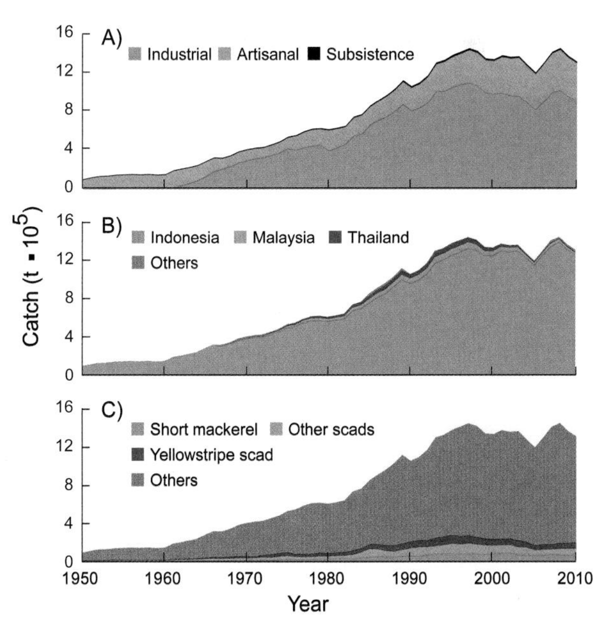

Figure 1. The Indonesian EEZ in the Indian Ocean subzone as defined here has a shelf of 174,000 km² and an EEZ of 1.41 million km². **Figure 2.** Domestic and foreign catches taken in the EEZ of Indonesia (Indian Ocean): (A) by sector (domestic scores: Ind. 2, 2, 2; Art. 2, 2, 2; Sub. 1, 1, 1; Rec. 1, 1, 1; Dis. 1, 1, 1); (B) by fishing country (foreign catches are very uncertain); (C) by taxon.

IRAN (ARABIAN SEA)[1]

Nardin Roshan Moniri, Nazanin Roshan Moniri, Dirk Zeller, Dalal Al-Abdulrazzak, Kyrstn Zylich, and Dyhia Belhabib

Sea Around Us, University of British Columbia, Vancouver, Canada

Iran has a coast along the Persian Gulf (see next page) and in the Northwestern Arabian Sea, or the Sea of Oman (figure 1). The catches were reconstructed by Roshan Moniri et al. (2013), based on various sources, notably the website of Sistan & Baluchestan Province of Iran (Anonymous 2012). Most of the catch is artisanal, in addition to a growing industrial and subsistence component (figure 2A). These catches amounted to about 35,000 t/year from 1950 to the onset of the Iran–Iraq war (1980–1988), increased rapidly after the Iran–Iraq war to a peak of 95,000 t in 2004, and slightly declined to the current level of about 90,000 t. These catches were 2.2 times the data reported by the FAO on behalf of Iran, with the discrepancy caused by underreported small-scale and industrial catches and unreported discards. Although most of the catches are domestic, catches by Pakistan and the United Arab Emirates also occur (figure 2B). Figure 2C presents the taxonomic breakdown for the catch of Iran in the Arabian Sea, based mainly on composition data from within the Gulf, because such data are not available from Iran's eastern coast. This includes longtail tuna (*Thunnus tonggol*), narrow-barred Spanish mackerel (*Scomberomorus commerson*), and ponyfishes (*Leiognathus* spp.). Valinassab et al. (2006) report on a trawl survey that yielded much higher densities of demersal fish in the Arabian Sea than in the Persian Gulf.

REFERENCES

Anonymous (2012) Sistan and Baluchestan province portal. Available at: http://www.sbportal.ir/en [Accessed: October 2, 2012].

Roshan Moniri N, Roshan Moniri N, Zeller D, Al-Abdulrazzak D, Zylich K and Belhabib D (2013) Fisheries catch reconstruction for Iran, 1950–2010. pp. 7–16 In Al-Abdulrazzak D and Pauly D (eds.), *From dhows to trawlers: a recent history of fisheries in the Gulf countries, 1950 to 2010*. Fisheries Centre Research Reports 21(2).

Valinassab T, Daryanabard R, Dehghani R and Pierce GJ (2006) Abundance of demersal fish resources in the Persian Gulf and Oman Sea. *Journal of the Marine Biological Association of the United Kingdom* 86(6): 1455–1462.

1. Cite as Roshan Moniri, N., N. Roshan Moniri, D. Zeller, D. Al-Abdulrazzak, K. Zylich, and D. Belhabib. 2016. Iran. P. 296 in D. Pauly and D. Zeller (eds.), *Global Atlas of Marine Fisheries: A Critical Appraisal of Catches and Ecosystem Impacts*. Island Press, Washington, DC.

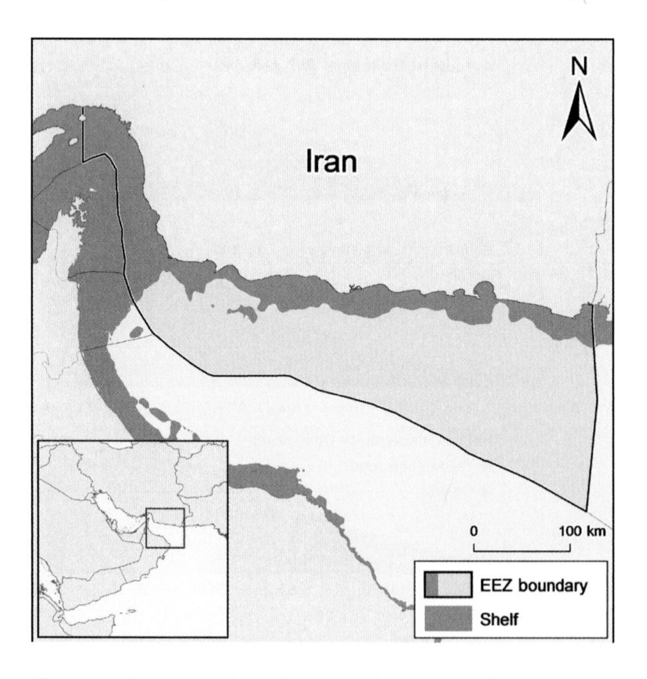

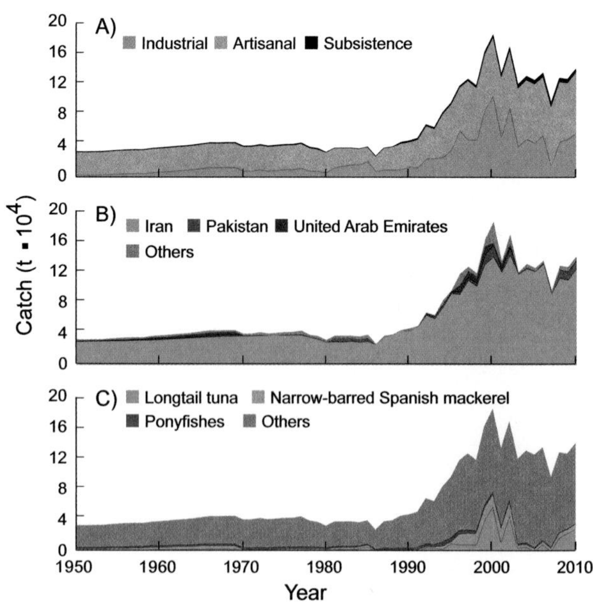

Figure 1. The coast of Iran in the Arabian Sea and the corresponding segment of its EEZ (65,800 km²) of which 17,500 km² is shelf. **Figure 2.** Domestic and foreign catches taken in the EEZ of Iran (Arabian Sea): (A) by sector (domestic scores: Ind. 2, 3, 2; Art. 2, 3, 2; Sub. 2, 2, 2; Dis. 1, 1, 1); (B) by fishing country (foreign catches are very uncertain); (C) by taxon.

IRAN (PERSIAN GULF)[1]

Nardin Roshan Moniri, Nazanin Roshan Moniri, Dirk Zeller, Dalal Al-Abdulrazzak, Kyrstn Zylich, and Dyhia Belhabib

Sea Around Us, University of British Columbia, Vancouver, Canada

Iran is the country with the longest coastline in the Persian Gulf (figure 1), hence the historic name of this water body. This account, adapted from Roshan Moniri et al. (2013), deals with Iran's fisheries in the Gulf (for the Iranian Arabian Sea EEZ, see previous page) and covers domestic industrial, artisanal (including weirs; Al-Abdulrazzak and Pauly 2013), subsistence, recreational, and discarded catches (figure 2A). Although the majority of catches in the EEZ are domestic, foreign catches by China, South Korea, and some of Iran's neighbors were also documented starting in the 1980s (figure 2B). The reconstructed domestic catch averaged 157,000 t/year in the 1950s, slowly increased to 192,000 t before declining during the Iran–Iraq war (1980–1988), rapidly recovered and peaked at 400,000 t in 1997, then declined to about 170,000 t/year in the late 2000s. This corresponded to a reconstructed catch 2.7 times the data reported by the FAO (and adjusted for Iranian catches from outside of the Gulf), largely because of substantial underreporting of artisanal catches. Catches were dominated by ponyfishes (*Leiognathus* spp.), green tiger prawn (*Penaeus semisulcatus*), and blue swimming crab (*Portunuss pelagicus*) but also included a huge variety of fish (figure 2C). Overall, Iran's nominal management of fisheries is hampered by lack of key data, as is the suppression of domestic and foreign illegal fishing.

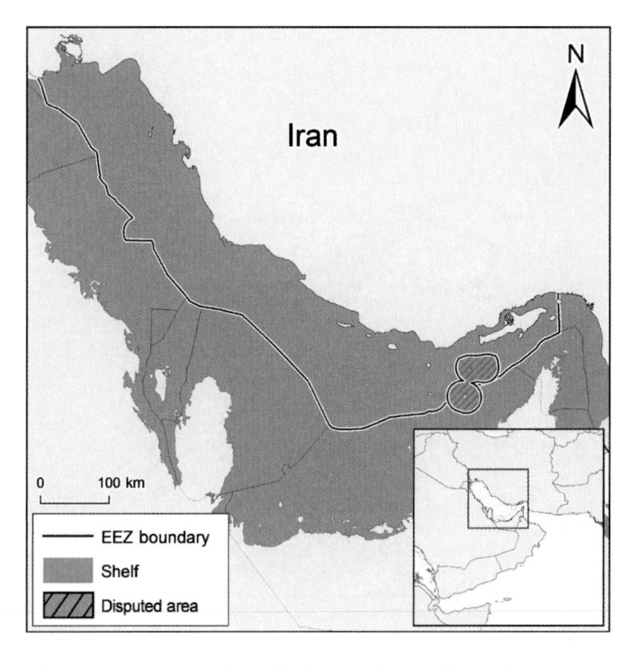

Figure 1. The coast of Iran in the Persian Gulf and the corresponding segment of its EEZ (97,900 km², including the contested zone with the United Arab Emirates), all of which is shelf.

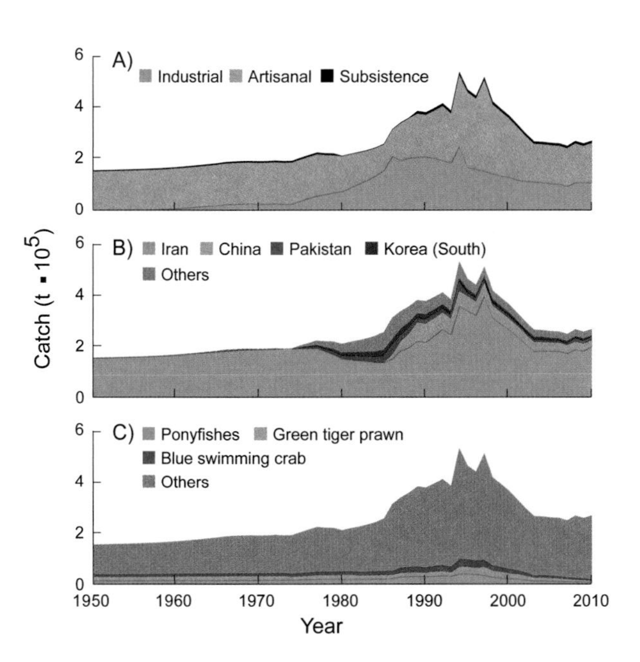

Figure 2. Domestic and foreign catches taken in the EEZ of Iran (Persian Gulf): (A) by sector (domestic scores: Ind. 2, 3, 2; Art. 2, 3, 2; Sub. 2, 2, 2; Rec. 1, 1, 1; Dis. 1, 1, 1); (B) by fishing country (foreign catches are very uncertain); (C) by taxon.

REFERENCES

Al-Abdulrazzak D and Pauly D (2013) Managing fisheries from space: Google Earth improves estimates of distant fish catches. *ICES Journal of Marine Science* 71(3): 450–455.

Roshan Moniri N, Roshan Moniri N, Zeller D, Al-Abdulrazzak D, Zylich K and Belhabib D (2013) Fisheries catch reconstruction for Iran, 1950–2010. pp. 7–16 In Al-Abdulrazzak D and Pauly D (eds.), *From dhows to trawlers: a recent history of fisheries in the Gulf countries, 1950 to 2010*. Fisheries Centre Research Reports 21(2).

1. Cite as Roshan Moniri, N., N. Roshan Moniri, D. Zeller, D. Al-Abdulrazzak, K. Zylich, and D. Belhabib. 2016. Iran (Persian Gulf). P. 297 in D. Pauly and D. Zeller (eds.), *Global Atlas of Marine Fisheries: A Critical Appraisal of Catches and Ecosystem Impacts*. Island Press, Washington, DC.

IRAQ[1]

Dalal Al-Abdulrazzak and Daniel Pauly

Sea Around Us, University of British Columbia, Vancouver, Canada

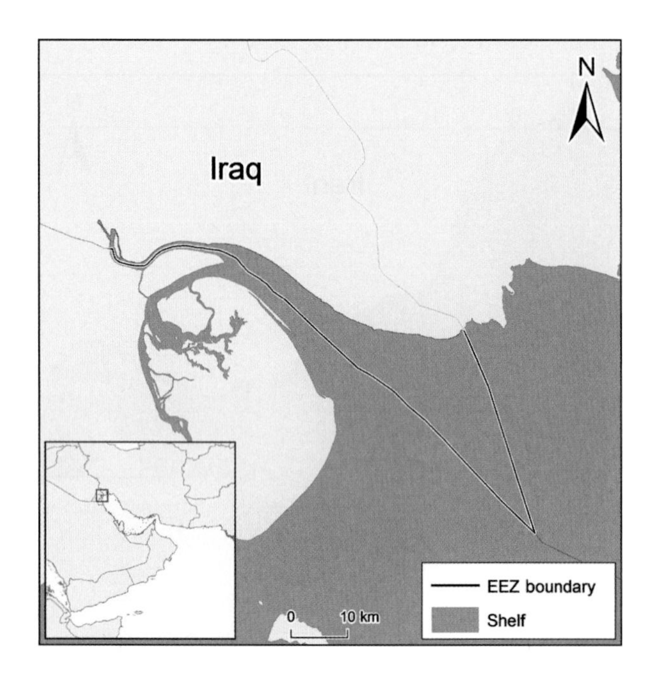

Figure 1. Iraq in the Persian Gulf, showing the extent of its small EEZ (541 km²), all of which is shelf.

Iraq has the smallest EEZ of the Gulf countries, at the mouth of the Shatt al-Arab River, formed by the confluence of the Euphrates and Tigris Rivers about 200 km upstream (figure 1). Consequently, Iraq's marine fisheries are of less importance than its freshwater fisheries, not considered here (see Jawad 2006). Catches were predominantly domestic and artisanal (figure 2A), although a subsistence sector does exist and was reconstructed by Al-Abdulrazzak and Pauly (2013), based on admittedly fragmentary evidence. Domestic catches were on the order of 1,000–3,000 t/year from 1950 to the early 1970s, then fluctuated between 10,000 and 30,000 t/year, as peace and war alternated in the Shatt-al-Arab region. Small foreign catches may have occurred within these waters before the EEZ declaration in 1985. Overall, the reconstructed catches are 1.6 times those reported by the FAO on behalf of Iraq and are dominated by previously unreported catches of hilsa shad (*Tenualosa ilisha*), an anadromous fish. The rest of the catch is likely to resemble that of neighboring Kuwait, whose taxonomic composition was used to disaggregate the non-hilsa marine catch of Iraq (figure 2B), and consisting of groups such as mullet (family Mugilidae), croakers (family Sciaenidae), and groupers (*Epinephelus* spp.).

REFERENCES

Al-Abdulrazzak D and Pauly D (2013) Reconstructing Iraq's fisheries: 1950–2010. pp. 17–22 In Al-Abdulrazzak D and Pauly D (eds.), *From dhows to trawlers: a recent history of fisheries in the Gulf countries, 1950 to 2010*. Fisheries Centre Research Reports 21(2).

Jawad LA (2006) Fishing gear and methods of the lower Mesopotamian Plain with reference to fishing management. *Marina Mesopotamica* 1(1): 1–37.

1. Cite as Al-Abdulrazzak, D., and D. Pauly. 2016. Iraq. P. 298 in D. Pauly and D. Zeller (eds.), *Global Atlas of Marine Fisheries: A Critical Appraisal of Catches and Ecosystem Impacts*. Island Press, Washington, DC.

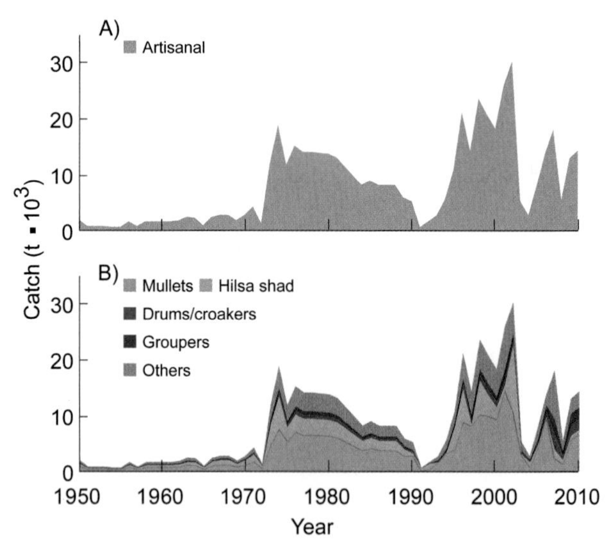

Figure 2. Domestic and foreign catches taken in the EEZ of Iraq: (A) by sector (domestic scores: Art. 2, 2, 2; Sub. 1, 1, 1; Dis. 1, 1, 1); (B) by taxon.

IRELAND[1]

Dana D. Miller[a] and Dirk Zeller[b]

[a]Fisheries Economics Research Unit, University of British Columbia, Vancouver, Canada
[b]*Sea Around Us*, University of British Columbia, Vancouver, Canada

Ireland (figure 1) has a long history of fishing. Irish fishery resources are managed under the European Union's Common Fisheries Policy, which has been criticized for failing to achieve social, economic, or environmental sustainability. The majority of the catches are industrial, in addition to the artisanal and recreational sectors (figure 2A). Reconstructed domestic catches in Ireland's EEZ, including discards and unreported landings, based on Miller and Zeller (2013), show an increase from about 20,000 t/year in the early 1950s to about 75,000 t/year in the 1970s. Then, in the early 1980s, catches sharply rose to nearly 200,000 t/year, eventually reaching a peak of 400,000 t in 1995, then declining to about 250,000 t/year in the late 2000s (Fahy 2013). These catches for 1950 to 2010 were 1.2 times the official landings from the Irish EEZ, as reported by the International Council for the Exploration of the Sea. Catches were mainly domestic; however, foreign fishing, notably by the United Kingdom, France, and the Netherlands occurs also (figure 2B). Catches were dominated by blue whiting (*Micromesistius poutassou*), Atlantic herring (*Clupea harengus*), Atlantic mackerel (*Scomber scombrus*), and Atlantic horse mackerel (*Trachurus trachurus*; figure 2C), and the highly vulnerable basking shark (*Cetorhinus maximus*; Pauly 2002) also contributed to catches in earlier years. The rapid increases in domestic catches over the last few years were driven by uncontrolled takes and not the result of any stock recoveries.

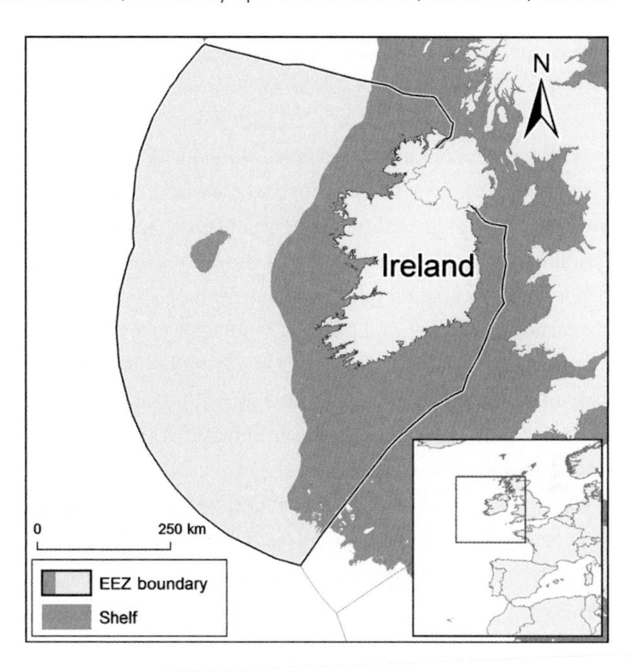

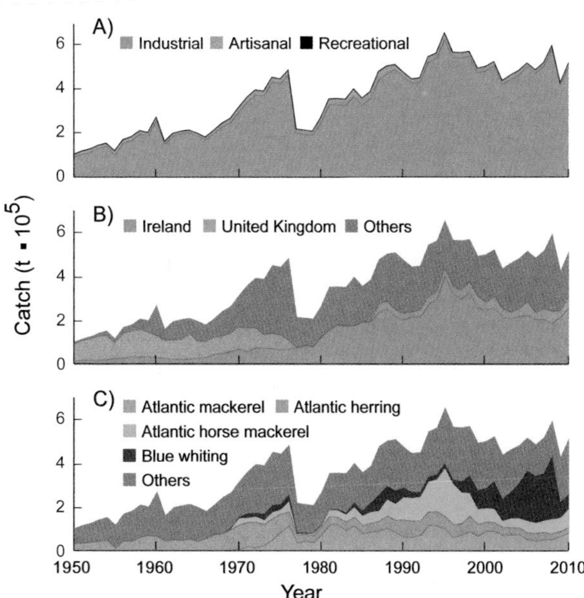

Figure 1. The Irish EEZ (410,000 km², of which 142,000 km² is shelf) and the ICES divisions around the Irish Coast.

Figure 2. Domestic and foreign catches taken in the EEZ of Ireland: (A) by sector (domestic scores: Ind. 3, 3, 4; Art. 2, 2, 3; Rec. 1, 1, 1; Dis. 2, 3, 4); (B) by fishing country (foreign catches are very uncertain); (C) by taxon.

REFERENCES

Fahy E (2013) Overkill! The euphoric rush to industrialise Ireland's sea fisheries and its unravelling sequel. Published by Edward Fahy (edwardfahy@eirecom.net). ISBN-13: 9780-95752180-3.

Miller D and Zeller D (2013) Reconstructing Ireland's marine fisheries catches: 1950–2010. Fisheries Centre Working Paper #2013–10, 48 p.

Pauly D (2002) Growth and mortality of basking shark *Cetorhinus maximus*, and their implications for whale shark *Rhincodon typus*, pp. 199–208 In Fowler SL, Reid T and Dipper FA (eds.), *Elasmobranch biodiversity: conservation and management*. Occasional Papers of the IUCN Survival Commission No. 25, Gland, Switzerland.

1. Cite as Miller, D., and D. Zeller. (2016). Ireland. P. 299 in D. Pauly and D. Zeller (eds.), *Global Atlas of Marine Fisheries: A Critical Appraisal of Catches and Ecosystem Impacts*. Island Press, Washington, DC.

ISRAEL (MEDITERRANEAN)[1]

Dori Edelist,[a] Aviad Scheinin,[a] Oren Sonin,[b] James Shapiro,[b] Pierre Salameh,[b] Gil Rilov,[c] Yehuda Benayahu,[d] Doron Schulz,[d] and Dirk Zeller[e]

[a]Department of Maritime Civilizations and The Leon Recanati Institute for Maritime Studies, University of Haifa, Haifa, Israel [b]Ministry of Agriculture and Rural Development, The Agricultural Center, Beit Dagan, Israel [c]National Institute of Oceanography (IORL), Haifa, Israel [d]Department of Zoology, Tel Aviv University, Ramat Aviv, Tel Aviv, Israel [e]*Sea Around Us*, University of British Columbia, Vancouver, Canada

Israel has a wide range of fisheries along its Mediterranean coast (figure 1): a small industrial trawler fleet; a small and shrinking artisanal sector of gillnetters, purse seiners, longliners, and SCUBA divers; and a growing recreational fishery (mostly angling but also spear-fishing). Over the past six decades, however, the FAO has reported Israel's commercial fisheries landings, not its total withdrawals from the Mediterranean. Thus, the FAO data do not include trawlers' discards (excluded by mandate) and the catch of recreational and subsistence fisheries. However, FAO included landings by Gaza fishers in the Gaza Strip in the 1960s and 1970s (Abudaya et al. 2013). Substantial artisanal and industrial fishing occurs, in addition to the subsistence and recreational sectors (figure 2A). Edelist et al. (2013) reconstructed total domestic removals from within the EEZ, which yielded an estimate of about 2,500 t/year in the early 1950s, a peak of 5,600 t in 1985, and 4,300 t/year in the 2000s. This was nearly 1.3 times the catch that Israel reported to the FAO after exclusion of data from the Gaza Strip. The domestic fishery constitutes the main component of the catch; however, foreign fishing by Spain, Greece, and Croatia, among others, was noticeable in the later time period (figure 2B). Figure 2C presents the composition of this catch, which saw the decline of sardines (*Sardinella* spp.) and the increase of invasive Red Sea fishes and invertebrates.

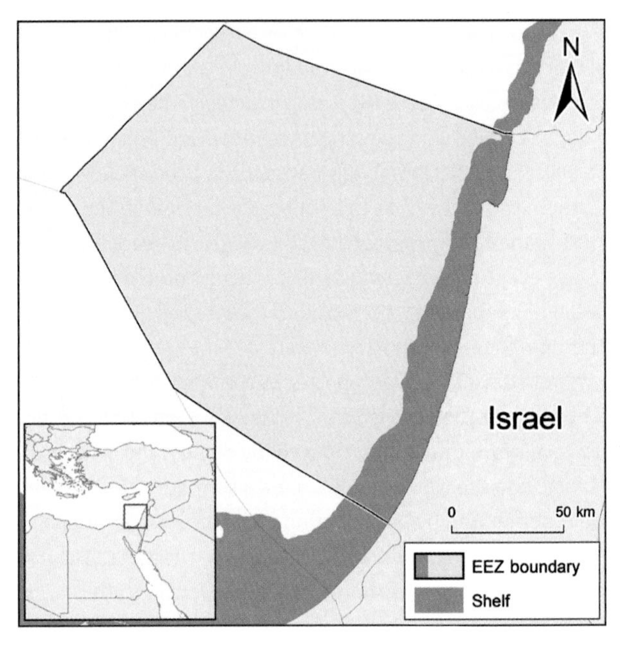

Figure 1. Israel and its Mediterranean EEZ of 25,100 km², of which 2,920 km² is shelf.

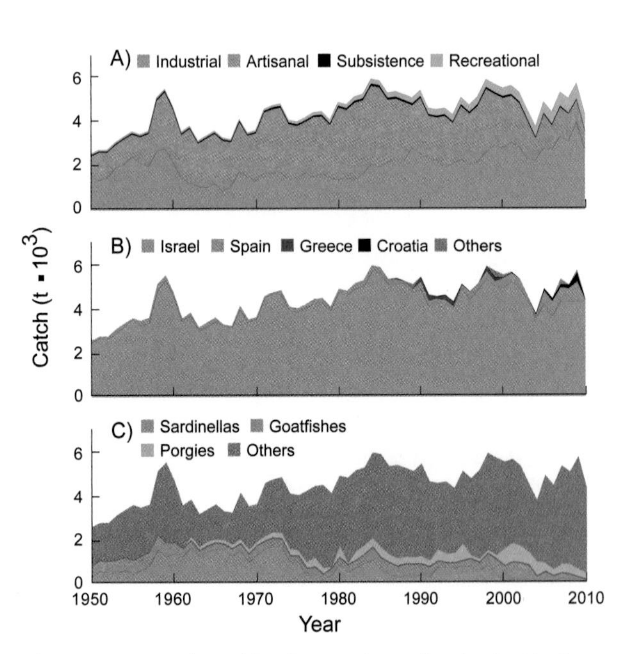

Figure 2. Domestic and foreign catches taken in the Mediterranean EEZ of Israel: (A) by sector (domestic scores: Ind. 1, 1, 2; Art. 4, 4, 4; Sub. 2, 2, 2; Rec. 1, 1, 2; Dis. 2, 2, 3); (B) by fishing country (foreign catches are very uncertain); (C) by taxon.

REFERENCES

Abudaya M, Harper S, Ulman A and Zeller D (2013) Correcting miss- and under-reported marine fisheries catches for the Gaza Strip: 1950–2010. *Acta Adriatica* 54(2): 241–252.

Edelist D, Scheinin A, Sonin O, Shapiro J, Salameh P, Benayahu Y, Schulz D and Zeller D (2013) Israel: Reconstructed estimates of total fisheries removals 1950–2010. *Acta Adriatica* 54(2): 252–264.

1. Cite as Edelist, D., A. Scheinin, O. Sonin, J. Shapiro, P. Salameh, G. Rilov, Y. Benayahu, D. Schulz, and D. Zeller. 2016. Israel. P. 300 in D. Pauly and D. Zeller (eds.), *Global Atlas of Marine Fisheries: A Critical Appraisal of Catches and Ecosystem Impacts*. Island Press, Washington, DC.

ISRAEL (RED SEA)[1]

Dawit Tesfamichael,[a,b] Rhona Govender,[a] and Daniel Pauly[a]

[a]*Sea Around Us*, University of British Columbia, Vancouver, Canada
[b]Department of Marine Sciences, University of Asmara, Asmara, Eritrea

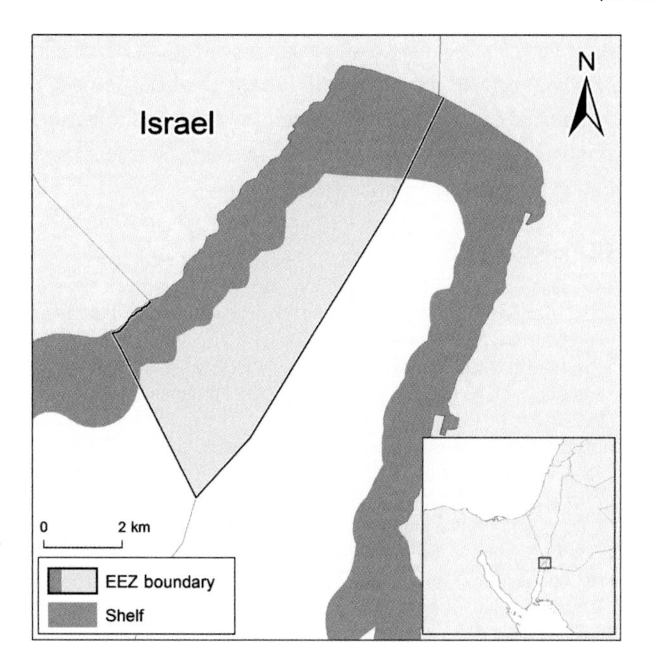

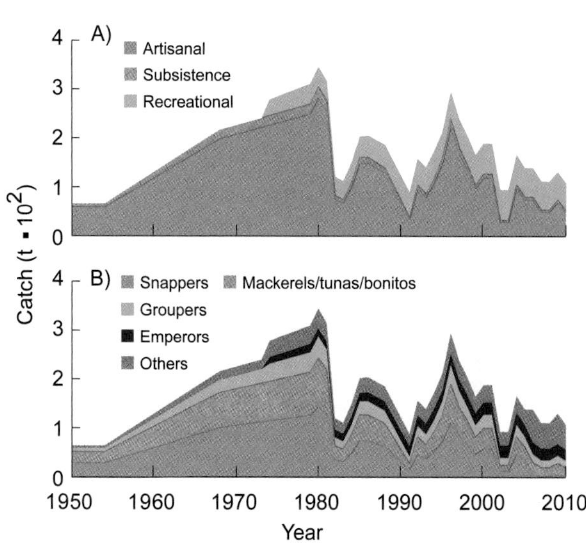

Figure 1. Israel has an EEZ of about 31 km² in the Red Sea, of which 14 km² is shelf. **Figure 2.** Domestic and foreign catches taken in the Red Sea EEZ of Israel: (A) by sector (domestic scores: Art. 2, 2, 2; Sub. 1, 1, 1; Rec. –, 1, 2); (B) by taxon.

Israel has a foothold in the Gulf of Aqaba (figure 1) but currently has no major Red Sea fishery. This was different from the late 1950s to the mid-1970s, when Israel had trawlers operating in the southern Red Sea, mainly in Eritrean (then Ethiopian) waters, with peak catches in 1960s. Israel's fisheries in the Gulf of Aqaba proper are predominantly artisanal, but a recreational fishery has recently emerged, which is now becoming more important than the artisanal fishery (Shapiro 2008; figure 2A). Israel's total catch in its Red Sea EEZ was less than 100 t/year in the early 1950s, then increased until it reached its peak of about 300 t/year in the early 1980s (figure 2A). In later years, the overall catch decreased, reaching values of about 100 t/year in the late 2000s (Tesfamichael et al. 2012). The cumulated reconstructed catch in Israel's waters from 1950 to 2010 was 2.3 times the catch reported by the FAO on behalf of Israel in the Red Sea after the catches taken in the southern Red Sea are excluded. This discrepancy is caused by unreported catch of the artisanal fishery; also, subsistence and recreational fisheries catches are not included in the FAO data. Snappers (family Lutjanidae) were the group with the highest contribution to the total catch (figure 2B), although large pelagics (Scombridae), groupers (Serranidae), and more recently emperors (Lethrinidae) also contributed.

REFERENCES

Shapiro J (2008) The Fisheries and Aquaculture of Israel 2007. Ministry of Agriculture, Department of Fisheries. 48 p.

Tesfamichael D, Govender R and Pauly D (2012) Preliminary reconstruction of fisheries catches of Jordan and Israel in the inner Gulf of Aqaba, Red Sea, 1950–2010. pp. 179–204 In Tesfamichael D and Pauly D (eds.), *Catch reconstruction for the Red Sea Large Marine ecosystem by countries (1950–2010)*. Fisheries Centre Research Reports 20(1).

1. Cite as Tesfamichael, D., R. Govender, and D. Pauly. 2016. Israel (Red Sea). P. 301 in D. Pauly and D. Zeller (eds.), *Global Atlas of Marine Fisheries: A Critical Appraisal of Catches and Ecosystem Impacts*. Island Press, Washington, DC.

ITALY (MAINLAND)[1]

Chiara Piroddi,[a] Michele Gristina,[b] Aylin Ulman,[c] Dirk Zeller,[c] and Daniel Pauly[c]

[a]Joint Research Centre, European Commission, Ispra, Italy [b]Centro Nazionale delle Ricerche di Mazaro del Vallo, Sicily, Italy [c]*Sea Around Us*, University of British Columbia, Vancouver, Canada

Italy (figure 1) has by far the highest fisheries catch in the Mediterranean, even if only the mainland is considered, that is, without Sardinia and Sicily (see following pages and Pauly et al. 2014). Most of the catches are attributed to the industrial sector. However, artisanal, subsistence, and recreational fisheries are known to occur in Italy's waters (figure 2A). Piroddi et al. (2014) performed a reconstruction that led to a mainland domestic catch of 350,000 t/year in the early 1950s, peaking at 850,000 t in 1982, which then declined and averaged about 300,000 t/year from 2006 to 2010. Overall, this was 2.6 times the data reported by the FAO, with the discrepancy driven largely by unreported industrial and artisanal landings and discards. Domestic fisheries compose the bulk of catches in the Italian EEZ, although foreign fishing by Spain and France, among others, has been documented (figure 2B). Since the 1950s, increasing global demand for fish in combination with a technological revolution prompted Italian fishing effort and capacity to increase exponentially until 1980 (Cataudella and Spagnolo 2011). Afterwards, catches rapidly decreased, primarily as a result of a decrease in the biomass of small pelagics, particularly European pilchard (*Sardina pilchardus*) and European anchovy (*Engraulis encrasicolus*) (Iborra Martin 2006; figure 2C). Unreported catches have never been assessed, and this study suggests that unreported landings account for at least one third of total fisheries removals.

REFERENCES

Cataudella S and Spagnolo M (2011) The state of Italian marine fisheries and aquaculture. Ministero delle Politiche Agricole, Alimentari e Forestali (MiPAAF), Rome (Italy), 620 p.

Iborra Martin J (2006) Fisheries in Italy. Policy Department Structural and Cohesion Policies. PE 369.027, European Parliament. IPOL/B/PECH/N/2006_01.

Pauly D, Ulman A, Piroddi C, Bultel E and Coll M (2014) "Reported" versus "likely" fisheries catches of four Mediterranean countries. pp. 11–17 In Lleonart J and Maynou F (eds.), *The Ecosystem approach to fisheries in the Mediterranean and Black Seas*. Scientia Marina 78S1.

Piroddi C, Gristina M, Ulman A, Zeller D and Pauly D (2014) Reconstruction of Italy's marine fisheries catches (1950–2010). Fisheries Centre Working Paper #2014–22, 42 p.

1. Cite as Piroddi, C., M. Gristina, A. Ulman, D. Zeller, and D. Pauly. 2016. Italy (Mainland). P. 302 in D. Pauly and D. Zeller (eds.), *Global Atlas of Marine Fisheries: A Critical Appraisal of Catches and Ecosystem Impacts*. Island Press, Washington, DC.

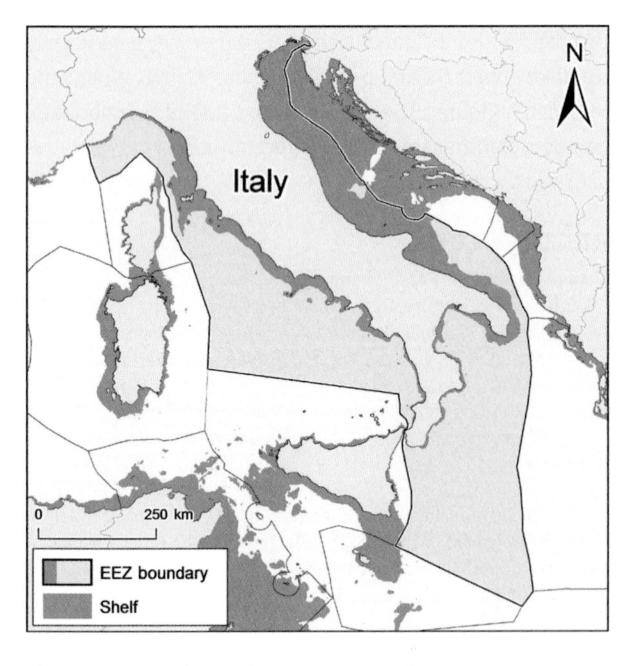

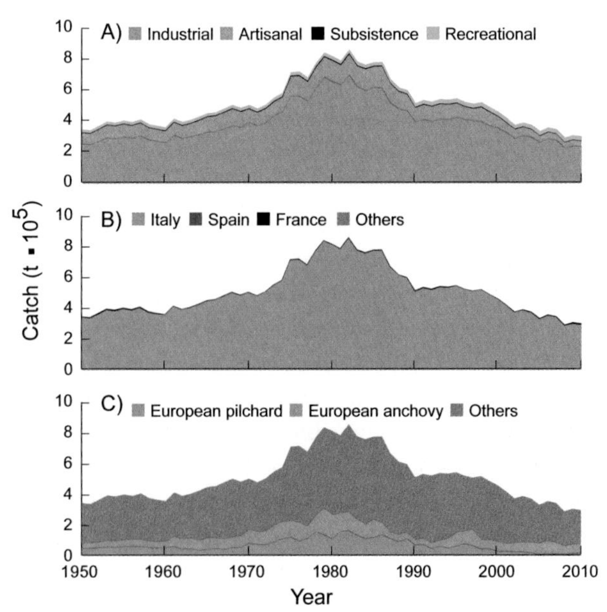

Figure 1. The Italian mainland has a shelf 77,000 km² and 316,000 km² of EEZ-equivalent waters. **Figure 2.** Domestic and foreign catches taken in the EEZ of Italy: (A) by sector (domestic scores: Ind. 2, 3, 4; Art. 2, 3, 4; Sub. 1, 1, 1; Rec. 1, 1, 2; Dis. 2, 2, 2); (B) by fishing country (foreign catches are very uncertain); (C) by taxon.

ITALY (SARDINIA)[1]

Chiara Piroddi,[a] Michele Gristina,[b] Aylin Ulman,[c] Dirk Zeller,[c] and Daniel Pauly[c]

[a]Joint Research Centre, European Commission, Ispra, Italy [b]Centro Nazionale delle Ricerche di Mazaro del Vallo, Sicily, Italy [c]*Sea Around Us*, University of British Columbia, Vancouver, Canada

Sardinia is the second largest island in the Mediterranean after Sicily (figure 1), with Cagliari as its capital. Industrial fisheries, mainly trawlers, multipurpose boats, and purse seiners, had the highest contribution to total landings (figure 2A). The artisanal catch, though substantial, appears to be declining, followed by the recreational and subsistence fisheries. The reconstruction by Piroddi et al. (2014) yielded domestic catches of about 13,000 t/year in the early 1950s, which increased to a short-lived peak of 45,000 t in 1988 and a subsequent steady decline to end at about 12,000 t in 2010, 2.7 times the official statistics. Although the catch is mostly domestic, foreign fishing by Spain, among others, is featured in the *Sea Around Us* catch allocation, as displayed in figure 2B. The industrial fisheries caught mainly European pilchard (*Sardina pilchardus*); however, the common octopus (*Octopus vulgaris*) and bogue (*Boops boops*) that dominated the artisanal and recreational sectors, respectively, were the main taxa overall (figure 2C). Other taxa fished include picarel (*Spicara smaris*) and European hake (*Merluccius merluccius*). Lorenzo et al. (2009) suggest that the artisanal fisheries of Sardinia are overcapitalized and suggest ways this could be overcome.

REFERENCES

Lorenzo I, Madau FA and Pulina P (2009) Capacity and economic efficiency in small-scale fisheries: evidence from the Mediterranean Sea. *Marine Policy* 33(5): 860–867.

Piroddi C, Gristina M, Ulman A, Zeller D and Pauly D (2014) Reconstruction of Italy's marine fisheries catches (1950–2010). Fisheries Centre Working Paper #2014-22, 42 p.

1. Cite as Piroddi, C., M. Gristina, A. Ulman, D. Zeller, and D. Pauly. 2016. Italy (Sardinia). P. 303 in D. Pauly and D. Zeller (eds.), *Global Atlas of Marine Fisheries: A Critical Appraisal of Catches and Ecosystem Impacts*. Island Press, Washington, DC.

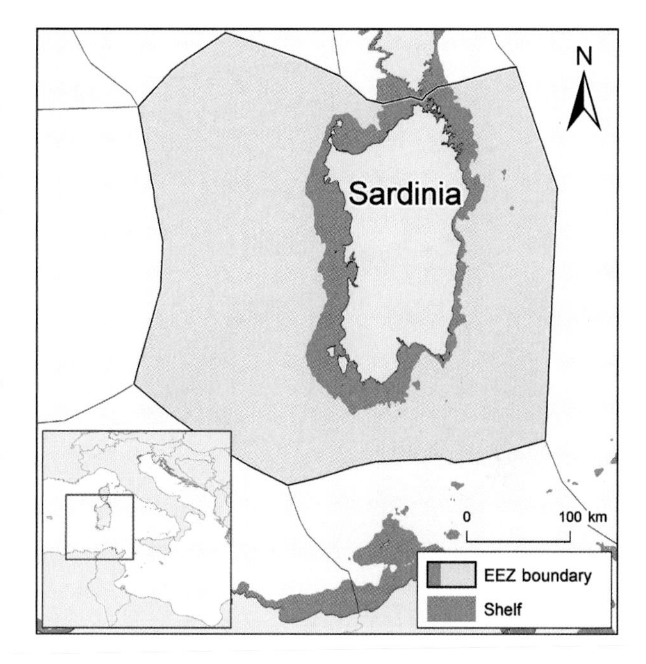

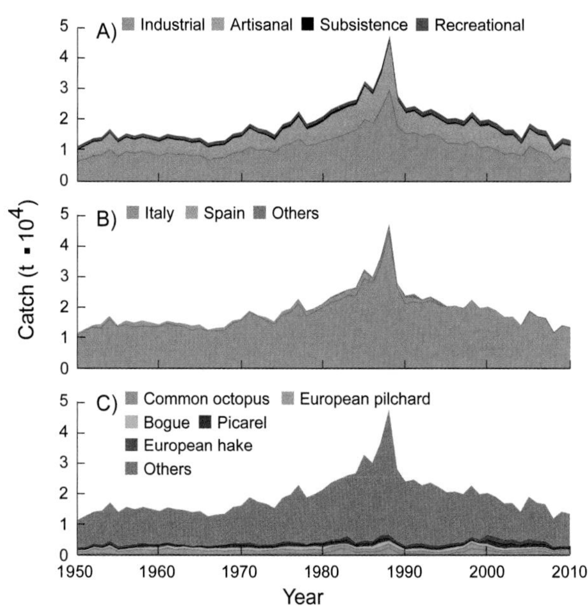

Figure 1. Sardinia has a land area of 24,000 km² and EEZ-equivalent waters of 117,000 km². **Figure 2.** Domestic and foreign catches taken in the EEZ of Sardinia: (A) by sector (domestic scores: Ind. 2, 3, 4; Art. 2, 3, 4; Sub. 1, 1, 1; Rec. 1, 1, 2; Dis. 1, 1, 1); (B) by fishing country (foreign catches are very uncertain); (C) by taxon.

ITALY (SICILY)[1]

Chiara Piroddi,[a] Michele Gristina,[b] Aylin Ulman,[c] Dirk Zeller,[c] and Daniel Pauly[c]

[a]Joint Research Centre, European Commission, Ispra, Italy [b]Centro Nazionale delle Ricerche di Mazaro del Vallo, Sicily, Italy [c]*Sea Around Us*, University of British Columbia, Vancouver, Canada

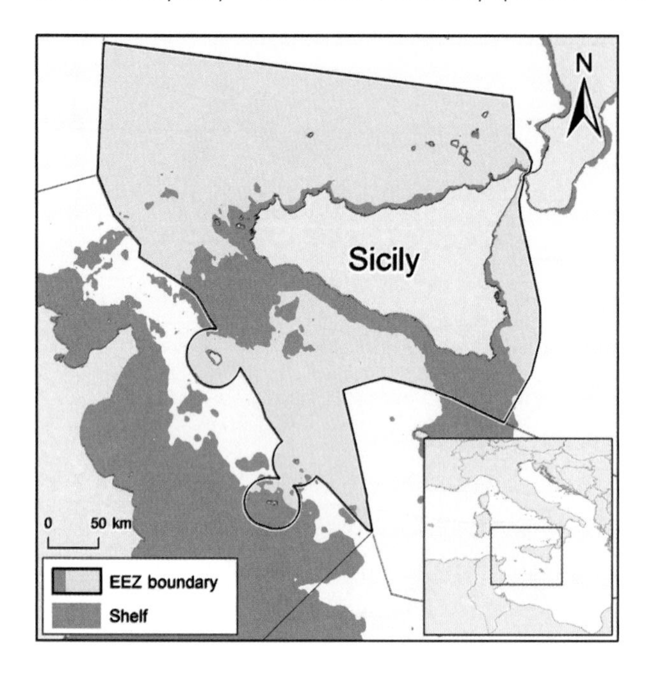

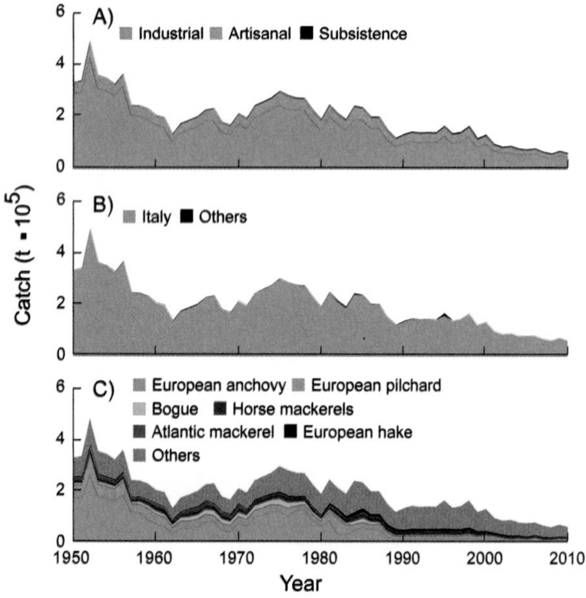

Figure 1. Sicily has a land area of 26,000 km² and EEZ-equivalent waters of 105,000 km². **Figure 2.** Domestic and foreign catches taken in the EEZ of Sicily: (A) by sector (domestic scores: Ind. 2, 3, 4; Art. 2, 3, 4; Sub. 1, 1, 1; Rec. 1, 1, 2; Dis. 1, 1, 1); (B) by fishing country (foreign catches are very uncertain); (C) by taxon.

The sun-drenched island of Sicily, separated from the Italian mainland by the Strait of Messina, is the largest island in the Mediterranean (figure 1). The industrial catch contributed by trawlers, multipurpose vessels, longliners, purse seiners, and midwater trawlers made up most of Sicilian total landings, followed by artisanal fisheries, whereas subsistence and recreational fisheries and discards contributed a small amount each (figure 2A). The domestic catch reconstruction by Piroddi et al. (2014) estimated 330,000 t in 1950, a peak at 488,000 t in 1952, and about 60,000 t/year in the late 2000s. This was 2.5 times the values officially reported. The bulk of the catch of Sicily's EEZ is domestic (figure 2B). The industrial catches consisted mainly of European anchovy (*Engraulis encrasicolus*) and European pilchard (*Sardina pilchardus*) for fish and deep-water rose shrimp (*Parapenaeus longirostris*) for crustaceans, whereas for artisanal fisheries bogue (*Boops boops*), picarel (*Spicara smaris*), common octopus (*Octopus vulgaris*), and cuttlefish (*Sepia officinalis*) dominated. Sharks and rays were commonly discarded. Overall, the major taxa caught in Sicily by all sectors were European anchovy (*Engraulis encrasicolus*), European pilchard (*Sardina pilchardus*), bogue (*Boops boops*), horse mackerel (*Trachurus* spp.), Atlantic mackerel (*Scomber scombrus*), and European hake (*Merluccius merluccius*; figure 2C). As elsewhere in the Mediterranean, artisanal fishers compete with trawlers (Pipitone et al. 2000; Himes 2003), and trawl bans could help.

REFERENCES

Himes AH (2003) Small-scale Sicilian fisheries: Opinions of artisanal fishers and sociocultural effects in two MPA case studies. *Coastal Management* 31(4): 389–408.

Pipitone C, Badalamenti F, D'Anna G and Patti B (2000) Fish biomass increase after a four-year trawl ban in the Gulf of Castellammare (NW Sicily, Mediterranean Sea). *Fisheries Research* 48(1): 23–30.

Piroddi C, Gristina M, Ulman A, Zeller D and Pauly D (2014) Reconstruction of Italy's marine fisheries catches (1950–2010). Fisheries Centre Working Paper #2014-22, 42 p.

1. Cite as Piroddi, C., M. Gristina, A. Ulman, D. Zeller, and D. Pauly. 2016. Italy (Sicily). P. 304 in D. Pauly and D. Zeller (eds.), *Global Atlas of Marine Fisheries: A Critical Appraisal of Catches and Ecosystem Impacts*. Island Press, Washington, DC.

Stephanie Lingard,[a] Sarah Harper,[a] Karl Aiken,[b] Nakhle Hado,[c] Stephen Smikle,[d] and Dirk Zeller[a]

[a]*Sea Around Us*, University of British Columbia, Vancouver, Canada
[b]Department of Life Sciences, University of the West Indies, Mona, Jamaica [c]Food for the Poor, Spanish Town, Jamaica
[d]Fisheries Division, Ministry of Agriculture and Fisheries, Kingston, Jamaica

Jamaica is a tropical island in the Caribbean (figure 1). There exists a large subsistence component, in addition to the artisanal, industrial, and recreational sectors (figure 2A). The reconstruction by Lingard et al. (2012) of the domestic catches of marine fisheries in the Jamaican EEZ starts with 50,000 t/year in the early 1950s, peaks at 62,000 t/year in the late 1970s, and, because of overfishing (Munro 1983), declines to 35,000–40,000 t/year in the 2000s. Overall, these catches are 4.3 times the figures reported by the FAO on behalf of Jamaica. The discrepancy between our figures and the reported data is attributable to large unmonitored subsistence fisheries and underreported artisanal catches, especially in earlier years. A large component of the catches is from the domestic fisheries; however, some foreign fishing by Cuba and Venezuela, among others, has been documented (figure 2B). Catches of jacks and pompanos (family Carangidae), the most important taxon caught throughout the study period, decreased through the time period, with similar trends noticeable for snappers (family Lutjanidae) and other groups (figure 2C). Also visible was a marked shift over time from top predators to taxa lower in the food web, such as parrotfishes (family Scaridae). Improved monitoring of, and public outreach to, subsistence and recreational fishers is imperative if recent management initiatives to create marine protected areas are to succeed.

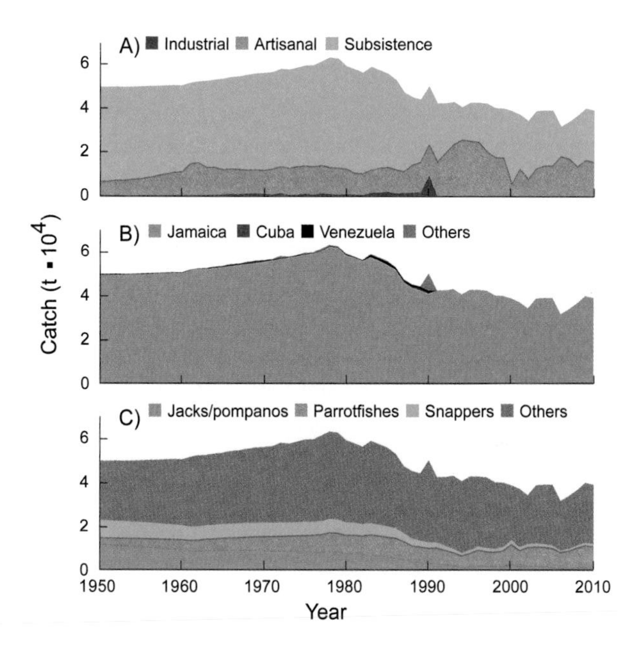

Figure 2. Domestic and foreign catches taken in the EEZ of Jamaica: (A) by sector (domestic scores: Art. 2, 2, 2; Sub. 1, 1, 1; Rec. 1, 1, 1; Dis. 1, 1, 1); (B) by fishing country (foreign catches are very uncertain); (C) by taxon.

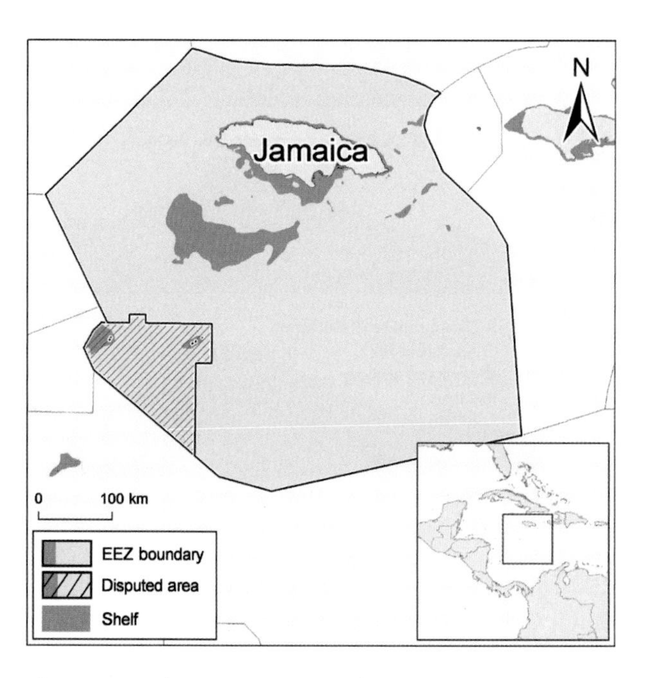

Figure 1. Jamaica has a land area of 11,000 km^2 and an EEZ of 263,000 km^2.

REFERENCES

Lingard S, Harper S, Aiken K, Hado N, Smikle S and Zeller D (2012) Marine fisheries of Jamaica: total reconstructed catch 1950–2010. pp. 47–59 In Harper S, Zylich K, Le Manach F, Pauly D and Zeller D (eds.), *Fisheries catch reconstructions: Islands, Part III*. Fisheries Centre Research Reports 20(5).

Munro JL (1983) Caribbean coral reef fishery resources: The biology, ecology, exploitation and management of Caribbean reef fishes: Scientific report of the ODA/UWI Fisheries ecology research project 1969–1973: University of the West Indies, Jamaica, 2nd edition. ICLARM Studies and Reviews 7, Manila, Philippines, 256 p.

1. Cite as Lingard, S., S. Harper, K. Aiken, N. Hado, S. Smikle, and D. Zeller. 2016. Jamaica. P. 305 in D. Pauly and D. Zeller (eds.), *Global Atlas of Marine Fisheries: A Critical Appraisal of Catches and Ecosystem Impacts*. Island Press, Washington, DC.

JAPAN[1]

Wilf Swartz[a] and Gakushi Ishimura[b]

[a]Nereus Program, University of British Columbia, Vancouver, Canada [b]Center for Sustainability Science, Hokkaido University, Sapporo, Japan

Japan is an archipelagic country (figure 1) with a long tradition of fishing. Most of the catches are industrial; however, there is a substantial artisanal component, in addition to the subsistence and recreational sectors (figure 2A). The extent of unreported fishery catches, including those of recreational fisheries, and discards has yet to be closely examined. Swartz and Ishimura (2014) estimated that the legal and illegal commercial catch, unreported recreational catch, and the associated discards amounted to 3.1 million t/year in the early 1950s, peaked in the mid-1980s with 9.5 million t/year, then declined to 3.8 million t in 2010. The reconstructed catch was about 11% higher than officially reported, a disparity not as large as observed in similar studies of other regions. Nonetheless, the reconstructed catches may provide a better baseline, particularly if Japan moves forward with further implementation of output control management regimes such as individual transferable quotas, which require close monitoring of catch levels to be effective. Removals from within Japan's EEZ are predominantly domestic; however, substantial catches by China and Russia are apparent in the later time period, as presented in figure 2B. Figure 2C gives a taxonomic breakdown of the reconstructed catch, dominated by the Pacific sardine (*Sardinops sagax*), which in the 1980s exhibited a tremendous and still largely unexplained expansion of their biomass (Chavez et al. 2003) and hence catches.

REFERENCES

Chavez FP, Ryan J, Lluch-Cota SE and Niquen M (2003) From anchovies to sardines and back: multidecadal change in the Pacific Ocean. *Science* 299(5604): 217–221.

Swartz W and Ishimura G (2014) Baseline assessment of total fisheries-related biomass removal from Japan's Exclusive Economic Zone: 1950–2010. *Fisheries Science* 80(4): 643–651.

1. Cite as Swartz, W., and G. Ishimura. 2016. Japan. P. 306 in D. Pauly and D. Zeller (eds.), *Global Atlas of Marine Fisheries: A Critical Appraisal of Catches and Ecosystem Impacts*. Island Press, Washington, DC.

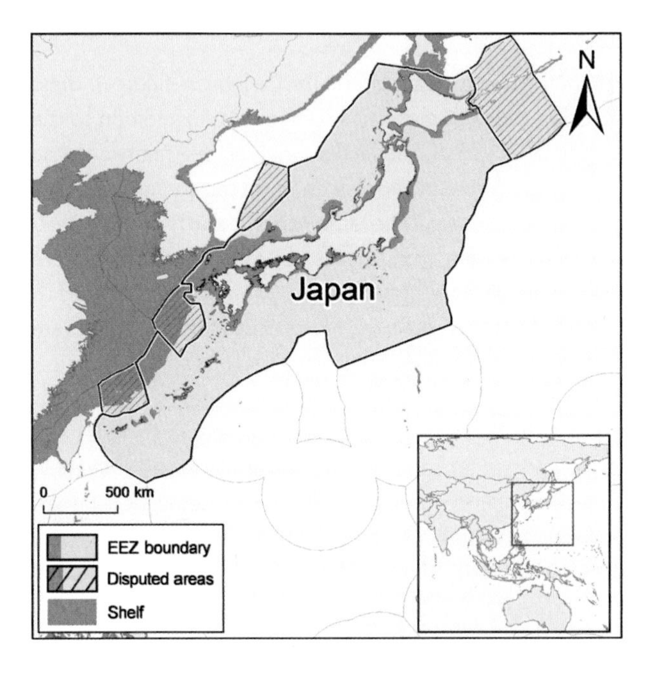

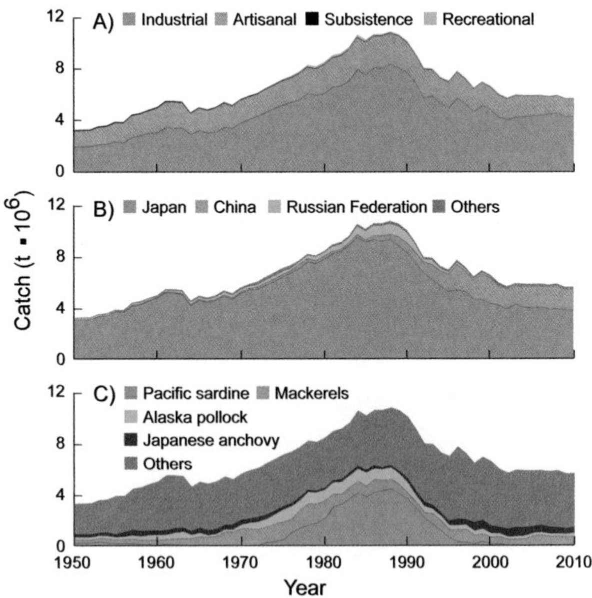

Figure 1. The main islands of Japan have a land area of 377,000 km² and an EEZ of 2.6 million km².

Figure 2. Domestic and foreign catches taken in the EEZ of Japan: (A) by sector (domestic scores: Ind. 3, 4, 4; Art. 4, 4, 4; Sub. 2, 2, 2; Rec. 1, 2, 4; Dis. 1, 2, 2); (B) by fishing country (foreign catches are very uncertain); (C) by taxon.

JAPAN (DAITO ISLANDS)[1]

Wilf Swartz

Nereus Program, University of British Columbia, Vancouver, Canada

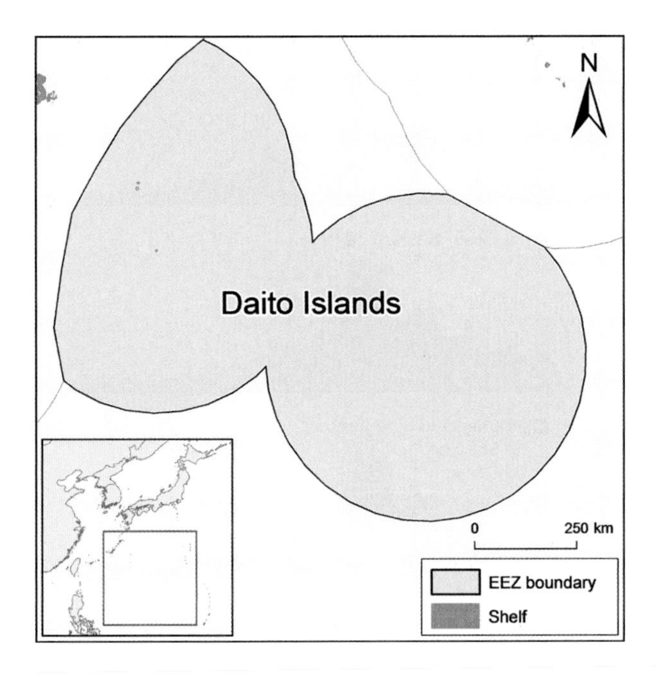

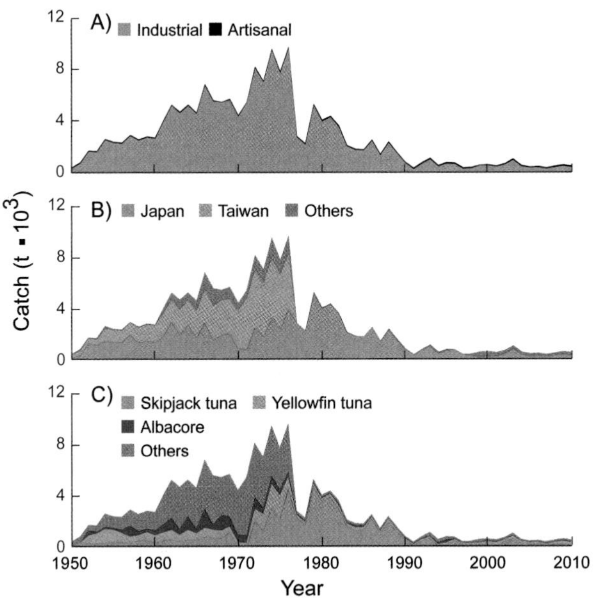

Figure 1. Japan's Daito Islands have a collective land area of 45 km² and an EEZ of 792,000 km². **Figure 2.** Domestic and foreign catches taken in the EEZ of Japan's Daito Islands: (A) by sector (domestic scores: Art. 3, 4, 4; Sub. 3, 4, 4); (B) by fishing country (foreign catches are very uncertain); (C) by taxon.

The Daito Islands are an archipelago consisting of three raised atolls (figure 1) southeast of Okinawa at the southeastern edge of the Kuroshio Current (Niiler et al. 2003; Long 2014). Two of the islands, Kita Daito-jima ("North Borodino") and Minami Daito-jima ("South Borodino"), have a total population of about 2,000, and the smallest island, Oki Daito-jima, remains uninhabited. The three islands were uninhabited until 1900, when pioneers from the Izu Islands south of Tokyo settled to cultivate sugar cane. The islands are administered by the Okinawa Prefecture. Catches allocated to this EEZ are predominantly from foreign industrial vessels, originating primarily from Hong Kong, China, Taiwan, and the Philippines (figure 2A, 2B) and their sudden departure after EEZ declaration in 1977 may not be realistic. Small-scale catches from the artisanal and subsistence sectors do occur, yielding average catches of about 100 t/year, as briefly described by Swartz (2015); however, they are dwarfed by foreign catches. Because of the remoteness of these islands, these small-scale catches are consumed exclusively by the local communities. As presented in figure 2C, overall, the fisheries in this EEZ target overwhelmingly large pelagic species, with skipjack tuna (*Katsuwonus pelamis*), yellowfin tuna (*Thunnus albacares*), and albacore (*Thunnus alalunga*) accounting for a sizable fraction of the total catch.

REFERENCES

Long D (2014) Shards of the shattered Japanese Empire that found themselves as temporary micronations. *Shima: The International Journal of Research into Island Cultures* 8(1): 104–108.

Niiler PP, Maximenko NA, Panteleev GG, Yamagata T and Olson DB (2003) Near-surface dynamical structure of the Kuroshio Extension. *Journal of Geophysical Research: Oceans (1978–2012)* 108(C6).

Swartz W (2015) Notes on the fisheries around Japan's so-called "outer-" or oceanic islands: Ogasawara (Bonin) and Daito Islands. Fisheries Centre Working Paper #2015–17, 5 p.

1. Cite as Swartz, W. 2016. Japan (Daito Islands). P. 307 in D. Pauly and D. Zeller (eds.), *Global Atlas of Marine Fisheries: A Critical Appraisal of Catches and Ecosystem Impacts*. Island Press, Washington, DC.

JAPAN (OGASAWARA ISLANDS)[1]

Wilf Swarz

Nereus Program, University of British Columbia, Vancouver, Canada

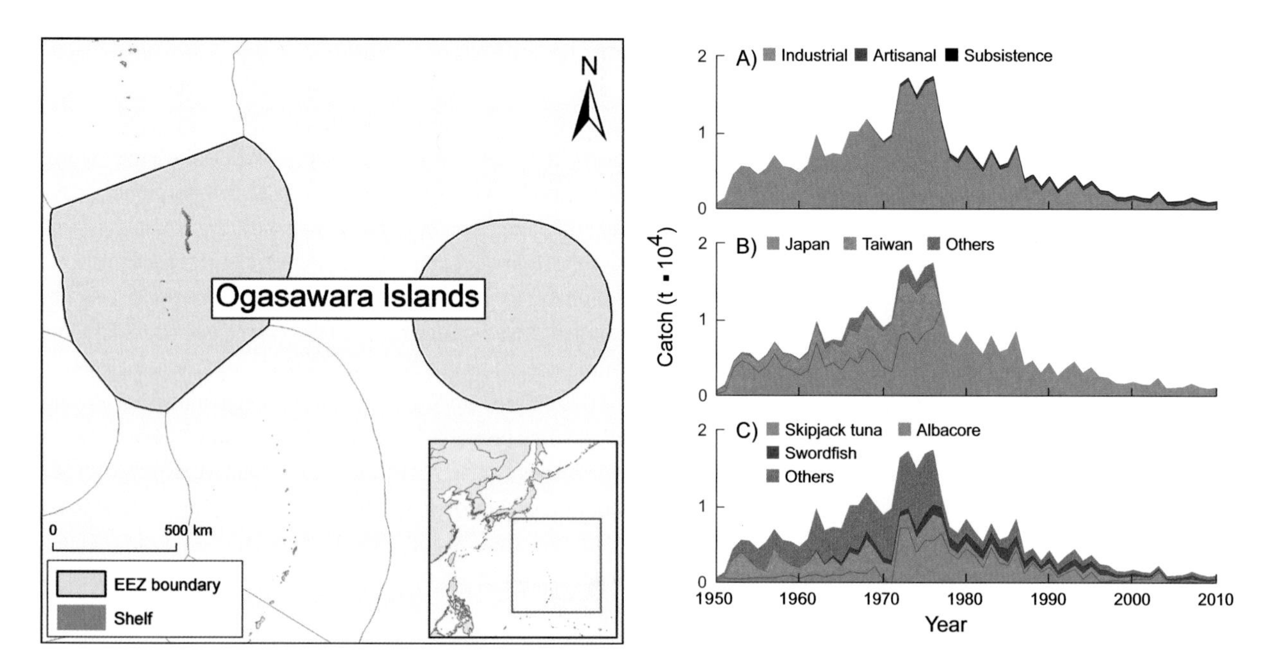

Figure 1. Japan's Ogasawara Islands have a collective land area of 110 km² and an EEZ of 1.08 million km². **Figure 2.** Domestic and foreign catches taken in the EEZ of Japan's Ogasawara Islands: (A) by sector (domestic scores Art. 2, 2, 4; Sub. 2, 2, 4); (B) by fishing country (foreign catches are very uncertain); (C) by taxon.

The Ogasawara Islands (or Bonin Islands), nominated a World Heritage Site, consist of more than 30 tropical and subtropical islands (figure 1), all but 2 of which (Chichi-jima and Haha-jima; jointly 2,600 people) are uninhabited, and are administered, along with Okinotori-shima (Parece Vela) and Minamitori-shima (Marcus Island), by the Tokyo Metropolitan Government. Although the artisanal and subsistence sectors are vital to the local population, the inclusion of the foreign industrial sector operating within the EEZ of Ogasawara Islands overshadows these small-scale catches (figure 2A). The local small-scale fisheries catches for 2004–2010 were used to infer a catch of about 500 t/year from the early 1970s onward, when Chichi-jima and Haha-jima were resettled (in 1968; Swartz 2015). Catches by foreign vessels, namely from China, Hong Kong, Taiwan, and the Philippines, among others, were estimated based on the approach in chapter 2 (this volume) and are thus very tentative (figure 2B). Their catches consist mainly of skipjack tuna (*Katsuwonus pelamis*), albacore (*Thunnus alalunga*), and swordfish (*Xiphias gladius*; figure 2C), although bigfin reef squid (*Sepioteuthis lessoniana*) and spiny lobsters (*Panilurus* spp.; Nishikiori and Sekiguchi 2001; Sekiguchi and George 2005) are also targeted. All fisheries operating in the EEZ (established in 1977) around the Ogasawara Islands are Japanese, with the possible exception of some illegal foreign fishing for large pelagics.

REFERENCES

Nishikiori K and Sekiguchi H (2001) Spiny lobster fishery in Ogasawara (Bonin) Islands, Japan. *Bulletin of the Japanese Society of Fisheries Oceanography* 65: 94–102.

Sekiguchi H and George RW (2005) Description of *Panulirus brunneiflagellum* new species with notes on its biology, evolution, and fisheries. *New Zealand Journal of Marine and Freshwater Research* 39(3): 563–570.

Swartz W (2015) Notes on the fisheries around Japan's so-called 'outer-' or oceanic islands: Ogasawara (Bonin) and Daito Islands. Fisheries Centre Working Paper #2015–17, 5 p.

1. Cite as Swartz, W. 2016. Japan (Ogasawara Islands). P. 308 in D. Pauly and D. Zeller (eds.), *Global Atlas of Marine Fisheries: A Critical Appraisal of Catches and Ecosystem Impacts*. Island Press, Washington, DC.

JORDAN[1]

Dawit Tesfamichael,[a,b] Rhona Govender,[a] and Daniel Pauly[a]

[a]*Sea Around Us*, Fisheries Centre, University of British Columbia, Vancouver, Canada [b]Department of Marine Sciences, University of Asmara, Asmara, Eritrea

Jordan has a very short coastline on the Gulf of Aqaba, its only access to the sea, which makes it all important for the country's maritime activities (figure 1). Because of the very small size of the Jordanian EEZ, the fishery of Jordan has always been one of the smallest in the Red Sea, even when Jordanian fishers fished outside their waters, mainly in Saudi Arabia, a practice that ceased in 1984 (PERSGA 2002). The fishery is mainly artisanal, although subsistence fishing is common and recreational fishing is increasing (figure 2A). The domestic catch of Jordan, according to the reconstruction by Tesfamichael et al. (2012), was about 150 t/year from 1950 to the mid-1960s. This catch then declined, because of conflicts in the area, and started to increase again in the mid-1980s, with some fluctuations, with highest catch of 330 t achieved in 2009. The total reconstructed catch, from 1950 to 2010, was 2 times the catch reported for Jordan by FAO, notably because subsistence and recreational fisheries catches are not included in the data for Jordan. The fishery changed from one dominated by coral reef–associated taxa (from 1950 to the mid-1980s) to one dominated by pelagic taxa. Overall, from 1950 to 2010, as presented in figure 2B, emperors (family Lethrinidae) were the taxon with the highest catch, with kawakawa (*Euthynnus affinis*) following.

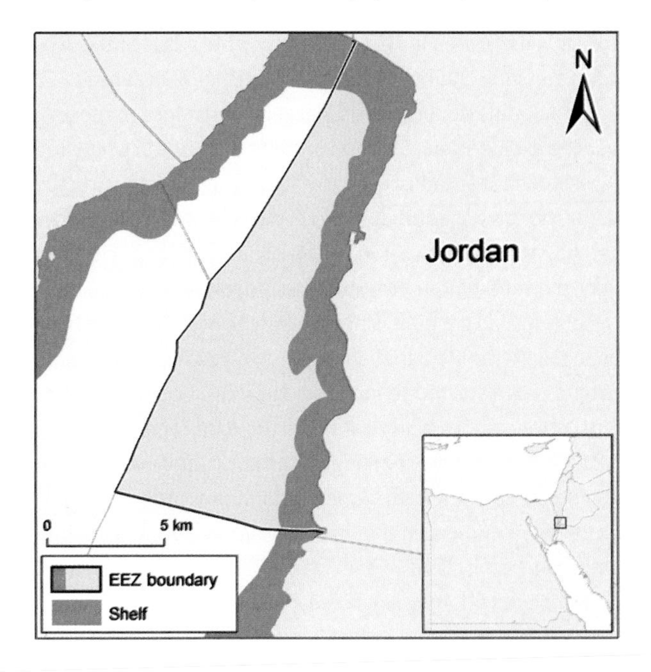

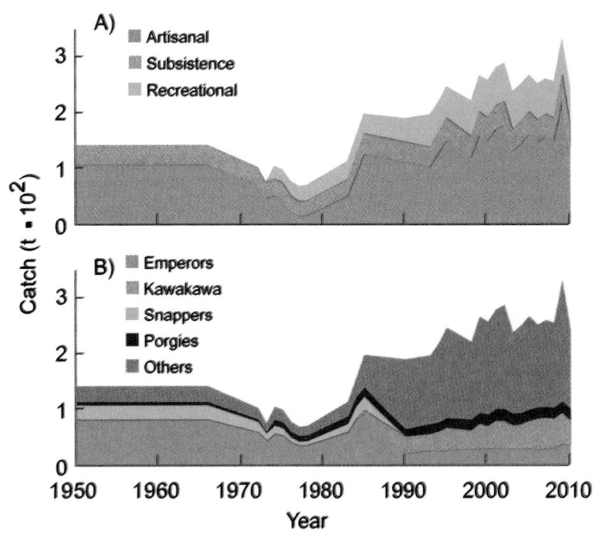

Figure 1. Jordan has a shelf of only 28 km² in an EEZ of about 97 km². **Figure 2.** Domestic and foreign catches taken in the EEZ of Jordan: (A) by sector (domestic scores: Art. 1, 1, 1; Sub. 1, 1, 2; Rec. –, 2, 2); (B) by taxon.

REFERENCES

PERSGA (2002) Status of the Living Marine Resources in the Red Sea and Gulf of Aden and Their Management. Strategic action programme for the Red Sea and Gulf of Aden, Regional Organization for the Conservation of the Environment of the Red Sea and Gulf of Aden (PERSGA), Jeddah, Saudi Arabia. 134 p.

Tesfamichael D, Govender R and Pauly D (2012) Preliminary reconstruction of fisheries catches of Jordan and Israel in the inner Gulf of Aqaba, Red Sea, 1950–2010. pp. 179–204 In Tesfamichael D and Pauly D (eds.), *Catch reconstruction for the Red Sea Large Marine Ecosystem by countries (1950–2010)*. Fisheries Centre Research Reports 20(1).

1. Cite as Tesfamichael, D., R. Govender, and D. Pauly. 2016. Jordan. P. 309 in D. Pauly and D. Zeller (eds.), *Global Atlas of Marine Fisheries: A Critical Appraisal of Catches and Ecosystem Impacts*. Island Press, Washington, DC.

KENYA[1]

Frédéric Le Manach,[a] Caroline A. Abunge,[b] Timothy R. McClanahan,[b] and Daniel Pauly[a]

[a]*Sea Around Us*, University of British Columbia, Vancouver, Canada [b]Wildlife Conservation Society, Mombasa, Kenya

Kenya is located in East Africa along the coast of the Mozambique Channel (figure 1). Most fisheries in Kenya's EEZ are small-scale (for both artisanal and subsistence purposes), but the exploitation of shrimp in the mouth of the Ungwana River is done at an industrial scale (figure 2A). The reconstructed domestic catches estimated by Le Manach et al. (2015) show catches at 10,000 t/year in the 1950s, peaking at 27,000 t in 1985, then declining in the 1990s, presumably because of unsustainable fishing pressure. Since the early 2000s, measures have been taken (see McClanahan et al. 2007, 2008), and domestic catches have started to increase again to 16,000 t by 2010, with total catches being 2.8 times the data reported to the FAO. Catches are predominantly domestic; however, foreign fishing by South Korea, Yemen, and Japan, among others, has been documented (figure 2B). Major groups caught are emperors (family Lethrinidae), parrotfishes (family Scaridae), and spinefoots (family Siganidae; figure 2C), besides valuable invertebrates that are exported (Marshall 2001). Kenyan waters also host both licensed and illegal foreign vessels targeting tunas, sharks, and billfishes (Le Manach et al. 2015), whereas foreign recreational fishers target large pelagic fishes (Abuoda 1999), and migrant fishers from Tanzania access Kenyan EEZ waters for artisanal livelihood purposes.

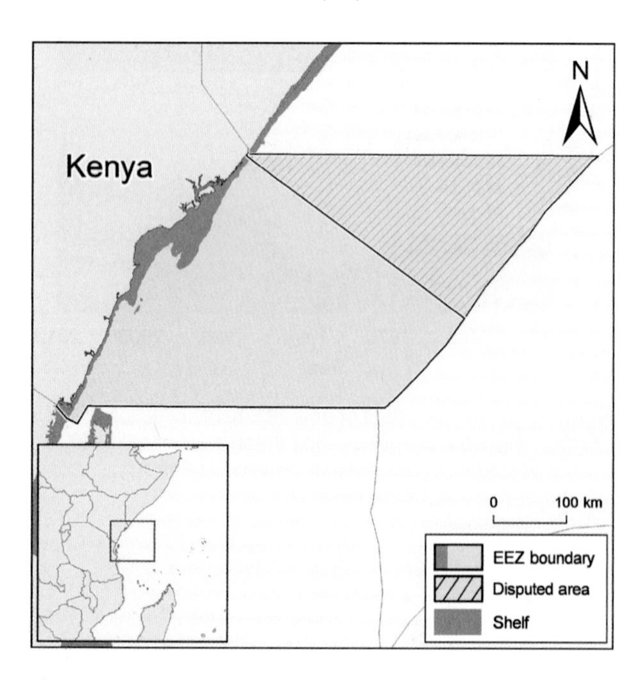

Figure 1. Kenya has a shelf of 8,440 km² in an EEZ of 163,000 km².

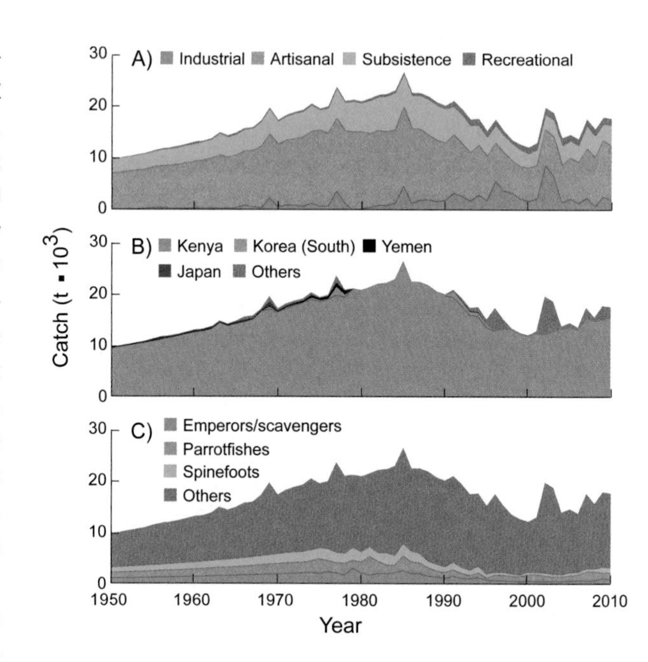

Figure 2. Domestic and foreign catches taken in the EEZ of Kenya: (A) by sector (domestic scores: Ind. –, 3, 4; Art. 2, 3, 4; Sub. 2, 3, 4; Rec. 2, 3, 3; Dis. –, 3, 3); (B) by fishing country (foreign catches are very uncertain); (C) by taxon.

REFERENCES

Abuoda P (1999) Status and trends in recreational marine fisheries in Kenya, pp. 46–50 in T. J. Pitcher (ed.), *Evaluating the benefits of recreational fishing*. Fisheries Centre Research Reports 7(2).

Le Manach F, Abunge CA, McClanahan TR and Pauly D (2015) Tentative Reconstruction of Kenya's Marine Fisheries Catch, 1950–2010. In Le Manach F and Pauly D (eds.), *Fisheries catch reconstructions in the Western Indian Ocean, 1950–2010*. Fisheries Centre Research Reports 23(2).

Marshall, NT (2001). *Stormy Seas for Marine Invertebrates: Trade in Sea Cucumbers, Sea Shells and Lobsters in Kenya, Tanzania and Mozambique*. Traffic East/Southern Africa.

McClanahan TR, Graham NAJ, Calnan JM and MacNeil MA (2007) Toward pristine biomass: reef fish recovery in coral reef marine protected areas in Kenya. *Ecological Applications* 17(4): 1055–1067.

McClanahan TR, Hicks CC and Darling ES (2008) Malthusian overfishing and efforts to overcome it on Kenyan coral reefs. *Ecological Applications* 18(6): 1516–1529.

1. Cite as Le Manach, F., C. A. Abunge, T. R. McClanahan, and D. Pauly. 2016. Kenya. P. 310 in D. Pauly and D. Zeller (eds.), *Global Atlas of Marine Fisheries: A Critical Appraisal of Catches and Ecosystem Impacts*. Island Press, Washington, DC.

KIRIBATI (GILBERT ISLANDS)[1]

Kyrstn Zylich, Sarah Harper, and Dirk Zeller

Sea Around Us, Fisheries Centre, University of British Columbia, Vancouver, Canada

Kiribati (pronounced "Kiri*bass*") consists of 33 islands scattered over a large area: the Gilbert Group (figure 1), the Phoenix Group, and the Line Islands (see next pages), ranging over 4,500 km and 2 FAO statistical areas. South Tarawa, the capital of Kiribati, is located on Tarawa Atoll, in the heavily populated Gilbert Group, and hosts 40% of Kiribati's population, which has one of the highest per capita seafood consumption rates in the world, >200 kg/person/year. Considerable industrial fishing occurs in these waters (figure 2A); however, most of it is by foreign vessels, notably from South Korea, Japan, and Taiwan (figure 2B). The reconstruction by Zylich et al. (2014) of catches in the Gilbert Islands, based on data from the FAO, the Forum Fisheries Agency (FFA), the Western and Central Pacific Fisheries Commission (WCPFC), and other sources suggests that the domestic (i.e., excluding largely foreign industrial tuna fisheries) catch was 8,000 t/year in the early 1950s, increased to about 30,000 t in 2002, and settled to 20,000 t by 2010. The reconstructed catch of the Gilbert Group overall was 1.2 times that reported by the FAO on behalf of Kiribati (as assigned to each of the 3 island groups). Catches were dominated by large pelagics such as skipjack (*Katsuwonus pelamis*), yellowfin (*Thunnus albacares*), and bigeye tuna (*Thunnus obesus*; figure 2C). These industrial catches of large pelagics, which are not truly domestic (see also Zeller et al. 2015), are well reported via the FFA and WCPFC, with only discards being unreported.

REFERENCES

Zeller D, Harper S, Zylich K and Pauly D (2015) Synthesis of under-reported small-scale fisheries catch in Pacific-island waters. *Coral Reefs* 34(1): 25–39.

Zylich K, Harper S and Zeller D (2014) Reconstruction of marine fisheries catches for the Republic of Kiribati (1950–2010). pp. 89–106 In Zylich K, Zeller D, Ang M and Pauly D (eds.), *Fisheries catch reconstructions: Islands, Part IV*. Fisheries Centre Research Reports 22(2).

1. Cite as Zylich, K., S. Harper, and D. Zeller. 2016. Kiribati (Gilbert Islands). P. 311 in D. Pauly and D. Zeller (eds.), *Global Atlas of Marine Fisheries: A Critical Appraisal of Catches and Ecosystem Impacts*. Island Press, Washington, DC.

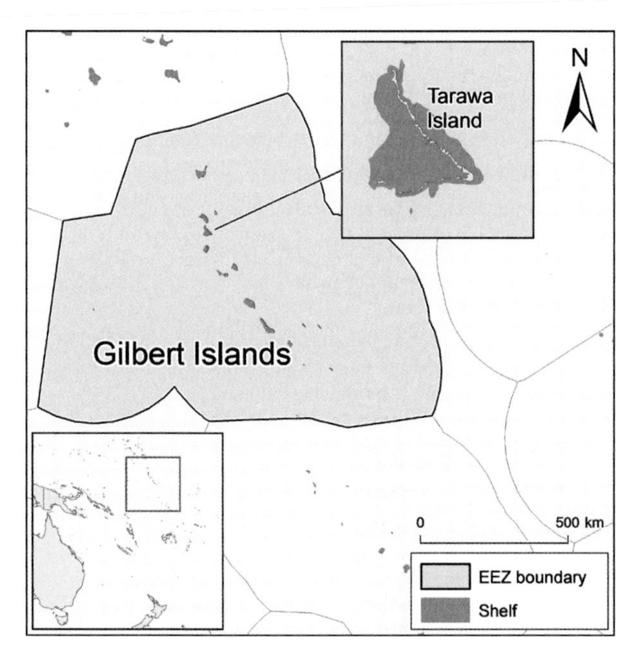

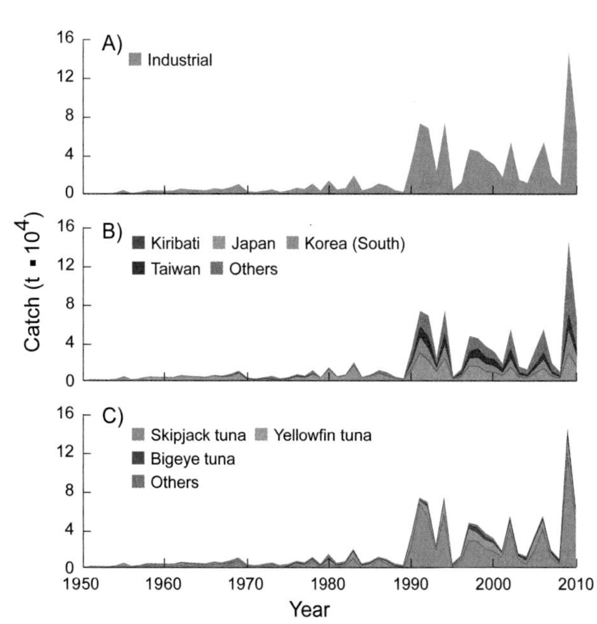

Figure 1. Kiribati's Gilbert Islands have a total land area of only 280 km² but an EEZ of 1.05 million km².

Figure 2. Domestic and foreign catches taken in the EEZ of Kiribati's Gilbert Islands: (A) by sector (domestic scores: Ind. –, 4, 4; Art. 2, 2, 3; Sub. 2, 2, 3); (B) by fishing country (foreign catches are very uncertain); (C) by taxon.

KIRIBATI (LINE ISLANDS)[1]

Kyrstn Zylich, Sarah Harper, Dirk Zeller, and Daniel Pauly

Sea Around Us, University of British Columbia, Vancouver, Canada

The Line Islands, named by Yankee whalers fishing "on the line" (Pala 2013), are part of a chain of atolls of which 8 belong to the easternmost part of the 3 Kiribati areas (pronounced "Kiri*bass*," figure 1; see previous and following page), and the other 3 belong to the United States. Zylich et al. (2014) performed a domestic catch reconstruction. This suggested that artisanal and subsistence catches were 300 t/year in the early 1950s and 1,700 t/year in the late 2000s and were 1.1 times the data reported to the FAO (as assigned to each of the 3 island groups). However, overall within the EEZ of Kiribati Line Islands, foreign industrial vessels, notably from South Korea, Japan, and Taiwan, overshadow the reconstructed small-scale catches (figure 2A, 2B). Reef taxa such as snappers (family Lutjanidae), emperors (Lethrinidae), and marine molluscs are important to the small-scale fisheries, although pelagics (especially skipjack tuna, *Katsuwonus pelamis*, and yellowfin tuna, *Thunnus albacares*) are prominent in the overall catches (figure 2C). Even low catches were shown by Sandin et al. (2008) to strongly modify the trophic functioning of the reef ecosystems, notably by inverting the top-heavy biomass structure that they exhibit when unfished. Thus, the reef fish biomass is dominated by sharks and other top predators in the uninhabited U.S. Line Islands, whereas zooplanktivorous fish biomass dominates on the exploited reefs of Kiritimati and Tabuaeran.

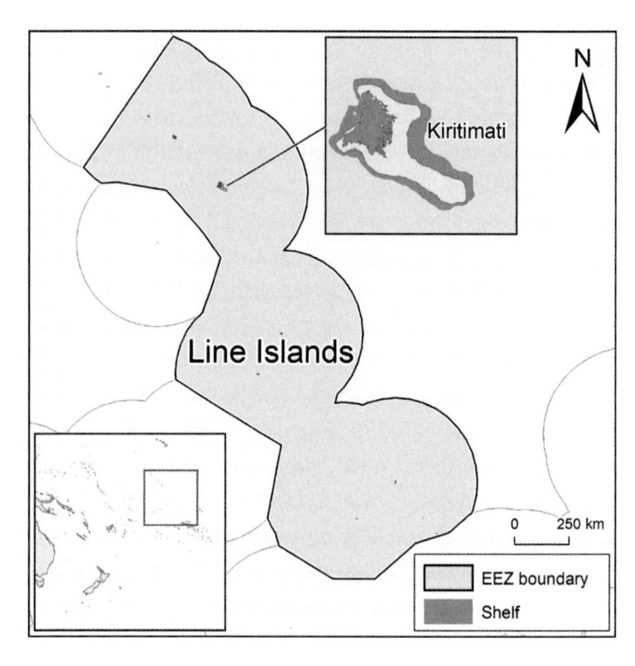

Figure 1. The Republic of Kiribati's Line Islands have a total land area of 490 km² but an EEZ of 1.05 million km².

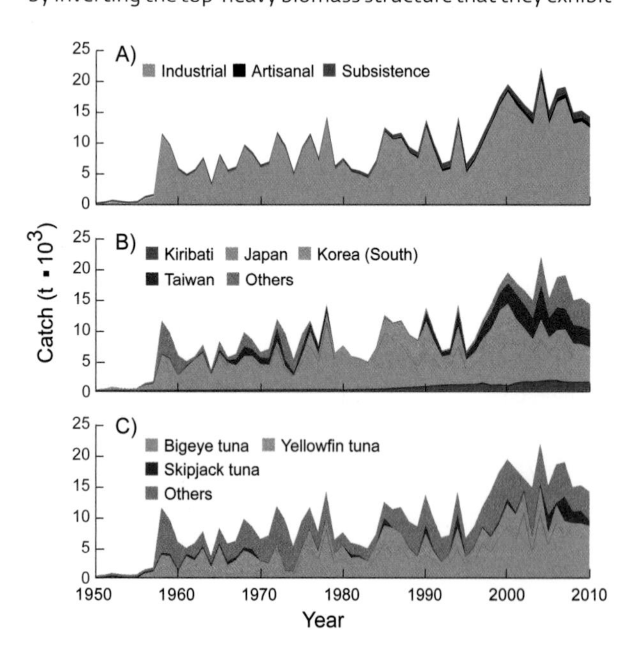

Figure 2. Domestic and foreign catches taken in the EEZ of Kiribati's Line Islands: (A) by sector (domestic scores: Ind. –, –, 4; Art. 2, 2, 3; Sub. 2, 2, 3); (B) by fishing country (foreign catches are very uncertain); (C) by taxon.

REFERENCES

Pala C (2013) History through a watery lens. pp. 15–23 In Stone GS and Obura D (eds.), *Underwater Eden: Saving the Last Coral Wilderness on Earth*. The University of Chicago Press, Chicago, IL.

Sandin SA, Smith JE, DeMartini EE, Dinsdale EA, Donner SD, Friedlander AM, Konotchick T, Malay M, Maragos JE, Obura D, Pantos O, Paulay G, Richie M, Rohwer F, Schoeder RE, Walsh S, Jackson JBC, Knowlton N and Sala E (2008) Baselines and Degradation of Coral Reefs in the Northern Line Islands. *PLoS ONE* 3(2): e1548. doi:10.1371/journal.pone.0001548.

Zylich K, Harper S and Zeller D (2014) Reconstruction of marine fisheries catches for the Republic of Kiribati (1950–2010). pp. 89–106 In Zylich K, Zeller D, Ang M and Pauly D (eds.), *Fisheries catch reconstructions: Islands, Part IV*. Fisheries Centre Research Reports 22(2).

1. Cite as Zylich, K., S. Harper, D. Zeller, and D. Pauly. 2016. Kiribati (Line Islands). P. 312 in D. Pauly and D. Zeller (eds.), *Global Atlas of Marine Fisheries: A Critical Appraisal of Catches and Ecosystem Impacts*. Island Press, Washington, DC.

KIRIBATI (PHOENIX ISLANDS)[1]

Kyrstn Zylich, Sarah Harper, Dirk Zeller, and Daniel Pauly

Sea Around Us, University of British Columbia, Vancouver, Canada

The Phoenix Islands (figure 1) are at the center of the wide-ranging Republic of Kiribati (still pronounced "Kiri*bass*"). The Phoenix Islands' population of about 720 in the late 1930s (from the Gilbert Islands) peaked at 1,300 in the mid-1950s, but lack of freshwater forced the evacuation of the entire population in the early 1960s, except for a few functionaries to service an airport and maintain Kiribati's claim to these islands (Pala 2013). In the early 2000s, a number of i-Kiribati families settled again in the Phoenix Islands. In 2006, the creation of the Phoenix Islands Protected Area (PIPA) was announced by Kiribati's government, although the extent of effective protection was a controversy for several years. Zylich et al. (2014) provide estimates of domestic catches, largely subsistence catches from 1950 to the early 1960s, virtually nonexistent until the mid-1980s, then consisting of occasional legal or illegal distant-water fleets. However, figure 2A and figure 2B illustrates allocated catch within these waters to be overwhelmingly industrial and foreign. Major taxa were large pelagics, encompassing skipjack tuna (*Katsuwonus pelamis*), yellowfin tuna (*Thunnus albacares*), and bigeye tuna (*Thunnus obesus*; figure 2C). Overfishing has been documented, in one instance from 2001, when a single longlining vessel wiped out "almost the entire adult population of shark from half of the Phoenix Islands" (Stone 2013). Shark finning (Biery and Pauly 2012) is definitely a nasty business.

REFERENCES

Biery L and Pauly D (2012) A global review of species-specific shark-fin-to-body-mass ratios and relevant legislation. *Journal of Fish Biology* 80(5): 1643–1677.

Pala C (2013) History through a watery lens. pp. 15–23 In Stone GS and Obura D (eds.), *Underwater Eden: Saving the Last Coral Wilderness on Earth*. The University of Chicago Press, Chicago, IL.

Stone GS (2013) In search of paradise. pp. 1–14 In Stone GS and Obura D (eds.), *Underwater Eden: Saving the Last Coral Wilderness on Earth*. The University of Chicago Press, Chicago, IL.

Zylich K, Harper S and Zeller D (2014) Reconstruction of marine fisheries catches for the Republic of Kiribati (1950–2010). pp. 89–106 In Zylich K, Zeller D, Ang M and Pauly D (eds.), *Fisheries catch reconstructions: Islands, Part IV*. Fisheries Centre Research Reports 22(2).

1. Cite as Zylich, K., S. Harper, D. Zeller, and D. Pauly. 2016. Kiribati (Phoenix Islands). P. 313 in D. Pauly and D. Zeller (eds.), *Global Atlas of Marine Fisheries: A Critical Appraisal of Catches and Ecosystem Impacts*. Island Press, Washington, DC.

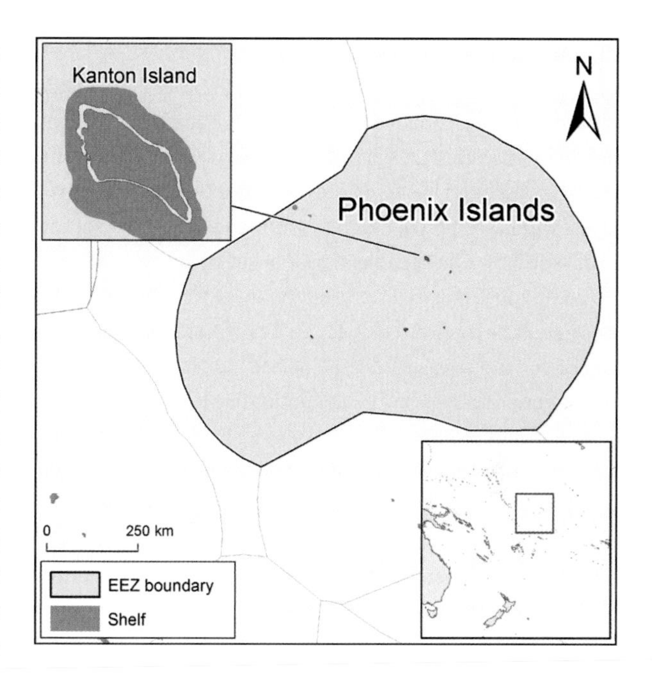

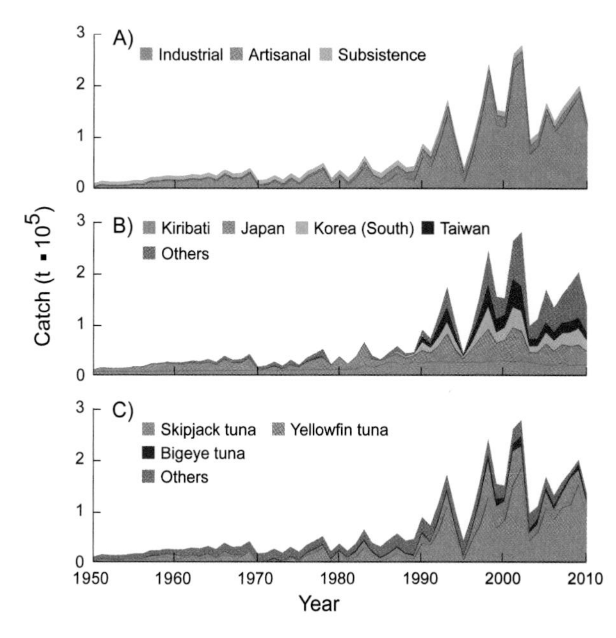

Figure 1. The Republic of Kiribati's Phoenix Islands have a total land area of 28 km² but an EEZ of 743,000 km².

Figure 2. Domestic and foreign catches taken in the EEZ of Kiribati's Phoenix Islands: (A) by sector (domestic scores: Art. 2, –, –; Sub. 2, 2, 3); (B) by fishing country (foreign catches are very uncertain); (C) by taxon.

KOREA (NORTH)[1]

Soohyun Shon, Sarah Harper, and Dirk Zeller

Sea Around Us, University of British Columbia, Vancouver, Canada

The Democratic People's Republic of Korea (or North Korea) is located on the northern portion of the Korean Peninsula (figure 1). Unlike most coastal countries, North Korea does not regularly report its annual marine fisheries catches to the FAO, although it is a member. Catches from within North Korea's EEZ are predominantly from the industrial sector, in addition to a noticeable artisanal sector and a gradually declining subsistence component (figure 2A). Shon et al. (2014) estimated the marine fisheries catches of North Korea within its EEZ for the 1950–2010 time period, based on household consumption figures, scattered information from the literature (e.g., Park and Hong 2012), and statistics from a South Korean government agency tasked with tracking the development of the North Korean economy. This yielded a domestic catch that increased from 430,000 t in 1950 to a peak of 1.28 million t in 1978, before collapsing to 200,000 t by 2010. Overall, this was 1.6 times the cumulative total catch inferred by the FAO from scattered information they receive from North Korea and consultant reports. Catches were largely domestic, with some small foreign removals by South Korea and China, allocated to these waters (figure 2B). Figure 2C presents top taxa as Alaska pollock (*Theragra chalcogramma*), Japanese flying squids (*Todarodes pacificus*), chub mackerel (*Scomber japonicus*), and Pacific herring (*Clupea pallasii*).

REFERENCES

Park SJ and Hong SG (2012) Revisiting changing patterns of North Korea's fisheries production: 1990s–2000s. *International Journal of Maritime Affairs and Fisheries* 4(1): 107–125.

Shon S, Harper S and Zeller D (2014) Reconstruction of marine fisheries catches from the Democratic People's Republic of Korea (North Korea) from 1950–2010. Fisheries Centre Working Paper #2014–20, 13 p.

1. Cite as Shon, S., S. Harper, and D. Zeller. 2016. Korea (North). P. 314 in D. Pauly and D. Zeller (eds.), *Global Atlas of Marine Fisheries: A Critical Appraisal of Catches and Ecosystem Impacts*. Island Press, Washington, DC.

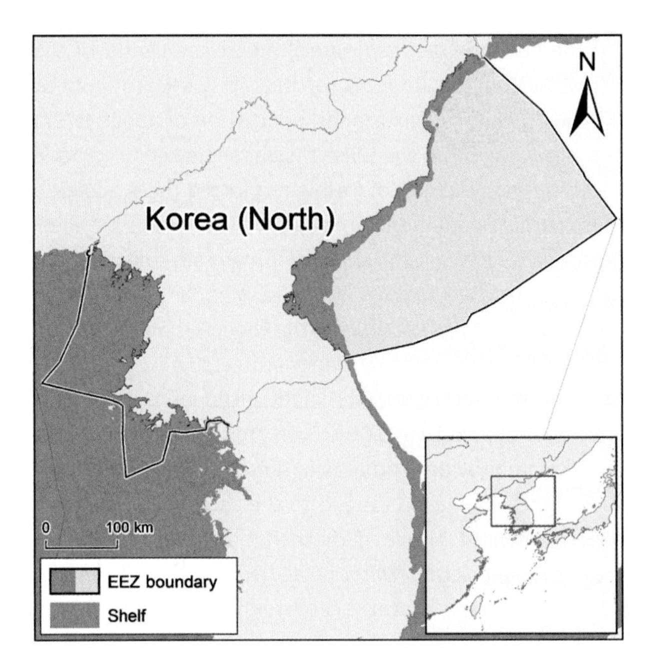

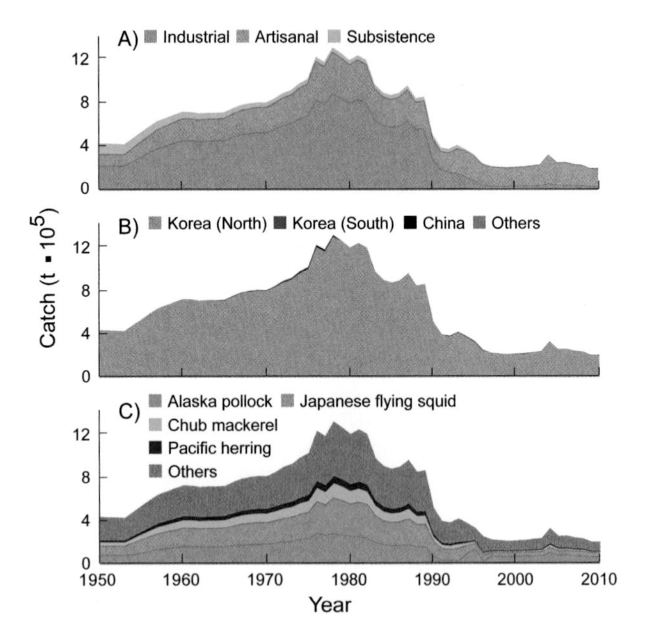

Figure 1. North Korea has a shelf of 35,900 km² and an EEZ of 116,000 km². **Figure 2.** Domestic and foreign catches taken in the EEZ of North Korea: (A) by sector (domestic scores: Ind. 2, 2, 2; Art. 1, 1, 1; Sub. 1, 1, 1; Dis. 1, 1, 1); (B) by fishing country (foreign catches are very uncertain); (C) by taxon.

KOREA (SOUTH)[1]

Soohyun Shon, Sarah Harper, and Dirk Zeller

Sea Around Us, University of British Columbia, Vancouver, Canada

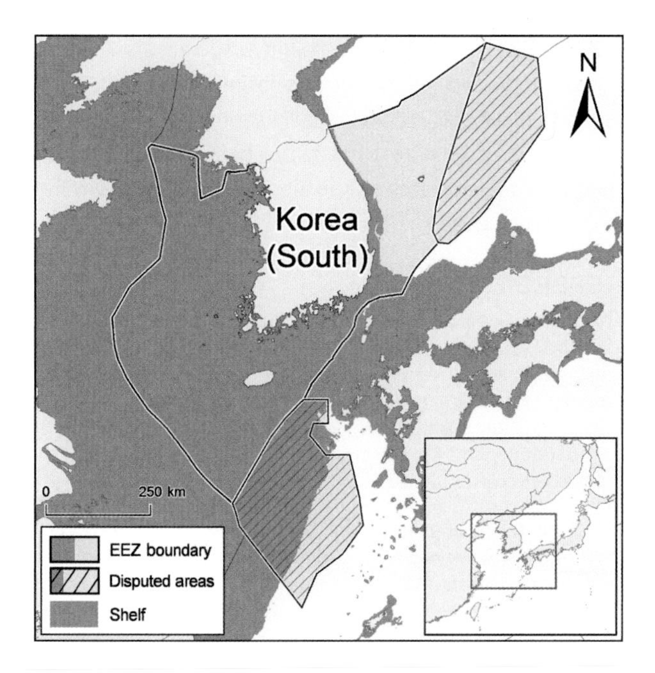

Figure 1. South Korea has 3,440 islands surrounding its mainland, a shelf of 290,000 km², and a claimed EEZ of 473,000 km².

The Republic of Korea is located on the southern half of the Korean Peninsula (figure 1) and is generally referred to as South Korea. After the devastating Korean war (1950–1953), South Korea transformed itself into an economic powerhouse, which included the development of ubiquitous distant-water fleets, in addition to smaller artisanal, subsistence, and recreational components (figure 2A). The reconstruction performed by Shon et al. (2014) relied strongly on Korean literature (e.g., Anonymous 2011) and dealt only with domestic fisheries in its EEZ. These catches increased, partially because of spatial expansion (Zhang and Kim 1999), from 0.7 million t in 1950 to a peak of 2.5 million t in 1986, before declining to 1.4 million t by 2010, 1.6 times the data reported by the FAO on behalf of South Korea. This occurred because of the addition of unreported subsistence, recreational, and domestic illegal catches, and discards, to the reported commercial landings. Although foreign catches by China have been steadily increasing in the 2000s, overall the catches are predominantly domestic (figure 2B). The changing composition of this catch (figure 2C) consists of a multitude of fish and invertebrate species dominated by seastars (Asteroidea), Japanese anchovy (*Engraulis japonicus*), and largehead hairtail (*Trichiurus lepturus*). Marine resources are valuable to the economy and food security of South Korea, so improved catch recording systems and estimation approaches are needed.

REFERENCES

Anonymous (2011). [Korean fisheries yearbook 2011]. Korea Fisheries Association, Seoul (South Korea). 589 p.

Shon S, Harper S and Zeller D (2014) Reconstruction of Marine Fisheries Catches for the Republic of Korea (South Korea) from 1950–2010. Fisheries Centre Working Paper #2014–19, 13 p.

Zhang, C. I. and S. Kim. 1999. Living marine resources in the Yellow Sea ecosystem in Korean waters: status and perspectives. pp. 163–178 In K. Sherman and Q. Tang. (eds.), *Large Marine Ecosystems of the Pacific Rim*. Blackwell Science, Malen, Mass.

1. Cite as Shon, S., S. Harper, and D. Zeller. 2016. Korea (South). P. 315 in D. Pauly and D. Zeller (eds.), *Global Atlas of Marine Fisheries: A Critical Appraisal of Catches and Ecosystem Impacts*. Island Press, Washington, DC.

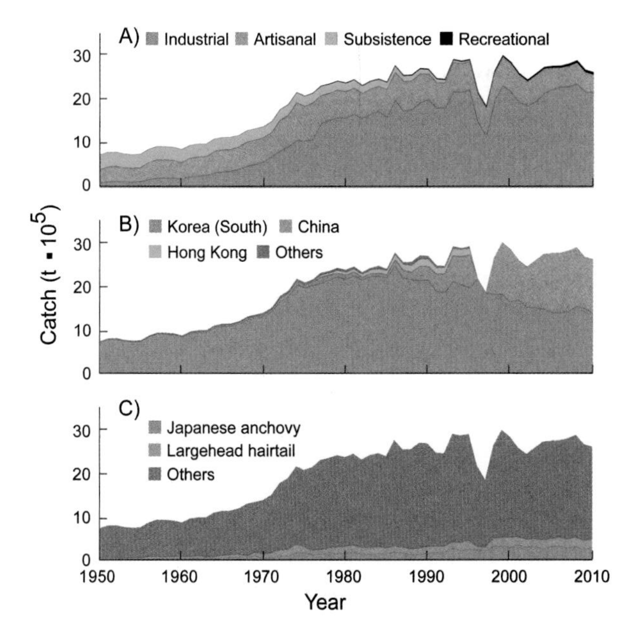

Figure 2. Domestic and foreign catches taken in the EEZ of South Korea: (A) by sector (domestic scores: Ind. 2, 2, 3; Art. 2, 2, 3; Sub. 2, 2, 2; Rec. –, 1, 2; Dis. 2, 2, 2); (B) by fishing country (foreign catches are very uncertain); (C) by taxon.

KUWAIT[1]

Dalal Al-Abdulrazzak

Sea Around Us, University of British Columbia, Vancouver, Canada

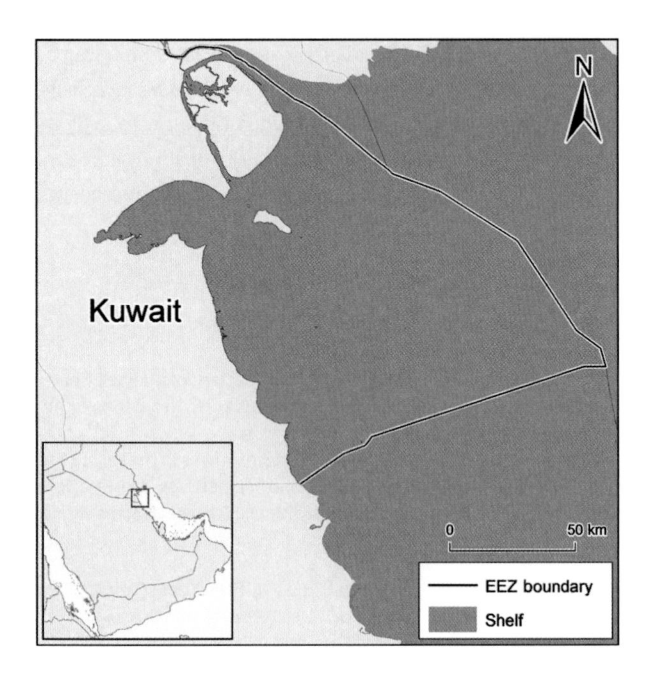

Figure 1. Kuwait has an EEZ of 11,800 km², all of which is shelf.

Kuwait is located in the northwest of the Persian Gulf (figure 1). Substantial artisanal and industrial fishing occurs in Kuwait's EEZ, in addition to the subsistence and recreational component (although not visible on graph; figure 2A). Kuwait's fishery catches have grown substantially over the past 60 years (Al-Sabbagh and Dashti 2009) and were reconstructed by Al-Abdulrazzak (2013). The result is a total domestic catch estimate of about 8,700 t/year for the early 1950s, which increased to a first peak of more than 40,000 t/year in the early 1970s, followed by decline in the late 1970s, caused by the Iran–Iraq war. The second peak, at more than 60,000 t, occurred in 1988 and was followed by a slow decline in total catches and their eventual stagnation at less than 40,000 t/year through the 2000s. Overall, this corresponded to a catch 6.4 times that reported by the FAO on behalf of Kuwait, mainly because of the discards from the trawl fishery being 10 times greater than the finfish landings. Foreign fishing, notably by Iraq, becomes visible only in the 2000s (figure 2B). Figure 2C shows a few of the major taxa caught in Kuwait: giant catfish (*Netuma thalassina*, whose catch is entirely discarded), shark and rays (Elasmobranchii), penaeid shrimps, and guitarfish (Rhinobatidae; discarded). It would be beneficial to find a way to use the huge bycatch of Kuwait's trawl fisheries, which is currently discarded.

REFERENCES

Al-Abdulrazzak D (2013) Reconstructing Kuwait's marine fishery catches: 1950–2010. In Al-Abdulrazzak D and Pauly D (eds.), *From dhows to trawlers: a recent history of fisheries in the Gulf countries, 1950 to 2010*. Fisheries Centre Research Reports 21(2).

Al-Sabbagh T and Dashti J (2009) Post-invasion status of Kuwait's fin-fish and shrimp fisheries (1991–1992). *World Journal of Fish and Marine Sciences* 1(2): 94–96.

1. Cite as Al-Abdulrazzak, D. 2016. Kuwait. P. 316 in D. Pauly and D. Zeller (eds.), *Global Atlas of Marine Fisheries: A Critical Appraisal of Catches and Ecosystem Impacts*. Island Press, Washington, DC.

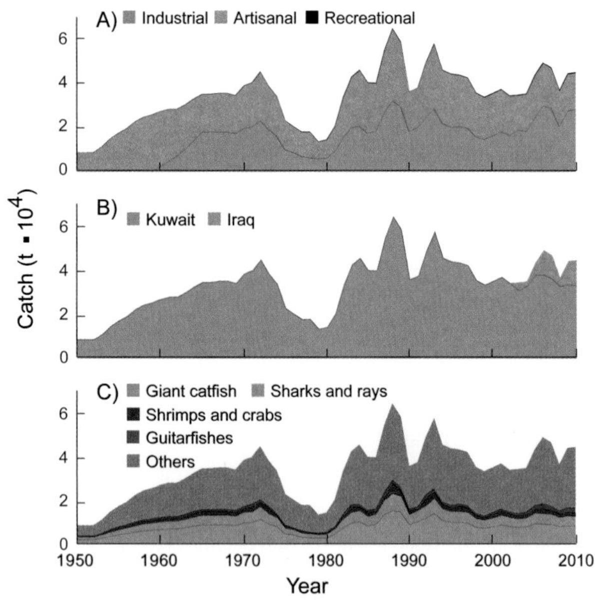

Figure 2. Domestic and foreign catches taken in the EEZ of Kuwait: (A) by sector (domestic scores: Ind. 2, 2, 2; Art. 2, 2, 3; Sub. 1, 1, 1; Rec. 2, 2, 2; Dis. 3, 3, 3); (B) by fishing country (foreign catches are very uncertain); (C) by taxon.

LATVIA[1]

Peter Rossing,[a] Maris Plikshs,[b] Shawn Booth,[a] Liane Veitch,[a] and Dirk Zeller[a]

[a]*Sea Around Us*, University of British Columbia, Vancouver, Canada [b]Latvian Fish Resource Agency (LATFRA), Riga, Latvia

Latvia lies on the eastern edge of the Baltic Sea (figure 1). The industrial sector contributes most of the catches allocated to Latvia's EEZ, in addition to the artisanal and recreational sectors (figure 2A). Total domestic marine fisheries catches by Latvia in the Baltic Sea (and its equivalent entity before separation from the former USSR in 1991) were reconstructed by Rossing et al. (2010), combining data from the International Council for the Exploration of the Sea (ICES) and the Latvian Fish Resource Agency for the pre-1991 period. The reconstructed domestic catch, as assigned to the Latvian EEZ, increased from 10,000–20,000 t/year in the 1950s to 60,000 t in 1974, decreased for 2 decades to a low of 20,000 t/year in the early 1990s, then bounced back to about 40,000 t/year into the 2000s. The reconstructed domestic catch was 1.2 times higher than reported by ICES. The discrepancy results mainly from unreported industrial landings, which makes Latvia the fourth highest in terms of unreported landings among all countries around the Baltic Sea (Zeller et al. 2011). Approximately half of the catches in Latvia's EEZ are domestic, and the remainder is taken by other Baltic countries, notably Sweden, Denmark, and Poland (figure 2B). Figure 2C presents top taxa as Atlantic herring (*Clupea harengus*), European sprat (*Sprattus sprattus*), and Atlantic cod (*Gadus morhua*).

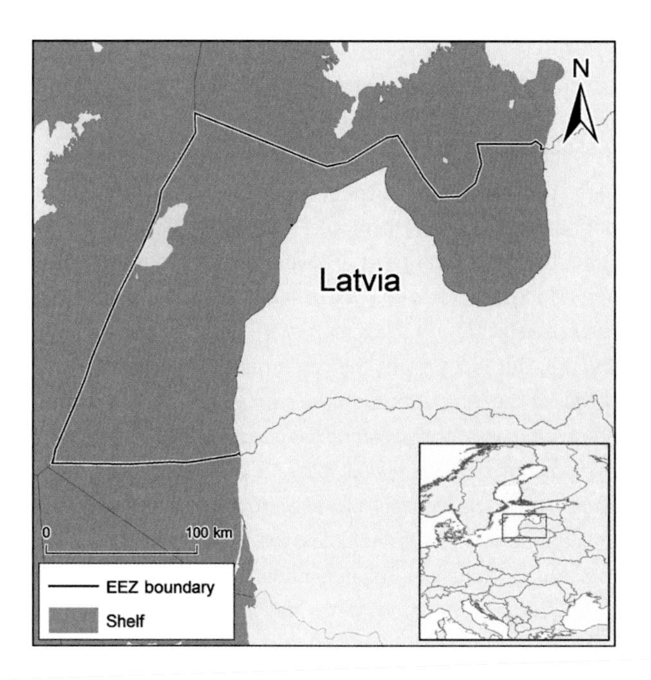

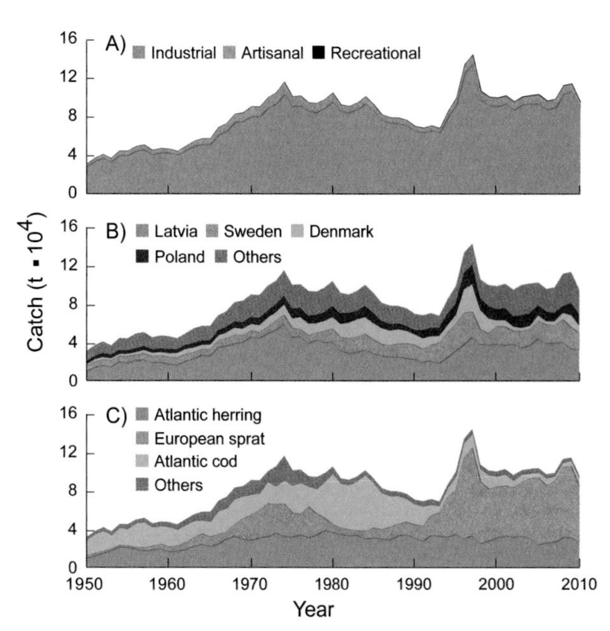

Figure 1. Latvia's Baltic Sea EEZ has an area of 29,000 km², of which 28,400 km² is shelf. **Figure 2.** Domestic and foreign catches taken in the EEZ of Latvia (Baltic): (A) by sector (domestic scores: Ind. 2, 3, 3; Art. 2, 3, 3; Rec. –, –, 2; Dis. 2, 2, 2); (B) by fishing country (foreign catches are very uncertain); (C) by taxon.

REFERENCES

Rossing P, Plikshs M, Booth S, Veitch L and Zeller D (2010) Catch reconstruction for Latvia in the Baltic Sea from 1950–2007. pp. 127–144 In Rossing R, Booth S and Zeller D (eds.), *Total marine fisheries extractions by country in the Baltic Sea: 1950–present*. Fisheries Centre Research Reports 18 (1). [Updated to 2010]

Zeller D, Rossing P, Harper S, Persson L, Booth S and Pauly D (2011) The Baltic Sea: estimates of total fisheries removals 1950–2007. *Fisheries Research* 108(356–363).

1. Cite as Rossing, P., M. Plikshs, S. Booth, L. Veitch, and D. Zeller. 2016. Latvia. P. 317 in D. Pauly and D. Zeller (eds.), *Global Atlas of Marine Fisheries: A Critical Appraisal of Catches and Ecosystem Impacts*. Island Press, Washington, DC.

LEBANON[1]

Manal R. Nader,[a] Shadi Indary,[a] Nazanin Roshan Moniri,[b] and Kyrstn Zylich[b]

[a]Institute of the Environment, University of Balamand, Al Kurah, Lebanon
[b]*Sea Around Us*, University of British Columbia, Vancouver, Canada

Lebanon has a coastline in the eastern Mediterranean (figure 1). Figure 2A presents a history of the marine fisheries catches in Lebanon (which are mainly artisanal), reconstructed by Nader et al. (2014) despite the destruction of the archives of the Lebanese Department of Fisheries and Wildlife in Beirut during the civil war (1975–1990). From 1950 to the early 1990s, the domestic catch appears to have fluctuated between 3,000 and 7,000 t/year, then doubled to about 9,000 t/year, at which they stayed through the 2000s (with a dip in 2008–2009). Overall, this is 2.5 times the catch reported by the FAO on behalf of Lebanon. Although the catches are predominantly domestic, some foreign fishing has occurred (figure 2B). Catch composition data were available only from punctual studies and through the *Sea Around Us* allocation process; figure 2C is indicative only and does not allow the computation of indicators relying on changes in species dominance, such as the mean temperature of the catch (MTC; Cheung et al. 2013). However, investigations conducted along the Lebanese coast (Carpentieri et al. 2009; Nader et al. 2012) suggest that warm-water fishes and hence the MTC should have increased. At present, top taxa include herring, shads, and sardines (family Clupeidae) and mackerels, tuna, and bonitos (family Scombridae). It will be useful to establish monitoring and management systems to document and minimize the impact of climate change on Lebanese fisheries.

REFERENCES

Carpentieri P, Lelli S, Colloca F, Mohanna C, Bartolino V, Moubayed S and Ardizzone GD (2009) Incidence of Lessepsian migrants on landings of the artisanal fishery of South Lebanon. *Marine Biodiversity Records* 2: e71.

Cheung WWL, Watson W and Pauly D (2013) Signature of ocean warming in global fisheries catch. *Nature* 497: 365–368.

Nader M, Indary S and Boustany L (2012) The puffer fish *Lagocephalus sceleratus* (Gmelin, 1789) in the eastern Mediterranean. EastMed Technical Documents 10, EastMed, Food and Agriculture Organization of the United Nations (FAO). vi + 33 p.

Nader MR, Indary S, Roshan Moniri Naz and Zylich K (2014) Historical fisheries catch reconstruction for Lebanon (GSA 27), 1950–2010. Fisheries Centre Working Paper #2014–11, 19 p.

1. Cite as Nader, M. R., S. Indary, N. Roshan Moniri, and K. Zylich. 2016. Lebanon. P. 318 in D. Pauly and D. Zeller (eds.), *Global Atlas of Marine Fisheries: A Critical Appraisal of Catches and Ecosystem Impacts*. Island Press, Washington, DC.

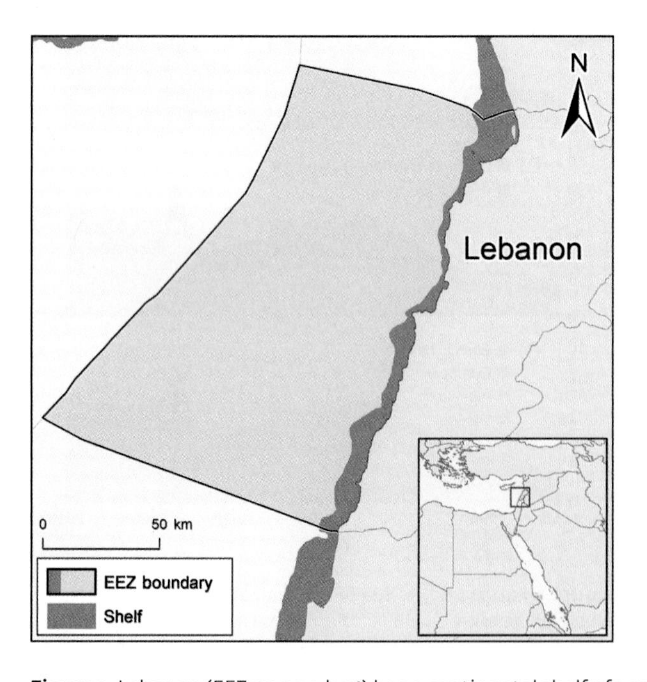

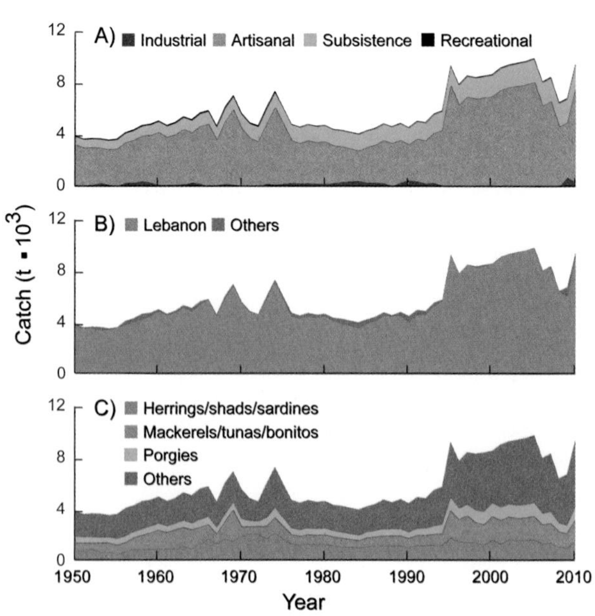

Figure 1. Lebanon (EEZ: 19,300 km²) has a continental shelf of 1,390 km². **Figure 2.** Domestic and foreign catches taken in the EEZ of Lebanon: (A) by sector (domestic scores: Art. 2, 2, 2; Sub. 1, 1, 1; Rec. 1, 1, 1; Dis. 1, 1, 1); (B) by fishing country (foreign catches are very uncertain); (C) by taxon.

LIBERIA[1]

Dyhia Belhabib,[a] Yevewuo Subah,[b] Nasi T. Broh,[b] Alvin S. Jueseah,[b] J. Nicolas Nipey,[b] William Y. Boeh,[b] Duncan Copeland,[c] Dirk Zeller,[a] and Daniel Pauly[a]

[a]*Sea Around Us*, University of British Columbia, Vancouver, Canada [b]Bureau of National Fisheries, Monrovia, Liberia [c]West Africa Regional Fisheries Project, Monrovia, Liberia

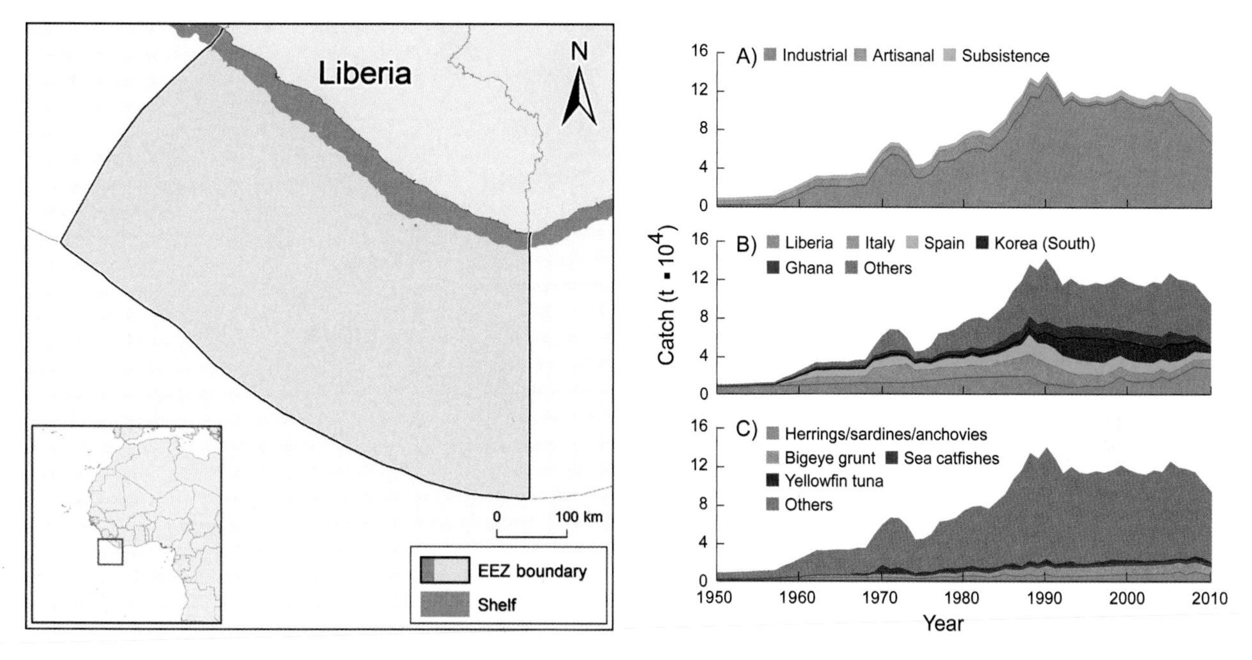

Figure 1. Liberia (EEZ: 246,000 km²) has a continental shelf of 18,500 km². **Figure 2.** Domestic and foreign catches taken in the EEZ of Liberia: (A) by sector (domestic scores: Ind. –, –, 4; Art. 3, 3, 3; Sub. 1, 3, 3; Dis. –, –, 4); (B) by fishing country (foreign catches are very uncertain); (C) by taxon.

Liberia, the only country in Africa that was never colonized, is located in Sub-Saharan West Africa (figure 1). The marine fisheries of Liberia could generate a sustainable catch sufficient to meet the animal protein needs of its people (Haakonson 1992), contrary to what the low official estimates suggest. Only part of the artisanal sector is covered by official statistics, and the industrial sector, consisting of foreign and "Liberian" vessels mostly owned by foreign companies, remains largely invisible in the data. Overall, catches in the EEZ are taken by industrial fleets (figure 2A), most of which are foreign fleets (figure 2B). Domestic catches were estimated by Belhabib et al. (2013) at 8,000 t in the 1950s, increased to 19,000 t/year in the late 1980s, and declined during the civil war period (1989 to mid-late 2000s), when catches consisted overwhelmingly of foreign-controlled industrial interests. Catches peaked at 29,000 t in 2009 and declined slightly to 28,000 t in 2010. Domestic catches were 2.5 times the data supplied to the FAO, almost 30% of which originated from coastal lagoons. Total catches were dominated by small pelagic species (figure 2C): various clupeoids such as herrings, sardines and anchovies, and bigeye grunt (*Brachydeuterus auritus*). Thus, Liberian fishers should benefit from the fact that illegal fishing, which previously made up to half the foreign catch, has declined because of improved monitoring and surveillance.

REFERENCES

Belhabib D, Subah Y, Broh NT, Jueseah AS, Nipey N, Boeh WY, Copeland D, Zeller D and Pauly D (2013) When "Reality leaves a lot to the imagination": Liberian fisheries from 1950 to 2010. Fisheries Centre Working Paper #2013-06, 18 p.

Haakonson JM (1992) Artisanal fisheries and Fishermen's migrations in Liberia. *Maritime Studies (MAST)* 1992(5): 75–87.

1. Cite as Belhabib, D., Y. Subah, N. T. Broh, A. S. Jueseah, J. N. Nipey, W. Y. Boeh, D. Copeland, D. Zeller, and D. Pauly. 2016. Liberia. P. 319 in D. Pauly and D. Zeller (eds.), *Global Atlas of Marine Fisheries: A Critical Appraisal of Catches and Ecosystem Impacts*. Island, Washington, DC.

LIBYA[1]

Myriam Khalfallah, Dyhia Belhabib, Dirk Zeller, and Daniel Pauly

Sea Around Us, University of British Columbia, Vancouver, Canada

Libya consists essentially of a heavily populated coastal strip along the Mediterranean (figure 1). In the 1960s, oil exports allowed Libya to develop different sectors, such as light industry and education, but this benefited only a minority. A revolution led by young colonel Al Gadhafi against King Idrisi brought deep changes in the country, some positive, but he quickly turned into a deluded dictator whose misrule demolished most institutions typical of functioning states. In terms of fisheries, total catches allocated to Libya's EEZ were predominantly artisanal, in addition to an industrial and subsistence component (figure 2A). The catch reconstruction by Khalfallah et al. (2015) was based on work published by Libyan colleagues, FAO publications (e.g., Lamboeuf et al. 2000; Reynolds et al. 1995), and inferences from neighboring countries. This yielded a domestic catch increasing from about 5,000 t in 1950 to 100,000 t/year in the late 2000s. Overall, this was about 2 times the catch reported by the FAO for Libya. Small foreign catches occurred before the EEZ declaration in 2009 (figure 2B). The catch composition, as assessed from Shakman and Kinzelbach (2007) and other sources, in addition to the *Sea Around Us* allocation, presents top taxa (figure 2C), consisting mainly of clupeids (e.g., round sardinella, *Sardinella aurita*), horse mackerels (*Trachurus* spp.), goatfishes (*Mullus* spp.), and spinefoots (*Siganus* spp.), which are Lessepsian migrants.

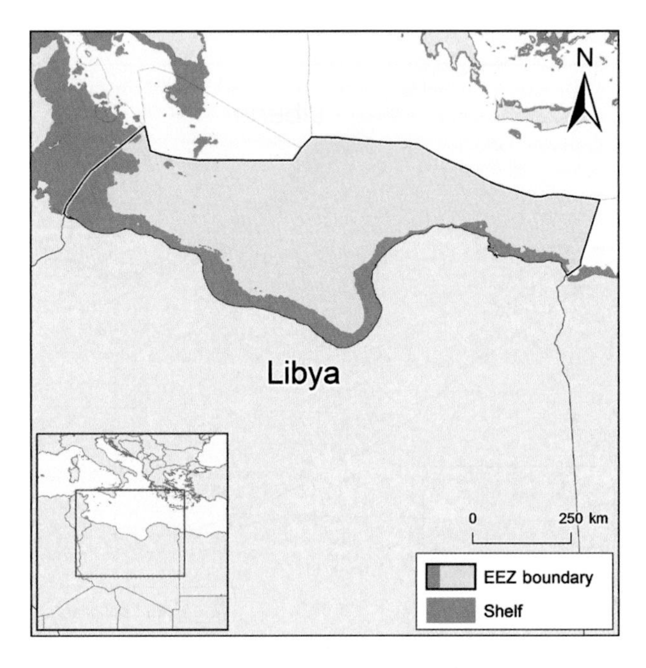

Figure 1. Libya has a shelf of 67,100 km² and an EEZ of 356,000 km².

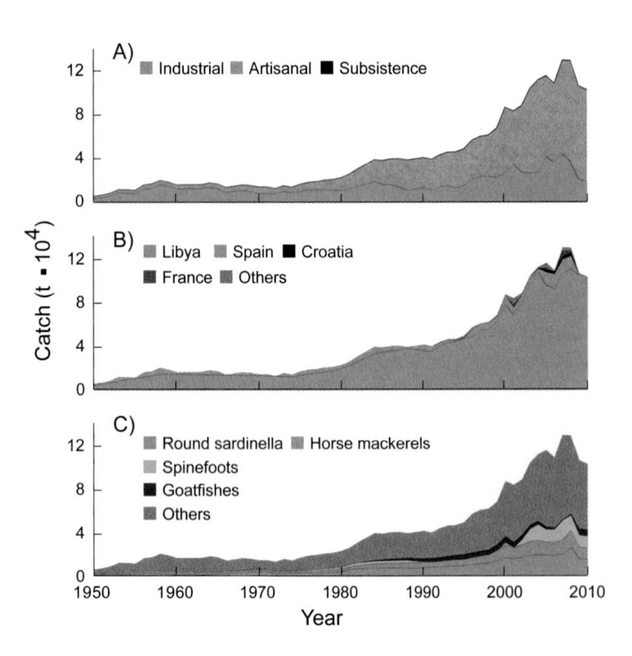

Figure 2. Domestic and foreign catches taken in the EEZ of Libya: (A) by sector (domestic scores: Ind. 1, 2, 2; Art. 1, 2, 2; Sub. 1, 2, 2; Dis. 1, 2, 2); (B) by fishing country (foreign catches are very uncertain); (C) by taxon.

REFERENCES

Khalfallah M, Belhabib D, Zeller D and Pauly D (2015) Reconstruction of marine fisheries catches for Libya (1950–2010). Fisheries Centre Working Paper #2015-47, 15 p.

Lamboeuf M, Abdallah AB, Coppola R, Germoni A and Spinelli M (2000) Artisanal Fisheries in Libya: Census of Fishing Vessels and Inventory of Artisanal Fishery Métiers. FAO-COPEMED-MBRC, 42 p.

Reynolds JE, Abukhader A and Abdallah AB (1995) The marine wealth sector of Libya: a development planning overview. Project reports. FAO, Tripoli/Rome. 122 p.

Shakman E and Kinzelbach R (2007) Commercial fishery and fish species composition in coastal waters of Libya. *Rostocker Meeresbiologische Beiträge* 18: 63–78.

1. Cite as Khalfallah, M., D. Belhabib, D. Zeller, and D. Pauly. 2016. Libya. P. 320 in D. Pauly and D. Zeller (eds.), *Global Atlas of Marine Fisheries: A Critical Appraisal of Catches and Ecosystem Impacts*. Island Press, Washington, DC.

LITHUANIA[1]

Liane Veitch,[a] Sarunas Toliusis,[b] Shawn Booth,[a] Peter Rossing,[a] Sarah Harper,[a] and Dirk Zeller[a]

[a]*Sea Around Us*, University of British Columbia, Vancouver, Canada [b]Fishery Research Laboratory, Klaipeda, Lithuania

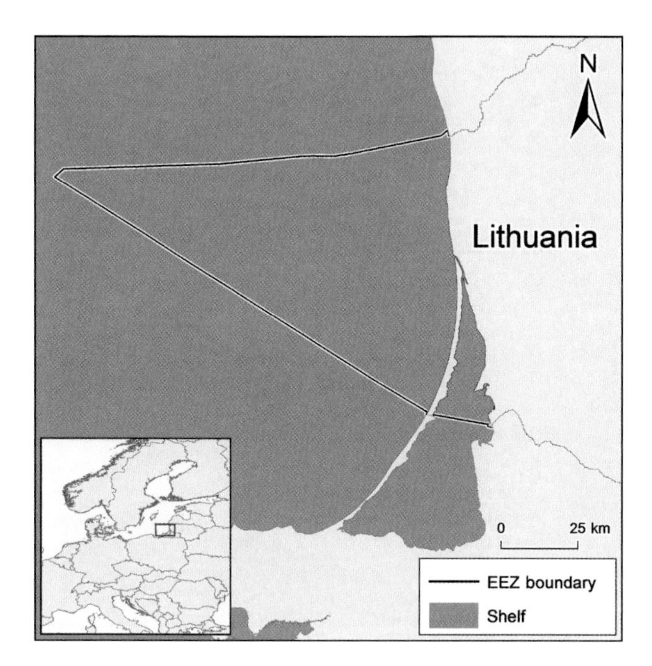

Figure 1. Lithuania's EEZ is 6,140 km², of which all is shelf.

Lithuania is the southernmost of the 3 Baltic republics (figure 1). Overall, catches in the EEZ are dominated by industrial fleets (figure 2A), mostly from the Russian Federation and Poland (figure 2B). The reconstructed domestic marine fisheries catches by Lithuania in the Baltic Sea (and its equivalent entity before separation from the USSR in 1991) were estimated by Veitch et al. (2010), but only its EEZ component is presented here. This was based on statistics of the International Council for the Exploration of the Sea (ICES) and data from ICES stock assessment working group reports since 1991, whereas the Lithuanian component of the reported catch of the ex-USSR was provided by the Latvian Fish Resource Agency (LATFRA), disaggregated by Lithuania, Latvia, Estonia, and Russia for 1950 to 1990 (see also Zeller et al. 2011). The reconstructed domestic catch as assigned to the Lithuanian EEZ increased from 1,500 t in 1950 to more than 4,000 t in 1979, decreased for two decades, then increased again to just under 4,000 t/year by the late 2000s. The reconstructed catch from 1992–2010 was 1.6 times the ICES reported landings. The main species targeted were European sprat (*Sprattus sprattus*), Atlantic cod (*Gadus morhua*), and Atlantic herring (*Clupea harengus*) (figure 2C). Although catches as assigned to its EEZ seem low, Lithuanian vessels are active throughout the Baltic Sea.

REFERENCES

Veitch L, Toliusis S, Booth S, Rossing P, Harper S and Zeller D (2010) Catch reconstruction for Lithuania in the Baltic Sea from 1950–2007. pp. 145–164 In Rossing R, Booth S and Zeller D (eds.), *Total marine fisheries extractions by country in the Baltic Sea: 1950–present*. Fisheries Centre Research Reports 18(1). [Updated to 2010]

Zeller D, Rossing P, Harper S, Persson L, Booth S and Pauly D (2011) The Baltic Sea: estimates of total fisheries removals 1950–2007. *Fisheries Research* 108: 356–363.

1. Cite as Veitch, L., S. Toliusis, S. Booth, P. Rossing, S. Harper, and D. Zeller. 2016. Lithuania. P. 321 in D. Pauly and D. Zeller (eds.), *Global Atlas of Marine Fisheries: A Critical Appraisal of Catches and Ecosystem Impacts*. Island Press, Washington, DC.

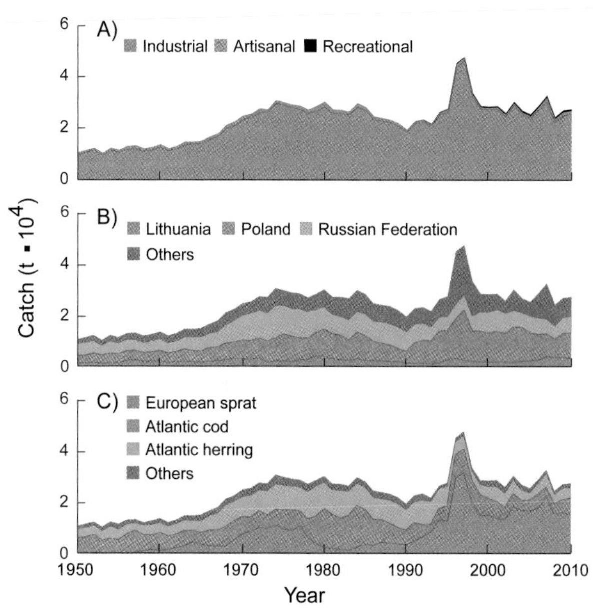

Figure 2. Domestic and foreign catches taken in the EEZ of Lithuania: (A) by sector (domestic scores: Ind. 3, 3, 4; Art. 3, 3, 4; Rec. –, –, 2; Dis. 2, 2, 3); (B) by fishing country (foreign catches are very uncertain); (C) by taxon.

MADAGASCAR[1]

Frédéric Le Manach,[a] Charlotte Gough,[b] Alasdair Harris,[b] Frances Humber,[b] Sarah Harper,[a] and Dirk Zeller[a]

[a]*Sea Around Us*, University of British Columbia, Vancouver, Canada
[b]Blue Ventures Conservation, Aberdeen Centre, London, United Kingdom

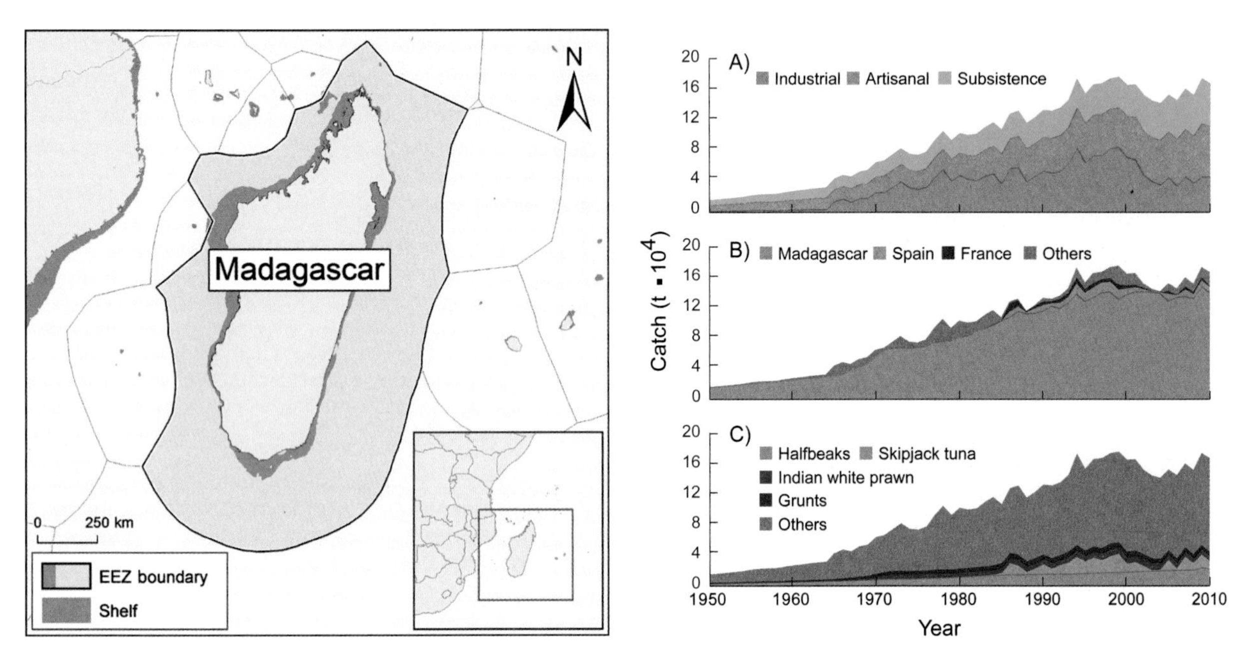

Figure 1. Madagascar has a land area of 587,000 km² and an EEZ of 1.2 million km². **Figure 2.** Domestic and foreign catches taken in the EEZ of Madagascar: (A) by sector (domestic scores: Ind. 3, 3, 4; Art. 1, 2, 2; Sub. 1, 2, 2; Dis. 3, 3, 3); (B) by fishing country (foreign catches are very uncertain); (C) by taxon.

Madagascar (figure 1) is one of the world's poorest developing countries, and its people depend heavily on marine resources for their livelihood (Barnes-Mauthe et al. 2013). Exports of these resources and foreign fishing access agreements are important economically. In recent years, concerns have been voiced among local fishers and industry groups, yet knowledge of Malagasy fisheries remains poor (Le Manach et al. 2012b). Unfortunately, fisheries legislation, management plans, and foreign fishing access agreements are often affected by incomplete data (Le Manach et al. 2012a). As presented in figure 2A, total catches allocated to Madagascar's EEZ are divided almost equally between industrial, artisanal, and subsistence fisheries. The catch reconstruction of all Malagasy fisheries sectors performed by Le Manach et al. (2011) suggests that domestic catches increased from 15,000 t/year in the early 1950s to 137,000 t/year from 2000 to 2010. Overall, this was about twice as high as officially reported, in part because of the subsistence sector, which is missing in the national statistics. A large component of the allocated catches originates from domestic fisheries; however, varying catches by Spain and France, among others, have been documented (figure 2B). This catch consists mainly of halfbeaks (family Hemiramphidae), skipjack tuna (*Katsuwonus pelamis*), Indian white prawn (*Fenneropenaeus incus*), and grunts (family Haemullidae; figure 2C). Signs of stock decline have been observed, suggesting that the current fishing pressure is excessive.

REFERENCES

Barnes-Mauthe M, Oleson KLL and Zafindrasilivonona B (2013) The total economic value of small-scale fisheries with a characterization of post-landing trends: An application in Madagascar with global relevance. *Fisheries Research* 147(0): 175–185.

Le Manach F, Andriamahefazafy M, Harper S, Harris A, Hosch G, Lange G-L, Zeller D and Sumaila UR (2012a) Who gets what? Developing a more equitable framework for EU fishing agreements. *Marine Policy* 38: 257–266.

Le Manach F, Gough C, Harris A, Humber F, Harper S and Zeller D (2012b) Unreported fishing, hungry people and political turmoil: The recipe for a food security crisis in Madagascar? *Marine Policy* 36: 218–225.

Le Manach F, Gough C, Humber F, Harper S and Zeller D (2011) Reconstruction of total marine fisheries catches for Madagascar (1950–2008). pp. 21–37 In Harper S and Zeller D (eds.), *Fisheries catch reconstructions: Islands, Part II*. Fisheries Centre Research Reports 19(4).

1. Cite as Le Manach, F, C. Gough, A. Harris, F. Humber, S. Harper, and D. Zeller. 2016. Madagascar. P. 322 in D. Pauly and D. Zeller (eds.), *Global Atlas of Marine Fisheries: A Critical Appraisal of Catches and Ecosystem Impacts*. Island Press, Washington, DC.

MALAYSIA (PENINSULAR)[1]

Lydia C. L. Teh and Louise S. L. Teh

Sea Around Us, University of British Columbia, Vancouver, Canada

The east coast of Peninsular Malaysia faces the South China Sea and is subject to the brunt of the northeast monsoon, whereas the west coast, which borders the Malacca Strait, is less exposed (figure 1). West coast fisheries have historically been more heavily capitalized and account for more than 65% of the reconstructed catch of Peninsular Malaysia (Teh and Teh 2014). Marine fisheries have long provided seafood and employment throughout the coastal areas in Malaysia (Firth 1946). After independence from British colonial rule in 1957, production-oriented national policies drove an uncontrolled fisheries expansion, leading to overexploitation of Peninsular Malaysia's inshore fisheries. Fish landings in Peninsular Malaysia increased by more than 300% between 1960 and 1980, but this did not account for catch from the nearly 2 million unlicensed fishers (Teh and Teh 2014). Overall, the industrial sector dominates total allocated catches, in addition to the artisanal, subsistence, and recreational components (figure 2A). Reconstructed domestic catches (Teh and Teh 2014) grew from 160,000 t/year in the early 1950s to 1.5 million t/year in the late 2000s. They were 1.8 times higher than reported statistics, a difference accounted for mostly by the industrial sector. Foreign fishing by Thailand contributed an important fraction of the allocated catches (figure 2B). Figure 2C presents catches dominated by low-value species such as threadfin, whiptail breams (family Nemipteridae), mackerels (*Rastrelliger* spp.), and jacks (family Carangidae).

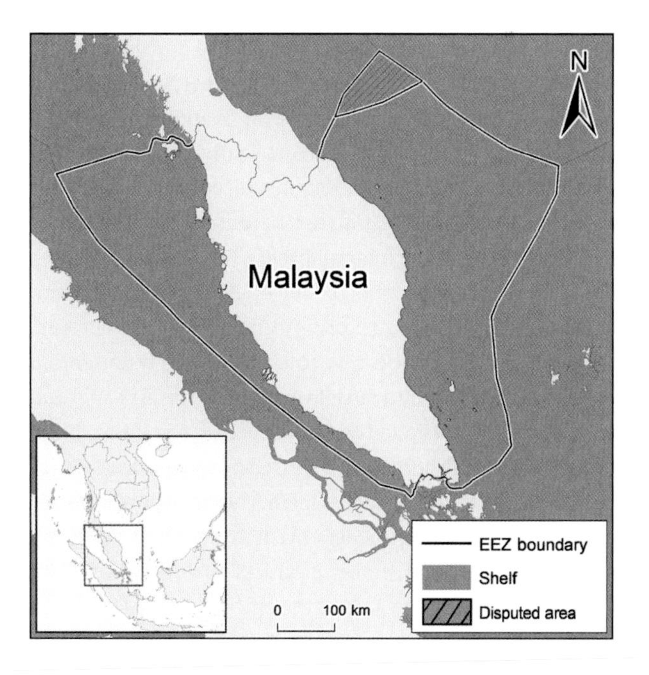

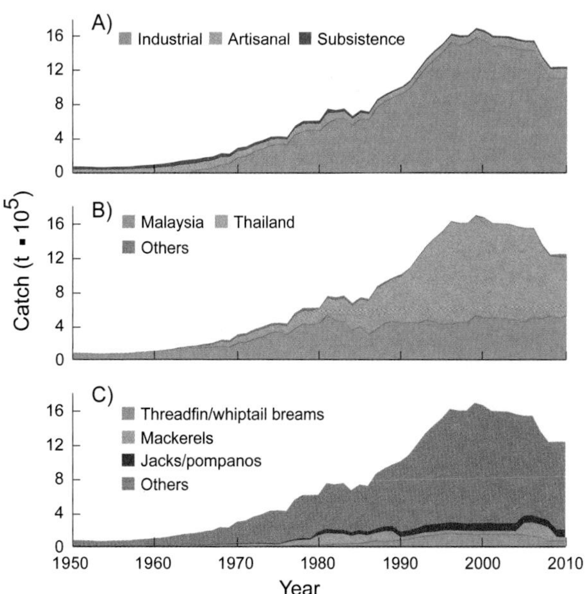

Figure 1. Peninsular Malaysia has shelves on its east coast (133,300 km²) and west coast (68,500 km²) within an EEZ totaling about 202,000 km². **Figure 2.** Domestic and foreign catches taken in the EEZ of Peninsular Malaysia: (A) by sector (domestic scores: Ind. 2, 2, 3; Art. 2, 2, 3; Sub. 2, 2, 2; Rec. –, 1, 1; Dis. 2, 2, 3); (B) by fishing country (foreign catches are very uncertain); (C) by taxon.

REFERENCES

Firth R (1946) *Malay Fishermen: Their Peasant Economy.* Archon Books, The Shoestring Press, Connecticut (US). 398 p.

Teh LCL and Teh LSL (2014) Reconstructing the marine fisheries catch of Peninsular Malaysia, Sarawak and Sabah, 1950–2010. Fisheries Centre Working Paper #2014–16, 20 p.

1. Cite as Teh, L. C. L., and L. S. L. Teh. 2016. Malaysia (Peninsular). P. 323 in D. Pauly and D. Zeller (eds.), *Global Atlas of Marine Fisheries: A Critical Appraisal of Catches and Ecosystem Impacts*. Island Press, Washington, DC.

MALAYSIA (SABAH)[1]

Louise S. L. Teh,[a] Lydia C. L. Teh,[a] Dirk Zeller,[a] and Annadel Cabanban[b]

[a]*Sea Around Us*, University of British Columbia, Vancouver, Canada [b]ASC Ecological and Engineering Solutions, Valencia Drive, Daro Dumaguete City 6200, Philippines

The Malaysian state of Sabah lies in the northeast of the island of Borneo (or Kalimantan; figure 1). Sabah's marine fishery resources are exploited mainly by industrial vessels, but small-scale fisheries are also active (figure 2A). Although landing statistics for both exist, it is recognized that small-scale catches are underestimated, because of a large number of unlicensed local and migrant fishers. The domestic catches based on the reconstruction by Teh et al. (2009) and Teh and Teh (2014) amounted to 20,000–30,000 t/year in the 1950s, which grew to 500,000 t by 2010. This more than tenfold increase was driven by a fivefold increase in the artisanal and subsistence sectors and the emergence, in the 1960s, of an industrial trawl fishery that, in the 2000s, generated about two thirds of the reconstructed catch. There is little discarding, however, because the trawlers' bycatch is used to make fish balls, fish cakes, and similar products. Overall, the reconstructed catches were 3.3 times the official reported landings. Catches by foreign fleets appear to be low (figure 2B), but this is probably an underestimate. Figure 2C shows the taxonomic composition to be dominated by jacks (family Carangidae) and Spanish mackerels (*Scomberomorus* spp.). Figure 2C is not detailed enough to show the effects of the live reef food fish trade (Sadovy 2005), which targets large fish such as coral groupers (*Plectropomus* and *Epinephelus* spp.) and humphead wrasse (*Cheilinus undulatus*).

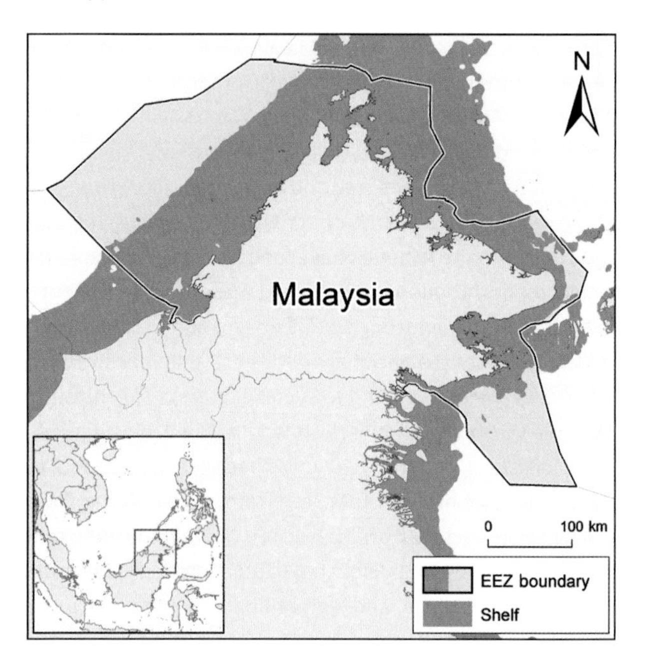

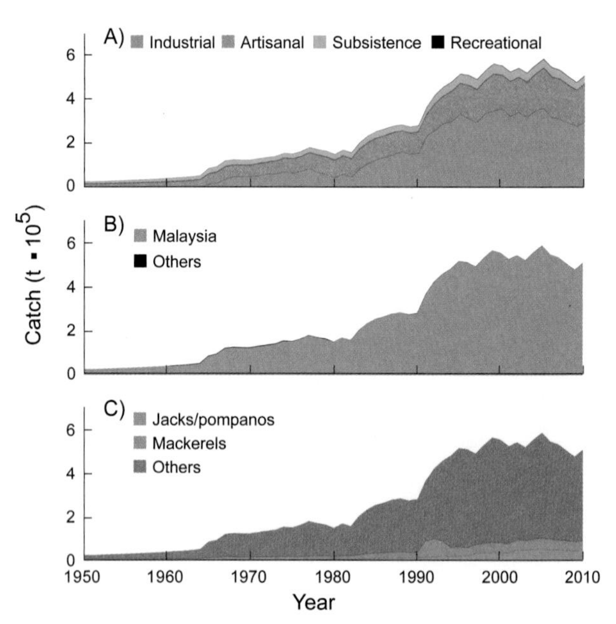

Figure 1. Sabah (Malaysia) occupies the northwest of Borneo, and its shelf is 54,200 km², within an EEZ covering about 91,700 km². **Figure 2.** Domestic and foreign catches taken in the EEZ of Sabah (Malaysia): (A) by sector (domestic scores: Ind. 2, 2, 3; Art. 2, 2, 2; Sub. 2, 2, 2; Rec. –, 2, 2; Dis. 3, 3, 3); (B) by fishing country (foreign catches are very uncertain); (C) by taxon.

REFERENCES

Teh LCL and Teh LSL (2014) Reconstructing the marine fisheries catch of Peninsular Malaysia, Sarawak and Sabah, 1950–2010. Fisheries Centre Working Paper #2014–16, 20 p.

Teh LSL, Teh LCL, Zeller D and Cabanban A (2009) Historical perspective of Sabah's marine fisheries. pp. 77–98 In Zeller D and Harper S (eds.), *Fisheries catch reconstructions: Islands, part 1*. Fisheries Centre Research Reports 17(5).

Sadovy Y (2005) Trouble on the reef: the imperative for managing vulnerable and valuable fisheries. *Fish and Fisheries* 6: 167–185.

1. Cite as Teh, L. S. L., L. C. L. Teh, D. Zeller, and A. Cabanban. 2016. Malaysia (Sabah). P. 324 in D. Pauly and D. Zeller (eds.), *Global Atlas of Marine Fisheries: A Critical Appraisal of Catches and Ecosystem Impacts*. Island Press, Washington, DC.

MALAYSIA (SARAWAK)[1]

Lydia C. L. Teh and Louise S. L. Teh

Sea Around Us, University of British Columbia, Vancouver, Canada

Sarawak is the western state of Malaysian Borneo facing the South China Sea (figure 1). Small-scale fishing is concentrated in inshore areas and is important for food security and employment in coastal communities. Local Chinese artisanal fishers actively engaged in the state's fish trade in the mid-20th century, and traditional Malay fishers operated mainly at a subsistence level. However, the industrial sector dominates total allocated catches in these waters (figure 2A). Teh and Teh (2014) estimated that national fisheries statistics were underreporting actual catches by a factor of 1.6, because of the high number of unlicensed fishers and unreported industrial catches. Reconstructed domestic catches increased from 7,000 t in 1950 to 235,000 t in 2003 before declining to 170,000 t by 2010. Rapid growth in the number of trawlers exploiting prawn stocks, many unlicensed, drove industrial fisheries expansion from the late 1960s onward. Discarding is minimal, because bycatch is used for fish balls, fish cakes, and the like. Government policies throughout the 1990s focused on modernizing the fisheries sector, but this aggravated the decline of biomass (Stobutzki et al. 2006). The bulk of the catches are domestic, although foreign catches by Indonesia, Taiwan, and Japan, among others, do occur and are underestimated on figure 2B. The composition of the reconstructed catch (figure 2C) showed top taxa to include shrimps (*Parapenaeopsis* spp.) and jellyfish (Cnidaria).

REFERENCES

Stobutzki IC, Silvestre GT and Garces LR (2006) Key issues in coastal fisheries in South and Southeast Asia, outcomes of a regional initiative. *Fisheries Research* 78: 109–118.

Teh LCL and Teh LSL (2014) Reconstructing the marine fisheries catch of Peninsular Malaysia, Sarawak and Sabah, 1950–2010. Fisheries Centre Working Paper #2014-16, 20 p.

1. Cite as Teh, L. C. L., and L. S. L. Teh. 2016. Malaysia (Sarawak). P. 325 in D. Pauly and D. Zeller (eds.), *Global Atlas of Marine Fisheries: A Critical Appraisal of Catches and Ecosystem Impacts*. Island Press, Washington, DC.

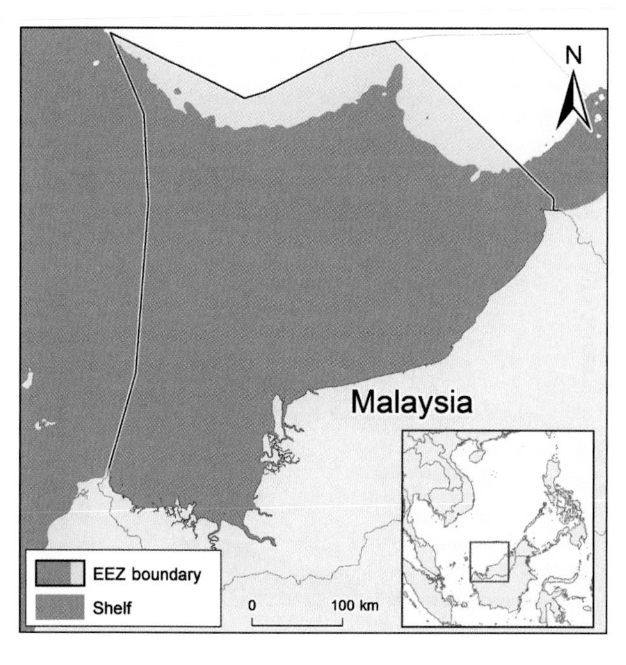

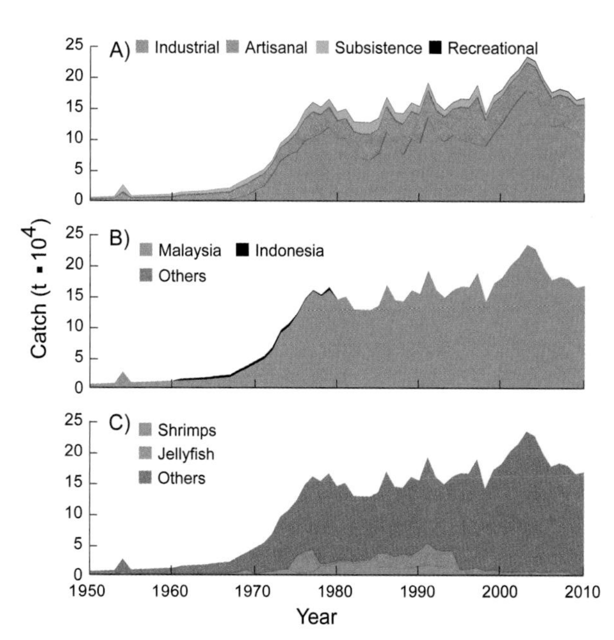

Figure 1. Sarawak has an EEZ of 156,000 km², of which 136,000 km² is shelf. **Figure 2.** Domestic and foreign catches taken in the EEZ of Sarawak: (A) by sector (domestic scores: Ind. 2, 2, 3; Art. 2, 2, 2; Sub. 2, 2, 2; Rec. –, 1, 1; Dis. –, 3, 3); (B) by fishing country (foreign catches are very uncertain); (C) by taxon.

MALDIVES[1]

Mark Hemmings,[a] Sarah Harper,[b] and Dirk Zeller[b]

[a]School of Marine Science and Engineering, Plymouth University, Plymouth, UK
[b]*Sea Around Us*, University of British Columbia, Vancouver, Canada

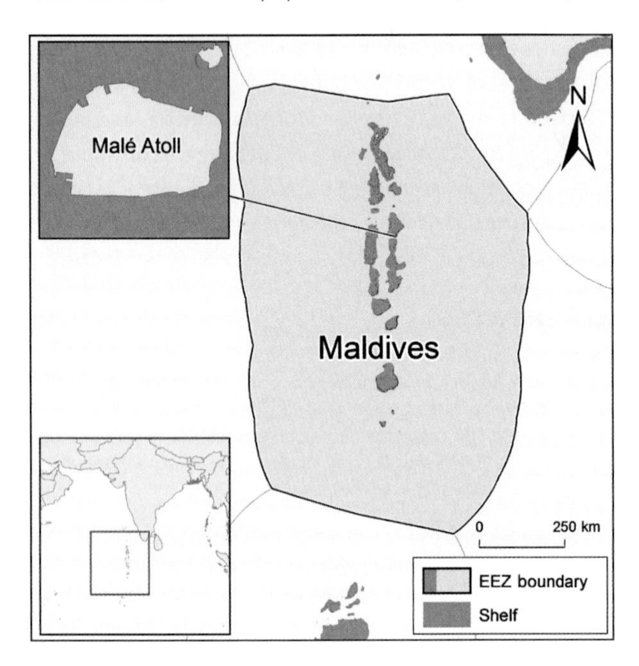

Figure 1. The Maldives, with a land area of 300 km², have an EEZ of about 916,000 km².

The Maldives, southwest of India, consist of 26 atolls with more than 1,000 islands, about 200 of which are inhabited, and another 80 have been developed into tourist resorts (figure 1). Catches have been recorded locally since 1959, but it is in the 1970s that the expansion and diversification of fisheries and tourism began. The existing enumeration system, initially focused on the pole-and-line tuna fishery, was expanded to other gears and species (Adam 2004), but there are still concerns about the catch data reported to the FAO. The majority of removals from within the Maldives' EEZ are deemed industrial (i.e., tuna), in addition to some artisanal and subsistence catches (figure 2A). Hemmings et al. (2014) present domestic reconstructed catches of 11,000 t/year in the 1950s, which grew to a peak of 91,000 t/year in 2006 before declining to about 60,000 t by 2010, 1.8 times the data reported by the FAO on behalf of the Maldives. Catches were predominantly domestic; however, removals by an unknown fishing country were allocated in noticeable quantities (figure 2B). Large catches of skipjack tuna (*Katsuwonus pelamis*), yellowfin tuna (*Thunnus albacares*), and kawakawa (*Euthynnus affinis*; figure 2C) mask the underreporting of reef-associated groups such as groupers (family Serranidae) and sharks, known to be susceptible to overfishing. The Maldives fishing and tourism industries are dependent on healthy marine ecosystems; reported catch statistics must better reflect total extractions.

REFERENCES

Adam SM (2004) Country review: Maldives. pp. 383–391 In De Young C (ed.), Review of the state of the world marine capture fisheries management: Indian Ocean. Fisheries Technical Paper 488. Food and Agriculture Organization of the United Nations (FAO), Rome, Italy.

Hemmings M, Harper S and Zeller D (2014) Reconstruction of total marine catches for the Maldives: 1950–2010. pp. 107–120 In Zylich K, Zeller D, Ang M and Pauly D (eds.), *Fisheries catch reconstructions: Islands, Part IV*. Fisheries Centre Research Reports 22(2).

1. Cite as Hemmings, M., S. Harper, and D. Zeller. 2016. Maldives. P. 326 in D. Pauly and D. Zeller (eds.), *Global Atlas of Marine Fisheries: A Critical Appraisal of Catches and Ecosystem Impacts*. Island Press, Washington, DC.

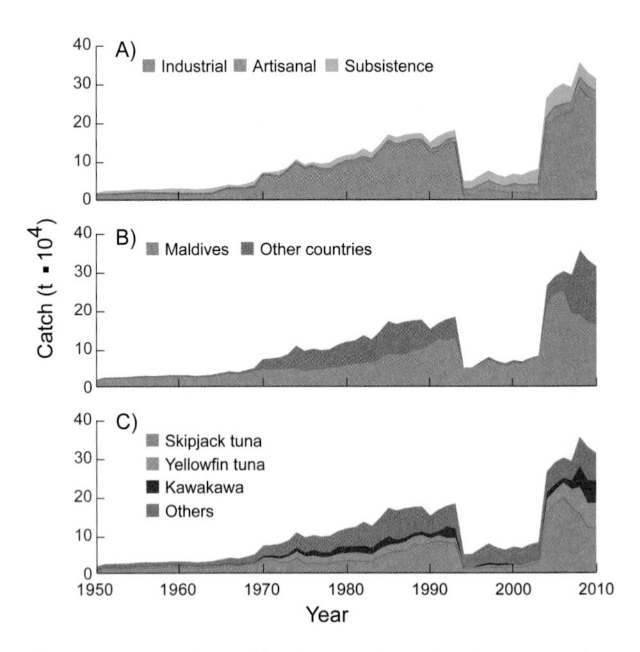

Figure 2. Domestic and foreign catches taken in the EEZ of Maldives: (A) by sector (domestic scores: Ind. 2, 3, 3; Art. 2, 3, 3; Sub. 2, 3, 3); (B) by fishing country (foreign catches are very uncertain); (C) by taxon.

MALTA[1]

Myriam Khalfallah,[a] Mark Dimech,[b] Aylin Ulman,[a] Dirk Zeller,[a] and Daniel Pauly[a]

[a]*Sea Around Us*, University of British Columbia, Vancouver, Canada [b]Malta Centre for Fisheries Sciences, Marsaxlokk, Malta

The Republic of Malta, a member of the European Union, is an archipelago in the center of the Mediterranean basin, which consists of 3 main inhabited islands: Malta proper (246 km²), where 90% of its 420,000 inhabitants reside; Gozo (67 km²) and Comino (3 km²); and a few limestone islets and rocks (figure 1). Although the artisanal sector has dominated catches through the time period, it appears to be declining, and the industrial sector has been rapidly expanding in the 2000s (figure 2A). Khalfallah et al. (2015) presents the reconstructed catches for Malta's domestic marine fisheries, which amounted to 1,100 t/year in the early 1950s, increased to 1,900 t in 1963, and declined to 840 t in 1992 before increasing to 2,200 t by 2010 (figure 2A). This excludes industrially caught large pelagics. Overall, the reconstructed catch of Malta was 1.5 times the amount presented by the FAO on behalf of Malta, mainly because of large catches by the recreational fishery. Catches are largely domestic, although foreign removals by Spain and Italy, among others, do occur, especially before the EEZ declaration in 1978 (figure 2B). The fishing sector in Malta differs from that of other Mediterranean countries in that it is very selective, targeting mainly high-value species such as *lampuki* (common dolphinfish, *Coryphaena hippurus*; figure 2C; De Leiva et al. 1998).

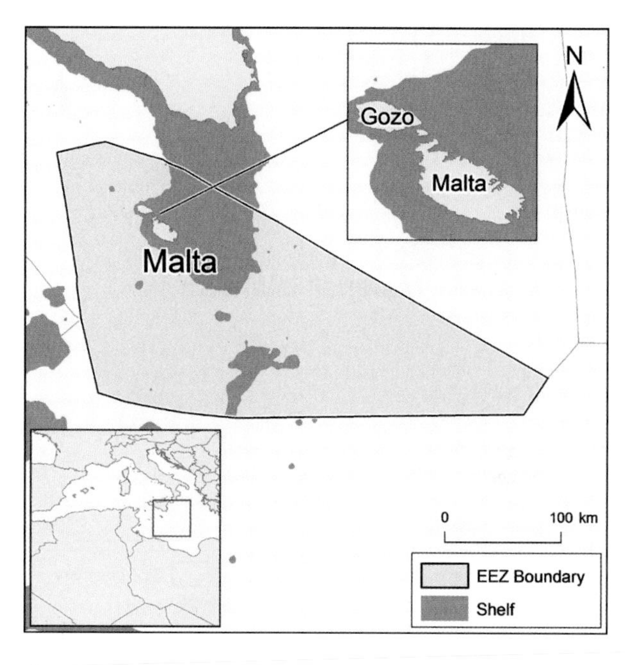

Figure 1. Malta has a total land area of 316 km² and an EEZ-equivalent of 55,500 km².

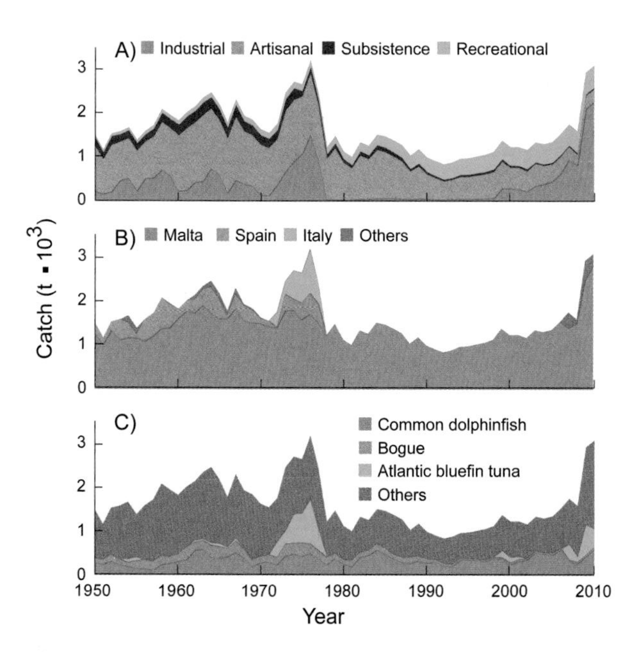

Figure 2. Domestic and foreign catches taken in the EEZ of Malta: (A) by sector (domestic scores: Ind. –, 3, 4; Art. 3, 3, 4; Sub. 2, 2, 2; Rec. 2, 2, 4; Dis. 1, 1, 1); (B) by fishing country (foreign catches are very uncertain); (C) by taxon.

REFERENCES

De Leiva, J I, C. Busuttil, M. Camilleri, and M. Darmanin (1998) Artisanal fisheries in the Western Mediterranean–Malta Fisheries. Copemed–FAO Sub-regional project, Malta, 21 p.

Khalfallah M, Dimech M, Ulman A and Zeller D (2015) Reconstruction of marine fisheries catches for the Republic of Malta (1950–2010). Fisheries Centre Working Paper #2015–43, 12 p.

1. Cite as Khalfallah, M., M. Dimech, A. Ulman, D. Zeller, and D. Pauly. 2016. Malta. P. 327 in D. Pauly and D. Zeller (eds.), *Global Atlas of Marine Fisheries: A Critical Appraisal of Catches and Ecosystem Impacts*. Island Press, Washington, DC.

MARSHALL ISLANDS[1]

Andrea Haas,[a] Sarah Harper,[a] Kyrstn Zylich,[a] James Hehre,[b] and Dirk Zeller[a]

[a]*Sea Around Us*, University of British Columbia, Vancouver, Canada [b]Centre for Marine Futures, The University of Western Australia, Perth, Australia

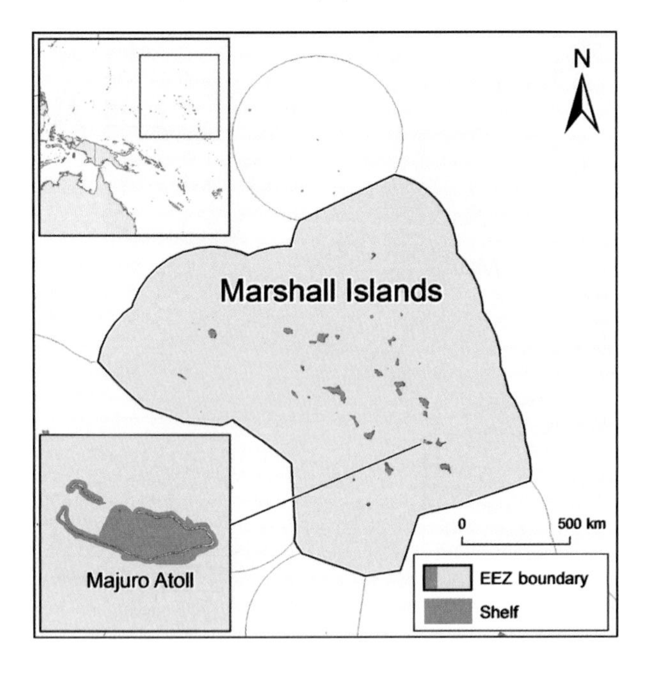

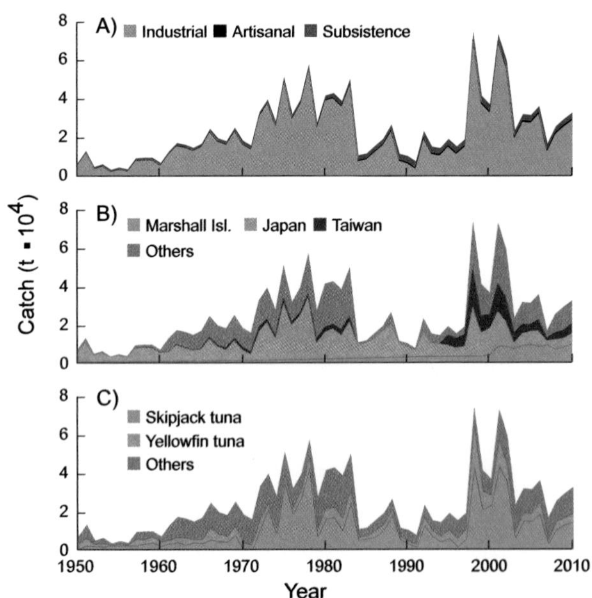

Figure 1. The Republic of the Marshall Islands covers a land area of 180 km² scattered over an EEZ of about 2 million km². **Figure 2.** Domestic and foreign catches taken in the EEZ of Republic of the Marshall Islands: (A) by sector (domestic scores: Ind. –, 3, 4; Art. 2, 2, 2; Sub. 2, 2, 2); (B) by fishing country (foreign catches are very uncertain); (C) by taxon.

The Republic of the Marshall Islands consists of 34 atolls and islands (24 of which are inhabited) located in the western Pacific (figure 1) and hosts former U.S. nuclear testing grounds (Cooke 2009). Two thirds of the country's population (70,000) live in the two major centers of Majuro and Ebeye. The outer islands have few inhabitants because of the scarcity of employment and economic development. The majority of catches in the Marshall Islands EEZ are industrial tuna fisheries, which overshadows the artisanal and subsistence sectors (figure 2A). Total domestic catches, as reconstructed by Haas et al. (2014) but excluding industrially caught tuna and billfishes (although including bycatch of sharks), were estimated at 1,000 t/year in the early 1950s and 9,000 t/year in the 2000s, and were 8 times the landings reported by the FAO. Industrially caught tuna added 30,000–60,000 t/year to the above data for the last decade (Haas et al. 2014). Catches are predominantly by foreign vessels, notably from Japan and Taiwan (figure 2B). Reef-based fisheries are important to Marshall Islanders; however, this is not reflected in the taxonomic composition in figure 2C, which is dominated by large pelagics such as skipjack tuna (*Katsuwonus pelamis*) and yellowfin tuna (*Thunnus albacares*). This highlights the substantial impact that large-scale fisheries can have on the marine resources and the economy of a small island state.

REFERENCES

Cooke S (2009) *In Mortal Hands: A Cautionary History of the Nuclear Age*. Bloomsbury, New York (US). 488 p.

Haas A, Harper S, Zylich K, Hehre J and Zeller D (2014) Reconstruction of the Republic of the Marshall Islands fisheries catches: 1950–2010. pp. 121–128 In Zylich K, Zeller D, Ang M and Pauly D (eds.), *Fisheries catch reconstructions: Islands, Part IV*. Fisheries Centre Research Reports 22(2).

1. Cite as Haas, A., S. Harper, K. Zylich, J. Hehre, and D. Zeller. 2016. Marshall Islands. P. 328 in D. Pauly and D. Zeller (eds.), *Global Atlas of Marine Fisheries: A Critical Appraisal of Catches and Ecosystem Impacts*. Island Press, Washington, DC.

MAURITANIA[1]

Dyhia Belhabib,[a] Didier Gascuel,[b] Elimane Abou Kane,[c] Sarah Harper,[a] Dirk Zeller,[a] and Daniel Pauly[a]

[a]*Sea Around Us*, University of British Columbia, Vancouver, Canada [b]Ecologie et Santé des Ecosystèmes, Université Européenne de Bretagne, Rennes, France [c]Institut Mauritanien de Recherches Océanographiques et des Pêches, Nouadhibou, Mauritania

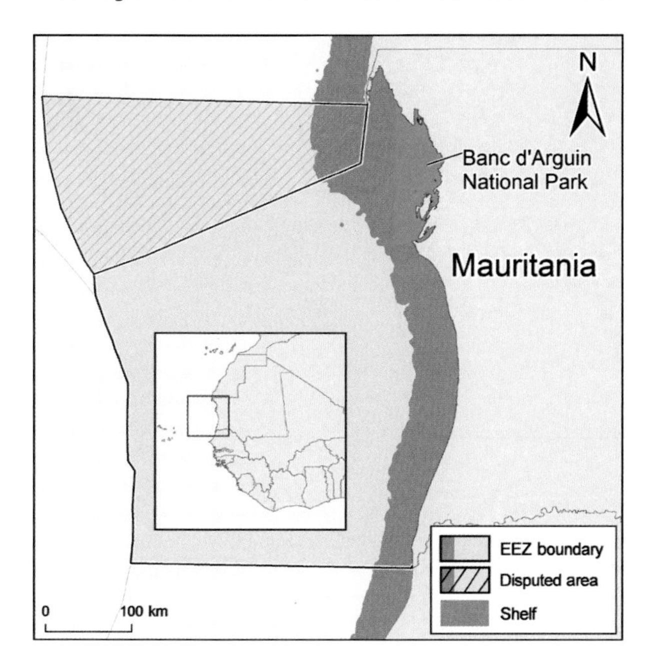

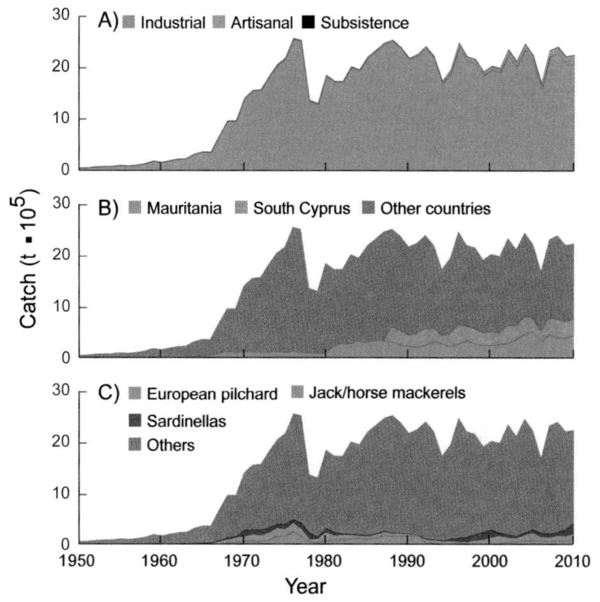

Figure 1. The Mauritanian EEZ (204,600 km²), of which 36,260 km² is shelf. **Figure 2.** Domestic and foreign catches taken in the EEZ of Mauritania: (A) by sector (domestic scores: Ind. 2, 2, 4; Art. 2, 3, 4; Sub. 2, 2, 2; Rec. –, 4, 4; Dis. 4, 4, 4); (B) by fishing country (foreign catches are very uncertain); (C) by taxon.

Mauritania, which boasts the largest marine protected area in West Africa, the Parc National du Banc d'Arguin (PNBA; figure 1), has productive fisheries due to its wide continental shelf and a seasonal upwelling. The domestic catch, as reconstructed by Belhabib et al. (2012), was 20,000 t/year in the 1950s, increased in 2 stages to more than 300,000 t/year in the late 1990s, and was about 440,000 t/year in the late 2000s. Overall, this is 3 times the data FAO reports on behalf of Mauritania. The majority of catches in this EEZ is industrial (figure 2A), exclusively by foreign fleets (figure 2B). The industrial sector is plagued by uncertainty about actual beneficial ownership; for example, about 60% of the "domestic" catch (and 40% of the discards) is generated by Mauritanian-flagged Chinese vessels. Foreign fishing is widespread in Mauritania, with 38% of foreign catch taken by Eastern European, 27% by Chinese, and 20% by European Union vessels. Part of the foreign catch was taken without authorization from Mauritania. Figure 2C presents top taxa, which include European pilchard (*Sardina pilchardus*), horse mackerels (*Trachurus* spp.), and sardinella (*Sardinella* spp.), which fluctuate less in Mauritania than further south in Senegal or north in Morocco (Samb and Pauly 2000). The only legal fishery inside the PNBA is conducted by the Imraguen people, whose traditional ways contrast with the massive industrial vessels operating outside (Picon 2002).

REFERENCES

Belhabib D, Gascuel D, Abou Kane E, Harper S, Zeller D and Pauly D (2012) Preliminary estimation of realistic fisheries removals from Mauritania: 1950–2010. pp. 61–78 In Belhabib D, Zeller D, Harper S and Pauly D (eds.), *Marine fisheries catches in West Africa, Part 1*. Fisheries Centre Research Reports 20(3).

Picon B (2002) Pêche et pêcheries du banc d'Arguin. Histoire d'une identité. Fondation internationale du banc d'Arguin (FIBA), Mauritania. 65 p.

Samb B and Pauly D (2000) On 'variability' as a sampling artefact: the case of Sardinella in North-western Africa. *Fish and Fisheries* 1: 206–210.

1. Cite as Belhabib, D., D. Gascuel, E. Abou Kane, S. Harper, D. Zeller, and D. Pauly. 2016. Mauritania. P. 329 in D. Pauly and D. Zeller (eds.), *Global Atlas of Marine Fisheries: A Critical Appraisal of Catches and Ecosystem Impacts*. Island Press, Washington, DC.

MAURITIUS[1]

Lea Boistol, Sarah Harper, Shawn Booth, and Dirk Zeller

Sea Around Us, University of British Columbia, Vancouver, Canada

Mauritius is a small island state in the western Indian Ocean, which because of its outer islands (Rodrigues, Agalega, and St. Brandon) has a large EEZ (figure 1). Mauritius, which recovered from postindependence social problems (Seetah 2010), now has a diversified economy, but its fisheries have major problems (Jehangeer 2006). Most of the catches are from the industrial tuna sector, in addition to the artisanal, subsistence, and recreational components (figure 2A). The marine catch of Mauritius was reconstructed by Boistol et al. (2011), including unreported catches from the small-scale fisheries, recreational fisheries, catch estimates for the Mauritian fleets fishing along the Mascarene Ridge, and discards of the tuna purse-seine fishery. The domestic reconstructed catch (excluding industrial tuna catches), of 5,500 t/year in the early 1950s, peaked at 13,000 t/year in the early 1990s and is now about 12,000 t/year. Total catches were 1.5 times the data reported by the FAO for Mauritius, largely because of better accounting of small-scale catches from around Mauritius and Rodrigues islands by part-time fishers. While domestic catches have remained steady, an increase in foreign fishing by Taiwan and Spain occurred, especially in the 1990s and 2000s (figure 2B). Figure 2C presents the major taxa caught in the EEZ of Mauritius, which were dominated by coral reef–associated taxa such as sky emperors (*Lethrinus mahsena*), as well as large pelagics (yellowfin tuna *Thunnus albacares*, and albacore, *T. alalunga*).

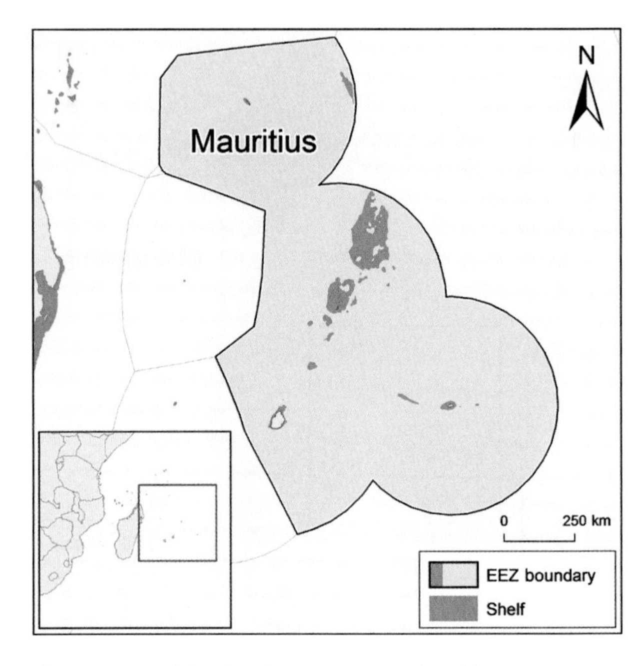

Figure 1. Mauritius (land area 2,040 km²) and its outer islands have an EEZ of about 1.3 million km².

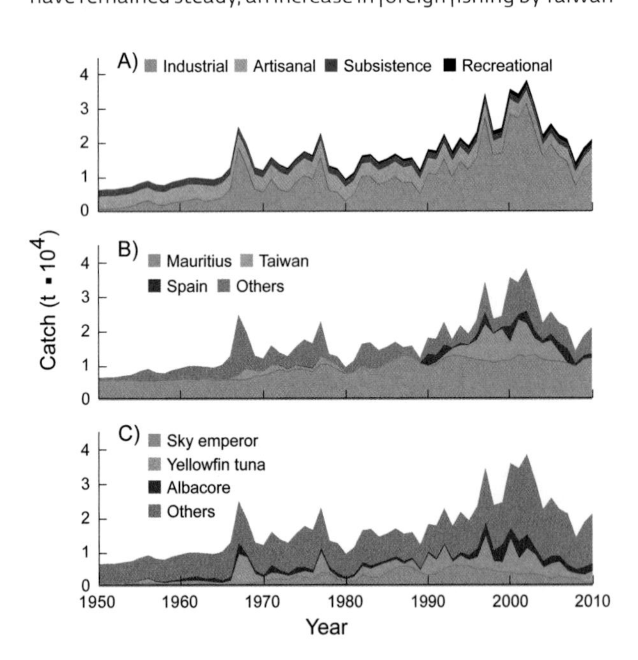

Figure 2. Domestic and foreign catches taken in the EEZ of Mauritius: (A) by sector (domestic scores: Ind. 2, 3, 3; Art. 2, 3, 3; Sub. 2, 3, 3; Rec. 1, 1, 1); (B) by fishing country (foreign catches are very uncertain); (C) by taxon.

REFERENCES

Boistol L, Harper S, Booth S and Zeller D (2011) Reconstruction of marine fisheries catches for Mauritius and its outer islands, 1950–2008. pp. 39–61 In Harper S and Zeller D (occurred eds.), *Fisheries catch reconstruction: Islands, part II*. Fisheries Centre Research Reports 19(4). [Updated to 2010]

Jehangeer I (2006) Country review: Mauritius. pp. 393–413 In C. De Young (ed.), Review of the state of world marine capture fisheries management: Indian Ocean. FAO Fisheries Technical Paper #488, Food and Agriculture Organization of the United Nations, Rome (Italy).

Seetah K (2010) "Our Struggle." Mauritius: an exploration of colonial legacies on an "island paradise." *Shima: The International Journal of Research into Island Cultures* 4(1): 99–112.

1. Cite as Boistol, L., S. Harper, S. Booth, and D. Zeller. 2016. Mauritius. P. 330 in D. Pauly and D. Zeller (eds.), *Global Atlas of Marine Fisheries: A Critical Appraisal of Catches and Ecosystem Impacts*. Island Press, Washington, DC.

MEXICO (ATLANTIC)[1]

Andrés M. Cisneros-Montemayor,[a] Miguel A. Cisneros-Mata,[b] Sarah Harper,[c] and Daniel Pauly[c]

[a]Fisheries Economics Research Unit, University of British Columbia, Vancouver, Canada [b]Centro Regional de Investigación Pesquera, Instituto Nacional de la Pesca, Guaymas, Mexico [c]*Sea Around Us*, University of British Columbia. Vancouver, Canada

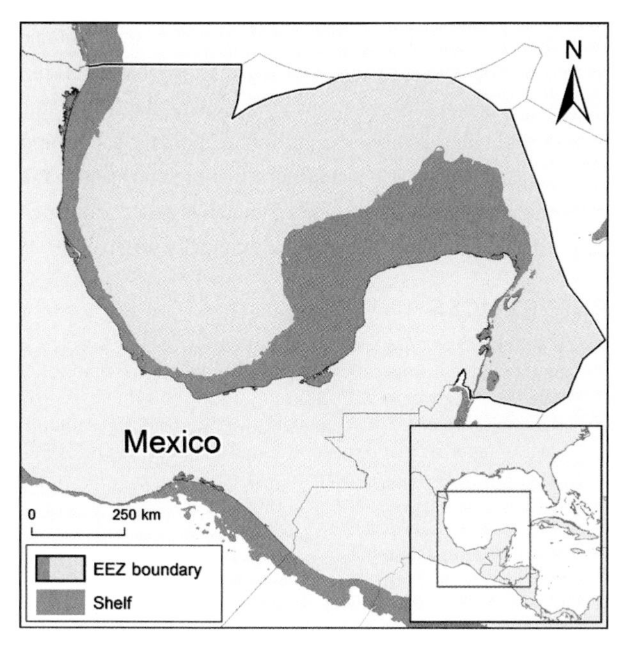

Figure 1. The EEZ of Mexico in the Atlantic covers 829,300 km², of which 235,000 km² is shelf.

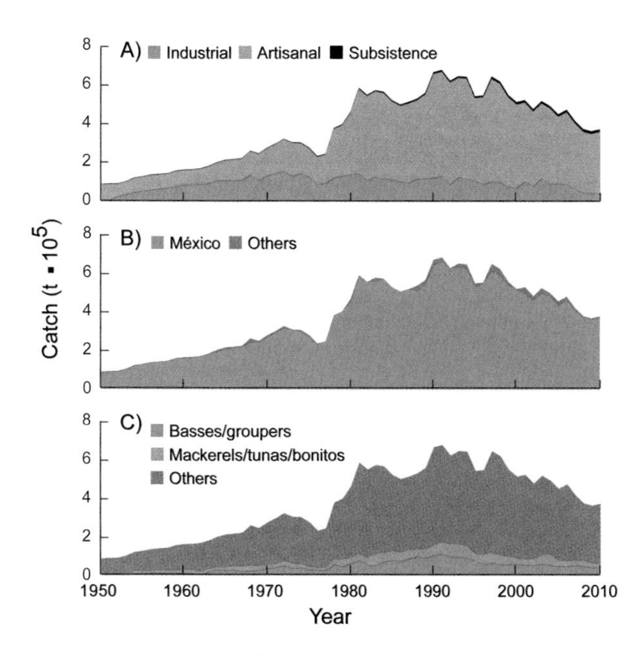

Figure 2. Domestic and foreign catches taken in the Atlantic EEZ of Mexico: (A) by sector (domestic scores: Ind. 3, 3, 4; Art. 1, 1, 2; Sub. 1, 1, 1; Rec. –, –, 2; Dis. 1, 1, 1); (B) by fishing country (foreign catches are very uncertain); (C) by taxon.

The Mexican Atlantic EEZ (Gulf of Mexico and Caribbean; figure 1), though less productive than its Pacific coast, provides important food and livelihoods. The coast consists of mangroves, mud flats, and coral reefs and supports clams, oysters, shrimps, and benthopelagic fishes. Atlantic fisheries have been neglected because of the more lucrative oil and tourism industries of the Gulf of Mexico and Caribbean coasts, respectively. The artisanal sector dominates total allocated catches to Mexico's EEZ, and the industrial ranks second (figure 2A). Reconstructed domestic catches in the Mexican Atlantic were about 85,000 t/year in the early 1950s, reached a peak of about 650,000 t in 1991, then declined to about 360,000 t/year in the late 2000s; overall, this was twice as much as reported by the FAO on behalf of Mexico (Cisneros-Montemayor et al. 2015). Although the fisheries are overwhelmingly domestic, there are some small amounts of foreign fishing in the 1990s, although the origin of this catch is unknown (figure 2B). Though underestimated, total catch is a serious issue in all of Mexico (Arreguín-Sánchez and Arcos-Huitrón 2007; Cisneros-Montemayor et al. 2013); the lack of attention placed on Atlantic fisheries, combined with continued harvesting of more valuable clams and shrimp, has obscured the clearly declining catches of groupers (family Serranidae) and mackerels, tunas, and bonitos (family Scombridae; figure 2C). Future management efforts must tackle issues of oversight and enforcement of policies.

REFERENCES

Arreguín-Sánchez F and Arcos-Huitrón E (2007) Fisheries catch statistics for Mexico. In Zeller D and Pauly D (eds.), *Reconstruction of marine fisheries catches for key countries and regions (1950–2005)*. Fisheries Centre Research Reports 15(2).

Cisneros-Montemayor AM, Cisneros-Mata MA, Harper S and Pauly D (2013) Extent and implications of IUU catch in Mexico's marine fisheries. *Marine Policy* 39: 283–288.

Cisneros-Montemayor AM, Cisneros-Mata MA, Harper S and Pauly D (2015) Unreported marine fisheries catch in Mexico, 1950–2010. Fisheries Centre Working Paper #2015–22, 9 p.

1. Cite as Cisneros-Montemayor, A. M., M. A. Cisneros-Mata, S. Harper, and D. Pauly. 2016. Mexico (Atlantic). P. 331 in D. Pauly and D. Zeller (eds.), *Global Atlas of Marine Fisheries: A Critical Appraisal of Catches and Ecosystem Impacts*. Island Press, Washington, DC.

MEXICO (PACIFIC)[1]

Andrés M. Cisneros-Montemayor,[a] Miguel A. Cisneros-Mata,[b] Sarah Harper,[c] and Daniel Pauly[c]

[a]Fisheries Economics Research Unit, University of British Columbia, Vancouver, Canada [b]Centro Regional de Investigación Pesquera, Instituto Nacional de la Pesca, Guaymas, Mexico [c]*Sea Around Us*, University of British Columbia, Vancouver, Canada

The Pacific coast of Mexico (figure 1) supports most of the country's fishing activity, including the largest stocks of small pelagic fishes and most of the valuable shrimp and tuna fisheries. Despite the political clout of these largely industrial fisheries (figure 2A), a significant factor in total catches are artisanal fishers, who catch any available species given seasonal and market conditions. Particularly since the 1970s, subsidized fishery expansion and operations have resulted in a large fleet, de facto open-access conditions, and poor industry oversight (Cisneros-Montemayor et al. 2013). Most of the catches are domestic (figure 2B). For the Mexican Pacific, reconstructed domestic catches totaled 400,000 t/year in the early 1950s, and they ranged between 1.5 and 1.7 million t/year in the late 2000s, which is twice the data reported by the FAO on behalf of Mexico (Cisneros-Montemayor et al. 2015). Given the high rates of malnutrition in Mexico, the large discards from industrial trawlers and artisanal gillnets are particularly troubling, despite having declined over time. Although total catch appears to be increasing, current large industrial catches of small pelagic fishes, mainly Pacific sardine (*Sardinops sagax*), have masked the concurrent declines of many species (e.g., benthopelagic fishes) that are probably much more important for fishing communities (figure 2C). In addition to improving knowledge of human impacts, future actions should increase compliance with existing policies while curtailing capacity expansion.

REFERENCES

Cisneros-Montemayor AM, Cisneros-Mata MA, Harper S and Pauly D (2013) Extent and implications of IUU catch in Mexico's marine fisheries. *Marine Policy* 39: 283–288.

Cisneros-Montemayor AM, Cisneros-Mata MA, Harper S and Pauly D (2015) Unreported marine fisheries catch in Mexico, 1950–2010. Fisheries Centre Working Paper #2015–22, 9 p.

1. Cite as Cisneros-Montemayor, A. M., M. A. Cisneros-Mata, S. Harper, and D. Pauly. 2016. Mexico (Pacific). P. 332 in D. Pauly and D. Zeller (eds.), *Global Atlas of Marine Fisheries: A Critical Appraisal of Catches and Ecosystem Impacts*. Island Press, Washington, DC.

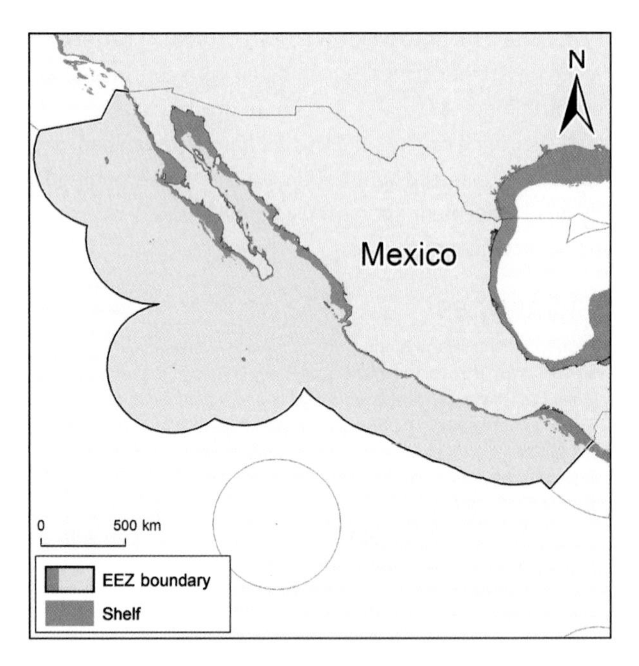

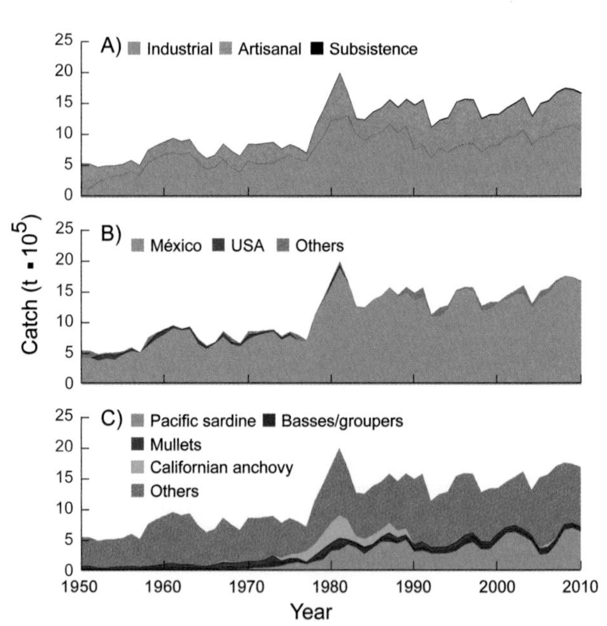

Figure 1. The EEZ of Mexico in the Pacific covers 2,440,000 km², of which 183,000 km² is shelf. **Figure 2.** Domestic and foreign catches taken in the Pacific EEZ of Mexico: (A) by sector (domestic scores: Ind. 3, 3, 4; Art. 1, 1, 2; Sub. 1, 1, 1; Rec. 1, 2, 2; Dis. 1, 1, 1); (B) by fishing country (foreign catches are very uncertain); (C) by taxon.

MICRONESIA (FEDERATED STATES OF)[1]

Sadiq Vali,[a] Kevin Rhodes,[b] Andrea Au,[a] Kyrstn Zylich,[a] Sarah Harper,[a] and Dirk Zeller[a]

[a]*Sea Around Us*, University of British Columbia, Vancouver, Canada [b]College of Aquaculture, Forestry and Natural Resource Management, University of Hawaii at Hilo, Hawaii

The Federated States of Micronesia (FSM) consist of more than 600 islands, divided into 4 states, located in the western Pacific (figure 1); many of these low-lying islands are at risk from climate change, although its perception varies between groups (Pam and Henry 2012). Vali et al. (2014) performed a reconstruction of the total domestic catches of the FSM, which yielded estimates of 5,000–6,000 t/year for the 1950s, increasing to almost 13,000 t/year in the 2000s. Over the 1950–2010 period, the reconstructed domestic catch (excluding industrially caught large pelagics) was 6.7 times the amount reported by the FAO on behalf of the FSM, mainly because of the subsistence sector being underrepresented in the officially reported data. Overall, fisheries in the EEZ were dominated by the industrial sector (figure 2A), which consists mainly of foreign fleets (figure 2B) targeting tuna such as skipjack tuna (*Katsuwonus pelamis*) and yellowfin tuna (*Thunnus albacares*; figure 2C). In the FSM, both women and men participate in subsistence and artisanal fishing, mainly using gillnets for fish and hooks for octopus. There are distinctions, however, with women collecting crabs and other invertebrates that inhabit intertidal areas and men catching fish and lobsters by spear and free-dive fishing (Lambeth 2000, 2001). Most of these catches are for subsistence purposes, although starting in the early 1990s, crabs were caught for domestic and foreign markets as well.

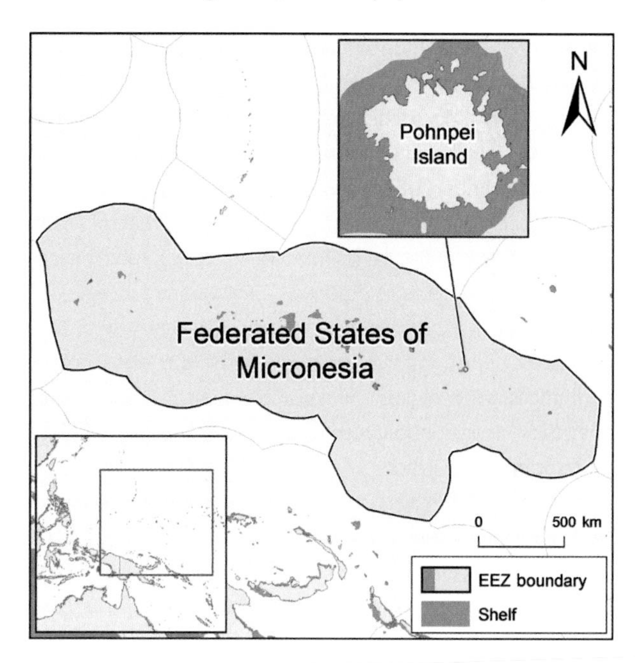

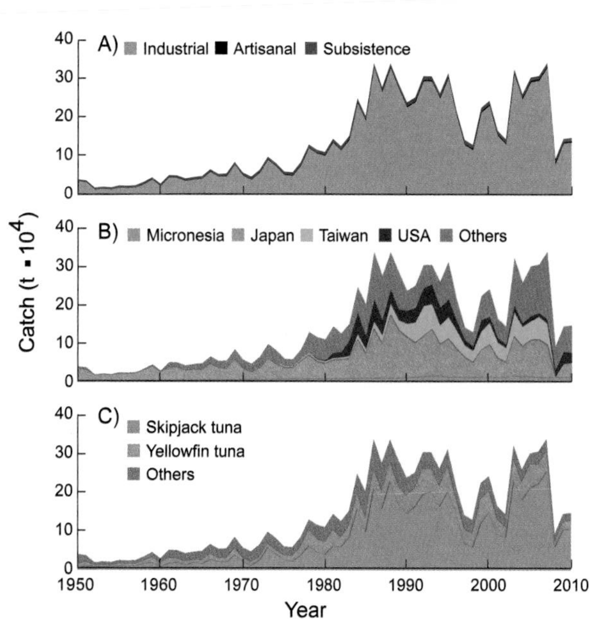

Figure 1. The Federated States of Micronesia (FSM) have a collective land area of 700 km², with an EEZ of almost 2.99 million km². **Figure 2.** Domestic and foreign catches taken in the EEZ of Federated States of Micronesia: (A) by sector (domestic scores: Art. 2, 2, 3; Sub. 2, 2, 3); (B) by fishing country (foreign catches are very uncertain); (C) by taxon.

REFERENCES

Lambeth L (2000) An assessment of the role of women in fisheries in Pohnpei, Federated States of Micronesia. Secretariat of the Pacific Community (SPC), Noumea, New Caledonia. vii + 37 p.

Lambeth L (2001) An assessment of the role of women in fisheries in Chuuk, Federated States of Micronesia. Secretariat of the Pacific Community (SPC), Noumea, New Caledonia. vii + 30 p. [Plus two 2001 reports of similar titles, in the same series and covering the states of Yap and Kosrae]

Pam C and Henry R (2012) Risky place: climate change discourse and the transformation of place on Moch (Federated States of Micronesia). *Shima: The International Journal of Research into Island Cultures* 6(1): 30–47.

Vali S, Rhodes K, Au A, Zylich K, Harper S and Zeller D (2014) Reconstruction of total fisheries catches for the Federated States of Micronesia (1950–2010). Fisheries Centre Working Paper #2014-06, 16 p.

1. Cite as Vali, S., K. Rhodes, A. Au, K. Zylich, S. Harper, and D. Zeller. 2016. Federated States of Micronesia. P. 333 in D. Pauly and D. Zeller (eds.), *Global Atlas of Marine Fisheries: A Critical Appraisal of Catches and Ecosystem Impacts*. Island Press, Washington, DC.

MONTENEGRO[1]

Çetin Keskin,[a] Davis Iritani,[b] and Dirk Zeller[b]

[a]Faculty of Fisheries, University of Istanbul, Laleli, Istanbul, Turkey
[b]*Sea Around Us*, University of British Colombia, Vancouver, Canada

The Republic of Montenegro, a component of the former Yugoslavia, lies on the Adriatic coast in the Mediterranean (figure 1). Domestic catches were reconstructed by Keskin et al. (2014), correcting and updating a previous disaggregation of Yugoslav landings from 1950 to 1991 by Rizzo and Zeller (2007). Total allocated catches in Montenegro's EEZ are predominantly industrial, with small catches from the artisanal, subsistence, and recreational sectors (figure 2A). This yielded a total domestic catch that grew from 500 t/year in the early 1950s to a peak catch of just under 2,000 t/year in the late 1980s, before declining to a low of 940 t in 1992 and ending at around 1,600 t by 2010. For the period from 1992, when Montenegro began to independently report its catch to the FAO, to 2010, the reconstructed catch was 3.1 times the catches presented by the FAO on behalf of Montenegro, with the difference attributed to unreported noncommercial catches and underreported commercial catches. However, the reconstructed total catches mirrored trends in reported landings. Although most of the catches are domestic, foreign fishing by Croatia and other countries has occurred (figure 2B). European pilchard (*Sardina pilchardus*) was a major contributor (figure 2C), in addition to other small pelagics (herrings, shads, and sardines; family Clupeidae), as did Atlantic bluefin tuna (*Thunnus thynnus*). The remainder of the catch is contributed by more than 50 other taxa.

REFERENCES

Keskin Ç, Ulman A, Iritani D and Zeller D (2014) Reconstruction of fisheries catches for Montenegro: 1950–2010. Fisheries Centre Working Paper #2014-27, 11 p.

Rizzo Y and Zeller D (2007) Country disaggregation of catches of former Yugoslavia. pp. 149–156 In Zeller D and Pauly D (eds.), *Reconstruction of marine fisheries catches for key countries and regions (1950–2005)*. Fisheries Centre Research Reports 15(2).

1. Cite as Keskin, Ç., D. Iritani, and D. Zeller. 2016. Montenegro. P. 334 in D. Pauly and D. Zeller (eds.), *Global Atlas of Marine Fisheries: A Critical Appraisal of Catches and Ecosystem Impacts*. Island Press, Washington, DC.

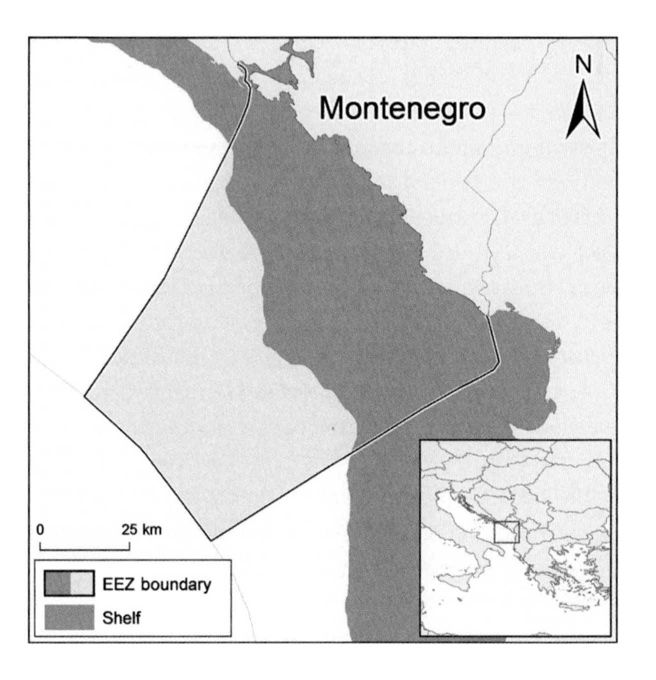

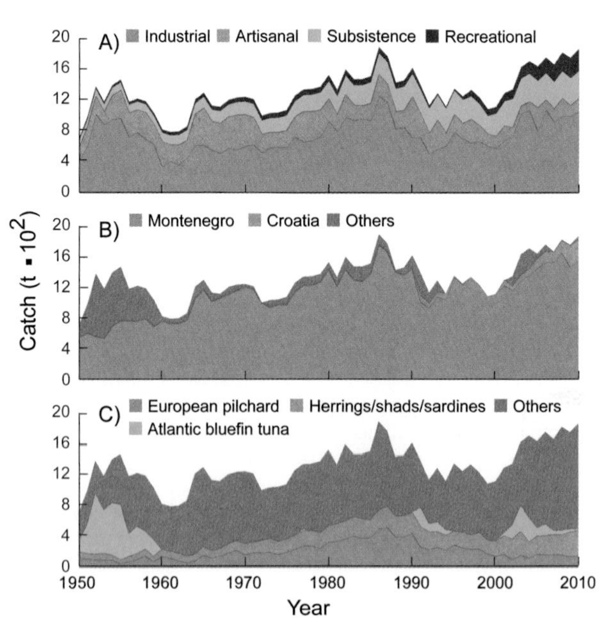

Figure 1. Montenegro's small coastline on the Adriatic Sea comes with a shelf of 3,850 km² and an EEZ of 7,460 km². **Figure 2.** Domestic and foreign catches taken in the EEZ of Montenegro: (A) by sector (domestic scores: Ind. 2, 3, 3; Art. 2, 3, 3; Sub. 1, 1, 1; Rec. 1, 1, 1; Dis. 2, 2, 2); (B) by fishing country (foreign catches are very uncertain); (C) by taxon.

MOROCCO (CENTRAL)[1]

Dyhia Belhabib, Sarah Harper, Dirk Zeller, and Daniel Pauly

Sea Around Us, University of British Columbia, Vancouver, Canada

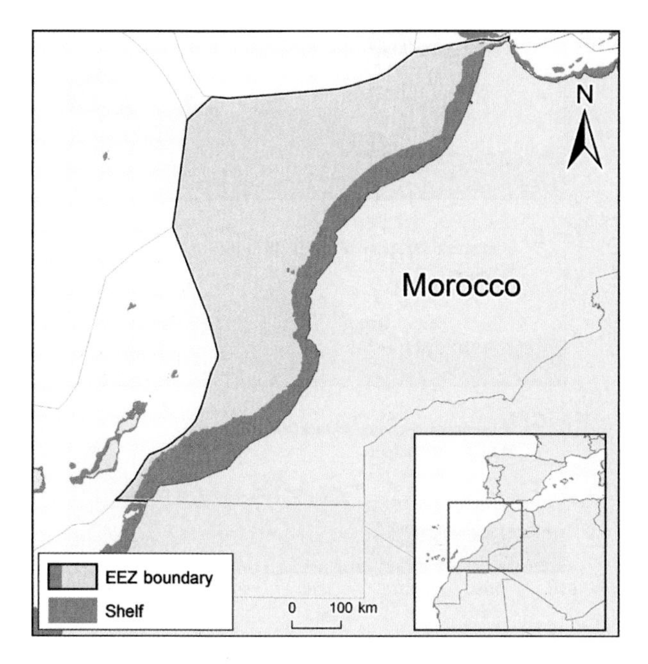

Figure 1. The central coast of Morocco has a shelf of 55,900 km² and an EEZ of 258,000 km².

"Central Morocco" refers here to the area between the Strait of Gibraltar and the former Spanish Sahara (figure 1; see following pages for other parts of Morocco). Morocco has some of the richest fishing grounds in the world, exploited largely by industrial fleets (figure 2A). Although domestic industrial fleets of West Africa do fish in these waters, catches are predominantly by foreign fleets such as Spain, Italy, and Japan (figure 2B; Baddyr and Guénette 2001). Other sectors, such as artisanal, subsistence, and recreational fisheries, exist but are hardly accounted for in official statistics. The domestic catch reconstructed by Belhabib et al. (2012a, 2012b) was 125,000 t in 1950 and reached a peak of 730,000 t in 2001 before declining slightly to 650,000 t by 2010. Domestic catches were 1.7 times the data supplied to the FAO, because of underreporting of industrial landings and a recently growing artisanal sector; discards, both industrial and artisanal, were small by global standards. Domestic catches were dominated by European pilchard (*Sardinella pilchardus*), as shown in figure 2C, and foreign catches were dominated by cephalopods and small pelagic species (Belhabib et al. 2012a). Total catches in Morocco have been stagnating since the mid-2000s despite an increasing fishing effort, which suggests overexploitation and threatens the sustainability of previously abundant resources. This is likely to intensify the competition between the different sectors and jeopardize the local economy.

REFERENCES

Baddyr M and Guénette S (2001) The fisheries off the Atlantic coast of Morocco 1950–1997. pp. 191–205 In Zeller D, Watson R and Pauly D (eds.), *Fisheries impacts on North Atlantic ecosystems: Catch, effort and national / regional data sets*. Fisheries Centre Research Reports 9(3).

Belhabib D, Harper S, Zeller D and Pauly D (2012a) An overview of fish removals from Western Sahara and Morocco by distant water fleets. In Belhabib D, Harper S, Zeller D and D Pauly (eds.), *Marine fisheries catches in West Africa, 1950–2010, Part I*. Fisheries Centre Research Reports 20(3).

Belhabib D, Harper S, Zeller D and Pauly D (2012b) Reconstruction of marine fisheries catches for Morocco (North, Central and South), 1950–2010. pp. 23–40 In Belhabib D, Zeller D, Harper S and Pauly D (eds.), *Marine fisheries catches in West Africa, Part I*. Fisheries Centre Research Reports 20(3).

1. Cite as Belhabib, D., S. Harper, D. Zeller, and D. Pauly. 2016. Morocco (Central). P. 335 in D. Pauly and D. Zeller (eds.), *Global Atlas of Marine Fisheries: A Critical Appraisal of Catches and Ecosystem Impacts*. Island Press, Washington, DC.

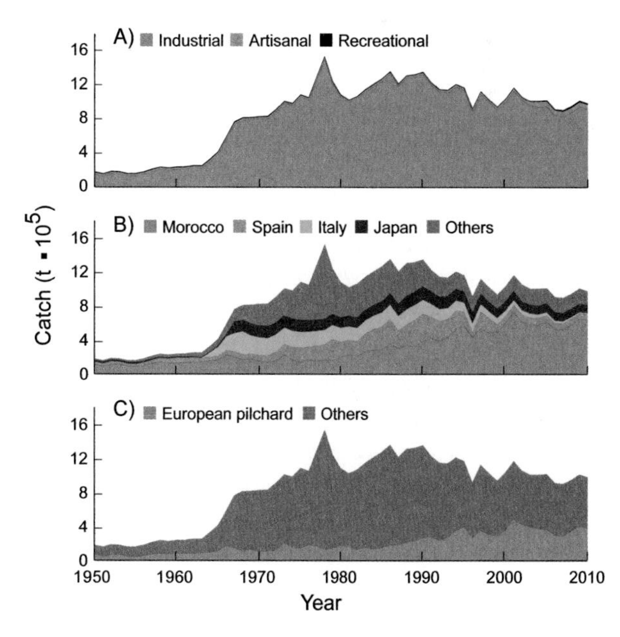

Figure 2. Domestic and foreign catches taken in the EEZ of Morocco (Central): (A) by sector (domestic scores: Ind. 3, 3, 3; Art. 3, 4, 4; Sub. 2, 2, 2; Rec. 2, 2, 4; Dis. 3, 4, 4); (B) by fishing country (foreign catches are very uncertain); (C) by taxon.

MOROCCO (MEDITERRANEAN)[1]

Dyhia Belhabib, Sarah Harper, Dirk Zeller, and Daniel Pauly

Sea Around Us, University of British Columbia, Vancouver, Canada

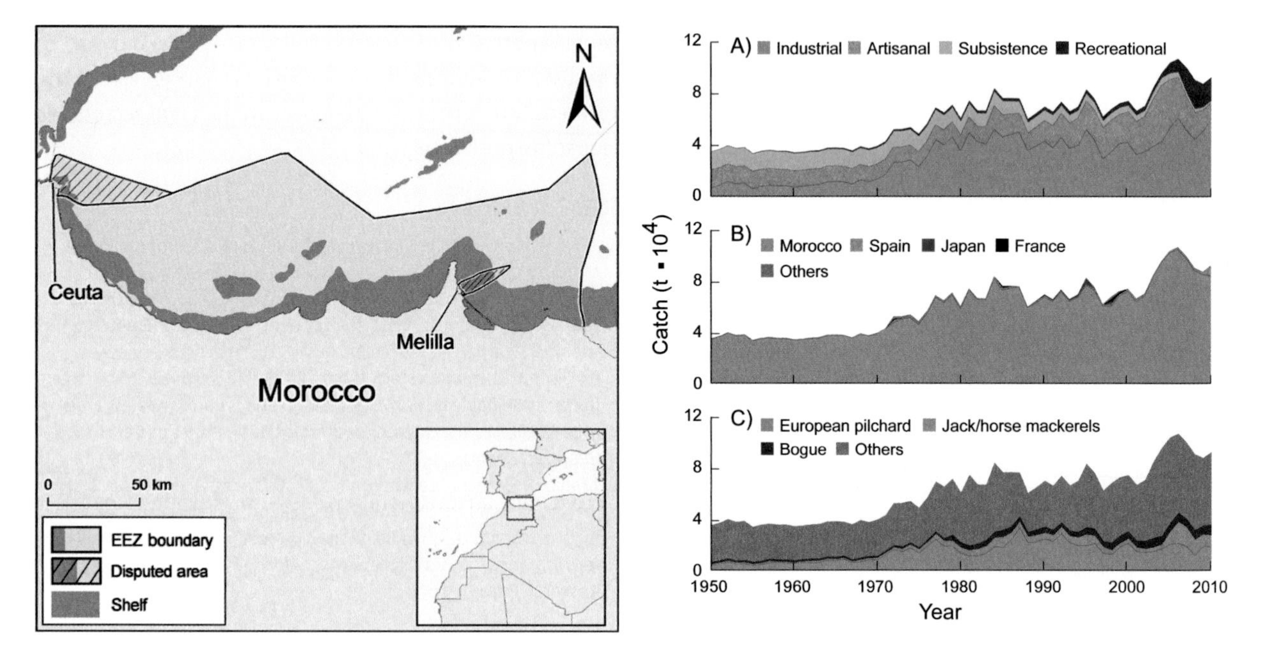

Figure 1. In the Mediterranean, Morocco has a continental shelf of 4,930 km² and an EEZ of 18,300 km²; the disputed area refers to the Spanish exclave of Ceuta and Melilla. **Figure 2.** Domestic and foreign catches taken in the Mediterranean EEZ of Morocco: (A) by sector (domestic scores: Ind. 3, 3, 3; Art. 3, 4, 4; Sub. 2, 2, 2; Rec. 2, 2, 4); (B) by fishing country (foreign catches are very uncertain); (C) by taxon.

Morocco is a North African country that borders both the Mediterranean and the Atlantic coasts (figure 1). The Mediterranean coast of Morocco immediately adjacent to the Strait of Gibraltar is influenced by the Atlantic Current that enters the Mediterranean, which contributes to the productivity of the fisheries, notably the artisanal fisheries (ArtFiMed 2009). Although the artisanal, subsistence, and recreational sectors contribute to the take from within Morocco's EEZ, the bulk of catches originate from the industrial sector (figure 2A). Belhabib et al. (2012) estimated a domestic catch of less than 40,000 t in the 1950s; a peak of about 110,000 t in 2006, due to increasing fishing effort, was followed by a decrease to about 83,000 t in 2010. Domestic catches were 2.5 times the data supplied to the FAO. Most of the domestic catches were unreported, as were catches of the subsistence and recreational fisheries. The majority of the catches were domestic, although some foreign fishing by Spain, Japan, and France has been documented (figure 2B). They were dominated by European pilchard (*Sardina pilchardus*), jack/horse mackerels (*Trachurus* spp.), and bogue (*Boops boops*; figure 2C). Declining catches, increased effort, and the continued use of destructive fishing techniques and gears such as driftnets and dynamite have created a situation where the resources of the western Mediterranean, on which Moroccan fishers and communities depend, are now severely depleted.

REFERENCES

ArtFiMed (2009) Diagnostique initial des sites de pêche artisanale du Maroc et de Tunisie. Développement durable de la pêche artisanale méditerranéenne au Maroc et en Tunisie. (GCP/INT/005/SPA). CopeMed II: ArtFiMed Technical Document No. 4, Malaga, 51 p.

Belhabib D, Harper S, Zeller D and Pauly D (2012) Reconstruction of marine fisheries catches for Morocco (North, Central and South), 1950–2010. pp. 23–40 In Belhabib D, Zeller D, Harper S and Pauly D (eds.), *Marine fisheries catches in West Africa, Part I*. Fisheries Centre Research Reports 20(3).

1. Cite as Belhabib, D., S. Harper, D. Zeller, and D. Pauly. 2016. Morocco (Mediterranean). P. 336 in D. Pauly and D. Zeller (eds.), *Global Atlas of Marine Fisheries: A Critical Appraisal of Catches and Ecosystem Impacts*. Isand Press, Washington, DC.

MOROCCO (SOUTH)[1]

Dyhia Belhabib, Sarah Harper, Dirk Zeller, and Daniel Pauly

Sea Around Us, University of British Columbia, Vancouver, Canada

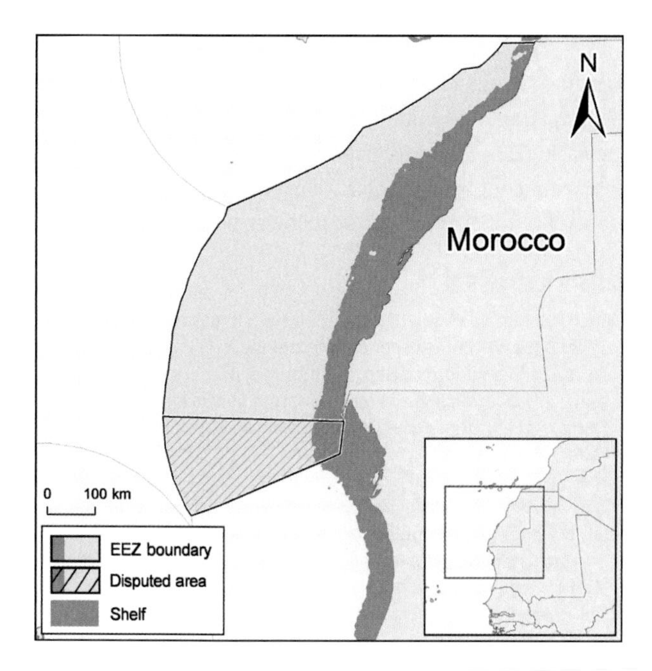

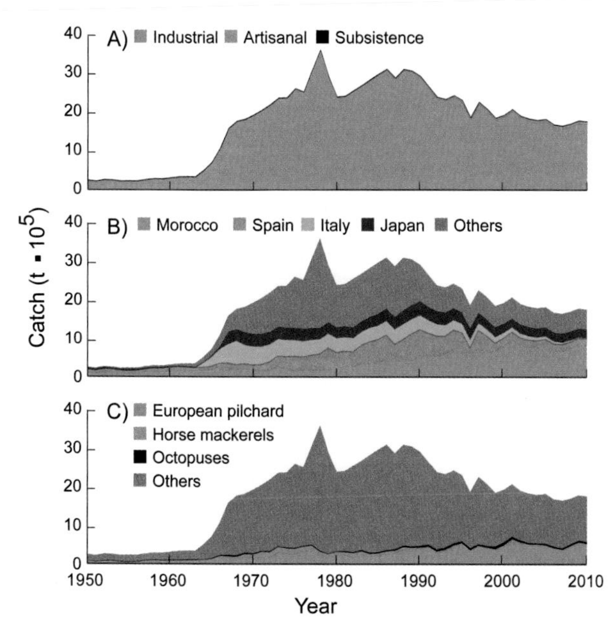

Figure 1. Morocco's southern region has a continental shelf of 63,800 km² and an EEZ of 301,000 km². **Figure 2.** Domestic and foreign catches taken in the EEZ of Morocco (South): (A) by sector (domestic scores: Ind. 3, 3, 3; Art. 3, 4, 4; Sub. 2, 2, 2; Rec. 2, 2, 4; Dis. 3, 4, 4); (B) by fishing country (foreign catches are very uncertain); (C) by taxon.

The south of Morocco corresponds to the former Spanish Sahara, in northwest Africa (figure 1). The waters of this region are extremely productive, with regard to valuable cephalopod (i.e., *Octopus vulgaris*) and demersal trawl fisheries (Baddyr and Guénette 2001). Total allocated catches from within Morocco South's EEZ are predominantly from the industrial sector, in addition to the artisanal, subsistence, and recreational components (figure 2A). The reconstructed domestic catch (excluding large pelagics) of this region (Belhabib et al. 2012b) was 140, 000 t/year in the early 1950s, reached a peak of 1.0 million t in 2001, and then decreased to about 860,000 t in 2010. Domestic catches were 1.7 times the data supplied to the FAO, largely because of underreported industrial catches. Foreign catches by the former USSR were dominated by cephalopods and demersal and small pelagic fish species (Belhabib et al. 2012a), in addition to catches by Spain, Italy, and Japan (figure 2B). As presented in figure 2C, the catch composition in these waters encompasses a wide variety of taxa; however, removals are dominated by European pilchard (*Sardina pilchardus*), horse mackerels (*Trachurus* spp.) and octopuses (Octopoda). Overall, the marine ecosystem exhibits signs of overexploitation, notably a sharp decline of trawler catch per unit of effort. Another sign is the catch abundance of octopus. Although it supports a lucrative fishery, it thrives when large predatory fishes have been removed.

REFERENCES

Baddyr M and Guénette S (2001) The fisheries off the Atlantic coast of Morocco 1950–1997. pp. 191–205 In Zeller D, Watson R and Pauly D (eds.), *Fisheries impacts on North Atlantic ecosystems: Catch, effort and national / regional data sets*. Fisheries Centre Research Reports 9(3).

Belhabib D, Harper S, Zeller D and Pauly D (2012a) An overview of fish removals from Western Sahara and Morocco by distant water fleets. In Belhabib D, Harper S, Zeller D and Daniel P (eds.), *Marine fisheries catches in West Africa, 1950–2010, Part I*. Fisheries Centre Research Reports 20(3).

Belhabib D, Harper S, Zeller D and Pauly D (2012b) Reconstruction of marine fisheries catches for Morocco (North, Central and South), 1950–2010. pp. 23–40 In Belhabib D, Zeller D, Harper S and Pauly D (eds.), *Marine fisheries catches in West Africa, Part I*. Fisheries Centre Research Reports 20(3).

1. Cite as Belhabib, D., S. Harper, D. Zeller, and D. Pauly. 2016. Morocco (South). P. 337 in D. Pauly and D. Zeller (eds.), *Global Atlas of Marine Fisheries: A Critical Appraisal of Catches and Ecosystem Impacts*. Island Press, Washington, DC.

MOZAMBIQUE[1]

Jennifer L. Jacquet,[a] Beau Doherty,[a] Margaret M. McBride,[b] Atanásio J. Brito,[c] Frédéric Le Manach,[a] Lizette Sousa,[c] Isabel Chauca,[c] and Dirk Zeller[a]

[a]*Sea Around Us*, University of British Columbia, Vancouver, Canada [b]Institute of Marine Research, Bergen, Norway
[c]Instituto Nacional de Investigação Pesqueira Maputo, Mozambique

Mozambique stretches along the coast of East Africa, between South Africa and Tanzania (figure 1), and is rich in marine resources (see contributions in Pauly 1992). However, Mozambique underreports its marine catches (Jacquet et al. 2010). Small-scale fisheries, including subsistence fishing by women, account for most marine fisheries landings in Mozambique's EEZ (figure 2A). Foreign fishing occurs but does not seem prominent (figure 2B). Doherty et al. (2015) reconstructed catches (including discards) between 55,000 and 64,000 t/year in the 1950s, which peaked at more than 200,000 t/year in the mid-1980s but were affected by the war of independence from Portugal, followed by a long civil war. By the late 2000s, catches were between 120,000 and 130,000 t/year. Between 1950 and 2010, the fishing sector is estimated to have caught 4.6 times the landings reported by the FAO on behalf of Mozambique. However, since 2003 annual catches reported to the FAO have increased because of substantial improvements in national data reporting systems, and the total reconstructed catches are only 1.6 times the statistics reported by the FAO. Figure 2C (based on Doherty et al. 2015), summarizes the taxonomic composition of the catch, to include herrings, shads and sardines (family Clupeidae), and anchovies (Engraulidae). Although globally Mozambique has a low (if increasing) per capita GDP, the country is demonstrating that this should not be a reason for inaccurate fisheries statistics. Hopefully, this example will be followed elsewhere.

REFERENCES

Doherty B, McBride MM, Brito AJ, Le Manach F, Sousa L, Chauca I and Zeller D (2015) Marine fisheries in Mozambique: catches updated to 2010 and taxonomic disaggregation. pp. 67–82 In Le Manach F and Pauly D (eds.), *Fisheries catch reconstructions in the Western Indian Ocean, 1950–2010*. Fisheries Centre Research Reports 23(2).

Jacquet J, Fox H, Motta H, Ngusaru A and Zeller D (2010) Few data but many fish: marine small-scale fisheries catches for Mozambique and Tanzania. *African Journal of Marine Science* 32(2): 197–206.

Pauly D, editor (1992) Population dynamics of exploited fishes and crustaceans in Mozambique: contributions from a course on the "Use of Computers for Fisheries Research", 23 February–15 March 1988, Maputo, Mozambique. Revista de Investigação Pesqueira 21. 135 p.

1. Cite as Jacquet, J., B. Doherty, M. M. McBride, A. Brito, and Dirk Zeller. 2016. Mozambique. P. 338 in D. Pauly and D. Zeller (eds.), *Global Atlas of Marine Fisheries: A Critical Appraisal of Catches and Ecosystem Impacts*. Island Press, Washington, DC.

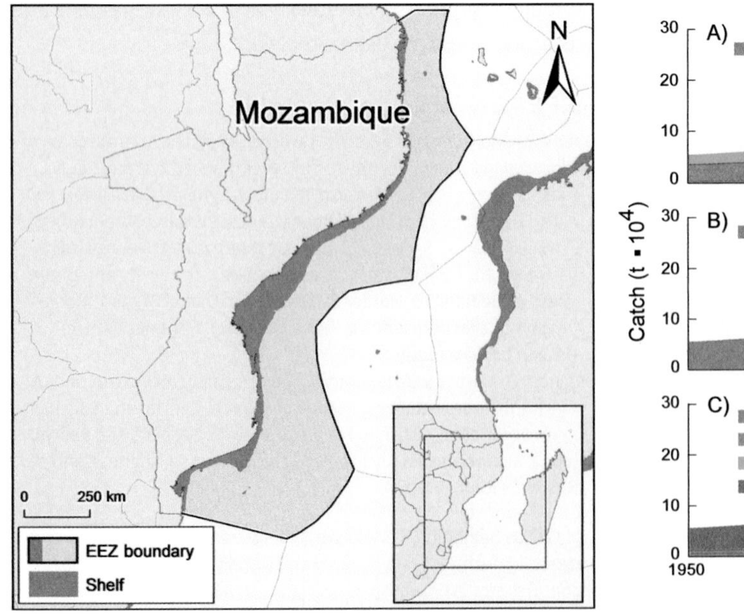

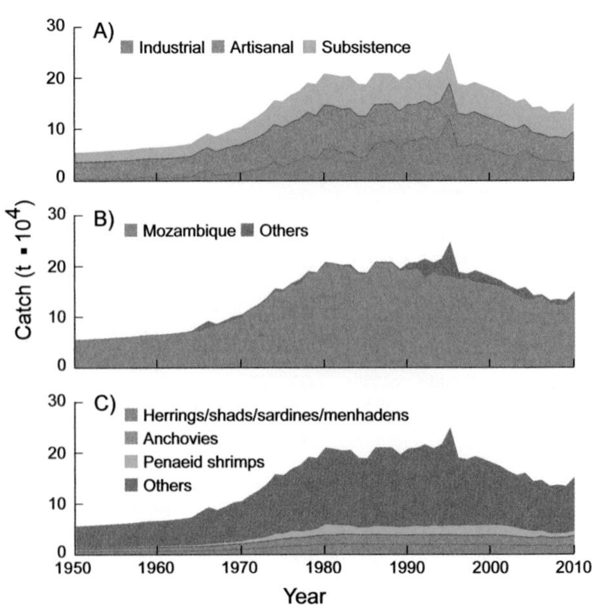

Figure 1. Mozambique has an EEZ of 571,000 km², with a shelf of 85,300 km². **Figure 2.** Domestic and foreign catches taken in the EEZ of Mozambique: (A) by sector (domestic scores: Ind. 1, 2, 3; Art. 1, 1, 2; Sub. 1, 1, 2; Dis. 1, 1, 1); (B) by fishing country (foreign catches are very uncertain); (C) by taxon.

MYANMAR[1]

Shawn Booth and Daniel Pauly

Sea Around Us, University of British Columbia, Vancouver, Canada

Myanmar (formerly Burma) is located in the east of the Bay of Bengal (figure 1), with formerly abundant marine resources (Pauly et al. 1984). Long an isolated country, Myanmar has recently begun to open up, which offers an opportunity to reform its fisheries monitoring systems, currently generating suspicious statistics. Notably, the catches, which are not taxonomically disaggregated (i.e., do not say what is caught), suggest that landings have been increasing exponentially for the last decades and were unaffected by cyclone Nargis in May 2008, which destroyed a large part of the Burmese fishing fleet. Here, based on Booth and Pauly (2011), we present estimates, not the actual catch of Myanmar but a possible alternative to official data, which lack credibility. Reconstructed domestic catches ranging from 200,000 to 260,000 t/year in the 1950s to 1.3 million t/year in the late 2000s, are 1.2 times the landing that FAO reports on behalf of Myanmar. The majority of allocated catches in Myanmar's EEZ are from the industrial sector, in addition to artisanal and subsistence components (figure 2A). Beginning in the 1990s, foreign fishing by Thailand increased substantially and became comparable in quantity to domestic catches (figure 2B). A likely catch composition for Myanmar's marine catch was inferred by interpolating between the well-documented catch composition of Thailand (Andaman Sea) and India (East Coast), yielding a simulacrum in figure 2C featuring threadfin, whiptail breams (family Nemipteridae), and drums and croakers (Sciaenidae).

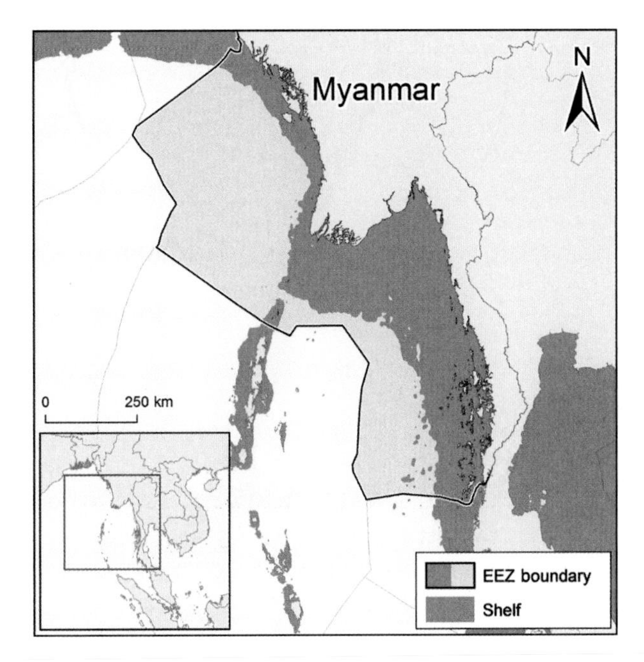

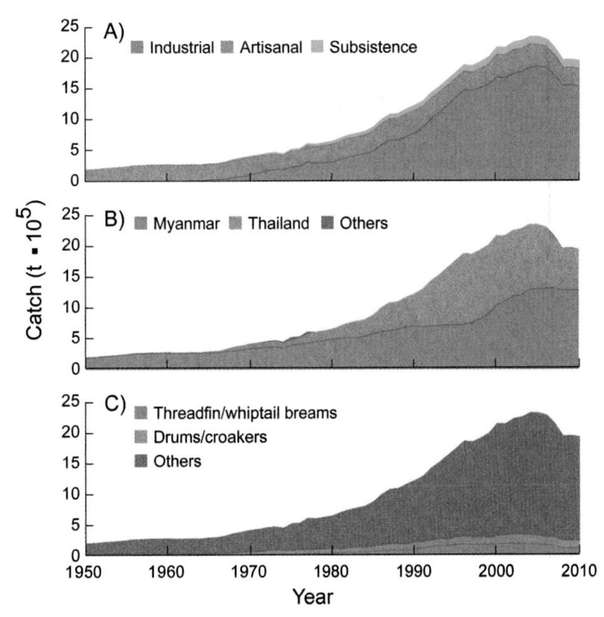

Figure 1. Myanmar and its EEZ (511,000 km², of which 220,000 km² is shelf), with the Rakhine Coastal Zone in the north, the Deltaic Coastal Zone in the center, and the Tanintharyi Coastal Zone in the south. **Figure 2.** Domestic and foreign catches taken in the EEZ of Myanmar: (A) by sector (domestic scores: Ind. 2, 3, 3; Art. 2, 3, 3; Sub. 2, 3, 3; Dis. 2, 2, 2); (B) by fishing country (foreign catches are very uncertain); (C) by taxon.

REFERENCES

Booth S and Pauly D (2011) Myanmar's marine capture fisheries 1950–2008: Expansion from the coast to the deep waters. pp. 101–134 In Harper S, O'Meara D, Booth S, Zeller D and Pauly D (eds.), *Fisheries Catches for the Bay of Bengal Large Marine Ecosystem since 1950*. Report to the Bay of Bengal Large Marine Ecosystem Project. BOBLME-2011-Ecology-16. [Updated to 2010]

Pauly D, Aung S, Rijavec L and Htein H (1984) The marine living resources of Burma: a short review. pp. 96–108 In Report of the 4th Session of the Standing Committee on Resources Research and Development of the Indo-Pacific Fishery Commission, 23–29 August 1984, Jakarta. FAO Fisheries Report 318.

1. Cite as Booth, S., and D. Pauly. 2016. Myanmar. P. 339 in D. Pauly and D. Zeller (eds.), *Global Atlas of Marine Fisheries: A Critical Appraisal of Catches and Ecosystem Impacts*. Island Press, Washington, DC.

NAMIBIA[1]

Dyhia Belhabib,[a] Nico E. Willemse,[b] and Daniel Pauly[a]

[a]*Sea Around Us*, University of British Columbia, Vancouver, Canada
[b]Benguela Current Commission Secretariat, Windhoek, Namibia

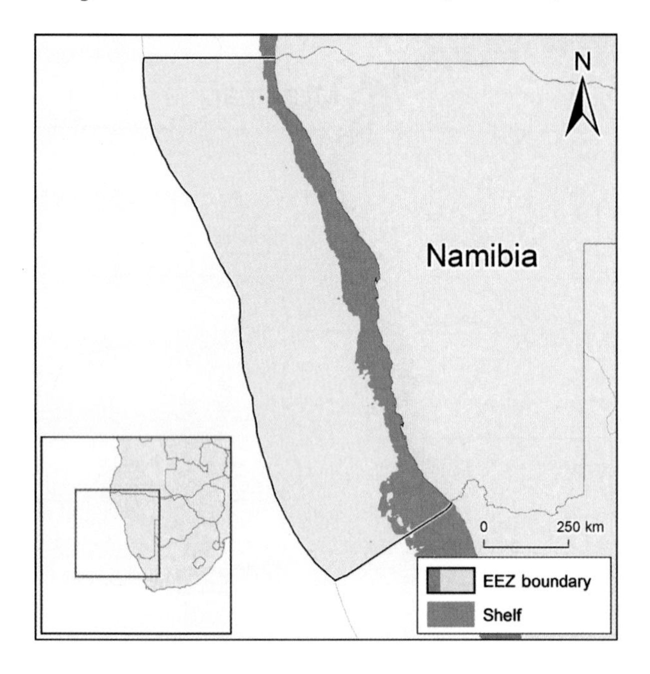

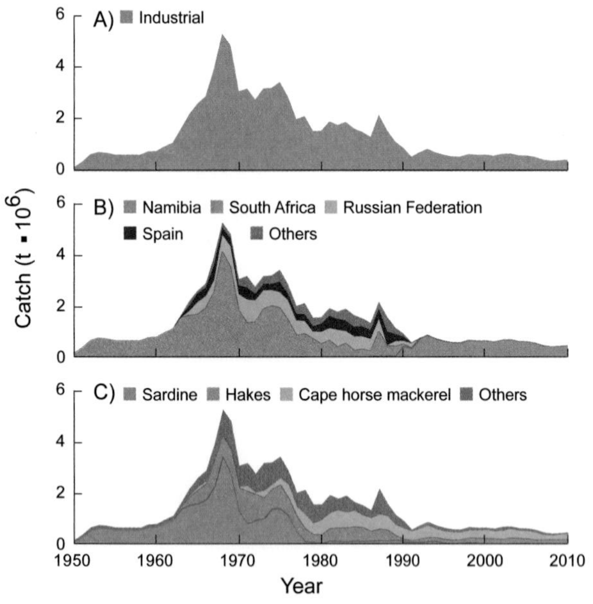

Figure 1. Namibia has a continental shelf of 94,900 km² and an EEZ of 560,000 km². **Figure 2.** Domestic and foreign catches taken in the EEZ of Namibia: (A) by sector (domestic scores: Ind. 4, 4, 4; Sub. 2, 2, 3; Rec. 3, 3, 3; Dis. –, 4, 4); (B) by fishing country (foreign catches are very uncertain); (C) by taxon.

Namibia's coast stretches along the northern part of the Benguela Current, in southwestern Africa (figure 1). Although its waters are extremely productive, the coast—the edge of the Namib desert—is largely uninhabitable, which led to the extreme dominance of industrial over small-scale fisheries (Willemse and Pauly 2004). Thus, the bulk of the catches are industrial, while the artisanal and recreational catches are not visible on figure 2A. Withdrawals by all fleets from the Namibian EEZ were reconstructed by Belhabib et al. (2015), yielding 90,000 t in 1950, a peak of 3.7 million t in 1968, and about 430,000 t in 2010. These estimates account for the catch of South African vessels off Namibia until independence in 1990 and by various foreign fleets, many illegal, a few years thereafter, until Namibia gained control of its EEZ. From 1991 to 2010, the reconstructed catches were about 10% higher than reported by the FAO on behalf of Namibia, mainly because of discards (recreational and subsistence fisheries jointly accounted for less than 0.2% of catches). Before 1991, the majority of the catches were from foreign vessels, notably from South Africa, Russia, and Spain (figure 2B); however, domestic catches came later to dominate. Earlier catches consisted mostly of Pacific sardine (*Sardinops sagax*) and hakes (*Merluccius* spp.), both of which declined in the 1980s and were partly replaced by Cape horse mackerel (*Trachurus capensis*; figure 2C). Namibia provides an example for Africa in that it wrested control of its fisheries resources from distant-water fleets, although it now must share them with the jellyfish that now dominate a large part of the Benguela ecosystem (Richardson et al. 2009).

REFERENCES

Belhabib D, Willemse N and Pauly D (2015) A fishery tale: Namibian fisheries between 1950–2010. Fisheries Centre Working Paper #2015–65, 17 p.

Richardson AJ, Bakun A, Hays GC and Gibbons MJ (2009) The jellyfish joyride: causes, consequences and management responses to a more gelatinous future. *Trends in Ecology & Evolution* 24(6): 312–322.

Willemse N and Pauly D (2004) Reconstruction and interpretation of marine fisheries catches from Namibian waters, 1950 to 2000. pp. 99–112 In Sumaila UR, Boyer DC, Skogen Morten D and Steinshamm I (eds.), *Namibia's Fisheries: Ecological, Economic and Social Aspects*. Eburon Academic Publishers, Amsterdam.

1. Cite as Belhabib, D., N. E. Willemse, and D. Pauly. 2016. Namibia. P. 340 in D. Pauly and D. Zeller (eds.), *Global Atlas of Marine Fisheries: A Critical Appraisal of Catches and Ecosystem Impacts*. Island Press, Washington, DC.

NAURU[1]

Pablo Trujillo, Sarah Harper, and Dirk Zeller

Sea Around Us, University of British Columbia, Vancouver, Canada

Nauru is a small island state in the western Pacific (figure 1) with an interior plateau that once held extensive deposits of phosphate-bearing rock (i.e., mineralized guano). Phosphate mining was the island's largest source of revenue, but it has now rendered 80% of the island uninhabitable and caused the substantial degradation of Nauru's adjacent reefs from silt and phosphate runoff (Vunisea et al. 2008). Most of the catches in Nauru's EEZ were from the industrial tuna sector (figure 2B), taken by foreign vessels such as Japan, Taiwan, and South Korea (figure 2A). The domestic catch reconstruction by Trujillo et al. (2011), excluding industrially caught large pelagics (see chapter 3), grew from 300 t/year in the early 1950s to 450 t/year in the 2000s and consists of small-scale inshore fisheries, including artisanal and subsistence components, as well as offshore domestic catch, and was 2.6 times the total landings reported to the FAO by Nauru for that period. Most of the total catch was taken by trolling gear, particularly tunas: skipjack (*Katsuwonus pelamis*), yellowfin (*Thunnus albacares*), and bigeye tuna (*T. obesus*; figure 2C). A significant portion was contributed by midwater handlines, almost exclusively catching fish of the family Carangidae such as rainbow runner (*Elagatis bipinnulata*) and demersal handlines targeting squirrelfish (family Holocentridae) and bluestripe snapper (*Lutjanus kasmira*). With the end of phosphate revenue in the late 1990s, subsistence fishing increased again in importance.

REFERENCES

Trujillo P, Harper S and Zeller D (2011) Reconstruction of Nauru's fisheries catches: 1950–2008. pp. 63–71 In Harper S and Zeller D (eds.), *Fisheries catch reconstruction: Islands, part II*. Fisheries Centre Research Reports 19(4). [Updated to 2010]

Vunisea A, Pinca S, Friedman K, Chapman L, Magron F, Sauni S, Pakoa K, Awira R and Lasi F (2008) Nauru country report: Profile and results from in-country survey work. Secretariat of the Pacific Community (SPC), Nouméa, New Caledonia, 68 p.

1. Cite as Trujillo, P., S. Harper, and D. Zeller. 2016. Nauru. P. 341 in D. Pauly and D. Zeller (eds.), *Global Atlas of Marine Fisheries: A Critical Appraisal of Catches and Ecosystem Impacts*. Island Press, Washington, DC.

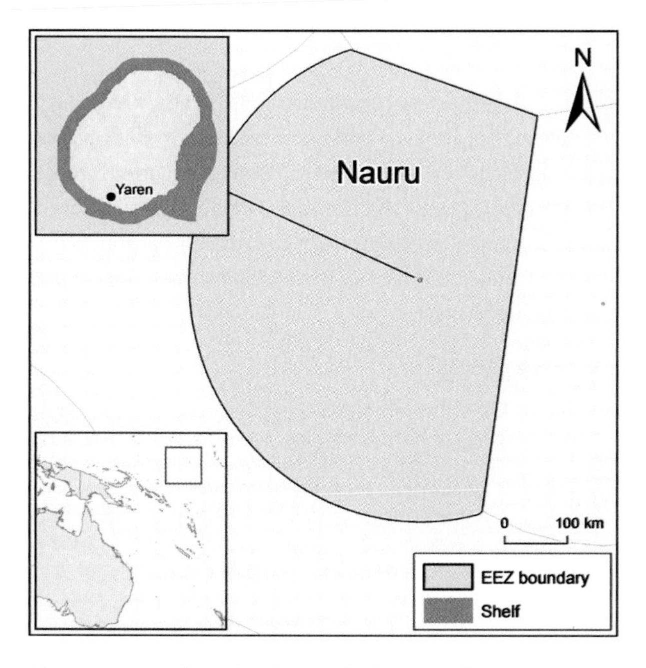

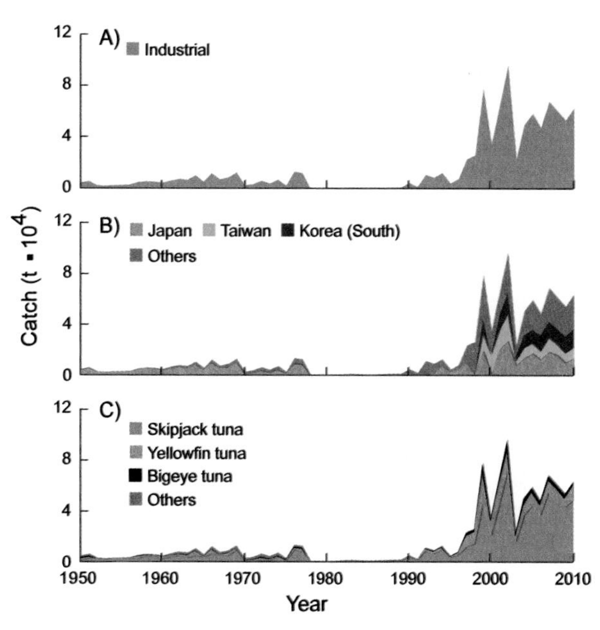

Figure 1. Nauru has a land area of 21 km², and its EEZ is 309,000 km².

Figure 2. Domestic and foreign catches taken in the EEZ of Nauru: (A) by sector (domestic scores: Art. 1, 1, 3; Sub. 1, 1, 3); (B) by fishing country (foreign catches are very uncertain); (C) by taxon.

NETHERLANDS[1]

Darah Gibson, Kyrstn Zylich, and Dirk Zeller

Sea Around Us, University of British Columbia, Vancouver, Canada

The Netherlands is a small country in Europe (figure 1), much of it lower than sea level and therefore protected from the sea by dikes. The Zuiderzee was closed off from the Wadden Sea and the Lauwerszee was turned into a freshwater lake, which probably affected the recruitment of many marine species (de Jonge et al. 1993; Wolff 2005). Most allocated catches to Netherlands' EEZ are industrial, with a small artisanal, subsistence, and recreational component (figure 2A). The Dutch marine fisheries catches within their EEZ-equivalent waters were reconstructed by Gibson et al. (2015), using data from the International Council for the Exploration of the Sea (ICES) as a reporting baseline, to which discards and unreported commercial, recreational, and subsistence catches were added from various sources. The reconstructed catch decreased from 234,000 t in 1950 to 176,000 t in 2010, with a peak of nearly 500,000 t in 1985. Over the 1950–2010 period, the reconstructed catch was about 2 times the catch reported to ICES, and discards from trawlers contributed most to the discrepancy. The bulk of the catches are domestic, although foreign fishing by Germany, the United Kingdom, and France occurred through the time period (figure 2B). Top taxa, as shown in figure 2C, are European plaice (*Pleuronectes platessa*), Atlantic herring (*Clupea harengus*), and common dab (*Limanda limanda*), which is the most discarded species.

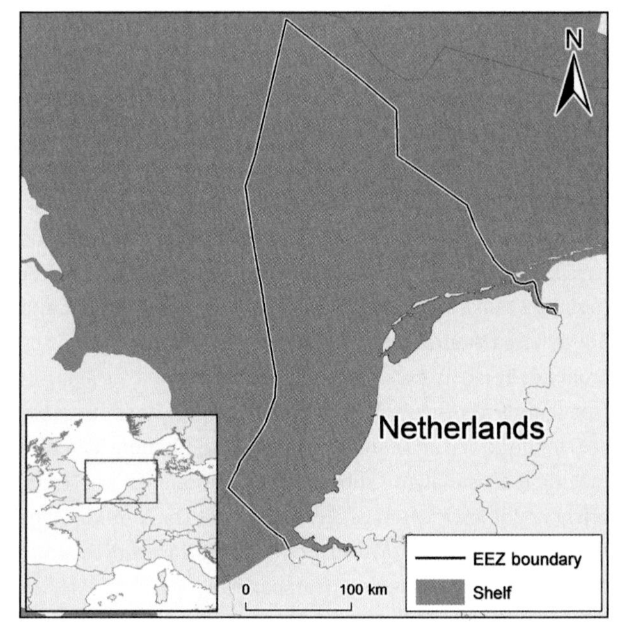

Figure 1. The Netherlands have EEZ-equivalent waters of 61,900 km², all of which is shelf.

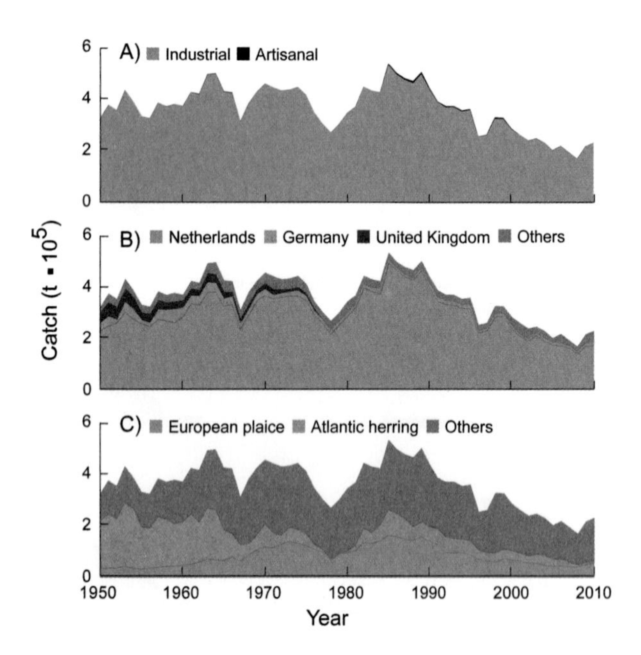

Figure 2. Domestic and foreign catches taken in the EEZ of The Netherlands: (A) by sector (domestic scores: Ind. 3, 3, 4; Art. 3, 3, 4; Sub. 2, 1, –; Rec. 1, 1, 2; Dis. 1, 1, 2); (B) by fishing country (foreign catches are very uncertain); (C) by taxon.

REFERENCES

de Jonge VN, Essink K and Boddeke R (1993) The Dutch Wadden Sea: a changed ecosystem. *Hydrobiologia* 265: 45–71.

Gibson D, Zylich K and Zeller D (2015) Preliminary reconstruction of total marine fisheries catches for the Netherlands in the North Sea (1950–2010). Fisheries Centre Working Paper #2015-46, 15 p.

Wolff WJ (2005) The exploitation of living resources in the Dutch Wadden Sea: a historical overview. *Helgoland Marine Research* 59: 31–38.

1. Cite as Gibson, D., K. Zylich, and D. Zeller. 2016. Netherlands. P. 342 in D. Pauly and D. Zeller (eds.), *Global Atlas of Marine Fisheries: A Critical Appraisal of Catches and Ecosystem Impacts*. Island Press, Washington, DC.

NETHERLANDS (ARUBA)[1]

Daniel Pauly, Sulan Ramdeen, and Aylin Ulman

Sea Around Us, University of British Columbia, Vancouver, Canada

The island of Aruba (figure 1) is part of the Leeward Netherlands Antilles island chain, which includes the other "ABC islands," Bonaire and Curaçao. The islands were part of the Netherlands Antilles, but Aruba went its own way in 1986 while maintaining ties to the Netherlands (Molenaar 2003). Given the absence of comprehensive statistics for Aruba, Pauly et al. (2015), based on various sources, presented a reconstruction of domestic fisheries catches that increased from 360 t/year in the 1950s to 700 t/year in the 2000s and, overall, were 1.8 times the amount deemed to have been reported by the FAO on behalf of Aruba. Overall, the allocated catches in Aruba's EEZ are predominantly industrial (figure 2A) and overwhelmingly feature catches by foreign fleets (figure 2B). The composition of the domestic catch changed from a dominance of reef fishes to large pelagics, because of the growing tourist-based recreational fishery and offshore expansion by the artisanal fishers. Figure 2C presents top taxa as king mackerel (*Scomberomorous cavalla*), southern red snapper (*Lutjanus purpureus*), yellowfin tuna (*Thunnus albacares*), and skipjack tuna (*Katsuwonus pelamis*). Note that, because of the berthing facilities available in Aruba, large oil refineries have been operating there for decades (Van Gelderen 1953), hiring a large number of workers. Therefore, small-scale fisheries did not have to absorb excess labor as they had in many other Caribbean countries.

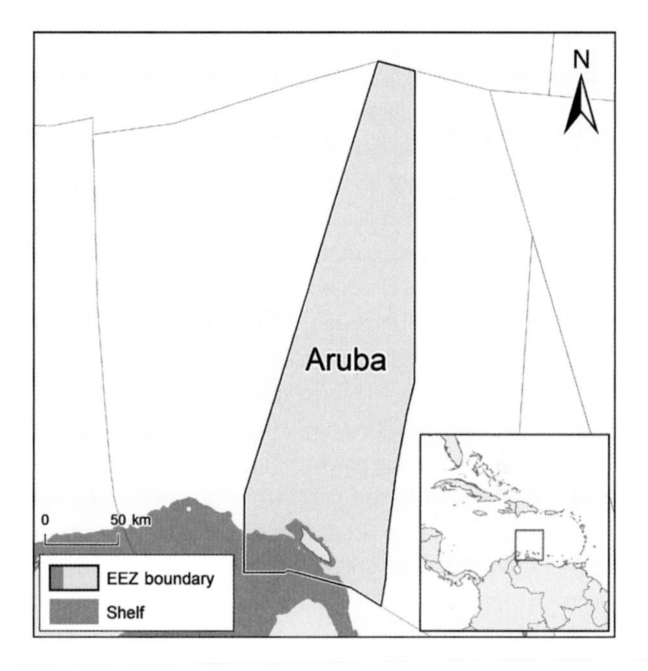

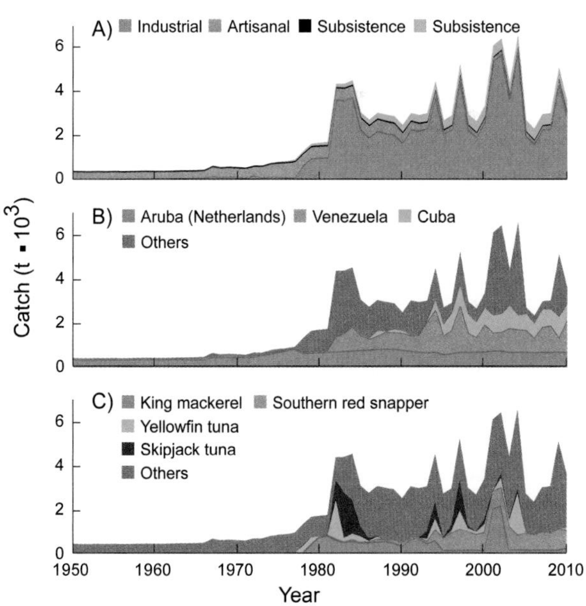

Figure 1. Aruba (land area 179 km²) and its EEZ (25,200 km²).
Figure 2. Domestic and foreign catches taken in the EEZ of Aruba: (A) by sector (domestic scores: Art. 2, 3, 4; Sub. 2, 2, 2; Rec. –, 2, 2; Dis. 2, 2, 2); (B) by fishing country (foreign catches are very uncertain); (C) by taxon.

REFERENCES

Molenaar EJ (2003) Current legal development: Marine fisheries in the Netherlands Antilles and Aruba in the context of international law. *The International Journal of Marine and Coastal Law* 18(1): 127–144.

Pauly D, Ramdeen S and Ulman A (2015) Reconstruction of total marine catches for Aruba, Southern Caribbean, 1950–2010. Fisheries Centre Working Paper #2015-10, 8 p.

van Gelderen P (1953) Fisheries in the Netherland Antilles. pp. 51–59 In Proceedings of the Fifth Annual Gulf and Caribbean Fisheries Institute, Coral Gables, Florida.

1. Cite as Pauly, D., S. Ramdeen, and A. Ulman. 2016. Netherlands (Aruba). P. 343 in D. Pauly and D. Zeller (eds.), *Global Atlas of Marine Fisheries: A Critical Appraisal of Catches and Ecosystem Impacts*. Island Press, Washington, DC.

NETHERLANDS (BONAIRE)[1]

Alasdair Lindop, Elise Bultel, Kyrstn Zylich, and Dirk Zeller

Sea Around Us, University of British Columbia, Vancouver, Canada

Bonaire, the second in the "ABC islands," is a municipality of the Netherlands 80 km from the northwest coast of Venezuela (figure 1). In principle, the reefs around Bonaire are protected by a marine park; however, this applies only to "nontraditional fishing," and traditional fishing, has reduced the biomass of large reef fish around Bonaire. The small-scale fisheries, though important to local livelihoods, contribute only a fraction to total allocated catches, as foreign industrial fleets, notably from France and Venezuela, dominate (figure 2A, 2B). Domestic catches, as reconstructed by Lindop et al. (2015) appear to have remained stable, decreasing slightly from 166 t in 1950 to 158 t in 2010, with the declining artisanal catch of reef fishes being partly compensated for by increasing subsistence and recreational catches. The reconstructed domestic catch was 2.9 times that reported by the FAO on behalf of Bonaire. Miscellaneous "marine fish nei" contributed a portion to allocated catches and consisted mostly of "potfish" (Dilrosun 2004). Pelagic species made up most of the rest of the catch, with yellowfin tuna (*Thunnus albacares*), skipjack tuna (*Katsuwonus pelamis*), and king mackerel (*Scomberomorus cavalla*) being particularly important (figure 2C). Snappers (Lutjanidae) were the most abundant demersal group, although they are not visible in figure 2C. Pauly (2002) describes how an attempt by Spanish industrial vessels to access Bonaire's EEZ was foiled; however, incursions by Venezuelan and other foreign vessels are frequent.

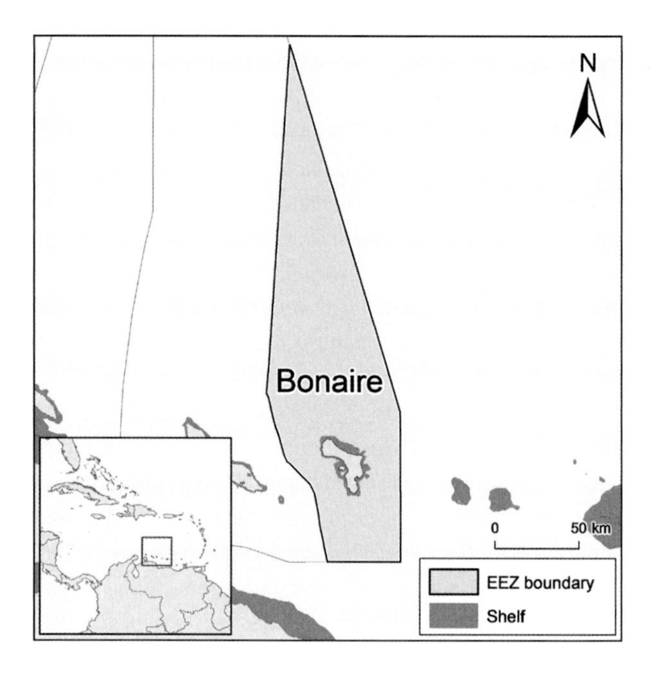

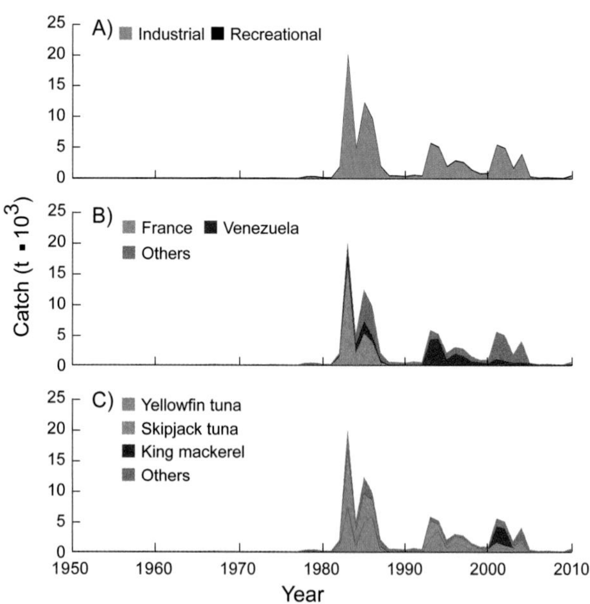

Figure 1. Bonaire (Netherlands) has a land area of 294 km² and an EEZ of 13,200 km². **Figure 2.** Domestic and foreign catches taken in the EEZ of Bonaire (Netherlands): (A) by sector (domestic scores: Art. 2, 3, 4; Sub. 2, 2, 2; Rec. 2, 2, 2); (B) by fishing country (foreign catches are very uncertain); (C) by taxon.

REFERENCES

Dilrosun F (2004) Korte inventarisatie van de visserijsector van Bonaire [Brief inventory of the fisheries sector in Bonaire]. Afdeling Milieu & Natuur (MINA) van de Directie Volksgezondheid. 12 p.

Lindop A, Bultel E, Zylich K and Zeller D (2015) Reconstructing the former Netherlands Antilles marine catches from 1950 to 2010. Fisheries Centre Working Paper #2015–69, 22 p

Pauly D (2002) Bonaire: 90 million years, plus a few days to think. *Sea Around Us Project Newsletter* (14): 1–3, 5.

1. Cite as Lindop, A., E. Bultel, K. Zylich, and D. Zeller. 2016. Netherlands (Bonaire). P. 344 in D. Pauly and D. Zeller (eds.), *Global Atlas of Marine Fisheries: A Critical Appraisal of Catches and Ecosystem Impacts*. Island Press, Washington, DC.

NETHERLANDS (CURAÇAO)[1]

Alasdair Lindop, Elise Bultel, Kyrstn Zylich, and Dirk Zeller

Sea Around Us, University of British Columbia, Vancouver, Canada

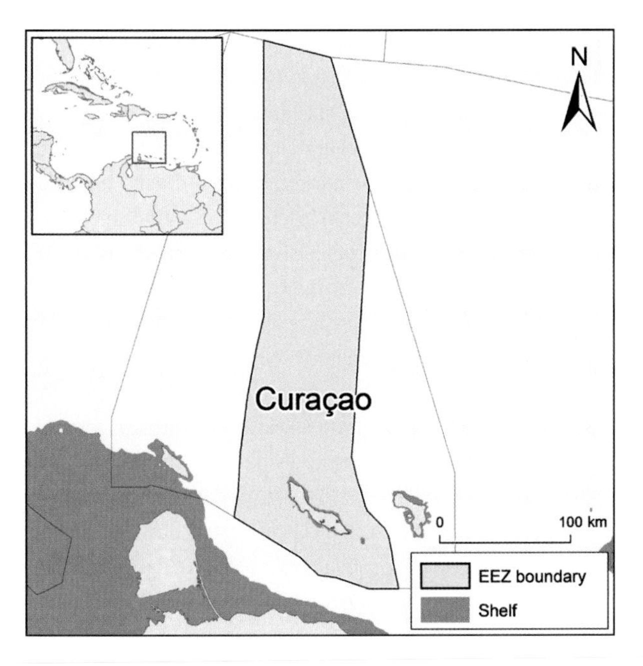

Figure 1. Curaçao (Netherlands) has a land area of 437 km² and an EEZ of 30,400 km².

Curaçao, the third "ABC Island" of the former Netherlands Antilles, is located north of Venezuela, between Aruba and Bonaire (figure 1), and has high per capita incomes because of a diversified economy, with oil refining and bunkering, shipping, financial services, and tourism being most important. The industrial sector dominated total allocated catches within Curaçao's EEZ; however, the artisanal sector remained constant and substantial through the time period (figure 2A). Domestic fisheries catches, as reconstructed by Lindop et al. (2015) based on very limited data (see Dilrosun 2002), increased from 330 t in 1950 to 1,000 t/year in the early 1990s, then declined to about 460 t by 2010. Some of the pelagic fish are caught using fish aggregating devices (FADs; Debrot and Nagelkerken 2000), and the artisanal fisheries uses chevron traps (Johnson 2010). Overall, reconstructed catch was 2.2 times the data reported by the FAO on behalf of Curaçao. Catches were dominated by foreign fishing by Venezuela, France, and the United States (figure 2B). Yellowfin tuna (*Thunnus albacares*) was the most prominent species caught, followed by skipjack tuna (*Katsuwonus pelamis*) and wahoo (*Acanthocybium solandri*; figure 2C). The data available to us do not allow assessing whether the recent sharp decline in Curaçao's catches is a result of overfishing or, perhaps more likely, a reduction of fishing effort.

REFERENCES

Debrot AO and Nagelkerken I (2000) User perceptions on coastal resource state and management options in Curaçao. *Revista de Biología Tropical* 48(Suppl. 1): 95–106.

Dilrosun F (2002) Progress report on Curaçao fishery monitoring programme from November 2000 to July 2001. pp. 9–20 in First meeting of the WECAFC Ad Hoc Working Group on the Development of Sustainable Moored Fish Aggregating Device Fishing in the Lesser Antilles. Le Robert, Martinique, 8–11 October 2001. Fisheries Report No. 683.

Johnson AE (2010) Reducing bycatch in coral reef trap fisheries: escape gaps as a step toward sustainability. *Marine Ecology Progress Series* 415: 201–209.

Lindop A, Bultel E, Zylich K and Zeller D (2015) Reconstructing the former Netherlands Antilles marine catches from 1950 to 2010. Fisheries Centre Working Paper #2015-69, 22 p.

1. Cite as Lindop, A., E. Bultel, K. Zylich, and D. Zeller. 2016. Netherlands (Curaçao). P. 345 in D. Pauly and D. Zeller (eds.), *Global Atlas of Marine Fisheries: A Critical Appraisal of Catches and Ecosystem Impacts*. Island Press, Washington, DC.

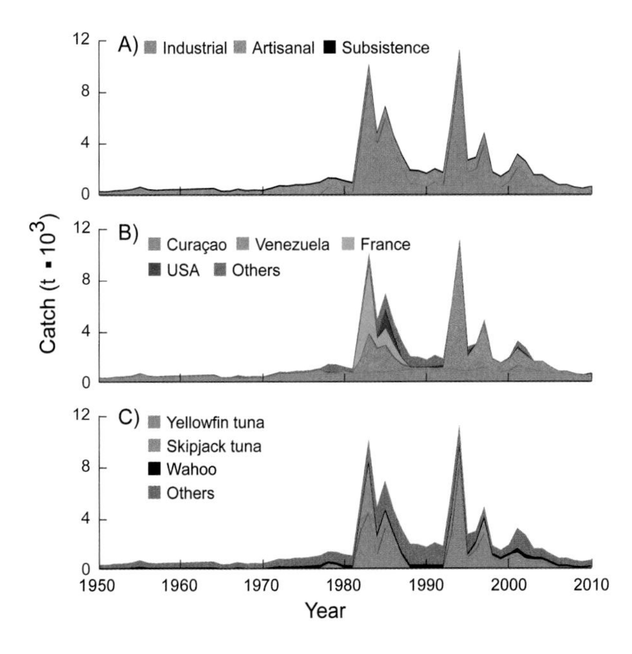

Figure 2. Domestic and foreign catches taken in the EEZ of Curaçao (Netherlands): (A) by sector (domestic scores: Art. 2, 3, 4; Sub. 2, 2, 2; Rec. 2, 2, 2); (B) by fishing country (foreign catches are very uncertain); (C) by taxon.

NETHERLANDS (SABA AND SINT EUSTATIUS)[1]

Alasdair Lindop, Elise Bultel, Kyrstn Zylich, and Dirk Zeller

Sea Around Us, University of British Columbia, Vancouver, Canada

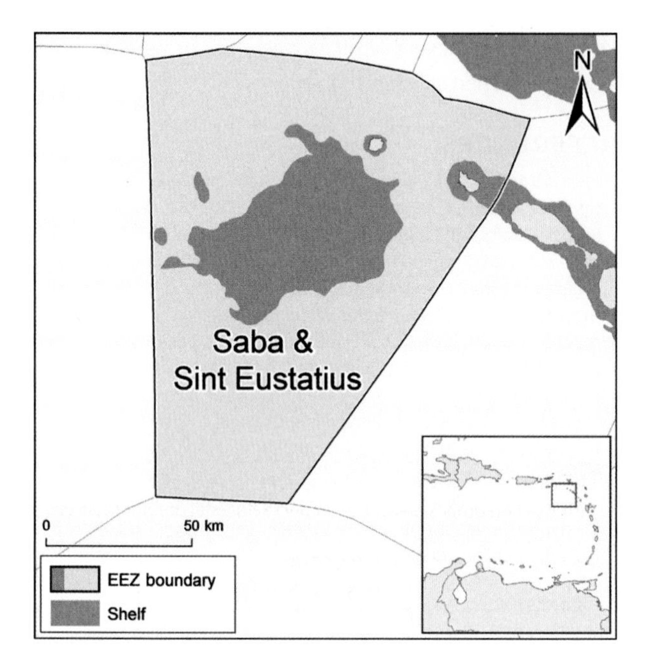

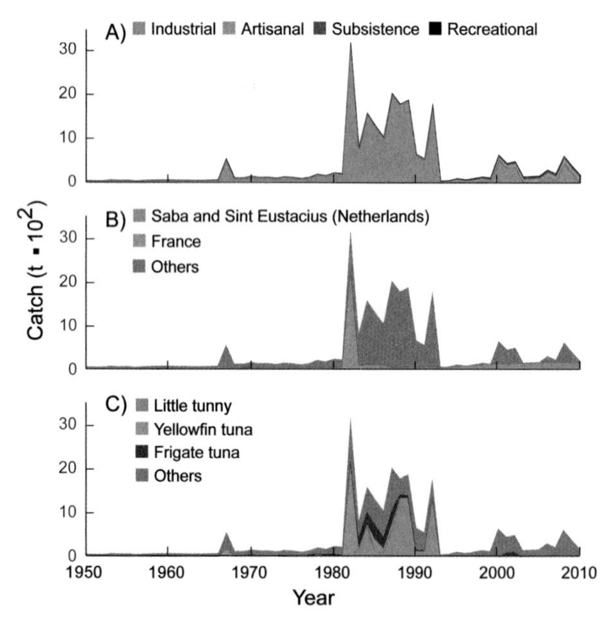

Figure 1. The islands of Saba and Sint Eustatius have a joint land area of 34 km² and EEZ of 11,600 km²; Saba Bank encompasses a very biodiverse area of 2,200 km² above the 200-m isobaths. **Figure 2.** Domestic and foreign catches taken in the EEZ of Saba and Sint Eustatius: (A) by sector (domestic scores: Art. 2, 3, 4; Sub. 2, 2, 2; Rec. 2, 2, 2); (B) by fishing country (foreign catches are very uncertain); (C) by taxon.

The Caribbean Islands of Saba and Sint Eustatius are in the northeastern Caribbean (Windward Islands; figure 1). The islands were first captured by the Dutch in 1630 but changed colonial masters several times, until the Netherlands claimed sovereignty. The economy, earlier based on trade and growing sugar, tobacco, and cotton is now based on tourism and financial services. Fish supply from the islands is insufficient to satisfy existing demand, necessitating seafood imports. Fishing in Saba consists of both near-shore fishing and offshore fishing on Saba Bank, whereas the Sint Eustatius fishing is conducted by about 24 artisanal fishers exploiting the narrow shelf surrounding the island (Dilrosun 2004b). However, the bulk of catches originate from the industrial sector, especially in the 1980s (figure 2A). This is predominantly from foreign vessels (figure 2B). The reconstruction by Lindop et al. (2015), based on Dilrosun (2000, 2004a, 2004b) and other sources, suggests a domestic catch of about 30 t in 1950, which increased to 150 t in 2010. Over the period from 1950 to 2010, this corresponds to 2.8 times the figures reported by the FAO on behalf of Saba and Sint Eustatius. This catch consists mainly of large pelagics such as little tunny (*Euthynnus alletteratus*), yellowfin tuna (*Thunnus albacares*), and frigate tuna (*Auxis thazard*; figure 2C). Also, increasingly fishers from Sint Maarten are fishing on Saba Bank.

REFERENCES

Dilrosun F (2000) Monitoring the Saba Bank fishery. Department of Public Health and Environmental Hygiene, Environmental Division, Curaçao, Netherlands Antilles. 56 p.

Dilrosun F (2004a) Inventory of the fishery sector of Saba. Island Territory of Curaçao, Department of Agriculture, Animal Husbandry and Fisheries. 14 p.

Dilrosun F (2004b) Inventory of the fishery sector of St. Eustatius. Island Territory of Curaçao, Department of Agriculture, Animal Husbandry and Fisheries. 14 p.

Lindop A, Bultel E, Zylich K and Zeller D (2015) Reconstructing the former Netherlands Antilles marine catches from 1950 to 2010. Fisheries Centre Working Paper #2015–69, 22 p.

1. Cite as Lindop, A., E. Bultel, K. Zylich, and D. Zeller. 2016. Netherlands (Saba and Sint Eustatius). P. 346 in D. Pauly and D. Zeller (eds.), *Global Atlas of Marine Fisheries: A Critical Appraisal of Catches and Ecosystem Impacts*. Island Press, Washington, DC.

NETHERLANDS (SINT MAARTEN)[1]

Alasdair Lindop, Elise Bultel, Kyrstn Zylich, and Daniel Pauly

Sea Around Us, University of British Columbia, Vancouver, Canada

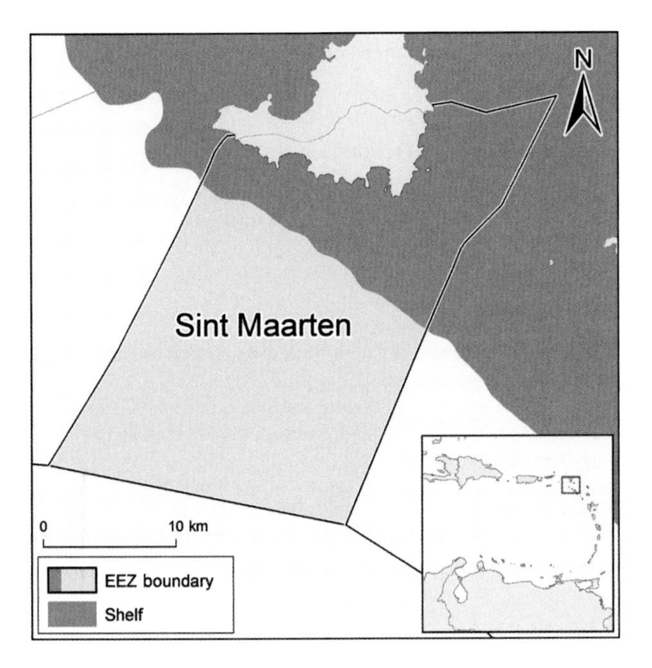

Figure 1. Sint Maarten (Netherlands) has a land area of 34 km² and EEZ of 500 km².

Sint Maarten is the Dutch part of a Caribbean island shared with the French territory of Saint Martin (Redon 2011; figure 1). There are two main landing sites on Sint Maarten, with about 50 fishers using 3 types of fishing vessels: small wooden boats, larger-decked fiberglass vessels equipped with diesel engines, and luxury yachts (Dilrosun 2004). The first deploy traps and hand-lines on the Sint Maarten shelf, the larger modern boats, which may be considered industrial, target snapper on Saba Bank (well over 50 km from Sint Maarten); while the luxury yachts perform day charters for tourists. Most of the catches around Sint Maarten are industrial (figure 2A), some attributable to French fleets (figure 2B). There is a large demand for fresh fish in Sint Maarten, with demersal species always available, and pelagic species are only occasionally available at market sites (Dilrosun 2004). The reconstruction by Lindop et al. (2015), based on Dilrosun (2004), NGMC (2003), and other sources, assessed that domestic catches from the waters around Sint Maarten remained stable, with 30 t/year in the 1950s to 60 t/year in 2010. This reconstructed domestic catch is, overall, 1.7 times the data deemed reported by the FAO on behalf of Sint Maarten. It consists mainly of finfish, Atlantic Spanish mackerel (*Scomberomorus maculatus*), little tunny (*Euthynnus alletteratus*), and silk snappers (Lutjanidae; figure 2C); however, the catch composition is tentative.

REFERENCES

Dilrosun F (2004) Inventory of the fishery sector of St. Maarten. Island Territory of Curaçao, Department of Agriculture, Animal Husbandry and Fisheries, 10 p.

Lindop A, Bultel E, Zylich K and Zeller D (2015) Reconstructing the former Netherlands Antilles marine catches from 1950 to 2010. Fisheries Centre Working Paper #2015–69, 22 p.

NGMC (2003) Market Profile: St. Maarten, A rapid reconnaissance survey. Guyana Office for Investment New Guyana Marketing Corporation Export Market Series Bulletin No. 10. 23 p.

Redon M (2011) One island, two landscapes: or does Otherness manifest itself on other sides of the border (Saint-Martin/Sint Maarten & Haiti/Dominican Republic). *Shima: The International Journal of Research into Island Cultures* 5(2): 68–85.

1. Cite as Lindop A., E. Butel, K. Zylich, and D. Pauly. 2016. Netherlands (Sint Maarten). P. 347 in D. Pauly and D. Zeller (eds.), *Global Atlas of Marine Fisheries: A Critical Appraisal of Catches and Ecosystem Impacts*. Island Press, Washington, DC.

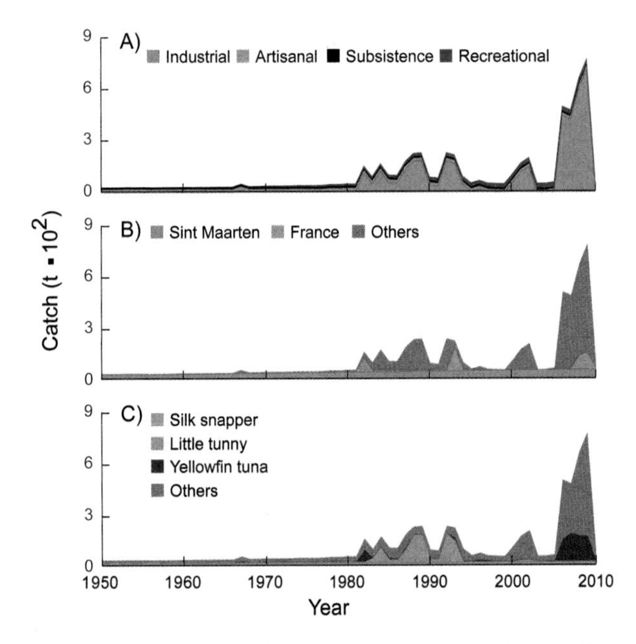

Figure 2. Domestic and foreign catches taken in the EEZ of Sint Maarten (Netherlands): (A) by sector (domestic scores: Art. 2, 3, 4; Sub. 2, 2, 2; Rec. 2, 2, 2); (B) by fishing country (foreign catches are very uncertain); (C) by taxon.

NEW ZEALAND[1]

Glenn Simmons,[a] Graeme Bremner,[b] Hugh Whittaker,[c] Philip Clarke,[b] Lydia Teh,[d] Kyrstn Zylich,[d] Dirk Zeller,[d] Daniel Pauly,[d] Christina Stringer,[e] Barry Torkington,[f] and Nigel Haworth[e]

[a]New Zealand Asia Institute, University of Auckland, Private Bag 92019, Auckland, New Zealand [b]Quadrat (NZ) Ltd, Otago, New Zealand [c]School of Interdisciplinary Area Studies, University of Oxford, UK [d]*Sea Around Us*, University of British Columbia, Vancouver, Canada [e]Department of Management and International Business, The University of Auckland, New Zealand [f]Warkworth, New Zealand

New Zealand, an island country in the southwestern Pacific Ocean, consists of two main landmasses (North and South Islands) and several smaller groups of islands, including the Antipodes, Auckland, Bounty, Campbell, Chatham, Kings, Kermadec, and Stewart Islands (figure 1). All but the Kermadec Islands (see next page) are covered in this account, which is based on the catch reconstruction by Simmons et al. (2015). The reconstructed domestic catch (i.e., excluding charter vessel catch) from the New Zealand EEZ was about 120,000 t/year in the early 1950s, increased to a peak of 900,000 t in 1977 and 890,000 t in 1988, then declined to about 400,000 t/year in the late 2000s. This is 2.6 times the data FAO reports on behalf of New Zealand, mainly because official data do not account for unreported commercial catch (i.e., "invisible" catch, Simmons et al. 2015), discards, recreational (Borch 2010), and customary (i.e., Māori; Bess 2001) catches (figure 2A). Although industrial catches dominate, small-scale artisanal fisheries continue to be important (figure 2A), although they have been a major victim of quota concentration in the New Zealand quota management system (Torkington 2016). Figure 2B summarizes the taxonomic composition of the catch, but does not do justice to its diversity, which grew from 36 species reported in the 1950s to 234 in 2010.

REFERENCES

Bess R (2001) New Zealand's indigenous people and their claims to fisheries resources. *Marine Policy* 25(1): 23–32.

Borch T (2010) Tangled lines in New Zealand's quota management system: The process of including recreational fisheries. *Marine Policy* 34: 655–662.

Simmons G, Bremner G, Whittaker H, Clarke P, Teh L, Zylich K, Zeller D, Pauly D, Stringer C, Torkington B, and Haworth N. (2015) Preliminary reconstruction of marine fisheries catches for New Zealand (1950–2010). Fisheries Centre Working Paper #2015–87, 33 p.

Torkington, B. 2016. New Zealand's quota management system: incoherent and conflicted. *Marine Policy* 63: 180–183.

1. Cite as Simmons, G., G. Bremner, H. Whittaker, P. Clarke, L. Teh, K. Zylich, D. Zeller, D. Pauly, C. Stringer, B. Torkington, and N. Haworth. 2016. New Zealand. P. 348 in D. Pauly and D. Zeller (eds.), *Global Atlas of Marine Fisheries: A Critical Appraisal of Catches and Ecosystem Impacts*. Island Press, Washington, DC.

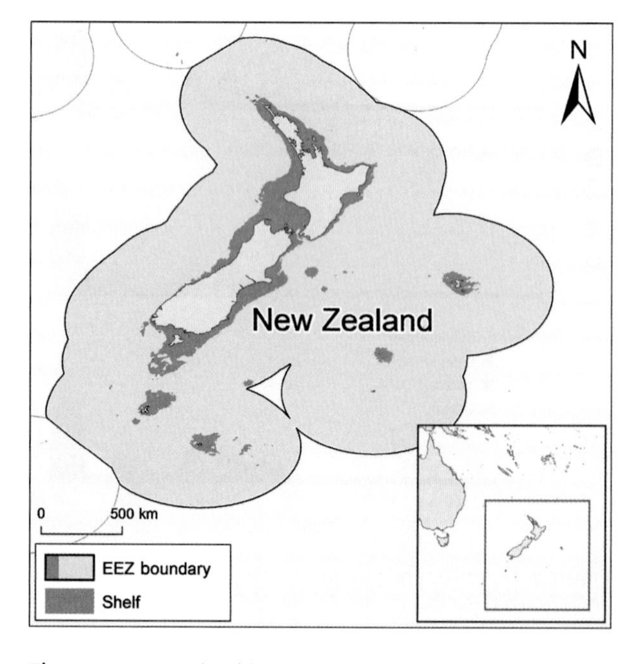

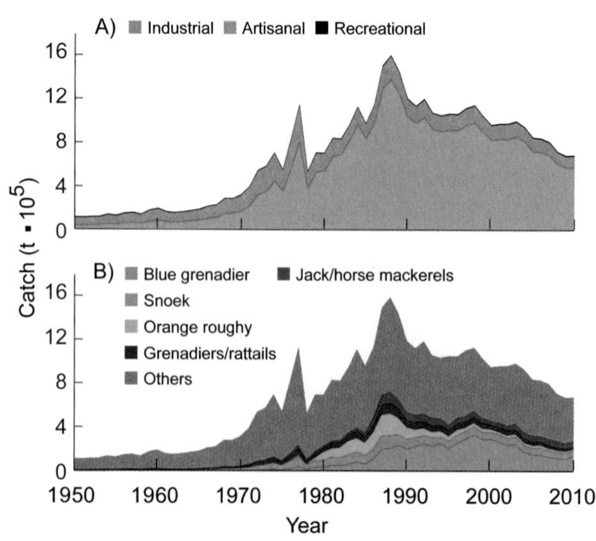

Figure 1. New Zealand (*sans* Kermadec Islands) has a land area of 269,400 km² and an EEZ of about 3,480,000 km².
Figure 2. Domestic catches in the EEZ of New Zealand: (A) by sector (domestic scores: Ind. 2, 2, 3; Art. 2, 2, 3; Sub. 2, 2, 2; Rec. 2, 2, 3; Dis. 1, 1, 1); (B) by taxon.

NEW ZEALAND (KERMADEC ISLANDS)[1]

Kyrstn Zylich, Sarah Harper, and Dirk Zeller

Sea Around Us, University of British Columbia, Vancouver, Canada

The Kermadec Islands are an isolated cluster of islands northeast of New Zealand's North Island, and are all uninhabited (except for the conservation staff on Raoul Island, the largest of the group). They represent the northernmost point of New Zealand, and their isolation has caused them to be exposed to little fishing. The isolation and subtropical location provides for a unique mix of tropical, subtropical, and temperate marine fauna. The EEZ surrounding the Kermadec Islands is nearly separated from New Zealand proper's EEZ (figure 1). Domestic (New Zealand) fishing started late, and total domestic catch from the Kermadec Islands' EEZ for the 1990–2010 period (Zylich et al. 2012) was less than 1,000 t (500 t excluding tuna and billfishes). Foreign fishing, predominantly by South Korea and Japan, began around the Kermadec Islands much earlier than domestic fishing and is bound to have had a greater impact on stocks (figure 2A). By 2010, there is apparently very little fishing around the Kermadec Islands (Zylich et al. 2012). Figure 2B presents top taxa as squids, albacore (*Thunnus alalunga*), and yellowfin tuna (*Thunnus albacares*). Indeed, the Kermadec EEZ has recently been declared a no-take marine reserve, complementing the terrestrial nature reserve and devoted to protecting its considerable marine biodiversity (Palomares et al. 2012), which figure 2B fails to reflect, given the selective nature of the few fisheries operating there.

REFERENCES

Palomares MLD, Harper S, Zeller D and Pauly D (2012) The marine biodiversity and fisheries catches of the Kermadec island group. A Report Prepared for the Global Ocean Project of the Pew Environment Group, University of British Columbia, Vancouver. 47 p.

Zylich K, Harper S and Zeller D (2012) Reconstruction of marine fisheries catches for the Kermadec Islands (1950–2010). pp. 61–67 In Harper S, Zylich K, Boonzaier L, Le Manach F, Pauly D and Zeller D (eds.), *Fisheries catch reconstructions: Islands, Part III*. Fisheries Centre Research Reports 20(5).

1. Cite as Zylich, K., S. Harper, and D. Zeller. 2016. New Zealand (Kermadec Islands). P. 349 in D. Pauly and D. Zeller (eds.), *Global Atlas of Marine Fisheries: A Critical Appraisal of Catches and Ecosystem Impacts*. Island Press, Washington, DC.

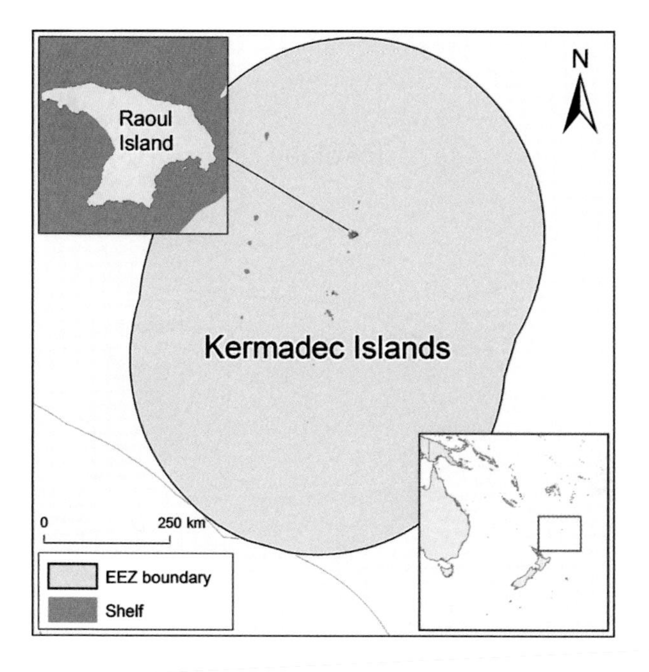

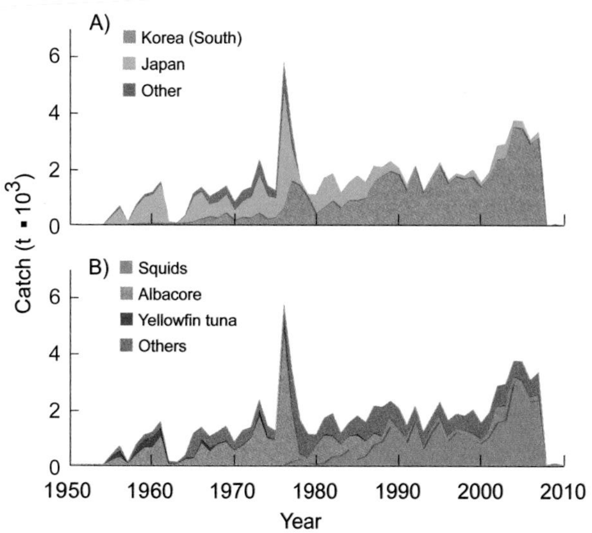

Figure 1. The Kermadec Islands have, jointly, a land area of 35 km² and an EEZ of about 622,000 km². **Figure 2.** Domestic and foreign catches taken in the EEZ of Kermadec Islands: (A) by fishing country (domestic scores: Ind. –, –, 3; Dis. –, –, 3; foreign catches are very uncertain); (B) by taxon.

NEW ZEALAND (TOKELAU)[1]

Kyrstn Zylich, Sarah Harper, and Dirk Zeller

Sea Around Us, University of British Columbia, Vancouver, Canada

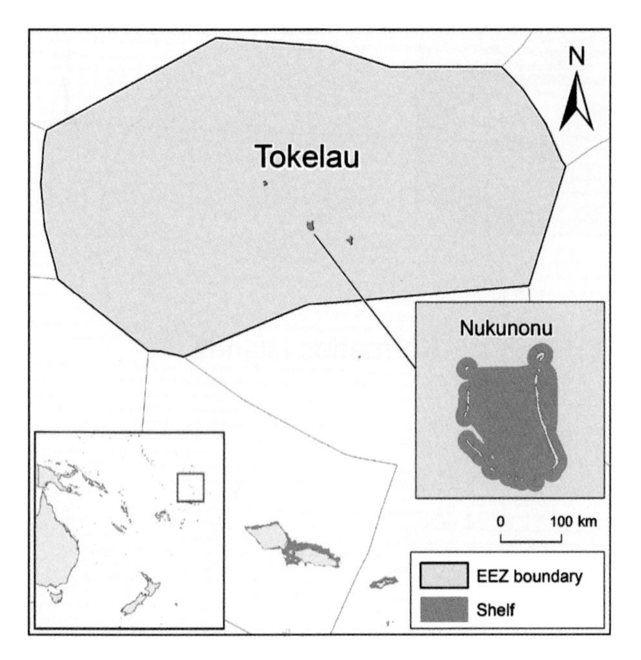

Figure 1. Tokelau (New Zealand) has a land area of only 10 km² but a large EEZ covering 319,000 km².

Tokelau, a territory of New Zealand, consists of 3 low-lying atolls, each consisting of many small islands and islets (figure 1). Because there are no domestic commercial fisheries (Chapman et al. 2005), the catch reconstruction by Zylich et al. (2011) represents the subsistence fisheries only, as estimated via per capita seafood consumption rates. These subsistence catches were consistently between 350 and 450 t/year from 1950 to 2010 and were 4.4 times that reported by the FAO on behalf of Tokelau. However, when all removals from within Tokelau's EEZ are considered, industrial fishing by foreign vessels, namely Japan and the United States, among others, strongly overshadows the small-scale sector (figure 2A, 2B). This catch is composed of large pelagic fishes, notably skipjack (*Katsuwonus pelamis*), albacore (*Thunnus alalunga*), yellowfin (*T. albacares*), and bigeye tuna (*T. obesus*), smaller pelagics often used as bait, flying fish (*Cypselurus* spp.), and a wide range of reef fishes and invertebrates (figure 2C). There is a general trend of human population decline from its peak in the 1960s. Countering the effect of this decline are the increased contacts between Tokelau and Samoa, which have led to an increase in informal exports. In 1980, Tokelauans began shipping frozen seafood to friends and family as gifts. The decrease in local demand, in combination with increasing amounts being sent overseas, has resulted in a nearly constant overall trend in catches.

REFERENCES

Chapman L, Des Rochers K and Pelasio M (2005) Survey of fishing activities in Tokelau. *SPC Fisheries Newsletter* 115. 36–40 p.

Zylich K, Harper S and Zeller D (2011) Reconstruction of fisheries catches for Tokelau (1950–2009). pp. 107–117 In Harper S and Zeller D (eds.), *Fisheries catch reconstructions: Islands, Part II*. Fisheries Centre Research Reports 19(4). [Updated to 2010]

1. Cite as Zylich, K., S. Harper, and D. Zeller. 2016. New Zealand (Tokelau). P. 350 in D. Pauly and D. Zeller (eds.), *Global Atlas of Marine Fisheries: A Critical Appraisal of Catches and Ecosystem Impacts*. Island Press, Washington, DC.

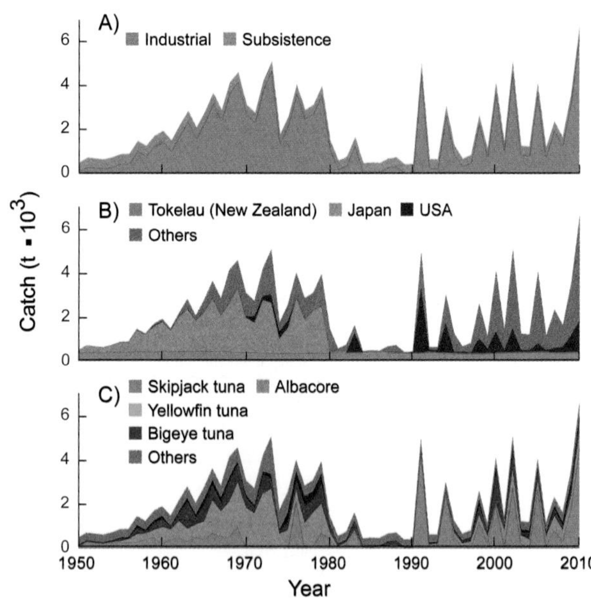

Figure 2. Domestic and foreign catches taken in the EEZ of Tokelau (New Zealand): (A) by sector (domestic scores: Sub. 1, 1, 2); (B) by fishing country (foreign catches are very uncertain); (C) by taxon.

NICARAGUA (CARIBBEAN)[1]

Andrea Haas, Sarah Harper, and Dirk Zeller

Sea Around Us, University of British Columbia, Vancouver, Canada

Nicaragua is the largest country in Central America and consists of 3 main geographic regions. The Pacific coastal region is the most economically developed, and the coastal plain's volcanic soils support the country's production of coffee, cotton, and sugar. Although the Caribbean lowlands make up more than half of the country's land area, they support less than 10% of its population. Despite high rainfall, rainforests, and swamps, the Mosquito Coast is named not after the insect but after the Miskito Indians (Nietschmann 1972). The domestic fisheries catch taken within this EEZ (figure 1) was reconstructed by Haas et al. (2015) as 4,800 t in 1950, which increased to a peak of 35,100 t in 1973, decreased to a low of 9,000 t in 1989, then increased to a second peak of 30,500 t in 1998 and declined to 24,000 t by 2010. Overall, catches were 3.6 times the data reported by the FAO on behalf of Nicaragua in the Caribbean (mainly because of discards, which are not considered by FAO). Catches were largely industrial, in addition to an artisanal and subsistence component (figure 2A). Total catches allocated to these waters are predominantly domestic, although some foreign fishing occurs (figure 2B). Figure 2C shows the catch composition to consist of snappers (family Lutjanidae), snooks (Centropomidae), and various invertebrates. Most of the fish catch originated as bycatch of the trawl fishery, targeting shrimps. Jameson et al. (2014) review the status of Miskitia's coral reefs.

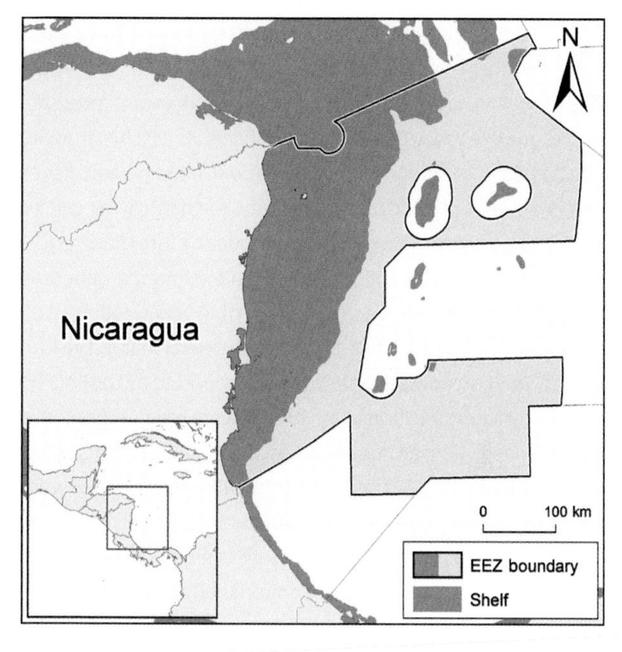

Figure 1. Nicaragua has a shelf of 61,300 km² and an EEZ of 161,000 km² on its Caribbean coast, or Mosquito Coast.

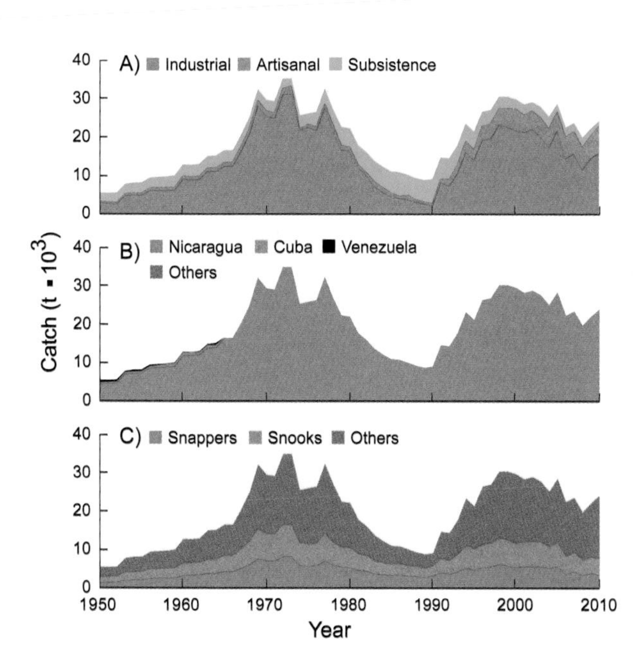

Figure 2. Domestic and foreign catches taken in the Caribbean EEZ of Nicaragua: (A) by sector (domestic scores: Ind. 2, 2, 2; Art. 2, 2, 2; Sub. 2, 1, 2; Dis. 3, 2, 3); (B) by taxon.

REFERENCES

Haas A, Harper S and Zeller D (2015) Reconstruction of Nicaragua's fisheries catches: 1950–2010. Fisheries Centre Working Paper #2015-23, 9 p.

Jameson S, AGGRA and CARICOMP (2014) Nicaragua. pp. 263–265 in J Jackson, M Donovan, K Kramer and VWY Lam (eds.), *Status and trends of Caribbean coral reefs: 1970–2012*. Global Coral Reef Monitoring Network, IUCN, Gland, Switzerland.

Nietschmann B (1972) Hunting and fishing focus among the Miskito Indians, eastern Nicaragua. *Human Ecology* 1(1): 41–67.

1. Cite as Hass, A., S. Harper, and D. Zeller. 2016. Nicaragua (Caribbean). P. 351 in D. Pauly and D. Zeller (eds.), *Global Atlas of Marine Fisheries: A Critical Appraisal of Catches and Ecosystem Impacts*. Island Press, Washington, DC.

NICARAGUA (PACIFIC)[1]

Andrea Haas, Sarah Harper, and Dirk Zeller

Sea Around Us, University of British Columbia, Vancouver, Canada

Nicaragua lies in the heart of Central America, with the Caribbean Sea to its east and the Pacific Ocean to its west (figure 1). This account, based on Haas et al. (2015), presents Nicaragua's reconstructed domestic fisheries catches from its Pacific EEZ. Excluding industrially caught large pelagic fishes (see chapter 3), catches increased from less than 2,500 t in 1950 to more than 7,800 t in 1965, then jumped to more than 15,000 t/year until the early 1980s. They then exhibited a generally lower pattern, but with a peak in 1999 of 19,200 t, followed by a decline, finally stabilizing at about 10,000 t/year in the late 2000s. This catch was 4 times the catch reported by the FAO on behalf of Nicaragua. Industrial catches were predominant, but generally declining, although artisanal and subsistence catches do exist (figure 2A). Furthermore, these allocated catches were mostly domestic, although substantial removals by the United States and others, were noted before the EEZ declaration in 1965 (figure 2B). Top taxa were snappers (family Lutjanidae) and grunts (Haemulidae; figure 2C). Many of the demersal fishes originated from the bycatch of the shrimp trawl fishery. Jentoft (1986) presents an illuminating review of Nicaraguan fisheries, with emphasis on attempts to overcome the legacy of the Somoza dictatorship (1936–1979) in the midst of a military conflict (Nicaragua's war against the U.S.-supported "contras"), given the limitations of foreign aid (here: Norwegian) under such conditions.

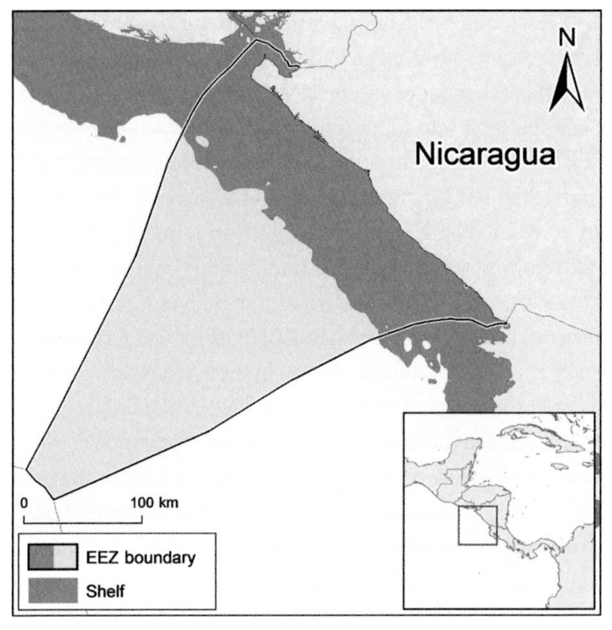

Figure 1. Nicaragua has a shelf of 21,900 km² and an EEZ of 61,500 km² on its Pacific coast.

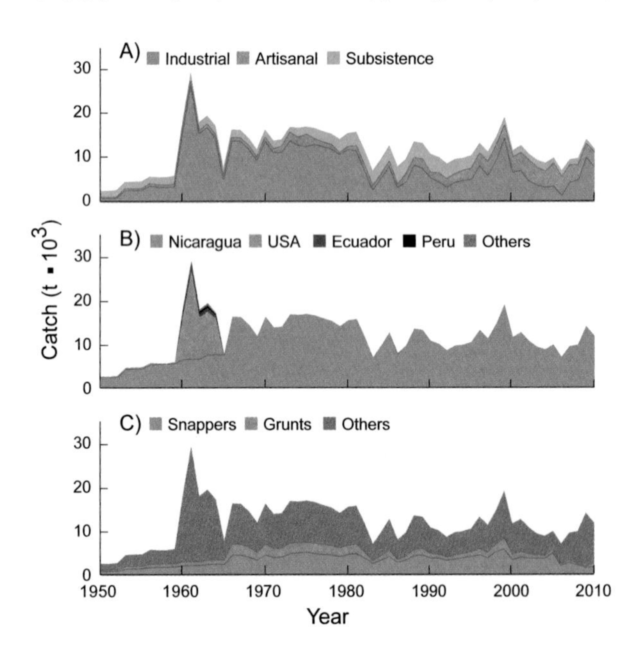

Figure 2. Domestic and foreign catches taken in the Pacific EEZ of Nicaragua: (A) by sector (domestic scores: Ind. 2, 2, 2; Art. 2, 2, 2; Sub. 2, 1, 2; Dis. 3, 2, 3); (B) by fishing country (foreign catches are very uncertain); (C) by taxon.

REFERENCES

Haas A, Harper S and Zeller D (2015) Reconstruction of Nicaragua's fisheries catches: 1950–2010. Fisheries Centre Working Paper #2015-23, 9 p.

Jentoft S (1986) Organizing fishery cooperatives: The case of Nicaragua. *Human Organization* 45(4): 353–358.

1. Cite as Hass, A., S. Harper, and D. Zeller. 2016. Nicaragua (Pacific). P. 352 in D. Pauly and D. Zeller (eds.), *Global Atlas of Marine Fisheries: A Critical Appraisal of Catches and Ecosystem Impacts*. Island Press, Washington, DC.

NIGERIA[1]

Lawrence Etim,[a] Dyhia Belhabib,[b] and Daniel Pauly[b]

[a]Department of Fisheries and Aquaculture, University of Uyo, Akwa Ibom State, Nigeria
[b]*Sea Around Us*, University of British Columbia, Vancouver, Canada

Nigeria, the most populous country in Africa (figure 1) is known for its diverse cultures and a controversial oil industry (see Ferguson 2012 for a fictional yet realistic account). The fisheries sector has grown since the 1950s (see Bayagbona 1965) and is now an important contributor to Nigeria's food security (Moses 2000). However, monitoring is highly variable between states. The artisanal and industrial sectors contribute most of the catches within Nigeria's EEZ; however, the subsistence component is notable, especially in the later time period (figure 2A). The reconstruction by Etim et al. (2015) estimated domestic catches of about 34,000 t in 1950, 540,000 t in 2005, and 490,000 t in 2010. Domestic catches were about 1.6 times the data supplied to the FAO, including some catches taken outside the Nigerian EEZ in the 1970s. Domestic catches form the majority of the removals, although foreign vessels, such as those from China, have been reported to fish in these waters (figure 2B). Increased illegal catches by foreign distant-water fleets endanger not only the food security of people in the region but also their livelihood, as local fishers, as occurred in Somalia, increasingly respond to this form of distant-water piracy with their own brand of piracy. Figure 2C presents croakers (*Pseudotolithus* spp.) and sardinellas (*Sardinella* spp.) as the largest contribution to allocated catches.

REFERENCES

Bayagbona EO (1965) The effect of fishing effort on croakers in the Lagos fishing ground. *Bulletin de l'Institut Français d'Afrique Noire XXVII*. 1(série A): 334–338.

Etim L, Belhabib D and Pauly D (2015) An overview of the Nigerian marine fisheries and a re-evaluation of its catch data for the years 1950–2010. pp. 66–76 In Belhabib D and Pauly D (eds.), *Fisheries catch reconstructions: West Africa, Part II*. Fisheries Centre Research Reports 23(3).

Ferguson W (2012) 419. Viking Canada, Toronto, 320 p.

Moses BS (2000) A review of artisanal marine and brackishwater fisheries of south-eastern Nigeria. *Fisheries Research* 47 (1): 81–92.

1. Cite as Etim, L., D. Belhabib, and D. Pauly. 2016. Nigeria. P. 353 in D. Pauly and D. Zeller (eds.), *Global Atlas of Marine Fisheries: A Critical Appraisal of Catches and Ecosystem Impacts*. Island Press, Washington, DC.

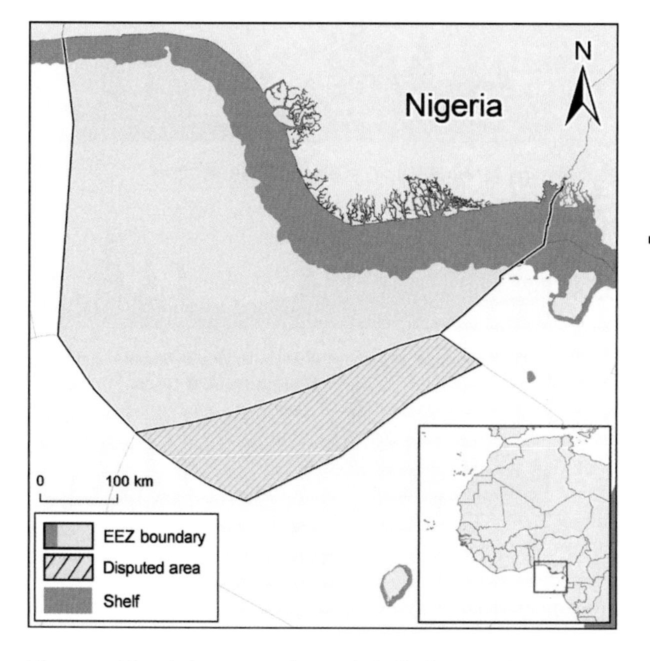

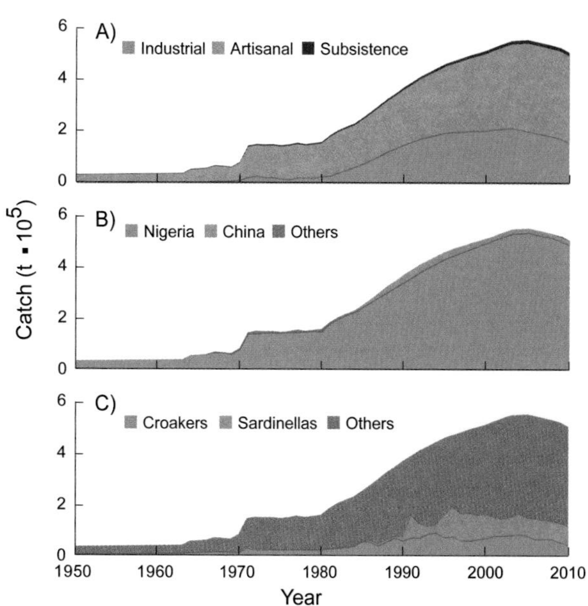

Figure 1. Nigeria has a continental shelf of 43,500 km² and an EEZ of 216,000 km². **Figure 2.** Domestic and foreign catches taken in the EEZ of Nigeria: (A) by sector (domestic scores: Ind. 1, 3, 3; Art. 2, 3, 3; Sub. 1, 3, 3; Dis. 1, 3, 3); (B) by fishing country (foreign catches are very uncertain); (C) by taxon.

NIUE[1]

Kyrstn Zylich, Sarah Harper, Nicolas Winkler, and Dirk Zeller

Sea Around Us, University of British Columbia, Vancouver, Canada

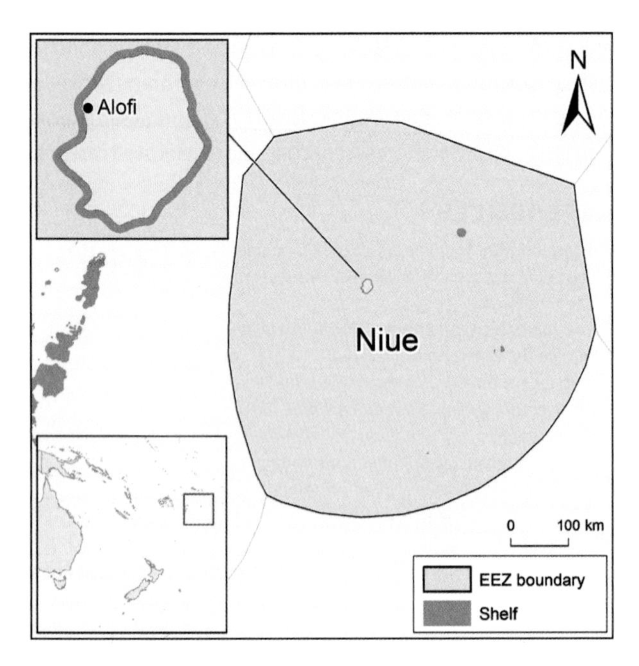

Figure 1. Niue has a land area of 260 km² and an EEZ of 317,000 km².

Niue, a single, uplifted atoll island surrounded by a narrow coral reef, which drops off to extreme depths within 3–5 km from shore, became in 1974 a self-governing entity in free association with New Zealand (Kronen et al. 2008). Because of emigration to New Zealand, the population of Niue declined substantially from its peak in 1960. This led to a decline of the subsistence fishery (figure 2A), in which women played a prominent role, at the same time that the data reported to the FAO, covering only the commercial fisheries, suggested increasing catches. The reconstruction of domestic catches by Zylich et al. (2012) suggested nonindustrial catches of more than 500 t/year in the 1950s and 1960s and 260 t in 2010, nearly 5 times the data FAO reports on behalf of Niue. However, this huge discrepancy appears to have been overcome in more recent years. Foreign industrial tuna fisheries feature large in Niue's EEZ waters (figure 2B, estimation uncertain). The total catches in the EEZ were dominated by large pelagic fishes (i.e., tuna), largely caught by foreign fleets, although taxa caught by small-scale sectors such as the rough turban (*Turbo setosus*, a sea snail) and soldierfish (family Holocentridae) also feature prominently, especially in earlier years (figure 2C). Measures are needed to manage the increasing strains caused by the recently developed industrial fishery and to ensure its sustainability.

REFERENCES

Kronen M, Fisk D, Pinca S, Magron F, Friedman K, Boblin P, Awira R and Chapman L (2008) Niue country report: Profile and results from in-country survey work. Secretariat of the Pacific Community (SPC), Nouméa (New Caledonia). xvi + 173 p.

Zylich K, Harper S, Winkler N and Zeller D (2012) Reconstruction of marine fisheries catches for Niue (1950–2010). pp. 77–86 In Harper S, Zylich K, Boonzaier L, Le Manach F, Pauly D and Zeller D (eds.), *Fisheries catch reconstructions: Islands, Part III*. Fisheries Centre Research Reports 20(5).

1. Cite as Zylich, K., S. Harper, N. Winkler, and D. Zeller. 2016. Niue. P. 354 in D. Pauly and D. Zeller (eds.), *Global Atlas of Marine Fisheries: A Critical Appraisal of Catches and Ecosystem Impacts*. Island Press, Washington, DC.

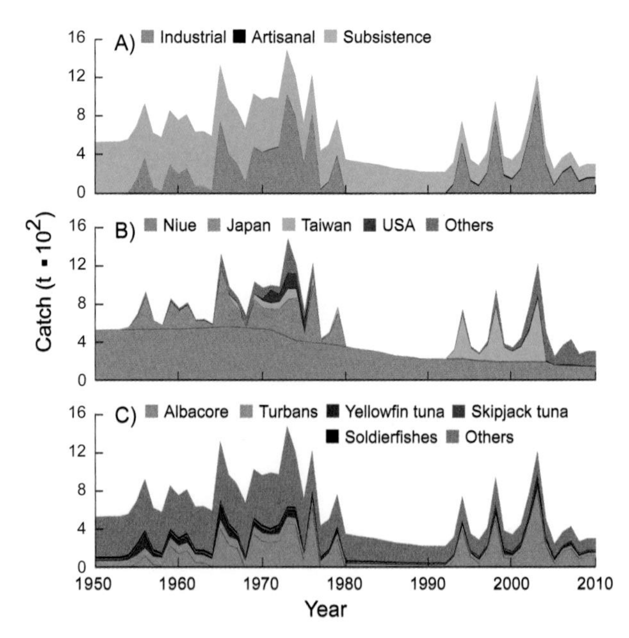

Figure 2. Domestic and foreign catches taken in the EEZ of Niue: (A) by sector (domestic scores: Art. 1, 1, 3; Sub. 1, 1, 2); (B) by fishing country (foreign catches are very uncertain); (C) by taxon.

NORWAY[1]

Kjell Nedreaas,[a] Svein A. Iversen,[a] and Grete Kuhnle[b]

[a]Institute of Marine Research, Bergen, Norway [b]Directorate of Fisheries, Bergen, Norway

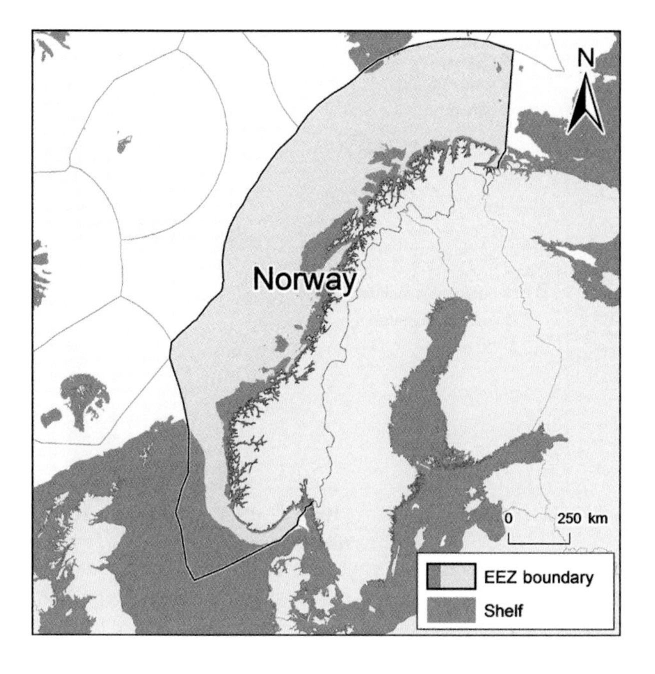

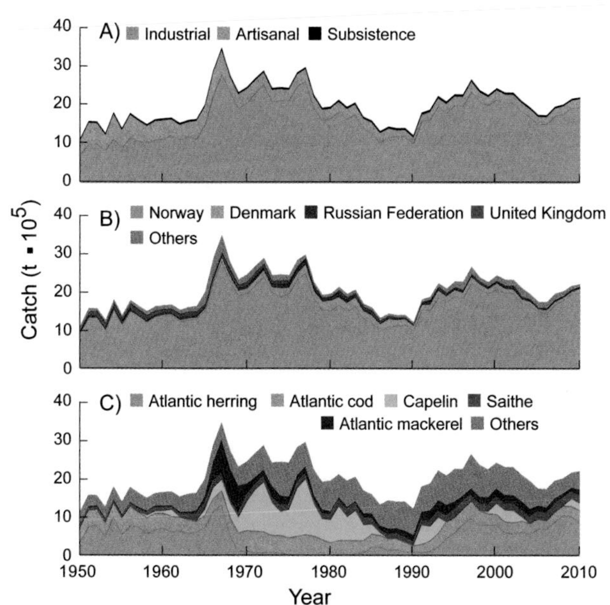

Figure 1. Norway's mainland has a shelf area of 211, 000 km², and its EEZ is 935,000 km². **Figure 2.** Domestic and foreign catches taken in the EEZ of Norway: (A) by sector (domestic scores: Ind. 2, 3, 4; Art. 2, 3, 4; Sub. 2, 2, 2; Rec. –, 1, 1; Dis. 2, 3, 3); (B) by fishing country (foreign catches are very uncertain); (C) by taxon.

Norway, which abuts an extremely productive marine ecosystem (Skjoldal 2004), is a pioneer in many aspects of fisheries, including their governance. Thus, a ban (established in 1987 and extended in 2008) makes it illegal for commercial and most other bycatch species to be discarded. A variety of other management and enforcement measures exist. Domestic and foreign catches in Norway's mainland EEZ (figure 1) are dominated by industrial fisheries, although artisanal catches were prominent in earlier years (figure 2A). The catch reconstruction of Nedreaas et al. (2015), which yielded domestic catches of just over 1 million t/year in the early 1950s and about 2 million t/year in the late 2000s, thus includes far less discard for the more recent time periods than one would expect for highly industrialized fisheries. These catches are only about 4% higher than those reported by the ICES for Norway, mainly because of earlier discards and because of the sport and subsistence fishery for cod (*Gadus morhua*), which is not considered in official statistics (Vølstad et al. 2011). Fisheries in this EEZ are dominated by Norway, although other countries fish also in these waters (figure 2B). Major species include herring (*Clupea harengus*), cod (*Gadus morhua*), capelin (*Mallotus villosus*), saithe (*Pollachius virens*), and Atlantic mackerel (*Scomber scombrus*; figure 2C). Another aspect of Norwegian fisheries policy is its success in rebuilding previously depleted stocks, which should be emulated worldwide.

REFERENCES

Nedreaas K, Iversen S and Kuhnle G (2015) Preliminary estimates of total removals by the Norwegian marine fisheries, 1950–2010. Fisheries Centre Working Paper #2015–94, 15 p.

Skjoldal HR (2004) *The Norwegian Sea Ecosystem*. Tapir Academic Press, Trondheim, 559 p.

Vølstad JH, Korsbrekke K, Nedreaas KH, Nilsen M, Nilsson GN, Pennington M, Subbey S, and Wienerroither R (2011) Probability-based surveying using self-sampling to estimate catch and effort in Norway's coastal tourist fishery. *ICES Journal of Marine Science*, 68(8): 1785–1791.

1. Cite as Nedreaas, K, S. Iversen, and G. Kuhnle. 2016. Norway. P. 355 in D. Pauly and D. Zeller (eds.), *Global Atlas of Marine Fisheries: A Critical Appraisal of Catches and Ecosystem Impacts*. Island Press, Washington, DC.

NORWAY (BOUVET ISLAND)[1]

Allan Padilla, Dirk Zeller, and Daniel Pauly

Sea Around Us, University of British Columbia, Vancouver, Canada

Bouvet Island is an isolated and uninhabited sub-Antarctic volcanic island in the South Atlantic (figure 1), covered to more than 90% by a glacier, and was officially declared a Norwegian dependency in 1930. It is deemed the most remote island in the world, being about 2,600 km southwest of South Africa and about 1,700 km north of Antarctica, and is known as the location for a cult movie (*Aliens vs. Predators*). The catch reconstruction by Padilla et al. (2015), based on data from the Commission for the Conservation of Antarctic Marine Living Resources (CCAMLR), Kock (1992), and other sources, includes the longline fisheries from 1977 to 2010, their discards, and unreported catches. According to CCAMLR, neither krill (*Euphausia suberba*) nor lanternfish (Myctophidae) are caught around Bouvet Island, and therefore they were not included in the reconstruction. Russia (as part of the former Soviet Union) was deemed to have a short-lived peak catch in this EEZ in 1970 (figure 2A), but otherwise, average total catches were negligible, rarely exceeding 1,000 t/year, although in recent years South Korea has increased their catches in these waters. Small as these catches were, they were largely unreported by the CCAMLR Statistical Bulletin for subarea 48.6. The reconstructed catch, which omits likely illegal catches, consisted mainly of marbled rockcod (*Notothenia rossii*) in the early years (figure 2B) and Antarctic toothfish (*Dissostichus mawsoni*) and Patagonian toothfish (*Dissostichus eleginoides*) in recent years.

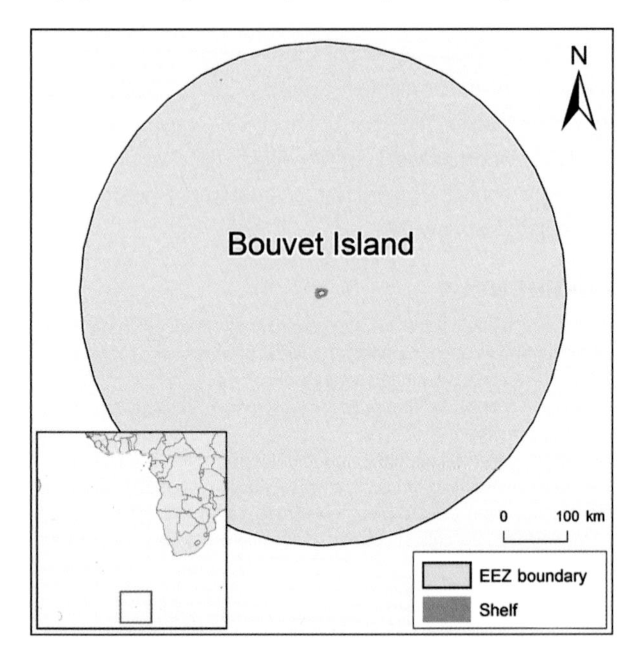

Figure 1. Bouvet's land area is 49 km² and is a nature reserve, although more than 90% of its surface is covered by ice; its EEZ is 441,000 km².

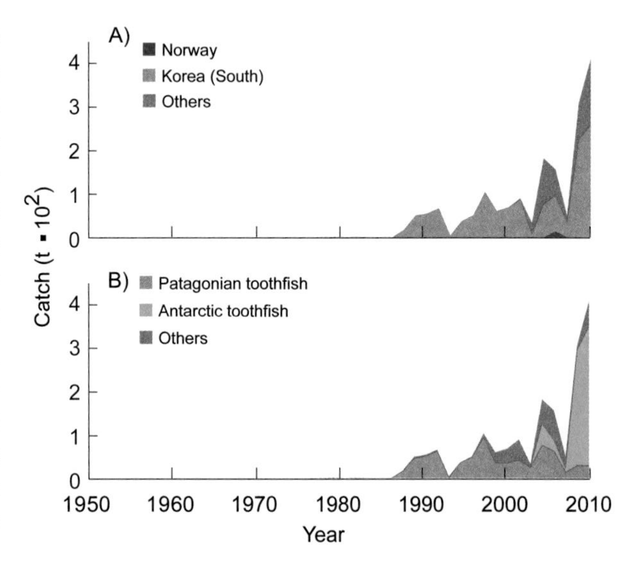

Figure 2. Domestic and foreign catches taken in the EEZ of Bouvet Island: (A) by fishing country (scores: Ind. –, –, 4; Dis. –, –, 2); (B) by taxon. Note that original data were for CCAMLR's "season" (from December 1 of a given year to November 30 of the next year), were extracted on a monthly basis, and were reaggregated to calendar years.

REFERENCES

Kock KH (1992) Antarctic fish and fisheries. p. 359 In *Studies in Polar Research*. Cambridge University Press, Cambridge and New York.

Padilla A, Zeller D and Pauly D (2015) The fish and fisheries of Bouvet Island. pp. 20–29 In Palomares MLD and Pauly D (eds.), *Marine fisheries catches of Sub-Antarctic islands, 1950 to 2010*. Fisheries Centre Research Reports 23(1).

1. Cite as Padilla, A., D. Zeller, and D. Pauly. 2016. Norway (Bouvet Island). P. 356 in D. Pauly and D. Zeller (eds.), *Global Atlas of Marine Fisheries: A Critical Appraisal of Catches and Ecosystem Impacts*. Island Press, Washington, DC.

NORWAY (JAN MAYEN)[1]

Kjell Nedreaas,[a] Svein A. Iversen,[a] and Grete Kuhnle[b]

[a]Institute of Marine Research, Bergen, Norway [b]Directorate of Fisheries, Bergen, Norway

Jan Mayen is a Norwegian volcanic island in the Arctic Ocean 600 km northeast of Iceland (figure 1). There are important fisheries resources in the waters around the island, and oil and gas deposits are suspected below the seafloor, which contributed to sovereignty disputes with Denmark, settled in 1988 largely in Denmark's favor. Nevertheless, the EEZ (more correctly a "fishery zone") around the island is large, allowing substantial fisheries to exist. The catch reconstruction for all Norwegian fisheries by Nedreaas et al. (2015) included Norwegian catches in the waters around Jan Mayen and was supplemented here by foreign catches from ICES by statistical areas (Jan Mayen falls mainly in ICES area IIa2 and XIVa, with minor areas in IIb2). Based on the reconstruction and *Sea Around Us* spatial catch allocation, catches around Jan Mayen, mainly by Danish vessels, were higher before the declaration of the EEZ ("fishery zone") in 1980, and in more recent years, mainly Iceland, Norway, and some EU vessels have fished in these waters (figure 2A). A wide range of taxa are caught, although cod (*Gadus morhua*) and haddock (*Melanogrammus aeglefinus*) featured prominently in early years, whereas Atlantic mackerel (*Scomber scombrus*) and Greenland halibut (*Reinhardtius hippoglossoides*) are important more recently (figure 2B). Expansion of fisheries, such as consideration of zooplankton fisheries (e.g., Tiller 2008) must be examined very carefully by Norway and others.

REFERENCES

Nedreaas K, Iversen S and Kuhnle G (2015) Preliminary estimates of total removals by the Norwegian marine fisheries, 1950–2010. Fisheries Centre Working Paper #2015-94, 15 p.

Tiller RG (2008) The Norwegian system and the distribution of claims to redfeed. *Marine Policy* 32(6): 928–940.

1. Cite as Nedreaas, K., S. Iversen, and G. Kuhnle. 2016. Norway (Jan Mayen). P. 357 in D. Pauly and D. Zeller (eds.), *Global Atlas of Marine Fisheries: A Critical Appraisal of Catches and Ecosystem Impacts*. Island Press, Washington, DC.

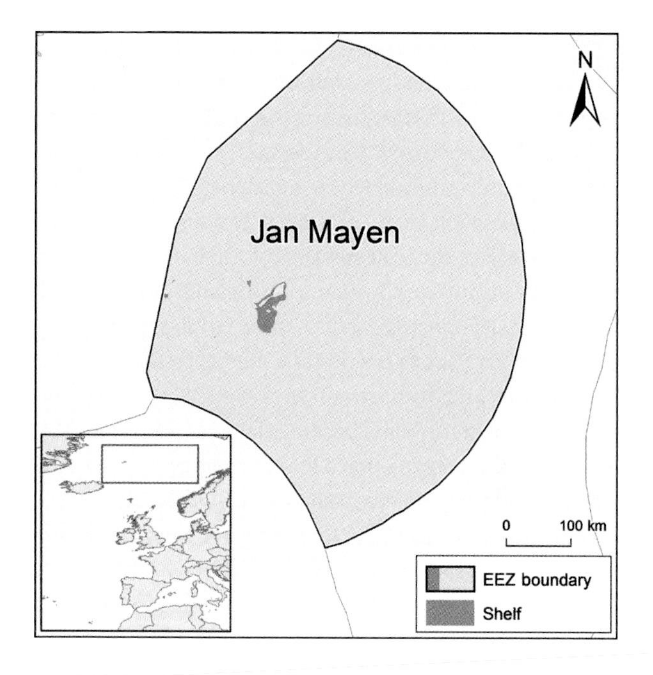

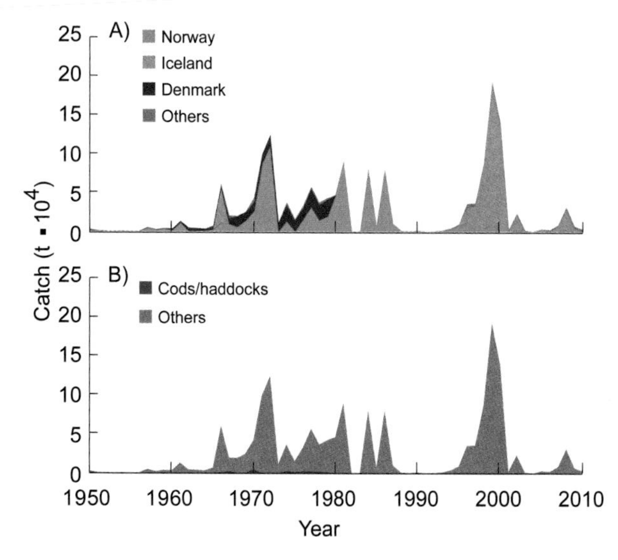

Figure 1. Jan Mayen's land area is 373 km²; its EEZ (fishery zone) is 292,000 km². **Figure 2.** Domestic and foreign catches taken in the EEZ (fishery zone) of Jan Mayen (Norway): (A) by fishing country (domestic scores: Ind. 2, 4, 4; Dis. 1, 2, 2; foreign catches are very uncertain); (B) by taxon.

NORWAY (SVALBARD)[1]

Kjell Nedreaas,[a] Svein A. Iversen,[a] and Grete Kuhnle[b]

[a]Institute of Marine Research, Bergen, Norway [b]Directorate of Fisheries, Bergen, Norway

The Archipelago of Svalbard, belonging to Norway and located between 74° and 81° north, about halfway between mainland Norway and the North Pole (figure 1), consists of 3 major islands (Spitsbergen, Nordaustlandet, and Edgeøya) and is surrounded by a Fisheries Protection Zone (closely equivalent to an EEZ). Norway maintains rigorous fisheries policies aimed at sustainable fisheries, although its claims are disputed by Russia. These islands were used as base by whalers since the 17th century, but this later shifted to coal mining (Russia maintains a mining settlement) and fishing. The catch reconstruction by Nedreaas et al. (2015) was supplemented by data from ICES by areas (ICES area IIb2). Based on the reconstruction and *Sea Around Us* spatial catch allocation, catches around Svalbard peaked in the early 1980s, with the majority taken by Norwegian and Russian vessels, although European vessels also fish in this zone (figure 2A). Although many taxa are caught, capelin (*Mallotus villosus*), cod (*Gadus morhua*), Atlantic herring (*Clupea harengus*), and haddock (*Melanogrammus aeglefinus*) contribute most to catches (figure 2B). Issues surrounding the potential and concerns of a zooplankton "redfeed" fishery (*Calanus finmarchicus*; Tiller 2011) are not considered here. Interestingly, Svalbard is also the location of the Global Seed Vault, which provides an environmentally safe location for protection of global plant diversity. It is funded by the Norwegian government, the Global Crop Diversity Trust, and philanthropic contributions.

REFERENCES

Nedreaas K, Iversen S and Kuhnle G (2015) Preliminary estimates of total removals by the Norwegian marine fisheries, 1950–2010. Fisheries Centre Working Paper #2015–94, 15 p.

Tiller R (2011) Institutionalizing the High North: Will the harvest of redfeed be a critical juncture for the solidification of the Svalbard Fisheries Protection Zone? *Ocean & Coastal Management* 54(5): 374–380.

1. Cite as Nedreaas, K., S. Iversen, and G. Kuhnle. 2016. Norway (Svalbard). P. 358 in D. Pauly and D. Zeller (eds.), *Global Atlas of Marine Fisheries: A Critical Appraisal of Catches and Ecosystem Impacts*. Island Press, Washington, DC.

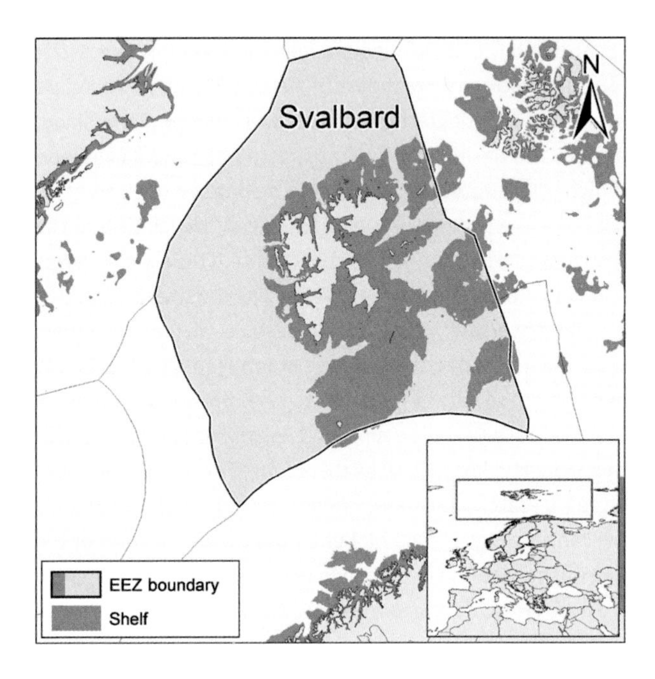

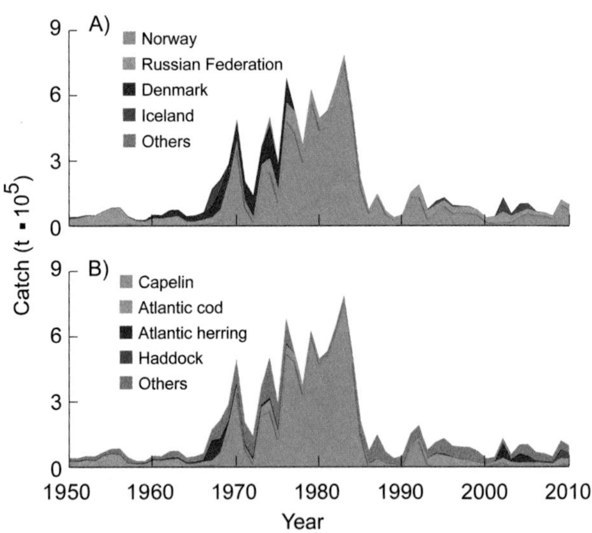

Figure 1. Svalbard's land area is 61,020 km², and its EEZ (Fisheries Protection Zone) is 805,000 km².

Figure 2. Domestic and foreign catches taken in the EEZ (Fisheries Protection Zone) of Svalbard (Norway): (A) by fishing country (domestic scores: Ind. 2, 3, 3; Dis. 2, 2, 2; foreign catches are very uncertain); (B) by taxon.

OMAN[1]

Myriam Khalfallah,[a] Hussein Al-Masroori,[b] Anesh Govender,[b]
Kyrstn Zylich,[a] Dirk Zeller,[a] and Daniel Pauly[a]

[a]*Sea Around Us*, University of British Columbia, Vancouver, Canada [b]Department of Marine Science & Fisheries
College of Agricultural and Marine Sciences, Sultan Qaboos University, Al-Khod, Sultanate of Oman

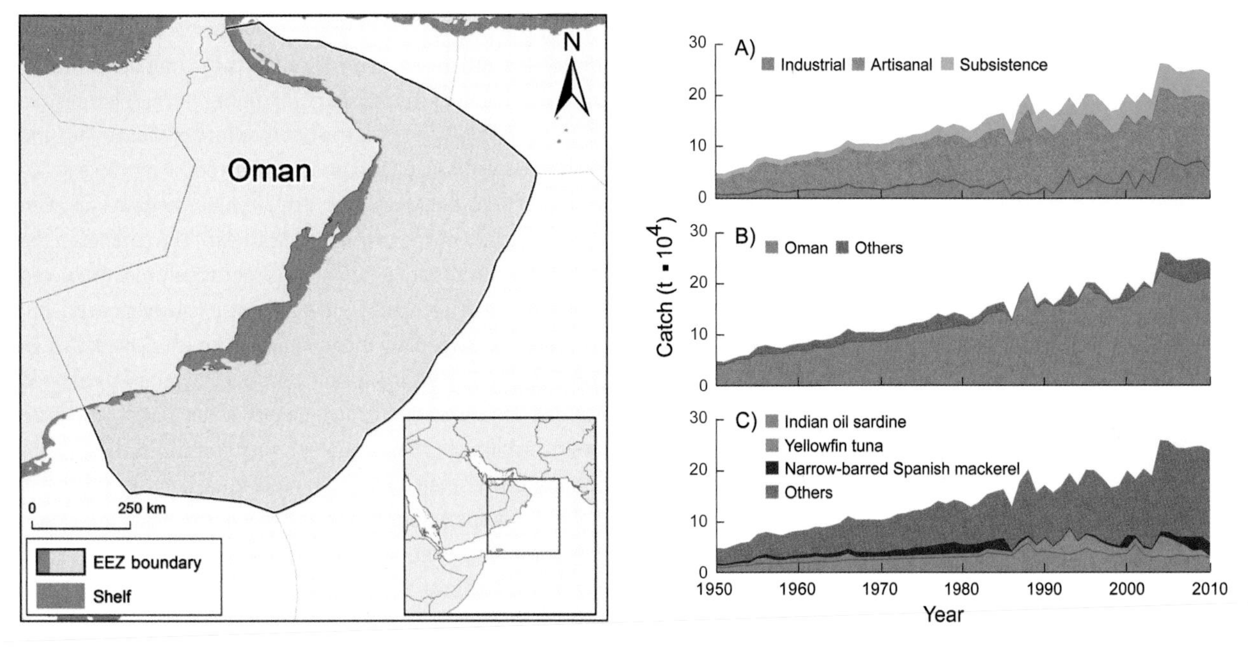

Figure 1. Oman (excluding the Musandam Peninsula) has 47,900 km² of shelf and an EEZ of 530,000 km². **Figure 2.** Domestic and foreign catches taken in the EEZ of Oman (excluding the Musandam Peninsula): (A) by sector (domestic scores: Ind. –, 3, 3; Art. 2, 3, 3; Sub. 2, 2, 2; Rec. 1, 1, 1; Dis. 1, 2, 2); (B) by fishing country (foreign catches are very uncertain); (C) by taxon.

The Sultanate of Oman is located in the southeast of the Arabian Peninsula (figure 1). Oman's neighbors are the United Arab Emirates (UAE), Saudi Arabia, and Yemen, and it fronts the Arabian Sea and the Gulf of Oman. Oman is divided into 11 governorates, including the Musandam Peninsula (see next page), an Omani exclave separated from the rest of the country by the United Arab Emirate of Fujairah. The total fisheries catch in Oman's EEZ (excluding Musandam), as reconstructed by Khalfallah et al. (2015), based on Fisheries Statistic Department (2014), Alhabsi and Mustapha (2011), and other sources and allocated by the *Sea Around Us*, increased steadily over the years and is dominated by the artisanal sector (figure 2A). Indian access to Oman's waters seems to have ceased in the early 1980s with the declaration of the EEZ, although foreign fishing seems to have increased more recently (figure 2B). The reconstructed domestic catch of Oman (including Musandam) is 1.3 times the catch reported by the FAO on behalf of Oman. Pelagic species, such as Indian oil sardine (*Sardinella longiceps*) and narrow-barred Spanish mackerel (*Scomberomorus commerson*), and tuna feature prominently in catches (figure 2C). Bottom trawling was banned in 2011, in an effort by the Omani authorities to privilege the artisanal fisheries. Wise decisions such as this are rare in the region and suggest that the Omani authorities are serious about attempting to ensure the sustainable use of their renewable resources.

REFERENCES

Alhabsi MS, and Mustapha NHN (2011) Fisheries sustainability in Oman. *Journal of Economics and Sustainable Development* 2(7): 35–48.

Fisheries Statistic Department (2014) Fisheries Statistics Book 2013. Ministry of Agriculture and Fisheries, Oman. 246 p.

Khalfallah M, Zylich K, Zeller D and Pauly D (2015) Reconstruction of marine fisheries catches for Oman (1950–2010). Fisheries Centre Working Paper #2015–89, 11 p.

1. Cite as Khalfallah, M., H. Al-Masroori, A. Govender, K. Zylich, D. Zeller, and D. Pauly. 2016. Oman. P. 359 in D. Pauly and D. Zeller (eds.), *Global Atlas of Marine Fisheries: A Critical Appraisal of Catches and Ecosystem Impacts*. Island Press, Washington, DC.

OMAN (MUSANDAM PENINSULA)[1]

Myriam Khalfallah and Daniel Pauly[a]

Sea Around Us, University of British Columbia, Vancouver, Canada

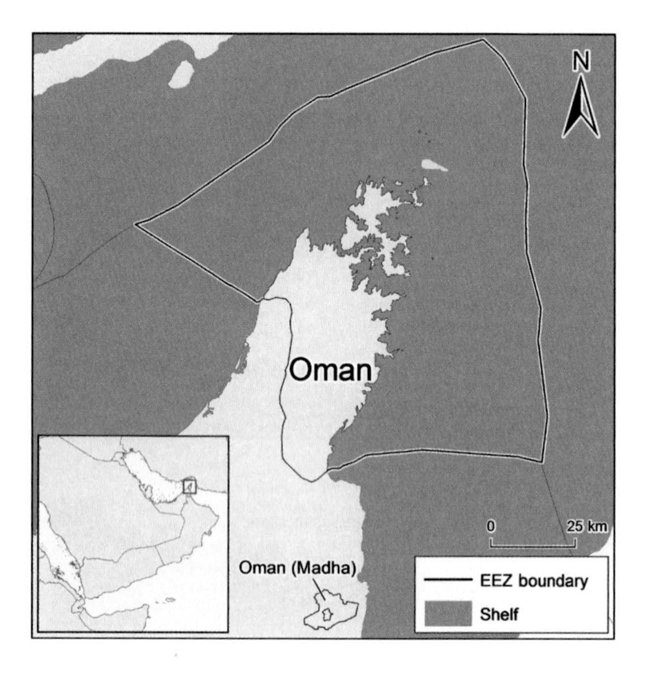

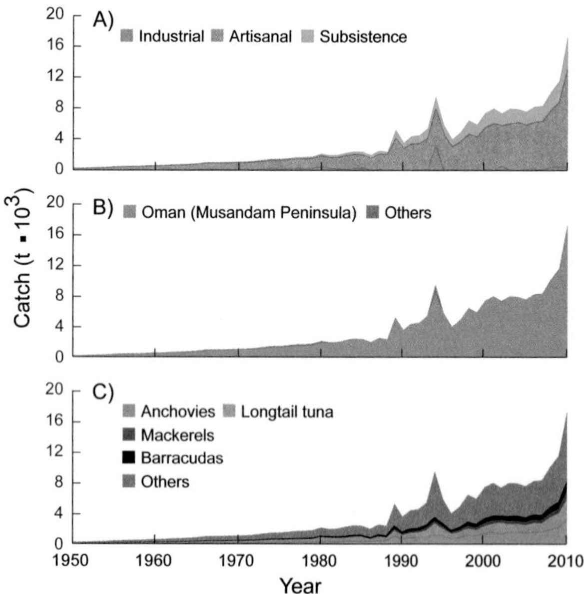

Figure 1. The Musandam Peninsula (Oman) has 6,680 km² of shelf and an EEZ of 6,680 km². **Figure 2.** Domestic and foreign catches taken in the EEZ of Musandam Peninsula (Oman): (A) by sector (domestic scores: Art. 2, 2, 3; Sub. 1, 1, 1; Rec. 1, 1, 1; Dis. 1, 1, 1); (B) by fishing country (foreign catches are very uncertain); (C) by taxon.

The Musandam Peninsula is an exclave of Oman surrounded by the United Arab Emirates (UAE) to the west and south and jutting into the strategically important Strait of Hormuz at the eastern end of the Persian Gulf (figure 1). This picturesque region, officially known as the Governorate of Musandam and inhabited by about 31,000 people, itself possesses an exclave in the UAE, called Madah (75 km²), which contains an even smaller exclave of the UAE called Mahwah. The catches in the EEZ of the Musandam Peninsula, as tentatively reconstructed by Khalfallah et al. (2015), are almost exclusively artisanal and seem to be rapidly increasing (figure 2A). The artisanal dominance is likely to continue, because trawling was banned from all Omani waters in 2011 (see previous page), and little foreign fishing seems to be occurring (figure 2B). Catches were dominated by small and medium pelagics, such as anchovies (*Stolephorus* spp.), longtail tuna (*Thunnus tonggol*), mackerel (e.g., narrow-barred Spanish mackerel, *Scomberomorus commerson*; McIlwain et al. 2005), and barracuda (*Sphyraena* spp.), among others (figure 2C). Cornelius et al. (1973) wrote that "the impression was gained that the area was not overfished, but it is not certain that large increases in fishing for inshore fish in the immediate vicinity of the Musandam peninsula could be supported." The large increases have occurred, but we are still uncertain about their sustainability.

REFERENCES

Cornelius PFS, Flacon NL, South D and Vita-Finzi C (1973) The Musandam Expedition 1971–2 scientific results. *The Geographical Journal* 139(3): 400–440.

Khalfallah M, Zylich K, Zeller D and Pauly D (2015) Reconstruction of marine fisheries catches for Oman (1950–2010). Fisheries Centre Working Paper #2015-89, 11 p.

McIlwain JL, Claereboudt MR, Al-Oufi HS, Zaki S and Goddard JS (2005) Spatial variation in age and growth of the kingfish *Scomberomorus commerson* in the coastal waters of the Sultanate of Oman. *Fisheries Research* 73(3): 283–298.

1. Cite as Khalfallah, M., and D. Pauly. 2016. Oman (Musandam Peninsula). P. 360 in D. Pauly and D. Zeller (eds.), *Global Atlas of Marine Fisheries: A Critical Appraisal of Catches and Ecosystem Impacts*. Island Press, Washington, DC.

PAKISTAN[1]

Claire Hornby,[a] M. Moazzam Khan,[b] Kyrstn Zylich,[a] and Dirk Zeller[a]

[a]*Sea Around Us*, University of British Columbia, Vancouver, Canada [b]Former Head of the Marine Fisheries Department, Government of Pakistan, Karachi Pakistan; currently WWF-Pakistan, Karachi, Pakistan

Pakistan is a South Asian country on the Arabian Sea, abutting India in the east and Iran in the west (figure 1). Perhaps because of its perennial political instability, little attention has been devoted to small-scale fishery development. Instead, industrial fishing became a dominant force that seems now difficult to control. Thus, the reconstruction by Hornby et al. (2014) of Pakistan's marine catches, complemented by spatially allocated foreign catches, suggest that artisanal fisheries contribute slightly more to total catches than the industrial sector (figure 2A; see also Siddiqi 1992). Apparent serious issues in the statistical data collection system, which largely underrepresents the small-scale sectors, contributed to the reconstructed catches being 2.6 times the landings reported by the FAO on behalf of Pakistan. Figure 2B suggests that although some foreign fishing occurs in Pakistan's EEZ, most catches are domestic. Overall, catches are dominated by small pelagics such as Indian oil sardines (*Sardinella longiceps*), although shrimp (family Penaeidae) featured prominently in earlier years and Indian mackerel (*Rastrelliger kanagurta*) more recently (figure 2C). The study by Hornby et al. (2014), performed with the support of the FAO regional representation in Pakistan, illustrates the need for improved estimation and reporting of catches for all fisheries sectors and greater consideration of the small-scale (artisanal and subsistence) sector, which is crucial for providing food security and employment to the fast-growing population of Pakistan.

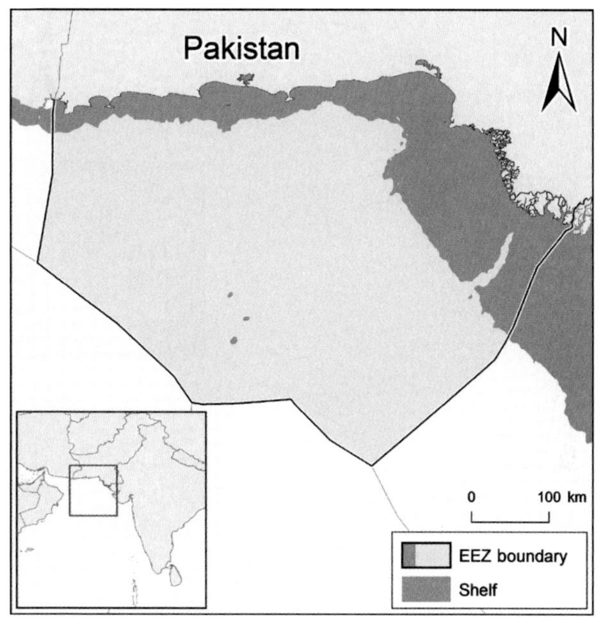

Figure 1. Pakistan and its EEZ (222,000 km²), of which 50,200 km² is shelf, with the Sind coast in the east and the Baluchistan coast in the west.

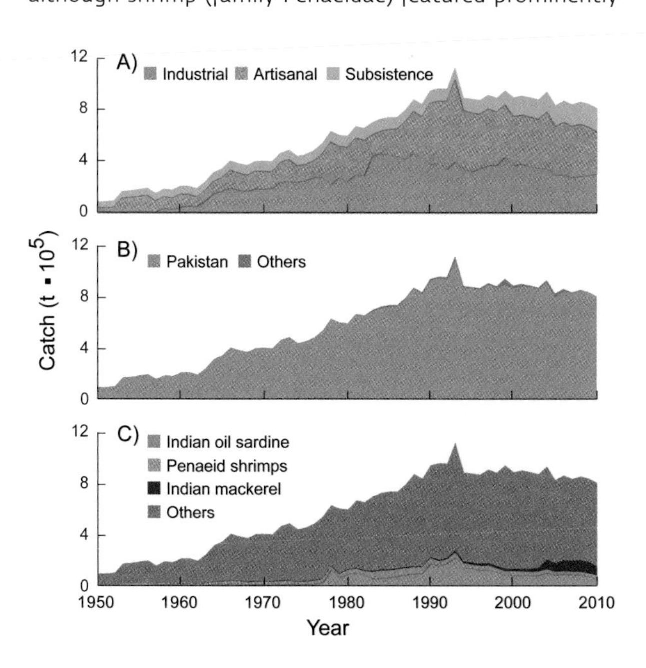

Figure 2. Domestic and foreign catches taken in the EEZ of Pakistan: (A) by sector (domestic scores: Ind. 2, 3, 3; Art. 2, 2, 2; Sub. 2, 2, 2; Rec. 1, 1, 1; Dis. 2, 2, 2); (B) by fishing country (foreign catches are very uncertain); (C) by taxon.

REFERENCES

Hornby C, Moazzam M, Zylich K and Zeller D (2014) Reconstruction of Pakistan's marine fisheries catches (1950–2010). Fisheries Centre Working Paper #2014–28, 54 p.

Siddiqi AH (1992) Fishery resources and development policy in Pakistan. *GeoJournal* 26(3): 395–411.

1. Cite as Hornby, C., M. Moazzam Khan, K. Zylich, and D. Zeller. 2016. Pakistan. P. 361 in D. Pauly and D. Zeller (eds.), *Global Atlas of Marine Fisheries: A Critical Appraisal of Catches and Ecosystem Impacts*. Island Press, Washington, DC.

PALAU[1]

Stephanie Lingard,[a] Sarah Harper,[a] Yoshi Ota,[b] and Dirk Zeller[a]

[a]*Sea Around Us*, University of British Columbia, Vancouver, Canada
[b]Nereus Program, University of British Columbia, Vancouver, Canada

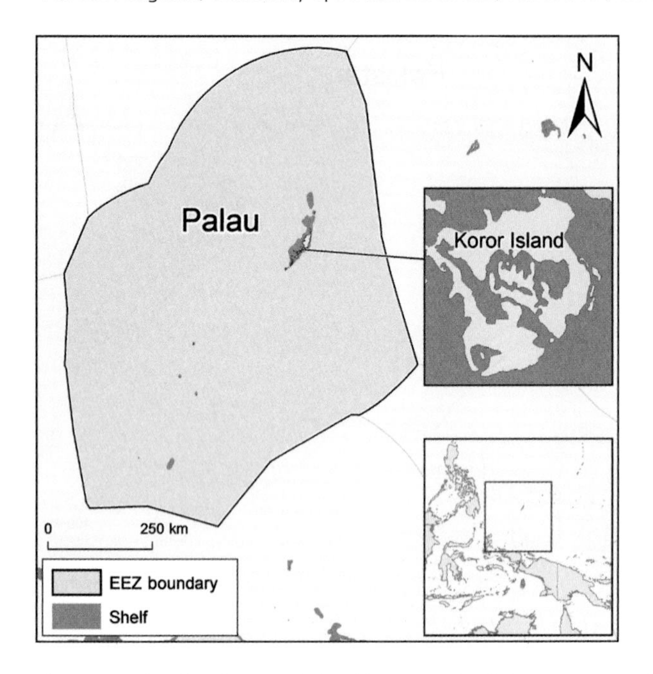

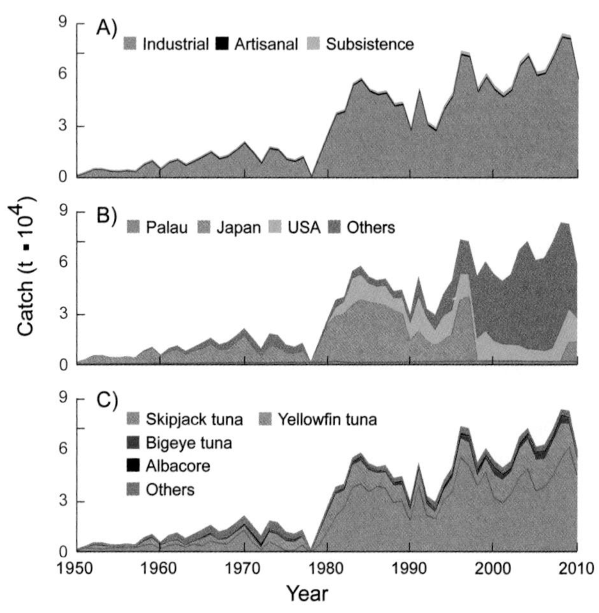

Figure 1. Palau (land area 466 km²) has an EEZ of about 604,200 km². **Figure 2.** Domestic and foreign catches taken in the EEZ of Palau: (A) by sector (domestic scores: Ind. 3, 3, 3; Art. 2, 2, 3; Sub. 1, 2, 2); (B) by fishing country (foreign catches are very uncertain); (C) by taxon.

Palau, an archipelagic country 500 km east of the Philippines (figure 1), consists of 340 islands, of which only the 5 major ones are inhabited. Palau maintained a predominantly traditional lifestyle, with subsistence fishing being a primary occupation for the majority of the population up until the post–World War II modernization of the 1950s. Lingard et al. (2011) estimated Palau's domestic marine fisheries catches, consisting of subsistence, artisanal, and some locally based tuna fisheries as less than 1,000 t in 1950, growing to a plateau of about 2,500 t/year in the 2000s. Overall, these catches were 1.4 times those reported by the FAO on behalf of Palau. This discrepancy was caused mainly by the subsistence catches (Ota 2006), which are underreported in the official statistics. Overall, however, industrial fisheries dominate total catches in the EEZ (figure 2A), driven exclusively by foreign fleets (figure 2B) targeting large pelagics such as skipjack (*Katsuwonus pelamis*), yellowfin (*Thunnus albacares*), bigeye (*T. obesus*), and albacore tuna (*T. alalunga*; figure 2C). However, domestic fisheries target a wide variety of reef fish, of which Johannes (1981) wrote eloquently, such as spinefoot (*Siganus* spp.), emperors (family Lethrinidae), and parrotfish (Scaridae) and invertebrates such as spiny lobsters and sea cucumbers. Palau has recently become a strong advocate of marine conservation, notably by banning the catching of sharks in its EEZ.

REFERENCES

Lingard S, Harper S, Ota Y and Zeller D (2011) Fisheries of Palau, 1950–2008: Total reconstructed catch. pp. 73–84 In Harper S and Zeller D (eds.), *Fisheries catch reconstruction: Islands, Part II*. Fisheries Centre Research Reports 19(4) [updated to 2010]

Johannes RE (1981) *Words of the Lagoon: Fishing and Marine Lore in the Palau District of Micronesia*. University of California Press, Berkeley (US). 245 p.

Ota Y (2006) An anthropologist in Palau. *Sea Around Us Newsletter*, May–June (35): 1–3.

1. Cite as Lingard, S., S. Harper, Y. Ota, and D. Zeller. 2016. Palau. P. 362 in D. Pauly and D. Zeller (eds.), *Global Atlas of Marine Fisheries: A Critical Appraisal of Catches and Ecosystem Impacts*. Island Press, Washington, DC.

PANAMA (CARIBBEAN)[1]

Sarah Harper,[a] Hector M. Guzman,[b] Kyrstn Zylich,[a] and Dirk Zeller[a]

[a]*Sea Around Us*, University of British Columbia, Vancouver, Canada
[b]Smithsonian Tropical Research Institute, Balboa, Republic of Panama

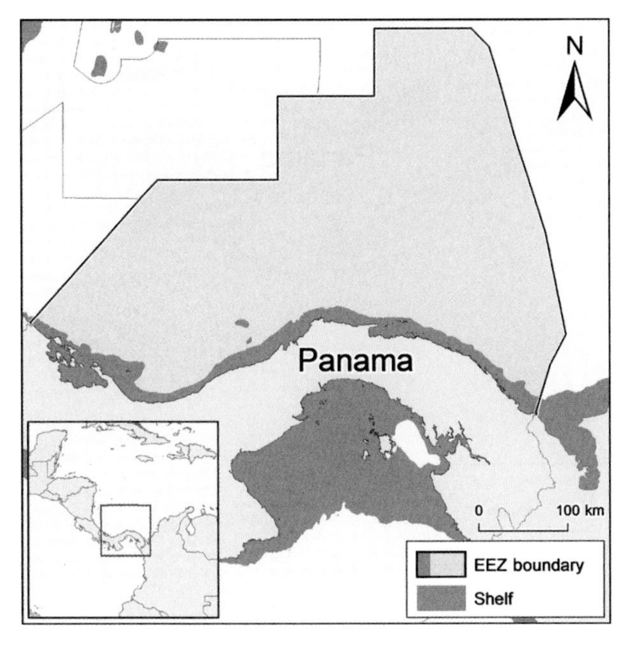

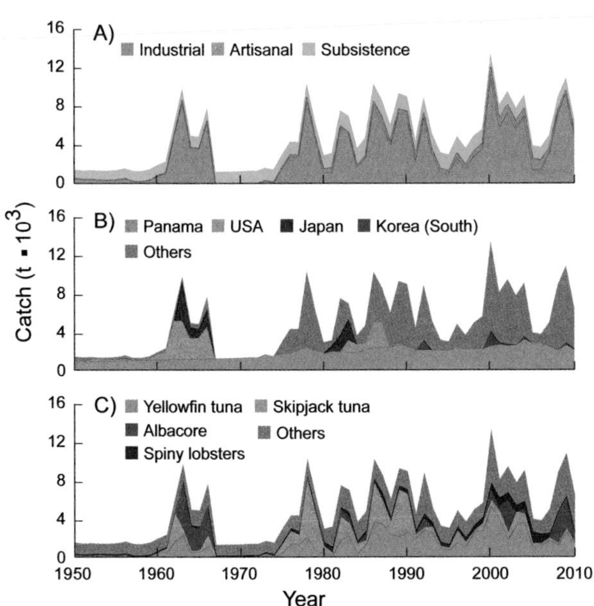

Figure 1. Panama's Caribbean coast has a shelf of 11,500 km² and an EEZ of 142,000 km². **Figure 2.** Domestic and foreign catches taken in the Caribbean EEZ of Panama: (A) by sector (domestic scores: Ind. –, 3, 3; Art. –, 2, 2; Sub. 1, 1, 1); (B) by fishing country (foreign catches are very uncertain); (C) by taxon.

Panama, the southernmost country in Central America, is located just north of Columbia and has coastlines on the Pacific and Atlantic Oceans, the latter being covered here (figure 1; see also Cramer et al. 2014). Based on Harper et al. (2014), reconstructed domestic marine catches for Panama's Caribbean coast were small compared with Pacific coast catches, with 800–1,000 t/year in the early 1950s, increasing to a plateau of about 2,000 t/year lasting into the 2000s, but with a gradual decline in recent years. However, this apparent stability masks the steady growth of an artisanal fishery that is slowly eroding the still dominant subsistence fishery along the Caribbean coast of Panama (figure 2A). Because the subsistence fishery is not officially monitored, the reconstructed domestic catches are as much as 8 times higher than the landings reported by the FAO on behalf of Panama in the Caribbean. Although small, these domestic catches play an important role in the food security and livelihoods of Panama's rural and small-island communities such as the Guna Yala. Overall, however, industrial catches (figure 2A), mainly by foreign fleets (figure 2B), dominate total catches in the EEZ. Figure 2C documents the composition of the catch, consisting of a mix of pelagic species (see Lasso and Zapata 1999), as dominated by large tuna catches by foreign fleets and reef fish species and various invertebrates, notably conch and lobster.

REFERENCES

Cramer K, Guzman H, Mate J, Shulman M, Weil E, AGGRA, CARICOMP, IUCN-CCCR and Reef Check (2014). pp. 266–268 In J Jackson, M Donovan, K Kramer and V Lam (eds.), *Status and trends of Caribbean coral reefs: 1970–2012*. Global Coral Reef Monitoring Network, IUCN, Gland, Switzerland.

Harper S, Guzman HM, Zylich K and Zeller D (2014) Reconstructing Panama's total fisheries catches from 1950 to 2010: highlighting data deficiencies and management needs. *Marine Fisheries Review* 76(1–2): 51–65.

Lasso J and Zapata L (1999) Fisheries and biology of *Coryphaena hippurus* (Pisces: Coryphaenidae) in the Pacific coast of Colombia and Panama. *Scientia Marina* 63(3–4): 387–399.

1. Cite as Harper, S., H. M. Guzman, K. Zylich, and D. Zeller. 2016. Panama (Caribbean). P. 363 in D. Pauly and D. Zeller (eds.), *Global Atlas of Marine Fisheries: A Critical Appraisal of Catches and Ecosystem Impacts*. Island Press, Washington, DC.

PANAMA (PACIFIC)[1]

Sarah Harper,[a] Hector M. Guzman,[b] Kyrstn Zylich,[a] and Dirk Zeller[a]

[a]*Sea Around Us*, University of British Columbia, Vancouver, Canada
[b]Smithsonian Tropical Research Institute, Balboa, Republic of Panama

Panama is located in Central America and is bisected, since the early 20th century, by a canal connecting the Atlantic and Pacific. Reconstructed domestic catches for Panama's Pacific coast (figure 1), based on Harper et al. (2014), were about 5,000 t/year in the early 1950s, increasing to an average of 200,000 t/year from the late 1970s to the mid-2000s and later declining. Overall, this is 1.4 times the landings reported to the FAO, the discrepancy being mainly discards (not considered in FAO data) from the shrimp trawl fishery. Overall, fisheries were dominated by the industrial sector, although the artisanal sector increased more recently (figure 2A). Although the majority of catches are taken by Panama, some foreign fishing is occurring (figure 2B). In the mid-1960s, Panama's most lucrative fishery developed: a fishmeal and oil reduction fishery for Pacific anchoveta (*Cetengraulis mysticetus*) and Pacific thread herring (*Opisthonema libertate*; figure 2C), which continues to the present. The artisanal fishery targets snappers (*Lutjanus* spp.) and croakers (Sciaenidae), as well as invertebrates, such as scallops (*Argopecten ventricosus*), conch (*Strombus* spp.), and lobster, notably *Panulirus gracilis* (see Guzman et al. 2008). The economically important trawl fishery has targeted seabob and shrimp (family Penaeidae) since the 1950s, mainly for export, but it has much diminished lately. Although management measures have been implemented to protect overexploited marine resources along Panama's Pacific coast (Harper et al. 2010), more are needed to protect threatened species.

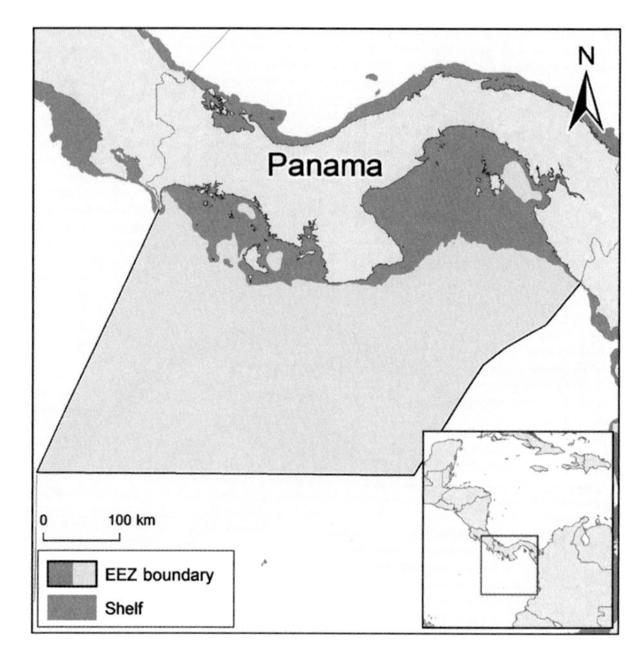

Figure 1. Panama's Pacific coast has a shelf of 40,700 km² and an EEZ of 189,000 km².

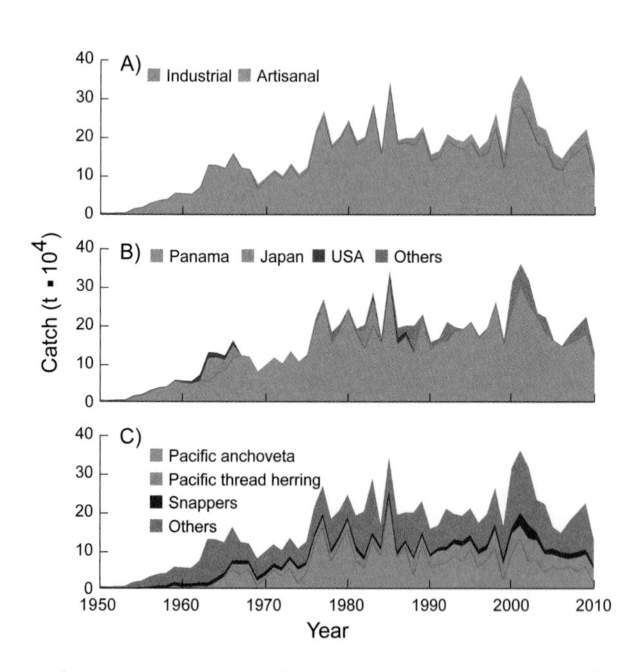

Figure 2. Domestic and foreign catches taken in the Pacific EEZ of Panama: (A) by sector (domestic scores: Ind. 3, 3, 3; Art. 2, 2, 2; Rec. 1, 2, 1; Dis. 1, 1, 2); (B) by fishing country (foreign catches are very uncertain); (C) by taxon.

REFERENCES

Guzman HM, Cipriani R, Vega AJ, Lopez M and Mair JM (2008) Population assessment of the Pacific green spiny lobster *Panulirus gracilis* in Pacific Panama. *Journal of Shellfish Research* 27: 907–915.

Harper S, Bates RC, Guzman HM and Mair JM (2010) Acoustic mapping of fish aggregation areas to improve fisheries management in Las Perlas Archipelago, Pacific Panama. *Ocean & Coastal Management* 54: 615–623.

Harper S, Guzman HM, Zylich K and Zeller D (2014) Reconstructing Panama's total fisheries catches from 1950 to 2010: highlighting data deficiencies and management needs. *Marine Fisheries Review* 76(1–2): 51–65.

1. Cite as Harper, S., H. M. Guzman, K. Zylich, and D. Zeller. 2016. Panama (Pacific). P. 364 in D. Pauly and D. Zeller (eds.), *Global Atlas of Marine Fisheries: A Critical Appraisal of Catches and Ecosystem Impacts*. Island Press, Washington, DC.

PAPUA NEW GUINEA[1]

Lydia C. L. Teh,[a] Jeff Kinch,[b] Kyrstn Zylich,[a], and Dirk Zeller[a]

[a]*Sea Around Us*, University of British Columbia, Vancouver, Canada [b]Papua New Guinea National Fisheries Authority

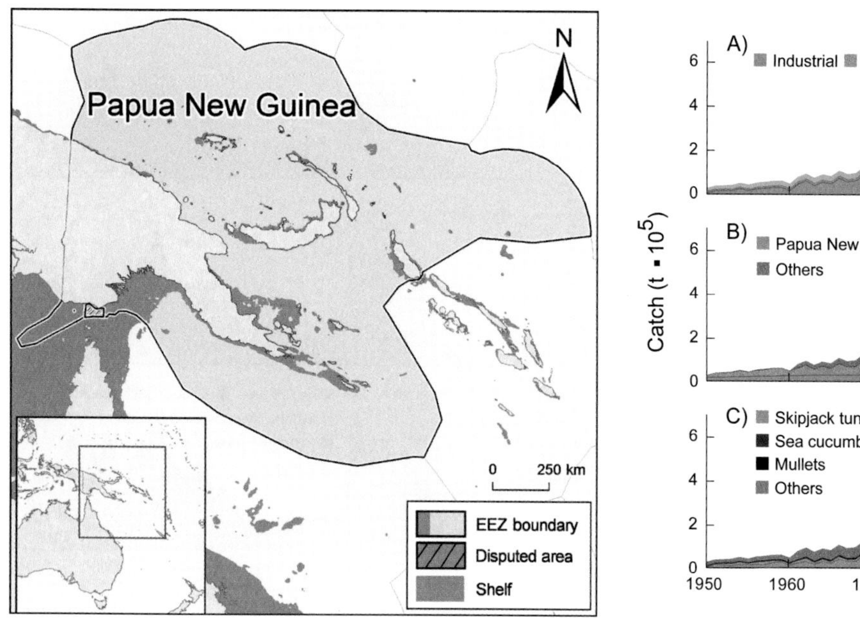

Figure 1. Papua New Guinea (land area 463,800 km²) has an EEZ of about 2.4 million km².

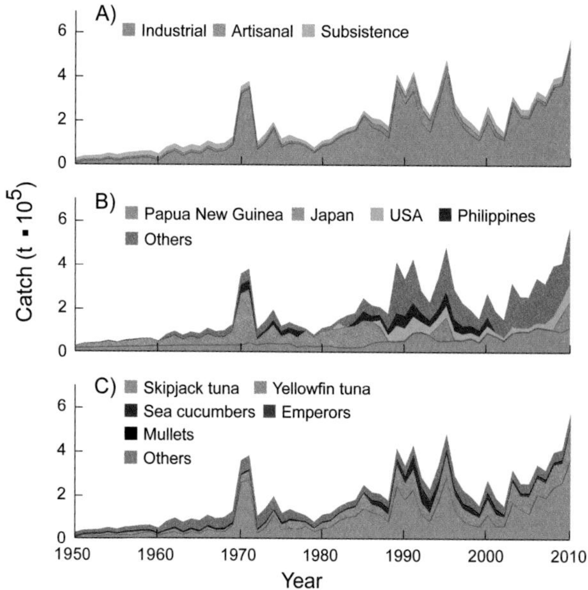

Figure 2. Domestic and foreign catches taken in the EEZ of Papua New Guinea: (A) by sector (domestic scores: Ind. 2, 2, 3; Art. 1, 2, 3; Sub. 2, 2, 2; Rec. –, 1, 1; Dis. 1, 2, 3); (B) by fishing country (foreign catches are very uncertain); (C) by taxon.

Papua New Guinea (PNG) in the western Pacific Ocean consists of the eastern half of the mountainous island of New Guinea and numerous smaller islands, notably New Britain, New Ireland, Bougainville, and Manus (figure 1). The coastal environments of PNG are diverse, with fringing and barrier coral reefs and extensive mangroves, and are surprisingly well studied (Kailola 2003). National landings statistics are dominated by tuna catches, which have been well documented since 2000. Nonetheless, PNG's fisheries statistics are still considered incomplete and underreported because of the omission of small-scale sector catches (Kuk and Tioti 2007). Teh et al. (2014) performed a catch reconstruction that focused on quantifying PNG's small-scale fisheries, leading to estimates of domestic nontuna catches of 22,000 t/year in the 1950s and 44,000 t/year in the 2000s. Reconstructed nontuna catch in PNG was about 4 times the nontuna catches reported by the FAO on behalf of PNG. Overall, fisheries in the EEZ were dominated by the industrial sector (figure 2A), which consists mainly of foreign fleets (figure 2B) targeting tuna such as skipjack (*Katsuwonus pelamis*) and yellowfin tuna (*Thunnus albacares*), whereas the domestic small-scale sectors focus on sea cucumbers (family Holothuriidae), emperors (Lethrinidae), and mullets (Mugilidae; figure 2C). There is high socioeconomic reliance on PNG's small-scale inshore fisheries, so steps should be taken to ensure the sustainability of these crucial food security resources.

REFERENCES

Kailola P (2003) Aquatic resources bibliography of Papua New Guinea. Secretariat of the Pacific Community, Noumea.

Kuk R and Tioti J (2007) Fisheries Policy and Management in Papua New Guinea. Australian Center for International Research, Project ASEM/2004/011, Project Paper 4, 41 p.

Teh LCL, Kinch J, Zylich K and Zeller D (2014) Reconstructing Papua New Guinea's Marine Fisheries Catch, 1950–2010. Fisheries Centre Working Paper #2014–09, 23 p.

1. Cite as Teh, L. C. L., J. Kinch, K. Zylich, and D. Zeller. 2016. Papua New Guinea. P. 365 In D. Pauly and D. Zeller (eds.), *Global Atlas of Marine Fisheries: A Critical Appraisal of Catches and Ecosystem Impacts*. Island Press, Washington, DC.

PERU[1]

Jaime Mendo[a] and Claudia Wosnitza-Mendo[b]

[a]Facultad de Pesqueria, Universidad Nacional Agraria La Molina, Lima, Peru [b]Calle B-190, San Miguel, Lima, Peru

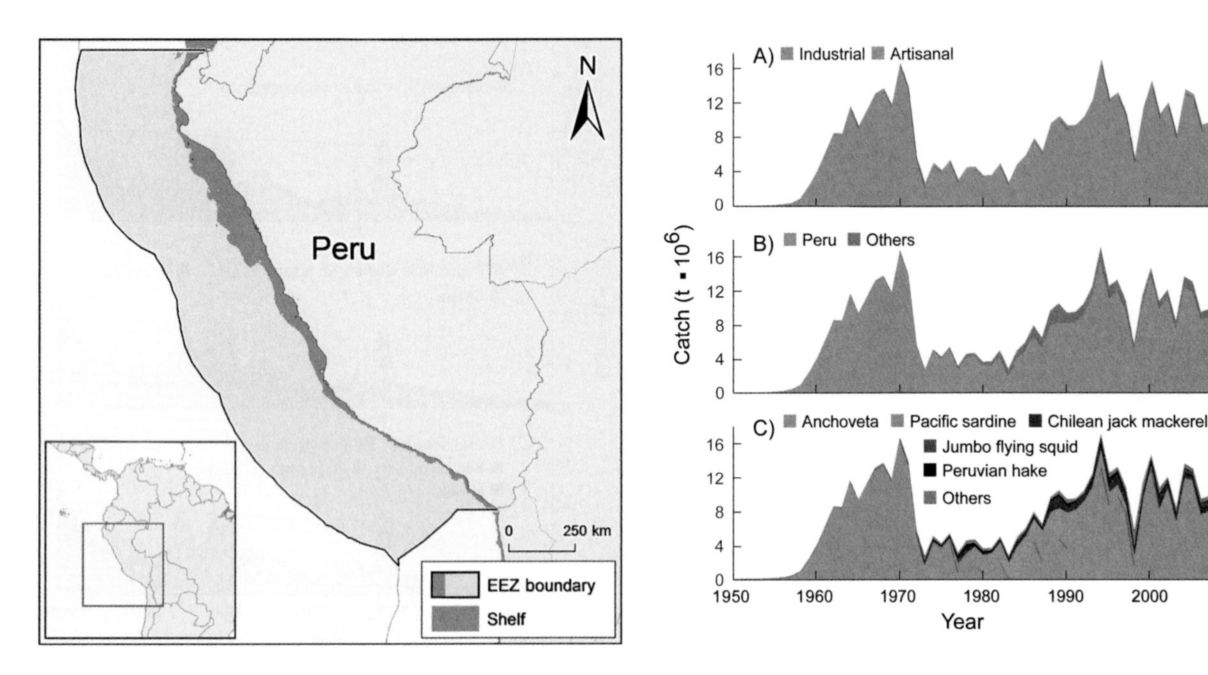

Figure 1. Peru has a narrow shelf, covering 88,560 km², and an EEZ of 857,000 km². **Figure 2.** Domestic and foreign catches taken in the EEZ of Peru: (A) by sector (domestic scores: Ind. 3, 4, 4; Art. 3, 3, 3; Sub. 3, 3, 3; Rec. 1, 2, 2; Dis. 3, 3, 3); (B) by fishing country (foreign catches are very uncertain); (C) by taxon.

The waters off Peru (figure 1) are probably the world's most productive, thanks to an intense coastal upwelling generating the planktonic food of a huge population of Peruvian anchoveta (*Engraulis ringens*), itself fed upon by higher-level consumers, including other fishes, seabirds, and marine mammals. The Peruvian industrial anchoveta fishery started in 1953 and is now a key component of Peru's economy and the fisheries sector with the largest nominal catch, although small-scale sectors, including subsistence, are gaining importance in more recent times with the rise in coastal populations (figure 2A). There seems to be only limited foreign fishing occurring in this EEZ (figure 2B). However, reported landings have always been lower than actual catches. Mendo and Wosnitza-Mendo (2014) reestimated total domestic catches based on sources of "losses" between the purse seining of anchoveta and the exported final fishmeal product (see Castillo and Mendo 1987). Sources of discrepancies between catch and landing pertaining to other species were also identified. From 1950 to 2010, the reconstructed catch is 1.2 times the landings reported by Peru to the FAO. Figure 2C illustrates that the non-anchoveta catch consists of a mix of other pelagic fishes (e.g., Pacific sardine, *Sardinops sagax*), Chilean jack mackerel (*Trachurus murphyi*), invertebrates (e.g., jumbo flying squid or Humboldt squid, *Dosidicus gigas*), and coastal demersals such as Peruvian hake (*Merluccius gayi peruanus*) and croakers (family Sciaenidae).

REFERENCES

Castillo S and Mendo J (1987) Estimation of unregistered Peruvian anchoveta (*Engraulis ringens*) in official catch statistics, 1951 to 1982. pp. 109–116 In Pauly D and Tsukayama, I (eds.), *The Peruvian anchoveta and its upwelling ecosystem: Three decades of change*. ICLARM Studies and Reviews 15.

Mendo J and Wosnitza-Mendo C (2014) Reconstruction of total marine fisheries catches for Peru: 1950–2010. Fisheries Centre Working Paper #2014-21, 23 p.

1. Cite as Mendo, J., and C. Wosnitza-Mendo. 2016. Peru. P. 366 In D. Pauly and D. Zeller (eds.), *Global Atlas of Marine Fisheries: A Critical Appraisal of Catches and Ecosystem Impacts*. Island Press, Washington, DC.

PHILIPPINES[1]

Maria Lourdes D. Palomares and Daniel Pauly

Sea Around Us, University of British Columbia, Vancouver, Canada

The Philippines are an archipelago in Southeast Asia (figure 1) that encompasses the global center of marine biodiversity (Carpenter and Springer 2005) and hosts equally diverse fisheries, with many articles on some of their aspects. However, there are few comprehensive accounts, and the country's statistics, though detailed and precise, are deemed inaccurate. A multiauthor reconstruction is presented in Palomares and Pauly (2014). Reconstructed domestic catches were about 250,000 t/year in the early 1950s and plateaued in the 1970s at about 900,000 t/year. Expansion into offshore stocks enabled further growth, leading to about 2.2 million t/year in 2010, with the majority (70%) being industrial (including so-called baby trawlers) but with substantial artisanal (20%) and subsistence components (10%; figure 2A). Overall, this is 0.96 times the catch reported by the FAO on behalf of the Philippines (i.e., national data overreport). Foreign fishing seems minimal (figure 2B) but is probably underestimated and either illegal or occurring in disputed waters (e.g., South China Sea). A very large number of taxa are caught in these high-diversity waters, although small pelagics (e.g., shortfin scad, *Decapterus macrosoma*; sardinellas, *Sardinella* spp.) and invertebrates seem to dominate, as do small tuna such as frigate tuna (*Auxis* spp.; figure 2C). Overfishing and "fishing down" are ubiquitous, but efforts are being made to counter these, notably through a multitude of marine protected areas (Alcala and Russ 2006), a tool pioneered in the Philippines.

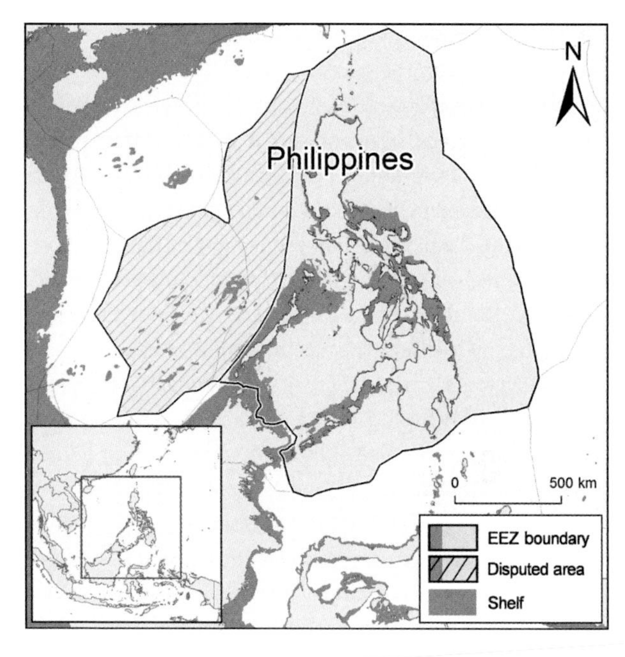

Figure 1. The Philippine Archipelago (land area 300,000 km²) and its EEZ (2.26 million km²).

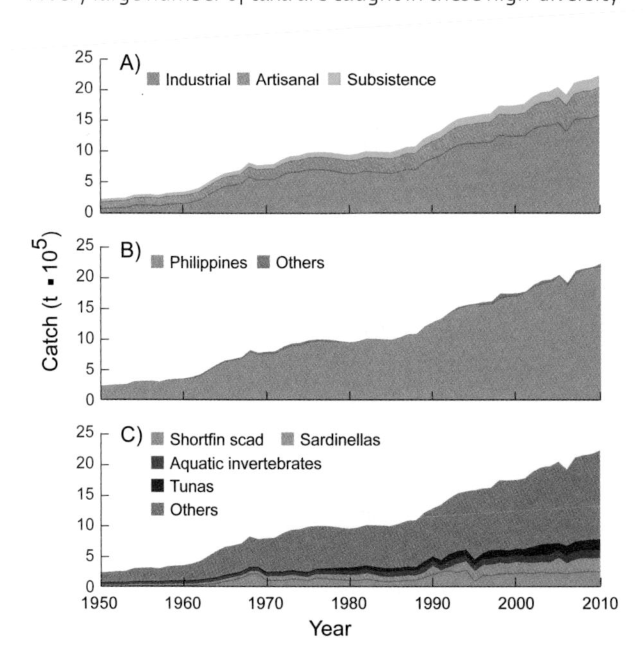

Figure 2. Domestic and foreign catches taken in the EEZ of the Philippines: (A) by sector (domestic scores: Ind. 2, 2, 2; Art. 3, 3, 3; Sub. 3, 3, 3; Dis. 2, 2, 2); (B) by fishing country (foreign catches are very uncertain); (C) by taxon.

REFERENCES

Alcala A and Russ GR (2006) No-take marine reserves and reef fisheries management in the Philippines: a new people power revolution. *AMBIO* 35(5): 245–254.

Carpenter KE and Springer VG (2005) The center of marine shore fish biodiversity: the Philippine Islands. *Environmental Biology of Fishes* 74(4): 467–480.

Palomares MLD and Pauly D, editors (2014) Philippine Marine Fisheries Catches: A Bottom-up Reconstruction, 1950 to 2010. *Fisheries Centre Research Reports* 22(1), 171 p.

1. Cite as Palomares, M. L. D., and D. Pauly. 2016. Philippines. P. 367 in D. Pauly and D. Zeller (eds.), *Global Atlas of Marine Fisheries: A Critical Appraisal of Catches and Ecosystem Impacts*. Island Press, Washington, DC.

POLAND[1]

Sarah Bale, Peter Rossing, Shawn Booth, Pawel Wowkonowicz, and Dirk Zeller

Sea Around Us, University of British Columbia, Vancouver, Canada

Poland, a member of the European Union (EU) since 2004, has a coast on the southern Baltic Sea (figure 1). Total fisheries catches of Poland in the Baltic Sea were reconstructed by Bale et al. (2010), based on statistics of the International Council for the Exploration of the Sea (ICES), ICES documents (e.g., stock assessment working group reports), and national statistics. The resulting estimates added 35% to Poland's reported data for the entire Baltic Sea, with most of the discrepancy since 1990. Total catches, as allocated to Poland's EEZ, were dominated by the industrial sector (figure 2A), with Poland accounting for the majority (43%), but Denmark and Sweden also fishing extensively in Polish waters (figure 2B). Catches focused heavily on three species: Atlantic herring (*Clupea harengus*), European sprad (*Sprattus sprattus*), and Atlantic cod (*Gadus morhua*), although other species (e.g., European flounder, *Platichthys flesus*) are also caught (figure 2C). The catch peak in the late 1990s is largely due to tremendous catches of European sprad and large catches of Atlantic cod, taken in excess of the EU-issued quota by Polish trawlers (figure 2B) with the tacit support of the Polish government at the time (Bale et al. 2010; Zeller et al. 2011). Despite estimates of illegal catches always being tentative, we believe that this reconstruction provides a conservative estimate of the extent of Polish catches in the Baltic Sea.

REFERENCES

Bale S, Rossing P, Booth S, Wowkonowicz P and Zeller D (2010) Poland's fisheries catches in the Baltic Sea (1950–2007). pp. 165–188 In Rossing R, Booth S and Zeller D (eds.), *Total marine fisheries extractions by country in the Baltic Sea: 1950–present*. Fisheries Centre Research Reports 18(1). [Updated to 2010]

Zeller D, Rossing P, Harper S, Persson L, Booth S and Pauly D (2011) The Baltic Sea: estimates of total fisheries removals 1950–2007. *Fisheries Research* 108: 356–363.

1. Cite as Bale, S., P. Rossing, S. Booth, P. Wowkonowicz, and D. Zeller. 2016. Poland. P. 368 in D. Pauly and D. Zeller (eds.), *Global Atlas of Marine Fisheries: A Critical Appraisal of Catches and Ecosystem Impacts*. Island Press, Washington, DC.

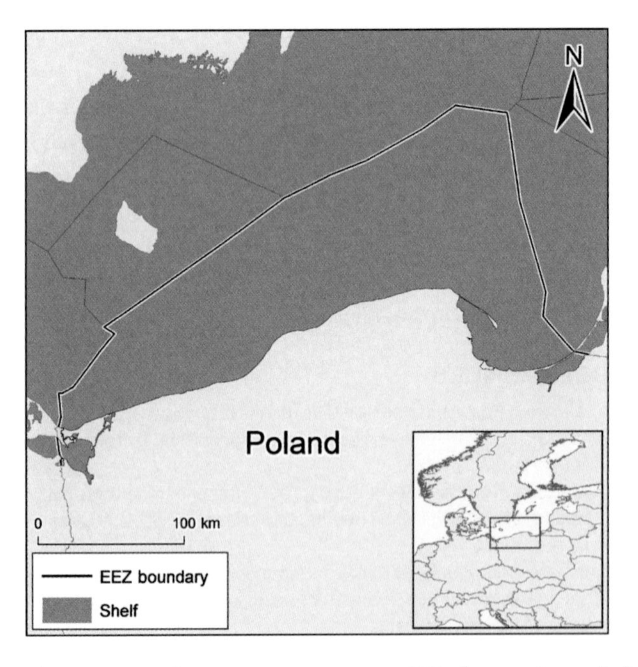

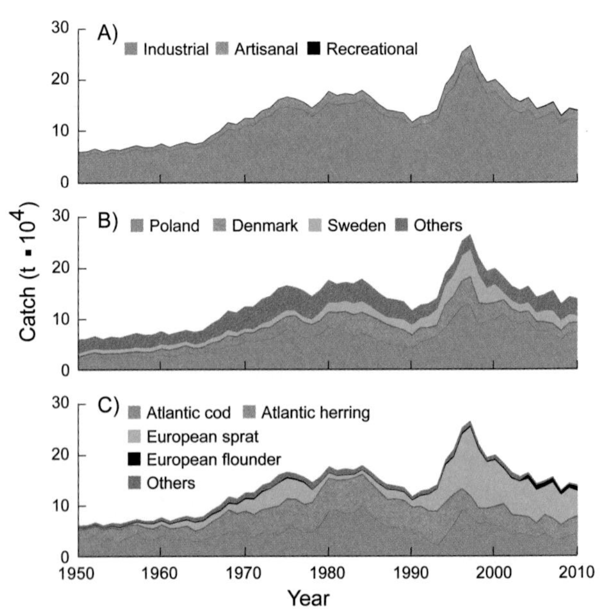

Figure 1. Poland's waters encompass an EEZ of 32,100 km², all of which is shelf. **Figure 2.** Domestic and foreign catches taken in the EEZ of Poland: (A) by sector (domestic scores: Ind. 2, 3, 2; Art. 2, 2, 2; Rec. –, 1, 1; Dis. 2, 2, 1); (B) by fishing country (foreign catches are very uncertain); (C) by taxon.

PORTUGAL (MAINLAND)[1]

Francisco Leitão,[a] Vania Baptista,[b] Karim Erzini,[a] Davis Iritani,[c] and Dirk Zeller[c]

[a]Centro de Ciências do Mar, Universidade do Algarve, Faro, Portugal [b]Instituto Ciências Biomédicas Abel Salazar, Universidade do Porto, Porto, Portugal [c]*Sea Around Us*, University of British Columbia, Vancouver, Canada

Portugal's mainland (figure 1), at the southwestern tip of Europe, has a long maritime history, including fishing, which has left deep marks on Portuguese culture. Portuguese consume about 60 kg/person/year, with the fisheries in the EEZ of the Portuguese mainland only partly meeting this demand (Leitão et al. 2014a, 2014b). The Portuguese reconstructed and total allocated catches increased from 280,000 t in 1950 to a plateau of about 500,000 t/year from the mid-1960s to the mid-1970s, then began a steady decline to 200,000–250,000 t/year in the 2000s, with the industrial sector (trawl and purse seine) accounting for 73%, the artisanal (multispecies) for 26%, and subsistence and recreational for 0.2–0.3% each (figure 2A). Overall, the reconstructed domestic catches were 2 times the amount reported by Portugal, mainly because of discards which are ignored in European statistics. Some limited foreign fishing seems to be occurring in the Portuguese EEZ (figure 2B). The Portuguese mainland fisheries are largely concentrated in near-shore waters, with European pilchard (*Sardina pilchardus*), Atlantic horse mackerel (*Trachurus trachurus*), chub mackerel (*Scomber japonicus*), and European hake (*Merluccius merluccius*) accounting for most of the catch (figure 2C). Efforts to conserve the biodiversity of coastal marine ecosystems in Portugal are constrained by political uncertainty and bureaucracy. However, the recent increase in public awareness has made discarding contentious, and it will eventually have to be reduced (e.g., within the framework of the recently revised EU Common Fisheries Policy).

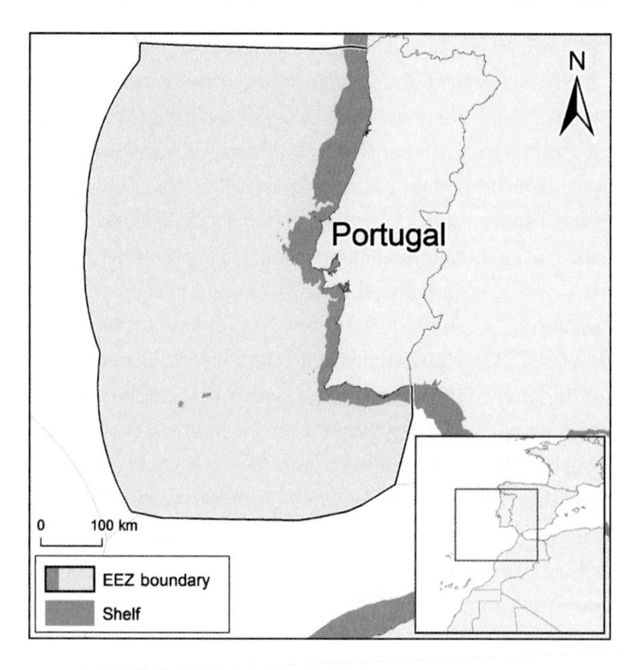

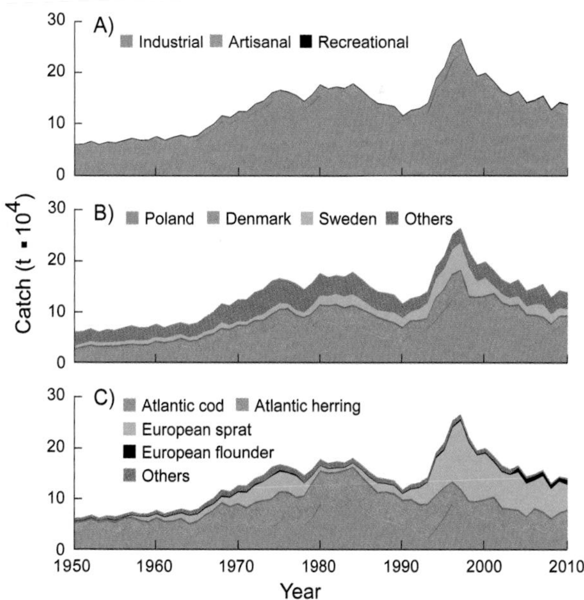

Figure 1. The Portuguese mainland has a shelf of 24,600 km² and an EEZ of 312,000 km². **Figure 2.** Domestic and foreign catches taken in the EEZ of Portugal: (A) by sector (domestic scores: Ind. 2, 3, 4; Art. 2, 3, 3; Sub. 2, 2, 2; Rec. 2, 2, 2; Dis. 2, 2, 2); (B) by fishing country (foreign catches are very uncertain); (C) by taxon.

REFERENCES

Leitão F, Baptista V, Erzini K, Iritani D and Zeller D (2014a) Reconstruction of mainland Portugal fisheries catches 1950–2010. Fisheries Centre Working Paper #2014-08, 29 p.

Leitão F, Baptista V, Zeller D and Erzini K (2014b) Reconstructed catches and trends for mainland Portugal fisheries between 1938 and 2009: implications for sustainability, domestic fish supply and imports. *Fisheries Research* 155: 33–50.

1. Cite as Leitão, F., V. Baptista, K. Erzini, D. Iritani, and D. Zeller. 2016. Portugal (Mainland). P. 369 In D. Pauly and D. Zeller (eds.), *Global Atlas of Marine Fisheries: A Critical Appraisal of Catches and Ecosystem Impacts*. Island Press, Washington, DC.

PORTUGAL (AZORES)[1]

Christopher Pham, Angela Canha, Hugo Diogo, João G. Pereira, and Telmo Morato

Departamento de Oceanografia e Pescas, Universidade dos Açores, Horta, Portugal

The Azores archipelago consists of 9 volcanic islands located in the North Atlantic Ocean (figure 1). As for most oceanic islands, fishing has always been a key driver of the local economy and currently, the fisheries of the Azores can be divided into four main components: tuna fishing using pole-and-line, demersal fishing using bottom longlines and handlines, swordfish fishing using pelagic longlines, and net fishing for small pelagic fishes (see also Carvalho et al. 2011). The catch reconstruction of Pham et al. (2013) for the Azores, together with *Sea Around Us* allocated foreign catches, with the exception of very high catches in the 1960s and 1970s, suggests that catches ranged between 15,000 and 25,000 t/year, with most taken by the industrial sector, although artisanal fishing accounted for 26% of total (including foreign) catches (figure 2A; see also Carvalho et al. 2011). Much of the industrial catch is taken by foreign or Portuguese mainland fleets (figure 2B). Overall, reconstructed domestic catches were 1.2 times the data officially reported by the Azores to the EU and FAO. Domestic catches from the subsistence and recreational fisheries explained 35% of the discrepancy, followed by discards from the bottom longline fishery (31%), baitfish catches for pole-and-line tuna fishing (15%), poaching of coastal invertebrates (9%), shark discards from the pelagic longline fleet (10% of total), and game fishing (0.1%). Figure 2C gives the composition of the Azorean catch.

REFERENCES

Carvalho N, Edwards-Jones G and Isidro E (2011) Defining scale in fisheries: Small versus large-scale fishing operations in the Azores. *Fisheries Research* 109: 360–369.

Pham CK, Canha A, Diogo H, Pereira JG, Prieto R and Morato T (2013) Total marine fishery catch for the Azores (1950–2010). *ICES Journal of Marine Science* 70(3): 564–577.

1. Cite as Pham, C., A. Canha, H. Diogo, J. G. Pereira, and T. Morato. 2016. Portugal (Azores). P. 370 in D. Pauly and D. Zeller (eds.), *Global Atlas of Marine Fisheries: A Critical Appraisal of Catches and Ecosystem Impacts*. Island Press, Washington, DC.

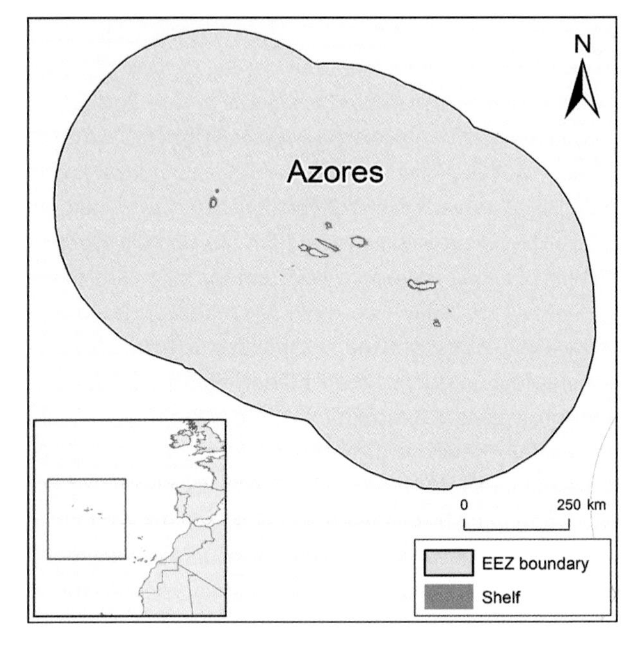

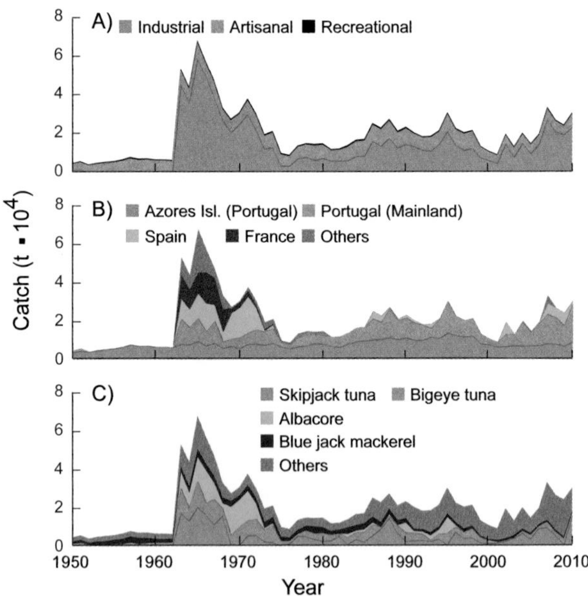

Figure 1. The Azores (Portugal) have a land area of 2,330 km² and an EEZ of 956,000 km². **Figure 2.** Domestic and foreign catches taken in the EEZ of Azores (Portugal): (A) by sector (domestic scores: Ind. 3, 3, 4; Art. 3, 3, 4; Sub. 2, 2, 3; Rec. 2, 3, 3; Dis. 1, 1, 3); (B) by fishing country (foreign catches are very uncertain); (C) by taxon.

PORTUGAL (MADEIRA)[1]

Soohyun Shon,[a] João Manuel Delgado,[b] Telmo Morato,[c] Christopher Kim Pham,[c] Dirk Zeller,[a] and Daniel Pauly[a]

[a]*Sea Around Us*, University of British Columbia, Vancouver, Canada
[b]Direcção de Serviços de Investigação das Pescas, 9004–365 Funchal, Madeira, Portugal
[c]Departamento de Oceanografia e Pescas, Universidade dos Açores, Horta, Azores, Portugal

The Portuguese Autonomous Region of Madeira (henceforth Madeira) consists of 4 islands (Madeira, Porta Santos, and the Desertas and Savage islands), of which only the first 2 are inhabited, with Madeira proper as the largest island. They are located northwest of Africa (figure 1) and are surrounded by extremely deep waters of low productivity, which sets limits for potential catches. This is also the reason why the main domestically exploited species are deep-water fishes (Bordalo-Machado et al. 2009). Time series of total domestic catch were reconstructed by Shon et al. (2015), resulting in domestic catches that were 16% higher than reported in official statistics, the discrepancy being caused by unreported subsistence and recreational catches, as well as discards from the commercial fisheries. Total catches consisted of 33% small-scale sectors (domestic artisanal, subsistence, and recreational), and the rest was industrial (figure 2A). More than 53% of total catch in the EEZ was taken by foreign, including Portuguese mainland, fleets (figure 2B). Given the limited shallow waters, the catch of coastal fish and invertebrates is very small, although diverse, whereas large pelagics, notably bigeye (*Thunnus obesus*) and skipjack (*Katsuwonus pelamis*), and a deep-sea fish, the black scabbardfish (*Aphanopus carbo*), contribute most to the catch from Madeiran waters (figure 2C). However, catches of black scabbardfish have shown a distinct decline in recent years, which may be a cause for concern.

REFERENCES

Bordalo-Machado P, Fernandes AC, Figueiredo I, Moura O, Reis S, Pestana G and Gordo LS (2009) The black scabbardfish (*Aphanopus carbo* Lowe, 1839) fisheries from the Portuguese mainland and Madeira Island. In Gordo LS (ed.), *Stock structure and quality of black scabbardfish in the southern NE Atlantic*. Scientia Marina, Barcelona, Spain.

Shon S, Delgado JM, Morato T, Pham CK, Zylich K, Zeller D and Pauly D (2015) Reconstruction of marine fisheries catches for Madeira Island, Portugal, from 1950–2010. Fisheries Centre Working Paper #2015–52, 13 p.

1. Cite as Shon, S., J. M. Delgado, T. Morato, C. K. Pham, D. Zeller, and D. Pauly. 2016. Portugal (Madeira). P. 371 in D. Pauly and D. Zeller (eds.), *Global Atlas of Marine Fisheries: A Critical Appraisal of Catches and Ecosystem Impacts*. Island Press, Washington, DC.

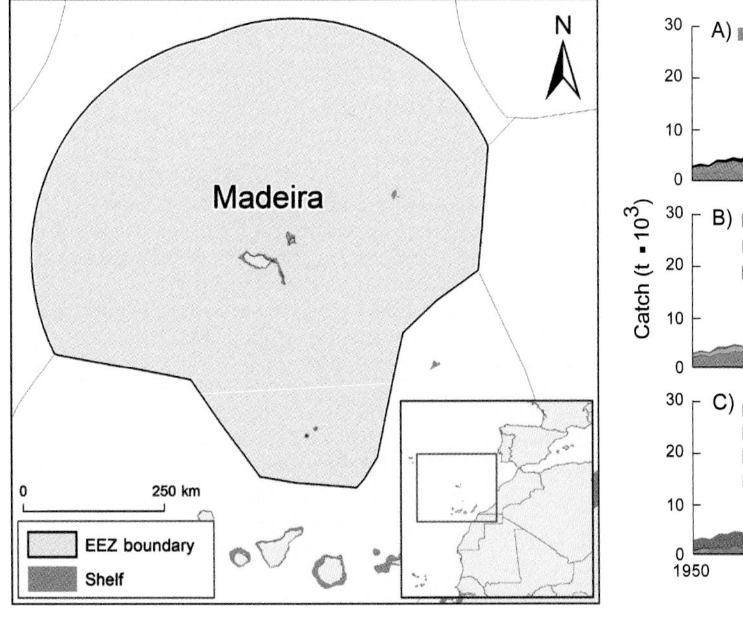

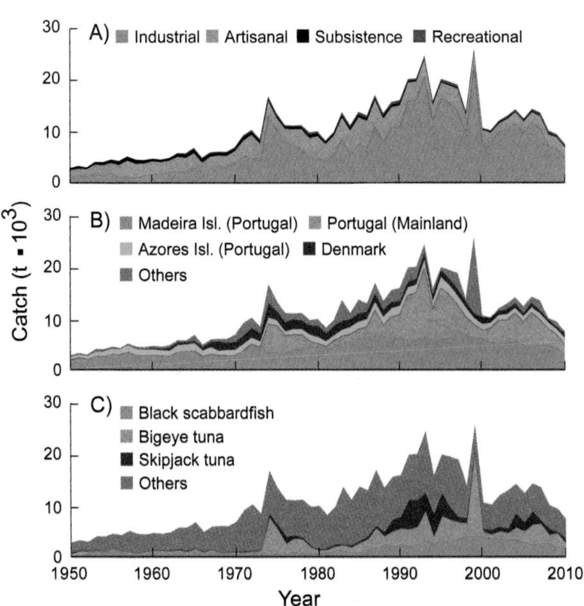

Figure 1. The Madeira Island group (land area: 801 km²) is surrounded by an EEZ of nearly 454,500 km².

Figure 2. Domestic and foreign catches taken in the EEZ of Madeira Island: (A) by sector (domestic scores: Ind. 2, 3, 4; Art. 2, 3, 3; Sub. 2, 2, 2; Rec. 2, 2, 2; Dis. 2, 2, 2); (B) by fishing country (foreign catches are very uncertain); (C) by taxon.

QATAR[1]

Dalal Al-Abdulrazzak

Sea Around Us, University of British Columbia, Vancouver, Canada

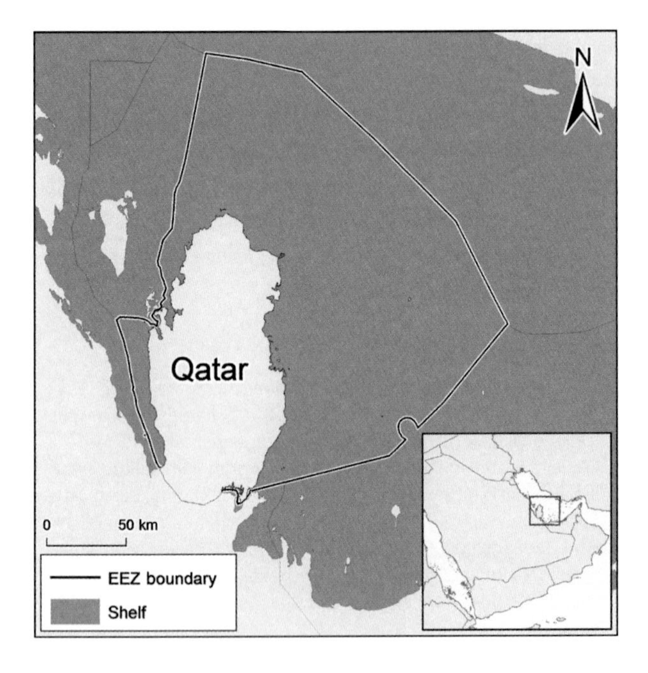

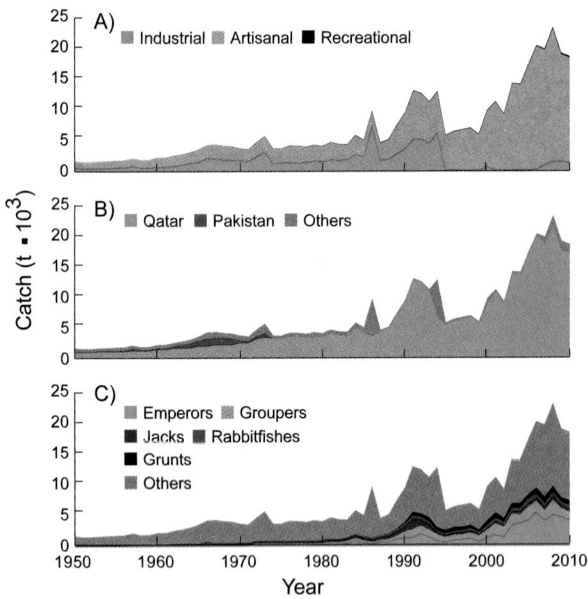

Figure 1. Qatar and its EEZ, 31,800 km², of which all is shelf. **Figure 2.** Domestic and foreign catches taken in the EEZ of Qatar: (A) by sector (domestic scores: Ind. –, 3, 2; Art. 3, 3, 3; Sub. 1, 1, 1; Rec. 1, 1, 1; Dis. 3, 3, 2); (B) by fishing country (foreign catches are very uncertain); (C) by taxon.

Qatar is a small Arab country located on a Persian Gulf peninsula abutting Saudi Arabia (figure 1). Qatar's catches have increased sharply over the past decade because of increased fishing effort, driven by increasing demand from a rapidly growing population. The peer-reviewed and gray literature was searched for qualitative and quantitative data on unreported catches from Qatar's fisheries. The resulting reconstruction by Al-Abdulrazzak (2013) suggested that domestic catches in the 1950s were about 1,000 t/year, increased to nearly 20,000 t/year in the 2000s, and were 38% higher than reported by the FAO on behalf of Qatar. In earlier years, catches by industrial vessels were more important (figure 2A), notably by foreign vessels before EEZ declaration in 1974 (figure 2B). More recently, domestic artisanal fleets have predominated (figure 2A). One of several reasons for the discrepancy between domestic reported and reconstructed catches is the neglect of discards from Qatar's bottom trawl fishery, which, between 1970 and 1993, were equivalent to 30% of the reported catch. The main taxa caught by Qatar are emperors (family Lethrinidae), groupers (Serranidae), jacks (Carangidae), rabbitfishes (Siganidae), and grunts (Haemulidae; figure 2C). The reconstruction also highlighted the extent of illegal domestic fishing. One example of this is that 14 tidal weirs (*hadrah*), which have been banned since 1994, were detected by Al-Abdulrazzak and Pauly (2014) along the Qatari coast in current Google Earth images.

REFERENCES

Al-Abdulrazzak D (2013) Total fishery extractions for Qatar: 1950–2010. pp. 31–37 In Al-Abdulrazzak D and Pauly D (eds.), *From dhows to trawlers: a recent history of fisheries in the Gulf countries, 1950 to 2010*. Fisheries Centre Research Reports 21(2).

Al-Abdulrazzak D and Pauly D (2014) Managing fisheries from space: Google Earth improves estimates of distant fish catches. *ICES Journal of Marine Science* 71(3): 450–455.

1. Cite as Al-Abdulrazzak, D. 2016. Qatar. P. 372 in D. Pauly and D. Zeller (eds.), *Global Atlas of Marine Fisheries: A Critical Appraisal of Catches and Ecosystem Impacts*. Island Press, Washington, DC.

ROMANIA[1]

Daniela Bănaru,[a] Frédéric Le Manach,[b] Leonie Färber,[b] Kyrstn Zylich,[b] and Daniel Pauly[b]

[a]Institut Méditerranéen d'Océanologie, Aix-Marseille Université, Marseille, France
[b]*Sea Around Us*, University of British Columbia, Vancouver, Canada

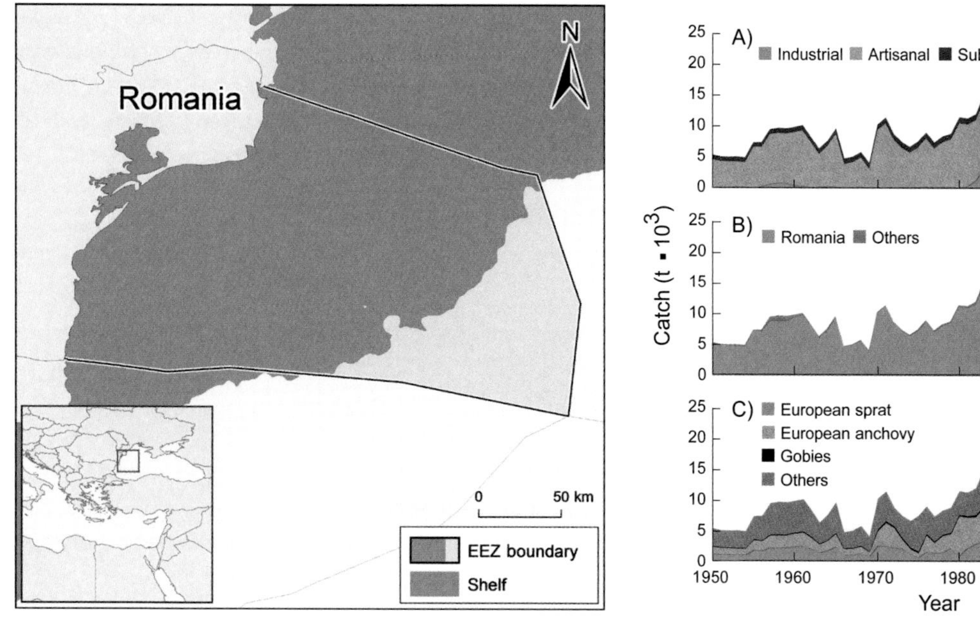

Figure 1. Romania has a shelf of 23,300 km² and an EEZ of 29,800 km² in the western Black Sea. **Figure 2.** Domestic and foreign catches taken in the EEZ of Romania: (A) by sector (domestic scores: Ind. –, 3, 3; Art. 2, 3, 3; Sub. 2, 2, 2; Dis. 2, 2, 2); (B) by fishing country (foreign catches are very uncertain); (C) by taxon.

Romania, along the western Black Sea (figure 1), once had vibrant fisheries. However, they have been reduced to small artisanal and large subsistence sectors, as the industrial trawler fleet collapsed after the end of the Ceauşescu dictatorship in 1989. Around that time, fish stocks in the western Black Sea had already started to collapse, because of decades of intense trawling, and they have not recovered despite much lower fishing pressures, possibly because of the increasing pollution, which contributes to the extension of the huge anoxic zone in the Black Sea (Petranu et al. 1999). Cociasu et al. (1996) commented that "it is evident that the principal cause of the long-term ecological change in Romanian coastal waters is the high nutrient input from the Danube and also, albeit to a minor extent, of municipal and industrial waste waters from along the coast." The reconstruction by Bănaru et al. (2015) led to catch estimates that increased from 5,000–10,000 t/year in the 1950s and 1960s to more than 18,000 t in 1990, because of the development of the industrial trawler fleets (figure 2A). Catches then collapsed and stagnated at slightly more than 1,000 t/year in the late 2000s (figure 2A), almost all domestic (figure 2B), although the foreign catch is almost certainly underestimated. Figure 2C shows that the once dominant European sprat (*Sprattus sprattus*) and anchovy (*Engraulis encrasicolus*) have been largely replaced with small, low-value fish such as gobies (family Gobiidae; figure 2B).

REFERENCES

Bănaru D, Le Manach F, Färber L, Zylich K and Pauly D (2015) From bluefin tuna to gobies: a reconstruction of the fisheries catch statistics in Romania, 1950–2010. Fisheries Centre Working Paper #2015-48, 10 p.

Cociasu A, Dorogan L, Humborg C and Popa L (1996) Long-term ecological changes in Romanian coastal waters of the Black Sea. *Marine Pollution Bulletin* 32(1): 32–38.

Petranu A, Apas M, Bodeanu N, Bologa AS, Dumitrache C, Moldoveanu M, Radu G and Tiganus V (1999) Status and evolution of the Romanian Black Sea coastal ecosystem. In Beşiktepe ST, Unluata U and Bologa AS (eds.), *Environmental Degradation of the Black Sea: Challenges and Remedies*. Springer.

1. Cite as Bănaru, D., F. Le Manach, L. Färber, K. Zylich, and D. Pauly. 2016. Romania. P. 373 in D. Pauly and D. Zeller (eds.), *Global Atlas of Marine Fisheries: A Critical Appraisal of Catches and Ecosystem Impacts*. Island Press, Washington, DC.

RUSSIA (BALTIC SEA)[1]

Sarah Harper,[a] Sergey Shibaev,[b] Olga Baryshnikova,[b] Peter Rossing,[a] Shawn Booth,[a] and Dirk Zeller[a]

[a]*Sea Around Us*, University of British Columbia, Vancouver, Canada [b]Department of Ichthyology and Ecology, Kaliningrad State Technical University, Kaliningrad, Russia

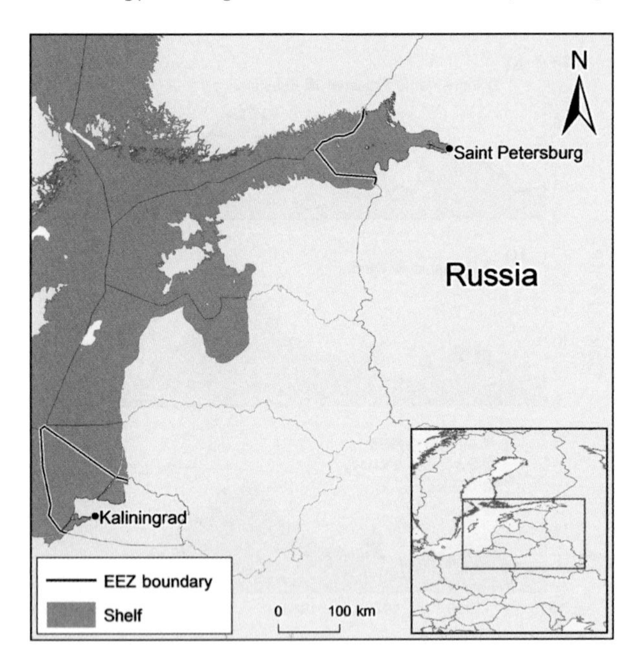

Figure 1. St. Petersburg oblast borders Finland to the north and Estonia to the west, and Kaliningrad oblast adjoins Poland to the west and Lithuania to the east. Jointly, this amounts to an EEZ area of 23,240 km², consisting entirely of shelf waters.

The Russian Federation (Russia), formed in 1991 after the collapse of the USSR, stretches from the Pacific Ocean to the Baltic Sea, but only 2 of its regions, St. Petersburg and Kaliningrad oblasts (the latter an exclave), have direct access to the Baltic Sea (figure 1). Harper et al. (2010) performed the catch reconstruction for Russian catches in the Baltic Sea (see also Zeller et al. 2011), based on statistics of the International Council for the Exploration of the Sea (ICES) and various ICES documents, such as stock assessment working group reports for the post-USSR period, and the Latvian Fish Resource Agency provided the former USSR landings data disaggregated by country for the pre-1991 period. Catches in the Russian EEZs of the Baltic Sea were largely industrial (figure 2A), and increased from about 40,000 t/year in the 1950s, to a peak of more than 120,000 t in 1997 (driven mainly by Polish catches, figure 2B) and about 90,000 t/year in the late 2000s. Between 1992 and 2010, the reconstructed Russian catches in the entire Baltic Sea were 28% higher than those reported by ICES for Russia. Total catches (figure 2C) were dominated by herring (*Clupea harengus*), sprat (*Sprattus sprattus*), and cod (*Gadus morhua*), but brackish-water species are also taken in coastal waters, notably European perch (*Perca fluviatilis*).

REFERENCES

Harper S, Shibaev SV, Baryshnikova O, Rossing P, Booth S and Zeller D (2010) Russian fisheries catches in the Baltic Sea from 1950–2007. pp. 189–224 In Rossing R, Booth S and Zeller D (eds.), *Total marine fisheries extractions by country in the Baltic Sea: 1950–present*. Fisheries Centre Research Reports 18(1). [Updated to 2010]

Zeller D, Rossing P, Harper S, Persson L, Booth S and Pauly D (2011) The Baltic Sea: estimates of total fisheries removals 1950–2007. *Fisheries Research* 108(356–363).

1. Cite as Harper, S., S. V. Shibaev, O. Baryshnikova, P. Rossing, S. Booth, and D. Zeller. 2016. Russia (Baltic Sea). P. 374 in D. Pauly and D. Zeller (eds.), *Global Atlas of Marine Fisheries: A Critical Appraisal of Catches and Ecosystem Impacts*. Island Press, Washington, DC.

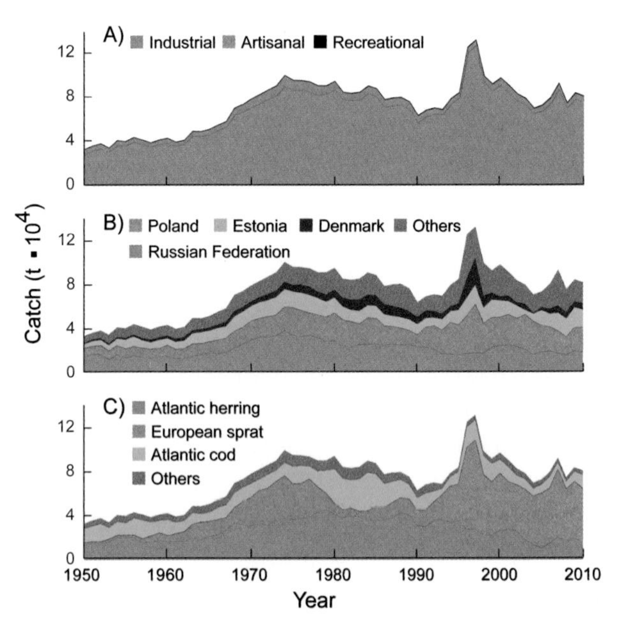

Figure 2. Domestic and foreign catches taken in the Baltic EEZs of Russia (St. Petersburg and Kaliningrad oblasts): (A) by sector (domestic scores: Ind. 2, 2, 2; Art. 2, 2, 2; Rec. 1, 1, 1; Dis. 1, 1, 1); (B) by fishing country (foreign catches are very uncertain); (C) by taxon.

RUSSIA (BARENTS SEA)[1]

Boris Jovanović, Esther Divovich, Sarah Harper, Dirk Zeller, and Daniel Pauly

Sea Around Us, University of British Columbia, Vancouver, Canada

Russia's Barents Sea region consists of the Barents Sea and the White Sea, bordered by the Norwegian Sea in the west, the Svalbard Archipelago in the northwest, Franz-Josef Land in the northeast, and Novaya Zemlya and the Kara Sea in the east (figure 1). The Barents Sea was among the first seas in the world to be exploited near-exclusively through large-scale industrial vessels, starting in 1906. Also, 1950 is the start of the trawling era in Russian fishing history, when the first large stern trawlers were deployed, eventually dominating the Barents fishery. The collapse of the ex-USSR led to interruptions of fisheries statistical time series, and this account, based on Jovanović et al. (2015), itself based on Rejwan et al. (2001), Spiridonov and Nikolaeva (2005), and other sources, conducted a catch reconstruction of Russian fisheries covering the years 1950 to 2010. The allocation of domestic and foreign catches to these waters (figure 2A) yielded catches of 200,000–300,000 t/year in the early 1950s, a peak of more than 1.3 million t in 1977, and 350,000 t/year in the late 2000s. For Russian catches alone, the reconstructed catches were 36% higher than data reported by the International Council for the Exploration of the Sea (ICES) on behalf of Russia, in part because of illegal operations. Figure 2B documents the taxonomic catch composition, dominated by Atlantic cod (*Gadus morhua*) and capelin (*Mallotus villosus*).

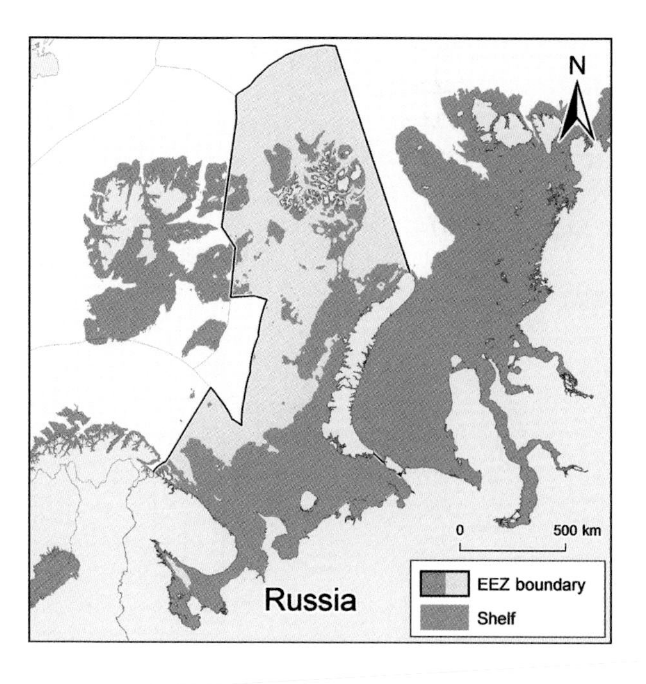

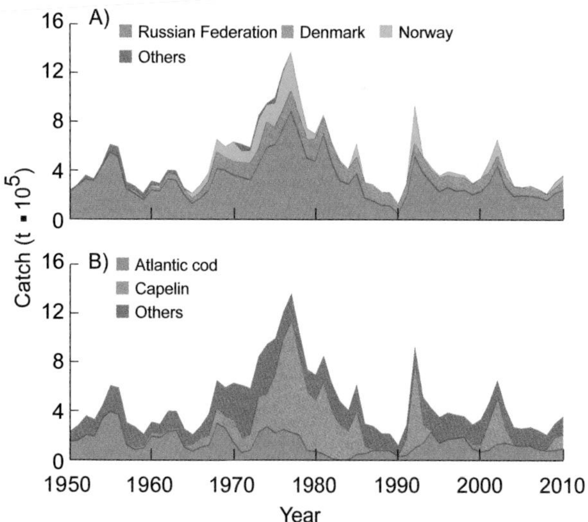

Figure 1. Russia's Barents Sea has a shelf of 606,000 km² and an EEZ of 1.2 million km². **Figure 2.** Domestic and foreign catches taken in the Barents Sea EEZ of Russia: (A) by fishing country (domestic scores: Ind. 3, 3, 2; Sub. 1, 1, 1; Rec. –, –, 1; Dis. 1, 1, 1; foreign catches are very uncertain); (B) by taxon.

REFERENCES

Jovanović B, Divovich E, Harper S, Zeller D and Pauly D (2015) Estimates of total Russian fisheries catches in the Barents Sea region (FAO 27 subarea I) between 1950 and 2010. Fisheries Centre Working Paper #2015–59, 16 p.

Rejwan C, Booth S and Zeller D (2001) Unreported catches in the Barents Sea and adjacent waters for periods from 1950 to 1998. pp. 99–106 In Zeller D and Pauly D (eds.), *Fisheries impacts on North Atlantic ecosystems: catch, effort and national/regional data sets*. Fisheries Centre Research Reports 9(3).

Spiridonov VA and Nikolaeva NG (2005) Fisheries in the Russian Barents Sea and the White Sea: Ecological challenges. WWF Barents Sea Ecoregion Programme, Oslo, 56 p.

1. Cite as Jovanović, B., S. Harper, D. Zeller, and D. Pauly. 2016. Russia (Barents Sea). P. 375 in D. Pauly and D. Zeller (eds.), *Global Atlas of Marine Fisheries: A Critical Appraisal of Catches and Ecosystem Impacts*. Island Press, Washington, DC.

RUSSIA (BLACK SEA)[1]

Esther Divovich, Boris Jovanović, Sarah Harper, Dirk Zeller, and Daniel Pauly

Sea Around Us, University of British Columbia, Vancouver, Canada

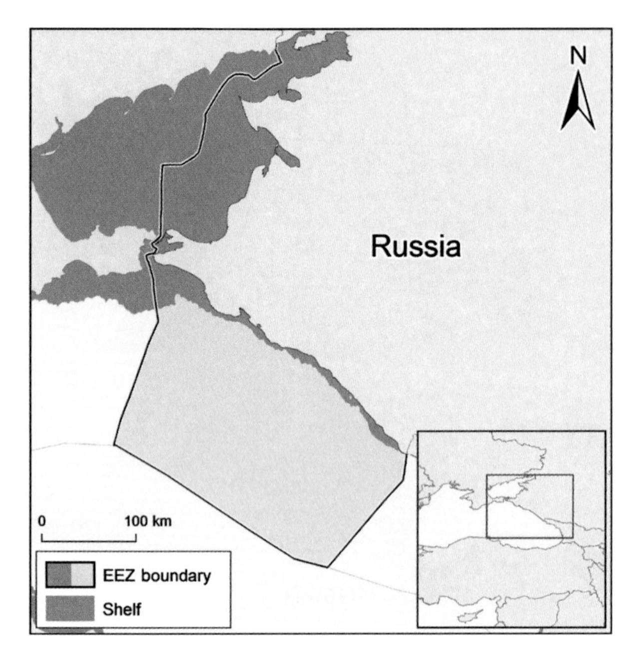

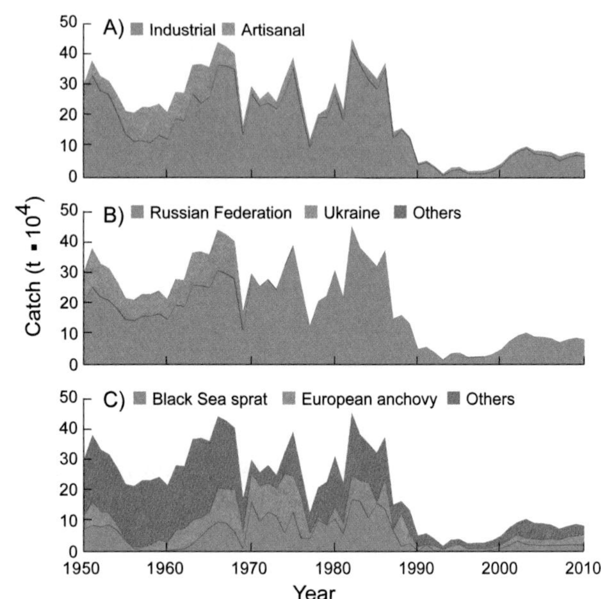

Figure 1. In 2010 Russia had a joint shelf area of 19,800 km² and an EEZ of 67,400 km² in the Black and Azov Seas (see Broad 2014 for an update). **Figure 2.** Domestic and foreign catches taken in the Black Sea EEZ of Russia (boundaries of 2010): (A) by sector (domestic scores: Ind. 2, 3, 2; Art. 2, 3, 2; Sub. 2, 2, 2; Rec. 1, 1, 2; Dis. 2, 2, 3); (B) by fishing country (foreign catches are very uncertain); (C) by taxon.

The shores of the Black Sea (including the Azov Sea; figure 1) have been important fishing grounds for the Russian Federation, the ex–Soviet Union, and Imperial Russia since the end of the Crimean War in 1856, when the Ottoman Empire signed a peace treaty allowing Russian fishing in the Black Sea. The first trawler was added to the Soviet fleet in 1950; during the period 1970–1990 the fish stocks in the Black Sea were reduced by overfishing and eutrophication, and the ecosystem gradually became dominated by gelatinous zooplankton. The final blow to the predominantly industrial fisheries (figure 2A) was the collapse of the Soviet Union, with the immediate cessation of state subsidies to the fleet, most of which has been disbanded; the fishery has not recovered since (Knudsen and Toje 2008). The reconstruction by Divovich et al. (2015), which is based on the 2010 borders of the Russian Federation (figure 1; but see Broad 2014), yielded Russian domestic catches of 225,000 t/year in the early 1950s, which peaked at more than 453,000 t in 1982 and decreased to 84,000 t/year in the 2000s (figure 2B). For 1950 to 2010, the reconstructed catch was 1.6 times the catch reported by the FAO on behalf of the Russian Federation (including the USSR catches assigned to Russia). The composition of the catch is given in figure 2C; in the 2000s, it consisted mainly of small pelagic species (i.e., anchovy and sprat).

REFERENCES

Broad WJ (2014) In Taking Crimea, Putin Gains a Sea of Fuel Reserves. *New York Times*, edition of May 17, 2014.

Divovich E, Jovanović B, Zylich K, Harper S, Zeller D and Pauly D (2015) Caviar and politics: A reconstruction of Russia's marine fisheries in the Black Sea and Sea of Azov from 1950 to 2010., Fisheries Centre Working Paper #2015–84, 24 p.

Knudsen S and Toje H (2008) Post-Soviet transformations in Russian and Ukrainian Black Sea fisheries: socio-economic dynamics and property relations. *Southeast European and Black Sea Studies* 8(1): 17–32.

1. Cite as Divovich, E., B. Jovanović, S. Harper, D. Zeller, and D. Pauly. 2014. Russia (Black Sea). P. 376 in D. Pauly and D. Zeller (eds.), *Global Atlas of Marine Fisheries: A Critical Appraisal of Catches and Ecosystem Impacts*. Island Press, Washington, DC.

RUSSIA (FAR EAST)[1]

Anna Sobolevskaya[a] and Esther Divovich[b]

[a]Institute of East Asian Studies, Duisburg–Essen University, Germany
[b]*Sea Around Us*, University of British Columbia, Vancouver, Canada

The Russian Far East (RFE; figure 1) is rich in fisheries resources, generating about two thirds of the marine catch of Russia (and ex-USSR) as a whole (Johnson 1996). The catches from this EEZ were reconstructed by Sobolevskaya and Divovich (2015), based on a variety of Russian- and English-language sources. The catches, overwhelmingly industrial (although artisanal catches increased more recently, figure 2A) by all countries were about 1 million t/year in the early 1950s, peaked at more than 10 million t in 1990, and declined to about 7 million t/year in the late 2000s. Although most catches were by Russian vessels, there was some foreign fishing, notably by Japan, China, and the two Koreas (figure 2B). Overall, the reconstructed domestic catches were 2.4 times the landings reported by the FAO since 1992 for Russia in these waters (FAO area 61), that is, since the distant-water fleets of the RFE returned to their home waters because of the declaration of EEZs by countries around the Pacific. It also accounts for the 1991 transition from a state-controlled economy, producing for domestic markets, to the current export-oriented fisheries and a small but growing share by indigenous and small-scale fisheries. Major targets are Alaska pollock (*Theragra chalcogramma*), Pacific herring (*Clupea palasii*), and pink salmon (*Oncorhynchus gorbuscha*; figure 2C). Illegal fishing is substantial, especially for various valuable species of king crab and snow and tanner crabs.

REFERENCES

Johnson T (1996) Changing Fisheries of the Russian Far East. *Marine Resource Economics* 11(2): 131–135.

Sobolevskaya A and Divovich E (2015) The Wall Street of fisheries: the Russian Far East, a catch reconstruction from 1950–2010. Fisheries Centre Working Paper #2015-45, 65 p.

1. Cite as Sobolevskaya, A., and E. Divovich. 2016. Russia (Far East). P. 377 in D. Pauly and D. Zeller (eds.), *Global Atlas of Marine Fisheries: A Critical Appraisal of Catches and Ecosystem Impacts*. Island Press, Washington, DC.

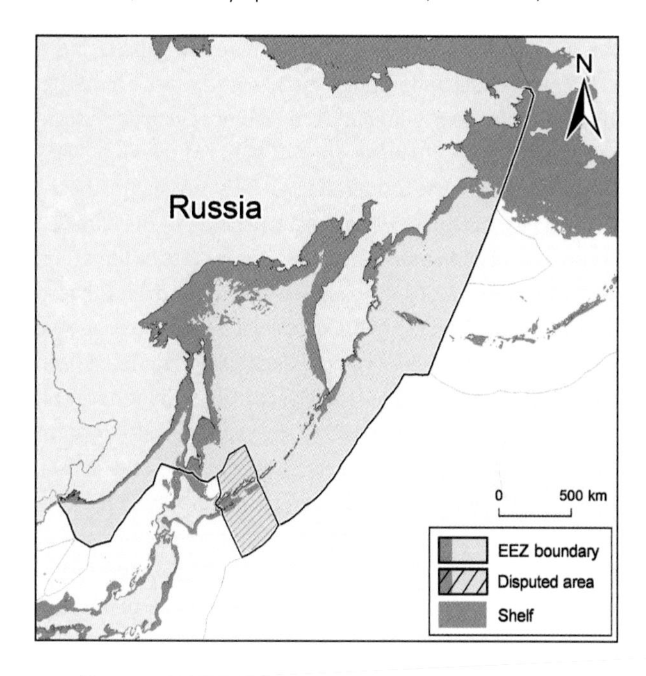

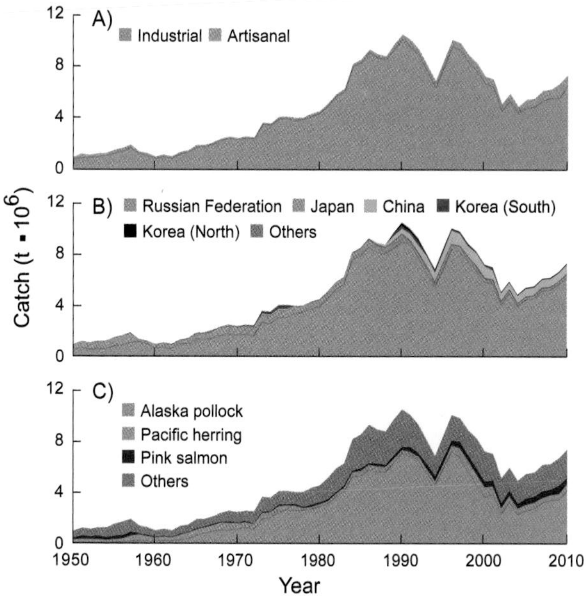

Figure 1. The Russian Far East, as defined here, has a shelf of 967,000 million km² and an EEZ of 3.45 million km².
Figure 2. Domestic and foreign catches taken in the EEZ of Russian Far East: (A) by sector (domestic scores: Ind. 4, 4, 4; Art. 4, 4, 4; Sub. 2, 3, 3; Rec. 2, 3, 3; Dis. 1, 1, 3); (B) by fishing country (foreign catches are very uncertain); (C) by taxon.

RUSSIA (KARA SEA)[1]

Daniel Pauly

Sea Around Us, University of British Columbia, Vancouver, Canada

Russia's Kara Sea (figure 1), in northwestern Siberia, has a complex oceanography. Notably, it receives occasional intrusions of "warm" water and the accompanying fauna from the Barents Sea. Otherwise, the fish fauna of the Kara Sea is as species poor as the Laptev and East Siberian Seas further to the east (see next page). The coastal fisheries of the Kara Sea are not well documented, at least in the non-Russian literature, and the reconstruction by Pauly and Swartz (2007) was based mainly on data in documents published by the International Northern Sea Route Programme (INSROP), conducted from 1993 to 1999. These included Larsen et al. (1996), itself based on internal Russian reports covering the years for 1980–1994 for Ob Bay and 1989 and 1991–1994 for three tributaries, the Yenisei, Pyasina, and Taimyskaya Rivers. From this and other information (Zeller et al. 2011), a subsistence catch of more than 14,000 t was estimated for 1950, which is very conservative. Other authors suggested much higher values, such as Slavin (1964), who wrote of an early catch of 30,000 t/year. This declined to about 700 t/year by 2004 but was thought to have increased slightly by 2010 (figure 2). Most of these catches (70%–90%) were contributed by several species of "whitefishes" or cisco (*Coregonus* spp., family Salmonidae), known as *sig* in Russian and caught in the lower reaches of rivers, their estuaries, and nearby coastal waters. The catch decline may be caused by the dwindling human population in Arctic Russia or, more ominously, by overfishing and pollution from mining, which Vilchek et al. (1996) describe as a major regional issue.

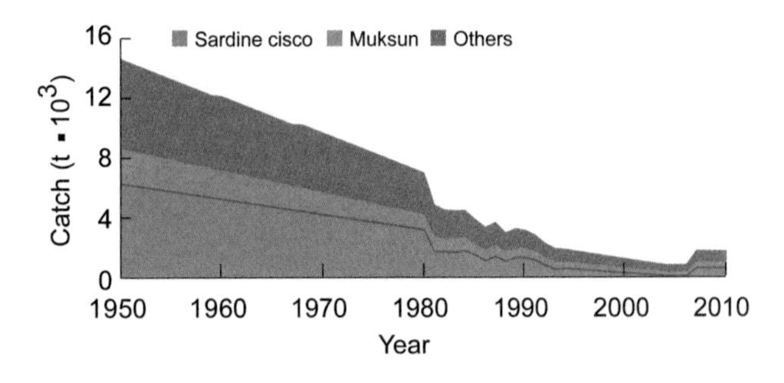

Figure 2. The subsistence fisheries catch in Russia's Kara Sea (scores: Sub. 1, 1, 1), by taxon.

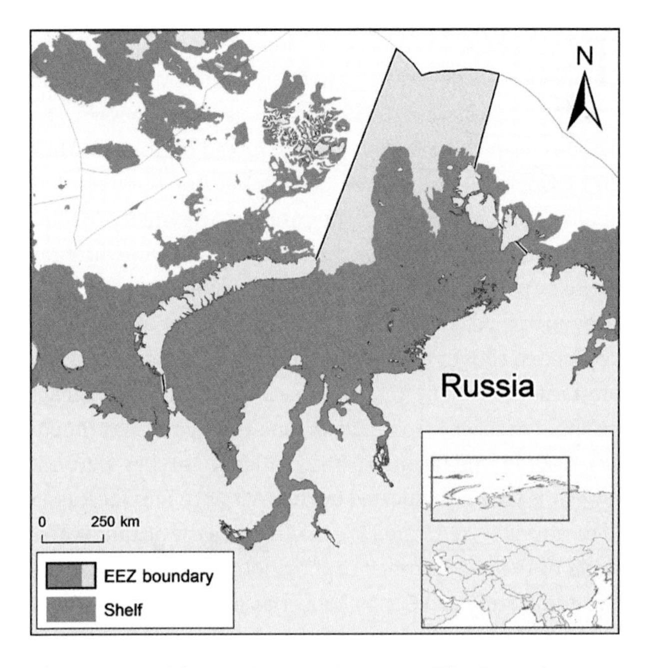

Figure 1. Russia's Kara Sea, northwestern Siberia, and corresponding EEZ of 1.06 million km², of which 801,000 km² consists of shelf.

REFERENCES

Larsen LH, Palerud R, Goodwin H and Sirenko B (1996) The marine invertebrates, fish and coastal zone features of the NSR area. INSROP Working Paper No 53, Fridtjof Nansen Institute. 42 p.

Pauly D and Swartz W (2007) Marine fish catches in North Siberia (Russia, FAO Area 18). pp. 17–33 In Zeller D and Pauly D (eds.), *Reconstruction of Marine Fisheries Catches for Key Countries and Regions* (1950–2005). Fisheries Centre Research Reports 15(2).

Slavin SV (1964) Economic development of the Siberian North. *Arctic* 17(2): 104–108.

Vilchek GE, Krasnovskaya TM and Chelyukanov VV (1996) The environment in the Russian Arctic: Status report. *Polar Geography* 20(1): 20–43.

Zeller D, Booth S, Pakhomov E, Swartz W and Pauly D (2011) Arctic fisheries catches in Russia, USA, and Canada: baselines for neglected ecosystems. *Polar Biology* 34(7): 955–973.

1. Cite as Pauly, D. 2016. Russia (Kara Sea). P. 378 in D. Pauly and D. Zeller (eds.), *Global Atlas of Marine Fisheries: A Critical Appraisal of Catches and Ecosystem Impacts*. Island Press, Washington, DC.

RUSSIA (LAPTEV TO CHUKCHI SEAS)[1]

Daniel Pauly,[a] Wilf Swartz,[a] Evgeny Pakhomov,[b] and Dirk Zeller[a]

[a]*Sea Around Us*, University of British Columbia, Vancouver, Canada
[b]Earth, Ocean and Atmospheric Sciences, University of British Columbia, Vancouver, Canada

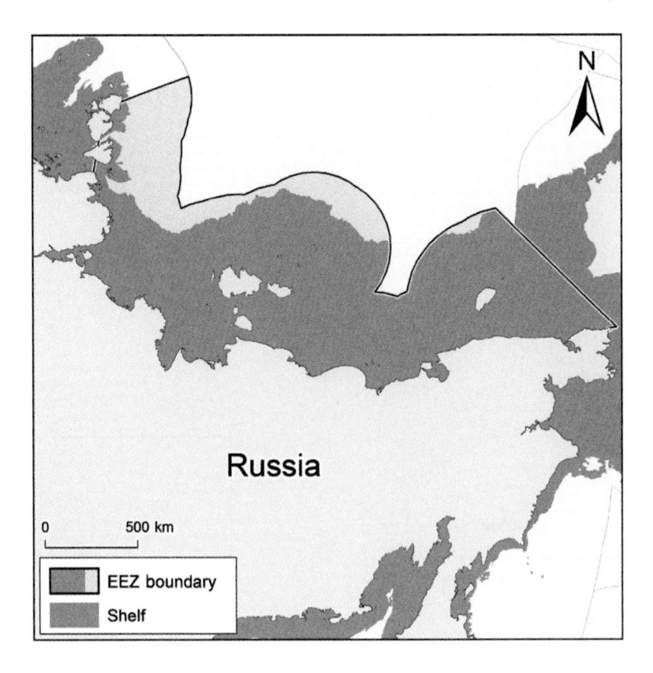

Figure 1. The Laptev, East Siberian and Chukchi Seas (Russia), and corresponding EEZ, of 2.09 million km², of which 1.71 million km² consists of shelf.

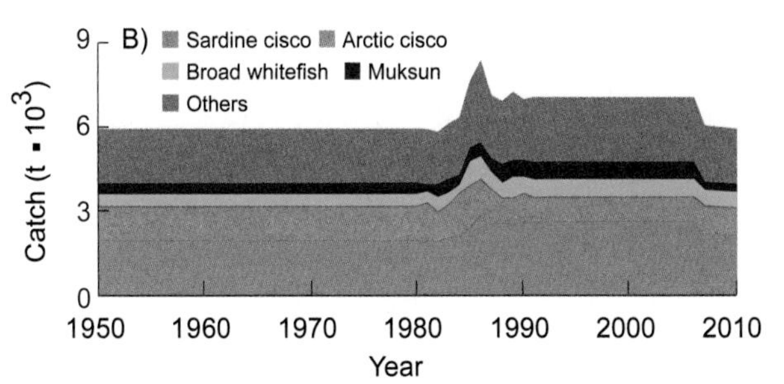

Figure 2. The subsistence fisheries catch in the Laptev, East Siberian and Chukchi Seas (Russia), by taxon (scores: Sub. 1, 1, 1).

There are four well-defined Large Marine Ecosystems (LMEs) along the northern coast of Siberia. This account covers 3 of those: the Laptev Sea, the East Siberian Sea, and the Russian part of the Chukchi Sea LMEs (see previous page for the Kara Sea LME-equivalent EEZ area). These 3 LMEs have in common a harsh and rapidly changing climate (Gordeev et al. 2006); an extreme paucity of fish fauna (see FishBase, www.fishbase.org), especially for the Laptev and East Siberian Seas; and limited subsistence fisheries, based mainly on cisco or "whitefish" (*Coregonus* spp.) in the latter 2 LME and on other species of diadromous salmonids in the Chukchi Sea (Newell 2004). These 3 LME-equivalent EEZ areas also share an extreme scarcity of fisheries-related data, and the very tentative catches in figure 2 are based mainly on Russian internal reports cited in Larsen et al. (1996), with a number of other, mainly anecdotal information sources documented in Pauly and Swartz (2007) and Zeller et al. (2011). No major catch trend was evident in the limited quantitative data that were available. This is perhaps compatible with the declining population of ethnic Russians in northern Siberia (who participated in the subsistence fishing along the coast) being compensated for by a more intense exploitation by the resurging aboriginal people of northern Siberia.

REFERENCES

Gordeev VV, Andreeva EN, Lisitzin AP, Kremer HH, Salomons W and Mashall-Crossland JI (2006) Russian Arctic basins: LOICZ global change assessment and synthesis of river catchment: Coastal areas interaction and human dimension. LOICZ Report and Studies No 29. 95 p.

Larsen LH, Palerud R, Goodwin H and Sirenko B (1996) The marine invertebrates, fish and coastal zone features of the NSR area. INSROP Working Paper No 53, Fridtjof Nansen Institute. 42 p.

Newell J (2004) The Russian far east: A reference guide for conservation and development. Daniel & Daniel Publishers, McKinleyville, U.S.

Pauly D and Swartz W (2007) Marine fish catches in North Siberia (Russia, FAO Area 18). pp. 17–33 In Zeller D and Pauly D (eds.), *Reconstruction of Marine Fisheries Catches for Key Countries and Regions (1950–2005)*. Fisheries Centre Research Reports 15(2).

Zeller D, Booth S, Pakhomov E, Swartz W and Pauly D (2011) Arctic fisheries catches in Russia, USA, and Canada: baselines for neglected ecosystems. *Polar Biology* 34(7): 955–973.

1. Cite as Pauly, D., W. Swartz, E. Pakhomov, and D. Zeller. 2016. Russia (Laptev to Chukchi Seas). P. 379 in D. Pauly and D. Zeller (eds.), *Global Atlas of Marine Fisheries: A Critical Appraisal of Catches and Ecosystem Impacts*. Island Press, Washington, DC.

SAMOA[1]

Stephanie Lingard, Sarah Harper, and Dirk Zeller

Sea Around Us, University of British Columbia, Vancouver, Canada

Samoa, located in the western South Pacific (figure 1), formerly known as Western Samoa, consists of 2 large islands and 7 small islets. Since the 1950s, artisanal (i.e., small-scale) commercial fisheries have developed, although noncommercial subsistence fishing dominates catches for the entire time period, accounting for more than 80% of total catches for 1950–2010 (figure 2A). In the 1970s, a local tuna fishery was initiated, and tuna has become a major export, complementing tourism as a source of foreign currency (Read 2006). Fisheries catches were reconstructed by Lingard et al. (2012), and domestic catches increased steadily from 1950 to a plateau in the 1980s, followed by a rapid and drastic decline in 1990 caused by the impacts of 2 successive cyclones that caused substantial damage to fishing vessels and the coral reef habitat, before catches increased again (figure 2B). Domestic reconstructed catches were 2.8 times the data submitted to the FAO, the disparity being caused mostly by underreported subsistence and artisanal catches, primarily in earlier years. Foreign fishing for large pelagics has increased in the last decade (figure 2B). Market surveys were used to improve resolution of the taxonomic composition of the small-scale (artisanal and subsistence) domestic catches (figure 2C). Emperors (family Lethrinidae) dominated earlier years, and small-scale octopus catches increased in recent years, as did large pelagics such as albacore (*Thunnus alalunga*) and yellowfin tuna (*T. albacares*).

REFERENCES

Lingard S, Harper S and Zeller D (2012) Reconstructed catches of Samoa 1950–2010. pp. 103–118 In Harper S, Zylich K, Boonzaier L, Le Manach F, Pauly D and Zeller D (eds.), *Fisheries catch reconstructions: Islands, Part III*. Fisheries Centre Research Reports 20(5).

Read R (2006) Sustainable natural resource use and economic development in small states: the tuna fisheries in Fiji and Samoa. *Sustainable Development* 14: 93–103.

1. Cite as Lingard, S., S. Harper, and D. Zeller. 2016. Samoa. P. 380 in D. Pauly and D. Zeller (eds.), *Global Atlas of Marine Fisheries: A Critical Appraisal of Catches and Ecosystem Impacts*. Island Press, Washington, DC.

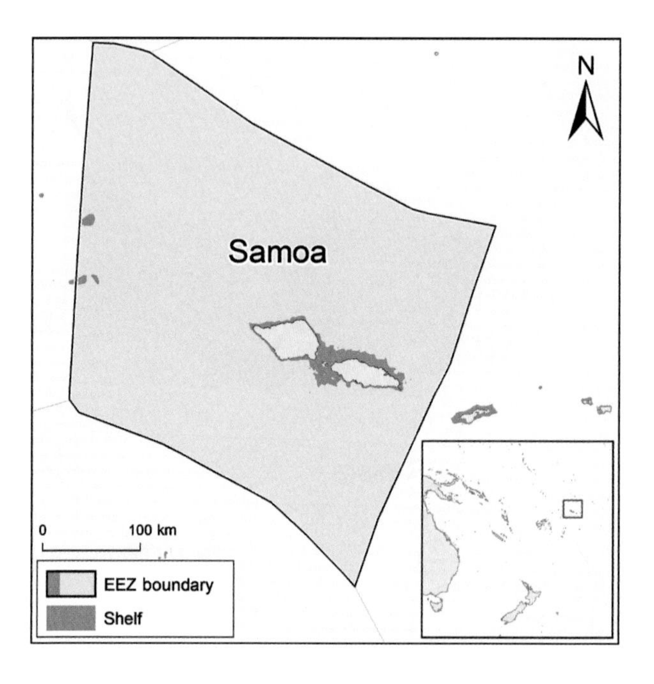

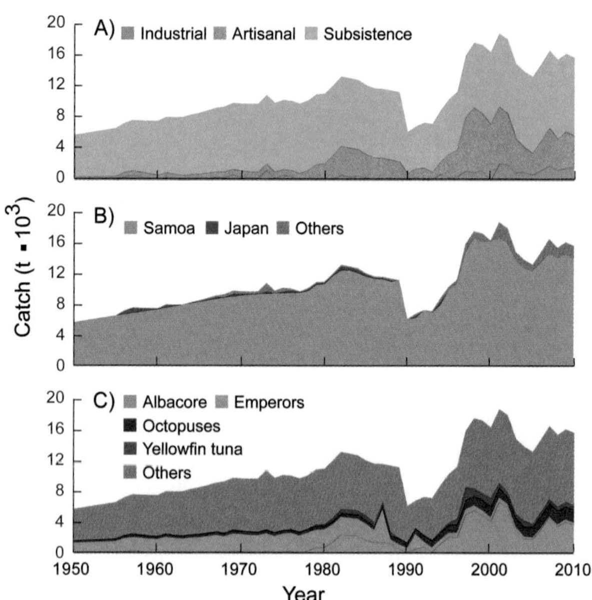

Figure 1. Samoa has a combined land area of 2,230 km² and an EEZ that does not extend 200 miles offshore and hence is the smallest (132,000 km²) in the region.

Figure 2. Domestic and foreign catches taken in the EEZ of Samoa: (A) by sector (domestic scores: Art. 1, 1, 3; Sub. 1, 1, 3; Dis. –, –, 1); (B) by fishing country (foreign catches are very uncertain); (C) by taxon.

SÃO TOMÉ AND PRÍNCIPE[1]

Dyhia Belhabib and Daniel Pauly

Sea Around Us, University of British Columbia, Vancouver, Canada

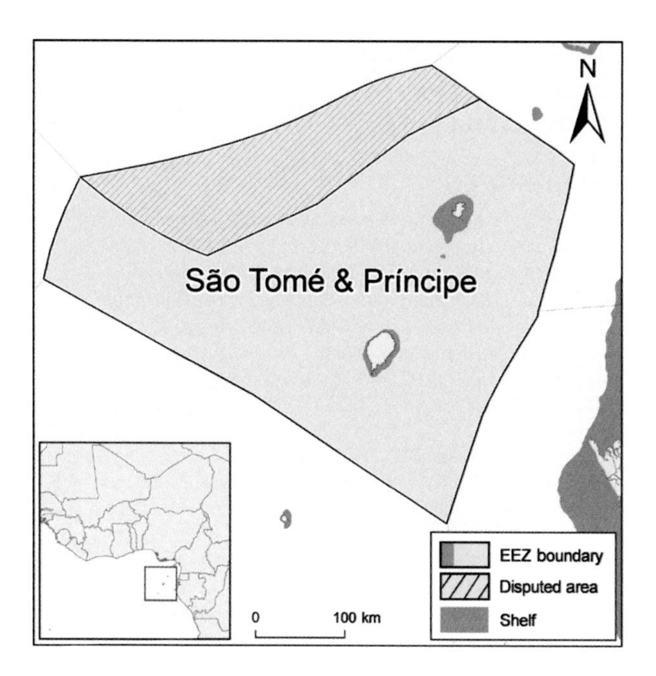

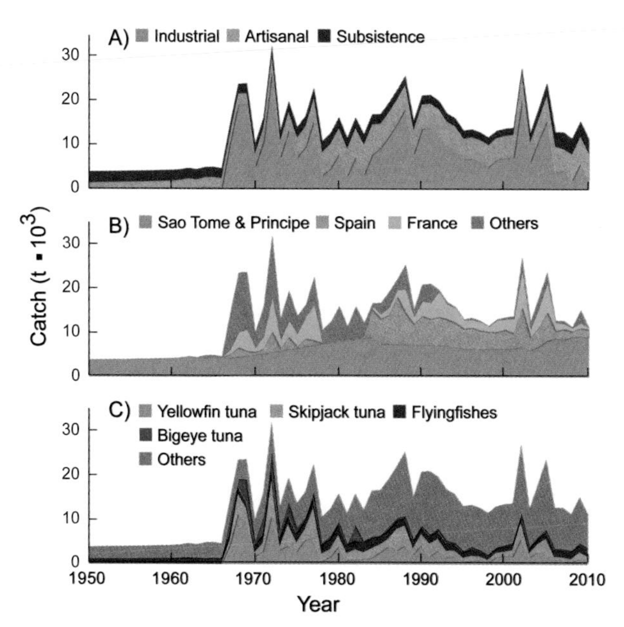

Figure 1. São Tomé and Príncipe has a land area of about 1,000 km² and an EEZ of 165,000 km². **Figure 2.** Domestic and foreign catches taken in the EEZ of São Tomé and Príncipe: (A) by sector (domestic scores: Art. 2, 3, 3; Sub. 2, 3, 3); (B) by fishing country (foreign catches are very uncertain); (C) by taxon.

São Tomé and Príncipe consists of two islands in the Gulf of Guinea (figure 1); it is also the least populated country of West Africa and one of the few with a functioning multiparty democracy. The population of São Tomé and Príncipe depends greatly on fisheries as a source of income and food, reflected in one of the highest consumption rates of West Africa. However, the domestic fisheries lack monitoring and management, and the foreign fisheries in the EEZ do what such fisheries generally do (Carneiro 2012). Although important artisanal and subsistence sectors exist domestically, total catches as allocated to this EEZ were dominated by industrial fleets (figure 2A), which are exclusively foreign (figure 2B). Total domestic catches from the EEZ of São Tomé and Príncipe were estimated by Belhabib (2015) to have grown from about 4,000 t in 1950 to almost 10,000 t by 2010 (figure 2B). Domestic catches were 4 times the data reported to the FAO, and foreign catches were higher, dominated by EU, Asian, and Russian fleets (figure 2B). Total catches in this EEZ were dominated by Atlantic flyingfish (*Cheilopogon melanurus*) for the domestic and tunas for the foreign sector (figure 2C). Declining catches despite increasing effort, and climate change, threaten the coastal communities of São Tomé and Príncipe, whose fish consumption has already begun to decline.

REFERENCES

Belhabib D (2015) Fisheries of São Tomé and Príncipe, a catch reconstruction 1950–2010. Fisheries Centre Working Paper #2015-67, 13 p.

Carneiro G (2012). They come, they fish, and they go. EC fisheries agreements with Cape Verde and São Tomé e Príncipe. *Marine Fisheries Review* 73(4): 1–25.

1. Cite as Belhabib, D., D. Pauly. 2016. São Tomé and Príncipe. P. 381 in D. Pauly and D. Zeller (eds.), *Global Atlas of Marine Fisheries: A Critical Appraisal of Catches and Ecosystem Impacts*. Island Press, Washington, DC.

SAUDI ARABIA (PERSIAN GULF)[1]

Dawit Tesfamichael and Daniel Pauly

Sea Around Us, University of British Columbia, Vancouver, Canada

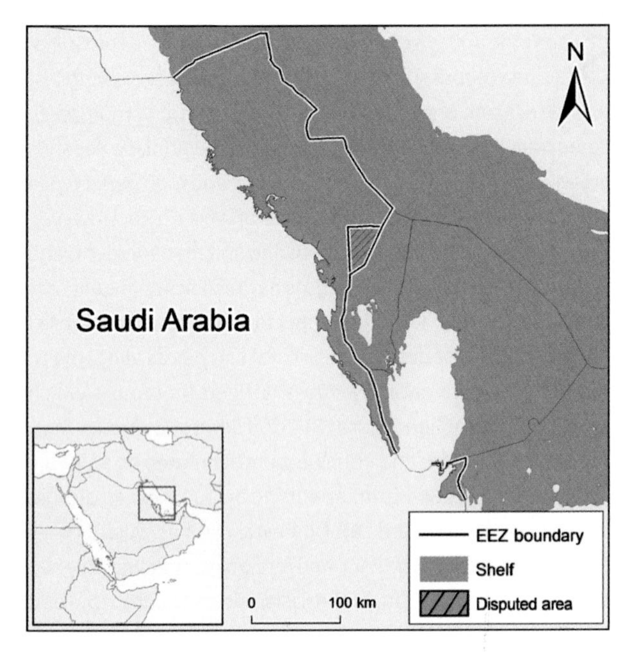

Figure 1. Map of Saudi Arabia, showing the extent of its EEZ in the Persian Gulf (33,800 km², all of which is shelf).

Saudi Arabia has a longer coastline in the Red Sea (see next page) than in the Gulf (figure 1), but its fish catch in the Gulf seems higher. The bulk of the Saudi fisheries in the Gulf are artisanal (figure 2A); their motorization started in the early 1960s and was completed in the late 1980s. Based on data from the Regional Commission for Fisheries (RECOFI), the domestic catches of Saudi Arabia in the Gulf were reconstructed (Tesfamichael and Pauly 2013) to increase from about 2,000 t/year in the early 1950s to 50,000 t/year in the late 2000s, with an intermediate phase, from the early 1960s to the early 1990s, with high trawl catches and especially discards, contributing substantially to overall catches from 1950 to 2010. The reconstructed Saudi catch is 2.4 times the data FAO reports on behalf of Saudi Arabia as deemed to relate to Gulf waters. The majority of catches were by the domestic fishery (figure 2B). Because of the nature of artisanal fisheries and Saudi Arabia's Gulf waters, which are generally shallow, with sandy and muddy bottoms covered by seagrass beds, the catch (figure 2C) consists mainly of a multitude of demersal fishes and shrimps, both reduced by oil pollution, notably in 1991 (Mathews et al. 1993).

REFERENCES

Mathews CP, Kedidi S, Fita NI, Al-Yahya A and Al-Rasheed K (1993) Preliminary assessment of the effects of the 1991 Gulf War on Saudi Arabian prawn stocks. *Marine Pollution Bulletin* 27: 251–271.

Tesfamichael D and Pauly D (2013) Catch reconstruction of the fisheries of Saudi Arabia in the Gulf, 1950–2010. pp. 39–52 In Al-Abdulrazzak D and Pauly D (eds.), *From dhows to trawlers: a recent history of fisheries in the Gulf countries, 1950 to 2010*. Fisheries Centre Research Reports 21(2).

1. Cite as Tesfamichael, D., and D. Pauly. 2016. Saudi Arabia (Persian Gulf). P. 382 in D. Pauly and D. Zeller (eds.), *Global Atlas of Marine Fisheries: A Critical Appraisal of Catches and Ecosystem Impacts*. Island Press, Washington, DC.

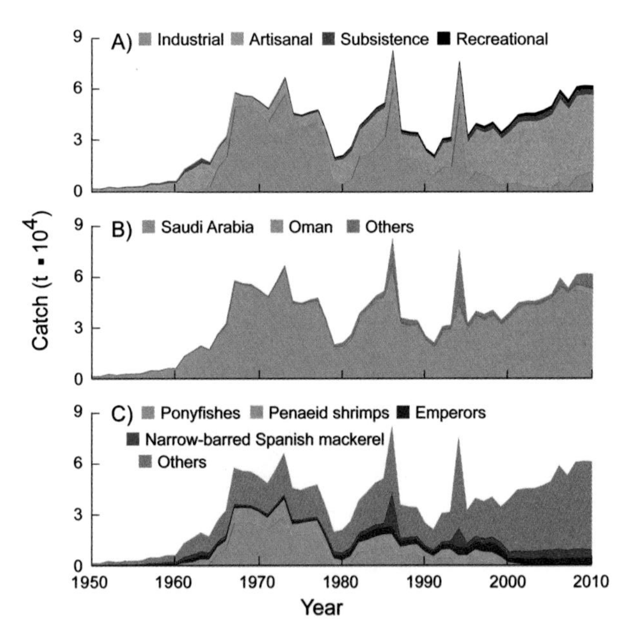

Figure 2. Domestic and foreign catches taken in the Persian Gulf EEZ of Saudi Arabia: (A) by sector (domestic scores: Ind. 2, 3, 3; Art. 2, 2, 2; Sub. 1, 1, 1; Rec. –, 2, 2; Dis. 2, 2, 2); (B) by fishing country (foreign catches are very uncertain); (C) by taxon.

SAUDI ARABIA (RED SEA)[1]

Dawit Tesfamichael[a,b] and Peter Rossing[a]

[a]*Sea Around Us*, University of British Columbia, Vancouver, Canada
[b]Department of Marine Sciences, University of Asmara, Eritrea

Saudi Arabia has coastlines on both the Red Sea (figure 1) and the Persian Gulf (see previous page), but the former is longer. Saudi Red Sea fisheries are dominated by the artisanal fishery (figure 2A), but the industrial sector has grown since the 1980s with economic development from oil, which also led to the development of a recreational fishery (Morgan 2006; Tesfamichael and Pitcher 2006). The reconstruction by Tesfamichael and Rossing (2012) generated estimates of the total catch of Saudi Arabia in its Red Sea EEZ of about 7,000 t/year in the 1950s, which rapidly increased in the mid-1980s, reaching its peak catch of about 50,000 t/year in 1994. It then declined slightly, to about 40,000 t/year by the end of the 2000s. The increase in the 1980s was caused mainly by motorization of artisanal boats and the introduction of an industrial fishery. The reconstructed domestic catch over the 1950–2010 period was 1.5 times the data officially reported, that is, the unreported catch consisted mainly of subsistence catches and industrial discards. The majority of catches were domestic and were dominated by Spanish mackerel (*Scomberomorus commerson*) followed by emperors (family Lethrinidae) and jacks (family Carangidae; figure 2C). The coal reefs and other marine habitats along the Saudi Arabian coast of the Red Sea have been mapped (Bruckner et al. 2012).

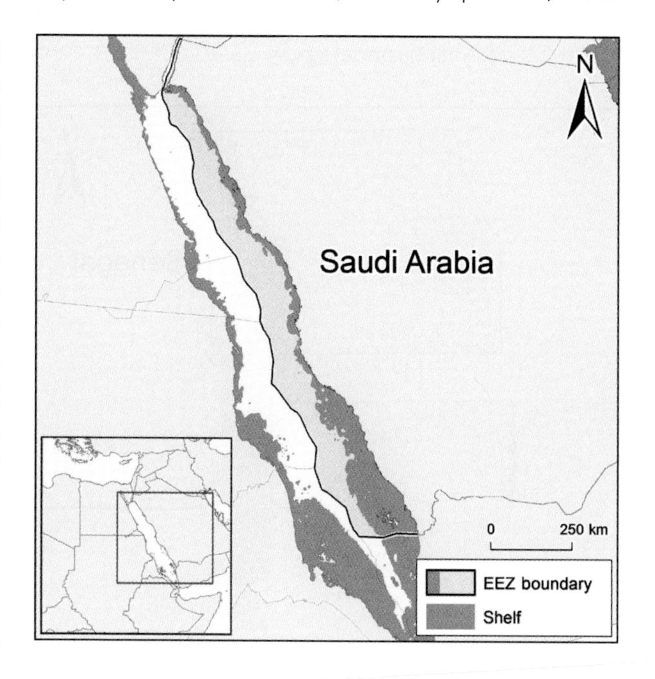

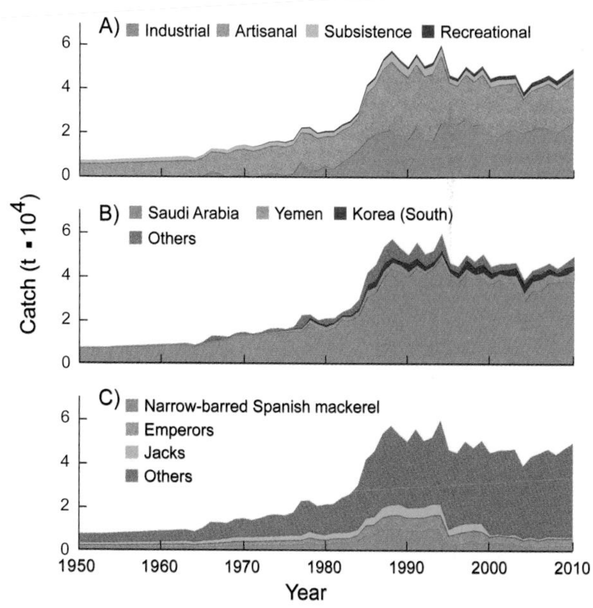

Figure 1. Saudi Arabia and its continental shelf (69,800 km²) and EEZ (186,000 km²) in the Red Sea. **Figure 2.** Domestic and foreign catches taken in the Red Sea EEZ of Saudi Arabia: (A) by sector (domestic scores: Ind. –, 3, 3; Art. 2, 2, 2; Sub. 1, 1, 1; Rec. –, 2, 2; Dis. –, 2, 2); (B) by fishing country (foreign catches are very uncertain); (C) by taxon.

REFERENCES

Bruckner A, Rowlands G, Riegl B, Purkis S, Williams A and Renaud P (2012) *Atlas of Saudi Arabian Red Sea Marine Habitats*. Khaled bin Sultan Living Ocean Foundation and Panoramic Press, Phoenix, AZ, 262 p.

Morgan G (2006) Country Review: Saudi Arabia, 303–314 In De Young C (ed.) *Review of the state of world marine capture fisheries management: Indian Ocean*. FAO Fisheries Technical Paper. No. 488.

Tesfamichael D and Pitcher TJ (2006) Multidisciplinary evaluation of the sustainability of Red Sea fisheries using Rapfish. *Fisheries Research* 78(2–3): 227–235.

Tesfamichael D and Rossing P (2012) Reconstructing Red Sea fisheries catches of Saudi Arabia: National wealth and fisheries transformation. In Tesfamichael D and Pauly D (eds.), *Catch reconstruction for the Red Sea large marine ecosystem by countries (1950–2010)*. Fisheries Centre Research Reports 20(1).

1. Cite as Tesfamichael, D., and P. Rossing. 2016. Saudi Arabia (Red Sea). P. 383 in D. Pauly and D. Zeller (eds.), *Global Atlas of Marine Fisheries: A Critical Appraisal of Catches and Ecosystem Impacts*. Island Press, Washington, DC.

SENEGAL[1]

Dyhia Belhabib,[a] Viviane Koutob,[a] Nassirou Gueye,[b] Lamine Mbaye,[c]
Christopher Mathews,[d] Vicky W. Y. Lam,[a] and Daniel Pauly[a]

[a]*Sea Around Us*, University of British Columbia, Vancouver, Canada [b]World Wide Fund for Nature, West African Marine Ecoregion, Dakar, Sénégal [c]Direction des Pêches Maritimes, Dakar, Sénégal
[d]Consultant; former Technical Leader, USAID/COMFISH, Dakar

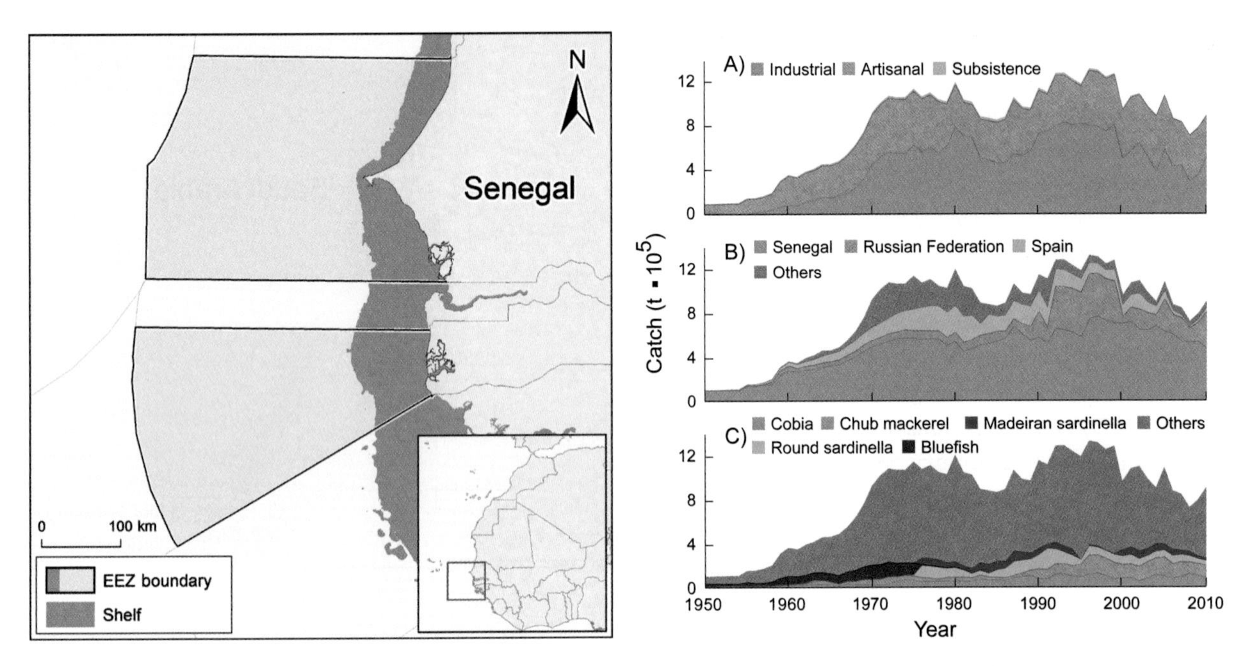

Figure 1. Senegal (EEZ: 158,000 km²) has a continental shelf of 23,900 km². **Figure 2.** Domestic and foreign catches taken in the EEZ of Senegal: (A) by sector (domestic scores: Ind. 3, 3, 4; Art. 2, 2, 3; Sub. 3, 3, 3; Rec. 2, 2, 4; Dis. 4, 4, 4); (B) by fishing country (foreign catches are very uncertain); (C) by taxon.

Senegal is located at the edge of 2 highly productive fishing zones: the Canary Current Large Marine Ecosystem and the Gulf of Guinea Large Marine Ecosystem (figure 1). Its colonial history contributed toward enriching fisheries data through surveys and analyses of the artisanal and industrial sectors. Official catches are lower than estimated maximum sustainable yield (MSY), but the pervasiveness of overexploitation (Cormier-Salem 1994; Thiao et al. 2013) suggests that this MSY will remain unattainable. Total catches are dominated by the industrial sector (figure 2A), although much of this is by foreign fleets (figure 2B). Reconstructed domestic catches were estimated by Belhabib et al. (2013, 2014) at 100,000 t in 1950, increasing to more than 700,000 t/year in the late 1990s before declining to about 500,000 t in 2010 (figure 2B). Domestic catches in the Senegalese EEZ were 2 times the data supplied to the FAO, once adjusted for its EEZ. Illegal foreign fishing appears to have peaked in the late 2000s, and legal foreign catches have continued to decline as Senegal reduces fishing opportunities by foreign fleets. Senegalese catches taken outside the Senegalese EEZ increased from low levels in the mid-1970s to about 250,000 t in 2010, much of which is taken from the Mauritanian EEZ but landed and reported in Senegal. Senegalese catches cover a multitude of taxa (figure 2C), but recently, the dominance of chub mackerel (*Scomber japonicus*) has increased.

REFERENCES

Belhabib D, Koutob V, Gueye N, Mbaye L, Mathews C, Lam VWY and Pauly D (2013) Lots of boats and fewer fishes: A preliminary catch reconstruction for Senegal, 1950–2010. Fisheries Centre Working Papers #2013-03, 34 p.

Belhabib D, Koutob V, Lam VWY, Sall A and Pauly D (2014) Fisheries catch misreporting and its implications: the case of Senegal. *Fisheries Research* 151: 1–11.

Cormier-Salem MC (1994) Environmental changes, agricultural crisis and small-scale fishing development in the Casamance region, Senegal. *Ocean & Coastal Management* 24(2): 109–124.

Thiao D, Chaboud C, Samba A, Laloë F and Cury PM (2012) Economic dimension of the collapse of the "false cod" *Epinephelus aeneus* in a context of ineffective management of the small-scale fisheries in Senegal. *African Journal of Marine Science* 34(3): 305–311.

1. Cite as Belhabib, D., V. Koutob, N. Gueye, L. Mbaye, C. Mathews, V. Lam, and D. Pauly. 2016. Senegal. P. 384 in D. Pauly and D. Zeller (eds.), *Global Atlas of Marine Fisheries: A Critical Appraisal of Catches and Ecosystem Impacts*. Island Press, Washington, DC.

SEYCHELLES[1]

Frédéric Le Manach,[a] Pascal Bach,[b] Léa Boistol,[a] Jan Robinson,[c] and Daniel Pauly[a]

[a]*Sea Around Us*, University of British Columbia, Vancouver, Canada [b]Institut de Recherche pour le Développement, Sète, France [c]Seychelles Fishing Authority, Victoria, Seychelles

The Seychelles is an archipelago located north of Madagascar and composed of 115 islands (figure 1). With about 15% of all formal jobs, the fisheries sector is the main pillar of the national economy, the second being tourism. The domestic fisheries sector is small-scale, with a fleet of small boats targeting demersal and small pelagics in and around the coral reefs (SFA 2014). Since the early 1990s, though, an expansion toward offshore waters has occurred, with a fleet of longliners targeting primarily swordfish. Total domestic catches were estimated to have increased from 3,100 t/year in the 1950s to about 20,000 t/year in the 2000s (of which about 5,000 t/year is artisanal; Le Manach et al. 2015), with reconstructed domestic catches being about 1.3 times the figures reported to the FAO. Overall, catches within the EEZ are dominated by industrial fleets (figure 2A), most of which is taken by foreign fleets (figure 2B). Because of the predominance of foreign industrial fleets, catches are dominated by large pelagics (figure 2C). A large sector contributing to the national economy is foreign owned: The large fleets of Spanish purse-seiners and Taiwanese and Japanese longliners are often Seychelles flagged and fish throughout the western Indian Ocean. A portion of their catch is landed in Victoria on Mahé (figure 1), the largest tuna hub in the Indian Ocean, and processed at the national cannery.

REFERENCES

Le Manach F, Bach P, Boistol L, Robinson J and Pauly D (2015) Artisanal fisheries in the world's second largest tuna fishing ground: Reconstruction of the Seychelles' marine fisheries catch, 1950–2010. pp. 99–109 In Le Manach F and Pauly D (eds.), *Fisheries catch reconstructions in the Western Indian Ocean, 1950–2010*. Fisheries Centre Research Reports 23(2).

SFA (2014) Annual report 2012. Seychelles Fishing Authority (SFA) Victoria (Seychelles). xiii + 75 p.

1. Cite as Le Manach, F., P. Bach, L. Boistol, J. Robinson, and D. Pauly. 2016. Seychelles. P. 385 In D. Pauly and D. Zeller (eds.), *Global Atlas of Marine Fisheries: A Critical Appraisal of Catches and Ecosystem Impacts*. Island Press, Washington, DC.

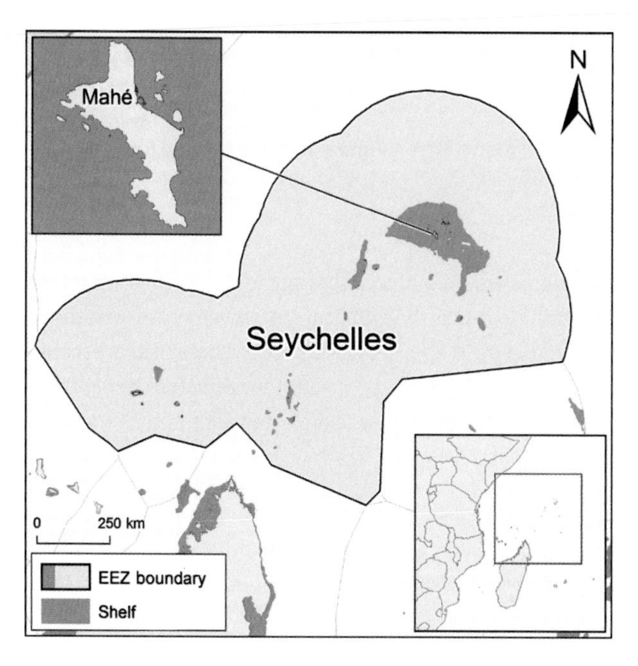

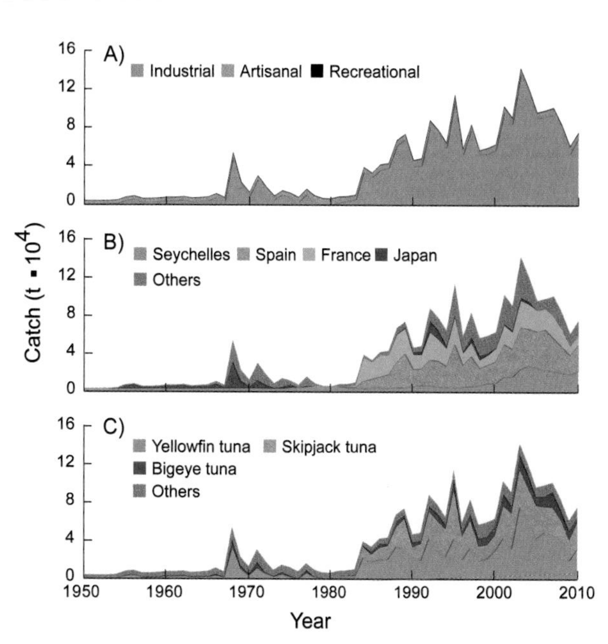

Figure 1. The Seychelles has a land area of 459 km² and an EEZ of 1.33 million km². **Figure 2.** Domestic and foreign catches taken in the EEZ of the Seychelles: (A) by sector (domestic scores: Ind. –, 3, 4; Art. 3, 3, 3; Rec. –, 1, 1; Dis. –, –, 3); (B) by fishing country (foreign catches are very uncertain); (C) by taxon.

SIERRA LEONE[1]

Katherine Seto,[a] Dyhia Belhabib,[b] Duncan Copeland,[b] Michael Vakily,[c] Heiko Seilert,[d] Salieu Sankoh,[e] Andrew Baio,[f] Ibrahim Turay,[g] Sarah Harper,[b] Dirk Zeller,[b] Kyrstn Zylich,[b] and Daniel Pauly[b]

[a]Department of Environmental Science, Policy, and Management, University of California at Berkeley, CA
[b]*Sea Around Us*, University of British Columbia, Vancouver, Canada [c]Deutsche Gesellschaft für Internationale Zusammenarbeit (GIZ) GmbH, Biodiversity Programme Office, New Delhi, India [d]GOPA Consultants, Bad Homburg, Germany
[e]West Africa Regional Fisheries Programme, Ministry of Fisheries and Marine Resources, Freetown, Sierra Leone
[f]Institute of Marine Biology and Oceanography, University of Sierra Leone, Freetown, Sierra Leone
[g]West African Regional Fisheries Programme Sub Regional Fisheries Commission, Dakar, Senegal

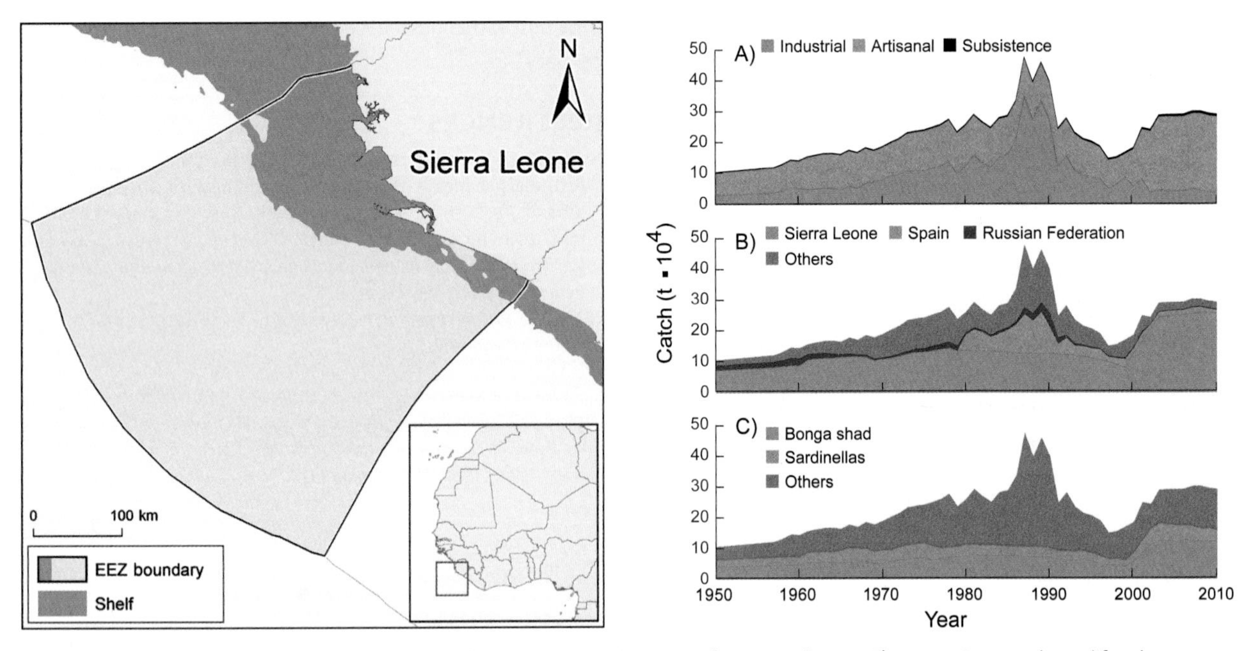

Figure 1. Sierra Leone has a continental shelf of 26,600 km² and an EEZ of 159,000 km². **Figure 2.** Domestic and foreign catches taken in the EEZ of Sierra Leone: (A) by sector (domestic scores: Ind. –, 4, 4; Art. 2, 3, 4; Sub. 1, 1, 1; Dis. –, 2, 4); (B) by fishing country (foreign catches are very uncertain); (C) by taxon.

Sierra Leone is located in northwest Africa (figure 1). After independence from British colonial rule in 1961, the country experienced nearly continuous strife over access to natural resources, culminating in a civil war (1991–2002) funded by blood diamonds. With deepening poverty, the people of Sierra Leone became increasingly dependent on marine fisheries as an alternative source of livelihood and food. But like blood diamonds, fisheries are usually dominated by both fleets large-scale operations and hard-to-monitor small-scale operations (Golley-Morgan 2012). The reconstruction by Seto et al. (2015) estimated domestic catches of 70,000 t in 1950, which peaked at 140,000 t in the late 1970s, declined to 80,000 t in 2000, and then increased again since the end of the civil war to 250,000 t in 2010. Domestic catches were about 30% higher than the data supplied to the FAO and were dominated by the artisanal sector (figure 2A). Although foreign fleets dominated in the 1980s, this has declined in recent years (figure 2B). Bonga shad (*Ethmalosa fimbriata*) had the largest contribution to catches, followed by sardinella (*Sardinella* spp.; figure 2C). Illegal foreign fisheries were ubiquitous during the civil war. Their recent decrease illustrates the significant improvement in monitoring and enforcement efforts by the country. The artisanal fisheries increased strongly in response to development aid and the opportunity to act as near-shore extensions of the remaining foreign industrial fleets.

REFERENCES

Golley-Morgan ETA (2012) The Fisheries Sector of Sierra Leone. pp. 17–21 In Vakily JM, Seto K and Pauly D (eds.), *The marine fisheries environment of Sierra Leone: Belated proceedings of a national seminar held in Freetown, November 1991*. Fisheries Center Research Reports 20(4).

Seto K, Belhabib D, Copeland D, Vakily M, Seilert H, Sankoh S, Baio A, Turay I, Harper S, Zeller D, Zylich K and Pauly D (2015) Colonialism, conflict, and fish: a reconstruction of marine fisheries catches for Sierra Leone, 1950–2010. Fisheries Centre Working Paper #2015-74, 23 p.

1. Cite as Seto, K., D. Belhabib, D. Copeland, M. Vakily, H. Seilert, S. Sankoh, A. Baio, I. Turay, S. Harper, D. Zeller, K. Zylich, and D. Pauly. 2016. Sierra Leone. P. 386 in D. Pauly and D. Zeller (eds.), *Global Atlas of Marine Fisheries: A Critical Appraisal of Catches and Ecosystem Impacts*. Island Press, Washington, DC.

SINGAPORE[1]

Loida Corpus

Sea Around Us, University of British Columbia, Vancouver, Canada

Singapore is a small island state at the tip of the Malaysian Peninsula (figure 1). Singapore started rebuilding its fisheries soon after World War II, and signs of overexploitation of the inshore waters (corresponding to the small EEZ it later acquired) became evident to its efficient administrators in the early 1950s. In response, Singapore fisheries expanded into the waters of nearby Malaysia and beyond, until declining catch rates made distant-water fishing unprofitable. The catch reconstruction by Corpus (2014) led to catch estimates of about 7,000 t in 1950, taken inshore by small-scale fisheries, followed by a rapid growth of the offshore (industrial) fishery (mainly outside EEZ waters), with a peak of about 30,000 t/year in the early to mid-1980s, while inshore EEZ catches stagnated in the early 1990s. The subsequent phasing out of the offshore fishery brought Singapore back to square one, with the inshore artisanal, subsistence, and recreational fisheries yielding about 3,000 t/year in the late 2000s (figure 2A). Overall, reconstructed domestic catches were 1.3 times the data reported to the FAO. The gross composition of catches from this EEZ is presented in figure 2B. Corpus (2014) presents more taxa, based on the detailed records and publications generated by Singapore's fisheries, which included the species composition of "trash fish" (Sinoda et al. 1978), eventually used for fish balls and other products.

REFERENCES

Corpus L (2014) Reconstructing Singapore's marine fisheries catch, 1950–2010. pp. 137–146 In Zylich K, Zeller D, Ang M and Pauly D (eds.), *Fisheries catch reconstructions: Islands, Part IV*. Fisheries Centre Research Reports 22(2).

Sinoda M, Lim PY and Tan SM (1978) Preliminary study of trash fish landed at Kangkar fish market in Singapore. *Bulletin of the Japanese Society for the Science of Fish* 44(6): 595–600.

1. Cite as Corpus, L. 2016. Singapore. P. 387 in D. Pauly and D. Zeller (eds.), *Global Atlas of Marine Fisheries: A Critical Appraisal of Catches and Ecosystem Impacts*. Island Press, Washington, DC.

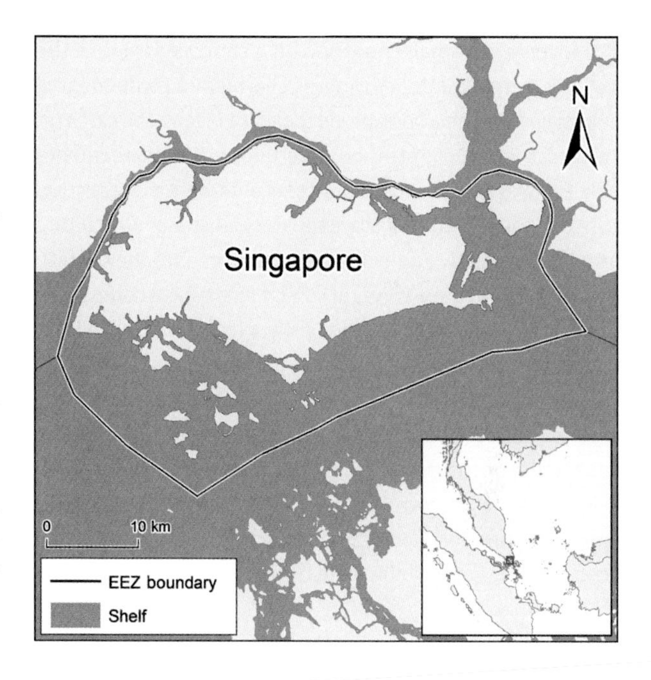

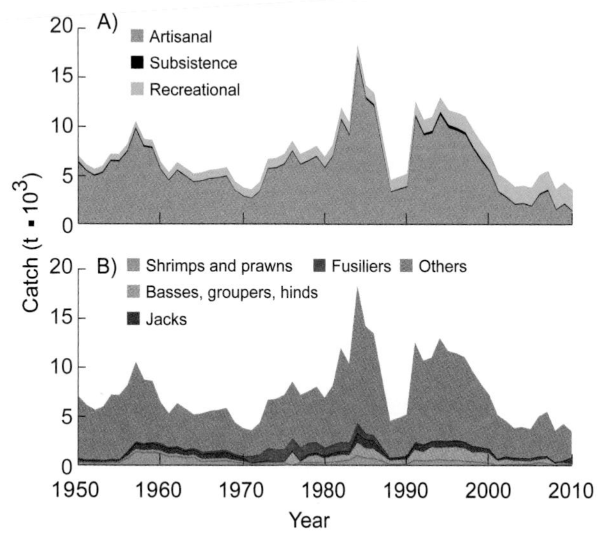

Figure 1. The total land area of Singapore Island is 718 km², and its small EEZ is 673 km². **Figure 2.** Domestic and foreign catches taken in the EEZ of Singapore: (A) by sector (domestic scores: Art. 4, 3, 3; Sub. 4, 3, 2; Rec. 3, 3, 4); (B) by taxon. Note that the domestic industrial sector was mainly a distant-water fishery, because it operated outside the inshore (i.e., EEZ) waters of Singapore.

SLOVENIA[1]

Aleš Bolje,[a] Bojan Marčeta,[a] Andrej Blejec,[b] and Alasdair Lindop[c]

[a]Fisheries Research Institute of Slovenia, Ljubljana-Šmartno, Slovenia [b]National Institute of Biology, Department of Entomology, Ljubljana, Slovenia [c]*Sea Around Us*, University of British Columbia, Vancouver, Canada

Slovenia is a small country with a short coastline in the northern Adriatic Sea (figure 1). Formerly a component of Yugoslavia, Slovenia became independent in 1991. The catch reconstruction by Bolje et al. (2015) pertains to Slovenian catches only, for 1950 to 2010. Major sources of data were the Statistical Office of the Republic of Slovenia, the Ministry of Agriculture and Environment, and data gathered from interviews with fishers and from the archives of commercial fishing companies. Landings from 2005 to 2012 and fishery monitoring data from 1995 to 2013 were used to determine the importance of different species to various fishing gear. Discards were estimated by species and gear from observer data, collected on board fishing vessels. Reconstructed catches, which started at about 2000 t/year in the 1950s, increased to a peak of 8,700 t in 1983, after which catches declined dramatically to just under 900 t by 2010. The majority of the industrial catch (figure 2A) before 1991 was taken in the waters of other former members of Yugoslavia (e.g., in Croatia), with little or no foreign fishing in the small Slovenian EEZ, Much of the catch consisted of European pilchard (*Sardina pilchardus*) and other small pelagics (figure 2B). Nastav et al. (2012) suggested that jellyfish (e.g., barrel jellyfish, *Rhizostoma pulmo*, and mauve stinger *Pelagia noctiluca*) affect the fisheries catches of Slovenia and hence its economy.

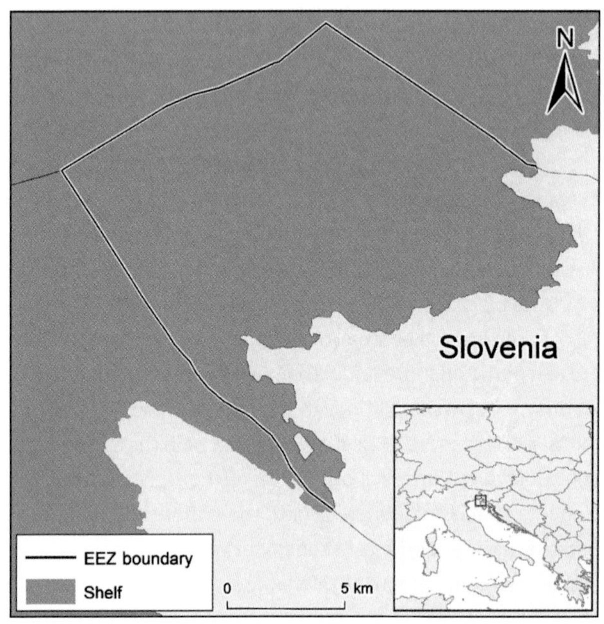

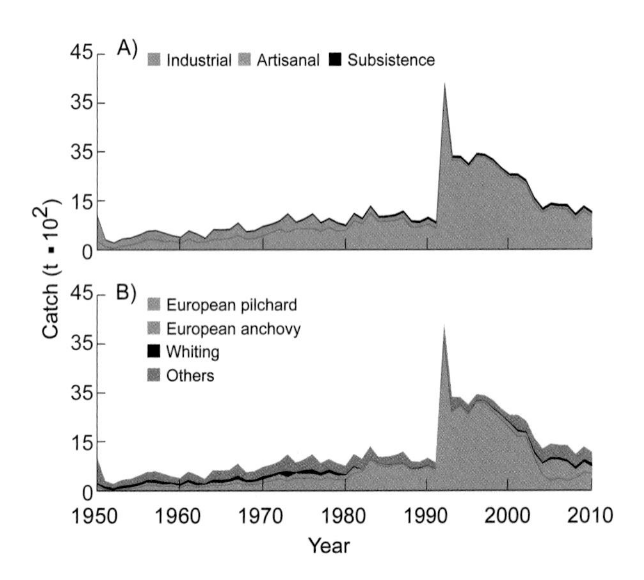

Figure 1. Slovenia has a shelf of 192 km², which fills all its EEZ. **Figure 2.** Domestic and foreign catches taken in the EEZ of Slovenia: (A) by sector (domestic scores: Ind. 2, 3, 3; Art. 2, 3, 3; Sub. 1, 1, 1; Rec. 1, 1, 1; Dis. 2, 2, 2); (B) by taxon.

REFERENCES

Bolje A, Marčeta B, Blejec A and Lindop A (2015) Marine fish catches in Slovenia between 1950 and 2010. Fisheries Centre Working Paper #2015-58, 13 p.

Nastav B, Malej M, Malej Jr A and Male A (2012) Is it possible to determine the economic impact of jellyfish outbreaks on fisheries? A case study–Slovenia. *Mediterranean Marine Science* 14(1): 214–223.

1. Cite as Bolje, A., B. Marčeta, A. Blejec, and A. Lindop. 2016. Slovenia. P. 388 In D. Pauly and D. Zeller (eds.), *Global Atlas of Marine Fisheries: A Critical Appraisal of Catches and Ecosystem Impacts*. Island Press, Washington, DC.

SOLOMON ISLANDS[1]

Bridget Doyle, Sarah Harper, Jennifer Jacquet, and Dirk Zeller

Sea Around Us, University of British Columbia, Vancouver, Canada

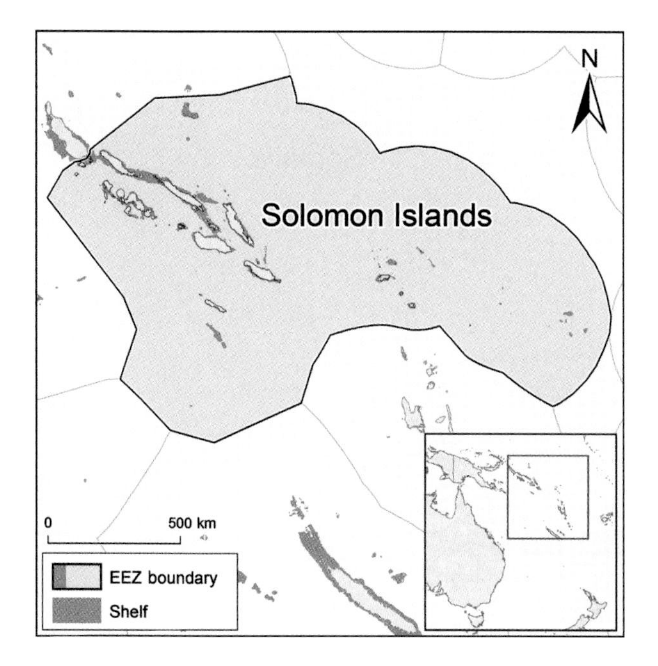

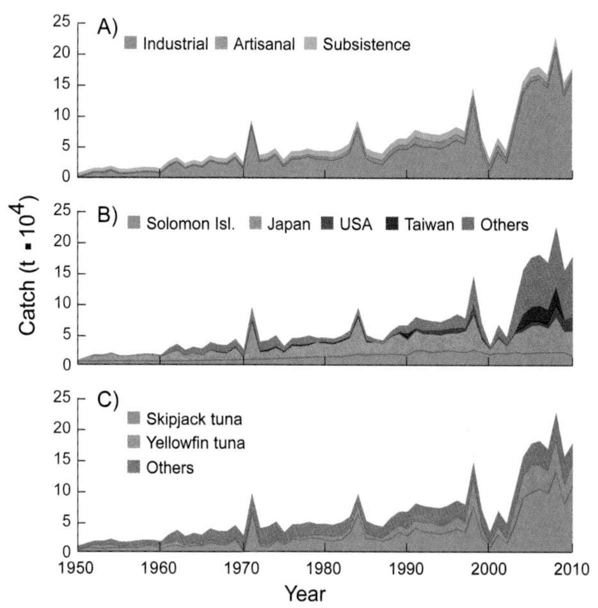

Figure 1. The total land area of the Solomon Islands is 28,400 km², and the EEZ is about 1.6 million km². **Figure 2.** Domestic and foreign catches taken in the EEZ of Solomon Islands: (A) by sector (domestic scores: Ind. –, 4, 4; Art. 2, 2, 3; Sub. 1, 2, 3); (B) by fishing country (foreign catches are very uncertain); (C) by taxon.

The Solomon Islands (figure 1) have some of the world's largest lagoons, notably Morovo Lagoon (Hviding and Baines 1994), and extensive coral reefs. The socioeconomic welfare and food security of the Solomon Islands rely heavily on its fisheries. Because the Solomon Islands are one of the few countries to report (at least partly) on the catch of its subsistence fisheries (Agassi 2005) and they do not have fisheries generating large amounts of discards, the catch reconstruction by Doyle et al. (2012) suggested that total reconstructed removals roughly corresponded to those presented by the FAO on behalf of the Solomon Islands. However, it included more than 200,000 t of unreported subsistence catch and 29,000 t of unreported artisanal shark catches. Catches within the EEZ were dominated by industrial fisheries (figure 2A), driven largely by foreign fleets (figure 2B). Figure 2C illustrates the dominance of skipjack (*Katsuwonus pelamis*) and yellowfin tunas (*Thunnus albacares*) in that fishery, whereas snappers (Lutjanidae), groupers (Serranidae), and barracuda (Sphyraenidae) dominate the near-shore fish catches, and spiny lobster (*Panulirus* spp.), trochus (*Trochus niloticus*), and sea cucumbers (Holothuridea) are important invertebrate resources. Global markets, population growth, increased migration to urban centers, and growing fishing pressure threaten to undermine many of the Solomon Islands' small-scale fisheries, and the presence of joint venture and foreign access commercial tuna fishing fleets is likely to expand with international demand and foreign exchange income opportunities.

REFERENCES

Agassi A (2005) Feast or famine? Fishing for a living in rural Solomon Islands. pp. 107–127 In Novaczek I, Mitchell J and Vietayaki J (eds.), *Pacific voices: Equity and sustainability in Pacific island fisheries*. Institute of Pacific Studies Publications, Suva (Fiji).

Doyle B, Harper S, Jacquet J and Zeller D (2012) Reconstructing marine fisheries catches in the Solomon Islands: 1950–2009. pp. 119–134 In Harper S, Zylich K, Boonzaier L, Le Manach F, Pauly D and D Zeller (eds.), *Fisheries catch reconstructions: Islands, Part III*. Fisheries Centre Research Reports 20(5). [Including on p. 129 an addendum which updates the dataset to 2010]

Hviding E and Baines G B K. (1994). Community-based fisheries management, tradition and the challenges of development in Marovo, Solomon Islands. *Development and Change* 25(1): 13–39.

1. Cite as Doyle, B., S. Harper, J. Jacquet, and D. Zeller. 2016. Solomon Islands. P. 389 in D. Pauly and D. Zeller (eds.), *Global Atlas of Marine Fisheries: A Critical Appraisal of Catches and Ecosystem Impacts*. Island Press, Washington, DC.

SOMALIA[1]

Lo Persson,[a,b] Alasdair Lindop,[b] Sarah Harper,[b] Kyrstn Zylich,[b] and Dirk Zeller[b]

[a]Sveriges Lantbruksuniversitet, Umeå, Sweden [b]*Sea Around Us*, University of British Columbia, Vancouver, Canada

Somalia is located on the Horn of Africa and has a long coastline, extending from Kenya in the southeast to Djibouti in the northwest and opening to the western Indian Ocean and the Gulf of Aden (figure 1). Somalia has suffered from political and social instability since the collapse of its last national government in 1991, resulting in unauthorized fishing by foreign industrial fleets. Persson et al. (2015) reconstructed domestic catches from a wide variety of sources and documented a Somali catch of about 20,000 t/year in the early decades, which increased slowly to the 1990s, then surged to 65,000 t/year in the late 2000s (figure 2A). The recent increase is probably the result of increased coastal settlement of previously nomadic people and a piracy-driven reduction in foreign illegal industrial fishing (Bahadur 2011; figure 2B) focusing heavily on tuna (figure 2C), which was itself a form of piracy. These recent trends enabled a profitable domestic artisanal sector to emerge into which Somali entrepreneurs, some connected to pirates, could safely invest. This recent development is the main reason why the reconstructed catch is nearly twice the catch reported by the FAO on behalf of Somalia from this productive ecosystem (Bakun et al. 1998). The revived domestic fisheries, since 2014 formally ensconced in an EEZ conforming to UNCLOS provisions, could thus contribute to the food security of a reborn Somalia.

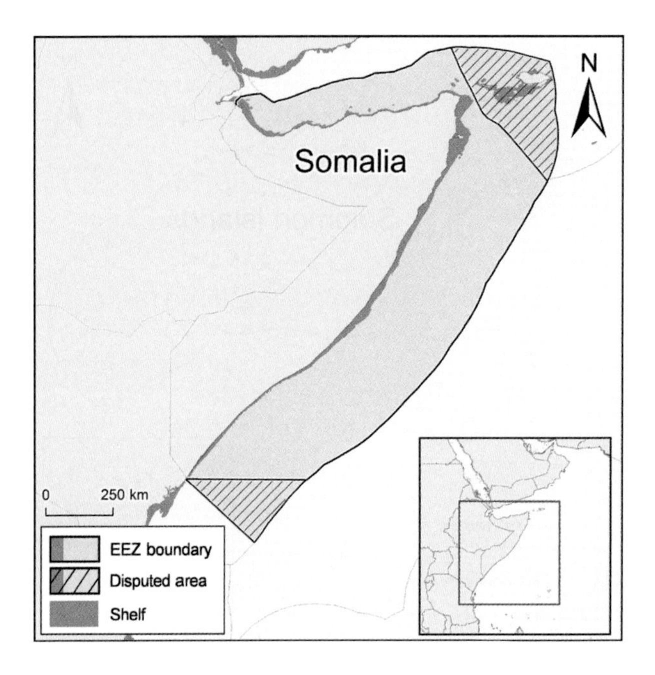

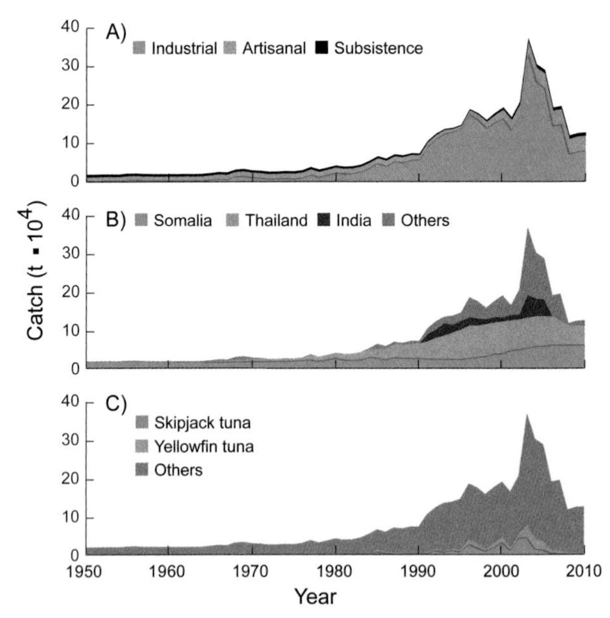

Figure 1. Somalia has a shelf of 50,800 km² and an EEZ of 831,000 km². **Figure 2.** Domestic and foreign catches taken in the EEZ of Somalia: (A) by sector (domestic scores: Ind. 1, 2, 2; Art. 1, 2, 2; Sub. 1, 1, 1; Dis. 1, 1, 1); (B) by fishing country (foreign catches are very uncertain); (C) by taxon.

REFERENCES

Bahadur J (2011) *The Pirates of Somalia: Inside Their Hidden World.* HarperCollins, Canada.

Bakun A, Claude R and Lluch-Cota S (1998). Coastal upwelling and other processes regulating ecosystem productivity and fish production in the Western Indian Ocean. pp. 103–142 In Sherman K, Okemwa EN, and Ntiba MJ (eds.), *Large marine ecosystems of the Indian Ocean: assessment, sustainability and management*. Blackwell Science, Oxford, U.K.

Persson L, Lindop A, Harper S, Zylich K and Zeller D (2015) Failed state: Reconstruction of domestic fisheries catches in Somalia 1950–2010. pp. 111–127 In Le Manach F and Pauly D (eds.), *Fisheries catch reconstructions in the Western Indian Ocean, 1950–2010*. Fisheries Centre Research Reports 23(2).

1. Cite as Persson, L., A. Lindop, S. Harper, K. Zylich, and D. Zeller. 2016. Somalia. P. 390 In D. Pauly and D. Zeller (eds.), *Global Atlas of Marine Fisheries: A Critical Appraisal of Catches and Ecosystem Impacts*. Island Press, Washington, DC.

SOUTH AFRICA (ATLANTIC AND CAPE)[1]

Sebastian Baust, Lydia C. L. Teh, Sarah Harper, and Dirk Zeller

Sea Around Us, University of British Columbia, Vancouver, Canada

South Africa, at the southern tip of the African continent, has coastlines facing both the southeast Atlantic and southwest Indian Oceans, but this account deals only with the former (figure 1; see next page for the latter). Because of the highly productive Benguela Current upwelling, the South African fisheries on its south and west coasts are mainly large-scale industrial (figure 2A) and pelagic, that is, they involve few species and fleets (Cury et al. 2000; Gasche et al. 2012), have very little if any foreign fleet participation (figure 2B), and, because of their size, justify an important management input. Still, the removals from what is now the South African EEZ were ill-defined, because of historic misreporting of catches taken by South African vessels in then-occupied Namibian waters. The reconstruction of South African fisheries catches by Baust et al. (2015) attempted to correct for this discrepancy and resulted in truly South African catches that were 0.74 times the data reported by the FAO on behalf of South Africa, which included huge quantities of Namibian fish being caught in the earlier decades (see Namibia chapter). This issue is now corrected, and in fact, South Africa's southwest Atlantic small pelagic fisheries for sardine (*Sardinops sagax*), anchovies (*Engraulis capensis*), and other species (figure 2C) are now well managed. Other important species include Cape horse mackerel (*Trachurus capensis*) and Cape hake (*Merluccius capensis* and *M. paradoxus*).

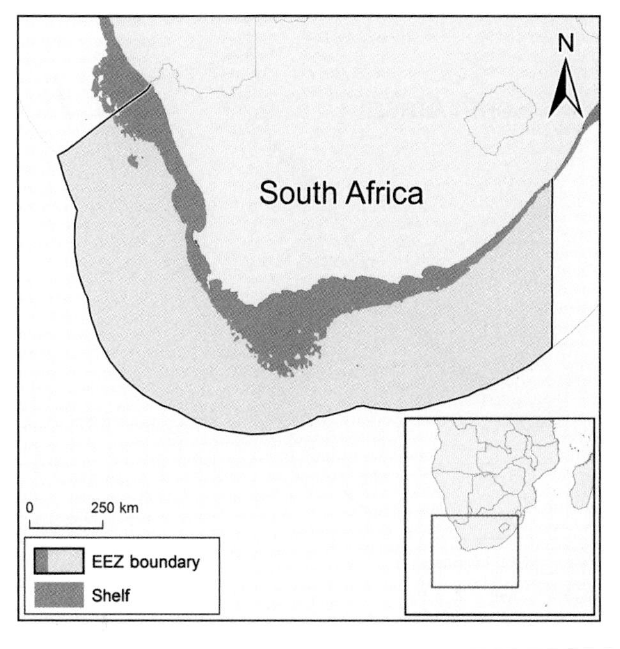

Figure 1. In the southeast Atlantic, South Africa as shelf of 64,200 km² and an EEZ of 375,000 km².

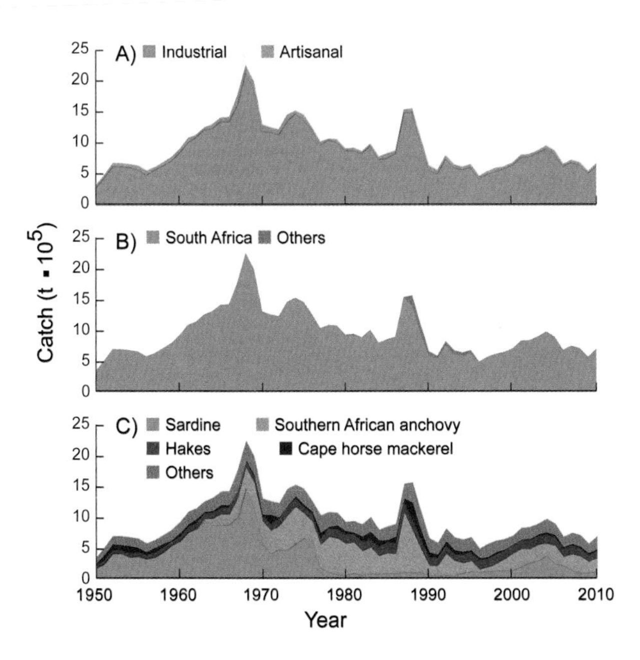

Figure 2. Domestic and foreign catches taken in the Southeast Atlantic EEZ of South Africa: (A) by sector (domestic scores: Ind. 2, 2, 2; Art. 2, 2, 2; Sub. 2, 2, 3; Rec. 2, 2, 3; Dis. 1, 2, 2); (B) by fishing country (foreign catches are very uncertain); (C) by taxon.

REFERENCES

Baust S, Teh LCL, Harper S and Zeller D (2015) Marine fisheries catches of South Africa's (1950–2010), including their recreational and subsistence components. pp. 129–150 In Le Manach F and Pauly D (eds.), *Fisheries catch reconstructions in the Western Indian Ocean, 1950–2010*. Fisheries Centre Research Reports 23(2).

Cury P, Bakun A, Crawford RJ, Jarre A, Quiñones RA, Shannon LJ and Verheye HM (2000) Small pelagics in upwelling systems: patterns of interaction and structural changes in "wasp-waist" ecosystems. *ICES Journal of Marine Science* 57(3): 603–618.

Gasche L, Gascuel D, Shannon L and Shin Y-J (2012) Global assessment of the fishing impacts on the Southern Benguela ecosystem using an EcoTroph modelling approach. *Journal of Marine Systems* 90(1): 1–12.

1. Cite as Baust, S., L. C. L. Teh, S. Harper, and D. Zeller. 2016. South Africa (Atlantic Coast). P. 391 In D. Pauly and D. Zeller (eds.), *Global Atlas of Marine Fisheries: A Critical Appraisal of Catches and Ecosystem Impacts*. Island Press, Washington, DC.

SOUTH AFRICA (INDIAN OCEAN COAST)[1]

Sebastian Baust, Lydia Teh, Sarah Harper, and Dirk Zeller

Sea Around Us, University of British Columbia, Vancouver, Canada

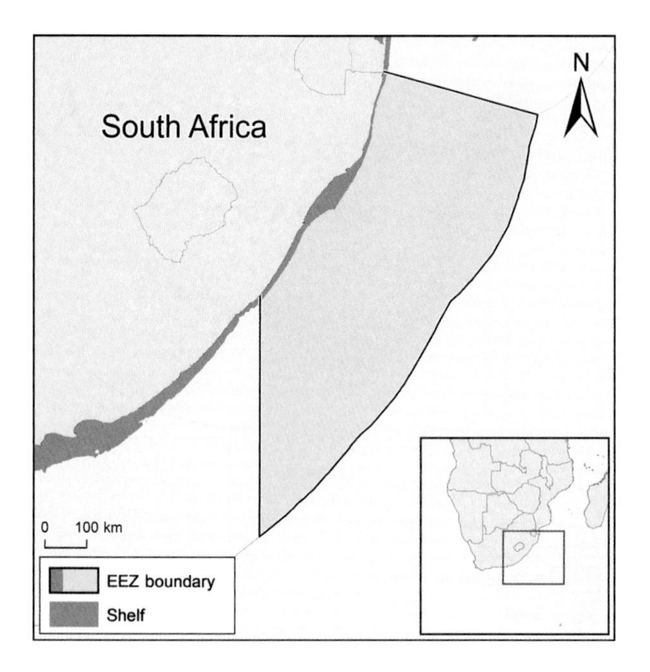

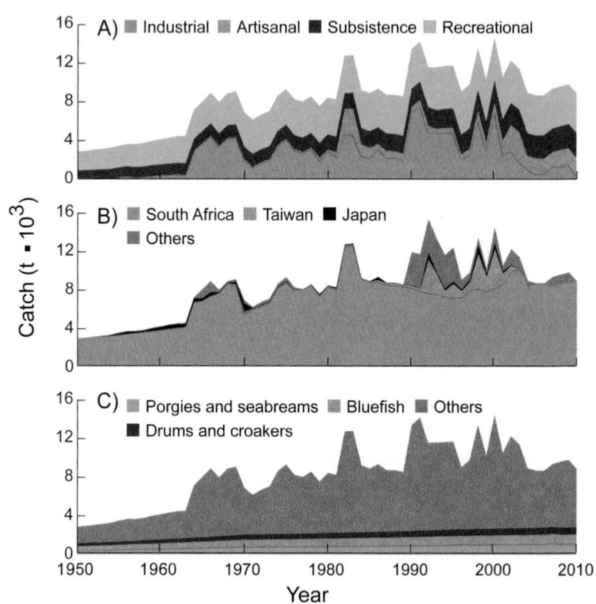

Figure 1. The Indian Ocean Coast of South Africa has a shelf of 96,000 km² and EEZ of 691,000 km². **Figure 2.** Domestic and foreign catches taken in the Indian Ocean Coast EEZ of South Africa: (A) by sector (domestic scores: Ind. 2, 2, 2; Art. 2, 2, 2; Sub. 2, 2, 3; Rec. 2, 2, 3; Dis. 1, 2, 2); (B) by fishing country (foreign catches are very uncertain); (C) by taxon.

South Africa has a long coastline along the southwestern Indian Ocean (figure 1), influenced by the eastward-flowing, cold Agulhas Current in the south (Lutjeharms 2006). This current becomes warmer northward and ends up in the northern province of Kwa-Zulu-Natal, with an Indo-Pacific coral reef fauna. The reconstruction of South African catches by Baust et al. (2015) estimated a domestic catch of 3,000 t/year for the early 1950s, which peaked at more than 12,000 t in 1982 and 1983 before settling at 8,500 t/year in the 2000s. Overall, the reconstructed domestic catch was 4.4 times that reported to the FAO by South Africa for these waters. The discrepancy results from underreported small-scale catches (figure 2A), especially the recreational fisheries (50% of total domestic catch) and subsistence (23%) and artisanal fisheries (9%), all usually not included in FAO statistics. About one third of the industrial catch was taken by foreign fleets especially in the later years (figure 2B). Figure 2C documents the taxonomic diversity of this catch. The dominance of small-scale fisheries along the South African Indian Ocean coast is due both to the popularity of recreational line fishing by the well-off segment of the population and the vital subsistence fishing (including invertebrate gathering by women) by the poorer segment of the population. Sowman (2006) cites the enormous social problems, most outside the reach of fisheries managers, that will have to be tackled when addressing this imbalance.

REFERENCES

Baust S, Teh LCL, Harper S and Zeller D (2015) Marine fisheries catches of South Africa's (1950–2010), including their recreational and subsistence components. pp. 129–150 In Le Manach F and Pauly D (eds.), *Fisheries catch reconstructions in the Western Indian Ocean, 1950–2010*. Fisheries Centre Research Reports 23(2).

Lutjeharms JRE (2006) *The Agulhas Current*. Springer, Berlin, Heidelberg, New York. xiv + 330 p.

Sowman M (2006) Subsistence and small-scale fisheries in South Africa: A ten-year review. *Marine Policy* 30(1): 60–73.

1. Cite as Baust, S., L. C. L. Teh, S. Harper, and D. Zeller. 2016. South Africa (Indian Ocean Coast). P. 392 In D. Pauly and D. Zeller (eds.), *Global Atlas of Marine Fisheries: A Critical Appraisal of Catches and Ecosystem Impacts*. Island Press, Washington, DC.

SOUTH AFRICA (PRINCE EDWARD ISLANDS)[1]

Lisa Boonzaier, Sarah Harper, Dirk Zeller, and Daniel Pauly

Sea Around Us, University of British Columbia, Vancouver, Canada

Located in the southwestern Indian Ocean, the Prince Edward Islands, which belong to South Africa, comprise two volcanic islands, Marion and Prince Edward. Most of the Prince Edward Islands' EEZ falls in subareas 58.6 and 58.7 of the Commission for the Conservation of Antarctic Marine Living Resources (CCAMLR), of which South Africa is a signatory (figure 1). Boonzaier et al. (2012) reconstructed the catch history in the archipelago, based on data obtained from the CCAMLR statistical database, CCAMLR stock assessment reports, and South African national commercial and observer datasets, re-expressed from "seasons" (southern summers) to calendar years and used to estimate landed and discarded catches. Numerous countries were known to fish in these waters (figure 2A). Catches of Patagonian toothfish (*Dissostichus eleginoides*), the only target species around the islands, show a sharp increase from 1994 on, peaking at nearly 24,000 t in 1997, an estimated 93% of which were taken by vessels operating illegally in the area. These large removals during the first years of the fishery had the effect of unsustainable "mining" of the stock (Agnew 2000); thereafter catches fell sharply. Grenadiers (family Macrouridae), notably Whitson's grenadier (*Macrourus whitsoni*), were a major bycatch (figure 2B). Currently, a small legal fishery remains operational in the area, but much of the EEZ is designated as a marine protected area whose complex ecosystem (Gurney et al. 2014) can now recover.

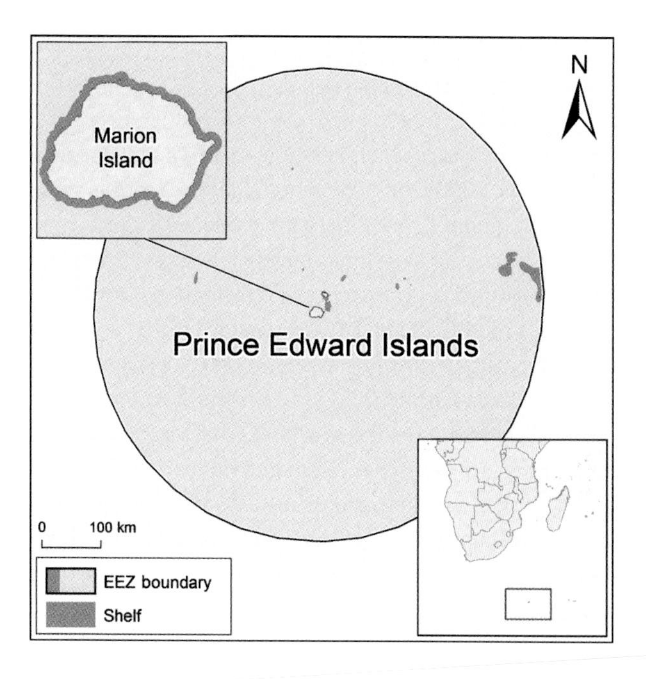

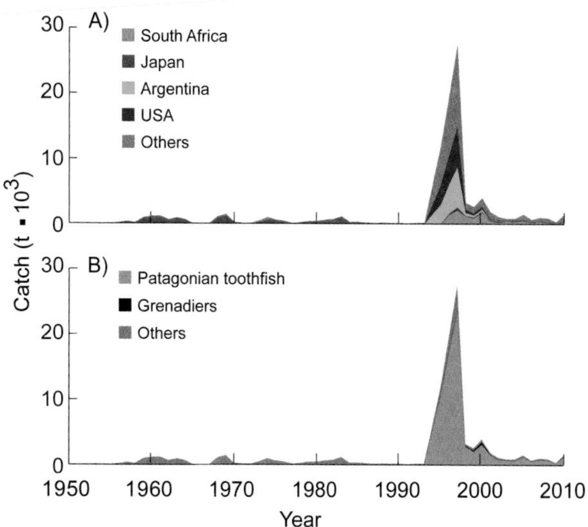

Figure 1. The Prince Edward Islands, consisting of Marion Island (290 km²) and Prince Edward (45 km²), have a joint EEZ of 473,400 km². **Figure 2.** Domestic and foreign catches taken in the EEZ of Prince Edward Islands (South Africa): (A) by fishing country (foreign catches are very uncertain); (B) by taxon. (scores: Ind. –, –, 4; Dis. –, –, 3).

REFERENCES

Agnew DJ (2000) The illegal and unregulated fishery for toothfish in the Southern Ocean, and the CCAMLR catch documentation scheme. *Marine Policy* 24: 361–374.

Boonzaier L, Harper S, Zeller D and Pauly D (2012) A brief history of fishing in the Prince Edward Islands, South Africa, 1950–2010. pp. 95–101 In Harper S, Zylich K, Boonzaier L, Le Manach F, Pauly D and Zeller D (eds.), *Fisheries catch reconstructions: Islands, Part III*. Fisheries Centre Research Reports 20(5).

Gurney LJ, Pakhomov EA and Christensen V (2014) An ecosystem model of the Prince Edward Island archipelago. *Ecological Modelling* 294: 117–136.

1. Cite as Boonzaier, L., S. Harper, D. Zeller, and D. Pauly. 2016. South Africa (Prince Edward Islands). P. 393 in D. Pauly and D. Zeller (eds.), *Global Atlas of Marine Fisheries: A Critical Appraisal of Catches and Ecosystem Impacts*. Island Press, Washington, DC.

SPAIN (BALEARIC ISLANDS)[1]

Marta Carreras,[a] Marta Coll,[b] Antoni Quetglas,[c] Raquel Goñi,[c] Xavier Pastor,[a] Maria-José Cornax,[a] Magdanena Iglesias, Enric Massutí,[c] Pere Oliver,[c] Ricardo Aguilar,[a] Andrea Au,[d] Kyrstn Zylich,[d] and Daniel Pauly[d]

[a]Oceana Europe, Madrid, Spain [b]Institut de Ciències del Mar (ICM-CSIC), Barcelona, Spain [c]Instituto Español de Oceanografía, Centre Oceanogràfic de les Balears, Palma, Spain [d]*Sea Around Us*, Institute for the Oceans and Fisheries. University of British Columbia, Vancouver, Canada

The Balearic Islands consist of four main islands, Mallorca, Menorca, Ibiza, and Formentera, in the western Mediterranean (figure 1). Artisanal fishing and bottom trawling (industrial) are the most important segment of the fisheries sector, followed by subsistence and recreational fishing by locals and tourists (figure 2A). Unreported catches are large, mainly because of discards and sales on the black market. Based on Carreras et al. (2014) and Carreras (2014), we present reconstructed catches in the Balearic EEZ that account for these deficiencies. These catches, estimated from official landings data (from national and regional agencies and daily sale bills), secondary sources (e.g., Massutí 1991), and interviews with fishers, amounted to about 9,000 t/year in the early 1950 and more than 15,000 t/year in the late 2000s (figure 2B). Overall, they were 2.3 times the reported landings. The targets of the trawlers are mullet (*Mullus surmuletus* and *M. barbatus*), hake (*Merluccius merluccius*), Norway lobster (*Nephrops norvegicus*), and red shrimp (*Aristeus antennatus*), whereas the artisanal fleet targets spiny lobster (*Palinurus elephas*), dolphinfish (*Coryphaena hippurus*), and transparent goby (*Aphia minuta*). The importance of picarel (*Spicara smaris*) has declined substantially over the years (figure 2C). Although the Spanish mainland fleet takes only 1% of the total Balearic catch, this amounts to one third of all Norway lobster and one tenth of the red shrimps from the Balearic Islands and thus should be considered when assessing their fisheries resources.

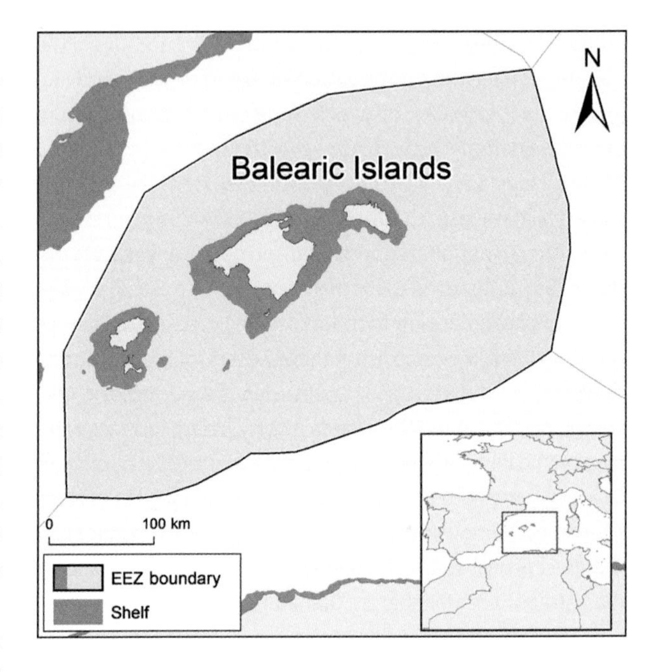

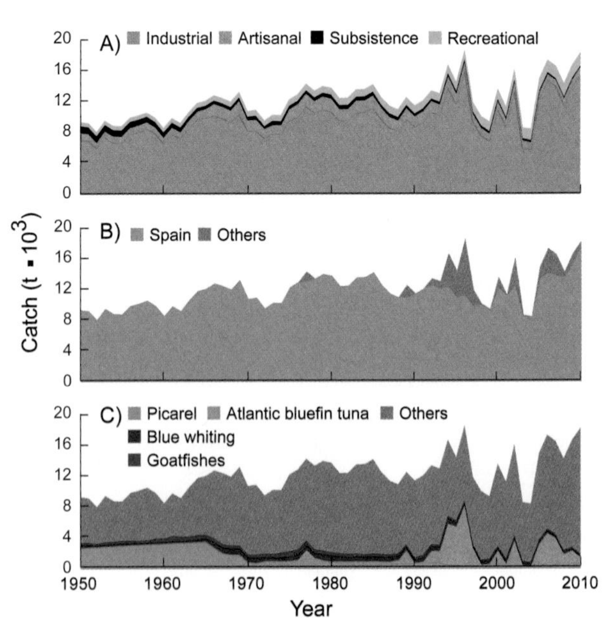

Figure 1. The Balearic Islands (land area: 5,000 km²) have an EEZ of 128,000 km². **Figure 2.** Domestic and foreign catches taken in the EEZ of Balearic Islands (Spain): (A) by sector (domestic scores: Ind. 2, 2, 4; Art. 2, 3, 3; Sub. 2, 2, 3; Rec. 2, 2, 3; Dis. 2, 3, 4); (B) by fishing country (foreign catches are very uncertain); (C) by taxon.

REFERENCES

Carrerras M (2014) Evolucion de la pesca Baleares. Oceana, Madrid, 32 p.

Carreras M, Coll M, Quetglas A, Goñi R, Pastor X, Cornax M, Iglesias M, Massutí E, Oliver P, Aguilar R, Au A, Zylich K and Pauly D (2014) Estimates of total fisheries removal for Balearic Islands (1950–2010). Fisheries Centre Working Paper #2015–19, 45 p.

Massutí M (1991) Les Illes Balears, un area de pesca individualitzada a la Mediterrània Occidental. *Quaderns de Pesca* 2: 1–62.

1. Cite as Carreras, M., M. Coll, A. Quetglas, R. Goñi, X. Pastor, M. Cornax, M. Iglesias, E. Massutí, P. Oliver, R. Aguilar, A. Au, K. Zylich, and D. Pauly. 2016. Spain (Balearic Islands). P. 394 in D. Pauly and D. Zeller (eds.), *Global Atlas of Marine Fisheries: A Critical Appraisal of Catches and Ecosystem Impacts*. Island Press, Washington, DC.

SPAIN (CANARY ISLANDS)[1]

José J. Castro,[a] Esther Divovich,[b] Alicia Delgado de Molina Acevedo,[c] and Antonio Barrera-Luján[a]

[a]Department of Biology, University of Las Palmas de Gran Canaria, Canary Islands, Spain [b]*Sea Around Us*, University of British Columbia, Vancouver, Canada [c]Instituto Español de Oceanografía, Tenerife, Canary Islands, Spain

The Canary Islands consist of 7 main islands approximately 100 km from the northwest African coast (figure 1). Being an autonomous community of Spain, the Canary Islands are the southernmost point of the European Union and thus form a maritime hub linking Europe, Africa, and the Americas. Total marine fishery catches within the EEZ of the Canary Islands, based on Melnychuck et al. (2001) and Castro et al. (2015), were dominated by artisanal fisheries in early decades and by recreational fisheries in recent years (figure 2A). Catches were almost exclusively domestic (figure 2B), peaked in the mid-1980s, declined into the early 2000s because of declining commercial catches (figure 2A), and increased again in the late 2000s because of growing recreational fishing (Pascual 2004). The commercial decline coincides with the depletion of fish stocks, caused by artisanal overcapacity, despite government attempts to limit effort. Only starting in 2006 were catches reported to the FAO by Spain; from 2006 to 2010 reconstructed catch was 7 times the reported catch. Figure 2C documents that this catch was dominated by the porgy family (Sparidae), parrotfish (*Sparisoma cretense*), Atlantic chub mackerel (*Scomber colias*), and various tuna species. Large catches, some taken illegally from the waters of northwest African countries, are also landed in Las Palmas on Gran Canaria, for reexport to Europe.

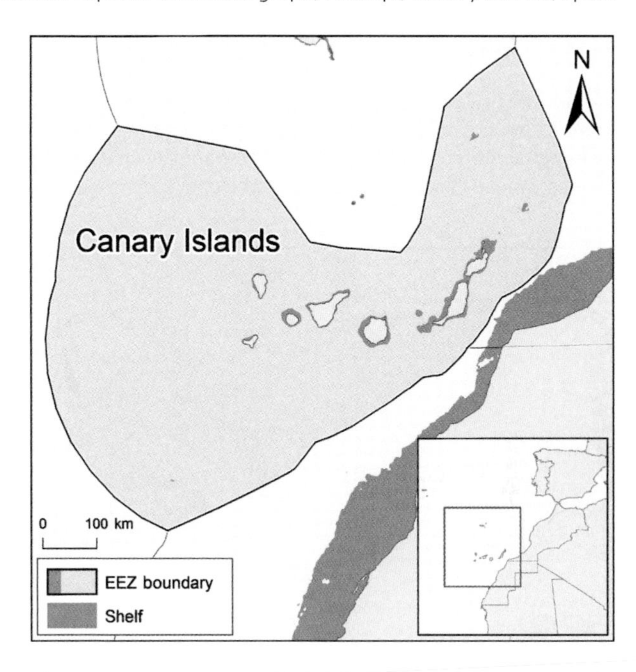

Figure 1. The Canary Islands (Spain) have a joint land area of 7,500 km² and an EEZ of 455,000 km².

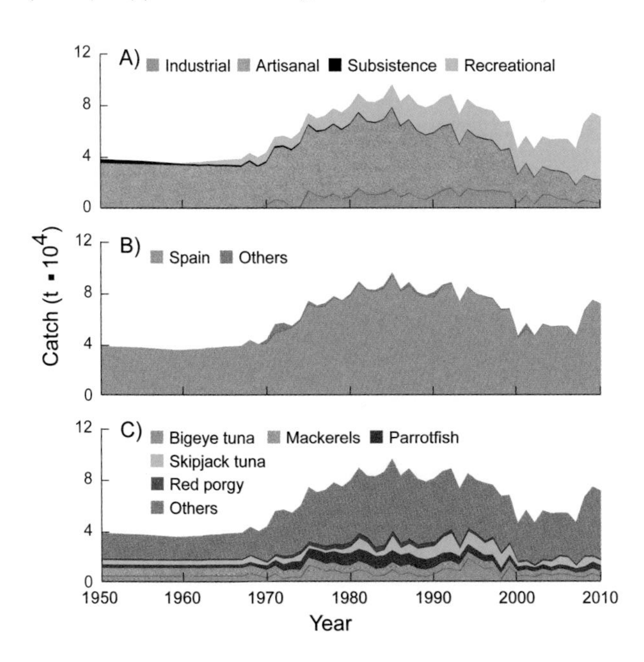

Figure 2. Domestic and foreign catches taken in the EEZ of Canary Islands (Spain): (A) by sector (domestic scores: Art. 3, 3, 4; Sub. 2, 1, 2; Rec. 2, 2, 3; Dis. 2, 2, 3); (B) by fishing country (foreign catches are very uncertain); (C) by taxon.

REFERENCES

Castro JJ, Divovich E, Delgado de Molina Acevedo A and Barrera-Luján A (2015) Over-looked and under-reported: A catch reconstruction of marine fisheries in the Canary Islands, Spain, 1950–2010. Fisheries Centre Working Paper #2015–26, 35 p.

Melnychuk M, Guénette S, Martín-Sosa P and Balguerias E (2001) Fisheries in the Canary Islands, Spain. pp. 221–224 In Zeller D, Watson R and Pauly D (eds.), *Fisheries Impacts on North Atlantic Ecosystems: Catch, Effort and National/Regional Data Sets*. Fisheries Centre Research Reports 9(3).

Pascual JJ (2004) Littoral fishermen, aquaculture and tourism in the Canary Islands: Attitudes and economic strategies. pp. 61–82 In J. Boissevain and T. Selwyn (eds.), *Contesting the foreshore: Tourism, society and politics on the coast*. Amsterdam University Press, Amsterdam.

1. Cite as Castro, J. J., E. Divovich, A. Delgado de Molina Acevedo, and A. Barrera-Luján. 2016. Spain (Canary Islands). P. 395 in D. Pauly and D. Zeller (eds.), *Global Atlas of Marine Fisheries: A Critical Appraisal of Catches and Ecosystem Impacts*. Island Press, Washington, DC.

SPAIN (MEDITERRANEAN AND GULF OF CADIZ)[1]

Marta Coll,[a] Marta Carreras,[b] Maria-Jose Cornax,[b] Enric Massutí,[c] Elvira Morote,[d] Xavier Pastor,[b] Antoni Quetglas,[b] Raquel Sáez,[a] Luis Silva,[e] Ignacio Sobrino,[e] Marian Torres,[e] Sergi Tudela,[f] Sarah Harper,[g] Dirk Zeller,[g] and Daniel Pauly[g]

[a]Institut de Ciències del Mar, Barcelona, Spain & Institut de Recherche pour le Développement, Sète, France [b]OCEANA, Madrid, Spain [c]Instituto Español de Oceanografía, Centre Oceanogràfic de les Balears, Palma, Spain [d]Universidad de Almería, Almeria, Spain [e]Instituto Español de Oceanografía, Centro Oceanográfico de Cádiz, Cádiz, Spain [f]WWF Mediterranean Programme Office, Barcelona, Spain [g]*Sea Around Us*, University of British Columbia, Vancouver, Canada

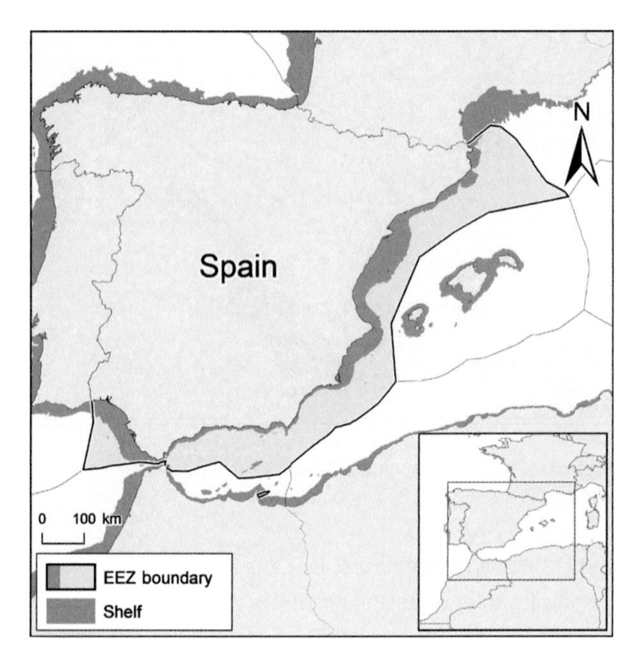

Figure 1. The Spanish Mediterranean and Gulf of Cadiz regions and their EEZ (148,000 km², of which 37,300 km² is shelf).

Underestimation of catches is often substantial where fishing fleets are highly diversified, the enforcement of fisheries management rules is low, and there is high demand for fish products. This is case for the Spanish Mediterranean and Gulf of Cadiz (SM&GC) regions, for which Coll et al. (2015), based on the literature and other sources of information that complemented officially reported data, estimated catches of about 400,000 t/year in the early 1950s, which increased to nearly 550,000 t/year by the late 1950s before beginning a steady decline to about 250,000 t/year in the late 2000s, and which were 1.7 times the data reported by the FAO on behalf of Spain for these waters. The fisheries were dominated by industrial gears, but small-scale fishing is occurring as well (figure 2A). Little if any foreign fishing seems to occur in SM&GC waters (figure 2B). Figure 2C documents the composition of the catch, with invertebrates and small and medium pelagic schooling species dominating. The results of the reconstructed catches are confirmed in fishers' interviews (Coll et al. 2014), suggesting that the degradation of resources in the SM&GC started earlier than previously assumed.

REFERENCES

Coll M, Carreras M, Ciércoles C, Cornax MJ, Morote E and Saez R (2014) Assessing fishing and marine biodiversity changes using fishers' perceptions: the Spanish Mediterranean and Gulf of Cadiz case study. *PLoS ONE* 9(1): e85670.

Coll M, Carreras M, Cornax M, Massuti E, Morote E, Pastor X, Quetglas A, Sáez R, Silva I, Sobrino I, Torres M, Tudela S, Harper S, Zeller D and Pauly D (2015) An estimate of the total catch in the Spanish Mediterranean Sea and Gulf of Cadiz regions (1950–2010). Fisheries Centre Working Paper #2015-60, 52 p.

1. Cite as Coll, M., M. Carreras, M. J. Cornax, E. Massutí, E. Morote, X. Pastor, A. Quetglas, R. Sáez, L. Silva, I. Sobrino, M. A. Torres, S. Tudela, S. Harper, D. Zeller, and D. Pauly. 2016. Spain (Mediterranean and Gulf of Cadiz). P. 396 in D. Pauly and D. Zeller (eds.), *Global Atlas of Marine Fisheries: A Critical Appraisal of Catches and Ecosystem Impacts.* Island Press, Washington, DC.

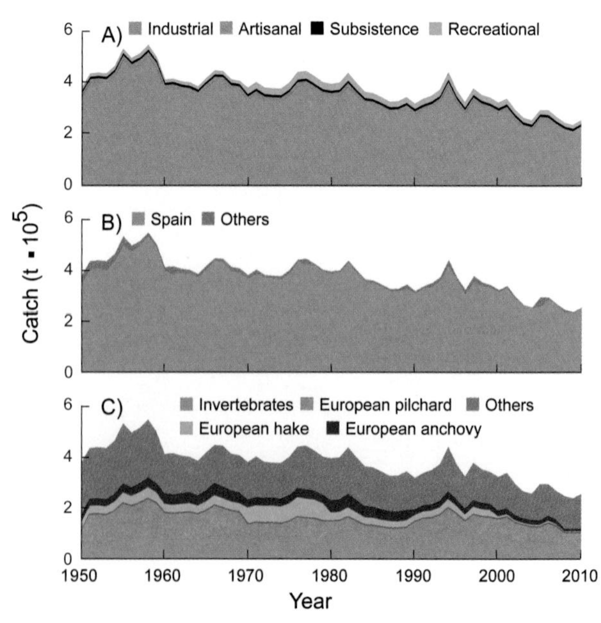

Figure 2. Domestic and foreign catches taken in the Spanish Mediterranean and Gulf of Cadiz EEZ: (A) by sector (domestic scores: Ind. 3, 3, 3; Art. 3, 3, 3; Sub. 1, 1, 1; Rec. 1, 1, 1; Dis. 2, 2, 2); (B) by fishing country (foreign catches are very uncertain); (C) by taxon.

SPAIN (NORTHWEST)[1]

Sebastian Villasante,[a] Gonzalo Macho,[b] Susana Rivero Rodriguez,[a,b] Josu Isusu de Rivero,[a,b] Esther Divovich,[c] Sarah Harper,[c] Dirk Zeller,[c] and Daniel Pauly[c]

[a]University Santiago de Compostela, Faculty of Political Sciences, A Coruña, Spain [b]Campus do Mar, International Campus of Excellence, Spain [c]*Sea Around Us*, University of British Columbia, Vancouver, Canada

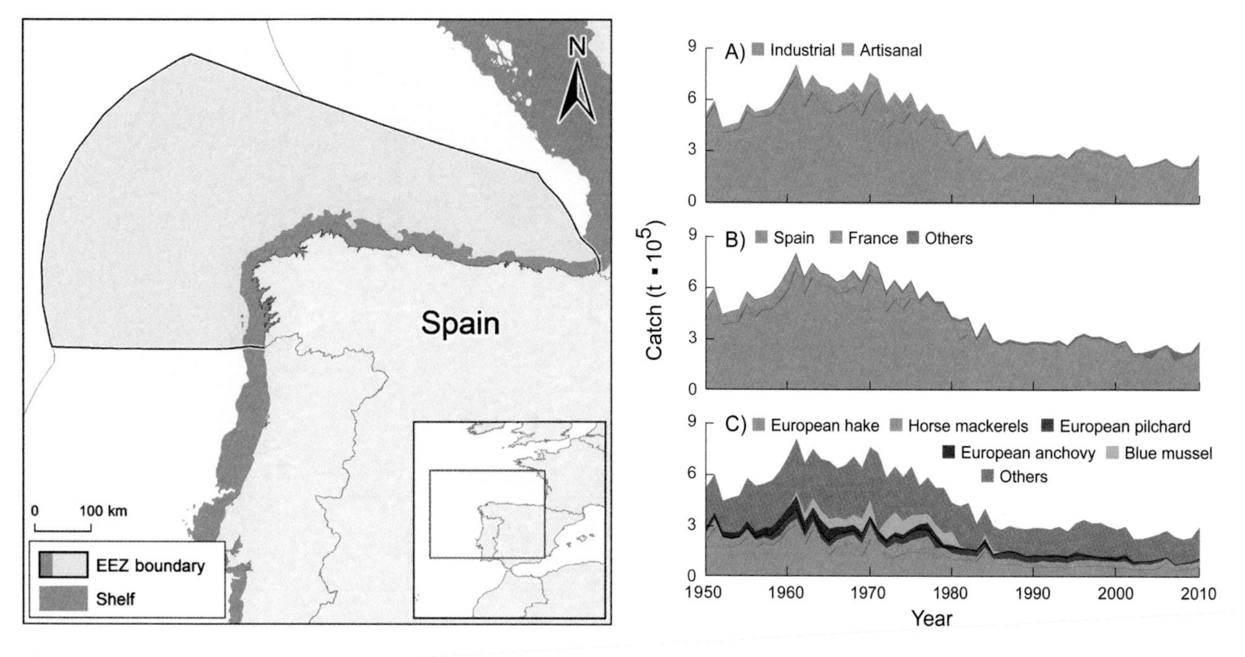

Figure 1. The northwest coast of Spain has a shelf of 22,900 km² and an EEZ of 289,000 km². **Figure 2.** Domestic and foreign catches taken in the EEZ of northwest Spain: (A) by sector (domestic scores: Ind. 1, 1, 2; Art. 1, 1, 2; Sub. 1, 1, 2; Rec. 1, 1, 2; Dis. 1, 1, 3); (B) by fishing country (foreign catches are very uncertain); (C) by taxon.

Spain has the largest fleet in Europe in terms of tonnage, and much of it is based in Galicia (but largely fishing in distant waters), in the northwest of Spain (figure 1). Galicia also is at the northern boundary of the Iberian–Canary Current upwelling system, where seasonal winds promote coastal upwelling and hence high marine productivity. However, the continental shelf is narrow. The fisheries catches from this region, as reconstructed by Villasante et al. (2015) were dominated by gears deemed industrial, although an important artisanal sector exists but seems to be in decline (figure 2A). Limited foreign fishing seems to take place (figure 2B). Total reconstructed catches in these waters increased from about 500,000 t/year in the early 1950s to a peak of about 700,000 t in the early 1960s before declining steadily to less than 300,000 t/year by the late 2000s (figure 2B). Reconstructed data highlight the importance and decline of European hake (*Merlucccius merluccius*), as well as artisanal resources such as blue mussel in catches (*Mytilus edulis*; figure 2C) and other invertebrates such as goose barnacle, *Pollicipes pollicipes* (Molares and Freire 2003) or common octopus, *Octopus vulgaris* (Otero et al. 2005). The Spanish and regional governments consider unregulated and black market catches as an important problem in the region, but no measures have been adopted to estimate these catches.

REFERENCES

Molares J and Freire J (2003) Development and perspectives for community-based management of the goose barnacle (*Pollicipes pollicipes*) fisheries in Galicia (NW Spain). *Fisheries Research* 65(1): 485–492.

Otero J, Rocha F, González ÁF, Gracia J and Guerra Á (2005) Modelling artisanal coastal fisheries of Galicia (NW Spain) based on data obtained from fishers: the case of *Octopus vulgaris*. *Scientia Marina* 69(4): 577–585.

Villasante S, Macho G, Rivero Rodriguez S, Isusu de Rivero J, Divovich E, Harper S, Zeller D and Pauly D (2015) Estimates of total fisheries removals from the Northwest of Spain (1950–2010). Fisheries Centre Working Paper #2015–51, 21 p.

1. Cite as Villasante, S., G. Macho, S. Rivero Rodriguez, J. Isusu de Rivero, E. Divovich, S. Harper, D. Zeller, and D. Pauly. 2016. Spain (Northwest). P. 397 in D. Pauly and D. Zeller (eds.), *Global Atlas of Marine Fisheries: A Critical Appraisal of Catches and Ecosystem Impacts*. Island Press, Washington, DC.

SRI LANKA[1]

Devon O'Meara,[a] Sarah Harper,[a] Nishan Perera,[b] and Dirk Zeller[a]

[a]*Sea Around Us*, University of British Columbia, Vancouver, Canada [b]Linnaeus University, Kalmar, Sweden

The Republic of Sri Lanka is located in the Bay of Bengal (figure 1), and it has an ancient reliance on the sea for the nutritional and economic well-being of its people. The reconstructed domestic marine catches of Sri Lanka (O'Meara et al. 2011) were estimated as 150,000 t/year in the 1950s, which increased to nearly 500,000 t/year in the 2000s, except for 2005, given the destruction caused by the tsunami of December 2004 (De Silva and Yamao 2007). This corresponds to twice the landings reported by the FAO on behalf of Sri Lanka, mainly because of unreported catches from the subsistence sector (40% of total domestic catch) and discarded bycatch from shrimp trawlers (13%). Substantial subsistence and artisanal catches exist in addition to industrial sector catches (figure 2A), and considerable foreign (mainly Indian) catches seem to occur according to the *Sea Around Us* allocation (figure 2B). However, it is possible that subsistence catches were underestimated, because several of the "countrywide" surveys we relied on included regions in the north affected by the long civil war in Sri Lanka. Consequently, it is likely that the household consumption estimate used to infer subsistence catches is conservative (details in O'Meara et al. 2011). Major contributing taxa (figure 2C) in the reconstructed catch included "silverbellies" or ponyfishes (family Leiognathidae, generally discarded), herrings and sardines (family Clupeidae), anchovies (Engraulidae), and croakers (Sciaenidae).

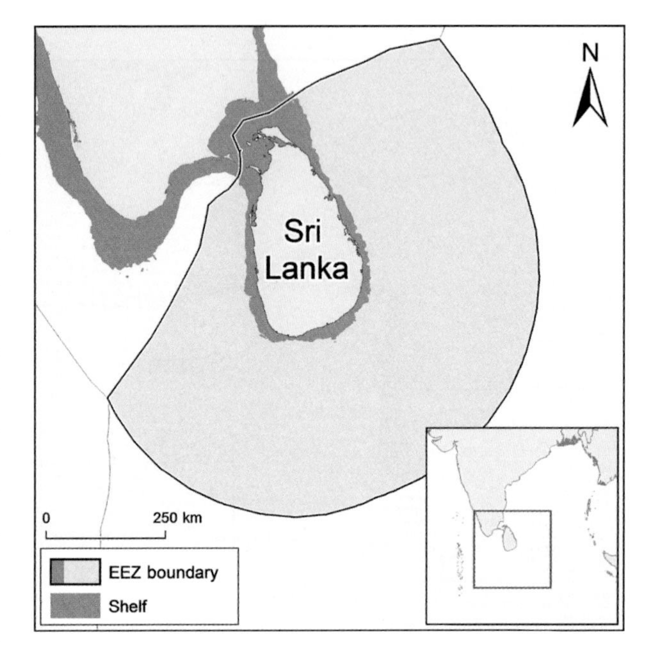

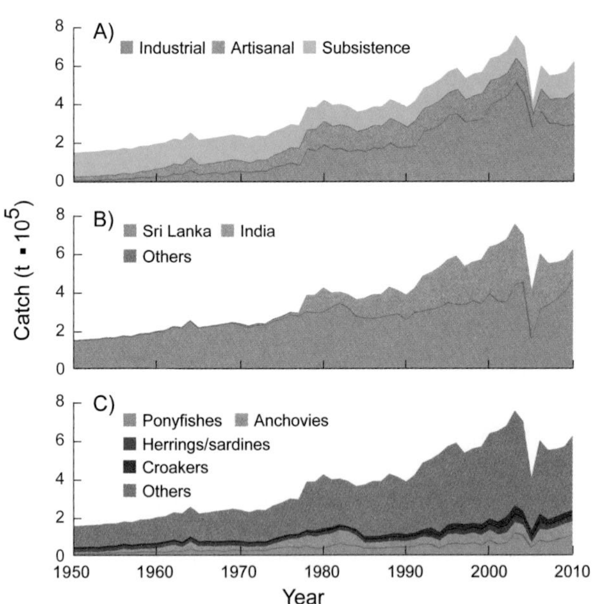

Figure 1. Sri Lanka, with a land area 65,600 km², has an EEZ of about 531,000 km². **Figure 2.** Domestic and foreign catches taken in the EEZ of Sri Lanka: (A) by sector (domestic scores: Ind. 1, 2, 2; Art. 1, 2, 2; Sub. 1, 1, 1; Dis. 1, 1, 1); (B) by fishing country (foreign catches are very uncertain); (C) by taxon.

REFERENCES

De Silva D and Yamao M (2007) Effects of the tsunami on fisheries and coastal livelihood: A case study of tsunami ravaged southern Sri Lanka. *Disasters* 3(4): 386–404.

O'Meara D, Harper S, Perera N and Zeller D (2011) Reconstruction of Sri Lanka's fisheries catches: 1950–2008. pp. 85–96 In Harper S and Zeller D (eds.), *Fisheries catch reconstruction: Islands, part II*. Fisheries Centre Research Reports 19(4). [Updated to 2010]

1. Cite as O'Meara, D., S. Harper, N. Perera, and D. Zeller. 2016. Sri Lanka. P. 398 in D. Pauly and D. Zeller (eds.), *Global Atlas of Marine Fisheries: A Critical Appraisal of Catches and Ecosystem Impacts*. Island Press, Washington, DC.

SAINT KITTS AND NEVIS[1]

Robin Ramdeen, Kyrstn Zylich, and Dirk Zeller

Sea Around Us, University of British Columbia, Vancouver, Canada

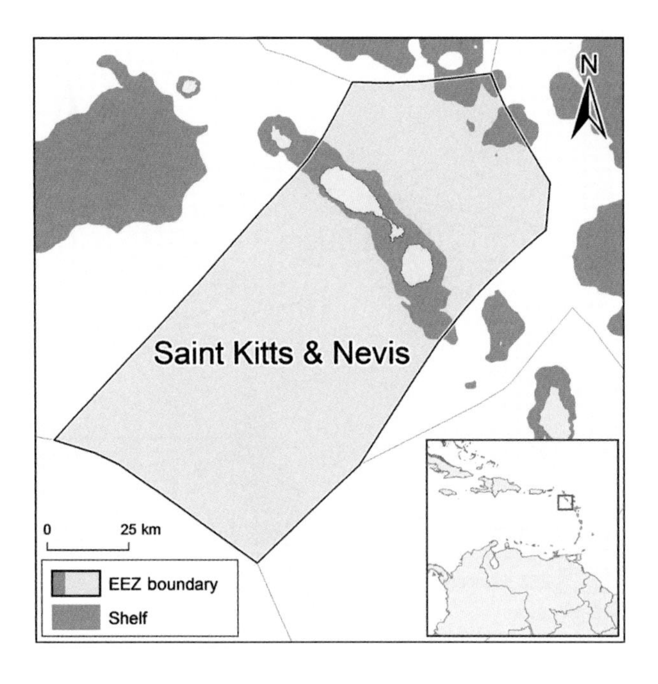

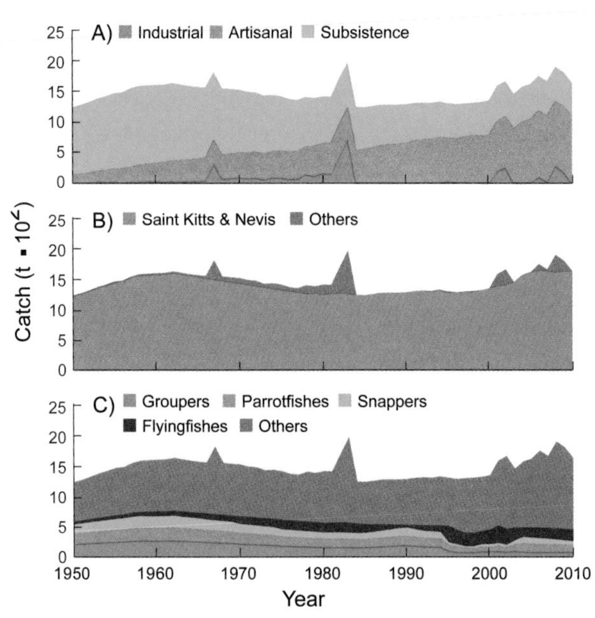

Figure 1. Saint Kitts and Nevis has a combined land area of 261 km², and an EEZ of 10,200 km². **Figure 2.** Domestic and foreign catches taken in the EEZ of Saint Kitts and Nevis: (A) by sector (domestic scores: Art. 2, 3, 3; Sub. 2, 2, 2; Rec. 1, 1, 1); (B) by fishing country (foreign catches are very uncertain); (C) by taxon.

Saint Kitts and Nevis are islands in the Caribbean Sea that have a combined population of 52,000. Both islands are of volcanic origin, with steep escarpments, hills in the interior, and gentle plains along the coasts. The islands are separated by a 3-km-wide channel, the Narrows (figure 1). This account, based on Ramdeen et al. (2014), presents the reconstruction of total marine fisheries catches for Saint Kitts and Nevis for the period 1950–2010. It includes estimates of unreported catches of conch and lobster for the early time period and underreported artisanal and subsistence catches for the entire time period (see also Wilkins 1984). Total catches range from 1,200 t in 1950 to 1,600 t/year in the late 2000s and consist of declining subsistence and increasing artisanal catches, plus a very small contribution from recreational fishing (figure 2A)—altogether about 2 times the landings reported by the FAO on behalf of Saint Kitts and Nevis. Fisheries are predominantly domestic, with little foreign fishing (figure 2B), and catches comprise a wide variety of fish ranging from groupers (family Serranidae), parrotfishes (Scaridae), and snappers (Lutjanidae) to flyingfishes (Exocoeditae), and invertebrates such as conchs and lobsters (figure 2C). These catches, accounting for total living marine resource extractions around Saint Kitts and Nevis, reflect the importance of small-scale fisheries in providing seafood to locals and visitors and livelihoods to fishers.

REFERENCES

Ramdeen R, Zylich K and Zeller D (2014) Reconstruction of total marine fisheries catches for St. Kitts and Nevis (1950–2010). pp. 129–136 In Zylich K, Zeller D, Ang M and Pauly D (eds.), *Fisheries catch reconstructions: Islands, Part IV*. Fisheries Centre Research Reports 22(2).

Wilkins R (1984) The Saint Kitts/Nevis fishery: A summary of the existing situation and constraints and requirements affecting development. Ministry of Agriculture, Lands, Housing, Labour and Tourism, Basseterre, St. Kitts. 3 p.

1. Cite as Ramdeen, R., K. Zylich, and D. Zeller. 2016. St. Kitts and Nevis. P. 399 in D. Pauly and D. Zeller (eds.), *Global Atlas of Marine Fisheries: A Critical Appraisal of Catches and Ecosystem Impacts*. Island Press, Washington, DC.

SAINT LUCIA[1]

Elizabeth Mohammed[a] and Alasdair Lindop[b]

[a]Research and Resource Assessment, Caribbean Regional Fisheries Mechanism Secretariat, Eastern Caribbean Office, St. Vincent and the Grenadines [b]*Sea Around Us*, University of British Columbia, Vancouver, Canada

Saint Lucia is a small island state in the eastern Caribbean Sea (figure 1). The island is volcanic in origin, resulting in a narrow shelf area relative to the extent of its EEZ. Its fisheries are largely artisanal and subsistence based, with an industrial element beginning in the 1990s with the introduction of longliners and larger boats capable of exploiting offshore pelagic fishes. Considerable industrial fishing exists in these waters (figure 2A); however, most of it is by foreign fleets (figure 2B). Based on the reconstruction by Mohammed and Lindop (2015), which built on the earlier work of Mohammed and Joseph (2003), total domestic catches remained stable from 1950 to 1990, averaging about 950 t/year in the early years but increasing to about 1,800 t/year in recent years. Overall, the reconstructed domestic catch is 1.15 times the data reported by the FAO on behalf Saint Lucia. Domestically, artisanal fisheries made up 59% of the total catch, with subsistence contributing 32% and the industrial sector 8.5%. Catches are strongly dominated by pelagic taxa, such as Atlantic bonito (*Sarda sarda*), tuna and mackerels (family Scombridae), dolphinfish (*Coryphaena hippurus*), jacks (Carangidae), and flyingfish (Exocoetidae), among others (figure 2C). Saint Lucia is home to a number of marine reserves (Roberts et al. 2001), some very small, which have all generated biomass increases from which adjacent fisheries have benefited.

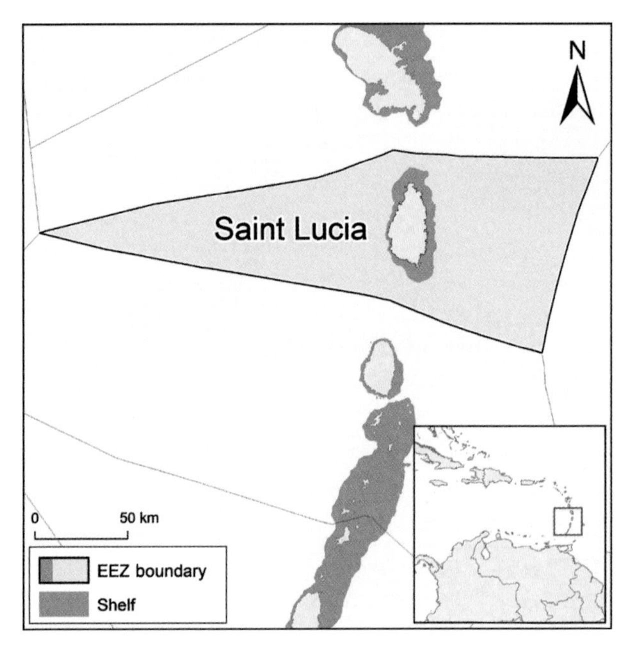

Figure 1. Saint Lucia has a land area 617 km² and an EEZ of about 15,500 km².

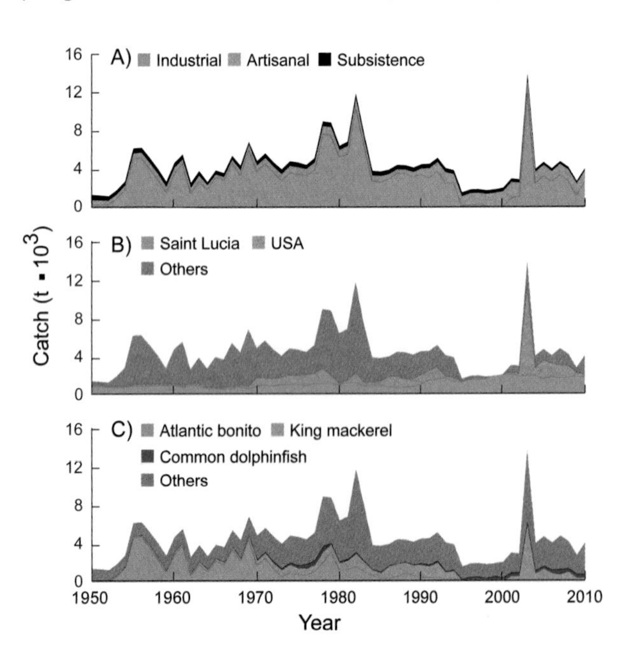

Figure 2. Domestic and foreign catches taken in the EEZ of Saint Lucia: (A) by sector (domestic scores: Ind. –, –, 3; Art. 2, 3, 3; Sub. 2, 3, 3; Rec. 1, 1, 1); (B) by fishing country (foreign catches are very uncertain); (C) by taxon.

REFERENCES

Mohammed E and Joseph W (2003) St. Lucia, eastern Caribbean: Reconstructed fisheries catches and fishing effort, 1942–2001. pp. 21–44 In Zeller D, Booth S, Mohammed E and Pauly D (eds.), *From Mexico to Brazil: Central Atlantic fisheries catch trends and ecosystem models*. Fisheries Centre Research Reports 11(6).

Mohammed E and Lindop A (2015) St. Lucia, reconstructed fisheries catches, 1950–2010. Fisheries Centre Working Paper #2015–53, 25 p.

Roberts CM, Bohnsack JA, Gell F, Hawkins JP and Goodridge R (2001) Effects of marine reserves on adjacent fisheries. *Science* 294(5548): 1920–1923.

1. Cite as Mohammed, E., and A. Lindop. 2016. St. Lucia. P. 400 In D. Pauly and D. Zeller (eds.), *Global Atlas of Marine Fisheries: A Critical Appraisal of Catches and Ecosystem Impacts*. Island Press, Washington, DC.

SAINT VINCENT AND THE GRENADINES[1]

Elizabeth Mohammed[a] and Alasdair Lindop[b]

[a]Research and Resource Assessment, Caribbean Regional Fisheries Mechanism Secretariat, Eastern Caribbean Office, St. Vincent and the Grenadines [b]*Sea Around Us*, University of British Columbia, Vancouver, Canada

Saint Vincent and the Grenadines is a small Caribbean island country in the southern part of the Windward group of the Lesser Antilles (figure 1). The local fisheries of Saint Vincent and the Grenadines are predominantly small-scale—some organized in cooperatives (Jentoft and Sandersen 1996)—with an industrial element beginning in the 1990s with the introduction of a multigear fleet. The domestic catch reconstructed by Mohammed and Lindop (2015), being an update of Mohammed et al. (2003), exhibits an overall increasing trend, although they remained largely static throughout the 1950s and 1960s (about 800 t/year), and increased to just over 2,000 t/year by the late 2000s. The reconstructed domestic catches for Saint Vincent and the Grenadines were 1.6 times the data reported to the FAO. The bulk (62%) of the domestic catch was artisanal, with subsistence fishing contributing 36%. Industrial and recreational fisheries made up only 1.6% and 0.1% of the domestic catch, respectively. However, total catches in the EEZ of Saint Vincent and the Grenadines were dominated by the industrial sector (figure 2A), almost all of it due to foreign fleets (figure 2B). Pelagic taxa dominated total catches (figure 2C), including Atlantic bonito (*Sarda sarda*), bullet tuna (*Auxis rochei*), scads (e.g., *Decapterus macarellus* and *Selar crumenophthalmus*), and various other tunas and mackerels (family Scombridae), whereas parrotfishes (Scaridae) and Caribbean spiny lobster (*Panulirus argus*) dominated reef catches.

REFERENCES

Jentoft S and Sandersen HT (1996) Cooperatives in fisheries management: the case of St. Vincent and the Grenadines. *Society & Natural Resources: An International Journal* 9(3): 295–305.

Mohammed E and Lindop A (2015) St. Vincent and the Grenadines: Reconstructed Fisheries Catches, 1950–2010. Fisheries Centre Working Paper #2015-54, 27 p.

Mohammed E, Straker LE and Jardine C (2003) St. Vincent and the Grenadines: Reconstructed fisheries catches and fishing effort, 1942–2001. pp. 95–116 In Zeller D, Booth S, Mohammed E and Pauly D (eds.), *From Mexico to Brazil: Central Atlantic fisheries catch trends and ecosystem models*. Fisheries Centre Research Reports 11(6).

1. Cite as Mohammed, E., and A. Lindop. 2016. Saint Vincent and the Grenadines. P. 401 In D. Pauly and D. Zeller (eds.), *Global Atlas of Marine Fisheries: A Critical Appraisal of Catches and Ecosystem Impacts*. Island Press, Washington, DC.

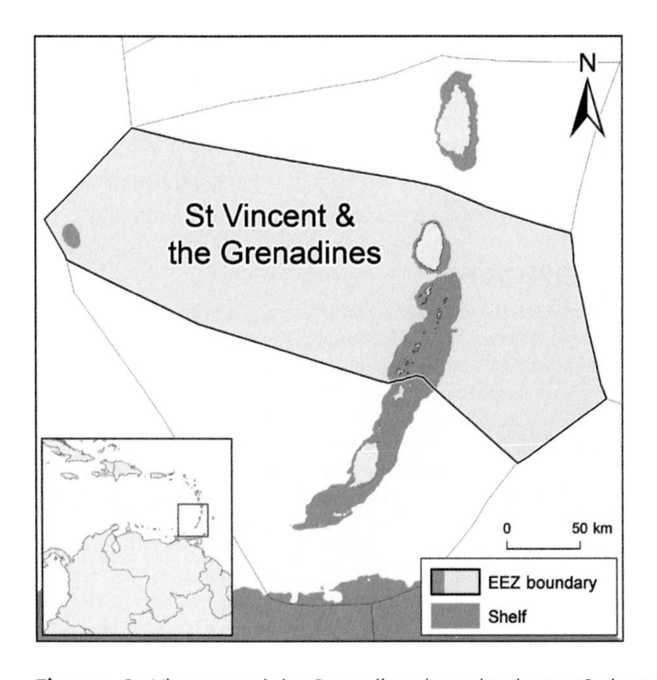

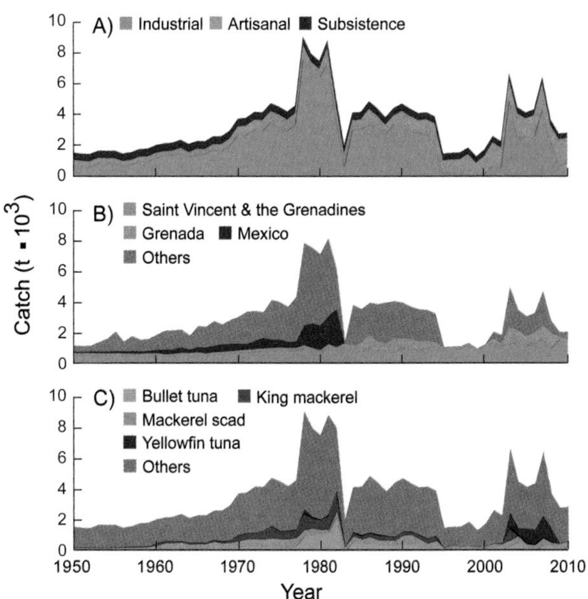

Figure 1. St. Vincent and the Grenadines has a land area 389 km² and an EEZ of about 36,300 km². **Figure 2.** Domestic and foreign catches taken in the EEZ of Saint Vincent and the Grenadines: (A) by sector (domestic scores: Ind. –, –, 3; Art. 3, 3, 3; Sub. 2, 2, 2; Rec. 2, 2, 2): by fishing country (foreign catches are very uncertain); (C) by taxon.

SUDAN[1]

Dawit Tesfamichael[a,b] and Abdella Nasser Elawad[c]

[a]*Sea Around Us*, University of British Columbia, Vancouver, Canada [b]Department of Marine Sciences, University of Asmara, Asmara, Eritrea [c]Fisheries Research Center, Port Sudan, Sudan

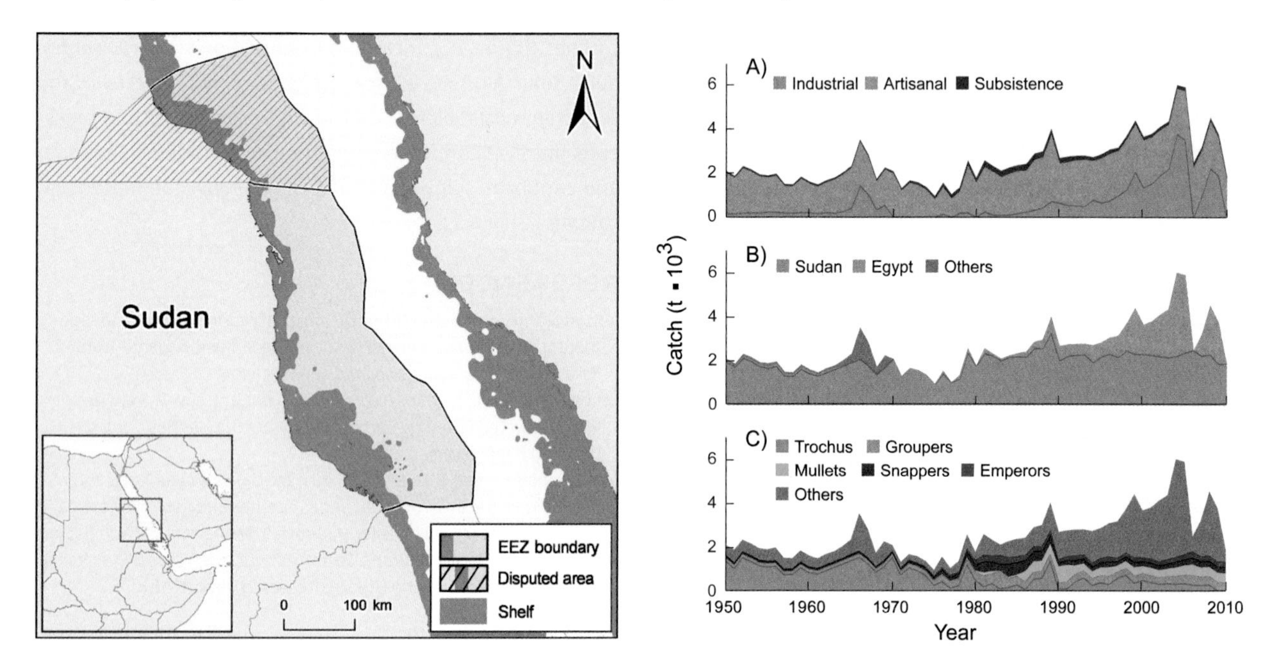

Figure 1. Sudan has a shelf of only 23,400 km² and an EEZ of about 92,500 km². **Figure 2.** Domestic and foreign catches taken in the EEZ of Sudan: (A) by sector (domestic scores: Art. 2, 3, 3; Sub. 1, 1, 1); (B) by fishing country (foreign catches are very uncertain); (C) by taxon.

Sudan borders the Red Sea and has a narrow shelf compared with other Red Sea countries (figure 1). Overall, Sudanese waters are deep and include the deepest part of the Red Sea. This account, adapted from Tesfamichael and Nasser Elawad (2012), presents the reconstructed catch of Sudan in the Red Sea, based on published articles and reports and unpublished data, but also on field interviews (see Tesfamichael et al. 2014). The results showed that domestic catches were about 2,000 t/year in the 1950s, declined throughout the 1970s to about 1,000 t/year, and increased again at the end of the 1970s, mainly because of development projects aided by foreign organizations, which led to a large increase in artisanal fishing effort (and artisanal catches, figure 2A), and ultimately, to domestic catches of about 2,000 t/year in the mid- to late 2000s. Starting in the 1980s, increasing volumes of industrial catches were also taken by Egyptian vessels in the waters of Sudan (figure 2B). The domestic fisheries also shifted from being dominated by shellfish (*Trochus* spp.) in the early years to an artisanal finfish fishery, exploiting a variety of fish species (figure 2C). For the first half of the time period from 1950 to 2010, the reconstructed catch was higher than the catch reported by the FAO on behalf of Sudan, but this was reversed in the second half, probably because Egyptian catches were mislabeled as Sudanese.

REFERENCES

Tesfamichael D and Nasser Elawad A (2012) Reconstructing the Red Sea fisheries of Sudan: foreign aid and fisheries. pp. 51–70 In Tesfamichael D and Pauly D (eds.), *Catch reconstruction for the Red Sea large marine ecosystem by countries (1950–2010)*. Fisheries Centre Research Reports 20(1).

Tesfamichael D, Pitcher TJ and Pauly D (2014) Assessing changes in fisheries using fishers' knowledge to generate long time series of catch rates: a case study from the Red Sea. *Ecology and Society* 19(1): 18.

1. Cite as Tesfamichael, D., and A. Nasser Elawad. 2016. Sudan. P. 402 In D. Pauly and D. Zeller (eds.), *Global Atlas of Marine Fisheries: A Critical Appraisal of Catches and Ecosystem Impacts*. Island Press, Washington, DC.

SURINAME[1]

Claire Hornby, Sarah Harper, Jessica MacDonald, and Dirk Zeller

Sea Around Us, University of British Columbia, Vancouver, Canada

Suriname, formally Dutch Guiana, is located between French Guiana and Guyana on the northeast coast of South America (figure 1). This region's continental shelf has historically been fished by numerous other countries (including the United States; see Dragovich and Coleman 1983) and once supported one of the most important shrimp fisheries in the world, targeting mainly brown shrimp (*Penaeus subtilis*), pink-spotted shrimp (*Penaeus brasiliensis*), and Atlantic seabob (*Xyphopenaeus kroyeri*). Thus, the industrial sector is substantial in these waters (figure 2A), and foreign fleets include South Korea, Japan, and Venezuela, plus the United States in earlier years (figure 2B). Based on the reconstruction by Hornby et al. (2012), Suriname's domestic catches were 5,000 t in 1950, increased toward a first peak of 46,000 t/year in the early 1970s, declined to about 20,000 t/year throughout the 1980s and early 1990s, then increased again to nearly 60,000 t/year in the early 2000s (figure 2B). For the 1950–2010 period, the reconstructed domestic catch was 3.4 times the landings reported by the FAO on behalf of Suriname, because of unreported artisanal and subsistence catch and high discards from the shrimp and seabob fisheries. Figure 2C shows that weakfishes and croakers (family Sciaenidae) contributed most to the catch, followed by penaeid shrimp (e.g., seabob in recent years) and sea catfishes (Ariidae). Hopefully, Suriname will acquire the management capability to control the exploitation of its marine fisheries resources, which are overexploited.

REFERENCES

Dragovich A and Coleman EM (1983) Participation of U.S. trawlers in the offshore shrimp fisheries of French Guiana, Suriname, and Guyana, 1978–79. *Marine Fisheries Review* 45(4–6): 1–9.

Hornby C, Harper S, MacDonald J and Zeller D (2012) Reconstruction of Suriname's marine fisheries catches from 1950–2010. Fisheries Centre Working Paper #2012-10, 24 p.

1. Cite as Hornby, C., S. Harper, J. MacDonald, and D. Zeller. 2016. Suriname. P. 403 in D. Pauly and D. Zeller (eds.), *Global Atlas of Marine Fisheries: A Critical Appraisal of Catches and Ecosystem Impacts*. Island Press, Washington, DC.

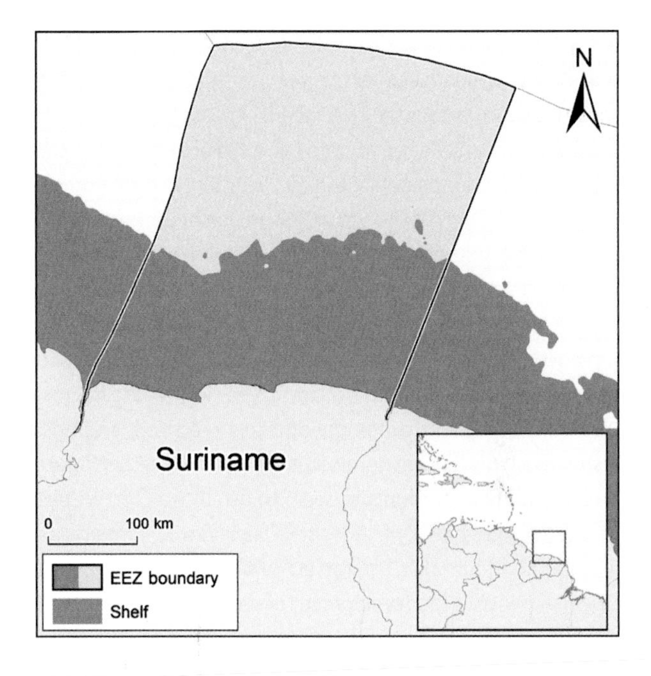

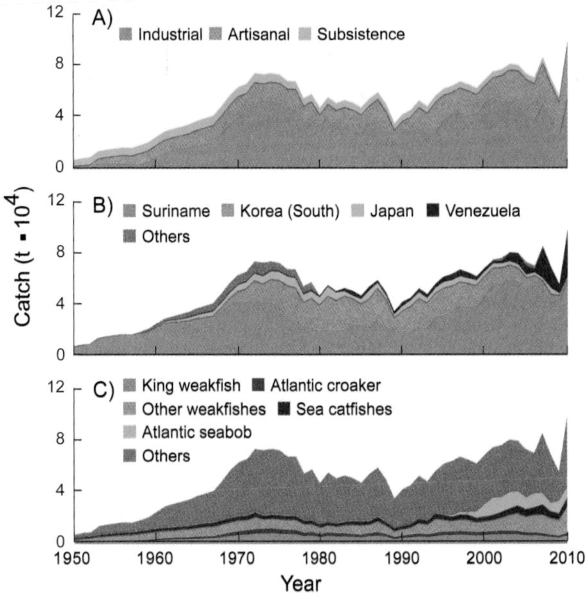

Figure 1. Suriname has 53,700 km² of continental shelf and an EEZ of 128,000 km². **Figure 2.** Domestic and foreign catches taken in the EEZ of Suriname: (A) by sector (domestic scores: Ind. 2, 3, 3; Art. 2, 3, 3; Sub. 1, 1, 1; Dis. 1, 1, 1); (B) by fishing country (foreign catches are very uncertain); (C) by taxon.

SWEDEN (BALTIC)[1]

Lo Persson

Sveriges Lantbruksuniversitet, Umeå, Sweden

Most of the coast of Sweden is in the Baltic Sea, but its West Coast (see next page) is not (figure 1). The catch reconstruction presented here applies to the Baltic Sea only (Persson 2010; Zeller et al. 2011) and was based on official reported landings as presented by the International Council for the Exploration of the Sea (ICES) on behalf of the Swedish government, data of the Swedish Board of Fisheries, and other sources. The reconstructed Swedish catch from the Baltic Sea, as allocated to the Swedish Baltic Sea EEZ, was about 40,000 t/year in the 1950s, increased to a peak of 230,000 t in 1998, then declined to 90,000 t/year in the late 2000s. Overall, this was 1.3 times the officially reported landings. Fisheries in this EEZ are dominated by industrial gears (figure 2A), and numerous countries seem to fish in Swedish waters (figure 2B). A tell-all book by Isabella Lövin (2012), formerly an investigative journalist, then an EU parliamentarian, and now Swedish minister for development assistance, describes how quota busting, high-grading, and other nefarious practices (largely accounted for in the reconstruction) contribute to this discrepancy and to similar divergences between official and actual catches in neighboring countries. Figure 2C presents the taxonomic composition of catches in the Swedish Baltic Sea EEZ, which is rather species poor, at least in contrast to the West Coast of Sweden.

REFERENCES

Lövin I (2012) *Silent Seas: The Fish Race to the Bottom*. Paragon Publishing. 244 p. [Swedish edition: 2007]

Persson L (2010) Sweden's fisheries catches in the Baltic Sea (1950–2007). pp. 225–263 In Rossing R, Booth S and Zeller D (eds.), *Total marine fisheries extractions by country in the Baltic Sea: 1950–present*. Fisheries Centre Research Reports 18(1) [Updated to 2010]

Zeller D, Rossing P, Harper S, Persson L, Booth S and Pauly D (2011) The Baltic Sea: estimates of total fisheries removals 1950–2007. *Fisheries Research* 108: 356–363.

1. Cite as Persson, L. 2016. Sweden (Baltic Sea). P. 404 in D. Pauly and D. Zeller (eds.), *Global Atlas of Marine Fisheries: A Critical Appraisal of Catches and Ecosystem Impacts*. Island Press, Washington, DC.

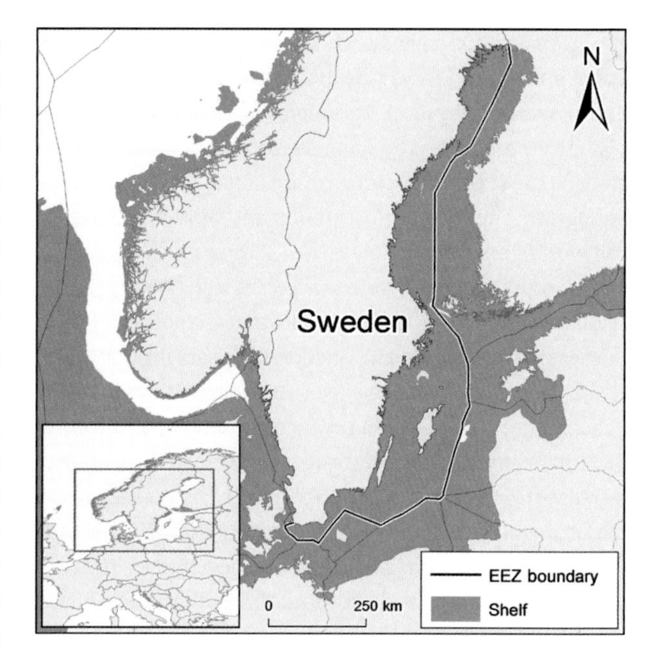

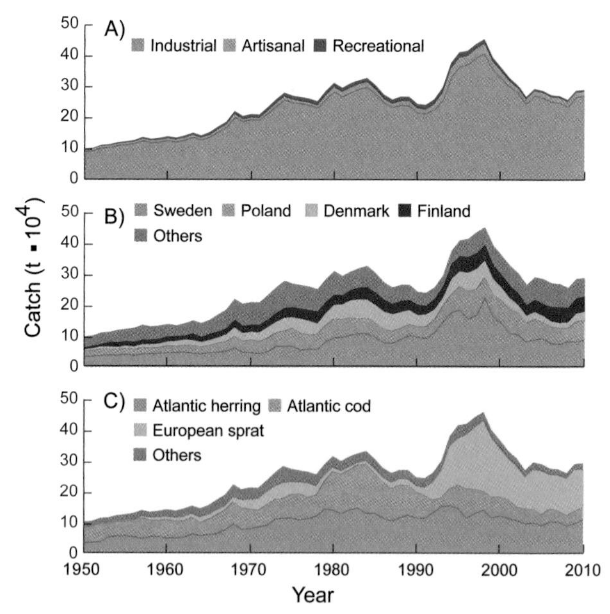

Figure 1. The Swedish Baltic Sea coast has a shelf of 142,000 km² and an EEZ of 142,000 km² in the Baltic. **Figure 2.** Domestic and foreign catches taken in the Baltic Sea EEZ of Sweden: (A) by sector (domestic scores: Ind. 2, 3, 3; Art. 2, 3, 3; Rec. 2, 2, 2; Dis. 2, 2, 2); (B) by fishing country (foreign catches are very uncertain); (C) by taxon.

SWEDEN (WEST COAST)[1]

Lo Persson

Sveriges Lantbruksuniversitet, Umeå, Sweden

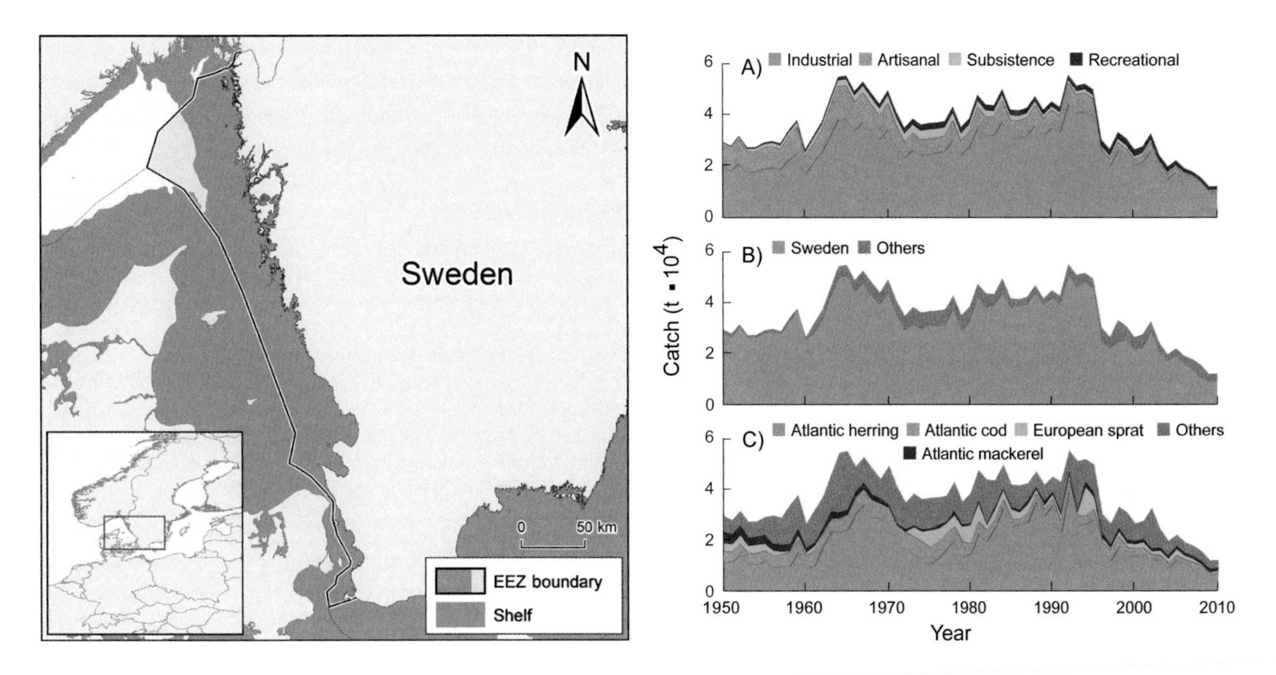

Figure 1. Sweden's West Coast has a shelf of 12,800 km² and an EEZ of 14,480 km². **Figure 2.** Domestic and foreign catches taken in the West Coast EEZ of Sweden: (A) by sector (domestic scores: Ind. 2, 3, 3; Art. 2, 3, 3; Sub. 2, 2, 2; Rec. 2, 2, 2; Dis. 2, 2, 2); (B) by fishing country (foreign catches are very uncertain); (C) by taxon.

The Swedish West Coast (figure 1), which has a long fishing tradition, was very productive compared with the Baltic Sea, because of favorable conditions in the vicinity of the North Sea. This has been reflected in past catches; for example, in the late 1940s, the Swedish West Coast accounted for 65% of the total Swedish marine catch (Andersson 1954). The reconstruction by Persson (2015) of Swedish West Coast catches distinguished catches from within the Swedish EEZ from those taken in the North Sea. These catches were more than 29,000 t/year in the 1950s, peaked in the mid-1960s and early 1990s with 50,000 t/year, and decreased drastically in the last decade, averaging 9,000 t/year in 2009–2010. Reconstructed domestic catch within the West Coast EEZ was 1.3 times the ICES reported landings for the same area. The discrepancy resulted from unreported industrial, recreational, and subsistence catches. Small-scale sectors are of considerable importance in these waters (figure 2A), and the majority of catches were Swedish (figure 2B). The taxonomic composition of catches is more diverse then in the Baltic Sea, but the 3 major contributors are the same (figure 2C). Cardinale and Svedang (2004) write on this area that "due to a prolonged period of high fishing pressure, local cod stocks in the study area, such as the Kattegat may be considered as severely depleted, or possibly, on the verge of total extinction."

REFERENCES

Andersson KA (1954) Swedish ocean fisheries. The West coast fisheries. p. 568 In Andersson KA (ed.), *Fish and fisheries in the Nordic Region*, Second edition. Fish and Fisheries in the Sea Volume 1. Bokförlaget Natur och Kultur, Stockholm.

Cardinale M and Svedang H (2004) Modelling recruitment and abundance of Atlantic cod, *Gadus morhua*, in the eastern Skagerrak-Kattegat (North Sea): evidence of severe depletion due to a prolonged period of high fishing pressure. *Fisheries Research* 69(2): 263–282.

Persson L (2015) Reconstructing total Swedish catches on the west coast of Sweden: 1950–2010. Fisheries Centre Working Paper #2015–24, 10 p.

1. Cite as Persson, L. 2016. Sweden (West Coast). P. 405 In D. Pauly and D. Zeller (eds.), *Global Atlas of Marine Fisheries: A Critical Appraisal of Catches and Ecosystem Impacts*. Island Press, Washington, DC.

SYRIA[1]

Aylin Ulman,[a] Adib Saad,[b] Kyrstn Zylich,[a] Daniel Pauly,[a] and Dirk Zeller[a]

[a]*Sea Around Us*, University of British Columbia, Vancouver, Canada [b]Faculty of Agriculture, Tishreen University, Lattakia, Syria

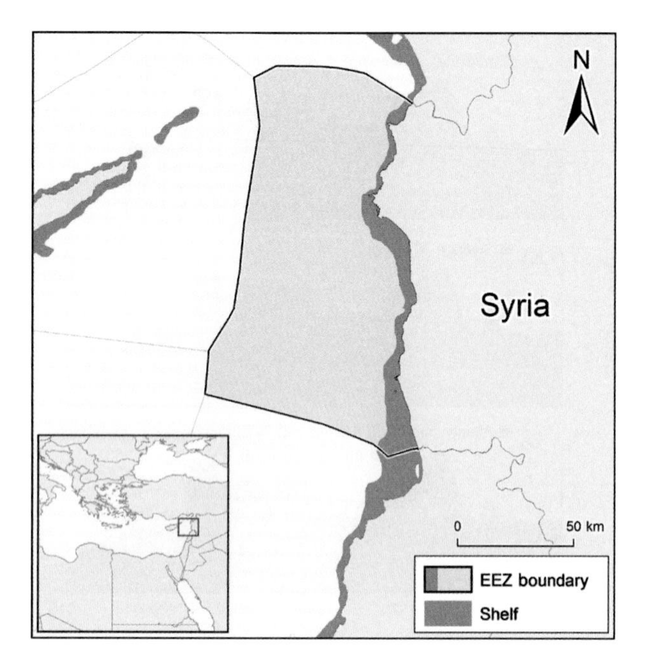

Figure 1. Syria has a shelf of 960 km² and an EEZ of 10,200 km².

Syria has a coastline on the Levantine Basin of the Mediterranean Sea (figure 1). It has a small fishery (Saad 1996, 2010) that is largely artisanal (figure 2A), which is rendered smaller by the fact that national data are collected only from major markets, resulting in underreporting. This was addressed through the catch reconstruction by Ulman et al. (2015), which, from 2004 onward, is based on samples of actual landings from fishing vessels. Reconstructed domestic fisheries catches (there appears to be little if any foreign fishing; figure 2B) were stable throughout the 1950s and 1960s, averaging 1,500 t/year, increased from the mid-1970s on, and averaged 6,500 t/year in the late 2000s. Overall, the reconstructed catches were 1.78 times the data reported by the FAO on behalf of Syria. The increase in catches since the 1990s is caused by increased industrial effort, coupled with spatial expansion of the heavily subsidized trawling fleet, whose catch/effort has strongly declined. The artisanal fisheries are also overexploited, as documented by declining catch rates, lots of juvenile fish in their landings, and minuscule profit margins despite subsidies. Figure 2C documents the major taxa caught, including hake (*Merluccius merluccius*), round sardinella (*Sardinella aurita*), little tunny (*Euthynnus alletteratus*), European barracuda (*Sphryraena sphryraena*), and red mullet (Mullidae). Seafood consumption in Syria is extremely low, about 1 kg/person/year, because of high fish prices and a nationwide fish import ban.

REFERENCES

Saad A (1996) Biology and life cycles of the small pelagic fish on the coast of Syria; landings and the catch profile for the Syria coastal fleet. Field document, Assistance to Artisanal Fleet Project, FAO. 47 p.

Saad A (2010) The fisheries and aquaculture in Syria, status and development perspective in Syria (in Arabic), Ed. Economic Technical Team, Prime Ministry, Syria. *Syrian Economic Bulletin* 1(1): 113–136.

Ulman A, Saad A, Zylich K, Pauly D and Zeller D (2015) Reconstruction of Syria's fisheries catches from 1950–2010: Signs of overexploitation. Fisheries Centre Working Paper #2015-80, 26 p.

1. Cite as Ulman, A., A. Saad, K. Zylich, D. Pauly, and D. Zeller. 2016. Syria. P. 406 In D. Pauly and D. Zeller (eds.), *Global Atlas of Marine Fisheries: A Critical Appraisal of Catches and Ecosystem Impacts*. Island Press, Washington, DC.

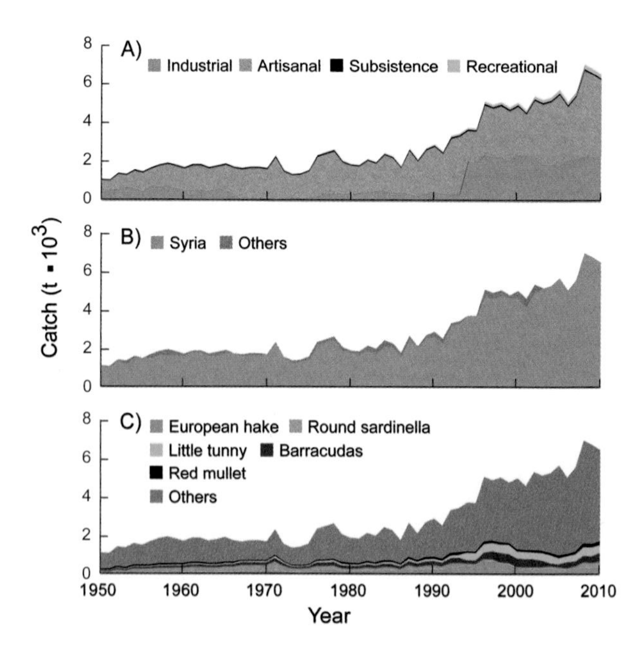

Figure 2. Domestic and foreign catches taken in the EEZ of Syria: (A) by sector (domestic scores: Ind. 1, 2, 3; Art. 1, 3, 4; Sub. 1, 1, 1; Rec. 1, 1, 1; Dis. 2, 3, 2); (B) by fishing country (foreign catches are very uncertain); (C) by taxon.

TAIWAN[1]

Daniel Kuo, Shawn Booth, Esther Divovich, Leonie Färber, Soohyun Shon, and Kyrstn Zylich

Sea Around Us, University of British Columbia, Vancouver, Canada

Taiwan is an island country off the southeast coast of the People's Republic of China in the South China Sea, and its claimed EEZ is disputed and challenged (figure 1). Taiwan's coastal fisheries were already considered overfished in the 1950s, leading to expansion of fisheries offshore and to distant waters beginning in 1959. The reconstruction of Taiwan's catches within its own EEZ by Kuo and Booth (2011), as updated and expanded by Divovich et al. (2015), suggested that industrial fisheries predominate but that the artisanal sector is also important (figure 2A). Foreign fishing occurred mainly in earlier years (figure 2B), and total domestic catches grew from 90,000 t/year in the 1950s to a peak of more than 600,000 t in 1980. Since 1980, total domestic catches within its EEZ have been declining to below 300,000 t/year in the 2000s (figure 2B), mainly because of excessive fishing effort (Huang and Chuang 2010). During this time, Taiwan's distant-water fleets have expanded into all major oceans, and by the 2000s they generated catches 3 to 5 times larger than those from its own EEZ. Figure 2C presents a rough taxonomic breakdown of Taiwan's catch in its own EEZ, documenting the former importance of shrimp and prawns and the recent reliance on chub mackerel (*Scomber japonicus*) and Pacific saury (*Cololabis saira*). Taiwan's diplomatic isolation has often caused problems during negotiations with neighboring countries, resulting in territorial disputes.

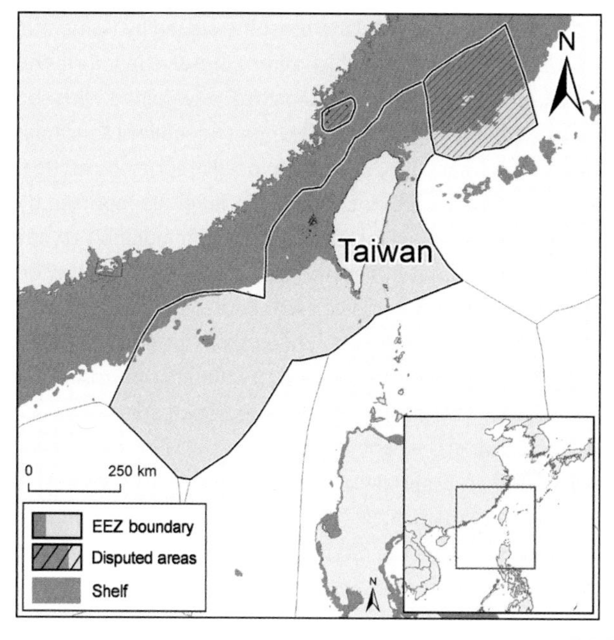

Figure 1. Taiwan (land area 36,200 km²) claims an EEZ of more than 1,150,000 km².

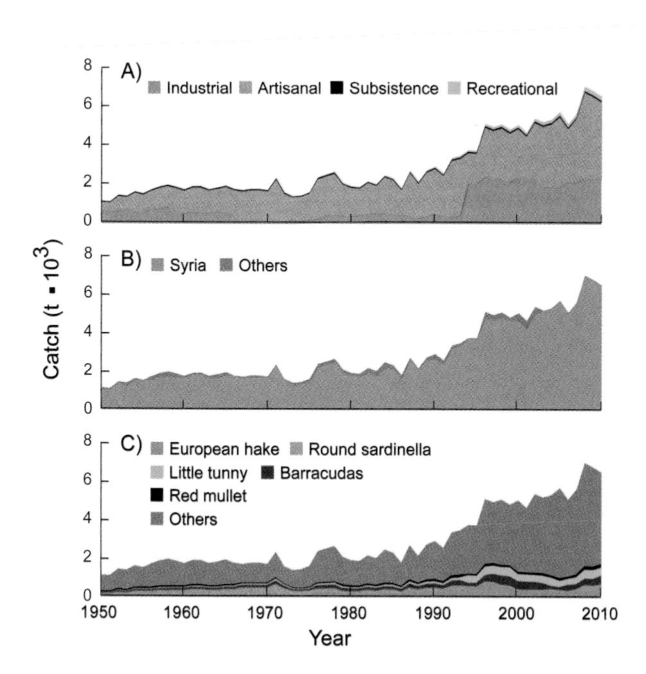

Figure 2. Domestic and foreign catches taken in the EEZ of Taiwan: (A) by sector (domestic scores: Ind. 2, 3, 3; Art. 2, 3, 3; Sub. 1, 2, 2; Rec. –, 1, 1; Dis. 2, 2, 2); (B) by fishing country (foreign catches are very uncertain); (C) by taxon.

REFERENCES

Divovich E, Färber L, Shon S and Zylich K (2015) An updated catch reconstruction of the marine fisheries of Taiwan from 1950–2010. Fisheries Centre Working Paper #2015–78, 7 p.

Huang, H.-W. and Chuang, C.-T. (2010) Fishing capacity management in Taiwan: Experiences and prospects. *Marine Policy* 34(1): 70–76.

Kuo D and Booth S (2011) From local to global: A catch reconstruction of Taiwan's fisheries from 1950–2007. pp. 97–106 In Harper S and Zeller D (eds.), *Fisheries catch reconstructions: Islands, part II*. Fisheries Centre Research Reports 19(4).

1. Cite as Kuo, D., S. Booth, E. Divovich, L. Färber, S. Shon, and K. Zylich. 2016. Taiwan. P. 407 in D. Pauly and D. Zeller (eds.), *Global Atlas of Marine Fisheries: A Critical Appraisal of Catches and Ecosystem Impacts*. Island Press, Washington, DC.

TANZANIA[1]

Jennifer L. Jacquet, Elise Bultel, Beau Doherty, Adam Herman, Frédéric Le Manach, and Dirk Zeller

Sea Around Us, University of British Columbia, Vancouver, Canada

Tanzania has a mainland coastline dotted by numerous islands (figure 1), of which Pemba and Zanzibar form the Zanzibar Region, previously separate, now joined with the mainland (Tanganyika) into the United Republic of Tanzania. Until a few years ago, Zanzibar's locally reported catches, although large, were not included in landings reported by the FAO on behalf of Tanzania. Also, the mainland catches were underestimated because of incomplete countrywide expansion of locally sampled catch data. Thus, a reconstruction of Tanzania catches was undertaken (Jacquet and Zeller 2007; Jacquet et al. 2010; updated by Bultel et al. 2015) that addressed these issues and estimated domestic catches of 18,000 t/year in the early 1950s and 110,000 t in 2010. Small-scale fisheries, especially artisanal, dominate catches in this EEZ (figure 2A), with little foreign fishing documented (figure 2B). This highlights the often neglected role of artisanal and subsistence fisheries (see Nakamura 2011), another reason why the reconstructed catches are 1.8 times the FAO catches for Tanzania, even after Zanzibar is taken into account. Figure 2C documents that these catches were dominated by emperors (Lethrinidae), sardine (*Sardinella* spp.), sharks and rays, jacks (Carangidae), rabbitfishes (Siganidae), and mackerels and tuna (Scombridae). The number of taxa reported to the FAO has increased in recent years, thus decreasing the amount of miscellaneous "marine fishes nei," a positive development also noted in many other countries.

REFERENCES

Bultel E, Doherty B, Herman A, Le Manach F and Zeller D (2015) An Update of the Reconstructed Marine Fisheries Catches of Tanzania with Taxonomic Breakdown. pp. 151–161 In Le Manach F and Pauly D (eds.), *Fisheries catch reconstructions in the Western Indian Ocean, 1950–2010*. Fisheries Centre Research Reports 23(2).

Jacquet JL and Zeller D (2007) Putting the 'United' in the United Republic of Tanzania: reconstructing marine fisheries catches for Tanzania. pp. 49–60 In Zeller D and Pauly D (eds.), *Reconstruction of marine fisheries catches for key countries and regions (1950–2005)*. Fisheries Centre Research Reports 15(2).

Jacquet J, Fox H, Motta H, Ngusaru A and Zeller D (2010) Few data but many fish: marine small-scale fisheries catches for Mozambique and Tanzania. *African Journal of Marine Science* 32(2): 197–206.

Nakamura R (2011) Multi-ethnic-coexistence in Kilwa Island, Tanzania. *Shima: The International Journal of Research into Island Cultures* 5(1): 44–68.

1. Cite as Jacquet, J., E. Bultel, B. Doherty, A. Herman, F. Le Manach, and D. Zeller. 2016. Tanzania. P. 408 In D. Pauly and D. Zeller (eds.), *Global Atlas of Marine Fisheries: A Critical Appraisal of Catches and Ecosystem Impacts*. Island Press, Washington, DC.

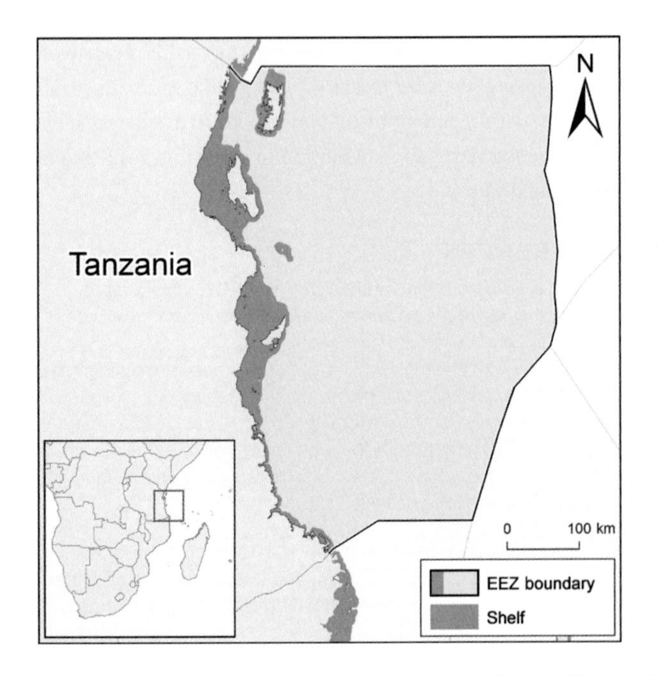

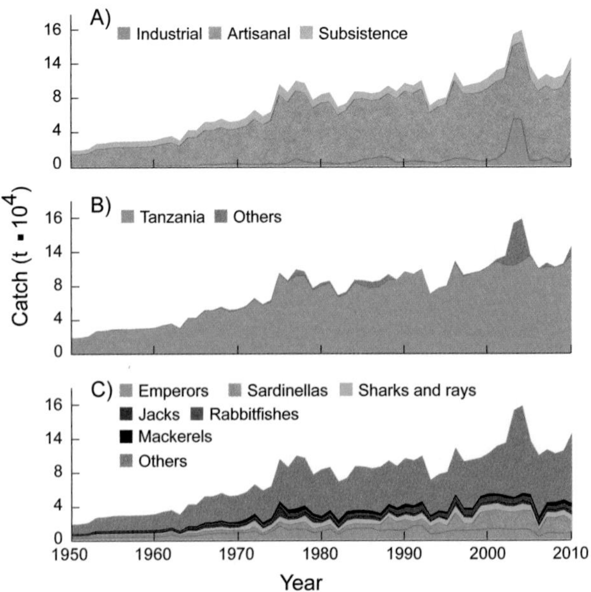

Figure 1. Tanzania, including the islands of Zanzibar, Mafia, and Pemba, with a shelf area of 19,000 km² and an EEZ of 241,000 km². **Figure 2.** Domestic and foreign catches taken in the EEZ of Tanzania: (A) by sector (domestic scores: Ind. 2, 3, 3; Art. 2, 3, 3; Sub. 1, 2, 2; Dis. 2, 2, 2); (B) by fishing country (foreign catches are very uncertain); (C) by taxon.

THAILAND (ANDAMAN SEA)[1]

Lydia C. L. Teh,[a] Ratana Chuenpagdee,[b] Dirk Zeller,[a] and Daniel Pauly[a]

[a]*Sea Around Us*, University of British Columbia, Vancouver, Canada [b]Department of Geography, Memorial University, St. John's, Newfoundland, Canada

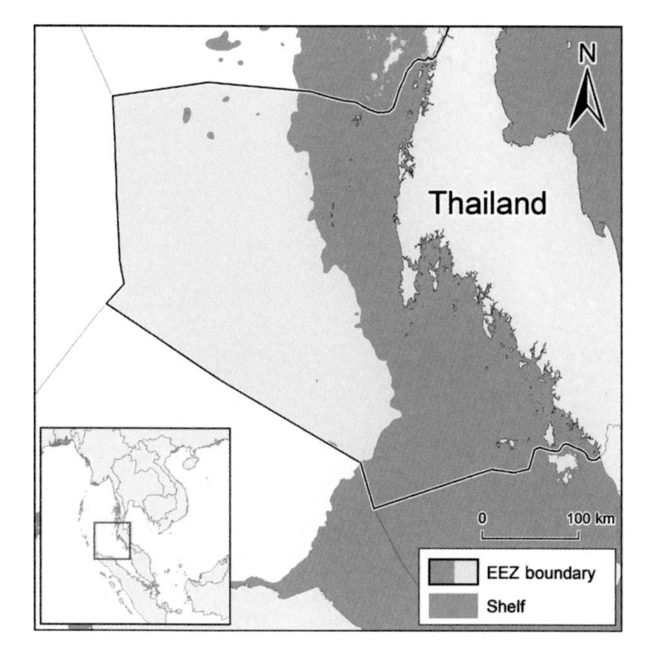

Figure 1. The Thai EEZ (119,000 km²) in the Andaman Sea, Indian Ocean, and the corresponding shelf (51,400 km²).

Thailand's coast on the Andaman Sea in the Indian Ocean, which ranges from Myanmar in the north to Malaysia in the south (figure 1), experienced a development of its marine fisheries similar to that in the better-documented Gulf of Thailand (see next page). Teh et al. (2015) showed that this involved a mainly subsistence and artisanal catch of about 20,000 t/year in the early 1950s, a boom in bottom trawling (industrial) and their associated catches starting in the 1960s, a peak of more than 800,000 t/year in the mid-1990s, and a domestic catch of about 700,000 t/year in the late 2000s, extracted from much-reduced stocks in the Thai Andaman Sea EEZ (figure 2A). Reconstructed domestic catches in this EEZ were 1.5 times the data reported by Thailand for these waters. There are also very large catches from ill-documented Thai distant-water fisheries active throughout the Indian Ocean (Butcher 1999). Virtually no foreign fishing occurs in Thailand's west coast waters (figure 2B). Figure 2C documents the taxonomic composition of the catch, which shows low-value demersal fish contributing most of the current catch in the late 2000s: threadfin breams (Nemipteridae), ponyfish (Leiognathidae), and lizardfish (Synodontidae). Panjarat (2008) suggests that demersal fisheries of the Thai portion of the Andaman Sea have been overexploited since the early 1970s, and pelagic catches were strongly overfished by the late 1980s.

REFERENCES

Butcher, J.G. 1999. Why do Thai trawlers get into so much trouble. Presentation at the Symposium on "The Indian Ocean: Past, Present and Future," November 1999, Western Australian Maritime Museum, Fremantle, Western Australia.

Panjarat S (2008) Sustainable fisheries in the Andaman Sea coast of Thailand, The United Nations-Nippon Foundation Fellowship Programme 2007–2008, Division for Ocean Affairs and the Law of the Sea, Office of Legal Affairs, United Nations, New York (US). 107 p.

Teh LCL, Zeller D and Pauly D (2015) Preliminary reconstruction of Thailand's fisheries catches: 1950–2010. Fisheries Centre Working Paper #2015–01, 15 p.

1. Cite as Teh, L. C. L., R. Chuenpagdee, D. Zeller, and D. Pauly. 2016 Thailand (Andaman Sea). P. 409 in D. Pauly and D. Zeller (eds.), *Global Atlas of Marine Fisheries: A Critical Appraisal of Catches and Ecosystem Impacts*. Island Press, Washington, DC.

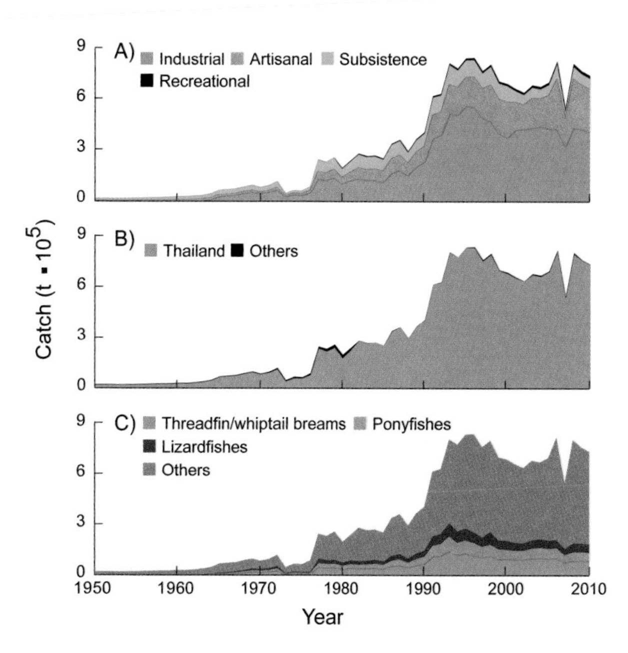

Figure 2. Domestic and foreign catches taken in the Andaman Sea EEZ of Thailand: (A) by sector (domestic scores: Ind. 2, 2, 3; Art. 2, 2, 3; Sub. 2, 2, 3; Rec. –, 1, 1; Dis. 2, 2, 3); (B) by fishing country (foreign catches are very uncertain); (C) by taxon.

THAILAND (GULF OF THAILAND)[1]

Ratana Chuenpagdee,[a] Lydia C. L. Teh,[b] Dirk Zeller,[b] and Daniel Pauly[b]

[a]Department of Geography, Memorial University, St. John's, Newfoundland, Canada
[b]*Sea Around Us*, University of British Columbia, Vancouver, Canada

Thailand is at the heart of Southeast Asia, with coastlines in the Pacific (Gulf of Thailand; figure 1) and the Indian Ocean (see previous page). Considerable literature exists on the development of Thai fisheries, from near-shore small-scale fisheries to a powerhouse of trawling, purse-seining, and other forms of industrial fishing, both locally and in distant waters (e.g., Panayotou and Jetanavanich 1987). However, this account, based on Pauly and Chuenpagdee (2003) and Teh et al. (2015), presents only reconstructed catches from the Gulf of Thailand EEZ, based on national fisheries statistics, estimates of small-scale and industrial fisheries using logbook data, and community surveys. This was about 400,000 t/year in the early 1950s, grew with the onset of trawling in the mid-1960s, peaked at 2 million t/year in 1995, and declined to about 1.2 million t/year in the late 2000s. Overall, this is 1.4 times the nationally reported catches from the Thai EEZ in the Gulf of Thailand, mainly because of underreported subsistence and artisanal fisheries (figure 2A), even though the nationally reported catches can be expected to include at least some of the large catches taken by Thai vessels in distant waters (Pauly and Chuenpagdee 2003). Almost no foreign fishing occurs in Thai waters. Figure 2B presents the Thai catch composition in the Gulf of Thailand, where fish biomass has been reduced by about 90% and shifted toward smaller forms.

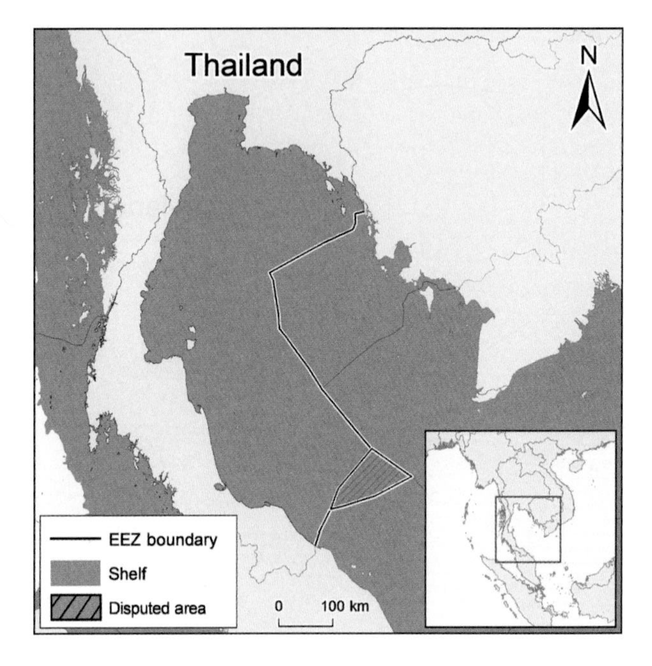

Figure 1. The Thai EEZ in the Gulf of Thailand is 187,000 km² and includes 187,000 km² of shelf.

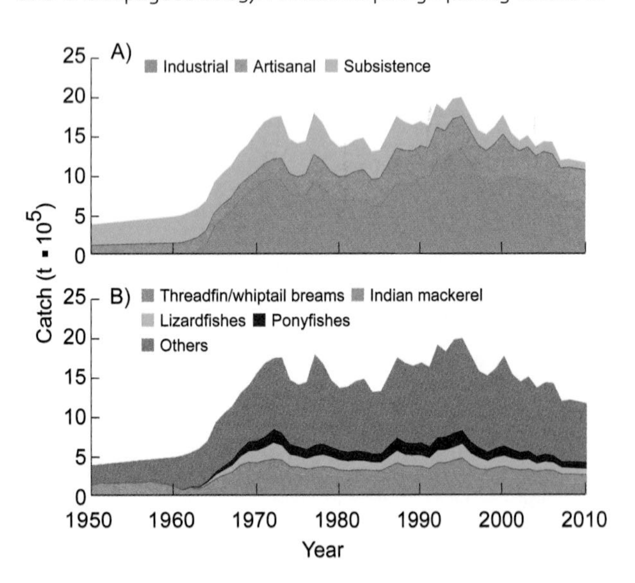

Figure 2. Domestic and foreign catches taken in the Gulf of Thailand EEZ of Thailand: (A) by sector (domestic scores: Ind. 2, 2, 3; Art. 2, 2, 3; Sub. 2, 2, 3; Dis. 2, 2, 3); (B) by taxon.

REFERENCES

Panayotou T and Jetanavanich S (1987) The economics and management of Thai marine fisheries. International Center for Living Aquatic Resources Management (ICLARM) and Winrock International Institute for Agricultural Development, Manila (Philippines) and Arkansas (US). 82 p.

Pauly D and Chuenpagdee R (2003) Development of fisheries in the Gulf of Thailand large marine ecosystem: Analysis of an unplanned experiment. pp. 337–354 in Hempel G and Sherman K (eds.), *Large marine ecosystems of the world: Trends in exploitation, protection, and research*. Elsevier Science.

Teh LCL, Zeller D and Pauly D (2015) Preliminary reconstruction of Thailand's fisheries catches: 1950–2010. Fisheries Centre Working Paper #2015-01, 15 p.

1. Cite as Chuenpagdee, R., L. C. L. Teh, D. Zeller, and D. Pauly. 2016. Thailand (Gulf of Thailand). P. 410 in D. Pauly and D. Zeller (eds.), *Global Atlas of Marine Fisheries: A Critical Appraisal of Catches and Ecosystem Impacts*. Island Press, Washington, DC.

TIMOR-LESTE[1]

Milton Barbosa,[a] Rui Pinto,[b] Shawn Booth,[a] and Daniel Pauly[a]

[a]*Sea Around Us*, University of British Columbia, Vancouver, Canada
[b]Timor-Leste Country Program, Conservation International, Dili, Timor-Leste

Timor-Leste (East Timor) is a small Southeast Asian country on the eastern half of Timor Island, including the islands of Atauro and Jaco and an exclave, Oecusse-Ambeno, within the Indonesian part of Timor Island (figure 1). A colony of Portugal until 1975, Timor-Leste, which never had much of a fishing industry (Felgas 1956; Cook 2000), experienced a short-lived independence before being forcibly annexed by Indonesia. The reconstructed domestic catch (dominated by subsistence and artisanal sectors; figure 2A) was estimated by Barbosa and Booth (2009) as about 1,600 t/year in the 1950s; it grew gradually, but this growth accelerated under Indonesian occupation (1975–1998, with extensive foreign fishing; figure 2B), which also brought skilled fishers from various areas of Indonesia, but declined steeply in 1999 as a result of the violence inflicted on Timor-Leste as it regained its independence. The fisheries sector rebounded and, in the late 2000s, contributed catches of 5,000–6,000 t/year (figure 2B). Large pelagics, such as tunas and tuna-like species (e.g., skipjack tuna, *Katsuwonus pelamis*) and sharks (e.g., silky shark, *Carcharhinus falciformes*), dominated the catch (figure 2C), although reef fishes were also important because of the dominant role of subsistence fisheries, notably snapper (family Lutjanidae), fusiliers (Caesionidae), and yellowfin surgeonfish (*Acanthurus xanthopterus*). Recent biomass and biodiversity studies (Erdmann and Mohan 2013) point to overfishing of reef fish species and thus show the need to strengthen coastal fisheries management.

REFERENCES

Barbosa M and Booth S (2009) Timor-Leste's fisheries catches (1950–2009): Fisheries under different regimes. pp. 39–51 In Zeller D and Harper S (eds.), *Fisheries catch reconstructions: Islands, part I*. Fisheries Centre Research Reports 17(5) [Updated to 2010]

Cook D (2000) Mission report: Rapid appraisal of the small-scale fisheries sector of East Timor. Food and Agriculture Organization of the United Nations (FAO), Rome (Italy).

Erdmann MV and Mohan C (2013) A Rapid Marine Biological Assessment of Timor-Leste. RAP Bulletin of Biological Assessment 66, Coral Triangle Support Partnership, Conservation International Timor-Leste, Dili. 166 p.

Felgas HAE (1956) Timor Português [Portuguese Timor]. Agência Geral do Ultramar. Divisão de Publicações e Biblioteca, Lisbon (Portugal). 570 p.

1. Cite as Barbosa, M., R. Pinto, S. Booth, and D. Pauly. 2016. Timor-Leste. P. 411 in D. Pauly and D. Zeller (eds.), *Global Atlas of Marine Fisheries: A Critical Appraisal of Catches and Ecosystem Impacts*. Island Press, Washington, DC.

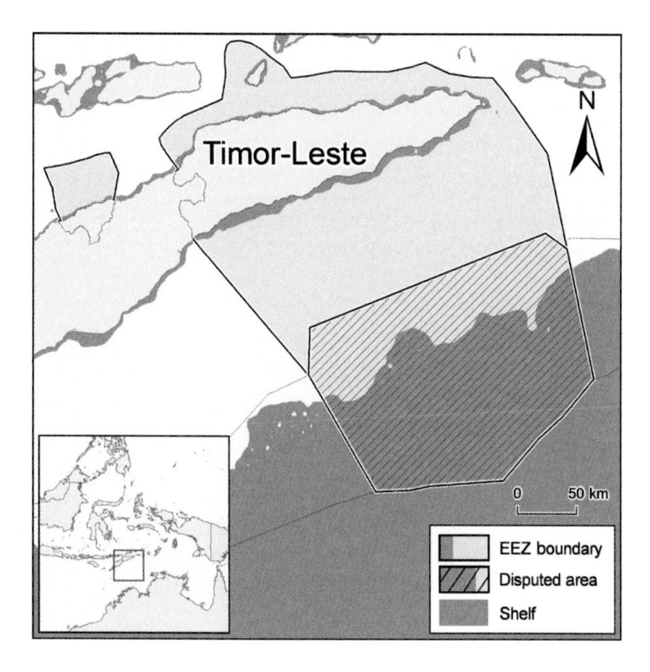

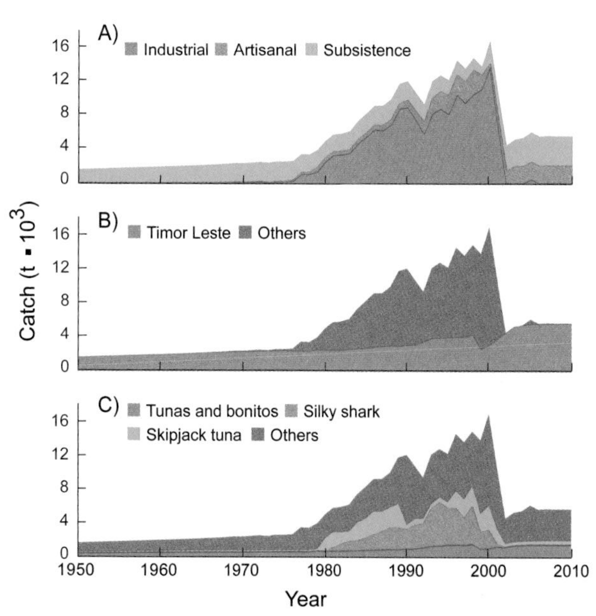

Figure 1. Timor-Leste (East Timor) has a total land area of 15,400 km² and an EEZ of 77,100 km². **Figure 2.** Domestic and foreign catches taken in the EEZ of Timor-Leste: (A) by sector (domestic scores: Art. 2, 2, 2; Sub. 2, 2, 2); (B) by fishing country (foreign catches are very uncertain); (C) by taxon.

TOGO[1]

Dyhia Belhabib, Viviane Koutob, Dirk Zeller, and Daniel Pauly

Sea Around Us, University of British Columbia, Vancouver, Canada

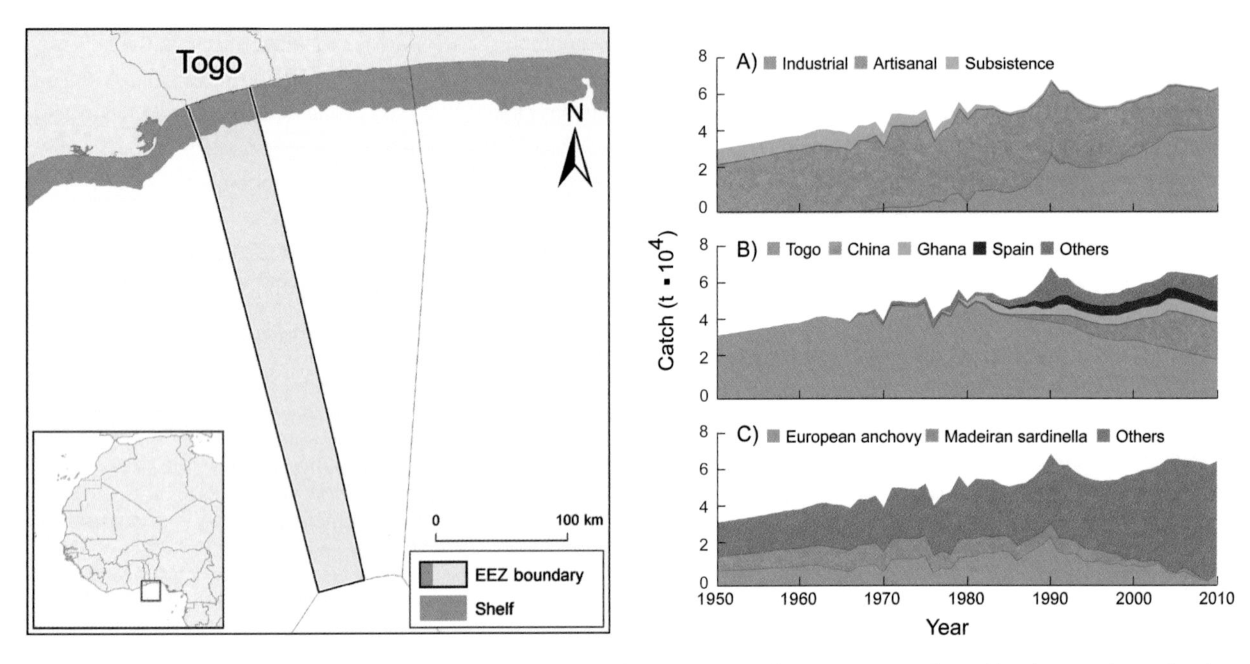

Figure 1. Togo has a continental shelf of 1,170 km² and an EEZ of 15,400 km². **Figure 2.** Domestic and foreign catches taken in the EEZ of Togo: (A) by sector (domestic scores: Ind. 3, 3, 3; Art. 3, 3, 3; Sub. 3, 3, 3; Rec. 2, 2, 2; Dis. –, 1, 2); (B) by fishing country (foreign catches are very uncertain); (C) by taxon.

Togo, formerly one of the few German colonies in Africa, stretches from the edge of the Sahel in the north to a narrow coast on the Gulf of Guinea in the south (figure 1). Although coastal lagoon fisheries remain understudied despite high yields (Laë 1997), marine fisheries are given higher importance in terms of monitoring. Yet catch data in Togo are largely anecdotal, and domestic fisheries are largely artisanal (figure 2A). Total catches reconstructed by Belhabib et al. (2015) were about 34,000 t in 1950, increased to 57,000 t in 1979 when foreign fleets began to operate in Togo, and then followed two separate trajectories: a steady decrease of domestic catches to about 20,000 t in 2010 and a drastic increase in foreign legal and illegal catches, with a total catch of nearly 65,000 t in 2010 (figure 2B). Domestic catches were 4 times the data supplied to the FAO, 12% of which originated from lagoon fisheries. Domestic catches were dominated by small pelagic species, notably anchovy (*Engraulis encrasicolus*) and Madeiran sardinella (*Sardinella maderensis*), whereas foreign catches consisted mainly of bigeye grunt (*Brachydeuterus auritus*), perch-like fishes, and porgies (family Sparidae; figure 2C). Despite the apparent nonoverlap of the species targeted by the domestic and foreign fleets, the recent decline in domestic small pelagic species is increasingly compensated by demersal species also targeted by foreign and mostly illegal fleets.

REFERENCES

Belhabib D, Koutob V, Zeller D and Pauly D (2015) The marine fisheries of Togo, the "Heart of West Africa," 1950 to 2010. Fisheries Centre Working Paper #2015-70 28 p.

Laë R (1997) Does overfishing lead to a decrease in catches and yields? An example of two West African coastal lagoons. *Fisheries Management and Ecology* 4(2): 149–164.

1. Cite as Belhabib, D., V. Koutob, D. Zeller, and D. Pauly. 2016. Togo. P. 412 in D. Pauly and D. Zeller (eds.), *Global Atlas of Marine Fisheries: A Critical Appraisal of Catches and Ecosystem Impacts*. Island Press, Washington, DC.

TONGA[1]

Patricia Sun, Sarah Harper, Shawn Booth, and Dirk Zeller

Sea Around Us, University of British Columbia, Vancouver, Canada

The Kingdom of Tonga is a South Pacific state consisting of approximately 170 islands, of which 37 are inhabited (figure 1). The reconstruction of domestic catches for Tonga (Sun et al. 2011) yielded values of 4,000 t/year in the early 1950s, a plateau at 6,000 t/year in the early 1970s, and about 5,000 t/year in the 2000s. Overall, this is more than 3.5 times the landings reported by the FAO on behalf of Tonga. Fisheries are dominated by subsistence and artisanal sectors (figure 2A), with industrial catches being largely from foreign fleets (figure 2B). Most of the domestic catches that were missing in the reported data were from the subsistence sector, which represented 70% of the reconstructed domestic catch and is based on the exploitation of reef and lagoon resources, with, e.g., flathead grey mullet (*Mugil cephalus*) being strongly overexploited. Figure 2C summarizes the taxonomic diversity of the catches taken in Tonga's waters, which include invertebrates gleaned on reefs, mostly by women, a wide range of reef and slope fishes, notably grouper and snapper caught by the subsistence and artisanal fisheries, and large offshore pelagic fish such as tuna and billfish (Kronen 2004). Although parts of Tonga have switched to a cash-based economy that relies less on subsistence fishing, bartering still exists on the outer islands, and therefore subsistence fisheries should be considered in policy development, particularly in terms of food security.

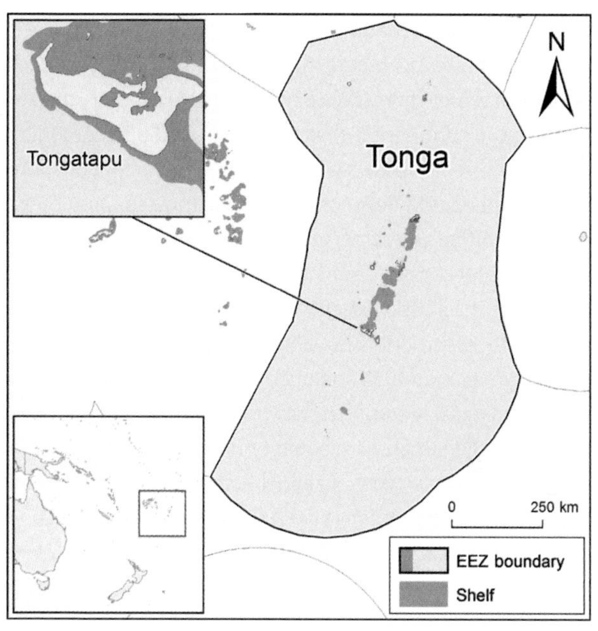

Figure 1. The islands of Tonga are clustered into 3 groups with a combined land area of about 748 km² and an EEZ of approximately 665,000 km².

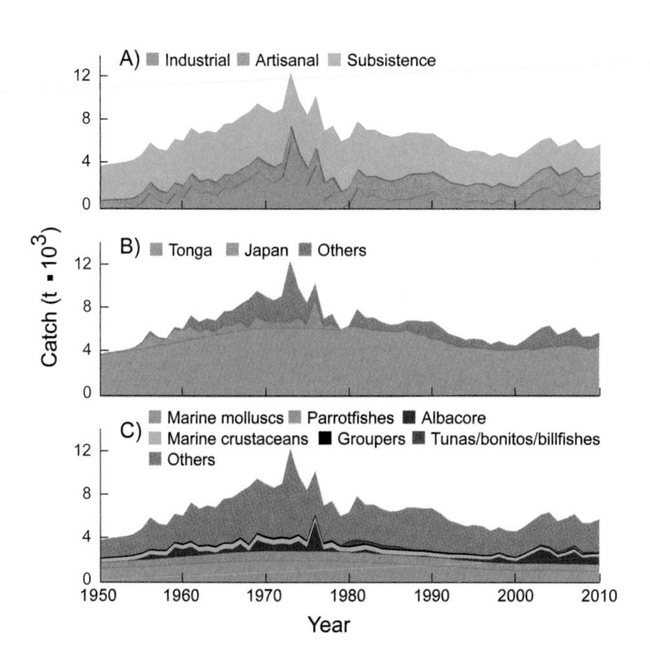

Figure 2. Domestic and foreign catches taken in the EEZ of Tonga: (A) by sector (domestic scores: Ind. –, 3, 3; Art. 2, 2, 2; Sub. 1, 2, 2); (B) by fishing country (foreign catches are very uncertain); (C) by taxon. Domestic and foreign catches taken in the EEZ of Tonga.

REFERENCES

Kronen M (2004) Fishing for fortunes? A socio-economic assessment of Tonga's artisanal fisheries. *Fisheries Research* 70(1): 121–134.

Sun P, Harper S, Booth S and Zeller D (2011) Reconstructing marine fisheries catches for the Kingdom of Tonga: 1950–2007. pp. 119–130 In Harper S and Zeller D (eds.), *Fisheries catch reconstruction: Islands, Part II*. Fisheries Centre Research Reports 19(4).

1. Cite as Sun, P., S. Harper, S. Booth, and D. Zeller. 2016. Tonga. P. 413 in D. Pauly and D. Zeller (eds.), *Global Atlas of Marine Fisheries: A Critical Appraisal of Catches and Ecosystem Impacts*. Island Press, Washington, DC.

TRINIDAD AND TOBAGO[1]

Elizabeth Mohammed[a] and Alasdair Lindop[b]

[a]Research and Resource Assessment, Caribbean Regional Fisheries Mechanism Secretariat, Eastern Caribbean Office, St. Vincent and the Grenadines [b]*Sea Around Us*, University of British Columbia, Vancouver, Canada

Trinidad and Tobago is a small country consisting of 2 islands and located at the southern end of the eastern Caribbean island chain (figure 1). The larger island of Trinidad rests on the continental shelf off northeast South America, close to the Venezuelan coast, and Tobago is further offshore and is surrounded by deep oceanic waters. Trinidad's shelf offers access to shrimp and demersal resources, absent in Tobago. Although the idea of performing catch reconstructions emerged in and was first illustrated with data from Trinidad (Pauly 1998), it is the reconstruction by Mohammed and Lindop (2015), building on Mohammed and Chan A Shing (2003), that led to the comprehensive domestic catch time series presented here. Fisheries in this EEZ are dominated by industrial fleets, although artisanal and subsistence fisheries also feature prominently (figure 2A). Foreign, mainly Venezuelan catches have declined in recent years (figure 2B), and domestic catches increased from 8,700 t in 1950 to 24,000 t in 1969, then declined to 16,000 t were in 1986. After large fluctuations, the catch eventually stabilized at 16,000 t/year by 2010. Overall, domestic catches were 2.6 times those reported by Trinidad and Tobago to the FAO. Tunas, mackerels, and bonitos (family Scombridae) were dominant (figure 2C), as were drums and croakers (Sciaenidae), snappers (Lutjanidae), jacks (Carangidae), and swimming crabs (Portunidae), while the economically valuable shrimps (Penaeidae) contribute less to catches.

REFERENCES

Mohammed E and Chan A Shing C (2003) Trinidad and Tobago: Preliminary reconstructions of fisheries catches and fishing effort, 1908–2002. pp. 117–132 In Zeller D, Booth S, Mohammed E and Pauly D (eds.), *From Mexico to Brazil: Central Atlantic fisheries catch trends and ecosystem models*. Fisheries Centre Research Reports 11(6).

Mohammed E and Lindop A (2015) Trinidad and Tobago: reconstructed fisheries catches, 1950–2010. Fisheries Centre Working Paper #2015-55, 42 p.

Pauly D (1998) Rationale for reconstructing catch time series. *EC Fisheries Cooperation Bulletin* 11(2): 4–7 [see also chapter 1, this volume]

1. Cite as Mohammed, E., and A. Lindop. 2016. Trinidad and Tobago. P. 414 in D. Pauly and D. Zeller (eds.), *Global Atlas of Marine Fisheries: A Critical Appraisal of Catches and Ecosystem Impacts*. Island Press, Washington, DC.

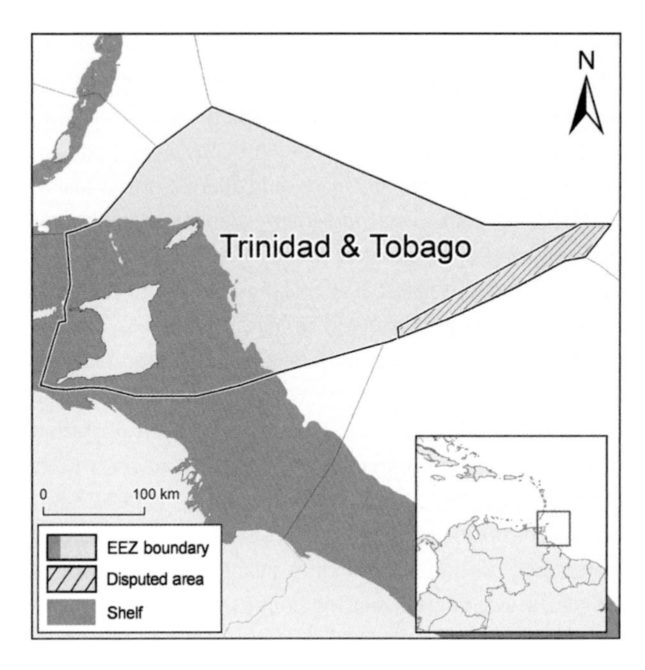

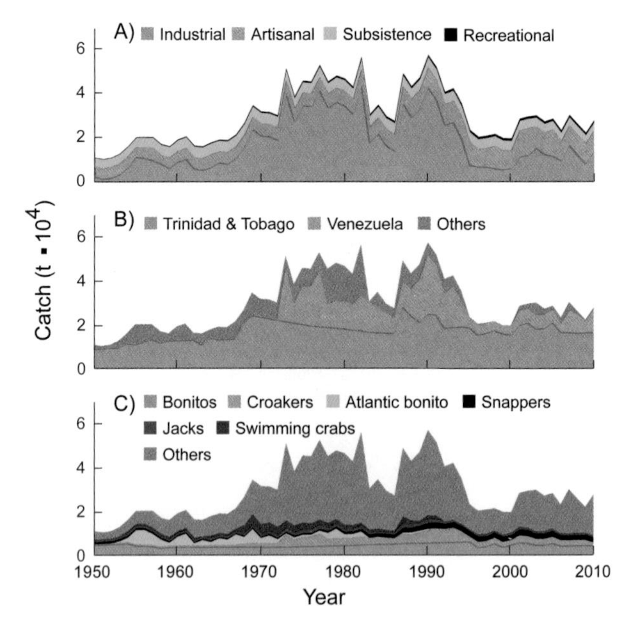

Figure 1. Trinidad and Tobago has a land area 5,130 km² and an EEZ of about 79,800 km². **Figure 2.** Domestic and foreign catches taken in the EEZ of Trinidad and Tobago: (A) by sector (domestic scores: Ind. 2, 2, 3; Art. 2, 2, 2; Sub. 1, 1, 1; Rec. 1, 1, 1; Dis. 1, 1, 1); (B) by fishing country (foreign catches are very uncertain); (C) by taxon.

TUNISIA[1]

Ghassen Halouani,[a] Frida Ben Rais Lasram,[a] Myriam Khalfallah,[a] Dirk Zeller,[b] and Daniel Pauly[b]

[a]Aquatic Resources and Ecosystems Laboratory, National Agronomic Institute, Tunis, Tunisia
[b]*Sea Around Us*, University of British Columbia, Vancouver, Canada

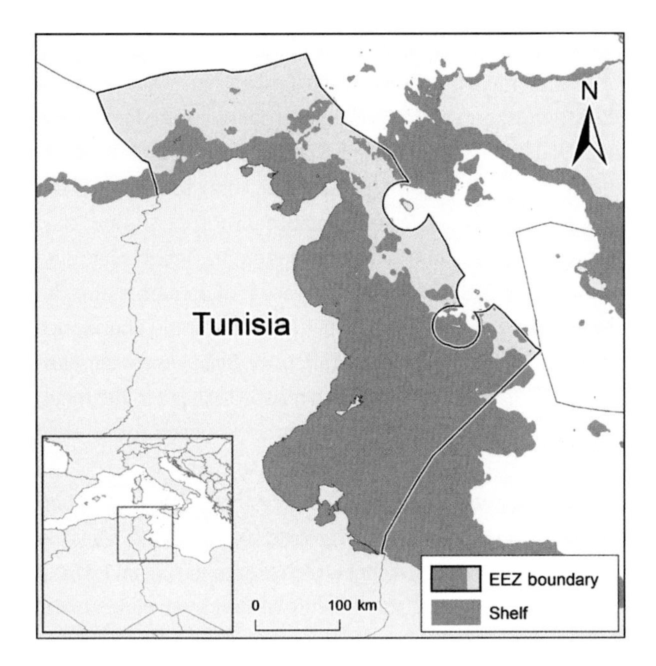

Figure 1. Tunisia has a large shelf (68,500 km²) for an EEZ of only 102,000 km².

Tunisia is situated in North Africa, at the juncture of the western and eastern Mediterranean basins (figure 1). The wide, soft-bottomed continental shelf has contributed to the development of artisanal and industrial fisheries in the south, in the Gulf of Gabes, where about 50% of total landings occur. In northern Tunisia, because of the narrower continental shelf, industrial fisheries target mainly pelagic species. Although industrial fisheries dominate in catches (figure 2A, and although some of these catches were foreign, figure 2B), artisanal fisheries play an important role in Tunisia. Halouandi et al. (2015) estimated total domestic removals from the EEZ by taking into account all reported and unreported catches from the subsistence, artisanal, recreational, and industrial fisheries and their discards, after reestimation by Khalfallah (2013) of raising factors used to derive artisanal catches from sampled boats. The results were domestic catches of about 15,000t/year in the 1950s, which peaked at 110,000 t in 1988 and declined to about 90,000 t in 2010. This is, overall, 1.13 times the data reported by the FAO on behalf of Tunisia. Figure 2C includes species caught by bottom trawling as well as pelagic fisheries, and some of the many species caught by the small-scale fisheries. That Tunisian fisheries catches seem to have reached a plateau while fishing effort keeps increasing is worrisome, because it suggests overexploitation.

REFERENCES

Halouani G, Lasram F, Khalfallah M, Zeller D and Pauly D (2015) Reconstruction of Marine Fisheries catches for Tunisia (1950–2010). Fisheries Centre Working Paper #2015–95, 11 p.

Khalfallah M (2013) Proposition d'amélioration du système de Statistiques de la Pêche en Tunisie : Cas du Golf de Tunis. Fisheries and Environmental Engineering thesis, University of Carthage, National Agronomic Institute of Tunisia, Department of Animal Resources, Fisheries and Food Technology, Tunisia. 62 p.

1. Cite as Halouani, G., F. Lasram, M. Khalfallah, D. Zeller, and D. Pauly. 2016. Tunisia. P. 415 In D. Pauly and D. Zeller (eds.), *Global Atlas of Marine Fisheries: A Critical Appraisal of Catches and Ecosystem Impacts*. Island Press, Washington, DC.

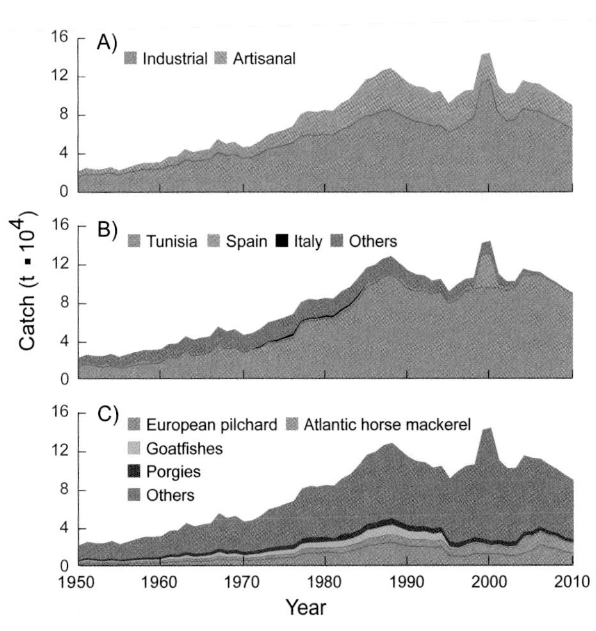

Figure 2. Domestic and foreign catches taken in the EEZ of Tunisia: (A) by sector (domestic scores: Ind. 2, 2, 3; Art. 2, 2, 3; Rec. 1, 1, 1; Dis. 2, 3, 3); (B) by fishing country (foreign catches are very uncertain); (C) by taxon.

TURKEY (BLACK SEA)[1]

Aylin Ulman,[a] Mustafa Zengin,[b] Ståle Knudsen[c]

[a]*Sea Around Us*, University of British Columbia, Vancouver, Canada [b]Central Fisheries Research Institute (MARA), Trabzon, Turkey [c]Department of Social Anthropology, University of Bergen, Bergen, Norway

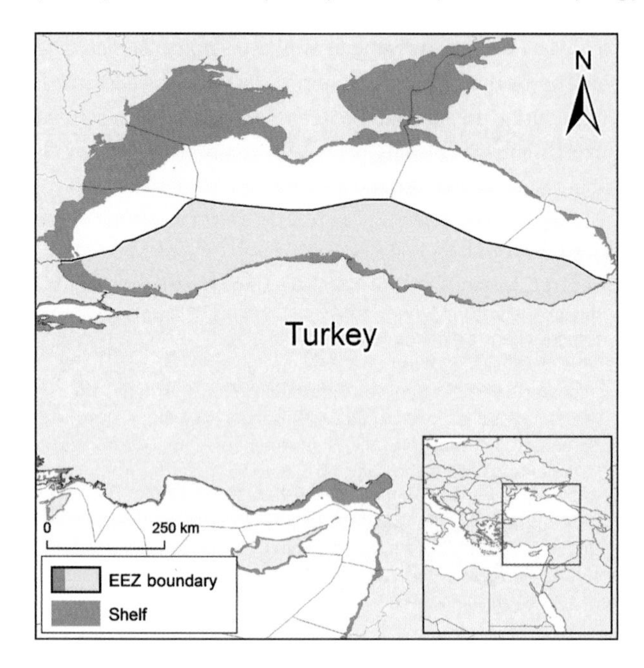

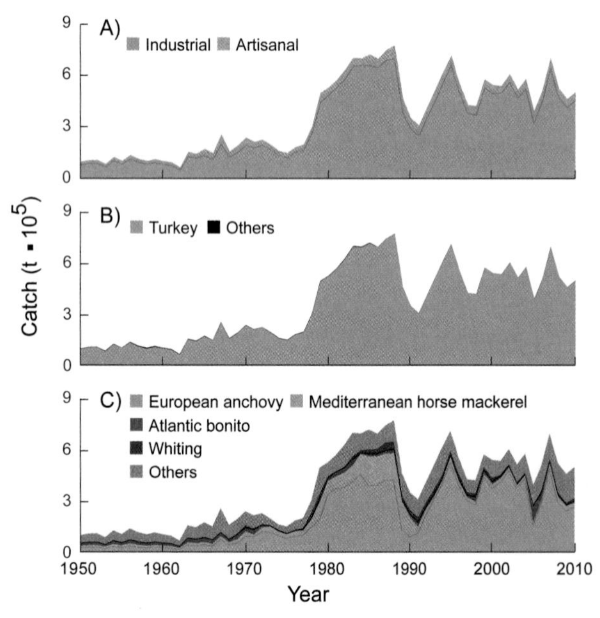

Figure 1. The Black Sea coast of Turkey and its shelf (18,900 km²) and EEZ (172,000 km²). **Figure 2.** Domestic and foreign catches taken in the Black Sea coast EEZ of Turkey: (A) by sector (domestic scores: Ind. 2, 3, 3; Art. 2, 3, 3; Sub. 1, 2, 2; Rec. 1, 2, 2; Dis. 3, 3, 4); (B) by fishing country (foreign catches are very uncertain); (C) by taxon. Note that the reconstructed catch is more than 1.5 times the FAO landings of Turkey in the Black Sea.

Turkey's northern coastline has a narrow shelf permitting only limited bottom trawling (figure 1), so most of its Black Sea fisheries target pelagic fishes. Fisheries are predominantly industrial (figure 2A), with almost no foreign fishing in this EEZ. The catch reconstruction (Ulman et al. 2013), based on Zengin et al. (1998), Knudsen (2009), and other sources, showed that domestic catches were stable throughout the 1950s, averaging 110,000 t/year. Because of the removal of predators and the development of the Turkish fleet (figure 2B), the abundance and catches of small pelagics (anchovy, *Engraulis encrasicolus*; and sprat, *Sprattus sprattus*, both mostly reduced to fishmeal) increased dramatically in the late 1970s (Knudsen 2009). They crashed in 1990 and 1991, partly because of an invasion of the carnivorous warty comb jelly (*Mnemiopsis leidyi*), but then recovered, varying between 170,000 and 500,000 t/year in the 2000s (figure 2C). Anchovy and sprat made up only 30% of Turkish Black Sea catches in the 1950s, but by the late 2000s they made up the majority of catches. Other species in the catch are Mediterranean and Atlantic horse mackerel (*Trachurus mediterraneus* and *T. trachurus*), bonito (*Sarda sarda*), and whiting (*Merlangius merlangus*). The Turkish reduction fisheries in the Black Sea are not well managed, which led to a massive overcapacity driving their expansion outside of Turkish waters and to the landing of undersized fish, often unsuitable even for processing into fishmeal.

REFERENCES

Knudsen S (2009) Fishers and scientists in modern Turkey: The management of natural resources, knowledge and identity on the eastern Black Sea coast. Studies in Environmental Anthropology and Ethnobiology 8. Bergahn Books, Oxford. xi + 273 p.

Ulman A, Bekişoğlu Ş, Zengin M, Knudsen S, Ünal V, Mathews C, Harper S, Zeller D and Pauly D (2013) From bonito to anchovy: a reconstruction of Turkey's marine fisheries catches (1950–2010). *Mediterranean Marine Science* 14(2): 309–342.

Zengin M, Genc Y and Duzgunes E (1998) Evaluation of the data from market samples on the commercial fish species in the Black Sea during 1990–1995. In Celikkale MS, Duzgunes E, Okumus I and Mutlu C (eds.), First International Symposium on Fisheries and Ecology Proceedings (FISHECO '98), September 2–4, 1998, Trabzon, Turkey. 9 p.

1. Cite as Ulman, A., M. Zengin, and S. Knudsen. 2016. Turkey (Black Sea). P. 416 In D. Pauly and D. Zeller (eds.), *Global Atlas of Marine Fisheries: A Critical Appraisal of Catches and Ecosystem Impacts*. Island Press, Washington, DC.

TURKEY (MARMARA SEA)[1]

Aylin Ulman,[a] Vahdet Ünal,[b] and Christopher Mathews[c]

[a]*Sea Around Us*, University of British Columbia, Vancouver, Canada [b]Faculty of Fisheries, Ege University, Izmir, Turkey [c]Fisheries Management and Planning Consultant, England

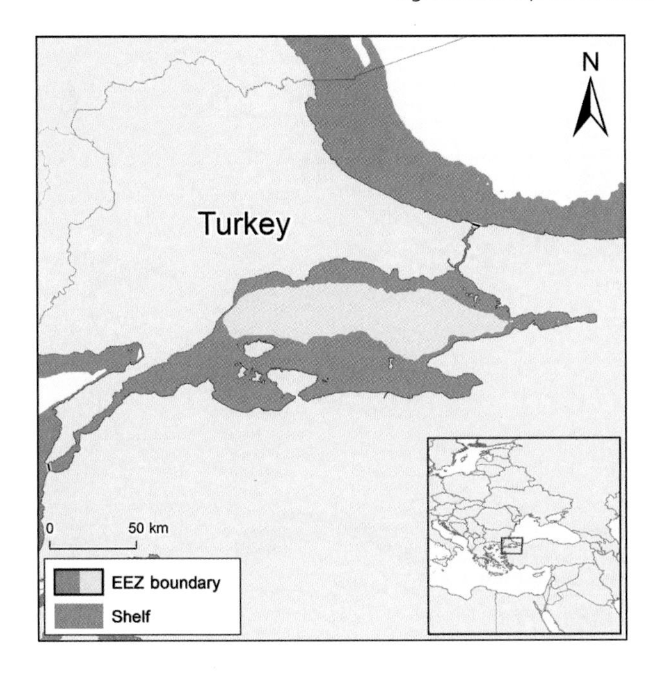

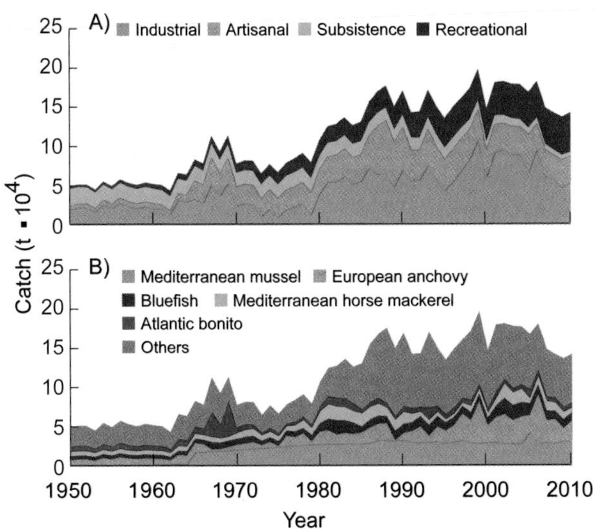

Figure 1. The Sea of Marmara (including the Bosphorus Strait and the Dardanelles) has an area of 11,700 km², of which more than 7,370 km² is shelf. **Figure 2.** Domestic and foreign catches taken in the Marmara Sea (including the Bosphorus Strait and the Dardanelles) EEZ of Turkey: (A) by sector (domestic scores: Ind. 2, 3, 3; Art. 2, 3, 3; Sub. 1, 2, 2; Rec. 1, 2, 2; Dis. 3, 3, 4); (B) by fishing country (foreign catches are very uncertain); (C) by taxon.

The Marmara Sea is a major part of the corridor connecting the Black Sea to the Mediterranean Sea. The catch reconstruction (Ulman et al. 2013), based mainly on Ünal et al. (2010) and Harlioğlu (2011), showed that total catches from the Marmara Sea (including the Bosphorus and Dardanelles straits) include substantial recreational and artisanal catches, in addition to industrial fisheries (figure 2A), with no foreign fishing. Catches were about 50,000 t/year throughout the 1950s, increased around 1980, peaked in the late 1990s at 190,000 t, and declined to about 140,000 t/year in the late 2000s. The Sea of Marmara lost several of its previous migratory species in the 1960s and 1970s, such as bluefin tuna (*Thunnus thynnus*), Atlantic mackerel (*Scomber scombrus*), and chub mackerel (*S. japonicus*). More recently, the main species caught are small pelagics, notably anchovy (*Engraulis encrasicolus*), and massive quantities of juveniles or small specimens of larger species, such as bluefish (*Pomatomus saltatrix*), horse mackerels (*Trachurus* spp.), and bonito (*Sarda sarda*), as well as invertebrates (figure 2B). Overall, the reconstructed catches were nearly 2.6 times the reported data, mainly because of previously undocumented large recreational and subsistence fisheries, notably in Istanbul. The state of the Sea of Marmara has been diminished by pollution and extensive bottom trawling, the latter being illegal since 1971 but ongoing because of weak enforcement. This highlights the serious governance problems besetting Turkish fisheries.

REFERENCES

Harlioğlu AG (2011) Present status of fisheries in Turkey. *Reviews in Fish Biology and Fisheries* 21: 667–680.

Ulman A, Bekişoğlu Ş, Zengin M, Knudsen S, Ünal V, Mathews C, Harper S, Zeller D and Pauly D (2013) From bonito to anchovy: a reconstruction of Turkey's marine fisheries catches (1950–2010). *Mediterranean Marine Science* 14(2): 309–342.

Ünal V, Acarli D and Gordoa A (2010) Characteristics of marine recreational fishing in the Çannakale strait. *Mediterranean Marine Science* 11(2): 315–330.

1. Cite as Ulman, A., V. Ünal, and C. Mathews. 2016. Turkey (Marmara Sea). P. 417 in D. Pauly and D. Zeller (eds.), *Global Atlas of Marine Fisheries: A Critical Appraisal of Catches and Ecosystem Impacts*. Island Press, Washington, DC.

TURKEY (MEDITERRANEAN)[1]

Aylin Ulman,[a] Vahdet Ünal,[b] and Şahin Bekişoğlu[c]

[a]*Sea Around Us*, University of British Columbia, Vancouver, Canada [b]Faculty of Fisheries, Ege University, Izmir, Turkey [c]DOLSAR Engineering, Ankara, Turkey

Turkish waters in the Mediterranean reach from the Aegean Sea to the Levantine Sea, and the shelf is mostly narrow, except in Iskenderun Bay in the east, where trawling was established as far back as the 1950s (figure 1). The waters of the Turkish Mediterranean are very clear, so that in contrast to the Black Sea, the Mediterranean is the "White Sea" in Turkish. The catch reconstruction of Turkey (Ulman et al. 2013), based mainly on Ünal and Erdem (2009) and Ünal et al. (2010), showed that although industrial catches dominate, an important and widespread artisanal sector exists (figure 2A). Reconstructed catches were stable and small throughout the 1950s, began to increase steadily after the mid-1970s, peaked at 180,000 t in 1993, and declined to about 110,000 t/year in the late 2000s; overall, this is 1.8 times the reported data. Major resources are European pilchards (*Sardina pilchardus*), grey mullet (Mugilidae), anchovy (*Engraulis encrasicolus*), gobies (Gobiidae), mussels (e.g., *Mytilus galloprovinvialis*), and whiting (*Micromestisius poutassou*) (figure 2B). A Red Sea migrant, the silver-cheeked toadfish (i.e., a pufferfish species, *Lagocephalus sceleratus*) is increasingly caught and discarded because of its toxicity. Seine nets and driftnets were popular but have been banned since 2001 and 2006, respectively. Longlines are now used to target swordfish (*Xiphias gladius*) and several species of tuna, and seine nets were replaced with gillnets and trammel nets.

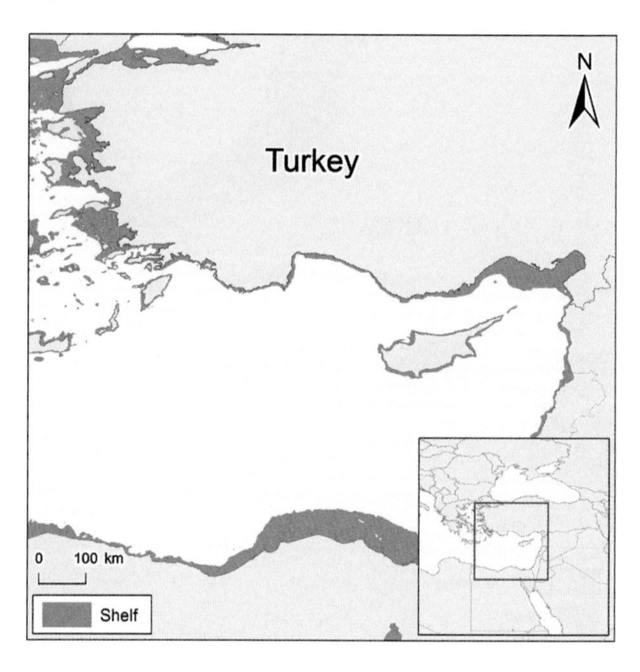

Figure 1. Turkish waters in the Mediterranean extend from south of the Dardanelles in the west to the Syrian border in the east; the corresponding shelf area is 26,000 km² and EEZ is 72,200 km².

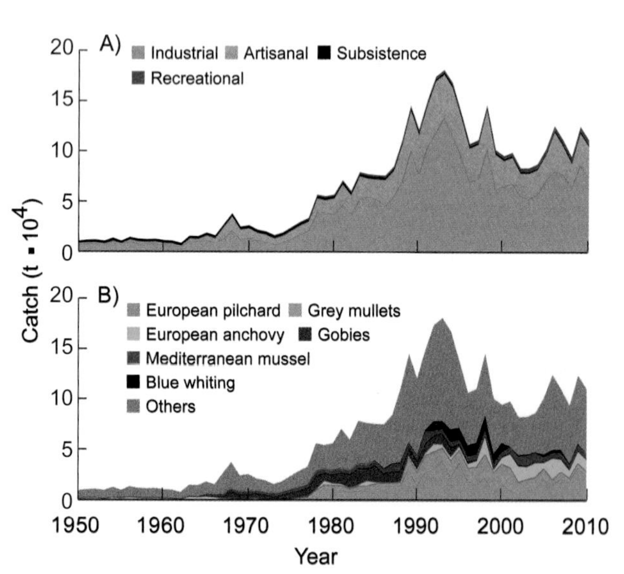

Figure 2. Domestic and foreign catches taken in the Mediterranean EEZ of Turkey: (A) by sector (domestic scores: Ind. 2, 3, 3; Art. 2, 3, 3; Sub. 1, 2, 2; Rec. 1, 2, 2; Dis. 3, 3, 4); (B) by taxon.

REFERENCES

Ulman A, Bekişoğlu Ş, Zengin M, Knudsen S, Ünal V, Mathews C, Harper S, Zeller D and Pauly D (2013) From bonito to anchovy: a reconstruction of Turkey's marine fisheries catches (1950–2010). *Mediterranean Marine Science* 14(2): 309–342.

Ünal, V., Acarli, D. and Gordoa, A., 2010. Characteristics of marine recreational fishing in the Çannakale Strait (Turkey). *Mediterranean Marine Science* 11 (2): 315–330.

Ünal V and Erdem M (2009) Combating illegal fishing in Gokova Bay (Aegean Sea), Turkey. pp. 80–86 In Proceedings of the 3rd International symposium on underwater research. Ege University, Izmir, Famagusta, Cyprus.

1. Cite as Ulman, A., V. Ünal, and Ş. Bekişoğlu. 2016. Turkey (Mediterranean Sea). P. 418 in D. Pauly and D. Zeller (eds.), *Global Atlas of Marine Fisheries: A Critical Appraisal of Catches and Ecosystem Impacts*. Island Press, Washington, DC.

TURKS AND CAICOS ISLANDS[1]

Aylin Ulman,[a] Lily Burke,[b] Edd Hind,[c] Robin Ramdeen,[a] and Dirk Zeller[a]

[a]*Sea Around Us*, University of British Columbia, Vancouver, Canada [b]Strawberry Isle Marine Research Society, Tofino, BC, Canada [c]The School for Field Studies, Center for Marine Resources Studies, South Caicos, Turks and Caicos Islands

The Turks and Caicos Islands (TCI) are a small chain of islands at the southern end of the Bahamian Archipelago and north of Hispaniola in the Caribbean Sea (figure 1). The TCI consist of 2 distinct island groups, the Turks Islands and the Caicos Islands, separated by the 35-km-wide and 2,134-m-deep Turks Island Passage, connecting the Atlantic Ocean to the Caribbean Sea. There are 3 main fisheries (Taylor and Medley 2003; Lockhart et al. 2007): queen conch (*Strombus gigas*), spiny lobster (*Panulirus argus*), and finfish. Queen conch and spiny lobster are a valuable export to the United States (Rudd 2003), and finfish, such as bone fish (*Albula vulpes*), grouper (family Serranidae), and snapper (Lutjanidae), are caught primarily for home consumption or local sale. Based on the reconstruction by Ulman et al. (2015), fisheries are predominantly artisanal and subsistence (figure 2A). The catches from the TCI appear to have been 16,000 t/year in the 1950s and about 12,000–13,000 t/year in the late 2000s (figure 2A). From 1950 to 2010, this catch, whose composition is shown in figure 2B, amounted to 2.8 times the data reported by the FAO on behalf of the TCI. The TCI are undergoing rapid tourist development because of their still pristine-looking environment and favorable location. They will need prudent management, because one of these assets can be easily lost.

REFERENCES

Lockhart K, De Fontaubert C and Clerveaux W (2007) Fisheries of the Turks and Caicos Islands: Status and Threats. *Proceedings of the Gulf and Caribbean Fisheries Institute* 58:66–71.

Rudd MA (2003) Fisheries Landings and Trade of the Turks and Caicos Islands. pp. 149–161 In D Zeller, S Booth, E Mohammed and D Pauly (eds.), *From Mexico to Brazil: Central Atlantic fisheries catch trends and models*. Fisheries Centre Research Reports 11(6).

Taylor O and Medley P (2003) Turks and Caicos Islands Field Report. Participatory Fisheries Stock Assessment project R7947. Marine Resource Assessment Group's (MRAG) Fisheries Management Science Programme (FMSP). 56 p.

Ulman A, Burke L, Hind E, Ramdeen R and Zeller D (2015) Reconstruction of total marine fisheries catches for the Turks and Caicos Islands (1950–2010). Fisheries Centre Working Paper #2015–63, 23 p.

1. Cite as Ulman, A., L. Burke, E. Hind, R. Ramdeen, and D. Zeller. 2016. Turks and Caicos Islands. P. 419 in D. Pauly and D. Zeller (eds.), *Global Atlas of Marine Fisheries: A Critical Appraisal of Catches and Ecosystem Impacts*. Island Press, Washington, DC.

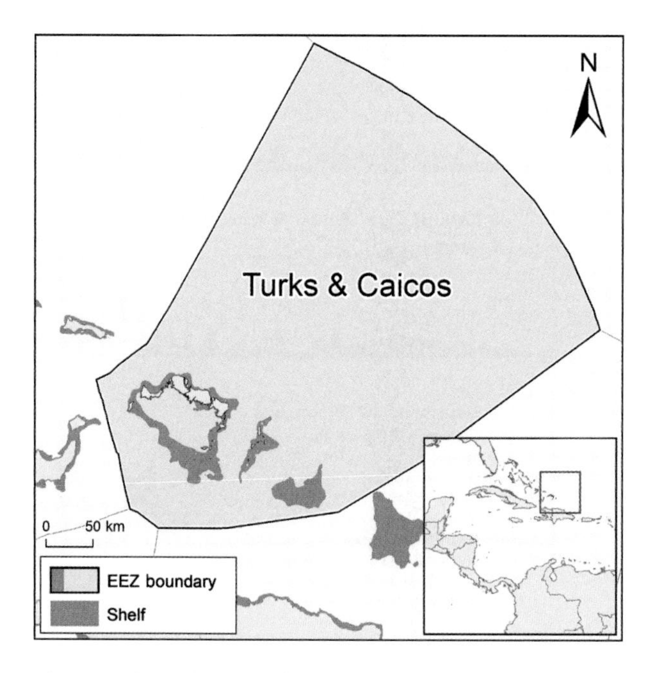

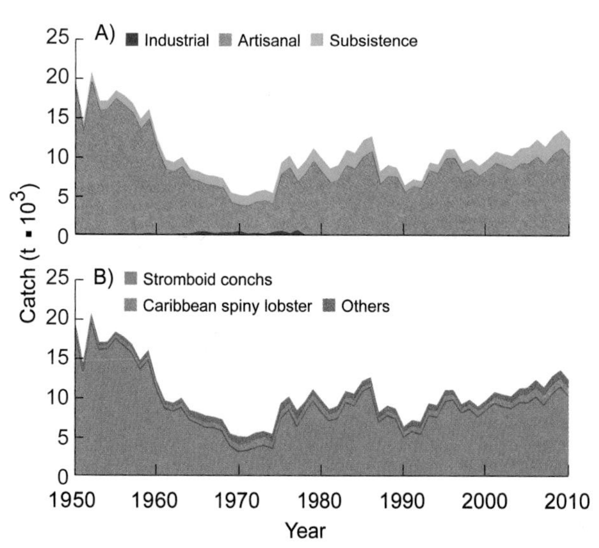

Figure 1. The Turks and Caicos Islands, with a land area of 616 km², have an EEZ of 154,000 km². **Figure 2.** Domestic and foreign catches taken in the EEZ of Turks and Caicos Islands: (A) by sector (domestic scores: Art. 2, 3, 3; Sub. 2, 3, 4; Rec. 2, 3, 4); (B) by taxon.

TUVALU[1]

Kendyl Crawford,[a] Sarah Harper,[b] and Dirk Zeller[b]

[a]Department of Marine and Environmental Science, Hampton University, Hampton, VA
[b]*Sea Around Us*, University of British Columbia, Vancouver, Canada

Tuvalu is an archipelago halfway between Australia and Hawaii in the south central Pacific, consisting of 9 atolls, of which 1, Funafuti, hosts the capital of the same name (figure 1). Most of the catches in this EEZ are industrial (figure 2A), exclusively because of foreign fleets targeting large pelagic species (figure 2B). The reconstruction of domestic catches by Crawford et al. (2011) was based on data reported by the FAO on behalf of Tuvalu, in combination with data from fish markets and household consumption. It suggested that domestic catches (mainly subsistence; figure 2A) increased from 800 t/year in the early 1950s to about 2,000 t/year in the late 2000s and were about 5 times the data reported by the FAO for Tuvalu. Eighty-seven percent of this catch was from the subsistence sector, much of it gathered by women (Lambeth 2000), the rest originating from a growing artisanal fishery, and the small take of a baitfish fishery, which operated from the 1950s to the early 1980s. Tuna (e.g., skipjack, *Katsuwonus pelamis*; yellowfin, *Thunnus albacares*; albacore, *T. alalunga*; and bigeye tuna, *T. obesus*) and billfishes (Indo-Pacific blue marlin, *Makaira mazara*) dominated total catches (figure 2C) because of the foreign fleets (figure 2B), whereas the most common families in the domestic inshore subsistence catch were groupers (family Serranidae) and emperors (Lethrinidae), although some tuna were also caught by domestic small-scale fisheries.

REFERENCES

Crawford K, Harper S and Zeller D (2011) Reconstruction of marine fisheries catches for Tuvalu (1950–2009), p. 131–143. In S Harper and D Zeller (eds.) *Fisheries catch reconstructions: Islands, Part II*. Fisheries Centre Research Reports 19(4). [Updated to 2010]

Lambeth L (2000) An Assessment of the Role of Women in Fishing Communities in Tuvalu. Secretariat of the Pacific Community, Nouméa, New Caledonia, 37 p.

1. Cite as Crawford, K., S. Harper, and D. Zeller. 2016. Tuvalu. P. 420 in D. Pauly and D. Zeller (eds.), *Global Atlas of Marine Fisheries: A Critical Appraisal of Catches and Ecosystem Impacts*. Island Press, Washington, DC.

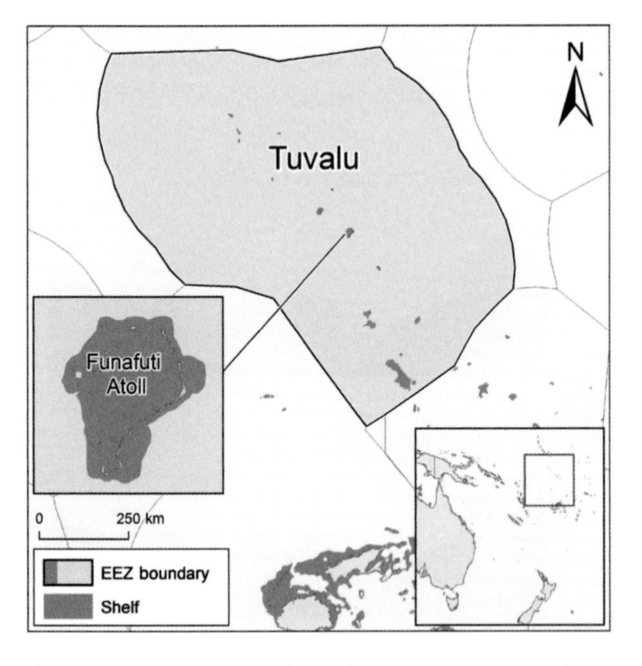

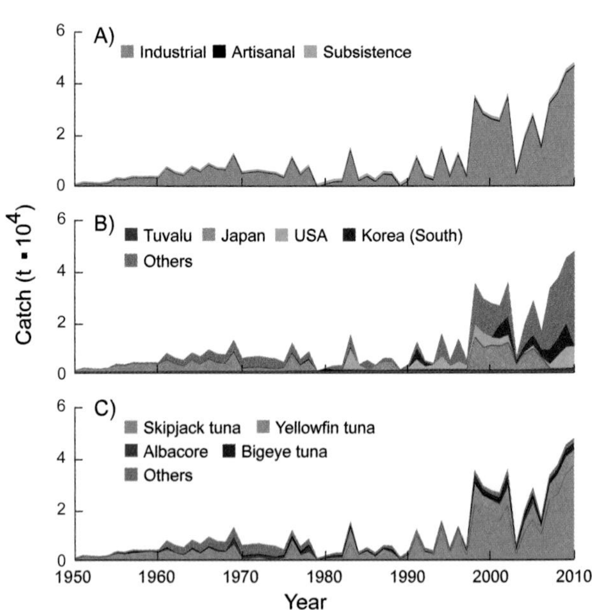

Figure 1. Tuvalu's land area is tiny (26 km²) compared with its EEZ of nearly 752,000 km².

Figure 2. Domestic and foreign catches taken in the EEZ of Tuvalu: (A) by sector (domestic scores: Art. 1, 1, 2; Sub. 1, 1, 2); (B) by fishing country (foreign catches are very uncertain); (C) by taxon.

UKRAINE[1]

Aylin Ulman,[a] Vladyslav Shlyakhov,[b] Sergei Jatsenko,[c] and Daniel Pauly[a]

[a]*Sea Around Us*, University of British Columbia, Vancouver, Canada [b]YugNIRO, Kerch, Crimea, Ukraine
[c]Head of Aquatic Resources, Derzhvodekologiya Research Institute, Ukraine

The Ukraine (including the Crimea) has a long coastline on the northern Black Sea and shares the Sea of Azov with Russia (figure 1). When the Ukraine was part of the USSR, there was strict monitoring of fishing activities, but since independence (1991), a large reduction in financial resources for the pertinent agencies no longer permitted this, which has increased the fraction of the catch that remains unreported. The catch reconstruction (Ulman et al. 2015), based on Shlyakhov and Charova (2003), Knudsen and Toje (2008), and other sources, showed that industrial catches dominated, although subsistence fisheries increased after independence (figure 2A), and total domestic catches (no or little foreign fishing seems to occur) averaged 55,000 t/year in the 1950s and increased dramatically in the 1970s because of the increased abundance of the anchovy (*Engraulis encrasicolus*) and sprat (*Sprattus sprattus*) populations. Catches peaked in 1988 at 175,000 t. In the early 1990s, these species were decimated by invasive comb jelly (*Mnemiopsis leidyi*), but they subsequently recovered, but not to their earlier abundances, and catches were about 100,000 t/year in the late 2000s. As figure 2B illustrates, Ukrainian catches in the Black Sea consisted of anchovy, sprat, goby (family Gobiidae), Mediterranean mussel (*Mytilus galloprovincialis*), Mediterranean horse mackerel (*Trachurus mediterraneus*), and so-iuy mullet (*Liza haematocheila*). Despite all its problems, Ukraine has apparently managed to create no-take marine reserves covering 40% of its territorial waters.

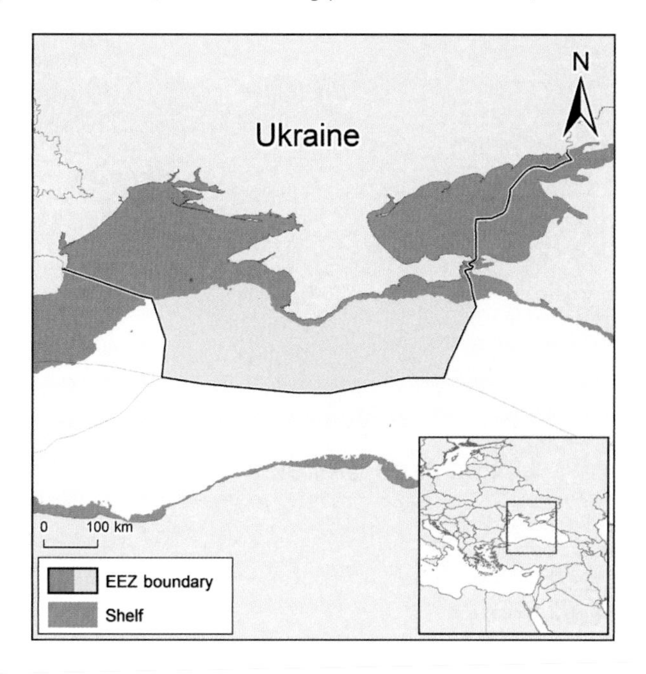

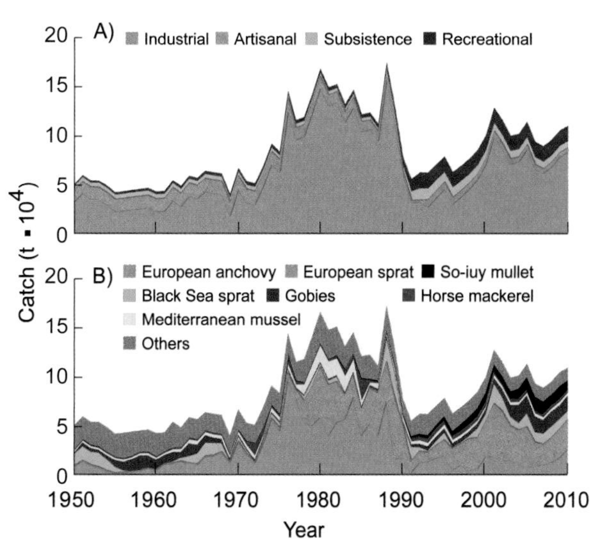

Figure 1. Ukraine and its Black Sea shelf (73,800 km²) and EEZ (132,000 km²). **Figure 2.** Domestic and foreign catches taken in the EEZ of Ukraine: (A) by sector (domestic scores: Ind. 3, 4, 2; Art. 3, 4, 2; Sub. 1, 1, 2; Rec. 1, 1, 2; Dis. 1, 1, 1); (B) by fishing country (foreign catches are very uncertain); (C) by taxon.

REFERENCES

Knudsen S and Toje H (2008) Post-Soviet transformations in Russian and Ukrainian Black Sea fisheries: Socio-economic dynamics and property relations. *Southeast European and Black Sea Studies* 8(1): 17–32.

Shlyakhov V and Charova I (2003) The status of the demersal fish populations along the Black Sea coast of Ukraine. p. 65–74 In Ozturk B and Karakulak FS (eds.) *Workshop on demersal resources in the Black and Azov Seas, 15–17 April 2003, Sile, Turkey*. Turkish Marine Research Foundation, Istanbul, Turkey.

Ulman A, Shlyakhov V, Jatsenko S and Pauly D (2015) A reconstruction of the Ukraine's marine fisheries catches, 1950–2010. Fisheries Centre Working Paper #2015-86, 23 p.

1. Cite as Ulman, A., V. Shlyakhov, S. Jatsenko, and D. Pauly. 2016. Ukraine. P. 421 in D. Pauly and D. Zeller (eds.), *Global Atlas of Marine Fisheries: A Critical Appraisal of Catches and Ecosystem Impacts*. Island Press, Washington, DC.

UNITED ARAB EMIRATES[1]

Dalal Al-Abdulrazzak

Sea Around Us, Fisheries Centre, University of British Columbia, Vancouver, Canada

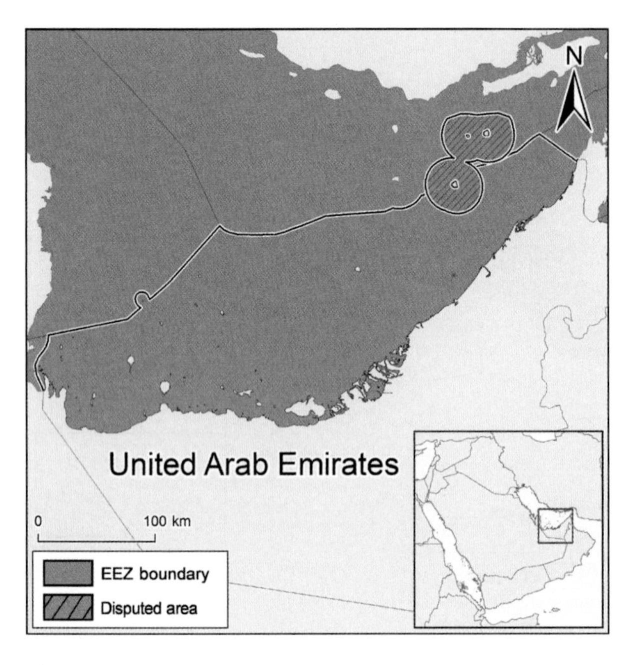

Figure 1. The United Arab Emirates (UAE), showing the extent of their EEZ in the Gulf (52,500 km², of which all is shelf), including the striped area, contested by Iran.

The United Arab Emirates (UAE) has coasts on the Persian Gulf (figure 1) and in the Gulf of Oman (next page). This account, based on Al-Abdulrazzak (2013), refers only to the UAE's Gulf coast, whose domestic fisheries (including a ban on trawling) are small in scale (figure 2A), with little foreign, industrial fishing (figure 2B), occurring mostly in Abu Dhabi, which makes up more than 60% of the UAE's Gulf EEZ. Because of its reliance on a market sampling program that did not differentiate between locally caught and imported seafood, the UAE systematically overreported its catches (Morgan 2004). Al-Abdulrazzak (2013), who considered this, estimated the UAE's catch adjusted for Gulf EEZ waters only as 8,000 t/year in the early 1950s, which increased until it reached a peak at 80,000 t in 1999, then declined to 50,000 t/year in the late 2000s. Overall, the figures reported by the FAO on behalf of the entire UAE *overestimated* reconstructed catches by 47%, despite the latter also accounting for subsistence and recreational catches, missed entirely by the market sampling program. This overestimation is reduced, however, when the UAE's catches from the Gulf of Oman are considered (see next page). The major taxa caught in the UAE (figure 2C) are mackerels and tuna (family Scombridae), herrings (Clupeidae), jacks (Carangidae), emperors (Lethrinidae), narrow-barred Spanish mackerel (*Scomberomorus commerson*), and groupers (Serranidae). Improving the UAE's catch reporting system appears essential if its fisheries are to be managed for sustainability.

REFERENCES

Al-Abdulrazzak D (2013) Estimating total fish extractions in the United Arab Emirates: 1950–2010. p. 53–59. In: Al-Abdulrazzak D and Pauly D (eds.) *From dhows to trawlers: a recent history of fisheries in the Gulf countries, 1950 to 2010*. Fisheries Centre Research Reports 21(2).

Morgan G (2004) Country review: United Arab Emirates. p. 327–335. In: C De Young (ed.), Review of the state of world marine capture fisheries management: Indian Ocean. FAO Fisheries Technical Paper (488), FAO, Rome.

1. Cite as Al-Abdulrazzak, D. 2016. United Arab Emirates (Persian Gulf). P. 422 in D. Pauly and D. Zeller (eds.), *Global Atlas of Marine Fisheries: A Critical Appraisal of Catches and Ecosystem Impacts*. Island Press, Washington, DC.

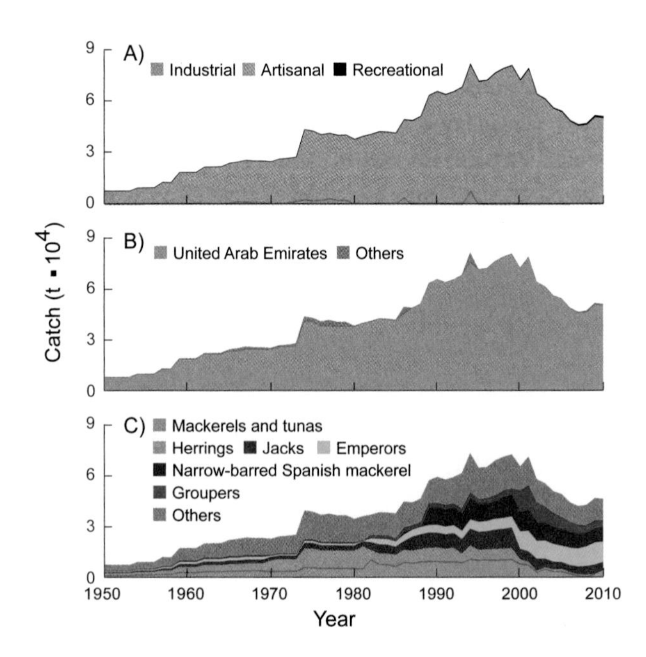

Figure 2. Domestic and foreign catches taken in the EEZ of United Arab Emirates: (A) by sector (domestic scores: Art. 3, 3, 3; Sub. 1, 1, 1; Rec. 1, 1, 1; Dis. 2, 2, 3); (B) by fishing country (foreign catches are very uncertain); (C) by taxon.

UNITED ARAB EMIRATES (FUJAIRAH)[1]

Myriam Khalfallah, Dirk Zeller, and Daniel Pauly

Sea Around Us, University of British Columbia, Vancouver, Canada

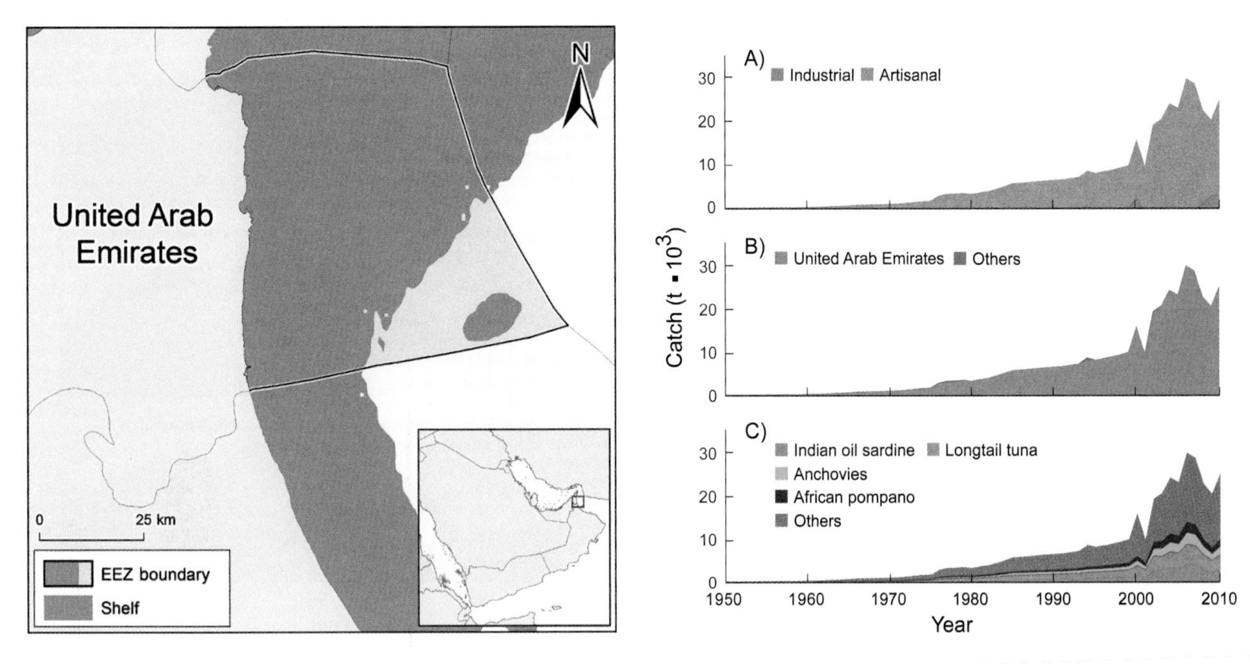

Figure 1. The Emirate of Fujairah (UAE), with its EEZ in the Gulf in the Gulf of Oman (4,370 km², of which 3,560 km² is shelf).
Figure 2. Domestic and foreign catches taken in the EEZ of Emirate of Fujairah (UAE): (A) by sector (domestic scores: Art. 2, 3, 3; Sub. 2, 2, 2; Rec. 1, 1, 1; Dis. 1, 2, 2); (B) by fishing country (foreign catches are very uncertain); (C) by taxon.

Fujairah is 1 of the 7 emirates constituting the United Arab Emirates (UAE). However, Fujairah has a coastline only on the Gulf of Oman and none on the Persian Gulf (figure 1). Total fisheries catches in the EEZ of Fujairah were estimated by Khalfallah et al. (2015), who reconstructed artisanal landings using catch rate information from the neighboring Omani provinces of Musandam and Al Batinah, combined with fishing effort (i.e., number of boats) in Fujairah and other sources (e.g., Pearson et al. 1998), including reconciling reported data with the rest of the UAE (Al-Abdulrazzak 2013). Subsistence catch estimates were obtained based on the number of foreign workers in Fujairah, who are assumed to fish for their own consumption, and recreational catches were estimated from recreational fishing participation rates. Discards were estimated based on the discard rate of the Omani artisanal fishery. The resulting, very tentative total catch (dominated by artisanal fleets; figure 2A), including very limited foreign catches (figure 2B), increased from less than 1,000 t/year in the 1950s to about 25,000 t/year in the late 2000s; the domestic catches correspond to 2.2 times the landings reported by the UAE on behalf of Fujairah. The taxonomic composition of the catch is presented in figure 2C. Much of this catch ends up, jointly with fish from Oman, in the markets of the large cities in the UAE (Al-Abdulrazzak 2013), where they complicate the estimation of catches from the Gulf part of that country.

REFERENCES

Al-Abdulrazzak D (2013) Estimating total fish extractions in the United Arab Emirates: 1950–2010. pp. 53–59 in Al-Abdulrazzak D and Pauly D (eds.). *From dhows to trawlers: a recent history of fisheries in the Gulf countries, 1950 to 2010*. Fisheries Centre Research Reports 21(2).

Khalfallah M, Zeller D and Pauly D (2015) Reconstruction of marine fisheries catches for Fujairah (UAE) (1950–2010). Fisheries Centre Working Paper #2015–57, 13 p.

Pearson WH, Al-Ghais SM, Neff JM, Brandt CJ, Wellman K and Green T (1998) Assessment of damages to commercial fisheries and marine environment of Fujairah, United Arab Emirates, resulting from the Seki oil spill of March 1994: A case study. *Bulletin Series, Yale School of Forestry and Environmental Studies* 103: 407–428.

1. Cite as Khalfallah, M., D. Zeller, and D. Pauly. 2016. United Arab Emirates (Gulf of Oman). P. 423 in D. Pauly and D. Zeller (eds.), *Global Atlas of Marine Fisheries: A Critical Appraisal of Catches and Ecosystem Impacts*. Island Press, Washington, DC.

UNITED KINGDOM[1]

Darah Gibson,[a] Emma Cardwell,[b] Kyrstn Zylich, and Dirk Zeller

[a]*Sea Around Us*, University of British Columbia, Vancouver, Canada
[b]School of Geography and Environment, Oxford University, England

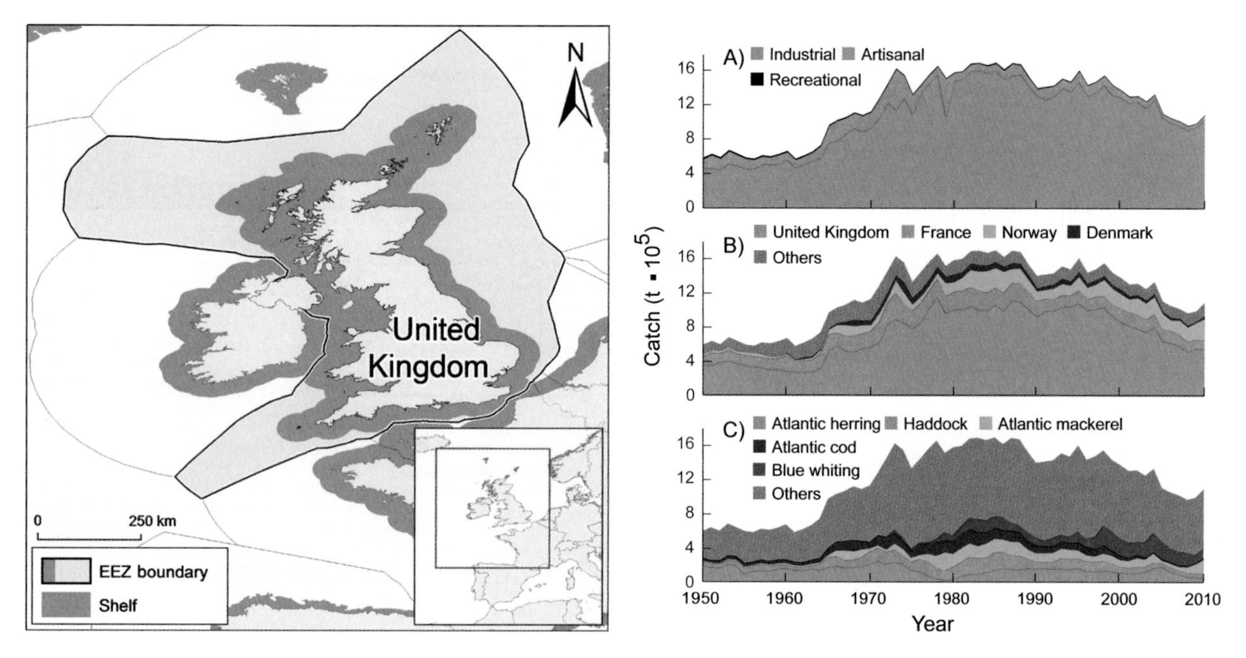

Figure 1. The United Kingdom has a land area of 242,500 km² and an EEZ of 757,000 km². **Figure 2.** Domestic and foreign catches taken in the EEZ of United Kingdom: (A) by sector (domestic scores: Ind. 2, 2, 3; Art. 2, 2, 3; Rec. 1, 1, 1; Dis. 2, 2, 3); (B) by fishing country (foreign catches are very uncertain); (C) by taxon.

The United Kingdom (figure 1), comprising England, Scotland, Northern Ireland, and Wales, is the home of the Industrial Revolution and also the first country to have deployed industrial trawlers in the 1880s (Roberts 2007). Powered by coal, these steam trawlers, representing the first application of fossil fuel to extract a living resource, were effective enough to rapidly reduce the accumulated biomass of fish around the British Isles. This triggered a spatial expansion, soon joined by other industrialized countries, which has continued to this date and engulfed the entire world ocean. Total catches taken in this EEZ, including the domestic reconstructed catches by Gibson et al. (2015), are clearly dominated by industrial fleets (figure 2A) and include considerable foreign catch (figure 2B). The domestic (UK) catch increased from about 350,000 t/year in the early 1950s to more than 1.0 million t/year in the 1980s before declining to about 550,000 t/year by the late 2000s. Not only did the biomass and the catch of traditionally exploited species (e.g., Atlantic herring, *Clupeus harengus*; Atlantic cod, *Gadus morhua*) decline (figure 2C), but the coastal ecosystems were radically transformed (Thurstan and Roberts 2010; Molfese et al. 2014), as in all of northwestern Europe (Gascuel et al. 2014), a region profoundly affected by the "fishing-down" phenomenon.

REFERENCES

Gascuel D, Coll M, Fox C, Guénette S, Guitton J, Kenny A, Knittweis L, Nielsen RJ, Piet G, Raid T, Travers-Trollet M, Shephard S (2014) Fishing impact and environmental status in European seas: a diagnosis from stock assessments and ecosystem indicators. *Fish and Fisheries* 17(1): 31–55.

Gibson D, Cardwell E, Zylich K and Zeller D (2015) Preliminary reconstruction of total marine fisheries catches for the United Kingdom and the Channel Islands in EEZ equivalent waters (1950–2010). Fisheries Centre Working Paper #2015-76, 20 p.

Molfese C, Beare D, Hall-Spencer JM (2014) Overfishing and the Replacement of Demersal Finfish by Shellfish: An Example from the English Channel. *PLoS ONE* 9(7): e101506.

Roberts C (2007) *The Unnatural History of the Sea*. Island Press, Washington, DC.

Thurstan RH and Roberts CM (2010) Ecological meltdown in the Firth of Clyde, Scotland: two centuries of change in a coastal marine ecosystem. *PloS One* 5(7): e11767.

1. Cite as Gibson, D., E. Cardwell, K. Zylich, and D. Zeller. 2016. United Kingdom. P. 424 in D. Pauly and D. Zeller (eds.), *Global Atlas of Marine Fisheries: A Critical Appraisal of Catches and Ecosystem Impacts*. Island Press, Washington, DC.

UNITED KINGDOM (ANGUILLA)[1]

Robin Ramdeen, Kyrstn Zylich, and Dirk Zeller

Sea Around Us, University of British Columbia, Vancouver, Canada

The British overseas territory of Anguilla is an arid, low-lying coralline island and the most northerly of the Leeward Islands in the eastern Caribbean (figure 1). Historically, salt production, lobster fishing, and overseas employment were the main sources of income. In the early 1980s, the government began positioning Anguilla as a luxury tourist destination. With its white sand beaches and turquoise seas, Anguilla has a tourism industry that currently contributes about 50% of GDP. The domestic fisheries of Anguilla are multigear (traditional Antillean arrowhead traps are common), multispecies, and mostly small-scale (figure 2A), with limited foreign industrial fishing (figure 2B). The reconstruction by Ramdeen et al. (2014), based on Lum Kong (2007), Wynne (2010), and other sources, generated estimates of domestic catches of 200 t in 1950, which increased to about 1,600 t/year by 2010. This catch is split mainly between subsistence and artisanal fisheries, and the unreported nature of the former is the main reason why the reconstructed catch is 2.75 times the official landings reported by the FAO on behalf of Anguilla. This catch consists of spiny lobsters, or "crayfish" (*Panulirus argus* and *P. guttatus*) and finfish such as groupers (family Serranidae), snappers (Lutjanidae), and grunts (Haemulidae; figure 2C). There is also a small fishery for conch (*Strombus* spp.), with most conch fishers using scuba gear, and a small recreational fishery targeting large pelagic fishes.

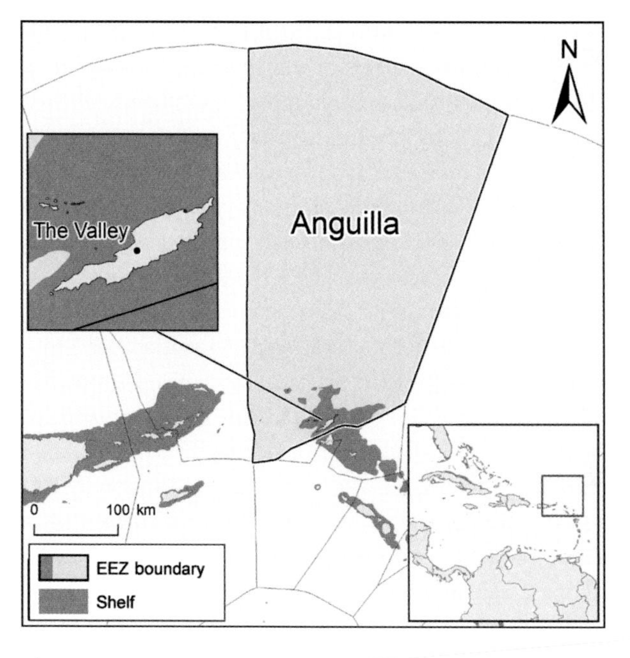

Figure 1. The UK overseas territory of Anguilla has a land area of 91 km² and an EEZ of 92,200 km².

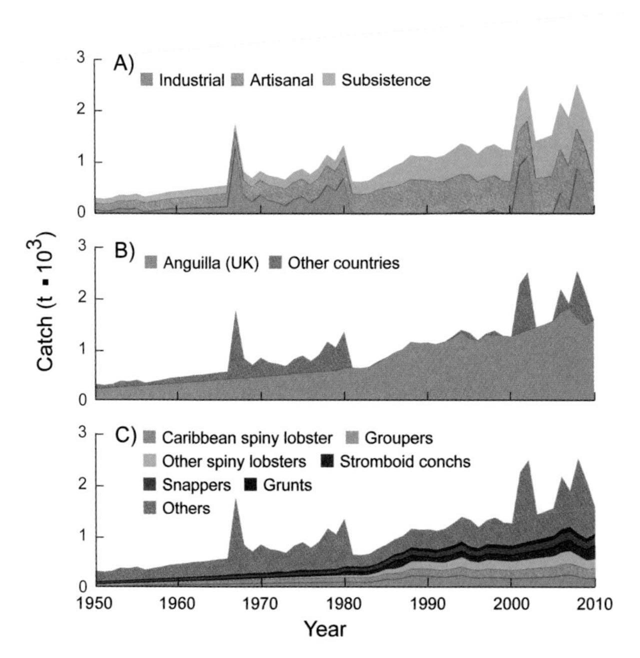

Figure 2. Domestic and foreign catches taken in the EEZ of Anguilla: (A) by sector (domestic scores: Art. 2, 3, 3; Sub. 2, 2, 2; Rec. 1, 1, 1); (B) by fishing country (foreign catches are very uncertain); (C) by taxon.

REFERENCES

Lum Kong A (2007) Report on the Anguilla Fisheries, Marine, Coastal Sector. Biodiversity Conservation Inc., Cannonball Complex, The Valley, Anguilla. 49 p.

Ramdeen R, Zylich K and Zeller D (2014) Reconstruction of total marine fisheries catches for Anguilla (1950–2010). pp. 1–8 In K Zylich, D Zeller, M Ang and D Pauly (eds.), *Fisheries catch reconstructions: Islands, Part IV*. Fisheries Centre Research Reports 22(2).

Wynne S (2010) Status of Anguilla's Marine Resources 2010. Department of Fisheries and Marine Resources for the Government of Anguilla, Anguilla. 46 p.

1. Cite as Ramdeen, R., K. Zylich, and D. Zeller. 2016. United Kingdom (Anguilla). P. 425 in D. Pauly and D. Zeller (eds.), *Global Atlas of Marine Fisheries: A Critical Appraisal of Catches and Ecosystem Impacts*. Island Press, Washington, DC.

UNITED KINGDOM (ASCENSION)[1]

Shawn Booth, Houman Azar, Danielle Knip, and Maria-Lourdes Palomares

Sea Around Us, University of British Columbia, Canada

Ascension Island is an isolated island in the South Atlantic, halfway between Africa and South America (figure 1). Along with the Tristan da Cunha group, it is administered from St. Helena as a British Overseas Territory. Until 2002, Ascension Island's main sources of funding were a military base and 2 major communication organizations (the BBC and Cable & Wireless), but now it is mainly personal income taxes that fund government activities, hence the limited fisheries data. The estimated per capita seafood demand of 83 kg/person/year, combined with other, scattered information (see Booth and Azar 2009), led to its artisanal catch being tentatively estimated at about 20 t/year in the early 1950s, increasing to catches fluctuating between 40 and 60 t/year in the 1990s and early 2000s, then decreasing to about 20–40 t/year in the late 2000s. However, highly intermittent industrial catches (figure 2A) by foreign fleets (figure 2B) dominate total catches taken from this EEZ. In the absence of better data, the taxonomic composition of this catch (figure 2C) was assumed, for the domestic component, to be the same as for the artisanal catch of St. Helena, the nearest island, whereas foreign catches focus on tuna and billfishes. However, using FishBase (www.fishbase.org, which includes all fish species listed in Lubbock 1980) we made sure that all taxa implied by this choice have been reported in the EEZ of Ascension.

REFERENCES

Booth S and Azar H (2009) The fisheries of St Helena and its dependencies. pp. 27–34. In: Zeller D and Harper S (eds.) *Fisheries catch reconstructions: Islands, Part I*. Fisheries Centre Research Reports 17(5). [Updated to 2010]

Lubbock R (1980) The shore fishes of Ascension Island. *Journal of Fish Biology* 17(3): 283–303.

1. Cite as Booth, S., H. Azar, D. Knip, and M. L. D. Palomares. 2016. United Kingdom (Ascension Island). P. 426 in D. Pauly and D. Zeller (eds.), *Global Atlas of Marine Fisheries: A Critical Appraisal of Catches and Ecosystem Impacts*. Island Press, Washington, DC.

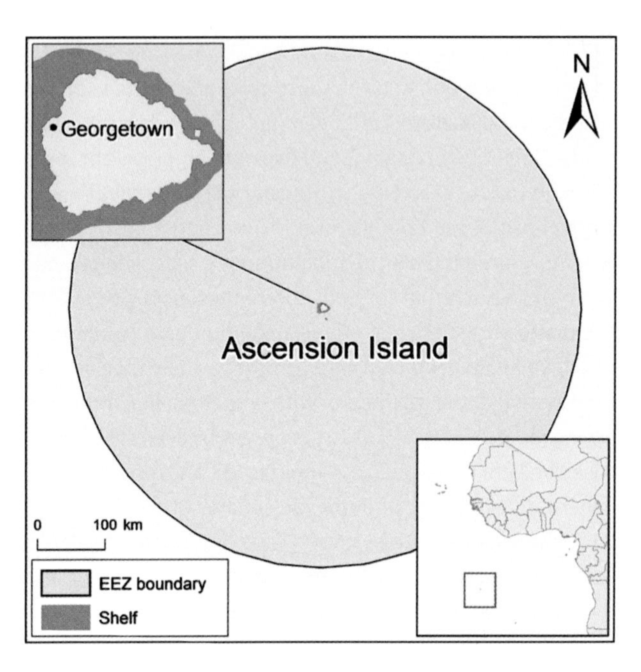

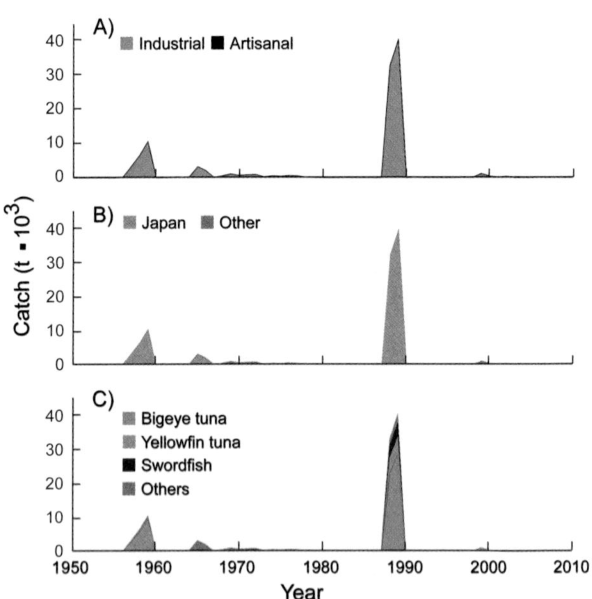

Figure 1. Ascension Island (United Kingdom) has a land area of 88 km² and an EEZ of 442,000 km². **Figure 2.** Domestic and foreign catches taken in the EEZ of Ascension Island (UK): (A) by sector (domestic scores: Art. 1, 2, 2); (B) by fishing country (foreign catches are very uncertain); (C) by taxon.

UNITED KINGDOM (BERMUDA)[1]

Esther Divovich, Lydia C. L. Teh, Kyrstn Zylich, and Dirk Zeller.

Sea Around Us, University of British Columbia, Vancouver, Canada

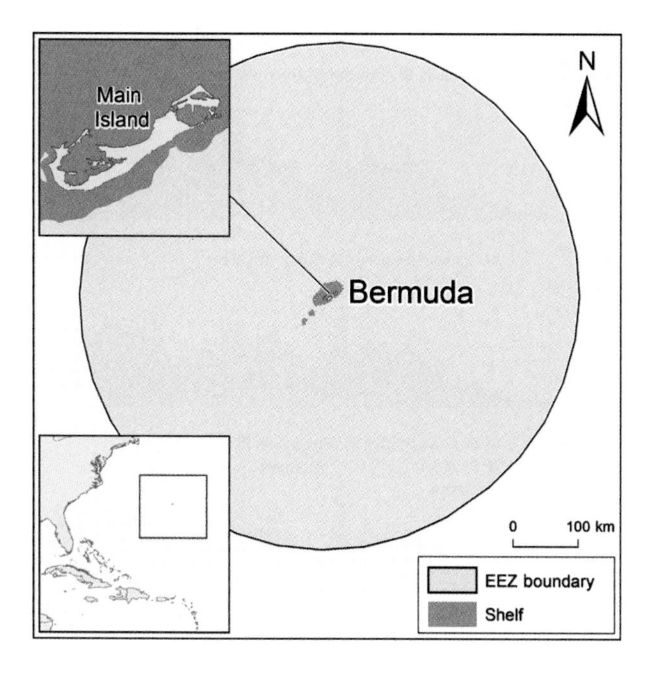

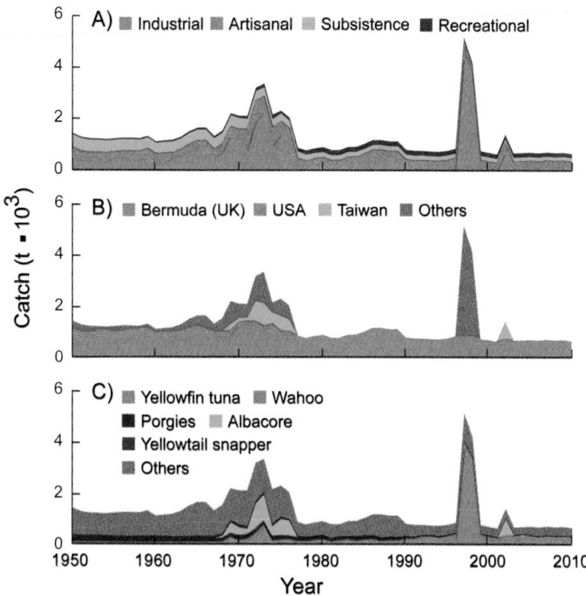

Figure 1. The Bermuda Islands (UK) have a land area of 53 km² and an EEZ of 450,000 km². **Figure 2.** Domestic and foreign catches taken in the EEZ of the Bermuda Islands (UK): (A) by sector (domestic scores: Art. 2, 3, 3; Sub. 2, 2, 2; Rec. 1, 2, 2); (B) by fishing country (foreign catches are very uncertain); (C) by taxon.

Bermuda is an archipelago of 7 main and numerous smaller islands located in the North Atlantic Ocean (figure 1). This British Overseas Territory has one of the world's highest per capita GDPs and an economy driven by financial services and tourism. Fishing is a minor economic activity, reflected by the small number of full-time licensed fishers (about 300 in 2010). Domestic fisheries are artisanal and subsistence, with the latter increasingly recreational in nature (figure 2A), and industrial catches are by foreign fleets (figure 2B). Reconstructed domestic catches averaged 850 t/year in the 1950s, peaked at about 1,500 t/year in 1972, and returned to about 800 t/year in the late 2000s (Divovich et al. 2015; Teh et al. 2014). Overall, this was 1.7 times the data reported to and by the FAO (see also Luckhurst et al. 2003). Artisanal catches made up 53% of total domestic catch, and subsistence and recreational fishing contributed 32% and 16%, respectively. The taxonomic composition (figure 2C) of EEZ-level catches is dominated by the foreign-caught tuna taxa, whereas domestic catches shifted from primarily high-trophic-level reef fishes such as groupers and snappers from 1950 to the 1980s (Ward 1988; Smith-Vanitz et al. 1999) to being dominated by pelagic fishes in latter years. The local fish catch is not sufficient to support demand from both residents and tourists to the island, and Bermuda imports about two thirds of its seafood.

REFERENCES

Divovich E, Teh LCL, Zylich K and Zeller D (2015) Preliminary reconstruction of Bermuda's marine fisheries catches, 1950–2010. Fisheries Centre Working Paper #2015-96, 18 p.

Luckhurst B, Booth S and Zeller D (2003) Brief history of Bermudian fisheries, and catch comparison between national sources and FAO records. pp. 163–169 In D Zeller, S Booth, E Mohammed and D Pauly (eds.), *From Mexico to Brazil: Central Atlantic fisheries catch trends and ecosystem models*. Fisheries Centre Research Reports 11(6).

Smith-Vaniz W, Collette BB, and Luckhurst BE (1999) Fisheries of Bermuda: History, Zoogeography, Annotated Checklist and Identification Keys. *American Society of Ichthyologists and Herpetologists Special Publication* 4: 424 p.

Teh LCL, Zylich K and Zeller D (2014) Preliminary reconstruction of Bermuda's marine fisheries catches, 1950–2010. Fisheries Centre Working Paper #2014-24, 17 p.

Ward J (1988) Mesh size selection in Antillean arrowhead fish traps. pp. 455–467 In S Venema, J Möller-Christensen and D Pauly (eds.), *Contributions to tropical fisheries biology*. FAO Fisheries Report No. 389.

1. Cite as Divovich, E., L. C. L. Teh, K. Zylich, and D. Zeller. 2016. United Kingdom (Bermuda). P. 427 in D. Pauly and D. Zeller (eds.), *Global Atlas of Marine Fisheries: A Critical Appraisal of Catches and Ecosystem Impacts*. Island Press, Washington, DC.

UNITED KINGDOM (CAYMAN ISLANDS)[1]

Sarah Harper,[a] John Bothwell,[b] Sarah Bale,[a] Shawn Booth,[a] and Dirk Zeller[a]

[a]*Sea Around Us*, Fisheries Centre, University of British Columbia, Vancouver, Canada
[b]Department of Environment, Grand Cayman, Cayman Islands

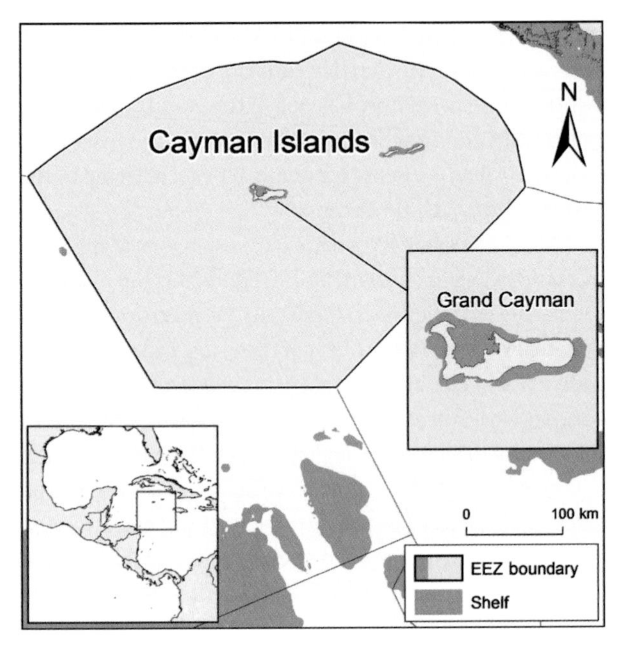

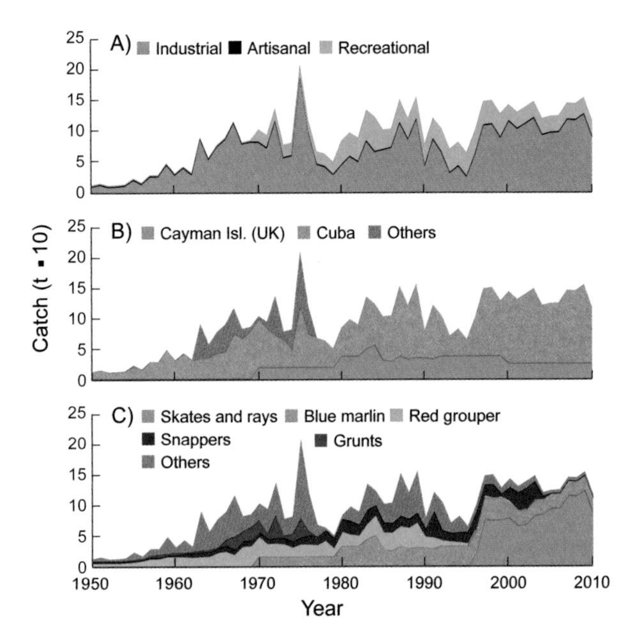

Figure 1. The Cayman Islands have a combined land area of 264 km² and an EEZ of 119,000 km². **Figure 2.** Domestic and foreign catches taken in the EEZ of Cayman Islands: (A) by sector (domestic scores: Art. 2, 2, 2; Sub. 2, 2, 2; Rec. 2, 2, 2); (B) by fishing country (foreign catches are very uncertain); (C) by taxon.

The Cayman Islands, a British overseas territory in the Caribbean Sea, famous for their financial service and tourism industries (Shackley 1998), consist of 3 islands, Grand Cayman, Cayman Brac, and Little Cayman, with Grand Cayman being the largest and most populated (figure 1). The coral reefs of the Cayman Islands are in bad shape, for a variety of reasons (including tourism) acting in concert (Austin et al. 2015). Total catches taken in this EEZ are dominated by the industrial sector (figure 2A), exclusively due to foreign fishing, mainly by Venezuela (early years) and more recently Cuba (figure 2B). The reconstructed domestic artisanal, subsistence, and recreational catches (Harper et al. 2009) within the EEZ of the Cayman Islands were low in the 1950s, increased in the 1980s and 1990s, then settled at 30 t/year in the 2000s, 3 times more than presented by the FAO on behalf of the Cayman Islands. These estimates omit very early shark catches by Cayman Island vessels outside of what is now the Cayman Island EEZ and earlier landings of tuna and "Natantian decapods" reported to the FAO as being caught by Cayman Island vessels in the eastern central Atlantic, but which were probably taken by non-Caymanian "flag-of-convenience" vessels, a practice suppressed since 1990. Figure 2C shows the major taxa caught in the Cayman EEZ, notably Atlantic blue marlin (*Makaira nigricans*).

REFERENCES

Austin T, Bush P, Fenner D, Manfrino C, McCoy C, Miller J, Nagelkerken I, Polunin N, Weil E, William I, AGRRA CORICOMP and Reef Check (2015) Cayman Islands, pp. 191–195 In J Jackson, M Donovan, K Kramer and VWY Lam (eds.) *Status and trends of Caribbean coral reefs: 1970–2012*. Global Coral Reef Monitoring Network, IUCN, Gland, Switzerland.

Harper S, Bothwell J, Bale S, Booth S and Zeller D (2009) Cayman Island fisheries catches: 1950–2007. pp. 3–11 In Zeller D and Harper S (eds.) *Fisheries catch reconstructions: Islands, Part I*. Fisheries Centre Research Reports 17(5) [Updated to 2010]

Shackley M (1998) "Stingray City": Managing the Impact of Underwater Tourism in the Cayman Islands. *Journal of Sustainable Tourism* 6(4): 328–338.

1. Cite as Harper, S., J. Bothwell, S. Bale, S. Booth, and D. Zeller. 2016. United Kingdom (Cayman Islands). P. 428 in D. Pauly and D. Zeller (eds.), *Global Atlas of Marine Fisheries: A Critical Appraisal of Catches and Ecosystem Impacts*. Island Press, Washington, DC.

UNITED KINGDOM (CHAGOS ARCHIPELAGO)[1]

Dirk Zeller and Daniel Pauly

Sea Around Us Project, University of British Columbia, Vancouver, Canada

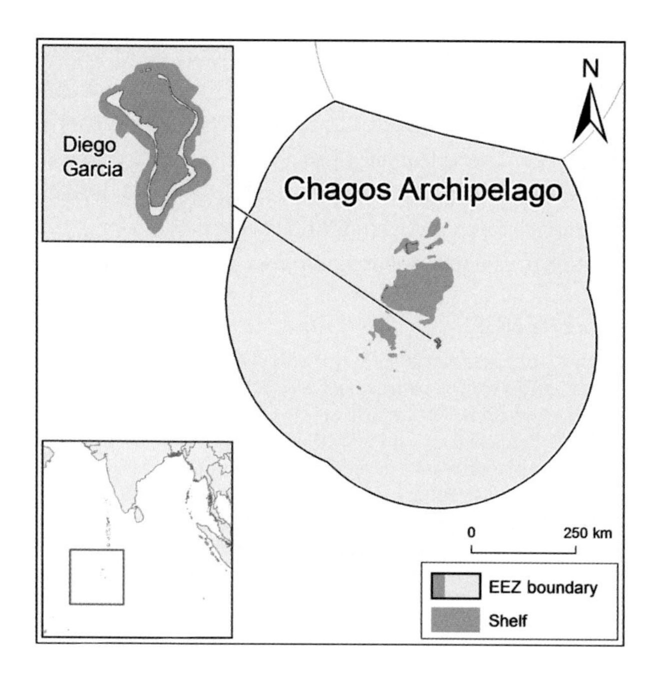

Figure 1. The Chagos Archipelago has a land area of 56 km² and the EEZ of 639,000 km² is, since April 2010, a no-take marine reserve.

The Chagos Archipelago is in the Indian Ocean, halfway between India and Madagascar (figure 1). The unfortunate inhabitants of the only remaining British Indian Ocean Territory were forcibly relocated to Mauritius and the Seychelles from 1967 to 1973 when part of the archipelago was turned into an American/British "Joint Defense and Naval Support Facility" located on Diego Garcia (Koldewey et al. 2010; Fischer 2012). In April 2010, the UK declared the entire EEZ of the Chagos Archipelago a no-take marine reserve. This history is reflected in the catch reconstruction by Zeller and Pauly (2014), who estimated a Chagossian subsistence catch of 90 t/year for the early 1950s, which declined to zero in 1972; and a recreational fishery by military personnel, which started in 1973 and may have grown to 100 t/year in the 2000s. This transition implied a radical change in targets, from medium-sized reef fish and invertebrates to trophy-type larger reef fishes and pelagic fishes. These catches are trivial compared with estimates of the industrial catches, taken by numerous foreign fleets (figure 2A).

The diversity of fish in the Chagos Archipelago, of which figure 2B represents but a minuscule fraction (see www.fishbase.org), is now in principle protected from fishing. Given the hardware available to the current residents of the Chagos Archipelago, there should be a good chance that the no-take part of the Chagos marine reserve can be enforced against illegal fishers, given enough goodwill.

REFERENCES

Fischer SR (2012) *Islands: from Atlantis to Zanzibar.* Reaktion Books, London, 336 p.

Koldewey H, Curnick D, Harding S, Harrison L and Gollock M (2010) Potential benefits to fisheries and biodiversity of the Chagos Archipelago/British Indian Ocean Territory as a no-take marine reserve. *Marine Pollution Bulletin* 60: 1906–1915.

Zeller D and Pauly D (2014) Reconstruction of domestic fisheries catches in the Chagos Archipelago: 1950–2010. pp. 17–24 In K Zylich, D Zeller, M Ang and D Pauly (eds.) *Fisheries catch reconstructions: Islands, Part IV.* Fisheries Centre Research Reports 22(2).

1. Cite as Zeller, D., and D. Pauly. 2016. United Kingdom (Chagos Archipelagos). P. 429 in D. Pauly and D. Zeller (eds.), *Global Atlas of Marine Fisheries: A Critical Appraisal of Catches and Ecosystem Impacts.* Island Press, Washington, DC.

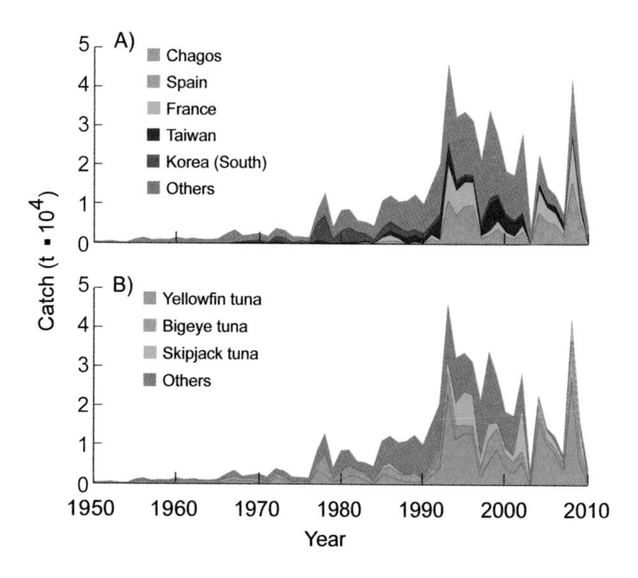

Figure 2. Domestic and foreign catches taken in the EEZ of Chagos Archipelago: (A) by fishing country (domestic scores: Sub. 1, 1, –; Rec. –, 1, 1; foreign catches are very uncertain); (B) by taxon.

UNITED KINGDOM (CHANNEL ISLANDS)[1]

Darah Gibson and Daniel Pauly

Sea Around Us, University of British Columbia, Vancouver, Canada

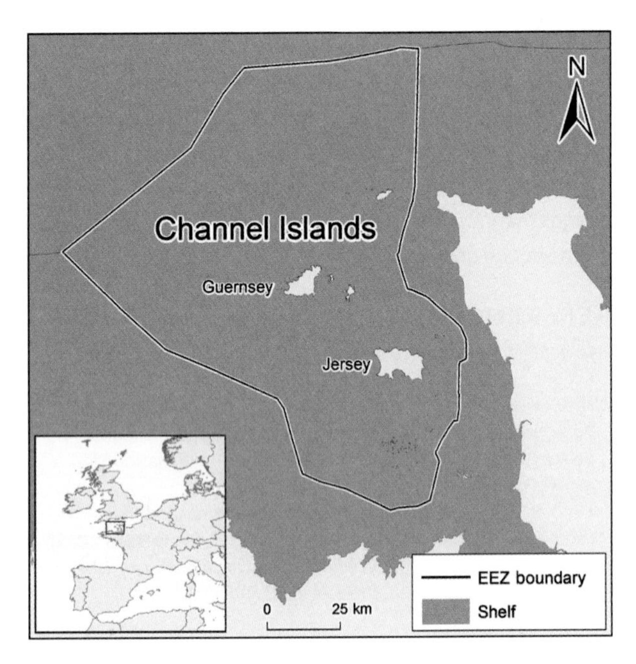

Figure 1. The Channel Islands (UK) have a land area of 194 km² and an EEZ of 11,600 km².

Jersey and Guernsey are small islands in the English Channel (figure 1) that are semi-independent from the UK. Their fisheries, governed by fiendishly complex arrangements (Fleury 2011), predominantly use dredges (deemed industrial; figure 2A) and target mainly edible crab (*Cancer pagurus*), spinous spider crab (*Maja squinado*), and great Atlantic scallop (*Pecten maximus*). As a complicating factor, the database of the International Council for the Exploration of the Sea (ICES), which documents European marine fisheries, appears to have double-counted their catch. Thus, from 1950 to 1972, the reported Channel Island catch was exactly equal to that of ICES areas VIIe (i.e., Channel Islands area) and VI (i.e., Scotland western area). These equal entries appear to be incorrect, because it is unlikely that vessels from the Channel Islands would be fishing in the waters off Scotland and report the exact same tonnage and species as they report from their home waters. This double accounting is mentioned here because ICES officials have repeatedly insisted that catch reconstructions cannot improve on ICES statistics. The conservative reconstruction by Gibson et al. (2015) yielded a domestic catch of 300 t/year in the early 1950s, which peaked at 4,300 t in 1997 and declined to 3,400 t by 2010, which is almost the same as reported by ICES on behalf of the Channel Islands. Considerable foreign fishing occurs in the context of the European Common Fisheries Policy (figure 2B), leading to a diverse taxonomic composition of catches (figure 2C), in addition to valuable molluscs and crustaceans.

REFERENCES

Fleury C (2011) Jersey and Guernsey: Two distinct approaches to cross-border fishery management. *Shima: The International Journal of Research into Island Cultures* 5(1): 24–43.

Gibson D, Cardwell E, Zylich K and Zeller D (2015) Preliminary reconstruction of total marine fisheries catches for the United Kingdom and the Channel Islands in EEZ equivalent waters (1950–2010). Fisheries Centre Working Paper #2015–76, 20 p.

1. Cite as Gibson, D., and D. Pauly. 2016. United Kingdom (Channel Islands). P. 430 in D. Pauly and D. Zeller (eds.), *Global Atlas of Marine Fisheries: A Critical Appraisal of Catches and Ecosystem Impacts*. Island Press, Washington, DC.

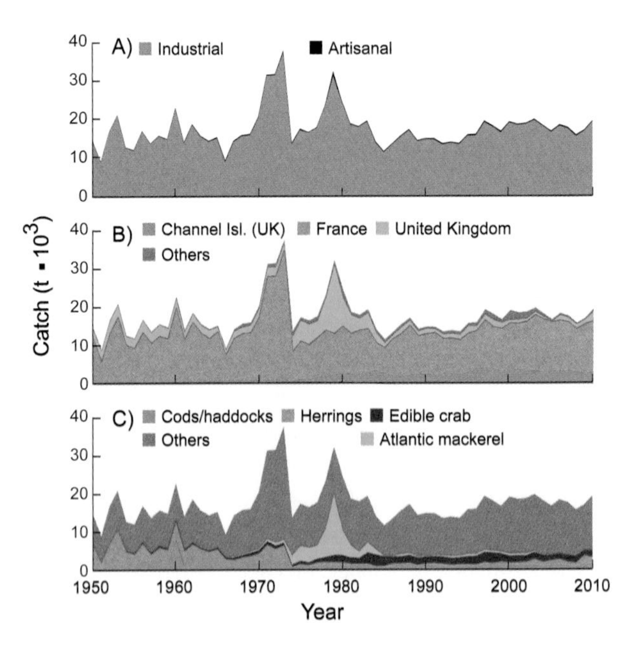

Figure 2. Domestic and foreign catches taken in the EEZ of the Channel Islands (UK): (A) by sector (domestic scores: Ind. 2, 3, 4; Art. 2, 3, 4; Dis. 2, 2, 2); (B) by fishing country (foreign catches are very uncertain); (C) by taxon.

UNITED KINGDOM (FALKLAND ISLANDS)[1]

Maria-Lourdes D. Palomares and Daniel Pauly

Sea Around Us, University of British Columbia, Vancouver, Canada

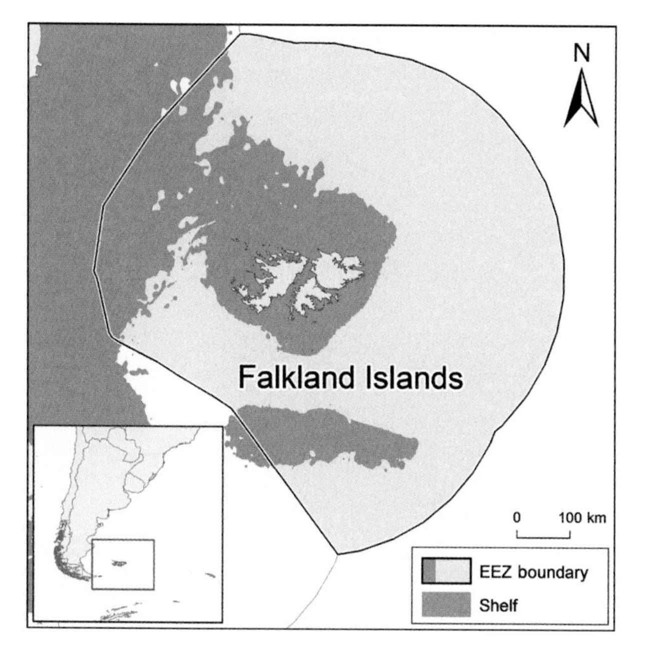

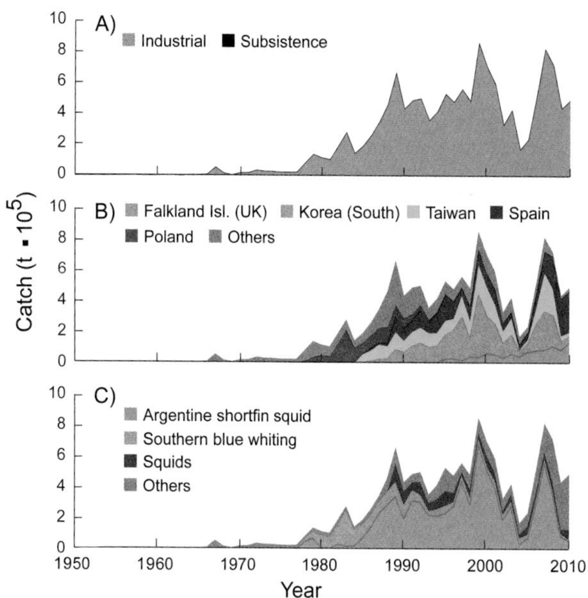

Figure 1. The Falklands Islands (U.K.) have a combined land area of about 12,200 km² and an EEZ of approximately 550,000 km².
Figure 2. Domestic and foreign catches taken in the EEZ of the Falklands Islands (UK): (A) by sector (scores: Ind. –, 3, 4; Art. 1, 1, 1; Sub. 2, 1, 1); (B) by fishing country (foreign catches are very uncertain); (C) by taxon.

The Falkland Islands, in the southwest Atlantic (figure 1), were first colonized by immigrants from Saint Malô in France (hence the name *Îles Malouines* in French and *Malvinas* in Spanish; Verne 1879). The islands were long contested between Spain and France, Spain and Britain, and Argentina and Britain, the last culminating in a brief but ferocious war in 1982. The Falkland Islands consist of 778 individual islands, of which the West and East Islands (with the capital city Stanley) are largest. Peopled by 3,000 inhabitants fiercely attached to Britain, the Falkland Islands are surrounded by a huge EEZ with valuable marine resources, notably Argentine shortfin squid (*Illex argentinus*) and Patagonian squid (*Loligo gahi*). Exploration for oil within the large EEZ remains controversial as a result of maritime disputes with Argentina. Although exploitation of the marine resources around these islands started more than 100 years ago and was initially focused on marine mammals (Palomares et al. 2006), this contribution covers only the years 1950–2010 and deals only with fish and invertebrates. The catch reconstruction of Palomares and Pauly (2015) yielded small domestic subsistence and artisanal catches (figure 2A) but documented considerable foreign industrial fisheries by a large number of countries in these waters, particularly since the mid-1980s (figure 2B). These fisheries targeted mainly industrially caught fish and invertebrates for export (figure 2C). The predominance of squid, especially the Argentine short-fin squid, is apparent.

REFERENCES

Palomares MLD and Pauly D (2015) Reconstruction of the marine fisheries catches of the Falkland Islands and the British Antarctic Territories 1950–2010. Pp. 1–19 In: Palomares MLD and Pauly D (eds.) *Marine Fisheries Catches of Sub-Antarctic Islands, 1950 to 2010*. Fisheries Centre Research Reports 23(1).

Palomares MLD, Mohammed E and Pauly D (2006) European expeditions as a source of historic abundance data on marine organisms. *Environmental History* 11(October): 835–847.

Verne J (1879) Les Grands Navigateurs du XVIIIe siècle. J Hetzel, Paris.

1. Cite as Palomares, M. L. D., and D. Pauly. 2016. United Kingdom (Falkland Islands). P. 431 in D. Pauly and D. Zeller (eds.), *Global Atlas of Marine Fisheries: A Critical Appraisal of Catches and Ecosystem Impacts*. Island Press, Washington, DC.

UNITED KINGDOM (MONTSERRAT)[1]

Robin Ramdeen,[a] Alwyn Ponteen,[b] Sarah Harper,[a] and Dirk Zeller[a]

[a]*Sea Around Us*, University of British Columbia, Vancouver, Canada [b]Montserrat Fisheries Division, Brades, Montserrat

The small eastern Caribbean island of Montserrat (figure 1) has been severely affected by natural disasters in its recent history. Notably, it was struck in 1989 by a devastating hurricane (Hugo) that damaged most of the island's infrastructure. Also, eruptions from the Soufrière Hills Volcano from 1995 until 2010 forced the evacuation of half of the residents to neighboring islands or beyond, and two thirds of the island, including the former capital of Plymouth, became uninhabitable (Kravtchenko and Fergus 2005). Montserrat's fisheries are small in scale (i.e., artisanal and subsistence; figure 2A), with the only industrial fishing being foreign (figure 2B). Total reconstructed domestic catches were estimated by Ramdeen et al. (2012) at about 300 t/year in the 1950s, declining to about 100 t/year in the late 1980s and only 50 t/year by 2010, which includes officially reported landings and estimates for unreported catches. This reconstruction shows that it is the subsistence catch, omitted from the statistics presented by the FAO on behalf of Montserrat, that caused the reconstructed catches, at least before Hurricane Hugo, to be about 3 times higher than FAO figures. Until 1989, Montserrat's catches consisted mainly of pelagic species (figure 2C), notably needlefishes (family Belonidae), various tuna (Scombridae) and targeted reef fish such as groupers and hinds (Serranidae), surgeonfish (Acanthuridae), squirrelfishes (Holocentridae), and grunts (Haemulidae). The current catch composition is assumed to be similar to that before 1989, although the amounts are much reduced.

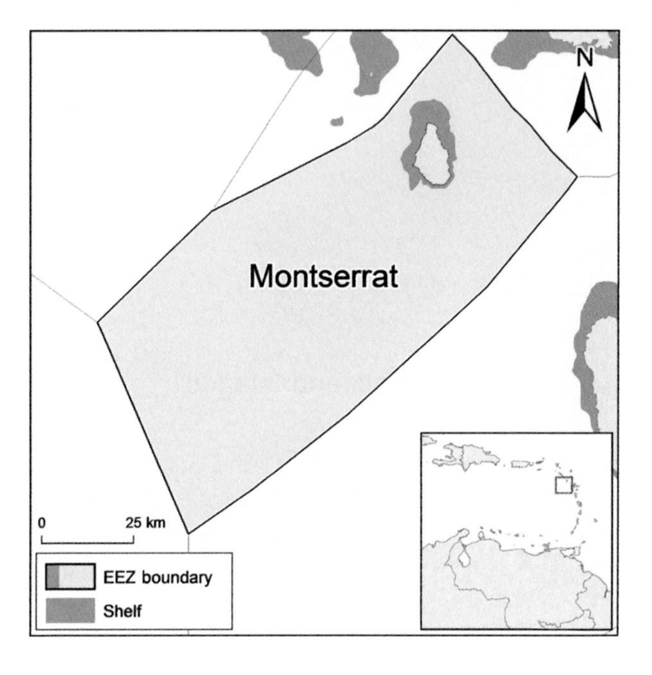

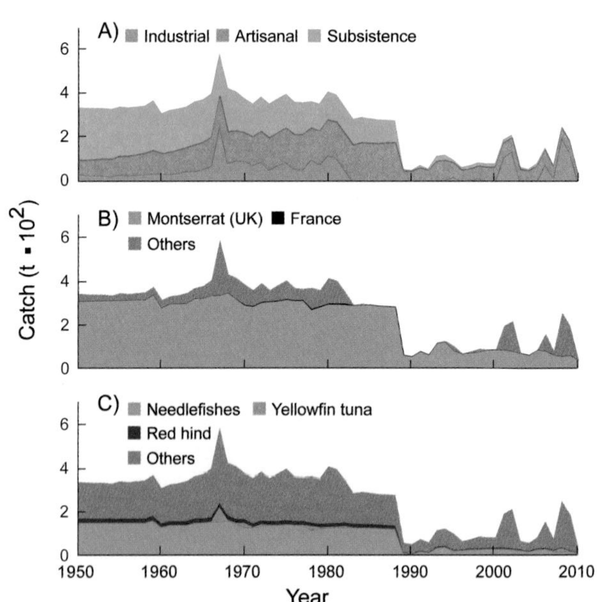

Figure 1. Montserrat has a land area of 102 km² and an EEZ of nearly 7,590 km². **Figure 2.** Domestic and foreign catches taken in the EEZ of Montserrat Island (UK): (A) by sector (domestic scores: Art. 1, 2, 3; Sub. 1, 2, 3); (B) by fishing country (foreign catches are very uncertain); (C) by taxon. The data illustrate the damage from Hurricane Hugo and the eruption from Soufrière Hills Volcano.

REFERENCES

Kravtchenko I and Fergus H (2005) Montserrat & Montserratians: Photo exploration. Commemorating ten years living with the volcano, 1995–2005. KiMAGIC, Canada.

Ramdeen R, Ponteen A, Harper S and Zeller D (2012) Reconstruction of total marine fisheries catches for Montserrat (1950–2010). pp. 69–76 In Harper S, Zylich K, Boonzaier L, Le Manach F, Pauly D and Zeller D (eds.), *Fisheries catch reconstructions: Islands, Part III*. Fisheries Centre Research Reports 20(5).

1. Cite as Ramdeen, R., A. Ponteen, S. Harper, and D. Zeller. 2016. Montserrat. P. 432 in D. Pauly and D. Zeller (eds.), *Global Atlas of Marine Fisheries: A Critical Appraisal of Catches and Ecosystem Impacts*. Island Press, Washington, DC.

UNITED KINGDOM (PITCAIRN ISLANDS)[1]

Devraj Chaitanya, Sarah Harper, and Dirk Zeller

Sea Around Us, University of British Columbia, Vancouver, Canada

Pitcairn, Henderson, Ducie, and Oeno are the 4 small islands that make up the Pitcairn Island group (figure 1), the last British Overseas Territory in the Pacific. The islands are located in the central South Pacific, roughly 5,300 km from New Zealand and 6,400 km from Chile (Amoamo 2011). The reconstruction by Chaitanya et al. (2012) suggests that Pitcairn Island's domestic catches were more than 6 times the landings reported by the FAO on behalf of the Pitcairn Islands, because artisanal catches were not accounted for. Overall, the reconstructed domestic catches for the Pitcairn Islands, which include a subsistence and a smaller artisanal component (figure 2A), totaled 25–30 t/year in the early 1950s but dropped to 10–15 t/year in the 2000s, because of a declining human population (descendants of *Bounty* mutineers and accompanying Tahitians). The small domestic catches are dwarfed by uncertain levels of industrial catches by largely illegal foreign fleets (figure 2B). The subsistence catches were dominated by blacktip grouper (*Epinephelus fasciatus*, Serranidae) and brown chub (*Kyphosus bigibbus*). Snappers (Lutjanidae), other Serranidae and Kyphosidae, and invertebrates, notably pronghorn spiny lobster (*Panulirus penicillatus*). Slipper lobsters (*Scyllarides* spp.) also provide substantial catch, while foreign fleets focus on a variety of taxa (figure 2C). The Pitcairn islanders expressed their unanimous wish to see their EEZ turned into a marine reserve, and the UK government granted their wish in March 2015.

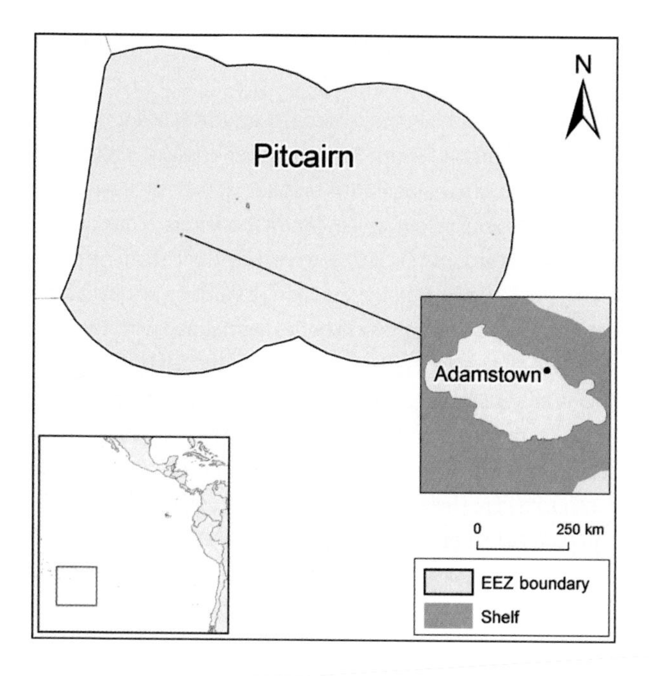

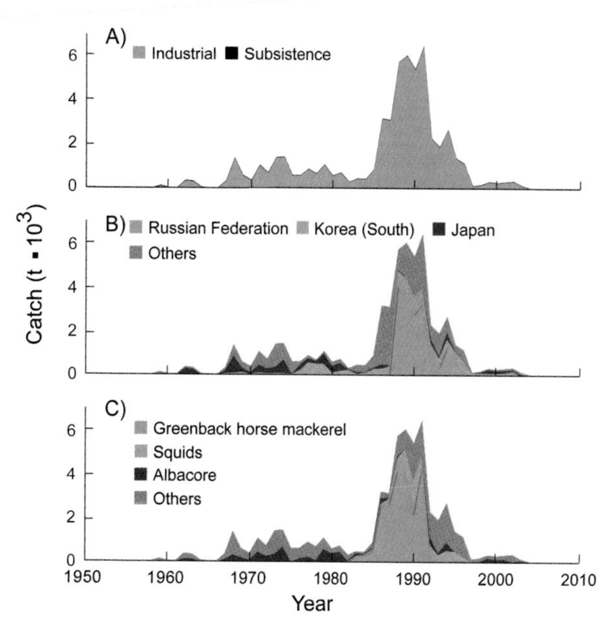

Figure 1. The 4 islands of the Pitcairn Islands, of which only Pitcairn proper is inhabited, have a combined land area of about 47 km² and an EEZ of approximately 836,000 km². **Figure 2.** Domestic and foreign catches taken in the EEZ of Pitcairn Islands (UK): (A) by sector (domestic scores: Art. 1, 1, 1; Sub. 2, 2, 2); (B) by fishing country (foreign catches are very uncertain); (C) by taxon.

REFERENCES

Amoamo M. (2011) Remoteness and Myth Making: Tourism Development on Pitcairn Island. *Tourism Planning & Development* 8(1): 1–19.

Chaitanya D, Harper S and Zeller D (2012) Reconstruction of total marine fisheries catches for the Pitcairn Islands (1950–2009), pp. 87–94. In: Harper S, Zylich K, Boonzaier L, Le Manach F, Pauly D and Zeller D (eds.) *Fisheries catch reconstructions: Islands, Part III*. Fisheries Centre Research Reports 20(5). [Incl. on p. 91 a short addendum which updates the dataset to 2010].

1. Cite as Chaitanya, D., S. Harper, and D. Zeller. 2016. United Kingdom (Pitcairn). P. 433 in D. Pauly and D. Zeller (eds.), *Global Atlas of Marine Fisheries: A Critical Appraisal of Catches and Ecosystem Impacts*. Island Press, Washington, DC.

UNITED KINGDOM (SOUTH GEORGIA/ SOUTH SANDWICH ISLANDS)[1]

Maria-Lourdes D. Palomares and Daniel Pauly

Sea Around Us, University of British Columbia, Canada

The uninhabited British Antarctic Territories of South Georgia Islands and South Sandwich Islands (SG&SSI; figure 1) are here treated as a single entity because of their similar history (Headland 1984), resource base, and fisheries governance. The first fisheries around the SG&SSI were subsistence fisheries supplying the crews of whaling vessels and stations, which lasted until the 1960s. Thus, these fisheries overlapped with the first Japanese and Soviet exploratory fisheries, which transformed into intensive bottom trawl fisheries exploiting mackerel icefish (*Chamsocephalus gunnari*) and other resources. These were succeeded by problematic longline fisheries for Patagonian toothfish (*Dissostichus eleginoides* ; Ashford et al. 1994), better known as Chilean seabass, managed, as are other fisheries, by the Commission for the Conservation of Antarctic Marine Living Resource (CCAMLR). The catches around SG&SSI, which are taken by a variety of countries (figure 2A) and reconstructed for the period from 1950 to 2010 by Palomares and Pauly (2015) from various historical records and CCAMLR statistics, were 2.5 t/year in the 1950s, peaked at more than 400,000 t/year in 1987, and declined to 30,000 t/year by 2010. Overall, this is 1.5 times the data officially reported by the CCAMLR. Figure 2B gives the composition of this catch, which was dominated by mackerel icefish and marbled rockcod (*Notothenia rossii*) in earlier decades but currently consists mainly of krill (*Euphausia superba*), Patagonian toothfish (*Dissostichus eleginoides*), and cod icefish (*Patagonotothen* spp.).

REFERENCES

Ashford JR, Croxall JP, Rubilar PS and Moreno CA (1994) Seabird interactions with longlining operations for *Dissostichus eleginoides* at the South Sandwich Islands and South Georgia. *CCAMLR Science* 1: 143–153.

Headland R (1984) The island of South Georgia. Cambridge University Press, Cambridge. 293 p.

Palomares MLD and Pauly D (2015) Reconstruction of the marine fisheries catches of the Falkland Islands and the British Antarctic Territories 1950–2010, p. 1–20. In: MLD Palomares and D Pauly (eds.) *Marine Fisheries Catches of the SubAntarctic Islands, 1950–2010*. Fisheries Centre Research Reports 23(1).

1. Cite as Palomares, M. L. D., and D. Pauly. 2016. United Kingdom (South Georgia and South Sandwich Islands). P. 434 in D. Pauly and D. Zeller (eds.), *Global Atlas of Marine Fisheries: A Critical Appraisal of Catches and Ecosystem Impacts*. Island Press, Washington, DC.

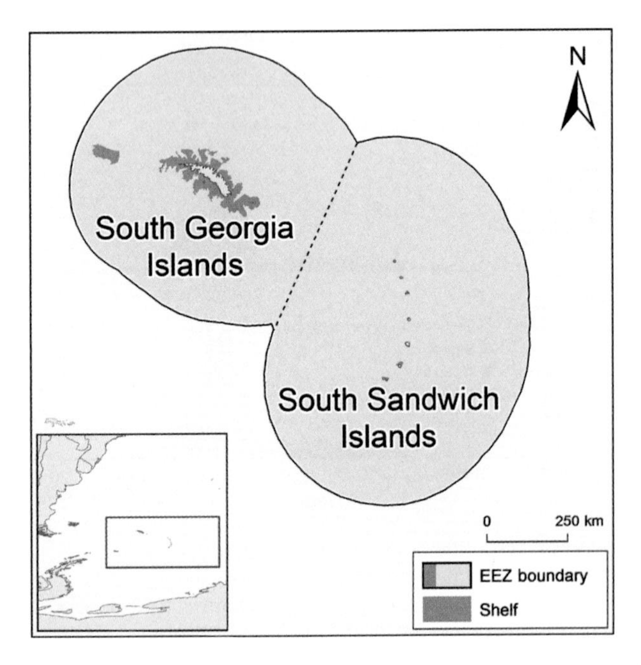

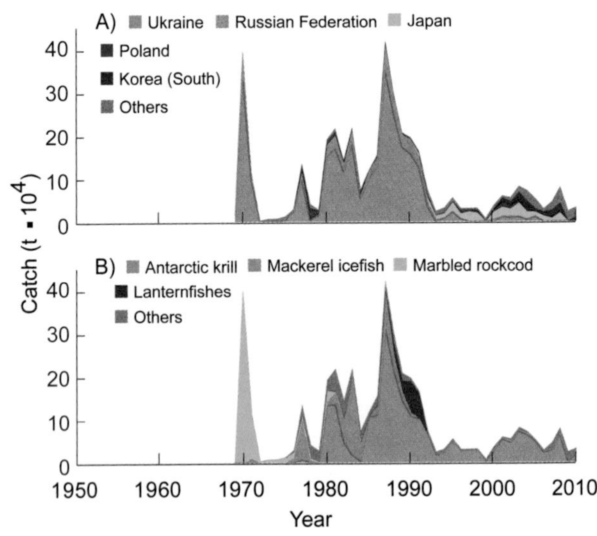

Figure 1. The South Georgia and South Sandwich Islands jointly have a land area of 3,900 km² and an EEZ of 1.2 million km².
Figure 2. Domestic and foreign catches taken in the EEZ of South Georgia and South Sandwich Islands (UK): (A) by fishing country (domestic scores: Ind. –, 3, 3; foreign catches are very uncertain); (B) by taxon.

UNITED KINGDOM (SOUTH ORKNEY ISLANDS)[1]

Maria-Lourdes Palomares and Daniel Pauly

Sea Around Us, University of British Columbia, Canada

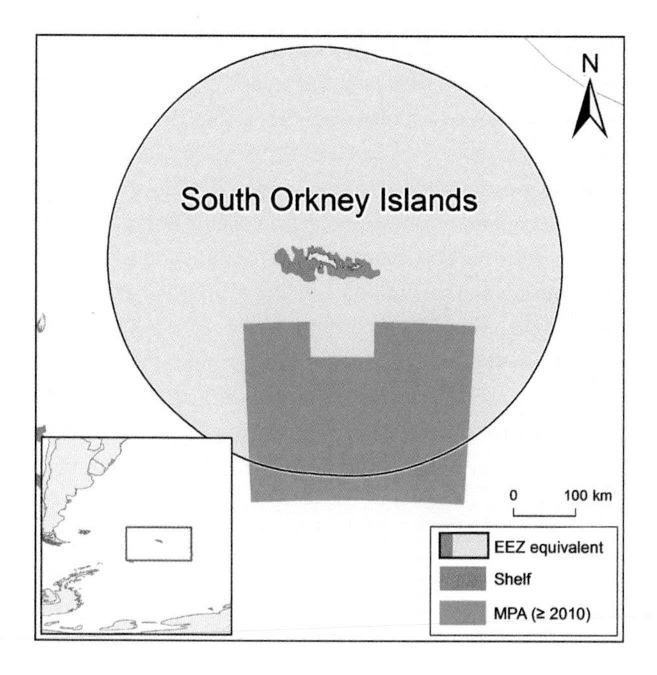

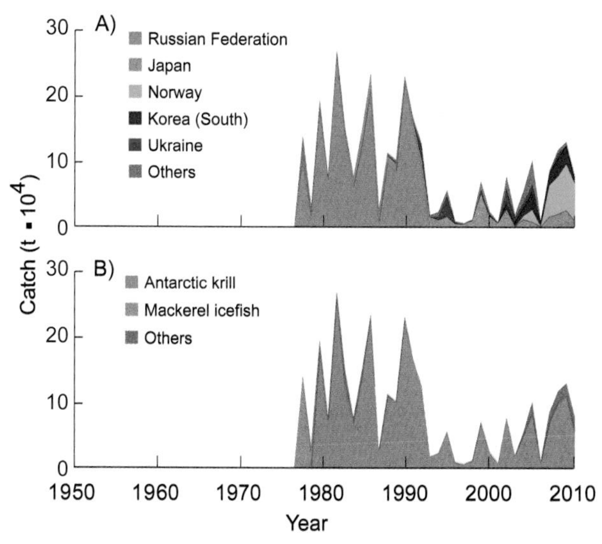

Figure 1. The South Orkney Islands (UK) jointly have a land area of 620 km² and would have an EEZ of about 390,000 km² if one were claimed. The "South Orkney Islands Southern Shelf MPA" has an area of 98,820 km². **Figure 2.** Domestic and foreign catches taken in the EEZ of South Orkney Islands (UK): (A) by fishing country (foreign catches are very uncertain); (B) by taxon. The lower catch in 2010 is meant to reflect the 2010 closure of part of the South Orkney waters (see also figure 1).

The South Orkney Islands, located near the tip of the Antarctic Peninsula, do not have a claimed EEZ, and the 200-nmi zone drawn around them in figure 1 is presented to allow comparison with other Antarctic territories. The corresponding catches, taken by a variety of countries (figure 2A) and reconstructed by Palomares and Pauly (2015) from historical sources and statistics of the Commission for the Conservation of Antarctic Marine Living Resource (CCAMLR), first consisted of a subsistence take by the crews of whaling stations and ships. Then, as the whales went, industrial fisheries started exploiting demersal fishes (mainly mackerel icefish, *Chamsocephalus gunnari*) and krill (*Euphausia superba*). Catches grew from zero in the 1950s to about 263,000 t in 1982 and down to less than 80,000 t in 2010 (figure 2B). This is probably the highest these catches will ever be, both because these resources were depleted (Jones et al. 2000) and because "upon approval in 2009, the South Orkney Islands Southern Shelf MPA became the first no-take marine reserve in the Commission for the Conservation of Antarctic Marine Living Resource's (CCAMLR's) network of Southern Ocean MPAs. The South Orkney Islands MPA designation was a positive first step as the world's first wholly high-seas MPA. The need to expand upon the area currently protected by the existing MPA is now a critical concern" (www.mpatlas.org/mpa/sites/7705283/).

REFERENCES

Jones CD, Kock K-H and Balguerias E (2000) Changes in biomass of eight species of finfish around the South Orkney Islands (Subarea 48.2) from three bottom trawl surveys. *CCAMLR Science* 7: 53–74.

Palomares MLD and Pauly D (2015) Reconstruction of the marine fisheries catches of the Falkland Islands and the British Antarctic Territories 1950–2010, pp. 1–20. in MLD Palomares and D Pauly (eds.) *Marine Fisheries Catches of the SubAntarctic Islands, 1950–2010*. Fisheries Centre Research Reports 23(1).

1. Cite as Palomares, M. L. D., and D. Pauly. 2016. United Kingdom (South Orkney Islands). P. 435 in D. Pauly and D. Zeller (eds.), *Global Atlas of Marine Fisheries: A Critical Appraisal of Catches and Ecosystem Impacts*. Island Press, Washington, DC.

UNITED KINGDOM (ST. HELENA)[1]

Shawn Booth, Houman Azar, and Danielle Knip

Sea Around Us, University of British Columbia, Canada

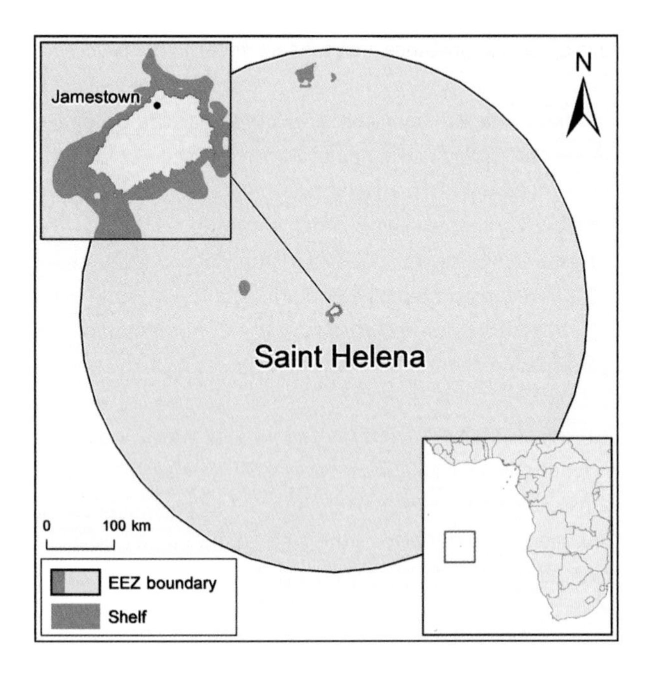

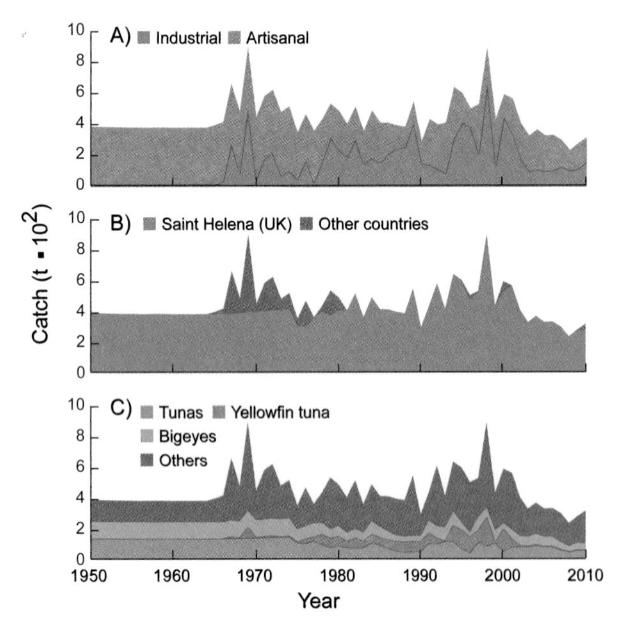

Figure 1. St. Helena Island has a land area of 120 km² and an EEZ of nearly 445,000 km². **Figure 2.** Domestic and foreign catches taken in the EEZ of St. Helena Island: (A) by sector (domestic scores: Ind. 1, 2, 2; Art. 1, 2, 2); (B) by fishing country (foreign catches are very uncertain); (C) by taxon.

St. Helena, a UK overseas territory most famous for being Napoleon's last place of exile, is a small, isolated island in the South Atlantic Ocean (figure 1). St. Helena has 2 dependencies, Ascension Island and Tristan da Cunha, both covered separately. Here, we estimate fisheries catches for the industrial and artisanal fisheries of St. Helena (figure 2A). Catches for St. Helena are estimated from data presented in Edwards (1990) and using a mass-balance approach detailed in Booth and Azar (2009), which used per capita seafood consumption and import and export data. The resulting estimate of total domestic catch (ignoring sporadic foreign fishing; figure 2B) was about 400 t/year from 1950 to 1974, when the artisanal catch began to decline, followed by strong fluctuations around an ascending trend caused by the growing industrial fishery, leading to catches of about 600 t/year in the early 2000s. The reconstructed domestic catch is 1.9 times the landings reported by the FAO on behalf of St. Helena, because much of the artisanal catch was not reported. Figure 2C shows the composition of this catch. St. Helena is heavily reliant on financial aid from the UK government and is currently experiencing a downward trend in population because of emigration due to a lack of economic opportunities. The anticipated opening of the first airport on St. Helena (scheduled for 2016) is expected to provide improved economic opportunities.

REFERENCES

Booth S and Azar H (2009) The fisheries of St Helena and its dependencies. pp. 27–34. In: Zeller D and Harper S (eds.) *Fisheries catch reconstructions: Islands, Part I*. Fisheries Centre Research Reports 17(5) [Updated to 2010]

Edwards A (1990) Fish and fisheries of Saint Helena Island. Centre for Tropical Coastal Management Studies, University of Newcastle upon Tyne, England, 152 p.

1. Cite as Booth, S., H. Azar, and D. Knip. 2016. United Kingdom (St. Helena). P. 436 in D. Pauly and D. Zeller (eds.), *Global Atlas of Marine Fisheries: A Critical Appraisal of Catches and Ecosystem Impacts*. Island Press, Washington, DC.

UNITED KINGDOM (TRISTAN DA CUNHA)[1]

Shawn Booth, Houman Azar, and Danielle Knip

Sea Around Us, University of British Columbia, Canada

The islands of Tristan da Cunha, in the South Atlantic (figure 1), include 5 neighboring islands, 2 of which are World Heritage Sites. Though administratively dependent on St. Helena, and thus on the UK, the people of Tristan da Cunha are economically self-reliant because of their exploitation of Tristan rock lobster (*Jasus tristani*), which began in 1949. Even when all the islanders were evacuated from 1961 to 1963 because of a volcanic eruption that temporarily ended the small-scale fishery, the large-scale fishery continued (figure 2A; Roscoe 1979; Cooper et al. 1992). Although other countries occasionally fish in these waters, most catch is domestic (figure 2B). Booth and Azar (2009) documented the reconstructed domestic catch in the Tristan da Cunha EEZ, consisting largely of rock lobster targeted by large vessels (about 400 t/year in the early 1950s and 420 t/year in the 2000s) and artisanal catches of lobster, baitfish, and fish for human consumption, about 500 t/year in the 1950s but less than 100 t/year from the mid-1980s on. Figure 2C provides details on the taxonomic composition, illustrating the predominance of lobster but also the occasional importance of large pelagics caught by foreign fleets. Other taxa of local importance are St. Paul's fingerfin (*Nemadactylus monodactylus*) and barrelfish (*Hyperoglyohe* sp.). There are concerns about illegal fishing around the islands, such as poaching for valuable lobster.

REFERENCES

Booth S and Azar H (2009) The fisheries of St Helena and its dependencies. pp. 27–34 In: Zeller D and Harper S (eds.) *Fisheries catch reconstructions: Islands, Part I*. Fisheries Centre Research Reports 17(5)[Updated to 2010].

Cooper J, Ryan PG and Andrew TG (1992) Conservation status of the Tristan da Cunha Islands, pp. 59–70 In: PR Dingwall (ed.) *Progress in conservation of the Subantarctic Islands*. Proceedings of the SCAR/IUCN Workshop on Protection, Research and Management of Subantarctic Islands, 27–29 April 1992. Paimpont, France.

Roscoe MJ (1979) Biology and exploitation of the rock lobster *Jasus tristani* at the Tristan da Cunha Islands, South Atlantic, 1949–1976. *South Africa, Sea Fisheries Branch Investigational Report* 118: 1–47.

1. Cite as Booth, S., H. Azar, and D. Knip. 2016. United Kingdom (Tristan de Cunha). P. 437 in D. Pauly and D. Zeller (eds.), *Global Atlas of Marine Fisheries: A Critical Appraisal of Catches and Ecosystem Impacts*. Island Press, Washington, DC.

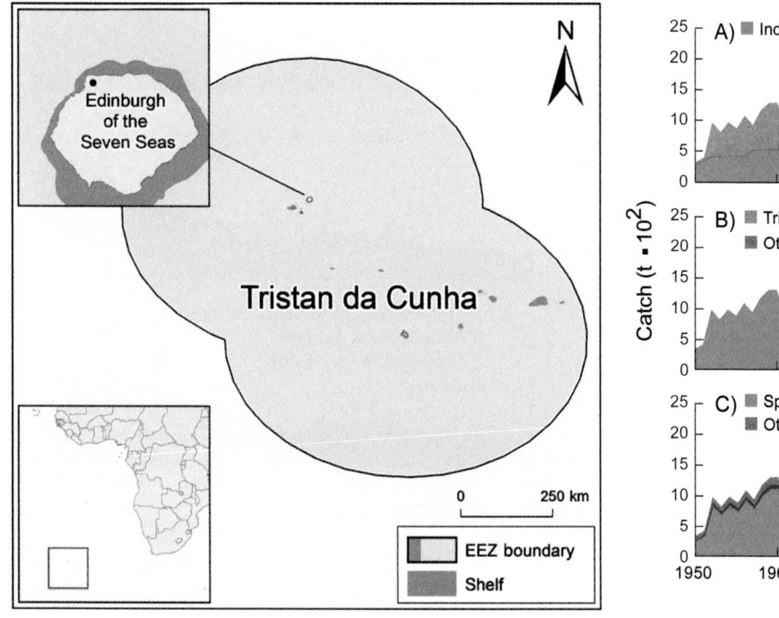

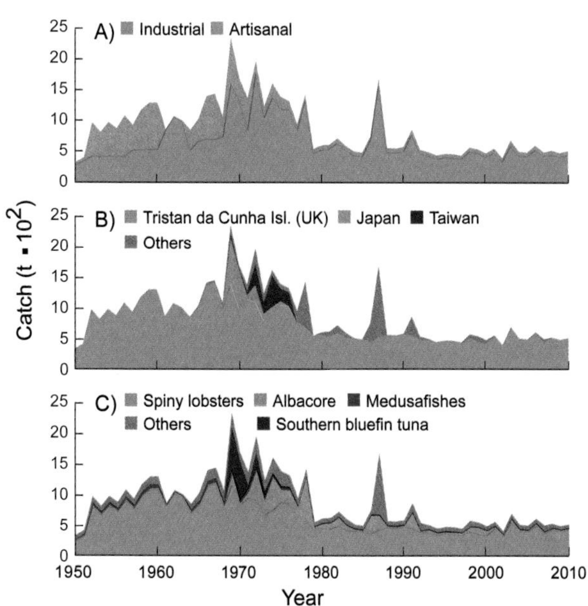

Figure 1. The Tristan de Cunha island group has a land area of 207 km² and an EEZ of 755,000 km². **Figure 2.** Domestic and foreign catches taken in the EEZ of Tristan de Cunha island group: (A) by sector (domestic scores: Ind. 1, 2, 2; Art. 1, 2, 2); (B) by fishing country (foreign catches are very uncertain); (C) by taxon.

UNITED KINGDOM (VIRGIN ISLANDS)[1]

Robin Ramdeen, Sarah Harper, and Dirk Zeller

Sea Around Us, University of British Columbia, Vancouver, Canada

The British Virgin Islands (BVI) consist of 60 islands, islets, and cays situated in the eastern Caribbean (figure 1), with an estimated total population of about 22,000 people. Sixteen of the islands are inhabited, but only 4 (Tortola, Jost Van Dyke, Virgin Gorda, and Anegada) have major settlements. The fishery catches, mostly artisanal and subsistence (figure 2A; Alimoso and Overing 1996), from the BVI's EEZ were reconstructed by Ramdeen et al. (2014). The domestic catches (some foreign fishing occurs in this EEZ; figure 2B) increased gradually from 500 t in 1950 to about 2,200 t in 2010, contrary to the figures reported by the FAO on behalf of the BVI, which suggest massive and unexplained short-term fluctuations. The reconstructed domestic catch, consisting of 72% artisanal, 25% subsistence, and 3% recreational contributions, was 2.3 times the reported catch overall. Catches were dominated (figure 2C) by reef species such yellowtail snapper (*Ocyurus chrysurus*) and silk snapper (*Lutjanus vivanus*), grunts (family Haemulidae), groupers (Serranidae), goatfish (Mullidae), and parrotfish (Scaridae). Catches of marine invertebrates such as spiny lobsters (*Panulirus argus*) and conch (*Strombus gigas*) were also very common. The "Others" category (figure 2C) consisted of 10 demersal families, 8 pelagic families, "miscellaneous marine fishes," and "miscellaneous marine invertebrates." With greater transparency from and some targeted investigations by the BVI Conservation and Fisheries Office, these estimates could all be improved.

REFERENCES

Alimoso S and Overing J (1996) Artisanal Fisheries and Resource Management in the British Virgin Islands, pp. 297–305 In *Proceedings of the Forty-Fourth Annual Gulf and Caribbean Fisheries Institute*. Charleston, South Carolina, USA.

Ramdeen R, Harper S and Zeller D (2014) Reconstruction of total marine fisheries catches for the British Virgin Islands (1950–2010). pp. 9–16. In: K Zylich, D Zeller, M Ang and D Pauly (eds.) *Fisheries catch reconstructions: Islands, Part IV*. Fisheries Centre Research Reports 22(2).

1. Cite as Ramdeen, R., S. Harper, and D. Zeller. 2016. United Kingdom (Virgin Islands. P. 438 in D. Pauly and D. Zeller (eds.), *Global Atlas of Marine Fisheries: A Critical Appraisal of Catches and Ecosystem Impacts*. Island Press, Washington, DC.

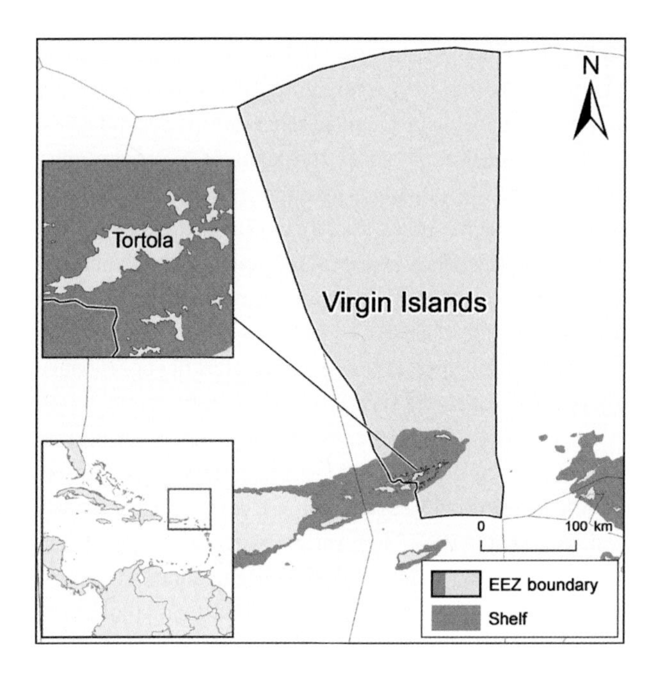

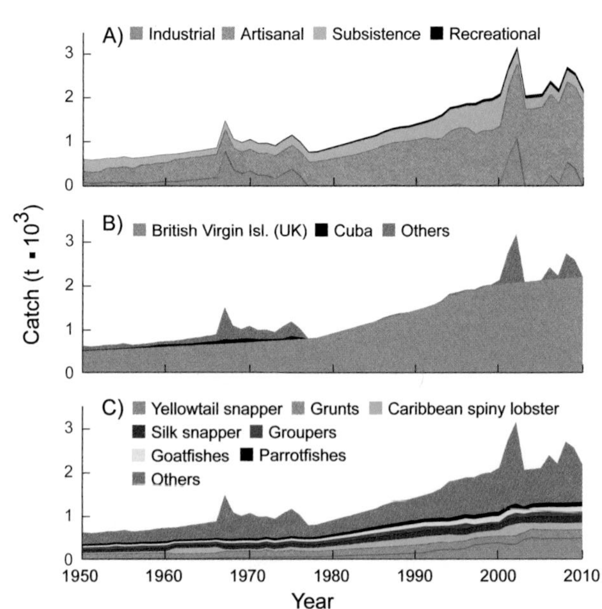

Figure 1. The British Virgin Islands have a land area of 153 km² and an EEZ of 80,100 km². **Figure 2.** Domestic and foreign catches taken in the EEZ of the British Virgin Islands: (A) by sector (domestic scores: Art. 2, 2, 1; Sub. 1, 1, 1; Rec. 1, 1, 1; Dis. –, –, 1); (B) by fishing country (foreign catches are very uncertain); (C) by taxon.

URUGUAY[1]

María Inés Lorenzo,[a] Omar Defeo,[a,b] Nazanin Roshan Moniri,[c] and Kyrstn Zylich

[a]Dirección Nacional de Recursos Acuáticos, Montevideo, Uruguay
[c]*Sea Around Us*, University of British Columbia, Vancouver, Canada

Uruguay is a South American country located between Brazil and Argentina (figure 1). Lorenzo et al. (2015) compiled reported catch statistics extracted from statistical yearbooks published by agencies of the Uruguayan government, complemented by new catch estimates for discards and recreational catches based on information from unpublished sources. This database leads to the conclusion that Uruguay has a growing artisanal sector (figure 2A). Total reconstructed catches by Uruguay in its EEZ increased from about 5,000 t/year in 1950 to about 160,000 t/year in the early 1980s and, after fluctuations, declined gradually to about 114,000 t by 2010. Overall, this was 1.25 times the catch reported by the FAO on behalf of Uruguay, mainly because of discards of the industrial fleet and the underestimation of the artisanal catch (see also Defeo et al. 2011), consisting of more than 50 fish and invertebrate species. In contrast, the industrial fishery (including foreign; figure 2B) mainly lands 3 species (figure 2C): Argentine hake (*Merluccius hubbsi*), whitemouth croaker (*Micropogonias furnieri*), and stripped weakfish (*Cynoscion guatucupa*), whose decline, shown by Milessi et al. (2005), has led to lower values of the mean trophic level of Uruguayan marine fisheries catches. Although industrial fisheries account for 90% of the total catch of Uruguay (including discards), artisanal (8%) and subsistence (2%) fisheries are important for the economic stability and food security of small-scale fishers and their families.

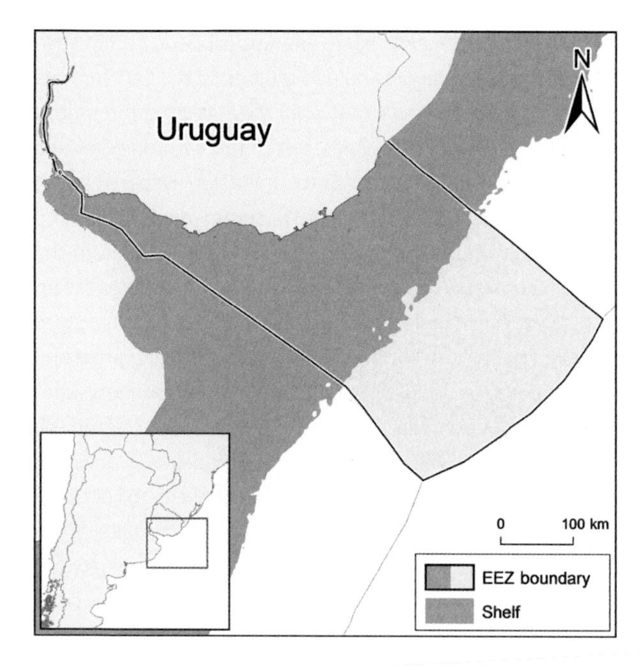

Figure 1. Uruguay has a shelf of 68,600 km² and an EEZ of 133,000 km².

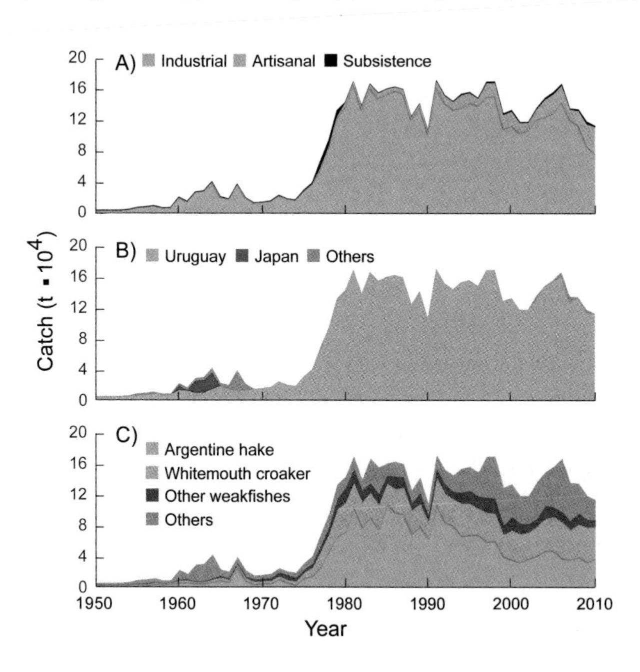

Figure 2. Domestic and foreign catches taken in the EEZ of Uruguay: (A) by sector (domestic scores: Ind. 2, 3, 4; Art. 2, 2, 3; Sub. 2, 3, 3; Rec. 1, 2, 2; Dis. 1, 1, 2); (B) by fishing country (foreign catches are very uncertain); (C) by taxon.

REFERENCES

Defeo O, Puig P, Horta S and de Álava A (2011) Coastal fisheries of Uruguay. pp. 357–384 In Salas S, Chuenpagdee R, Charles A and Seijo JC (eds.), *Coastal fisheries of Latin America and the Caribbean*. FAO Fisheries and Aquaculture Technical Paper No. 544, Rome.

Lorenzo MI, Defeo O, Roshan Moniri N, and Zylich K (2015) Fisheries catch statistics for Uruguay. Fisheries Centre Working Paper #2015–25, 6 p.

Milessi A, Arancibia H, Neira S and Defeo O (2005) The mean trophic level of Uruguayan landings during the period 1990–2001. *Fisheries Research* 74: 223–231.

1. Cite as Lorenzo, M. I., O. Defeo, N. Roshan Moniri, and K. Zylich. 2016. Uruguay. P. 439 in D. Pauly and D. Zeller (eds.), *Global Atlas of Marine Fisheries: A Critical Appraisal of Catches and Ecosystem Impacts*. Island Press, Washington, DC.

USA (ALASKA, ARCTIC)[1]

Shawn Booth, Dirk Zeller, and Daniel Pauly

Sea Around Us, University of British Columbia, Vancouver, Canada

The U.S. National Oceanic and Atmospheric Administration (NOAA) reports no fish catches to the FAO from its statistical Area 18 (Arctic), that is, for the Alaskan communities north of Cape Prince of Wales, which fall within FAO's Area 18 (figure 1). Thus, while the State of Alaska's Department of Fish and Game collects time series of commercial catches and occasionally undertakes subsistence fisheries studies, in the 15 main communities concerned, NOAA does not report on either of these fisheries, because they take place within state waters. The reconstruction of fisheries catches of marine and anadromous species of Booth and Zeller (2008) suggests that these fisheries generated catches of about 1,300 t/year in the 1950s and 1,500 t/year in the 2000s, with violent fluctuations since the 1970s, caused by the variability in the catch of chum salmon (*Oncorhychus keta*). This contrasts to the stability of the subsistence catch (figure 2A), which is apparent only because much of the "whitefish" catch, notably cisco (*Coregonus* spp.; figure 2B), was used as food for sled dogs (as in the Canadian Arctic) but went to human consumption after dog sleds were, in the 1960s, phased out as a major mean of long-distance transport. The baseline provided here (and in Zeller et al. 2011) is important given of the expected impact of global climate change on the food security of Arctic peoples.

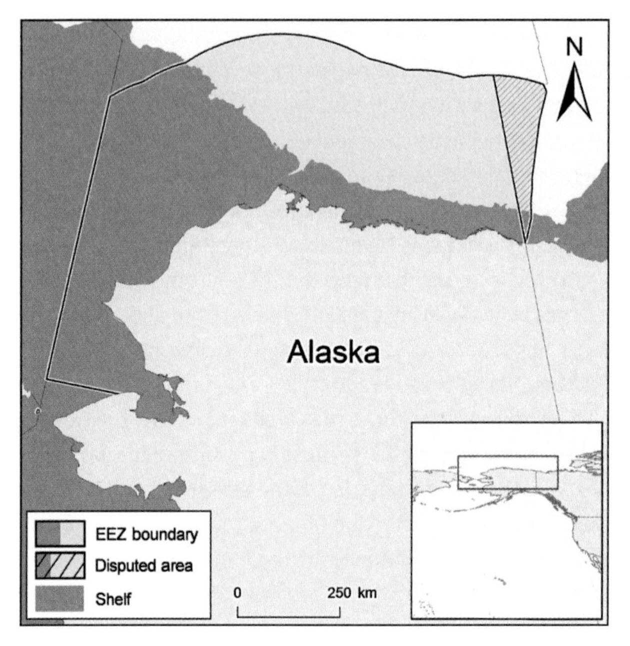

Figure 1. Arctic Alaska (USA) has an EEZ of about 508,810 km², encompassing a shelf of 283,230 million km², both within FAO area 18.

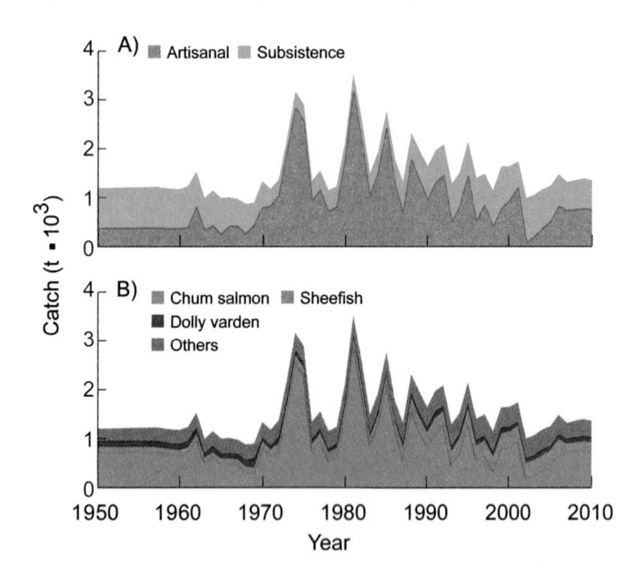

Figure 2. Fisheries catches in the EEZ of Arctic Alaska (USA): (A) by sector (domestic scores: Art. 3, 3, 4; Sub. 1, 1, 2); (B) by taxon.

REFERENCES

Booth S and Zeller D (2008) *Marine fisheries catches in Arctic Alaska*. Fisheries Centre Research Reports 16(9). [Updated to 2010]

Zeller D, Booth S, Pakhomov E, Swartz W and Pauly D (2011) Arctic fisheries catches in Russia, USA and Canada: baselines for neglected ecosystems. *Polar Biology* 34(7): 955–973.

1. Cite as Booth, S., D. Zeller, and D. Pauly. 2016. USA (Alaska, Arctic). P. 440 in D. Pauly and D. Zeller (eds.), *Global Atlas of Marine Fisheries: A Critical Appraisal of Catches and Ecosystem Impacts*. Island Press, Washington, DC.

USA (ALASKA, SUBARCTIC)[1]

Beau Doherty,[a] Darah Gibson,[a] and Yunlei Zhai

[a]*Sea Around Us*, University of British Columbia, Vancouver, Canada

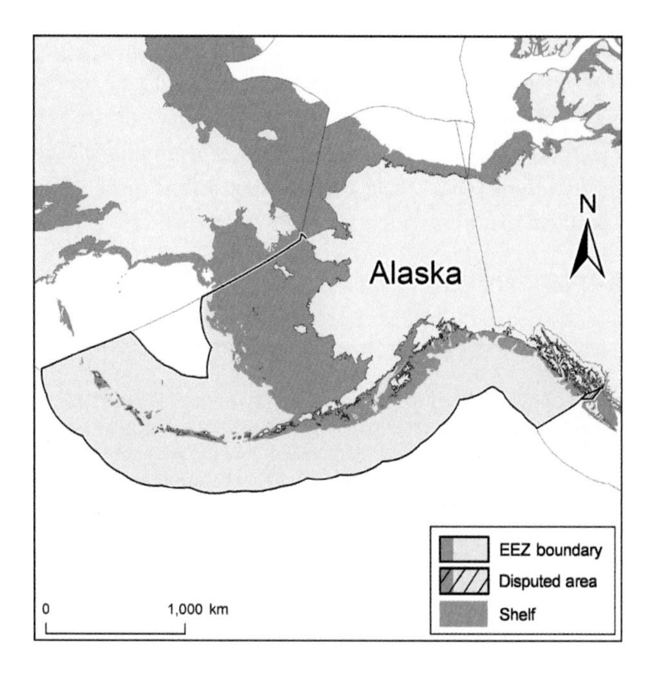

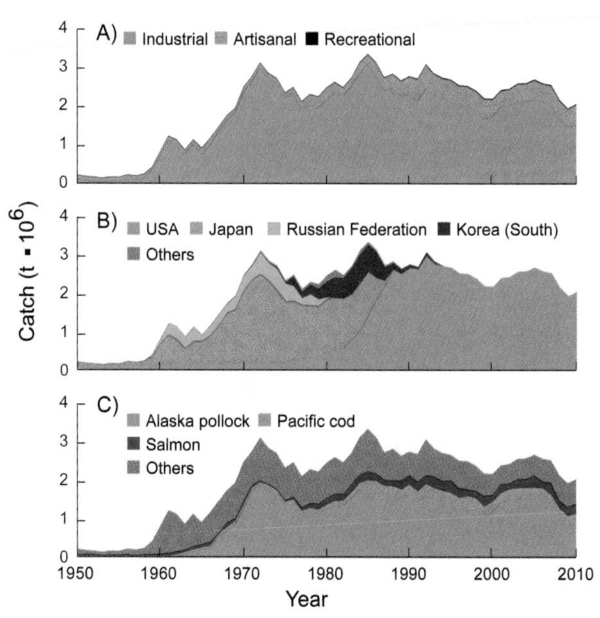

Figure 1. Subarctic Alaska (USA) has an EEZ of about 3,192,320 km², encompassing a shelf 1,087,270 km², within FAO areas 67 and 61. **Figure 2.** Total reconstructed catch within the Alaskan subarctic EEZ from 1950–2010: (A) by sector (the invisible subsistence and recreational catches make up 0.7% of the total; domestic scores: Ind. 3, 3, 4; Art. 3, 3, 4; Sub. 1, 1, 2; Rec. 1, 2, 3; Dis. 1, 1, 1); (B) by fishing country (foreign catches are more uncertain); (C) by taxon.

Alaska, the largest U.S. state, also has an immense EEZ (figure 1). Doherty et al. (2015) reconstructed U.S. catch for Alaska within its subarctic EEZ or equivalent waters (i.e., within FAO areas 67 and 61) from 1950 to 2010 using commercial landings data from the National Marine Fisheries Service (NMFS) and Queirolo et al. (1995) as a reporting baseline. Additional sources of catch in the form of recreational, subsistence, discards, and joint venture catches were compiled from historical data from the National Oceanic and Atmospheric Administration, the Alaska Department of Fish and Game, and the International Pacific Halibut Commission. Domestic catches from 1950 to 1975 averaged about 200,000 t/year, more than half of which were Pacific salmon. Catches increased sharply in the late 1970s and early 1980s, coinciding with the establishment of joint venture fisheries for groundfish between foreign processing ships and domestic fishing vessels (Queirolo et al. 1995). Although most catches are industrial, there is a substantial artisanal sector (figure 2A), and the heavy foreign fishing ended in the late 1980s (figure 2B). Since 1985, domestic catches have averaged 2.4 million t/year, peaking at nearly 3 million t in 1992 (figure 2B), with Alaska pollock (*Theragra chalcogramma*) accounting for 43%–63% of annual catch (figure 2C). Overall, reconstructed domestic catches are 1.1 times the commercial landings baseline, a discrepancy caused mostly by joint venture catch, accounting differences, and discards from trawlers. Catch and related data were available, transparent, and detailed, providing a sound basis for this reconstruction and, more importantly, for the prudent management of Alaska's fisheries, which should serve as a model elsewhere.

REFERENCES

Doherty B, Gibson D, Zhai Y, McCrea-Strub A, Zylich K, Zeller D and Pauly D (2015) Reconstruction of marine fisheries catches for Subarctic Alaska, 1950–2010. Fisheries Centre Working Paper #2015–82, University of British Columbia, Vancouver, 34 p.

Queirolo LE, Fritz LW, Livingston PA, Loefflad MR, Colpo DA and deReynier YL (1995) Bycatch, Utilization, and Discards in the Commercial Groundfish Fisheries of the Gulf of Alaska, Eastern Bering Sea, and Aleutian Islands. U.S. Department of Commerce. 148 p.

1. Cite as Doherty, B., D. Gibson, and Y. Zhai. 2016. USA (Alaska, Subarctic). P. 441 in D. Pauly and D. Zeller (eds.), *Global Atlas of Marine Fisheries: A Critical Appraisal of Catches and Ecosystem Impacts*. Island Press, Washington, DC.

USA (AMERICAN SAMOA)[1]

Dirk Zeller, Shawn Booth, and Daniel Pauly

Sea Around Us, University of British Columbia, Canada

American Samoa is the only U.S. territory south of the Equator (figure 1), and its small-scale fisheries consist of shore- and boat-based sectors (Zeller et al. 2006). A clear separation between commercial and noncommercial aspects in each fishery is difficult because fish from either sector can be sold or retained for personal consumption (figure 2A). The existing catch data on the predominantly commercial boat-based sector by the American Samoa Department of Marine and Wildlife Resources (DMWR) have been reported through the Western Pacific Fisheries Information Network (WPacFIN) since the early 1980s. The noncommercial sector, especially the shore-based fisheries, is not monitored, and catches are not reported on a regular basis. However, DMWR surveys of shore-based fisheries and other studies were conducted sporadically between 1980 and 2002. Although foreign industrial fishing (figure 2B) targeting tuna and billfishes (figure 2C) predominates in this EEZ, reconstructed domestic catches, based on Zeller et al. (2006, 2007a, 2015) amounted to 800 t/year in the 1950s, declining to 200 t/year in the 2000s. From 1950 to 2010, this corresponded to 17 times the catch officially reported and consisted nearly entirely of nearshore and reef fishes (figure 2C). For the 1982–2003 period, this suggested that the contribution of small-scale fisheries to GDP was underestimated by a factor of 5 (Zeller et al. 2007b). Such underestimation of the contribution of small-scale, especially subsistence fisheries, is a problem that needs addressing.

REFERENCES

Zeller D, Booth S, Craig P and Pauly D (2006) Reconstruction of coral reef fisheries catches in American Samoa, 1950–2002. *Coral Reefs* 25: 144–152.

Zeller D, Booth S, Davis G and Pauly D (2007a) Re-estimation of small-scale fisheries catches for U.S. flag-associated islands in the western Pacific: the last 50 years. *Fishery Bulletin* 105(2): 266–277.

Zeller D, Booth S and Pauly D (2007b) Fisheries contribution to the Gross Domestic Product: Underestimating small-scale fisheries in the Pacific. *Marine Resources Economics* 21: 355–374.

Zeller D, Harper S, Zylich K and Pauly D (2015) Synthesis of under-reported small-scale fisheries catch in Pacific island waters. *Coral Reefs* 34(1): 25–39.

1. Cite as Zeller, D., S. Booth, and D. Pauly. 2016. USA (American Samoa). P. 442 in D. Pauly and D. Zeller (eds.), *Global Atlas of Marine Fisheries: A Critical Appraisal of Catches and Ecosystem Impacts*. Island Press, Washington, DC.

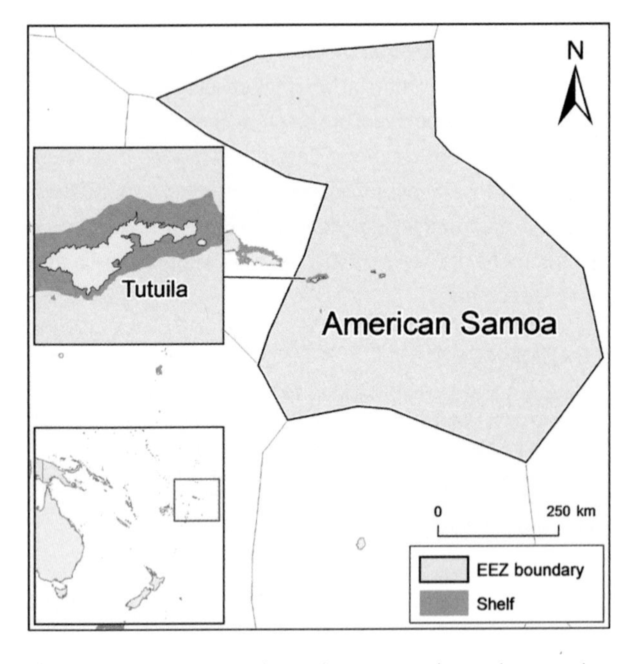

Figure 1. The land area of American Samoa is 200 km², against an EEZ of 404,000 km².

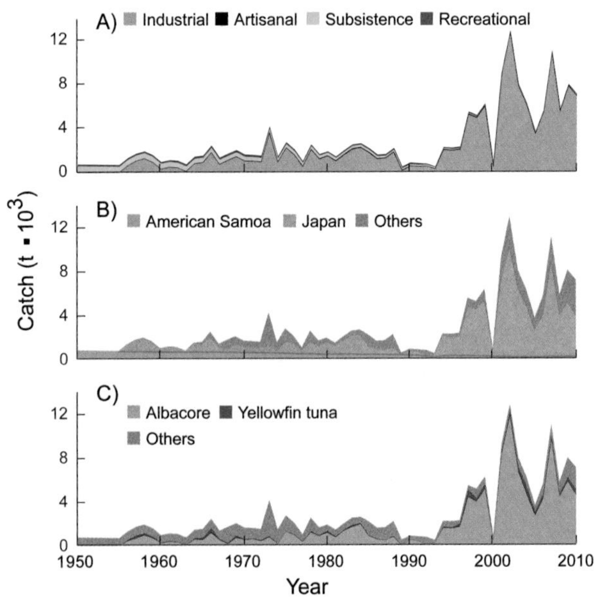

Figure 2. Domestic and foreign catches taken in the EEZ of American Samoa: (A) by sector (domestic scores: Art. 1, 2, 3; Sub. 1, 2, 2; Rec. 1, 1, 1); (B) by fishing country (foreign catches are very uncertain); (C) by taxon.

USA (EAST COAST)[1]

Ashley McCrea-Strub

Sea Around Us, University of British Columbia, Vancouver, Canada

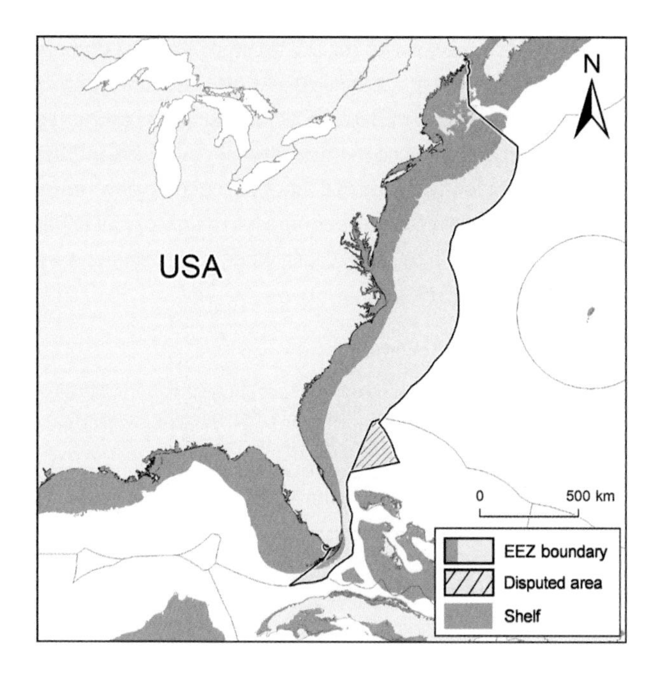

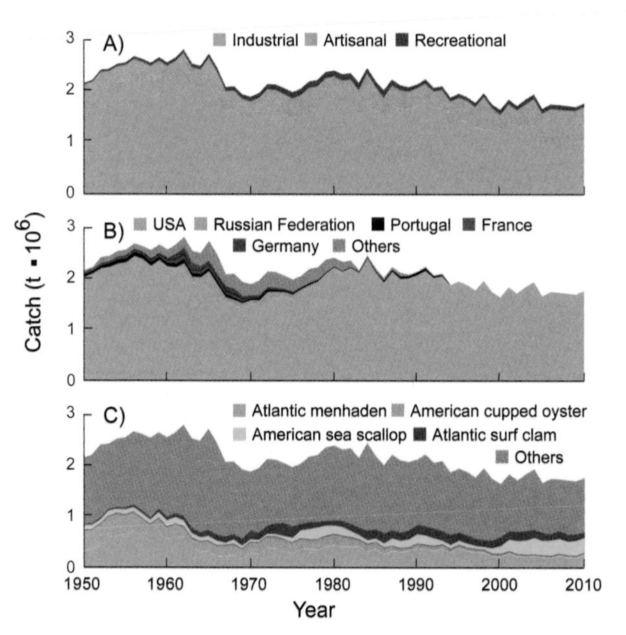

Figure 1. Map of USA (East Coast), with its maritime states, shelf of 361,000 km², and EEZ of 926,000 km². **Figure 2.** Domestic and foreign catches taken in the EEZ of USA (East Coast): (A) by sector (domestic scores: Ind. 4, 4, 4; Art. 4, 4, 4; Rec. 2, 3, 4; Dis. 2, 2, 3); (B) by fishing country (foreign catches are very uncertain); (C) by taxon.

The U.S. East Coast, from Maine to the tip of Florida (figure 1), has a long fishing tradition. This started with Native Americans (Steneck et al. 2004) but was intensified by the first colonists from Europe, notably in New England, where the fisheries were industrialized in the early 20th century. The resulting fishing pressure, by both industrial and artisanal fleets (figure 2A), jointly with that of fleets from Europe and East Asia, reduced the once abundant stocks of demersal fish until the United States claimed an EEZ in 1983 (figure 2B). Many formerly overexploited fish stocks are rebuilding, thanks to stringent regulations, which in part explain the recent low catches. The reconstruction (McCrea-Strub 2015), based on detailed landings data available from the National Oceanic and Atmospheric Administration (NOAA), individual states, and other sources, suggested that domestic catches were about 2.0 million t/year in 1950, decreased in the late 1960s to 1.4 million t/year before increasing to just under 2.4 million t in 1984, and settled at 1.7 million t/year in the late 2000s (figure 2B). Overall, this is about 60% more than the NOAA-reported data (as supplied by the FAO), mainly because of the noninclusion of recreational catches and discards in the NOAA database. However, this database was extremely detailed (including taxonomically; figure 2C), very transparent, and publicly available, in stark contrast to data from some other developed countries.

REFERENCES

McCrea-Strub A (2015) Reconstruction of total catch by U.S. fisheries in the Atlantic and Gulf of Mexico: 1950–2010. Fisheries Centre Working Paper #2015-79, 46 p.

Steneck RS, Vavrinec J and Leland AV (2004) Accelerating trophic-level dysfunction in kelp forest ecosystems of the western North Atlantic. *Ecosystems* 7(4): 323–332.

1. Cite as McCrea-Strub, A. 2016. USA (East Coast). P. 443 in D. Pauly and D. Zeller (eds.), *Global Atlas of Marine Fisheries: A Critical Appraisal of Catches and Ecosystem Impacts*. Island Press, Washington, DC.

USA (GUAM)[1]

Dirk Zeller and Daniel Pauly[a]

Sea Around Us, University of British Columbia, Vancouver, Canada

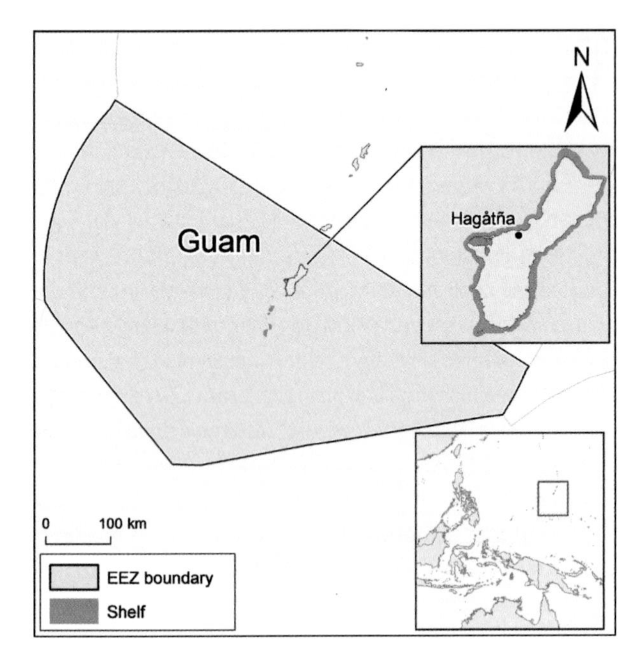

Figure 1. Guam has a land area of 540 km² and an EEZ of approximately 222,000 km².

Guam is the southernmost island in the Mariana Archipelago, in the western Pacific (figure 1). Guam's coral reef fisheries are both economically and culturally important. Reef fishes are a crucial component of the local diet. During the Japanese occupation, limitations were placed on local large-scale fisheries development, which together with the destruction of the Japanese fishing infrastructure at the end of World War II led to continued reliance on subsistence fisheries into the postwar years. The coral reefs around Guam were heavily fished and degraded, and concerns about overfishing were raised as early as 1970. Although considerable foreign industrial fishing occurred in these waters in early years (figure 2A, 2B), it essentially ceased with the EEZ declaration in 1983 (figure 2B). Guam's domestic fisheries have 2 sectors (ignoring transshipment and distant-water fleet catches of tuna): shore-based fisheries and small boat-based fisheries. There are few full-time commercial fishers, and many fishing trips contribute to all 3 segments, making the distinction between commercial, subsistence, and recreational catches challenging. The catch reconstruction performed for the Western Pacific Regional Fishery Management Council (WPRFMC), documented in Zeller et al. (2007) and updated in Zeller et al. (2015) estimated total domestic catch of 900–1,000 t/year in the 1950s and less than 50 t/year in the 2000s, 3.6 times as much as reported by NOAA/WPacFIN. This catch is composed mainly of reef fishes (figure 2C), except in the early years, when large pelagics were targeted by foreign fleets.

REFERENCES

Zeller D, Booth S, Davis G and Pauly D (2007) Re-estimation of small-scale fishery catches for U.S. flag-associated islands in the western Pacific: the last 50 years. *Fishery Bulletin* 105: 266–277.

Zeller D, Harper S, Zylich K and Pauly D (2015) Synthesis of under-reported small-scale fisheries catch in Pacific island waters. *Coral Reefs* 34(1): 25–39.

1. Cite as Zeller, and D. Pauly. 2016. USA (Guam). P. 444 in D. Pauly and D. Zeller (eds.), *Global Atlas of Marine Fisheries: A Critical Appraisal of Catches and Ecosystem Impacts*. Island Press, Washington, DC.

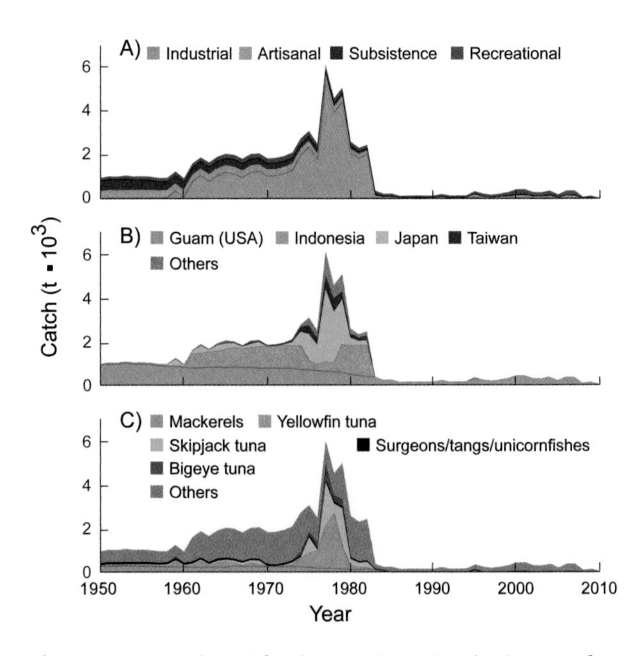

Figure 2. Domestic and foreign catches taken in the EEZ of Guam: (A) by sector (domestic scores: Art. 1, 2, 3; Sub. 1, 2, 3; Rec. 1, 2, 3); (B) by fishing country (foreign catches are very uncertain); (C) by taxon.

USA (GULF OF MEXICO)[1]

Ashley McCrea-Strub

Sea Around Us, University of British Columbia, Vancouver, Canada

U.S. waters in the Gulf of Mexico (GoM) extend from the tip of Florida in the east to Texas and its border with Mexico in the west (figure 1), a productive zone for both fish and offshore oil, which sometimes interact (McCrea-Strub et al. 2011). The total domestic catch from the US GoM EEZ was reconstructed by McCrea-Strub (2015) from landings data kindly made available by the National Oceanic and Atmospheric Administration (NOAA); the fisheries departments of the states of Florida, Alabama, Louisiana, Mississippi, and Texas; and other agencies. Emphasis was given to the recreational catch (figure 2A) and to discards, both of which are important in the GoM. This yielded a total domestic (no noticeable foreign fishing occurred in these waters) catch of 1.0 million t/year in the early 1950s, which peaked at 2.0 million t in 1984 and settled at about 1.0 million t/year in the late 2000s. This is 60% higher than the catch reported by NOAA for the GoM, mainly because of recreational catches and discards, omitted from official statistics. The catch from the U.S. GoM waters is dominated by Gulf menhaden (*Brevoortia patronus*), but other groups are also prominent (figure 2B). The GoM has been frequently used as a laboratory for testing ecosystem models (Browder 1993; Walters et al. 2008). The catch data herein should be useful for the generation of such ecosystem models.

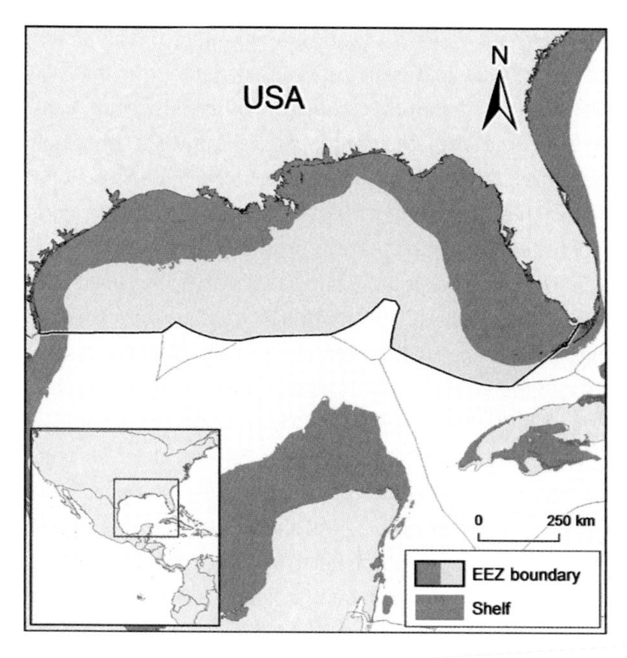

Figure 1. The USA (Gulf of Mexico) a shelf area of 336,000 km² and an EEZ of 696,000 km².

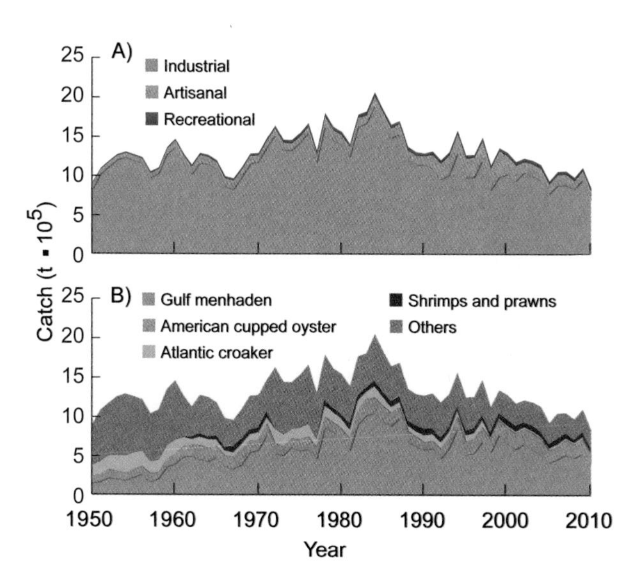

Figure 2. Domestic and foreign catches taken in the EEZ of the USA (Gulf of Mexico): (A) by sector (domestic scores: Ind. 4, 4, 4; Art. 4, 4, 4; Rec. 2, 3, 4; Dis. 3, 4, 4); (B) by taxon.

REFERENCES

Browder, JA (1993) A pilot model of the Gulf of Mexico continental shelf. pp. 279–284 In V Christensen and D Pauly (ed.) *Trophic models of aquatic ecosystems*. ICLARM Conference Proceedings 26, Manila.

McCrea-Strub A (2015) Reconstruction of total catch by U.S. fisheries in the Atlantic and Gulf of Mexico: 1950–2010. Fisheries Centre Working Paper #2015-79, 46 p.

McCrea-Strub A, Kleisner K, Sumaila UR, Swartz W, Watson R, Zeller D and Pauly D (2011) Potential Impact of the Deepwater Horizon Oil Spill on Commercial Fisheries in the Gulf of Mexico. *Fisheries* 37(7): 332–336.

Walters C, Martell S, Christensen V and Mahmoudi B (2008) An Ecosim model for exploring Gulf of Mexico ecosystem management options: implications of including multistanza life-history models for policy predictions. *Bulletin of Marine Science* 83(1): 251–271.

1. Cite as McCrea-Strub, A. 2016. USA (Gulf of Mexico). P. 445 in D. Pauly and D. Zeller (eds.), *Global Atlas of Marine Fisheries: A Critical Appraisal of Catches and Ecosystem Impacts*. Island Press, Washington, DC.

USA (MAIN HAWAIIAN ISLANDS)[1]

Dirk Zeller and Daniel Pauly[a]

Sea Around Us, University of British Columbia, Vancouver, Canada

The Main Hawaiian Islands (MHI; figure 1) have a high human population density and a major tourism industry. Although the majority of commercial fisheries revenue (80%–90%) lies with the large pelagic species, the MHI's nonpelagic resources have high cultural, subsistence, and recreational value. These resources consist of both shallow-water coral reef fishes and deeper-water "bottomfish," notably snappers (family Lutjanidae), jacks (Carangidae), and Hawaiian grouper (*Epinephelus quernus*). The nonpelagic catches, reconstructed by Zeller et al. (2008) and featuring substantial recreational catches (figure 2A), based on data from the Hawaii Division of Aquatic Resources (HDAR) and other sources (augmented by U.S. pelagic catches) and updated in Zeller et al. (2015), were 3,000 t/year in the early 1950s, declined to 1,600 t/year in the early 1960s, and grew to a peak of 6,600 t by the late 2000s. Overall, the official (commercial) statistics may have under-reported the actual catch by a factor of 2.24, mainly because they did not take subsistence and recreational catches into account. Foreign fishing, present in the 1970s, has essentially ceased (figure 2B). Figure 2C illustrates the composition of this catch. McClenachan and Kittinger (2013) reconstructed the catch of the well-managed reef and other coastal fisheries of ancient Hawaii and estimated figures of 15,000 to well above 20,000 t/year from 1400 to 1800, about 2–3 times the catch estimated here.

REFERENCES

McClenachan L and Kittinger JN (2013) Multicentury trends and the sustainability of coral reef fisheries in Hawai'i and Florida. *Fish and Fisheries* 14(3): 239–255.

Zeller D, Darcy M, Booth S, Lowe MK and Martell S (2008) What about recreational catch? Potential impact on stock assessment for Hawaii's bottomfish fisheries. *Fisheries Research* 91(1): 88–97.

Zeller D, Harper S, Zylich K and Pauly D (2015) Synthesis of under-reported small-scale fisheries catch in Pacific island waters. *Coral Reefs* 34(1): 25–39.

1. Cite as Zeller, D., and D. Pauly. 2016. USA (Main Hawaiian Islands). P. 446 in D. Pauly and D. Zeller (eds.), *Global Atlas of Marine Fisheries: A Critical Appraisal of Catches and Ecosystem Impacts*. Island Press, Washington, DC.

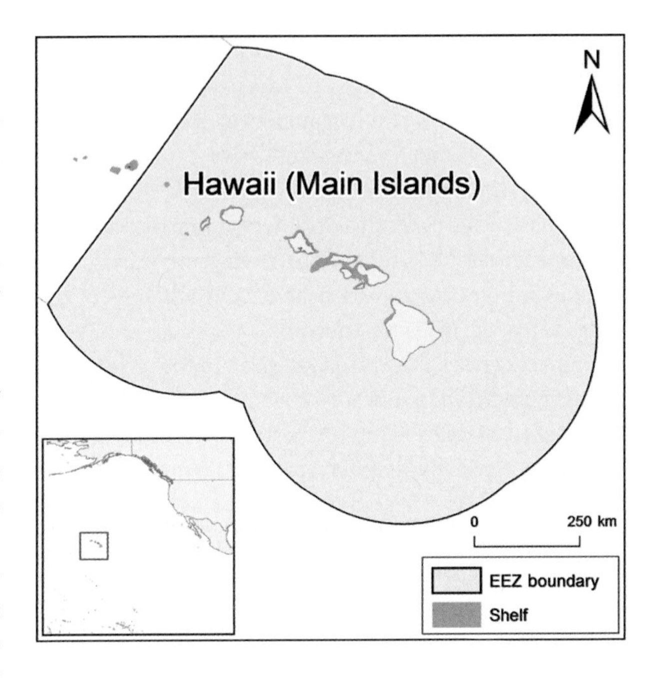

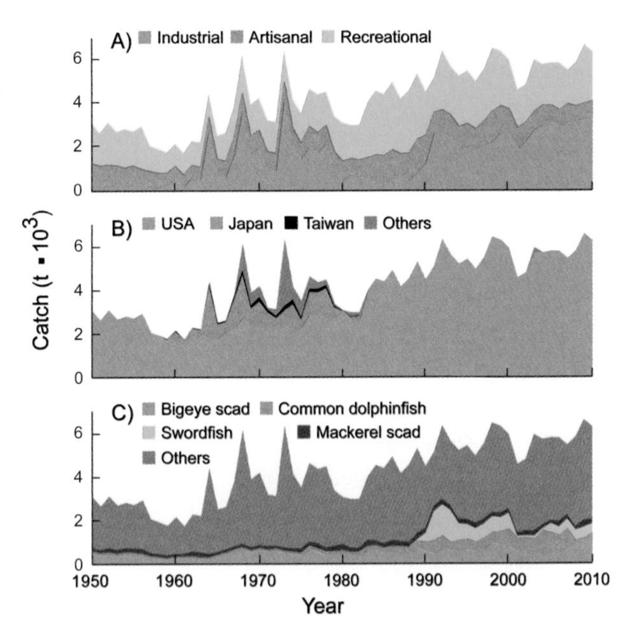

Figure 1. The Main Hawaiian Islands have a cumulative land area of 16,700 km² and an EEZ of 896,000 km². **Figure 2.** Domestic and foreign catches taken in the EEZ of Main Hawaiian Islands: (A) by sector (domestic scores: Ind. 3, 3, 4; Art. 3, 3, 4; Rec. 2, 2, 3; Dis. 2, 2, 3); (B) by fishing country (foreign catches are very uncertain); (C) by taxon.

USA (NORTHERN MARIANA ISLANDS)[1]

Dirk Zeller and Daniel Pauly[a]

Sea Around Us, University of British Columbia, Vancouver, Canada

The Commonwealth of the Northern Mariana Islands (CNMI) consists of 14 volcanic islands over a 680-km stretch in the Pacific (figure 1). More than 99% of the small population lives on the 3 southern islands of Saipan, Tinian, and Rota, and much of it is occupied in tourism and garment manufacturing. After World War II, the local fisheries reverted to subsistence fisheries well into the 1970s; the development of commercial and recreational fisheries did not start until the 1960s. The Division of Fish and Wildlife of CNMI ran a commercial catch collection system since the mid-1970s, but reported data have been available only since the early 1980s (through WPacFIN). Based on these and other data, Zeller et al. (2007a, 2015) estimated the domestic catch as 400 t/year in the 1950s, which decreased to 160 t/year in the 2000s (but see Cuetos-Bueno and Houk 2014). For 1950 to 2010, this was 3.9 times higher than reported by WPacFIN, and the official underestimation of catches affected the estimation of CNMI's GDP (Zeller et al. 2007b). Figure 2A presents the catch in this EEZ by sector, although sectoral boundaries are blurred because the same fishing trip can have commercial, subsistence, and recreational aspects. Furthermore, these data are highly skewed by the substantial foreign industrial fishing for tuna that occurred especially in pre-EEZ years (figure 2B). Figure 2C presents the taxonomic composition of these catches.

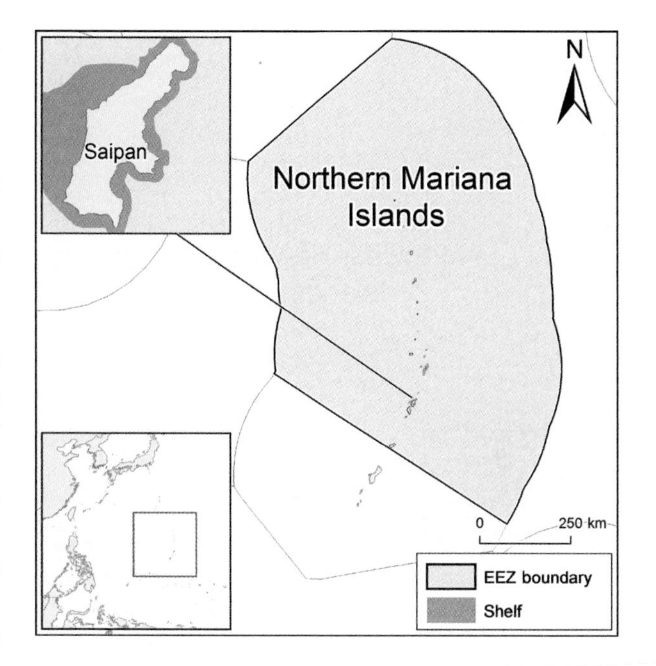

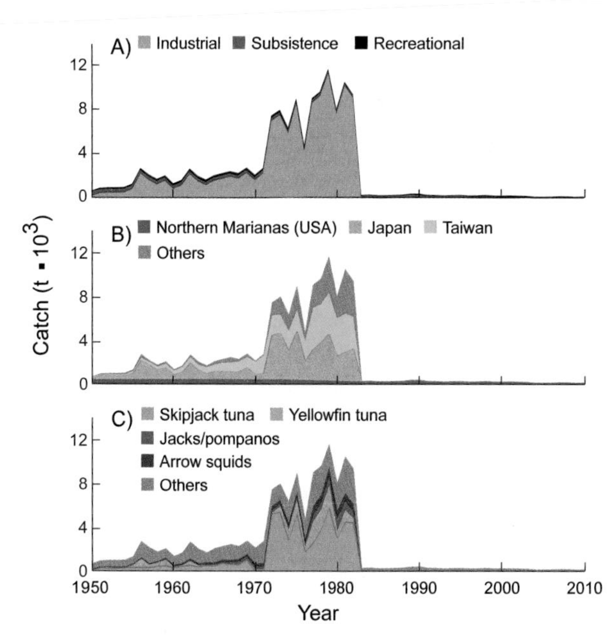

Figure 1. The Commonwealth of the Northern Mariana Islands (CNMI) has a land area of 475 km² and an EEZ of 749,000 km². **Figure 2.** Domestic and foreign catches taken in the EEZ of Northern Mariana Islands: (A) by sector (domestic scores: Art. 1, 2, 2; Sub. 1, 1, 2; Rec. 1, 1, 1); (B) by fishing country (foreign catches are very uncertain); (C) by taxon.

REFERENCES

Cuetos-Bueno J and Houk P (2014) Re-estimation and synthesis of coral-reef fishery landings in the Commonwealth of the Northern Mariana Islands since the 1950s suggests the decline of a common resource. *Reviews in Fish Biology and Fisheries* 25(1): 179–194.

Zeller D, Booth S, Davis G and Pauly D (2007a) Re-estimation of small-scale fishery catches for U.S. flag-associated islands in the western Pacific: the last 50 years. *Fishery Bulletin* 105: 266–277.

Zeller D, Booth S and Pauly D (2007b) Fisheries contribution to the Gross Domestic Product: Underestimating small-scale fisheries in the Pacific. *Marine Resources Economics* 21: 355–374.

Zeller D, Harper S, Zylich K and Pauly D (2015) Synthesis of under-reported small-scale fisheries catch in Pacific island waters. *Coral Reefs* 34(1): 25–39.

1. Cite as Zeller, D., and D. Pauly. 2016. USA (Northern Mariana Islands). P. 447 in D. Pauly and D. Zeller (eds.), *Global Atlas of Marine Fisheries: A Critical Appraisal of Catches and Ecosystem Impacts*. Island Press, Washington, DC.

USA (NORTHWEST HAWAIIAN ISLANDS)[1]

Daniel Pauly and Dirk Zeller

Sea Around Us, University of British Columbia, Vancouver, Canada

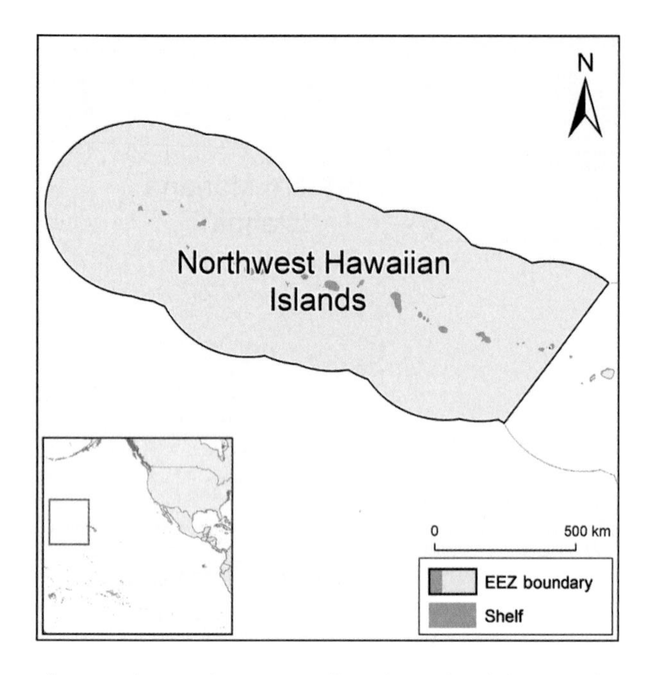

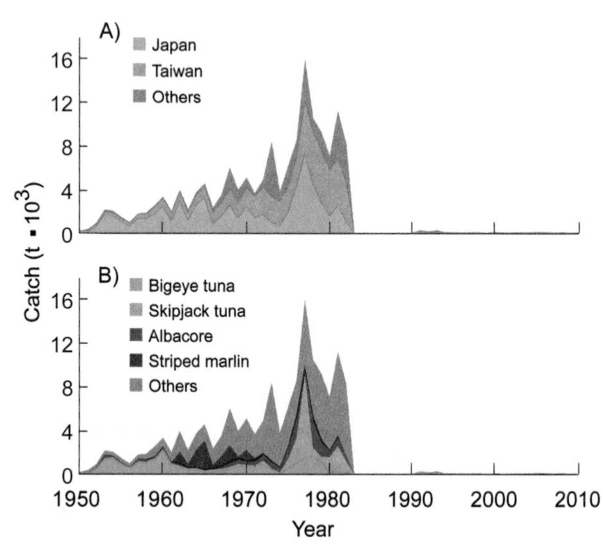

Figure 1. The Northwest Hawaiian Islands (USA), here also including Midway Atoll, have a cumulative land area of 8 km² and an EEZ of 1.58 million km². **Figure 2.** Domestic and foreign catches taken in the EEZ of Northwest Hawaiian Islands: (A) by fishing country (foreign catches are very uncertain); (B) by taxon. (Domestic scores: Ind. 1, 2, 3)

The uninhabited Northwest Hawaiian Islands (NWHI; figure 1), a part of the U.S. state of Hawaii, include about 80% of the Hawaiian coral reefs, most in a reasonable state, in contrast to those around the Main Hawaiian Islands. The NWHI include the French Frigate Island ecosystem, for which Polovina (1984) developed the first Ecopath model; see chapter 9. Because of their distance from inhabited lands, the NWHI were frequented mainly by larger fishing vessels from the Main Islands pursuing lobster and tuna fisheries (Zeller et al. 2005, 2015), and Asian fleets for tuna and shark fins (figure 2A). Given the process that turned the first 50 miles of the NWHI EEZ into the Papahānaumokuākea Marine National Monument and closed all fishing therein from 2011 on (thus creating a no-take marine reserve of 363,000 km²), it can be expected that fishing pressure in the NWHI EEZ will decline. The total catches taken from these waters by domestic and (formerly) foreign fleets amounted to 800 t in 1950 and reached values above 20,000 t before the closure of these waters to foreign fleets with the declaration of the U.S. EEZ in 1983, after which catches dropped dramatically to about 50 t by 2010. Catch composition is presented in figure 2B. More recent illegal catches are not quantified, although they occurred (see Mueller 2003).

REFERENCES

Mueller M (2003) Coast Guard seizes illegal nets: Two Chinese fishing vessels are caught with driftnets during a three-week span. Star Bulletin (Honolulu), July 7, http://archives.starbulletin.com/2003/07/25/news/story14.html

Polovina JJ (1984) Model of a coral reef ecosystem. *Coral Reefs* 3(1): 1–11.

Zeller D, Booth S and Pauly D (2005) *Reconstruction of coral reef and bottom fisheries catches for U.S. Flag Islands in the Western Pacific, 1950–2002*. Western Pacific Regional Fishery Management Council, Honolulu, 113 p.

Zeller D, Harper S, Zylich K and Pauly D (2015) Synthesis of under-reported small-scale fisheries catch in Pacific island waters. *Coral Reefs* 34(1): 25–39.

1. Cite as Pauly, D., and D. Zeller. 2016. USA (Northwestern Hawaiian Islands). P. 448 in D. Pauly and D. Zeller (eds.), *Global Atlas of Marine Fisheries: A Critical Appraisal of Catches and Ecosystem Impacts*. Island Press, Washington, DC.

USA (PACIFIC SMALL ISLAND TERRITORIES)[1]

Dirk Zeller and Daniel Pauly

Sea Around Us, University of British Columbia, Vancouver, Canada

The U.S. small island territories in the Pacific (also called U.S. Minor Outlying Islands), which are managed by the U.S. federal government, comprise several widely separated atolls and islands: the Johnston and Palmyra atolls and Howland & Baker, Jarvis, and Wake islands (figure 1). Midway Atoll, which is officially part of the small island territories, are here treated as part of the U.S. Northwest Hawaiian Islands (see previous page). These islands have no permanent resident population, except for Wake Island and Johnston Atoll, which have some government personnel and contractors (installations on Johnston Atoll have been closed since the late 2000s) and Palmyra Atoll, which has a small caretaker staff. Thus, it can be assumed that no domestic fisheries catches occurred (figure 2A) except for some recreational fishing on Wake Island and Johnston Atoll, as estimated by Zeller et al. (2005), and here extended to 2010, based on Zeller et al. (2015). The EEZ surrounding these islands, parts of which are currently protected as National Wildlife Refuges, are currently being evaluated as candidates for marine reserves, a development that would make the United States the country with the largest no-take marine space in the world. However, this would require that sizable resources be devoted to monitoring, control, and surveillance because their fish, notably tuna (figure 2B) and sharks, though still generally abundant, are frequently raided by illegal fishing vessels (Maragos et al. 2008), notwithstanding the 1983 declaration of EEZs by the United States.

REFERENCES

Maragos J, Friedlander AM, Godwin S, Musburger C, Tsuda R, Flint E, Pantos O, Ayotte P, Sandin ES, McTee S, Siciliano D and Oburl D (2008) US coral reefs in the Line and Phoenix Islands, Central Pacific Ocean: Status, threats and significance, p. 643–654 In: B. M. Riegl and R. E. Dodge (eds.) *Coral Reefs of the USA*, Springer, Amsterdam, the Netherlands.

Zeller D, Booth S and Pauly D (2005) *Reconstruction of coral reef- and bottom fisheries catches for U.S. Flag Islands in the Western Pacific, 1950–2002*. Western Pacific Regional Fishery Management Council, Honolulu, 113 p.

Zeller D, Harper S, Zylich K and Pauly D (2015) Synthesis of underreported small-scale fisheries catch in Pacific island waters. *Coral Reefs* 34(1): 25–39.

1. Cite as Zeller, D., and D. Pauly. 2016. USA (Pacific Small Island Territories). P. 449 in D. Pauly and D. Zeller (eds.), *Global Atlas of Marine Fisheries: A Critical Appraisal of Catches and Ecosystem Impacts*. Island Press, Washington, DC.

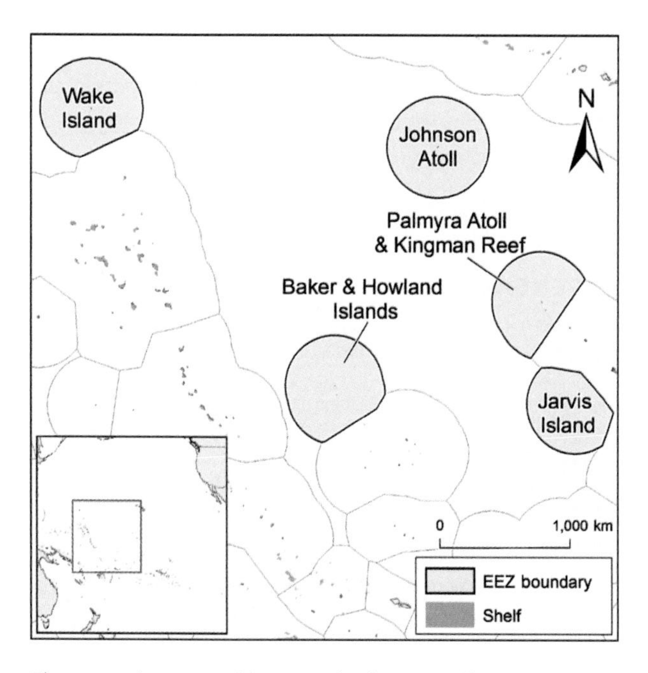

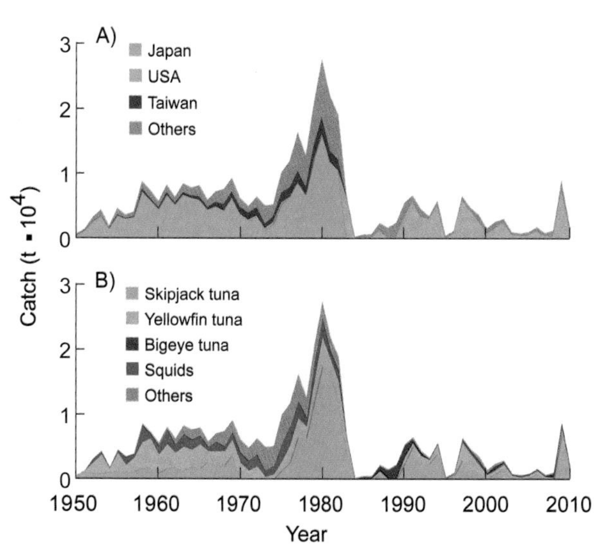

Figure 1. Johnston, Midway, and Palmyra atolls and Baker & Howland, Jarvis, and Wake islands (USA) have a cumulative land area of 29 km² (24 km² without Midway) and EEZs totaling 1,516,000 km². **Figure 2.** Domestic and foreign catches taken in the EEZ of the U.S. Pacific Small Island Territories: (A) by fishing country (foreign catches are very uncertain); (B) by taxon. (Domestic scores: Rec. 1, 1, 2.)

USA (PUERTO RICO)[1]

Richard S. Appeldoorn[a], Ilse M. Sanders[a], and Leonie Färber[b]

[a]Department of Marine Sciences, University of Puerto Rico, Mayagüez, Puerto Rico, USA
[b]*Sea Around Us*, University of British Columbia, Vancouver, Canada

Puerto Rico is the easternmost island of the Great Antilles (figure 1). The coast is lined with coral and rock reefs, seagrass beds, fringing mangroves, lagoons, and sandy beaches (Appeldoorn and Meyers 1993). To appreciate the extent of fishing impacts on ecosystems and the economic status of coastal communities requires a clear baseline of the timing and magnitude of fishery extractions. This account, based on the reconstruction by Appeldoorn et al. (2015), summarizes their comprehensive baseline, which included bait fish catches and growing recreational catches in addition to artisanal fishing (figure 2A). The domestic reconstructed catch was about 5,000 t/year in the early 1950s, grew to 6,500–6,800 t/year in the 1960s and 1970s, then gradually declined to about 1,500 t/year in the late 2000s (figure 2A). Overall, this was 1.9 times the data reported by the FAO on behalf of Puerto Rico. Some foreign fishing was deemed to have occurred in these waters, mainly before the EEZ declaration (figure 2B). Figure 2C does not reflect all species composition changes known to have occurred (Valdes-Pizzini 2007). For example, Nassau grouper (*Epinephelus striatus*) was once one of the most important commercial species, but it was largely replaced by the smaller red hind (*E. guttatus*). These changes suggest that fishery policy should be reevaluated and that potential cascading impacts of fishing on the ecosystem be considered when setting catch limits.

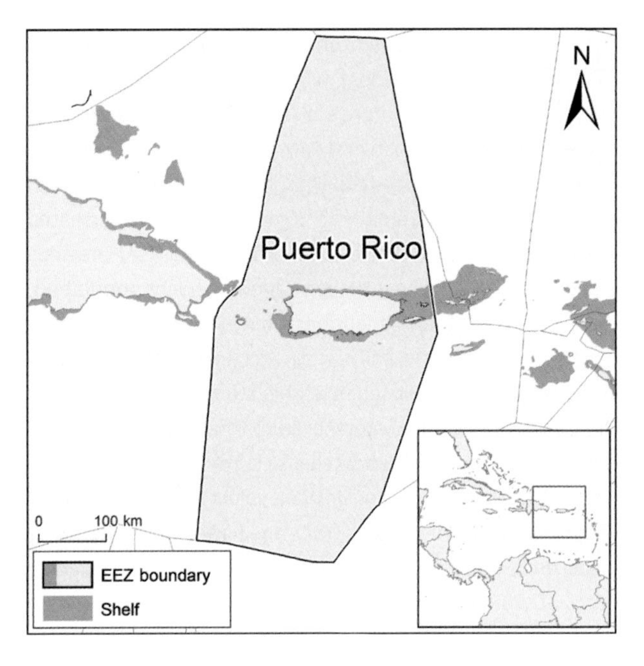

Figure 1. Puerto Rico, including the adjacent islands of Vieques and Culebra off the east coast, and Mona and Desecheo off the west coast, has a land area of 9,100 km² and an EEZ of 177,000 km².

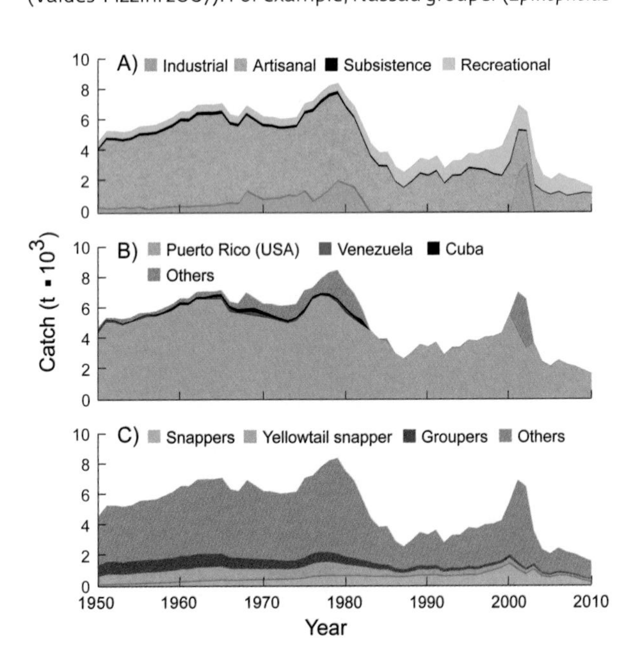

Figure 2. Domestic and foreign catches taken in the EEZ of Puerto Rico: (A) by sector (domestic scores: Art. 3, 3, 3; Sub. 2, 2, 2; Rec. 2, 2, 2; Dis. 2, 2, 2); (B) by fishing country (foreign catches are very uncertain); (C) by taxon.

REFERENCES

Appeldoorn RS and Meyers S (1993) Puerto Rico and Hispaniola, pp. 99–158 In Marine fishery resources of the Antilles. FAO Fish. Tech. Pap. 326.

Appeldoorn R, Sanders I and Färber L (2015) A 61 year reconstruction of fisheries catch in Puerto Rico. Fisheries Centre Working Paper #2015–44, 15 p.

Valdes-Pizzini M 2007. Reflections on the way life used to be: Anthropology, history and the decline of fish stocks in Puerto Rico, p. 37–47. Proc. 56th Gulf and Caribbean Fisheries Institute 59: 37–47.

1. Cite as Appeldoorn, R., I. Sanders, and L. Färber. 2016. USA (Puerto Rico). P. 450 in D. Pauly and D. Zeller (eds.), *Global Atlas of Marine Fisheries: A Critical Appraisal of Catches and Ecosystem Impacts*. Island Press, Washington, DC.

USA (VIRGIN ISLANDS)[1]

Robin Ramdeen, Kyrstn Zylich, and Dirk Zeller

Sea Around Us, University of British Columbia, Vancouver, Canada

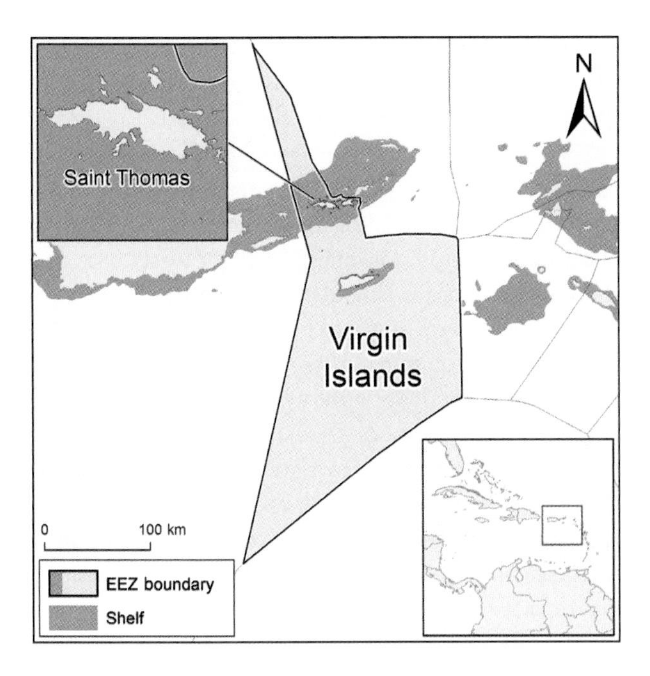

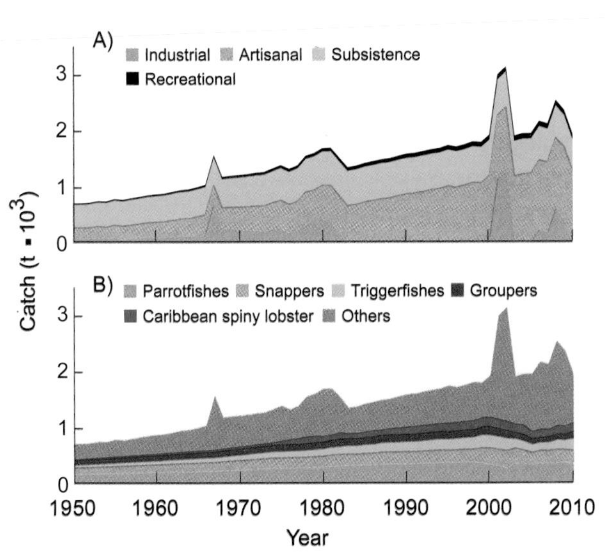

Figure 1. The U.S. Virgin Islands have a joint land area of 362 km², and their EEZ is 33,730 km². **Figure 2.** Marine fisheries catches from the U.S. Virgin Islands: (A) by sector (domestic scores: Art. 2, 3, 3; Sub. 2, 2, 2; Rec. 1, 1, 1); B) by taxon.

The U.S. Virgin Islands (USVI) is an overseas territory of the United States, located in the northeast Caribbean east of Puerto Rico (figure 1). It consists of 3 major islands—Thomas, St. John, and St. Croix—and about 50 cays. Originally inhabited by Ciboney, Caribs, and Arawaks, the islands were claimed for Spain by Christopher Columbus in 1493, then alternated between rulers from Spain, England, France, The Netherlands, and Denmark, with the last of these selling these islands to the United States in 1916. The fisheries of the USVI consist of three sectors: artisanal, subsistence, and recreational/sport. The sport fishery for billfishes in the USVI began in the 1950s. Annually 40–70 fishing boats from the southeast United States and a number of Caribbean countries travel to St. Thomas to fish for blue marlin and other billfishes. Three internationally recognized billfish tournaments are held annually in the USVI, along with several local tournaments (Kojis 2004; Mateo Rabelo 2004). Ramdeen et al. (2015) reconstructed the catch from the USVI as about 600 t in 1950, which steadily increased to 1,900 t/year in the mid-2000s, where it remained (figure 2A). This catch, consisting of artisanal (52%), subsistence (44%), and recreational/sport fishing (4%), is about 1.8 times that reported by the FAO, with omits subsistence and recreational/sport catches. Figure 2B gives the taxonomic composition of this catch.

REFERENCES

Kojis B (2004) Census of the marine commercial fishers of the U.S. Virgin Islands. Caribbean Fishery Management Council, San Juan, Puerto Rico, 91 p.

Mateo Rabelo I (2004).Survey of Resident Participants In Recreational Fisheries Activities In The Us Virgin Islands, p. 205–222 In: *Proceedings of The Fifty-Fifth Annual Gulf And Caribbean Fisheries Institute*. Xel Ha, Mexico.

Ramdeen R, Zylich K and Zeller D (2015) Reconstruction of total marine catches for the US Virgin Islands (1950–2010). Fisheries Centre Working Paper #2015–64, University of British Columbia, Vancouver, 9 p.

1. Cite as Ramdeen, R., K. Zylich, and D. Zeller. 2016. USA (Virgin Islands). P. 451 in D. Pauly and D. Zeller (eds.), *Global Atlas of Marine Fisheries: A Critical Appraisal of Catches and Ecosystem Impacts*. Island Press, Washington, DC.

USA (WEST COAST)[1]

Beau Doherty,[a] Haley Harguth,[b] Ashley McCrea-Strub,[a] LeKelia D. Jenkins,[b] and Will Figueira[c]

[a]*Sea Around Us*, University of British Columbia, Vancouver, Canada [b]School of Marine and Environmental Affairs, University of Washington, Seattle, WA [c]School of Biological Sciences, University of Sydney, NSW, Australia

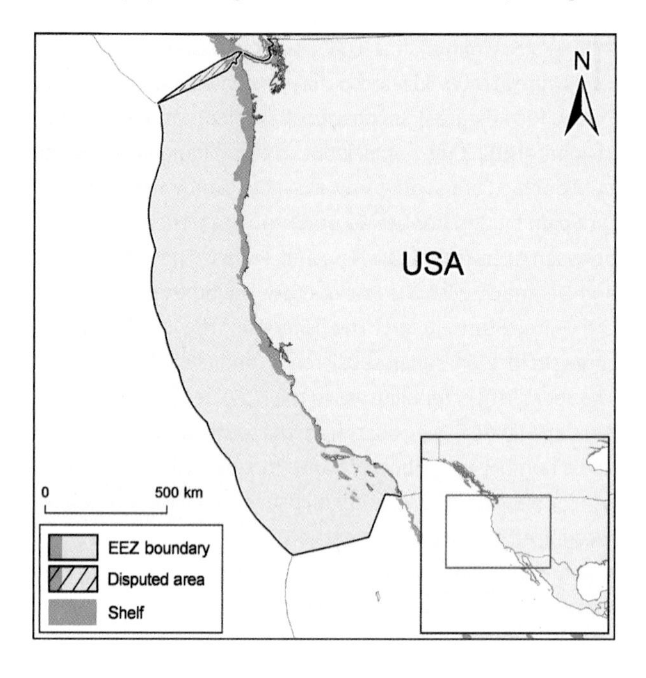

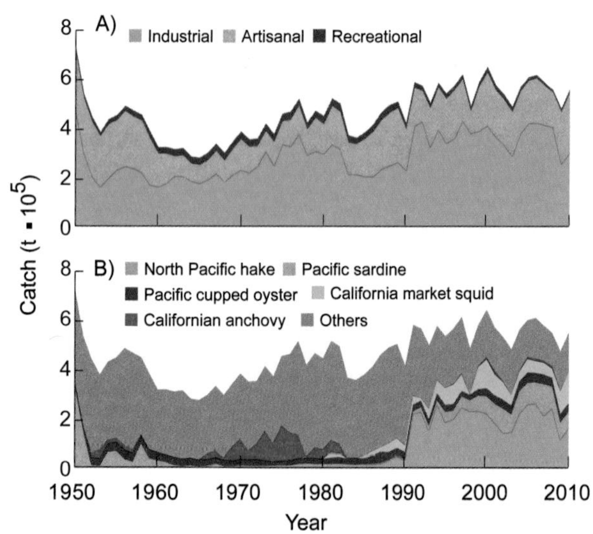

Figure 1. The U.S. West Coast and its EEZ (822,000 km²), with a shelf (66,500 km²) encompassing the coasts of the states of California, Oregon, and Washington. **Figure 2.** Domestic and foreign catches taken in the EEZ of U.S. West Coast: (A) by sector (domestic scores: Ind. 4, 4, 4; Art. 4, 4, 4; Rec. 2, 3, 3; Dis. 2, 1, 2); (B) by taxon.

The U.S. West Coast (figure 1) has some of the best documented fisheries in the world. Doherty et al. (2015) derived a reconstruction of total catch from the area of the EEZ of the U.S. West Coast from 1950 to 2010. Annual catches peaked at 750,000 t in 1950, just before the waning of the California sardine (*Sardinops sagax*) fishery. Both industrial and artisanal sectors are important on the West Coast, and the recreational sector, although smaller in catch, is socioeconomically important (figure 2A). With relatively little foreign fishing having occurred in these waters, catches declined to about 270,000 t/year in the mid-1960s before increasing again to 550,000–600,000 t/year by the mid 2000s. Overall, total reconstructed catches consisted of commercial landings (87.4%), followed by discards (8.2%; see also Bellman and Heery 2013) and recreational catch (4.4%; see also Figueira and Coleman 2010), and were 1.2 times the commercial landings reported by NOAA for the West Coast. It is evident that there remain sources of unreported catch in U.S. fisheries, namely discards and recreational fisheries before 1980. The NOAA commercial data also provide an excellent taxonomic resolution (simplified here, figure 2B), with most catch being reported to the species level. In this context, the United States is a global model of transparency, because fishery data are either publicly available on websites or readily shared by the agencies that generated them.

REFERENCES

Bellman MA and Heery E (2013) Discarding and fishing mortality trends in the U.S. West Coast groundfish demersal trawl fishery. *Fisheries Research* 147: 115–126.

Doherty B, Harguth H, McCrea-Strub A, Jenkins LD and Figueira W (2015) Reconstructing catches along Highway 101: Historic catch estimates for marine fisheries in California, Oregon and Washington from 1950–2010. Fisheries Centre Working Paper #2015-81, 66 p.

Figueira WF and Coleman CF (2010) Comparing landings of the United States recreational fishery sectors. *Bulletin of Marine Science* 86(3): 499–514.

1. Cite as Doherty, B., H. Harguth, A. McCrea-Strub, L. Jenkins, and W. Figueira. 2016. USA (West Coast). P. 452 in D. Pauly and D. Zeller (eds.), *Global Atlas of Marine Fisheries: A Critical Appraisal of Catches and Ecosystem Impacts*. Island Press, Washington, DC.

VANUATU[1]

Kyrstn Zylich, Soohyun Shon, Sarah Harper, and Dirk Zeller

Sea Around Us, University of British Columbia, Vancouver, Canada

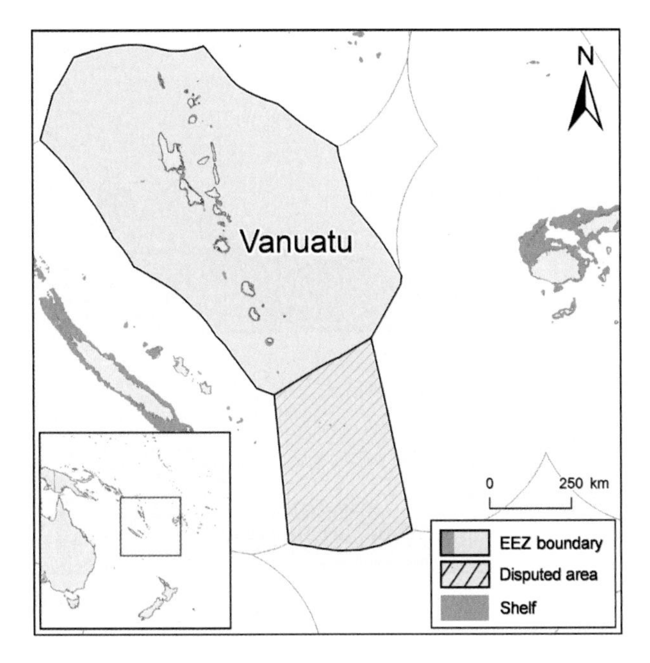

Figure 1. The republic of Vanuatu (land area 12,190 km²) and its claimed EEZ (828,000 km²), which includes the waters surrounding Matthew and Hunter Islands, also claimed by France for New Caledonia.

The Republic of Vanuatu (here Vanuatu, formerly New Hebrides when still under an unusual "condominium" by the UK and France) is an archipelagic country consisting of 83 islands (63 permanently inhabited by 267,000 people) in the southwestern Pacific (figure 1). Vanuatu has a low per capita seafood consumption, but seafood is still important to the local economy. The reconstruction of fisheries catch of Vanuatu by Zylich et al. (2014) suggested that domestic catches dominated by the subsistence sector (figure 2A, although substantial foreign industrial fishing occurs, figure 2B), with domestic catches of 1,750 t/year in the 1950s and 4,800 t/year in the 2000s. Overall, the domestic reconstructed catches are 1.1 times those reported by the FAO on behalf of Vanuatu for 1950 to 2010. However, if only small-scale catches are considered (i.e., excluding industrial tuna and shark fisheries), these domestic catches are 64% higher than the reported catches assigned to the small-scale sector. The subsistence sector was found to be most important among small-scale fisheries (David and Cillauren 1992), contributing almost 84% of their catches.

As illustrated by figure 2C, albacore (*Thunnus alalunga*) and yellowfin tuna (*T. albacares*) dominate total catches taken in this EEZ. Small-scale catches were dominated by invertebrates, including gastropods, pronghorn spiny lobster (*Panulirus penicillatus*), giant clams (*Tridacna* spp.), octopus, and other bivalves. The most important reef fish species was crimson jobfish (*Pristipomoides filamentosus*).

REFERENCES

David G and Cillaurren E (1992) National fisheries development policy for coastal waters, small-scale village fishing, and food self-reliance in Vanuatu. *Man and Culture in Oceania* 8: 35–58.

Zylich K, Shon S, Harper S and Zeller D (2014) Reconstruction of total marine fisheries catches for the Republic of Vanuatu, 1950–2010. pp. 147–156. In K Zylich, D Zeller, M Ang and D Pauly (eds.) *Fisheries catch reconstructions: Islands, Part IV.* Fisheries Centre Research Reports 22(2).

1. Cite as Zylich, K., S. Shon, S. Harper, and D. Zeller. 2016. Vanuatu. P. 453 in D. Pauly and D. Zeller (eds.), *Global Atlas of Marine Fisheries: A Critical Appraisal of Catches and Ecosystem Impacts*. Island Press, Washington, DC.

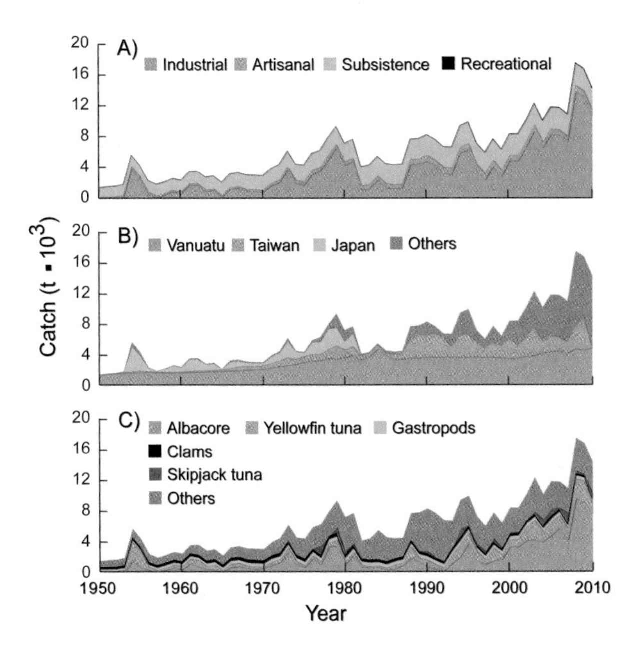

Figure 2. Domestic and foreign catches taken in the EEZ of Vanuatu: (A) by sector (domestic scores: Ind. –, 4, 4; Art. 2, 3, 3; Sub. 2, 3, 3; Rec. –, 2, 2; Dis. –, 3, 3); (B) by fishing country (foreign catches are very uncertain); (C) by taxon.

VENEZUELA[1]

Jeremy J. Mendoza

Instituto Oceanográfico de Venezuela, Universidad de Oriente, Cumaná, Venezuela

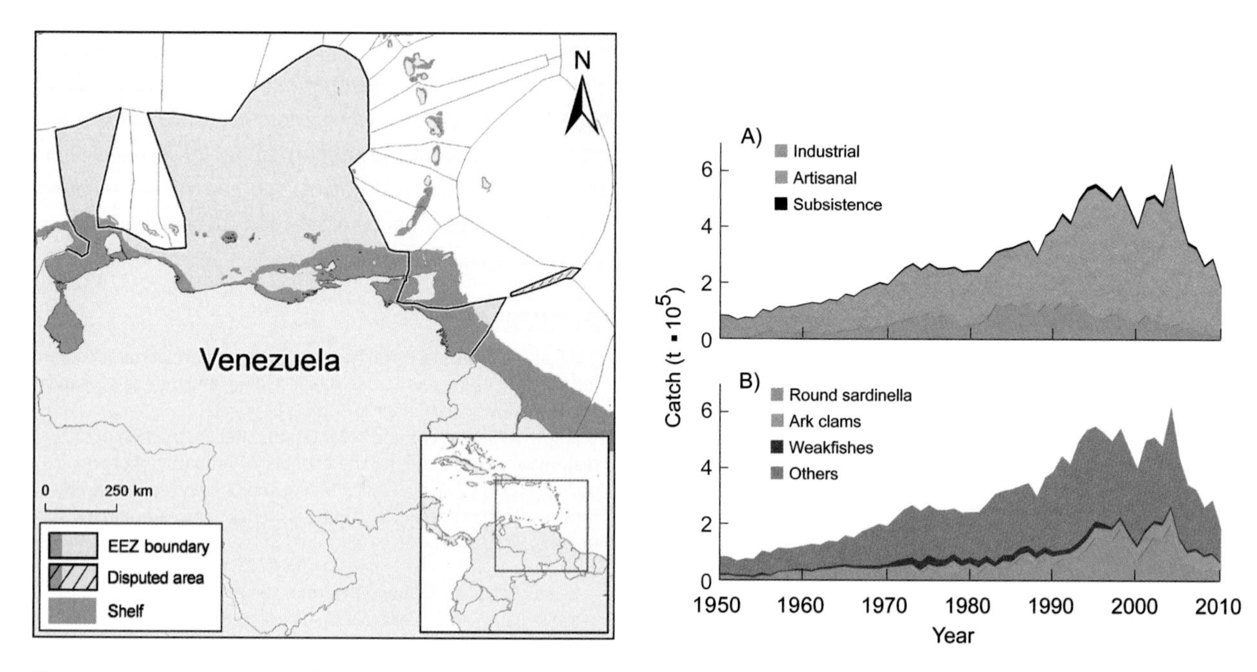

Figure 1. Venezuela has a shelf of 108,000 km² and an EEZ of about 475,000 km². **Figure 2.** Domestic and foreign catches taken in the EEZ of Venezuela: (A) by sector (domestic scores: Ind. 3, 3, 3; Art. 1, 2, 2; Sub. 1, 1, 1; Rec. 2, 3, 3; Dis. 1, 2, 3); (B) by taxon.

The Venezuelan coastline occupies most of the southern margin of the Caribbean Sea (figure 1); consequently, the Venezuelan fisheries, operating on a large continental shelf whose productivity is heightened by seasonal upwelling and river runoff, contribute a substantial proportion of catches from the Caribbean Sea. Statistics from national and international sources, publications, gray literature, and expert opinions were used by Mendoza (2015) to reconstruct Venezuelan marine fisheries catches, which are overwhelmingly artisanal (figure 2A), by estimating landings, unreported catches, and discards at the lowest taxonomic resolution possible. Total catches by all fisheries increased from just over 80,000 t in 1950 to a peak of more than 600,000 t in 2004, then declined to 178,000 t by 2010. Discards reached a maximum of 74,000 t/year in 1989 at the height of the trawl fishery but later declined significantly to just under 1,000 t in 2010, after the closure of the trawl fishery in early 2009 (Mendoza et al. 2010). Overall, reconstructed catches were 1.45 times the data reported by the FAO on behalf of Venezuela. Figure 2B presents the taxonomic composition. The data of Mendoza (2015) showed a sequential rise and decline of the trawl and then artisanal fisheries during the study period, a result of sequential overfishing. The most pressing problem for the Venezuelan fisheries administration is how to regulate the diverse artisanal sector in order to recover overfished populations and prevent further overfishing.

REFERENCES

Mendoza J (2015) Rise and fall of Venezuelan industrial and artisanal marine fisheries: 1950–2010. Fisheries Centre Working Paper #2015–27, 15 p.

Mendoza J, Marcano L, Alió J and Arocha F (2010) Autopsia de la Pesquería de Arrastre del Oriente de Venezuela: Análisis de los Datos de Desembarques y Esfuerzo de Pesca. pp. 69–76 In *Proceedings of the 62nd Gulf and Caribbean Fisheries Institute*.

1. Cite as Mendoza, J. 2016. Venezuela. P. 454 in D. Pauly and D. Zeller (eds.), *Global Atlas of Marine Fisheries: A Critical Appraisal of Catches and Ecosystem Impacts*. Island Press, Washington, DC.

VIETNAM[1]

Lydia C. L. Teh, Dirk Zeller, Kyrstn Zylich, George Nguyen and Sarah Harper

Sea Around Us, University of British Columbia, Vancouver, Canada

Vietnam stretches along the eastern margin of the Indochina Peninsula (figure 1). Fishing and aquaculture have a long tradition in Vietnam, and they have developed into some of the country's top economic performers, despite long-lasting social and political conflicts. As assessed by Teh et al. (2014), marine catches grew slowly from subsistence and artisanal fisheries in the 1950s (figure 2A), when they amounted to about 95,100 t/year. After the transition to a market-oriented economy in the mid-1980s, both inshore and offshore fishing intensified, and catches grew to 1.9 million t/year in the 1990s and to about 2.9 million t/year in the late 2000s, most of it industrial (figure 2A). The decline of reconstructed catches between 1965 and 1975 reflects reduced fishing during the peak of the Second Indochina War, whereas the statistics reported by Vietnam to the FAO (reconstructed catches were 75% higher than reported data) featured a steady and suspiciously regular increase of catches, as they did for the 2000s (and probably requiring stronger downward adjustments than performed here). Yet while Vietnam reported ever-increasing catches to the FAO, the declining catch per unit effort of its trawlers and other industrial vessels suggested a strong decline in fish biomass (Thuoc and Long 1997). The composition of fisheries catches (figure 2B) has shifted from consisting mainly of small and medium inshore pelagics in the 1950s to being dominated by small, low-value fish such as lizardfishes (family Synodontidae) and ponyfishes (Leiognathidae) in 2010.

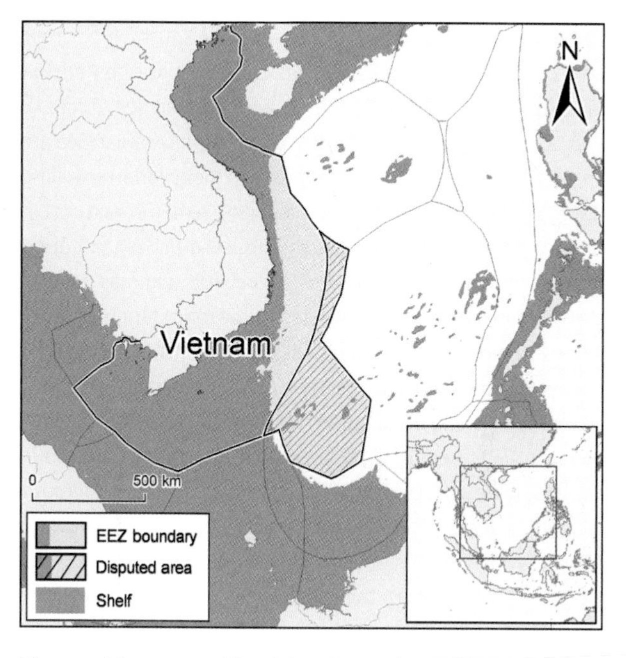

Figure 1. Vietnam and its claimed EEZ of 1.4 million km², of which 432,000 km² is shelf.

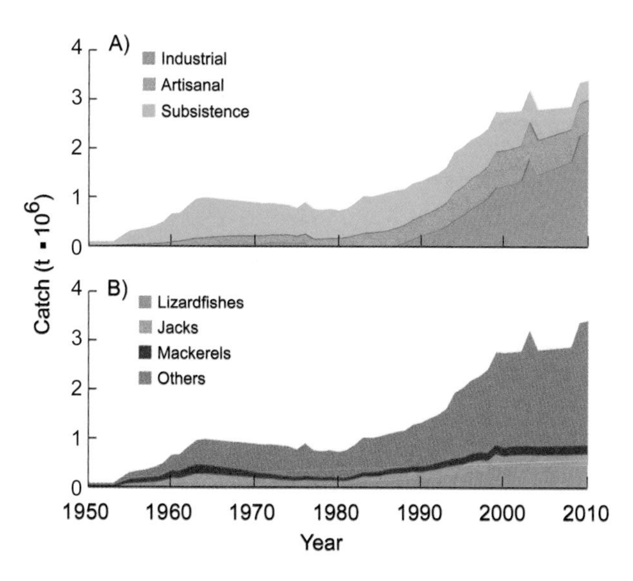

Figure 2. Domestic and foreign catches taken in the EEZ of Vietnam: (A) by sector (domestic scores: Ind. –, 2, 2; Art. 2, 2, 1; Sub. 2, 2, 1; Dis. –, 2, 2); (B) by taxon.

REFERENCES

Teh LCL, Zeller D, Zylich K, Nguyen G and Harper S (2014) Reconstructing Vietnam's Marine Fisheries Catch, 1950–2010. Fisheries Centre Working Paper #2014–17, 11 p.

Thuoc P and Long N (1997) Overview of the coastal fisheries of Vietnam. p. 96–106. In: G. Silvestre and D. Pauly (eds.) *Status and Management of tropical coastal fisheries in Asia*. ICLARM Conference Proceedings 53.

1. Cite as Teh, L. C. L., D. Zeller, K. Zylich, G. Nguyen, and S. Harper. 2016. Vietnam. P. 455 in D. Pauly and D. Zeller (eds.), *Global Atlas of Marine Fisheries: A Critical Appraisal of Catches and Ecosystem Impacts*. Island Press, Washington, DC.

YEMEN (ARABIAN SEA)[1]

Dawit Tesfamichael,[a,b] Peter Rossing,[a] and Hesham Saeed[c]

[a]*Sea Around Us*, University of British Columbia, Vancouver, Canada [b]Department of Marine Sciences, University of Asmara, Eritrea [c]Marine Research and Resource Center, Hodeidah, Yemen

The Republic of Yemen, located on the southwest corner of the Arabian Peninsula, has coastlines on the Red Sea (see next page) and the Gulf of Aden (or Arabian Sea; figure 1). The Gulf of Aden and the adjacent region of the Arabian Sea are characterized by high productivity due to monsoonal upwelling (Morgan 2006; Sanders and Morgan 1989). Yemen's main port in the Gulf is Aden, used mainly by the industrial fishery, targeting mainly the pharaoh cuttle (*Sepia pharaonis*), whereas its substantial artisanal fisheries (figure 2A) use many landing places along its coast. Although substantial foreign fishing (mainly by former Soviet Union republics) occurred in the 1970s and 1980s, less foreign fishing occurs today (figure 2B). The reconstruction by Tesfamichael et al. (2012) suggests that domestic catches in this EEZ were 40,000 t/year in the 1950s, increased to about 100,000 t/year in the early 1990s, mainly because of a trawling fishery (which did not survive the reunification of South and North Yemen in 1990), then massively increased, with peak catches of 430,000 t/year in the mid-2000s and 260,000 t by 2010 (figure 2B). Overall, reconstructed catches are about 2.1 times the officially reported catches, mainly because of unreported subsistence catches and discards of the industrial fishery. Figure 2C presents the taxonomic composition of the catch, consisting mainly of small pelagic fishes and some tuna and sharks.

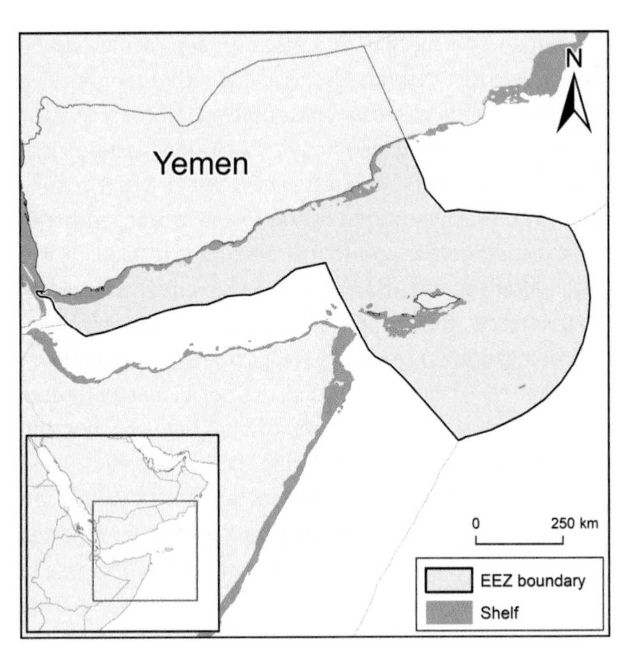

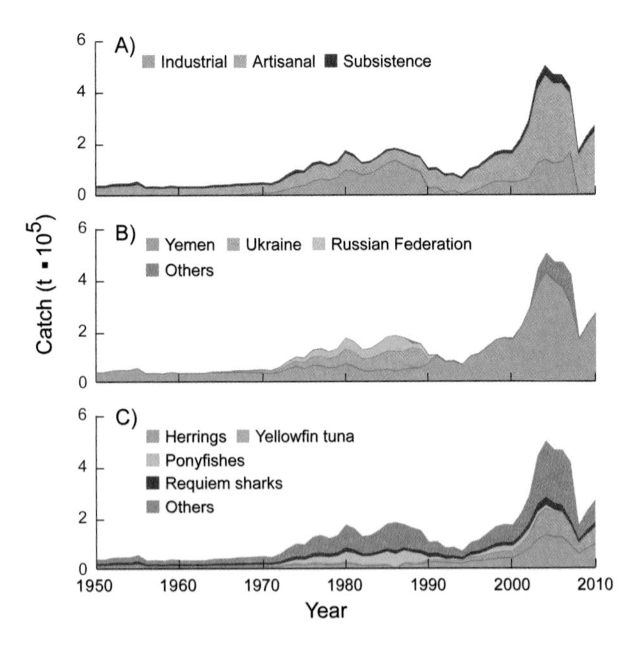

Figure 1. Yemen and its continental shelf (28,600 km²) and EEZ (509,000 km²) in the Gulf of Aden (Arabian Sea). **Figure 2.** Domestic and foreign catches taken in the EEZ of Gulf of Aden (Arabian Sea) EEZ: (A) by sector (domestic scores: Ind. –, –, 3; Art. 3, 3, 3; Sub. 1, 1, 1; Dis. –, –, 2); (B) by fishing country (foreign catches are very uncertain); (C) by taxon.

REFERENCES

Morgan G (2006) Country Review: Yemen, p. 458 In De Young C (ed.) *Review of the state of world marine capture fisheries management: Indian Ocean*. FAO Fisheries Technical Paper. No. 488. FAO, Rome.

Sanders MJ and Morgan GR (1989) *Review of the fisheries resources of the Red Sea and Gulf of Aden*. FAO Fisheries Technical Paper 304, FAO, Rome, 138 p.

Tesfamichael D, Rossing P and Awadh H (2012) The marine fisheries of Yemen with emphasis on the Red Sea and cooperatives. pp. 337–348 In Tesfamichael D and Pauly D (eds.) *Catch reconstruction for the Red Sea large marine ecosystem by countries (1950–2010)*. Fisheries Centre Research Reports 20(1).

1. Cite as Tesfamichael, D., P. Rossing, and H. Saeed. 2016 Yemen (Arabian Sea). P. 456 in D. Pauly and D. Zeller (eds.), *Global Atlas of Marine Fisheries: A Critical Appraisal of Catches and Ecosystem Impacts*. Island Press, Washington, DC.

YEMEN (RED SEA)[1]

Dawit Tesfamichael,[a,b] Peter Rossing,[a] and Hesham Saeed[c]

[a]*Sea Around Us*, University of British Columbia, Vancouver, Canada [b]Department of Marine Sciences, University of Asmara, Eritrea [c]Marine Research and Resource Center, Hodeidah, Yemen

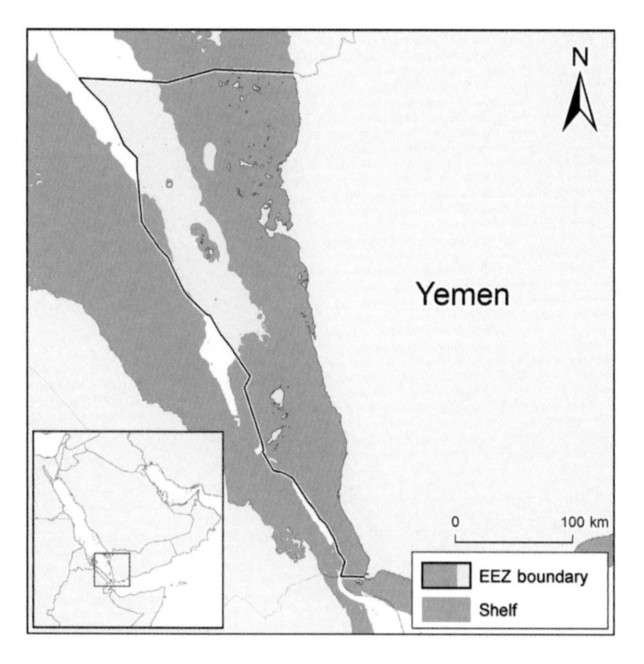

Figure 1. Yemen and its continental shelf (25,900 km²) and EEZ (35,900 km²) in the Red Sea.

The Republic of Yemen, at the southwestern end of the Arabian Peninsula, has access to both the Gulf of Aden (or Arabian Sea, see previous page) and the Red Sea (figure 1). The flow of nutrient-rich water from the Gulf of Aden boosts the productivity of the southern part of the Red Sea but also limits the development of coral reefs, and the fisheries tend to be mainly pelagic (Morgan 2006; Sanders and Morgan 1989). The Yemeni Red Sea fishing fleets, which often stray into neighboring countries, are predominantly artisanal (figure 2A) and organized in cooperatives scattered along the coast. Foreign fishing (mainly by Egypt) may have been common in recent years (figure 2B). The reconstructed domestic catch for Yemen in its Red Sea EEZ (Tesfamichael et al. 2012) was about 10,000 t/year in the 1950s, then increased, especially in the 1990s, until reaching a peak of about 75,000 t/year in 1993 (figure 2B), then declining to about 28,000 t/year by the end of the 2000s. Overall, this was about 1.9 times the officially reported catches, mainly because of discarding by the industrial fishery and the nonreporting of the subsistence fishery. The catch is dominated by pelagic species: Indian mackerel (*Rastrelliger kanagurta*) and narrow-barred Spanish mackerel (*Scomberomorus commerson*), followed by the demersal ponyfish (family Leiognathidae; figure 2C), one of the many groups discarded by the trawl fishery.

REFERENCES

Morgan G (2006) Country Review: Yemen, p. 458 In De Young C (ed.) *Review of the state of world marine capture fisheries management: Indian Ocean*. FAO Fisheries Technical Paper. No. 488. FAO, Rome.

Sanders MJ and Morgan GR (1989) *Review of the fisheries resources of the Red Sea and Gulf of Aden*. FAO Fisheries Technical Paper 304, FAO, Rome, 138 p.

Tesfamichael D, Rossing P and Awadh H (2012) The marine fisheries of Yemen with emphasis on the Red Sea and cooperatives. pp. 337–348 In Tesfamichael D and Pauly D (eds.) *Catch reconstruction for the Red Sea large marine ecosystem by countries (1950–2010)*. Fisheries Centre Research Reports 20(1).

1. Cite as Tesfamichael, D., P. Rossing, and H. Saeed. 2016. Yemen (Red Sea). P. 457 in D. Pauly and D. Zeller (eds.), *Global Atlas of Marine Fisheries: A Critical Appraisal of Catches and Ecosystem Impacts*. Island Press, Washington, DC.

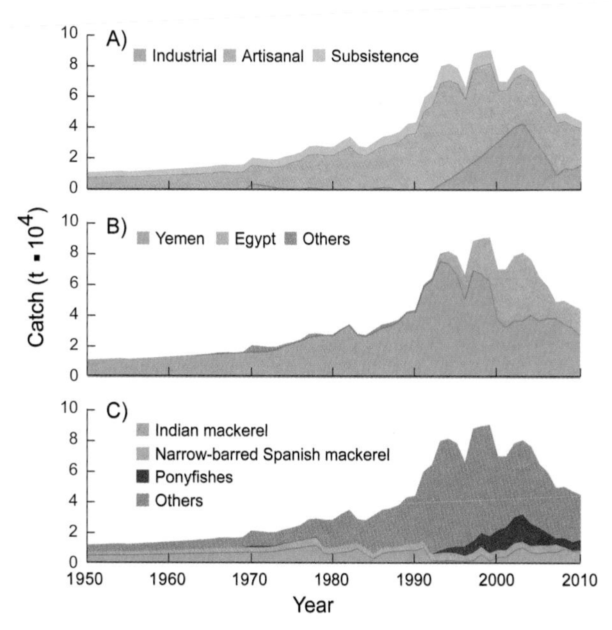

Figure 2. Domestic and foreign catches taken in the EEZ of Yemen (Red Sea): (A) by sector (domestic scores: Art. 3, 3, 3; Sub. 1, 1, 1); (B) by fishing country (foreign catches are very uncertain); (C) by taxon.

ACRONYMS AND GLOSSARY

Adapted from the glossary in FishBase (www.fishbase.org), Wikipedia, and other sources, including Holt, S. J. 1960. *Multilingual Vocabulary for Fishery Dynamics*. FAO, Rome. The symbol * refers to an entry elsewhere in this list.

Acadja: A simple enclosed fish culture technique (also referred to as "brush park") commonly used in West African coastal lagoons (particularly in Benin) but also in Madagascar, Sri Lanka, Bangladesh, Cambodia, China, Ecuador, and Mexico in very shallow waters with little or no tidal change. They can be more than a hectare in size and consist of large sticks being driven into the lagoon floor, with dense bundles of branches (often from *mangroves) bound to the sticks. They serve both as a shelter from predation and for increasing local food production for fish as the nutrient release from the branches increases local primary production. This culture system involves no artificial stocking or feeding and thus is rather a natural fish aggregation and shelter device that is harvested once or twice a year.

Acoustic survey: Surveys of fishing grounds with a vessel equipped with a sound emitting device (i.e., an echosounder), producing sounds that are reflected by fish schools and even single fish (especially if they have a gas bladder), and whose abundance can thus be estimated.

AFMA: *Australian Fisheries Management Authority.

AGRRA: *Atlantic and Gulf Rapid Reef Assessment.

Aquaculture: According to *FAO, aquaculture is the farming of aquatic organisms including fish, molluscs, crustaceans, and aquatic plants, with some sort of intervention in the rearing process to increase production, such as regular stocking, feeding, or protection from predators. See also *Mariculture.

Aquatic Sciences and Fisheries Abstracts (ASFA): ASFA is an extremely useful International Cooperative Information System that comprises a world-leading abstracting and indexing service covering the world's literature on the science, technology, management, and conservation of marine, brackish water, and freshwater resources and environments, including their socioeconomic and legal aspects. It is maintained by and based at *FAO. ASFA is produced as a cooperative effort by the international network of ASFA Partners, which consists of United Nations Co-sponsoring Partners, National and International Partners, and the Publishing Partner. The objective is to disseminate bibliographic information to the world community.

Arctic: The region of the earth defined as north of 66°33′44″N, i.e., north of the Arctic Circle.

Areas beyond national jurisdiction (ABNJ): See *High sea(s).

Arrowhead trap: A type of fish trap widespread in the Caribbean.

Artisanal: Referring to small-scale fishers who are catching fish that is predominantly sold, i.e., small-scale commercial fishers. The distinction between small-scale and large-scale (i.e., *industrial) in this atlas is, for each country, the definition prevailing in that country or a regional equivalent. "Artisanal" roughly correspond to "traditional" in Malaysia, "municipal" in the Philippines, and *petits métiers* in France. Also see chapter 2.

ASFA: *Aquatic Sciences and Fisheries Abstracts.

Atlantic and Gulf Rapid Reef Assessment (AGRRA): The AGRRA Program is an international collaboration of scientists and managers aimed at determining the regional condition of reefs in the western Atlantic and Gulf of Mexico. It is the first and only program that has developed an extensive regional database on Caribbean coral reef condition. Using an innovative regional approach to examine the condition of reef-building corals, algae, and fishes, AGRRA's teams of reef scientists have assessed more than 800 reef sites throughout the Caribbean region. Preliminary findings have already provided valuable baseline data for scientists and government officials responsible for selecting marine protected areas and maintaining their condition. The goals of the AGRRA Project are to complete the regional assessment of the health of coral reefs throughout the western Atlantic, analyze the results and develop a database so as to establish a practical scale of comparative reef condition, and promote the transfer of this information to a wider audience including the general public, resource managers, government officials, policy makers, tourist operators, and students.

Atoll: An island (or group of islands), mostly surrounded by deep water, and consisting of a ring-like perimeter of shallow coral reefs enclosing a shallow *lagoon, the entire structure sitting on the top of a sunken volcano.

Australian Fisheries Management Authority (AFMA): The Australian federal government agency responsible for the efficient management and sustainable use of commonwealth fish resources on behalf of the Australian community. AFMA deals with commercial fisheries from 3 nmi out to the extent of the Australian Fishing Zone (i.e., EEZ). The individual states and the Northern Territory look after recreational, commercial coastal, and inland fishing and aquaculture. Through AFMA's foreign compliance functions, they work with other Australian government agencies and international counterparts to deter illegal fishing in the Australian Fishing Zone.

Bait/baitfish: Fish used to catch other fishes, e.g., in pole and line fishing, or in longlining for tuna. Here, unreported baitfish catches were mostly taken into account in reconstructing fisheries catches.

Bay of Bengal Large Marine Ecosystem Project (BOBLME): Bangladesh, India, Indonesia, Malaysia, Maldives, Myanmar, Sri Lanka, and Thailand are collaborating through the BOBLME Project to better the lives of their coastal populations by improving regional management of the Bay of Bengal environment and its fisheries. The project promotes ecosystem-based management approaches for sustaining some of the most important shared fish stocks, including hilsa shad, Indian mackerel, and sharks.

Beach seining: A fishing method where a net and a length of rope are laid out from and back to the shore and retrieved by hauling the net onto the shore. Often, the hauling is performed by a large group of people (e.g., from a village community), with the fish that are caught shared between them. Beach seines are problematic in that they catch juvenile fish and thus contribute to *growth overfishing.

Benthos: The community of organisms that live on the bottom of a water body, in it, or near it.

Benthopelagic: Community of organisms that live and opportunistically feed either near the bottom on benthic (see *benthos) or on free-swimming organisms in midwater (see *pelagic). Also used as term describing the depth zone of 100 m above seafloor for waters beyond (deeper than) the continental *shelf.

Bioaccumulation: The increase of a concentration of substances such as pesticides, or other organic chemicals in organisms, due to absorption by the organism occurring at rates greater than that at which the substances are lost by the organism. Thus, the longer the biological half-life of the substance, the greater the risk of chronic poisoning, even if environmental levels of the toxin are not very high (see chapter 13).

Biodiversity: The full range of living organisms, including terrestrial, marine, and other aquatic ecosystems and the ecological complexes of which they are part; this includes diversity within species, between species, and of ecosystems.

Biomagnification: Also known as bioamplification, this process occurs when the concentration of a substance, such as *dioxin, *PCB, or mercury in an organism exceeds the background concentration of the substance in its diet. This increase can occur as a result of persistence (i.e., the substance cannot be broken down by environmental processes), food chain energetics (i.e., the substance concentration increases progressively as it moves up a food web), or a low or nonexistent rate of internal degradation or excretion of the substance (e.g., due to water insolubility). See chapter 13.

Biomass: Weight of a *stock or of one of its components; e.g., "spawning biomass" is the combined weight of all sexually mature animals in a stock. **Standing stock* is an alternative term for biomass. Also, the mass of living tissues across organisms in a *population or *ecosystem. Used as a measure of population abundance.

Black fish: A term used in some countries for the illegally caught fish that are sold on the *black market.

Black market: The (virtual) place where illegally caught fish (*black fish) are sold.

B_{MSY}: Biomass level at which an exploited stock generates *maximum sustainable yield (*MSY), i.e., generally half the unexploited biomass (see box 1.1, chapter 1).

BOBLME: *Bay of Bengal Large Marine Ecosystem Project.

Brackish/brackishwater: Water or water bodies with a salinity higher than freshwater but lower than seawater, as occurs e.g., in *estuaries or *lagoons, when the salinity of the water can vary considerably over space or time. Brackish water contains between 0.5 and 30 grams of salt per liter, often expressed as 0.5 to 30 parts per thousand (ppt or ‰), but for which the correct current expression is *Practical Salinity Unit, or *psu.

Bycatch: The part of a fish catch that is caught in addition to the *target species because the fishing gear (e.g., a *trawl) is not selective. Bycatch may be retained, landed and sold or used, or dumped at sea (see *discard).

Capacity (fleet): In input terms, fleet capacity can be considered the minimum fleet size and *effort needed to generate a given catch. In output terms, capacity can be considered the maximum catch that a fisher or a fleet can produce with given levels of inputs, such as fuel, amount of fishing gear, ice, bait, engine horsepower, and vessel size.

Caribbean Coastal Marine Productivity Program (CARICOMP): The CARICOMP Program is a Caribbean-wide research and monitoring network of 27 marine laboratories, parks, and reserves in 17 countries. It consists of datasets collected from 42 stations at 29 sites in the Caribbean from 1993 to 1998. Line transects were used to determine the abundance of hard and soft corals, algae, sponges, urchins, and abiotic material such as substrate type. Data sets are archived through the Caribbean Coastal Data Centre at the University of the West Indies at Mona, Jamaica.

CARICOMP: *Caribbean Coastal Marine Productivity Program.

Cascade (trophic): Trophic cascades occur when predators in a *food web suppress the abundance or alter the behavior of their prey, such that the next lower *trophic level (i.e., the prey of the prey) is released from predation (or *herbivory if the intermediate trophic level is an herbivore). For example, if the abundance of large piscivorous fish is increased, the abundance of their prey, zooplanktivorous fish, should decrease, large *zooplankton abundance should increase, and *phytoplankton biomass should decrease. This concept has stimulated research in many areas of marine ecology and fisheries biology.

Catch: The number or weight of fish or other animals caught or killed by a fishery, including fishes that are landed (*landings, whether reported in statistics or not), discarded at sea (*discard), or killed by lost gear (ghost fishing).

Catch composition: The different taxa (species, genera, family; see *taxon) making up the catch of a fishery. The more detailed a catch composition is, the more useful it is. When their composition is unavailable, catches are often labeled as "miscellaneous fish" or similarly uninformative labels.

Catch per unit of effort (CPUE; or catch/effort): A measure of relative abundance, obtained by dividing the *catch by a measure of the fishing *effort needed to realize this catch. CPUE is generally proportional to *biomass.

Cay: A low bank or reef of rock, sand or coral; see also *key.

CCAMLR: *Commission for the Conservation of Antarctic Marine Living Resources.

CCCR: *Climate Change and Coral Reefs Marine Working Group.

CFP: *Common Fisheries Policy (of the European Union).

Ciguatera: An illness resulting from the consumption of reef fishes contaminated by *Gambierdiscus toxicus* and related dinoflagellates, which tend to appear in reef fishes after coral reefs have been stressed (e.g., by typhoons or fisheries). Ciguatoxicity is reported from reef-associated fishes of the Indian and Pacific Oceans and from the Caribbean; see "France (Saint Barthélemy)" in Part II, this volume, for more information.

Climate change: A lasting change in the statistical distribution of weather patterns over periods ranging from decades to millions of years. It may be a change in average weather conditions or in the distribution of weather around the average conditions (i.e., more or fewer extreme weather events such as storms or heat waves). Climate change is caused by factors such as variations in solar radiation received by Earth, plate tectonics, and volcanic eruptions. The release of greenhouse gases (notably carbon dioxide and methane) by human activities has been identified as the cause of recent climate changes, often referred to as global warming.

Climate Change and Coral Reefs Marine Working Group (CCCR): The objective of the CCCR *working group of the *IUCN is to use coral reefs as a model *ecosystem to identify priority information gaps and issues to be addressed through parallel workshop and research tracks, provide a mechanism to focus scientific contributions from different leading research groups, and synthesize the relevance of resilience to coral reefs and *climate change. The working group thus aims to bridge gaps between theoretical science and management application in order to fast-track the development and use of tools that would improve the protection of coral reefs under the threat of climate change and interacting or synergistic human threats. It provides a forum to facilitate the reciprocal flow of information between scientists and managers to continually update and improve recommended management practices for mitigating climate change threats to coral reefs. At government and intergovernmental levels, policy outputs from the working group will seek to inform at the highest levels the possibilities of mitigating climate change impacts on coral reefs and thereby empower management efforts at local, national, and regional levels.

Coastal zone: The region where interactions of the sea and land processes occur.

Coastline (length of): Never mentioned in this atlas, even if our sources mentioned such lengths, because coastline lengths can be evaluated only if the length of the "stick" used to measure coastline length is given—which it never is, because the authors are not aware of the "coastline paradox" (see Wikipedia).

Collapse(d): Rapid decline in the abundance (*biomass) of a stock, generally reflected in a rapid decline of *catches from that *stock because there are fewer fish to be caught than previously. In *SSPs, collapsed stock are defined as stocks with catches of less than 10% of the historically maximum catch.

Commercial: Refers to a fishery whose catch is sold. This means that both *large-scale (or *industrial) and *small-scale fisheries (i.e., *artisanal, or *petit métiers*) are commercial fisheries and that the term *commercial fisheries* should not be considered synonymous with industrial or large-scale fisheries.

Commercial fisher: A person who fishes for a living and sells most of his or her catch.

Commission for the Conservation of Antarctic Marine Living Resources (CCAMLR): CCAMLR was established by international convention in 1982 with the objective of conserving Antarctic marine life. This was in response to increasing commercial interest in Antarctic krill resources, a keystone component of the Antarctic ecosystem and a history of overexploitation of several other marine resources in the Southern Ocean. CCAMLR is an international commission with 25 members, and another 11 countries have acceded to the convention. Based on the best available scientific information, the commission agrees a set of conservation measures that determine the use of marine living resources in the Antarctic.

Common Fisheries Policy (CFP): The policy governing the marine fisheries in the 25 million km^2 of EEZs jointly held by the member states of the European Union (EU). The CFP covers a range of activities, of which the main one is probably the setting of annual *quotas (suggested by *ICES), and allocating them to the fleets of its maritime member states. The CFP, whose policy outcome had been widely criticized, was reformed in 2013 through an act of the EU Parliament, with the result that more emphasis will be given to stock *rebuilding and the gradual banning of *discards.

Compliance: Adherence of individual fishers to the harvest strategies mandated by national or international fisheries management bodies.

Continental shelf: see *shelf.

CPI: The Consumer Price Index is an indicator of changes in consumer prices experienced over time. It can be used to adjust for the effect of inflation to obtain the "real" values of economic indicators, i.e., values not distorted by inflation.

CPUE: *Catch per unit of effort, or catch/effort.

Demersal: Organisms swimming just above or lying on the seafloor and usually feeding on *benthic organisms.

Dioxin: A group of 17 chemical forms with varying toxicological effects, with the most toxic form 2,3,7,8-tetrachlorodibenzo-*p*-dioxin being classified as a human carcinogen (see also chapter 13).

Discard: Portion of catch that is thrown overboard, but which may be of important ecological or commercial value. Discards typically consist of nontarget species or undersized specimens of the target species. *High-grading is a special form of (mostly illegal) discarding where a catch of target species is thrown overboard to make space in the hull for (or accommodate under a *quota) fresher, larger, or otherwise more valuable catch of the same species. See also chapters 2 and 14.

Distant-water fleet/fishery: The fleet of a country that is fishing in the *EEZ of another country (or the EEZs of other countries), or in *high seas regions not adjacent to its own EEZ. Under *UNCLOS, a distant-water fishery can be conducted in the EEZ of a coastal state only with its explicit access permission, generally in exchange for compensation.

Domestic: Here, pertaining to a country's or territory's own *EEZ.

Driftnet: Net hanging vertically in the water column, without being anchored to the bottom. The nets are kept vertical in the water by floats attached to a rope along the top of the net and weights attached to another rope along the bottom of the net. Driftnets generally rely on the entanglement properties of loosely affixed netting. Folds of loose netting, much like a window drapery, snag on a fish's fins and tail and wrap it up in loose netting as it struggles to escape. However, driftnets can also function as gill nets if fish are captured when their head gets stuck in its mesh. Driftnets are unselective and thus kill thousands of marine mammals, turtles and *seabirds, beside the fish *bycatch. Before the 1960s, the size of driftnets was not limited, and they grew to lengths of more than 50 km. In 1992, the UN banned the use of driftnets longer than 2.5 km long in the *high seas.

DWF: *Distant-water fleet or fishery.

Ecopath: An approach and software package allowing the straightforward construction of a mass-balance models (quantified representations) of the trophic linkages in (aquatic) ecosystems at a given time or during a given period (see chapters 9 and 13).

Ecopath with Ecosim: An ecosystem modeling software currently integrating *Ecopath, *Ecosim, and *Ecospace (see chapters 9 and 13).

Ecosim: An add-on to *Ecopath that uses its parametrization to define a system of differential equations allowing changes in, e.g., fishing tactics or environmental forcing to be evaluated in terms of their effects on the time series dynamics of an ecosystem as a whole (see chapter 9).

Ecospace: An add-on to *Ecosim that allows the processes it simulates to be represented spatially.

Ecosystem: Community of plants, animals, and other living organisms, together with the nonliving components of their environment, found in a particular habitat and interacting with each other.

Ectotherms: Animals whose body temperature is determined largely by the temperature of their habitat. Most fishes and all invertebrates are ectotherms, whereas (*marine) mammals and (*sea)birds are endotherms.

EEZ: *Exclusive Economic Zone.

Effort (fishing): Any activity or devices deployed to catch fish that can be quantified. Thus, the number of nets of a certain type deployed in a set time period is a measure of effort, as is the amount of fuel used by a fishing fleet or the days fished per time period.

Endemic: Native and restricted to a particular area, e.g., an island, a country, a continent, an ocean.

Estuary: Wide *brackish mouth of a river, resulting in transitional environments between freshwater and saltwater.

EU: *European Union.

European Union: A unique political and economic union and partnership of 28 states in Europe that functions through intergovernmental negotiated decisions and supranational institutions (e.g., EU Parliament, EU Commission). The EU encompasses more than 4 million km^2 and has more than 500 million inhabitants, and as a single market is a major world trading power.

Eutrophication: A condition caused by the oversupply of nutrients such as nitrates or phosphates, inducing growth of *phytoplankton in excess of what can be grazed by zooplankton. Upon dying, these excess phytoplankton support bacterial growth, which consumes all the oxygen in a body of water, thereby creating "dead zones."

EwE: *Ecopath with *Ecosim (and *Ecospace).

Exclave: A part of a country that is physically separated from its main territory by the territory of another country. Examples are Kaliningrad Oblast (Russia), Musandam (Oman), and Oecusse (Timor-Leste).

Exclusive Economic Zone (EEZ): Generally, all waters within 200 nautical miles (370 km) of a country and its outlying islands, unless such areas would overlap because neighboring countries are less than 400 nautical miles (740 km) apart. If an overlap exists, it is up to countries to negotiate a delineation of the actual maritime boundary. Under *UNCLOS, a country has special rights regarding the exploration and use of marine resources inside its EEZ, such as the power to control and manage all fishery resources in this zone. Not until 1982, with the adoption of UNCLOS, did 200-nmi EEZs become formally adopted, and a country must formally declare its EEZ.

Ex-vessel price: The price fishers get for a unit weight of fish landed at the dock or beach, i.e., at the first point of sale, corresponding to farmgate prices in agriculture.

FADs: *Fish aggregating devices.

FAO: *Food and Agriculture Organization of the United Nations.

Feeding (or trophic) interactions: Linkages between (groups of) species due to grazing (in herbivores) or predation (in other animals), whose strength defines the dependencies of these species on each other. Feeding interactions are stronger within *ecosystems than between them, and this largely defines their borders.

FFA: *Forum Fisheries Agency.

Finfish: Members of the taxonomic (*taxon) class *Pisces*, i.e., aquatic animals with fins, as distinguished from *shellfish.

Finning: Removing (by cutting) the valuable fins off freshly caught sharks, often with subsequent *discarding of the carcasses.

Fish/Fishes: The term *fish* sensu stricto refers to the taxonomic (*taxon) class *Pisces* (*finfish) in the subphylum Vertebrata, phylum Chordata. In the wider sense, *fish* refer to aquatic animals sought by fisheries, i.e., *finfish + invertebrate *shellfish; the plural *fishes* is used when explicitly referring to more than one species of *finfish.

Fish aggregating devices (FADs): Floating objects made of vegetable matter (e.g., palm fronds) or artificial materials (e.g., plastic, steel, and even concrete), some anchored on the seafloor, which attract *pelagic fishes, mainly tuna, based on their propensity to congregate under floating debris (see also chapter 3). Some FADs are now equipped with sensors (linked to satellites), to assess when they can be fished. FADs attract juvenile fish and thus can contribute to *growth overfishing.

Fisheries-independent data: Information about fishery resources not based on catches and derived statistics such as the *catch/effort (or *CPUE) of fishing vessels. Typically, fisheries-independent data are obtained from dedicated research vessels performing *trawling or *acoustic surveys and from fish *tagging operations. However, it can involve remote sensing (satellite) data or shore-based surveys, e.g., of human fish consumption.

Fisherman: A person who fishes, usually professionally (or for *subsistence or recreation). The term *fisherman* is commonly replaced by *fisher* to recognize the fact that many women are also engaged in fishing.

Fishery: A set of people and gear interacting with an aquatic resource (one or several species of *fish) for the purpose of generating a *catch.

Fishing down (marine food webs): The process whereby fisheries in a given ecosystem, having depleted the large predatory fish on top of the food web, turn to increasingly smaller species, finally ending up with previously spurned small fish and invertebrates, as demonstrated for multiple countries in this atlas. See also Wikipedia entry on this, and www.fishingdown.org.

Fishing effort: *Effort (fishing).

Fishmeal: Protein-rich animal feed product based on ground-up fish, usually small *pelagic fishes such as anchovies and sardines, which are also directly consumed by people.

Fixed cost: Expenses incurred by a fishery that are not dependent on the level of fishing that takes place, e.g., the cost of owning a fishing vessel; see also *variable cost.

F_{MSY}: The value of fishing mortality *F* that produces the maximum yield in the long-term, i.e., *MSY.

Food and Agriculture Organization of the United Nations (FAO): The only United Nations agency in the world tasked with annually assembling global fisheries statistics and generally assisting member countries in managing their fisheries. FAO leads international efforts to defeat hunger.

Food security: According to the *FAO, this occurs "when all people, at all times, have physical and economic access to sufficient, safe and nutritious food to meet their dietary needs and food preferences for an active and healthy life." Seafood contributes crucially to food security in numerous countries where alternative sources of animal protein and micronutrients are lacking.

Food web: The ensemble of *feeding (or trophic) interactions connecting the elements of (and defining) an *ecosystem.

Forage fish: Small pelagic fish, often also called bait fish or prey fish, which are a major food item for larger predators, such as larger fish, *seabirds, and *marine mammals. Forage fish usually feed near the base of the food chain, on *plankton.

Forum Fisheries Agency: The Pacific Islands Forum Fisheries Agency (FFA), based in Honiara, Solomon Islands, was established in 1979 to help its 17 Pacific member countries (Australia, Cook Islands, Federated States of Micronesia, Fiji, Kiribati, Marshall Islands, Nauru, New Zealand, Niue, Palau, Papua New Guinea, Samoa, Solomon Islands, Tokelau, Tonga, Tuvalu, and Vanuatu) sustainably manage their offshore (tuna) resources that fall within their 200-nmi EEZs. FFA is an advisory body providing expertise, technical assistance, and other support to its members who make sovereign decisions about their tuna resources and participate in regional decision making on tuna management through agencies such as the *Western and Central Pacific Fisheries Commission (WCPFC).

Game fishing: A form of *recreational fishing wherein large fish (e.g., tuna, marlins) are targeted, usually by deploying an array of high technologies against an animal with a brain the size of a pea. Such game fishing is the equivalent of playing 3-D chess against a ground sloth.

Growth overfishing: The catching of *fish that are too small relative to their growth potential, even when account is taken of their natural mortality. Growth overfishing is caused by the excessive deployment of gear catching undersize fish.

Herbivory: Feeding only on organisms with a *trophic level of 1, i.e., plants.

High-grading: See *discards.

High sea(s): The areas of the world ocean that is outside of the 200-nmi *EEZs of coastal states; the high seas cover about 60% of the world ocean.

IATTC: *Inter-American Tropical Tuna Commission.

ICES: *International Council for the Exploration of the Sea.

IFA: *Inshore Fishing Area.

IFREMER: *Institut Français de Recherche pour l'Exploitation de la Mer.

Illegal: Fishing in violation of the laws of a fishery, either under the jurisdiction of a coastal state (i.e., within an *EEZ) or in high seas fisheries regulated by *regional fisheries management organizations (RFMOs). The *Sea Around Us* defines "illegal" as fishing in the *EEZ waters of another country without explicit or implicit permission and thus does not include noncompliance with domestic fishing laws by domestic fleets in their own *EEZ (i.e., poaching) under this definition, nor does it include violations of RFMO regulations in high seas fisheries.

Illegal, Unreported, and Unregulated: *IUU.

Individual transferable quota (ITQ): ITQs are a type of catch share used by many governments to regulate commercial fishing. The regulatory agency sets a species-specific *total allowable catch (TAC), typically by weight and for a certain time period (e.g., annually). A percentage of the TAC, called quota "shares", is then allocated to individual fishers or fishing entities (e.g., companies, communities). ITQ are transferable, i.e., they can typically be sold, bought, or leased, and thus often end up being concentrated in the hands of financial operators (banks, hedge funds), with the fishers becoming hired (and fired) hands.

Industrial: Referring to *large-scale fisheries that are catching fish for commercial marketing or global export, i.e., large-scale commercial fisheries. The distinction between large-scale and *small-scale (i.e., *artisanal) in this atlas is, for each country, the definition prevailing in that country and is usually related to vessel size and gear type used. Note, however, that we define any gear type that is actively moved across the seafloor (e.g., bottom and shrimp trawl) or through the water column (e.g., midwater trawl) as industrial, irrespective of vessel size or local definition (see also chapter 2).

Inshore Fishing Area (IFA): The Inshore Fishing Area (IFA) represents the area between the shoreline and either 200 m depth or 50 km distance from shore, whichever comes first. We restrict *small-scale (i.e., *artisanal, *subsistence, and *recreational) fisheries to the IFA waters of each country.

Institut Français de Recherche pour l'Exploitation de la Mer (IFREMER): *IFREMER is an oceanographic institution in France (English: French Research Institute for Exploitation of the Sea) that focuses its research activities on monitoring, use and enhancement of coastal seas, aquaculture, fishery resources, ocean biodiversity, and oceanography.

Institut de Recherche pour le Développement (IRD): IRD, the Institute of Research for Development, is a French public science and technology research institute under the joint authority of the French ministries in charge of research and overseas development. It is the successor of the *Office de la Recherche Scientifique et Technique Outre-Mer (ORSTOM).

Inter-American Tropical Tuna Commission (IATTC): An international commission (a *regional fisheries management organization) responsible for the conservation and management of tuna and other marine resources in the eastern Pacific Ocean. The IATTC was created by the Convention for the Establishment of an Inter-American Tropical Tuna Commission, signed between the United States and Costa Rica on May 31, 1949. In 2003, the members of the IATTC signed the Antigua Convention, which strengthened the commission's powers. Most members of the commission ratified the Antigua Convention between 2004 and 2009, but as of 2011, the United States had not ratified the Antigua Convention. The headquarters of the IATTC are located in La Jolla, San Diego, California. It has more than 20 member countries.

International Council for the Exploration of the Sea (ICES): The International Council for the Exploration of the Sea (*ICES, founded in 1902) is the oldest intergovernmental organization in the world concerned with marine and fisheries science. It conducts science and develops advice to support the sustainable use of the oceans in Europe, for the European Union and associated partner countries. ICES is headquartered in Copenhagen, Denmark, and consists of a network of more than 350 marine institutes in 20 member countries and beyond. Their work also extends into the Arctic, the Mediterranean Sea, the Black Sea, and the North Pacific Ocean.

International Criminal Police Organization (Interpol): An intergovernmental organization facilitating international police cooperation. It was established as the International Criminal Police Commission (ICPC) in 1923. Interpol is funded mostly through annual contributions by 190 member countries. The organization's headquarters is in Lyon, France. It is the second largest international organization after the United Nations in terms of international representation. To keep Interpol as politically neutral as possible, its charter forbids it, at least in theory, from undertaking interventions or activities of a political, military, religious, or racial nature or involving itself in disputes over such matters. Its work focuses primarily on public safety and battling terrorism, crimes against humanity, environmental crime, genocide, war crimes, organized crime, piracy, illicit traffic in works of art, illicit drug production, drug trafficking, weapon smuggling, human trafficking, money laundering, child pornography, white-collar crime, computer crime, intellectual property crime, and corruption.

International Union for Conservation of Nature (IUCN): IUCN helps the world find pragmatic solutions to the most pressing environment and development challenges and focuses on valuing and conserving nature, ensuring effective and equitable governance of its use, and deploying nature-based solutions to global challenges in climate, food, and development. IUCN is the world's oldest and largest global environmental organization, with more than 1,200 government and nongovernment members and almost 11,000 volunteer experts in some 160 countries. IUCN's work is supported by more than 1,000 staff in 45 offices and hundreds of partners in public and private sectors around the world.

Interpol: *International Criminal Police Organization.

IRD: *Institut de Recherche pour le Développement.

Isobath: An imaginary line on a map or chart that connects all points having the same depth below the water surface, also called depth contour.

ITQ: *Individual Transferable Quota.

IUCN: *International Union for Conservation of Nature.

IUU: *Illegal, Unreported and Unregulated, an acronym proposed by *FAO to describe problematic fisheries and catches. This acronym is not used in this atlas because it has become synonymous with *illegal* in practice, and thus confuses people.

Key: A low-lying island or reef (especially in the Caribbean); see also *cay.

Lagoon: Smaller water body, associated with a coastline or coral reef. Coastal lagoons are formed behind permanent or occasionally (or seasonally) occurring sand bars, and their salinity (and hence the flora and fauna of costal lagoons) depend on their water exchange with the land interior (rivers) and the sea (breaks in the sand bars). Coastal lagoons can be very productive, e.g., along the coast of the Gulf of Guinea. The shallow water bodies whose periphery is defined by *atolls are also called lagoons, and their productivity, particularly if the water exchanges with the outside, can also be high. Both lagoon types are vulnerable to rain-induced, sudden drops in salinity, which can kill the resident organisms.

Landings: Weight of the catch landed at a wharf or beach. Also, the number or weight of fish unloaded at a dock by *commercial fishers or brought to shore by *subsistence and *recreational fishers for personal use. Landings are reported at the points at which fish are brought to shore. Note that the catch = landings + *discards.

Large Marine Ecosystem: Large Marine Ecosystems (*LMEs) are large areas of ocean space (often about 200,000 km^2 or greater), adjacent to the continents in coastal waters where *primary productivity is generally higher than in open ocean areas. The delineation and definition of individual LMEs and their boundaries are based on four ecological, rather than political (see *EEZ) or economic, criteria. These are bathymetry, hydrography, productivity, and trophic relationships. Based on these ecological criteria, as of Spring 2015, 66 distinct LMEs have been delineated around the coastal margins of the Atlantic, Pacific, and Indian Oceans.

Large-scale (fishery): *Industrial.

Leeward: The side (e.g., of an island) facing the direction opposite to that of the prevailing winds; the opposite of *windward.

Lessepsian migrants: Organisms that have migrated from the Red Sea into the Mediterranean through the Suez Canal or, much more rarely, from the Mediterranean into the Red Sea.

LME: *Large Marine Ecosystem.

Longline/Longlining: A line of great length, bearing numerous baited hooks, which is usually set horizontally in the water column, used, e.g., in snapper, grouper, ling, and tuna fisheries. The line is set for varying periods up to several hours on the seafloor, or in the case of tuna, in midwater at various depths. Also, a fishing line with baited hooks set at intervals on branch lines; it may be 150 km long and have several thousand hooks and can be on the seabed or above it supported by floats. It may be anchored or drifting free and is marked by floats.

Mangrove: Trees and shrubs that grow in saline coastal sediment habitats in the tropics and subtropics, mainly between latitudes 25°N and 25°S. Mangroves are saline woodland or shrubland habitats characterized by depositional coastal environments, where fine sediments (often with high organic content) collect in areas protected from high-energy wave action. The saline conditions tolerated by various mangrove species range from brackish water, through pure seawater, to water concentrated by evaporation to more than twice the salinity of ocean seawater. Mangroves are crucial juvenile nursery habitats for many fisheries species and also fulfill a very important coastal protection function, where they, similar to healthy coral reefs, shield coastlines from the impacts of high-energy ocean surges and tropical storm damage.

Mariculture: The farming of aquatic organisms in seawater, such as fjords, inshore, and open waters in which the salinity generally exceeds 20‰. Earlier stages in the life cycle of these aquatic organisms may be spent in *brackish water or freshwater. See also chapter 12.

Marine mammals: A diverse group of more than 100 species that rely on the ocean for their existence and include seals, whales, dolphins, porpoises, manatees, dugongs, otters, walruses, and polar bears. They do not represent a distinct biological grouping but rather are unified by their reliance on the aquatic environment for feeding, although their dependence on the aquatic ecosystems varies widely between species. Marine mammals can be subdivided into four major groups: cetaceans (whales, dolphins, and porpoises), pinnipeds (seals, sea lions, and walruses), sirenians (manatees and dugongs), and fissipeds, which are the group of carnivores with separate digits (the polar bear and two species of otters). Both cetaceans and sirenians are fully and obligate ocean dwellers. Pinnipeds are semi-aquatic, i.e., they spend most of their time in the water but need to return to land for mating and breeding. Otters and the polar bear are much less modified for ocean living. Although the number of marine mammals is small compared with land mammals, their total biomass is large. The hunting of marine mammals has both an artisanal or subsistence component and a (highly destructive) industrial component. Marine mammals are not included in our reconstructed catch data.

Marine protected area (MPA): Area of the ocean within which fishing and other extractive activities are limited. Often used to mean *no-take area*, where no fishing is allowed, but for which the term *marine reserve is more appropriate.

Marine reserve: A form of *marine protected area that includes legal protection against fishing (i.e., is designated a no-take area), i.e., an area of ocean space where all fishing is prohibited. Benefits include increases in the biodiversity, abundance, biomass, body size, and reproductive output of fisheries populations.

Maximum economic yield (MEY): The catch level that maximizes the economic "rent" (profits over the opportunity cost of labor and capital) from fisheries. Given conventional fisheries models, MEY is slightly lower than *MSY but requires a much lower level of fishing *effort.

Maximum sustainable yield (MSY): The maximum amount that can be taken (caught) over the long term from a fisheries resource. MSY is now considered an upper limit for fishery management, as opposed to a *target level.

MCS: *Monitoring, control, and surveillance.

Midwater: Trawling, net fishing, or line fishing at a water depth that is higher in the water column than the bottom of the ocean. It is contrasted with bottom (or *benthic) fishing. Also known as *pelagic fishing.

Monitoring, control, and surveillance (MCS): The broadening of traditional means of enforcing national rules over fishing, to the support of the broader problem of fisheries management. MCS has aspects distinct from fisheries management, although there is overlap, and increasingly management cannot be effective without MCS. Although in the traditional basic definition MCS does not include enforcement, that category will need to be included as part of successful implementation of MCS operations. There is a strong emphasis that the success of MCS is not to be measured in number of arrests but in the level of compliance with reasonable management frameworks.

MPA: *Marine protected area.

MSY: *Maximum sustainable yield.

MTC: Mean temperature of the catch; see chapter 8.

NAFO: *Northwest Atlantic Fisheries Organization.

Natural mortality: A mathematical expression of the part of the total rate of deaths of fish from all causes except fishing (e.g., predation, cannibalism, disease, and senescence).

Nei: FAO acronym for "not elsewhere included"; a synonym for "other fish species" or "other invertebrate species" not specifically identified.

Neritic: The shallow part of *pelagic zone over the continental shelf, down to a depth of 200 m.

Northwest Atlantic Fisheries Organization (NAFO): An intergovernmental fisheries science and management organization that ensures the long-term conservation and sustainable use of the fishery resources in the northwest Atlantic. NAFO manages the fisheries outside the *EEZs of the coastal states in the international waters of the northwest Atlantic.

No-take area: *Marine reserve.

NPP: Net primary *production; the fraction of primary production that is available to grazers, e.g., *zooplankton.

Nursery: The area where fish larvae metamorphose into juveniles and where the latter remain until they "recruit" to the adult stock. *Mangroves, in the tropics, often serve as nurseries, as do *estuaries, coastal *lagoons, and generally, shallow areas. However, contrary to a widespread belief, mangroves and the other nearshore habitats are *not* the places where most marine fish spawn (which generally occurs offshore).

Nutrients: The nutritional constituents organisms need for body maintenance and growth; for marine primary production, the key nutrients are nitrates, silicates, and phosphates, generally supplied by *tidal mixing and *upwellings.

Office de la Recherche Scientifique et Technique Outre-Mer (ORSTOM): ORSTOM, now the *Institut de Recherche pour le Développement (IRD, Institute of Research for Development) is a French public science and technology research institute under the joint authority of the French ministries in charge of research and overseas development.

Omnivory: This occurs when consumers feed at several trophic levels, such as a bear feeding on both berries (TL = 1) and salmon (TL = 3.5–4.0); also see *Herbivory.

Open-access fisheries: The common situation wherein those exploiting a resource do not have exclusive rights to it, i.e., anyone can enter the fishery.

ORSTOM: *Office de la Recherche Scientifique et Technique Outre-Mer, now *Institut de Recherche pour le Développement (*IRD).

Outgroup: An item useful when classifying objects, organisms, or institutions that provides a contrast, by virtue of being more closely related to other objects, organisms, or institutions than to all those that are being classified.

Overfishing: Applying a level of fishing *effort beyond that which will generate a desirable, *sustainable, or "safe" population or *stock level. The level of effort can be in excess of that needed to generate *maximum sustainable yield (biological overfishing), *maximum economic yield (economic overfishing), maximum yield per recruit (growth overfishing), or maximum *recruitment (recruitment overfishing).

P/B ratio: The ratio of *production to biomass of an organism (or roughly, its "turnover rate"), per unit of time, is a measure of its productivity. It is usually equivalent to its observed instantaneous rate of total mortality, because any population of organisms, if it is going to persist, must compensate the losses it experiences by *recruitment and growth, which jointly define productivity.

PCB: *Polychlorinated biphenyls.

Pelagic: Living and feeding in the open sea; associated with the surface or middle depths of a body of water; free swimming in the seas, oceans, or open waters; not in association with the bottom (*benthic). Many pelagic fish feed on *plankton.

Phytoplankton: The smallest components of the *plankton community, consisting of microscopic plants, which are a key component of marine and freshwater ecosystems. They are the major transformer of solar energy into biomass that feeds the aquatic food chain.

Plankton: Community of living plants (microscopic *phytoplankton) and animals (*zooplankton) whose lack of powerful propulsive organs leaves them to drift with the water body in which the vagaries of turbulence, currents, or *upwelling have placed them.

Polychlorinated biphenyls: Synthetic organic chemicals related to *dioxin, consisting of a compound of chlorine attached to biphenyl, i.e., a molecule composed of two benzene rings. Polychlorinated biphenyls were widely used as dielectric and coolant fluids, for example in electrical apparatus, and as cutting fluids for machining operations. However, PCBs have been shown to cause cancer in animals and humans, and their production was banned by the 2001 Stockholm Convention on Persistent Organic Pollutants.

Population: A set of interacting organisms of the same species that live in the same geographic area and have the capability of interbreeding. Roughly corresponds to the concept of *stock, used by fishery scientists.

PPR: Primary production required.

Practical Salinity Unit: The ratio *K* of the electrical conductivity of a seawater sample of 15°C and the pressure of 1 standard atmosphere, to that of a potassium chloride (KCl) solution in which the mass fraction of KCl is 0.0324356, at the same temperature and pressure. The *K* value exactly equal to 1 corresponds, by definition, to a practical salinity equal to 35. In this definition, salinity is a ratio, and parts per thousand (‰) is therefore no longer used, but an old value of 35‰ corresponds to a value of 35 practical salinity. Practical salinity is a ratio, and strictly no units should be used; however, "psu" is often added to the value.

Precautionary approach: Also known as the precautionary principle, the notion that supports taking prudent and protective action before there is sufficient scientific evidence for consensus. In short, protective action should not be delayed only because scientific information is lacking. This principle implies that there is a social responsibility to protect from harm when science suggests a plausible risk. This principle is now incorporated into several international environmental agreements, and it is now recognized as a general principle of international environmental law.

Primary production required (PPR): The primary production (i.e., the phytoplankton *production) that is embodied in the catch taken by fisheries in a given ecosystem, usually expressed as a percentage of the overall primary production of that ecosystem. PPR can be estimated given the *trophic level of the species caught, an assumed transfer efficiency, and an estimate of primary production.

Privateer: Primarily in the 16th to 19th centuries, a pirate with a government license to attack ships belonging to its declared enemy.

Production: In ecology and fisheries biology, production is the sum of all growth increments of the animals or plants of a *population over a given time period, including the growth of individuals that may not have survived to the end of that period. Most of the *primary production of the ocean is due to *phytoplankton, and secondary production is due to herbivorous *zooplankton. The term *production*, which is appropriate in agriculture, may be applied to *aquaculture (including *mariculture) but should never be applied to fisheries *catches (or even *landings), which are not "produced" by fishing.

Psu: *Practical Salinity Unit.

Purse-seine: A fishing net used to encircle *pelagic fish. The net may be up to 1 km long and extend 300 m deep and is used to encircle surface schooling fish such as anchovies, mackerel, or tuna. It is usually set from a larger vessel in cooperation with a small boat. During retrieval, the bottom of the net is closed or pursed by drawing a purse line through a series of rings to prevent the fish from escaping.

Quota: The amount of fish that a country, enterprise, or fisher is allowed to take in a given year. Also refers to a constant fraction of a variable *total allowable catch (TAC).

Rebuilding: Reducing fishing (e.g., via low to zero *quotas) until the natural processes of *recruitment and individual growth cause the biomass of a *stock to increase to some preset level, e.g., *B_{MSY}.

Reconstruction (catch): A set of procedures documented in chapter 2 to derive a coherent time series of likely total catches for all fisheries of a country or area from various sources, not necessarily including only official catch statistics; also, the product of these procedures. The word and concept are derived from the science of linguistics, which "reconstructs" extinct words (or languages) from daughter languages.

Recreational fishing: This form of fishing, also called *"sport" fishing, is fishing for pleasure or in competition. This differs from *commercial fishing (both *artisanal and *industrial), where the main motivation is to catch fish for eventual sale, and from *subsistence fishing, where fish is mainly caught for personal or family and friends' consumption.

Recruitment: The process by which young fish enter a fish *population. Recruitment is distinguished from reproduction because the high mortality experienced by fish eggs and larvae usually precludes the prediction of population sizes from the abundance of these early stages (whereas the number of recruits can be used to predict the number of adults).

Reduction fisheries: Fisheries whose catch is used for making *fishmeal (often with fish oil as a byproduct), which is then used to feed animals, e.g., pigs, chickens, or farm-raised salmon. An example of a fish that is usually reduced to fishmeal is the Peruvian anchoveta (*Engraulis ringens*). Most fish used for fishmeal are perfectly edible, and thus reduction fisheries are usually competing with fisheries whose catch is used for direct human consumption.

Reference point: A fishery indicator corresponding to a situation considered desirable (target reference point, TRP) or undesirable (limit reference point, LRP).

Regional fisheries management organization (RFMO): International government organization tasked with managing the fisheries of a region of the ocean (including the *high seas) for the benefit of member states (see chapter 7).

Republic of Cyprus: Cyprus, officially the Republic of Cyprus, is an island country in the eastern Mediterranean Sea. Cyprus is the third largest and third most populous island in the Mediterranean and a member state of the European Union. The Republic of Cyprus has de jure sovereignty over the whole island of Cyprus and its surrounding waters, according to international law, except for the British Overseas Territory of Akrotiri and Dhekelia, administered as British Sovereign Base Areas. However, the Republic of Cyprus is de facto partitioned into 2 main parts: the area in the South under the effective control of the republic, comprising about 59% of the island's area, and the Turkish-controlled area in the north, calling itself the *Turkish Republic of Northern Cyprus (*TRNC) and recognized only by Turkey, covering about 36% of the island's area. The international community considers the northern part of the island as territory of the Republic of Cyprus.

Resilience/resilient: The capacity of a system to tolerate impacts without irreversible change in its outputs or structure. In a species or *population, resilience is usually understood as the capacity to withstand exploitation.

RFMO: *Regional fisheries management organization.

***Sea Around Us (The)*:** Title of a 1951 bestselling book by Rachel Carson, who inspired the fisheries research activity of the same name, launched in July 1999.

Seabirds: Birds that have adapted to life in the marine environment, although they vary greatly in lifestyle, behavior, and physiology. Usually, seabirds have a longer lifespan, breed later, and have fewer young than other birds, but they invest a great deal of time in their young. Most seabirds nest in colonies and may make long annual migrations, crossing the Equator or circumnavigating Earth in some cases. They feed both at the ocean's surface and below it. Seabirds and humans have a long history together; seabirds have provided food for hunters, guided fishers to fishing stocks, and led sailors to land. Many species are threatened by human activities, and conservation efforts are under way. "Catches" of seabirds (see chapter 11) are not included in our reconstructed catch data.

SEAFDEC: *South East Asian Fisheries Development Center.

Secretariat of the Pacific Community (SPC): The "South Pacific Commission", as the SPC was formerly called, was founded in Australia in 1947 by the 6 "participating governments" that then administered territories in the Pacific: Australia, France, New Zealand, the Netherlands, the United Kingdom, and the United States of America. They established the organization to restore stability to a region that had experienced the trauma of World War II, to assist in administering their dependent territories and to benefit the people of the Pacific. In 1962, Samoa was the first island nation to become an independent state and in 1965 was the first to become a full member of SPC. Other island nations in turn became independent or largely self-governing, and in 1983 all 22 Pacific Island member countries and territories were recognized as full voting and contributing members of SPC. The name "South Pacific Commission" was changed to the Pacific Community at the 50th anniversary conference in 1997 to reflect the organization's Pacific-wide membership. By 2010, SPC's 26 members included the 22 Pacific Island countries and territories along with 4 of the original founders (the Netherlands and United Kingdom withdrew in 1962 and 2004, respectively, when they relinquished their Pacific interests).

Shelf: The seafloor between the coast and the 200 m *isobath around the continents ("continental shelf"), and less commonly, around islands. Shelves are the most productive parts of the oceans and support their most important fisheries.

Shellfish: An aquatic invertebrate animal, such as a mollusc or crustacean, with a shell or shell-like external skeleton.

Slope (continental): Region of the outer edge of a continent between the generally shallow continental *shelf and the deep ocean floor, i.e., from 200 to 2,000 m; often steep.

Small-scale: *Artisanal, *subsistence, and *recreational.

SOFIA: *State of the World Fisheries and Aquaculture*, a report issued every 2 years by *FAO, whose content is based mainly on analyses of recent fisheries and aquaculture statistics by FAO staff.

South East Asian Fisheries Development Center (SEAFDEC): An autonomous intergovernmental body established in 1967 with a mandate to develop and manage the fisheries potential of the region by rational use of the resources for providing food security and safety to the people and alleviating poverty through transfer of new technologies, research, and information dissemination. SEAFDEC consists of 11 member countries: Brunei Darussalam, Cambodia, Indonesia, Japan, Laos, Malaysia, Myanmar, Philippines, Singapore, Thailand, and Vietnam.

Soviet: Pertaining to the former USSR, i.e., the now-defunct Union of Socialist Soviet Republics.

SPC: *Secretariat of the Pacific Community.

Spear fishing: Fishing with devices functioning like crossbows or airguns. Spear fishing using scuba gear is widely prohibited. Spearfishing by *small-scale fishers is widely used in parts of the Caribbean and Asia-Pacific, where it can be highly selective.

Sport fishing: *Recreational fishing.

SSP: *Stock-status plot.

SST: Sea surface temperature.

Standing stock: Synonym for *biomass.

Stern trawlers: A term used formerly to distinguish *trawlers in which the trawl is pulled on deck via a slipway at the back (stern) from others, i.e., side trawlers. Not used much today because most trawlers are stern trawlers.

Stock: The exploited part of a fish population.

Stock assessment: A set of mathematical procedures through which the current and probable future abundance (or biomass) of commercial fish stocks can be estimated, using data from life history studies, environmental surveys, and catch statistics. Generally forms the basis for setting *total allowable catches.

Stock-status plot (SSP): A manner of summarizing time series of fisheries catches such that trends in the status of the fisheries that generated these catches are highlighted (see chapter 1).

Subsidies: Government funds made available to a segment of the population of a country or a sector of its economy. When given to a well-developed fishery, subsidies tend to encourage *overfishing (see chapter 6).

Subsistence fishing: A form of *small-scale (inshore) fishing (or "gleaning"), often practiced by women and children, where the catch (often small fish and invertebrates, particularly bivalves) is caught mainly for consumption by oneself or one's family or bartered for other commodities.

Surplus: (1) In *surplus production models, the surplus is the biomass that is produced by a *stock in excess of what is needed to maintain the stock at its current abundance level, which can thus be taken by a fishery without the stock declining further. *MSY is the highest surplus level, and in the Schaefer model, it occurs when the biomass is reduced, by fishing, to half its unexploited biomass. (2) *Surplus* is also a term used in the context of *UNCLOS, wherein the vessels of one or several *distant-water fishing countries should be given access to the *EEZ of a coastal state (against payment of an "access fee") if it is not itself fully exploiting the fishery resources therein.

Surplus production model: One of the simplest analytical *stock assessment models, which pools *recruitment, mortality, and growth of a *stock into a single *production function.

Surveillance: A key element of *"monitoring, control, and surveillance"' (MCS), necessary to support fisheries management in the context of extended jurisdiction, i.e., in larger *EEZs. Surveillance is the degree and types of observations needed to maintain compliance with the regulatory controls imposed on fishing activities. Radar, including coastal, airborne, and spaceborne systems, even if intended for national security or law enforcement, can provide information to fisheries managers and environmental protection authorities.

Sustainable: An activity or process whose properties are such that it could last indefinitely (or at least into the long-term future). By this definition, "sustainable growth" is an oxymoron.

t: *Tonne, or metric ton, being 1,000 kg.

TAC: *total allowable catch.

Target: This term has two meanings in fisheries. One refers to the fish (species or group) that are meant to be caught; this contrasts to *bycatch, which is caught because the gear targeting a certain resource type is not sufficiently selective. The other meaning of *target* refers to fisheries management, which often defines *MSY and its associated fishing mortality (*F_{MSY}) as a limit, with the target being slightly more conservative.

Taxon (plural: taxa): According to the International Code of Zoological Nomenclature, any formal unit or category of organisms (species, genus, family, order, class, etc.). Derived terms include *taxonomist*, *taxonomic*, and *taxonomically*.

Technological coefficient: In input–output analysis, this identifies the percentage or portion of the total inputs of a sector needed to be purchased from another sector irrespective of the geographic origin of this purchase.

Territorial waters: The area beyond the tidal baseline of the open coasts of a country over which that country exercises full control and sovereignty except for innocent passage of foreign vessels. Set at a maximum of 12 nmi by the 1982 *UNCLOS. The United States and a few other countries claim territorial waters of only 3 nmi (see also *EEZ, often mistakenly labeled "territorial waters"). For the purposes of the *Sea Around Us* spatial definitions, territorial waters are treated as part of the EEZ.

Thermocline: The distinct interface between surface waters and cooler, deeper waters; region between the warm upper layer and the lower cold layer of the sea or lake, where temperature declines abruptly (1°C/m or more) with increasing depth.

Threatened: Species of animals, plants, fungi, and so on that are vulnerable to endangerment. The International Union for Conservation of Nature (IUCN) is the foremost authority on threatened species, and classifies threatened species (in their "Red List") as Vulnerable, Endangered, or Critically Endangered species. Less-than-threatened categories are Near Threatened and Least Concern. Species that have not been evaluated (NE) or do not have sufficient data (Data Deficient) also are not considered threatened by the IUCN.

Tidal mixing: The mixing of water layers due to the action of alternating high and low tides.

Tonne: A weight unit corresponding to 1,000 kg; used in this atlas as unit of catches. Equivalent to a metric ton.

Total allowable catch (TAC): The amount of fish (or *quota) that can be taken legally by a given fishery in a given period (usually a year, or a fishing season), as determined by a fishery management agency.

Traditional: An adjective misleadingly used in some countries to describe *artisanal fishers and fisheries.

Trammel net: A set-net consisting of three layers of netting, designed so that a fish entering through one of the large-meshed outer sections will push part of the finer-meshed central section through the large meshes on the further side, forming a pocket in which the fish is trapped.

Transhipment (at sea): The transfer of goods (here: fish) from one boat to the other while at sea, often to avoid controls available in ports.

Trash fish: The earlier and badly misleading name for the fraction of the *bycatch for which no market had been identified and which was therefore *discarded (see also account for Guyana).

Trawl/Trawler/Trawling: Fishing methods where a vessel--a trawler--tows a large bag-shaped trawl net. A wide range of *benthic (also called demersal or bottom) or *pelagic (midwater) species of fish are taken by this fishing method. The trawl net usually features a buoyed head (top) rope, a weighted foot (bottom) rope, and two heavy "otter" doors to keep the net mouth open. Variation include beam trawls that use a horizontal beam instead of otter doors and foot rope to keep the net open, or pair-trawls in which two vessels are used to tow a single, often much larger net. Bottom trawling is unselective and destructive of habitats and is gradually being banned from areas that people care about. All trawls are here considered *industrial gear, whatever the size of the vessel pulling them.

TRNC: *Turkish Republic of Northern Cyprus.

Trolling: A fishing method where baited hooks or lures are towed behind a boat; not to be confused with *trawling.

Trophic level: Numbers expressing the relative "height" of an organism within the food web, with plants having a trophic level (TL) of 1 (by definition), herbivores 2, their predators 3, etc. Because fishes have mixed diets, they tend to have intermediate TL values, e.g., 2.5 for an omnivore feeding half on plants and half on herbivores.

Tropics/Tropical: A climate zone ranging north and south from the Equator to the limits of the Subtropical Zone, generally limited to *SST above 20°C.

Turkish Republic of Northern Cyprus (TRNC): The Turkish Republic of Northern Cyprus (TRNC) is a self-declared state that comprises the northeastern portion of the island of Cyprus. It is recognized only by Turkey, whereas the international community considers northern Cyprus as part of the *Republic of Cyprus.

UN Fish Stocks Agreement: The UN Fish Stocks Agreement (UNFSA), whose full title is "The United Nations Agreement for the Implementation of the Provisions of the United Nations Convention on the Law of the Sea of 10 December 1982 relating to the Conservation and Management of Straddling Fish Stocks and Highly Migratory Fish Stocks" came into force in December 2001 and provides principles for state cooperation on the conservation and optimal use of highly migratory fish stocks both within and beyond *EEZs. It emphasizes the *precautionary approach and the need to use best available scientific information.

UNCLOS: *United Nations Convention on the Law of the Sea.

UNFSA: *UN Fish Stocks Agreement.

United Nations Convention on the Law of the Sea (UNCLOS): Also called the Law of the Sea Convention or the Law of the Sea treaty, it is the international agreement that came into force in 1994 and defines the rights and responsibilities of nations with respect to their use of the world's oceans, establishing guidelines for businesses, the environment, and the management of marine natural resources. Among other things, UNCLOS enabled countries to declare an *EEZ out to a maximum of 200 nmi.

Upwelling: An oceanographic phenomenon involving the wind-driven rise of dense, cooler, and usually nutrient-rich water toward the ocean surface, where it replaces (and pushes offshore) warmer, usually nutrient-depleted surface water. The cold but nutrient-rich upwelled water stimulates the growth of primary producers (mainly *phytoplankton) and *secondary producers (mainly *zooplankton), upon which the rest of the *ecosystem (*forage fish, other fishes, *seabirds, and *marine mammals) depends.

USSR: The former Union of Soviet Socialist Republics, or Soviet Union for short, was a state on the Eurasian continent that existed between 1922 and 1991, governed by the Communist Party of the USSR, with Moscow as its capital. Nominally a union of "Soviet Republics," with a highly centralized government and economy. It has since separated into its constituent republics, as largely independent states. For maritime issues, the constituents are Estonia, Georgia, Latvia, Lithuania, the Russian Federation (Russia), and Ukraine.

Variable cost: Expenses incurred by a fishery that depend on the amount of fishing that takes place, e.g., fuel costs; see also *fixed cost.

Vessel monitoring system (VMS): Used by agencies tasked with monitoring the position, time at a position, and course and speed of fishing vessels. VMSs are a key part of *monitoring, control, and surveillance (MCS) programs. VMSs may be used to monitor vessels in the territorial waters of a country or a subdivision of a country, or in their EEZs.

VMS: *Vessel monitoring system.

WCPFC: *Western and Central Pacific Fisheries Commission.

Western and Central Pacific Fisheries Commission: The Western and Central Pacific Fisheries Commission (WCPFC) was established in 2004. The WCPF Convention draws on many of the provisions of the *UN Fish Stocks Agreement (UNFSA) while reflecting the special political, socioeconomic, geographic, and environmental characteristics of the western and central Pacific Ocean region. The WCPFC seeks to address problems in the management of high seas fisheries resulting from unregulated fishing, overcapitalization, excessive fleet capacity, vessel reflagging to escape controls, insufficiently selective gear, unreliable databases, and insufficient multilateral cooperation in respect to conservation and management of highly migratory fish stocks.

It has 26 members (Australia, China, Canada, Cook Islands, European Union, Federated States of Micronesia, Fiji, France, Indonesia, Japan, Kiribati, Republic of Korea, Republic of Marshall Islands, Nauru, New Zealand, Niue, Palau, Papua New Guinea, Philippines, Samoa, Solomon Islands, Chinese Taipei, Tonga, Tuvalu, United States of America, and Vanuatu), 7 participating territories (American Samoa, Commonwealth of the Northern Mariana Islands, French Polynesia, Guam, New Caledonia, Tokelau, Wallis and Futuna), and 8 cooperating nonmembers (Ecuador, El Salvador, Mexico, Panama, Senegal, Liberia, Thailand, Vietnam).

Western Pacific Fisheries Information Network: The Western Pacific Fisheries Information Network (*WPacFIN), established in 1981 and hosted by the (U.S.) National Oceanic and Atmospheric Administration. (Hawaii), updates and maintains a central database that provides access to the publicly available fisheries data from the Western Pacific Region under U.S. management consideration to support fisheries management in that region. It maintains cooperative agreements with participating state and territorial fisheries agencies in American Samoa, the Commonwealth of the Northern Mariana Islands (CNMI), Guam, and Hawaii. It also works closely with the relevant U.S. Fishery Management Council. Its vision is to provide data on a timely basis for managing fisheries resources in the U.S. Pacific EEZ.

Western Pacific Regional Fishery Management Council: The WPRFMC is 1 of 8 regional U.S. fishery management councils established by the Magnuson Fishery Conservation and Management Act in the USA (now the Magnuson–Stevens Fishery Conservation and Management Act). This council consists of Hawaii, American Samoa, Guam, and the Northern Mariana Islands and has authority over marine fisheries in the Pacific Ocean of these entities and all territories and possession of the United States in the Pacific.

Windward: Facing the wind or on the side facing the wind; the opposite of *leeward.

Withdrawal(s): Synonym of *catch or catches when applied to fish; same as "removals".

Working group: An ad hoc or permanent group of experts with a specific task, e.g., assessing the state of an exploited *stock of fish. Extensively used by, e.g., *ICES.

World Heritage Site: Area identified by UNESCO as being of great cultural or historical importance.

WPacFIN: *Western Pacific Fisheries Information Network.

WPRFMC: *Western Pacific Regional Fishery Management Council.

Yield: *Catch in weight during a conventional period, e.g., a year; see also *maximum sustainable yield or *MSY.

Zooplankton: Small components of the *plankton community, consisting of small to microscopic animals, feeding on either *phytoplankton or other zooplankton. They form part of the bottom of the food chain and are often a major food source for small or medium-sized fishes and invertebrates.

GEOGRAPHIC INDEX

(The page where a country or territory is treated in depth is **bold;** note that this index does not include countries from which foreign fleets originate, nor countries that were actors in negotiations)

SUBJECT INDEX

NOTE: This is the SUBJECT INDEX. Please also see the GEOGRAPHIC INDEX starting on page 479.
NOTE: Page numbers followed by *b*, *f*, *t*, or *n* refer to boxes, figures, tables, or endnotes, respectively.